Lecture Notes in Computer Science 3889

Commenced Publication in 1973
Founding and Former Series Editors:
Gerhard Goos, Juris Hartmanis, and Jan van Leeuwen

Editorial Board

David Hutchison
Lancaster University, UK

Takeo Kanade
Carnegie Mellon University, Pittsburgh, PA, USA

Josef Kittler
University of Surrey, Guildford, UK

Jon M. Kleinberg
Cornell University, Ithaca, NY, USA

Friedemann Mattern
ETH Zurich, Switzerland

John C. Mitchell
Stanford University, CA, USA

Moni Naor
Weizmann Institute of Science, Rehovot, Israel

Oscar Nierstrasz
University of Bern, Switzerland

C. Pandu Rangan
Indian Institute of Technology, Madras, India

Bernhard Steffen
University of Dortmund, Germany

Madhu Sudan
Massachusetts Institute of Technology, MA, USA

Demetri Terzopoulos
New York University, NY, USA

Doug Tygar
University of California, Berkeley, CA, USA

Moshe Y. Vardi
Rice University, Houston, TX, USA

Gerhard Weikum
Max-Planck Institute of Computer Science, Saarbruecken, Germany

D1769107

Justinian Rosca Deniz Erdogmus
José C. Príncipe Simon Haykin (Eds.)

Independent Component Analysis and Blind Signal Separation

6th International Conference, ICA 2006
Charleston, SC, USA, March 5-8, 2006
Proceedings

Springer

Volume Editors

Justinian Rosca
Siemens Corporate Research
Real-Time Vision and Modeling Department
755 College Road East, Princeton, New Jersey 08540, USA
E-mail: justinian.rosca@siemens.com

Deniz Erdogmus
Oregon Health and Science University
CSE Department
20000 NW Walker Road, Beaverton, Oregon, 97006, USA
E-mail: derdogmus@ieee.org

José C. Príncipe
University of Florida
Computational NeuroEngineering Laboratory
Gainesville, Florida, 32611, USA
E-mail: principe@cnel.ufl.edu

Simon Haykin
McMaster University
1280 Main Street West, Hamilton, Ontario, L8S 4K1, Canada
E-mail: haykin@mcmaster.ca

Library of Congress Control Number: 2006920910

CR Subject Classification (1998): C.3, F.1.1, E.4, F.2.1, G.3, H.1.1, H.5.1, I.2.7

LNCS Sublibrary: SL 3 – Information Systems and Application, incl. Internet/Web and HCI

ISSN 0302-9743
ISBN-10 3-540-32630-8 Springer Berlin Heidelberg New York
ISBN-13 978-3-540-32630-4 Springer Berlin Heidelberg New York

Springer is a part of Springer Science+Business Media

springer.com

Printed in Germany

Typesetting: Camera-ready by author, data conversion by Scientific Publishing Services, Chennai, India
Printed on acid-free paper SPIN: 11679363 06/3142 5 4 3 2 1 0

Preface

This volume contains the papers presented at the 6th International Conference on Independent Component Analysis (ICA) and Blind Source Separation (BSS) organized in historic Charleston, SC, USA, March 5-8, 2006.

The sixth edition of the conference has brought the latest developments in one of the most exciting areas of statistical signal processing/unsupervised machine learning. ICA theory has received attention from several research communities including machine learning, neural networks, statistical signal processing and Bayesian modeling. ICA/BSS has applications at the intersection of many science and engineering disciplines concerned with understanding and extracting useful information from data as diverse as neuronal activity and brain images, bioinformatics, communications, the world wide-web, audio, video, sensor signals, or time series.

Papers were solicited in all areas of independent component analysis and blind source separation, including the following: algorithms and architectures (e.g. statistical learning algorithms based on ICA and BSS using linear/nonlinear mixture models, convolutive and noisy models, extensions of basic models, combinatorial optimization, kernel methods, graphical models), applications (innovative applications or fielded systems that use ICA and BSS, including systems for acoustic signal separation, time series prediction, data mining, multimedia processing, telecommunications), medical applications (e.g., bioinformatics, neuroimaging, processing of electrocardiograms, electroencephalograms, magnetoencephalograms, and functional magnetic resonance imaging), speech and signal processing (e.g., computational auditory speech analysis, source separation, auditory perception, coding, recognition, synthesis, denoising, segmentation, dynamic and temporal models), theory (e.g., information theory, estimation methods, complex methods, time/frequency representations, optimization, sparse representations, asymptotic analysis, unsupervised learning, coding), visual and sensory processing (e.g., image processing and coding, segmentation, object detection and recognition, motion detection and tracking, visual scene analysis and interpretation).

Accepted papers covered these topics well, and as a result this volume has a simple organization based on the six sections: Algorithms and Architectures, Applications, Medical Applications, Speech and Signal Processing, Theory, and Visual and Sensory Processing. Within each section, papers were organized alphabetically by the first author's last name. Several topics are widely represented in the present volume such as audio source separation, bioinformatics, convolutive models of ICA, denoising, estimation methods, linear/nonlinear mixture models, optimization in ICA/BSS, time/frequency representations, sparse representations, and statistical learning.

The 2006 event introduced several innovations compared to previous meetings. The paper review/acceptance system relied on the Program Committee members' responsibility in assigning papers for review and drawing acceptance decisions. For the first time two tutorials were included in the program about outstanding developments in the area: "Neural theory and neural analysis using ICA," lectured by Tony Bell of the University of California at Berkeley, and "Bayesian machine learning for signal processing," lectured by Hagai T. Attias of Golden Metallic, Inc. The conference offered Student Best Paper Awards and travel support to participating students.

The interest in the conference was demonstrated by the large number of author registrations and the healthy submission rate. The conference database included 183 submissions. The review process was more selective than at the previous conferences and many meritourious submissions could not be accepted for the final program. In the end, the Program Committee selected 64% of the papers for inclusion in this volume. The vast majority of papers benefited from at least four reviews. The authors of accepted papers had the opportunity to upgrade their manuscripts based on the peer review feedback.

The conference had a combination of high-quality tutorials, research papers, applications papers, posters, and invited talks, which demonstrated that ICA has become a mature conference and the main venue for researchers and practitioners in this area.

Many people deserve credit for their hard work on behalf of the conference. Thanks go to all paper authors in this volume. In addition we thank the members of the Program Committee and the reviewers for their efforts in organizing the reviews, and for reviewing and selecting the papers to be included in this volume. We are also grateful to the organizers of the special sessions for their work in inviting, selecting presentations, and putting together the sessions. All these efforts have been essential in compiling a high-quality scientific program.

Special acknowledgements go to many people whose effort and dedication contributed to the success of the conference. We thank Jose Principe for his efforts in organizing the conference, the staff of the University of Florida for the support with various phases of the process, Thomas Preuss for designing and helping with the excellent web submission and conference database engine ConfMaster, and Antonio Paiva for acting as webmaster of the conference. We thank the members of the ICA Steering Committee for their advice and for assigning the job to the present team.

Last but not least, the cooperation with Springer in preparation of this volume and the CD-ROM proceedings was outstanding. We hope you will find the proceedings interesting and stimulating.

January 2006

Justinian Rosca
Deniz Erdogmus
Jose Principe
Simon Haykin

ICA 2006 Committee Listings

Organizing Committee

General Chairs:	Jose C. Principe, University of Florida, USA Simon Haykin, McMaster University, Canada
Program Chair:	Justinian Rosca, Siemens Corporate Research, USA
Technical Chair:	Deniz Erdogmus, Oregon Health and Science University, USA
Special Sessions and Publicity:	Te-Won Lee, University of California San Diego, USA
European Liaison:	Erkki Oja, Helsinki University of Technology, Finland
Far East Liaison:	Andrzej Cichocki, RIKEN BSI, Japan
Americas Liaison:	Allan K. Barros, UFMA, Brazil

Program Committee

Chong-Yung Chi, National Tsing Hua University, Taiwan
Andrzej Cichocki, RIKEN BSI, Japan
Deniz Erdogmus, Oregon Health and Science University, USA
Simon Haykin, McMaster University, Canada
Aapo Hyvarinen, University of Helsinki, Finland
Yujiro Inouye, Shimane University, Japan
Christian Jutten, INPG-LIS, France
Juha Karhunen, Helsinki University of Technology, Finland
Te-Won Lee, University of California San Diego, USA
Shoji Makino, NTT Communication Science Laboratories, Japan
Klaus Muller, First.Ida/Fraunhofer Society, Germany
Erkki Oja, Helsinki University of Technology, Finland
Jose C. Principe, University of Florida, USA
Justinian Rosca, Siemens Corporate Research, USA
Michael Zibulevsky, Technion, Israel

Reviewers

Karim Abed-Meraim
Tulay Adali

Luis Almeida
Miroslav Andrle
Jrn Anemller
Shoko Araki
Francis Bach
Radu Balan
Adel Belouchrani
Maria Carmen Carrion
Jonathon Chambers
Zhe Chen
Chong-Yung Chi
Seungjin Choi
Andrzej Cichocki
Sergio Cruces-Alvarez
Adriana Dapena
Mike Davies
Lieven De Lathauwer
Yannick Deville
Konstantinos Diamantaras
Scott Douglas
Shinto Eguchi
Deniz Erdogmus
Jan Eriksson
Da-Zheng Feng
Cdric Fvotte
Simone Fiori
Mark Girolami
Pedro Gomez Vilda
Juan Manuel Gorriz Saez
Rmi Gribonval
Ulas Gunturkun
Jiucang Hao
Zhaoshui He
Tom Heskes
Kenneth Hild
Susana Hornillo Mellado
Patrik Hoyer
Yingbo Hua
Jarmo Hurri
Aapo Hyvarinen
Shiro Ikeda
Mika Inki
Yujiro Inouye
Tzyy-Ping Jung
Christian Jutten

Juha Karhunen
Wlodzimierz Kasprzak
Mitsuru Kawamoto
Taesu Kim
Visa Koivunen
Tian Lan
Elmar Lang
Soo-Young Lee
Intae Lee
Ta-Sung Lee
Yuanqing Li
Philippe Loubaton
Hui Luo
Wing-Kin Ma
Shoji Makino
Dmitry Malioutov
Ali Mansour
Ruben Martin-Clemente
Oleg Michailovich
Dmitri Model
Enric Monte
Eric Moreau
Ryo Mukai
Klaus Muller
Noboru Murata
Juan Jose Murillo-Fuentes
Wakako Nakamura
Noboru Nakasako
Asoke Nandi
Klaus Obermayer
Erkki Oja
Kazunori Okada
Stanislaw Osowski
Umut Ozertem
Dinh-Tuan Antoine Pham
Mark Plumbley
Kenneth Pope
Carlos Puntonet
Duangmanee Putthividhya
Laura Rebollo-Neira
Scott Rickard
Stephen Roberts
Justinian Rosca
Diego Ruiz
Tomasz Rutkowski

Fathi Salem
Jaakko Srel
Hiroshi Saruwatari
Hiroshi Sawada
Santiago Sazo
Odelia Schwartz
Sarit Shwartz
Jordi Sol-Casals
Toshihisa Tanaka
Fabian Theis
Kari Torkkola
Harri Valpola
Alle-Jan Van Der Veen
Jean-Marc Vesin
Ricardo Vigrio
Vincent Vigneron
Emmanuel Vincent
Ben Vincent
Tuomas Virtanen
Frederic Vrins
Deliang Wang
Yoshikazu Washizawa
Karl Wiklund
Lei Xu
Isao Yamada
Kehu Yang
Arie Yeredor
Vicente Zarzoso
Liqing Zhang
Yimin Zhang
Michael Zibulevsky
Andreas Ziehe

Table of Contents

Algorithms and Architectures

Applications

Medical Applications

Speech and Signal Processing

Theory

Visual and Sensory Processing

Simple LU and QR Based Non-orthogonal Matrix Joint Diagonalization

Bijan Afsari

Institute for Systems Research and Department of Applied Mathematics,
University of Maryland, College Park,
20742 MD,USA
bijan@glue.umd.edu

Abstract. A class of simple Jacobi-type algorithms for non-orthogonal matrix joint diagonalization based on the LU or QR factorization is introduced. By appropriate parametrization of the underlying manifolds, i.e. using triangular and orthogonal Jacobi matrices we replace a high dimensional minimization problem by a sequence of simple one dimensional minimization problems. In addition, a new scale-invariant cost function for non-orthogonal joint diagonalization is employed. These algorithms are step-size free. Numerical simulations demonstrate the efficiency of the methods.

1 Introduction

The problem of matrix (approximate) Joint Diagonalization (JD) has found applications in many blind signal processing algorithms, see for example [4,6]. In one formulation it can be presented as: given a set of $n \times n$ symmetric matrices $\{C_i\}_{i=1}^N$ find a non-singular B such that the matrices $\{BC_iB^T\}_{i=1}^N$ are "as diagonal as possible". We call such a B a joint diagonalizer. In general diagonalization can happen only approximately. If B is restricted to the set of orthogonal $n \times n$ matrices $\mathrm{O}(n)$, the problem is referred to as orthogonal JD. Here, we are interested in non-orthogonal JD or NOJD, i.e. where B is in the set of non-singular $n \times n$ matrices $\mathrm{GL}(n)$. The reader is referred to [2,8] for further references on this subject. We remind that in [7] the NOJD problem is formulated differently.

A natural and effective cost function for orthogonal JD is [4]:

$$J_1(\Theta) = \sum_{i=1}^{n} \left\| \Theta C_i \Theta^T - \mathrm{diag}(\Theta C_i \Theta^T) \right\|_F^2 \tag{1}$$

where $\mathrm{diag}(X)$ is the diagonal part of matrix X, $\|.\|_F$ is the Frobenius norm and $\Theta \in \mathrm{O}(n)$. The algorithm introduced in [4], which is a part of the JADE algorithm, to minimize $J_1(\Theta)$ is an iterative minimization method using orthogonal Jacobi matrices. This algorithm breaks the $\frac{n(n-1)}{2}$ dimensional minimization problem to a sequence of one dimensional minimization problems and also uses the group structure of $\mathrm{O}(n)$ by using multiplicative updates. Here, we extend this idea to the NOJD problem.

J. Rosca et al. (Eds.): ICA 2006, LNCS 3889, pp. 1–7, 2006.

In many cases, such as noisy ICA, the joint diagonalizer sought can not assumed to be orthogonal. The NOJD problem is more challenging than orthogonal JD. It is natural to consider the NOJD as a minimization problem. Motivating physical problems such as ICA and BSS suggest that a good cost function J for NOJD should be invariant under permutation Π and under scaling by a non-singular diagonal matrix Λ, i.e. $J(\Lambda\Pi B) = J(B)$[1]. If we extend J_1 to $\mathrm{GL}(n)$ then clearly $J_1(\Lambda B) \neq J_1(B)$. In fact by reducing the norm of B we can reduce $J_1(B)$ arbitrarily. In order to still use J_1 we can extend J_1 to a smaller subgroup of $\mathrm{GL}(n)$ such as $\mathrm{SL}(n)$ [3] or as in [3,8] we can restrict the "reduction" of J_1 only to the directions that do not correspond to multiplication by diagonal matrices. The latter results in updates of the form:

$$B_{k+1} = (I + \Delta_k)B_k \tag{2}$$

where I is the $n \times n$ identity matrix, $\mathrm{diag}(\Delta_k) = 0$ and Δ_k is found such that $J_1(B_{k+1})$ is reduced at each step. This can be done, for example, from a gradient descent step as in [3]. One consequence of the update in (2) is that if the norm of Δ_k is small enough [8] we can guarantee invertibility of B_{k+1}. Also if we choose Δ_k to be a triangular matrix with $\mathrm{diag}(\Delta_k) = 0$ and if $B_0 = I$ then $\det B_{k+1} = 1$ for all k and hence $\|B_{k+1}\|_2 \geq 1$. The significance of the latter is that it ensures that the cost J_1 is not reduced merely due to reducing the norm of B_k. In this article we consider triangular Δ_k with only one non-zero element and we refer to $I + \Delta_k$ as a Jacobi triangular matrix. In Section 2, we describe a class of NOJD methods using orthogonal and triangular Jacobi matrices which are based on the LU or QR factorization of the sought diagonalizer.

Another idea in devising NOJD algorithms is to use cost functions other than J_1. In [8,2] some different cost functions are mentioned. In [1] a scale-invariant cost function is used for NOJD which has the form:

$$J_2(B) = \sum_{i=1}^{N} \|C_i - B^{-1}\mathrm{diag}(BC_iB^T)B^{-T}\|_F^2 \tag{3}$$

Note that $J_2(\Lambda B) = J_2(B)$ for diagonal Λ and that $J_2(\Theta) = J_1(\Theta)$ for $\Theta \in \mathrm{O}(n)$. J_2 is the normalized version of J_1 in the sense that:

$$\frac{J_1(B)}{n\|B\|_F^2} \leq J_2(B) \leq n\|B^{-1}\|_F^2 J_1(B) \tag{4}$$

A drawback of J_2 is that in its calculation we need to compute the inverse of B. In Section 3 we propose a simple algorithm for minimization of J_2, too. In Section 4 we test the developed methods numerically and provide a comparison with one existing efficient NOJD method.

2 Use of LU and QR Factorizations in Minimization of J_1

Any non-singular matrix B admits the LU factorization:

$$B = \Pi\Lambda LU \tag{5}$$

[1] Intuitively, we do not expect that $\Lambda\Pi C_i\Pi^T\Lambda$ can become more diagonal than C_i.

where Π is a permutation matrix, Λ is a non-singular diagonal matrix and L and U are $n \times n$ unit lower and upper triangular matrices, respectively. By a unit triangular matrix we mean a triangular matrix with diagonal elements of one [5]. The factorization in (5) exactly matches the invariances in NOJD. On the other hand the SVD factorization, for example, can not match this. Unit lower and upper triangular matrices of dimension n, form Lie groups denoted by $\mathcal{L}(n)$ and $\mathcal{U}(n)$, respectively. This fact simplifies the minimization process a lot. B also admits the QR factorization:

$$B = \Lambda L \Theta \tag{6}$$

where $\Theta \in$O(n) and $L \in \mathcal{L}(n)$. The idea is to find L and U separately in the LU form or L and Θ in the QR form such that J_1 is reduced at each step and repeat this till convergence. If the initial condition is the identity matrix, by construction, the solution's determinant will remain unity. Furthermore, we replace each of these $\frac{n(n-1)}{2}$ dimensional minimization problems by a sequence of simple one-dimensional problems by using triangular and orthogonal Jacobi matrices.

2.1 Triangular and Orthogonal Jacobi Matrices

A lower triangular Jacobi matrix with parameter a corresponding to the position $(i,j), i > j$ is denoted by $L_{ij}(a)$. $L_{ij}(a)$ is an element of $\mathcal{L}(n)$ whose $(i,j)^{th}$ entry is a and the rest of its off-diagonal entries are zero. In a similar fashion we define the upper triangular Jacobi matrix with parameter a corresponding to the position $(i,j), i < j$ and denote it by $U_{ij}(a)$. Any element of $\mathcal{L}(n)$ ($\mathcal{U}(n)$) can be represented as a product of lower (upper) triangular Jacobi matrices. We replace the problem of minimization of $J_1(L)$ with $L \in \mathcal{L}(n)$ which is a high dimensional problem with a sequence of simple one-dimensional quadratic problems of finding the parameter of triangular Jacobi matrices for minimizing J_1. The following simple proposition solves the one-dimensional problem. For brevity the proof is omitted.

Notation: (MATLAB's indexing) For matrix A, $A(k, index)$ where $index$ is a row vector denotes a row-vector whose elements are from the kth row of A indexed by $index$. $A(index, l)$ is defined similarly. Specificality we are interested in vectors like $A(l, [1:i-1, i+1:n])$.

Proposition 1. *If $\hat{a}$ is such that:*

$$\hat{a} = -\frac{\sum_{i=1}^{N} C_i(k,[1:l-1,l+1:n])C_i(l,[1:l-1,l+1:n])^T}{\sum_{i=1}^{N} \|C_i(k,[1:l-1,l+1,:n])\|_F^2} \tag{7}$$

then: with $k < l$, $\hat{a}$ minimizes $J_1(L_{lk}(a))$ and with $k > l$, $\hat{a}$ minimizes $J_1(U_{lk}(a))$. If $\sum_{i=1}^{N} \|C_i(k,[1:l-1,l+1,:n])\|_F^2 = 0$ set $\hat{a} = 0$, i.e. J_1 can not be reduced by that particular L_{lk} or U_{lk}.

Similarly, if $\Theta_{kl}(\theta)$ is the Jacobi rotation matrix corresponding to the position (k,l) and a counter-clockwise rotation by θ, then we have that[4]:

Proposition 2. *If θ_{kl} is such that $\vec{v} = [\cos 2\theta_{kl} \quad \sin 2\theta_{kl}]^T$ is a unit-norm eigen vector corresponding to the larger eigen value of the matrix $G_{kl}^T G_{kl}$ where G_{kl} is an $N \times 2$ matrix defined as $G_{kl}(i,1) = C_i(k,k) - C_i(l,l)$ and $G_{kl}(i,2) = 2C_i(k,l)$ for $1 \leq i \leq N$, then θ_{kl} minimizes $J_1(\Theta_{kl}(\theta))$.*

Based on these two propositions we can have two algorithms LUJ1D and QRJ1D. We juxtapose the two algorithms together:

Algorithm LUJ1D (QRJ1D)

1. `set` $B = I$. `set` ϵ.
2. U-phase (Q-Phase): `set` $U = I(\Theta = I)$. `for` $1 \leq l < k \leq n$:
   - Find $a_{lk} = \arg\min_a J_1(U_{lk}(a))$ ($\theta_{lk} = \arg\min_\theta J_1(\Theta_{lk}(\theta))$) from Proposition 1 (Proposition 2)
   - $C_i \leftarrow U_{lk}(a_{lk}) C_i U_{lk}(a_{lk})^T$ ($C_i \leftarrow \Theta_{lk}(\theta_{lk}) C_i \Theta_{lk}(\theta_{lk})^T$) `and` $U \leftarrow U_{lk}(a_{lk})U$ ($\Theta \leftarrow \Theta_{lk}(\theta_{lk})\Theta$)
3. L-phase (R-Phase): `set` $L = I$. `for` $1 \leq l < k \leq n$:
   - Find $a_{kl} = \arg\min_a J_1(L_{kl}(a))$ from Proposition 1
   - Update $C_i \leftarrow L_{kl}(a_{kl}) C_i L_{kl}(a_{kl})^T$ `and` $L \leftarrow L_{kl}(a_{kl})L$
4. `if` $\|LU - I\|_F > \epsilon$ ($\|L\Theta - I\|_F > \epsilon$), `then` $B \leftarrow LUB$ ($B \leftarrow L\Theta B$) `and goto` 2, `else end`

We could use other stoping criteria such as keeping track of J_1 or J_2. The LUJ1D (as well as QRJ1D) algorithm is iterative in the sense that we find the L and U matrices repetitively, and it is sequential in the sense that the problem of finding a triangular matrix minimizing J_1 has been replaced (or approximated) by a finite sequence of one dimensional problems. Note that for updating C_i, the matrix multiplications can be realized by few vector scalings and vector additions. We also mention that, as other Jacobi methods, these methods are suitable for parallel implementation. For parallel implementation we may combine (multiply) all the lower triangular matrices corresponding to the same column and find the minimizing parameters of this new matrix at one shot [2].

2.2 Row Balancing

In practice, if the rows of the large matrix $C = [C_1, \ldots C_N]$ are not balanced in their norms, especially when n and N are large, the value found for a can be inaccurate (see for example (7)). To alleviate this, after every few iterations, we use updates $C_i \leftarrow DC_iD$ and $B \leftarrow DBD$ where D is a diagonal matrix that approximately balances the rows of C. We choose $D(k,k) = \frac{1}{\sqrt{\|C(k,:)\|_F}}$ where $C(k,:)$ is the kth row of C. With this modification, the algorithms perform desirably. As mentioned we could keep track of the values of a cost function (either J_1 or J_2) as a stopping criterion. Since J_1 is not scale invariant and it can change dramatically as a result of row balancing, J_2 is more preferable in this case.

[2] This unit triangular matrix is also known as the Gauss transformation [5].

3 Minimization of J_2 for Joint Diagonalization

Now, we introduce LU and QR based algorithms using Jacobi matrices for minimization of J_2. The inverse of a Jacobi matrix is a linear function of its elements. For example, the inverse of $L_{ij}(a) \in \mathcal{L}$ is $L_{ij}(-a)$. This fact can mitigate the effect of the presence of B^{-1} in J_2. We, again, replace the high dimensional minimization problem with a sequence of one dimensional problems involving parameters of Jacobi matrices in LU or QR factorizations. The difference is that here $J_2(L_{ij}(a))$ is a quadric function of a and in order to minimize it we need to employ either an iterative scheme or the known formulae to find the roots of the cubic polynomial $\frac{\partial J_2(L_{ij}(a))}{\partial a}$. Proposition 3 gives $J_2(L_{lk}(a))$ and $J_2(U_{lk}(a))$ in terms of the elements of C_i's:

Proposition 3. *If $k < l$ then: $J_2(L_{lk}(a)) = a_4a^4 + a_3a^3 + a_2a^2 + a_1a + a_0$ and if $k > l$ then $J_2(U_{lk}(a)) = a_4a^4 + a_3a^3 + a_2a^2 + a_1a + a_0$, where:*

$$a_4 = 4\sum_{i=1}^{N} C_i(k,k)^2, \quad a_3 = 8\sum_{i=1}^{N} C_i(k,k)C_i(k,l), \quad a_2 = 2\sum_{i=1}^{N} C_i(k,k)^2 + 2C_i(k,l)^2$$

and:

$$a_1 = 4\sum_{i=1}^{N} C_i(k,l)C_i(k,k), \quad a_0 = 2\sum_{i=1}^{N} C_i(k,l)^2 \tag{8}$$

As mentioned the corresponding minimization is a straight forward task. Similar to QRJ1D and LUJ1D we can have QRJ2D and LUJ2D algorithms by replacing steps referring to Proposition 1 with steps referring to Proposition 3. As it can be seen from the above formula the value of a in minimization of $J_2(L_{lk}(a))$ depends only on the elements of the matrices $\{C_i\}_{i=1}^{N}$ at positions (k,k) and (k,l). Note that a for minimization of $J_1(L_{lk}(a))$ depends on the elements of $\{C_i\}$ at other positions too. As a result, assuming the computation cost in minimization of $J_2(L_{lk}(a))$ is mainly due to calculating the coefficients, we can see that the complexity of calculating a is of the order $\mathcal{O}(N)$, whereas for $J_1(L_{lk}(a))$ it is of the order $\mathcal{O}(Nn)$. However, the complexity of one iteration (including the costly update of the C_i's) for all the methods is of the order $\mathcal{O}(Nn^3)$. We mention that here also row balancing proves to be useful.

4 Numerical Experiments

We examine the performance of the developed methods by joint diagonalization of a set of matrices that are generated as:

$$C_i = A\Lambda_i A^T + tN_i, \quad \Lambda_i = \text{diag}(\text{randperm}(n))$$

where $\text{diag}(x)$ for a vector x denotes a diagonal matrix whose diagonal is x, $\text{randperm}(n)$ denotes a random permutation of the set $\{1, 2, ..., n\}$, N_i is the symmetric part of a matrix whose elements are i.i.d standard normal random variables and t measures the noise contribution. We try $n = 10$, $N = 100$ with values for $t = 0$ and $t = 0.1$. A is randomly generated. We apply QRJ1D, LUJ1D, QRJ2D and LUJ2D methods [3] with row balancing to find B. The row balancing is performed once per each three iterations. The index:

$$\text{Index}(P) = \sum_{i=1}^{n}(\sum_{j=1}^{n} \frac{|p_{ij}|}{\max_k |p_{ik}|} - 1) + \sum_{j=1}^{n}(\sum_{i=1}^{n} \frac{|p_{ij}|}{\max_k |p_{kj}|} - 1) \tag{9}$$

which measures how far $P = BA$ is from being permuted diagonal is used to measure the performance. Plots (1.a) and (1.b) show the result. Note that for

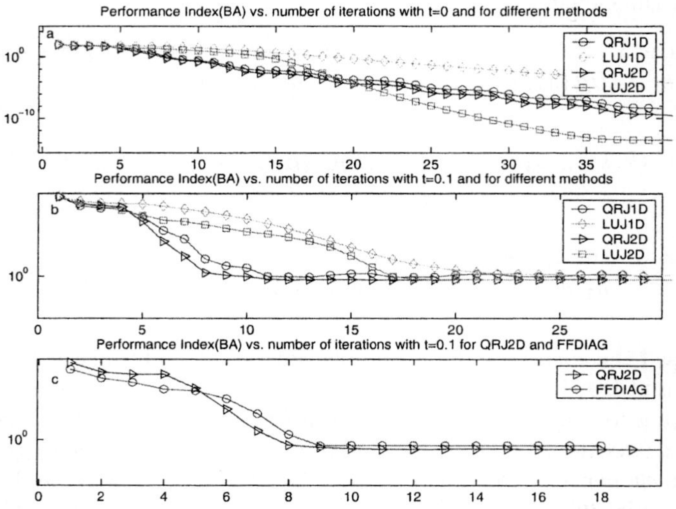

Fig. 1. (a), (b) The performance index Index(BA) for different methods with two noise levels $t = 0$ and $t = 0.1$, respectively. (c) Performance index vs. number of iterations for QRJ2D and FFDIAG with noise level $t = 0.1$.

$t = 0$ the index values are very small. Of course, $t = 0.1$ is a more realistic case for which the convergence is faster. For both $t = 0$ and $t = 0.1$ the QRJ2D and LUJ2D outperform the J_1 based methods. Yet, since in simulations this has not been consistently observed we refrain from any comparison of the methods. In another experiment we compare the QRJ2D method and the FFDIAG [8] algorithm for which the available MATLAB code has been used. With $t = 0.1$ we repeat the previous example and apply both the algorithms. Plot (1.c) shows the index for the two methods. QRJ2D outperform FFDIAG little bit, both in terms of speed and performance. Again, this situation may vary in different experiments. However, we can confirm comparable performance for FFDIAG and the developed methods.

[3] Matlab code is available at http://www.isr.umd.edu/Labs/ISL/ICA2006/

5 Conclusion

We presented simple NOJD algorithms based on the QR and LU factorizations. Using Jacobi matrices we replaced high dimensional minimization problems with a sequence of simple one-dimensional problems. Also a new scale invariant cost function has been introduced and used for developing NOJD algorithms. A comparison with one efficient existing method shows the competence of the developed methods. The idea of resorting to a matrix factorization and solving a sequence of minimization sub-problems over one-parameter subgroups can be useful in other minimization problems over matrix groups.

Acknowledgments

This research was supported in part by Army Research Office under ODDR&E MURI01 Program Grant No. DAAD19-01-1-0465 to the Center for Communicating Networked Control Systems (through Boston University). The author is grateful to Dr. P.S. Krishnaprasad for his support as well as his comments on this paper. The author would also like to thank Dr. U. Helmke for his helpful hints and guidance during his September 2004 visit in College Park. The author is also greatly indebted to the anonymous reviewers for their useful comments on this work.

References

1. B. Afsari, P.S .Krishnaprasad: A Novel Non-orthogonal Joint Diagonalization Cost Function for ICA, ISR technical report, 2005(Available at:`http://techreports.isr.umd.edu/reports/2005/TR_2005-106.pdf`)
2. B. Afsari: Gradient Flow Based Matrix Joint Diagonalization for Independent Componenet Analysis, MS Thesis, ECE Department, University of Maryland, College Park, May 2004.(Available at: `http://techreports.isr.umd.edu/reports/2004/MS_2004-4.pdf`)
3. B. Afsari, P.S. Krishnaprasad: Some Gradient Based Joint Diagonalization Methods for ICA, in C. G.Puntonet and A. Prieto(ed's), Proceedings of ICA 2004, Springer LNCS series, Vol. 3195, pp 437-444, 2004
4. J.F. Cardoso and A. Soulumiac:Blind Beamforming For Non-Gauusian Signals, IEE-Proceedings, Vol.140, No 6, Dec 1993
5. G. H. Golub, C. F. Van Loan: Matrix Computations, third eddition, Johns Hopkins University Press, 1996
6. D.T. Pham and J.F. Cardoso: Blind separation of instantaneous mixtures of non stationary sources, IEEE Trans. Signal Processing, pp 1837-1848, vol 49, no 9, 2001
7. A.Yeredor: Non-Orthogonal Joint Diagonalization in the Least-Squares Sense With Application in Blind Source Separation, IEEE Transactions on Signal Processing, Vol 50, No.7.July 2002.
8. A. Ziehe, M. Kawanabe, S. Harmeling, and K.-R. Mller: A Fast Algorithm for Joint Diagonalization with Non-orthogonal Transformations and its Application to Blind Source Separation. Journal of Machine Learning Research; 5(Jul):801–818, 2004

Separation of Nonlinear Image Mixtures by Denoising Source Separation

Mariana S.C. Almeida[1], Harri Valpola[2], and Jaakko Särelä[3]

[1] Instituto de Telecomunicações,
Av. Rovisco Pais - 1, 1049-001 Lisboa, Portugal
mariana.almeida@lx.it.pt
[2] Laboratory of Computational Engineering,
Helsinki University of Technology,
P.O. Box 9203, FI-02015 HUT, Espoo, Finland
harri.valpola@tkk.fi
http://www.lce.hut.fi/~harri/
[3] Adaptive Informatics Research Centre,
Laboratory of Computer and Information Science,
Helsinki University of Technology,
P.O. Box 5400, FI-02015 HUT, Espoo, Finland
jaakko.sarela@tkk.fi
http://www.cis.hut.fi/jaakkos/

Abstract. The denoising source separation framework is extended to nonlinear separation of image mixtures. MLP networks are used to model the nonlinear unmixing mapping. Learning is guided by a denoising function which uses prior knowledge about the sparsity of the edges in images. The main benefit of the method is that it is simple and computationally efficient. Separation results on a real-world image mixture proved to be comparable to those achieved with MISEP.

1 Introduction

Nonlinear source separation refers to separation of sources from their nonlinear mixtures (for reviews, see [1, 2]). It is much harder than linear source separation because the problem is highly ill-posed. In practice, some type of regularisation is needed. It is, for instance, possible to require that the nonlinear mixing or unmixing mapping is smooth or belongs to a restricted class of nonlinear functions. Alternatively, it is possible to impose restrictions on the extracted sources. In any case, it is important to reduce the number of available degrees of freedom.

Denoising source separation (DSS, [3]) has been introduced as a framework for source separation algorithms, where separation is constructed around denoising procedures. DSS algorithms can range from almost blind to highly tuned separation with detailed prior knowledge. The framework has already been successful in several applications such as biomedical signal processing [3], CDMA signal recovery [4] and climatology [5].

J. Rosca et al. (Eds.): ICA 2006, LNCS 3889, pp. 8–15, 2006.

So far, DSS has been applied to linear separation only, but in this paper we show that nonlinear separation is possible, too. In the DSS framework, it is easy to use detailed prior information. This means that separation becomes possible even if the nonlinear mappings are not carefully regularised. This is a significant benefit because this translates to significant savings in computational complexity, particularly in large problems with many sources and mixtures.

The rest of the paper is organised as follows. The nonlinear DSS method is introduced in Sec. 2. In many respects the separation procedure is exactly like linear separation except that decorrelation and scaling of the sources need to be embedded into the denoising whereas in linear separation this can be implemented by orthogonalising the mixing matrix.

In the rest of the paper, we demonstrate the nonlinear DSS in a real-world nonlinear separation problem introduced by [6]. The problem is to separate two images which have been printed on opposite sides of a paper. Due to partial transparency of the paper, both images are visible from each side, corresponding to two nonlinear mixtures of the source images. In the DSS framework, separation is built around a denoising procedure which can be tailored to the problem at hand. A suitable denoising function which utilises the sparsity of image edges is introduced in Sec. 3 and separation results are reported in Sec. 4.

Finally, in Sec. 5, we discuss the relation of the proposed nonlinear DSS framework with other nonlinear separation methods and also discuss possible future research directions.

2 Nonlinear DSS Method

In DSS, separation consists of the following steps:

1. estimation of the current sources using current mapping,
2. denoising of the sources and
3. adaptation of the mapping to match the denoised sources.

Note that the procedure bears resemblance to the EM algorithm: the first two steps correspond roughly to the E-step and the last step to the M-step. The main difference is that the EM algorithm is a generative approach where the mixing mapping is estimated. With generative models assuming uncorrelated sources, the sources will automatically become approximately uncorrelated due to the so-called explaining-away phenomenon. This needs to be emulated in DSS using some type of competition mechanism (see, e.g., [7] for discussion about emulating explaining away by lateral inhibition).

In linear separation, decorrelation and scaling can be realised by prewhitening the data and orthogonalising and scaling the projection vectors in the last step. In nonlinear DSS, this option is not available as there is, in general, no easy way to make sure that the outputs of a nonlinear mapping are orthogonal and suitably scaled. Instead, the decorrelation and scaling must be embedded in the denoising step. Besides this, the basic principle in nonlinear DSS is exactly the same as in linear DSS.

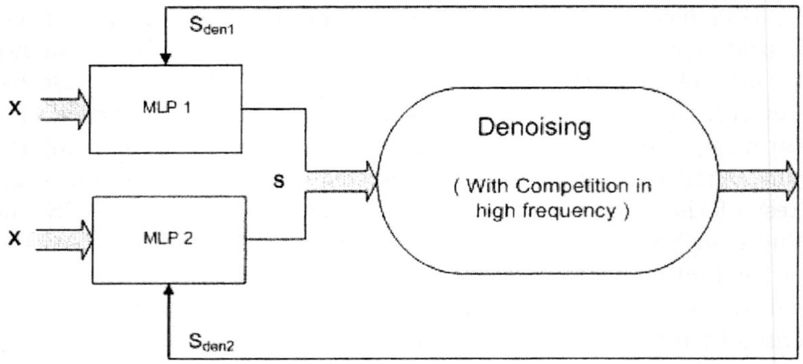

Fig. 1. Schematic representation of the nonlinear DSS method

The method that we have used for nonlinear separation is illustrated in Fig. 1, for the case of separation of a two-source mixture.

The principle of operation is as follows. The mixture vector $\boldsymbol{X}$ is fed into two multilayer perceptrons (MLP1 and MLP2), which yield the current estimates of the sources $\boldsymbol{S}$ as their outputs (step 1). The estimates are then denoised (step 2). Finally the MLP networks are adapted to the denoised source estimates $\boldsymbol{S}_{den1}$ and $\boldsymbol{S}_{den2}$ (step 3). Provided that the denoising step is well chosen, iterating these three steps will result in the separation of the mixed sources.

3 Denoising for Image Separation

The crucial element in DSS, with linear or nonlinear mapping, is the choice of the denoising function. A lengthy discussion of denoising functions and their properties can be found in [3]. In brief, removing noise helps identify the signal subspace and removing the interference from other sources promotes separation. In this paper, we focus on the case where there is an equal number of sources and mixtures. Therefore, the most important thing is to reduce the interference from other sources.

3.1 Mixing Process

The image mixtures that were studied correspond to a well known practical situation: when an image of a paper document is acquired, the back page sometimes shows through. The paper that was used was onion skin, which leads to a strong mixture, which is significantly nonlinear. This separation problem has first been introduced by [6]. We show the effectiveness of the proposed DSS method using the first, the second and the fifth mixtures from that paper. The source images, which were printed on the onion skin paper, are shown in Fig. 3a. The acquired images (mixtures) are shown in Fig. 3b. For more detailed description of the data aquisition, see [6].

3.2 Edge Denoising

Looking at the last two pair of mixtures in Fig. 3b, it is evident that despite strong nonlinear mixtures, a human being can easily separate the images, i.e., can tell which features or objects belong in which image, even without knowing the original images. What features could be used for separating, i.e., denoising, the images?

A characteristic feature of most natural images is the sparsity of edges. When an edge is found in the same place in both mixtures, it probably originates from only one of the source images. Decision about which source the edge belongs to can be based on the relative strength of the edges in the mixtures. Hence, we suggest the following denoising scheme:

1. Represent each of the source estimates by their edges.
2. Induce a competition between the edges in different images in such a way that stronger edges tend to eliminate weaker ones.

Note that the edge features in different natural images are usually almost independent, which is not necessarily true for low-frequency features. Consider for instance natural images of faces.

Edge detection in images. A crude approach for edge detection that already leads to somewhat acceptable results, is to use simple high-pass filtering to extract the edges. Another, more advanced possibility is to use wavelet analysis. We decided to use a wavelet family that forms a spatio-frequency representation of an image separately with horizontal, vertical and diagonal components (H, V and D). The representation results in a hierarchy of increasing frequencies. A schematic illustration of the wavelet transform that was used, is depicted on the left side of Fig. 2.

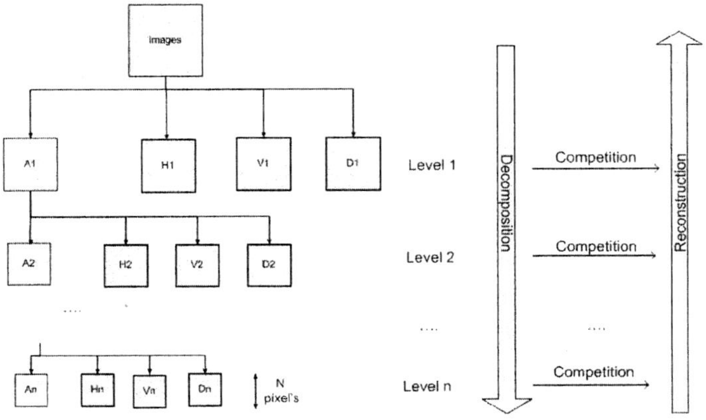

Fig. 2. Diagram of the wavelet-based denoising operation

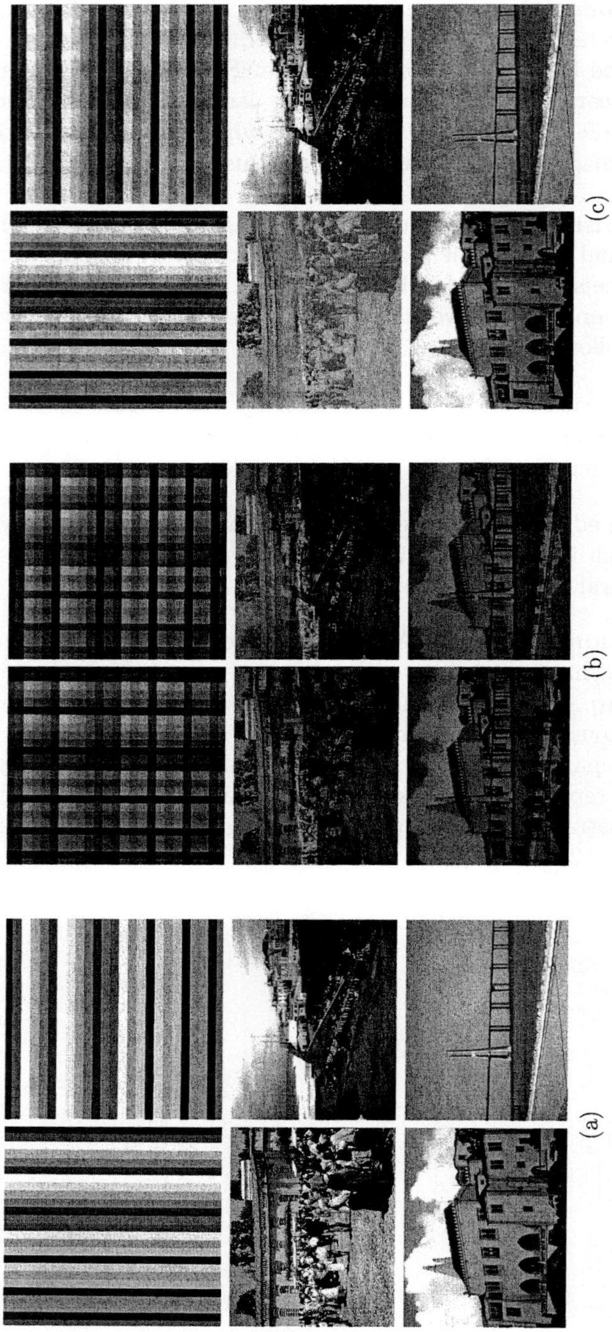

Fig. 3. a) Source images b) mixture (acquired) images c) separation results

Competition between the edges. Once the edges of both source estimates have been extracted, one should decide which edge belongs to which image. On average, edges of the foreground image appear stronger on the foreground mixture. Hence strong edges on the foreground images should be privileged for the foreground source estimate. This has been achieved by using a soft winner-take-all operation, which assigned most of the energy to the stronger component. The competition was induced in each level of the wavelet transform, except for the first one that represents the slowest frequencies (see the right side of Fig. 2).

Additionally, the artificial nature of the first pair of mixtures (Fig. 3c, top-row), was taken into account. Since one of the source images contained only vertical and the other one only horizontal edges, the horizontal (H) components were set to zero in one of the images, prior to reconstruction, and the vertical (V) ones in the other image.

4 Results

The multilayer perceptrons that were used had a hidden layer with five sigmoidal units. They also had direct "shortcut" connections between inputs and output, and their output units were linear. With this structure they were able to implement linear operations. These perceptrons were initialised to perform an approximate linear whitening (also called sphering) of the mixture data, subject to the restriction of being symmetrical (processing the two input components equally). Training was performed with the adaptive step sizes speedup method [8]. Fifty training epochs were performed, within each iteration of the global nonlinear DSS procedure. Two-level description was used in the wavelet decomposition.

Figure 3c shows the results obtained after 10 iterations of the nonlinear DSS. For comparison, the results obtained with the MISEP technique of nonlinear ICA can be consulted in [6].

For an objective quality assessment, the four quality measures defined in [6] were computed. Q_1 is simply the signal-to-noise ratio (SNR). Q_2 is also an SNR measure, but with a correction for possible nonlinear distortions of the intensity scale of the separated images. Q_3 is the mutual information between each separated component and the corresponding source. Finally, Q_4 is the mutual information between each separated component and the opposite source. For Q_1, Q_2 and Q_3, higher values are better, while for Q_4 lower values are better. See [6] for more details. Table 1 shows the results, together with the results obtained with the MISEP method, for comparison (the latter were obtained from [6]).

In the first pair, nonlinear DSS performed better than MISEP. This is probably due to the specific denoising operation that was used, which is very well suited to this pair of sources. In the second image pair, nonlinear DSS and MISEP performed approximately equally on the right-hand image, and MISEP performed better on the left-hand image. In the third pair, nonlinear DSS performed globally better. This pair of sources is not independent (see [6]), and therefore nonlinear DSS is probably more suited to handle it than MISEP, which is an independence based method.

Table 1. Quality measures. For each pair (Nonlinear DSS and MISEP, for the same source), the best result is shown in bold. For Q_1, Q_2 and Q_3 higher results are better, while for Q_4 lower results are better.

		Nonlinear DSS		MISEP	
Image pair	**Quality measure**	source 1	source 2	source 1	source 2
1	Q_1 (dB)	**14.6**	**14.1**	13.8	13.1
	Q_2 (dB)	**15.3**	**14.7**	14.7	14.2
	Q_3 (bit)	**2.57**	**2.50**	2.45	2.39
	Q_4 (bit)	0.29	0.27	**0.23**	**0.26**
2	Q_1 (dB)	6.4	13.6	**9.3**	**13.9**
	Q_2 (dB)	9.5	**15.1**	**11.0**	15.0
	Q_3 (bit)	1.62	1.93	**1.83**	**1.95**
	Q_4 (bit)	0.44	**0.39**	**0.24**	0.40
3	Q_1 (dB)	13.5	**9.2**	**14.2**	6.4
	Q_2 (dB)	**15.5**	**9.9**	15.3	7.8
	Q_3 (bit)	**2.23**	**1.62**	2.19	1.29
	Q_4 (bit)	0.74	0.56	**0.56**	**0.49**

5 Discussion

In this paper, we reported the first results about nonlinear separation with DSS. As the results show, separation was relatively successful but still far from perfect. For instance, from the extracted image pair in the middle of Fig. 3c, it is evident that the contrast on the image on the left depends on the intensity of the image on the right (lighter on the right implies better contrast on the left). Furthermore, we had to resort to early stopping in the separation of the mixtures of natural images. Such problems could be avoided by improving the denoising function, for example by introducing a local normalisation of image contrast, or by using more prior information about the mixing process to restrict the parametric form of the unmixing mapping.

Of the existing nonlinear separation techniques, MISEP is similar to the one proposed here in that it, too, estimates a separating MLP network. The main advantage over MISEP is that the learning procedure is simpler and computationally more efficient. In MISEP, the Jacobian matrix of the nonlinear mapping needs to be computed for every sample, inverted and then propagated back through the MLP network. For two-dimensional case this is not of importance and MISEP was actually faster in these simulations. However, it means that MISEP cannot be extended to problems with a large number of sources.

Slow-feature analysis (SFA, [9]) resembles nonlinear DSS in its use of denoising for guiding separation. In SFA, the denoising is implemented by low-pass filtering (see [3] for details) and therefore assumes that the sources have slowly changing temporal or spatial structure. In DSS, the denoising can be more general and tuned to the problem at hand, such as the presented edge-based denoising for separating images. Interestingly, SFA has been shown to be applicable to very large problems when the set of nonlinearities is fixed and only a linear mapping is learned [9]. It should therefore be possible to apply nonlinear DSS in very large problems using a similar restricted mapping.

6 Conclusion

We have presented a nonlinear separation method based on the denoising source separation framework. The method uses a competition-based denoising stage which performs a partial separation of the sources, the partially separated components being used to iteratively re-train a set of nonlinear separators. The method was applied to real-life nonlinear mixtures of images, and proved to be competitive with ICA-based nonlinear separation.

Acknowledgements

We are grateful to L. B. Almeida for helpful comments on this manuscript and for providing the images and related information.

References

1. Jutten, C., Karhunen, J.: Advances in blind source separation (BSS) and independent component analysis (ICA) for nonlinear mixtures. International Journal of Neural Systems **14**(5) (2004) 267–292
2. Almeida, L.: Nonlinear Source Separation. Morgan and Claypool (2005) in press.
3. Särelä, J., Valpola, H.: Denoising source separation. Journal of Machine Learning Research **6** (2005) 233–272
4. Raju, K., Särelä, J.: A denoising source separation based approach to interference cancellation for DS-CDMA array systems. In: Proceedings of the 38th Asilomar Conference on Signals, Systems, and Computers, Pacific grove, CA, USA (2004) 1111 – 1114
5. Ilin, A., Valpola, H., Oja., E.: Semiblind source separation of climate data detects El Niño as the component with the highest interannual variability. In: Proceedings of the International Joint Conference on Neural Networks (IJCNN 2005), Montréal, Québec, Canada (2005) 1722 – 1727
6. Almeida, L.: Separating a real-life nonlinear image mixture. Journal of Machine Learning Research **6** (2005) 1199–1229
7. Ghahramani, Z., Hinton, G.E.: Hierarchical non-linear factor analysis and topographic maps. In Jordan, M.I., Kearns, M.J., Solla, S.A., eds.: Advances in Neural Information Processing Systems 10. The MIT Press, Cambridge, MA, USA (1998) 486–492
8. Almeida, L.: Multilayer perceptrons. In Fiesler, E., Beale, R., eds.: Handbook of Neural Computation, Institute of Physics, Oxford University Press (1997)
9. Wiskott, L., Sejnowski, T.: Slow feature analysis: Unsupervised learning of invariances. Neural Computation **14** (2002) 715 – 770

Second-Order Separation of Multidimensional Sources with Constrained Mixing System

Jörn Anemüller

Medical Physics Section,
Dept. of Physics, University of Oldenburg,
26111 Oldenburg, Germany

Abstract. The case of sources that generate multidimensional signals, filling a subspace of dimensionality K, is considered. Different coordinate axes of the subspace ("subspace channels") correspond to different signal portions generated by each source, e.g., data from different spectral bands or different modalities may be assigned to different subspace channels. The mixing system that generates observed signals from the underlying sources is modeled as superimposing within each subspace channel the contributions of the different sources. This mixing system is constrained as it allows *no* mixing of data that occurs in *different* subspace channels. An algorithm based on second order statistics is given which leads to a solution in closed form for the separating system. Correlations across different subspace channels are utilized by the algorithm, whereas properties such as higher-order statistics or spectral characteristics within subspace channels are not considered. A permutation problem of aligning different sources' subspace channels is solved based on ordering of eigenvalues derived from the separating system. Effectiveness of the algorithm is demonstrated by application to multidimensional temporally i.i.d. Gaussian signals.

1 Introduction

The notion of multi-dimensional or subspace ICA has been developed in [3] and [6] to account for the fact that not all sources may reasonably be modeled as one-dimensional processes with mutual independence. Rather, some sources may generate signals that fill a multi-dimensional subspace that resists decomposition into one-dimensional mutually independent sources. This can occur both in situations where underlying sources are unknown and rather a plausible model of the observed data is sought for, and in situations where analytical reasons dictate a multi-dimensional character of the sources, such as separation of spectral domain speech, which has originally motivated this work.

The present work suggests a second-order approach to the separation of multidimensional sources and considers a constrained version of the general linear mixing system.

2 Multidimensional Sources and Constrained Mixing

The multidimensional signal generated by source i ($i = 1, \ldots, N$) is denoted by $s_i^f(t)$. The source is regarded as stationary and ergodic with respect to parameter t, $t =$

J. Rosca et al. (Eds.): ICA 2006, LNCS 3889, pp. 16–23, 2006.

$t_1, \ldots, t_L$, i.e., expectation values can be estimated as sample means w.r.t. t. E.g., t may denote time or spatial position in an image, provided stationarity may be assumed w.r.t. these variables. Without loss of generality, we limit our treatment to zero mean sources.

Index f spans the subspace of dimensionality K that is filled by the source, with $f = 1, \ldots, K$ denoting the "subspace channels" (or "channels") of the source. Source statistics w.r.t. individual channels $f \neq f'$ may differ, hence, expectations cannot be computed as sample averages across f. Subspace channels, e.g., may correspond to different frequency bands for which data of an audio source or multispectral image data has been collected. More generally, subspace channels may also correspond to different modalities of recorded data, e.g., audio and video data may be stored in different channels. Another example would be data with non-constant mixing, e.g., image data with position-dependent mixing parameters; here, different subspace channels could correspond to different spatial positions.

As they are generated by the same source, data in different subspace channels $f \neq f'$ of a single source may in general covary,

$$E\left\{s_i^f(t)\,(s_i^{f'}(t))^*\right\} \neq 0. \tag{1}$$

Data from different sources $i \neq j$ must be uncorrelated since the sources are assumed to be independent systems. If correlations are computed from source components at two subspace channels, the result is zero for all pairs of channels (f, f'),

$$E\left\{s_i^f(t)\,(s_j^{f'}(t))^*\right\} = 0 \qquad \forall i \neq j,\, \forall f, f'. \tag{2}$$

Mixing of sources is assumed to be separable in the sense that the mixing system only mixes data from corresponding subspace channels of different sources, but does *not* mix data "across" different channels. Gathering data from the f-th subspace channel of all N sources into a single vector $\mathbf{s}^f(t) = [s_1^f(t), \ldots, s_N^f(t)]^T$, mixing is written as

$$x_i^f(t) = \sum_{j=1}^{N} a_{ij}^f\, s_j^f(t) \qquad \Longleftrightarrow \qquad \mathbf{x}^f(t) = \mathbf{A}^f\, \mathbf{s}^f(t) \tag{3}$$

This model is compatible with the mixing scenarios in the examples mentioned above. E.g., multimodal data may plausibly be explained by superposition of basis-patterns within each modality. A-priori, intermingling of data from different subspace channels may be regarded as a less significant process and may be ruled out completely in some applications (e.g., frequency-domain separation of convolutive audio mixtures) on the grounds of known physics.

From knowledge of the mixed signals $\mathbf{x}^f(t)$, only, it is aimed to find an estimate $\hat{\mathbf{A}}^f$ of the mixing matrix so that unmixed signals

$$\mathbf{u}^f(t) = [\hat{\mathbf{A}}^f]^{-1}\, \mathbf{x}^f(t) \tag{4}$$

can be obtained which resemble the source signals.

The simplest approach to solve system (3) would be to perform ICA or second-order source separation separately for each subspace channel f. For two reasons this approach might not be optimal. First, by neglecting information present in the across-channel correlations (Eq.s 1, 2), the obtained separation quality may not be optimal, e.g., because the data in individual channels might be lacking sufficient higher-order or spectral cues. In the evaluation (Sec. 4), we demonstrate that across-channel correlations make it possible to separate data that have no spectral, higher-order or non-stationarity cues.

Second, care has to be taken to reconstruct coherent subspaces, each pertaining to one source process. When treating Eq. 3 as N individual source separation problems, the permutation invariance inherent to blind source separation algorithm applies individually to each one, so that gathering the subspace channels for each source process is a non-trivial issue. A similar problem is frequently encountered in frequency-domain approaches to convolutive blind source separation. It is shown that across-channel correlations as in Eq.s 1, 2 can be exploited to this end.

3 Solution Based on Correlations Across Subspace Channels

Defining the sources' cross-covariance matrix $\mathbf{R}_s^{ff'}$ computed from channels f and f' as

$$\mathbf{R}_s^{ff'} = E\left\{\mathbf{s}^f(t)\,(\mathbf{s}^{f'}(t))^H\right\}, \tag{5}$$

equations (2) and (1) can be restated such that $\mathbf{R}_s^{ff'}$ is diagonal for all (ff'),

$$\left[\mathbf{R}_s^{ff'}\right]_{ij} = \delta_{ij} E\left\{s_i^f(t)\,(s_i^{f'}(t))^*\right\}, \tag{6}$$

where δ_{ij} is the Kronecker symbol.

Since the mixed signals are not independent, their covariance matrix $\mathbf{R}_x^{ff'}$,

$$\mathbf{R}_x^{ff'} = E\left\{\mathbf{x}^f(t)\,(\mathbf{x}^{f'}(t))^H\right\}, \tag{7}$$

is not diagonal. It can be expressed in terms of the sources' covariance matrix as

$$\mathbf{R}_x^{ff'} = \mathbf{A}^f\,\mathbf{R}_s^{ff'}\left(\mathbf{A}^{f'}\right)^H. \tag{8}$$

If the mixing system was identical in both subspace channels, $\mathbf{A}^f = \mathbf{A}^{f'}$, then an eigenvalue equation could be derived in exactly the same manner as presented by [7]. However, since in general $\mathbf{A}^f \neq \mathbf{A}^{f'}$, the analog derivation is not possible.

It is observed that by forming the products

$$\mathbf{Q}_s^{ff'} = \mathbf{R}_s^{ff'}\,[\mathbf{R}_s^{f'f'}]^{-1}\,\mathbf{R}_s^{f'f} \tag{9}$$

$$\mathbf{Q}_x^{ff'} = \mathbf{R}_x^{ff'}\,[\mathbf{R}_x^{f'f'}]^{-1}\,\mathbf{R}_x^{f'f} \tag{10}$$

the algebraic relation between the sources' $\mathbf{Q}_s^{ff'}$ and the mixed signals' $\mathbf{Q}_x^{ff'}$ involves matrix $\mathbf{A}^f$, but not $\mathbf{A}^{f'}$,

$$\mathbf{Q}_s^{ff'} = [\mathbf{A}^f]^{-1}\,\mathbf{Q}_x^{ff'}\,[\mathbf{A}^f]^{-H} \tag{11}$$

Hence, $[\mathbf{A}^f]^{-1}$ diagonalizes $\mathbf{Q}_x^{ff'}$ for all f'.

An eigenvalue equation for $\mathbf{A}^f$ can be derived from (11) by forming the product

$$\mathbf{Q}_x^{ff'}\,[\mathbf{Q}_x^{ff}]^{-1}, \tag{12}$$

yielding

$$\mathbf{A}^f\,\mathbf{\Lambda}^{ff'} = \mathbf{Q}_x^{ff'}\,[\mathbf{Q}_x^{ff}]^{-1}\,\mathbf{A}^f, \tag{13}$$

where

$$\mathbf{\Lambda}^{ff'} = \mathbf{Q}_s^{ff'}\,[\mathbf{Q}_s^{ff}]^{-1} \tag{14}$$

is diagonal and contains the eigenvalues of $\mathbf{Q}_x^{ff'}\,[\mathbf{Q}_x^{ff}]^{-1}$.

Similarly, $\mathbf{A}^{f'}$ is obtained from the Eigenvalue equation

$$\mathbf{A}^{f'}\,\mathbf{\Lambda}^{f'f} = \mathbf{Q}_x^{f'f}\,[\mathbf{Q}_x^{f'f'}]^{-1}\,\mathbf{A}^{f'}. \tag{15}$$

3.1 Conditions for Identifiability

Equation (13) has a unique solution if all eigenvalues on the diagonal of $\mathbf{\Lambda}^{ff'}$ are different. Similarly, for (15) it must hold that the diagonal elements of $\mathbf{\Lambda}^{f'f}$ are different. Since $\mathbf{R}_s^{ff'}$ is diagonal and $\mathbf{R}_s^{ff'} = [\mathbf{R}_s^{f'f}]^H$, we obtain

$$\begin{aligned}\mathbf{\Lambda}^{ff'} &= \mathbf{\Lambda}^{f'f} = \\ &\mathbf{R}_s^{ff'}\,[\mathbf{R}_s^{ff'}]^H\,[\mathbf{R}_s^{ff}]^{-1}\,[\mathbf{R}_s^{f'f'}]^{-1}.\end{aligned} \tag{16}$$

Hence, together with (6) it follows that for $\mathbf{A}^f$ and $\mathbf{A}^{f'}$ to be identifiable it must be fulfilled that $\forall i \neq j$

$$\frac{\left|E\{s_i^f(t)\,(s_i^{f'}(t))^*\}\right|^2}{E\{\left|s_i^f(t)\right|^2\}\,E\{\left|s_i^{f'}(t)\right|^2\}} \neq \frac{\left|E\{s_j^f(t)\,(s_j^{f'}(t))^*\}\right|^2}{E\{\left|s_j^f(t)\right|^2\}\,E\{\left|s_j^{f'}(t)\right|^2\}}. \tag{17}$$

3.2 Solving the Permutation Problem

Since the eigenvectors corresponding to the solution of (13) are unambiguous only upto their order and a scale factor, the mixing matrix $\mathbf{A}^f$ cannot be determined uniquely. Rather, any matrix $\tilde{\mathbf{A}}^f$ which can be expressed as

$$\tilde{\mathbf{A}}^f = \mathbf{A}^f\,\mathbf{D}^f\,\mathbf{P}^f, \tag{18}$$

where $\mathbf{D}^f$ is a diagonal matrix and $\mathbf{P}^f$ a permutation matrix, represents a solution of (13). Hence, it is only possible to determine $\mathbf{A}^f$ upto an unknown rescaling and permutation of its columns by $\mathbf{D}^f$ and $\mathbf{P}^f$, respectively. This corresponds to the well-known invariances inherent to all blind source separation algorithms.

For one-dimensional source signals this is usually not a problem. With multidimensional sources, the components belonging to a single source are reconstructed with disparate (unknown) order and scale in different subspace channels $f \neq f'$ if the corresponding channel-specific permutation and diagonal matrices differ, i.e.,

$$\mathbf{P}^f \neq \mathbf{P}^{f'} \qquad \mathbf{D}^f \neq \mathbf{D}^{f'}. \tag{19}$$

Thus, a coherent picture of each source's activity cannot be obtained.

No solution is given for the invariance with respect to varied scaling in different channels. Instead, each row of the estimated unmixing matrix $[\hat{\mathbf{A}}^f]^{-1}$ is rescaled to have unit norm.

The solution to the permutation problem is based on the observation that transformation (18) results in rearranged eigenvalues $\tilde{\mathbf{\Lambda}}^{ff'}$,

$$\tilde{\mathbf{\Lambda}}^{ff'} = [\mathbf{P}^f]^T \mathbf{\Lambda}^{ff'} \mathbf{P}^f. \tag{20}$$

That is, the column permutation of $\mathbf{A}^f$ results in a corresponding permutation of the eigenvalues' order on the diagonal of $\tilde{\mathbf{\Lambda}}^{ff'}$.

Denote by $\hat{\mathbf{A}}^f$ and $\hat{\mathbf{A}}^{f'}$ the estimates of the true mixing matrices $\mathbf{A}^f$ and $\mathbf{A}^{f'}$, respectively. Without loss of generality, we assume

$$\hat{\mathbf{A}}^f = \mathbf{A}^f \qquad \hat{\mathbf{A}}^{f'} = \mathbf{A}^{f'}\,\mathbf{P}, \tag{21}$$

so that the estimates $\hat{\mathbf{\Lambda}}^{ff'}$ and $\hat{\mathbf{\Lambda}}^{f'f}$ of the true eigenvalue matrices $\mathbf{\Lambda}^{ff'}$ and $\mathbf{\Lambda}^{f'f}$, respectively, are

$$\hat{\mathbf{\Lambda}}^{ff'} = \mathbf{\Lambda}^{ff'} \tag{22}$$

$$\hat{\mathbf{\Lambda}}^{f'f} = \mathbf{P}^T \mathbf{\Lambda}^{f'f} \mathbf{P} \tag{23}$$

Since, according to (16) we have $\mathbf{\Lambda}^{f'f} = \mathbf{\Lambda}^{ff'}$, it follows

$$\hat{\mathbf{\Lambda}}^{f'f} = \mathbf{P}^T \mathbf{\Lambda}^{ff'} \mathbf{P} = \mathbf{P}^T \hat{\mathbf{\Lambda}}^{ff'} \mathbf{P}. \tag{24}$$

Therefore, the permutation matrix $\mathbf{P}$ can be directly read from the relative ordering of the eigenvalues on the diagonals of $\hat{\mathbf{\Lambda}}^{ff'}$ and $\hat{\mathbf{\Lambda}}^{f'f}$. Permutations are corrected by replacing $\hat{\mathbf{A}}^{f'}$ by $\hat{\mathbf{A}}^{f'}\,\mathbf{P}^T$ whose columns are ordered in accordance with $\hat{\mathbf{A}}^f$.

3.3 More Than Two Subspace Channels

Separation. If channels $f = 1, \ldots, K$, $K \geq 2$, are to be used for separation, the mixing matrix $\mathbf{A}^f$ is obtained as the matrix which simultaneously solves the K diagonalization equations

$$\begin{aligned}
\mathbf{Q}_s^{f,1} &= [\mathbf{A}^f]^{-1}\,\mathbf{Q}_x^{f,1}\,[\mathbf{A}^f]^{-H} \\
\mathbf{Q}_s^{f,2} &= [\mathbf{A}^f]^{-1}\,\mathbf{Q}_x^{f,2}\,[\mathbf{A}^f]^{-H} \\
&\ \ \vdots \\
\mathbf{Q}_s^{f,K} &= [\mathbf{A}^f]^{-1}\,\mathbf{Q}_x^{f,K}\,[\mathbf{A}^f]^{-H}.
\end{aligned} \tag{25}$$

The solution can be obtained by using numerical techniques for simultaneous diagonalization [4].

Identifiability. Equations (25) have a unique solution (up to rescaling and permutation) if, analogous to Eq. (17), for each $f = 1, \ldots, K$ there exists at least one subspace channel f' for which it is fulfilled that $\forall i \neq j$

$$\frac{\left|E\{s_i^f(t)\,(s_i^{f'}(t))^*\}\right|^2}{E\{\left|s_i^f(t)\right|^2\}\,E\{\left|s_i^{f'}(t)\right|^2\}} \neq \frac{\left|E\{s_j^f(t)\,(s_j^{f'}(t))^*\}\right|^2}{E\{\left|s_j^f(t)\right|^2\}\,E\{\left|s_j^{f'}(t)\right|^2\}}. \tag{26}$$

Permutations. The permutations must be sorted for each pair of subspace channels (f, f') by using the method outlined in section 3.2.

4 Evaluation

A synthetic data set of Gaussian i.i.d. noise in two channels is separated. Since the data in each subspace channel is purely Gaussian, these data cannot be separated by looking at a single channel only.

The data consisted of four sources $s_1^f(t), \ldots, s_4^f(t)$, each containing a two-dimensional subspace with channels, $f = 1, 2$, and time-points $t = 1, \ldots, 10000$. Within each subspace channel of each source, the data was chosen to be i.i.d. noise with Gaussian distribution. To enable separation by the proposed algorithm, correlations were introduced between the data in different channels of each source by composing the signals as the sum

$$s_i^f(t) = \xi_i^f(t) + \zeta_i(t) \tag{27}$$

of channel-dependent and channel-independent Gaussian random variables $\xi_i^f(t)$ and $\zeta_i(t)$, respectively.

Since the data within each subspace channel contained neither cues related to higher-order statistics, nor cues related to auto-correlation information or non-stationarity, it is inseparable for any algorithm looking at isolated channels. Only integrating information across different channels makes separation feasible.

The correlations within each source and the independence of the different sources are reflected by the covariance matrices $\mathbf{R}_s^{ff'}$,

$$\mathbf{R}_s^{1,1} = \begin{pmatrix} 1.99 & 0.00 & 0.00 & 0.00 \\ 0.00 & 0.89 & 0.00 & 0.00 \\ 0.00 & 0.00 & 0.20 & 0.00 \\ 0.00 & 0.00 & 0.00 & 0.04 \end{pmatrix} \quad \mathbf{R}_s^{1,2} = \begin{pmatrix} 1.00 & 0.00 & 0.00 & 0.00 \\ 0.00 & 0.64 & 0.00 & 0.00 \\ 0.00 & 0.00 & 0.16 & 0.00 \\ 0.00 & 0.00 & 0.00 & 0.04 \end{pmatrix} \tag{28}$$

$$\mathbf{R}_s^{2,1} = \begin{pmatrix} 1.00 & 0.00 & 0.00 & 0.00 \\ 0.00 & 0.64 & 0.00 & 0.00 \\ 0.00 & 0.00 & 0.16 & 0.00 \\ 0.00 & 0.00 & 0.00 & 0.04 \end{pmatrix} \quad \mathbf{R}_s^{2,2} = \begin{pmatrix} 2.00 & 0.00 & 0.00 & 0.00 \\ 0.00 & 0.89 & 0.00 & 0.00 \\ 0.00 & 0.00 & 0.20 & 0.00 \\ 0.00 & 0.00 & 0.00 & 0.04 \end{pmatrix}. \tag{29}$$

Since the different sources are independent, the off-diagonal terms of all covariance matrices are zero. The diagonals of $\mathbf{R}_s^{1,2}$ and $\mathbf{R}_s^{2,1}$ are non-zero due to the correlations across channels within each source subspace.

The eigenvalues of equation (16) are computed as

$$\operatorname{diag}\mathbf{\Lambda}^{1,2} = (\Lambda_1^{1,2}, \ldots, \Lambda_4^{1,2}) = (0.25, 0.51, 0.64, 1.00). \tag{30}$$

Since all eigenvalues are different, the condition for identifiability (17) is fulfilled.

The 4×4 mixing matrices $\mathbf{A}^1$ and $\mathbf{A}^2$ were chosen at random. Covariance matrices of the mixed signals were processed by the proposed algorithm using the eigenvalue method, yielding the combined mixing-unmixing system $[\hat{\mathbf{A}}^f]^{-1}\mathbf{A}^f$

$$[\hat{\mathbf{A}}^1]^{-1}\mathbf{A}^1 = \begin{pmatrix} \mathbf{-3.11} & -0.03 & 0.01 & -0.02 \\ 0.00 & 0.01 & \mathbf{-2.43} & -0.02 \\ -0.01 & \mathbf{2.39} & 0.04 & -0.01 \\ 0.00 & 0.00 & 0.00 & \mathbf{2.23} \end{pmatrix} \quad [\hat{\mathbf{A}}^2]^{-1}\mathbf{A}^2 = \begin{pmatrix} 0.00 & 0.00 & 0.00 & \mathbf{-1.50} \\ \mathbf{1.69} & 0.01 & -0.03 & -0.02 \\ -0.01 & \mathbf{1.49} & 0.02 & 0.02 \\ 0.00 & 0.00 & \mathbf{-1.43} & -0.01 \end{pmatrix}$$

Since each row of the combined system contains only one significant non-zero element, the algorithm has successfully separated the signals. The increase in signal-to-interference ratio from before to after separation amounts to 37.8 dB.

Sources' components are reconstructed in a different order in the two frequency channels, as can be seen from the different positions of the non-zero elements of $(\hat{\mathbf{A}}^{-1}\mathbf{A})(1)$ and $(\hat{\mathbf{A}}^{-1}\mathbf{A})(2)$. Therefore, the method for sorting permutations described in section 3.2 must be employed. To this end, the estimated eigenvalue matrices $\hat{\mathbf{\Lambda}}(1,2)$ and $\hat{\mathbf{\Lambda}}(2,1)$ obtained from solving the eigenvalue problems (13) and (15), respectively, are

$$\hat{\mathbf{\Lambda}}(1,2) = \begin{pmatrix} \mathbf{0.25} & 0.00 & 0.00 & 0.00 \\ 0.00 & \mathbf{0.64} & 0.00 & 0.00 \\ 0.00 & 0.00 & \mathbf{0.51} & 0.00 \\ 0.00 & 0.00 & 0.00 & \mathbf{1.00} \end{pmatrix} \quad \hat{\mathbf{\Lambda}}(2,1) = \begin{pmatrix} \mathbf{1.00} & 0.00 & 0.00 & 0.00 \\ 0.00 & \mathbf{0.25} & 0.00 & 0.00 \\ 0.00 & 0.00 & \mathbf{0.51} & 0.00 \\ 0.00 & 0.00 & 0.00 & \mathbf{0.64} \end{pmatrix}. \tag{31}$$

By permuting the eigenvalues on the diagonals of $\hat{\mathbf{\Lambda}}(1,2)$ and $\hat{\mathbf{\Lambda}}(2,1)$ to occur in the same order in both matrices, and by performing the same permutations for the rows of $\hat{\mathbf{A}}^{-1}(1)$ and $\hat{\mathbf{A}}^{-1}(2)$, respectively, it is ensured that the sources' components are reconstructed in the same order in both frequencies.

5 Discussion

We have proposed a solution to the BSS problem when sources generate subspaces with second-order dependencies within each source subspace. Under the assumption of a constrained mixing system, that can be separated into one linear instantaneous mixing system per subspace channel, an eigenvalue/joint diagonalization based approach has been developed for source identification and correct assignment of subspace dimensions across different sources.

Under additional assumptions, existing second-order separation methods are recovered as special cases of our method. If different subspace channels are derived from underlying one-dimensional sources by temporal shifting, approaches like SOBI [2], Molgedey-Schuster [7] and TDSEP [8] are recovered. In this case, the signal $s_i^f(t)$ would be constructed from a one-dimensional signal $s_i(t)$ as $s_i^f(t) = s_i(t + \tau^f)$, for time-shifts $\tau^1, \ldots, \tau^K$, and constant mixing matrices $\mathbf{A}^f = \mathbf{A}$ would be assumed.

In the two-input-two-output (TITO) case with a whitening preprocessing step, the separation equations of our algorithm boil down to the TITO identification of FIR channels proposed by [5] (while the permutation alignment step of both algorithms remains different).

It is straight-forward to combine the techniques outlined here with standard second-order separation techniques that can employ spectral cues or non-stationarity of variance within each source subspace channel. Such a combination approach yields a large number of equations for simultaneous diagonalization that are expected to lead to decent signal separation.

The developed method may be useful in two applications. For the separation of data with multiple spectral bands, e.g., spectrogram sound data or spectral image data, correlations across different frequency-channels constitute a criterion for source separation that can be used on its own, or in addition to existing methods of decorrelation with respect to time- or spatial shifts. By using this additional source of information, it should be possible to improve on the performance of source separation algorithms in a similar way as, e.g., decorrelation with multiple time-delays can improve over decorrelation with only a single time-delay.

Concerning separation of time-varying mixtures, present approaches average over short time segments to estimate the averaged unmixing system. The presented method may improve the quality of separation since it allows to estimate the unmixing system for time t taking into account data from time $t + \tau$ even though the unmixing system at both times is different, and without necessarily averaging over the entire temporal range $t \dots t + \tau$.

References

1. J. Anemüller. *Across-Frequency Processing in Convolutive Blind Source Separation.* PhD thesis, Dept. of Physics, University of Oldenburg, Oldenburg, Germany, 2001.
2. A. Belouchrani, K. Abed-Meraim, J. F. Cardoso, and E. Moulines. A blind source separation technique using second order statistics. *IEEE Transactions on Speech and Audio Processing*, 45(2):434–444, 1997.
3. Jean-Francois Cardoso. Multidimensional independent component analysis. In *Proceedings of the IEEE International Conference on Acoustics, Speech, and Signal Processing*, Seattle, USA, 1998.
4. Jean-François Cardoso and Antoine Souloumiac. Jacobi angles for simultaneous diagonalization. *SIAM Journal of Matrix Analysis and Applications*, 17(1):161–164, January 1996.
5. Konstantinos I. Diamantaras, Athina P. Petropulu, and Binning Chen. Blind two-input-two-output FIR channel identification based on frequency domain second-order statistics. *IEEE Transactions on Signal Processing*, 48:534–542, 2000.
6. Aapo Hyvärinen and Patrik Hoyer. Emergence of phase- and shift-invariant features by decomposition of natural images into independent feature subspaces. *Neural Computation*, 12(7):1705–1720, 2000.
7. L. Molgedey and H. G. Schuster. Separation of a mixture of independent signals using time delayed correlations. *Physical Review Letters*, 72:3634–3637, 1994.
8. A. Ziehe and K.-R. Müller. Tdsep – an efficient algorithm for blind separation using time structure. In L. Niklasson, M. Boden, and T. Ziemke, editors, *ICANN'98*, pages 675–680, Skövde, Sweden, 1998. Springer.

Fast Kernel Density Independent Component Analysis

Aiyou Chen

Bell Labs, Lucent Technologies
aychen@research.bell-labs.com

Abstract. We develop a super-fast kernel density estimation algorithm (FastKDE) and based on this a fast kernel independent component analysis algorithm (KDICA). FastKDE calculates the kernel density estimator exactly and its computation only requires sorting n numbers plus roughly $2n$ evaluations of the exponential function, where n is the sample size. KDICA converges as quickly as parametric ICA algorithms such as FastICA. By comparing with state-of-the-art ICA algorithms, simulation studies show that KDICA is promising for practical usages due to its computational efficiency as well as statistical efficiency. Some statistical properties of KDICA are analyzed.

Keywords: independent component analysis, kernel density estimation, nonparametric methods.

1 Introduction

Independent component analysis (ICA) has been a powerful tool for blind source separation in many applications such as image and acoustic signal processing, brain imaging analysis (Hyvarinen, Karhunen and Oja 2001). Suppose that an observable signal, say $\mathbf{X}$, can be modeled as an unknown linear mixture of m mutually independent hidden sources $(S_1, \cdots, S_m)$. Denote $\mathbf{S} \equiv (S_1, \cdots, S_m)^T$, so

$$\mathbf{X} = \mathbf{AS} \tag{1}$$

for some matrix $\mathbf{A}$. Assume that $\{\mathbf{X}(t) : 1 \leq t \leq n\}$ are n i.i.d. observations of $\mathbf{X}$, where t is the time index. That is, at time t the hidden sources produce signals $\mathbf{S}(t) \equiv (S_1(t), \cdots, S_m(t))^T$ that are observed as $\mathbf{X}(t) = \mathbf{AS}(t)$. The problem is to recover $\{\mathbf{S}(t) : 1 \leq t \leq T\}$ without knowing either $\mathbf{A}$ or the distributions of $\mathbf{S}$. In order to solve this problem, it is necessary that $\dim(\mathbf{X}) \geq m$. Without loss of generality, we may assume that the dimension of $\mathbf{X}$ is the same as $\mathbf{S}$ and that A is an $m \times m$ nonsingular matrix. It is well-known that $\mathbf{W} = \mathbf{A}^{-1}$ (called the unmixing matrix) is identifiable up to permutation and scale transformations of the rows of $\mathbf{A}$ if $\mathbf{S}$ has at most one Gaussian component (Comon, 1994). The order and scale can be controlled such that $\mathbf{W}$ is unique. The ICA problem becomes to estimate $\mathbf{W}$.

Classical ICA algorithms such as FastICA fit parametric models for the hidden sources and thus are limited to particular families of hidden sources

J. Rosca et al. (Eds.): ICA 2006, LNCS 3889, pp. 24–31, 2006.

(Cardoso 1998). It has been realized that the unknown distributions of hidden sources can be estimated by nonparametric methods, which can be applied to a wide range of distribution families. For example, Hastie and Tibshirani (2002) proposed penalized maximal likelihood based on log-spline density estimation. Miller and Fisher (2003) proposed the RADICAL algorithm based on the neighborhood density estimator. Vlassis & Motomura (2001), Boscolo et al. (2004) and recently Shwartz et al. (2005) used kernel density estimation to deal with the unknown source distributions. These nonparametric algorithms are in general more accurate and more robust but on the other side are computationally much heavier than classical parametric ICA algorithms such as FastICA. The computational bottleneck is the nonparametric density estimators[1]. There exists other nonparametric ICA algorithms such as KCCA, KGV (Bach & Jordan 2002), CFICA (Eriksson & Koivunen 2003), PCFICA (Chen & Bickel 2005) and *kernel mutual information* (Gretten et al 2005), which do not deal with the source density functions directly. Among different nonparametric density estimators, the kernel density estimator (KDE) is most popular. But naive implementation requires $O(n^2)$ complexity, where n is the sample size. In the statistical literature, the binning and clustering techniques have been used to reduce the complexity, see Silverman (1986). For example, Pham (2004) applied the binning technique in the ICA literature. Fast Gauss transform (Greengard & Strain 1991) and the dual-tree algorithm by Gray & Moore (2003) are alternative fast algorithms for KDE. All these KDE algorithms are based on different approximation techniques and are faster than $O(n^2)$. But these techniques require careful choices of certain tuning parameters in order to balance computational speed-up and approximation errors, and occasionally are as slow as $O(n^2)$ in order to achieve good performance.

In this paper, we develop a super-fast kernel density estimation algorithm (FastICA) and based on this a fast kernel ICA algorithm (KDICA). The remaining of the paper is structured as follows. In Section 2, the FastKDE algorithm is developed. In Section 3, the KDICA algorithm is described. In Section 4, some simulation studies are used to show both computational and statistical efficiency of KDICA. In Section 4, some statistical properties of KDICA are analyzed. Section 5 concludes the paper. From now on, vectors and matrices are in bold and capital. W_k denotes the kth row vector of $\mathbf{W}$.

2 The FastKDE Algorithm

Let $\{x_i : 1 \leq i \leq n\} \subset \mathcal{R}$ be from a density function $p(\cdot)$. The kernel density estimator of $p(\cdot)$ is defined by

$$\hat{p}(x) = \frac{1}{nh} \sum_{i=1}^{n} K(\frac{x_i - x}{h}), \tag{2}$$

[1] The neighborhood density estimator used by RADICAL only requires $n \log n$ complexity, but it does not produce a continuous objective function w.r.t. W.

where $K(\cdot)$ is a kernel density function and h is the bin width, usually $h = O(n^{-1/5})$. Popular choices of $K(\cdot)$ are symmetric density functions such as Gaussian kernel, Laplacian kernel, Uniform, Epanechnikov, etc. We need to evaluate $\hat{p}(x)$ for $x \in \{x_i : i = 1, \cdots, n\}$. Direct evaluation requires $O(n^2)$ complexity, and alternative algorithms based on approximation are available with complexity less than $O(n^2)$, but are *not* fast enough for ICA.

It is known that the choice of K is not crucial for KDE. Here we use the Laplacian kernel and develop a simple fast algorithm. The Laplacian kernel is $K(x) = \frac{1}{2}e^{-|x|}$, $x \in \mathcal{R}$. Although $K(x)$ is not differentiable at $x = 0$, $\hat{p}(x) \approx \int p(x + th)K(t)dt$ is differentiable wherever $p(x)$ is.

First the sample points $\{x_i\}$ are sorted. Sorting n numbers can be performed very quickly, for example the *quick sort* algorithm has complexity in the worst case $O(n \log n)$ and the *bucket sort* algorithm requires linear time only. Without loss of generality, let $x_1 \leq \cdots \leq x_n$. It is not hard to show that for $k = 1, \cdots, n$,

$$\hat{p}(x_k) = \frac{1}{2nh}\{\exp(\frac{x_k}{h}) \sum_{i=k+1}^{n} \exp(-\frac{x_i}{h}) + \exp(-\frac{x_k}{h}) \sum_{i=1}^{k} \exp(\frac{x_i}{h})\}.$$

Then FastKDE can now be described as follows.

Algorithm. FastKDE (given h and $x_1 \leq \cdots \leq x_n$)

1. Initialize $\underline{\mathbf{s}}_1 = e^{x_1/h}$ and $\bar{\mathbf{s}}_n = 0$, then calculate for $i = 2, \cdots, n$,

$$\underline{\mathbf{s}}_i = \underline{\mathbf{s}}_{i-1} + \exp(\frac{x_i}{h}) \text{ and } \bar{\mathbf{s}}_{n-i+1} = \bar{\mathbf{s}}_{n-i+2} + \exp(-\frac{x_{n-i+2}}{h}).$$

2. For $i = 1, \cdots, n$, compute

$$\hat{p}(x_i) = \frac{1}{2nh}\{\underline{\mathbf{s}}_i \exp(-\frac{x_i}{h}) + \bar{\mathbf{s}}_i \exp(\frac{x_i}{h})\}.$$

The exponential values $\{(\exp(x_i/h), \exp(-x_i/h)) : 1 \leq i \leq n\}$ only need to be computed once and saved for both Step 1 and Step 2. Then Step 1 and Step 2 require about $3n$ summations in total. Thus the total complexity of FastKDE is about $2n$ exponential evaluations. The bin width h is chosen for simplicity by the reference method which minimizes $\int(\hat{p}(x) - p(x))^2 dx$ and gives $h = O(n^{-1/5})$ (Silverman 1986). We recommend to use

$$\hat{h} = 0.6\hat{\sigma}n^{-1/5} \tag{3}$$

where $\hat{\sigma}$ is the sample standard deviation of $\{x_i\}$.

3 The KDICA Algorithms

In this section we develop the KDICA algorithm, for which the FastKDE algorithm as the key technology is implemented. We use the maximum profile likelihood and later establish its relationship with criteria derived from information theory in Section 5.

3.1 Maximum Profile Likelihood

Suppose each S_k has a density function $r_k(\cdot)$, for $k = 1, \cdots, m$. Then the density function of $\mathbf{X}$ can be expressed as $p_{\mathbf{X}}(\mathbf{x}) = |\det(\mathbf{W})| \prod_{k=1}^{m} r_k(W_k \mathbf{x})$, where W_k is the kth row of $\mathbf{W}$. The classical maximum likelihood estimator (MLE) maximizes the likelihood of observations of X with respect to all the parameters $(\mathbf{W}, r_1, \cdots, r_m)$. However, since $(r_1, \cdots, r_m)$ are unknown functions, model (1) is called *semiparametric* (Bickel et al. 1993) and direct implementation of MLE does not work by using finite samples. In this scenario, maximum profile likelihood (MPLE) can serve as an alternative of MLE (see Murphy and van der Vaart 2000). If $\mathbf{W}$ is known, then r_k is identical to the density function of $W_k\mathbf{X}$. Thus r_k can be estimated by the kernel density estimator $\hat{r}_{W_k}(x) = (nh)^{-1} \sum_{t=1}^{n} K((W_k \mathbf{X}(t) - x)/h)$, where for the KDICA algorithm the Laplacian kernel is used for K. The profile likelihood, say l_p, is to modify the likelihood function by replacing r_k by $\hat{r}_{W_k}$, that is,

$$l_p(\mathbf{W}) = \frac{1}{n} \sum_{t=1}^{n} \sum_{k=1}^{m} \log \hat{r}_{W_k}(W_k \mathbf{X}(t)) + \log |\det(\mathbf{W})|. \tag{4}$$

Since $l_p(\mathbf{W})$ is just a function of $\mathbf{W}$, the maximum profile likelihood estimator (MPLE) is defined by

$$\hat{\mathbf{W}} = \arg\max l_p(\mathbf{W}). \tag{5}$$

Obviously the computational bottleneck of MPLE is to evaluate $\{\hat{r}_{W_k} (W_k\mathbf{X}(t)) : t = 1, \cdots, n\}_{k=1}^{m}$. By using the FastKDE algorithm developed above, the complexity of MPLE is reduced to $O(mn)$.

3.2 Algorithm

This subsection describes the KDICA algorithm which implements the estimator (5). Since prewhitening can reduce computational complexity while keeps statistical consistency (Chen & Bickel 2005), we use this technique to preprocess the data. That is, let $\tilde{\mathbf{X}}(t) = \hat{\Sigma}_{\mathbf{X}}^{-1/2} \mathbf{X}(t)$ for $t = 1, \cdots, n$, where $\hat{\Sigma}_{\mathbf{X}}$ is the sample variance-covariance matrix of $\mathbf{X}(t)$. By assuming unitary variances for $\mathbf{S}$, $\tilde{\mathbf{X}}$ can be separated by a rotation matrix. Then we seek for a rotation matrix $\hat{\mathbf{O}}$, such that

$$\hat{\mathbf{O}} = \arg \min_{O \in \mathcal{O}(m)} F(\mathbf{O}), \tag{6}$$

where $F(\mathbf{O}) = -\sum_{k=1}^{m} \frac{1}{n} \sum_{t=1}^{n} \log \tilde{r}_{O_k}(O_k \tilde{\mathbf{X}}(t))$, and $\tilde{r}_{O_k}(s) = \frac{1}{nh} \sum_{t=1}^{n} K((O_k \tilde{\mathbf{X}}(t) - s)/h)$ is the Laplacian kernel density estimator for $O_k \tilde{\mathbf{X}}$. $\mathcal{O}(m)$ is the set of $m \times m$ rotation matrices. Since $O_k \tilde{\mathbf{X}}$ has unitary variance, by (3), $h = 0.6 n^{-1/5}$.

The optimization of (6) can be done efficiently by using the gradient algorithm on the Stiefel manifold (Edelman, Arias & Smith 1999). We refer to Bach & Jordan (2002) for how to implement it. The KDICA algorithm then has three steps as follows. Note that the KDICA does not need any tuning parameters.

Algorithm. KDICA ($h_n = 0.6n^{-1/5}$)

1. Prewhiten : $\tilde{\mathbf{X}}(t) = \hat{\Sigma}_{\mathbf{X}}^{-1/2}\mathbf{X}(t)$ for $t = 1, \cdots, n$, where $\hat{\Sigma}_{\mathbf{X}}$ is the sample variance-covariance matrix of $\{\mathbf{X}(t) : 1 \le t \le n\}$.
2. Optimize $\hat{\mathbf{O}} = \arg\max_{\mathbf{O} \in \mathcal{O}(m)} F(\mathbf{O})$ using the gradient algorithm.
3. Output $\hat{\mathbf{W}} = \hat{\mathbf{O}}\hat{\Sigma}_{\mathbf{X}}^{-1/2}$.

4 Simulation Studies

We compare KDICA with several well-known ICA algorithms such as the generalized FastICA (Hyvarinen 1999), JADE (Cardoso 1999) and KGV (Bach and Jordan 2002). Some recent algorithms such as NPICA (Boscolo et al 2004) and EFFICA (Chen 2004) are also included for comparison. FastICA is used to initialize KGV, NPICA, EFFICA and KDICA. We used $m = 4$ and $m = 8$ sources with different sample sizes 1000 and 4000. The 8 sources were generated from: $\mathcal{N}(0,1)$, exp(1), t(3), lognorm(1,1), t(5), logistic(0,1), Weibull(3,1), and exp(10)+$\mathcal{N}(0,1)$. When $m = 4$, the first four distributions were used for hidden sources. Each experiment was replicated 100 times and the boxplots of Amari errors were reported. Figure 1 shows that KDICA is comparable to EFFICA which has been proven to be asymptotically efficient under mild conditions, and like other nonparametric algorithms, KDICA performs much better than FastICA and JADE. The right panel of Figure 1 reports the average running time of all algorithms. The plot shows that KDICA is more than 20 times faster than NPICA which uses the FFT based KDE algorithm and 50 times faster than KGV. KDICA is about 10 times slower than but comparable to FastICA and JADE. The KDICA algorithm exhibits very good simulation performance. But due to space limitation, we are not allowed to report further simulation studies.

We next apply the KDICA algorithm for blind separation of mixtures of images. Two natural images and a Gaussian noise image are given in the first row of Figure 2, each of size 80×70 pixels (black/white). First, each pixel matrix is reshaped into a column and each column is normalized by its sample standard deviation. Second, a random 3×3 matrix $W \in \Omega$ is inverted to obtain three columns $\{\mathbf{X}(t) \in \mathcal{R}^3 : 1 \le t \le 5600\}$ and each column is reshaped into a matrix of size 80×70. This gives three contaminated images, as shown in the second row of Figure 2. Third, $\{\mathbf{X}(t)\}$ is separated into three vectors by using KDICA, and each vector is reshaped into an image with 80×70 pixels. Three random restarting points were used in KDICA. It is surprising that human eyes can hardly tell the difference between natural images and separated images. This type of experiments have also been done by several different researchers in the ICA literature (e.g. Yang and Amari 1997).

We ran this experiment 10 times with random **W** by using KDICA and several other ICA algorithms. The average running times for the generalized FastICA, JADE, and KDICA are 0.05, 0.03 and 1.82 seconds separately. Other nonparametric algorithms such as NPICA and KGV take more than one minute.

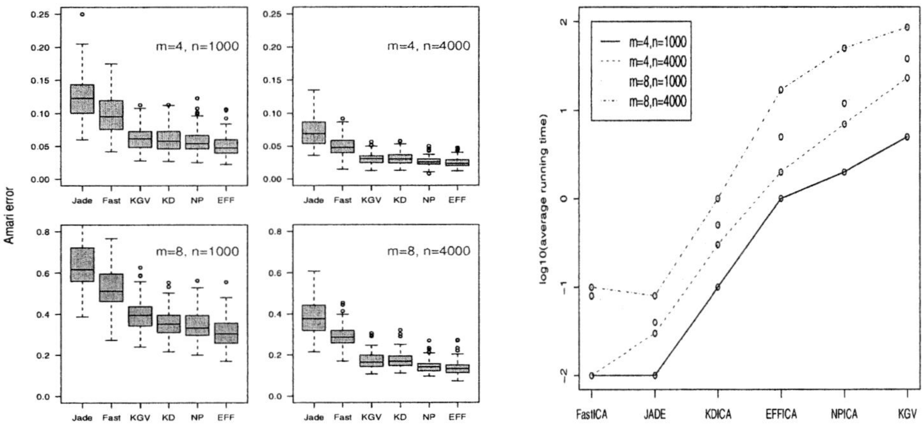

Fig. 1. Left panel: Comparison of KDICA and other ICA algorithms in terms of the Amari errors, where the numbers below the x-labels are the average running time (seconds/per experiment) of the corresponding algorithms. Right panel: Comparison of running time of different ICA algorithms.

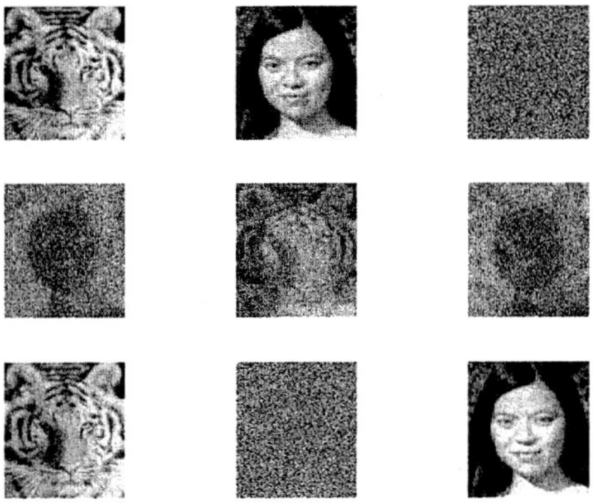

Fig. 2. Face identification by KDICA, where the three original, mixed and separated images are given in the three rows separately

5 Statistical Consistency and Efficiency of KDICA

In this Section, we study the statistical properties of the estimator (5). Obviously as $n \uparrow \infty$, $\hat{r}_{W_k} \to r_{W_k}$, the density function of $W_k X$. Thus for $n = \infty$, the profile likelihood is equal to $l_p(\mathbf{W}) = E\sum_{k=1}^{m} \log r_{W_k}(W_k X) + \log|\det(\mathbf{W})|$. Let $p_{\mathbf{W}}(\cdot)$ be the joint density function of $(W_1\mathbf{X}, \cdots, W_m\mathbf{X})$, then $p_{\mathbf{W}}(\mathbf{Wx}) = p_{\mathbf{X}}(\mathbf{x})/|\det(\mathbf{W})|$. Thus the mutual information of $(W_1\mathbf{X}, \cdots, W_m\mathbf{X})$ is equal to

$$\mathbf{I}(\mathbf{W}) = E \log \frac{p_{\mathbf{W}}(\mathbf{W}\mathbf{X})}{\prod_{k=1}^{m} r_{W_k}(W_k\mathbf{X})} = E \log p_{\mathbf{X}}(\mathbf{X}) - l_p(\mathbf{W}).$$

Notice that $E \log p_{\mathbf{X}}(\mathbf{X})$ does not depend on the parameter $\mathbf{W}$. The above equation implies that the profile likelihood criteria is equivalent to the mutual information criteria which has been popularly used in the ICA literature. Thus we would expect the statistical performance of KDICA to be similar to or better than other nonparametric ICA algorithms. General connection between likelihood inference and information theory criteria has been studied by Lee, Girolami, Bell & Sejnowski (2000). We obtain statistical consistency of the KDICA algorithm as summarized in Theorem 1, whose technical conditions and proof are omitted here due to space limitation but refer to Chen (2004).

Theorem 1. *Suppose that* $\mathbf{W}$ *is identifiable and the density functions of the hidden sources are continuous and satisfy mild smoothness conditions. If* $h_n = O(n^{-1/5})$. *Then the estimator* $\hat{\mathbf{W}}$ *given by (5) is consistent, that is,* $||\hat{\mathbf{W}} - \mathbf{W}_P|| = o_P(1)$, *where* $\mathbf{W}_P$ *is the true unmixing matrix.*

6 Concluding Remarks

In this paper, we have presented the FastKDE and KDICA algorithms. Due to its computational and statistical efficiency, KDICA makes nonparametric ICA applicable for large size problems of blind source separation. We conjecture that FastKDE will make it convenient to deal with nonlinear independent component analysis (Jutten et. al, 2004) in a truly nonparametric manner.

Acknowledgment

The paper is extracted from Chapter 4 of the author's Ph.D. thesis. The author would like to thank his advisor Peter J. Bickel for support and many helpful discussions, and thank Michael I. Jordan, Anant Sahai and Diane Lambert for helpful comments.

References

1. Bach, F. and Jordan, M. (2002). Kernel independent component analysis. *Journal of Machine Learning Research* **3** 1-48.
2. Bickel, P., Klaassen, C. , Ritov, Y. and Wellner, J. (1993). *Efficient and Adaptive Estimation for Semiparametric Models.* Springer Verlag, New York, NY.
3. Boscolo, R., Pan, H. and Roychowdhury V. P. (2004). Independent component analysis based on nonparametric density estimation. *IEEE Trans. Neural Networks,* **15** (1): 55-65.
4. Cardoso, J. F. (1998). Blind signal separation: statistical principles. *Proceedings of the IEEE,* **9**(10) 2009-2025.
5. Cardoso, J. F. (1999). High-order contrasts for independent component analysis. *Neural Computation* **11**(1) 157-192.

6. Chen, A. (2004). Semiparametric inference for independent component analysis. *Ph.D Thesis, Advisor: Peter J. Bickel, Department of Statistics, University of California, Berkeley, 2004.*
7. Chen, A. and Bickel, P. J. (2005). Consistent independent component analysis and prewhitening. *IEEE Trans. on Signal Processing*, Vol. 53, 10, page 3625-3632.
8. Comon, P. (1994). Independent component analysis, a new concept? *Signal Processing* **36**(3):287-314.
9. Edelman, A., Arias, T. and Smith, S. (1999). The geometry of algorithms with orthogonality constraints. *SIAM journal on Matrix Analysis and Applications*, **20**(2): 303-353.
10. Eriksson, J. and Visa Koivunen (2003). Characteristic-function based independent component analysis. *Signal Processing*, Vol. 83, pp2195-2208.
11. Greengard, L. and Strain, J. (1991). The fast Gauss transform. *SIAM J. Sci. Stat. Comput.*, 12, page 79-94.
12. Gray, A. and Moore, A. (2003). Very fast multivariate kernel density estimation via computational geometry. *Proceedings of the Joint Statistical Meeting*, San Francisco, CA, 2003.
13. Gretton, A., Herbrich, R., Smola, A., Bousquet, O. and Scholkopf, B. (2005). Kernel methods for testing independence. Submitted.
14. Hastie, T. and Tibshirani, R. (2002). Independent component analysis through product density estimation, *Technical report*, Department of Statistics, Stanford University.
15. Hyvarinen, A. (1999). Fast and robust fixed-point algorithms for independent component analysis. *IEEE Trans. on Neural Networks* **10**(3) 626-634.
16. Hyvarinen, A., Karhunen, J. and Oja, E. (2001). *Independent Component Analysis.* John Wiley & Sons, New York, NY.
17. Jutten, C., Babaie-Zadeh, M, and Hosseini, S. (2004). Three easy ways for separating nonlinear mixtures? *Signal Processing*, 84, pp.217-229.
18. Lee, T. W., Girolami, M., Bell, A. and Sejnowski, T. (2000). A unifying information-theoretic framework for independent component analysis. *Computers and Mathematics with Applications* **39** 1-21.
19. Miller, E. and Fisher, J. (2003). ICA using spacings estimates of entropy. *Journal of Machine Learning Research*, **4**, pp. 1271-1295.
20. Murphy, S. and van der Vaart, A. (2000). On profile likelihood. *Journal of the American Statistical Association* **95** 449-485.
21. Pham, D. T. (2004). Fast algorithms for mutual information based independent component analysis. *IEEE Trans. on Signal Processing*, Vol. 52, 10, page 2690-2700.
22. Shwartz, S., Zibulevsky, M. and Schechner, Y. (2005). Fast kernel entropy estimation and optimization. *Signal Processing*, **85**, 1045-1058.
23. Silverman, B. W. (1986). *Density Estimation for Statistics and Data Analysis.* Chapman Hall: London.
24. Vlassis, N. and Motomura, Y. (2001). Efficient source adaptivity in independent component analysis. *IEEE Trans. Neural Networks* **12**(3) 559-565.
25. Yang, H. H. and Amari, S. (1997). Adaptive on-line learning algorithms for blind separation - maximum entropy and minimum mutual information. *Neural Computation*, **9**(7): 1457-1482.

Csiszár's Divergences for Non-negative Matrix Factorization: Family of New Algorithms

Andrzej Cichocki[1,*], Rafal Zdunek[1,**], and Shun-ichi Amari[2]

[1] Laboratory for Advanced Brain Signal Processing
[2] Laboratory for Mathematical Neuroscience,
RIKEN BSI, Wako-shi Japan

Abstract. In this paper we discus a wide class of loss (cost) functions for non-negative matrix factorization (NMF) and derive several novel algorithms with improved efficiency and robustness to noise and outliers. We review several approaches which allow us to obtain generalized forms of multiplicative NMF algorithms and unify some existing algorithms. We give also the flexible and relaxed form of the NMF algorithms to increase convergence speed and impose some desired constraints such as sparsity and smoothness of components. Moreover, the effects of various regularization terms and constraints are clearly shown. The scope of these results is vast since the proposed generalized divergence functions include quite large number of useful loss functions such as the squared Euclidean distance,Kulback-Leibler divergence, Itakura-Saito, Hellinger, Pearson's chi-square, and Neyman's chi-square distances, etc. We have applied successfully the developed algorithms to blind (or semi blind) source separation (BSS) where sources can be generally statistically dependent, however they satisfy some other conditions or additional constraints such as nonnegativity, sparsity and/or smoothness.

1 Introduction and Problem Formulation

The non-negative matrix factorization (NMF approach is promising in many applications from engineering to neuroscience since it is designed to capture alternative structures inherent in the data and, possibly to provide more biological insight [1, 2, 3, 4, 5, 6]. Lee and Seung introduced NMF in its modern formulation as a method to decompose patterns or images [3, 7].

In this paper we impose nonnegativity constraints and other penalties such as sparseness and/or smoothness. The NMF decomposes the data matrix $\boldsymbol{Y} = [\boldsymbol{y}(1), \boldsymbol{y}(2), \ldots, \boldsymbol{y}(N)] \in \mathbb{R}^{m \times N}$ as a product of two matrices $\boldsymbol{A} \in \mathbb{R}^{m \times n}$ and $\boldsymbol{X} = [\boldsymbol{x}(1), \boldsymbol{x}(2), \ldots, \boldsymbol{x}(N)] \in \mathbb{R}^{n \times N}$ having only non-negative elements. Although some decompositions or matrix factorizations provide an exact reconstruction of the data (i.e., $\boldsymbol{Y} = \boldsymbol{A}\boldsymbol{X}$), we shall consider here decompositions which are approximative in nature, i.e.,

* On leave from Warsaw University of Technology, Poland.
** On leave from Institute of Telecommunications, Teleinformatics and Acoustics, Wroclaw University of Technology, Poland.

J. Rosca et al. (Eds.): ICA 2006, LNCS 3889, pp. 32–39, 2006.

$$\boldsymbol{Y} = \boldsymbol{A}\boldsymbol{X} + \boldsymbol{V}, \qquad \boldsymbol{A} \geq 0, \ \ \boldsymbol{X} \geq 0 \tag{1}$$

or equivalently $\boldsymbol{y}(k) = \boldsymbol{A}\boldsymbol{x}(k) + \boldsymbol{v}(k)$, $k = 1, 2, \ldots, N$ or in a scalar form as $y_i(k) = \sum_{j=1}^{n} a_{ij} x_j(k) + \nu_i(k)$, $i = 1, \ldots, m$, where $\boldsymbol{V} \in \mathbb{R}^{m \times N}$ represents noise or error matrix, $\boldsymbol{y}(k) = [y_1(k), \ldots, y_m(k)]^T$ is a vector of the observed signals (typically nonnegative) at the discrete time instants k while $\boldsymbol{x}(k) = [x_1(k), \ldots, x_n(k)]^T$ is a vector of components or source signals at the same time instant [8]. Our objective is to estimate the mixing (basis) matrix $\boldsymbol{A}$ and sources $\boldsymbol{X}$ subject to nonnegativity constraints on all entries. Usually, in BSS applications it is assumed that $N >> m \geq n$ and n is known or can be relatively easily estimated using SVD or PCA. Throughout this paper, we use the following notations: $x_j(k) = x_{jk}$, $y_i(k) = y_{ik}$ and $z_{ik} = [\boldsymbol{A}\boldsymbol{X}]_{ik}$ means ik-element of the matrix $(\boldsymbol{A}\boldsymbol{X})$, the ij-th element of the matrix $\boldsymbol{A}$ is denoted by a_{ij}.

The basic approach to NMF is alternating minimization or alternating projection: the specified loss function is alternately minimized with respect to two sets of parameters $\{x_{jk}\}$ and $\{a_{ij}\}$, each time optimizing one set of arguments while keeping the other one fixed [2, 3, 8].

The most popular adaptive multiplicative algorithms for NMF are based on two loss functions: 1. square Euclidean distance expressed by the Frobenius norm:

$$D_F(\boldsymbol{A}, \boldsymbol{X}) = \frac{1}{2}\|\boldsymbol{Y} - \boldsymbol{A}\boldsymbol{X}\|_F^2 = \frac{1}{2}\sum_{i=1}^{m}\sum_{k=1}^{N} |y_{ik} - [\boldsymbol{A}\boldsymbol{X}]_{ik}|^2$$
$$\text{s. t. } \ a_{ij} \geq 0, \quad x_j(k) = x_{jk} \geq 0 \quad \forall\, i, j, k, \tag{2}$$

which is optimal for a Gaussian distributed noise). Based on of this cost function Lee and Seung proposed the following multiplicative algorithm:

$$a_{ij} \leftarrow a_{ij} \frac{[\boldsymbol{Y}\,\boldsymbol{X}^T]_{ij}}{[\boldsymbol{A}\,\boldsymbol{X}\,\boldsymbol{X}^T]_{ij}}, \qquad x_{jk} \leftarrow x_{jk} \frac{[\boldsymbol{A}^T\,\boldsymbol{Y}]_{jk}}{[\boldsymbol{A}^T\boldsymbol{A}\,\boldsymbol{X}]_{jk}}. \tag{3}$$

which is an extension of the well known ISRA (Image Space Reconstruction Algorithm) algorithm [9]. Alternative mostly used loss function that intrinsically ensures non-negativity constraints and it is related to the Poisson likelihood is a functional based on the Kullback-Leibler divergence [3, 5]:

$$D_{KL}(\boldsymbol{Y}\|[\boldsymbol{A}\boldsymbol{X}]) = \sum_{ik}\left(y_{ik}\log\frac{y_{ik}}{[\boldsymbol{A}\boldsymbol{X}]_{ik}} + [\boldsymbol{A}\boldsymbol{X}]_{ik} - y_{ik}\right) \tag{4}$$
$$\text{s. t. } \ x_{jk} \geq 0, \quad a_{ij} \geq 0, \quad \|\boldsymbol{a}_j\|_1 = \sum_{i=1}^{m} a_{ij} = 1.$$

Using the alternating minimization approach, Lee and Seung derived the following multiplicative learning rules:

$$x_{jk} \leftarrow x_{jk} \frac{\sum_{i=1}^{m} a_{ij}\,(y_{ik}/[\boldsymbol{A}\,\boldsymbol{X}]_{ik})}{\sum_{q=1}^{m} a_{qj}}, \quad a_{ij} \leftarrow a_{ij} \frac{\sum_{k=1}^{N} x_{jk}\,(y_{ik}/[\boldsymbol{A}\,\boldsymbol{X}]_{ik})}{\sum_{p=1}^{N} x_{jp}}, \tag{5}$$

which are extensions (by alternating minimization) of the well known EMML or Richardson-Lucy algorithm (RLA) [9].

It should be noted that he most existing NMF algorithms perform blind source separation rather very poorly due to the non-uniqueness of solution and/or the lack of additional constraints which should be satisfied. The main objective of this contribution is to propose flexible and improved NMF algorithms that generalize or combine several different criteria in order to extract physically meaningful sources, especially for biomedical signal applications such as EEG and MEG.

2 Generalized Divergences for NMF

There are three large classes of generalized divergences which can be potentially useful for developing new flexible algorithms for NMF: the Bregman divergences, Amari's alpha divergence [1] and the Csiszár's φ-divergences [10]. In this contribution we limit our discussion to the Csiszár's divergences and as the special case the alpha divergence. The Csiszár's φ-divergence s defined as

$$D_C(\boldsymbol{z}||\boldsymbol{y}) = \sum_{k=1}^{N} z_k \varphi(\frac{y_k}{z_k}) \tag{6}$$

where $y_k \geq 0, z_k \geq 0$ and $\varphi: \ [0,\infty) \to (-\infty,\infty)$ is a function which is convex on $(0,\infty)$ and continuous at zero. Depending on the application, we can impose different restrictions on φ. In order to use the Csiszár's divergence as a distance measure, we assume that $\varphi(1) = 0$ and that it is strictly convex at 1.

Several basic examples include ($u_{ik} = y_{ik}/z_{ik}$):

1. If $\varphi(u) = (\sqrt{u}-1)^2$, then $D_{C-H} = \sum_{ik}(\sqrt{y}_{ik} - \sqrt{z}_{ik})^2$ -Hellinger distance;
2. If $\varphi(u) = (u-1)^2$, then $D_{C-P} = \sum_{ik}(y_{ik} - z_{ik})^2/z_{ik}$ -Pearson's distance;
3. For $\varphi(u) = u(u^{\beta-1}-1)/(\beta^2-\beta) + (1-u)/\beta$ we have a family of Amari's alpha divergences:

$$D_A^{(\beta)}(\boldsymbol{AX}||\boldsymbol{Y}) = \sum_{ik} y_{ik}\frac{(y_{ik}/z_{ik})^{\beta-1}-1}{\beta(\beta-1)} + \frac{z_{ik}-y_{ik}}{\beta}, \qquad z_{ik} = [\boldsymbol{AX}]_{ik}, \tag{7}$$

where $\beta = (1+\alpha)/2$ [1] (see also Ali-Sllvey, Liese & Vajda, Cressie-Read disparity, Eguchi beta divergence,Kompass) [11, 12]. It is interesting to note that in the special cases for $\beta = 2, 0.5, -1$, we obtain Pearson's, Hellinger and Neyman's chi-square distances, respectively (while for the cases $\beta = 1$ and $\beta = 0$ the divergences have to be defined as limiting cases as $\beta \to 1$ and $\beta \to 0$, respectively). When these limits are evaluated one gets for $\beta \to 1$ the generalized Kullback-Leibler divergence (called I-divergence) defined by equations (4) and for $\beta \to 0$ the dual generalized KL divergence:

$$D_{KL}(\boldsymbol{AX}||\boldsymbol{Y}) = \sum_{ik}\left([\boldsymbol{AX}]_{ik}\log\frac{[\boldsymbol{AX}]_{ik}}{y_{ik}} - [\boldsymbol{AX}]_{ik} + y_{ik}\right) \tag{8}$$

As an illustrative example, let us derive a new multiplicative learning rule for the loss function (8). By applying multiplicative exponentiated gradient (EG) descent updates:

$$x_{jk} \leftarrow x_{jk} \exp\left(-\eta_j \frac{\partial D_{KL}}{\partial x_{jk}}\right), \quad a_{ij} \leftarrow a_{ij} \exp\left(-\tilde{\eta}_j \frac{\partial D_{KL}}{\partial a_{ij}}\right), \tag{9}$$

we obtain new simple multiplicative learning rules for NMF

$$x_{jk} \leftarrow x_{jk} \exp\left(\sum_{i=1}^{m} \eta_j a_{ij} \log(\frac{y_{ik}}{[\boldsymbol{A}\boldsymbol{X}]_{ik}})\right) = x_{jk} \prod_{i=1}^{m} \left(\frac{y_{ik}}{[\boldsymbol{A}\boldsymbol{X}]_{ik}}\right)^{\eta_j a_{ij}}, \tag{10}$$

$$a_{ij} \leftarrow a_{ij} \exp\left(\sum_{k=1}^{N} \tilde{\eta}_j x_{jk} \log(\frac{y_{ik}}{[\boldsymbol{A}\boldsymbol{X}]_{ik}})\right) = a_{ij} \prod_{k=1}^{N} \left(\frac{y_{ik}}{[\boldsymbol{A}\boldsymbol{X}]_{ik}}\right)^{\tilde{\eta}_j x_{jk}}, \tag{11}$$

The nonnegative learning rates $\eta_j, \tilde{\eta}_j$ can take different forms. Typically, in order to guarantee stability of the algorithm we assume that $\eta_j = \omega\,(\sum_{i=1}^{m} a_{ij})^{-1}$, $\tilde{\eta}_j = \omega\,(\sum_{k=1}^{N} x_{jk})^{-1}$, where $\omega \in (0,2)$ is an over-relaxation parameter. The above algorithm can be considered as an alternating minimization/projection extension of the well known SMART (Simultaneous Multiplicative Algebraic Reconstruction Technique) [9].

Similarly, for $\beta \neq 0$ we have developed the following new algorithm (the proof is omitted due to the lack of space)

$$x_{jk} \leftarrow x_{jk} \left(\sum_{i=1}^{m} a_{ij}\ (y_{ik}/[\boldsymbol{A}\,\boldsymbol{X}]_{ik})^{\beta}\right)^{1/\beta}, \; a_{ij} \leftarrow a_{ij} \left(\sum_{k=1}^{N} (y_{ik}/[\boldsymbol{A}\,\boldsymbol{X}]_{ik})^{\beta}\, x_{jk}\right)^{1/\beta}$$

with normalization of columns of $\boldsymbol{A}$ in each iteration to unit length: $a_{ij} \leftarrow a_{ij}/\sum_p a_{pj}$. The algorithm can be written in a compact matrix form using some MATLAB notations:

$$\boldsymbol{X} \leftarrow \boldsymbol{X}\ .*\ \left(\boldsymbol{A}^T\,((\boldsymbol{Y}+\varepsilon)./\,(\boldsymbol{A}\boldsymbol{X}+\varepsilon)).^{\beta}\right).^{1/\beta} \tag{12}$$

$$\boldsymbol{A} \leftarrow \boldsymbol{A}\ .*\ \left(((\boldsymbol{Y}+\varepsilon)./\,(\boldsymbol{A}\boldsymbol{X}+\varepsilon)).^{\beta}\boldsymbol{X}^T\right).^{1/\beta}, \quad \boldsymbol{A} \leftarrow \boldsymbol{A}\,\mathrm{diag}\{1./sum(\boldsymbol{A},1)\},$$

where in practice a small constant $\varepsilon = 10^{-9}$ is introduced in order to ensure non-negativity constraints and avoid possible division by zero.

3 Modified Multiplicative NMF Algorithms with Regularization, Sparsity and/or Smoothing

Although the standard NMF (without any auxiliary constraints) provides sparseness of its components, we can achieve some control of this sparsity by imposing additional constraints in addition to non-negativity constraints. In fact, we can incorporate smoothness or sparsity constraints in several ways. As an illustrative example, let us consider a modified alpha divergence with regularization terms (which is an extension of the generalized divergence proposed recently by Raul Kompass [12]):

$$D_{Ko}(\boldsymbol{Y}||\boldsymbol{AX}) = \sum_{ik} \left(y_{ik} \, \frac{y_{ik}^{\beta-1} - [\boldsymbol{AX}]_{ik}^{\beta-1}}{\beta(\beta-1)} + [\boldsymbol{AX}]_{ik}^{\beta-1} \frac{[\boldsymbol{AX}]_{ik} - y_{ik}}{\beta} \right)$$
$$+\alpha_X f_X(\boldsymbol{X}) + \alpha_A f_A(\boldsymbol{A}), \tag{13}$$

where regularization parameters and terms $f_A(\boldsymbol{A})$ and $f_X(\boldsymbol{X})$ are introduced to enforce a certain application-dependent characteristic of solutions such as smoothness or sparsity. For example, in order to achieve sparse representation we usually choose $f_X(x_j) = x_j$ with constraints $x_j \geq 0$.

It is interesting to note that such defined divergence for $\alpha_X = \alpha_A = 0$ and $\beta = 2$ simplifies to the Frobenius norm (2); for $\beta \to 0$ it tends to Itakura -Saito distance, and for $\beta \to 1$ reduces to the Kullback-Leibler divergence (4).

Applying the standard gradient descent to (13) we have

$$x_{jk} \leftarrow x_{jk} - \eta_{jk} \left(\sum_{i=1}^{m} a_{ij} \left([\boldsymbol{AX}]_{ik}^{\beta-1} - y_{ik}/[\boldsymbol{AX}]_{ik}^{2-\beta} \right) - \alpha_X \psi_X(\boldsymbol{X}) \right) \tag{14}$$

$$a_{ij} \leftarrow a_{ij} - \eta_{ij} \left(\sum_{k=1}^{N} \left([\boldsymbol{AX}]_{ik}^{\beta-1} - y_{ik}/[\boldsymbol{AX}]_{ik}^{2-\beta} \right) x_{jk} - \alpha_A \psi_A(\boldsymbol{A}) \right), \tag{15}$$

where the functions $\psi_A(\boldsymbol{A})$ and $\psi_X(\boldsymbol{X})$ are defined as

$$\psi_A(\boldsymbol{A}) = \frac{\partial f_A(\boldsymbol{A})}{\partial a_{ij}}, \quad \psi_X(\boldsymbol{X}) = \frac{\partial f_X(\boldsymbol{X})}{\partial x_{jk}}. \tag{16}$$

Similar to the Lee and Seung approach, by choosing suitable learning rates:

$$\eta_{jk} = \frac{x_{jk}}{\sum_{i=1}^{m} a_{ij} [\boldsymbol{AX}]_{ik}^{\beta-1}}, \qquad \eta_{ij} = \frac{a_{ij}}{\sum_{k=1}^{N} [\boldsymbol{AX}]_{ik}^{\beta-1} \, x_{jk}}, \tag{17}$$

we obtain multiplicative update rules:

$$x_{jk} \leftarrow x_{jk} \frac{[\sum_{i=1}^{m} a_{ij} \, (y_{ik}/[\boldsymbol{AX}]_{ik}^{2-\beta}) - \alpha_X \psi_X(\boldsymbol{X})]_{\varepsilon}}{\sum_{i=1}^{m} a_{ij} \, [\boldsymbol{AX}]_{ik}^{\beta-1}}, \tag{18}$$

$$a_{ij} \leftarrow a_{ij} \frac{[\sum_{k=1}^{N} (y_{ik}/[\boldsymbol{AX}]_{ik}^{2-\beta}) \, x_{jk} - \alpha_A \psi_A(\boldsymbol{A})]_{\varepsilon}}{\sum_{k=1}^{N} [\boldsymbol{AX}]_{ik}^{\beta-1} \, x_{jk}}, \tag{19}$$

where the additional nonlinear operator is introduced in practice defined as $[x]_\varepsilon = \max\{\varepsilon, x\}$ with a small ε in order to avoid zero and negative values.

Another simple approach which can be used for control of sparsification of estimated variables is to apply nonlinear projections via suitable nonlinear monotonic functions which increase or decrease the sparseness. In this paper we have applied a very simple nonlinear transformation $x_{jk} \leftarrow (x_{jk})^{1+\alpha_{sX}}$, $\forall k$, where α_{sX} is a small coefficient typically, from 0.001 to 0.005 and it is positive if we want to increase sparseness of an estimated component and negative if we want to decrease the sparseness (see Table 1 for practical implementations).

Table 1. New Multiplicative NMF algorithms with regularization and/or sparsity constraints

Minimization of loss function subject to $a_{ij} \geq 0$ and $x_{jk} \geq 0$	Iterative Learning Algorithm Relaxation parameter $\omega \in (0,2)$
Alpha divergence, $\beta \neq 0, \beta \neq 1$ $\sum_{ik} \left\{ \frac{y_{ik}^{\beta} z_{ik}^{1-\beta} - \beta y_{ik} + (\beta - 1) z_{ik}}{\beta(\beta - 1)} \right\}$	$x_{jk} \leftarrow \left(x_{jk} \left(\sum_{i=1}^{m} a_{ij} \, (\frac{y_{ik}}{[\boldsymbol{A}\,\boldsymbol{X}]_{ik}})^{\beta} \right)^{\omega/\beta} \right)^{1+\alpha_{sX}}$ $a_{ij} \leftarrow \left(a_{ij} \left(\sum_{k=1}^{N} x_{jk} \, (\frac{y_{ik}}{[\boldsymbol{A}\,\boldsymbol{X}]_{ik}})^{\beta} \right)^{\omega/\beta} \right)^{1+\alpha_{sA}}$ $a_{ij} \leftarrow a_{ij} / \sum_{p} a_{pj},$
Pearson and Hellinger distances $\sum_{ik} \frac{(y_{ik} - [\boldsymbol{AX}]_{ik})^2}{[\boldsymbol{AX}]_{ik}},$ $\sum_{ik} \left(\sqrt{[\boldsymbol{AX}]_{ik}} - \sqrt{y_{ik}} \right)^2,$	 $(\beta = 2)$ $(\beta = 0.5)$
Kulback-Leibler divergence $\sum_{ik} (y_{ik} \log \frac{y_{ik}}{[\boldsymbol{AX}]_{ik}} - y_{ik} + [\boldsymbol{AX}]_{ik})$ $(\beta = 1)$	$x_{jk} \leftarrow \left(x_{jk} \left(\sum_{i=1}^{m} a_{ij} \, \frac{y_{ik}}{[\boldsymbol{A}\,\boldsymbol{X}]_{ik}} \right)^{\omega} \right)^{1+\alpha_{sX}}$ $a_{ij} \leftarrow \left(a_{ij} \left(\sum_{k=1}^{N} x_{jk} \, \frac{y_{ik}}{[\boldsymbol{A}\,\boldsymbol{X}]_{ik}} \right)^{\omega} \right)^{1+\alpha_{sA}}$ $a_{ij} \leftarrow a_{ij} / (\sum_{p} a_{pj})$
K-L divergence (dual) $\sum_{ik} ([\boldsymbol{AX}]_{ik} \log \frac{[\boldsymbol{AX}]_{ik}}{y_{ik}} + y_{ik} - [\boldsymbol{AX}]_{ik})$ $(\beta = 0)$	$x_{jk} \leftarrow \left(x_{jk} \prod_{i=1}^{m} \left(\frac{y_{ik}}{[\boldsymbol{AX}]_{ik}} \right)^{\omega \, a_{ij}} \right)^{1+\alpha_{sX}}$ $a_{ij} \leftarrow \left(a_{ij} \prod_{k=1}^{N} \left(\frac{y_{ik}}{[\boldsymbol{AX}]_{ik}} \right)^{\tilde{\eta}_j x_{jk}} \right)^{1+\alpha_{sA}}$ $a_{ij} \leftarrow a_{ij} / (\sum_{p} a_{pj}), \; \tilde{\eta}_j = \omega \, (\sum_{k} x_{jk})^{-1}$
Euclidean distance $\Vert \boldsymbol{Y} - [\boldsymbol{AX}] \Vert_F^2 + \alpha_X f_X(\boldsymbol{X}) + \alpha_A f_A(\boldsymbol{A})$	$x_{jk} \leftarrow x_{jk} \frac{\left[[\boldsymbol{A}^T \boldsymbol{Y}]_{ik} - \alpha_X \, \psi_X(\boldsymbol{X}) \right]_{\varepsilon}}{[\boldsymbol{A}^T \boldsymbol{A}\,\boldsymbol{X}]_{ik} + \varepsilon}$ $a_{ij} \leftarrow a_{ij} \frac{\left[[\boldsymbol{Y}\,\boldsymbol{X}^T]_{ij} - \alpha_A \, \psi_A(\boldsymbol{A}) \right]_{\varepsilon}}{[\boldsymbol{A}\,\boldsymbol{X}\,\boldsymbol{X}^T]_{ij} + \varepsilon}$
Kompass generalized divergence $\sum_{ik} (y_{ik} \frac{y_{ik}^{\beta-1} - [\boldsymbol{AX}]_{ik}^{\beta-1}}{\beta(\beta-1)} +$ $+ [\boldsymbol{AX}]_{ik}^{\beta-1} \frac{[\boldsymbol{AX}]_{ik} - y_{ik}}{\beta}) + \alpha_X f_X(\boldsymbol{X})$	$x_{jk} \leftarrow x_{jk} \frac{[\sum_{i=1}^{m} a_{ij} \, (y_{ik} / [\boldsymbol{AX}]_{ik}^{2-\beta}) - \alpha_X \psi_X(\boldsymbol{X})]_{\varepsilon}}{\sum_{i=1}^{m} a_{ij} \, [\boldsymbol{AX}]_{ik}^{\beta-1} + \varepsilon}$ $a_{ij} \leftarrow \left(a_{ij} \frac{[\sum_{k=1}^{N} x_{jk} \, (y_{ik} / [\boldsymbol{AX}]_{ik}^{2-\beta})]_{\varepsilon}}{\sum_{k=1}^{N} x_{jk} \, [\boldsymbol{AX}]_{ik}^{\beta-1} + \varepsilon} \right)^{1+\alpha_{sA}}$ $a_{ij} \leftarrow a_{ij} / (\sum_{p} a_{pj}), \;\; \beta \in [0, 2]$

4 Simulation Results

All the NMF algorithms discussed in this paper (see Table 1) have been extensively tested for many difficult benchmarks for sparse signals and images with various statistical distributions. The simulation results confirmed that the developed algorithms are stable, efficient and provide consistent results for a wide set of parameters. Due to the limit of space we give here only one illustrative example: Nine nonnegative sparse signals (some of them are statistically dependent) shown in Fig.1 (a) have been mixed by randomly generated nonnegative matrix $\boldsymbol{A} \in \mathbb{R}^{18 \times 9}$. To the mixture we added an uniform distributed noise with SNR=20 dB. The mixed signals are shown in Fig.1 (b). Using the known standard NMF algorithm (5) we failed to estimate the original sources (see Fig.1 (c)). However, for the new algorithms we reconstructed successfully all the sources. Fig. 1 (d) illustrates the results obtained by using algorithm (12) with $\beta = 2$ and the nonlinear projection with $\alpha_{sX} = \alpha_{sA} = 0.002$ (see also Table 1). Similar

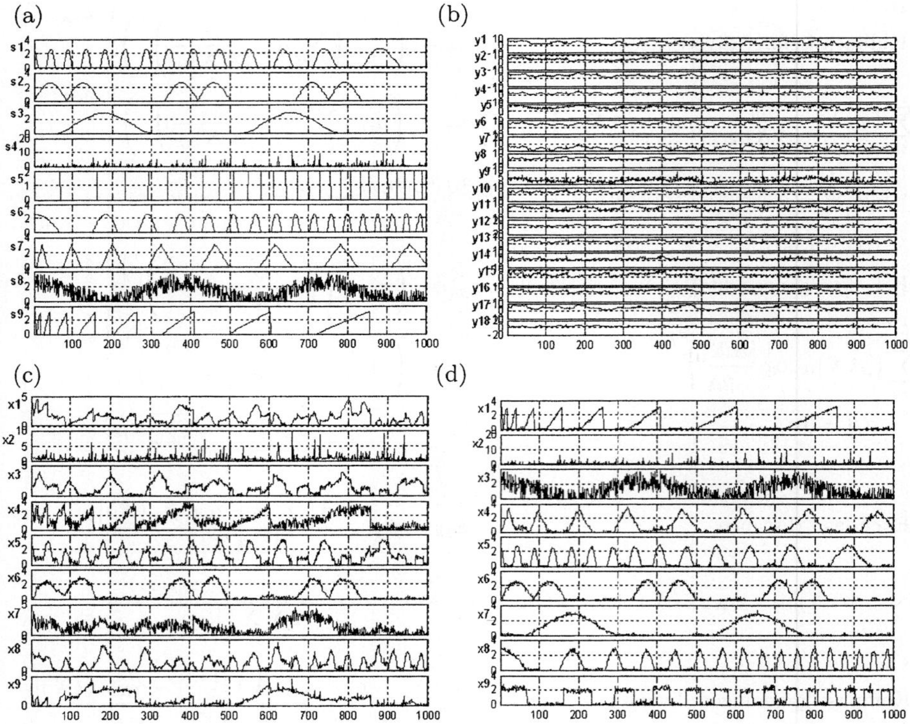

Fig. 1. Example 1: (a) The original 9 source signals, (b) observed 18 mixed signals, (c) Estimated sources using standard Lee-Seung algorithm (5) (d) Estimated source signals using the new algorithm (12) for $\beta = 2$ with nonlinear projection $\alpha_{sX} = \alpha_{sA} = 0.002$ with SIR=32dB, 20dB, 19dB, 18dB, 23dB, 25dB, 27dB, 26dB, 19dB, for individual sources respectively

or even slightly better performance we achieved by applying the other proposed algorithms with regularization/projection (Table 1).

5 Conclusions and Discussion

In this paper we discuss loss functions which allow to derive us a very large class of flexible, robust and efficient NMF adaptive algorithms. The optimal choice of β depends on the distribution of data and *a priori* knowledge about noise. If such knowledge is not available, we may run NMF algorithms for various sets of parameters to find an optimal solution. For some tasks and distributions there are particular divergence measures that are uniquely suited. On the other hand, if the approximating model fits the true distribution well, then it does not matter which divergence measure is used, since all of them will give similar results. The discussed loss functions are jointly convex. This property is stronger than the individual convexity in $\{y_{ik}\}$ and $\{z_{ik}\}$. However, it very difficult to prove the global convergence of the derived NMF algorithms. Our simulation experiments indicate that for $m >> n$, typically $m > 5n$ and $N = 10^3 \sim 10^4$, we usually avoid stucking in poor local minima. We found by extensive simulations that regularization/projections techniques play a key role in improving the performance of blind source separation by using the NMF approach.

References

1. Amari, S.: Differential-Geometrical Methods in Statistics. Springer Verlag (1985)
2. Amari, S.: Information geometry of the EM and em algorithms for neural networks. Neural Networks **8** (1995) 1379–1408.
3. Lee, D.D., Seung, H.S.: Learning of the parts of objects by non-negative matrix factorization. Nature **401** (1999) 788–791.
4. Hoyer, P.: Non-negative matrix factorization with sparseness constraints. Journal of Machine Learning Research **5** (2004) 1457–1469.
5. Sajda, P., Du, S., Parra, L.: Recovery of constituent spectra using non-negative matrix factorization. In: Proceedings of SPIE – Volume 5207, Wavelets: Applications in Signal and Image Processing (2003) 321–331.
6. Cho, Y.C., Choi, S.: Nonnegative features of spectro-temporal sounds for classification. Pattern Recognition Letters **26** (2005) 1327–1336.
7. Lee, D.D., Seung, H.S.: Algorithms for nonnegative matrix factorization. Volume 13. NIPS, MIT Press (2001)
8. Cichocki, A., Amari, S.: Adaptive Blind Signal And Image Processing (New revised and improved edition). John Wiley, New York (2003)
9. Byrne, C.: Accelerating the EMML algorithm and related iterative algorithms by rescaled block-iterative (RBI) methods. IEEE Transactions on Image Processing **7** (1998) 100 - 109.
10. Csiszár, I.: Information measures: A critical survey. In: Prague Conference on Information Theory, Academia Prague. Volume A. (1974) 73–86.
11. Cressie, N.A., Read, T.: Goodness-of-Fit Statistics for Discrete Multivariate Data. Springer, New York (1988)
12. Kompass, R.: A generalized divergence measure for nonnegative matrix factorization, Neuroinfomatics Workshop, Torun, Poland (September, 2005)

Second-Order Blind Identification of Underdetermined Mixtures*

Lieven De Lathauwer and Joséphine Castaing

ETIS (CNRS, ENSEA, UCP), UMR 8051, Cergy-Pontoise, France
{delathau, castaing}@ensea.fr

Abstract. In this paper we show that the underdetermined ICA problem can be solved using a set of spatial covariance matrices, in case the sources have sufficiently different temporal autocovariance functions. The result is based on a link with the decomposition of higher-order tensors in rank-one terms. We discuss two algorithms and present theoretical bounds on the number of sources that can be allowed.

1 Introduction

Let us use the following notation for the basic Independent Component Analysis (ICA) model:

$$\mathbf{x}_t = \mathbf{A}\mathbf{s}_t + \mathbf{n}_t, \tag{1}$$

in which the observation vector $\mathbf{x}_t \in \mathbb{C}^J$, the noise vector $\mathbf{n}_t \in \mathbb{C}^J$ and the source vector $\mathbf{s}_t \in \mathbb{C}^R$ are zero-mean. The mixing matrix $\mathbf{A} \in \mathbb{C}^{J \times R}$. The goal is to exploit the assumed mutual statistical independence of the source components to estimate the mixing matrix and/or the source signals from the observations. The literature on ICA addresses for the most part the so-called *overdetermined* case, where $J \geqslant R$. Here, we consider the *underdetermined* or *overcomplete* case, where $J < R$.

A large class of algorithms for underdetermined ICA starts from the assumption that the sources are (quite) sparse [2, 12, 15, 22]. In this case, the scatter plot typically shows high signal values in the directions of the mixing vectors. These extrema may be localized by maximization of some clustering measure [2, 12]. Some of these techniques are based on an exhaustive search in the mixing vector space, and are therefore very expensive when there are more than two observation channels. In a preprocessing step a linear transform may be applied such that the new representation of the data is sparser (e.g. short-time Fourier transform in the case of audio signals) [2].

* L. De Lathauwer holds a permanent research position with the French CNRS; he also holds a honorary position with the K.U.Leuven. This work is supported in part by the Research Council K.U.Leuven under Grant GOA-AMBioRICS, in part by the Flemish Government under F.W.O. Project G.0321.06, Tournesol 2005 - Project T20013, and F.W.O. research communities ICCoS, ANMMM, and in part by the Belgian Federal Science Policy Office under IUAP P5/22.

J. Rosca et al. (Eds.): ICA 2006, LNCS 3889, pp. 40–47, 2006.

There are two aspects to ICA: estimation of the mixing matrix and separation of the sources. In the overdetermined case, sources are usually separated by multiplying the observations with the pseudo-inverse of the mixing matrix estimate. This is no longer possible in the case of underdetermined mixtures: for each sample $\mathbf{x}_t$, the corresponding source sample $\mathbf{s}_t$ that satisfies $\mathbf{x}_t = \mathbf{A}\mathbf{s}_t$ is only known to belong to an affine variety of dimension $J - R$ — hence the term "underdetermined". One could estimate the sources and the mixing matrix simultaneously by exploiting prior knowledge on the sources [15, 22]. An other approach is to estimate the mixing matrix first, and then estimate the sources. In this paper we will show that the estimation of the mixing matrix is actually an overdetermined subproblem, even in the case of underdetermined mixtures. If the source densities are known, then $\mathbf{s}_t$ may subsequently be estimated by maximizing the log posterior likelihood [15]. In the case of Laplacian probability densities, modelling sparse sources, this can be formulated in terms of a linear programming problem [2, 15]. In the case of finite alphabet signals in telecommunication, one may perform an exhaustive search over all possible combinations. In this paper we will focus on the estimation of the mixing matrix.

This paper presents new contributions to the class of algebraic algorithms for underdetermined ICA. In [6, 7, 8] algorithms are derived for the case of two mixtures and three sources. An arbitrary number of mixing vectors can be estimated from two observation channels by sampling derivatives of sufficiently high order of the second characteristic function [19]. Algebraic underdetermined ICA is based on the decomposition of a higher-order tensor in a sum of rank-1 terms. Some links with the literature on homogeneous polynomials are discussed in [5].

In this paper we assume that the sources are individually correlated in time. The spatial covariance matrices of the observations then satisfy [1]:

$$\begin{aligned} \mathbf{C}_1 \equiv E\{\mathbf{x}_t \mathbf{x}_{t+\tau_1}^H\} &= \mathbf{A} \cdot \mathbf{D}_1 \cdot \mathbf{A}^H \\ &\vdots \\ \mathbf{C}_K \equiv E\{\mathbf{x}_t \mathbf{x}_{t+\tau_K}^H\} &= \mathbf{A} \cdot \mathbf{D}_K \cdot \mathbf{A}^H \end{aligned} \tag{2}$$

in which $\mathbf{D}_k \equiv E\{\mathbf{s}_t \mathbf{s}_{t+\tau_k}^H\}$ is diagonal, $k = 1, \ldots, K$. For simplicity, we have dropped the noise terms; they can be considered as a perturbation of (2).

Let us stack the matrices $\mathbf{C}_1, \ldots, \mathbf{C}_K$ in Eq. (2) in a tensor $\mathcal{C} \in \mathbb{C}^{J \times J \times K}$. Define a matrix $\mathbf{D} \in \mathbb{C}^{K \times R}$ by $(\mathbf{D})_{kr} \equiv (\mathbf{D}_k)_{rr}$, $k = 1, \ldots, K$, $r = 1, \ldots, R$. Then we have

$$c_{ijk} = \sum_{r=1}^{R} a_{ir} a_{jr}^* d_{kr}, \tag{3}$$

which we write as

$$\mathcal{C} = \sum_{r=1}^{R} \mathbf{a}_r \circ \mathbf{a}_r^* \circ \mathbf{d}_r, \tag{4}$$

in which $\circ$ denotes the tensor outer product and in which $\{\mathbf{a}_r\}$ and $\{\mathbf{d}_r\}$ are the columns of $\mathbf{A}$ and $\mathbf{D}$, respectively. Eq. (4) is a decomposition of tensor $\mathcal{C}$ in a sum of R rank-1 terms. In the literature, this is called a "Canonical Decomposition" (CANDECOMP) [4] or a "Parallel Factors Model" (PARAFAC) [13]. The minimal number of rank-1 tensors in which a higher-order tensor can be decomposed, is called its rank. Note that each rank-1 term in (4) consists of the contribution of one distinct source to $\mathcal{C}$. Hence, in terms of this tensor, "source separation" amounts to the computation of decomposition (4), provided it is unique. In contrast to the matrix case, PARAFAC can be unique (up to some trivial indeterminacies) even when (i) the rank-1 terms are not mutually orthogonal and (ii) the rank is greater than the smallest tensor dimension. This allows for the determination of the mixing matrix (up to a scaling and permutation of its columns) in the overcomplete case.

Uniqueness issues are discussed in Section 2. Section 3 and 4 present algorithms for the computation of decomposition (4). Section 3 deals with the case where $R > K$. More powerful results are obtained for the case where $R \leqslant K$ in Section 4. Section 5 shows the results of some simulations. The presentation is in terms of complex signals. Whenever the results cannot be directly applied to real data, this will be explicitly mentioned.

This paper is a short version of [11]. The foundations of Section 4 were laid in [3]. Some mathematical aspects are developed in more detail in [10].

2 PARAFAC Uniqueness

PARAFAC can only be unique up to permutation of the rank-1 terms and scaling and counterscaling of the factors of the rank-1 terms. We call the decomposition in (4) essentially unique if any other matrix pair $\mathbf{A}'$ and $\mathbf{D}'$ that satisfies (4), is related to $\mathbf{A}$ and $\mathbf{D}$ via

$$\mathbf{A} = \mathbf{A}' \cdot \mathbf{P} \cdot \boldsymbol{\Omega}_1 \qquad \mathbf{D} = \mathbf{D}' \cdot \mathbf{P} \cdot \boldsymbol{\Omega}_2, \tag{5}$$

with $\boldsymbol{\Omega}_1$, $\boldsymbol{\Omega}_2$ diagonal matrices, satisfying $\boldsymbol{\Omega}_1 \cdot \boldsymbol{\Omega}_1^* \cdot \boldsymbol{\Omega}_2 = \mathbf{I}$, and $\mathbf{P}$ a permutation matrix.

A first uniqueness result requires the notion of Kruskal-rank or k-rank $k(\mathbf{A})$ of a matrix $\mathbf{A}$ [14]. It is defined as the maximal number k such that any set of k columns of $\mathbf{A}$ is linearly independent. From [14, 18] we have then immediately that decomposition (4) is essentially unique when

$$2k(\mathbf{A}) + k(\mathbf{D}) \geqslant 2(R+1). \tag{6}$$

For a second uniqueness condition we assume that the number of sources R does not exceed the number of covariance matrices K. We call decomposition (4) generic when the (noiseless) entries of $\mathbf{A}$ and $\mathbf{D}$ can be considered drawn from continuous probability densities. It turns out that in the complex case the generic decomposition is essentially unique when $2R(R-1) \leqslant J^2(J-1)^2$ [10, 11]. For real-valued tensors, we have uniqueness if $R \leqslant R_{max}$, given by [18]:

J	2	3	4	5	6	7	8
R_{max}	2	4	6	10	15	20	26

3 Case 1: $R > K$

Generically, a matrix is full rank and full k-rank. Hence, in practice, $k(\mathbf{A}) = \min(J, R) = J$ and $k(\mathbf{D}) = \min(K, R) = K$ if $R > K$. Eq. (6) then guarantees identifiability if $2J + K \geqslant 2R + 2$, i.e., when the number of sources $R \leqslant J - 1 + K/2$.

The standard way to compute PARAFAC, is by means of an "Alternating Least Squares" (ALS) algorithm [13]. The aim is to minimize the (squared) Frobenius norm of the difference between $\mathcal{C}$ and its estimated decomposition in rank-1 terms by means of an iteration in which each step consists of fixing a subset of unknown parameters to their current estimates, and optimizing w.r.t. the remaining unknowns, followed by fixing an other subset of parameters, and optimizing w.r.t. the complimentary set, etc. (Like for matrices, the squared Frobenius norm of a tensor is the sum of the squared moduli of its entries.) More specifically, one optimizes the cost function

$$f(\mathbf{U}, \mathbf{V}, \mathbf{D}) = \|\mathcal{C} - \sum_{r=1}^{R} \mathbf{u}_r \circ \mathbf{v}_r^* \circ \mathbf{d}_r\|^2. \tag{7}$$

Due to the multi-linearity of the model, estimation of one of the arguments, given the other two, is a classical linear least squares problem. One alternates between updates of $\mathbf{U}$, $\mathbf{V}$ and $\mathbf{D}$. After updating $\mathbf{U}$ and $\mathbf{V}$, their columns are rescaled to unit length, to avoid under- and overflow. Although during the iteration the symmetry of the problem is broken, one supposes that eventually $\mathbf{U}$ and $\mathbf{V}$ both converge to $\mathbf{A}$. If some difference remains, then the mixing vector $\mathbf{a}_r$ can be estimated as the dominant left singular vector of the matrix $[\mathbf{u}_r \, \mathbf{v}_r]$, $r = 1, \ldots, R$. The rank of $\mathcal{C}$ is estimated by trial-and-error. In [16] an exact line search is proposed to enhance the convergence of the ALS algorithm.

4 Case 2: $R \leqslant K$

In this case, one can still work as in the previous section. However, more powerful results can be derived. We assume that the second uniqueness condition in Section 2 is satisfied. This implies in particular that $R < J^2$.

We stack the entries of tensor $\mathcal{C}$ in a $(J^2 \times K)$ matrix $\mathbf{C}$ as follows:

$$(\mathbf{C})_{(i-1)J+j,k} = c_{ijk}, \quad i \in [1, J], j \in [1, J], k \in [1, K].$$

Eq. (4) can be written in a matrix format as:

$$\mathbf{C} = (\mathbf{A} \odot \mathbf{A}^*) \cdot \mathbf{D}^T, \tag{8}$$

in which $\odot$ denotes the Khatri-Rao or column-wise Kronecker product, i.e., $\mathbf{A} \odot \mathbf{A}^* \equiv [\mathbf{a}_1 \otimes \mathbf{a}_1^*, \ldots, \mathbf{a}_R \otimes \mathbf{a}_R^*]$. If $R \leqslant \min(J^2, K)$, then $\mathbf{A} \odot \mathbf{A}^*$ and $\mathbf{S}$ are generically full rank [10]. This implies that the number of sources R is simply equal to the rank of $\mathbf{C}$. Instead of determining it by trial-and-error, as in the previous section, it can be estimated as the number of significant singular values of $\mathbf{C}$. Let the "economy size" Singular Value Decomposition (SVD) of $\mathbf{C}$ be given by:

$$\mathbf{C} = \mathbf{U} \cdot \mathbf{\Sigma} \cdot \mathbf{V}^H, \tag{9}$$

in which $\mathbf{U} \in \mathbb{C}^{J^2 \times R}$ and $\mathbf{V} \in \mathbb{C}^{K \times R}$ are column-wise orthonormal matrices and in which $\mathbf{\Sigma} \in \mathbb{R}^{R \times R}$ is positive diagonal. We deduce from (8) and (9) that there exists an a priori unknown matrix $\mathbf{F} \in \mathbb{C}^{R \times R}$ that satisfies:

$$\mathbf{A} \odot \mathbf{A}^* = \mathbf{U} \cdot \mathbf{\Sigma} \cdot \mathbf{F}. \tag{10}$$

If we would know $\mathbf{F}$, then the mixing matrix $\mathbf{A}$ would immediately follow. Stack the entries of the columns $\mathbf{m}_r$ of $\mathbf{A} \odot \mathbf{A}^*$ in $(J \times J)$ matrices $\mathbf{M}_r$ as follows: $(\mathbf{M}_r)_{ij} \equiv (\mathbf{m}_r)_{(i-1)J+j}$, $i, j = 1, \ldots, J$. Then $\mathbf{M}_r$ is theoretically a rank-one matrix: $\mathbf{M}_r = \mathbf{a}_r \mathbf{a}_r^H$. This means that $\mathbf{a}_r$ can, up to an irrelevant scaling factor, be determined as the left singular vector associated with the largest singular value of $\mathbf{M}_r$, $r = 1, \ldots, R$.

We will now explain how the matrix $\mathbf{F}$ in (10) can be found. Let $\mathbf{E}_r$ be a $(J \times J)$ matrix in which the rth column of matrix $\mathbf{U\Sigma}$ is stacked as above. We have

$$\mathbf{E}_r = \sum_{k=1}^{R} \left(\mathbf{a}_k \mathbf{a}_k^H\right) (\mathbf{F}^{-1})_{kr}. \tag{11}$$

This means that the matrices $\mathbf{E}_r$ consist of linear combinations of the rank-one matrices $\mathbf{a}_k \mathbf{a}_k^H$ and that the linear combinations are the entries of the nonsingular matrix $\mathbf{F}^{-1}$. It would be helpful to have a tool that allows us to determine whether a matrix is rank-one or not. Such a tool is offered by the following theorem [3, 10].

Theorem 1. *Consider the mapping* Φ: $(\mathbf{X}, \mathbf{Y}) \in \mathbb{C}^{J \times J} \times \mathbb{C}^{J \times J} \longmapsto \Phi(\mathbf{X}, \mathbf{Y}) = \mathcal{P} \in \mathbb{C}^{J \times J \times J \times J}$ *defined by:*

$$p_{ijkl} = x_{ij} y_{kl} + y_{ij} x_{kl} - x_{il} y_{kj} - y_{il} x_{kj}$$

for all index values. Given $\mathbf{X} \in \mathbb{C}^{J \times J}$, $\Phi(\mathbf{X}, \mathbf{X}) = 0$ *if and only if the rank of* $\mathbf{X}$ *is at most one.*

From the matrix $\mathbf{U\Sigma}$ in the SVD (9) we construct the set of R^2 tensors $\{\mathcal{P}_{rs} \equiv \Phi(\mathbf{E}_r, \mathbf{E}_s)\}_{r,s \in [1,R]}$. It can now be proved [3, 10, 11] that generically there exist exactly R linearly independent symmetric matrices $\mathbf{B}_r \in \mathbb{C}^{R \times R}$ that satisfy:

$$\sum_{t,u=1}^{R} \mathcal{P}_{tu} (\mathbf{B}_r)_{tu} = 0. \tag{12}$$

Moreover, these matrices can all be diagonalized by $\mathbf{F}$:

$$\begin{aligned} \mathbf{B}_1 &= \mathbf{F} \cdot \mathbf{\Lambda}_1 \cdot \mathbf{F}^T \\ &\vdots \\ \mathbf{B}_R &= \mathbf{F} \cdot \mathbf{\Lambda}_R \cdot \mathbf{F}^T \end{aligned} \tag{13}$$

in which $\mathbf{\Lambda}_1, \ldots, \mathbf{\Lambda}_R$ are diagonal. Eqs. (12) and (13) provide the means to find $\mathbf{F}$. Eq. (12) is just a linear set of equations, from which the matrices $\mathbf{B}_r$ may be computed. Note that, in the absence of noise, $\mathbf{F}$ already follows from the Eigenvalue Decomposition (EVD)

$$\mathbf{B}_1 \cdot \mathbf{B}_2^{-1} = \mathbf{F} \cdot (\mathbf{\Lambda}_1 \cdot \mathbf{\Lambda}_2^{-1}) \cdot \mathbf{F}^{-1}.$$

In practice, it is more reliable to take all matrices in (13) into account. The set can be simultaneously diagonalized by means of the algorithms presented in [9, 13, 20, 21].

5 Simulations

We conduct a Monte-Carlo experiment consisting of 100 runs. In each run, signal values are drawn from a standardized complex Gaussian distribution and subsequently passed through a filter of which the coefficients are rows of a (16×16) Hadamard matrix. (More specifically, rows 1, 2, 4, 7, 8 and 13 are considered.) The resulting sources are mixed by means of a matrix of which the entries are also drawn from a standardized complex Gaussian distribution and additive Gaussian noise is added.

In the first experiment, we assume $J = 4$ observation channels and $R = 5$ sources. Covariance matrices are computed for $\tau = 0, \ldots, 3$. This problem is quite difficult since two of the (4×4) submatrices of $\mathbf{D}$ have a condition number of about 30, which indicates some lack of "diversity" for these submatrices. The number of samples $T = 10000$. The mixing matrix is computed by means of the ALS algorithm described in Section 3. In Fig. 1 we plot the mean relative error $E\{\|\mathbf{A} - \hat{\mathbf{A}}\|/\|\mathbf{A}\|\}$, in which the norm is the Frobenius-norm. (The columns of $\mathbf{A}$ are normalized to unit length and $\hat{\mathbf{A}}$ represents the optimally ordered and scaled estimate.)

In the second experiment, we compute 12 covariance matrices ($\tau = 0, \ldots, 11$). This makes the problem better conditioned. We consider $R = 5$ or 6 sources.

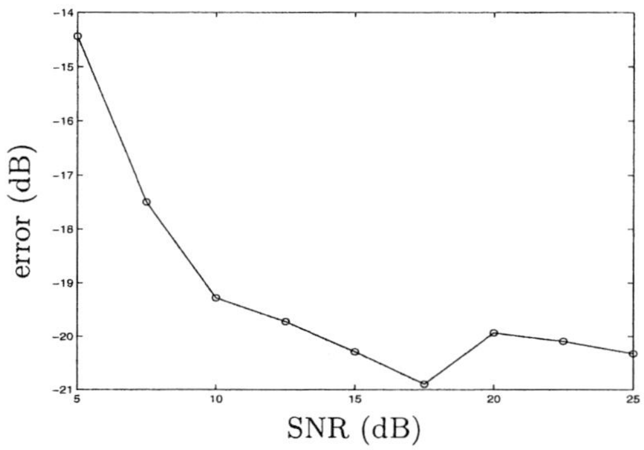

Fig. 1. Relative error in the first experiment ($K = 4$)

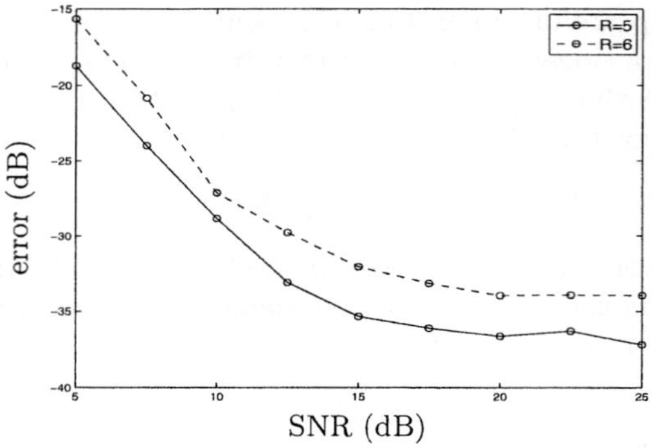

Fig. 2. Relative error in the second experiment ($K = 12$)

The number of samples $T = 5000$. The mixing matrix is computed by means of the algorithm described in Section 4, where we used the algorithm derived in [20] for the simultaneous diagonalization. The mean relative error is shown in Fig. 2. Note that the mixing matrix is estimated with an accuracy of two digits.

To have a reference, we also computed the solution by means of the AC-DC algorithm proposed in [21]. In neither experiment, AC-DC yielded a reliable estimate of the mixing matrix.

6 Conclusion

In this paper we exploited differences in autocovariance to solve the underdetermined ICA problem. The joint decomposition of a set of spatial covariance matrices was interpreted as the decomposition in rank-one terms of the third-order tensor in which these matrices can be stacked. We distinguished between two cases, depending on whether the number of covariance matrices K exceeds the number of sources R or not. For both cases, we presented theoretical bounds on the number of sources that can be allowed and discussed algebraic algorithms. We explained that, in the case $K > R$, the noise-free solution can be obtained by means of an EVD. The same approach can be used for nonstationary sources, by considering spatial covariance matrices at different time instances, sets of spatial time-frequency distributions, etc.

References

1. A. Belouchrani, et al., "A Blind Source Separation Technique using Second Order Statistics," *IEEE Trans. Signal Processing*, Vol. 45, No. 2, Feb. 1997, pp. 434–444.
2. P. Bofill, M. Zibulevsky, "Underdetermined Blind Source Separation Using Sparse Representations," *Signal Process.*, Vol. 81, 2001, pp. 2353–2362.
3. J.-F. Cardoso, "Super-symmetric Decomposition of the Fourth-Order Cumulant Tensor. Blind Identification of More Sources than Sensors," *Proc. ICASSP-91*, Toronto, Canada, 1991, pp. 3109–3112.

4. J. Carroll, J. Chang, "Analysis of Individual Differences in Multidimensional Scaling via an N-way Generalization of "Eckart-Young" Decomposition, *Psychometrika*, Vol. 9, 1970, pp. 267–283.
5. P. Comon, B. Mourrain, "Decomposition of Quantics in Sums of Powers of Linear Forms," *Signal Process.*, Vol. 53, No. 2, Sept. 1996, pp. 93–107.
6. P. Comon, "Blind Identification and Source Separation in 2×3 Under-Determined Mixtures," *IEEE Trans. Signal Processing*, Vol. 52, No. 1, Jan. 2004, pp. 11–22.
7. L. De Lathauwer, P. Comon, B. De Moor, "ICA Algorithms for 3 Sources and 2 Sensors," *Proc. Sixth Sig. Proc. Workshop on Higher Order Statistics*, Caesarea, Israel, June 14–16, 1999, pp. 116–120.
8. L. De Lathauwer, B. De Moor, J. Vandewalle, "An Algebraic ICA Algorithm for 3 Sources and 2 Sensors," *Proc. Xth European Signal Processing Conference (EUSIPCO 2000)*, Tampere, Finland, Sept. 5–8, 2000.
9. L. De Lathauwer, B. De Moor, J. Vandewalle, "Computation of the Canonical Decomposition by Means of Simultaneous Generalized Schur Decomposition", *SIAM J. Matrix Anal. Appl.*, Vol. 26, 2004, pp. 295–327.
10. L. De Lathauwer, "A Link between the Canonical Decomposition in Multilinear Algebra and Simultaneous Matrix Diagonalization," submitted.
11. L. De Lathauwer, J. Castaing, "Independent Component Analysis Based on Simultaneous Matrix Diagonalization: the Underdetermined Case," in preparation.
12. D. Erdogmus, L. Vielva, J.C. Principe, "Nonparametric Estimation and Tracking of the Mixing Matrix for Underdetermined Blind Source Separation", *Proc. ICA'01*, San Diego, CA, Dec. 2001, pp. 189–194.
13. R. Harshman, "Foundations of the PARAFAC Procedure: Model and Conditions for an "Explanatory" Multi-mode Factor Analysis, *UCLA Working Papers in Phonetics*, Vol. 16, 1970, pp. 1–84.
14. J.B. Kruskal, "Three-Way Arrays: Rank and Uniqueness of Trilinear Decompositions, with Application to Arithmetic Complexity and Statistics", *Linear Algebra and its Applications*, No. 18, 1977, pp. 95–138.
15. M. Lewicki, T.J. Sejnowski, "Learning Overcomplete Representations", *Neural Computation*, Vol. 12, 2000, pp. 337–365.
16. M. Rajih, P. Comon, "Enhanced Line Search: A Novel Method to Accelerate Parafac," *Proc. Eusipco'05*, Antalya, Turkey, Sept. 4–8, 2005.
17. A. Stegeman, J.M.F. ten Berge, L. De Lathauwer, "Sufficient Conditions for Uniqueness in Candecomp/Parafac and Indscal with Random Component Matrices," *Psychometrika*, to appear.
18. A. Stegeman, N.D. Sidiropoulos, "On Kruskal's Uniqueness Condition for the Candecomp/Parafac Decomposition," Tech. Report, Heijmans Inst., Univ. Groningen, submitted.
19. A. Taleb, "An Algorithm for the Blind Identification of N Independent Signals with 2 Sensors," *Proc. 16th Int. Symp. on Signal Processing and Its Applications (ISSPA'01)*, Kuala-Lumpur, Malaysia, Aug. 13–16, 2001, pp. 5–8.
20. A.-J. van der Veen, A. Paulraj, "An Analytical Constant Modulus Algorithm", *IEEE Trans. Signal Processing*, Vol. 44, 1996, pp. 1136–1155.
21. A. Yeredor, "Non-orthogonal Joint Diagonalization in the Least-Squares Sense with Application in Blind Source Separation," *IEEE Trans. Signal Processing*, Vol. 50, 2002, pp. 1545–1553.
22. M. Zibulevsky, B.A. Pearlmutter, "Blind Source Separation by Sparse Decomposition in a Signal Dictionary," in: S.J. Robert, R.M.Everson (Eds.), *Independent Component Analysis: Principles and Practice*, Cambridge University Press, Cambridge, 2000.

Differential Fast Fixed-Point BSS for Underdetermined Linear Instantaneous Mixtures

Yannick Deville, Johanna Chappuis, Shahram Hosseini, and Johan Thomas

Laboratoire d'Astrophysique de Toulouse-Tarbes,
Observatoire Midi-Pyrénées - Université Paul Sabatier,
14 Av. Edouard Belin, 31400 Toulouse, France
{ydeville, shosseini, jthomas}@ast.obs-mip.fr,
johanna.chappuis@airbus.com

Abstract. This paper concerns underdetermined linear instantaneous blind source separation (BSS), i.e. the case when the number P of observed mixed signals is lower than the number N of sources. We propose a partial BSS method, which separates P supposedly non-stationary sources of interest one from the others (while keeping residual components for the other $N-P$, supposedly stationary, "noise" sources). This method is based on the general differential BSS concept that we introduced before. Unlike our previous basic application of that concept, this improved method consists of a differential extension of the FastICA method (which does not apply to underdetermined mixtures), thus keeping the attractive features of the latter algorithm. Our approach is therefore based on a differential sphering, followed by the optimization of the differential kurtosis that we introduce in this paper. Experimental tests show that this differential method is much more robust to noise than standard FastICA.

1 Introduction

Blind source separation (BSS) methods [9] aim at restoring a set of N unknown source signals $s_j(n)$ from a set of P observed signals $x_i(n)$. The latter signals are linear instantaneous mixtures of the source signals in the basic case, i.e.

$$x(n) = As(n) \tag{1}$$

where $s(n) = [s_1(n) \dots s_N(n)]^T$ and $x(n) = [x_1(n) \dots x_P(n)]^T$ are the source and observation vectors, and A is a constant mixing matrix. We here assume that the signals and mixing matrix are real-valued and that the sources are centered and statistically independent. Moreover, we consider the underdetermined case, i.e. $P < N$, and we require that $P \geq 2$. Some analyses and statistical BSS methods have been reported for this difficult case (see e.g. [2],[3],[4],[7],[10]). However, they set major restrictions on the source properties (discrete sources are especially considered) and/or on the mixing conditions. Other reported approaches use in several ways the assumed sparsity of the sources (see e.g. [1] and references

J. Rosca et al. (Eds.): ICA 2006, LNCS 3889, pp. 48–56, 2006.

therein). In [6], we introduced a general differential BSS concept for processing underdetermined mixtures. In its standard version, we consider the situation when (at most) P of the N mixed sources are non-stationary while the other $N - P$ sources (at least) are stationary. The P non-stationary sources are the signals of interest in this approach, while the $N - P$ stationary sources are considered as "noise sources". Our differential BSS concept then achieves the "partial BSS" of the P sources of interest, i.e. it yields output signals which each contain contributions from only one of these P sources, still superimposed with some residual components from the noise sources (this is described in [6]).

Although we first defined this differential BSS concept in a quite general framework in [6], we then only applied it to a simple but restrictive BSS method, which is especially limited to $P = 2$ mixtures and based on slow-convergence algorithms. We here introduce a much more powerful BSS criterion and associated algorithms, based on differential BSS. This method is obtained by extending to underdetermined mixtures the kurtotic separation criterion [5] and the associated, fast converging, fixed-point, FastICA algorithm [8], thus keeping the attractive features of the latter algorithm.

2 Proposed Differential BSS Method

2.1 A New BSS Criterion Based on Differential Kurtosis

The standard FastICA method [8], which is only applicable to the case when $P = N$ (or $P > N$), extracts a source by means of a two-stage procedure. The first stage consists in transferring the observation vector $x(n)$ through a real PxP matrix M, which yields the vector

$$z(n) = Mx(n). \tag{2}$$

In the standard FastICA method, M is selected so as to sphere the observations, i.e. so as to spatially whiten and normalize them. The second stage of that standard method then consists in deriving an output signal $y_i(n)$ as a linear instantaneous combination of the signals contained by $z(n)$, i.e

$$y_i(n) = w^T z(n) \tag{3}$$

where w is a vector, which is constrained so that $\| w \| = 1$. This vector w is selected so as to optimize the (non-normalized) kurtosis of $y_i(n)$, defined as its zero-lag 4th-order cumulant

$$K_{y_i}(n) = cum(y_i(n), y_i(n), y_i(n), y_i(n)). \tag{4}$$

Now consider the underdetermined case, i.e. $P < N$. We again derive an output signal $y_i(n)$ according to (2) and (3). We aim at defining how to select M and w, in order to achieve the above-defined partial BSS of the P sources of interest. To this end, we apply the general differential BSS concept that we described in [6] to the specific kurtotic criterion used in the standard FastICA method.

We therefore consider two times n_1 and n_2. We then introduce the differential (non-normalized) kurtosis that we associate to (4) for these times. We define this parameter as

$$DK_{y_i}(n_1, n_2) = K_{y_i}(n_2) - K_{y_i}(n_1). \tag{5}$$

Let us show that, whereas the standard parameter $K_{y_i}(n)$ depends on all sources, its differential version $DK_{y_i}(n_1, n_2)$ only depends on the non-stationary sources. Eq. (1), (2) and (3) yield

$$y_i(n) = v^T s(n) \tag{6}$$

where the vector

$$v = (MA)^T w \tag{7}$$

includes the effects of the mixing and separating stages. Denoting v_q, with $q = 1 \ldots N$, the entries of v, (6) implies that the output signal $y_i(n)$ may be expressed with respect to all sources as

$$y_i(n) = \sum_{q=1}^{N} v_q s_q(n). \tag{8}$$

Using cumulant properties and the assumed independence of all sources, one derives easily

$$K_{y_i}(n) = \sum_{q=1}^{N} v_q^4 K_{s_q}(n) \tag{9}$$

where $K_{s_q}(n)$ is the kurtosis of source $s_q(n)$, again defined according to (4). The standard output kurtosis (9) therefore actually depends on the kurtoses of all sources. The corresponding differential output kurtosis, defined in (5), may then be expressed as

$$DK_{y_i}(n_1, n_2) = \sum_{q=1}^{N} v_q^4 DK_{s_q}(n_1, n_2) \tag{10}$$

where we define the differential kurtosis $DK_{s_q}(n_1, n_2)$ of source $s_q(n)$ in the same way as in (5). Let us now take into account the assumption that P sources are non-stationary, while the other sources are stationary. We denote by $\mathcal{I}$ the set containing the P unknown indices of the non-stationary sources. The standard kurtosis $K_{s_q}(n)$ of any source $s_q(n)$ with $q \notin \mathcal{I}$ then takes the same values for $n = n_1$ and $n = n_2$, so that $DK_{s_q}(n_1, n_2) = 0$ [1]. Eq. (10) then reduces to

$$DK_{y_i}(n_1, n_2) = \sum_{q \in \mathcal{I}} v_q^4 DK_{s_q}(n_1, n_2). \tag{11}$$

[1] Note that the "complete" stationarity of the sources $s_q(n)$ with $q \notin \mathcal{I}$ is sufficient for, but not required by, our method: we only need their differential kurtoses (and their differential powers below) to be zero for the considered times.

This shows explicitly that this differential parameter only depends on the non-stationary sources. Moreover, for given sources and times n_1 and n_2, it may be seen as a function $f(.)$ of the set of variables $\{v_q,\ q \in \mathcal{I}\}$, i.e $DK_{y_i}(n_1, n_2)$ is equal to

$$f(v_q,\ q \in \mathcal{I}) = \sum_{q \in \mathcal{I}} v_q^4 \alpha_q \tag{12}$$

where the parameters α_q are here equal to the differential kurtoses $DK_{s_q}(n_1, n_2)$ of the non-stationary sources. The type of function defined in (12) has been widely studied in the framework of standard kurtotic BSS methods, i.e. methods for the case when $P = N$, because the standard kurtosis used as a BSS criterion in that case may also be expressed according to (12) [2]. The following result has been established (see [9] p. 173 for the basic 2-source configuration and [5] for a general proof). Assume that all parameters α_q with $q \in \mathcal{I}$ are non-zero, i.e. that all non-stationary sources have non-zero differential kurtoses for the considered times n_1 and n_2. Consider the variations of the function in (12) on the P- dimensional unit sphere, i.e. for $\{v_q,\ q \in \mathcal{I}\}$ such that

$$\sum_{q \in \mathcal{I}} v_q^2 = 1. \tag{13}$$

The results obtained in [5],[9] imply in our case that the maxima of the absolute value of $f(v_q,\ q \in \mathcal{I})$ on the unit sphere are all the points such that only one of the variables v_q, with $q \in \mathcal{I}$, is non zero. Eq. (8) shows that the output signal $y_i(n)$ then contains a contribution from only one non-stationary source (and contributions from all stationary sources). We thus reach the target partial BSS for one of the non-stationary sources.

The last aspect of our method that must be defined is how to select the matrix M and to constrain the vector w (which is the parameter controlled in practice, unlike v) so that the variables $\{v_q,\ q \in \mathcal{I}\}$ meet condition (13). To this end, we define the differential correlation matrix of $z(n)$ as

$$DR_z(n_1, n_2) = R_z(n_2) - R_z(n_1) \tag{14}$$

where $R_z(n) = E\{z(n)z(n)^T\}$ is its standard correlation matrix. The differential correlation matrix $DR_s(n_1, n_2)$ of the sources is defined in the same way. It is diagonal, since the sources are assumed to be uncorrelated and centered, and its non-zero entries are the differential powers of the non-stationary sources, i.e.

$$DP_{s_q}(n_1, n_2) = E\{s_q^2(n_2)\} - E\{s_q^2(n_1)\}. \tag{15}$$

The BSS scale indeterminacy makes it possible to rescale these differential powers up to *positive* factors. Therefore, provided the diagonal elements of $DR_s(n_1, n_2)$

[2] In standard approaches, the summation for $q \in \mathcal{I}$ in (12) is performed over all $P = N$ sources and the parameters α_q are equal to the standard kurtoses $K_{s_q}(n)$ of all these sources. However, this has no influence on the discussion below, which is based on the general properties of the type of functions defined by (12).

corresponding to the P sources of interest are strictly positive for the considered times n_1 and n_2, they may be assumed to be equal to 1 without loss of generality. We then select the matrix M so that

$$DR_z(n_1, n_2) = I \tag{16}$$

and we control w so as to meet $\| w \| = 1$. This method is the differential extension of the sphering stage of the FastICA approach. As shown in Appendix A, these conditions on M and w guarantee that the constraint (13) is satisfied.

2.2 Summary of Proposed Method

The practical method which results from the above analysis operates as follows:

Step 1 Select two non-overlapping bounded time intervals (which correspond to n_1 and n_2 in the above theoretical analysis) such that all non-stationary[3] sources have non-zero differential kurtoses and positive differential powers (15). These intervals may be derived by only resorting to the observed signals, $x_i(n)$, e.g. as explained in [6].

Step 2 Compute an estimate $\widehat{DR_x}(n_1, n_2)$ of the differential correlation matrix of the observations, defined in the same way as in (14). Then perform the eigendecomposition of that matrix. This yields a matrix Ω whose columns are the unit-norm eigenvectors of $\widehat{DR_x}(n_1, n_2)$ and a diagonal matrix Λ which contains the eigenvalues of $\widehat{DR_x}(n_1, n_2)$. Then derive the matrix $M = \Lambda^{-1/2}\Omega^T$. This matrix performs a "differential sphering" of the observations, i.e. it yields a vector $z(n)$ defined by (2) which meets (16).

Step 3 Create an output signal $y_i(n)$ defined by (3), where w is a vector which satisfies $\| w \| = 1$ and which is adapted so as to maximize the absolute value of the differential kurtosis of $y_i(n)$, defined by (5). Various algorithms may be used to achieve this optimization, especially by developing differential versions of algorithms which were previously proposed for the case when $P = N$. The most classical approach is based on gradient ascent [9]. We here preferably derive an improved method from the standard fixed-point FastICA algorithm [8], which yields several advantages with respect to the gradient-based approach, i.e. fast convergence and no tunable parameters. Our differential fast fixed-point algorithm then consists in iteratively performing the following couple of operations:

---

[3] "non-stationary" here means "long-term non-stationary". More precisely, all sources should be stationary inside each of the two time intervals considered here, so that their statistics may be estimated for each of these intervals, by time averaging. This corresponds to "short-term stationarity". The above-mentioned "sources of interest" (resp. "noise sources") then consist of source signals whose statistics are requested to vary (resp. not to vary) from one of the considered time intervals to the other one, i.e. sources which are "long-term non-stationary" (resp. "long-term stationary").

1) Differential update of w

$$w = \left[E\{z(w^T z)^3\} - 3w\right]_{n_2} - \left[E\{z(w^T z)^3\} - 3w\right]_{n_1} \quad (17)$$

$$= \left[E\{z(w^T z)^3\}\right]_{n_2} - \left[E\{z(w^T z)^3\}\right]_{n_1} \quad (18)$$

where the expressions $\left[E\{z(w^T z)^3\}\right]_{n_i}$ are resp. estimated over the two considered time intervals.
2) Normalization of w, to meet condition $\| w \| = 1$, i.e

$$w = w/||w||. \quad (19)$$

Step 4 The non-stationary source signal extracted as $y_i(n)$ in Step 3 is then used to subtract its contributions from all observed signals. The resulting signals are then processed by using again the above complete procedure, thus extracting another source, and so on until all non-stationary sources have been extracted. This corresponds to a deflation procedure, as in the standard FastICA method [8], except that a *differential* version of this procedure is required here again. This differential deflation operates in the same way as the standard deflation, except that the statistical parameters are replaced by their differential versions, as in (17). Here again, a parallel (differential) approach [8] could be considered instead of deflation.

3 Experimental Results

We now illustrate the performance of the proposed method for a configuration involving 2 linear instantaneous mixtures of 3 artificial sources. Each of the 2 non-stationary sources $s_1(n)$ and $s_2(n)$ consists of two 5000-sample time windows. Both sources have a Laplacian distribution $p(x) = 1/2 \exp(-|x|)$ in the first window and a uniform distribution over $[-0.5, 0.5]$ in the second window. The "noise" source $s_3(n)$ has the same distribution over all 10000 samples.

The overall relationship between the original sources and the outputs of our BSS system reads $y(n) = Cs(n)$, where $C = [c_{ij}]$ is here a 2x3 matrix. If $s_1(n)$ and $s_2(n)$ appear without permutation in $y(n)$, c_{12} and c_{21} correspond to the undesired residual components of $s_2(n)$ and $s_1(n)$ resp. in $y_1(n)$ and $y_2(n)$ and should ideally be equal to zero. The "error" associated to the partial BSS of $s_1(n)$ and $s_2(n)$ may then be measured by the parameter $(E\{c_{12}^2\} + E\{c_{21}^2\})$, where the expectations $E\{.\}$ are estimated over a set of 100 tests hereafter. Equivalently, the quality of this partial BSS may be measured by the inverse of the above error criterion, i.e

$$Q = \frac{1}{E\{c_{12}^2\} + E\{c_{21}^2\}}. \quad (20)$$

We investigated the evolution of this criterion with respect to the input Signal to Noise Ratio (SNR) associated to the observed mixed signals, defined as

$$SNR_{in} = \sqrt{SNR_{in}^1 \ . \ SNR_{in}^2} \quad (21)$$

where the input SNR associated to each mixed signal $x_i(n)$ reads

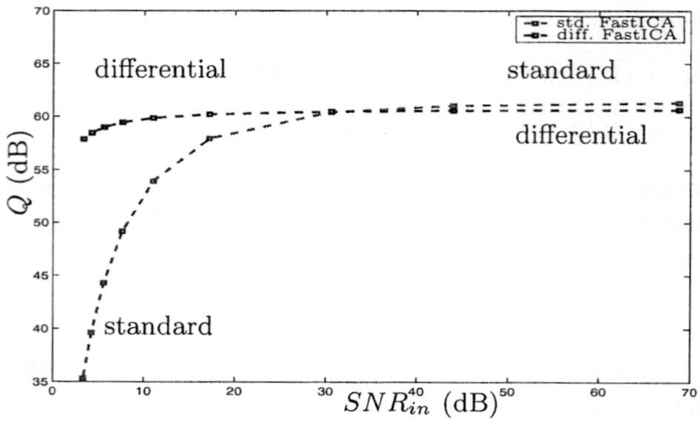

Fig. 1. Separation quality criterion Q of standard and differential FastICA methods vs input SNR

$$SNR_{in}^{i} = \frac{a_{i1}E\{s_1^2\} + a_{i2}E\{s_2^2\}}{a_{i3}E\{s_3^2\}} \quad i \in \{1, 2\}. \tag{22}$$

The input SNR was varied in our tests by changing the magnitude of the noise source $s_3(n)$. Fig. 1 shows the performance of the proposed differential BSS method and of the standard FastICA algorithm. This proves the effectiveness of our differential approach, since its quality criterion Q remains almost unchanged down to quite low input SNRs, i.e less than 5 dB, whereas the performance of standard FastICA already starts to degrade around 30 dB input SNR[4].

4 Conclusion

In this paper, we considered underdermined BSS. By using our differential BSS concept, we proposed a partial BSS method which has the same general structure as the kurtotic methods (especially FastICA) which have been developed for the case when $P = N$: it consists of a first stage which uses the second-order statistics of the signals, followed by a second stage which takes advantage of their fourth-order statistics. However, these stages are here based on new statistical parameters, that we introduce as the differential versions of the standard parameters. The proposed BSS method thus basically consists of a differential sphering, followed by the optimization of the differential kurtosis of an output signal. This optimization may especially be performed by using our differential version of the fast fixed-point algorithm which has been introduced in the standard FastICA approach, thus keeping the advantages of the latter algorithm. This has been

[4] For very high input SNRs (which is not the target situation for our approach !) standard FastICA performs slightly better than its differential counterpart. This probably occurs because the differential statistical parameters involved in the latter approach are estimated with a slightly lower accuracy than their standard version, partly because each expectation in the differential parameters is only estimated over one half of the available signal realization.

confirmed by our experimental tests, which show that our method is much more robust to noise than standard FastICA. Our future investigations will especially aim at extending our differential BSS method to convolutive mixtures.

References

1. F. Abrard, Y. Deville, "A time-frequency blind signal separation method applicable to underdetermined mixtures of dependent sources", Signal Processing, Vol. 85, Issue 7, pp. 1389-1403, July 2005,
2. X-R Cao and R-W Liu, "General approach to blind source separation", IEEE Trans. Signal Process., Vol. 44, No. 3, pp. 562-571, March 1996.
3. P. Comon and O. Grellier, "Non-linear inversion of underdetermined mixtures", Proc. ICA'99, pp. 461-465, Aussois, France, 11-15 January 1999.
4. P. Comon, "Blind identification and source separation in 2x3 under-determined mixtures", IEEE Trans. on Sig. Proc, vol. 52, no. 1, pp. 11-22, Jan. 2004.
5. N. Delfosse, P. Loubaton, "Adaptive blind separation of independent sources: a deflation approach", Signal Processing, vol. 45, no. 1, pp. 59-84, July 1995.
6. Y. Deville, M. Benali, F. Abrard, "Differential source separation for underdetermined instantaneous or convolutive mixtures: concept and algorithms", Signal Processing, Vol. 84, Issue 10, pp. 1759-1776, Oct. 2004.
7. F. Gamboa, "Separation of sources having unknown discrete supports", IEEE Workshop Higher-Order Stat., pp. 56-60, Begur, Spain, 12-14 June 1995.
8. A. Hyvarinen, E. Oja, "A fast fixed-point algorithm for Independent Component Analysis", Neural Computation, vol. 9, pp. 1483-1492, 1997.
9. A. Hyvarinen, J. Karhunen, E. Oja, "Independent Component Analysis", Wiley, New York, 2001.
10. A. Taleb, C. Jutten, "On underdetermined source separation", Proc. ICASSP-99, pp. 1445-1448, Phoenix, Arizona, 15-19 March 1999.

A Proof of Condition (13)

We first introduce the matrix H, defined as the diagonal matrix with entries equal to 1 for indices $q \in \mathcal{I}$ and 0 otherwise. We then define the vector

$$\tilde{v} = Hv \tag{23}$$

which is such that

$$||\tilde{v}||^2 = \sum_{q \in \mathcal{I}} v_q^2. \tag{24}$$

Besides, Eq. (23) and (7) yield

$$||\tilde{v}||^2 = w^T (MA) H (MA)^T w. \tag{25}$$

Moreover, Eq. (1) and (2) yield

$$DR_z(n_1, n_2) = (MA) DR_s(n_1, n_2) (MA)^T. \tag{26}$$

In addition, the properties of $DR_s(n_1, n_2)$ provided in Section 2.1 mean that $DR_s(n_1, n_2) = H$. Eq. (26) and (25) then yield

$$||\tilde{v}||^2 = w^T DR_z(n_1, n_2)w. \tag{27}$$

Therefore, (27) and (24) show that, if M is selected so that (16) is met and w is controlled so as to meet $\parallel w \parallel = 1$, then the requested condition (13) is met.

Equivariant Algorithms for Estimating the Strong-Uncorrelating Transform in Complex Independent Component Analysis

Scott C. Douglas[1], Jan Eriksson[2], and Visa Koivunen[2]

[1] Department of Electrical Engineering,
Southern Methodist University, Dallas, Texas 75275, USA
[2] Signal Processing Laboratory, SMARAD CoE,
Helsinki University of Technology,
Espoo 02015, Finland

Abstract. For complex-valued multidimensional signals, conventional decorrelation methods do not completely specify the covariance structure of the whitened measurements. In recent work [1,2], the concept of strong-uncorrelation and its importance for complex-valued independent component analysis has been identified. Few algorithms for estimating the strong-uncorrelating transform currently exist. This paper presents two novel algorithms for estimating and computing the strong uncorrelating transform. The first algorithm uses estimated covariance and pseudo-covariance matrices, and the second algorithm estimates the strong uncorrelating transform directly from measurements. An analysis shows that the only stable stationary point of both algorithms produces the strong uncorrelating transform when the circularity coefficients of the sources are distinct and positive. Simulations show the efficacy of the approach in a source clustering task for wireless communications.

1 Introduction

In most treatments of blind source separation and independent component analysis, the signals are assumed to be real-valued. In a number of practical applications, however, measurements are naturally represented using complex linear models. In wireless communications, multiantenna or multiple-input, multiple-output systems can be conveniently described using a complex-valued mixture model. Multiple-sensor recordings in various biological signal processing applications are also well-represented in complex form [3]. These applications motivate the study of m-dimensional complex-valued signal mixtures of the form

$$\mathbf{x}(k) = \mathbf{A}\mathbf{s}(k), \tag{1}$$

where $\mathbf{A}$ is an arbitrary complex-valued $(m \times m)$ matrix and the source signal vector sequence $\mathbf{s}(k)$ contains statistically-independent complex-valued elements.

Recently, work in complex ICA has uncovered a statistical structure that is unlike the real-valued case [1,2]. In particular, it is possible in some cases to

J. Rosca et al. (Eds.): ICA 2006, LNCS 3889, pp. 57–65, 2006.

identify $\mathbf{A}$ in (1) using only second-order statistics from $\mathbf{x}(k)$ at time k, a situation that is distinct from the real-valued case. The key construct in these results is the *strong-uncorrelating transform,* which we now describe. Without loss of generality, assume that the source covariance and pseudo-covariance matrices are $E\{\mathbf{s}(k)\mathbf{s}^H(k)\} = \mathbf{I}$ and $E\{\mathbf{s}(k)\mathbf{s}^T(k)\} = \boldsymbol{\Lambda}$, respectively, where $\boldsymbol{\Lambda}$ is a diagonal matrix of ordered real-valued entries between zero and one called *circularity coefficients* $\{\lambda_i\}$, $i \in \{1, \ldots, m\}$. Define the covariance and pseudo-covariance matrices of $\mathbf{x}(k)$ as

$$\mathbf{R} = E\{\mathbf{x}(k)\mathbf{x}^H(k)\} = \mathbf{A}\mathbf{A}^H \qquad \text{and} \qquad \mathbf{P} = E\{\mathbf{x}(k)\mathbf{x}^T(k)\} = \mathbf{A}\boldsymbol{\Lambda}\mathbf{A}^T, \quad (2)$$

respectively. Then, the strong-uncorrelating transform $\underline{\mathbf{W}}$ is a matrix satisfying

$$\underline{\mathbf{W}}\mathbf{R}\underline{\mathbf{W}}^H = \mathbf{I} \qquad \text{and} \qquad \underline{\mathbf{W}}\mathbf{P}\underline{\mathbf{W}}^T = \boldsymbol{\Lambda}. \quad (3)$$

If the $\{\lambda_i\}$ values are distinct, then a matrix $\mathbf{W}$ satisfying (3) is also a separating matrix for the mixing model in (1). Additional results for the strong-uncorrelating transform are in [1,2], and [9] uses the transform to derive kurtosis-based fixed-point algorithms for complex signal mixtures.

In [1], a technique for computing the strong uncorrelating transform for given values of $\mathbf{R}$ and $\mathbf{P}$ is described. This technique employs both an eigenvalue decomposition of a Hermitian symmetric matrix and the *Takagi factorization* of a complex symmetric matrix, the latter of which requires specialized numerical code [5]. A Jacobi-type rotation method for the Takagi factorization is outlined in [6], but its numerical and convergence properties are not established. Both of these methods are computationally-complex and not amenable to situations in which the second-order data statistics are slowly-varying. Since few methods for computing the strong-uncorrelation transform currently exist, it is of great interest to derive simple algorithms for the strong-uncorrelating transform that could be employed in adaptive estimation and tracking tasks.

This paper describes two simple iterative procedures for computing the strong uncorrelating transform adaptively. Both procedures can be viewed as extensions of the method in [7]. The first procedure employs sample estimates of the covariance and pseudo-covariance matrices and is equivariant with respect to the mixing system $\mathbf{A}$ when sample-based averages of these matrices are used. The second equivariant procedure estimates the strong-uncorrelating transform directly from measurements. Both techniques have the significant advantage of not requiring estimates of the circularity coefficients $\{\lambda_i\}$ for their successful operation. Simulations show the abilities of the methods to perform strong-uncorrelation in a source clustering task for wireless communications.

2 An Adaptive Algorithm for the Strong Uncorrelating Transform

The simple algorithms described in this paper adapt a row-scaled version of $\underline{\mathbf{W}}$, termed $\mathbf{W}(k)$, to compute the strong uncorrelating transform. In the interest of algorithm simplicity, and because overall output signal scaling is often not an issue, we define the space of allowable solutions for $\mathbf{W}(k)$ as

$$\lim_{k\to\infty} \mathbf{W}(k)\mathbf{R}\mathbf{W}^H(k) = \widetilde{\mathbf{D}} \qquad \text{and} \qquad \lim_{k\to\infty} \mathbf{W}(k)\mathbf{P}\mathbf{W}^T(k) = \widetilde{\mathbf{D}}\boldsymbol{\Lambda}, \tag{4}$$

where $\widetilde{\mathbf{D}}$ is an arbitrary diagonal matrix of positively-valued diagonal entries. If $\mathbf{R}$ is available or can be estimated, then a $\mathbf{W}(k)$ satisfying (4) can be turned into a $\underline{\mathbf{W}}$ satisfying (3) using $\underline{\mathbf{W}} = \widetilde{\mathbf{D}}^{-\frac{1}{2}}\mathbf{W}(k)$.

Our first proposed algorithm for adaptively computing the strong-uncorrelating transform is

$$\mathbf{W}(k+1) = \mathbf{W}(k) + \mu\Big(\mathbf{I} - \mathbf{W}(k)\widehat{\mathbf{R}}(k)\mathbf{W}^H(k) - \text{tri}[\mathbf{W}(k)\widehat{\mathbf{P}}(k)\mathbf{W}^T(k)]\Big)\mathbf{W}(k), \tag{5}$$

where $\widehat{\mathbf{R}}(k)$ and $\widehat{\mathbf{P}}(k)$ are sample estimates of $\mathbf{R}$ and $\mathbf{P}$ and $\text{tri}[\mathbf{M}]$ denotes a matrix whose lower triangular portion is identical to that of $\mathbf{M}$ and whose strictly-upper triangular portion is zero.

The following three theorems describe important theoretical and practical convergence properties of this algorithm, the proofs of which are in the Appendix.

Theorem 1. *The algorithm in (5) is equivariant with respect to the mixing matrix* $\mathbf{A}$ *under the data model in (1).*

Remark. Although the algorithm is equivariant with respect to the mixing matrix $\mathbf{A}$, its performance is affected by the values in $\boldsymbol{\Lambda}$ that depend on the sources. Thus, convergence of the algorithm may be fast or slow depending on $\boldsymbol{\Lambda}$.

Theorem 2. *The space of stationary points for the algorithm in (5) are* $\mathbf{W} = \mathbf{0}$ *and the set of matrices that satisfy*

$$\mathbf{W}\mathbf{R}\mathbf{W}^H = \mathbf{I} - \mathbf{D} \;\; \textit{and} \;\; \mathbf{W}\mathbf{P}\mathbf{W}^T = \mathbf{D}, \tag{6}$$

where $\mathbf{D}$ *is a diagonal matrix of real-valued unordered entries that are all less than or equal to one.*

Theorem 3. *Suppose the diagonal entries of* $\boldsymbol{\Lambda}$ *are distinct and positive. Then, the only locally-stable stationary point of the algorithm in (5) is the unique matrix* $\mathbf{W}$ *that yields the solution*

$$\mathbf{W}\mathbf{R}\mathbf{W}^H = (\mathbf{I} + \boldsymbol{\Lambda})^{-1} \;\; \text{and} \;\; \mathbf{W}\mathbf{P}\mathbf{W}^T = (\mathbf{I} + \boldsymbol{\Lambda})^{-1}\boldsymbol{\Lambda}. \tag{7}$$

Remark. We could have $\lambda_i = 0$ or $\lambda_i = \lambda_j$ for some diagonal entries of $\boldsymbol{\Lambda}$. In such cases, there is not one unique stationary point for the algorithm. This situation is similar to that for the strong uncorrelated transform, in which a unique solution is not guaranteed. Experience shows that the algorithm still accurately computes a strong uncorrelating transform satisfying (4) despite the fact that this transform may not be unique.

3 A Simple Algorithm for Tracking the Strong Uncorrelating Transform

In many applications, tracking versions of algorithms are desired. We seek a simpler version of (5) for tracking a strong-uncorrelating transform solution given a measured sequence $\mathbf{x}(k)$. Our second proposed algorithm replaces $\widehat{\mathbf{R}}(k)$ and $\widehat{\mathbf{P}}(k)$ in (5) with their instantaneous values $\mathbf{x}(k)\mathbf{x}^H(k)$ and $\mathbf{x}(k)\mathbf{x}^T(k)$ to yield

$$\mathbf{y}(k) = \mathbf{W}(k)\mathbf{x}(k) \tag{8}$$

$$\mathbf{W}(k+1) = \mathbf{W}(k) + \mu(k)[\mathbf{W}(k) - \mathbf{y}(k)\mathbf{y}^H(k)\mathbf{W}(k) - \text{tri}[\mathbf{y}(k)\mathbf{y}^T(k)]\mathbf{W}(k)]. \tag{9}$$

This algorithm is particularly simple, requiring approximately $5m^2$ complex-valued multiply/adds at each iteration if an order-recursive procedure is used to compute $\text{tri}[\mathbf{y}(k)\mathbf{y}^T(k)]\mathbf{W}(k)$. As in all similar adaptive algorithms, the step size sequence $\mu(k)$ controls both the data-averaging of the $\mathbf{x}(k)$ terms and the convergence performance of $\mathbf{W}(k)$. Care must be taken in choosing $\mu(k)$.

The algorithm in (8)–(9) is equivariant with respect to the mixing matrix $\mathbf{A}$ in (1). Moreover, because the discrete-time and differential averaged versions of (8)–(9) are the same as those for the updates in (5) and (17), respectively, Theorems 2 and 3 also apply to (8)–(9). Provided a suitably small step size is chosen and $\mathbf{x}(k)$ is a stationary input signal with distinct non-zero circularity coefficients, the only stable stationary point of (8)–(9) satisfies (7).

Eqns. (8)–(9) are closely related to simple decorrelation methods for real-valued signals [8]. One could view (8)–(9) as the complex extension of the natural gradient method in [8], with the additional feature that it computes the strong uncorrelating transform if $\mathbf{P} \neq \mathbf{0}$. In situations where $\mathbf{x}(k)$ is circularly-symmetric (*i.e.* $\mathbf{P} = \boldsymbol{\Lambda} = \mathbf{0}$), then $E\{\text{tri}[\mathbf{y}(k)\mathbf{y}^T(k)]\} \approx \mathbf{0}$, such that (8)–(9) becomes a natural gradient algorithm for ordinary whitening of complex signals.

For source separation or clustering based on non-circularity, both (5) and (8)–(9) have the nice property that the sources $\{s_i(k)\}$ are grouped in $\mathbf{y}(k)$ in the order of their decreasing circularity coefficients. This property is maintained despite the fact that *the algorithm does not estimate the circularity coefficients of the sources explicitly.* A similar feature was noted for the algorithm in [7].

4 Simulations

We now explore the behaviors of the two proposed algorithms via simulations. The first set of simulations illustrate the algorithms' convergence behaviors when $\mathbf{A}$ is identifiable through the strong-uncorrelating transform. Let $\mathbf{s}(k)$ contain six zero-mean, unit-variance, uncorrelated, and non-circular Gaussian sources with distinct circularity coefficients $\{\lambda_1, \lambda_2, \lambda_3, \lambda_4, \lambda_5, \lambda_6\} = \{1, 0.8, 0.6, 0.4, 0.2, 0.1\}$. One hundred simulations are run, in which $\mathbf{A}$ is generated as a random mixing matrix with jointly-Gaussian real and imaginary elements. Both exponential ($\alpha = 0.999$), denoted by 'exp', and growing-window, denoted by 'lin', averaging of the sequences $\mathbf{x}(k)\mathbf{x}^H(k)$ and $\mathbf{x}(k)\mathbf{x}^T(k)$ with $\widehat{\mathbf{R}}(0) = \widehat{\mathbf{P}}(0) = 0.01\mathbf{I}$ were

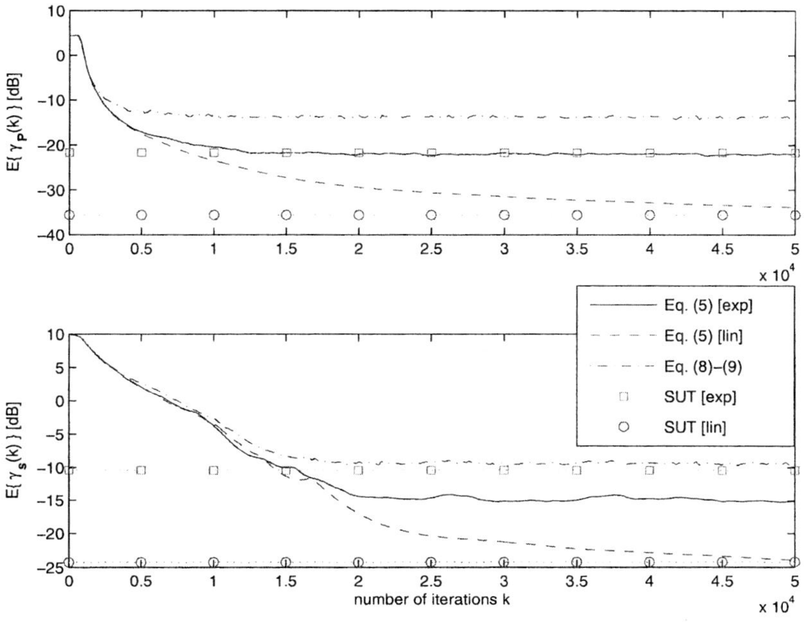

Fig. 1. Convergence of $E\{\gamma_{\mathbf{P}}(k)\}$ and $E\{\gamma_{\mathbf{s}}(k)\}$ in the first simulation example showing the proposed algorithms' successful estimation of the strong-uncorrelating transform

used to estimate $\widehat{\mathbf{R}}(k)$ and $\widehat{\mathbf{P}}(k)$ for two versions of (5). The combined system coefficient vector $\mathbf{C}(k) = \mathbf{W}(k)\mathbf{A}$ is computed and used to evaluate two metrics:

1. Pseudo-covariance Diagonalization: This cost verifies that the algorithms diagonalize the pseudo-covariance and is given by

$$\gamma_{\mathbf{P}}(k) = \frac{||\mathbf{C}(k)\boldsymbol{\Lambda}\mathbf{C}^T(k) - \text{diag}[\mathbf{C}(k)\boldsymbol{\Lambda}\mathbf{C}^T(k)]||_F^2}{||\text{diag}[\mathbf{C}(k)\boldsymbol{\Lambda}\mathbf{C}^T(k)]||_F^2}. \qquad (10)$$

2. Source Separation Without De-rotation: This cost is the average of the inter-channel interferences of the combined system matrices $\mathbf{C}(k)$ and $\mathbf{C}^T(k)$, as

$$\gamma_{\mathbf{s}}(k) = \frac{1}{2m}\left(\sum_{i=1}^{n}\sum_{l=1}^{n}\frac{|c_{il}(k)|^2}{\max_{1\leq i\leq n}|c_{il}(k)|^2} + \frac{|c_{il}(k)|^2}{\max_{1\leq l\leq n}|c_{li}(k)|^2}\right) - 1. \qquad (11)$$

Shown in Figure 1(a) and (b) are the evolutions of $E\{\gamma_{\mathbf{P}}(k)\}$ and $E\{\gamma_{\mathbf{s}}(k)\}$ for the various algorithms with their associated data averaging methods, where $\mu = \mu(k) = 0.007$ for (5) and (8)–(9). As can be seen, all versions of the algorithms diagonalize the pseudo-covariance matrix over time, and they also perform source separation for this scenario.

We now illustrate the behaviors of the simple algorithm in (8)–(9) in a more-practical setting. Let $\mathbf{s}(k)$ contain two BPSK and one 16-QAM source signals.

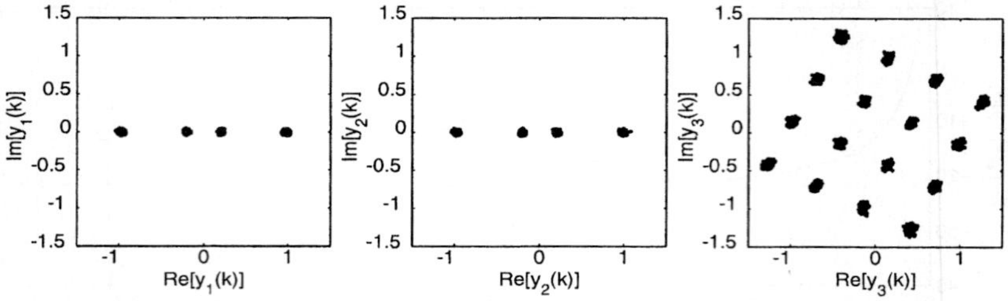

Fig. 2. Output signal constellations obtained by (8)–(9) for a source clustering task in wireless communications

The circularity coefficients in this situation are $\{\lambda_1, \lambda_2, \lambda_3\} = \{1, 1, 0\}$. The strong-uncorrelating transform applied to mixtures of these sources creates a combined system matrix $\mathbf{C}(k) = \mathbf{W}(k)\mathbf{A}$ in which the first two rows (resp. columns) are nearly orthogonal to the third row (resp. column). Thus, $y_1(k)$ and $y_2(k)$ largely contain mixtures of the two real-valued BPSK sources, and $y_3(k)$ largely contains the 16-QAM source. Shown in Figure 2 are the output signal constellations from $y_i(k)$, $i \in \{1, 2, 3\}$, $20000 \leq n \leq 25000$, obtained by applying (8)–(9) with $\mu = 0.0001$ to noisy mixtures of these sources, in which $\mathbf{A}$ contains jointly circular Gaussian entries with variance 2 and the (complex circular Gaussian) additive noise has variance 0.001. The first two outputs clearly show mixtures of the two real BPSK sources, whereas the last output contains the 16-QAM source.

5 Conclusions

The strong-uncorrelating transform is an important linear transform in complex independent component analysis. This paper describes two simple algorithms for adaptively estimating the strong-uncorrelating transform from known covariance and pseudo-covariance matrices and from measured signals, respectively. The algorithms are equivariant to the mixing system, and local stability analyses verify that they perform strong-uncorrelation reliably. Simulations illustrate their performances in separation and source clustering tasks.

References

1. J. Eriksson and V. Koivunen, "Complex-valued ICA using second order statistics," *Proc. IEEE Workshop Machine Learning Signal Processing,* Sao Luis, Brazil, pp. 183-191, Oct. 2004.
2. J. Eriksson and V. Koivunen, "Complex random vectors and ICA models: Identifiability, uniqueness, and separability," *IEEE Trans. Inform. Theory,* accepted.
3. V. D. Calhoun, T. Adali, G. D. Pearlson, P. C. van Zijl, and J. J. Pekar, "Independent component analysis of fMRI data in the complex domain," *Magn Reson. Med.*, vol. 48, pp. 180-192, 2002.

4. G.H. Golub and C.F. Van Loan, *Matrix Computations,* 3rd. ed. (Baltimore: Johns Hopkins Press, 1996).
5. C. Guo and S. Qiao, "A stable Lanczos tridiagonalization of complex symmetric matrices," Tech. Rep. CAS 03-08-SQ, Dept. of Comput. Software Engr., McMaster Univ., June 2003.
6. L. De Lathauwer and B. De Moor, "On the blind separation of non-circular sources," *Proc. XI European Signal Processing Conf.,* Toulouse, France, vol. II, pp. 99-102, Sept. 2002.
7. S.C. Douglas, "Simple algorithms for decorrelation-based blind source separation," *Proc. IEEE Workshop Neural Networks Signal Processing,* Martingy, Switzerland, pp. 545-554, Sept. 2002.
8. S.C. Douglas and A. Cichocki, "Neural networks for blind decorrelation of signals," *IEEE Trans. Signal Processing,* vol. 45, pp. 2829-2842, Nov. 1997.
9. S.C. Douglas, "Fixed-point FastICA algorithms for the blind separation of complex-valued signal mixtures," *Proc. 39th Asilomar Conf. Signals, Syst., Comput.*, Pacific Grove, CA, Oct. 2005.

Appendix

Proof of Theorem 1. Substituting the expressions for $\mathbf{R}$ and $\mathbf{P}$ in (2) for $\widehat{\mathbf{R}}(k)$ and $\widehat{\mathbf{P}}(k)$ in (5) and defining $\mathbf{C}(k) = \mathbf{W}(k)\mathbf{A}$, an equivalent expression for (5) is

$$\mathbf{C}(k+1) = \mathbf{C}(k) + \mu\left(\mathbf{I} - \mathbf{C}(k)\mathbf{C}^H(k) - \mathrm{tri}[\mathbf{C}(k)\boldsymbol{\Lambda}\mathbf{C}^T(k)]\right)\mathbf{C}(k), \qquad (12)$$

which does not depend on $\mathbf{W}(k)$ or $\mathbf{A}$ individually.

Proof of Theorem 2. The stationary points of the algorithm are defined by

$$\left(\mathbf{I} - \mathbf{W}\mathbf{R}\mathbf{W}^H - \mathrm{tri}[\mathbf{W}\mathbf{P}\mathbf{W}^T]\right)\mathbf{W} = \mathbf{0}. \qquad (13)$$

Clearly, $\mathbf{W} = \mathbf{0}$ defines one stationary point. The other stationary points are determined by the solutions of $\mathbf{M} = \mathbf{0}$, where

$$\mathbf{M} = \mathrm{tri}[\mathbf{W}\mathbf{P}\mathbf{W}^T] + \mathbf{W}\mathbf{R}\mathbf{W}^H - \mathbf{I}. \qquad (14)$$

Consider the symmetric and anti-symmetric parts of $\mathbf{M}$ separately. The anti-symmetric part of $\mathbf{M}$ is

$$\mathbf{M}_a = \frac{1}{2}(\mathbf{M} - \mathbf{M}^H) = \frac{1}{2}\left(\mathrm{tri}[\mathbf{W}\mathbf{P}\mathbf{W}^T] - \mathrm{tri}[\mathbf{W}\mathbf{P}\mathbf{W}^T]^H\right). \qquad (15)$$

For $\mathbf{M}_a = \mathbf{0}$, we must have that $\mathbf{W}\mathbf{P}\mathbf{W}^T = \mathbf{D}$, where $\mathbf{D}$ has real-valued but potentially-unordered entries. Under this condition, the symmetric part of $\mathbf{M}$ is

$$\mathbf{M}_s = \frac{1}{2}(\mathbf{M} + \mathbf{M}^H) = \mathbf{W}\mathbf{R}\mathbf{W}^H - \mathbf{I} + \mathbf{D}. \qquad (16)$$

For $\mathbf{M}_s = \mathbf{0}$ to hold, we must have $\mathbf{W}\mathbf{R}\mathbf{W}^H = \mathbf{I}-\mathbf{D}$, which verifies (6). Moreover, since $\mathbf{R}$ is non-negative definite, the diagonal entries of $\mathbf{I} - \mathbf{D}$ are non-negative, and the diagonal entries of $\mathbf{D}$ must satisfy $0 < d_i \leq 1$.

Proof of Theorem 3. Consider the differential form of the update in (5):

$$\frac{d\mathbf{W}}{dt} = \mathbf{W} - \mathbf{W}\mathbf{R}\mathbf{W}^H\mathbf{W} - \mathrm{tri}[\mathbf{W}\mathbf{P}\mathbf{W}^T]\mathbf{W}. \tag{17}$$

Substituting the expressions for $\mathbf{R}$ and $\mathbf{P}$ in (2) into (17) and post-multiplying both sides of (17) by $\mathbf{A}$, we re-write (17) in the combined matrix $\mathbf{C} = \mathbf{W}\mathbf{A}$ as

$$\frac{d\mathbf{C}}{dt} = \mathbf{C} - \mathbf{C}\mathbf{C}^H\mathbf{C} - \mathrm{tri}[\mathbf{C}\boldsymbol{\Lambda}\mathbf{C}^T]\mathbf{C}. \tag{18}$$

First, assume that $\mathbf{C}$ is near a stationary point corresponding to $\mathbf{W} = 0$, and let $\mathbf{C} = \boldsymbol{\Delta}$, where $\boldsymbol{\Delta}$ is a matrix of small complex-valued entries. Then, we can rewrite the update in (18) in the entries of $\boldsymbol{\Delta}$ as

$$\frac{d\boldsymbol{\Delta}}{dt} = \boldsymbol{\Delta} + \mathcal{O}(\Delta_{ij}^2) \tag{19}$$

where $\mathcal{O}(\Delta_{ij}^2)$ denotes terms that are second and higher-order in the entries of $\boldsymbol{\Delta}$. Eq. (19) is exponentially unstable; $\mathbf{W} = \mathbf{0}$ is not a stable stationary point.

Now, assume that $\mathbf{C}$ is near a stationary point such that $\mathbf{C}_s\mathbf{C}_s^H = \mathbf{I} - \mathbf{D}$ and $\mathbf{C}_s\boldsymbol{\Lambda}\mathbf{C}_s^T = \mathbf{D}$, where $\mathbf{D}$ is a diagonal matrix of real-valued scaling factors $\{d_i\}$ satisfying $0 < d_i \leq 1$, and let $\mathbf{C} = (\mathbf{I} + \boldsymbol{\Delta})\mathbf{C}_s$, where $\boldsymbol{\Delta}$ is a matrix of small complex-valued entries. Then, we can rewrite the update in (18) in the entries of $\boldsymbol{\Delta}$ as

$$\frac{d\boldsymbol{\Delta}}{dt} = -\boldsymbol{\Delta}(\mathbf{I} - \mathbf{D}) - (\mathbf{I} - \mathbf{D})\boldsymbol{\Delta}^H - \mathrm{tri}[\boldsymbol{\Delta}\mathbf{D} + \mathbf{D}\boldsymbol{\Delta}^T] + \mathcal{O}(\Delta_{ij}^2). \tag{20}$$

Ignoring second and higher-order terms, the diagonal entries of $\boldsymbol{\Delta}$ evolve as

$$\frac{d\Delta_{ii}}{dt} = -2\Delta_{ii}, \tag{21}$$

and they are exponentially convergent. The off-diagonal entries of $\boldsymbol{\Delta}$ evolve in a pairwise coupled manner and for $i < j$ satisfy

$$\frac{d\Delta_{ij}}{dt} = (-1 + d_j)\Delta_{ij} + (-1 + d_i)\Delta_{ji}^* \tag{22}$$

$$\frac{d\Delta_{ji}}{dt} = -\Delta_{ij} + (-1 + d_i)\Delta_{ji}^* - d_i\Delta_{ji} \tag{23}$$

Considering the real and imaginary parts of $\Delta_{ij} = \Delta_{R,ij} + j\Delta_{I,ij}$ and $\Delta_{ji} = \Delta_{R,ji} + j\Delta_{I,ji}$ jointly, we have

$$\frac{d}{dt}\begin{bmatrix} \Delta_{R,ij} \\ \Delta_{R,ji} \\ \Delta_{I,ij} \\ \Delta_{I,ji} \end{bmatrix} = \begin{bmatrix} -1 + d_j & -1 + d_i & 0 & 0 \\ -1 & -1 & 0 & 0 \\ 0 & 0 & -1 + d_j & 1 - d_i \\ 0 & 0 & -1 & 1 - 2d_i \end{bmatrix} \begin{bmatrix} \Delta_{R,ij} \\ \Delta_{R,ji} \\ \Delta_{I,ij} \\ \Delta_{I,ji} \end{bmatrix}. \tag{24}$$

For these terms to be convergent, the (2×2) dominant sub-matrices in the above transition matrix must have negative real parts. Recall that $0 < d_l \leq 1$ for all

$1 \leq l \leq m$ by the stationary point condition. Then, the eigenvalue of the first dominant (2×2) matrix with the largest real part is

$$r_{max}^{(1)} = \frac{2 - d_j}{2}\left(-1 + \sqrt{1 - 4\frac{d_i - d_j}{(2 - d_j)^2}}\right). \tag{25}$$

For $\Re\mathrm{e}[r_{max}^{(1)}] < 0$, we require that $d_i > d_j$. With this result, the eigenvalue of the second dominant (2×2) matrix with the largest real part is

$$r_{max}^{(2)} = \frac{2d_i - d_j}{2}\left(-1 + \sqrt{1 - 4\frac{d_i + d_j - 2d_i d_j}{(2d_i - d_j)^2}}\right), \tag{26}$$

which for $d_i > d_j$ is guaranteed to satisfy $\Re\mathrm{e}[r_{max}^{(2)}] < 0$. Thus, the only stable stationary point of the algorithm is when $d_1 > d_2 > \cdots > d_m$.

Now, consider the only stable stationary point solution in (6). Define $\underline{\mathbf{W}} = (\mathbf{I} - \mathbf{D})^{-1/2}\mathbf{W}$ such that

$$\underline{\mathbf{W}}\mathbf{R}\underline{\mathbf{W}}^H = \mathbf{I} \text{ and } \underline{\mathbf{W}}\mathbf{P}\underline{\mathbf{W}}^T = (\mathbf{I} - \mathbf{D})^{-1}\mathbf{D}. \tag{27}$$

It is straightforward to show that $d_i > d_j$ implies $d_i/(1 - d_i) > d_j/(1 - d_j)$, such that $(\mathbf{I} - \mathbf{D})^{-1}\mathbf{D}$ has ordered entries. Eqn. (27) is exactly the strong uncorrelating transform, such that $(\mathbf{I} - \mathbf{D})^{-1}\mathbf{D} = \boldsymbol{\Lambda}$, or $\mathbf{D} = (\mathbf{I} + \boldsymbol{\Lambda})^{-1}\boldsymbol{\Lambda}$ and $\mathbf{I} - \mathbf{D} = (\mathbf{I} + \boldsymbol{\Lambda})^{-1}$. This proves the theorem.

Blind Source Separation of Post-nonlinear Mixtures Using Evolutionary Computation and Order Statistics

Leonardo Tomazeli Duarte, Ricardo Suyama, Romis Ribeiro de Faissol Attux, Fernando José Von Zuben, and João Marcos Travassos Romano

DSPCOM/LBiC, School of Electrical and Computer Engineering, University of Campinas (UNICAMP), C.P. 6101, CEP 13083-970, Campinas - SP, Brazil
{ltduarte, rsuyama, romisri, romano}@decom.fee.unicamp.br, vonzuben@dca.fee.unicamp.br

Abstract. In this work, we address the problem of source separation of post-nonlinear mixtures based on mutual information minimization. There are two main problems related to the training of separating systems in this case: the requirement of entropy estimation and the risk of local convergence. In order to overcome both difficulties, we propose a training paradigm based on entropy estimation through order statistics and on an evolutionary-based learning algorithm. Simulation results indicate the validity of the novel approach.

1 Introduction

The problem of blind source separation (BSS) is related to the idea of recovering a set of source signals from samples that are mixtures of these original signals. Until the end of the last decade, the majority of the proposed techniques [1] were designed to solve the standard linear and instantaneous mixture problem. However, in some applications [2], as a consequence of the nonlinear character of the sensors, the use of linear BSS algorithms may lead to unsatisfactory results, which motivates the use of nonlinear mixing models.

An inherent difficulty associated with the separation of nonlinear mixtures comes from the fact that, in contrast to the linear case, there is no guarantee that it be always possible to recover the sources solely by means of independent component analysis (ICA). Nonetheless, the ICA framework still holds in the so-called Post-Nonlinear (PNL) model as pointed out in [2], and further analyzed in [3].

In [2], Taleb and Jutten proposed a solid paradigm for inverting the action of a PNL mixture system that was based on the minimization of the mutual information between the source estimates. Despite its theoretical solidness, this approach suffers from two major practical drawbacks. The first one comes from the fact that the evaluation of the mutual information demands estimation of the marginal entropies, which may be a rather complex task. The second one is

J. Rosca et al. (Eds.): ICA 2006, LNCS 3889, pp. 66–73, 2006.

related with the presence of local minima in the mutual information-based cost function [4], which poses a problem to the adaptation of the separating system via gradient-based algorithms.

In view of interesting results obtained in entropy estimation using order statistics [6, 7] and of the inherent ability of evolutionary algorithms to perform multimodal optimization tasks, in this work we propose a novel paradigm for training separating systems for PNL mixtures based on these approaches. Related efforts can be found in [4, 5], in which standard genetic algorithms (GA) are employed in the learning process. The present proposal differs from them in two aspects: firstly, we deal with a class of evolutionary techniques that are more robust to suboptimal convergence than the GA, as observed, for instance, in [8]; secondly, a significantly different mutual information estimator is adopted in this work.

The work is structured as follows. In Section 2, the fundamentals of the problem of separating PNL mixtures are discussed. In Section 3, we discuss the major problems present in the treatment of PNL mixture model and expose the training algorithm. The simulation results are shown and discussed in Section 4. Finally, in Section 5, we present the concluding remarks.

2 Problem Statement

Let $\mathbf{s}(t) = [s_1(t), s_2(t), \ldots, s_N(t)]^T$ denote N mutually independent sources and $\mathbf{x}(t) = [x_1(t), x_2(t), \ldots, x_N(t)]^T$ be the N mixtures of the source signals, i.e., $\mathbf{x}(t) = \phi(\mathbf{s}(t))$. The aim of a BSS technique is to recover the source signals based solely on the observed samples of the mixtures.

The simplest form of BSS problem takes place when the mixture process is modeled as a linear and instantaneous system, i.e., $\mathbf{x}(t) = \mathbf{As}(t)$, where $\mathbf{A}$ denotes the mixing matrix. In this case, separation can be achieved by multiplying the mixture vector by a separating matrix $\mathbf{W}$, i.e. $\mathbf{y}(t) = \mathbf{Wx}(t)$, so that the elements of $\mathbf{y}(t)$ be mutually statistically independent. This approach, known as *Independent Component Analysis* (ICA), allows the recovery of the sources up to scaling and permutation ambiguities [1].

A natural extension of the standard BSS problem is to consider a nonlinear mixture process. In such case, the independence hypothesis may no longer be enough to obtain the original sources, indicating that, in general, the solution of the nonlinear BSS problem goes beyond the scope of ICA, in the sense that *a priori* information about the sources and/or the mixture model is necessary. Thus, one possible approach to deal with nonlinear BSS would be to restrain the nonlinear mixing model to a class of separable models, i.e., mixing systems in which statistical independence of the outputs leads to a perfect recovery of the sources.

The most representative example of nonlinear separable mixture model is the PNL system (Fig. 1), where the mixture process is given by $\mathbf{x}(t) = \mathbf{f}(\mathbf{As}(t))$, where $\mathbf{f}(\cdot) = [f_1(\cdot), f_2(\cdot), \ldots, f_n(\cdot)]^T$ denotes the nonlinearities applied to each output of the linear mixing stage.

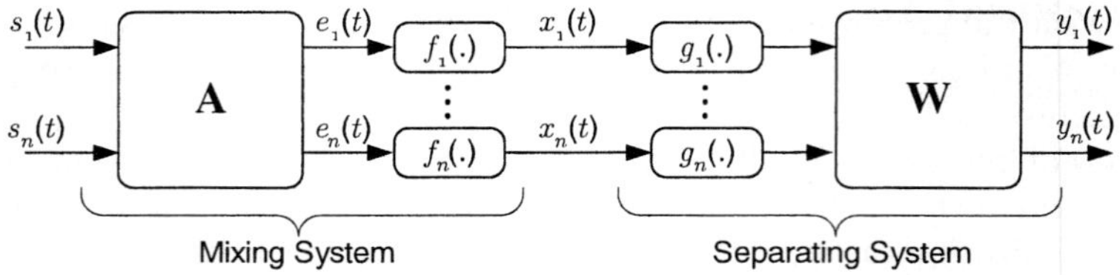

Fig. 1. The PNL problem structure

In [2], source separation of PNL mixtures was achieved by considering the separating system $\mathbf{y}(t) = \mathbf{W}\mathbf{g}(\mathbf{x}(t))$, where $\mathbf{g}(\cdot) = [g_1(\cdot), g_2(\cdot), \ldots, g_n(\cdot)]^T$ is a set of nonlinear functions that must be carefully adjusted to invert the action of $\mathbf{f}(\cdot)$, and $\mathbf{A}$ corresponds to the linear separating matrix. In [2] it has been shown that it is possible, under some mild conditions over $\mathbf{A}$, $\mathbf{W}$, $\mathbf{f}(\cdot)$ and $\mathbf{g}(\cdot)$, to perform source separation in this sort of system relying exclusively on the ICA framework.

2.1 Source Separation Based on the Minimization of Mutual Information

According to the previous discussion, independence between the components of the estimated vector $\mathbf{y}$ leads to source separation. Consequently, one possible criterion to recover the source signals is to minimize the mutual information between the components of $\mathbf{y}$, given by

$$I(\mathbf{y}) = \sum\nolimits_i H(y_i) - H(\mathbf{y}), \tag{1}$$

where $H(\mathbf{y})$ represents the joint entropy of $\mathbf{y}$ and $H(y_i)$ the entropy of each one of its components. Considering the separating structure depicted in Fig. 1, it is possible to express the mutual information as [2]

$$I(\mathbf{y}) = \sum\nolimits_i H(y_i) - H(\mathbf{x}) - \log|\det \mathbf{W}| - E\left\{\log \prod\nolimits_i |g_i'(x_i)|\right\}, \tag{2}$$

with $g_i'(\cdot)$ denoting the first derivative of the nonlinearity $g_i(\cdot)$. It is important to note that Eq. (2) holds *only if the functions $g_i(\cdot)$ are invertible*, a restriction that must be taken into account in the development of the learning algorithm.

Eq. (2) states that accurate estimation of the mutual information relies on proper estimation of both $H(y_i)$ and $H(\mathbf{x})$. However, it should be observed that $H(\mathbf{x})$ does not depend on the parameters of the separating system, and, moreover, is constant for static mixing systems. As a consequence, it can be ignored in the learning process, since our goal is to minimize Eq.(2) with respect to the parameters of the separating system.

On the other hand, $H(y_i)$ is closely related with the system under adaptation, and therefore must be efficiently calculated. In this context, an attractive method for entropy estimation is that based on order statistics [6], which will be employed in our proposal.

3 Proposed Technique

As discussed in the previous section, the mutual information between the estimates of the sources is a consistent cost function. In nonlinear BSS, however, the problem of convergence to local minima becomes more patent, as we are adapting a nonlinear system to separate the sources. Hence, the use of more robust optimization tools is required. In addition to that, the evaluation of the cost function demands effective methods for estimating the marginal entropies of each output y_i.

In the present work, we employ an evolutionary algorithm, the opt-aiNet, to locate the global optimum of the cost function. The opt-aiNet has a good performance in the optimization of multimodal cost functions, as indicated in [8]. With respect to the entropy estimation task, we adopted a solution based on order statistics, which is a good compromise between accuracy and complexity.

In the following, we describe the fundamentals of entropy estimation based on order statistics and the employed search algorithm.

3.1 Entropy Estimation Using Order Statistics

Entropy estimation based on order statistics has been successfully used in blind source separation of instantaneous mixtures [6, 7]. An attractive feature associated with this approach is its low computational complexity when compared, for instance, to density estimation methods. However, the use of order statistics does not easily yield gradient-based algorithms [6], and therefore other optimization tools, such as the evolutionary algorithm presented in section 3.2, must be employed.

Consider a set of T samples of the variable Y organized as

$$y_{(1:T)} \leq y_{(2:T)} \leq \cdots \leq y_{(T:T)}, \tag{3}$$

where $y_{(k:T)}$, called *kth order statistic* [6], is the kth value, in ascending order, among the T available samples.

In order clarify the applicability of order statistics to the problem of entropy estimation, let us rewrite the entropy of a random variable Y in terms of its quantile function $Q_Y(u) = \inf\{y \in \Re : P(Y \leq y) \geq u\}$, which is, in fact, the inverse of the cumulative distribution function $F_Y(y) = P(Y \leq y)$. Using this definition, it is possible to show that [6]

$$H(y) = \int_{-\infty}^{\infty} f_Y(\tau) \log Q_Y^{'}[F_Y(\tau)]\, d\tau = \int_0^1 \log Q_Y^{\prime}(u) du, \tag{4}$$

where $f_Y(y)$ and $Q_Y^{'}(y)$ denote the probability density function and the derivative of the quantile function of y, respectively.

For practical reasons, in order to evaluate the entropy of a given signal y_i, it is necessary to obtain a discretized form of (4), which is given by

$$H(y_i) \approx \sum_{k=2}^{L} \log \left[\frac{Q_{Y_i}(u_k) - Q_{Y_i}(u_{k-1})}{u_k - u_{k-1}} \right] \frac{u_k - u_{k-1}}{u_L - u_1} \tag{5}$$

with $\{u_1, u_2, \ldots, u_L\}$ denoting a set of increasing number in the interval $[0, 1]$.

The link between entropy estimation and order statistics relies on the close relationship between order statistics and the quantile function. In fact, an estimate of the value of $Q_Y(\frac{k}{T+1})$, called *empirical quantile function*, is given by kth order statistic $y_{(k:T)}$ [6]. Therefore, we approximate the value of $Q_Y(\cdot)$ in Eq. (5) by $Q_{Y_i}(u) \approx y_{(k:T)}$, for k such that $\frac{k}{T+1}$ is the closest point to u. This simplification results in a fast algorithm for entropy estimation, which is a desirable characteristic when dealing with optimization using evolutionary algorithms.

3.2 Evolutionary Optimization Technique

The research field of evolutionary computation encompasses a number of optimization techniques whose modus operandi is based on the idea of evolution. In this work, we shall concentrate our attention on a member of the class of artificial immune systems (AIS): the opt-aiNet (Optimization version of an Artificial Immune Network) [9], which can be defined as a multimodal optimization algorithm founded on two theoretical constructs, viz., the notions of *clonal selection and affinity maturation* and the idea of *immune network*.

Under the conceptual framework of clonal selection and affinity maturation, the immune system is understood as being composed of cells and molecules that carry receptors for antigens (disease-causing agents). In simple terms, when these receptors recognize a given antigen, they are stimulated to proliferate. During the process, controlled mutation takes place, and, thereafter, the individuals are subjected to a natural selection mechanism that tends to preserve the most adapted.

A different view on the defense system is provided by the immune network theory, which states that it is possible for the immune cells and molecules to interact with each other in a way that engenders "eigen-behaviors" even in the absence of antigens. As a consequence, the "invasion" could be thought of as a sort of "perturbation" of a well-organized state of things.

In order to transform these ideas into an efficient optimization technique, it is imperative that some parallels be established: the function to be optimized represents a measure of affinity between antibody and antigen (fitness), each solution corresponds to the information contained in a given receptor (network cell), and, finally, the affinity between cells is quantified with the aid of a simple Euclidean distance measure. Having these concepts in mind, let us expose the algorithm.

1. Initialization: randomly create initial network cells;
2. Local search: while stopping criterion is not met, do:
   (a) Clonal expansion: for each network cell, determine its fitness (an objective function to be optimized). Generate a set of N_c antibodies, named clones, which are the exact copies of their parent cell;
   (b) Affinity maturation: mutate each clone with a rate that is inversely proportional to the fitness of its parent antibody, which itself is kept unmutated. The mutation follows

$$c' = c + \alpha N(0,1), \text{ with } \alpha = \beta^{-1} \exp(-f^*) \tag{6}$$

where c' and c represent the mutated and the original individual, respectively; β is a free parameter that controls the decay of the inverse exponential function, and f^* is the fitness of an individual. For each mutated clone, select the one with highest fitness and calculate the average fitness of the selected cells;

(c) Local convergence: if the average fitness of the population does not vary significantly from one iteration to the other, go to the next step; else, return to Step 2;

3. Network interactions: determine the similarity between each pair of network antibodies;
4. Network suppression: eliminate all but one of the network cells whose affinity with each other is lower than a pre-specified threshold σ_s, and determine the number of remaining cells in the network;
5. Diversity: introduce a number of new randomly generated cells into the network and return to Step 2.

After initialization, Step 2 is responsible for performing a process of local search that is based on a fitness-dependent mutation operator; Steps 3 and 4 account for the "immune network character" of the technique, a consequence of which is the emergence of a method for controlling the population size; at last, Step 5 allows that new features be introduced in the population. The combination of these stages produces an algorithm that allies a good balance between exploration and exploitation with the notion of seeking a parsimonious use of the available resources.

4 Results

In order to evaluate our technique, we conducted simulations under two different scenarios. In the first one, two sources with uniform distribution between $[-1, 1]$ were mixed through a system with

$$\mathbf{A} = \begin{bmatrix} 1 & 0.6 \\ 0.5 & 1 \end{bmatrix} \text{ and } \begin{array}{l} f_1(e_1) = \tanh(2e_1) \\ f_2(e_2) = 2\sqrt[5]{e_2} \end{array} . \tag{7}$$

The second scenario is composed of three uniform sources and a mixing system defined by

$$\mathbf{A} = \begin{bmatrix} 1 & 0.6 & 0.5 \\ 0.5 & 1 & 0.4 \\ 0.4 & 0.6 & 1 \end{bmatrix} \text{ and } \begin{array}{l} f_1(e_1) = 2\sqrt[3]{e_1} \\ f_2(e_2) = 2\sqrt[3]{e_2} \\ f_3(e_3) = 2\sqrt[3]{e_3} \end{array} . \tag{8}$$

In both cases, the separating system to be optimized consists of a square matrix $\mathbf{W}$ and a polynomial of order 5, only with odd powers, i.e., $y = ax^5 + bx^3 + cx$. In view of the fact that the separability property [2] of the PNL model requires that $\mathbf{g}(\mathbf{f}(\cdot))$ be a monotonic function, the coefficients of each polynomial were restricted to be positive. The parameters of the opt-aiNet were adjusted after several experiments, and for both cases we considered the following set: $N = 10$, $N_c = 7$, $\beta = 60$ and $\sigma_s = 2$.

In Fig. 2(a), the joint distribution of the mixture signals in the first case is shown. For this situation, 2000 samples of the mixtures were considered in the training stage. The joint distribution of the recovered is depicted in Fig. 2(b), where it can be noted that, despite some residual nonlinear distortion (which is expected, given that it is impossible to invert the hyperbolic tangent using a polynomial), the obtained distribution is also uniform, indicating that the separation task was fulfilled.

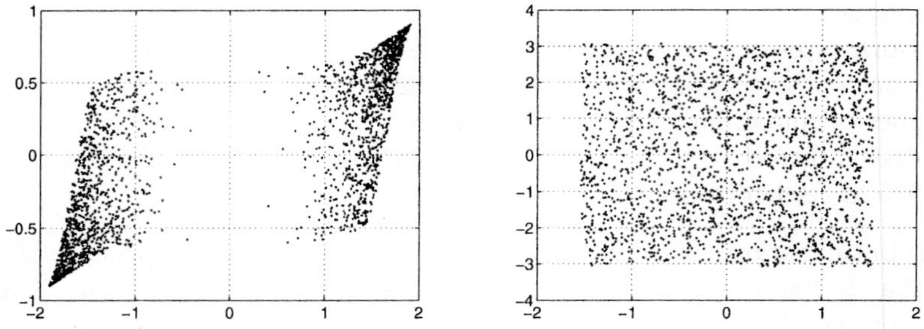

(a) Mixed signals. (b) Recovered signals.

Fig. 2. First scenario

The application of our proposal has also led to accurate source estimates in the second scenario, as it can be seen in Fig. 3, in which one source and its estimated version are depicted. In this case, the mean-square error (MSE) between these signals (after variance normalization) was 0.0335. For the other two sources, the obtained MSEs were 0.011 and 0.015.

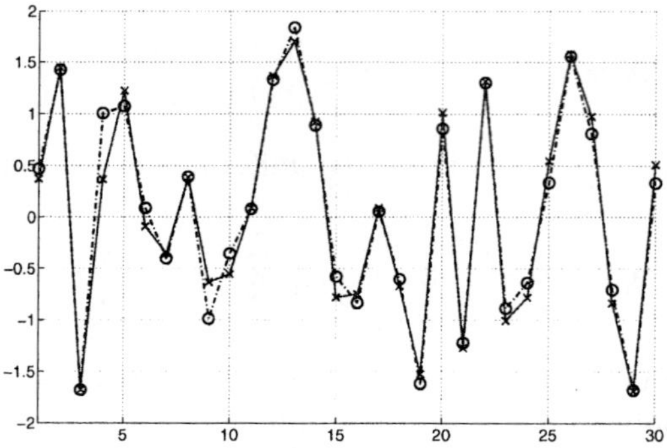

Fig. 3. Second scenario - one of the recovered sources:(−) original source; (· − ·) estimated source

5 Final Comments and Conclusions

In this work, a new training method for PNL separating systems was presented. The proposal has a twofold motivation: 1) to avoid local convergence, a menace originated by nonlinear BSS cost functions; and 2) to obtain a fast and effective method for evaluating the chosen function. In order to meet these requirements, the proposed technique was founded on two pillars: 1) an evolutionary optimization tool, the opt-aiNet, specially-tailored to solve multimodal problems; and 2) an entropy estimation method based on order statistics. Two sets of simulation results attest the efficacy of the proposed methodology in representative PNL scenarios. Finally, a possible extension of this work is to analyze the possibility of employing more flexible functions in the separating system.

Acknowledgements

The authors would like to thank FAPESP and CNPq for the financial support.

References

1. Hyvrinen, A., Karhunen, J., and Oja, E.: Independent Component Analysis. John Wiley & Sons, New York, NY (2001)
2. Taleb, A., Jutten, C.: Source separation in postnonlinear mixtures. IEEE Trans. Signal Processing, Vol. 47, Issue 10, October 1999, 2807–2820
3. Achard S., Jutten, C.: Identifiability of Post-Nonlinear Mixtures. IEEE Signal Processing Letters, Vol. 12, Issue 5, May 2005, 423–426
4. Rojas, F.,Rojas, I., Clemente, R. M., Puntonet, C. G.: Nonlinear Blind Source Separation Using Genetic Algorithms, Proc. of the 3rd Int. Conf. on Independent Component Analysis and Blind Signal Separation (ICA2001). December 9-12, San Diego, California, USA (2001) 400–405
5. Tan, Y.,Wang, J.: Nonlinear blind source separation using higher-order statistics and a genetic algorithm, IEEE Trans. on Evolutionary Computation, 5, 6, (2001) 600–612
6. Pham, D.-T.: Blind Separation of Instantenaous Mixtures of Sources Based on Order Statistics. IEEE Trans. Signal Processing, 48, 2, (2000) 363–375
7. Even, J., Moisan, E.: Blind source separation using order statistics. Signal Processing 85 (2005) 1744–1758
8. Attux, R. R. F., Loiola, M. B., Suyama, R., de Castro, L. N., Von Zuben, F. J., Romano, J. M. T., Blind Search for Optimal Wiener Equalizers Using an Artificial Immune Network Model. EURASIP Journal on Applied Signal Processing Special Issue on Genetic and Evolutionary Computation for Signal Processing and Image Analysis, vol. 2003, No. 8, (2003) 740–747
9. de Castro, L. N., Von Zuben, F. J.: Learning and Optimization Using the Clonal Selection Principle, IEEE Trans. on Evolutionary Computation, Special Issue on Artificial Immune Systems, 6, 3, (2002) 239–251

Model Structure Selection in Convolutive Mixtures

Mads Dyrholm[1], Scott Makeig[2], and Lars Kai Hansen[1]

[1] Informatics and Mathematical Modelling,
Technical University of Denmark, 2800 Lyngby, Denmark
{mad, lkh}@imm.dtu.dk
[2] Swartz Center for Computational Neuroscience,
University of California, San Diego 0961, La Jolla, CA 92093-0961
scott@sccn.ucsd.edu

Abstract. The CICAAR algorithm (convolutive independent component analysis with an auto-regressive inverse model) allows separation of white (i.i.d) source signals from convolutive mixtures. We introduce a source color model as a simple extension to the CICAAR which allows for a more parsimonious representation in many practical mixtures. The new filter-CICAAR allows Bayesian model selection and can help answer questions like: 'Are we actually dealing with a convolutive mixture?'. We try to answer this question for EEG data.

1 Introduction

Convolutive ICA (CICA) is a topic of high current interest and several schemes are now available for recovering mixing matrices and sources signals from convolutive mixtures, see e.g., [7]. Convolutive models are more complex than conventional instantaneous models, hence, the issue of model optimization is important. Convolutive ICA in its basic form concerns reconstruction of the L+1 mixing matrices $\mathbf{A}_\tau$ and the N source signal vectors $\mathbf{s}_t$ of dimension K, from a D-dimensional convolutive mixture

$$\mathbf{x}_t = \sum_{\tau=0}^{L} \mathbf{A}_\tau \mathbf{s}_{t-\tau} \tag{1}$$

Here we focus, for simplicity, on the case where the number of sources equals the number of sensors, $D = K$.

We have earlier proposed the CICAAR approach for convolutive ICA [4] as a generalization of Infomax [3] to convolutive mixtures. The CICAAR exploits the relatively simple structure of the un-mixing system resulting when the inverse mixing is represented as an autoregressive process. In the original derivation we were forced to assume white (i.d.d) sources, i.e., that all temporal correlation in the mixture signals appeared through the convolutive mixing process. A more economic representation is obtained, however, if we explicitly introduce filters to represent possible auto-correlation of sources. This added degree of freedom also

J. Rosca et al. (Eds.): ICA 2006, LNCS 3889, pp. 74–81, 2006.

carries another benefit, it allows for optimizing the model structure: How much correlation should be accounted for by the source filters, and how much should be accounted for by the convolutive mixture? Explicit source auto-correlation modeling using filtered white noise has been proposed earlier by several authors, see e.g., [12,11,2].

2 Modelling Convolutive ICA with Auto-correlated Sources

We introduce a model for each of the sources

$$s_k(t) = \sum_{\lambda=0}^{M} h_k(\lambda) z_k(t-\lambda) \tag{2}$$

where $z_k(t)$ represents a whitened version of the source signal. The negative log likelihood for the model combining (1) and (2) is given by

$$\mathcal{L} = N \log|\det \mathbf{A}_0| + N \sum_k \log|h_k(0)| - \sum_{t=1}^{N} \log \mathrm{p}(\hat{\mathbf{z}}_t) \tag{3}$$

where $\hat{\mathbf{z}}_t$ is a vector of whitened source signal estimates at time t using an operator that represents the inverse of (2). We can without loss of generality set $h_k(0) = 1$, then

$$\mathcal{L} = N \log|\det \mathbf{A}_0| - \sum_{t=1}^{N} \log \mathrm{p}(\hat{\mathbf{z}}_t) \tag{4}$$

The number of parameters in this model is $D^2(L+1) + DM$, and it can thus be minimized if M is increased so as to explain the source auto-correlations allowing L to be reduced in return. An algorithm for convolutive ICA which includes the source model can be derived by making a relative straight forward modification to the equations of the CICAAR algorithm found in [4], see appendix A.

3 Model Selection Protocol

Let $\mathcal{M}$ represent a specific choice of model structure (L, M). The Bayes Information Criterion (BIC) is given by $\log \mathrm{p}(\mathcal{M}|\mathbf{X}) \approx \log \mathrm{p}(\mathbf{X}|\boldsymbol{\theta}_0, \mathcal{M}) - \frac{\dim \boldsymbol{\theta}}{2} \log N$ where $\dim \boldsymbol{\theta}$ is the number of parameters in the model, and $\boldsymbol{\theta}_0$ are the maximum likelihood parameters [13]. BIC has previously been used in context of ICA, see e.g. [5,8,6].

We propose a simple protocol for the dimensions (L, M) of the convolutional- and source-filters. First, expand the convolution length L without a source model (i.e. keeping $M = 0$). This will model the total temporal dependency structure of the system. The optimal L, denote it $L_{\max}$, is found by monitoring BIC. Next, expand the dimensions M of the source model filters while keeping the temporal dependency constant, i.e. keeping $(L + M) = L_{\max}$.

3.1 Simulation Example

The first experiment is designed to illustrate the protocol for determining the dimensions of the convolution and the source filters. We create a 2×2 system with known source filters $M = 15$ and known convolution $L = 10$...

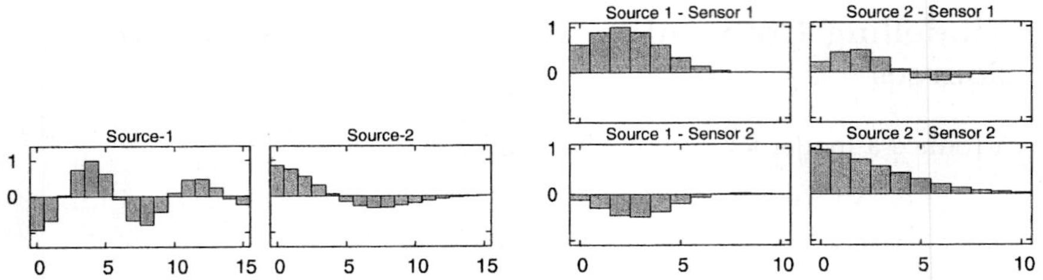

(a) Source color filters, $M = 15$. (b) Convolutive mixing system, $L = 10$.

Fig. 1. Filters for generating synthetic data. First, two i.i.d. signals are colored through their respective filters (a). Then, the colored signals are convolutively mixed using a distinct filter for each source-sensor path (b).

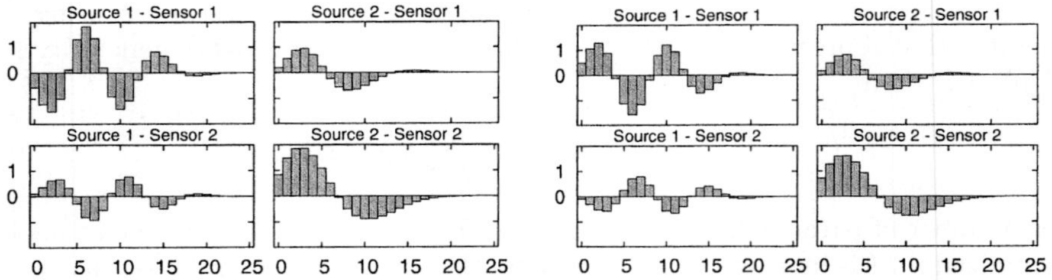

(a) Generating model $(L, M) = (10, 15)$ (b) Estimated by algorithm $(L, M) = (10, 15)$

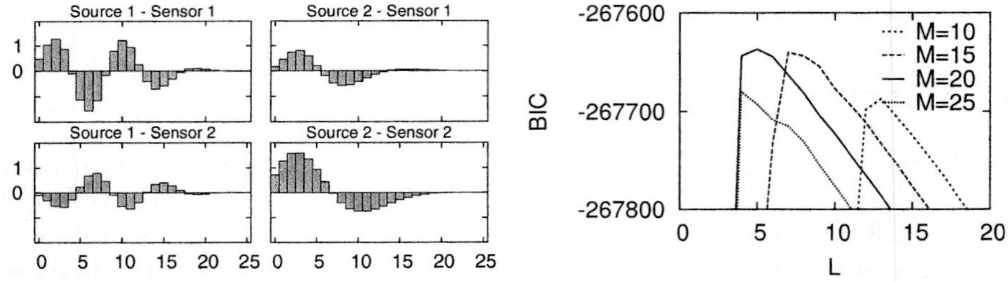

(c) Estimated by algorithm $(L, M) = (5, 20)$ (d) BIC optimal is $(L, M) = (5, 20)$.

Fig. 2. Mixing filters convolved with respective color filters. (a) for the generating model. (b) for an estimated model with the 'true' L and M. (c) for the Bayes optimal model with $(L, M) = (5, 20)$. (d) shows the BIC for various models, and (L,M)=(5,20) is found optimal.

Data — Two signals are generated by filtering temporally white signals using the filters shown on Figure-1(a). The signals are then mixed using the $2 \times 2 \times 10$ system shown on Figure-1(b). The generating model has thus $(L, M) = (10, 15)$.

Result — First we note, the model is in itself ambiguous; an arbitrary filter can be applied to a color filter if the inverse filter is applied to the respective column of mixing filters. Therefore, to compare results we inspect the system as a whole, i.e. source color convolved with a column of mixing filters.

Figure-2 displays convolutive mixing systems where each mixing channel has been convolved with the respective color filter; (a) for the true generating model; (b) a run with the algorithm using $N = 300000$ training samples and using the (L, M) of the generating model. The result is perfect up to sign and scaling ICA ambiguities; (c) shows a run with the algorithm using $N = 100000$ and the Bayes optimal choice of $(L, M) = (5, 20)$ c.f. (d), in the finite data the protocol has found a parsimonious model with similar overall transfer function. We first study the learning curves, i.e., how does the training set dimension N, influence learning. We use the likelihood evaluated on a test set to measure the learning of different models. We now compare learning curves for three models; one which is the generating model $(L, M) = (10, 15)$, one $(L, M) = (25, 0)$ which is more complex but fully capable of imitating the first model, and $(L, M) = (5, 20)$ which is optimal according to BIC. Figure-3 shows learning curves of the three models, the test set is $N_{\text{test}} = 300000$ samples. The uniform improvements in generalization of the 'optimal model' further underlines the importance of model selection in the context of convolutive mixing.

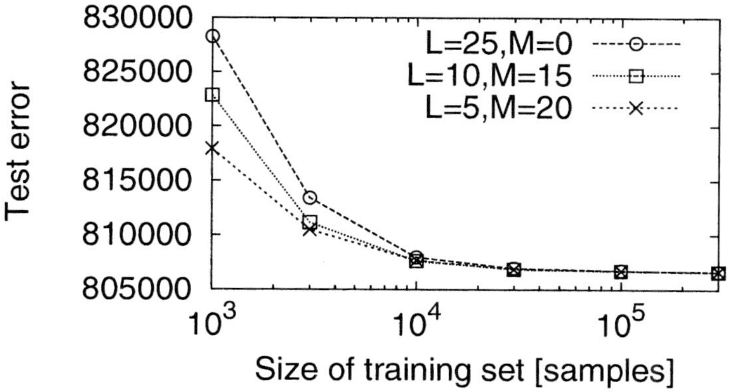

Fig. 3. Learning curves for three models: The generating model $(L, M) = (10, 15)$, a model with $(L, M) = (25, 0)$ which is more complex but fully capable of 'imitating' the first model, and the model $(L, M) = (5, 20)$ which was found Bayes optimal according to BIC. The generalization error is estimated as the likelihood of a test set ($N_{\text{test}} = 300000$). The uniform improvements in generalization of the 'optimal model' further underlines the importance of model selection in the context.

3.2 Rejecting Convolution in an Instantaneous Mixture

We will now illustrate the importance of the source color filters when dealing with the following fundamental question: 'Do we learn anything by using Convolutive ICA instead of instantaneous ICA?'—or put in another way: 'should L be larger than zero?'.

Data — To produce an instantaneous mixture we now mix the two colored sources from before using a random matrix.

Result — Figure-4(a) shows the result of using Bayesian model selection without allowing for a filter ($M = 0$). This corresponds to model selection in a conventional convolutive model. Since the signals are non-white L is detected and the model BIC simply increases as function of L up to the maximum which is attained at a value of $L = 15$. Next, in Figure-4(b) we fix $L + M = 15$. Models with a greater L have at least the same capability as a model with a lower L; but as expected lower L are preferable because the models has fewer parameters. Thus, thanks to the filters, we now get the correct answer: 'There is no evidence of convolutive ICA'.

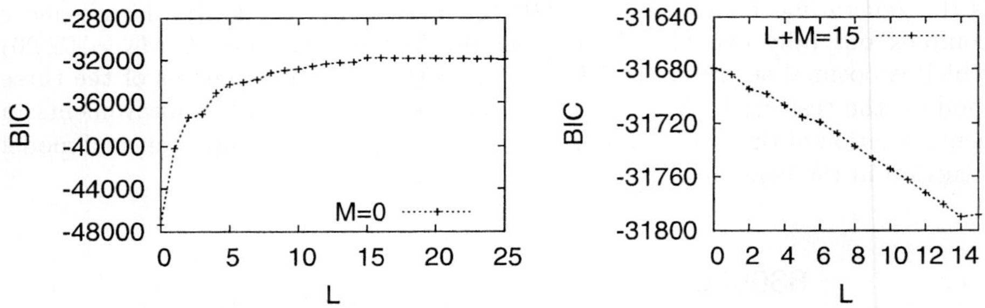

(a) $M = 0$ (b) $M + L = 15$

Fig. 4. (a) the result of using Bayesian model selection without allowing for a filter ($M = 0$). Since the signals are non-white L is detected at a value of $L = 15$. (b) We fix $L + M = 15$, and now get the correct answer: $L = 0$ — 'There is no evidence of convolutive ICA'.

4 Is Convolutive ICA Relevant for EEG?

The EEG signals from the entire brain superimpose onto every EEG electrode instantaneously; there are no delays or echoes, hence, the mixing of the electromagnetic activity is definitely not a convolutive process. However, the question is whether the convolutive mixing model is relevant as a model for the brain activity itself, see also [1]. It is well known that EEG activity exhibits rich spatio-temporal dynamics and that different tasks of the brain combine different regions in different frequency bands, and so, we expect the Bayes optimal model to potentially include some convolutive mixing $L > 0$.

Data — 20 minutes of a 71-channel human EEG recording downsampled to a 50-Hz sampling rate after filtering between 1 and 25 Hz with phase-indifferent FIR filters. First, the recorded (channels-by-times) data matrix ($\mathbf{X}$) was decomposed using extended infomax ICA [3,9] into 71 maximally independent components whose ('activation') time series were contained in (components-by-times) matrix $\mathbf{S}^{\mathrm{ICA}}$ and whose ('scalp map') projections to the sensors were specified in (channels-by-components) mixing matrix $\mathbf{A}^{\mathrm{ICA}}$, assuming instantaneous linear mixing $\mathbf{X} = \mathbf{A}^{\mathrm{ICA}}\mathbf{S}^{\mathrm{ICA}}$. Three of the resulting independent components were selected for further analysis on the basis of event-related coherence results that showed a transient partial collapse of component independence following the subject button presses [10]. Their scalp maps (the relevant three columns of $\mathbf{A}^{\mathrm{ICA}}$) are shown on Figure 5(a).

Convolutive ICA analysis — Next, convolutive ICA decomposition was applied to the three component activation time series (relevant three rows of $\mathbf{S}^{\mathrm{ICA}}$) which we shall refer to as channels ch_1, ch_2 and ch_3. Following our proposed protocol, we find $L_{\max} = 110$, then $L = 9$ as shown on Figure-5(c) — so, we are in fact dealing with a convolutive mixture. Figure-5(b) shows, for one of the resulting convolutive ICA components, cross correlation functions between its contribution to the channels (with each a scalp map associated). Clearly, there are delayed correlation between the different brain regions, and this is not possible to model with an instantaneous ICA model, hence the need for convolutive mixing.

(a)

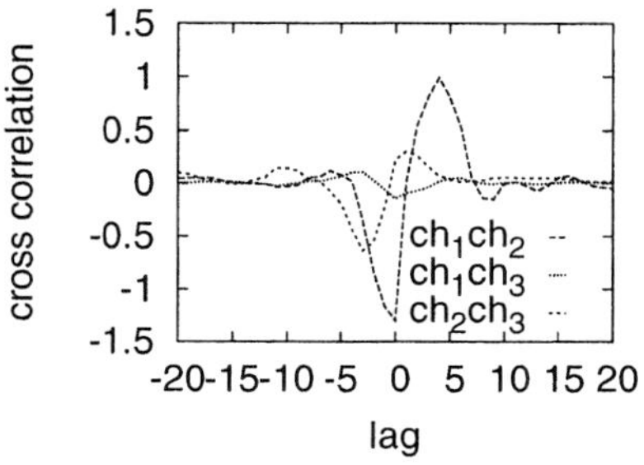

(b)

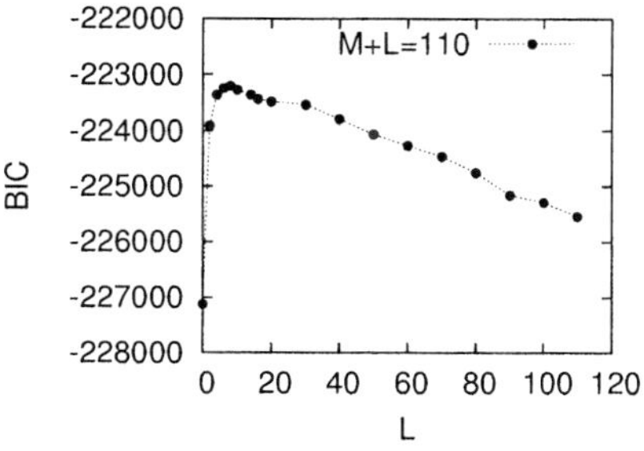

(c) We find $L = 9$

Fig. 5. (a) Scalp maps for the three ICA components. (b) For one of the resulting convolutive ICA components, cross correlation functions between its contribution to the channels. (c) Finding $L = 9$ for the EEG data.

5 Conclusion

We have incorporated filters for modelling possible source auto-correlations into an existing algorithm for convolutive ICA. We have proposed a protocol for determining the dimension L of a convolutive mixture utilizing the filters. We have shown that convolutive ICA is relevant for real EEG data.

References

1. J. Anemüller, T. Sejnowski, and S. Makeig. Complex independent component analysis of frequency-domain eeg data. *Neural Networks*, 16:1313–1325, 2003.
2. H. Attias and C. E. Schreiner. Blind source separation and deconvolution: the dynamic component analysis algorithm. *Neural Computation*, 10(6):1373–1424, 1998.
3. A. J. Bell and T. Sejnowski. An information maximisation approach to blind separation and blind deconvolution. *Neural Computation*, 7:1129–1159, 1995.
4. M. Dyrholm and L. K. Hansen. CICAAR: Convolutive ICA with an auto-regressive inverse model. In Carlos G. Puntonet and Alberto Prieto, editors, *Independent Component Analysis and Blind Signal Separation*, volume 3195, pages 594–601, sep 2004.
5. L. K. Hansen, J. Larsen, and T. Kolenda. Blind detection of independent dynamic components. *In proc. IEEE ICASSP'2001*, 5:3197–3200, 2001.
6. P. Højen-Sørensen, L. K. Hansen, and O. Winther. Mean field implementation of bayesian ICA. In *In proceedings of 3rd International Conference on Independent Component Analysis and Blind Signal Separation (ICA2001).* Institute of Neural Computation, 2001.
7. A. Hyvarinen, J. Karhunen, and E. Oja. *Independent Component Analysis.* John Wiley & Sons., 2001.
8. T. Kolenda, L. Hansen, and J. Larsen. Signal detection using ica: application to chat room topic spotting. In Lee, Jung, Makeig, and Sejnowski, editors, *Proc. of the Third International Conference on Independent Component Analysis and Signal Separation (ICA2001), San Diego, CA*, pages 540–545, 2001.
9. S. Makeig, A. J. Bell, T.-P. Jung, and T. J. Sejnowski. Independent component analysis of electroencephalographic data. *Advances in Neural Information Processing Systems*, 8:145–151, 1996.
10. S. Makeig, A. Delorme, M. Westerfield, T.-P. Jung, J. Townsend, E. Courchesne, and T. J. Sejnowski. Electroencephalographic brain dynamics following manually responded visual targets. *PLoS Biology*, 2004.
11. L. Parra, C. Spence, and B. Vries. Convolutive source separation and signal modeling with ml. In *International Symposioum on Inteligent Systems*, 1997.
12. B. A. Pearlmutter and L. C. Parra. Maximum likelihood blind source separation: A context-sensitive generalization of ICA. In Michael C. Mozer, Michael I. Jordan, and Thomas Petsche, editors, *Advances in Neural Information Processing Systems*, volume 9, page 613. The MIT Press, 1997.
13. G. Schwarz. Estimating the dimension of a model. *Annals of Statistics*, 6:461–464, 1978.

Appendix A: Source Modeling with the CICAAR Algorithm

For notational convenience we introduce the following matrix notation instead of (2), handling all sources in one matrix equation

$$\mathbf{s}_t = \sum_{\lambda=0}^{M} \mathbf{H}_\lambda \mathbf{z}_{t-\lambda} \tag{5}$$

where the $\mathbf{H}_\lambda$'s are diagonal matrices defined by $(\mathbf{H}_\lambda)_{ii} = h_i(\lambda)$.

Given a current estimate of the mixing matrices $\mathbf{A}_\tau$ and the source filter coefficients $h_k(\lambda)$, First apply equation 7 of [4] to obtain $\hat{\mathbf{s}}_t$. Then apply the inverse source coloring operator

$$\hat{\mathbf{z}}_t = \hat{\mathbf{s}}_t - \sum_{\lambda=1}^{M} \mathbf{H}_\lambda \hat{\mathbf{z}}_{t-\lambda} \tag{6}$$

which must replace $\hat{\mathbf{s}}_t$ in [4] (in equations 6,8,9 and 11). This involves the following partial derivatives which in turn uses the result from [4] (from equations 7,10,12)

$$\frac{\partial(\hat{\mathbf{z}}_t)_k}{\partial(\mathbf{B}_\tau)_{ij}} = \frac{\partial(\hat{\mathbf{s}}_t)_k}{\partial(\mathbf{B}_\tau)_{ij}} - \sum_{\lambda=1}^{M} \mathbf{H}_\lambda \frac{\partial(\hat{\mathbf{z}}_{t-\lambda})_k}{\partial(\mathbf{B}_\tau)_{ij}} \tag{7}$$

where $\mathbf{B}_\tau = \mathbf{A}_\tau$ for $\tau > 0$ and $\mathbf{B}_0 = \mathbf{A}_0^{-1}$. Furthermore

$$\frac{\partial(\hat{\mathbf{z}}_t)_k}{\partial(\mathbf{H}_\lambda)_{ii}} = -\delta(k-i)(\hat{\mathbf{z}}_{t-\lambda})_i - \left(\sum_{\lambda'=1}^{M} \mathbf{H}_{\lambda'} \frac{\partial \hat{\mathbf{z}}_{t-\lambda'}}{\partial(\mathbf{H}_\lambda)_{ii}} \right)_k . \tag{8}$$

The work involved in this plug-in is minimal due to the diagonal structure of the $\mathbf{H}_\lambda$ matrices. Finally,

$$\frac{\partial \mathcal{L}}{\partial(\mathbf{H}_\lambda)_{ii}} = -\sum_{t=1}^{N} \psi_t^T \frac{\partial \hat{\mathbf{z}}_t}{\partial(\mathbf{H}_\lambda)_{ii}} \tag{9}$$

where $(\psi_t)_k = \mathrm{p}'((\hat{z}_t)_k)/\,\mathrm{p}((\hat{z}_t)_k)$.

Estimating the Information Potential with the Fast Gauss Transform

Seungju Han, Sudhir Rao, and Jose Principe

CNEL, Department of Electrical and Computer Engineering, University of Florida, Gainesville, USA
{han, sudhir, principe}@cnel.ufl.edu
http://www.cnel.ufl.edu

Abstract. In this paper, we propose a fast and accurate approximation to the information potential of Information Theoretic Learning (ITL) using the Fast Gauss Transform (FGT). We exemplify here the case of the Minimum Error Entropy criterion to train adaptive systems. The FGT reduces the complexity of the estimation from $O(N^2)$ to $O(pkN)$ where p is the order of the Hermite approximation and k the number of clusters utilized in FGT. Further, we show that FGT converges to the actual entropy value rapidly with increasing order p unlike the Stochastic Information Gradient, the present $O(pN)$ approximation to reduce the computational complexity in ITL. We test the performance of these FGT methods on System Identification with encouraging results.

1 Introduction

Information Theoretic Learning (ITL) is a methodology to non-parametrically estimate entropy and divergence directly from data, with direct applications to adaptive systems training [1]. The centerpiece of the theory is a new estimator for Renyi's quadratic entropy that avoids the explicit estimation of the probability density function. The argument of the logarithm of Renyi's entropy is called the Information Potential (IP), and since the logarithm is a monotonic function, it is sufficient to use the IP in training [2]. ITL has been used in ICA [3], blind equalization [4], clustering [5], and projections that preserve discriminability [6]. One of the difficulties of ITL is that the calculation of the IP is $O(N^2)$, which may become prohibitive for large data sets. A stochastic approximation of the IP called the Stochastic Information Gradient (SIG) [7] decreases the complexity to $O(N)$, but slows down training due to the noise in the estimate. This paper presents an effort to make the estimation faster and more accurate using the Fast Gauss Transform (FGT). The FGT is one of a class of very interesting and important new families of fast evaluation algorithms that have been developed over the past dozen years to enable rapid calculation of approximations at arbitrary accuracy to matrix-vector products of the form Ad where $a_{ij} = \Phi(|x_i - x_j|)$ and Φ is a particular special function. These sums first arose in astrophysical observations where the function Φ was the gravitational field. The basic idea is to cluster the sources and target points using appropriate data structures, and to replace the sums

J. Rosca et al. (Eds.): ICA 2006, LNCS 3889, pp. 82 – 89, 2006.

with smaller summations that are equivalent to a given level of precision. We will use here the FGT algorithm proposed by Greengard and Strain [8] and the *farthest-point clustering* proposed by Gonzalez [9] for evaluating Gaussian sums.

The paper will be organized as follows. First we will briefly describe one of the simplest ITL algorithms that minimize the error entropy between a desired response and the adaptive filter output. Next, we present the FGT algorithm and its interaction with the MEE criterion, followed by some simulation results and conclusions.

2 Minimum Error Entropy (MEE)

Suppose that the adaptive system is an FIR structure with a weight vector $\mathbf{w}$. The error samples are $e_k = d_k - \mathbf{w}_k^T \mathbf{u}_k$, where d_k is the desired response, and $\mathbf{u}_k$ is the input vector. The error PDF is estimated using Parzen windows as

$$\hat{f}_e(e) = \frac{1}{N} \sum_{i=1}^{N} k_\sigma (e - e_i) \tag{1}$$

where $k_\sigma(\cdot)$ is kernel function with a kernel size σ. So, Renyi's quadratic entropy estimator for a set of discrete data samples becomes:

$$H_{R2}(e) = -\log \int f^2(e)\, de = -\log V(e) \tag{2}$$

$$V(e) = \frac{1}{N^2} \sum_{j=1}^{N} \sum_{i=1}^{N} k_{\sigma\sqrt{2}} (e_j - e_i) \cdot \tag{3}$$

Minimizing the entropy in (2) is equivalent to maximizing the information potential since the log is a monotonic function. Thus, the weight update of MEE is

$$\mathbf{w}_{k+1} = \mathbf{w}_k + \mu \nabla V(e) \tag{4}$$

where for a Gaussian kernel the gradient is,

$$\nabla V(e) = \frac{1}{2\sigma^2 N^2} \sum_{j=1}^{N} \sum_{i=1}^{N} G_{\sigma\sqrt{2}} (e_j - e_i)\{e_j - e_i\}\{\mathbf{u}_j - \mathbf{u}_i\} \cdot \tag{5}$$

For online training methods, the information potential can be estimated using the Stochastic Information Gradient (SIG) as shown in (6), where the sum is over the most recent L samples at time k. Thus for a filter order of length M, the complexity of MEE is equal to O(ML) per weight update,

$$V(e) \approx \frac{1}{L} \sum_{i=k-L}^{k-1} k_{\sigma\sqrt{2}} (e_k - e_i) \tag{6}$$

where $e_i = d_i - \mathbf{w}_k^T \mathbf{u}_i$, for $k - L \le i \le k$.

3 MEE Using the Fast Gauss Transform

For efficient computation of information potential, we use the principle of Fast Gauss Transform. Direct evaluation of the information potential (3) requires O(N^2). We apply the FGT idea by using the following expansions for the Gaussian in one dimension (the method can be easily extended to multiple dimensions):

$$\exp\left(-\frac{(e_j-e_i)^2}{4\sigma^2}\right)=\sum_{n=0}^{p-1}\frac{1}{n!}\left(\frac{e_i-s}{2\sigma}\right)^n h_n\left(\frac{e_j-s}{2\sigma}\right)+\varepsilon(p) \tag{7}$$

where the Hermite function $h_n(x)$ is defined by

$$h_n(x)=(-1)^n\frac{d^n}{dx^n}\left(\exp(-x^2)\right). \tag{8}$$

In practice a single expansion about one center is not always valid or accurate over the entire domain. A space subdivision scheme is applied in the FGT and the Gaussian functions are expanded at multiple centers. To efficiently subdivide the space, we use a very simple greedy algorithm, called *farthest-point clustering* that computes a data partition with a maximum radius at most twice the optimum. The direct implementation of farthest-point clustering has running time O(*kN*), which *k* is the number of clusters. Thus, the information potential $V(e)$ is given as

$$V(e)\approx\frac{1}{2\sigma N^2\sqrt{\pi}}\sum_{j=1}^{N}\sum_{B}\sum_{n=0}^{p-1}\frac{1}{n!}h_n\left(\frac{e_j-s_B}{2\sigma}\right)C_n(B) \tag{9}$$

where B is a cluster with center s_B and $C_n(B)$ is defined by

$$C_n(B)=\sum_{e_i\in B}\left(\frac{e_i-s_B}{2\sigma}\right)^n. \tag{10}$$

From the above equation, we can see that the total number of operations required is O(*pkN*) per data dimension. The truncation order *p* depends on the desired accuracy alone, and is independent of *N*.

The gradient of the information potential with respect to the weights is given as

$$\nabla V(e)=\frac{1}{2\sigma N^2\sqrt{\pi}}\sum_{j=1}^{N}\sum_{B}\sum_{n=0}^{p-1}\frac{1}{n!}\left[h_{n+1}\left(\frac{e_j-s_B}{2\sigma}\right)\left[\frac{\mathbf{u}_j}{2\sigma}\right]\cdot C_n(B)+h_n\left(\frac{e_j-s_B}{2\sigma}\right)\cdot\nabla C_n(B)\right] \tag{11}$$

where $\nabla C_n(B)$ is defined by

$$\nabla C_n(B)=\sum_{e_i\in B}n\left(\frac{e_i-s_B}{2\sigma}\right)^{n-1}\left[-\frac{\mathbf{u}_i}{2\sigma}\right]. \tag{12}$$

4 Simulations

4.1 Entropy Estimation Using Fast Gauss Transform

We start by analyzing the accuracy of the FGT in the calculation of the IP for the Gaussian and Uniform distributions, using the original definition (3), the SIG (6) and the FGT approximation (9) for two sample sizes (100 and 1,000 samples). For a comparison between SIG and FGT we use $p = L$ in all our simulations. We fix the radius of the farthest point clustering algorithm at $r = \sigma$. This radius is related to the number of clusters, i.e., as the radius increases, the number of clusters (hence the computation time) decreases, but the approximation accuracy may suffer. Results are depicted in Fig. 1 and 2.

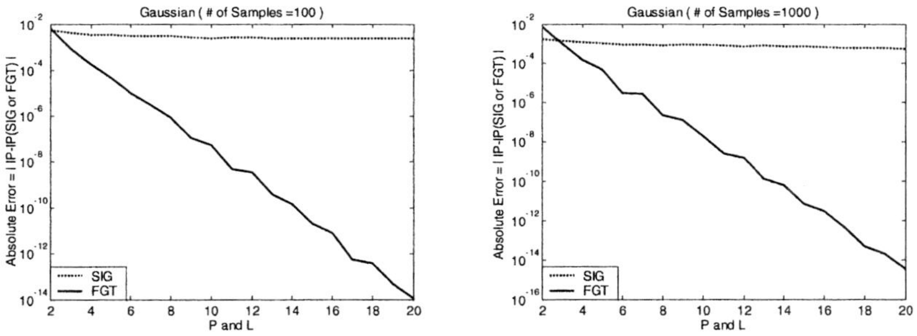

Fig. 1. Plot of the absolute error for SIG and FGT with respect to the IP estimated using Parzen window for a Gaussian distribution with 100 and 1000 samples

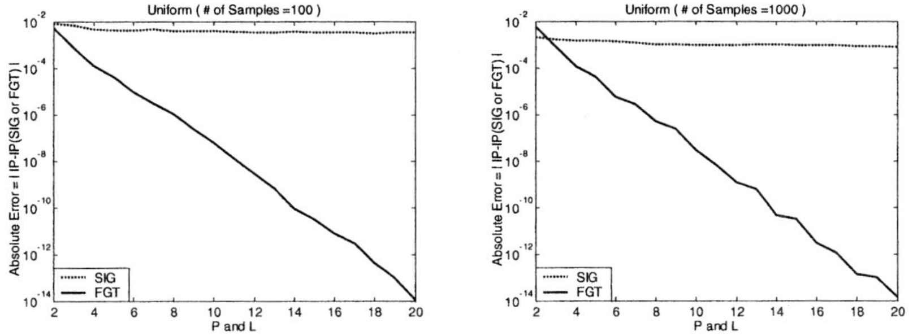

Fig. 2. Plot of the absolute error for SIG and FGT with respect to the IP estimated using Parzen window for a uniform distribution with 100 and 1000 samples

As can be observed in Fig.1 and 2, the absolute error between the IP and the FGT estimation decreases with the order p of the Hermite expansion to very small values, while that of the SIG fluctuates around 0.005 (100 samples) and 0.001 (1000 samples). We can conclude that from a strictly absolute error point of view, a FGT with order $p > 3$ outperforms the SIG method for all cases.

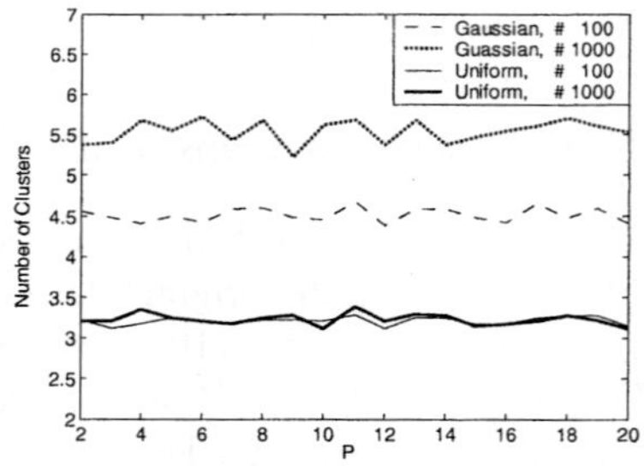

Fig. 3. Plot of the average number of clusters in FGT when estimating the IP for the Gaussian and uniform distribution with 100 and 1000 samples (40 times Monte Carlo)

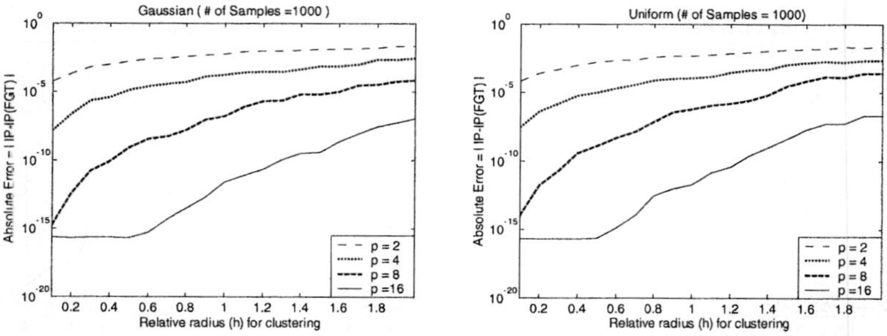

Fig. 4. Plot of the absolute error for a given p (=2, 4, 8 and 16) as the radius of the farthest-point clustering algorithm ($r = h \times \sigma$) for Gaussian and uniform distribution with 1000 samples

Fig. 3 shows the relation between FGT estimation and the number of clusters. According to data size, the number of clusters does not vary for the uniform distribution, while for the Gaussian distribution the number of cluster is larger as the number of data samples increases.

We also fix the number of points to N=1000 and vary the radius r for clustering from 0.1σ to 2σ and plot the absolute error for a given p (= 2, 4, 8 and 16) in Fig. 4. The results show that the error of the FGT is reduced as the radius decreases, as expected such that the user can control the approximation error to IP.

However, for our ITL application, the accuracy of the IP is not the primary objective. Indeed, in ITL we would like to train adaptive systems using gradient information, so the smoothness of the cost function is perhaps more important.

4.2 System Identification

We next consider the system identification of a moving-average model with a 9th order transfer function given by

$$\begin{aligned} H(z) = 0.1 + 0.2z^{-1} + 0.3z^{-2} + 0.4z^{-3} + 0.5z^{-4} \\ + 0.4z^{-5} + 0.3z^{-6} + 0.2z^{-7} + 0.1z^{-8} \end{aligned} \tag{13}$$

using the minimization of the error entropy [10]. Although the true advantage of MEE is for nonlinear system identification with nonlinear filters, here the goal is to compare adaptation accuracy and speed so we elected to use a linear plant and a FIR adaptive filter with the same plant order (zero achievable error). A standard method of comparing the performance in system identification problems is to plot the weight error norm since this is directly related to misadjustment. In each case the power of the weight noise was plotted versus the number of epochs performed. In this simulation, the inputs to both the plant and the adaptive filter are also white Gaussian or uniform noise. We choose a proper kernel size by using Silverman's rule ($\sigma = 0.707$) the radius of the farthest point clustering algorithm $r=\sigma$.

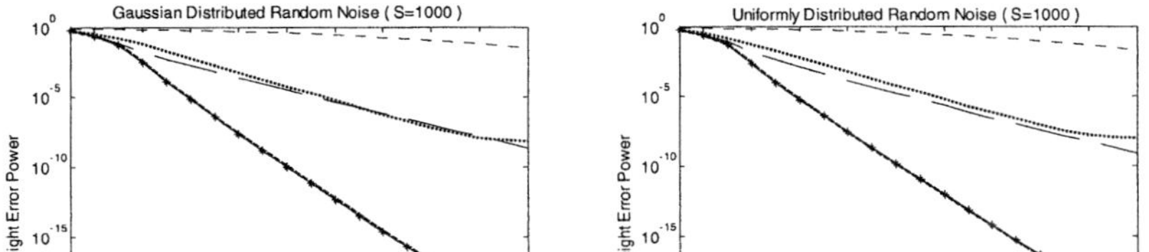

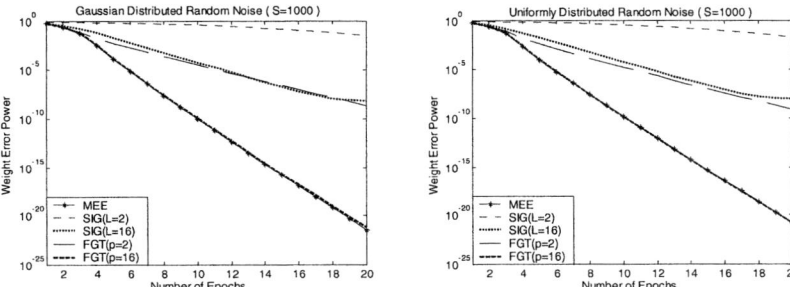

Fig. 5. Comparison of different methods for system identification with Gaussian and uniform noise, using (S) = 1000 samples

As can be observed in Fig. 5, all the versions of IP produce converging filters. However, the speed of convergence and the actual value of the final error are different. The FGT method performs better in training the adaptive system as compared to SIG. A SIG with 16 samples approaches the FGT with $p=2$, and the FGT with $p=16$ is virtually identical to the true IP. The case of the uniform input noise does not change the conclusions.

Fig. 6. shows the plot of the number of clusters during adaptation. Since the error is decreasing at each epoch, the number of clusters gets progressively smaller. In this case, where the achievable error is zero, the number reduces to one cluster after 5 epochs.

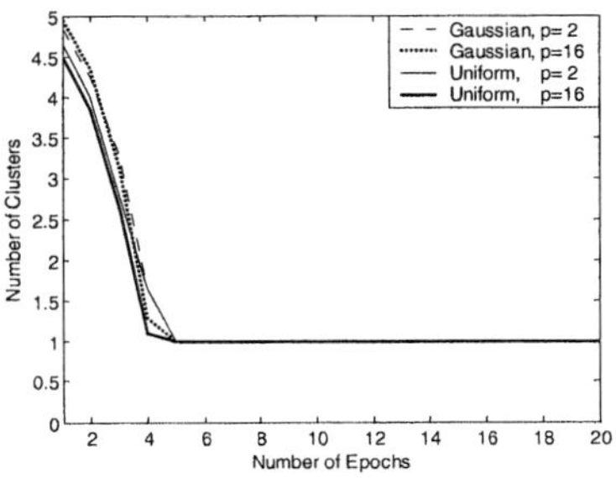

Fig. 6. Plot of the average number of clusters during adaptation in system identification

5 Conclusions

Information Theoretic Learning and in particular the Minimum Error Entropy criterion has been recently proposed as a more principled approach for training adaptive systems. But, a major bottleneck in this method is the high computational complexity of $O(N^2)$ per epoch, thus limiting its use for many practical applications in signal processing, communications and machine learning. The method of the Fast Gaussian Transform helps alleviate this problem by accurate and efficient computation of entropy using the Hermite series expansion in $O(pN)$ operations. Furthermore, since this series converges rapidly, a small order p gives a very good approximation of the IP and can therefore provide accurate and fast converging optimal filters. Indeed we have shown that the FGT has a performance virtually identical to the exact information potential for p=16. The FGT seems therefore to be preferable to the SIG algorithm we have been using.

We still need to quantify the performance of FGT for training MIMO (multiple input multiple output) systems such in ICA or discriminative projections. In these cases ITL algorithms will be applied to multidimensional signals and the computation becomes prohibitive. A straight application of the algorithm presented in this paper will raise p to the number of dimensions in the complexity calculation. However, recent results show that it is possible to avoid the multiplicative factor in complexity brought by the dimensionality of the space of interactions [11]. If further testing corroborates these initial results, the class of FGT algorithms may very well take away the computational drawback of ITL versus the MSE criterion to adapt nonlinear models both in Adaptive Systems and Pattern Recognition applications.

Acknowledgement. This work was partially supported by NSF grant ECS-0300340.

References

1. Principe, J.C., Xu, D., Fisher, J.: Information Theoretic Learning. In: Haykin, S., Unsupervised Adaptive Filtering. Wiley, New York, Vol.I, (2000) 265-319
2. Principe, J.C., Xu, D.: Information-Theoretic Learning Using Renyi's Quadratic Entropy. In Proceedings of the 1st Int. Workshop on Independent Component Analysis and Signal Separation, Aussois, France, (January 11-15 1999) 407-412
3. Hild, K.E., Erdogmus, D., Principe, J.C.: Blind Source Separation using Renyi's Mutual Information. IEEE Signal Processing Letters, Vol. 8, No. 6, (2001) 174-176
4. Lazaro, M., Santamaria, I., Erdogmus, D., Hild, K.E., Pantaleon, C., Principe, J.C.: Stochastic Blind Equalization Based on PDF Fitting Using Parzen Estimator. IEEE Transactions on Signal Processing, Vol. 53, No. 2, (Feb 2005) 696-704
5. Jenssen, R., Eltoft, T. Principe, J.C.: Information Theoretic Spectral Clustering. In Proc. Int. Joint Conference on Neural Networks, Budapest, Hungary, (Jul 2004) 111-116
6. Torkkola, K.: Learning discriminative feature transforms to low dimensions in low dimensions. In advances in neural information processing systems 14, Vancouver, BC, Canada, (Dec 3-8 2001a) MIT Press
7. Erdogmus, D., Principe, J.C., Hild, K.E.: Online entropy manipulation: stochastic Information Gradient. IEEE Signal Processing Letters, Vol. 10, No. 8, (Aug 2003) 242-245

8. Greengard, L., Strain, J.: The fast Gauss transform. SIAM J. Sci. Statist. Comput. 12 (1991) 79-94
9. Gonzalez. T.: Clustering to minimize the maximum intercluster distance. Theoretical Computer Science, 38 (1985) 293-306
10. Erdogmus, D., Principe, J.C.: An Entropy Minimization algorithm for Supervised Training of Nonlinear Systems. IEEE trans. of Signal Processing, Vol. 50, No. 7, (Jul 2002) 1780-1786
11. Yang, C., Duraiswami, R., Gumerov, N., Davis, L.: Improved fast gauss transform and efficient kernel density estimation. In 9th Int. Conference on Computer Vision, Vol. 1, (2003) 464-471

K-EVD Clustering and Its Applications to Sparse Component Analysis

Zhaoshui He[1,2] and Andrzej Cichocki[1]

[1] Laboratory for Advanced Brain Signal Processing,
RIKEN Brain Science Institute, Wako-shi, Saitama 351-0198, Japan
{he_shui, cia}@brain.riken.jp
[2] School of Electronic and Information Engineering,
South China University of Technology, Guangzhou 510640, China

Abstract. In this paper we introduce a simple distance measure from a m-dimensional point a hyper-line in the complex-valued domain. Based on this distance measure, the K-EVD clustering algorithm is proposed for estimating the basis matrix $\boldsymbol{A}$ in sparse representation model $\boldsymbol{X} = \boldsymbol{AS} + \boldsymbol{N}$. Compared to existing clustering algorithms, the proposed one has advantages in two aspects: it is very fast; furthermore, when the number of basis vectors is overestimated, this algorithm can estimate and identify the significant basis vectors which represent a basis matrix, thus the number of sources can be also precisely estimated. We have applied the proposed approach for blind source separation. The simulations show that the proposed algorithm is reliable and of high accuracy, even when the number of sources is unknown and/or overestimated.

1 Introduction

Clustering is one of the most widely used techniques for exploratory data analysis. For example, it can be used for signal compression, blind source separation, feature extraction, regularization in inverse problems, and so on. Especially in the applications of sparse representation and undetermined blind source separation, the basis matrix usually needs to be identified in advance using line orientation clustering algorithms such as K-SVD method [1], K-means method [10], Georgiev, Theis and Cichocki method [2][14], potential-function method [9], time-frequency mask method [13], the extension method of DUET and TIFROM [3][4], soft-LOST [7], hard-LOST [8]. After the basis matrix is estimated, some algorithms, such as FOCUSS algorithms [11], shortest-path-decomposition [9] or its extension [12], linear programming [3][4][10] etc, can be employed to estimate the matrix $\boldsymbol{S}$ representing unknown source signals. In our approach the hyper-line orientation clustering plays a key role in sparse representation under assumption that all sources are sufficiently reach represented and they are all sparse, i.e., each source for many samples achieves dominant value, while other sources at the same time are negligibly small [2][14]. If the basis matrix is not well estimated, it's impossible to find the source signals.

For above mentioned clustering algorithms, it is usually assumed that the number of sources is known. However, in some applications such as BSS, we do not know the

J. Rosca et al. (Eds.): ICA 2006, LNCS 3889, pp. 90 – 97, 2006.

number of source signals in advance and their number is usually overestimated. It's very difficult to accurately estimate the number of source signals. On the other side, if one underestimates the number of source signals, the estimated source signals could be completely different from the true sources. One possible approach is to overestimate the number of source signals and in the next stage to remove possibly all redundant or spurious basis vectors $\hat{\boldsymbol{a}}_i$ of the overestimated basis matrix $\hat{\boldsymbol{A}}$.

For this purpose, in this paper we discuss K-EVD clustering method for complex valued data based on the distance measure presented in Section 2. In contrast to the K-SVD which estimates the clustering centroid matrix by Singular Value Decomposition (SVD) for long matrix, our approach achieve the same goal by Eigen-Value Decomposition (EVD) for smaller dimension matrices.

2 The Distance Formula from a Point to a Hyper-line

Consider a point $P(p_1,\cdots,p_m)$ (see Figure 1) and hyper-line in the m-dimensional complex domain. We attempt to calculate the distance from the point P to the hyper-line L, whose direction vector is $\boldsymbol{l}=(l_1,\cdots,l_m)^T$. This problem can be converted to searching an optimal point Z_* (located on the hyper-line L), which is closest to P, i.e., we can formulate this problem as the following optimization problem:

$$\begin{cases} \min\limits_{z} d(z)=\|\boldsymbol{p}-z\|^2, \\ \text{subject to}: \dfrac{z_1}{l_1}=\cdots=\dfrac{z_m}{l_m}, \end{cases} \tag{1}$$

where $\boldsymbol{z}=(z_1,z_2,\cdots,z_m)^T\in C^m$, $\boldsymbol{p}=(p_1,p_2,\cdots,p_m)^T\in C^m$, C denotes complex valued number and $\min\limits_{z} d(\boldsymbol{z})$ denotes minimizing with respect to vector argument $\boldsymbol{z}$.

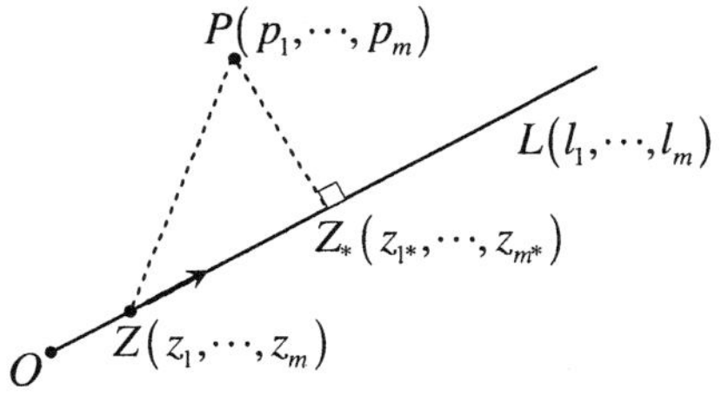

Fig. 1. The distance from a point P to a line L

Let $\frac{z_1}{l_1}=\cdots=\frac{z_m}{l_m}=k\in C$. So we have

$$\boldsymbol{z}=(z_1,z_2,\cdots,z_m)^T=k(l_1,l_2,\cdots,l_m)^T=k\boldsymbol{l}, \tag{2}$$

From equation (2), optimization problem (1) can be described as

$$\min_{k\in C} d(k) = \|p - kl\|^2 = \langle p - kl, p - kl\rangle$$
$$= \langle p, p\rangle - \langle p, kl\rangle - \langle kl, p\rangle + \langle kl, kl\rangle \tag{3}$$
$$= \langle p, p\rangle - k\langle p, l\rangle - k^*\langle l, p\rangle + kk^*\langle l, l\rangle,$$

where $\langle \bullet, \bullet \rangle$ stands for inner product. For any real valued function $f(z)$ of a complex valued variable z the gradients with respect to the real and imaginary part are obtained by taking derivatives formally with respect to the conjugate quantities z^*, ignoring the nonconjugate occurrences of z [5][6], i.e.,

$$\frac{\partial f(z)}{\partial R(z)} + i\frac{\partial f(z)}{\partial I(z)} = 2\frac{\partial f(z)}{\partial z^*}. \tag{4}$$

Therefore the derivative of cost function $d(\bullet)$ with respect to k^* is

$$\frac{\partial d(\bullet)}{\partial k^*} = -\langle l, p\rangle + k\langle l, l\rangle. \tag{5}$$

Let $\frac{\partial d(\bullet)}{\partial k^*} = 0$, we get

$$k = \frac{\langle l, p\rangle}{\langle l, l\rangle}. \tag{6}$$

Substitute equation (6) and equation (2) into (1), we have

$$d(p, l) = \langle p, p\rangle - \frac{\langle l, p\rangle \cdot \langle p, l\rangle}{\langle l, l\rangle}. \tag{7}$$

3 K-EVD Clustering for Estimating the Basis Matrix *A*

In this paper, we assume that sources are very sparse in the sense that each source for many samples is dominant. In such case the observed data $x(t)$ builds up hyper-lines. Of course not all observed datum belongs to any hyper-line. So, our objective is to detect and estimate directions of all hyper-lines where there are many outliers and noise. Roughly speaking, we obtain easily some preliminary K clusters representing vectors $\hat{a}_k, k = 1\cdots, K$. Usually, due to noise and many outliers the number of clusters is much larger than the number of sparse sources.

K-EVD clustering algorithm for estimating A (where $X = AS$ or equivalently $x(t) = As(t), t = 1,\cdots,T$) can be outlined as follows:

(1) Initialize the clustering matrix A as $\hat{A} \in C^{m\times K}$, where $K \geq n$, we assume that only matrix $X \in C^{m\times T}$ is known.

(2) **Partition stage:** Assign the sample points in observation matrix $X = [x(1), \cdots, x(T)] \in C^{m\times T}$ into K different clusters $\theta(\hat{a}_i), i = 1, \cdots, K$, where $\theta(\hat{a}_i)$ is a

data vector set. The estimation of line orientation of set $\theta(\hat{\boldsymbol{a}}_i)$ is $\hat{\boldsymbol{a}}_i, i=1,\cdots,K$ (here $\hat{A}=[\hat{\boldsymbol{a}}_1,\cdots,\hat{\boldsymbol{a}}_K]$). According to distance formula (7), we can compute each distance $d(\boldsymbol{x}(t),\hat{\boldsymbol{a}}_i)$ from observation point $\boldsymbol{x}(t), t=1,\cdots,T$ to cluster center line $\hat{\boldsymbol{a}}_i, i=1,\cdots,K$. The observation point $\boldsymbol{x}(t)\in\theta(\hat{\boldsymbol{a}}_i)$ if and only if it satisfies $d(\boldsymbol{x}(t),\hat{\boldsymbol{a}}_i)=\min\{d(\boldsymbol{x}(t),\hat{\boldsymbol{a}}_j), j=1,\cdots,K\}$.

(3) **Update the cluster centroid matrix stage:** for $i=1,\cdots,K$

Consider each cluster $\theta(\hat{\boldsymbol{a}}_i)$, assume it contains $T^{(i)}$ entries $\boldsymbol{x}^{(i)}(1),\cdots,\boldsymbol{x}^{(i)}(T^i)$, which compose a matrix $\boldsymbol{X}^{(i)}=\left[\boldsymbol{x}^{(i)}(1),\cdots,\boldsymbol{x}^{(i)}\left(T^{(i)}\right)\right]$. For the entries of cluster $\theta(\hat{\boldsymbol{a}}_i)$, apply EVD decomposition $\frac{1}{T}\left(\boldsymbol{X}^{(i)}\right)^H \boldsymbol{X}^{(i)}=\boldsymbol{VDV}^H$. Suppose $d_1^{(i)}$ is the largest eigenvalue, we update the line orientation of cluster $\theta(\hat{\boldsymbol{a}}_i)$ by $\boldsymbol{v}_1^{(i)}$, i.e., $\hat{\boldsymbol{a}}_i=\boldsymbol{v}_1^{(i)}$, $\hat{A}=[\hat{\boldsymbol{a}}_1,\cdots,\hat{\boldsymbol{a}}_K]$.

(4) Return to step (2) and repeat until $\hat{A}$ converges.

(5) Without the loss of generality, we assume $d_1^{(1)},\dots,d_1^{(K)}$ such that $d_1^{(l1)}\geq\cdots\geq d_1^{(lK)}$. We rank column vectors $\hat{\boldsymbol{a}}_1,\cdots,\hat{\boldsymbol{a}}_K$ by the order $l1,\cdots,lK$.

(6) Output the clustering centroid matrix $A_*=[\hat{\boldsymbol{a}}_{l1},\cdots,\hat{\boldsymbol{a}}_{lK}]$.

Remark. Similar to the K-SVD [1], we call this algorithm "K-EVD" in analogy to the standard clustering K-means. In above algorithm we at most need to perform the EVD decomposition for a set of $m\times m$ symmetrical matrices. In contrast sometimes the K-SVD algorithm even needs to do SVD decomposition for a set of large $m\times T$ matrice. Since in BSS applications $T>>m$, the proposed K-EVD algorithm is much faster than the K-SVD. Additionally for the K-EVD, sorting eigenvalues are used to speed up the convergence.

As mentioned in section 1, usually we have no information about how many clusters we should discriminate from the observation $\boldsymbol{X}$. So we usually overestimate the number of clusters. To reduce the overestimated basis vectors, it's necessary to evaluate which basis vectors are significant, i.e. which represent true vectors. The K maximum eigenvalues $d_1^{(1)},\dots,d_1^{(K)}$ can be used to identify such vectors. We can rank vectors $\hat{\boldsymbol{a}}_1,\cdots,\hat{\boldsymbol{a}}_K$ by the values of corresponding eigenvalues. In other words, we choose only those K basis vectors which correspond to the largest $d_1^{(\cdot)}$. Our algorithm is especially useful for such applications where the number of source is unknown and we can overestimate their number.

The full description of sparse representation algorithm is given as follows:

(1) Estimate the extended basis matrix A_* (with spurious vectors) and corresponding eigenvalues $d_1^{(l1)},\dots,d_1^{(lK)}$ using K-EVD algorithm and next, we can only choose the first n basis vectors $\hat{A}_*=[\hat{\boldsymbol{a}}_{l1},\cdots,\hat{\boldsymbol{a}}_{ln}]$ corresponding to n largest eigenvalues $d_1^{(l1)},\cdots,d_1^{(ln)}$. The optimal threshold is still an open problem, but some information criteria can be used here.

(2) Estimate the coefficients of matrix $\boldsymbol{S}$ using *shortest-path-decomposition method* [9].

4 Numerical Experiments and Comparison

In order to demonstrate the validity and very good performance of the K-EVD algorithm, here we give an example how sparse representation is used for BSS. The basis matrix $\boldsymbol{A}$ is corresponding to the mixing matrix of the BSS and the rows of matrix $\boldsymbol{S}$ are corresponding to the source signals.

To check how well the mixing matrix is estimated, we introduce the following Biased Angle (BA) as a performance index (i.e., the angle between the column vector $\boldsymbol{a}_i$ (of mixing matrix) and its corresponding estimation $\hat{\boldsymbol{a}}_i$):

$$BA(\boldsymbol{a}_i,\hat{\boldsymbol{a}}_i)=acos\langle \boldsymbol{a}_i,\hat{\boldsymbol{a}}_i\rangle, \tag{8}$$

where $acos(\bullet)$ denotes the inverse cosine function, $\langle \bullet,\bullet\rangle$ denotes the inner product and $\boldsymbol{A}=[\boldsymbol{a}_1,\cdots,\boldsymbol{a}_n]$.

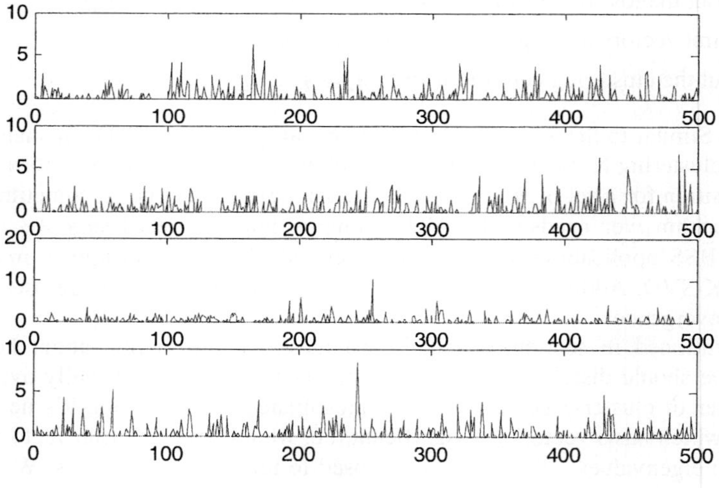

Fig. 2. Source signals

In addition, the Signal to Interference Ratio (SIR) is employed to measure the accuracy of the estimated source signals.

$$SIR(s,\hat{s})=10\log\frac{\sum_{t=1}^{T}(s)^2}{\sum_{t=1}^{T}(s-\hat{s})^2}\ \text{(dB)}. \tag{9}$$

In the BSS a separated signal $\hat{s}$ may have an arbitrary nonzero scale constant factor $c\,(c\neq 0)$, we suitably rescaled the estimated sources in order to optimally match them to the original sources. Usually, when SIR is larger than 18dB, the source signal is considered to be successfully estimated.

In the following two experiments, the 3 by 4 (full row rank) mixing matrix is generated randomly and normalized to unit length vectors. The sparse source matrix $S \in R^{4\times 5000}$ is generated by the following MATLAB command:

$$S = -\log\left(rand\left(n,T\right)\right).*\max\left(0, sign\left(rand\left(n,T\right)-p\right)\right)\cdot \quad (10)$$

Table 1. The results of K-SVD clustering tests

K	$\hat{A}$	$d_1^{(l1)},\cdots,d_1^{(lK)}$	$BA(\boldsymbol{a}_i,\hat{\boldsymbol{a}}_i)$
4	$\begin{pmatrix} 0.3351 & 0.1164 & 0.7361 & -0.4133 \\ -0.8367 & 0.7810 & -0.1214 & -0.4917 \\ 0.4333 & 0.6136 & 0.6659 & 0.7665 \end{pmatrix}$	0.6172 0.5309 0.5304 0.3974	0.2067 0.0405 0.2202 0.1882
5	$\begin{pmatrix} 0.3449 & 0.1122 & 0.7574 & -0.4208 & 0.2310 \\ -0.8286 & 0.7856 & -0.0999 & -0.4931 & -0.2486 \\ 0.4410 & 0.6085 & 0.6453 & 0.7614 & 0.9407 \end{pmatrix}$	0.6205 0.5339 0.5100 0.3942 0.1500(small)	0.1785 0.0373 0.2278 0.1827
6	$\begin{pmatrix} 0.3465 & 0.0955 & 0.7573 & -0.4203 & 0.2310 & 0.4435 \\ -0.8305 & 0.7662 & -0.1000 & -0.4924 & -0.2486 & 0.8863 \\ 0.4361 & 0.6355 & 0.6453 & 0.7622 & 0.9407 & 0.1331 \end{pmatrix}$	0.6221 0.5176 0.5101 0.3940 0.1500(small) 0.1485(small)	0.1786 0.0326 0.2013 0.1828

In Table 1, the "small" $d_1^{(\cdot)}$ mean that the corresponding basis vectors are spurious.

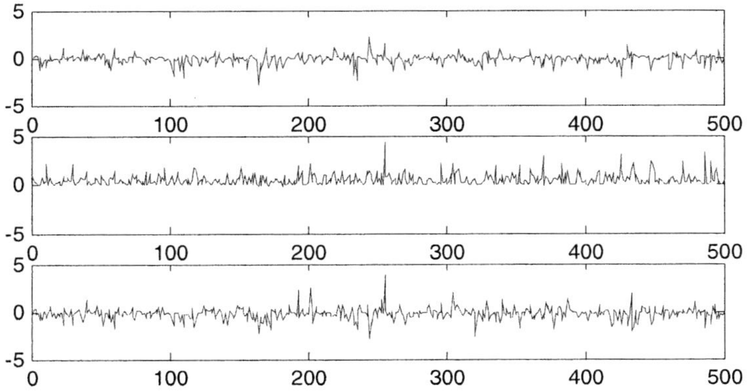

Fig. 3. Observed signals

In expression (10), by choosing different parameter $p(0 \le p \le 1)$, we can obtain the source signals with different sparseness degree. The larger the parameter p is, the sparser the sources signals generated by equation (10) are. Here $n = 4$, $T = 5000$, $p = 0.55$. It should be noted that the source signals are not very sparse, but we still can estimate matrix $\boldsymbol{A}$ and recover the original sources.

Example 1. When the number of source signals is overestimated, in this example we demonstrate how to prior choose those basis vectors with large reliability.

We choose $K = 4,5,6$, respectively, to perform the tests. All tests were convergent for less than 20 iterations. The detailed results are shown in Table 1.

Example 2. We compare the K-EVD algorithm with K-SVD algorithm [1] in this example. Here we mainly compare the speed of these two algorithms. For the $K = 4$ case of example 1, in the same simulation environment, K-EVD took about 0.6000 seconds, while K-SVD cost about 26.8890 seconds. Obviously, K-EVD is much faster than K-SVD.

In addition, in this case, all SIRs of the estimated signals are larger than 18dB, and are 18.8872dB, 21.3923dB, 21.8182dB, 19.6349dB, respectively.

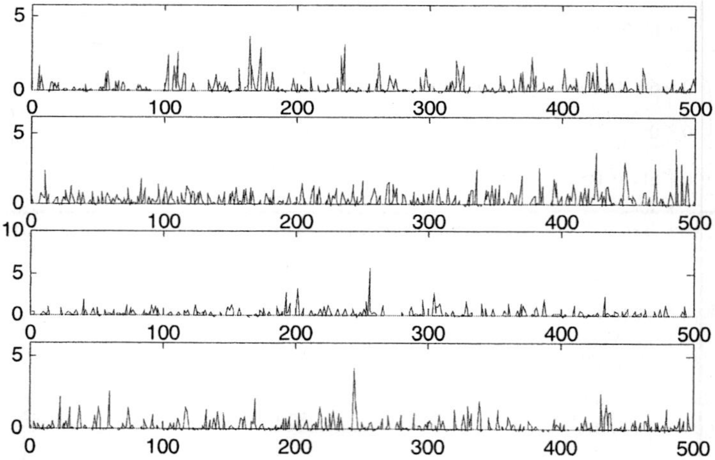

Fig. 4. Separated signals

5 Conclusions

We proposed a simple distance measure. Based on this distance formula, the K-EVD algorithm is presented for estimating the basis matrix $\boldsymbol{A}$ in very sparse representation. When the number of basis vectors is overestimated, K-EVD can evaluate which basis vectors are most reliable by their corresponding principal eigenvalues $d_1^{(\cdot)}$. The simulation experiments illustrate the validity and some advantages of the proposed algorithm. In fact the K-EVD algorithm can be applied not only for real but also to sparse representation for the complex valued data. For signals that are not sparse in the time domain but sparse in the frequency domain, the proposed algorithm can work. Here we limited our considerations to real valued data due to the space limit.

References

[1] M. Aharon, M. Elad, A.M. Bruckstein. K-SVD and its non-negative variant for dictionary design. *Proceedings of the SPIE conference wavelets*, Vol. 5914, July 2005.

[2] P. G. Georgiev, F. Theis, and A. Cichocki, Sparse component analysis and blind source separation of underdetermined mixtures, *IEEE Trans. on Neural Networks*, July 2005, Vol. **16**, No.4, pp. 992-996.

[3] Y.Q. Li, S. Amari, A. Cichocki, and D. W. C. Ho: Underdetermined Blind Source Separation Based on Sparse Representation, *IEEE Trans. on Signal Processing* (in print).

[4] Y. Q. Li, A. Cichocki, and S. Amari, Blind estimation of channel parameters and source components for EEG signals: A sparse factorization approach, *IEEE Transactions on Neural Networks*, 2006, (accepted for publication).

[5] D. Brandwood, A complex gradient operator and its application in adaptive array theory. *Proc. Inst. Elect. Eng.*, vol **130**, pp.11-16, Feb. 1983.

[6] L. Parra, C. Spence. Convolutive blind separation of non-stationary sources. *IEEE Trans. Speech and audio processing*, vol. **8**(3), pp. 320-327, May 2000.

[7] P.D. O'Grady and B. A. Pearlmutter, Soft-LOST: EM on a mixture of oriented lines, *International Conference on Independent Component Analysis 2004*, pages 428-435.

[8] P.D. O'Grady and B. A. Pearlmutter, Hard-LOST: Modified K-means for oriented lines, *Proceedings of the Irish Signals and Systems Conference 2004*, pages 247-252.

[9] P. Bofill, M. Zibulevsky. Undetermined blind source separation using sparse representations. *Signal processing*, vol.**81**, 2353-2362, 2001.

[10] Y.Q. Li, A. Cichocki, S. Amari. Analysis of sparse representation and blind source separation. *Neural computation*, vol.**16**, 1193-1234, 2004.

[11] K.K. Delgado, J.F. Murry, B.D. Rao, *et al.* Dictionary learning algorithms for sparse representation. *Neural computation*, vol. **15**, pp. 349-396, 2003.

[12] I. Takigawa, M. Kudo, J. Toyama, Performance analysis of Minimum l_1 -norm solutions for underdetermined source separation, *IEEE Trans. Signal processing*, vol.**52**, no. 3, pp.582-591, March 2004.

[13] O. Yilmaz, S. Rickard. Blind separation of speech mixtures via time-frequency masking. IEEE Trans. Signal Processing, vol.**52**(7), pp.1830-1847, 2004.

[14] F.J. Theis, P.G. Georgiev, and A. Cichocki, Robust overcomplete matrix recovery for sparse sources using a generalized Hough transform, in Proceedings of 12th European Symposium on Artificial Neural Networks (ESANN2004), (Bruges, Belgium), pp. 343-348, Apr. 2004.

An EM Method for Spatio-temporal Blind Source Separation Using an AR-MOG Source Model

Kenneth E. Hild II[1], Hagai T. Attias[2], and Srikantan S. Nagarajan[1]

[1] Dept. of Radiology, University of California at San Francisco, CA, USA 94122
k.hild@ieee.org, Srikantan.Nagarajan@radiology.ucsf.edu
[2] Golden Metallic, San Francisco CA, USA 94147
htattias@goldenmetallic.com

Abstract. A maximum likelihood blind source separation algorithm is developed. The temporal dependencies are explained by assuming that each source is an AR process and the distribution of the associated i.i.d. innovations process is described using a Mixture of Gaussians (MOG). Unlike most maximum likelihood methods the proposed algorithm takes into account both spatial and temporal information, optimization is performed using the Expectation-Maximization method, and the source model is learned along with the demixing parameters.

1 Introduction

Blind source separation (BSS) involves the application of a linear transformation to an observed set of M mixtures, $\boldsymbol{x}$, in an attempt to extract the original M (unmixed) sources, $\boldsymbol{s}$. Two of the main types of BSS methods for stationary data include decorrelation approaches and approaches based on Independent Components Analysis (ICA). Methods based on decorrelation minimize the squared cross-correlation between all possible pairs of source estimates at two or more lags [1], [2], [3]. Methods based on ICA attempt to make the source estimates statistically independent at lag 0 [4], [5], [6]. Herein it is assumed that the sources are mutually statistically independent, the mixing matrix is invertible, and there are as many sensors as there are sources. If, in addition, at most one source has a Gaussian probability density function (pdf) then ICA methods are appropriate for BSS even if all the sources have identical spectra, whereas this is not the case for decorrelation methods. Similarly, if the M sources possess sufficient spectral diversity then decorrelation methods are appropriate for BSS even if all the sources are Gaussian-distributed, whereas this is not the case for ICA methods. Consequently, the appropriate BSS algorithm for a given application depends on the spatial and temporal structure of the sources in question.

The approach presented here, AR-MOG, differs from most ML methods [7], [8], [9] in three important ways. First, the proposed criterion makes use of both the spatial and temporal structure of the sources. Consequently, AR-MOG may be used in situations for which either of the above two types of BSS algorithms

J. Rosca et al. (Eds.): ICA 2006, LNCS 3889, pp. 98–105, 2006.

are appropriate. Second, AR-MOG is formulated in terms of latent variables so that it can be optimized using the Expectation-Maximization (EM) method. Third, instead of assuming the target distributions are known, the proposed method learns the target distributions directly from the observations.

2 Generative Model

It is assumed that there are M mutually statistically independent sources, each of which are N samples in length. The variable $\boldsymbol{s}$ represents the $(M \times N)$ source matrix, $\boldsymbol{s}_{m,1:\mathrm{N}}$ represents the $(1 \times N)$ vector of the mth row of $\boldsymbol{s}$, and $\boldsymbol{s}_{1:\mathrm{M},n}$ represents the $(M \times 1)$ vector of the nth column of $\boldsymbol{s}$. Each source, $s_{m,n}$, is assumed to be an autoregressive (AR) process that is generated from a temporally i.i.d. innovations process, $u_{m,n}$. The relationship between a given source and the associated innovations process is assumed to be $u_{m,n} = \sum_{k=0}^{K_g} g_{m,k} s_{m,n-k}$, where $g_{m,0} = 1$ $m \in \{1, 2, \ldots, M\}$, $g_{m,k}$ is an element of the $(M \times K_g + 1)$ matrix $\boldsymbol{g}$ of AR coefficients, and K_g is the order of each of the AR filters. The sources are therefore given by

$$s_{m,n} = -\sum_{k=1}^{K_g} g_{m,k} s_{m,n-k} + u_{m,n} \quad . \tag{1}$$

The M observations at time n are assumed to be generated from the sources by means of a linear, memory-less $(M \times M)$ mixing matrix, i.e., $\boldsymbol{x}_n = \boldsymbol{A}\boldsymbol{s}_n$.

The pdf of each innovations process is assumed to be parameterized by a Mixture of Gaussians (MOG),

$$\begin{aligned} p_{\mathrm{U}_{m,n}}(u_{m,n}) &= \sum_{q=1}^{K_Q} p_{\mathrm{U}_{m,n}|\mathrm{Q}_{m,n}}(u_{m,n}|Q_{m,n}=q) p_{\mathrm{Q}_{m,n}}(Q_{m,n}=q) \\ &= \sum_{q=1}^{K_Q} \mathcal{N}(u_{m,n}|\mu_{m,q}, \nu_{m,q})\, \pi_{m,q} \quad , \end{aligned} \tag{2}$$

which should not be confused with $p_{\bar{\mathrm{U}}_{m,n}}(u_{m,n})$ (the target pdf of each innovations process) or $p_{\hat{\mathrm{U}}_{m,n}}(\hat{u}_{m,n})$ (the actual pdf of the estimate of the innovations), and where $p_{\mathrm{U}_{m,n}|\mathrm{Q}_{m,n}}(u_{m,n}|Q_{m,n} = q)$ has a normal distribution, $\mu_{m,q}$ is the mean of the qth component (or state) of the mth source, $\nu_{m,q}$ is the corresponding precision, $\pi_{m,q} \equiv p_{\mathrm{Q}_{m,n}}(Q_{m,n} = q)$ is the corresponding prior probability (constrained such that $\sum_{q=1}^{K_Q} \pi_{m,q} = 1\ \forall m$), and $Q_{m,n} \in \{1, 2, \ldots, K_Q\}$ represents the state (latent variable) of the mth source at the nth time point. This particular generative model is able to describe both the non-Gaussianity and the temporal dependencies of the sources.

3 Criterion

Let $p_{\mathrm{U}_{m,n}}(u_{m,n})$ denote the marginal pdf of a particular innovations process and let $p_{\mathrm{U}}(\boldsymbol{u})$ and $p_{\mathrm{U}_{1:\mathrm{M},n}}(\boldsymbol{u}_{1:\mathrm{M},n})$ denote the order-MN and order-M joint pdf's of the innovations, respectively. It is assumed that all variables are identically distributed in time (although this is not valid for the outputs of the IIR filter $s_{m,n}$ until after the transients have died out). Using this notation and the preceding generative model the data likelihood is given by

$$p_{\mathbf{X}}(\boldsymbol{x}) = \prod_{n=1}^{N} p_{\mathbf{X}_{1:\mathrm{M},n}|\mathbf{X}_{1:\mathrm{M},1:n-1}}(\boldsymbol{x}_{1:\mathrm{M},n}|\boldsymbol{x}_{1:\mathrm{M},1:n-1}) = |\boldsymbol{W}|^N \prod_{m=1}^{M}\prod_{n=1}^{N} p_{\mathrm{U}_{m,n}}(u_{m,n}) \;, \tag{3}$$

where $\boldsymbol{W} = \boldsymbol{A}^{-1}$ (hence, $\boldsymbol{s}_n = \boldsymbol{W}\boldsymbol{x}_n$), it is understood that the set of all parameters, $\{\boldsymbol{W}, \boldsymbol{g}, \boldsymbol{\mu}, \boldsymbol{\nu}, \boldsymbol{\pi}, K_Q, K_g\}$, is given for each pdf, and where $u_{m,n} = \sum_{k=0}^{K_g} \sum_{l=1}^{M} W_{m,l} g_{m,k} x_{l,n-k}$. Hence, the log likelihood is given by

$$\begin{aligned} \mathcal{L} &= N \ln|\boldsymbol{W}| + \sum_{m=1}^{M}\sum_{n=1}^{N} \ln p_{\mathrm{U}_{m,n}}(u_{m,n}) \\ &= N \ln|\boldsymbol{W}| + \sum_{m=1}^{M}\sum_{n=1}^{N}\sum_{q=1}^{K_Q} \gamma_{m,n,q} \ln \frac{p_{\mathrm{U}_{m,n},\mathrm{Q}_{m,n}}(u_{m,n}, Q_{m,n}=q)}{\gamma_{m,n,q}} \;, \end{aligned} \tag{4}$$

where the latter expression is given as a function of the posterior state probabilities, $\gamma_{m,n,q} \equiv p_{\mathrm{Q}_{m,n}|\mathbf{X}}(Q_{m,n}=q|\boldsymbol{x})$. Adaptation using EM, which is guaranteed to converge (possibly to a local maximum), involves maximizing (4) by alternating between the E-step and the M-step.

4 EM Algorithm for AR-MOG

In this section we present an EM algorithm for inferring the model from the data and extracting independent sources.

4.1 E-Step

The E-step maximizes the log likelihood w.r.t. the posteriors, $\gamma_{m,n,q}$, while keeping the parameters fixed. The estimates of the posteriors are given by

$$\hat{\gamma}_{m,n,q} = \frac{p_{\bar{\mathrm{U}}_{m,n}|\hat{\mathrm{Q}}_{m,n}}(\hat{u}_{m,n}|\hat{Q}_{m,n}=q)\hat{\pi}_{m,q}}{\xi_{m,n}} \;, \tag{5}$$

where $\xi_{m,n}$ ensures that $\sum_{q=1}^{K_Q} \hat{\gamma}_{m,n,q} = 1 \; \forall m, n$, the true pdf's (conditioned on the state) have been replaced with the target pdf's, and all other quantities have been replaced with their estimates (denoted using the hat symbol).

4.2 M-Step

The M-step maximizes the log likelihood w.r.t. the parameter estimates $\{\hat{\boldsymbol{W}}, \hat{\boldsymbol{g}}, \hat{\boldsymbol{\mu}}, \hat{\boldsymbol{\nu}}, \hat{\boldsymbol{\pi}}\}$ while keeping the posteriors fixed. The two parameters that are not learned by AR-MOG, $\{\hat{K}_Q, \hat{K}_g\}$, are assumed to be known. The update of $\hat{\boldsymbol{W}}$ is performed using multiple iterations of gradient ascent where

$$\frac{\partial \mathcal{L}}{\partial \hat{W}_{m,l}} = \sum_{i=1}^{M} \Big(N I_{m,i} - \sum_{n=1}^{N} \sum_{q=1}^{\hat{K}_Q} \sum_{k=0}^{\hat{K}_g} \hat{\gamma}_{m,n,q} (\hat{u}_{m,n} - \hat{\mu}_{m,q}) \hat{\nu}_{m,q} \hat{g}_{m,k} \hat{s}_{i,n-k} \Big) \hat{W}_{i,l} \ , \tag{6}$$

which makes use of the natural gradient [10] (also known as the relative gradient [11]). The solution for the matrix of AR coefficients is

$$\begin{aligned} \Phi_{m,k} &= \sum_{n=1}^{N} \sum_{q=1}^{K_Q} \hat{\gamma}_{m,n,q} \hat{\nu}_{m,q} (\hat{\mu}_{m,q} - \hat{s}_{m,n}) \hat{s}_{m,n-k} \\ \Psi_{m,k,k'} &= \sum_{n=1}^{N} \sum_{q=1}^{K_Q} \hat{\gamma}_{m,n,q} \hat{\nu}_{m,q} \hat{s}_{m,n-k} \hat{s}_{m,n-k'} \\ \hat{\boldsymbol{g}}_{m,1:\mathrm{N}} &= \boldsymbol{\Phi}_{m,1:\mathrm{M}} (\boldsymbol{\Psi}_{m,1:\mathrm{M},1:\mathrm{M}})^{-1} \end{aligned} \tag{7}$$

for $m \in \{1, \ldots, M\}$. The solutions for the parameters that constitute the target distributions are

$$\begin{aligned} \hat{\mu}_{m,q} &= \frac{\sum_{n=1}^{N} \hat{u}_{m,n} \hat{\gamma}_{m,n,q}}{\sum_{n=1}^{N} \hat{\gamma}_{m,n,q}} \\ \hat{\nu}_{m,q} &= \frac{\sum_{n=1}^{N} \hat{\gamma}_{m,n,q}}{\sum_{n=1}^{N} (\hat{u}_{m,n} - \hat{\mu}_{m,q})^2 \hat{\gamma}_{m,n,q}} \\ \hat{\pi}_{m,q} &= \frac{\sum_{n=1}^{N} \hat{\gamma}_{m,n,q}}{\sum_{n=1}^{N} \sum_{q'=1}^{\hat{K}_Q} \hat{\gamma}_{m,n,q'}} \ . \end{aligned} \tag{8}$$

5 Experiments

Several different experiments are performed in order to assess the separation performance of AR-MOG. Separation performance is gauged using the signal-to-interference ratio (SIR), which is defined by

$$\mathrm{SIR} = \frac{1}{M} \sum_{m=1}^{M} 10 \log_{10} \frac{\sum_{n=1}^{N} (\hat{\boldsymbol{W}}_{m,1:\mathrm{M}} \boldsymbol{A}_{1:\mathrm{M},m} s_{m,n})^2}{\sum_{\substack{m'=1 \\ (m' \neq m)}}^{M} \sum_{n=1}^{N} (\hat{\boldsymbol{W}}_{m,1:\mathrm{M}} \boldsymbol{A}_{1:\mathrm{M},m'} s_{m',n})^2} \ (\mathrm{dB}) \ .$$

Unless otherwise specified the data is drawn from the same model that is used by AR-MOG, the innovations are assumed to have the same distribution, $M = 2$, and $N = 10^4$. The error bars represent one standard error. When they are included the mean results represent the average of 10 Monte Carlo trials. Results from JADE [12], which does not use temporal dependencies ($\hat{K}_g = 0$), and MRMI-SIG [6], which essentially uses $\hat{K}_g = 1$, are also included as benchmarks.

Figure 1a shows the mean separation performance of AR-MOG as a function of $\hat{K}_g$, where $K_g = 10$ and $\hat{K}_Q = K_Q = 4$. The means, precisions, and priors are not adapted in this experiment or the next experiment so that the change in performance due to the addition of the AR filters may be better quantified. For $\hat{K}_g = 0$

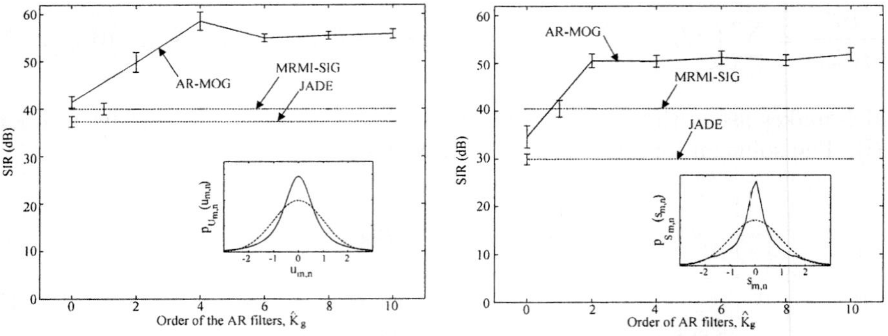

(a) Experiment 1 (b) Experiment 2

Fig. 1. Separation performance as a function of $\hat{K}_g$. (a) Experiment 1. The inset shows $p_{\mathrm{U}_{m,n}}(u_{m,n})$ and a Gaussian distribution (dashed line) having the same mean and variance. (b) Experiment 2. The inset shows $p_{\mathrm{S}_{m,n}}(s_{m,n})$ and a Gaussian distribution (dashed line) having the same mean and variance.

AR-MOG defaults to the case where no AR model is used, i.e., $\hat{\boldsymbol{u}} = \hat{\boldsymbol{s}}$. When no temporal dependencies are used the AR-MOG method performs similarly, but slightly better, than JADE and MRMI-SIG. For $\hat{K}_g \geq 4$ the performance improvement of AR-MOG is approximately 15-20 dB.

The separation results in Fig. 1a represent a best case scenario for AR-MOG since the data are drawn from the model. The results shown in Fig. 1b use (artificially-mixed) real speech data not drawn from the AR-MOG model. Performance is shown as a function of $\hat{K}_g$ where $\hat{K}_Q = 3$. The target distribution $p_{\bar{\mathrm{U}}_{m,n}}(u_{m,n})$ is chosen to be unimodal and super-Gaussian since speech is known to be approximately Laplacian. For the speech data the performance of both AR-MOG and JADE are reduced by approximately 5-8 dB with respect to the first experiment. Figure 1b shows that it is not strictly necessary for the sources to be stationary processes for AR-MOG to perform well (speech is commonly assumed to be stationary, but only for very short segments [13]).

The third experiment shows the sensitivity of the three BSS algorithms to an increase in the temporal correlation of the sources. For this experiment $N = 3 * 10^4$, $\hat{K}_Q = K_Q = 3$, $\hat{K}_g = 4$, and each $\boldsymbol{s}_{m,1:\mathrm{N}}$ is related to the associated $\boldsymbol{u}_{m,1:\mathrm{N}}$ by means of a moving average (MA) filter, $\boldsymbol{h}_{m,1:K_h+1}$. Performance is shown in Fig. 2a as the order of this filter, K_h, is varied (increasing K_h increases the overall correlation at an exponentially decreasing rate). Unlike the previous experiments the means and variances are adapted. For this dataset increasing the temporal correlation (i.e., K_h) causes the separation performance of JADE and MRMI-SIG to decrease by roughly 30 dB and 6 dB, respectively. The performance of AR-MOG is not affected by the change in temporal correlation.

The fourth experiment attempts to measure the separation performance as a function of the initialization of $p_{\bar{\mathrm{U}}_{m,n}}(u_{m,n})$. For each case considered the separation performance is given when the parameters that constitute $p_{\bar{\mathrm{U}}_{m,n}}(u_{m,n})$ are

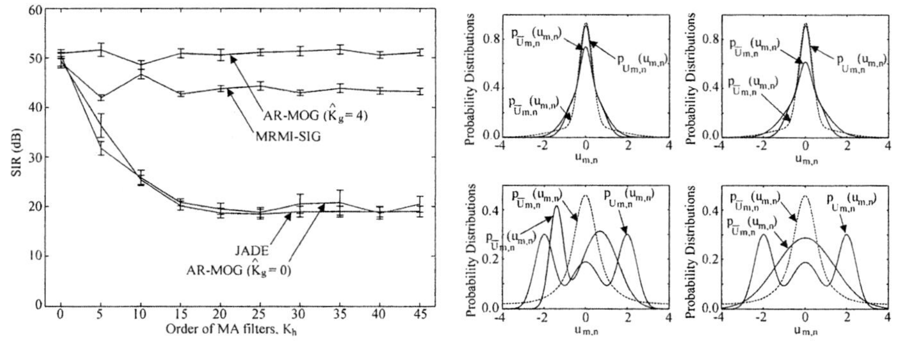

(a) Experiment 3

(b) Experiment 4

Fig. 2. (a) Separation performance as a function of K_h (the length of each h_m) for Experiment 3. (b) Initial (dashed line) and final $p_{\bar{U}_{m,n}}(u_{m,n})$ distributions and $p_{U_{m,n}}(u_{m,n})$ for Experiment 4. Upper-left: Case 1. Upper-right: Case 2. Lower-left: Case 3. Lower-right: Case 4.

adapted and when they are fixed. The resulting SIR values are shown in Table I, where the left column corresponds to when $p_{\bar{U}_{m,n}}(u_{m,n})$ is adapted, whereas the right column keeps $p_{\bar{U}_{m,n}}(u_{m,n})$ fixed at the distribution used for initialization of the left column results. The initial and final $p_{\bar{U}_{m,n}}(u_{m,n})$ distributions and the true distribution, $p_{U_{m,n}}(u_{m,n})$, are shown in Fig. 2b. For Cases 1 & 2 the initial innovations distribution and the true distribution are similar and for Cases 3 & 4 the assumed (initial) innovations distribution is far from correct. Likewise, for Cases 1 & 3 $\hat{K}_g = K_g = 0$ and for Cases 2 & 4 $\hat{K}_g = K_g = 4$. When the initial innovations distribution is similar to the true distribution the separation performance is excellent independent of whether or not $p_{\bar{U}_{m,n}}(u_{m,n})$ is adapted. When they are not similar, based on these results, it is advantageous to adapt $p_{\bar{U}_{m,n}}(u_{m,n})$. Notice that $p_{\bar{U}_{m,n}}(u_{m,n})$ gets trapped in a local maximum for Cases 3 and 4. This is indicated by the fact that the target distribution converges to a bimodal solution for Case 3 and a unimodal solution for Case 4. If AR-MOG is initialized with the true distribution the final SIR is 62.9 and 67.7 dB, respectively, and the target distributions for both cases converge to a trimodal solution. The fact that the final target distribution is incorrect does not necessarily preclude the possibility of

Table 1. Final SIR separation performance for Experiment 4

Case	Adapt $p_{\bar{U}_{m,n}}(u_{m,n})$	Fixed $p_{\bar{U}_{m,n}}(u_{m,n})$
1	45.6 dB	46.3 dB
2	59.5 dB	47.1 dB
3	44.2 dB	0.0 dB
4	51.0 dB	39.5 dB

achieving good separation performance, as indicated in Table I, because it is neither sufficient nor necessary for good separation performance that the final $p_{\bar{\mathrm{U}}_{m,n}}(u_{m,n})$ approximates $p_{\mathrm{U}_{m,n}}(u_{m,n})$. What is necessarily required (but is not sufficient, e.g., for Gaussian distributions) is that $p_{\hat{S}_{m,n}}(s_{m,n})$ approximates $p_{\mathrm{S}_{m,n}}(s_{m,n})$ for each m (and allowing for possible permutations). The ability of AR-MOG to separate sources even if $p_{\bar{\mathrm{U}}_{m,n}}(u_{m,n})$ is incorrect is identical to the well-known fact that ML methods that assume the cumulative density function (cdf) is sigmoidal are oftentimes able to separate sources even if the cdf of each source is not sigmoidal [4], [11], [14], [15], [16], [17]. There is no assurance that AR-MOG will be able to find a solution for $p_{\bar{\mathrm{U}}_{m,n}}(u_{m,n})$ that allows for good separation, but Table I indicates that it may be advantageous to try to improve on the original assumptions.

6 Conclusions

This paper develops a BSS algorithm that is based on maximizing the data likelihood where each source is assumed to be an AR process and the innovations are described using a MOG distribution. It differs from most ML methods in that it uses both spatial and temporal information, the EM algorithm is used as the optimization method, and the parameters that constitute the source model are adapted to maximize the criterion. Due to the combination of the AR process and the MOG model, the update equations for each parameter has a very simple form. The separation performance was compared to several other methods, one that does not take into account temporal information and one that does. The proposed method outperforms both. Future work will focus on incorporating noise directly into the model in a manner similar to that used for the Independent Factor Analysis method [18].

Acknowledgments

This work was supported by National Institutes of Health grants 1 F32 NS 52048-01 and RO1 DC 4855.

References

1. Van Gerven, S., Van Compernolle, D.: Signal separation by symmetric adaptive decorrelation: stability, convergence, and uniqueness. IEEE Trans. on Signal Proc. **43** (1995) 1602–1612
2. Weinstein, E., Feder, M., Oppenheim, A.: Multi-channel signal separation by decorrelation. IEEE Trans. on Speech and Audio Proc. **1** (1993) 405–413
3. Wu, H.C., Principe, J.C.: A unifying criterion for blind source separation and decorrelation: simultaneous diagonalization of correlation matrices. Neural Networks for Signal Proc. (1997) 496–505
4. Bell, A.J., Sejnowski, T.J.: An information-maximization approach to blind separation and blind deconvolution. Neural Computation **7** (1995) 1129–1159

5. Muller, K.R., Philips, P., Ziehe A.: Jade$_{TD}$: Combining higher-order statistics and temporal information for blind source separation (with noise). Intl. Workshop on Independent Component Analysis and Signal Separation (1999) 87–92
6. Hild II, K.E., Erdogmus, D., Principe, J.C.: An Analysis of Entropy Estimators for Blind Source Separation. Signal Processing **86** (2006) 182–194
7. Moulines, E., Cardoso, J.F., Gassiat, E.: Maximum Likelihood for blind source separation and deconvolution of noisy signals using mixture models. Intl. Conf. on Acoustics, Speech, and Signal Processing **5** (1997) 3617–3620
8. Pham, D.T., Garat, P.: Blind separation of mixture of independent sources through a quasi-maximum likelihood approach. IEEE Trans. on Signal Proc. **45** (1997) 1712–1725
9. Pearlmutter, B.A., Parra, L.C.: Maximum likelihood blind source separation: A context-sensitive generalization of ICA. Advances in Neural Information Proc. Systems **9** (1996) 613–619
10. Amari, S.: Neural learning in structured parameter spaces - Natural Riemannian gradient. Advances in Neural Information Proc. Systems **9** (1996) 127–133
11. Cardoso, J.F., Laheld, B.H.: Equivariant adaptive source separation. IEEE Trans. on Signal Proc. **44** (1996) 3017–3030
12. Cardoso, J.F., Souloumiac, A.: Blind beamforming for non-Gaussian signals. IEE Proceedings-F **140** (1993) 362–370
13. Rabiner, L.R., Schafer, R.W.: Digital Processing of Speech Signals (1978)
14. Hosseini, S., Jutten, C., Pham, D.T.: Markovian source separation. IEEE Trans. on Signal Proc. **51** (2003) 3009–3019
15. Amari, S.I., Cardoso, J.F.: Blind source separation-Semiparameteric statistical approach. IEEE Trans. on Signal Proc. **45** (1997) 2692–2700
16. Cruces-Alvarez, S.A., Cichoki, A., Amari, S.I.: On a new blind signal extraction algorithm: Different criteria and stability analysis. IEEE Signal Proc. Letters **9** (2002) 233–236
17. Cardoso, J.F.: Infomax and maximum likelihood for blind source separation. IEEE Signal Proc. Letters **4** (1997) 112–114
18. Attias, H.: Independent factor analysis. Neural Computation **11** (1999) 803–851

Markovian Blind Image Separation

Shahram Hosseini[1], Rima Guidara[1], Yannick Deville[1], and Christian Jutten[2]

[1] Laboratoire Astrophysique de Toulouse-Tarbes (LATT), Observatoire Midi-Pyrénées - Université Paul Sabatier, 14 Avenue Edouard Belin, 31400 Toulouse, France
{shosseini, rguidara, ydeville}@ast.obs-mip.fr
[2] Laboratoire des Images et des Signaux (LIS), UMR CNRS-INPG-UJF, Grenoble, France
Christian.Jutten@inpg.fr

Abstract. We recently proposed a markovian image separation method. The proposed algorithm is however very time consuming so that it cannot be applied to large-size real-world images. In this paper, we propose two major modifications *i.e.* utilization of a low-cost parametric score function estimator and derivation of a modified equivariant version of Newton-Raphson algorithm for solving the estimating equations. These modifications make the algorithm much faster and allow us to perform more experiments with artificial and real data which are presented in the paper.

1 Introduction

We recently proposed [1] a quasi-efficient Maximum Likelihood (ML) approach for blindly separating mixtures of temporally correlated, mono-dimensional independent sources where a Markov model was used to simplify the joint Probability Density Functions (PDF) of successive samples of each source. This approach exploits both source non-gaussianity and autocorrelation in a quasi-optimal manner.

In [2], we extended this idea to bi-dimensional sources (in particular images), where the spatial autocorrelation of each source was described using a second-order Markov Random Field (MRF). The idea of using MRF for image separation has recently been exploited by other authors [3], where the source PDF are supposed to be known, and are used to choose the Gibbs priors. In [2], however, we made no assumption about the source PDF so that the method remains quasi-efficient whatever the source distributions. The first experimental results reported in [2] confirmed the better performance of our method with respect to the ML methods which ignore the source autocorrelation [4] and the autocorrelation-based methods which ignore the source non-gaussianity [5], [6].

The algorithm used in [2] is however very slow: its implementation requires the estimation of some 5-dimensional conditional score functions using a non-parametric estimator and the maximization of a likelihood function using a time

J. Rosca et al. (Eds.): ICA 2006, LNCS 3889, pp. 106–114, 2006.

consuming gradient method. In the present paper, we propose a parametric polynomial estimator of the conditional score functions which is much faster than the non-parametric estimator. We also derive a modified equivariant Newton-Raphson algorithm which considerably reduces the computational cost of the optimization procedure. Using this fast algorithm, we performed more simulations with artificial and real-world data to compare our method with classical approaches.

2 ML Method for Separating Markovian Images

Assume we have $N = N_1 \times N_2$ samples of a K-dimensional vector $\mathbf{x}(n_1, n_2)$ resulting from a linear transformation $\mathbf{x}(n_1, n_2) = \mathbf{A}\mathbf{s}(n_1, n_2)$, where $\mathbf{s}(n_1, n_2)$ is the vector of independent image sources $s_i(n_1, n_2)$, each one of dimension $N_1 \times N_2$ and possibly spatially autocorrelated, and $\mathbf{A}$ is a $K \times K$ invertible matrix. Our objective is to estimate the separating matrix $\mathbf{B} = \mathbf{A}^{-1}$ up to a diagonal matrix and a permutation matrix.

The ML method consists in maximizing the joint PDF of all the samples of all the components of the vector $\mathbf{x}$ (all the observations), with respect to the separating matrix $\mathbf{B}$. We denote this PDF

$$f_{\mathbf{x}}(x_1(1,1), \cdots, x_K(1,1), \cdots, x_1(N_1, N_2), \cdots, x_K(N_1, N_2)) \tag{1}$$

Under the assumption of independence of the sources, this function is equal to

$$(\frac{1}{|\det(\mathbf{B}^{-1})|})^N \prod_{i=1}^{K} f_{s_i}(s_i(1,1), \cdots, s_i(N_1, N_2)) \tag{2}$$

where $f_{s_i}(.)$ represents the joint PDF of N samples of the source s_i. Each joint PDF can be decomposed using Bayes rule in many different manners following different sweeping trajectories within the image corresponding to source s_i (Fig. 1). These schemes being essentially equivalent, we chose the horizontal sweeping. Then, the joint PDF of source s_i can be decomposed using Bayes rule to obtain

$$\begin{aligned}&f_{s_i}(s_i(1,1))f_{s_i}(s_i(1,2)|s_i(1,1))f_{s_i}(s_i(1,3)|s_i(1,2), s_i(1,1)) \cdots\cdots \\ &f_{s_i}(s_i(1,N_2)|s_i(1,N_2-1), \cdots, s_i(1,1))f_{s_i}(s_i(2,1)|s_i(1,N_2), \cdots, s_i(1,1)) \cdots\cdots \\ &f_{s_i}(s_i(N_1,N_2)|s_i(N_1,N_2-1), \cdots, s_i(1,1))\end{aligned} \tag{3}$$

This equation may be simplified by assuming a Markov model for the sources. We suppose hereafter that the sources are second-order Markov random fields, *i.e.* the conditional PDF of a pixel $s(n_1, n_2)$ given all the other pixels is equal to its conditional PDF given its 8 nearest neighbors (Fig. 2). From this assumption, it is clear that the conditional PDF of a pixel not situated on the boundaries, given all its predecessors (in the sense of sweeping trajectory) is equal to its conditional PDF given its three top neighbors and its left neighbor (squares in Fig. 2). In other words, if D_{n_1,n_2} is the set of pixel values $s_i(k,l)$ such that $\{k < n_1\}$ or $\{k = n_1, l < n_2\}$, then

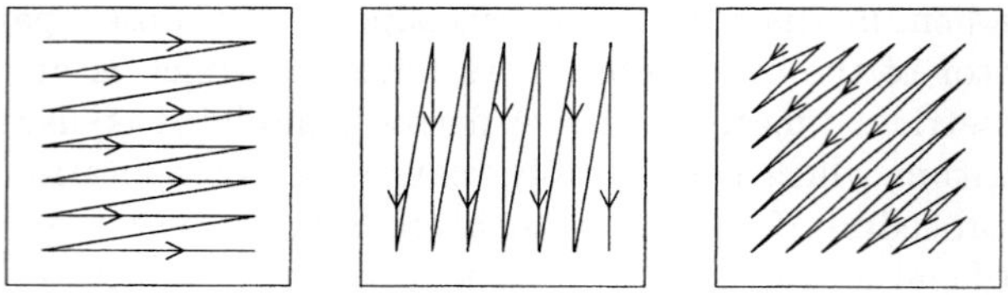

Fig. 1. Different sweeping possibilities

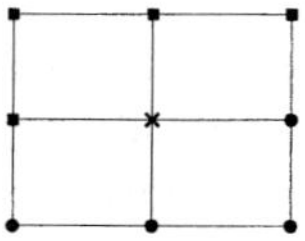

Fig. 2. Second-order Markov random field

$$f_{s_i}(s_i(n_1,n_2)|D_{n_1,n_2}) = f_{s_i}(s_i(n_1,n_2)|s_i(n_1,n_2-1), s_i(n_1-1,n_2+1), s_i(n_1-1,n_2), s_i(n_1-1,n_2-1)) \quad (4)$$

If N is sufficiently large, the conditional PDF of the pixels located on the left, top and right image boundaries (for which, the 4 mentioned neighbors are not available) may be neglected in (3). Supposing that the sources are stationary so that the conditional PDF (4) does not depend on n_1 and n_2, it follows from (4) that the decomposed joint PDF (3) can be rewritten as

$$f_{s_i}(s_i(1,1), s_i(1,2), \cdots, s_i(1,N_2), s_i(2,1), \cdots, s_i(N_1,N_2)) \simeq \prod_{n_1=2}^{N_1} \prod_{n_2=2}^{N_2-1} f_{s_i}(s_i(n_1,n_2)|s_i(n_1,n_2-1), s_i(n_1-1,n_2+1), s_i(n_1-1,n_2), s_i(n_1-1,n_2-1))$$

The log-likelihood function may be obtained by replacing the above PDF in (2) and taking the logarithm:

$$N\log(|\det(\mathbf{B})|) + \sum_{i=1}^{K}\sum_{n_1=2}^{N_1}\sum_{n_2=2}^{N_2-1} \log f_{s_i}(s_i(n_1,n_2)|s_i(n_1,n_2-1), s_i(n_1-1,n_2+1), s_i(n_1-1,n_2), s_i(n_1-1,n_2-1)) \quad (5)$$

Dividing the above cost function by N and defining the spatial average operator $E_N[.] = \frac{1}{N}\sum_{n_1=2}^{N_1}\sum_{n_2=2}^{N_2-1}[.]$, Equation (5) may be rewritten in the following simpler form

$$L_1 = \log(|\det(\mathbf{B})|) + E_N[\sum_{i=1}^{K} \log f_{s_i}(s_i(n_1,n_2)|s_i(n_1,n_2-1), s_i(n_1-1,n_2+1), s_i(n_1-1,n_2), s_i(n_1-1,n_2-1))]$$

In [2], the separating matrix $\mathbf{B}$ was obtained by maximizing the above cost function using a relative gradient ascent algorithm which is very time consuming. Here, we choose another approach which consists in solving the equation $\frac{\partial L_1}{\partial \mathbf{B}} = 0$ using a modified equivariant Newton-Raphson algorithm.

3 Estimating Equations and Their Solution

As shown in [2], the gradient of the cost function L_1 is equal to

$$\frac{\partial L_1}{\partial \mathbf{B}} = \mathbf{B}^{-T} - E_N[\sum_{(k,l)\in \Upsilon} \boldsymbol{\Psi}_{\mathbf{s}}^{(k,l)}(n_1,n_2).\mathbf{x}^T(n_1-k,n_2-l)] \tag{6}$$

where $\Upsilon = \{(0,0),(0,1),(1,-1),(1,0),(1,1)\}$ and the vector $\boldsymbol{\Psi}_{\mathbf{s}}^{(k,l)}(n_1,n_2)$ contains the conditional score functions of the K sources, which are denoted $\psi_{s_i}^{(k,l)}(n_1,n_2)$ hereafter for simplicity, and which read explicitly

$$\begin{aligned}&\psi_{s_i}^{(k,l)}(n_1,n_2) = \psi_{s_i}^{(k,l)}(s_i(n_1,n_2)|s_i(n_1,n_2-1),s_i(n_1-1,n_2+1),\\ &s_i(n_1-1,n_2),s_i(n_1-1,n_2-1)) = \frac{-\partial}{\partial s_i(n_1-k,n_2-l)}\log f_{s_i}(s_i(n_1,n_2)|\\ &s_i(n_1,n_2-1),s_i(n_1-1,n_2+1),s_i(n_1-1,n_2),s_i(n_1-1,n_2-1))\end{aligned} \tag{7}$$

Setting (6) to zero, then post-multiplying by $\mathbf{B}^T$ we obtain

$$E_N[\sum_{(k,l)\in \Upsilon} \boldsymbol{\Psi}_{\mathbf{s}}^{(k,l)}(n_1,n_2).\mathbf{s}^T(n_1-k,n_2-l)] = \mathbf{I} \tag{8}$$

This yields the $K(K-1)$ estimating equations

$$E_N[\sum_{(k,l)\in \Upsilon} \psi_{s_i}^{(k,l)}(n_1,n_2).s_j(n_1-k,n_2-l)] = 0 \qquad i \neq j = 1,\cdots,K \tag{9}$$

which determine $\mathbf{B}$ up to a diagonal and a permutation matrix. The other K equations $E_N[\sum_{(k,l)\in \Upsilon} \psi_{s_i}^{(k,l)}(n_1,n_2).s_i(n_1-k,n_2-l)] = 1 \quad i = 1,\cdots,K$ are not important and can be replaced by any other scaling convention.

The system of equations (9) may be solved using the Newton-Raphson algorithm. We propose a modified version of this algorithm which has the equivariance property, *i.e.* its performance does not depend on the mixing matrix.

To ensure the equivariance property, the adaptation gain must be proportional to the previous value of $\mathbf{B}$. Let $\tilde{\mathbf{B}}$ be an initial estimation of $\mathbf{B}$. We want to find a matrix $\boldsymbol{\Delta}$ so that the estimation $\hat{\mathbf{B}} = (\mathbf{I}+\boldsymbol{\Delta})\tilde{\mathbf{B}}$ be a solution of (9). To simplify the notations, we here only consider the case $K = 2$ but the same approach may be used for higher values of K. In the appendix, we show that the off-diagonal entries of $\boldsymbol{\Delta}$, δ_{12} and δ_{21}, are the solutions of the following linear system of equations

$$E_N[\sum_{(k,l)\in\Upsilon} \psi_{\hat{s}_1}^{(k,l)}(n_1,n_2).\tilde{s}_1(n_1-k,n_2-l)]\delta_{21}$$

$$+E_N[\sum_{(k,l)\in\Upsilon}\{\sum_{(i,j)\in\Upsilon} \frac{\partial\psi_{\hat{s}_1}^{(k,l)}(n_1,n_2)}{\partial s_1(n_1-i,n_2-j)}\tilde{s}_2(n_1-i,n_2-j)\}.\tilde{s}_2(n_1-k,n_2-l)]\delta_{12}$$

$$= -E_N[\sum_{(k,l)\in\Upsilon} \psi_{\hat{s}_1}^{(k,l)}(n_1,n_2).\tilde{s}_2(n_1-k,n_2-l)]$$

$$E_N[\sum_{(k,l)\in\Upsilon} \psi_{\hat{s}_2}^{(k,l)}(n_1,n_2).\tilde{s}_2(n_1-k,n_2-l)]\delta_{12}$$

$$+E_N[\sum_{(k,l)\in\Upsilon}\{\sum_{(i,j)\in\Upsilon} \frac{\partial\psi_{\hat{s}_2}^{(k,l)}(n_1,n_2)}{\partial s_2(n_1-i,n_2-j)}\tilde{s}_1(n_1-i,n_2-j)\}.\tilde{s}_1(n_1-k,n_2-l)]\delta_{21}$$

$$= -E_N[\sum_{(k,l)\in\Upsilon} \psi_{\hat{s}_2}^{(k,l)}(n_1,n_2).\tilde{s}_1(n_1-k,n_2-l)] \tag{10}$$

The computation of the coefficients δ_{12} and δ_{21} requires the estimation of the conditional score functions and their derivatives. In [2], we used a non-parametric method proposed in [7] involving the estimation of joint entropies using a discrete Riemann sum and third-order cardinal spline kernels. This estimator is very time and memory consuming and does not provide the derivatives of the score functions required for Newton-Raphson algorithm. In the following section, we propose another solution based on a third order polynomial parametric estimation of the score functions which is very fast and directly provides the derivatives of the score functions. Then, the terms δ_{12} and δ_{21} can be obtained by solving (10). The diagonal entries of $\boldsymbol{\Delta}$ are not important because they influence only the scale factors. Thus, we can fix them arbitrarily to zero.

4 Parametric Estimation of the Score Functions

Our parametric estimator of the conditional score functions is based on the following theorem, proved in [8] in the scalar case:

Theorem. If $\lim_{y_i\to\pm\infty} f_y(y_0,\cdots,y_q)g(y_0,\cdots,y_q) = 0$ [1] where f_y is the joint PDF of $y_0,\cdots.y_q$ and g is an arbitrary function of these variables, then

$$E[-\frac{\partial \log f_y(y_0,\cdots,y_q)}{\partial y_i}g(y_0,\cdots,y_q)] = E[\frac{\partial g(y_0,\cdots,y_q)}{\partial y_i}] \tag{11}$$

Following this theorem, if $g(y_0,\cdots,y_q,\mathbf{W})$ is a mean-square parametric estimator of the joint score function $\psi_{y_i}(y_0,\cdots,y_q) = -\frac{\partial \log f_y(y_0,\cdots,y_q)}{\partial y_i}$, its parameter vector $\mathbf{W}$, can be found from

[1] When $g(.)$ is bounded, this condition is satisfied for every real-world signal because its joint PDF tends to zero at infinity.

$$\mathbf{W} = argmin\{E[g^2(y_0, \cdots, y_q, \mathbf{W})] - 2E[\frac{\partial g(y_0, \cdots, y_q, \mathbf{W})}{\partial y_i}]\} \qquad (12)$$

Note that the function to be minimized does not explicitly depend on the score function itself. In our problem, we want to estimate the conditional score functions. Each conditional score function can be written as the difference between two joint score functions:

$$\begin{aligned}\psi_{y_i}(y_0|y_1, \cdots, y_q) &= -\frac{\partial \log f_y(y_0, \cdots, y_q)}{\partial y_i} + \frac{\partial \log f_y(y_1, \cdots, y_q)}{\partial y_i} \\ &= \psi_{y_i}(y_0, \cdots, y_q) - \psi_{y_i}(y_1, \cdots, y_q) \qquad (13)\end{aligned}$$

Each of two joint score functions in the above equation can be estimated using a parametric estimator which may be realized in different manners. In our work, we used the polynomial functions because of their linearity with respect to the parameters which simplifies the computations.

The conditional score functions used in our work being of dimension 5, they may be written as the difference between two joint score functions of dimensions 5 and 4 respectively. We used third-order polynomial functions for estimating them. The polynomial function modeling the 5-dimensional joint score function must contain all the possible terms in $\{1, (y_0, y_1, y_2, y_3, y_4), (y_0, y_1, y_2, y_3, y_4)^2, (y_0, y_1, y_2, y_3, y_4)^3\}$. Hence, it contains $\sum_{k=0}^{3}\binom{5+k-1}{k} = 56$ coefficients. In the same manner, the polynomial function modeling the 4-dimensional joint score function contains $\sum_{k=0}^{3}\binom{4+k-1}{k} = 35$ coefficients.

Our tests confirm that the above parametric estimator is much more faster, roughly 100 times, than the non-parametric estimator used in [2] and leads to the same performance.

5 Simulation Results

In the following experiments, we compare our method with two well-known algorithms: SOBI [6] and Pham-Garat [4]. SOBI is a second-order method which consists in jointly diagonalizing several covariance matrices evaluated at different lags. The Pham-Garat algorithm is based on a maximum likelihood approach which supposes that the sources are i.i.d. and therefore does not take into account their possible autocorrelation. For each experiment, the output Signal to Interference Ratio (in dB) was computed by $SIR = 0.5\sum_{i=1}^{2} 10\log_{10}\frac{E[s_i^2]}{E[(\hat{s}_i - s_i)^2]}$, after normalizing the estimated sources, $\hat{s}_i(n_1, n_2)$, so that they have the same variances and signs as the source signals, $s_i(n_1, n_2)$.

In the first experiment, we use artificial image sources of size 100×100 which satisfy exactly the considered Markov model. Two independent non-autocorrelated and uniformly distributed image noises, $e_1(n_1, n_2)$ and $e_2(n_1, n_2)$, are filtered by two autoregressive (AR) filters using the following formula:

$$\begin{aligned}s_i(n_1, n_2) = e_i(n_1, n_2) &+ \rho_{0,1}s_i(n_1, n_2 - 1) + \rho_{1,-1}s_i(n_1 - 1, n_2 + 1) \\ &+ \rho_{1,0}s_i(n_1 - 1, n_2) + \rho_{1,1}s_i(n_1 - 1, n_2 - 1) \qquad (14)\end{aligned}$$

The coefficients $\rho_{i,j}$ are chosen to guarantee a sufficient stability condition. Thus, the coefficients of the first and the second filters are respectively fixed to $\{-0.5, 0.4, 0.5, 0.3\}$ and $\{-0.5, \rho_{1,-1}, 0.5, 0.3\}$. The coefficient $\rho_{1,-1}$ of the second filter may change in its stability interval, *i.e.* $[0.2, 0.6]$. Then, the source images $s_i(n_1, n_2)$ are mixed by the mixing matrix $\mathbf{A} = \begin{pmatrix} 1 & 0.99 \\ 0.99 & 1 \end{pmatrix}$. The mean of SIR over 100 Monte Carlo simulations as a function of the coefficient $\rho_{1,-1}$ of the second AR filter is shown in Fig. 3-a. Our algorithm outperforms the other two, whatever $\rho_{1,-1}$.

In the second experiment, the same non-autocorrelated and uniformly distributed image noises, $e_1(n_1, n_2)$ and $e_2(n_1, n_2)$, were generated and one of them was filtered by a symmetrical FIR filter. It is evident that the filtered signal is no longer a 2nd-order MRF. Then, the signals were mixed by the same matrix as in the first experiment. The mean of SIR as a function of the selectivity of the FIR filter is shown in Fig. 3-b. The performance of our method is always better than SOBI. It also outperforms Pham-Garat unless the filter selectivity is small so that the filtered signal is nearly uncorrelated. In the last experiment, the two real images of dimension 230×270 pixels, shown in Fig. 4, were mixed by the same matrix. It is clear that the working hypotheses are no longer true because the images are not stationary and cannot be described by 2nd-order MRFs. However, the images are autocorrelated and nearly cyclostationary because the correlation profiles on different circles are similar. Thus, the conditional score functions on different circles are nearly similar. Once more, the three mentioned algorithms were used for separating the sources. Our algorithm led to 57-dB SIR while SOBI and Pham-Garat led to 23-dB and 12-dB SIR respectively.

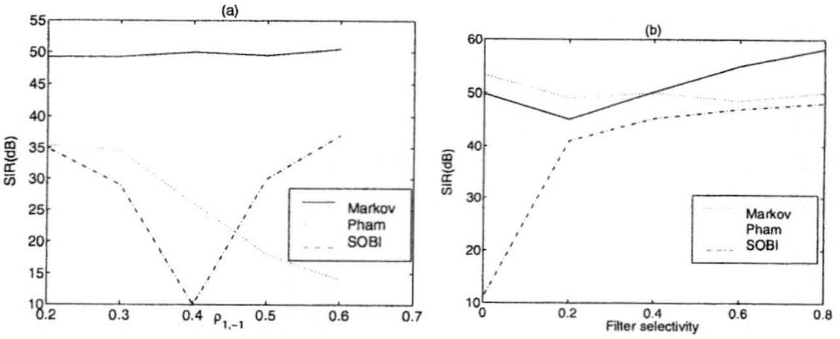

Fig. 3. Simulation results using (a) IIR and (b) FIR filters

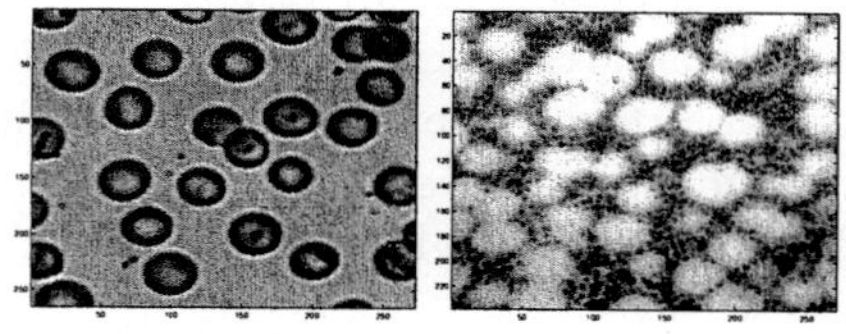

Fig. 4. Real-world images used in the experiment

6 Conclusion

In this paper, we made two major modifications in our markovian blind image separation algorithm *i.e.* utilization of a low-cost parametric score function estimator instead of the non-parametric estimator, and derivation of a modified equivariant Newton-Raphson algorithm for solving the estimating equations instead of maximizing the log-likelihood function by a relative gradient algorithm. These modifications led to a much faster algorithm and allowed us to perform more experiments using artificial and real-world data. These experiments confirmed the better performance of our method in comparison to the classical methods which ignore spatial autocorrelation or non-gaussianity of data.

References

1. S. Hosseini, C. Jutten and D.-T. Pham, Markovian source separation, *IEEE Trans. on Signal Processing*, vol. 51, pp. 3009-3019, 2003.
2. S. Hosseini, R. Guidara, Y. Deville and C. Jutten, Maximum likelihood separation of spatially autocorrelated images using a Markov model, in *Proc. of MAXENT'05*, San Jose, USA, August 2005.
3. E. E. Kuruoglu, A. Tonazzini and L. Bianchi, Source separation in noisy astrophysical images modelled by markov random fields, in *Proc. ICIP'04*, pp. 2701-2704.
4. D.-T. Pham and P. Garat, Blind separation of mixture of independent sources through a quasi-maximum likelihood approach, *IEEE Trans. on Signal Processing*, vol. 45, pp. 1712-1725, July 1997.
5. L. Tong, R. Liu, V. Soon and Y. Huang, Indeterminacy and identifiability of blind identification, *IEEE Trans. on Circuits Syst.*, vol. 38, pp. 499-509, May 1991.
6. A. Belouchrani, K. Abed Meraim, J.-F. Cardoso and E. Moulines, A blind source separation technique based on second order statistics, *IEEE Trans. on Signal Processing*, vol. 45, pp. 434-444, Feb. 1997.
7. D.-T. Pham, Fast algorithms for mutual information based independent component analysis, *IEEE Trans. on Signal Processing*, vol. 52, no. 10, Oct. 2004.
8. D.-T. Pham, P. Garat and C. Jutten, Separation of mixture of independent sources through a quasi-maximum likelihood approach, in *Proc. of EUSIPCO'92*, pp. 771-774, Brussels, August 1992.

Appendix. Derivation of Equations (10)

Post-multiplying $\hat{\mathbf{B}} = (\mathbf{I} + \boldsymbol{\Delta})\tilde{\mathbf{B}}$ by $\mathbf{x}$ we obtain $\hat{\mathbf{s}} = (\mathbf{I} + \boldsymbol{\Delta})\tilde{\mathbf{s}}$. Denoting $\boldsymbol{\Delta} = \begin{pmatrix} \delta_{11} & \delta_{12} \\ \delta_{21} & \delta_{22} \end{pmatrix}$, it implies that $\widehat{s}_1(n_1, n_2) = \widetilde{s}_1(n_1, n_2) + \delta_{11}\widetilde{s}_1(n_1, n_2) + \delta_{12}\widetilde{s}_2(n_1, n_2)$ and $\widehat{s}_2(n_1, n_2) = \widetilde{s}_2(n_1, n_2) + \delta_{21}\widetilde{s}_1(n_1, n_2) + \delta_{22}\widetilde{s}_2(n_1, n_2)$. Since $\hat{s}_1$ and $\hat{s}_2$ must satisfy the estimating equations (9), by replacing the above relations in the first estimating equation and considering (7) we obtain

$$E\Big[\sum_{(k,l)\in\Upsilon}\{\psi_{\widetilde{s}_1}^{(k,l)}(\widetilde{s}_1(n_1,n_2)+\delta_{11}\widetilde{s}_1(n_1,n_2)+\delta_{12}\widetilde{s}_2(n_1,n_2)|\widetilde{s}_1(n_1,n_2-1)$$
$$+\delta_{11}\widetilde{s}_1(n_1,n_2-1)+\delta_{12}\widetilde{s}_2(n_1,n_2-1),\ldots,\widetilde{s}_1(n_1-1,n_2-1)$$
$$+\delta_{11}\widetilde{s}_1(n_1-1,n_2-1)+\delta_{12}\widetilde{s}_2(n_1-1,n_2-1))\}.\{\widetilde{s}_2(n_1-k,n_2-l)$$
$$+\delta_{21}\widetilde{s}_1(n_1-k,n_2-l)+\delta_{22}\widetilde{s}_2(n_1-k,n_2-l)\}\Big]=0 \quad (15)$$

Using a first-order Taylor development of the score function, noting that the separated sources are independent at the vicinity of the solution, neglecting the terms containing the products of δ_{ij}, and neglecting δ_{22} with respect to 1, we obtain by some simple calculus the first equation in (10). The second equation can be derived by symmetry.

New Permutation Algorithms for Causal Discovery Using ICA

Patrik O. Hoyer[1], Shohei Shimizu[1,2], Aapo Hyvärinen[1], Yutaka Kano[2], and Antti J. Kerminen[1]

[1] HIIT Basic Research Unit, Dept. of Comp. Science, University of Helsinki, Finland
[2] Graduate School of Engineering Science, Osaka University, Japan
http://www.cs.helsinki.fi/group/neuroinf/lingam/

Abstract. Causal discovery is the task of finding plausible causal relationships from statistical data [1,2]. Such methods rely on various assumptions about the data generating process to identify it from uncontrolled observations. We have recently proposed a causal discovery method based on independent component analysis (ICA) called LiNGAM [3], showing how to completely identify the data generating process under the assumptions of linearity, non-gaussianity, and no hidden variables. In this paper, after briefly recapitulating this approach, we focus on the algorithmic problems encountered when the number of variables considered is large. Thus we extend the applicability of the method to data sets with tens of variables or more. Experiments confirm the performance of the proposed algorithms, implemented as part of the latest version of our freely available Matlab/Octave LiNGAM package.

1 Introduction

Several authors [1,2] have recently formalized concepts related to causality using probability distributions defined on directed acyclic graphs. This line of research emphasizes the importance of understanding the process which generated the data, rather than only characterizing the joint distribution of the observed variables. The reasoning is that a causal understanding of the data is essential to be able to predict the consequences of interventions, such as setting a given variable to some specified value.

An interesting question within this theoretical framework is: 'Under what circumstances and in what way can one determine causal structure on the basis of observational data alone?'. In many cases it is impossible or too expensive to perform controlled experiments, and hence methods for discovering likely causal relations from uncontrolled data would be very valuable.

For continuous-valued data the main approach has been based on assumptions of linearity and gaussianity [1,2]. Those assumptions generally lead only to a *set* of possible models equivalent in their conditional correlation structure. We have recently showed [3] that an assumption of *non-gaussianity* in fact allows the full model to be identified using a method based on independent component

J. Rosca et al. (Eds.): ICA 2006, LNCS 3889, pp. 115–122, 2006.

analysis (ICA). However, this new method poses some challenging computational problems. In this paper we describe and solve these problems, allowing the application of the method to problems of high dimensionality.

The paper is structured as follows. In Section 2 we briefly describe the basics of LiNGAM, before focusing on the computational problems in Section 3. The proposed algorithms are empirically evaluated in Section 4. Conclusions are given in Section 5.

2 LiNGAM

Assume that we observe data generated from a process with the following properties:

1. The observed variables x_i, $i = \{1 \dots n\}$ can be arranged in a causal order $k(i)$, defined to be an ordering of the variables such that no later variable in the order participates in generating the value of any earlier variable. That is, the generating process is *recursive* [4], meaning it can be represented graphically by a *directed acyclic graph* (DAG) [2,1].
2. The value assigned to each variable x_i is a *linear function* of the values already assigned to the earlier variables, plus a 'disturbance' (noise) term e_i, and plus an optional constant term c_i, that is

$$x_i = \sum_{k(j)<k(i)} b_{ij} x_j + e_i + c_i. \tag{1}$$

3. The disturbances e_i are all continuous random variables having *non-gaussian* distributions with non-zero variances, and the e_i are independent of each other, i.e. $p(e_1, \dots, e_n) = \prod_i p_i(e_i)$.

A model with these three properties we call a *Linear, Non-Gaussian, Acyclic Model*, abbreviated LiNGAM.

We assume that we observe a large number of data vectors $\mathbf{x}$ (containing the components x_i), and each is generated according to the above described process, with the same causal order $k(i)$, same coefficients b_{ij}, same constants c_i, and the disturbances e_i sampled independently from the same distributions. Note that the above assumptions imply that there are *no unobserved confounders* [2] (hidden variables). Spirtes et al. [1] call this the *causally sufficient* case.

To see how we can identify the parameters of the model from the set of data vectors $\mathbf{x}$, we start by subtracting out the mean of each variable x_i, leaving us with the following system of equations:

$$\mathbf{x} = \mathbf{B}\mathbf{x} + \mathbf{e}, \tag{2}$$

where $\mathbf{B}$ is a matrix that contains the coefficients b_{ij} and that could be permuted (by simultaneous equal row and column permutations) to strict lower triangularity if one knew a causal ordering $k(i)$ of the variables. (Strict lower triangularity is here defined as lower triangular with all zeros on the diagonal.) Solving for $\mathbf{x}$ one obtains

$$\mathbf{x} = \mathbf{A}\mathbf{e}, \tag{3}$$

where $\mathbf{A} = (\mathbf{I} - \mathbf{B})^{-1}$. Again, $\mathbf{A}$ could be permuted to lower triangularity (although not *strict* lower triangularity, actually in this case all diagonal elements will be *non-zero*) with an appropriate permutation $k(i)$. Taken together, equation (3) and the independence and non-gaussianity of the components of $\mathbf{e}$ define the standard linear independent component analysis (ICA) model [5,6], which is known to be identifiable.

While ICA is essentially able to estimate $\mathbf{A}$ (and $\mathbf{W} = \mathbf{A}^{-1}$), there are two important indeterminacies that ICA cannot solve: First and foremost, the order of the independent components is in no way defined or fixed. Thus, we could reorder the independent components and, correspondingly, the columns of $\mathbf{A}$ (and rows of $\mathbf{W}$) and get an equivalent ICA model (the same probability density for the data). In most applications of ICA, this indeterminacy is of no significance and can be ignored, but in LiNGAM, we can and we have to find the correct permutation as described in Section 3 below.

The second indeterminancy of ICA concerns the scaling of the independent components. In ICA, this is usually handled by assuming all independent components to have unit variance, and scaling $\mathbf{W}$ and $\mathbf{A}$ appropriately. On the other hand, in LiNGAM (as in structural equation modeling, SEM [4]) we allow the disturbance variables to have arbitrary (non-zero) variances, but fix their weight (connection strength) to their corresponding observed variable to unity. This requires us to re-normalize the rows of $\mathbf{W}$ so that all the diagonal elements equal unity, before computing $\mathbf{B}$.

Our LiNGAM discovery algorithm [3] can thus be briefly summarized: First, use a standard ICA algorithm to obtain an estimate of the demixing matrix $\mathbf{W}$, permute its rows such that there are no zeros on its diagonal, rescale each row by dividing by the element on the diagonal, and finally compute $\mathbf{B} = \mathbf{I} - \mathbf{W}'$, where $\mathbf{W}'$ denotes the permuted and rescaled $\mathbf{W}$.

To find a causal order $k(i)$ we must subsequently find a second permutation, to be applied equally both to the rows and columns of $\mathbf{B}$, which yields strict lower triangularity.

3 Algorithms for Solving the Permutation Problems

3.1 Permuting the Rows of W

As pointed out above, because of the permutation indeterminancy of ICA, the rows of $\mathbf{W}$ will be in random order. This means that we do not yet have the correct correspondence between the disturbance variables e_i and the observed variables x_i. The former correspond to the rows of $\mathbf{W}$ while the latter correspond to the columns of $\mathbf{W}$. Thus, our first task is to permute the rows to obtain a correspondence between the rows and columns. If $\mathbf{W}$ were estimated exactly, there would exist one (and only one!) row permutation that would give a matrix with no zeros on the diagonal, and this permutation gives the correct correspondence [3]. Furthermore, finding the correct permutation would be trivial.

In practice, however, ICA algorithms applied on finite data sets will yield estimates which are only approximately zero for those elements which should be exactly zero. Thus, we need to search for the correct permutation by minimizing a cost function which heavily penalizes small absolute values in the diagonal, such as $\sum_i 1/|\widetilde{W}_{ii}|$, where $\widetilde{\mathbf{W}}$ denotes the row-permuted $\mathbf{W}$.

An exhaustive search over all possible row-permutations is feasible only in relatively small dimensions [3]. For larger problems other optimization methods are needed. Fortunately, it turns out that the optimization problem can be written in the form of the classical *linear assignment problem*. To see this set $C_{ij} = 1/|W_{ij}|$, in which case the problem can be written as the minimization of

$$\sum_{i=1}^{n} C_{\phi(i),i} \tag{4}$$

where ϕ denotes the permutation to be optimized over. A great number of algorithms exist for this problem, with the best achieving worst-case complexity of $O(n^3)$ where n is the number of variables, see e.g. [7]. In our current implementation though, we simply use general-purpose linear programming software to find the optimum, which is good enough to solve problems involving tens of variables. Future implementations will use the more efficient algorithms.

3.2 Permuting B to Get a Causal Order

Once we have obtained the correct correspondence between rows and columns of the ICA decomposition, calculating estimates of the b_{ij} is straightforward. First, we normalize the rows of the permuted matrix to yield $\mathbf{W}'$, and then calculate $\mathbf{B} = \mathbf{I} - \mathbf{W}'$ as described in Section 2 [3].

Although we now have initial estimates of all coefficients b_{ij} we do not yet have available a causal ordering $k(i)$ of the variables. Such an ordering (in general there may exist many if the generating network is not fully connected) is needed to achieve a directed acyclic graph, thus completing the estimation process. Essentially, after the ordering we can force half of the coefficients to equal zero such that the resulting network has no directed cycles.

A causal ordering can be found by permuting both rows and columns (using the same permutation) of the matrix $\mathbf{B}$ (containing the initial estimated connection strengths) to yield a strictly lower triangular matrix. If the estimates were exact, this would be a trivial task, using the following algorithm:

Algorithm A: Testing for DAGness, and returning a causal order if true

1. Initialize the permutation p to be an empty list
2. Repeat until $\mathbf{B}$ contains no more elements:
   (a) Find a row i of $\mathbf{B}$ containing all zeros, if not possible return **false**
   (b) Append i to the end of the list p
   (c) Remove the i:th row and the i:th column from $\mathbf{B}$
3. Return **true** and the found permutation p

However, since our estimates will not contain exact zeros, we will have to find a permutation such that setting the upper triangular elements to zero changes the matrix as little as possible. For instance, we could define our objective to be to minimize the sum of squares of elements on and above the diagonal, that is $\sum_{i \le j} \widetilde{\mathbf{B}}_{ij}^2$ where $\widetilde{\mathbf{B}} = \mathbf{P}\mathbf{B}\mathbf{P}^T$ denotes the permuted $\mathbf{B}$, and $\mathbf{P}$ denotes the permutation matrix representing the sought permutation. In low dimensions, the optimal permutation can be found by exhaustive search. However, for larger problems this is obviously infeasible. Since we are not aware of any efficient method for exactly solving this combinatorial problem, we have taken another approach to handling the high-dimensional case.

Our approach is based on setting small (absolute) valued elements to zero, and testing whether the resulting matrix can be permuted to strict lower triangularity. Thus, the algorithm is:

Algorithm B: Finding a permutation of $\mathbf{B}$ by iterative pruning and testing

1. Set the $n(n+1)/2$ smallest (in absolute value) elements of $\mathbf{B}$ to zero
2. Repeat
   (a) Test if $\mathbf{B}$ can be permuted to strict lower triangularity (using Algorithm A above). If the answer is yes, stop and return the permuted $\mathbf{B}$
   (b) Additionally set the next smallest (in absolute value) element of $\mathbf{B}$ to zero

---

If in the problem, all the true zeros resulted in estimates smaller than all of the true non-zeros, this algorithm finds the optimal permutation. In general, however, the result is not optimal in terms of the above proposed objective; more elements are usually set to zero than would be needed. Fortunately, this is not a big problem because in sparse networks there are many more zeros in the coefficients than required by the acyclicity of the model, hence we would nevertheless like to prune out the small values from the estimated coefficients. See [3] for some discussion on pruning the estimated coefficients.

4 Experiments

In [3] we empirically verified the basic concept of LiNGAM by generating data from such models and estimating them using our method. However, because of the lack of efficient permutation algorithms we were limited to problems with small numbers of variables (8 variables or less). In the present paper we demonstrate that the method also works well in high dimensions by employing the permutation algorithms discussed in Section 3. All experimental code (including the precise code to produce Figure 1) is included in the LiNGAM code package, available at:

`http://www.cs.helsinki.fi/group/neuroinf/lingam/`

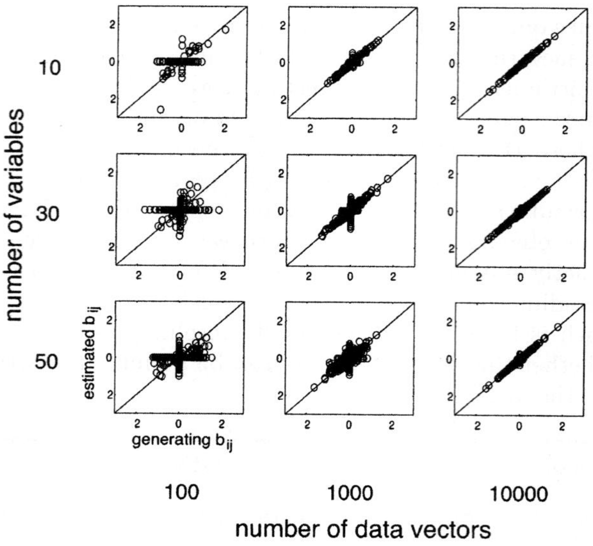

Fig. 1. Scatterplots of the estimated b_{ij} versus the original (generating) values. The different plots correspond to different numbers of variables and different numbers of data vectors. Although for small data sizes the estimation often fails, when there is sufficient data the estimation works essentially flawlessly, as evidenced by the grouping of the points along the diagonal.

We repeatedly performed the following experiment:

1. First, we randomly constructed a strictly lower-triangular matrix $\mathbf{B}$. Various dimensionalities (10, 30, and 50) were used. Both fully connected (no zeros in the strictly lower triangular part) and sparse networks (many zeros) were tested. We also randomly selected variances of the disturbance variables and values for the constants c_i.
2. Next, we generated data by independently drawing the disturbance variables e_i from gaussian distributions and subsequently passing them through a power non-linearity (raising the absolute value to an exponent in the interval [0.5, 0.8] or [1.2, 2.0], but keeping the original sign) to make them non-gaussian. Various data set sizes were tested. The e_i were then scaled to yield the desired variances, and the observed data $\mathbf{X}$ was generated according to the assumed recursive process (1).
3. Before feeding the data to the LiNGAM algorithm, we randomly permuted the rows of the data matrix $\mathbf{X}$ to hide the causal order with which the data was generated. At this point, we also permuted the generating coefficients, the c_i, as well as the variances of the disturbance variables to match the new order in the data.
4. Finally, we fed the data to our discovery algorithm, and compared the estimated parameters to the generating parameters. In particular, we made a scatterplot of the entries in the estimated matrix $\mathbf{B}$ against the corresponding generating coefficients.

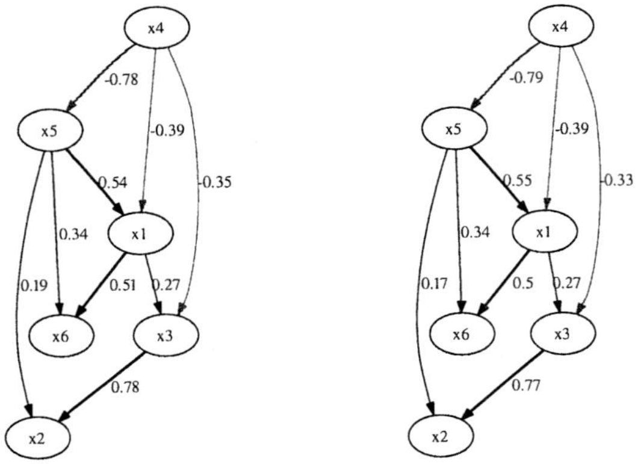

Fig. 2. Left: example original network. Right: estimated network. Graphs plotted using the latest version of the LiNGAM package which connects seamlessly to the free Graphviz software, a sophisticated tool for plotting graphs.

Since the number of different possible parameter configurations is limitless, we feel that the reader is best convinced by personally running the simulations using various settings. This can be easily done by anyone by downloading our software and running it using Matlab or the freely available Octave software. Nevertheless, we here show some representative results.

Figure 1 gives combined scatterplots of the elements of **B** versus the generating coefficients. The different plots correspond to different dimensionalities (numbers of variables) and different data sizes (numbers of data vectors), where each plot combines the data for a number of different network sparseness levels and non-linearities. Although for very small data sizes the estimation often fails, when the data size grows the estimation works practically flawlessly, as evidenced by the grouping of the datapoints onto the main diagonal.

In summary, the experiments verify that the new algorithms are able to find the appropriate permutations even in high dimensions, and demonstrate that reliable estimation is possible even when the number of variables is large. Comparing with the experiments in [3] we note that for larger dimensions we clearly need more data, but the amounts of data required are still reasonable.

5 Conclusions

Developing methods for causal inference from non-experimental data (data which does not come from controlled, randomized experiments) is a fundamental problem with a very large number of potential applications. Although one can never fully prove the validity of a causal model from observational data alone, such methods are nevertheless crucial in cases where it is impossible or very costly to perform experiments.

The estimation of linear causal models can be based purely on the covariance structure of the data [4,1,2] but such methods cannot in most cases distinguish between multiple equally possible causal models that all imply the same conditional correlation structure. We have recently shown [3] that an assumption of non-gaussianity of the disturbance variables allows the full causal model to be identified, and provided an algorithm for this estimation. The method is essentially a post-processing method of ICA results.

In this paper we have shown how to solve one of the main remaining problems with our LiNGAM method, that of finding the appropriate permutations when the number of variables is large. The proposed algorithms have been implemented in our freely available software package, and tested in thorough experiments. The code package has also been extended to include graph plotting capability (in combination with Graphviz), as Figure 2 demonstrates.

How well real-world causal processes fit our assumptions, in particular that of linearity, will be crucial to the success or failure of applications of LiNGAM. We are currently involved in testing the method on real-world data and comparing its power and usefulness with other causal discovery methods, such as those based purely on conditional correlation structure. For the most recent developments, please see the project webpage.

Acknowledgements. The authors would like to thank Aristides Gionis, Heikki Mannila, and Alex Pothen for discussions relating to algorithms for solving the permutation problems. P.O.H. was supported by the Academy of Finland project #204826. S.S. was supported by Grant-in-Aid for Scientific Research from the Ministry of Education, Culture and Sports, Japan. A.H. was supported by the Academy of Finland through an Academy Research Fellow Position and project #203344.

References

1. P. Spirtes, C. Glymour, and R. Scheines. *Causation, Prediction, and Search.* MIT Press, 2000.
2. J. Pearl. *Causality: Models, Reasoning, and Inference.* Cambridge University Press, 2000.
3. S. Shimizu, A. Hyvärinen, Y. Kano, and P. O. Hoyer. Discovery of non-gaussian linear causal models using ICA. In *Proc. 21st Conference on Uncertainty in Artificial Intelligence (UAI-2005)*, pages 526–533, Edinburgh, Scotland, 2005.
4. K. A. Bollen. *Structural Equations with Latent Variables.* John Wiley & Sons, 1989.
5. P. Comon. Independent component analysis – a new concept? *Signal Processing*, 36:287–314, 1994.
6. A. Hyvärinen, J. Karhunen, and E. Oja. *Independent Component Analysis.* Wiley Interscience, 2001.
7. R. E. Burkard and E. Cela. Linear assignment problems and extensions. In P. M. Pardalos and D.-Z. Du, editors, *Handbook of Combinatorial Optimization - Supplement Volume A*, pages 75–149. Kluwer, 1999.

Eigenvector Algorithms with Reference Signals for Frequency Domain BSS

Masanori Ito[1], Mitsuru Kawamoto[2,3], Noboru Ohnishi[1], and Yujiro Inouye[4]

[1] Graduate School of Information Engineering, Nagoya University, Furo-cho Chikusa-ku Nagoya, 464-8603 Japan
{ito, ohnishi}@ohnishi.m.is.nagoya-u.ac.jp
[2] Advanced Industrial Science and Technology, Tsukuba, Japan
[3] Bio-mimetic Control Research Center, RIKEN, Nagoya, Japan
[4] Department of Electronic and Control Systems Engineering, Shimane University, Matsue, Japan

Abstract. This paper describes blind source separation (BSS) problems in the frequency domain using an eigenvector algorithm (EVA) with reference signals. The proposed EVA has such an attractive feature that all source signals are separated simultaneously from their mixtures. This is an advantage against the methods using deflation process (e.g., super-exponential method), because those methods sometimes do not work so as to converge to desired solutions, due to deflation failure. Computer simulation demonstrates the validity of the proposed EVA.

1 Introduction

This paper deals with the blind source separation (BSS) problem for a multiple-input multiple-output (MIMO) static system driven by independent source signals. To solve this problem, we draw on the ideas of eigenvector algorithms (EVAs) with reference signals. Jelonnek et al. have proposed EVAs derived from a criterion using a reference signal, in order to solve blind equalization of single-input single-output (SISO) systems [1,2]. They have shown that the equalizer can be derived from the eigenvectors of a fourth-order cumulant matrix. In this paper, the EVA derived from a criterion with reference signals is used for solving the BSS problem of MIMO static systems. The proposed EVA has such an attractive feature that all source signals are separated simultaneously from their mixtures, while the other methods using deflation process extract signals one by one. If the deflation process fails, all the signals cannot be separated. However, the EVA with reference signals enables us to extract all the sources without any deflation methods.

Through computer simulations and real environment experiments, we show the effectiveness of the proposed methods.

J. Rosca et al. (Eds.): ICA 2006, LNCS 3889, pp. 123–131, 2006.

2 Problem Formulation

Throughout this paper, let us consider the following MIMO static system with n inputs and m outputs, a convolutive mixture model with additive noise (See figure 1.);

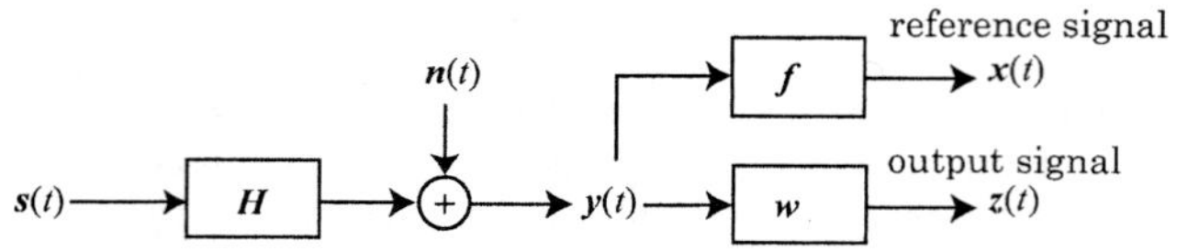

Fig. 1. The composite system of an unknown system and a filter, and reference system

$$\boldsymbol{y}(t) = \sum_k \boldsymbol{H}(k)\boldsymbol{s}(t-k) + \boldsymbol{n}(t), \tag{1}$$

where $\boldsymbol{y}(t)$ represents an m-column output vector called the *observed signal*, $\boldsymbol{s}(t)$ represents an n-column input vector called the *source signal*, $\boldsymbol{n}(t)$ represents an m-column noise vector and $\boldsymbol{H}(t)$ is an $m \times n (m \geq n)$ mixing matrix.

To achieve the blind source separation for the system (1), a convolutive mixture in the time domain is converted into instantaneous mixtures in the frequency domain with the short-time Fourier transform (STFT),

$$\mathbf{Y}(f,t) = \mathbf{H}(f)\mathbf{S}(f,t) + \mathbf{N}(f,t). \tag{2}$$

The following n filters, which are m-input single-output (MISO) static systems driven by the observed signals, are used for each frequency bin:

$$\mathrm{Z}_l(f,t) = \mathbf{w}_l^H(f)\mathbf{Y}(f,t), \quad l = 1,2,\ldots,n, \tag{3}$$

where superscript ^{H} denotes the conjugate transpose (Hermitian) of a matrix or a vector and $\mathrm{Z}_l(f,t)$ is the lth output of the filter, $\mathbf{w}_l(f) = [\mathrm{w}_{l1}(f), \mathrm{w}_{l2}(f), \ldots, \mathrm{w}_{lm}]^H$ is an m-column vector representing the m coefficients of the filter in frequency bin f. Substituting (2) into (3), we obtain

$$\begin{aligned}\mathrm{Z}_l(f,t) &= \mathbf{w}_l^H(f)\mathbf{H}(f)\mathbf{S}(f,t) + \mathbf{w}_l^H(f)\mathbf{N}(f,t),\\ &= \mathbf{g}_l^H(f)\mathbf{S}(f,t) + \mathbf{w}_l^H(f)\mathbf{N}(f,t), \quad l = 1,2,\ldots,n,\end{aligned} \tag{4}$$

where $\mathbf{g}_l(f) = [\mathrm{g}_{l1}(f), \mathrm{g}_{l2}(f), \ldots, \mathrm{g}_{ln}(f)]^H = \mathbf{H}^H(f)\mathbf{w}_l(f)$ is an n-column vector. The BSS problem considered in this paper can be formulated as follows: Find n filters $\mathbf{w}_l(f)$'s denoted by $\tilde{\mathbf{w}}_l(f)$'s satisfying the following condition, without the knowledge of $\mathbf{H}(f)$, even if the Gaussian noise $\mathbf{N}(f,t)$ is added to the observed signal $\mathbf{Y}(f,t)$,

$$\tilde{\mathbf{g}}_l(f) = \mathbf{H}^H(f)\tilde{\mathbf{w}}_l(f) = \tilde{\boldsymbol{\delta}}_l(f), \quad l = 1,2,\ldots,n, \tag{5}$$

where $\tilde{\boldsymbol{\delta}}_l(f)$ is an n-column vector whose elements $\tilde{\delta}_{lr}(f)(r = 1,2,\ldots,n)$ are equal to zero except for $\rho_l(f)$th element.

To solve the blind separation problem, we put the following assumptions on the system and the source signals.

A1) The matrix $\mathbf{H}(f)$ in (2) has full column rank.
A2) The input sequence $\{\mathbf{S}(f,t)\}$ is a zero-mean, non-Gaussian vector whose element processes $\{\mathrm{S}_i(f,t)\}, i = 1, 2, \ldots, n$, are mutually statistically independent and have nonzero variance, $\sigma^2_{s_i}(f)$ and nonzero fourth-order cumulants, $\gamma_i(f)$, $i = 1, 2, \ldots, n$.
A3) The noise sequence $\{\mathbf{N}(f,t)\}$ is stationary process vector, whose elements, $\{N_i(f,t)\}, i = 1, 2, \ldots, m$ are Gaussian processes with zero mean.
A4) The two vector sequences $\{\mathbf{N}(f,t)\}$ and $\{\mathbf{S}(f,t)\}$ are mutually independent.

3 Eigenvector Algorithms (EVAs)

3.1 Analysis of EVAs with the Reference Signal for MIMO Static Systems

In this subsection we assume that there is no noise $\boldsymbol{n}(t)$ in the output $\boldsymbol{y}(t)$. Next we propose the eigenvector algorithm with the reference signal. To solve the BSS problem, the following cross-cumulant between $\mathrm{Z}_l(f,t)$ and the reference signal $\mathrm{X}(f,t)$ is defined:

$$\mathrm{C}_{\mathrm{ZX}}(f) = \mathrm{cum}\{\mathrm{Z}_l(f,t), \mathrm{Z}_l^*(f,t), \mathrm{X}(f,t), \mathrm{X}^*(f,t)\}, \tag{6}$$

where * denotes the complex conjugate and the reference signal $\mathrm{X}(f,t)$ is given by $\mathbf{f}^H(f)\mathbf{Y}(f,t) = \mathbf{f}^H(f)\mathbf{H}(f)\mathbf{S}(f,t) = \mathbf{a}^H(f)\mathbf{S}(f,t)$ ($\mathbf{a}^H(f) = \mathbf{f}^H(f)\mathbf{H}(f)$ is a vector whose elements are $\mathrm{a}_1(f), \mathrm{a}_2(f), \ldots, \mathrm{a}_n(f)$), using an appropriate filter $\mathbf{f}(f)$. The filter $\mathbf{f}(f)$ is called a *reference system*. Moreover we define the constrain $\sigma^2_{\mathrm{Z}_l}(f) = \sigma^2_{S_{\rho_l}}(f)$, where $\sigma^2_{\mathrm{Z}_l}(f)$ and $\sigma^2_{S_{\rho_l}}(f)$ denote the variance of the output $\mathrm{Z}_l(f,t)$ and a source signal $S_{\rho_l}(f,t)$, respectively. In the case of SISO systems, Jelonnek et al. [1,2] have shown that the maximization of $|\mathrm{C}_{\mathrm{ZX}}(f)|$ under $\sigma^2_{\mathrm{Z}_l}(f) = \sigma^2_{S_{\rho_l}}(f)$ leads to a closed-form expression as the following generalized eigenvector problem:

$$\mathbf{C}_{\mathrm{YX}}(f)\mathbf{w}_l(f) = \lambda\mathbf{R}(f)\mathbf{w}_l(f). \tag{7}$$

Then they utilize the facts that $\mathrm{C}_{\mathrm{ZX}}(f)$ and $\sigma^2_{\mathrm{Z}_l}(f)$ can be expressed in terms of the vector $\mathbf{w}_l(f)$ as, respectively,

$$\mathrm{C}_{\mathrm{ZX}}(f) = \mathbf{w}_l^H(f)\mathbf{C}_{\mathrm{YX}}(f)\mathbf{w}_l(f), \tag{8}$$
$$\sigma^2_{\mathrm{Z}_l}(f) = \mathbf{w}_l^H(f)\mathbf{R}(f)\mathbf{w}_l(f), \tag{9}$$

where $\mathbf{C}_{\mathrm{YX}}(f)$ is a matrix whose (i,j)th element is calculated by $\mathrm{cum}\{\mathrm{Y}_i(f,t), \mathrm{Y}_j^*(f,t), \mathrm{X}(f,t), \mathrm{X}^*(f,t)\}$, $\mathbf{R}(f) = E[\mathbf{Y}(f,t)\mathbf{Y}^H(f,t)]$ is the covariance matrix of m-column vector $\mathbf{Y}(f,t)$ and λ is an eigenvalue of $\mathbf{R}^\dagger(f)\mathbf{C}_{\mathrm{YX}}(f)$, where $\dagger$ denotes the pseudo-inverse operation of a matrix. Moreover they have shown that the eigenvector corresponding to the maximum eigenvalue of $\mathbf{R}^\dagger(f)\mathbf{C}_{\mathrm{YX}}(f)$ becomes the solution of the blind equalization problem in

[1,2], which is referred to as an *eigenvector algorithm* (EVA). However, the algorithm proposed by Jelonnek et al. is for SISO or SIMO infinite impulse response channel. Therefore, we want to show how the eigenvector algorithm (7) works for the BSS in the case of the MIMO static system in the frequency domain. To this end, we use following equalities:

$$\mathbf{R}(f) = \mathbf{H}(f)\mathbf{\Sigma}(f)\mathbf{H}^H(f), \tag{10}$$

$$\mathbf{C}_{\mathrm{YX}}(f) = \mathbf{H}(f)\mathbf{\Lambda}(f)\mathbf{H}^H(f), \tag{11}$$

where $\mathbf{\Sigma}(f)$ is a diagonal matrix whose elements are $\sigma_{s_i}^2(f), i = 1, 2, \ldots, n$ and $\mathbf{\Lambda}(f)$ is a diagonal matrix whose elements are $|\mathrm{a}_i(f)|^2\gamma_i(f), i = 1, 2, \ldots, n$. Then we obtain the following theorem.

Theorem 1. *Suppose the values* $|\mathrm{a}_i(f)|^2\gamma_i(f)/\sigma_{s_i}^2(f), i = 1, 2, \ldots, n$ *are all nonzero and distinct. If the noise* $\mathbf{N}(f,t)$ *is absent in (2), the* n *eigenvectors corresponding to* n *nonzero eigenvalues of* $\mathbf{R}^\dagger(f)\mathbf{C}_{\mathrm{YX}}(f)$ *become the the vectors* $\tilde{\mathbf{w}}_l(f)$*'s satisfying (5).*

Proof. Based on (7), we consider the following eigenvector problem:

$$\mathbf{R}^\dagger(f)\mathbf{C}_{\mathrm{YX}}(f)\mathbf{w}_l(f) = \lambda\mathbf{w}_l(f). \tag{12}$$

Then substituting (10) and (11) into (12), we obtain

$$\mathbf{H}^{H\dagger}(f)\mathbf{\Sigma}^{-1}(f)\mathbf{H}^\dagger(f)\mathbf{H}(f)\mathbf{\Lambda}(f)\mathbf{H}^H(f)\mathbf{w}_l(f) = \lambda\mathbf{w}_l(f). \tag{13}$$

Since $\mathbf{H}(f)$ has full column rank, using a property of the pseudo-inverse operation([3], p.433),

$$\mathbf{H}^{H\dagger}(f)\mathbf{\Sigma}^{-1}(f)\mathbf{\Lambda}(f)\mathbf{H}^H(f)\mathbf{w}_l(f) = \lambda\mathbf{w}_l(f). \tag{14}$$

Multiplying (14) by $\mathbf{H}^H(f)$ from left side and using a property of the pseudo-inverse operation again, (14) becomes

$$\mathbf{\Sigma}^{-1}(f)\mathbf{\Lambda}(f)\mathbf{H}^H(f)\mathbf{w}_l(f) = \lambda\mathbf{H}^H(f)\mathbf{w}_l(f). \tag{15}$$

By noting that $\mathbf{\Sigma}^{-1}(f)\mathbf{\Lambda}(f)$ is a diagonal matrix whose elements, $|\mathrm{a}_i(f)|^2\gamma_i(f)/\sigma_{s_i}^2(f), i = 1, 2, \ldots, n$ are all nonzero and distinct, if $\mathbf{g}_l(f) = \mathbf{H}^H(f)\mathbf{w}_l(f) \neq \mathbf{0}$, then the eigenvector $\mathbf{g}_l(f)$ obtained from (15) becomes the vector $\tilde{\mathbf{g}}_l(f)$ satisfying (5). Namely, the n eigenvectors $\mathbf{w}_l(f)$ corresponding to n nonzero eigenvalues of $\mathbf{R}^\dagger(f)\mathbf{C}_{\mathrm{YX}}(f)$ obtained from (12) become the vectors $\tilde{\mathbf{w}}_l(f)$ satisfying (5). □

3.2 Robust Eigenvector Algorithm (REVA)

In the previous subsection, we assume that there are no noises in the output signals. In this subsection, we shall show such an eigenvector algorithm that the solutions (5) can be obtained, even if the noise $\boldsymbol{n}(t)$ is present in the output

$\boldsymbol{y}(t)$. To this end, we introduce fourth-order cumulants matrices of m-column vector process $\{\mathbf{Y}(f,t)\}$, which constitute a set of $m \times m$ matrices $\mathbf{C}_{\mathbf{Y},i}(f), i = 1, 2, \ldots, m$. The matrix $\mathbf{C}_{\mathbf{Y},i}(f)$ is defined by

$$[\mathbf{C}_{\mathbf{Y},i}(f)]_{q,r} = \text{cum}\left\{\mathrm{Y}_q(f,t), \mathrm{Y}_r^*(f,t), \mathrm{Y}_i(f,t), \mathrm{Y}_i^*(f,t)\right\}, \tag{16}$$

where $[\cdot]_{q,r}$ denotes the (q,r)th element of the matrix $\mathbf{C}_{\mathbf{Y},i}(f)$. Then we consider an $m \times m$ matrix $\mathbf{Q}(f)$ expressed by

$$\mathbf{Q}(f) = \sum_{i=1}^{m} \mathbf{C}_{\mathbf{Y},i}(f). \tag{17}$$

It is shown by a simple calculation (see [4]) that (17) becomes

$$\mathbf{Q}(f) = \mathbf{H}(f)\tilde{\mathbf{\Lambda}}(f)\mathbf{H}^H(f), \tag{18}$$

where $\tilde{\mathbf{\Lambda}}(f)$ is a diagonal matrix defined by

$$\tilde{\mathbf{\Lambda}}(f) = \text{diag}\{\gamma_1(f)\tilde{\mathrm{a}}_1(f), \gamma_2(f)\tilde{\mathrm{a}}_2(f), \ldots, \gamma_n(f)\tilde{\mathrm{a}}_n(f)\}, \tag{19}$$

$$\tilde{\mathrm{a}}_r(f) = \sum_{i=1}^{m} \mathrm{h}_{ir}(f)\mathrm{h}_{ir}^*(f), \quad r = 1, 2, \ldots, n, \tag{20}$$

and diag$\{\cdots\}$ denotes a diagonal matrix with the diagonal elements built from its arguments, $\mathrm{h}_{ir}(f)$ is the (i,r)th element of $\mathbf{H}(f)$.

Here, as a constraint, we take the following value:

$$\begin{aligned}|\mathrm{C}_{\mathrm{ZY}}(f)| &= \left|\sum_{i=1}^{m} \text{cum}\{\mathrm{Z}_l(f,t), \mathrm{Z}_l^*(f,t), \mathrm{Y}_i(f,t), \mathrm{Y}_i^*(f,t)\}\right| = \left|\mathbf{w}_l^H(f)\mathbf{Q}(f)\mathbf{w}_l(f)\right| \\ &= \left|\sum_{i=1}^{m} \tilde{\mathrm{a}}_i(f)\gamma_i(f)\mathrm{g}_{li}(f)\mathrm{g}_{li}^*(f)\right|. \end{aligned} \tag{21}$$

Then, we consider solving the problem that the fourth-order cumulants $|\mathrm{C}_{\mathrm{ZX}}(f)|$ is maximized under the condition that $|\mathrm{C}_{\mathrm{ZY}}(f)| = |\tilde{\mathrm{a}}_{\rho_l}(f)\gamma_{\rho_l}(f)|$. Then by the Lagrangian method, the following generalized eigenvector problem is derived from the problem:

$$\mathbf{C}_{\mathrm{YX}}(f)\mathbf{w}_l(f) = \tilde{\lambda}\mathbf{Q}(f)\mathbf{w}_l. \tag{22}$$

From the following theorem, by solving the eigenvector problem of the matrix $\mathbf{Q}^\dagger(f)\mathbf{C}_{\mathrm{YX}}(f)$, the n eigenvectors $\mathbf{w}_l(f) (l = 1, 2, \ldots, n)$ correspond to the vectors $\tilde{\mathbf{w}}_l(f) (l = 1, 2, \ldots, n)$ in (5).

Theorem 2. *Suppose the values* $|\mathrm{a}_i(f)|^2/\tilde{\mathrm{a}}_i(f), i = 1, 2, \ldots, n$ *are all nonzero and distinct. The* n *eigenvectors corresponding to* n *nonzero eigenvalues of* $\mathbf{Q}^\dagger(f)\mathbf{C}_{\mathrm{YX}}(f)$ *become the the vectors* $\tilde{\mathbf{w}}_l(f)$*'s satisfying (5).*

Proof. We omit the proof because it is easily proved as well as Theorem 1.

Remark 1. Since the matrix $\mathbf{Q}^{\dagger}(f)\mathbf{C}_{\mathrm{YX}}(f)$ consists of only the fourth-order cumulants, the eigenvector derived from the matrix can be obtained with as little influence of Gaussian noises as possible, which is referred as a *robust eigenvector algorithm* (REVA).

4 Adaptive Version of REVA

REVA can be implemented adaptively. To this end we must specify the dependency of each time t and omit frequency bin f for simplicity. We show the update procedure in the case of 2-input 2-output static system.

$\tilde{\mathbf{Q}}(t)$, which is the estimator of $\mathbf{Q}$ at time t is calculated by

$$\begin{aligned}\tilde{\mathbf{Q}}(t) &= \alpha\tilde{\mathbf{Q}}(t-1)\\ &+(1-\alpha)\left\{\left(\mathbf{C}_1(t)-\tilde{\mathbf{C}}_1(t)-\mathrm{tr}\{\tilde{\mathbf{C}}_1(t)\}\right)\mathbf{C}_1(t)-\tilde{\mathbf{C}}_2(t)\mathbf{C}_2^*(t)\right\},\end{aligned} \tag{23}$$

where α is a forgetting factor close to, but less than 1 and $\mathrm{tr}\{X\}$ denotes the trace of matrix X.

Here $\mathbf{C}_1(t)$ and $\mathbf{C}_2(t)$ in (23) are defined by $\mathbf{C}_1(t) = \mathbf{Y}(t)\mathbf{Y}^H(t)$ and $\mathbf{C}_2(t) = \mathbf{Y}^*(t)\mathbf{Y}^H(t)$, respectively. $\tilde{\mathbf{C}}_1(t)$ and $\tilde{\mathbf{C}}_2(t)$ are the moving averages of $\mathbf{C}_1(t)$ and $\mathbf{C}_2(t)$, respectively, which are calculated by

$$\tilde{\mathbf{C}}_1(t) = \beta\tilde{\mathbf{C}}_1(t-1) + (1-\beta)\mathbf{C}_1(t), \tag{24}$$
$$\tilde{\mathbf{C}}_2(t) = \beta\tilde{\mathbf{C}}_2(t-1) + (1-\beta)\mathbf{C}_2(t), \tag{25}$$

where β is also a forgetting factor close to, but less than 1 and $\alpha > \beta$.

$\tilde{\mathbf{C}}_{\mathrm{YX}}(t)$, which is the estimator of $\mathbf{C}_{\mathrm{YX}}$ at time t is calculated by

$$\begin{aligned}\tilde{\mathbf{C}}_{\mathrm{YX}}(t) &= \alpha\tilde{\mathbf{C}}_{\mathrm{YX}}(t-1) + (1-\alpha)\{\mathbf{Y}(t)\mathbf{Y}^H(t)\mathrm{X}(t)\mathrm{X}^*(t) - \mathbf{Y}(t)\mathbf{Y}^H(t)\tilde{\mathrm{V}}_{\mathrm{X}}(t)\\ &-\mathbf{Y}(t)\mathrm{X}(t)\tilde{\mathbf{V}}_{\mathrm{Y}_1}(t) - \mathbf{Y}(t)\mathrm{X}^*(t)\tilde{\mathbf{V}}_{\mathrm{Y}_2}(t)\},\end{aligned} \tag{26}$$

where $\tilde{\mathrm{V}}_{\mathrm{X}}(t)$ and $\tilde{\mathbf{V}}_{\mathrm{Y}_i}(t), i = 1,2$ are the moving averages of $\mathrm{V}_{\mathrm{X}}(t)$ and $\mathbf{V}_{\mathrm{Y}_i}(t)$ defined by

$$\tilde{\mathrm{V}}_{\mathrm{X}}(t) = \beta\tilde{\mathrm{V}}_{\mathrm{X}}(t-1) + (1-\beta)\mathrm{V}_{\mathrm{X}}(t), \tag{27}$$
$$\tilde{\mathbf{V}}_{\mathrm{Y}_i}(t) = \beta\tilde{\mathbf{V}}_{\mathrm{Y}_i}(t-1) + (1-\beta)\mathbf{V}_{\mathrm{Y}_i}(t), \quad i = 1,2, \tag{28}$$

where $\mathrm{V}_{\mathrm{X}}(t) = \mathrm{X}(t)\mathrm{X}^*(t)$, $\mathbf{V}_{\mathrm{Y}_1}(t) = \mathrm{Y}^H(t)\mathrm{X}^*(t)$ and $\mathbf{V}_{\mathrm{Y}_2}(t) = \mathrm{Y}^H(t)\mathrm{X}(t)$.

Then the separator $\mathbf{w}_l(t)$ is calculated by solving eigenvector problem (7).

5 Experiments

5.1 Simulation

We conducted a simulation experiment. $\mathbf{H}(z)$, which is z-transform of the mixing matrix $\boldsymbol{H}(t)$, is defined as:

$$\mathbf{H}(z) = \begin{pmatrix} 1-0.4z^{-1} & 0.5z^{-1}-0.2z^{-2} \\ 0.5z^{-1}-0.2z^{-2} & 1-0.4z^{-1} \end{pmatrix}. \tag{29}$$

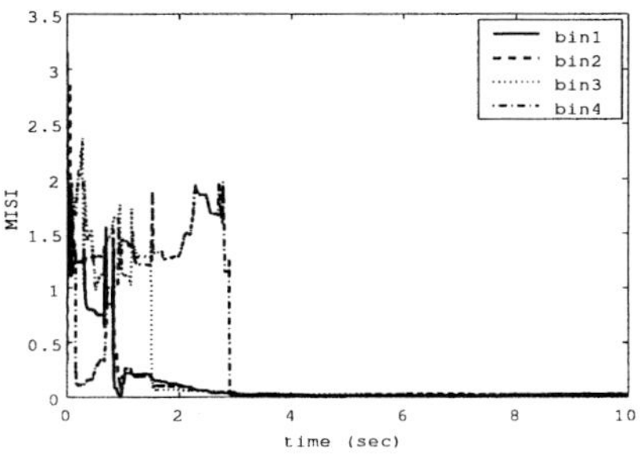

Fig. 2. MISIs of EVA

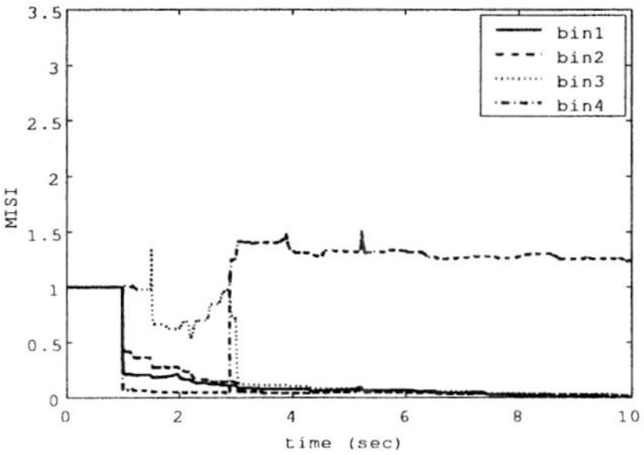

Fig. 3. MISIs of SEM

The BSS problem is solved by adaptive REVA. To measure the separation performance, multichannel intersymbol-interference (MISI) was used, which is defined as

$$\mathrm{MISI} = \sum_{i=1}^{2}\left(\frac{\sum_{j=1}^{2}|g_{ij}|^2}{\max_j |g_{ij}|^2} - 1\right) + \sum_{j=1}^{2}\left(\frac{\sum_{i=1}^{2}|g_{ij}|^2}{\max_i |g_{ij}|^2} - 1\right). \tag{30}$$

The MISI becomes zero if $\tilde{\mathbf{g}}_l$'s satisfying (5) are obtained. The smaller the MISI value is, the closer the obtained solution is to the desired one. Figure 2 shows the MISIs of some frequency bins using EVA with the reference signal and Figure 3 shows those of SEM [5], which uses the deflation process. Obviously

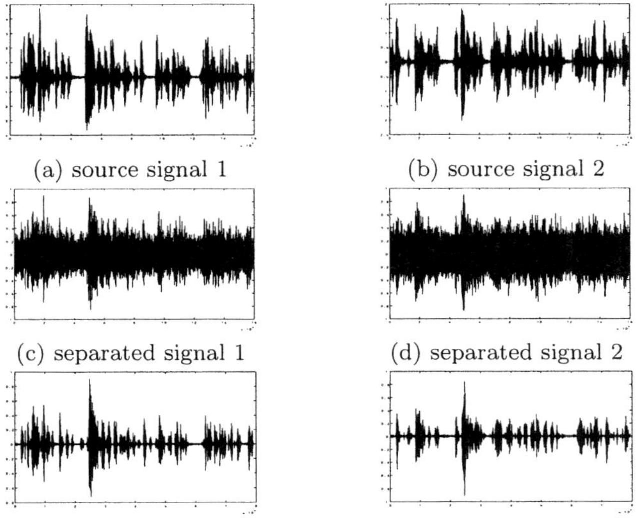

(a) source signal 1 (b) source signal 2

(c) separated signal 1 (d) separated signal 2

(e) enhanced signal 1 (f) enhanced signal 2

Fig. 4. Waveforms of source, separated and enhanced signals using SEM the deflation process failed in a frequency bin, while EVA with the reference signal could converge to the desired solution in all frequency bins.

Remark 2. REVA utilizes the fourth-order cumulants. To estimate the fourth-order cumulants accurately a large number of samples are generally needed. Therefore it takes a rather long time to converge using REVA.

5.2 Real Environment

In an office room, we conducted separation experiments using REVA. Because the reference signal is needed, the number of microphones is three, while the number of source is two, one of the observed signals is used as a reference signal. Manually 5dB Gaussian noises are added to the observed signals to show that the proposed REVA works in a noisy environment. Figure 4 shows a set of waveforms of the source signals, the separated signals and the enhanced signals which were given by the ES 202 050 software [6]. In the enhanced signals additive Gaussian noises were reduced. We can see that REVA can extract independent but distorted source signals.

6 Conclusion

We described the BSS problem in the frequency domain. We proposed the eigenvector algorithm (EVA) with reference signals. The proposed method has such an attractive property that all source signal are extracted simultaneously without the deflation process. EVA can be robust to Gaussian noises using only the higher-order cumulants (REVA). We have also shown the adaptive version of REVA.

The computer simulations and real environment experiments have clarified the validity of the proposed methods.

Acknowledgement

A part of this work is supported by Grant-in-Aid for JSPS Fellow (17-7606) and 21st Century COE Program, Intelligent Media Integration for Socail Infromation Infrastructure, Nagoya University. The authors would like to thanks Dr. Masashi Ohata and Dr. Toshiharu Mukai for helping our experiments.

References

1. Jelonnek, B., Kammeyer, K.D.: A closed-form solution to blind equalization. Signal Processing **36** (1994) 251–259
2. Jelonnek, B., Boss, D., Kammeyer, K.D.: Generalized eigenvector algorithm for blind equalization. Signal Processing **61** (1997) 237–264

3. Lancaster, P., Tismenetsky, M.: The Theory of Matrices. second edition. Academic Press. (1985)
4. Kawamoto, M., Kohno, K., Inouye, Y.: Robust super-exponential methods for deflationary blind source separation of instantaneous mixtures. IEEE Trans. on Singal Processing **53** (2005) 1933–1937
5. Ito, M., Ohata, M., Kawamoto, M., Mukai, T., Ohnishi, N., Inouye, Y.: Blind source separation in a noisy environment using super-exponential algorithm. In: Proceedings of ISSPIT2005. (2005) To appear.
6. European Telecommunications Standards Institute (ETSI) Standard ES 202 050: (Speech Processing, Transmission and Quality Aspects (STQ); Distributed Speech Recognition; Advanced Front-end Feature Extraction Algorithm; Compression Algorithms)

Sparse Coding for Convolutive Blind Audio Source Separation*

Maria G. Jafari, Samer A. Abdallah, Mark D. Plumbley, and Mike E. Davies

Centre for Digital Music,
Queen Mary University of London, UK
{maria.jafari, samer.abdallah}@elec.qmul.ac.uk
http://www.elec.qmul.ac.uk

Abstract. In this paper, we address the convolutive blind source separation (BSS) problem with a sparse independent component analysis (ICA) method, which uses ICA to find a set of basis vectors from the observed data, followed by clustering to identify the original sources. We show that, thanks to the temporally localised basis vectors that result, phase information is easily exploited to determine the clusters, using an unsupervised clustering method. Experimental results show that good performance is obtained with the proposed approach, even for short basis vectors.

1 Introduction

The convolutive blind audio source separation problem arises when an array of microphones records mixtures of a set of sound sources that are convolved with the impulse response between each source and sensor. The problem is often addressed in the frequency domain, through the short-time fourier transform (STFT), where the statistics of the sources are sparser, so that ICA algorithms achieve better performance [1], and the approximations of convolutions by multiplications yield reduced computational complexity. Source separation is then performed separately at each frequency bin, resulting in the introduction of the well-known problem of frequency permutations [2], whose solution amounts to clustering the frequency components of the recovered sources, using additional information about the mixing system or the sources. The most successful methods in this context have perhaps been beamforming approaches [2-5], which exploit phase information contained in the de-mixing filters identified by the source separation algorithm, but suffer from phase ambiguities in the upper frequencies, since phase is defined exclusively up to 2π. An alternative approach to convolutive BSS was proposed in [7], and is based on the use of sparse coding to identify the mixing matrix from the observed data. No assumptions are required on the number of microphones, or the type of mixing (eg. instantaneous or convolutive) in the underlying model, but the recovered matrix implicitly encodes

* This work was funded by EPSRC grants GR/S85900/01, GR/R54620/01, and GR/S82213/01.

J. Rosca et al. (Eds.): ICA 2006, LNCS 3889, pp. 132–139, 2006.

these characteristics of the system. Thus, it could even potentially deal with the more sources than sensors case. The subspaces corresponding to the original sources are then identified using clustering techniques. In this paper we investigate the performance of the frequency domain ICA (FD-ICA) and sparse coding approaches. We find that the latter yields mostly temporally localised basis vectors, that do not suffer from the phase ambiguity encountered in the frequency domain. Hence, in contrast to the approach in [7], which uses manual clustering, we propose an unsupervised clustering method that exploits phase information to separate the sources. The structure of this paper is as follows: the convolutive BSS problem is described in section 2, together with an overview of FD-ICA; the sparse coding method is summarised in section 3. The clustering technique proposed is discussed in section 4, where the performance of the sparse coding and FD-ICA methods are also compared. Conclusions are drawn in section 5.

2 Problem Formulation

We consider the problem of separating 2 sampled real-valued speech signals, $\mathbf{s}(n)$, from 2 convolutive mixtures, $\mathbf{x}(n)$, recorded from an array of microphones, so that the signal recorded at the q-th microphone, $x_q(n)$, is

$$x_q(n) = \sum_{p=1}^{2} \sum_{l=1}^{L} a_{qp}(l) s_p(n-l), \quad q = 1,2 \tag{1}$$

where $s_p(n)$ is the p-th source signal, $a_{qp}(l)$ denotes the impulse response from source p to sensor q, and L is the maximum length of all impulse responses [2]. The aim of blind source separation is to find estimates for the unmixing filters $w_{qp}(l)$, using only the sensor measurements, and to reconstruct the sources from

$$y_p(n) = \sum_{q=1}^{2} \sum_{l=1}^{L} w_{qp}(l) x_q(n-l), \quad p = 1,2 \tag{2}$$

where $y_p(n)$ is the p-th recovered source. Typically, the N-point STFT is evaluated, and the mixing and separating models in (1) and (2) become, respectively $\mathbf{X}(f,t) = \mathbf{A}(f)\mathbf{S}(f,t)$ and $\mathbf{Y}(f,t) = \mathbf{W}(f)\mathbf{X}(f,t)$ where t denotes the STFT block index. The resulting N instantaneous BSS problems, are addressed independently in each subband with an ICA algorithm, and the problem of frequency permutations that is introduced is solved essentially by clustering the frequency components of the recovered sources. This is often done using beamforming techniques, such as in [2-5], where the direction of arrival (DOA) of the sources are evaluated from the beamformer directivity patterns

$$F_p(f,\theta) = \sum_{q=1}^{2} W_{qp}^{\mathrm{ICA}}(f) e^{j2\pi f d \sin\theta_p / c}, \quad p = 1,2 \tag{3}$$

where W_{qp}^{ICA} is the ICA de-mixing filter from the q-th sensor to the p-th output, d is the spacing between two sensors, θ_p is the angle of arrival of the p-th source signal, and $c \approx 340\text{m/s}$ is the speed of sound in air. The frequency permutations are then determined by ensuring that the directivity pattern for each beamformer is approximately aligned along the frequency axis. There exists, however, an ambiguity in the DOA estimation, due to the restriction on the phase difference to lie between $-\pi$ and π, which results in the creation of additional nulls in the directivity pattern of magnitude similar to that corresponding to the angle of arrival [5]. The distance between two microphones should satisfy $d \leq c/2f_{\max}$, in order to avoid spatial aliasing [2]; when this condition is not met, ambiguities in the position of the nulls are introduced, resulting in inaccurate DOA estimates, and the frequency, f_{M}, above which multiple nulls are expected is $f_{\mathrm{M}} = c/2d$.

3 Overview of Sparse ICA

The aim of sparse coding is to find sparse dictionaries from the mixtures, so that only a small number of coefficients, $\mathbf{s}(n)$, are needed to encode the observed data, $\mathbf{x}(n)$ [6]. The convolutive BSS problem was first addressed within this framework in [7], by finding a set of basis vectors for the observed data, followed by clustering to identify the subspaces corresponding to the original sources. The approach does not explicitly model the mixing process nor the number of mixtures, but is based on the assumption that the recordings are generated by signals that are sparse in the dictionary domain. Prior to estimating the basis vectors, the observed vector is reshaped into a $K \times k_{max}$ matrix on which learning is performed. A frame of $K/2$ samples is taken from each mixture, with an overlap of T samples. Thus, the (i,k)-th element of the new matrix, $\tilde{\mathbf{X}}$, is

$$\tilde{\mathbf{X}}_{i,k} = \begin{cases} x_1\left[(k-1)Z + \frac{i+1}{2}\right] & : i \text{ odd} \\ x_2\left[(k-1)Z + \frac{i}{2}\right] & : i \text{ even} \end{cases} \tag{4}$$

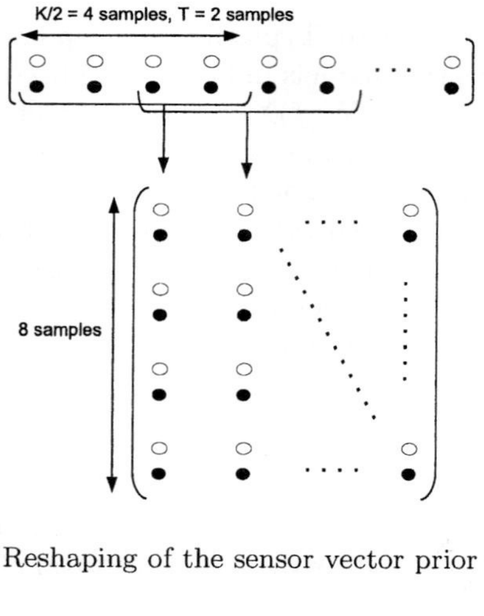

Fig. 1. Reshaping of the sensor vector prior to training with ICA

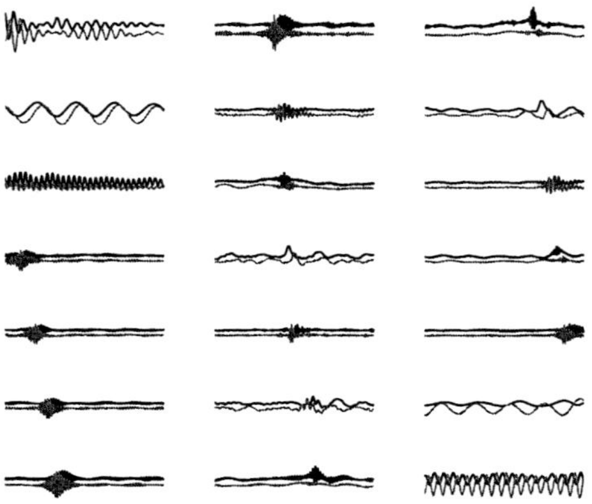

Fig. 2. Examples of basis vectors extracted with the sparse ICA algorithm

where $Z = K/2 - T$, and $i \in \{1, \ldots, K/2\}$, and $k \in \{1, \ldots, k_{max}\}$. The reshaping of the sensor vector $\mathbf{x}(n)$ is illustrated in figure 1. The basis vectors are learned from the resulting matrix, and with any ICA algorithm using a sparse prior. Here we use [7]

$$\mathbf{\Delta W} = \eta \left(\mathbf{I} - E\{\mathbf{f}(\mathbf{y})\mathbf{y}^T\}\right) \mathbf{W} \tag{5}$$

where η is the learning rate, and $\mathbf{f}(\mathbf{y})$ is the activation function. Details for its choice can be found in [7]. The algorithm (5) operates upon $\mathbf{y} = \mathbf{W}\tilde{\mathbf{X}}$, where the time index n has been dropped for the sake of clarity, and $\mathbf{W} \in \mathbb{R}^{K \times K}$. The reshaping of $\mathbf{x}(n)$ into the matrix $\tilde{\mathbf{X}}(n)$ emphasises the correlations between the sources at the two microphones. Stacking the columns of $\mathbf{x}(n)$ ensures that features relating to temporally correlated signals from each recording are extracted, leading to basis pairs that encode information about the mixing channel, as can be seen from the basis pairs plotted in figure 2, where a time-delay is clearly visible in several of the vector pairs. The strong directionality observed indicates that each basis pair relates to a particular source, and thus the proposed method is based on the property of spatial diversity. However, should the sources be aligned along the same DOA, the technique cannot be used.

4 Frequency Domain Versus Sparse ICA

In this section, we consider the separation of two speech signals, one each from a male and a female speaker, from two mixtures recorded in a university lecture room, and sampled at 16kHz. Further details of the experimental set up can be found in [10]. The sources were also recorded separately, so that they could be used for performance evaluation. The performance of the sparse ICA approach is compared to a representative FD-ICA method [10] (MD2003) since, due to their inherent similarities, we expect other FD-ICA algorithms to have comparable performance. The sparse ICA approach was first used to learn the basis vectors from the real data; the mixtures were buffered into frames of 512 samples, so that 256 samples from each mixture were taken, as shown in figure 1, and the algorithm (5) was used for training. The learned basis vector pairs are found in the columns of $\mathbf{W}^{-1}$, examples of which are shown in figure 2. This figure illustrates that the basis vector pairs encode how the extracted features are received at the microphones and, therefore, they capture information about time-delay and amplitude differences that characterise the mixing channel. Moreover, most of the basis vectors have the additional property of being localised in time, which implies that time delays can be estimated more accurately. Reconstruction of the two original signals is achieved with $\hat{\mathbf{s}}_1 = \mathbf{W}^{-1}\mathbf{H}^{(1)}\mathbf{y}$, and $\hat{\mathbf{s}}_2 = \mathbf{W}^{-1}\mathbf{H}^{(2)}\mathbf{y}$ where $\mathbf{H}^{(1)}$ is a diagonal matrix whose diagonal elements are ones or zeros depending on whether a component belongs to the first source, and similarly for $\mathbf{H}^{(2)}$ [7]. We propose clustering the basis vector pairs, and therefore determine the diagonal elements of $\mathbf{H}^{(1)}$ and $\mathbf{H}^{(2)}$, according to the following algorithm:

1 For each basis pair k find the time delay τ_k between the vectors
2 Form the histogram of τ_k, and use k-means to find the peaks, τ_{k_1} and τ_{k_2} corresponding to the sources
3 $h_{kk}^{(p)} = \begin{cases} 1, \text{ if } (\tau_{k_p} - \tau_\delta) \leq \tau_k \leq (\tau_{k_p} + \tau_\delta), \\ 0, \text{ otherwise} \end{cases}$

for $p = 1, 2$, where $h_{kk}^{(p)}$ is the kk-th element of $\mathbf{H}^{(p)}$. The inclusion of the τ_δ allows the algorithm to perform a degree of de-noising. We estimate the time delay between sensor pairs using the popular generalised cross-correlation with phase transform (GCC-PHAT) algorithm, originally proposed in [9], $R_{a_1 a_2}(\tau) = \int_{-\infty}^{\infty} A_1(\omega)A_2^*(\omega)/(|A_1(\omega)A_2^*(\omega)|)e^{j\omega\tau}d\omega$, where $A_1(\omega)$, $A_2(\omega)$ are the Fourier transforms of the basis vectors. The function $R_{a_1 a_2}(\tau)$, typically exhibits a sharp peak at the lag corresponding to the time delay between the two signals. Figure 3(a) depicts the time-delay estimates obtained with GCC-PHAT, for all basis vector pairs, and figure 3(b) shows their histogram; values of τ_{k_1} and τ_{k_2} were obtained with k-means as 10.04 and -9.03 samples, and τ_δ was set to 2 samples.

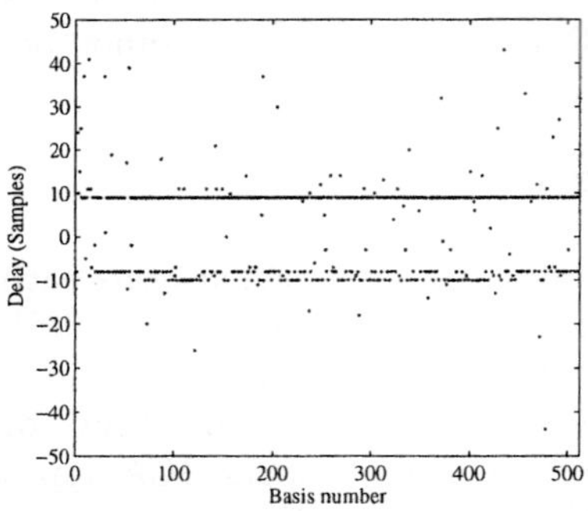

(a) Delays for all basis vectors

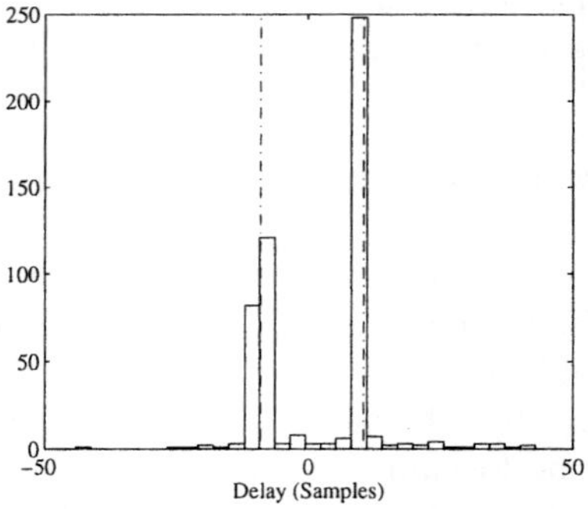

(b) Histogram of delays

Fig. 3. Plot of the time delays estimated for all basis vectors, and its histogram

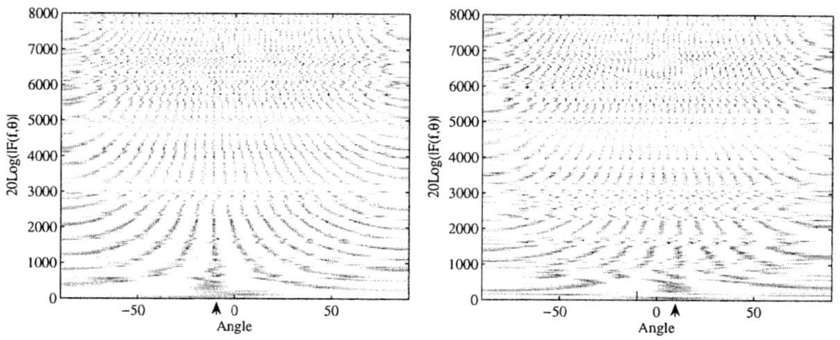

(a) Directivity pattern for source $\hat{s}_1$. (b) Directivity pattern for source $\hat{s}_2$.

Fig. 4. Directivity patterns for the outputs of FD-ICA, after permutation alignment

Frequency domain separation was performed with the algorithm in [10] using the 256 and 2048-point STFT, and permutations were aligned as in [3]. MD2003 with the latter frame length has been shown to successfully achieve separation on this data in [10]. Figure 4 shows a plot of the directivity pattern of the outputs evaluated at all frequencies with (3), following permutation alignment. The plots show that permutations are correctly aligned in the low frequency bands, while the behaviour of the algorithm is less clear in the higher frequencies, where time delay estimation is less accurate. The DOAs estimated from the plots were found to be 12° and −11°, corresponding to time delays of approximately 10 and −9 samples. The sample delay at a frequency f is estimated from the directions of arrival by $\tau = 2\pi f \sin\theta f_s/c$, where f_s is the sampling frequency.

Tables 1 and 2 show the global performance of the two methods, as evaluated from [11]. The evaluation criteria allows for the recovered sources to be modified by a permitted distortion. In Table 1, we consider a time-invariant gain distortion, and the sources recovered at the two channels are compared to the

Table 1. Global performance measures when a gain distortion is allowed. SDR, SIR, and SAR measures are respectively the signal-to-distortion, signal-to-interference, and signal-to-artifact ratios.

Method	SDR (dB)		SIR (dB)		SAR (dB)	
	$\hat{s}^{\mathrm{M}}$	$\hat{s}^{\mathrm{F}}$	$\hat{s}^{\mathrm{M}}$	$\hat{s}^{\mathrm{F}}$	$\hat{s}^{\mathrm{M}}$	$\hat{s}^{\mathrm{F}}$
	Channel 1					
$\mathrm{MD2003}_{256}$	−5.13	−8.61	2.73	1.83	−2.49	−6.01
$\mathrm{MD2003}_{2048}$	−5.24	−6.26	4.69	6.17	−3.50	−5.07
Sparse ICA	−8.69	−10.09	1.59	2.74	−5.98	−8.00
	Channel 2					
$\mathrm{MD2003}_{256}$	−8.29	−6.09	−0.81	2.64	−4.00	−3.58
$\mathrm{MD2003}_{2048}$	−6.94	−3.42	3.86	7.24	−5.06	−2.28
Sparse ICA	−6.76	−11.61	−0.41	7.40	−2.40	−10.83

Table 2. Global performance measures when a filter distortion is allowed

Method	Channel 1					
	SDR (dB)		SIR (dB)		SAR (dB)	
	$\hat{s}^{M}$	$\hat{s}^{F}$	$\hat{s}^{M}$	$\hat{s}^{F}$	$\hat{s}^{M}$	$\hat{s}^{F}$
$MD2003_{256}$	−7.12	−8.25	2.47	2.01	−4.66	−5.69
$MD2003_{2048}$	−6.27	−7.07	7.82	9.26	−5.43	−6.48
Sparse ICA	−7.77	−10.50	3.94	3.55	−6.00	−8.74
	Channel 2					
$MD2003_{256}$	−10.33	−8.03	−0.37	2.21	−6.67	−5.55
$MD2003_{2048}$	−8.15	−5.62	6.52	9.99	−7.12	−5.08
Sparse ICA	−8.27	−9.40	1.29	12.00	−5.35	−9.10

original sources recorded at the microphones. Negative SIR values indicate that the interfering source is larger than the target source, and the algorithm has failed to recover the target. Large negative SAR values, with SAR $\approx$ SDR, indicate that large artifacts are present, and dominate distortion [11]. The results suggest that sparse ICA and $MD2003_{256}$ have similar performance, and both fail to recover the source $\hat{s}^{M}$ at channel 2. An informal listening test indicates, however, that the objective assessment in Table 1 is not a good guide to the audible performance. The test reveals that MD2003 separates the sources with a frame of 2048 samples, while it fails with a short frame of 256 samples. This is in contrast to sparse ICA which uses a frame of 256 samples, and whose outputs are clearly separated, although the interfering source is still audible. Interestingly, the algorithm seems also to have performed some de-reverberation, which is particularly audible for the female source, $\hat{s}^{F}$, at the second channel. Moreover, the outputs sound quite natural and large artifacts do not appear to be present. This is in disagreement with the large negative SAR values which suggest that sparse ICA introduces large artifacts. To obtain a more meaningful objective assessment, a time-invariant filter distortion is allowed, with a 64 taps filter. The results are shown in Table 2, where the recovered sources are compared to the original signals at the speakers. In this case, it was found that the objective assessment is more closely in agreement with the informal listening test, but still overcritical of sparse ICA. The results in this section also show how the STFT length is a crucial parameter for FD-ICA. Since modeling of real room transfer functions typically requires long frame sizes, better separation is achieved with a frame size of 2048 samples. Sparse ICA, on the other hand, provides good separation even with a very short frame size.

5 Conclusions

In this paper, we have shown that most of the basis vectors extracted with sparse coding are temporally localised functions that do not suffer from phase ambiguities encountered in the frequency domain. A simple unsupervised clustering technique that exploits this property has been proposed. The performance of the algorithm with real data has been investigated, and informal listening tests have suggested that it separates the signals with short basis vectors, in contrast to FD-ICA, which requires long basis vectors. Currently available objective testing methods fail to verify this, so further subjective listening tests are planned to formally substantiate this performance.

References

1. J.F. Cardoso, "Blind signal separation: statistical principles," *Proceedings of the IEEE*, vol. 86, pp. 2009–2025, 1998.
2. H. Sawada, R. Mukai, S. Araki, and S. Makino, "A robust and precise method for solving the permutation problem of frequency-domain blind source separation," *IEEE Trans. on Speech and Audio Processing*, vol. 12, pp. 530–538, 2004.
3. N. Mitianoudis and M. Davies, "Permutation alignment for frequency domain ICA using subspace beamforming methods," in *Proc. ICA*, 2004, pp. 669–676.
4. H. Saruwatari, S. Kurita, and K. Takeda, "Blind source separation combining frequency-domain ICA and beamforming," in *Proc. ICASSP*, 2001, vol. 5, pp. 2733–2736.
5. M. Ikram and D. Morgan, "A beamforming approach to permutation alignment for multichannel frequency-domain blind speech separation," in *Proc. ICASSP*, 2002, vol. 1, pp. 881–884.
6. J.-H. Lee, T.-W. Lee, H.-Y. Jung, and S.-Y. Lee, "On the efficient speech feature extraction based on independent component analysis," *Neural Processing Letters*, vol. 15, pp. 235–245, 2002.
7. S. Adballah and M. Plumbley, "Application of geometric dependency analysis to the separation of convolved mixtures," in *Proc. ICA*, 2004, pp. 22–24.
8. J.F. Cardoso and B. Laheld, "Equivariant adaptive source separation," *IEEE Trans. Signal Processing*, vol. 44, pp. 3017–3030, 1996.
9. C. Knapp and G. Carter, "The generalized correlation method for estimation of time delay," *IEEE Trans. Acoustic, Speech, and Signal Processing*, vol. 24, pp. 320–327, 1976.
10. N. Mitianoudis and M. Davies, "Audio source separation of convolutive mixtures," *IEEE Trans. on Audio and Speech Processing*, vol. 11, pp. 489–497, 2003.
11. C. Févotte, R. Gribonval and E. Vincent, "BSS_EVAL Toolbox User Guide," *IRISA Technical Report 1706*, April 2005. http://www.irisa.fr/metiss/bss_eval/.

ICA-Based Binary Feature Construction*

Ata Kabán[1] and Ella Bingham[2]

[1] School of Computer Science, The University of Birmingham, Birmingham, B15 2TT, UK
a.kaban@cs.bham.ac.uk
[2] HIIT BRU, University of Helsinki, Finland
ella@iki.fi

Abstract. We address the problem of interactive feature construction and denoising of binary data. To this end, we develop a variational ICA method, employing a multivariate Bernoulli likelihood and independent Beta source densities. We relate this to other binary data models, demonstrating its advantages in two application domains.

1 Introduction

Binary data becomes more and more abundant, arising from areas as diverse as bioinformatics, e-businesses and paleontological research. The processing of binary data requires appropriate tools and methods for tasks such as exploratory analysis, feature construction and denoising. These necessarily must follow the specific distributional characteristics of the data and cannot be accomplished with tools that exist for continuous valued data analysis.

Previous successes of Independent Component Analysis (ICA) [5] make it an important statistical principle worthy of investigation for tackling such problems. However, contrarily to continuous-valued signals, work on ICA methods for binary data has been very scarce [4,3]. A few methods exist, though, that seek binary sources [9,10] from continuous data. Due to the discrete combinatorial nature of the problem, these latter works resort to search heuristics [10] or indeed an exhaustive search [9], that are, at best, computationally intensive.

In this paper we develop a linear ICA model for binary data. We employ a probabilistic framework and make use of the variational methodology to alleviate the computational demand. Application examples will demonstrate the workings of our approach and its advantages over other binary data models.

2 Binary ICA with Beta Sources

Consider an independent factor model for binary data $\boldsymbol{x}$, having a Bernoulli likelihood model and independent Beta latent priors.

$$P(\boldsymbol{x}_n) = \int P(\boldsymbol{x}_n|\boldsymbol{b}) \prod_k p(b_k) db_k \tag{1}$$

* Part of this work has been done while visiting HIIT BRU, Helsinki, Finland.

J. Rosca et al. (Eds.): ICA 2006, LNCS 3889, pp. 140–148, 2006.

$$= \int \prod_t (\sum_k a_{tk} b_k)^{x_{tn}} (1 - \sum_k a_{tk} b_k)^{1-x_{tn}} \prod_k B(b_k | \alpha_k^0, \beta_k^0) db_k \quad (2)$$

where $B(b|\alpha, \beta) = \frac{\Gamma(\alpha+\beta)}{\Gamma(\alpha)\Gamma(\beta)}(1-b)^{\beta-1}b^{\alpha-1}$ is the Beta density [1]. This is defined on the $[0, 1]$ domain, which is desirable for our purposes, since we may be able to interpret the components as grey-scale representations of the binary data. In addition, the particularly flexible shape of the Beta density is advantageous.

Further, a linear-convex mixing process will be assumed, so that the mixing coefficients are all non-negative and satisfy $\sum_k a_{tk} = 1, \forall t = 1 : T$. This is mainly due to computational convenience, since then it follows that $\sum_k a_{tk} b_k$ will necessarily fall into $[0, 1]$ so that we do not need any further nonlinear transformation to obtain the mean parameter of the Bernoulli likelihood. While nonlinear models are also of interest, here we seek the 0-s and 1-s to be exchangeable within the model, and this would not be possible if a nonlinearity is applied to non-negative variables. Thus, a Dirichlet prior may be assumed for the mixing coefficients, to make the specification fully Bayesian.

2.1 Inference and Estimation

In order to make the problem tractable, we will employ the well-known Jensen's inequality to lower bound the data probability, and we make use of the factorial posterior approximation to simplify the computations:

$$\log \int P(\boldsymbol{x}_n | \boldsymbol{b}) \prod_k B(b_k | \alpha_k^0, \beta_k^0) db_k \geq \int \prod_k q_n(b_k) \log \frac{\prod_t P(x_{tn} | \boldsymbol{b}) \prod_k B(b_k | \alpha_k^0, \beta_k^0)}{\prod_k q_n(b_k)} db_k$$

where $\prod_k q_n(b_k)$ is the factorial variational posterior.

Due to the Bernoulli likelihood term $P(x_{tn}|\boldsymbol{b})$, this integral is still intractable, therefore the ultimate lower bound will be obtained by a further application of Jensen's inequality. The convexity constraint imposed on the mixing proportions comes in useful, as the likelihood term can be rewritten and lower bounded:

$$\log P(x_{tn}|\boldsymbol{b}) = \log \left\{ (\sum_k a_{tk} b_k)^{x_{tn}} (1 - \sum_k a_{tk} b_k)^{1-x_{tn}} \right\}$$

$$= \log \left\{ \sum_k a_{tk} b_k^{x_{tn}} (1 - b_k)^{1-x_{tn}} \right\} \geq \sum_k Q_{k|t,n,x_{tn}} \log \frac{a_{tk} b_k^{x_{tn}} (1 - b_k)^{1-x_{tn}}}{Q_{k|t,n,x_{tn}}} \quad (3)$$

Here $Q_{k|t,n,x_{tn}} \geq 0, \sum_k Q_{k|t,n,x_{tn}} = 1$ is a discrete variational distribution with values in $\{1, ..K\}$, where K denotes the number of components.

Using (3) we obtain a lower bound on the log likelihood, which is tractable and will be referred to as $\mathcal{L}^{bound}$.

2.2 Variational Solution

Let $q_n(b_k) = B(b_k|\alpha_{kn}, \beta_{kn})$ be parameterised Beta variational posteriors with variational parameters α_{kn}, β_{kn}. Then, maximising $\mathcal{L}^{bound}$ yields the following update equations for the variational parameters

$$\alpha_{kn} = \alpha_k^0 + \sum_t x_{tn} Q_{k|t,n,x_{tn}=0} = \alpha_k^0 + e^{\langle \log b_{kn} \rangle} \sum_t \frac{x_{tn} a_{tk}}{\sum_k a_{tk} e^{\langle \log b_{kn} \rangle}} \quad (4)$$

$$\beta_{kn} = \beta_k^0 + \sum_t (1 - x_{tn}) Q_{k|t,n,x_{tn}=1} = \beta_k^0 + e^{\langle \log(1-b_{kn}) \rangle} \sum_t \frac{(1 - x_{tn}) a_{tk}}{\sum_k a_{tk} e^{\langle \log(1-b_{kn}) \rangle}} \quad (5)$$

where

$$Q_{k|t,n,x_{tn}} \propto a_{tk} (e^{\langle \log b_{kn} \rangle})^{x_{tn}} (e^{\langle \log(1-b_{kn}) \rangle})^{1-x_{tn}} \quad (6)$$

is obtained by maximising $\mathcal{L}^{bound}$ w.r.t. $Q_{k|t,n,x_{tn}}$ and this has been replaced into the expressions of all variational parameter estimates above.

The required variational posterior expectations are easily evaluated as $\langle \log b_{kn} \rangle \equiv E_{q_n(b_k)}[\log b_k] = \psi(\alpha_{kn}) - \psi(\alpha_{kn} + \beta_{kn})$ and $\langle \log(1 - b_{kn}) \rangle \equiv E_{q_n(b_k)}[\log(1 - b_k)] = \psi(\beta_{kn}) - \psi(\alpha_{kn} + \beta_{kn})$.

Maximising $\mathcal{L}^{bound}$ in a_{tk} under the constraint that $\sum_k a_{tk} = 1$ and replacing the expression of $Q_{k|t,n,x_{tn}}$ as before, yields the update equation below.

$$a_{tk} \propto a_{tk} \left\{ \sum_n \frac{x_{tn}}{\sum_k a_{tk} e^{\langle \log b_{kn} \rangle}} e^{\langle \log b_{kn} \rangle} + \frac{1 - x_{tn}}{\sum_k a_{tk} e^{\langle \log(1-b_{kn}) \rangle}} e^{\langle \log(1-b_{kn}) \rangle} \right\} \quad (7)$$

Finally, the prior parameters α_k^0 and β_k^0 will both be set to one, in order to express a uniform prior.

To make some connections with earlier work, it can easily be shown that a maximum likelihood estimator for our model (2) would yield equations that (after some algebra) are identical to the aspect Bernoulli (AB) algorithm in [7]. Vice-versa, the above construction offers an interpretation of AB as an ICA model. By analogy, other popular aspect models [2,3] may also be related to ICA in a similar manner, and this is different from, and complementary to the connection initially envisaged in [3].

2.3 Bayesian Model Selection

As already mentioned, a prior may also be naturally specified for the mixing coefficients, and due to the imposed convexity constraint, a Dirichlet is appropriate. As a result, the optimal number of components can determined simply by choosing the model order that maximises the log of the data evidence bound

$$E_{q_t(\boldsymbol{a})}[\mathcal{L}^{bound}] + E_{q_t(\boldsymbol{a})}[\log Dir(\boldsymbol{a}|\gamma^0)] - E_{q_t(\boldsymbol{a})}[\log q_t(\boldsymbol{a})] \quad (8)$$

where $q_t(\boldsymbol{a}) = Dir(\boldsymbol{a}|\gamma_t)$ is the variational posterior of the mixing variable.

The modification this brings to the previously presented estimation procedure is minimal — denoting by γ_{tk} the additional variational parameters associated with a_{tk} and omitting the straightforward algebra, the parameters a_{tk} in (4) will need to be replaced by $e^{\langle \log \gamma_{tk} \rangle}$ and instead of eq (5) we will have:

$$\gamma_{tk} = \gamma_k^0 + e^{\langle \log a_{tk} \rangle} \left\{ \sum_n \frac{x_{tn} e^{\langle \log b_{kn} \rangle}}{\sum_k e^{\langle \log a_{tk} \rangle} e^{\langle \log b_{kn} \rangle}} + \frac{(1 - x_{tn}) e^{\langle \log(1-b_{kn}) \rangle}}{\sum_k e^{\langle \log a_{tk} \rangle} e^{\langle \log(1-b_{kn}) \rangle}} \right\} \quad (9)$$

The parameter of the Dirichlet prior, γ_{tk}^0 will again be set to 1 to express a uniform prior, and the remaining posterior expectation in (9) is computed as $\langle \log a_{tk} \rangle \equiv E_{q_t(\boldsymbol{a})}[\log a_k] = \psi(\gamma_{tk}) - \psi(\sum_{k'} \gamma_{tk'})$.

3 Analyst Input and Posterior Data Reconstruction

Perhaps the greatest reason for the popularity of ICA methods for exploratory data analysis is that the independent components are often easier to comprehend and interpret by humans separately, rather than in their mixture. This has been exploited in numerous applications, most notably for signal denoising [6]. Once the independent signals of different genuine and artifact sources are separated from the data, artifact-corrected signals may be derived by eliminating the contributions of the artifact sources. Our methodology is conceptually similar, although the formalism differs according to our probabilistic framework.

Let us denote the posterior means obtained from our algorithm by $\langle a_{tk} \rangle$ and $\langle b_{kn} \rangle$: $\langle b_{kn} \rangle = E_{q_n(b_k)}[b_k] = \int db_k b_k B(b_k|\alpha_{kn}, \beta_{kn}) = \frac{\alpha_{kn}}{\alpha_{kn}+\beta_{kn}}$ and analogously $\langle a_{tk} \rangle = \frac{\gamma_{tk}}{\sum_{k'} \gamma_{tk'}}$. These are themselves discrete probabilities, so that $\sum_k \langle a_{tk} \rangle = 1$. After inspecting the independent components $\langle \boldsymbol{b}_k \rangle$, the elimination of undesired components may now be accomplished by specifying a probability value, $P(u|k)$, for each component and using these to modify our unsupervised estimates. Denoting by $P_t(k)$ the posterior expectations $\langle a_{tk} \rangle$, for each t, the Bayes rule will provide us the post-processed data representation.

$$\langle a_{tk} \rangle_{postproc} := P_t(k|u) = \frac{P_t(k)P(u|k)}{\sum_{k'} P_t(k')P(u|k')} \tag{10}$$

Typically a 0 probability will be specified for components that are capturing undesirable noise, while 1 will specify a clearly meaningful component. Clearly, if for a component k a value of $p(u|k) = 0$ was specified, then $\langle a_{tk} \rangle_{postproc} = 0$ will become zero for all t. Naturally, the formalism straightforwardly permits the specification of analyst inputs at more detailed levels. E.g. nothing prevents us from specifying a separate set of probabilities, $P(u|k,t)$, for each t. However, we may typically expect human experts to feed back on the components' level, since those are hoped to provide some interpretable representations.

For computing the posterior data reconstruction, we re-express the above in terms of a conditional posterior: $q_t(\boldsymbol{a}|\boldsymbol{u}) := Dir(\boldsymbol{a}|\boldsymbol{\gamma}_t \circ P(u|.))$, whose expectation is exactly $\langle a_{tk}|\boldsymbol{u} \rangle = \langle a_{tk} \rangle_{postproc}$. Here, $\circ$ denotes element-wise product and $\boldsymbol{u}$ is the random vector of $u|k$ when $k = 1 : K$. Then the posterior post-processed data reconstruction can be computed as follows (omitting the algebra):

$$P(\hat{x}_{tn}|\boldsymbol{X}, \boldsymbol{u}) = \int d\boldsymbol{a} d\boldsymbol{b} P(\hat{x}_{tn}|\boldsymbol{a}, \boldsymbol{b}) q_t(\boldsymbol{a}|\boldsymbol{u}) \prod_k q_n(b_k) \tag{11}$$

$$= (\sum_k \langle a_{tk}|\boldsymbol{u} \rangle \langle b_{kn} \rangle)^{\hat{x}_{tn}} (1 - \sum_k \langle a_{tk}|\boldsymbol{u} \rangle \langle b_{kn} \rangle)^{1-\hat{x}_{tn}} \tag{12}$$

In consequence, the grey-scale posterior reconstruction of the (t, n)-th data entry is

$$\langle \hat{x}_{tn} | \boldsymbol{u} \rangle = p(\hat{x}_{tn} = 1 | \boldsymbol{X}, \boldsymbol{u}) = \sum_k \langle a_{tk} | \boldsymbol{u} \rangle \langle b_{kn} \rangle \quad (13)$$

and so the binary reconstruction is given by thresholding this value.

4 Experiments

4.1 Restoration of Corrupted Binary Images

For the first set of experiments we use a data set of handwritten digit images[1]. The subset of the first five digits were taken, each having 200 examples, which totals 1,000 image instances. We artificially created a corrupted version of this data set, by simulating a uniformly varying process of degradation, which turns off some of the pixels that were initially 'on'. Fifteen randomly chosen examples are shown from the initial data set, along with their corrupted version, on Figure 1. Figure 2 then shows the ICA representation obtained: several components can clearly be recognised as typical digits, and one other – completely blank – separates out the corruption factor. Inspecting the mixing proportions for the data instances shown earlier, it is clear that the white component is indeed present in those images that suffered a degradation. To remove the noise component, we apply the procedure described earlier. The results can be followed on Figure 3: The grey-scale posterior reconstruction of the data has indeed filtered out the degradation source and presents a smoothed reconstruction of the initial clean data. The grey levels correspond to probabilities of pixels being 'on'. Thresholding these probabilities at 0.5 gives us the binary reconstruction of the data shown on the right-hand plot. The degradation has now been eliminated.

A comparative set of experiments has then been conducted in order to objectively and quantitatively assess the performance of our method in reconstructing the clean data from its corrupted version. We included a comprehensive set of binary data analysis methods in this comparison: mixtures of Bernoulli (MB), Bernoulli (logistic) PCA [11] (LPCA), our binary ICA with and without post-processing (BICA-postproc and BICA respectively), and a Bernoulli version of non-negative matrix factorisation [8], that we created for the purpose of this comparison (BNMF). (For the latter, a shifted and rescaled sigmoid nonlinearity was used, which transforms the non-negatively constrained factors and mixing proportions into the [0,1] interval.) None of the methods except BICA was able to separate out the noise factor. In consequence no obvious correction post-processing is applicable to the other methods. In this experiment, 500 corrupted images were used for training and another 500 corrupted images formed an independent test set. For the previously unseen data instances, the required posteriors were first estimated. In the case of BNMF we just implemented a Maximum Likelihood estimation method and in this case the required parameter matrix was

[1] http://www.ics.uci.edu/mlearn/MLSummary.html

Fig. 1. Examples of clean (left) and corrupted (right) images

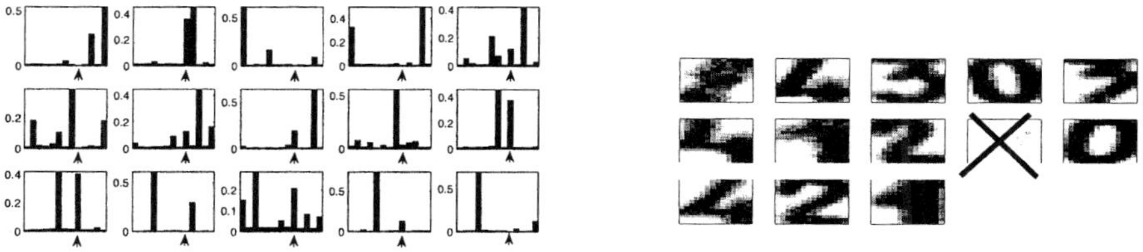

Fig. 2. Right: Independent components estimated from the corrupted binary image data set; Left: The mixing coefficients associated with the examples shown on the right hand plot of Fig.1. Small arrow heads point to the mixing coefficients associated with the noise component.

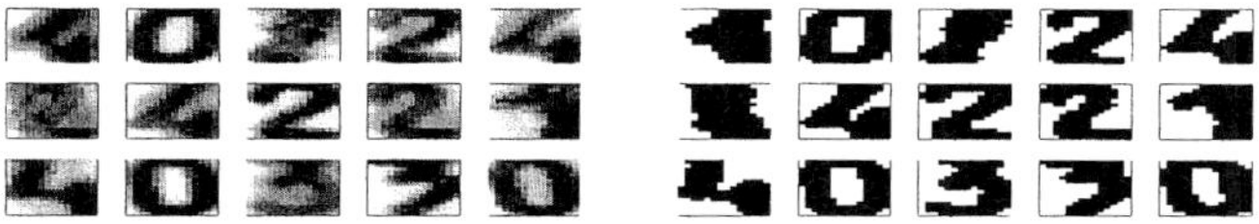

Fig. 3. Reconstructed grey-scale (left) and binary (right) images after the post-processing

estimated anew for the previously unseen test data. The upper plots of Figure 4 show the areas under the ROC curve of the posterior data reconstruction (both grey-scale and black&white, using a threshold of 0.5), averaged over all pixels of the corrupted test set. LPCA is the overall winner in reconstructing the corrupted test data set. The lower plots of the same figure, in turn, show the AUC values averaged over the blank pixels of the test images, but computed against the pixel values of the true, uncorrupted test set (not used anywhere else). As we can see, the proposed post-processing, by the removal of the automatically separated noise component, BICA becomes the most successful in this exercise – comparable with the nonlinear and time-consuming LPCA at grey-scale reconstruction and net superior at binary reconstruction.

4.2 Age Discovery and Missingness Detection in Paleontological Data

We now demonstrate our method in paleontological data[2]. The data consists of findings of 139 mammals among 501 sites of excavation and is seen in Figure 5

[2] NOW database, http://www.helsinki.fi/science/now/, a public resource based on collaboration between mammal paleontologists.

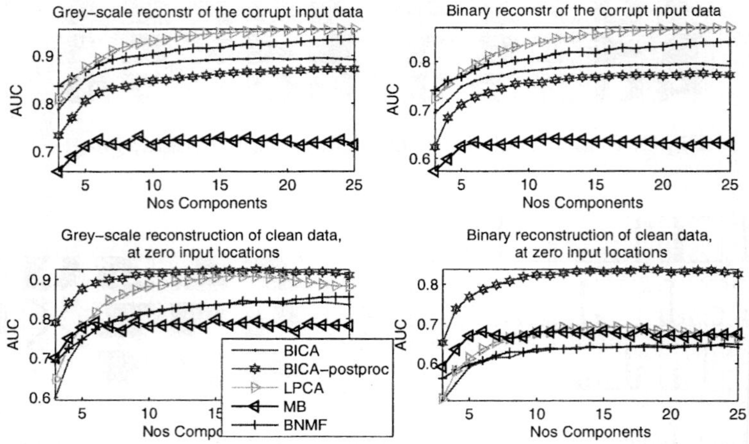

Fig. 4. Comparison of BICA with other binary data models on test inputs. Since both the training and the test data sets are corrupted, all methods try to reconstruct the data including the corruptions, LPCA being the best (upper plots). However, by the described post-processing, BICA stops reconstructing the corrupted regions, instead it becomes net superior in terms of restoration of the uncorrupted images (lower plots).

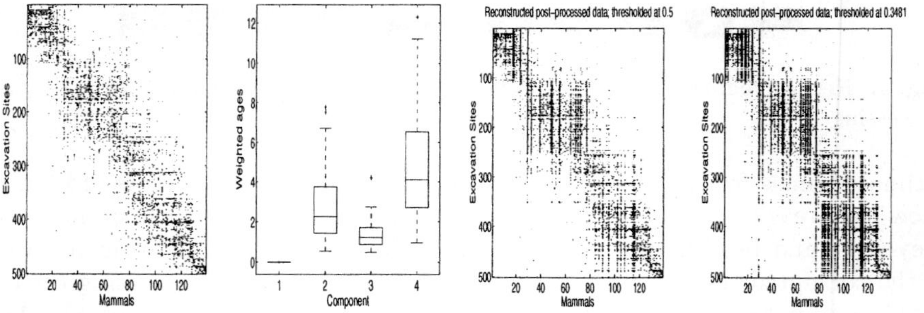

Fig. 5. From left to right: The palaeontological data, both the sites and the remains of mammals are ordered by age, for the ease of visual analysis of the results; Distributions of ages of mammals, weighted by $\langle b_{kn} \rangle$, for each component; Binary reconstruction of the absences in the data after having removed the noise component, using a threshold of 0.5 – these are superimposed with the observed presences; Binary reconstruction, when using an estimated threshold

(leftmost plot). Four components have been estimated, out of which three turned out to capture contiguous disjoint time periods. The fourth component in turn is completely blank — having all elements nearly zero. The second left plot of Figure 5 shows the box plots of the ages of remains[3], weighted by b_{kn}. The Kolmogorov-Smirnov test indicates that these distributions are indeed distinct:

[3] The age information is auxiliary and it is not used during the parameter estimation.

the P values range between $5 \cdot 10^{-13}$ and $3 \cdot 10^{-4}$. The blank component is the one shrunken to zero on this figure – clearly it does not contribute to the age discovery. In turn, its presence indicates that not all zero observations are due to age, but another reason for absence of remains exists.

Often, remains of a mammal are not observed at a site even though it probably lived there, as the preservation, recovery and identification of fossils are subject to random effects. According to palaeontologists[4], an indication of missingness can be derived from the age order of the sites: if a mammal is observed at two sites but not at an intermediate site, it is possible (although not certain) that an observation at the intermediate site is missing. This may be the additional independent noise factor that our method has separated out, and in order to verify this, we will now remove this noise factor from the data. Employing the probabilistic post-processing procedure described previously, and thresholding at 0.5 (see Figure 5, third plot from the left), we obtain a significant decrease in such intermediate, "probably missing" values: 1369 of them will be filled in. Furthermore, by thresholding at a smaller value of 0.3481 (obtained by considering all such intermediate values as missing, and dividing the number of 1s plus missing values by the size of the data) the decrease in "probably missing" values raises to 3642. The continuity of mammals as recovered by our binary ICA is now quite apparent on the rightmost plot of Figure 5.

5 Conclusions

We have devised a variational ICA method for binary data, employing independent Beta latent densities. This turned out to be a flexible model and has allowed us to include human input in a principled manner. We demonstrated the use of our approach on two application examples.

References

1. J.M Bernardo and A.F.M Smith. Bayesian Theory. Wiley, 2001.
2. D.M. Blei, A.Y.Ng and M.I. Jordan, Latent Dirichlet Allocation. *Journal of Machine Learning Research*, 3(5):993–1022, 2003.
3. W Buntine and A Jakulin. Applying Discrete PCA in Data Analysis. Proc. 20th Conference on Uncertainty in Artificial Intelligence, pp. 59 - 66, 2004.
4. J Himberg and A Hyvärinen. Independent Component Analysis for Binary Data: an Experimental Study. Proc. ICA2001, pp. 552-556, 2001.
5. A Hyvärinen, J Karhunen, E Oja. Independent Component Analysis. Wiley, 2001.
6. T.P Jung, S Makeig, C Humphries, T.W Lee, M.J McKeown, V Iragui, T.J Sejnowski. Removing Electroencephalographic Artifacts by Blind Source Separation, Psychophysiology, 37:163-78, 2000.
7. A Kabán, E Bingham, T Hirsimäki, Learning to Read Between the Lines: The Aspect Bernoulli Model, Proc. SIAM Int Conf on Data Mining, 2004, pp. 462–466.
8. D.D Lee and H.S Seung. Algorithms for non-negative matrix factorization, Advances in Neural Information Processing Systems, 2000, pp. 556–562.

[4] Professor Mikael Fortelius, University of Helsinki, personal communication.

9. P. Pajunen. Blind Separation of Binary Sources With Less Sensors Than Sources. Proc Int. Conf on Neural Networks (ICNN-97), 1997, pp. 1994-1997.
10. E Segal, A Battle, D Koller. Decomposing gene expression into cellular components. Proc. Pacific Symposium on Biocomputing, 2003, pp. 89–100.
11. M.E Tipping. Probabilistic visualisation of high dimensional data, Advances in Neural Information Processing Systems, 1999, pp. 592–598.

A Novel Dimension Reduction Procedure for Searching Non-Gaussian Subspaces

Motoaki Kawanabe[1], Gilles Blanchard[1], Masashi Sugiyama[2], Vladimir Spokoiny[3], and Klaus-Robert Müller[1,4]

[1] Fraunhofer FIRST.IDA, Germany
[2] Department of Computer Science, Tokyo Institute of Technology, Japan
[3] Weierstrass Institute and Humboldt University, Germany
[4] Department of Computer Science, University of Potsdam, Germany
{blanchar, nabe}@first.fhg.de, sugi@cs.titech.ac.jp, spokoiny@wias-berlin.de, klaus@first.fhg.de

Abstract. In this article, we consider high-dimensional data which contains a low-dimensional non-Gaussian structure contaminated with Gaussian noise and propose a new *linear* method to identify the non-Gaussian subspace. Our method NGCA (Non-Gaussian Component Analysis) is based on a very general semi-parametric framework and has a theoretical guarantee that the estimation error of finding the non-Gaussian components tends to zero at a parametric rate. NGCA can be used not only as preprocessing for ICA, but also for extracting and visualizing more general structures like clusters. A numerical study demonstrates the usefulness of our method.

1 Introduction

Suppose that we are given a set of i.i.d. observations $\boldsymbol{x}_i \in \mathbb{R}^d$, $(i = 1, \ldots, n)$ obtained as a sum of a signal $\boldsymbol{s} \in \mathbb{R}^m$ $(m \leq d)$ with an unknown non-Gaussian distribution and an independent Gaussian noise component $\boldsymbol{n} \in \mathbb{R}^d$:

$$\boldsymbol{x} = A\boldsymbol{s} + \boldsymbol{n}, \tag{1}$$

where A is a $d \times m$ matrix and $\boldsymbol{n} \sim N(\boldsymbol{0}, \Gamma)$. The rationale behind this model is that in most real-world applications the 'signal' or 'information' contained in the high-dimensional data is essentially non-Gaussian while the 'rest' can be interpreted as high-dimensional Gaussian noise. We want to emphasize that we do *not* assume the Gaussian components to be of *smaller* order of magnitude than the signal components. This setting therefore excludes the use of common (nonlinear) dimensionality reduction methods such as PCA, Isomap [12] and LLE [11] that are based on the assumption that the data lies, say, on a lower dimensional manifold, up to some small noise distortion.

If the non-Gaussian components s_i's are mutually independent, the model turns out to be the under-complete noisy ICA [9]. Although some algorithms have been proposed, combinations of dimension reduction like PCA or Factor Analysis and noise-free ICA methods are often used, when the number m of the sources is relatively small. However, the classical methods for dimension reduction are based on second order statistics

J. Rosca et al. (Eds.): ICA 2006, LNCS 3889, pp. 149–156, 2006.

and do not consider non-Gaussianity of the sources. In this research, we will construct a dimension reduction procedure called NGCA (Non-Gaussian Component Analysis) which extracts the non-Gaussian subspace by higher order statistics. Since mutual independence of the sources is not assumed, our NGCA method can be used not only as preprocessing for ICA, but also for searching more general and dependent non-Gaussian structures (cf. [10]).

The NGCA approach is built upon a very general semi-parametric framework where the density of the sources is not specified at all. We will present an implementation here which is close in spirit to *Projection Pursuit (PP)* [5, 7, 8, 9] for visualization of interesting structures in high-dimensional data. However, the philosophy that we would like to promote in this paper is in a sense different: in fact we do not specify what we are interested in, but we rather define what is *not interesting*. To be more precise, in PP methods, a *single* index which measures the non-Gaussianity (or 'interestingness') of a projection direction has to be fixed and optimized , while NGCA takes many various indices into account at the same time. Therefore it can outperform PP algorithms, if the data contains say, both super- and sub-Gaussian components.

In the following section we will outline a novel semi-parametric theory for *linear* dimension reduction and theoretical guarantees of the NGCA procedure. The algorithm will be presented in Section 3 and simulation results underline the usefulness of NGCA; finally a brief conclusion is given.

2 Theoretical Framework

The probability density function $p(\boldsymbol{x})$ of the observations defined by the mixing model (1) can be put under the following semi-parametric form:

$$p(\boldsymbol{x}) = g(T\boldsymbol{x})\phi_{\Gamma}(\boldsymbol{x}), \tag{2}$$

where T is an unknown linear mapping from $\mathbb{R}^d$ to another subspace $\mathbb{R}^m$, g is an unknown function on $\mathbb{R}^m$ related to the distribution of the source $\boldsymbol{s}$ and ϕ_{Γ} is a centered Gaussian density with unknown covariance matrix Γ. The model (2) includes as particular cases both the pure parametric ($m = 0$) and pure non-parametric ($m = d$) models. In practice we are interested in an intermediate case where d is large and m is rather small.

Note that the decomposition (2) is non-unique, but we will show that the following m-dimensional *linear* subspace $\mathcal{I}$ of $\mathbb{R}^d$ is *identifiable*:

$$\mathcal{I} = \mathrm{Ker}(T)^{\perp} = \mathrm{Range}(T^{\top}) .$$

We call $\mathcal{I}$ the *non-Gaussian index space*. Its geometrical meaning is the following: in the model (1), the noise term can be decomposed into two components, $\boldsymbol{n} = \boldsymbol{n}_1 + \boldsymbol{n}_2$, where $\boldsymbol{n}_1 = A\boldsymbol{\eta} \in \mathrm{Range}(A)$ and $\boldsymbol{n}_2$ is restricted in the $(d-m)$-dimensional complementary subspace s.t. $\mathrm{Cov}(\boldsymbol{n}_1, \boldsymbol{n}_2) = 0$ (i.e. $\boldsymbol{n}_1$ and $\boldsymbol{n}_2$ are independent). Thus, we have the representation

$$\boldsymbol{x} = A\widetilde{\boldsymbol{s}} + \boldsymbol{n}_2, \tag{3}$$

where $\widetilde{\boldsymbol{s}} := \boldsymbol{s} + \boldsymbol{\eta}$ and the noise term $\boldsymbol{n}_2$ distributes with a $(d-m)$-dimensional degenerated Gaussian independent of $\widetilde{\boldsymbol{s}}$. The subspace $\mathcal{I}$ is then the orthogonal complement of the $(d-m)$-dimensional subspace containing the independent Gaussian component $\boldsymbol{n}_2$. Once we can estimate the index space $\mathcal{I}$, we can project out the noise $\boldsymbol{n}_2$ by projecting the data $\boldsymbol{x}$ onto $\mathcal{I}$. In the representation (2) we can assume that $TA = I_m$ and $T\boldsymbol{x} = \widetilde{\boldsymbol{s}}$ without loss of generality, in which case T corresponds to the demixing matrix in under-complete ICA, but here we are not interested in the individual directions of the components $\widetilde{\boldsymbol{s}}_i$ (which are not assumed to be independent).

The main idea underlying our approach is summed up in the following Proposition (proof in Appendix). Whenever the variable $\boldsymbol{x}$ has covariance matrix identity, this result allows, from an *arbitrary* smooth real function h on $\mathbb{R}^d$, to find a vector $\boldsymbol{\beta}(h) \in \mathcal{I}$.

Proposition 1. *Let $\boldsymbol{x}$ be a random variable whose density function $p(\boldsymbol{x})$ satisfies* (2) *and suppose that $h(\boldsymbol{x})$ is a smooth real function on $\mathbb{R}^d$. Assume furthermore that $\Sigma = \mathbb{E}\left[\boldsymbol{x}\boldsymbol{x}^\top\right] = I_d$. Then under mild regularity conditions the following vector belongs to the target space $\mathcal{I}$:*

$$\boldsymbol{\beta}(h) = \mathbb{E}_{\boldsymbol{x}}\left[\nabla h(\boldsymbol{x}) - \boldsymbol{x}h(\boldsymbol{x})\right] . \tag{4}$$

Since an expectation over the unknown density $p(\boldsymbol{x})$ is used to define $\boldsymbol{\beta}$ by Eq.(4), in practice, it must be approximated using empirical expectation over the available data:

$$\widehat{\boldsymbol{\beta}}(h) = \frac{1}{n}\sum_{i=1}^{n}\{\nabla h(\boldsymbol{x}_i) - \boldsymbol{x}_i h(\boldsymbol{x}_i)\,.\} \tag{5}$$

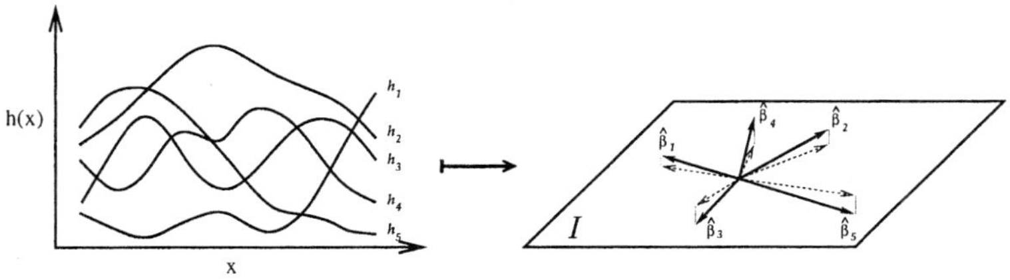

Fig. 1. The NGCA principle idea: from a varied family of real functions h, compute a family of vectors $\widehat{\boldsymbol{\beta}}$ belonging to the target space up to small estimation error

In the extended version of this paper [3], we show a probabilistic confidence bound of estimation error of our NGCA method under certain regularity conditions.

- If we assume $\mathbb{E}\left[\boldsymbol{x}\boldsymbol{x}^\top\right] = I_d$, the empirical estimator $\widehat{\boldsymbol{\beta}}(h)$ converges at a rate $\mathcal{O}(n^{-1/2})$ to a vector in the index space $\mathcal{I}$.
- In the general case where $\mathbb{E}\left[\boldsymbol{x}\boldsymbol{x}^\top\right]$ is an arbitrary positive definite matrix, we consider a "whitening" step, computing $\widehat{\boldsymbol{y}}_i = \widehat{\Sigma}^{-1/2}\boldsymbol{x}_i$ beforehand, where $\widehat{\Sigma} := \frac{1}{n}\sum_{i=1}^{n}\boldsymbol{x}_i\boldsymbol{x}_i^\top$. Taking into account the extra error introduced by this step, we can bound the the convergence rate of $\gamma(h) := \widehat{\Sigma}^{-1/2}\widehat{\boldsymbol{\beta}}_{\boldsymbol{y}}(h)$ to the index space $\mathcal{I}$ by $\mathcal{O}(\sqrt{d\log n/n})$.
- The entire index space $\mathcal{I}$ can be estimated from a family of vectors $\widehat{\boldsymbol{\beta}}_k$ (see Fig. 1) for a large set of functions $\{h_k\}_{k=1}^{L}$ and applying PCA to the set $\{\widehat{\boldsymbol{\beta}}_k\}_{k=1}^{L}$.

- Thanks to an exponential deviation inequality for the convergence rate of single functions, a union bound over L functions leads to a *uniform* convergence bound over the whole set of functions with rate of order $\mathcal{O}(\sqrt{d \log n / n} + \sqrt{\log L / n})$. Therefore, taking, e.g., $L = O(n^d)$ we still have insurance that convergence holds.

3 The NGCA Algorithm

As is briefly mentioned in the last section, in our NGCA procedure, basically we calculate a family of vectors $\widehat{\beta}_k$ for a large family of such functions $\{h_k\}_{k=1}^L$ and apply PCA to the set $\{\widehat{\beta}_k\}_{k=1}^L$ to find out the m-dimensional subspace $\widehat{\mathcal{I}}$ which gives the least approximation error. Although the principle of NGCA is very simple, there are some implementation issues.

- The theoretical results guarantee that the convergence order is achieved for any smooth functions $\{h_k\}_{k=1}^L$ with mild regularity conditions. However, in practice, it is important to find out good functions which provide a lot of information on the index space $\mathcal{I}$ and make the estimator $\widehat{\mathcal{I}}$ more accurate, because there exist many uninformative functions.
- Since the mapping $h \mapsto \beta(h)$ is linear, we need an appropriate renormalization of h or $\beta(h)$, otherwise it is meaningless to combine many vectors $\{\beta_k\}$ from various functions $\{h_k\}$ by PCA. Here we propose renormalizing by the trace of the variance $\mathrm{Var}\{\widehat{\beta}(h)\}$. Under this condition the norm of each vector is proportional to its signal-to-noise ratio so that longer vectors are more informative, while vectors with too small a norm are uninformative and can be discarded.

In the proposed algorithm we will restrict our attention to functions of the form $h_{f,\omega}(\boldsymbol{x}) = f(\langle \boldsymbol{\omega}, \boldsymbol{x} \rangle)$, where $\boldsymbol{\omega} \in \mathbb{R}^d$, $\|\boldsymbol{\omega}\| = 1$, and f belongs to a finite family $\mathcal{F}$ of smooth real functions of real variable. Our theoretical setting allows to ensure that the approximation error remains small uniformly over $\mathcal{F}$ and $\boldsymbol{\omega}$. However it is not feasible in practice to sample the whole parameter space for $\boldsymbol{\omega}$ as soon as it has more than a few dimensions. To overcome this difficulty we advocate using a well-known PP algorithm, FastICA [8], as a heuristic to find good candidates for $\boldsymbol{\omega}_f$ for a fixed f. We remark that FastICA, as a standalone procedure, requires to fix the "index function" f beforehand. The new point of our method is that we provide a theoretical setting and a methodology which allows to *combine* the results of this Projection Pursuit method when used over a possibly large spectrum of arbitrary index functions f.

Summing up, the NGCA algorithm then consists of the following steps: (1) Data whitening, (2) Applying FastICA to each function $f \in \mathcal{F}$ to find a promising candidate value for $\boldsymbol{\omega}_f$, (3) Computing the corresponding family of vectors $(\widehat{\beta}(h_{f,\boldsymbol{\omega}_f}))_{f \in \mathcal{F}}$ (using Eq. (5)), (4) Normalize the vectors appropriately; threshold and throw out uninformative ones, (5) apply PCA, (6) Pull back in original space (cf. Pseudocode). Note that the PCA step could be replaced by other, more refined principal directions extraction methods. In the implementation tested, we have used the following forms of the functions f_k: $f_\sigma^{(1)}(z) = z^3 \exp(-z^2/2\sigma^2)$ (Gauss-Pow3), $f_b^{(2)}(z) = \tanh(bz)$ (Hyperbolic Tangent), $f_a^{(3)}(z) = \{\sin, \cos\}(az)$ (Fourier). More precisely, we consider

PSEUDOCODE FOR THE NGCA ALGORITHM

Input: Data points $(\boldsymbol{x}_i) \in \mathbb{R}^d$, dimension m of target subspace.

Parameters: Number $T_{\max}$ of FastICA iterations; threshold ε; family of real functions (f_k).

Whitening.

The data $\boldsymbol{x}_i$ is recentered by subtracting the empirical mean.

Let $\widehat{\Sigma}$ denote the empirical covariance matrix of the data sample $(\boldsymbol{x}_i)$;

put $\widehat{\boldsymbol{y}}_i = \widehat{\Sigma}^{-\frac{1}{2}} \boldsymbol{x}_i$ the empirically whitened data.

Main Procedure.

Loop on $k = 1, \ldots, L$:

Draw $\boldsymbol{\omega}_0$ at random on the unit sphere of $\mathbb{R}^d$.

Loop on $t = 1, \ldots, T_{\max}$: *[FastICA loop]*

Put $\widehat{\boldsymbol{\beta}}_t \leftarrow \frac{1}{n} \sum_{i=1}^{n} \left(\widehat{\boldsymbol{y}}_i f_k(\langle \boldsymbol{\omega}_{t-1}, \widehat{\boldsymbol{y}}_i \rangle) - f_k'(\langle \boldsymbol{\omega}_{t-1}, \widehat{\boldsymbol{y}}_i \rangle) \boldsymbol{\omega}_{t-1} \right)$.

Put $\boldsymbol{\omega}_t \leftarrow \widehat{\boldsymbol{\beta}}_t / \| \widehat{\boldsymbol{\beta}}_t \|$.

End Loop on t

Let N_i be the trace of the empirical covariance matrix of $\widehat{\boldsymbol{\beta}}_{T_{\max}}$:

$$N_i = \frac{1}{n} \sum_{i=1}^{n} \left\| \widehat{\boldsymbol{y}}_i f_k(\langle \boldsymbol{\omega}_{T_{\max}-1}, \widehat{\boldsymbol{y}}_i \rangle) - f_k'(\langle \boldsymbol{\omega}_{T_{\max}-1}, \widehat{\boldsymbol{y}}_i \rangle) \boldsymbol{\omega}_{T_{\max}-1} \right\|^2 - \left\| \widehat{\boldsymbol{\beta}}_{T_{\max}} \right\|^2 .$$

Store $v^{(k)} \leftarrow \widehat{\boldsymbol{\beta}}_{T_{\max}} * \sqrt{n/N_i}$. *[Normalization]*

End Loop on k

Thresholding.

From the family $\boldsymbol{v}^{(k)}$, throw away vectors having norm smaller than threshold ε.

PCA step.

Perform PCA on the set of remaining $\boldsymbol{v}^{(k)}$.

Let V_m be the space spanned by the first m principal directions.

Pull back in original space.

Output: $W_m = \widehat{\Sigma}^{-\frac{1}{2}} V_m$.

discretized ranges for $a \in [0, A], b \in [0, B], \sigma \in [\sigma_{\min}, \sigma_{\max}]$, which gives rise to a finite family (f_k) (which includes *simultaneously* functions of the three different above families).

4 Numerical Results

All the experiments presented where obtained with exactly the same set of parameters: $a \in [0, 4]$ for the Fourier functions; $b \in [0, 5]$ for the Hyperbolic Tangent functions; $\sigma^2 \in [0.5, 5]$ for the Gauss-pow3 functions. Each of these ranges was divided into 1000 equispaced values, thus yielding a family (f_k) of size 4000 (Fourier functions count twice because of the sine and cosine parts). Some preliminary calibration suggested to take $\varepsilon = 1.5$ as the threshold under which vectors are not informative. Finally we fixed the number of FastICA iterations $T_{\max} = 10$. With this choice of parameters, with 1000 points of data the computation time is typically of the order of 10 seconds on a modern PC under a Matlab implementation.

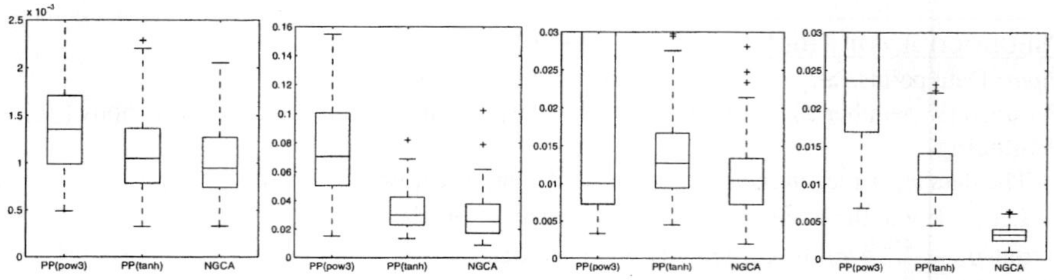

(A) (B) (C) (D)

Fig. 2. Boxplots of the error criterion $\mathcal{E}(\widehat{\mathcal{I}}, \mathcal{I})$ over 100 training samples of size 1000

Tests in a controlled setting. We performed numerical experiments using various synthetic data. We report exemplary results using 4 data sets. Each data set includes 1000 samples in 10 dimensions, and consists of 8-dimensional independent standard Gaussian and 2 non-Gaussian components as follows:

(A) Simple Gaussian Mixture: 2-dimensional independent bimodal Gaussian mixtures;
(B) Dependent super-Gaussian: 2-dimensional density is proportional to $\exp(-\|x\|)$;
(C) Dependent sub-Gaussian: 2-dimensional uniform on the unit circle;
(D) Dependent super- and sub-Gaussian: 1-dimensional Laplacian with density proportional to $\exp(-|x_{Lap}|)$ and 1-dimensional dependent uniform $U(c, c+1)$, where $c = 0$ for $|x_{Lap}| \leq \log 2$ and $c = -1$ otherwise.

We compare the NGCA method against standalone FastICA with two different index functions. Figure 2 shows boxplots, over 100 samples, of the error criterion $\mathcal{E}(\widehat{\mathcal{I}}, \mathcal{I}) = m^{-1} \sum_{i=1}^{m} \|(I_d - \Pi_{\mathcal{I}})\widehat{v}_i\|^2$, where $\{\widehat{v}_i\}_{i=1}^{m}$ is an orthonormal basis of $\widehat{\mathcal{I}}$, I_d is the identity matrix, and $\Pi_{\mathcal{I}}$ denotes the orthogonal projection on $\mathcal{I}$. In datasets (A),(B),(C), NGCA appears to be on par with the best FastICA method. As expected the best index for FastICA is data-dependent: the 'tanh' index is more suited to the super-Gaussian data (B) while the 'pow3' index works best with the sub-Gaussian data (C) (although in this case FastICA with this index has a tendency to get caught in local minima, leading to a disastrous result for about 25% of the samples. Note that NGCA does *not* suffer from this problem). Finally, the advantage of the implicit index adaptation feature of NGCA can be clearly observed in the data set (D), which includes both sub- and super-Gaussian components. In this case neither of the two FastICA index functions taken alone does well and NGCA gives significantly lower error than either FastICA flavor.

Example of application for realistic data: visualization and clustering. We now give an example of application of NGCA to visualization and clustering of realistic data. We consider here "oil flow" data which has been obtained by numerical simulation of a complex physical model. This data was already used before for testing techniques of dimension reduction [2]. The data is 12-dimensional and our goal is to visualize the data and possibly exhibit a clustered structure. We compared results obtained with the NGCA methodology, regular PCA, FastICA with tanh index and Isomap. The results are shown on Figure 3. A 3D projection of the data was first computed using these methods, which was in turn projected in 2D to draw the figure; this last projection

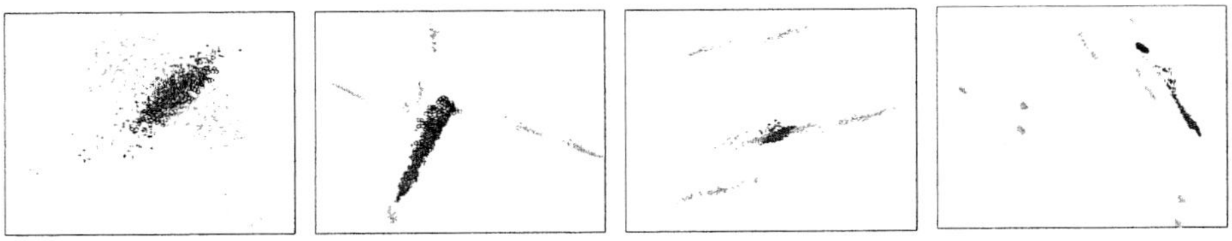

Fig. 3. 2D projection of the "oil flow" (12-dimensional) data obtained by different algorithms, from left two right: PCA, Isomap, FastICA (tanh index), NGCA. In each case, the data was first projected in 3D using the respective methods, from which a 2D projection was chosen visually so as to yield the clearest cluster structure. Colors indicate label information (*not used to determine the projections*).

was chosen manually so as to make the cluster structure as visible as possible in each case. The NGCA result appears better with a clearer clustered structure appearing. This structure is only partly visible in the Isomap result; the NGCA method additionally has the advantage of a clear geometrical interpretation (linear orthogonal projection). Finally, datapoints in this dataset are distributed in 3 classes. This information was not used in the different procedures, but we can see *a posteriori* that only NGCA clearly separates the classes in distinct clusters.

5 Conclusion

We proposed a new semi-parametric framework for constructing a linear projection to separate an uninteresting, possibly of large amplitude multivariate Gaussian 'noise' subspace from the 'signal-of-interest' subspace. We also provided generic consistency results on how well the non-Gaussian directions can be identified (an extended version of this paper). Once the low-dimensional 'signal' part is extracted, we can use it for a variety of applications such as data visualization, clustering, denoising or classification. Numerically we found comparable or superior performance to, e.g., FastICA in deflation mode as a generic representative of the family of ICA/PP algorithms. Note that in general, PP methods need to pre-specify a projection index with which they search non-Gaussian components. By contrast, an important advantage of our method is that we are able to simultaneously use several families of nonlinear functions; moreover, also inside a same function family we are able to use an entire range of parameters (such as frequency for Fourier functions). Thus, NGCA provides higher flexibility, and less restricting assumptions *a priori* on the data. In a sense, the functional indices that are the most relevant for the data at hand are automatically selected.

Future research will adapt the theory to simultaneously estimate the dimension of the non-Gaussian subspace. Extending the proposed framework to non-linear projection scenarios [11, 12, 1, 6] and to finding the most discriminative directions using labels are examples for which the current theory could be taken as a basis.

Acknowledgements. This work was supported in part by the IST Programme of the European Community, under the PASCAL Network of Excellence, IST-2002-506778. This publication only reflects the authors' views.

References

1. M. Belkin and P. Niyogi. Laplacian eigenmaps for dimensionality reduction and data representation. *Neural Computation*, 15(6):1373–1396, 2003.
2. C.M. Bishop, M. Svensen and C.K.I. Wiliams. GTM: The generative topographic mapping. *Neural Computation*, 10(1):215–234, 1998.
3. G. Blanchard, M. Kawanabe, M. Sugiyama, V. Spokoiny and K.-R. Müller. In search of non-Gaussian components of a high-dimensional distribution. (Preprint available at http://www.cs.titech.ac.jp/)
4. P. Comon. Independent component analysis—a new concept? *Signal Processing*, 36:287–314, 1994.
5. J.H. Friedman and J.W. Tukey. A projection pursuit algorithm for exploratory data analysis. *IEEE Transactions on Computers*, 23(9):881–890, 1975.
6. S. Harmeling, A. Ziehe, M. Kawanabe and K.-R. Müller. Kernel-based nonlinear blind source separation. *Neural Computation*, 15(5):1089–1124, 2003.
7. P.J. Huber. Projection pursuit. *The Annals of Statistics*, 13:435–475, 1985.
8. A. Hyvärinen. Fast and robust fixed-point algorithms for independent component analysis. *IEEE Transactions on Neural Networks*, 10(3):626–634, 1999.
9. A. Hyvärinen, J. Karhunen and E. Oja. *Independent Component Analysis*. Wiley, 2001.
10. M. Kawanabe and K.-R. Müller. Estimating functions for blind separation when sources have variance dependencies. *Journal of Machine Learning Research*, 6:453–482, 2005.
11. S. Roweis and L. Saul. Nonlinear dimensionality reduction by locally linear embedding. *Science*, 290(5500):2323–2326, 2000.
12. J.B. Tenenbaum, V. de Silva and J.C. Langford. A global geometric framework for nonlinear dimensionality reduction. *Science*, 290(5500):2319–2323, 2000.

Proof of Proposition 1

Put $\boldsymbol{\alpha} = \mathbb{E}_{\boldsymbol{x}}[\boldsymbol{x}h(\boldsymbol{x})]$ and $\psi(\boldsymbol{x}) = h(\boldsymbol{x}) - \boldsymbol{\alpha}^\top \boldsymbol{x}$. Note that $\nabla\psi = \nabla h - \boldsymbol{\alpha}$, hence $\beta(h) = \mathbb{E}_{\boldsymbol{x}}[\nabla\psi(\boldsymbol{x})]$. Furthermore, it holds by change of variable that

$$\int \psi(\boldsymbol{x}+\boldsymbol{u})p(\boldsymbol{x})d\boldsymbol{x} = \int \psi(\boldsymbol{x})p(\boldsymbol{x}-\boldsymbol{u})d\boldsymbol{x}.$$

Under mild regularity conditions on $p(\boldsymbol{x})$ and $h(\boldsymbol{x})$, differentiating this with respect to u gives

$$\mathbb{E}_{\boldsymbol{x}}[\nabla\psi(\boldsymbol{x})] = \int \nabla\psi(\boldsymbol{x})p(\boldsymbol{x})d\boldsymbol{x} = -\int \psi(\boldsymbol{x})\nabla p(\boldsymbol{x})d\boldsymbol{x} = -\mathbb{E}_{\boldsymbol{x}}[\psi(\boldsymbol{x})\nabla \log p(\boldsymbol{x})],$$

where we have used $\nabla p(\boldsymbol{x}) = \nabla \log p(\boldsymbol{x})\, p(\boldsymbol{x})$. Eq.(2) now implies $\nabla \log p(\boldsymbol{x}) = \nabla \log g(T\boldsymbol{x}) - \Gamma^{-1}\boldsymbol{x}$, hence

$$\begin{aligned}\beta(\psi) &= -\mathbb{E}_{\boldsymbol{x}}[\psi(\boldsymbol{x})\nabla \log g(T\boldsymbol{x})] + \mathbb{E}_{\boldsymbol{x}}\left[\psi(\boldsymbol{x})\Gamma^{-1}\boldsymbol{x}\right] \\ &= -T^\top \mathbb{E}_{\boldsymbol{x}}[\psi(\boldsymbol{x})\nabla g(T\boldsymbol{x})/g(T\boldsymbol{x})] + \Gamma^{-1}\mathbb{E}_{\boldsymbol{x}}\left[\boldsymbol{x}h(\boldsymbol{x}) - \boldsymbol{x}\boldsymbol{x}^\top \mathbb{E}[\boldsymbol{x}h(\boldsymbol{x})]\right].\end{aligned}$$

The last term above vanishes because we assumed $\mathbb{E}_{\boldsymbol{x}}\left[\boldsymbol{x}\boldsymbol{x}^\top\right] = Id$. The first term belongs to $\mathcal{I}$ by definition. This concludes the proof. □

Estimating Non-Gaussian Subspaces by Characteristic Functions

Motoaki Kawanabe[1] and Fabian J. Theis[2]

[1] Fraunhofer FIRST.IDA, Germany
[2] Institute of Biophysics, University of Regensburg, Germany
nabe@first.fhg.de, fabian@theis.name

Abstract. In this article, we consider high-dimensional data which contains a low-dimensional non-Gaussian structure contaminated with Gaussian noise and propose a new method to identify the non-Gaussian subspace. A *linear dimension reduction* algorithm based on the fourth-order cumulant tensor was proposed in our previous work [4]. Although it works well for sub-Gaussian structures, the performance is not satisfactory for super-Gaussian data due to outliers. To overcome this problem, we construct an alternative by using Hessian of characteristic functions which was applied to (multidimensional) independent component analysis [10,11]. A numerical study demonstrates the validity of our method.

1 Introduction

Recently enormous amount of data with a huge number of features have been stored and are to be analyzed. In most real-world applications, the 'signal' or 'information' is typically contained only in a low-dimensional subspace of the high-dimensional data, thus dimensionality reduction is a useful preprocessing for further data analysis. Here we make an assumption on the data: the high-dimensional data $\boldsymbol{x} \in \mathbb{R}^d$ is a sum of low-dimensional non-Gaussian components ('signal') $\boldsymbol{s} \in \mathbb{R}^m$ $(m < d)$ and a Gaussian noise $\boldsymbol{n} \sim N(\boldsymbol{0}, \Gamma)$,

$$\boldsymbol{x} = A\boldsymbol{s} + \boldsymbol{n} \tag{1}$$

where A is a $d \times m$ full rank matrix indicating the non-Gaussian subspace and $\boldsymbol{s}$ and $\boldsymbol{n}$ are assumed to be independent. Under this modeling assumption, therefore, the tasks are to estimate the relevant non-Gaussian subspace and to recover the low-dimensional non-Gaussian structures by *linear dimension reduction*. Although our goal is dimension reduction, we want to emphasize that we do *not* assume the Gaussian components to be of *smaller* order of magnitude than the signal components. This setting therefore excludes the use of common linear and non-linear dimensionality reduction methods such as PCA, Isomap [9] and LLE [8].

If the non-Gaussian components s_i's are mutually independent, the model turns out to be the under-complete noisy ICA, and there exist algorithms to extract the independent components in the presence of Gaussian noise [7]. However, this is often a too strict assumption on the practical data.

In contrast, Projection Pursuit (PP) [3,5] or FastICA in the deflation mode [6,7] can also extract dependent non-Gaussian structures by maximizing a prefixed non-Gaussianity index which contains higher order information. Recently two procedures

J. Rosca et al. (Eds.): ICA 2006, LNCS 3889, pp. 157–164, 2006.

have been developped in the same spirit of PP/FastICA. Non-Gaussian Component Analysis (NGCA) [1] was built upon a general semi-parametric framework in mathematical statistics, while the other [4] is a modification of an ICA algorithm (JADE [2]) to the dimension reduction problem. In this paper, we will propose an alternative of the second algorithm with Hessian of characteristic functions which was applied to (multidimensional) independent component analysis [10,11]. In comparison with the fourth-order cumulant, characteristic functions yield more robust and efficient method when data contain super-Gaussian structures.

2 Mathematical Preliminaries

Since the decomposition (1) is not uniquely determined, we will further transform the model to reduce indeterminacies. The noise term $\boldsymbol{n}$ can be decomposed into two independent parts as $\boldsymbol{n} = \boldsymbol{n}_1 + \boldsymbol{n}_2$, where $\boldsymbol{n}_1 = A\boldsymbol{\eta} \in \mathrm{Range}(A)$ and $\boldsymbol{n}_2$ is restricted in the $(d-m)$-dimensional complementary subspace s.t. $\mathrm{Cov}(\boldsymbol{n}_1, \boldsymbol{n}_2) = 0$. Thus, we get a representation with less indeterminacies

$$\boldsymbol{x} = A\widetilde{\boldsymbol{s}} + \boldsymbol{n}_2, \tag{2}$$

where $\widetilde{\boldsymbol{s}} := \boldsymbol{s} + \boldsymbol{\eta}$ and the noise term $\boldsymbol{n}_2$ distributes with a $(d-m)$-dimensional degenerated Gaussian. We remark that we can only recover $\widetilde{\boldsymbol{s}}$, the signal with contaminated noise in the non-Gaussian subspace Range(A). By changing the symbols as $A \to A_N$, $\widetilde{\boldsymbol{s}} \to \boldsymbol{s}_N$ and $\boldsymbol{n}_2 \to A_G\boldsymbol{s}_G$, we will consider

$$\boldsymbol{x} = A_N\boldsymbol{s}_N + A_G\boldsymbol{s}_G \tag{3}$$

as our model fomulation, where A_G indicates the subspace and $\boldsymbol{s}_G$ denotes a $(d-m)$-dimensional Gaussian random vector. Independence of $\boldsymbol{s}_N$ and $\boldsymbol{s}_G$ implies that the non-Gaussian subspace and the Gaussian noise components are orthogonal with the metric Σ^{-1}, i.e. $A_N^\top \Sigma^{-1} A_G = 0$, where $\Sigma := \mathrm{Cov}(\boldsymbol{x})$.

Let $(B_N^\top, B_G^\top)^\top$ be the inverse matrix of (A_N, A_G). Then, the submatrices B_N and B_G extract the non-Gaussian and the Gaussian parts of the data $\boldsymbol{x}$, i.e. $B_N\boldsymbol{x} = \boldsymbol{s}_N$ and $B_G\boldsymbol{x} = \boldsymbol{s}_G$. The primal goal of dimension reduction in this paper is estimating the linear mapping B_N onto the non-Gaussian subspace in order to project out the irrelevant Gaussian components $\boldsymbol{s}_G$ and obtain the non-Gaussian signals $\boldsymbol{s}_N = B_N\boldsymbol{x}$. We remark that other matrices B_G, A_N and A_G can also be determined automatically, once B_N is derived. From independence of $\boldsymbol{s}_N$ and $\boldsymbol{s}_G$, the density function of $\boldsymbol{x}$ can be expressed as a product of the non-Gaussian and the Gaussian components

$$p(\boldsymbol{x}) = g(B_N\boldsymbol{x})\phi_L(B_G\boldsymbol{x}), \tag{4}$$

where g is an unknown function describing the density of $\boldsymbol{s}_N$ and ϕ_L is the Gaussian density with covariance L.

There still remain trivial indeterminacies in the model (3)

$$\boldsymbol{x} = (A_NC_1)(C_1^{-1}\boldsymbol{s}_N) + (A_GC_2)(C_2^{-1}\boldsymbol{s}_G), \tag{5}$$

where C_1 and C_2 are m- and $(d-m)$-dimensional square invertible matrices, respectively. Because of the indeterminacies (5) we should evaluate the results by $\mathcal{I} = \mathrm{Range}(B_N^\top)$ (called non-Gaussian index space here) rather than B_N itself. We recently proved that the decomposition (3) is unique up to this indeterminacies, if we assume that the dimension m of the non-Gaussian subspace is correct [12].

3 Joint Low-Rank Approximation Method

3.1 Dimension Reduction by Using Fourth-Order Cumulant Tensor

In our previous work [4], we propose a procedure for estimating the non-Gaussian subspace $\mathcal{I}$ based on the fourth-order cumulant tensor

$$\begin{aligned}&\mathrm{cum}(x_i, x_j, x_k, x_l)\\&:= \mathbb{E}[x_i x_j x_k x_l] - \mathbb{E}[x_i x_j]\mathbb{E}[x_k x_l] - \mathbb{E}[x_i x_k]\mathbb{E}[x_j x_l] - \mathbb{E}[x_i x_l]\mathbb{E}[x_j x_k].\end{aligned}$$

The method was inspired by the JADE algorithm [2] for ICA which uses this tensor.

As is used in the JADE algorithm, we also apply the whitening transformation $\boldsymbol{z} = V^{-1/2}\boldsymbol{x}$ as preprocessing, where $V = \mathrm{Cov}[\boldsymbol{x}]$. Let us define the matrices

$$W_N := B_N V^{1/2}, \qquad W_G := B_G V^{1/2},$$

which are the linear transformations from the sphered data to the factors $\boldsymbol{s} = (\boldsymbol{s}_N^\top, \boldsymbol{s}_G^\top)^\top$. We remark that the non-Gaussian index space can be expressed as

$$\mathcal{I} = \mathrm{Range}(B_N^\top) = V^{-1/2}\mathrm{Range}(W_N^\top).$$

and therefore, it is enough to estimate the matrix W_N. Without loss of generality, we can assume that $\mathrm{Cov}[\boldsymbol{s}] = I$. Then, $(W_N^\top, W_G^\top)$ becomes an orthogonal matrix.

The method proposed in [4] rests on the fact that the cumulant tensor of the sources $(\boldsymbol{s}_N, \boldsymbol{s}_G)$ has simple structure. Let us order the sources as $\boldsymbol{s}_N = (s_1, \ldots, s_m)$ and $\boldsymbol{s}_G = (s_{m+1}, \ldots, s_d)$. The cumulant tensor $\mathrm{cum}(s_i, s_j, s_k, s_l)$ takes 0, unless $1 \le i, j, k, l \le m$ (i.e. all components should belong to the non-Gaussian part). Let $Q^{(kl)}$ be the matrix whose (i,j) element is $\mathrm{cum}(z_i, z_j, z_k, z_l)$ for all $1 \le k, l \le d$ and W° be a d-dimensional orthogonal matrix which recovers the sources, i.e. $\boldsymbol{s} = W^\circ \boldsymbol{z}$. Then, it can be proven that, for all (k,l),

$$W^\circ Q^{(kl)} (W^\circ)^\top = \begin{pmatrix} * & 0 \\ 0 & 0 \end{pmatrix}$$

holds, that is, all components which are not contained in the $m \times m$ submatrix $*$ vanish after the similar transformation by W°. This fact implies that we can estimate the transformation W_N° to the non-Gaussian components $\boldsymbol{s}_N$ by maximizing the Frobenius norms of the $m \times m$ submatrices corresponding to the non-Gaussian subspace

$$\mathcal{L}(W_N) = \sum_{k,l=1}^{d} \|W_N Q^{(kl)} W_N^\top\|_{\mathrm{Fro}}^2 = \sum_{k,l=1}^{d} \sum_{i',j'=1}^{m} \mathrm{cum}(y_{i'}, y_{j'}, z_k, z_l)^2 \quad (6)$$

w.r.t. W_N s.t. $W_N W_N^\top = I_m$, where $\boldsymbol{y}_N = (y_1, \ldots, y_m)^\top = W_N \boldsymbol{z}$ denotes the reconstructed non-Gaussian components by W_N and $\|\cdot\|_{\mathrm{Fro}}^2$ is Frobenius norm of matrices. The contrast function (6) was optimized by iterative eigenvalue decomposition in [4],

$$W_N^{(t+1)} \sum_{k,l=1}^{d} \widehat{Q}^{(kl)} \{W_N^{(t)}\}^\top W_N^{(t)} \widehat{Q}^{(kl)} = \Lambda W_N^{(t+1)}, \tag{7}$$

where $\widehat{Q}^{(kl)}$ is the empirical correspondent of the matrix $Q^{(kl)}$ and $W_N^{(t)}$ is the t-step estimator.

The algorithm works well for sub-Gaussian structures. However, due to outliers it performs worse when the data contains heavy-tailed structures. In the remaining of this section, we will introduce joint low-rank approximation (JLA) of matrices by generalizing this method and show its global consistency. A novel algorithm using Hessian of the characteristic function will be proposed as an example.

3.2 Joint Low-Rank Approximation of Matrices

Let us consider the ideal situation as the discussion with the expected cumulant tensor in the previous section. Suppose that K complex matrices $M_1, \ldots, M_K$ can be simultaneously transformed into

$$W^\circ M_k (W^\circ)^\top = \begin{pmatrix} * & 0 \\ 0 & 0 \end{pmatrix}, \qquad k = 1, \ldots, K, \tag{8}$$

that is, all components of all the transformed matrices vanish except for those in $m \times m$ submatrices indicated by $*$, where W° is a d-dimensional orthogonal matrix. Let W_N° be the $m \times d$ matrix composed of the first m rows of W°. We remark that $W_N^\circ (W_N^\circ)^\top = I_m$. The goal here is to estimate the mapping W_N° as before.

Let us consider the contrast function

$$\mathcal{L}(W_N) = \sum_{k=1}^{K} \|W_N M_k W_N^\top\|_{\mathrm{Fro}}^2, \tag{9}$$

where Frobenius norm $\|C\|_{\mathrm{Fro}}^2 = \mathrm{tr}(CC^*)$ in complex case. We can show that the desired mapping W_N° can be obtained up to an orthogonal matrix by maximizing the contrast function $\mathcal{L}(W_N)$.

Theorem 1. *The objective function $\mathcal{L}(W_N)$ is maximal at $W_N = UW_N^\circ$, where W_N° is the first $m \times d$ submatrix of W° defined by Eq. (8) and U is an m-dimensional orthogonal matrix.*

Proof. We remark that Frobenius norm $\|WM_kW^\top\|_{\mathrm{Fro}}^2$ is unchanged for any orthogonal matrix W, i.e. $\|WM_kW^\top\|_{\mathrm{Fro}}^2 = \|M_k\|_{\mathrm{Fro}}^2 = \|W^\circ M_k (W^\circ)^\top\|_{\mathrm{Fro}}^2$. From the property (8) of the matrix W°, we get

$$\|W^\circ M_k (W^\circ)^\top\|_{\mathrm{Fro}}^2 = \left\| \begin{pmatrix} W_N^\circ M_k (W_N^\circ)^\top & 0 \\ 0 & 0 \end{pmatrix} \right\|_{\mathrm{Fro}}^2 = \|W_N^\circ M_k (W_N^\circ)^\top\|_{\mathrm{Fro}}^2,$$

where we divided W° into two submatrices W_N° and W_G°. On the other hand, for a general orthogonal matrix W,

$$\begin{aligned}\|WM_kW^\top\|^2_{\text{Fro}} &= \left\|\begin{pmatrix} W_N M_k W_N^\top & W_N M_k W_G^\top \\ W_G M_k W_N^\top & W_G M_k W_G^\top \end{pmatrix}\right\|^2_{\text{Fro}} \\ &= \|W_N M_k W_N^\top\|^2_{\text{Fro}} + \|W_N M_k W_G^\top\|^2_{\text{Fro}} \\ &\qquad + \|W_G M_k W_N^\top\|^2_{\text{Fro}} + \|W_G M_k W_G^\top\|^2_{\text{Fro}} \\ &\geq \|W_N M_k W_N^\top\|^2_{\text{Fro}},\end{aligned}$$

where W was also divided into two submatrices W_N and W_G. Since the inequality is strict if and only if one of the three terms is non-zero, we notice that all global maxima were already found. Therefore, for all orthogonal matrices U, finally we get

$$\|UW_N^\circ M_k (W_N^\circ)^\top U^\top\|^2_{\text{Fro}} = \|W_N^\circ M_k (W_N^\circ)^\top\|^2_{\text{Fro}} \geq \|W_N M_k W_N^\top\|^2_{\text{Fro}}. \qquad \square$$

For simplicity, we further assume that $M_k^\top = M_k$, as is the case with our algorithms. By differentiating the criterion $\mathcal{L}$ under the constraint $W_N W_N^\top = I_m$, we get

$$W_N \sum_{k=1}^{K} \mathcal{M}_k(W_N) = \Lambda W_N, \tag{10}$$

where

$$\mathcal{M}_k(W_N) := M_k W_N^\top W_N M_k^* + M_k^* W_N^\top W_N M_k \tag{11}$$

is a $d \times d$ matrix depeding on W_N and Lagrange multipliers Λ is assumed to be diagonal without loss of generality. As the algorithm with the cumulant tensor, the maximization of the contrast function (9) can be solved by iterating the eigenvalue problem

$$W_N^{(t+1)} \sum_{k=1}^{K} \widehat{\mathcal{M}}_k(W_N^{(t)}) = \Lambda W_N^{(t+1)} \tag{12}$$

where $\widehat{\mathcal{M}}_k$ is the empirical correspondent of the matrix $\mathcal{M}_k$ and $W_N^{(t)}$ is the t-step estimator.

3.3 Dimension Reduction by Using Characteristic Functions

In [10] and [11], Hessians of the characteristic function were used for (multidimensional) independent component analysis. Since they satisfy the property (8) under our model assumption as we will show, they can also be used as the matrices M_k in the joint low-rank approximation procedure. The characteristic function of the random variable $\boldsymbol{z}$ can be defined by $\widehat{Z}(\boldsymbol{\zeta}) := \mathbb{E}[\exp(i\boldsymbol{\zeta}^\top \boldsymbol{z})]$. Let $W^\circ = ((W_N^\circ)^\top, (W_G^\circ)^\top)^\top$ be an orthogonal matrix s.t. $\boldsymbol{s} = W^\circ \boldsymbol{z}$. Then, the characteristic function can be expressed as

$$\widehat{Z}(\boldsymbol{\zeta}) = \widehat{S}(W^\circ \boldsymbol{\zeta}) = \widehat{S}_N(W_N^\circ \boldsymbol{\zeta}) \exp\left(-\frac{1}{2}\|W_G^\circ \boldsymbol{\zeta}\|^2\right), \tag{13}$$

where $\widehat{S}$ and $\widehat{S}_N$ are the characteristic functions of s and s_N, respectively. Therefore, if $\log \widehat{Z}(\zeta)$ exists, the Hessian of $\log \widehat{Z}(\zeta)$ becomes

$$H_{\log \widehat{Z}}(\zeta) := \frac{\partial^2}{\partial \zeta \partial \zeta^\top} \log \widehat{Z}(\zeta)$$
$$= (W^\circ)^\top \begin{pmatrix} \frac{\partial^2}{\partial \xi_N \partial \xi_N^\top} \log \widehat{S}_N(W_N^\circ \zeta) & 0 \\ 0 & -I_{d-m} \end{pmatrix} W^\circ, \tag{14}$$

where $\xi_N = W_N^\circ \zeta$. For K selected vectors $\zeta_1, \ldots, \zeta_K \in \mathbb{R}^d$, each matrix $M_k := H_{\log \widehat{Z}}(\zeta_k) + I_d$ satisfies the property (8).

Suppose that samples $\boldsymbol{x}_1, \ldots, \boldsymbol{x}_n$ are given. The algorithm with Hessians of the characteristic function is summarized as follows.

Algorithm

1. Sphere the data $\{\boldsymbol{x}_i\}_{j=1}^n$ by $\widehat{\boldsymbol{z}}_j = \widehat{V}^{-1/2}\boldsymbol{x}_j$, where $\widehat{V} = \widehat{\mathrm{Cov}}[\boldsymbol{x}]$.
2. Calculate the Hessian $\widehat{M}_k := \widehat{H}_{\log \widehat{Z}}(\zeta_k) + I_d$ at selected vectors ζ_k from the empirical characteristic function $\widehat{Z}_{\mathrm{emp}}(\zeta) = \frac{1}{n}\sum_{j=1}^n \exp(i\zeta^\top \widehat{\boldsymbol{z}}_j)$.
3. Compute m eigenvectors with largest absolute eigenvalues.

$$W_N^{(0)} \sum_{k=1}^K \left\{ \mathrm{Re}(\widehat{M}_k) + \mathrm{Im}(\widehat{M}_k) \right\} = \Lambda W_N^{(0)}$$

4. Solve the following eigenvalue problem until the matrix $W_N^{(t)}$ converges.

$$W_N^{(t+1)} \sum_{k=1}^K \widehat{\mathcal{M}}_k(W_N^{(t)}) = \Lambda W_N^{(t+1)}$$

The symbols with hat denote the empirical versions of the corresponding quantities, for example, $\widehat{\mathrm{Cov}}$ is the sample covariance. Re(M) and Im(M) are the real and the imaginary parts of a matrix M. The matrix $\widehat{\mathcal{M}}_k(W_N)$ is calsulated from $\widehat{M}_k$ by Eq. (11).

4 Numerical Results

For testing our algorithm, we performed numerical experiments using various synthetic data used in [1]. Each data set includes 1000 samples in 10 dimension. Each sample consists of 8-dimensional independent standard Gaussian and 2 non-Gaussian components as follows.

(A) Simple: 2-dimensional independent Gaussian mixtures with density of each component given by $\frac{1}{2}\phi_{-3,1}(x) + \frac{1}{2}\phi_{3,1}(x)$.

(B) Dependent super-Gaussian: 2-dimensional isotropic distribution with density proportional to $\exp(-\|x\|)$.

(C) Dependent sub-Gaussian: 2-dimensional isotropic uniform with constant positive density for $\|x\| \leq 1$ and 0 otherwise.

(D) Dependent super- and sub-Gaussian: 1-dimensional Laplacian with density proportional to $\exp(-|x_{Lap}|)$ and 1-dimensional dependent uniform $U(c, c+1)$, where $c = 0$ for $|x_{Lap}| \leq \log 2$ and $c = -1$ otherwise.

The profiles of the density functions of the non-Gaussian components in the above data sets are described in Fig. 1. The mean and standard deviation of samples are normalized to zero and one in a component-wise manner.

Besides the proposed algorithm, we applied for reference the following four methods in the experiments: FastICA with 'pow3' or 'tanh' index (denoted by FIC3 and FICt, respectively), JADE and joint low-rank approximation (JLA) algorithms with the fourth-order cumulant tensor and Hessian of the characteristic function (denoted by JLA4 and JLAH, respectively). In JLA with Hessian, 1000 vectors ζ were randomly chosen and 10% of them with high norm $\|M_k\|_{\mathrm{Fro}}$ were taken in the contrast function. We remark that we did not include the better method [1], because the main purpose of the experiments is compareing the two JLA algorithm. Further research is necessary to improve the algorithm. In FastICA and JLA with the cumulant tensor, additionally 9 runs from random initial matrices were also carried out and the optimum among these 10 solutions were chosen to avoid local optima.

Fig. 2 shows boxplots of the error criterion

$$\mathcal{E}(\widehat{\mathcal{I}}, \mathcal{I}) = \frac{1}{m} \|(I_d - \Pi_{\mathcal{I}})\Pi_{\widehat{\mathcal{I}}}\|^2_{\mathrm{Fro}}, \tag{15}$$

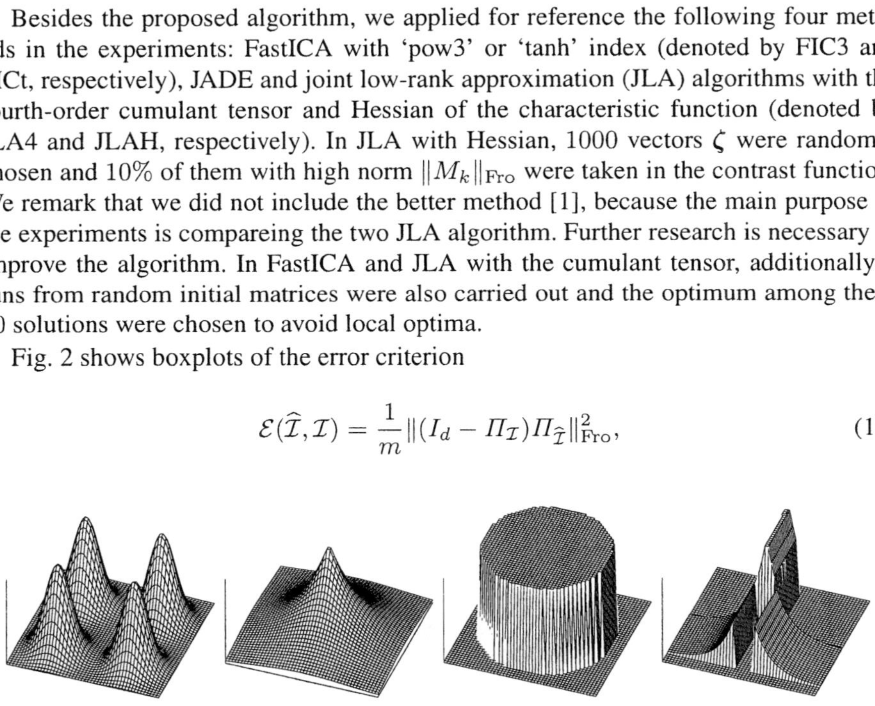

(A) (B) (C) (D)

Fig. 1. Densities of non-Gaussian components. The datasets are: (a) 2D independent Gaussian mixtures, (b) 2D isotropic super-Gaussian, (c) 2D isotropic uniform and (d) dependent 1D Laplacian + 1D uniform.

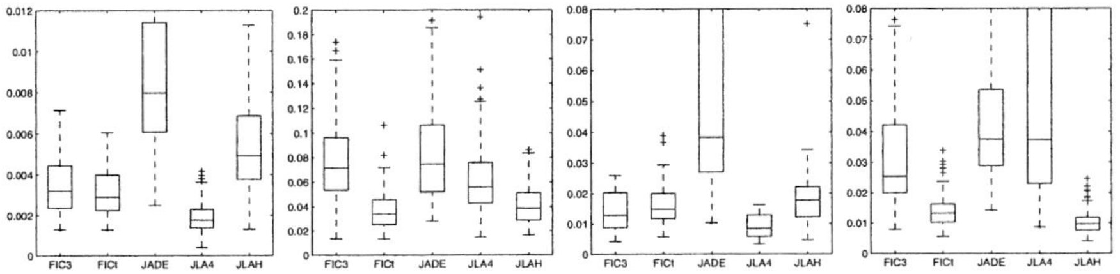

(A) (B) (C) (D)

Fig. 2. Boxplots of the error criterion $\mathcal{E}(\widehat{\mathcal{I}}, \mathcal{I})$. Algorithms are FIC3, FICt, JADE, JLA4 and JLAH (from left to right).

obtained from 100 runs, where $\Pi_{\mathcal{I}}$ (resp. $\Pi_{\widehat{\mathcal{I}}}$) is the projection matrix onto the true non-Gaussian subspace $\mathcal{I}$ (resp. the estimated one $\widehat{\mathcal{I}}$).

Although we did not prove theoretically, JADE could find the non-Gaussian subspace $\mathcal{I}$ in all these examples. Unfortunately, the performance of the proposed algorithm JLAH was worse than that of the previous version JLA4 for the simple data (A) and was on par for the sub-Gaussian data (C). However, for data (B) and (D) which contain super-Gaussian structures, the Hessian version JLAH outperformed the cumulant one JLA4. Moreover, JLAH was much more robust than JLA4. The proposed algorithm (JLAH) missed only one case, while the latter (JLA4) failed to estimate the index space $\mathcal{I}$ many times ((B)7%, (C)21% and (D)30%).

5 Conclusions

In this paper, we proposes a new *linear* method to identify the non-Gaussian subspace based on Hessian of the characteristic function. In our numerical experiments, the proposed algorithm was more robust and efficient than the previous version with the cumulant tensor when data contain super-Gaussian structures. Global consistency of the method was also proved in a more general framework. Further research should be done on selection of the vectors ζ_k to improve its performance. Other examples of joint low-rank approximation procedure can also be interesting.

References

1. G. Blanchard, M. Kawanabe, M. Sugiyama, V. Spokoiny and K.-R. Müller. In search of non-Gaussian components of a high-dimensional distribution. submitted to *Journal of Machine Learning Research*.
2. J.-F. Cardoso and A. Souloumiac. Blind beamforming for non Gaussian signals. *IEE Proceedings-F*, 140(6):362–370, 1993.
3. J.H. Friedman and J.W. Tukey. A projection pursuit algorithm for exploratory data analysis. *IEEE Transactions on Computers*, 23(9):881–890, 1975.
4. M. Kawanabe. Linear dimension reduction based on the fourth-order cumulant tensor. *Proc. ICANN 2005*, LNCS 3697: 151-156, Warsaw, Poland, 2005.
5. P.J. Huber. Projection pursuit. *The Annals of Statistics*, 13:435–475, 1985.
6. A. Hyvärinen. Fast and robust fixed-point algorithms for independent component analysis. *IEEE Transactions on Neural Networks*, 10(3):626–634, 1999.
7. A. Hyvärinen, J. Karhunen and E. Oja. *Independent Component Analysis*. Wiley, 2001.
8. S. Roweis and L. Saul. Nonlinear dimensionality reduction by locally linear embedding. *Science*, 290(5500):2323–2326, 2000.
9. J.B. Tenenbaum, V. de Silva and J.C. Langford. A global geometric framework for nonlinear dimensionality reduction. *Science*, 290(5500):2319–2323, 2000.
10. F.J. Theis. A new concept for separability problems in blind source separation. *Neural Computation*, 16: 1827-1850, 2004
11. F.J. Theis. Multidimensional independent component analysis using characte ristic functions. *Proc. EUSIPCO 2005*, Antalya, Turkey, 2005.
12. F.J. Theis and M. Kawanabe. Uniqueness of non-Gaussian component analysis. submitted to ICA 2006.

Independent Vector Analysis: An Extension of ICA to Multivariate Components

Taesu Kim[1,2], Torbjørn Eltoft[3], and Te-Won Lee[1]

[1] Institute for Neural Computation, UCSD, USA
{taesu, tewon}@ucsd.edu
[2] Department of BioSystems, KAIST, Korea
[3] Department of Physics, University of Tromsø, Norway
torbjorn.eltoft@phys.uit.no

Abstract. In this paper, we solve an ICA problem where both source and observation signals are multivariate, thus, vectorized signals. To derive the algorithm, we define dependence between vectors as Kullback-Leibler divergence between joint probability and the product of marginal probabilities, and propose a vector density model that has a variance dependency within a source vector. The example shows that the algorithm successfully recovers the sources and it does not cause any permutation ambiguities within the sources. Finally, we propose the frequency domain blind source separation (BSS) for convolutive mixtures as an application of IVA, which separates 6 speeches with 6 microphones in a reverberant room environment.

1 Introduction

Independent component analysis (ICA) is proposed as a method to find statistically independent sources from mixture observations by utilizing higher-order statistics [1]. In its simplest form, the ICA model assumes linear, instantaneous mixing without sensor noise, the number of sources being equal to the number of sensors, and so on. Before considering these assumptions, there is more fundamental assumption, which is that *every component is independent of the others.* Of course, it is. However, what if the sources are multivariate or vectorized signal? Let's consider some examples such as complex-valued signal, time-frequency representation of audio signal, color image signal, etc. Are the components still independent? Usually not. Elements within a source vector are sometimes correlated or sometimes uncorrelated but dependent.

In this paper, we consider an algorithm for solving the following problem.

Independent Vector Analysis (IVA)
Given observations $\mathbf{x}_i$,

$$\mathbf{x}_i = \sum_{j}^{L} \mathbf{a}_{ij} \circ \mathbf{s}_j \tag{1}$$

finding source vectors $\mathbf{s}_j$ by

J. Rosca et al. (Eds.): ICA 2006, LNCS 3889, pp. 165–172, 2006.

$$\mathbf{s}_i \approx \hat{\mathbf{s}}_i = \sum_j^M \mathbf{w}_{ij} \circ \mathbf{x}_j \tag{2}$$

where $\circ$ denotes element-wise product, and L and M is the number of sources and observations, respectively. Notation used in this paper is defined in the footnote.[1]

Assumptions
1. Elements of a source vector are mutually independent of elements of the other source vectors.
2. Within a source vector, the elements are highly dependent on the others.
3. The number of sources is less than or equal to the number of observations.

Easily, one can treat this problem as several numbers of ICA problems, because (1) can be rewritten as

$$\mathbf{x}^{(1)} = A^{(1)}\mathbf{s}^{(1)}, \qquad \mathbf{x}^{(2)} = A^{(2)}\mathbf{s}^{(2)}, \qquad \cdots, \qquad \mathbf{x}^{(D)} = A^{(D)}\mathbf{s}^{(D)} \tag{3}$$

However, once the ICA algorithm is separately applied to each element of a vector, the elements of the recovered source vectors would be randomly ordered. In this case, afterwards, it should be decided which component belongs to which source vector. It causes another clustering problem, which is not easy to solve when the number of sources is large. Instead of applying ICA separately, we tackle the problem by defining dependence between multivariate components and deriving an algorithm for the IVA problem directly.

2 Method

2.1 Objective Function

In order to separate multivariate components from multivariate observations, we need to define the objective function for multivariate random variables. Here, we define Kullback-Leibler divergence between two functions as the measure of dependence. One is an exact joint probability density function, $p(\mathbf{s}_1, \cdots, \mathbf{s}_L)$ and the other is a nonlinear function which is the product of approximated marginal probability distribution functions, $\prod_i q(\mathbf{s}_i)$.

$$\begin{aligned}\mathcal{C} &= \mathcal{KL}\left(p(\mathbf{s}_1, \cdots, \mathbf{s}_L) \,\|\, \prod_i q(\mathbf{s}_i)\right) \\ &= const. + \sum_d \log|\det A^{(d)}| - \sum_i E_{\mathbf{s}_i} \log q(\mathbf{s}_i)\end{aligned} \tag{4}$$

[1] **Notation.** We use lower-cased, bold-faced letters to denote vector variables, upper cased letters to denote matrix variables, e.g. $\mathbf{s}_i = [s_i^{(1)}, \cdots, s_i^{(D)}]^\mathsf{T}$. $\mathbf{x}_i = [x_i^{(1)}, \cdots, x_i^{(D)}]^\mathsf{T}$, and $\mathbf{a}_{ij} = [a_{ij}^{(1)}, \cdots, a_{ij}^{(D)}]^\mathsf{T}$, where $a_{ij}^{(d)}$ is the ith row, jth column element of the dth mixing matrix $A^{(d)}$.

Note that the random variables in above equations are multivariate. The interesting parts of this objective function are that each source is multivariate and it would be minimized when dependency between the source vectors is removed, but dependency between the components of each vector does not need to be removed. Therefore, the objective function preserves the inherent dependency within each source vector, although it removes dependency between the source vectors.

2.2 Learning Algorithm: A Gradient Descent Method

Now that we have defined the objective function for IVA, derivation of the learning algorithm is straightforward. Here, we are using a gradient descent method to minimize the objective function. By differentiating the objective function $\mathcal{C}$ with respect to the coefficients of unmixing matrices $w_{ij}^{(d)}$, we can derive the learning rule as follows.

$$\begin{aligned} \Delta w_{ij}^{(d)} &= -\frac{\partial \mathcal{C}}{\partial w_{ij}^{(d)}} \\ &= a_{ji}^{(d)} - E\varphi^{(d)}\left(\hat{\mathbf{s}}_i^{(1)}, \cdots, \hat{\mathbf{s}}_i^{(D)}\right)\mathbf{x}_j^{(d)} \end{aligned} \tag{5}$$

By multiplying scaling matrices, $W^{(d)^{\mathsf{T}}}W^{(d)}$, the natural gradient learning rule [2], which is well known as a fast convergence method, can be obtained as

$$\Delta w_{ij}^{(d)} = \sum_{l=1}^{L}\left(I_{il} - E\varphi^{(d)}\left(\hat{\mathbf{s}}_i^{(1)}, \cdots, \hat{\mathbf{s}}_i^{(D)}\right)\hat{\mathbf{s}}_l^{(d)}\right) w_{lj}^{(d)} \tag{6}$$

where I_{il} is one when $i = l$, otherwise zero, and a multivariate score function is given by

$$\varphi^{(d)}\left(\hat{\mathbf{s}}_i^{(1)}, \cdots, \hat{\mathbf{s}}_i^{(D)}\right) = -\frac{\partial \log q\left(\hat{\mathbf{s}}_i^{(1)}, \cdots, \hat{\mathbf{s}}_i^{(D)}\right)}{\partial \hat{s}_i^{(d)}} \tag{7}$$

3 Vector Density Model

In order to minimize the objective function, defining an optimal form of the function $q(\cdot)$ as an approximated marginal probability density function is the most critical part. Here, the function $q(\cdot)$ has to be characterized as a vector density model that has dependency within a source vector. We define a vector density model as a scale mixture of Gaussians distribution.

3.1 Scale Mixture of Gaussians Distribution

Suppose that there is a D-dimensional random variable, which is defined by

$$\mathbf{s} = \sqrt{v}\,\mathbf{z} + \mu, \tag{8}$$

where v is a scalar random variable, $\mathbf{z}$ is a D-dimensional random variable, and μ is a deterministic bias. Here, the random variable, $\mathbf{z}$, has a Gaussian distribution with mean 0 and covariance matrix Σ.

$$\mathbf{z} \sim \mathcal{N}(0, \Sigma) = \frac{1}{(2\pi)^{D/2}|\Sigma|^{1/2}} \exp\left(-\frac{\mathbf{z}^{\mathsf{T}}\Sigma^{-1}\mathbf{z}}{2}\right) \tag{9}$$

Obviously, the random variable, v, is non-negative. We assume that v has a Gamma distribution, which is a commonly used distribution for non-negative random variables.

$$v \sim \mathcal{G}(\alpha, \lambda) = \frac{\lambda^{\alpha} v^{\alpha-1}}{\Gamma(\alpha)} \exp(-\lambda v), \tag{10}$$

where α and λ are the parameters of a Gamma distribution, and $\Gamma(\cdot)$ is a complete Gamma function. Then, the random variable $\mathbf{s}$ given v has a Gaussian distribution. The mean and variance of this distribution are $E\mathbf{s} = \sqrt{v}E\mathbf{z} + \mu = \mu$ and $E(\mathbf{s}-\mu)(\mathbf{s}-\mu)^{\mathsf{T}} = \sqrt{v}Ezz^{\mathsf{T}}\sqrt{v} = v\Sigma$, respectively.

$$\mathbf{s}|v \sim \mathcal{N}(\mu, v\Sigma) = \frac{1}{(2\pi v)^{D/2}|\Sigma|^{1/2}} \exp\left(-\frac{(\mathbf{s}-\mu)^{\mathsf{T}}\Sigma^{-1}(\mathbf{s}-\mu)}{2v}\right) \tag{11}$$

In this model, each component of $\mathbf{s}$ is not only correlated to others, but also has variance dependency generated by v. Even though we assume the covariance matrix Σ is identity, that is, each element of $\mathbf{s}$ is uncorrelated, it is dependent on the others. We can obtain probability distribution function of variance dependent random variable $\mathbf{s}$, by integrating joint distribution of $\mathbf{s}$ and v over v.

$$p(\mathbf{s}) = \int_0^{\infty} p(\mathbf{s}|v)p(v)dv \tag{12}$$

Let $\delta = \sqrt{((\mathbf{s}-\mu)^{\mathsf{T}}\Sigma^{-1}(\mathbf{s}-\mu))}$ and $\gamma = \sqrt{2\lambda}$. Now, we rearrange the joint p.d.f as a form of Inverse Gaussian distribution [3] as follows.

$$\begin{aligned}(12) &= \frac{\lambda^{\alpha}}{(2\pi)^{D/2}\Gamma(\alpha)|\Sigma|^{1/2}} \frac{(2\pi)^{1/2}}{\delta} \exp(-\gamma\delta) \\ &\times \int_0^{\infty} v^{\alpha-(D-1)/2} \underbrace{\frac{\delta}{(2\pi)^{1/2}} \exp(\gamma\delta)\, v^{-3/2} \exp\left(-\frac{1}{2}\left(\frac{\delta^2}{v} + \gamma^2 v\right)\right)}_{\text{Inverse Gaussian p.d.f.}} dv \end{aligned} \tag{13}$$

Then, the integral in (13) is the $(\alpha - (D-1)/2)$-th order moment of Inverse Gaussian. Therefore, the variance dependent source p.d.f is obtained as

$$p(\mathbf{s}) = c\left((\mathbf{s}-\mu)^{\mathsf{T}}\Sigma^{-1}(\mathbf{s}-\mu)\right)^{\alpha/2-D/4} \mathcal{K}_{\alpha-D/2}\left(\sqrt{2\lambda(\mathbf{s}-\mu)^{\mathsf{T}}\Sigma^{-1}(\mathbf{s}-\mu)}\right), \tag{14}$$

where c is a normalization term and $\mathcal{K}_{\nu}(z)$ is the modified Bessel function of the second kind, which is approximated as

$$\mathcal{K}_{\nu}(z) \approx \sqrt{\frac{\pi}{2z}} e^{-z} \left(1 + \frac{4\nu^2 - 1}{8z} + \frac{(4\nu^2-1)(4\nu^2-9)}{2!(8z)^2} + \cdots\right) \tag{15}$$

3.2 Multivariate Score Function

So far, we have derived an algorithm and defined a vector density model. Finally in the algorithm, one can notice that the only difference between IVA and the conventional ICA is caused by the form of a score function. If we define the multivariate score function given in (7) as a single-variate score function, $\varphi^{(d)}\left(\hat{s}_i^{(d)}\right)$, which is a function of only one variable, the algorithm is converted to the same as the conventional ICA such as InfoMax algorithm. According to the density model we defined, we can obtain a form of a multivariate score function by differentiating log prior (14) with respect to each element of a source vector, because $q(\hat{\mathbf{s}}_i)$ in the objective function is an approximated probability density function of a source vector, that is, $q(\mathbf{s}_i) \approx p(\mathbf{s}_i)$. Therefore, we can obtain following form of a multivariate score function.

$$\varphi^{(k)}\left(\hat{s}_i^{(1)}, \cdots, \hat{s}_i^{(D)}\right) = \frac{\mathcal{K}_{\alpha-D/2-1}(\delta)}{\mathcal{K}_{\alpha-D/2}(\delta)} \frac{\hat{s}_i^{(d)}}{\delta} = \xi(\delta)\frac{\hat{s}_i^{(d)}}{\delta} \tag{16}$$

where $\xi(\delta) \approx 1$ for large δ. To obtain a simplified score function, we may approximate the Bessel function in (14) up to the 1st order, which results the following function.

$$\varphi^{(k)}\left(\hat{s}_i^{(1)}, \cdots, \hat{s}_i^{(D)}\right) \approx \left(\frac{D+1-2\alpha}{2\delta} + 1\right)\frac{\hat{s}_i^{(d)}}{\delta} \tag{17}$$

Although it is possible to estimate the mean vector μ and the covariance matrix Σ while the algorithm learns. We would, in this paper, fix them to zero mean and unit variance, and assume that the elements in a source vector are uncorrelated. Thus, simply $\delta = \sqrt{\sum_d \left|\hat{s}_i^{(d)}\right|^2}$. Although we propose above 2 forms of multivariate score functions, we believe that another form of a multivariate score function will be still possible by choosing a different vector density model that has different dependencies.

4 Example

We verified our algorithm with artificially generated signals. First, we generated 3 i.i.d. Gaussian random vector signals, which were 4 dimensional vectors. Then, the same amplitude modulation was applied to the elements of each vector signal as follows.

$$\mathbf{s}_2(t) = \cos(2\pi t/3)\,\mathbf{z}_1(t) \tag{18}$$

$$\mathbf{s}_1(t) = \sin(2\pi t)\,\mathbf{z}_2(t) \tag{19}$$

$$\mathbf{s}_3(t) = \mathbf{U}(\sin(2\pi t/3))\,\mathbf{z}_3(t), \tag{20}$$

where $\mathbf{z}_i$ is 4 dimensional i.i.d. Gaussian random vector, and $\mathbf{U}(\cdot)$ denotes a unit step function. Mixing matrices were randomly generated. Fig. 1 shows the original sources, observations signals, and recovered sources by both of ICA and IVA.

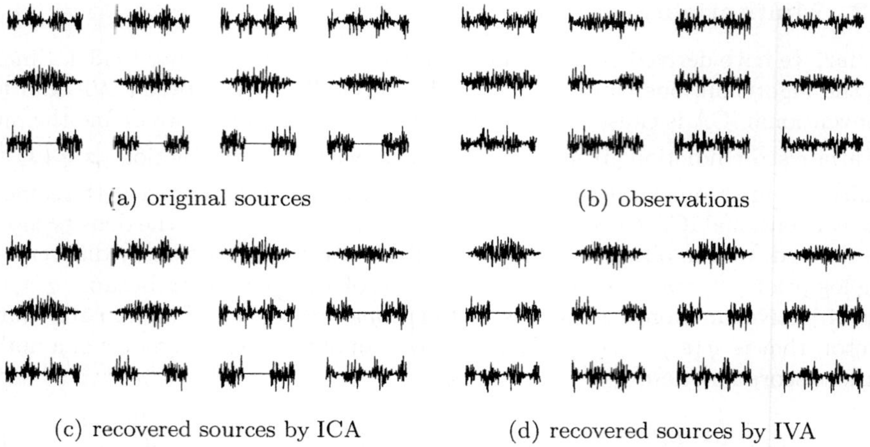

Fig. 1. The original sources, observations, and recovered sources by ICA and IVA. Each row is corresponding to a single source vector, which is 4 dimensional in the example. In contrast to ICA, IVA does not suffer the inter-element permutation problem as well as it separates sources properly.

Each row is corresponding to a single source vector, which is 4 dimensional in the example. As shown in the figure, ICA solution disorders elements in a source vector, whereas IVA does not suffer the inter-element permutation problem as well as it separates the sources properly. Following matrices show the product of the unmixing matrix and the mixing matrix, which should be identity matrix with permutations. Those obtained by ICA was

$$W^{(1)}A^{(1)} = \begin{bmatrix} 0.031 & -0.034 & \boxed{1.557} \\ 0.012 & \boxed{1.3695} & 0.039 \\ \boxed{1.395} & -0.081 & 0.019 \end{bmatrix} \quad W^{(2)}A^{(2)} = \begin{bmatrix} \boxed{1.360} & -0.065 & 0.060 \\ 0.054 & \boxed{-1.399} & 0.013 \\ -0.005 & 0.021 & \boxed{-1.536} \end{bmatrix}$$

$$W^{(3)}A^{(3)} = \begin{bmatrix} 0.018 & \boxed{1.422} & -0.019 \\ \boxed{-1.391} & 0.058 & 0.075 \\ 0.013 & -0.054 & \boxed{-1.538} \end{bmatrix} \quad W^{(4)}A^{(4)} = \begin{bmatrix} 0.011 & \boxed{1.357} & 0.018 \\ -0.002 & 0.001 & \boxed{-1.557} \\ \boxed{-1.428} & -0.047 & 0.029 \end{bmatrix}$$

In contrast to ICA, IVA provided a well-ordered solution, which has the same permutations in a source vector as follows.

$$W^{(1)}A^{(1)} = \begin{bmatrix} -0.006 & \boxed{-2.388} & -0.048 \\ -0.011 & 0.099 & \boxed{-2.592} \\ \boxed{2.370} & -0.079 & 0.092 \end{bmatrix} \quad W^{(2)}A^{(2)} = \begin{bmatrix} 0.022 & \boxed{-2.395} & 0.036 \\ 0.015 & 0.066 & \boxed{2.610} \\ \boxed{-2.306} & 0.1270 & -0.076 \end{bmatrix}$$

$$W^{(3)}A^{(3)} = \begin{bmatrix} -0.009 & \boxed{2.409} & 0.007 \\ -0.011 & -0.077 & \boxed{-2.609} \\ \boxed{-2.386} & 0.033 & 0.103 \end{bmatrix} \quad W^{(4)}A^{(4)} = \begin{bmatrix} -0.003 & \boxed{-2.338} & -0.009 \\ 0.027 & -0.083 & \boxed{2.587} \\ \boxed{2.421} & -0.012 & 0.013 \end{bmatrix}$$

In the above matrices, the values covered by rectangles to the other values ratio was used to calculate the performance measure. ICA and IVA result 28.5dB and 30dB, respectively.

5 Application to the Frequency Domain BSS

We applied the proposed IVA algorithm to separate convolutive mixture in the frequency domain, because the convolution is equivalent to multiplication at each frequency bin, which is the same as the model given by (1). Although one can use the conventional ICA algorithm to separate each frequency bin separately, it causes another problem which is called the frequency permutation problem. Thus, the permutations of separating matrices at each frequency should be corrected so that the separated signal in the time domain is reconstructed properly. Various algorithms have been proposed to solve the permutation problem, e.g. method that limits the filter length in the time domain [4], uses direction of arrival estimation [5], and uses inter-frequency correlation [6]. Although these algorithms perform well in some cases, they are sometimes very sensitive to the parameters or mixing conditions. However, IVA algorithm we proposed in this paper does not suffer the permutation problem at all as well as it separates sources properly.

We tested the proposed algorithm to separate 6 speeches with 6 microphones in a reverberant room environment. In this experiment, we used 8kHz sampling rate, a 2048 point FFT and a hanning window to convert time domain signal to the frequency domain. The length of window was 2048 samples and shift size was 512 samples. The condition of the room was illustrated in Fig. 2(a), and the separated sources are shown in Fig. 2(b). The improvement of signal to interference ratio (SIR) was 18dB. More intensive experiments are included in our web site [2] and another work [7].

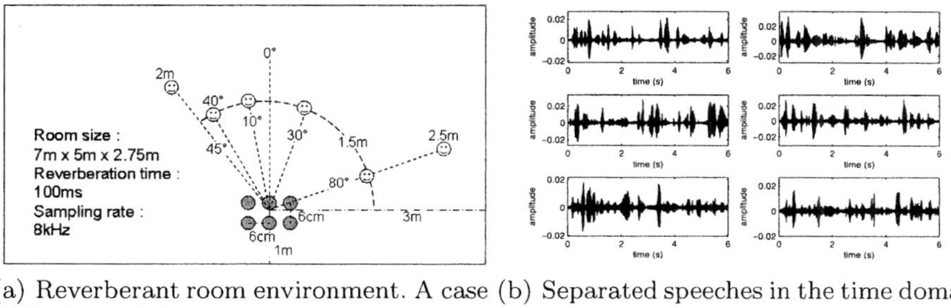

(a) Reverberant room environment. A case of 6 mics and 6 sources (b) Separated speeches in the time domain

Fig. 2. Room environment and the separated speeches. 2048 sample sized hanning window and 2048 FFT point was used. SIR improvement was approximately 18dB.

[2] *http://ergo.ucsd.edu/~taesu/source_separation.html*

6 Conclusions

We have extended the conventional ICA problem to multivariate components, which we termed IVA. While ICA algorithm has a single-variate score function, IVA algorithm has a multivariate score function, which is caused by higher-order dependency within source vectors. To model a vector density, we have used scale mixture of Gaussians distribution, which models variance dependency. The results have shown that the proposed algorithm successfully recovers the sources not only in a simple example but also real world problem such as frequency domain BSS. Further, researches on various kinds of higher-order dependency models and multivariate score functions would be important to separate multivariate components.

References

1. Hyvärinen, A., Oja, E.: Independent Component Analysis. John Wiley and Sons (2002)
2. Amari, S.I., Cichocki, A., Yang, H.H.: A new learning algorithm for blind signal separation. In: Adv. Neural information Processing Systems. Volume 8. (1996)
3. Barndorff-Nielsen, O.E.: Normal inverse gaussian distributions and stochastic volatility modeling. Scand. J. Statist. **24** (1997) 1–13
4. Parra, L., Spence, C.: Convolutive blind separation of non-stationary sources. IEEE Trans. Speech Audio Processing **8** (2000) 320–327
5. Ikram, M.Z., Morgan, D.R.: A beamforming approach to permutation alignment for multichannel frequency-domain blind speech separation. In: Proc. IEEE Int. Conf. on Acoustics, Speech, and Signal Processing. (2002) 881–884
6. Murata, N., Ikeda, S., Ziehe, A.: An approach to blind source separation based on temporal structure of speech signals. Neurocomputing **41** (2001) 1–24
7. Kim, T., Attias, H., Lee, S.Y., Lee, T.W.: Frequency domain blind source separation based on variance dependencies. In: Proc. IEEE Int. Conf. on Acoustics, Speech, and Signal Processing. (2006)

A One-Bit-Matching Learning Algorithm for Independent Component Analysis*

Jinwen Ma[1,2], Dengpan Gao[1], Fei Ge[1], and Shun-ichi Amari[2]

[1] Department of Information Science, School of Mathematical Sciences and LMAM, Peking University, Beijing 100871, China
[2] Laboratory of Mathematical Neuroscience, RIKEN Brain Science Institute, Wako-shi, Saitama 351-0198, Japan
jwma@math.pku.edu.cn

Abstract. Independent component analysis (ICA) has many practical applications in the fields of signal and image processing and several ICA learning algorithms have been constructed via the selection of model probability density functions. However, there is still a lack of deep mathematical theory to validate these ICA algorithms, especially for the general case that super- and sub-Gaussian sources coexist. In this paper, according to the one-bit-matching principle and by turning the de-mixing matrix into an orthogonal matrix via certain normalization, we propose a one-bit-matching ICA learning algorithm on the Stiefel manifold. It is shown by the simulated and audio experiments that our proposed learning algorithm works efficiently on the ICA problem with both super- and sub-Gaussian sources and outperforms the extended Infomax and Fast-ICA algorithms.

1 Introduction

Independent component analysis (ICA) [1]-[2] aims to blindly separate some independent sources $\mathbf{s}$ from their linear mixture $\mathbf{x} = \mathbf{A}\mathbf{s}$ via

$$\mathbf{y} = \mathbf{W}\mathbf{x}, \quad \mathbf{x} \in \mathbb{R}^m, \quad \mathbf{y} \in \mathbb{R}^n, \quad \mathbf{W} \in \mathbb{R}^{m \times n}, \tag{1}$$

where $\mathbf{A}$ is a mixing matrix, and $\mathbf{W}$ is the de-mixing matrix to be estimated. For simplicity of analysis, the number of mixed signals is required to be equal to the number of source signals, i.e., $m = n$, and $\mathbf{A}$ is an $n \times n$ nonsingular matrix. Although the ICA problem has been studied from different perspectives [3]-[5], it can be typically solved by minimizing the following objective function:

$$D = -H(\mathbf{y}) - \sum_{i=1}^{n} \int p_{\mathbf{W}}(y_i; \mathbf{W}) \log p_i(y_i) dy_i, \tag{2}$$

where $H(\mathbf{y}) = -\int p(\mathbf{y}) \log p(\mathbf{y}) d\mathbf{y}$ denotes the entropy of $\mathbf{y}$, $p_i(y_i)$ denotes the pre-determined model probability density function (pdf), and $p_{\mathbf{W}}(y_i; \mathbf{W})$ denotes the probability distribution on $\mathbf{y} = \mathbf{W}\mathbf{x}$.

* This work was supported by the Natural Science Foundation of China for Project 60471054.

J. Rosca et al. (Eds.): ICA 2006, LNCS 3889, pp. 173–180, 2006.

In the literature, how to choose the model pdf's $p_i(y_i)$ remains a key issue for the ICA algorithms using the objective function Eq.(2). In general, any gradient descent learning algorithm, such as the relative or natural gradient algorithms [3]-[4], can work only in the cases that the components of $\mathbf{s}$ are either all super-Gaussians or all sub-Gaussians. Recently, many new algorithms (e.g., the extended Infomax algorithm [6] and the Fast-ICA algorithm [7]) have been proposed to solve the general ICA problem, but their theoretical foundations are yet unclear. In order to solve the general ICA problem, Xu et al. [8] proposed the one-bit-matching conjecture which states that "all the sources can be separated as long as there is a one-to-one same-sign-correspondence between the kurtosis signs of all source pdf's and the kurtosis signs of all model pdf's". Recently, Liu et al. [9] proved this conjecture by globally minimizing the objective function under certain assumptions on the model pdf's. Ma et al. [10] further proved the conjecture by locally minimizing the same objective function on the two-source ICA problems. It is generally believed that the one-bit-matching condition can serve as a reasonable principle for the design of the model pdf's. On the other hand, if the observed $\mathbf{x}$ and the output $\mathbf{y}$ are both normalized with zero mean and unit covariance matrix, the de-mixing matrix becomes orthogonal, which can be learned on the Stiefel manifold.

In this paper, under the condition that the model pdf's are designed according to the one-bit-matching principle, with the observed $\mathbf{x}$ and the output $\mathbf{y}$ being properly normalized, we propose a gradient-type ICA learning algorithm on the Stiefel manifold, which we call as one-bit-matching learning algorithm. It is shown by the simulated and audio experiments that the proposed algorithm works efficiently on the general blind source separation problems and outperforms the typical existing ICA algorithms.

2 The One-Bit-Matching Learning Algorithm

We start to introduce the Stiefel manifold. Roughly, the Stiefel manifold $V_{n,p}$ consists of n-by-p "tall skinny" orthogonal matrices. That is, the p column vectors of each matrix in $V_{n,p}$ are pair-wised orthogonal in $\mathbb{R}^n$. Here, we need only to consider the special Stiefel manifold $V_{n,n}$, i.e., the orthogonal group O_n consisting of n-by-n orthogonal matrices. For a smooth function $F(\mathbf{Z})$ on the Stiefel manifold O_n, i.e., $\mathbf{Z} \in O_n$, with the canonical Euclidean metric, its gradient on the manifold is computed by

$$\nabla F = F_{\mathbf{Z}} - \mathbf{Z} F_{\mathbf{Z}}^T \mathbf{Z}, \tag{3}$$

where $F_{\mathbf{Z}}$ is the conventional gradient of $F(\mathbf{Z})$ with respect to the matrix $\mathbf{Z}$. This gradient is consistent with the natural Riemannian gradient on the Stiefel manifold from information geometry.

We further pre-whiten the observed $\mathbf{x}$ and the output $\mathbf{y}$ so that the de-mixing matrix $\mathbf{W}$ can only be orthogonal, i.e, on the Stiefel manifold O_n. Clearly, we can easily pre-whiten the observed $\mathbf{x}$ such that

$$E(\mathbf{x}) = 0, \qquad\qquad E\mathbf{x}\mathbf{x}^T = \mathbf{I}_n, \tag{4}$$

where $\mathbf{I}_n$ is the $n \times n$ identity matrix. With each matrix $\mathbf{W}$, we can also pre-whiten the output $\mathbf{y} = \mathbf{W}\mathbf{x}$ such that

$$E(\mathbf{y}) = 0, \qquad E\mathbf{y}\mathbf{y}^T = \mathbf{I}_n. \tag{5}$$

In this way, we have

$$\mathbf{I}_n = E(\mathbf{y}\mathbf{y}^T) = \mathbf{W}E(\mathbf{x}\mathbf{x}^T)\mathbf{W}^T = \mathbf{W}\mathbf{W}^T. \tag{6}$$

Thus, $\mathbf{W}\mathbf{W}^T = \mathbf{I}_n$. That is, $\mathbf{W}$ must be an orthogonal matrix. Therefore, if we can pre-whiten or normalize the observed $\mathbf{x}$ and the output $\mathbf{y}$ during each phase of the learning process, the resulted $\mathbf{W}$ should be an orthogonal matrix. Therefore, we can solve it on the Stiefel manifold.

We now revisit the objective function defined in Eq.(2). Suppose that the observed $\mathbf{x}$ and the output $\mathbf{y}$ are both pre-whitened. Then, the required de-mixing $\mathbf{W}$ should be orthogonal. So, we can search the feasible solution $\mathbf{W}$ of the ICA problem via minimizing the objective function of $\mathbf{W}$ on the Stiefel manifold O_n. For the design of model pdf's, we use the one-bit-matching condition. Suppose that p is the number of super-Gaussian sources in the ICA problem. The model pdfs of sub- and super-Gaussians are selected as

$$p_{super}(u) = \frac{1}{\pi}\text{sech}(u), \qquad p_{sub}(u) = \frac{1}{2}[p_{N(1,1)}(u) + p_{N(-1,1)}(u)],$$

respectively, where $p_{N(\mu,\sigma^2)}(u)$ is the Gaussian probability density with mean μ and variance σ^2, and sech$(\cdot)$ is the hyperbolic secant function. That is, the first p model pdf's are selected as $p_{super}(u)$, while the rest $n-p$ model pdf's are selected as $p_{sub}(u)$. In this way, the conventional gradient of the objective function can be computed as follows.

We let $\mathbf{V} = (v_1, v_2, \cdots, v_n)^T$ be an n-dim vector. For each observed $\mathbf{x}$ and the corresponding output $\mathbf{y}$ via the relation $\mathbf{y} = \mathbf{W}\mathbf{x}$, we define

$$v_i = -\tanh(y_i), \quad \text{for} \quad i = 1, \cdots, p;$$
$$v_i = \tanh(y_i) - y_i, \quad \text{for} \quad i = p+1, \cdots, n.$$

Then, the adaptive gradient $J_{\mathbf{W}}$ of the objective function Eq. (2) with respect to $\mathbf{W}$ is simplified as

$$J_{\mathbf{W}} = -\mathbf{W} - \mathbf{V}\mathbf{x}^T. \tag{7}$$

Given Eq.(7), we can construct the one-bit-matching learning algorithm as a local gradient-descent learning algorithm of $\mathbf{W}$ on the Stiefel manifold O_n as follows.

$$\triangle\mathbf{W} = -\eta(J_{\mathbf{W}} - \mathbf{W}J_{\mathbf{W}}^T\mathbf{W}) = \eta(\mathbf{V}\mathbf{x}^T - \mathbf{W}\mathbf{x}\mathbf{V}^T\mathbf{W}), \tag{8}$$

where $\eta > 0$ is the learning rate parameter which is generally selected as a small positive constant.

Since $J_{\mathbf{W}}$ is just the adaptive gradient of the objective function, this algorithm is adaptive. As $\mathbf{W}$ keeps an orthogonal matrix during the learning process, the output $\mathbf{y}$ will be always normalized or whitened. Therefore, we need only to pre-whiten the observed $\mathbf{x}$ at the beginning of the algorithm. Certainly, we can establish the batch gradient learning algorithm on the Stiefel manifold with the batch gradient of the objective function.

3 Experimental Results

In order to test our one-bit-matching learning algorithm, we conducted several simulated and audio experiments on three source separation problems: (i). mixed super-Gaussian and sub-Gaussian ones; (ii). uniform noises which are all sub-Gaussian; (iii). audio samples which are all super-Gaussian. We also compared it with the extended Informax and Fast-ICA algorithms.

3.1 On Separating Mixed Super-Gaussian and Sub-Gaussian Sources

We began to consider the ICA problem of seven independent sources in which there are four super-Gaussian sources generated from one Exponential distribution $E(0.5)$, one Chi-square distribution $\chi^2(6)$, one Gamma distribution $\gamma(1,4)$ and one F-distribution $F(10,50)$, respectively, and three sub-Gaussian sources generated from two β distributions $\beta(2,2)$, $\beta(0.5,0.5)$, and one Uniform distribution $U([0,1])$, respectively. From each source or distribution, 100000 i.i.d. samples were generated to form a source. Accordingly, these samples were further pre-whitened.

The first set of linearly mixed signals was generated from these seven source signals via a random orthogonal mixing matrix $\mathbf{A}_1$. We implemented the one-bit-matching learning algorithm ($p = 4, n = 7$) on the first set of linearly mixed signals with the learning rate being selected as $\eta = 0.001$ and the initial $\mathbf{W}$ being set as a randomly generated orthogonal matrix. The one-bit-matching learning algorithm operated adaptively and was stopped after 100000 iterations to ensure the fulfilment of separation.

The result of the one-bit-matching learning algorithm on the first linearly mixed signal set is given by Eq. (9). As a feasible solution of the ICA problem, the obtained $\mathbf{W}$ will make $\mathbf{WA} = \mathbf{\Lambda P}$ be satisfied or approximately satisfied to a certain extent, where $\mathbf{\Lambda P} = diag[\lambda_1, \lambda_2, \cdots, \lambda_n]$ with each $\lambda_i \neq 0$, and $\mathbf{P}$ is a permutation matrix. Since $\mathbf{A}$ was selected as an orthogonal matrix, $\mathbf{WA}$ should be just a permutation matrix up to sign indeterminacy.

$$\mathbf{WA_1} = \begin{bmatrix} \mathbf{1.0000} & 0.0033 & 0.0027 & -0.0043 & -0.0020 & -0.0044 & -0.0043 \\ 0.0026 & 0.0058 & -\mathbf{0.9998} & -0.0156 & 0.0031 & -0.0007 & -0.0032 \\ 0.0044 & 0.0032 & -0.0156 & \mathbf{0.9998} & -0.0003 & 0.0079 & 0.0054 \\ 0.0032 & -\mathbf{0.9999} & -0.0058 & 0.0032 & 0.0006 & -0.0128 & 0.0008 \\ -0.0020 & -0.0006 & -0.0031 & -0.0004 & -\mathbf{1.0000} & 0.0027 & -0.0015 \\ -0.0044 & 0.0128 & 0.0006 & 0.0079 & -0.0027 & -\mathbf{0.9999} & -0.0007 \\ -0.0043 & -0.0008 & 0.0031 & 0.0055 & 0.0015 & 0.0007 & -\mathbf{1.0000} \end{bmatrix} \quad (9)$$

For comparison, we also ran the extended Infomax algorithm [5] (a kind of natural or relative gradient learning with a switch criterion) on this set of linearly mixed signals and obtained the separation result given by Eq. (10).

$$\mathbf{WA}_1 = \begin{bmatrix} 0.0148 & -\mathbf{0.7588} & 0.0085 & -0.0005 & -0.0241 & -0.0189 & 0.0088 \\ 0.0222 & 0.0167 & -0.0109 & 0.0135 & -\mathbf{1.4220} & -0.0111 & 0.0093 \\ 0.0088 & -0.0042 & -\mathbf{0.7532} & -0.0197 & -0.0133 & 0.0336 & 0.0103 \\ -0.0144 & -0.0141 & 0.0037 & -0.0333 & -0.0280 & -0.0141 & \mathbf{1.4943} \\ -\mathbf{0.8065} & 0.0161 & -0.0018 & -0.0146 & -0.0581 & -0.0465 & 0.0777 \\ 0.0176 & -0.0197 & -0.0057 & 0.0288 & -0.0210 & -\mathbf{1.4393} & 0.0343 \\ 0.0001 & -0.0353 & 0.0284 & \mathbf{0.7675} & 0.0004 & -0.0537 & -0.0017 \end{bmatrix} \quad (10)$$

From the above two tables, it can be found that the one-bit-matching learning algorithm is much better than that of the extended Infomax algorithm. Precisely, we calculated the performance index (introduced in [3]) defined by

$$PI = \sum_{i=1}^{n}(\sum_{j=1}^{n} \frac{|r_{ij}|}{\max_k |r_{ik}|} - 1) + \sum_{j=1}^{n}(\sum_{i=1}^{n} \frac{|r_{ij}|}{\max_k |r_{kj}|} - 1),$$

where $\mathbf{R} = (r_{ij})_{n \times n} = \mathbf{WA}$. For a perfect separation, this index should be zero. Actually, the performance indexes of the one-bit-matching learning and extended Infomax algorithms are 0.3411 and 1.6399, respectively, which quantitatively shows that the one-bit-matching learning algorithm is much better than the extended Infomax. Moreover, we implemented the Fast-ICA algorithm on this linearly mixed signal set and obtained the separation result with the performance index being 0.3028, which is slightly better than that of the one-bit-matching learning algorithm.

3.2 On Separating Uniform Noises

Next, we considered the ICA problem of separating eight independent uniform noises. That is, each source was sampled from a uniform distribution and contains 100000 samples. These sources are all sub-Gaussian, being recognized as the uniform noises. Our second set of linearly mixed signals was generated from these eight uniform noises via another random orthogonal mixing matrix $\mathbf{A}_2$. The signals were further pre-whitened. On this set of linearly mixed signals, we implemented the one-bit-matching learning and extended Infomax algorithms and their results are given by Eq. (11) and Eq. (12), respectively. It was found that their performance indices are 0.1713 and 2.2776, respectively, which also shows that the one-bit-matching learning algorithm also outperforms the extended Infomax. Moreover, it was also found that the the performance index of the separation results via the Fast-ICA algorithm on this set is 0.2342, which is considerably larger than 0.1713. That is, the one-bit-matching learning algorithm also outperforms the Fast-ICA algorithm in this case.

$$\mathbf{WA_2} = \begin{bmatrix} -0.003 & 0.000 & 0.001 & -0.001 & \mathbf{1.000} & 0.000 & -0.002 & -0.003 \\ -0.002 & 0.000 & -0.002 & 0.001 & -0.002 & 0.000 & -\mathbf{1.000} & -0.001 \\ 0.000 & 0.002 & 0.000 & -\mathbf{1.000} & -0.001 & -0.004 & -0.001 & -0.002 \\ 0.001 & -0.003 & -\mathbf{1.000} & 0.000 & 0.001 & 0.000 & 0.002 & 0.002 \\ -0.003 & -\mathbf{1.000} & 0.003 & -0.002 & 0.000 & 0.001 & 0.000 & 0.001 \\ -0.002 & -0.001 & -0.002 & 0.002 & -0.003 & 0.002 & 0.001 & -\mathbf{1.000} \\ -0.001 & 0.001 & 0.000 & -0.004 & 0.000 & \mathbf{1.000} & 0.000 & 0.002 \\ \mathbf{1.000} & -0.003 & 0.001 & 0.000 & 0.003 & 0.001 & -0.002 & -0.003 \end{bmatrix} \quad (11)$$

$$\mathbf{WA_2} = \begin{bmatrix} 0.021 & -0.002 & -\mathbf{1.449} & 0.029 & -0.014 & 0.038 & -0.017 & -0.053 \\ 0.009 & -0.057 & -0.039 & -0.033 & 0.053 & \mathbf{1.430} & -0.008 & -0.022 \\ \mathbf{1.450} & -0.040 & -0.014 & -0.010 & 0.033 & 0.052 & 0.009 & -0.076 \\ 0.050 & -0.035 & -0.047 & 0.045 & -0.014 & 0.025 & -0.008 & -\mathbf{1.499} \\ -0.037 & -\mathbf{1.452} & 0.031 & -0.040 & -0.048 & 0.037 & -0.038 & -0.023 \\ -0.031 & -0.046 & -0.031 & 0.007 & -\mathbf{1.444} & -0.047 & -0.005 & -0.023 \\ -0.014 & 0.015 & 0.037 & 0.022 & 0.037 & -0.016 & \mathbf{1.443} & 0.032 \\ -0.013 & 0.056 & -0.006 & \mathbf{1.404} & 0.014 & -0.018 & 0.057 & -0.013 \end{bmatrix} \quad (12)$$

3.3 On Separating Audio Sources

Finally, we considered the ICA problem of separating 8 independent real-life audio recordings (downloaded from Barak Pearlmutter's homepage: `http://www- bcl.cs.may.ie/~bap/demos.html`). Each audio source consists of 100000 data sampled at 22050 Hz. We pre-whitened these audio sources and then linearly mixed them via an 8×8 random orthogonal mixing matrix $\mathbf{A}_3$ to form the third set of linearly mixed signals. On such a data set, we implemented the one-bit-matching algorithm, obtaining a successful separation result shown in Fig. 1.

For comparison, we also implemented the extended Infomax and Fast-ICA algorithms on the data set. It was found by the experiments that the performance indices of the one-bit-matching learning, extended Infomax and Fast-ICA algorithms are 1.2979, 2.2746, and 1.3288, respectively, which again shows that the one-bit-matching learning algorithm outperforms the extended Infomax and Fast-ICA algorithms in this case.

For further comparison, we calculated the signal-to-noise ratios (SNRs) of the output signals of the one-bit-matching, extended Infomax, and Fast-ICA learning algorithms on the data set. Their results are listed in Table 1, which again shows our proposed one-bit-matching learning algorithm outperforms the other two popular ICA learning algorithms.

In addition, extensive experiments on the different ICA problems with mixed super- and sub-Gaussian sources also showed that the one-bit-matching learning algorithm always reaches an accurate feasible solution. It was even found that as the number of sources increases, the one-bit-matching learning algorithm can still maintain a similarly good performance on the source separation problems. By comparison, we have found that the one-bit-matching learning algorithm considerably outperforms the extended Infomax algorithm in the general case. As

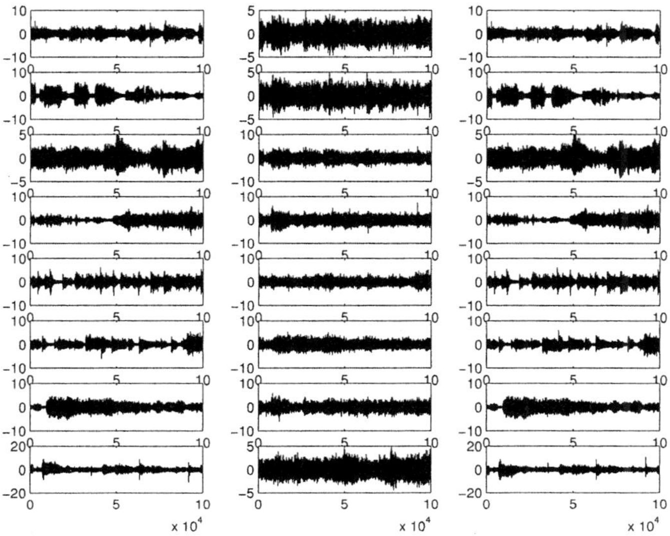

Fig. 1. The waveforms of sources signals (left column), linearly mixed signals (middle column), and output signals (right column) of the one-bit-matching learning algorithm

Table 1. The SNRs of the recovered sources of the three algorithms

	Signal-to-Noise Ratio (dB)									
Audio Source	1	2	3	4	5	6	7	8	Med.	Avg.
One-bit-matching	23.06	37.25	21.84	25.69	28.71	31.19	36.64	30.15	29.43	29.32
Extended Infomax	18.82	23.96	22.56	26.97	22.39	26.97	25.71	23.67	23.81	23.88
Fast-ICA	22.21	34.37	22.25	25.74	29.96	31.97	36.33	28.07	29.02	28.86

compared with the Fast-ICA algorithm, the one-bi-matching learning algorithm leads to a similar result in the case of mixed sub- and Super-Gaussian sources, but a better result in the case of all the sub- or super-Gaussian sources.

In practice, the number of super-Gaussian sources, p, may not be available in ceratin cases. In this situation, the one-bit-matching learning algorithm cannot work directly. However, we can implement the one-bit-matching learning algorithm on the pre-whitened observed $\mathbf{x}$ with p varying from zero to n, then there must be a feasible solution $\mathbf{W}$ with which the components of the output $\mathbf{y} = \mathbf{W}\mathbf{x}$ are independent, which can be checked by certain statistical independence test method. That is, for each p, we can check whether the n components of the output $\mathbf{y}$ by the resulted $\mathbf{W}$ are mutually independent. If they are, this $\mathbf{W}$ is just a feasible solution for the ICA problem. Otherwise, it is not a feasible solution for the ICA problem. Since the independence between the components of the output $\mathbf{y}$ is sufficient for the feasible solution of the ICA problem, we can find out the feasible solution of the ICA problem by this test and checking procedure with the one-bit-matching learning algorithm. In fact, with a certain independence test criterion, we can use the one-bit-matching learning algorithm to obtain the feasible solution for all the above three cases without knowing the number of super-Gaussian sources.

4 Conclusions

In this paper, we have investigated the ICA problem from the point of view of the one-bit-matching principle, and established an efficient one-bit-matching ICA learning algorithm based on the Stiefel manifold gradient under the condition that the number of super-Gaussian sources is known and the observed signals are pre-whitened. It is demonstrated by the simulated and audio experiments that the proposed one-bit-matching learning algorithm can solve the source separation problem of mixed super- and sub-Gaussian sources efficiently and even outperforms the existing extended Infomax and Fast-ICA learning algorithms. Moreover, with certain independence test criterion, the one-bit-matching learning algorithm can be used to solve the source separation problem without knowing the number of super-Gaussians sources.

References

1. Tong L., Inouye Y., and Liu R.: Waveform-preserving blind estimation of multiple independent sources. IEEE Trans. Signal Processing, 41(7)(1993) 2461-2470
2. Comon P.: Independent component analysis–a new concept?. Signal Processing, 36(3)(1994) 287-314
3. Bell A. and Sejnowski T.: An information-maximization approach to blind separation and blind deconvolution. Neural Computation, 7(6)(1995) 1129-1159
4. Amari S. I., Cichocki A., and Yang H.: A new learning algorithm for blind separation of sources. Advances in Neural Information Processing, 8(1996) 757-763
5. Cardoso J. F. and Laheld B.: Equivalent adaptive source separation. IEEE Trans. Signal Processing, 44(12)(1996) 3017-3030
6. Lee T. W., Girolami M., and Sejnowski T.: Independent component analysis using an extended infomax algorithm for mixed subgaussian and supergaussian sources. Neural Computation, 11(2)(1999) 417-441
7. Hyvarinen A.: Fast and robust fixed-point algorithms for independent component analysis. IEEE Trans. Neural Networks, 10(3)(1999) 626-634
8. Xu L., Cheung C. C., and Amari S. I.: Learned parametric mixture based ica algorithm. Neurocomputing, 22(1998) 69-80
9. Liu Z. Y., Chiu K. C., and Xu L.: One-bit-matching conjecture for independent component analysis. Neural Computation, 16(2)(2004) 383-399
10. Ma J., Liu Z., and Xu L.: A further result on the ICA one-bit-matching conjecture. Neural Computation, 17(2)(2005) 331-334

Blind Separation of Underwater Acoustic Signals*

Ali Mansour, Nabih Benchekroun, and Cedric Gervaise

Lab. E^3I^2, ENSIETA, 29806 Brest cedex 09, France
mansour@ieee.org, n.bench@gmail.com, gervaice@ensieta.fr
http://www.ensieta.fr, http://ali.mansour.free.fr

Abstract. In last two decades, many researchers have been involved in acoustic tomography applications. Recently, few algorithms have been dedicated to the passive acoustic tomography applications in a single input single output channel. Unfortunately, most of these algorithms can not be applied in a real situation when we have a Multi-Input Multi-Output channel. In this paper, we propose at first a realistic model of an underwater acoustic channel, then a general structure to separate acoustic signals crossing an underwater channel is proposed. Concerning ICA algorithms, many algorithms have been implemented and tested but only two algorithms give us good results. The latter algorithms minimize two different second order statistic criteria in the frequency domain. Finally, some simulations have been presented and discussed.

Keywords: Underwater acoustic applications, passive acoustic Tomography, second order statistics in frequency domain, multipath channel, sparseness or non-stationary signals.

1 Introduction

Acoustic oceanic tomography are used in many civil or military applications such as: Mapping underwater surfaces, meteorological applications (to measure the temperature, the salinity, the motion and the depth of the water), to improve sonar technology, so on. Many algorithms [1,2] have been developed to deal with active acoustic tomography. Recently, the Passive Acoustic Tomography (PAT) [3] has taken an increased importance mainly for the three following reasons: related to submarine acoustic warfare, ecological reasons (the underwater ecological system isn't disturbed since no signal is emitted) and economical and logistical reasons because there is no need for emitters.

The main drawbacks of PAT are the lack of information about the number, the positions and the natures of the emitted signals. With more than two sources many

* The authors are grateful for sustained funds provided by the French Military Center for Hydrographic & Oceanographic Studies (SHOM i.e. Service Hydrographique et Ocanographique de la Marine, Centre Militaire d'Ocanographie) under research contract CA/2003/06/CMO. The authors are grateful to Dr. M. Legris for discussions and comments.

J. Rosca et al. (Eds.): ICA 2006, LNCS 3889, pp. 181–188, 2006.

actual tomography algorithms can't give satisfactory results. Many other don't work well or at all when the emitted signals are wide band signals [4]. Some algorithms take into consideration the position of the acoustic sound emitters [5]. Typically, in real world PAT applications, underwater acoustic signals are generated by various moving sources whose number and positions are hardly (impossible to be) identified (as in the case of shoal of fish or wave noises).

Since the early of the ninetieth, Independent Component Analysis (ICA) has been considered as a set of many important signal processing tools [6]. By assuming that the unknown p emitted signals (i.e sources) are statistically independent from each other, ICA consists on retrieving a set of independent signals (output signals) from the observation of unknown mixtures of the p sources. It was proved that the output signals can be the sources up to a factor (or filter) scale and up to a permutation [7].

This paper deals with the application of ICA algorithms in PAT in order to improve and simplified the PAT algorithms as well as the processing of the received signals.

2 Channel Model, Assumptions and Background

In passive acoustic tomography (PAT) applications, the sources are obviously some signals of opportunities. Therefore, an extensive experimental study has been conducted by a research engineer in our laboratory to classify and characterize the divers recorded artificial signals (made by human activities as boats, ships or submarine noises, *etc.*) and natural signals (mainly animals sounds or noises) signals in our data base. A part of his study was of extreme important to us. In fact, according to that study, one can conclude the following facts:

- Each signal in our data base corresponds to a well identified source. These recorded signals are affected by a background ocean noise which can be considered as an Additive White Gaussian Noise (AWGN).
- Many signals are Gaussian ones or they have a very weak kurtosis.
- Almost all of the signals are non-stationary signals, however some of them have more or less periodic components as boat noises.
- Natural signals are very sparse ones and artificial ones are very noisy.

The above mentioned properties have been considered to select appropriate ICA algorithms. Once the appropriate sources (non gaussian signals) are identified and characterized, an underwater acoustic channel should be simulated in order to conduct our experimental studies.

According to [8], the sound speed in the ocean is an increasing function of temperature, salinity, and pressure, the latter being a function of depth. Since most of these later parameters depend on time as well as geographic positions and hydrographic properties of the sea, we consider a simplified model where the sound propagation speed is assumed to be a constant.

It is well known [8] that the underwater sound is produced by natural or artificial phenomena through forced mass injection leading to inhomogeneous wave equations which can be converted to frequency domain. The frequency-domain wave equation is called the Helmholtz equation. The solutions of the Helmholtz equation give us an underwater sound propagation model. A general solution of the Helmholtz equation is very difficult to be obtained. Therefore researchers use some simplified models (such as the ray theory, the mode theory, the parabolic model, the hybrid model, *etc*) according to their applications. The choice of a propagation model depends on many parameters such as wave frequency, the depth of the sea, *etc.* In our case, the ray theory was the more appropriate propagation model.

The reflected acoustic waves on the bottom of the propagation channel depend on many parameters such as the constitution and the geometrical properties of the bottom. In our model a standard sand bottom has been considered and random coefficients have been added to characterize the other unknown parameters.

The reflected acoustic waves on the top of the propagation channel, i.e. the water surface, depend on many parameters such as the wind, the wave frequency as well as the swell properties. For this reason, the water surface cann't be considered as a flat surface. Therefore the direction of the reflected acoustic wave is dispersed in the space. However in average term, the reflected acoustic wave can be considered as obtained by a flat surface with some absorption coefficients. In our model a flat surface has been considered and random coefficients have been added to characterize the other unknown parameters.

Finally to consider the acoustic propagation effect, an acoustic model proposed by Schulkin [9] was considered. According to that model, the received signal should be multiplied by a corrective coefficient p given by the following equation:

$$p = \frac{\exp\left(-\frac{\alpha r}{20}\right)}{r} \tag{1}$$

here r is the propagation distance and α stands for the Rayleigh absorption coefficient which it can be approximated by the following equation, [9]:

$$\alpha = (1 - 6.54 * 10^{-4} * P_w)\left(\frac{SAf_T f^2}{f^2 + f_T^2} + \frac{Bf^2}{f_T}\right) \tag{2}$$

where $f_T = 21.9 * 10^{\left(6 - \frac{1520}{T+273}\right)}$ (in kHz), T is the water temperature (°C), $S = 3.5\%$ is the water salinity (in the ocean $S \approx 35g/l$), P_w is the water pressure (in kg/m^2), $A = 2.34 * 10^{-6}$ and $B = 3.38 * 10^{-6}$.

3 Mathematical Model

Under some mild assumptions [2], acoustic underwater channel can be considered as a multiple paths which, in frequency domain, each of them can be defined by a complex constant gain. Let $S(n)$ denotes the p unknown sources which are statistically independent from each other. $X(n)$ is the $q \times 1$ observed vector. The relationship between $S(n)$ and $X(n)$ is given by:

$$X(n) = [\mathcal{H}(z)]S(n) + N(n) \tag{3}$$

where $\mathcal{H}(z)$ stands for the channel effect. In the case of convolutive mixture, $\mathbf{H}(z) = (h_{ij}(z))$ becomes a $q \times p$ complex polynomial matrix. In the following, we consider that the channel is a linear and causal one and that the coefficients $h_{ij}(z)$ are RIF filter. Let M denotes the degree of the channel which is the highest degree of $h_{ij}(z)$. The previous equation (3) can be rewritten as:

$$X(n) = \sum_{i=0}^{M} \mathbf{H}(i) S(n-i) + N(n) \tag{4}$$

Here $\mathbf{H}(i)$ denotes the $q \times p$ real constant matrix corresponding to the impulse response of the channel at time i and $S(n-i)$ is the source vector at time $(n-i)$.

4 Pre- and Post-processing

Many ICA algorithms have been implemented and tested during this project. Each of these algorithms has been tested using the following three steps:

- At first, we use the same (or similar) signals used by the authors of the algorithm, and we try to obtain same (or similar) results shown by the authors.
- After that, the same algorithm should be tested on simple mixture of acoustic signals.
- At the end, we try the algorithm on real signals which cross our simulated underwater acoustic channel.

Using the three above mentioned steps, we found that at the third step none of the tested algorithms can unfortunately achieve a satisfactory separation according to a set of performance indexes [10]. For this reason, a complete separation structure has been implemented using pre- and post-processing modules of the signals.

Most of our sources are bounded in frequency domain. Therefore, a low-pass filter was of great helpful for us to reduce the impact of the AWGN and then achieve better performances. Using this filter, we found that among the tested algorithms, only three ones have given satisfactory results. These three algorithms [11-13] were dedicated to separate non-stationary sources (audio or music signals). The last two algorithms [12,13], which be called in the following SOS [12] and Parra [13] algorithms, are implemented in frequency domain using discrete frequency adapted filter. Experimental studies showed that best results can be obtained by applying the SOS algorithm over the signals mainly divided in three frequency bands. Once the separation in each frequency bound are achieved, than a reconstitution module should be used to recover the original sources. Our reconstitution module is based on the second order statistics (but it can easily by generalized to use other statistical features) and it uses the correlation of the signals in time or frequency domain.

Finally, we should mention that best results have been obtained when both algorithms Parra and SOS are used and the number of sensors is strict great than the number of sources, as shown in Fig. 1.

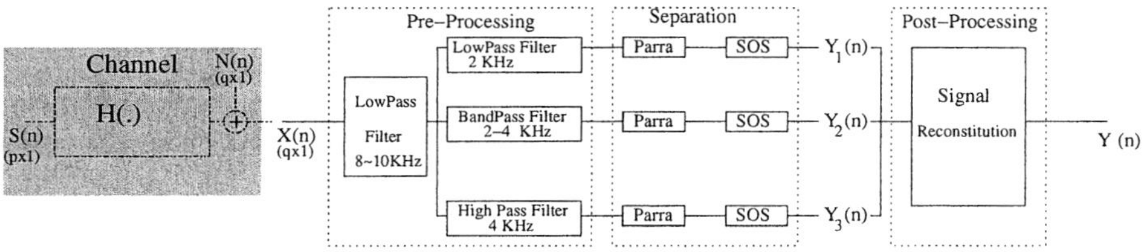

Fig. 1. General Structure

5 Frequency Domain Approach Applied to Acoustic Signals

As it was mentioned before that best experimental results were obtained using two algorithms [12,13]. These two algorithms are minimized second order statistics criteria in frequency-domain. In the following, we describe briefly both of them, for more details please refer to the cited references.

5.1 A Frequency Domain Method for Blind Source Separation of Convolutive Audio Mixture (SOS)

K. Rahbar *et al.* in [12,14] propose an algorithm which minimize a criterion Γ based on the cross-spectral density matrix of the observed signals. For non-stationary signals, the latter matrix depends of frequency and time epoch m:

$$\Gamma = \int_0^{\pi} \sum_{m=0}^{M-1} \|F(w,m)\|_F^2 dw \tag{5}$$

$$F(w,m) = \hat{P}_m(w) - \sum_{\alpha=0,\beta=0}^{L} \hat{H}_\alpha \hat{D}_m(w) \hat{H}_\beta^T \exp\left(-j(\alpha-\beta)w\right) \tag{6}$$

where $\|F(w,m)\|_F^2$ is the Frobenius norm of $F(w,m)$, L is an estimation of channel degree $H(z) = \sum_{i=0}^{L} H(i)z^{-i} = \sum_{i=0}^{L} H_i z^{-i}$, $\hat{H}_\alpha$ is an estimation of the channel response at time α, and $\hat{D}_m(w)$ are diagonal matrices as estimated cross-spectral density matrix of the sources. To estimate the cross-spectral density matrix of the signals, the authors use M estimation windows with L_m samples each:

$$\hat{P}_m(w) = \frac{1}{J} \sum_{i=0}^{J-1} X_{im}(w) X_{im}^H(w) \tag{7}$$

where $X_{im}(w)$ is the Fourier transform of the observed signals, and J is the number of estimated windows such that $L_J < L_m$ and $JL_J > L_m$.

It is clear that the minimization of (5) needs a continues variable w which it is very difficult to be implemented. To solve that problem, the authors proposed the minimization of another criterion over K frequency points such that $w_k = \frac{\pi k}{K}$:

$$\Gamma = \sum_{k=0}^{K-1} \sum_{m=0}^{M-1} \mathrm{Tr}\left(F_R(w_k, m) F_R^H(w_k, m) + F_I(w_k, m) F_I^H(w_k, m)\right) \quad (8)$$

where $F_R(w, m)$ and $F_I(w, m)$ are the real and the imaginary parts of equation 6. Finally, the minimization is done using a conjugate gradient algorithm.

5.2 Convolutive Blind Separation of Non-stationary Sources

The approach proposed by Parra *et al.* [13] is similar to the previous one proposed by Rahbar *et al.* Using the spectral density of different signals, the authors propose the minimization of the following criterion by using a gradient algorithm:

$$\hat{G}, \hat{R}_S, \hat{R}_N = \mathrm{argmin} \sum_k \sum_w \|G(w) \left[\hat{R}_X(w, k) - R_N(w, k)\right] W^H(w) - R_S(w, k)\|^2 \quad (9)$$

where $\hat{R}_X(w, k)$ is the estimated cross-power spectra of X. To improve the performance of their algorithm, the authors propose the minimization using a joint diagonalization algorithm of the following criterion $J(w)$ and subject to a constraint in time domain concerning the filter size which aims to solve the permutation indeterminacy in frequency domain:

$$J(w) = \sum_{t,w} \left(\sum_t \|R_X(w, t)\|^{-2} \right) \|R_X(t, w) - \mathrm{diag}(R_X(t, w))\|_F^2 \quad (10)$$

6 Experimental Results

Using the structure proposed in Fig. 1, many simulations have been conducted. Generally, over 500000-1000000 samples are needs to achieve the separation. The original sources are sampled at 44KHz. In almost all the simulations, The separation of artificial or natural signals have been successfully achieved. In these simulations, we have set the channel depth between 100 to 500m, the distances among the sources or the sensors are from 30 to 100 m, the distances among the different sources and the divers sensors are from 1500 to 2500 m, the number of sensors is strictly great to the number of sources.

Fig. 2 represents the experimental results obtained by only applying SOS algorithm to separate a mixture of acoustic signals (Ship and Whale).

We should mention here, that good results have been obtained by only applying SOS algorithm except for some configurations notably when the sources are close to the water surface. For the latter cases, we found that the Parra algorithm before SOS algorithm could improve the overall results. Fig. 3 shows us different experimental results obtained by the different algorithms (Parra, SOS or Parra + SOS), each point corresponds to results of random simulations using Parra, SOS or Parra & SOS algorithms. In this figure, a normalized positive performance index based on a nonlinear decorrelation is used [10]. The normalized performance index is forced to be zero for the mixture values and 1 for the sources.

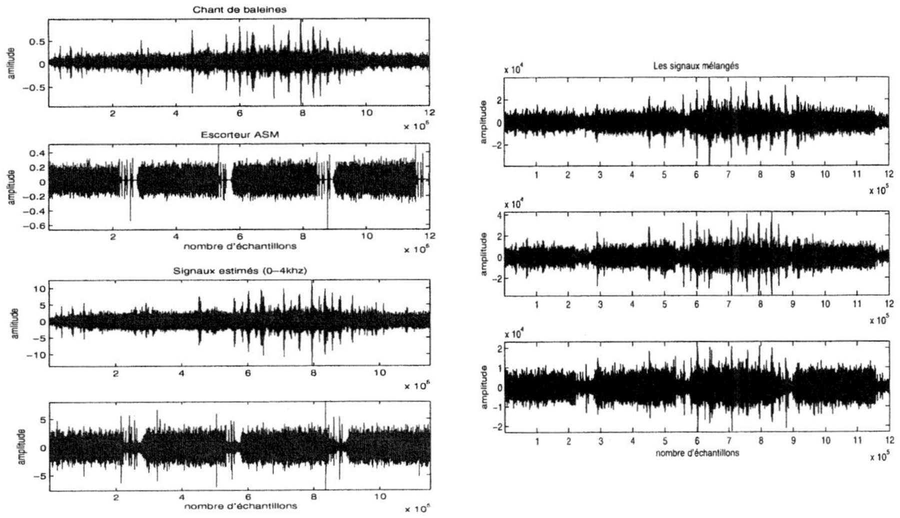

Fig. 2. Experimental results: First column contains the original and the estimated sources, and second column contains the observed signals (the sources are: Whale sound and a boat noise)

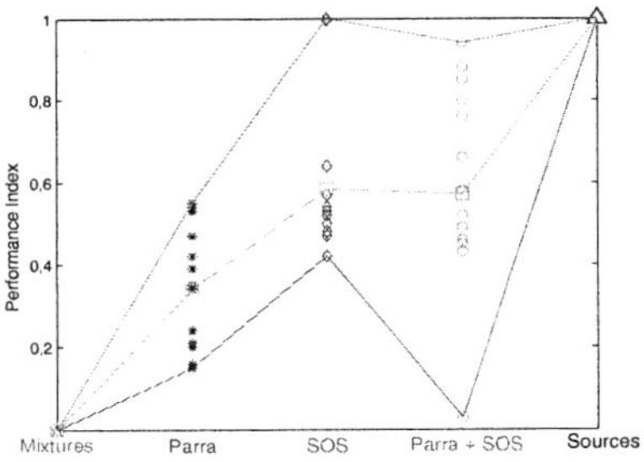

Fig. 3. Experimental results obtained by the different algorithms (Parra, SOS or SOS + Parra) on divers configuration and using a normalized performance index

7 Conclusion

In this paper, a general structure for applying ICA algorithms on real world application such the Passive Acoustic Tomography (PAT) has been presented. Many simulations have been conducted and experimental studies show the necessity of considering pre-processing and post processing of the observed signals in order to achieve properly the separation of the sources.

Many algorithms have been implemented and tested on our application. However, few algorithms which are dedicated to the separation of non-stationary signals, give us satisfactory results. Our future work consists on developing an ICA algorithm which can use other features of acoustic signals such as sparseness along with non-stationarity, *etc.*

References

1. C. Gervaise, A. Quinquis, and I. Luzin, "High resolution identification of an underwater channel from unknown transient stimuli," in *18eme Colloque Gretsi*, Toulouse, France, Sept. 2001.
2. C. Gervaise, A. Quinquis, and N. Martins, "Time frequency approach of blind study of acoustic submarine channel and source recognition," in *Physics in Signal and Imane Processing, PSIP 2001*, Marseille, France, January 2001.
3. D. Gaucher, C. Gervaise, and G. Jourdain, "Feasibility of passive oceanic acoustic tomography in shallow water context: Optimal design of experiments," in *European Conference on UNDERWATER ACOUSTICS ECUA 2004*, Delft, Netherlands, 5-8 July 2004, pp. 56–60.
4. N. Martins, S. Jesus, C. Gervaise, and A. Quinquis, "A time-frequency approach to blind deconvolution in multipath underwater channels," in *Proceedings of ICASSP 2002*, Orlando, Florida, U.S.A, May 2002.
5. D. Gaucher and C. Gervaise, "Feasibility of passive oceanic acoustic tomography: a cramer rao bounds approach," in *Oceans 2003 Marine Technology and Ocean Science Conference*, San Diego, USA, Sept. 2003, pp. 56–60.
6. A. Mansour and M. Kawamoto, "Ica papers classified according to their applications & performances.," *IEICE Trans. on Fundamentals of Electronics, Communications and Computer Sciences*, vol. E86-A, no. 3, pp. 620–633, 2003.
7. P. Comon, "Independent component analysis, a new concept?," *Signal Processing*, vol. 36, no. 3, pp. 287–314, April 1994.
8. F. B. Jensen, W. A. Kuperman, M.B. Porter, and H. Schmidt, *Computational ocean acoustics*, Springer-Verlag New York, Inc., 2000.
9. M. Shulkin and H. W. Marsh, "Sound absorption in sea water," *Journal of the Acoustical Society of America*, vol. 134, pp. 864–865, 1962.
10. A. Mansour, "A survey of real world performance indexes of ICA algorithms," in preparation 2006.
11. M. Kawamoto, A. Kardec Barros, A. Mansour, K. Matsuoka, and N. Ohnishi, "Real world blind separation of convolved non-stationary signals.," in *ICA99*, Aussois, France, January 1999, pp. 347–352.
12. K. Rahbar and J. Reilly, "Blind separation of convolved sources by joint approximate diagonalization of cross-spectral density matrices," in *Proceedings of ICASSP 2001*, Salt Lake City, Utah, USA, May 2001.
13. L. Parra and C. Alvino, "Convolutive blind separation of non-stationnary sources," *IEEE Trans. on Speech and Audio Processing*, vol. 8, no. 3, pp. 320–327, May 2000.
14. K. Rahbar, J. Reilly, and J. H. Manton, "Blind identification of mimo fir systems driven by quasistationary sources using second order statistics: A frequency domain approach," *IEEE Trans. on Signal Processing*, vol. 52, no. 2, pp. 406–417, 2004.

Partitioned Factor Analysis for Interference Suppression and Source Extraction

Srikantan S. Nagarajan[1], Hagai T. Attias[2], Kensuke Sekihara[3], and Kenneth E. Hild II[1]

[1] Dept. of Radiology, University of California at San Francisco, CA 94122, USA
Srikantan.Nagarajan@radiology.ucsf.edu, k.hild@ieee.org
[2] Golden Metallic Inc., San Francisco, CA 94147, USA
htattias@goldenmetallic.com
[3] Dept. of Electronic Systems and Engineering, Tokyo Metropolitan Institute of Technology, 191-0065, Japan
ksekiha@cc.tmit.ac.jp

Abstract. It is common for data to be contaminated with artifacts, interference, and noise. Several methods including independent components analysis (ICA) and principal components analysis (PCA) have been used to suppress these undesired signals and/or to extract the underlying (desired) source waveforms. For some data it is known, or can be extracted post hoc, how to partition the data into periods of source activity and source inactivity. Two examples include cardiac data and data collected using the stimulus-evoked paradigm. However, neither ICA nor PCA are able to take full advantage of the knowledge of the partition. Here we introduce an interference suppression method, partitioned factor analysis (PFA), that takes into account the data partition.

1 Introduction

Raw data are corrupted by artifacts, interference, and sensor noise. When the power of these undesired signals is large one of several signal processing techniques may be applied to either reduce the level of interference or to extract the underlying source waveforms directly (we use "denoising" and "interference suppression" interchangeably). Linear denoising methods, including independent components analysis (ICA), attempt to find the source subspace and produce denoised signals by projecting intermediate lower-dimensional data back into the space of the observations. Denoising is useful for spatio-temporal visualization and for source localization [1]. Furthermore, the intermediate data produced by denoising methods can be used as the input to an ICA algorithm when the desire is to extract the source waveforms. While ICA can be applied directly to the observations to perform denoising and source extraction simultaneously, many ICA algorithms are too computationally intensive to be used in this manner when there are many sensors. An alternative is to reduce the dimensionality with a (non-ICA) denoising method and then use ICA.

J. Rosca et al. (Eds.): ICA 2006, LNCS 3889, pp. 189–197, 2006.

Principal component analysis (PCA) is the most widely used method of denoising. PCA is an ideal choice when either the power of the sources is large relative to the power of all undesired signals or when the undesired signals are spatially uncorrelated and isotropic in sensor space [2]. Likewise, ICA is ideal for denoising when the sources of interest are statistically independent of all undesired signals, there are at least as many sensors as sources (desired and undesired), and the subspace that contains the sources of interest can be robustly determined automatically or with the aid of a human expert. However, these requisite conditions may not always be met. In addition, both PCA and ICA are unable to take full advantage of additional information that is available for some data, i.e., knowledge of when the sources are active and when they are inactive. Data for which the timing of the source activity is known or can be estimated include, e.g., cardiac data collected using a magnetocardiogram (MCG) and stimulus-evoked data collected using a magnetoencephalogram (MEG).

Here we introduce a denoising method, partitioned factor analysis (PFA), that is based on what is thought to be a more realistic set of assumptions than PCA or ICA and that is able to incorporate knowledge of the data partition.

2 Partitioned Factor Analysis

Generative model. The proposed method, PFA, is based on the following generative model,

$$\boldsymbol{y}_n = \begin{cases} \boldsymbol{B}\boldsymbol{u}_n + \boldsymbol{v}_n & n=1,\ldots,N_0-1 \\ \boldsymbol{A}\boldsymbol{x}_n + \boldsymbol{B}\boldsymbol{u}_n + \boldsymbol{v}_n & n=N_0,\ldots,N \end{cases}, \tag{1}$$

where the $(M_y \times 1)$ vector $\boldsymbol{v}_n$ is used to represent all signals that are spatially uncorrelated in sensor space and that exist in the active and inactive periods, the $(M_u \times 1)$ vector $\boldsymbol{u}_n$ represents all signals that are spatially correlated in sensor space and that exist throughout, and the $(M_x \times 1)$ vector $\boldsymbol{x}_n$ represents all signals that exist only during the active period. We refer to $\boldsymbol{x}_n$ as the factors (which are an arbitrary linear combination of the sources of interest, $\boldsymbol{x}_n = \boldsymbol{W}\boldsymbol{s}_n$), $\boldsymbol{u}_n$ as the interference, and $\boldsymbol{v}_n$ as the noise. The inclusion of the interference signals allow us to model signals of no interest that, unlike the model commonly used for sensor noise, are spatially correlated in sensor space, e.g., respiration, muscle artifacts, eye blinks, and ongoing neural activity. Also, there is a common assumption that the spatial auto-correlation matrix of the post-stimulus equals the auto-correlation of the pre-stimulus plus the auto-correlation of the evoked response. This structure is directly reflected by the model of (1), which provides a more complete representation of the data than what is inherently assumed in PCA and the vast majority of ICA algorithms (which combine $\boldsymbol{x}_n$ and $\boldsymbol{u}_n$ into a single vector and assume that $\boldsymbol{v}_n$ is zero).

The proper choice for the active and inactive periods is problem-dependent. For example, if the goal is to recover a cardiac signal the active period should correspond to all portions of the data that are near a QRS complex. Likewise, if the goal is to recover an evoked response for data collected using the stimulus-evoked paradigm, the active period corresponds to the post-stimulus period. To simplify the notation the data are assumed to ordered so that the first $N_0 - 1$ samples of the $(M_y \times N)$ matrix of observations $\boldsymbol{y}$ (where $\boldsymbol{y}$ corresponds to the collection of $\boldsymbol{y}_n \forall n$) correspond to the concatenation of all inactive periods and the remaining samples correspond to the concatenation of all active periods.

PFA probabilistic graphical model. Each factor $x_{m,n}$ and each interference $u_{m,n}$ is modeled as having a Gaussian probability density function (pdf) with zero mean and unit precision, where the precision is defined as the inverse variance and $x_{m,n}$ and $u_{m,n}$ are the mth element of vectors $\boldsymbol{x}_n$ and $\boldsymbol{u}_n$, respectively. Likewise, the noise at sensor m is modeled as having a Gaussian pdf with zero mean and precision λ_m. We model the factors, interferences, and noises as mutually statistically independent,

$$p(\boldsymbol{x}_n) = \mathcal{N}(\boldsymbol{x}_n|\boldsymbol{0}, \boldsymbol{I}), \quad p(\boldsymbol{u}_n) = \mathcal{N}(\boldsymbol{u}_n|\boldsymbol{0}, \boldsymbol{I}), \quad p(\boldsymbol{v}_n) = \mathcal{N}(\boldsymbol{v}_n|\boldsymbol{0}, \boldsymbol{\Lambda}) \ , \tag{2}$$

where $\boldsymbol{0}$ is a column vector of zeros, $\boldsymbol{I}$ is an identity matrix, and $\boldsymbol{\Lambda}$ is a diagonal matrix. By inspection the data likelihood is

$$p(\boldsymbol{y}_n|\boldsymbol{x}_n,\boldsymbol{u}_n,\boldsymbol{A},\boldsymbol{B}) = \begin{cases} \mathcal{N}(\boldsymbol{y}_n|\boldsymbol{B}\boldsymbol{u}_n, \boldsymbol{\Lambda}) & n=1,\ldots,N_0{-}1 \\ \mathcal{N}(\boldsymbol{y}_n|\boldsymbol{A}\boldsymbol{x}_n{+}\boldsymbol{B}\boldsymbol{u}_n, \boldsymbol{\Lambda}) & n=N_0,\ldots,N \end{cases} . \tag{3}$$

We also assume that the signals are temporally i.i.d. so that

$$p(\boldsymbol{y}|\boldsymbol{x},\boldsymbol{u},\boldsymbol{A},\boldsymbol{B}) = \prod_{n=1}^{N} p(\boldsymbol{y}_n|\boldsymbol{x}_n,\boldsymbol{u}_n,\boldsymbol{A},\boldsymbol{B}), \ p(\boldsymbol{x}) = \prod_{n=N_0}^{N} p(\boldsymbol{x}_n), \ p(\boldsymbol{u}) = \prod_{n=1}^{N} p(\boldsymbol{u}_n) \ . \tag{4}$$

The elements of the two mixing matrices are assumed to be independent zero-mean Gaussians that have a precision that is proportional to the noise precision of the corresponding sensor,

$$p(\boldsymbol{A}) = \prod_{m=1}^{M_y} \prod_{k=1}^{M_x} \mathcal{N}(A_{m,k}|0, \lambda_m \alpha_k), \qquad p(\boldsymbol{B}) = \prod_{m=1}^{M_y} \prod_{k=1}^{M_u} \mathcal{N}(B_{m,k}|0, \lambda_m \beta_k) \ , \tag{5}$$

where the proportionality constants, α_k, β_k, are referred to as hyperparameters. These priors are chosen so that they have the same functional form as the posterior distribution (when this is true the prior is referred to as a conjugate prior).

Inferring the PFA model from data. All three types of signals, $\boldsymbol{x}_n, \boldsymbol{u}_n, \boldsymbol{v}_n$, are unobserved, as are the $(M_y \times M_x)$ matrix $\boldsymbol{A}$ and the $(M_y \times M_u)$ matrix $\boldsymbol{B}$. Hence, PFA must use only $\boldsymbol{y}$ to infer the quantities of interest, which are $\tilde{\boldsymbol{y}}_n = \boldsymbol{A}\boldsymbol{x}_n$ for denoising and $\boldsymbol{x}_n$ for subsequent source extraction. To infer the model from $\boldsymbol{y}$ we use an extended version of the Expectation-Maximization (EM) algorithm, which is known as the variational Bayesian EM method (VB-EM) [3]. Whereas standard EM computes the most likely parameter value given the data, i.e., the maximum a posteriori (MAP) estimate, VB-EM computes full posterior distributions. Furthermore, VB-EM provides a natural mechanism for inferring the model order through hyper-parameter optimization [3], whereas standard EM requires ad-hoc methods for model order selection.

Standard EM maximizes the log likelihood, which can be written as

$$\log p(\boldsymbol{y}) = \log p(\boldsymbol{y}) \int p(\boldsymbol{\theta}|\boldsymbol{y}) d\boldsymbol{\theta} = \int p(\boldsymbol{\theta}|\boldsymbol{y}) \log \frac{p(\boldsymbol{\theta}, \boldsymbol{y})}{p(\boldsymbol{\theta}|\boldsymbol{y})} d\boldsymbol{\theta} \ , \tag{6}$$

where $\boldsymbol{\theta} = \{\boldsymbol{x},\boldsymbol{u},\boldsymbol{A},\boldsymbol{B}\}$. Since the exact posterior distribution is computationally intractable, we approximate the posterior with a function that factorizes the hidden variables given the data from the parameters given the data,

$$p(\boldsymbol{\theta}|\boldsymbol{y}) \approx q(\boldsymbol{\theta}|\boldsymbol{y}) = q(\boldsymbol{x},\boldsymbol{u}|\boldsymbol{y}) q(\boldsymbol{A},\boldsymbol{B}|\boldsymbol{y}) \ . \tag{7}$$

The result is that VB-EM adapts $q(\boldsymbol{x},\boldsymbol{u}|\boldsymbol{y})$ and $q(\boldsymbol{A},\boldsymbol{B}|\boldsymbol{y})$ to maximize an approximation of the log likelihood, which can be written as

$$\mathcal{F} = \int q(\boldsymbol{x},\boldsymbol{u}|\boldsymbol{y}) q(\boldsymbol{A},\boldsymbol{B}|\boldsymbol{y}) \log \frac{p(\boldsymbol{x},\boldsymbol{u},\boldsymbol{A},\boldsymbol{B},\boldsymbol{y})}{q(\boldsymbol{x},\boldsymbol{u}|\boldsymbol{y}) q(\boldsymbol{A},\boldsymbol{B}|\boldsymbol{y})} d\boldsymbol{x} d\boldsymbol{u} d\boldsymbol{A} d\boldsymbol{B} \ . \tag{8}$$

It can be shown that maximizing $\mathcal{F}$ w.r.t. $q(\boldsymbol{x},\boldsymbol{u}|\boldsymbol{y})$, $q(\boldsymbol{A},\boldsymbol{B}|\boldsymbol{y})$ is equivalent to minimizing the Kullback-Leibler divergence [4] between $p(\boldsymbol{\theta}|\boldsymbol{y})$ and $q(\boldsymbol{\theta}|\boldsymbol{y})$. Like standard EM the VB-EM optimization method is an iterative algorithm where each iteration is composed of an E-step and an M-step.

E-step. Maximization of $\mathcal{F}$ with respect to the posterior over hidden variables is accomplished by setting the derivative of $\mathcal{F}$ to zero and solving for $q(\boldsymbol{x},\boldsymbol{u}|\boldsymbol{y})$ while keeping $q(\boldsymbol{A},\boldsymbol{B}|\boldsymbol{y})$ fixed. This produces

$$q(\boldsymbol{x},\boldsymbol{u}|\boldsymbol{y}) = \frac{1}{z_1} \exp\left[\int q(\boldsymbol{A},\boldsymbol{B}|\boldsymbol{y}) \log p(\boldsymbol{x},\boldsymbol{u},\boldsymbol{A},\boldsymbol{B},\boldsymbol{y}) d\boldsymbol{A} d\boldsymbol{B}\right] \ , \tag{9}$$

where the joint pdf, due to the previous assumptions, simplifies to

$$p(\boldsymbol{x},\boldsymbol{u},\boldsymbol{A},\boldsymbol{B},\boldsymbol{y}) = p(\boldsymbol{y}|\boldsymbol{x},\boldsymbol{u},\boldsymbol{A},\boldsymbol{B}) p(\boldsymbol{x}) p(\boldsymbol{u}) p(\boldsymbol{A}) p(\boldsymbol{B}) \ , \tag{10}$$

and where z_1 is the normalizing constant (normalization of this posterior and the posterior over parameters is enforced by adding two Lagrange multipliers to $\mathcal{F}$). The quantities in (10) are given by (2)-(5). The posterior over hidden variables factorizes over time so that

$$q(\boldsymbol{x},\boldsymbol{u}|\boldsymbol{y}) = \prod_{n=1}^{N_0-1} q(\boldsymbol{u}_n|\boldsymbol{y}_n) \prod_{n=N_0}^{N} q(\boldsymbol{x}_n,\boldsymbol{u}_n|\boldsymbol{y}_n) \ , \tag{11}$$

where

$$
\begin{aligned}
q(\boldsymbol{u}_n|\boldsymbol{y}_n) &= \mathcal{N}(\boldsymbol{u}_n|\bar{\boldsymbol{u}}_n, \boldsymbol{\Phi}^{-1}) \\
q(\boldsymbol{x}_n,\boldsymbol{u}_n|\boldsymbol{y}_n) &= \mathcal{N}(\begin{bmatrix}\boldsymbol{x}_n\\ \boldsymbol{u}_n\end{bmatrix} | \begin{bmatrix}\bar{\boldsymbol{x}}_n\\ \bar{\boldsymbol{u}}_n\end{bmatrix}, \boldsymbol{\Gamma}^{-1}) \\
\bar{\boldsymbol{x}}_n &= (\boldsymbol{\Gamma}_{xx}\bar{\boldsymbol{A}}^T + \boldsymbol{\Gamma}_{xu}\bar{\boldsymbol{B}}^T)\boldsymbol{\Lambda}\boldsymbol{y}_n \\
\bar{\boldsymbol{u}}_n &= \boldsymbol{\Phi}\bar{\boldsymbol{B}}^T\boldsymbol{\lambda}\boldsymbol{y}_n \qquad\qquad n \in \{1,\dots,N_0-1\} \\
\bar{\boldsymbol{u}}_n &= (\boldsymbol{\Gamma}_{xu}^T\bar{\boldsymbol{A}}^T + \boldsymbol{\Gamma}_{uu}\bar{\boldsymbol{B}}^T)\boldsymbol{\Lambda}\boldsymbol{y}_n \quad n \in \{N_0,\dots,N\} \\
\boldsymbol{\Phi} &= (\bar{\boldsymbol{B}}^T\boldsymbol{\Lambda}\bar{\boldsymbol{B}} + \boldsymbol{I} + M_y\boldsymbol{\Psi}_{BB})^{-1} \\
\boldsymbol{\Gamma} &= (\begin{bmatrix}\bar{\boldsymbol{A}}\\ \bar{\boldsymbol{B}}\end{bmatrix}\boldsymbol{\Lambda}\,[\bar{\boldsymbol{A}}\ \bar{\boldsymbol{B}}] + \boldsymbol{I} + M_y\boldsymbol{\Psi})^{-1} = \begin{bmatrix}\boldsymbol{\Gamma}_{xx} & \boldsymbol{\Gamma}_{xu}\\ \boldsymbol{\Gamma}_{xu}^T & \boldsymbol{\Gamma}_{uu}\end{bmatrix} ,
\end{aligned} \tag{12}
$$

and where $\bar{\boldsymbol{A}}, \bar{\boldsymbol{B}}, \boldsymbol{\lambda}, \boldsymbol{\Psi}$ are computed in the M-step.

M-step. Similarly, maximization of $\mathcal{F}$ with respect to the posterior over parameters is accomplished by setting the derivative of $\mathcal{F}$ to zero and solving for $q(\boldsymbol{A},\boldsymbol{B}|\boldsymbol{y})$ while keeping $q(\boldsymbol{x},\boldsymbol{u}|\boldsymbol{y})$ fixed. This produces

$$
q(\boldsymbol{A},\boldsymbol{B}|\boldsymbol{y}) = \frac{1}{z_2}\exp[\int q(\boldsymbol{x},\boldsymbol{u}|\boldsymbol{y})\log p(\boldsymbol{x},\boldsymbol{u},\boldsymbol{A},\boldsymbol{B},\boldsymbol{y})d\boldsymbol{x}d\boldsymbol{u}] , \tag{13}
$$

where z_2 is the normalizing constant.

It follows from (13) that the posterior over parameters factorizes over the rows of the two mixing matrices. Hence,

$$
q(\boldsymbol{A},\boldsymbol{B}|\boldsymbol{y}) = \prod_{m=1}^{M_y}\mathcal{N}(\begin{bmatrix}\boldsymbol{A}_m^T\\ \boldsymbol{B}_m^T\end{bmatrix} | \begin{bmatrix}\bar{\boldsymbol{A}}_m^T\\ \bar{\boldsymbol{B}}_m^T\end{bmatrix}, \lambda_m\boldsymbol{\Psi}^{-1}) , \tag{14}
$$

where $\boldsymbol{A}_m$ is the mth row of $\boldsymbol{A}$ and

$$
\begin{aligned}
\bar{\boldsymbol{A}} &= \left(\sum_{n=N_0}^{N}\boldsymbol{y}_n\bar{\boldsymbol{x}}_n^T\right)\boldsymbol{\Psi}, \quad \bar{\boldsymbol{B}} = \left(\sum_{n=1}^{N}\boldsymbol{y}_n\bar{\boldsymbol{u}}_n^T\right)\boldsymbol{\Psi} \\
\boldsymbol{\Psi} &= \begin{bmatrix}\boldsymbol{R}_{xx}+\alpha & \boldsymbol{R}_{xu}\\ \boldsymbol{R}_{xu}^T & \boldsymbol{R}_{uu}+\beta\end{bmatrix}^{-1} = \begin{bmatrix}\boldsymbol{\Psi}_{AA} & \boldsymbol{\Psi}_{AB}\\ \boldsymbol{\Psi}_{AB}^T & \boldsymbol{\Psi}_{BB}\end{bmatrix} \\
\boldsymbol{R}_{\boldsymbol{xx}} &= \sum_{n=N_0}^{N}\left(\bar{\boldsymbol{x}}_n\bar{\boldsymbol{x}}_n^T + \boldsymbol{\Gamma}_{\boldsymbol{xx}}\right), \quad \boldsymbol{R}_{\boldsymbol{xu}} = \sum_{n=N_0}^{N}\left(\bar{\boldsymbol{x}}_n\bar{\boldsymbol{u}}_n^T + \boldsymbol{\Gamma}_{\boldsymbol{xu}}\right) \\
\boldsymbol{R}_{\boldsymbol{uu}} &= \sum_{n=1}^{N_0-1}\left(\bar{\boldsymbol{u}}_n\bar{\boldsymbol{u}}_n^T + \boldsymbol{\Phi}\right) + \sum_{n=N_0}^{N}\left(\bar{\boldsymbol{u}}_n\bar{\boldsymbol{u}}_n^T + \boldsymbol{\Gamma}_{\boldsymbol{uu}}\right) ,
\end{aligned} \tag{15}
$$

and $\boldsymbol{\alpha}, \boldsymbol{\beta}$ are diagonal matrices that contain α_k, β_k, respectively.

The solutions for the noise precision matrix and the hyperparameters are found by computing the derivative of $\mathcal{F}$ and equating the result with zero,

$$\begin{aligned} \boldsymbol{\alpha}^{-1} &= \mathrm{diag}\left(\frac{1}{M_y}\bar{\boldsymbol{A}}^T\boldsymbol{\Lambda}\bar{\boldsymbol{A}} + \boldsymbol{\Psi}_{AA}\right) \\ \boldsymbol{\beta}^{-1} &= \mathrm{diag}\left(\frac{1}{M_y}\bar{\boldsymbol{B}}^T\boldsymbol{\Lambda}\bar{\boldsymbol{B}} + \boldsymbol{\Psi}_{BB}\right) \\ \boldsymbol{\Lambda}^{-1} &= \frac{1}{N}\mathrm{diag}\left(\boldsymbol{R}_{yy} - \bar{\boldsymbol{A}}\boldsymbol{R}_{yx}^T - \bar{\boldsymbol{B}}\boldsymbol{R}_{yu}^T\right) \ , \end{aligned} \tag{16}$$

where

$$\boldsymbol{R}_{yy} = \sum_{n=1}^{N} \boldsymbol{y}_n\boldsymbol{y}_n^T, \quad \boldsymbol{R}_{yx} = \sum_{n=N_0}^{N} \boldsymbol{y}_n\bar{\boldsymbol{x}}_n^T, \quad \boldsymbol{R}_{yu} = \sum_{n=1}^{N} \boldsymbol{y}_n\bar{\boldsymbol{u}}_n^T \ . \tag{17}$$

3 Results

Denoising performance is measured using the output signal-to-noise/interference ratio (SNIR),

$$\mathrm{SNIR} = \frac{1}{M}\sum_{m=1}^{M} 10\log_{10}\frac{\sum_{n=1}^{N}(\boldsymbol{A}\boldsymbol{x})_{m,n}^2}{\sum_{n=1}^{N}((\boldsymbol{A}\boldsymbol{x})_{m,n} - (\bar{\boldsymbol{A}}\bar{\boldsymbol{x}})_{m,n})^2} \ (\mathrm{dB}) \ .$$

For real data we replace $\boldsymbol{A}\boldsymbol{x}_n$ with the sensor signals due to the 5 principal components (representing 97% of the energy) of the average sensor data, where the average is taken over 525 trials. The metric for source extraction performance is the source-to-distortion ratio (SDR),

$$\mathrm{SDR} = \frac{1}{M_s}\sum_{m=1}^{M_s} 10\log_{10}\frac{1}{M_s}\sum_{m'=1}^{M_s}\left(\frac{1}{2 - \frac{2}{N-N_0+1}\left|\sum_{n=N_0}^{N} s_{m,n}\bar{s}_{m',n}\right|}\right) \ (\mathrm{dB}) \ ,$$

where the distortion for source estimate m includes noise, interference, and all sources other than source m, $\boldsymbol{s}_n = \boldsymbol{W}^{-1}\boldsymbol{x}_n$, $\boldsymbol{W}$ is found using ICA, and both $s_{m,n}$ and $\bar{s}_{m,n}$ (the estimate of source m at time n) are normalized to have unit variance. For simulated data the SNIR and SDR are shown as a function of the input signal-to-interference ratio (SIR) for a fixed value of input signal-to-noise ratio (SNR). The former is defined as the ratio of the power of the factors to the power of the interferences measured in sensor space. The latter is defined in a similar fashion.

Simulated data. For the simulations $N = 1000$ data points/trial, $N_0 = 631$, $M_y = 132$, $M_x = M_s = 2$, $M_u = 1000$, SNR$= 0$ dB, and the number of trials is 10. The results represent the mean over 10 Monte Carlo experiments (per trial) and error

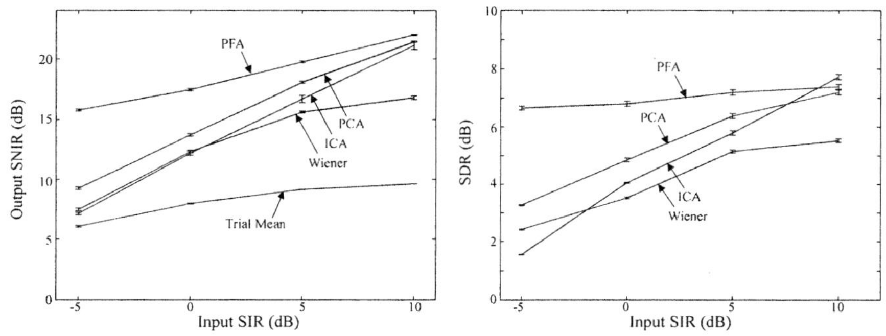

Fig. 1. Left subplot (1a): Output SNIR as a function of input SIR for 10 trials, input SNR = 0 dB, and the ICA method is TDSEP. Right subplot (1b): SDR as a function of input SIR for 10 trials, input SNR = 0 dB, and the ICA method is FastICA.

bars are used to indicate one standard error. The comparison includes the proposed method (PFA), PCA [2], Wiener [5], ICA (TDSEP [6] or FastICA [7]), and the mean over trials. For ICA, the source subspace is determined as the components having the largest ratio of active power to inactive power.

Figure 1a shows the denoising performance as a function of input SIR. All of the methods perform better than the trial mean. PFA performs the best across all values of input SIR. The performances of both PCA and Wiener approach that of PFA as the input SIR increases.

Figure 1b shows the SDR as a function of input SIR. In this figure PFA, PCA, and Wiener are all combined with ICA. Also shown is the result for ICA without dimension reduction. PFA produces the best overall results and is the least sensitive to input SIR. The results for ICA (with no dimension reduction) indicate that for this dataset dimension reduction should be used when the input SIR is low and is not needed if the input SIR is > 10 dB.

Real data. Figure 2a shows the denoising performance as a function of the number of trials for a real MEG dataset, which uses a somatosensory stimulus ($M_y = 274$, M_x is assumed to be 2, M_u is assumed to be 50, $N = 361$, $N_0 = 121$). PFA performs the best and both PCA and Wiener (which performs almost identically to PCA and is not shown here) outperform the trial mean.

Figure 2b shows the sensor signals before and after PFA denoising is applied to real fetal MCG data, which is a mixture of both fetal and maternal cardiac sources ($M_y = 51$, M_x is assumed to be 10, M_u is assumed to be 50, $N = 4000$, $N_0 = 501$). Since the goal is to recover the fetal cardiac factors, the inactive period is chosen such that it contains minimal activity of the fetal heart (two 250-length portions) and the active period is chosen such that it contains both maternal and fetal cardiac activity. Notice that QRS complexes of the mother are effectively suppressed and several fetal QRS complexes that were previously obscured, e.g., at 1.3 sec and 3.1 sec, are now visible.

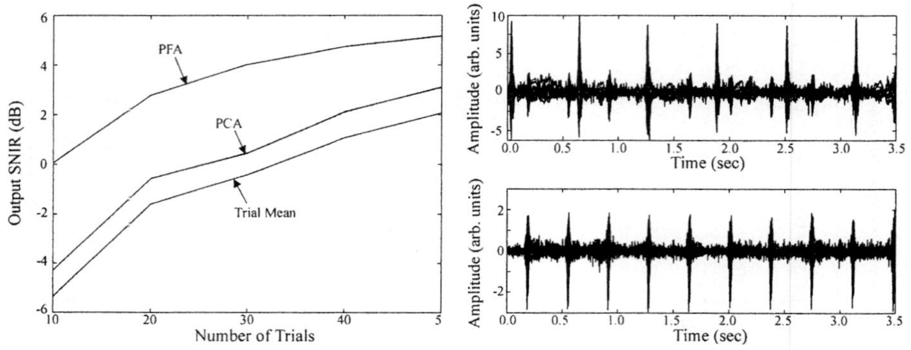

Fig. 2. Left subplot (2a):Output SNIR as a function of number of trials for real MEG data. Right subplot (2b): Observations before (upper subplot) and after (lower subplot) PFA denoising for real fetal MCG data.

4 Conclusions

The PFA graphical model for denoising and dimension reduction is introduced. This model takes into account additional information that is available for several types of data, including cardiac data and data collected using the evoked-response paradigm. The results of simulated and real data indicate that PFA may be a viable alternative to ICA for interference suppression and may, when used as a preprocessor, improve the performance of ICA for source extraction. This appears to be true especially when the power of the noise/interference is large or there are only a few trials available.

Acknowledgments

This work supported by National Institutes of Health grants 1 F32 NS 52048-01 and RO1 DC 4855.

References

1. Ossadtchi, A., Baillet, S., Mosher, J.C., Thyerlei, D., Sutherling, W., Leahy, R.M.: Automated interictal spike detection and source localization in magnetoencephalography using independent components analysis and spatio-temporal clustering. Clinical Neurophysiology **115** (2004) 508–522
2. Jackson, J.E.: A User's Guide to Principal Components (2003)
3. Attias, H.: A variational Bayesian framework for graphical models. Advances in Neural Information Proc. Systems **13** (2000) 209–215
4. Cover, T.M., Thomas, J.A.: Elements of Information Theory (1991)

5. Ungan, P., Basar, E.: Comparison of Wiener filtering and selective averaging of evoked potentials. Electroencephalography and Clinical Neurophysiology **40** (1976) 516–520
6. Ziehe, A., Mller, K.R.: TDSEP - an efficient algorithm for blind separation using time structure. Intl. Conf. on Artificial Neural Networks (1998) 675–680
7. Hyvarinen, A.: Fast and robust fixed-point algorithms for Independent Component Analysis. IEEE Trans. on Neural Networks **10** (1999) 626–634

Recursive Generalized Eigendecomposition for Independent Component Analysis

Umut Ozertem[1], Deniz Erdogmus[1,2], and Tian Lan[2]

[1] CSEE Department, OGI, Oregon Health & Science University, Portland, OR, USA
{ozertemu, deniz}@csee.ogi.edu
[2] BME Department, OGI, Oregon Health & Science University, Portland, OR, USA
lantian@bme.ogi.edu

Abstract. Independent component analysis is an important statistical tool in machine learning, pattern recognition, and signal processing. Most of these applications require on-line learning algorithms. Current on-line ICA algorithms use the stochastic gradient concept, drawbacks of which include difficulties in selecting the step size and generating suboptimal estimates. In this paper a recursive generalized eigendecomposition algorithm is proposed that tracks the optimal solution that one would obtain using all the data observed.

1 Introduction

Independent component analysis (ICA) has now established itself as an essential statistical tool in signal processing and machine learning, both as a solution to problems (such as blind source separation) [1,2] and as a preprocessing instrument that complements other pieces of a more comprehensive solution (such as dimensionality reduction and feature extraction) [3-5]. All of these applications of ICA require on-line learning algorithms that can operate in real-time on contemporary digital signal processors (DSP).

Currently, the on-line ICA solutions are obtained using algorithms designed using the stochastic gradient concept (e.g., Infomax [6]), similar to the well-known least-mean squares (LMS) algorithm [7]. The drawbacks of stochastic gradient algorithms in on-line learning include difficulty in selecting the step size for optimal speed misadjustment trade-off and suboptimal estimates of the weights given the information contained in all the samples seen at any given iteration.

Recursive least squares (RLS) is an on-line algorithm for supervised adaptive filter training, which has the desirable property that the estimated weights correspond to the optimal least squares solution that one would obtain using all the data observed so far, provided that initialization is done *properly* [7]. This benefit, of course comes at the cost of additional computational requirements compared to LMS. Nevertheless, certain applications where an on-line ICA algorithm that tracks the optimal solution one would have obtained using all samples observed up to that point in time would be beneficial. To this end, we derive a recursive generalized eigendecomposition (GED) based ICA algorithm that is similar to RLS in principle, but solves the simultaneous diagonalization problem using second and fourth order joint statistics of the observed mixtures.

J. Rosca et al. (Eds.): ICA 2006, LNCS 3889, pp. 198 – 205, 2006.

The joint diagonalization of higher order statistics have been known to solve the ICA problem under the assumed linear mixing model and have led to popular algorithms (e.g., JADE [8]). The joint diagonalization problem in ICA is essentially a GED problem, a connection which has been nicely summarized in a recent paper by Parra and Sajda [9] for various signal models in linear instantaneous BSS; others have pointed out this connection earlier as well. The algorithm we develop here is based on the non-Gaussian independent sources assumption, with independent and identically distributed samples of mixtures (the latter assumption eliminates the need for weighted averaging for minimum bias estimation of the expectations).

In Section 2, we derive the recursive update equations for the required higher order statistics and the corresponding *optimal* ICA solution. In Section 3, we demonstrate using Monte Carlo simulations that the algorithm tracks the optimal ICA solution exactly when all matrices are initialized to their ideal values and that the algorithm converges to the optimal ICA solution when the matrices are initialized to arbitrary small matrices (whose bias on the solution should diminish as more samples are observed and utilized).

2 Recursive ICA Algorithm

The square linear ICA problem is expressed in (1), where $\mathbf{X}$ is the $n \times N$ observation matrix, $\mathbf{A}$ is the $n \times n$ mixing matrix, and $\mathbf{S}$ is the $n \times N$ independent source matrix.

$$\mathbf{X} = \mathbf{AS} \tag{1}$$

Each column of $\mathbf{X}$ and $\mathbf{S}$ represents one sample of data. If we consider each column as a sample in time, (1) becomes:

$$\mathbf{x}_t = \mathbf{A}\mathbf{s}_t \tag{2}$$

The joint diagonalization of higher order cumulant matrices can be compactly formulated in the form of a generalized eigendecomposition problem that gives the ICA solution in an analytical form [9]. According to this formulation, under the assumption of independent non-Gaussian source distributions the separation matrix $\mathbf{W}$ is the solution to the following generalized eigendecomposition problem:

$$\mathbf{R_x W} = \mathbf{Q_x W \Lambda} \tag{3}$$

where $\mathbf{R_x}$ is the covariance matrix and $\mathbf{Q_x}$ is the cumulant matrix estimated using sample averages. While any order of cumulants could have been employed, lower orders are more robust to outliers and small sample sizes, therefore we focus on the fourth order cumulant matrix: $\mathbf{Q_x}$=E[$\mathbf{x^T x x x^T}$]-$\mathbf{R_x}$tr($\mathbf{R_x}$)-E[$\mathbf{xx^T}$]E[$\mathbf{xx^T}$]-$\mathbf{R_x R_x}$. Given the sample estimates for these matrices, the ICA solution can be easily determined using efficient generalized eigendecomposition algorithms (or the *eig* command in Matlab®). With the assumption of iid samples, expectations reduce to simple sample averages, and the estimates of covariance and cumulant matrices are given by (for real-valued mixtures)

$$\mathbf{R_x} = \sum_i \mathbf{x}_i \mathbf{x}_i^T \qquad \mathbf{Q_x} = \sum_i \left(\mathbf{x}_i^T \mathbf{x}_i \mathbf{x}_i \mathbf{x}_i^T\right) - \mathbf{R_x} tr(\mathbf{R_x}) - 2\mathbf{R_x^2} \tag{4}$$

2.1 The Update Equations

Given the estimates in (4), one can define recursive update rules for the estimates of the covariance and cumulant matrices **R** and **Q**, as well as $\mathbf{R}^{-1}$ and $\mathbf{R}^{-1}\mathbf{Q}$ for further computational savings. The recursive update for the covariance matrix is

$$\mathbf{R}_t = \frac{t-1}{t}\mathbf{R}_{t-1} + \frac{1}{t}\mathbf{x}_t\mathbf{x}_t^T \tag{5}$$

and the update rule for the cumulant matrix is given by

$$\mathbf{Q}_t = \mathbf{C}_t - \mathbf{R}_t tr(\mathbf{R}_t) - 2\mathbf{R}_t^2 \tag{6}$$

where the matrix **C** is defined as $\mathbf{C} = E\{\mathbf{x}^T\mathbf{x}\,\mathbf{x}\,\mathbf{x}^T\}$, and estimating the expectation using sample averages as before, it becomes

$$\mathbf{C} = \sum_i \left(\mathbf{x}_i^T\mathbf{x}_i\ \mathbf{x}_i\ \mathbf{x}_i^T\right) \tag{7}$$

Now, we can define the update rules for **C** and $\mathbf{R}^2$ to obtain the recursive update for the cumulant matrix. The update rule for **C** is given by

$$\mathbf{C}_t = \frac{t-1}{t}\mathbf{C}_{t-1} + \frac{1}{t}\left(\mathbf{x}_t^T\mathbf{x}_t\right)\mathbf{x}_t\mathbf{x}_t^T \tag{8}$$

The recursive update of $\mathbf{R}^2$ can be derived from (5) by squaring both sides. Hence, the update rule for $\mathbf{R}^2$ becomes

$$\mathbf{R}_t^2 = \frac{(t-1)^2}{t^2}\mathbf{R}_{t-1}^2 + \frac{1}{t^2}\left(\mathbf{x}_t^T\mathbf{x}_t\right)\mathbf{x}_t\mathbf{x}_t^T + \frac{(t-1)}{t^2}[\mathbf{v}_t\mathbf{x}_t^T + \mathbf{x}_t\mathbf{v}_t^T] \tag{9}$$

where for further computational savings we introduce the vector $\mathbf{v}_t$ as

$$\mathbf{v}_t = \mathbf{R}_{t-1}\mathbf{x}_t \tag{10}$$

Finally, the update rule for the cumulant matrix **Q** can be obtained by substituting (5), (8), and (9) into (6). Further computational savings can be obtained by iterating $\mathbf{R}^{-1}$ and $\mathbf{R}^{-1}\mathbf{Q}$ to avoid matrix multiplications and inversions, each having an $O(n^3)$ computational load. The reason why we need these two matrices will be clear as we proceed to the fixed-point algorithm that solves for the generalized eigendecomposition. Employing the matrix inversion lemma, the recursion rule for $\mathbf{R}^{-1}$ becomes

$$\mathbf{R}_t^{-1} = \frac{t}{t-1}\mathbf{R}_{t-1}^{-1} - \frac{t}{(t-1)\alpha_t}\mathbf{u}_t\mathbf{u}_t^T \tag{11}$$

where α_t and $\mathbf{u}_t$ are defined as

$$\alpha_t = (t-1) + \mathbf{x}_t^t\mathbf{u}_t \qquad \mathbf{u}_t = \mathbf{R}_{t-1}^{-1}\mathbf{x}_t \tag{12}$$

Here we define the matrix **D**, the update equation of whom can easily be defined by substituting the previously given update equations for $\mathbf{R}^{-1}$ and **Q**, using (11) and (6).

$$\mathbf{D}_t = \mathbf{R}_t^{-1}\mathbf{Q}_t \tag{13}$$

2.2 Deflation Procedure

Having the update equations, the aim is to find the optimal solution for the eigendecomposition for the updated correlation and cumulant matrices in each iteration. Recall the original problem given in (3); we need to solve for the weight matrix **W**. We will employ the deflation procedure to determine each generalized eigenvector sequentially. Every generalized eigenvector $\mathbf{w}_d$ that is a column of the weight matrix **W** is a stationary point of the function

$$J(\mathbf{w}) = \frac{\mathbf{w}^T \mathbf{R}\, \mathbf{w}}{\mathbf{w}^T \mathbf{Q}\, \mathbf{w}}. \tag{14}$$

This fact can be easily seen by taking the derivative and equating it to zero:

$$\begin{aligned} \frac{\partial J(\mathbf{w})}{\partial \mathbf{w}} &= \frac{\mathbf{w}^T \mathbf{Q}\, \mathbf{w}(2\,\mathbf{R}\,\mathbf{w}) - \mathbf{w}^T \mathbf{R}\, \mathbf{w}(2\,\mathbf{Q}\,\mathbf{w})}{\left(\mathbf{w}^T \mathbf{Q}\, \mathbf{w}\right)^2} = 0 \\ \mathbf{R}\mathbf{w} &= \frac{\mathbf{w}^T \mathbf{R}\, \mathbf{w}}{\mathbf{w}^T \mathbf{Q}\, \mathbf{w}} \mathbf{Q}\mathbf{w} \end{aligned} \tag{15}$$

This is nothing but the generalized eigenvalues equation where the eigenvalues are the values of the objective criterion $J(\mathbf{w})$ given in (14) evaluated at its stationary points. Hence, the fixed-point algorithm becomes

$$\mathbf{w} \leftarrow \frac{\mathbf{w}^T \mathbf{R}\, \mathbf{w}}{\mathbf{w}^T \mathbf{Q}\, \mathbf{w}} \mathbf{R}^{-1}\mathbf{Q}\mathbf{w}. \tag{16}$$

This fixed-point optimization procedure converges to the largest generalized eigenvector[1] of **R** and **Q**, and the deflation procedure is employed to manipulate the matrices such that the obtained matrices have the same generalized eigenvalue and eigenvector pairs except the ones that have been determined previously. The larger eigenvalues are replaced by zeros in each deflation step. Note that in this subsection the time index is implicit and omitted for notational convenience. While d represents the dimension index, the deflation procedure employed while iterating the dimensions is given by

$$\mathbf{Q}_d = \left[\mathbf{I} - \frac{\mathbf{Q}_{d-1}\mathbf{w}_{d-1}\mathbf{w}_{d-1}^T}{\mathbf{w}_{d-1}^T\mathbf{Q}_{d-1}\mathbf{w}_{d-1}}\right]\mathbf{Q}_{d-1} \qquad \mathbf{R}_d = \mathbf{R}_{d-1}. \tag{17}$$

The deflated matrices are initialized to $\mathbf{Q}_1$=**Q** and $\mathbf{R}_1$=**R**. Obtaining the new matrices using deflation, we will employ the same fixed-point iteration procedure given in (16) to find the corresponding eigenvector.

Investigating the fixed-point algorithm in (16), it is clear that iterating $\mathbf{R}^{-1}$ and **D** as suggested earlier will result in computational savings. The deflation rules for these matrices can be deduced from (17) easily. The deflation of $\mathbf{R}^{-1}$ is

$$\mathbf{R}_d^{-1} = \mathbf{R}_{d-1}^{-1}. \tag{18}$$

[1] The largest eigenvector is the one that corresponds to the greatest eigenvalue.

Similarly, the deflation rule for **D** can be obtained by combining (17) and (18) as

$$\mathbf{D}_d = [\mathbf{I} - \mathbf{w}_{d-1}\mathbf{w}_{d-1}^T \mathbf{Q}_{d-1}]\,\mathbf{D}_{d-1} \tag{19}$$

For each generalized eigenvector, the corresponding fixed-point update rule then becomes as follows:

$$\mathbf{w}_d \leftarrow \frac{\mathbf{w}_d^T \mathbf{R}_d\, \mathbf{w}_d}{\mathbf{w}_d^T \mathbf{Q}_d \mathbf{w}_d} \mathbf{D}_d\, \mathbf{w}_d \tag{20}$$

Employing this fixed-point algorithm for each dimension and solving for the eigenvectors sequentially, one can update the **W** matrix and proceed to the next time update step. In the following section we will present results comparing the original GED-ICA algorithm [9] with the results of the proposed recursive GED-ICA algorithm.

3 Experimental Results

In this section, the results provided by the proposed recursive algorithm will be compared with those of the original GED-ICA algorithm. The experiments are done on a synthetic dataset, which is simply generated by a linear mixture of independent uniform sources. Experiments using mixing matrices with varying condition numbers are employed to test the dependency of the tracking performance on the mixture conditioning.

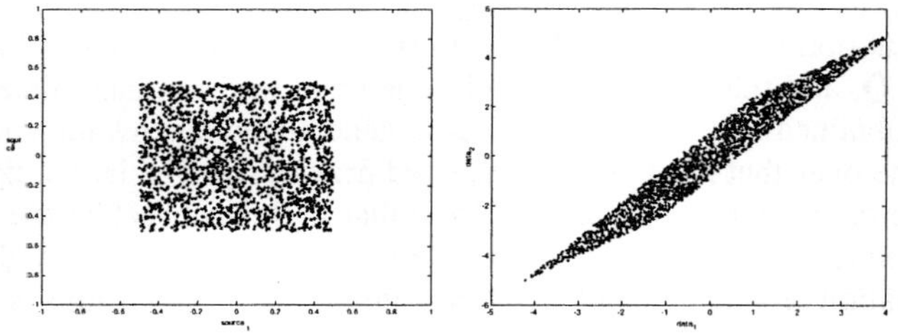

(a) (b)

Fig. 1. An illustration of the samples (a) from independent uniform sources (b) after the linear mixing

The joint distribution of the sources is presented in Figure 1 for a two-dimensional case. Mixing matrices with condition numbers 10 and 100 are employed for the mixing and the corresponding results are presented for two cases. In the first case, the original GED-ICA is employed on a small initialization data set to obtain ideal initial values for all matrices involved, including the eigenvectors. The expected result for the first simulation is to observe that the recursive algorithm is capable of tracking the result of the original algorithm within a range of small numerical error. In second case, these values are initialized to arbitrary small matrices. As increasing number of samples are utilized for the matrix updates, the bias of these arbitrary initial conditions is expected to decay. The second experiment will allow us to investigate this decay process by comparing the *biased* solution to that of the original GED-ICA procedure.

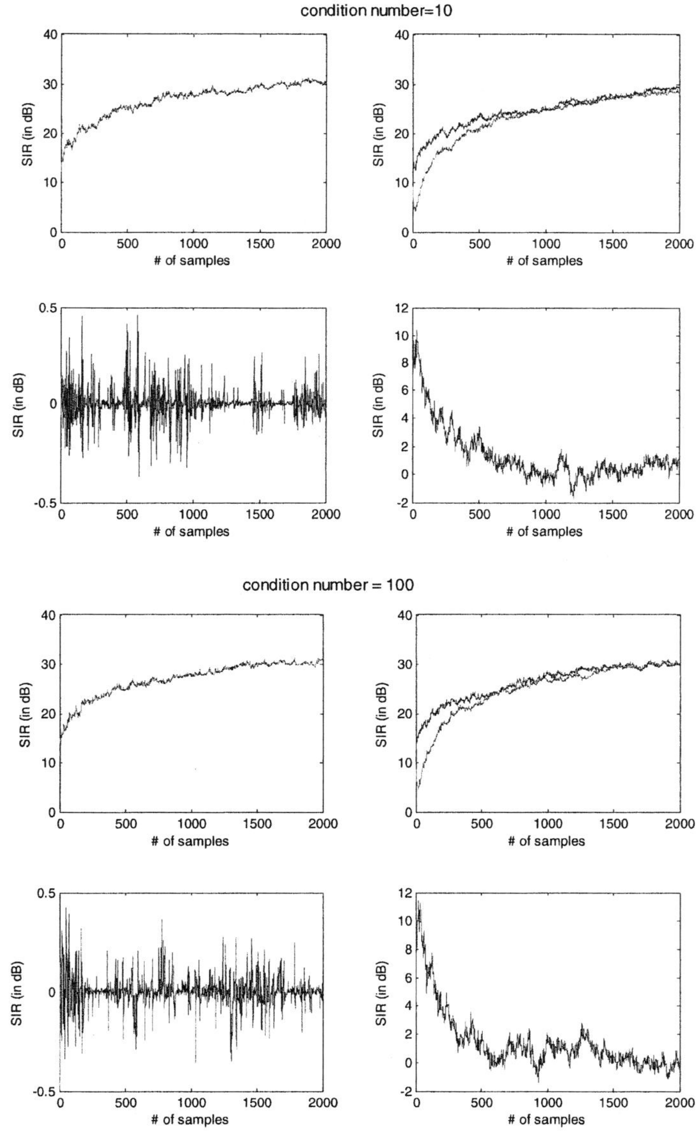

Fig. 2. The performances of recursive and the original methods are compared for a mixing of condition numbers 10 and 100. Performances and performance differences for exact initialization (top and bottom left, accordingly) and random initialization (top and bottom right) are shown.

In the simulations, the 20 samples have been used for initialization the ideal solution, and for the arbitrary initialization identity matrices with a small variances are employed (note that once **R** and **C** are initialized to values at the order of 10^{-6} all other matrices can be determined consistently with the equations). The corresponding average tracking results for 2000 samples are shown in Figure 2 for mixing matrix condition numbers of 10 and 100. These results are averaged over 100 Monte Carlo runs keeping the condition number of the mixture and the joint source distribution fixed and randomizing the right and left eigenvectors of the mixing matrix as well as the actual source samples using in the sample averages.

4 Conclusions

On-line ICA algorithms are essential for may signal processing and machine learning applications, where the ICA solution acts as a front-end preprocessor, a feature extractor, or a major portion of the solution itself. Stochastic gradient based algorithm motivated by various ICA criteria have been utilized successfully in such situations and they have the advantage of yielding computationally simple weight update rules. On the other hand, they are not able to offer optimal solutions at every iteration.

In this paper, we derived a recursive ICA algorithm based on the joint diagonalization of covariance and fourth order cumulants. The derivation employs the use of the matrix inversion lemma and the sample update rules for expectations approximated by sample averages. Since the proposed method is the recursive version of the algorithm proposed in [9], and it is tracking the optimal solution given by this algorithm in a recursive manner, the experimental results section is limited to the comparisons between the proposed recursive method and the original algorithm.

The resulting algorithm, of course, is computationally more expensive than its stochastic gradient counterpart. However, it has the ability to converge to and track the optimal solution based on this separation criterion in a small number of samples, even when initialized to arbitrary matrices.

Acknowledgments

This work was supported by DARPA under contract DAAD-16-03-C-0054 and by NSF under grant ECS-0524835.

References

1. Hyvarinen A., Karhunen J., Oja E.: Independent Component Analysis, Wiley, New York (2001)
2. Cichocki A., Amari S.I.: Adaptive Blind Signal and Image Processing: Learning Algorithms and Applications, Wiley, 2002.
3. Hyvärinen A., Oja E., Hoyer P., Hurri J.: Image Feature Extraction by Sparse Coding and Independent Component Analysis, Proceedings of ICPR'98, (1998) 1268-1273

4. Lan T., Erdogmus D., Adami A., Pavel M.: Feature Selection by Independent Component Analysis and Mutual Information Maximization in EEG Signal Classification, Proceedings IJCNN 2005, Montreal, (2005) 3011-3016
5. Everson R., Roberts S.: Independent Component Analysis: A Flexible Nonlinearity and Decorrelating Manifold Approach, Neural Computation, vol. 11. (2003) 1957-1983
6. Bell A., Sejnowski T.: An Information-Maximization Approach to Blind Separation and Blind Deconvolution, Neural Computation, vol. 7. (1195) 1129-1159
7. Haykin S.: Adaptive Filter Theory, Prentice Hall, Upper Saddle River, New Jersey, (1996)
8. Cardoso J.: Bind signal separation: Statistical principles, Proc. of the IEEE, vol. 86. (1998)
9. Parra L., Sajda P.: Blind Source Separation via Generalized Eigenvalue Decomposition, Journal of Machine Learning Research, vol. 4. (2003) 1261-1269

Recovery of Sparse Representations by Polytope Faces Pursuit

Mark D. Plumbley

Department of Electronic Engineering, Queen Mary University of London,
Mile End Road, London E1 4NS, United Kingdom
mark.plumbley@elec.qmul.ac.uk

Abstract. We introduce a new greedy algorithm to find approximate sparse representations $\mathbf{s}$ of $\mathbf{x} = \mathbf{A}\mathbf{s}$ by finding the Basis Pursuit (BP) solution of the linear program $\min\{\|\mathbf{s}\|_1 \mid \mathbf{x} = \mathbf{A}\mathbf{s}\}$. The proposed algorithm is based on the geometry of the polar polytope $P^* = \{\mathbf{c} \mid \tilde{\mathbf{A}}^T\mathbf{c} \leq \mathbf{1}\}$ where $\tilde{\mathbf{A}} = [\mathbf{A}, -\mathbf{A}]$ and searches for the vertex $\mathbf{c}^* \in P^*$ which maximizes $\mathbf{x}^T\mathbf{c}$ using a path following method. The resulting algorithm is in the style of Matching Pursuits (MP), in that it adds new basis vectors one at a time, but it uses a different correlation criterion to determine which basis vector to add and can switch out basis vectors as necessary. The algorithm complexity is of a similar order to Orthogonal Matching Pursuits (OMP). Experimental results show that this algorithm, which we call *Polytope Faces Pursuit*, produces good results on examples that are known to be hard for MP, and it is faster than the interior point method for BP on the experiments presented.

1 Introduction

Suppose are given a sequence of observations $\mathbf{X} = [\mathbf{x}_1, \ldots, \mathbf{x}_T]$, $\mathbf{x}_t \in \mathbb{R}^d$, which we wish to decompose according to the usual independent component analysis (ICA) generative model $\mathbf{X} = \mathbf{A}\mathbf{S}$, where $\mathbf{A} = [\mathbf{a}_1, \ldots, \mathbf{a}_n]$ is a $d \times n$ mixing matrix of real basis vectors $\mathbf{a}_i$ and sequence $\mathbf{S} = [\mathbf{s}_1, \ldots, \mathbf{s}_T]$ of source coefficient vectors $\mathbf{s}_t \in \mathbb{R}^n$. If we have $n > d$, i.e. the number of basis vectors is larger than the dimensionality of the basis space, then the system is *overcomplete*. This means in particular that if we have identified the mixing matrix $\mathbf{A}$, using some dictionary learning method (see e.g. [1]), then the equation $\mathbf{x}_t = \mathbf{A}\mathbf{s}_t$ at a particular t still has multiple solutions in general for $\mathbf{s}_t$ given $\mathbf{A}$ and $\mathbf{x}_t$. In *sparse coding*, we favour the minimum ℓ_0-norm solution $\mathbf{s}$ to $\mathbf{x} = \mathbf{A}\mathbf{s}$ which has the smallest number of non-zero elements $\|\mathbf{s}\|_0$,

$$\min_{\mathbf{s}} \|\mathbf{s}\|_0 \quad \text{such that} \quad \mathbf{x} = \mathbf{A}\mathbf{s}\,. \tag{1}$$

However, finding the ℓ_0 solution (1) is known to be a hard problem. Instead, the method of *Basis Pursuit* (BP) [2] proposes to find the minimum ℓ_1 norm solution

$$\min_{\mathbf{s}} \|\mathbf{s}\|_1 \quad \text{such that} \quad \mathbf{x} = \mathbf{A}\mathbf{s} \tag{2}$$

J. Rosca et al. (Eds.): ICA 2006, LNCS 3889, pp. 206–213, 2006.

which is equivalent to a linear programming (LP) problem and can be solved with e.g. interior point methods. However these methods can still be slow, so in practice greedy algorithms such as *Matching Pursuits* (MP) [3] or *Orthogonal Matching Pursuit* (OMP) [4] have been used as an efficient way to find approximate solutions to (2). Recently there has been much investigation into the conditions under which the solutions to (1) and (2) coincide (ℓ_1/ℓ_0 equivalence) and will be found by MP/OMP (exact recovery condition). For discussions see e.g. [5,6,7,8] and references therein.

In an interesting new direction, Donoho [9] has shown that by considering the d-dimensional *polytope* (the d-dimensional generalization of a polygon) whose vertices are the $2n$ signed basis vectors $\pm\mathbf{a}_i$, $\mathbf{a}_i \in \mathbf{A}$, results from the theory of polytopes can give us insight into the question of ℓ_1/ℓ_0 equivalence. Using a similar geometric approach, the present author [10] has explored the geometry of the problem (2), giving a visualization for the conditions discussed by Fuchs [7] and Tropp [6] for a unique optimal solution to (2), and hence ℓ_1/ℓ_0 equivalence.

In this paper we build on this geometrical visualization to propose a new greedy algorithm that performs Basis Pursuit (2). The algorithm adopts a path-following approach through the relative interior of the faces of the polar (dual) polytope associated with the dual LP problem. We refer to this as the *Polytope Faces Pursuit* algorithm.

2 Dual Linear Programs and Polar Polytopes

It is sometimes convenient to convert (2) into its standard form [2,11]

$$\min_{\tilde{\mathbf{s}}} \mathbf{1}^T\tilde{\mathbf{s}} \quad \text{such that} \quad \mathbf{x} = \tilde{\mathbf{A}}\tilde{\mathbf{s}},\ \tilde{\mathbf{s}} \geq 0 \tag{3}$$

where $\mathbf{1}$ is a column vector of ones, $\tilde{\mathbf{A}} = [\mathbf{A}, -\mathbf{A}]$ and $\tilde{\mathbf{s}}$ has $2n$ nonnegative components $\tilde{s}_i = \max(s_i, 0)$ for $1 \leq i \leq n$ and $\tilde{s}_i = \max(-s_{i-n}, 0)$ for $n+1 \leq i \leq 2n$. Clearly $s_i = \tilde{s}_i - \tilde{s}_{i+n}$. The new linear program (3) has a corresponding *dual* linear program [2]

$$\max_{\mathbf{c}} \mathbf{x}^T\mathbf{c} \quad \text{such that} \quad \tilde{\mathbf{A}}^T\mathbf{c} \leq \mathbf{1} \tag{4}$$

which has an optimum $\mathbf{c}^*$ associated with any optimum $\tilde{\mathbf{s}}^*$ of (3). Thus to perform BP we can search for the optimum $\mathbf{c}$ in (4) and solve the resulting (determined) system for $\tilde{\mathbf{s}}$.

To help us visualize this search space, we introduce some geometric concepts. A d-dimensional *polytope* P is a bounded subset of $\mathbb{R}^d$ defined by a finite set of inequalities, or alternatively as the convex hull of a finite set of extreme points, its *vertices*. If the inequality $\mathbf{a}^T\mathbf{x} \leq b$ is *valid* for P, i.e. $\mathbf{a}^T\mathbf{x} \leq b$ for all $\mathbf{x} \in P$, then $F = \{\mathbf{x} \in P \mid \mathbf{a}^T\mathbf{x} = b\}$ is a *face* of P. Examples of faces include the *improper* faces $\emptyset$ and P itself, as well as the vertices (0-dimensional faces) and facets ($(d-1)$-dimensional faces) of P. For more definitions and notation see e.g. [9,12].

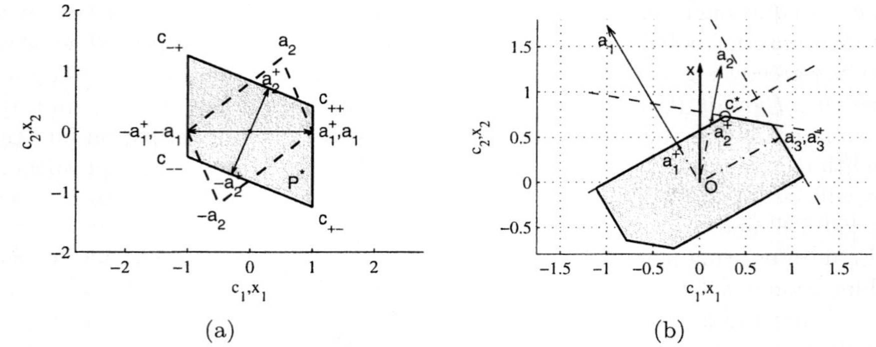

Fig. 1. Examples of polytopes in 2-D: (a) primal (dashed) and polar (solid) polytopes; (b) polar polytope showing ideal basis vertex c^*

The dashed polygon in Fig. 1(a) shows the *primal polytope* P with vertices given by the signed basis vectors $\pm\mathbf{a}_i$. Of more interest to us is the *polar polytope* P^* defined by the inequalities $P^* = \{\mathbf{c} \mid \tilde{\mathbf{a}}_i^T\mathbf{c} \leq 1\}$ where $\tilde{\mathbf{a}}_i \in \tilde{\mathbf{A}}$, i.e. $\tilde{\mathbf{a}}_i \in \{\mathbf{a}_1, \mathbf{a}_2, -\mathbf{a}_1, -\mathbf{a}_2\}$. The scaled vectors $\mathbf{a}_i^+ = \mathbf{a}_i/|\mathbf{a}_i|^2$, satisfy $\mathbf{a}_i^T\mathbf{a}_i^+ = 1$ so touch the faces, extended if necessary, defined by $\mathbf{a}_i^T\mathbf{c} = 1$ ($\mathbf{a}_i^+$ is also the transpose of the Moore-Penrose pseudoinverse of $\mathbf{a}_i$). Clearly this polar polytope P^* is the feasible region for $\mathbf{c}$ in (4). The vertices $\mathbf{c}_{++}$ etc. of P^* correspond to particular sets of selected vertices. If we let $\mathbf{A}_{+-} = [\mathbf{a}_1, -\mathbf{a}_2]$ then $\mathbf{c}_{+-}$ is the vertex such that $\mathbf{A}_{+-}^T\mathbf{c}_{+-} = \mathbf{1}$, i.e. $\mathbf{c}_{+-} = \mathbf{A}_{+-}^{\dagger\ T}\mathbf{1}$ where $\mathbf{A}^\dagger$ is the Moore-Penrose pseudoinverse of $\mathbf{A}$ (used so that we can define $\mathbf{c}$ for an $\mathbf{A}$ with less than d columns). It is a standard result from linear programming that the optimum of (4) will be achieved at one of the vertices [11]. Therefore it remains for us to identify which is the optimal vertex which maximizes $\mathbf{x}^T\mathbf{c}$, and use that to find the optimal vector $\tilde{\mathbf{s}}$ (and hence $\mathbf{s}$).

As an example, consider the shaded polytope in Fig. 1(b), which is defined by three basis vector pairs $\mathbf{A} = [\mathbf{a}_1, \mathbf{a}_2, \mathbf{a}_3]$. The scaled vectors $\mathbf{a}_i^+$, $i = 1, 2, 3$ are also shown. The dual vector $\mathbf{c}$ that maximizes the inner product $\mathbf{x}^T\mathbf{c}$ with $\mathbf{x}$ is the one furthest along the direction parallel to $\mathbf{x}$, and is marked $\mathbf{c}^*$. In this particular case $\mathbf{c}^*$ corresponds to the optimal basis set $\tilde{\mathbf{A}}_{\text{opt}} = [\mathbf{a}_1, \mathbf{a}_2]$, which has the corresponding primal solution $\tilde{\mathbf{s}} = \tilde{\mathbf{A}}_{\text{opt}}^\dagger\mathbf{x}$, or in original form $\mathbf{s} = \mathbf{A}_{\text{opt}}^\dagger\mathbf{x}$ where $\mathbf{A}_{\text{opt}} = [\mathbf{a}_1, \mathbf{a}_2]$.

However, consider what happens if we apply either MP or OMP, suitably adjusted for non-unit-norm basis vectors, to the situation pictured in Fig. 1(b). The first vector selected is $\arg\max_{\mathbf{a}_i} \mathbf{a}_i^T\mathbf{x} = \mathbf{a}_1$, so after the first step, $\mathbf{A}^1 = [\mathbf{a}_1]$. This produces a residual $\mathbf{r}^1$ perpendicular to $\mathbf{a}_1$, i.e. along $\mathbf{a}_3$, so on the second step both MP and OMP will select $\mathbf{a}_3$. The basis set $\mathbf{A}^2 = [\mathbf{a}_1, \mathbf{a}_3]$ now spans the space, the residual $\mathbf{r}^2$ is zero, and both MP and OMP stop after 2 steps. MP and OMP have the same behaviour in this case because $\mathbf{a}_1$ and $\mathbf{a}_3$ are orthogonal. Thus both MP and OMP have failed to find the optimal solution to (2). In Natarajan's algorithm [13], sometimes called *Order Recursive*

Matching Pursuit, there would be an arbitrary choice between $\mathbf{a}_2$ and $\mathbf{a}_3$ in step 2, since both would give zero residual after selection, so this can also fail on this example.

3 Deriving the Faces Pursuit Algorithm

Let us now derive an algorithm to find the true optimal LP solution of (2), but will build up its solution in a similar way to MP/OMP. Our first insight is that if we project from the origin O along $\mathbf{h} = \alpha\mathbf{x}$ for $\alpha \geq 0$ the first polytope face we encounter is at

$$\alpha_1 = \min\{\alpha > 0 \mid \mathbf{a}_i^T(\alpha\mathbf{x}) = 1\} = \min\{\alpha > 0 \mid \mathbf{a}_i^T\mathbf{x} = 1/\alpha\} = 1/\max\{\mathbf{a}_i^T\mathbf{x}\}$$

which is our normal MP/OMP maximum correlation condition. Let us select $\mathbf{a}_1$ as our first vector, and then continue our path 'towards' $\mathbf{x}$, but with our path now constrained to be within the polytope face $F^1 = \{\mathbf{h} \in P^* \mid \mathbf{a}_1^T\mathbf{h} = 1\}$. We can achieve this by projecting $\mathbf{x}$ into the subspace parallel to F^1, to give $\mathbf{r}^1 = (\mathbf{I} - \mathbf{Q}_1)\mathbf{x}$ where $\mathbf{Q}_1 = \mathbf{a}_1\mathbf{a}_1^{+T} = \mathbf{a}_1\mathbf{a}_1^T/|\mathbf{a}_1|^2$. Since $\tilde{\mathbf{s}}^1 = \mathbf{a}_1^{+T}\mathbf{x}$ and $\hat{\mathbf{x}}^1 = \mathbf{a}_1\tilde{\mathbf{s}}^1$ we have $\mathbf{r}^1 = \mathbf{x} - \mathbf{a}_1\tilde{\mathbf{s}}^1 = \mathbf{x} - \hat{\mathbf{x}}^1$ so $\mathbf{r}^1$ is therefore the residual from the approximation $\hat{\mathbf{x}}^1$ to $\mathbf{x}$ obtained after Step 1 (Fig. 2).

The second step is where the difference from MP/OMP arises. These would find $\max_i \mathbf{a}_i^T\mathbf{r}^1$, but this is the first face encountered projecting along the residual $\mathbf{r}^1$ *from the origin*, not within face F^1. Instead, to correctly determine the next face we encounter along the face F^1 we project along the residual starting at

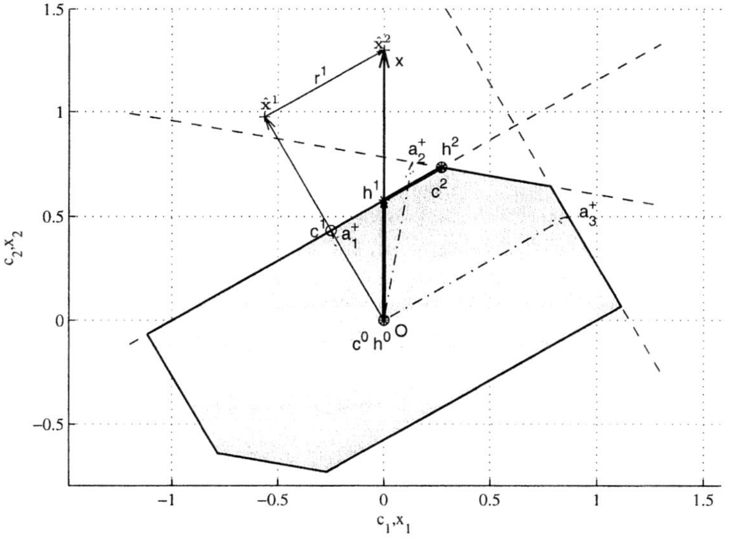

Fig. 2. Path of the Faces Pursuit Algorithm, starting at $\mathbf{c}^0 = \mathbf{h}^0 = \mathbf{0}$

$\mathbf{h}_1 = \alpha_1 \mathbf{x}$. A little manipulation will confirm that the next face is encountered at $\min\{\alpha > \alpha^1 \mid \mathbf{a}_i^T(\mathbf{c}^1 + \alpha \mathbf{r}^1) = 1\} = 1/\max\{(\mathbf{a}_i^T \mathbf{r}^1)/(1 - \mathbf{a}_i^T \mathbf{c}^1) \mid \mathbf{a}_i^T \mathbf{r}^1 > 0\}$ where $\mathbf{c}^1 = \mathbf{a}_1^+ \cdot 1$ and we exclude faces already encountered. Further consideration along these lines shows that this generalizes so at each step k we do not simply want the maximum correlation $\mathbf{a}_i^T \mathbf{r}^{k-1}$, but the maximum scaled correlation

$$\mathbf{a}^k = \arg\max_{\mathbf{a}_i} \frac{\mathbf{a}_i^T \mathbf{r}^{k-1}}{1 - \mathbf{a}_i^T \mathbf{c}^{k-1}} \tag{5}$$

where we consider only vectors $\mathbf{a}_i$ for which $\mathbf{a}_i^T \mathbf{r}^{k-1} > 0$. Note that if we have a fast method to compute $\mathbf{a}_i^T \mathbf{r}^{k-1}$, such as a fast Wavelet transform if $\mathbf{A}$ is a wavelet basis, we can use the same method to compute $\mathbf{a}_i^T \mathbf{c}^{k-1}$.

In the complete algorithm (Algorithm 1) we also need to be able to optionally switch out certain constraints once we have encountered a new face (consider 'climbing up' the face corresponding to $\mathbf{a}_3$ instead of $\mathbf{a}_1$ in Fig. 2: we leave that constraint once we encounter the face corresponding to $\mathbf{a}_2$). A little manipulation will show that constraints should be switched out after step k if $\tilde{\mathbf{s}}^k$ contains any negative entries: the negative entries corresponding to the constraints to be removed. We also have to take practical steps to ensure that $\mathbf{a}_i^T \mathbf{c}^k \approx 1$ does not give divide-by-zero errors, and to ensure the same bases are not considered candidates again.

Algorithm 1. Polytope Faces Pursuit

Input: $\tilde{\mathbf{A}} = [\tilde{\mathbf{a}}_i]$, $\mathbf{x}$ {If required, set $\tilde{\mathbf{A}} \leftarrow [\mathbf{A}, -\mathbf{A}]$}
Set stopping conditions $k_{\max}$ and $\theta_{\min}$
Initialize: $k \leftarrow 0$, $\mathcal{I}^k \leftarrow \emptyset$, $\tilde{\mathbf{A}}^k \leftarrow \emptyset$, $\mathbf{c}^k \leftarrow \mathbf{0}$, $\tilde{\mathbf{s}}^k \leftarrow \emptyset$, $\hat{\mathbf{x}}^k \leftarrow \mathbf{0}$, $\mathbf{r}^k \leftarrow \mathbf{x}$
while $k < k_{\max}$ and $\max_i \tilde{\mathbf{a}}_i^T \mathbf{r}^{k-1} > \theta_{\min}$ **do** {Find next face}
  $k \leftarrow k + 1$
  Find face: $i^k \leftarrow \arg\max_{i \notin \mathcal{I}^{k-1}} \{(\tilde{\mathbf{a}}_i^T \mathbf{r}^{k-1})/(1 - \tilde{\mathbf{a}}_i^T \mathbf{c}^{k-1}) \mid \tilde{\mathbf{a}}_i^T \mathbf{r}^{k-1} > 0\}$
  Optionally: $\alpha^k \leftarrow (1 - \tilde{\mathbf{a}}_{i^k}^T \mathbf{c}^{k-1})/(\tilde{\mathbf{a}}_{i^k}^T \mathbf{r}^{k-1})$, $\mathbf{h}^k \leftarrow \mathbf{c}^{k-1} + \alpha^k \mathbf{r}^{k-1}$
  Add constraint: $\tilde{\mathbf{A}}^k \leftarrow [\tilde{\mathbf{A}}^{k-1}, \mathbf{a}_{i^k}]$, $\mathcal{I}^k \leftarrow \mathcal{I}^{k-1} \cup \{i^k\}$, $\mathbf{B}^k \leftarrow (\tilde{\mathbf{A}}^k)^\dagger$, $\tilde{\mathbf{s}}^k \leftarrow \mathbf{B}^k \mathbf{x}$
  while $\tilde{\mathbf{s}}^k \not\geq \mathbf{0}$ **do** {Release retarding constraints}
    Select some $j \in \mathcal{I}^k$ such that $\tilde{s}_j^k < 0$; remove column $\mathbf{a}_j$ from $\tilde{\mathbf{A}}^k$
    Update: $\mathcal{I}^k \leftarrow \mathcal{I}^k \setminus \{j\}$, $\mathbf{B}^k \leftarrow (\tilde{\mathbf{A}}^k)^\dagger$, $\tilde{\mathbf{s}}^k \leftarrow \mathbf{B}^k \mathbf{x}$
  end while
  $\mathbf{c}^k \leftarrow (\mathbf{B}^k)^T \mathbf{1}$, $\hat{\mathbf{x}}^k \leftarrow \tilde{\mathbf{A}}^k \tilde{\mathbf{s}}^k$, $\mathbf{r}^k \leftarrow \mathbf{x} - \hat{\mathbf{x}}^k$
end while
Output: $\mathbf{c}^* = \mathbf{c}^k$, $\tilde{\mathbf{s}}^* \leftarrow \mathbf{0} +$ corresponding entries from $\tilde{\mathbf{s}}^k$
{If required, get $s_i^* \leftarrow (\tilde{s}_i^* - \tilde{s}_{i+n}^*)$, $1 \leq i \leq n$}

The most expensive operations in the algorithm are the dictionary analysis calculations $\tilde{\mathbf{a}}_i^T \mathbf{r}^{k-1}$ and $\tilde{\mathbf{a}}_i^T \mathbf{c}^{k-1}$ (two per step k instead of one per step for MP/OMP) and the pseudoinverse calculation $(\tilde{\mathbf{A}}^k)^\dagger$ (one per step k as for OMP, plus one each retarding constraint release in the inner loop). Thus the complexity is similar to OMP but with an additional dictionary probe each loop. Constraint releases appear to be relatively infrequent: on the Gong example in Section 4 constraint releases occur on average every 5–6 steps. Note that the Algorithm 1 does not require **A** to be a unit norm dictionary.

4 Experiments

To confirm the operation of the Polytope Faces Pursuit algorithm and compare it with MP and the interior point method for BP, we applied the algorithm to some examples where MP and BP have already been compared [2]. Fig. 3 shows MP, OMP, BP and the proposed Polytope Faces Pursuit algorithm (FP) applied to the signal 'TwinSine', a superposition of two sinusoids separated by less than the Rayleigh Distance $2\pi/n$. The analysis is performed in a 4-fold overcomplete

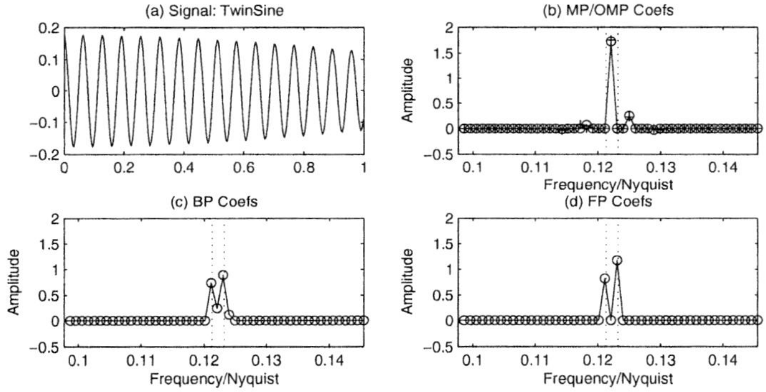

Fig. 3. (a) The 'TwinSine' signal (see [2]) with decompositions by (b) MP ('o') and OMP ('+'), (c) BP using the interior point method, and (d) Polytope Faces Pursuit

discrete cosine dictionary [2]. The stopping condition $\theta_{\min}$ was set to 10^{-2} times the signal norm for MP and Polytope Faces Pursuit.

As is already known, MP and OMP fail to resolve this signal, in that they both initially select the middle frequency atom (atom 126) and subsequently do not remove it, while BP (using the interior point method) does resolve the signal correctly. Faces Pursuit quickly produces a clean sparse decomposition: atom 126 is initially selected, the same as for MP/OMP as we would expect, but following the addition of atoms 125 and 127 atom 126 it is removed at step 3 as a retarding constraint. Thus Faces Pursuit stops in 3 steps, yielding a sparse representation consisting of only 2 basis atoms (Fig. 3(d)). Remaining differences with BP appear to be due to algorithm tolerances.

Fig. 4 shows MP, BP and Polytope Faces Pursuit applied to the signal 'Gong', which is zero until t_0 and then follows a decaying sinusoid. The analysis is performed using a cosine packet dictionary [2]. As already known, BP using the interior point method produces a 'cleaner' decomposition than MP, although it is approximately an order of magnitude slower than MP on this example. We

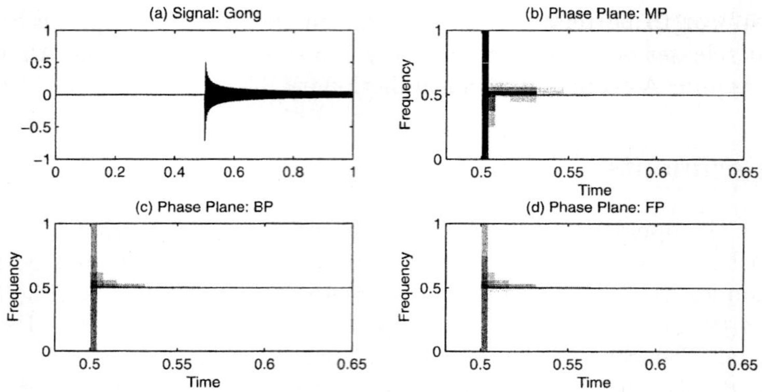

Fig. 4. (a) The 'Gong' signal (see [2]) with decompositions by (b) MP, (c) BP using the interior point method, and (d) Polytope Faces Pursuit

can confirm that Polytope Faces Pursuit produces a similar result to BP, taking a time between MP and BP for the same accuracy. A typical run on the three algorithms using Matlab 7.0 (R14) under Windows XP on a 1.5GHz Intel Pentium M laptop takes $t_{\mathrm{MP}} = 28$s, $t_{\mathrm{BP}} = 224$s, $t_{\mathrm{FP}} = 74$s. We expect to improve the speed of the Faces Pursuit algorithm in future through the use of matrix and vector updating and downdating formulae in place of the expensive pseudo-inverse calculation each step (see e.g. [14]).

5 Conclusions

We have introduced a new greedy algorithm to find approximate sparse representations $\mathbf{s}$ of $\mathbf{x} = \mathbf{A}\mathbf{s}$ given $\mathbf{A}$ and $\mathbf{x}$. The algorithm is based on the geometry of the polar polytope $P^* = \{\mathbf{c} \mid \tilde{\mathbf{A}}^T\mathbf{c} \leq \mathbf{1}\}$ where $\tilde{\mathbf{A}} = [\mathbf{A}, -\mathbf{A}]$ which defines the feasible region of the dual linear program $\max\{\mathbf{x}^T\mathbf{c} \mid \tilde{\mathbf{A}}^T\mathbf{c} \leq \mathbf{1}\}$. The algorithm searches for the vertex $\mathbf{c}^* \in P^*$ which maximizes $\mathbf{x}^T\mathbf{c}$ using a path following method through the relative interior of faces of P. We call this method *Polytope Faces Pursuit.*

The resulting algorithm is in the style of Matching Pursuits (MP), in that it adds new basis vectors one at a time based in a correlation criterion, but has two major differences: (1) the correlation criterion depends on the current vertex $\mathbf{c}^k$ at step k as well as the residual $\mathbf{r}^k$; and (2) basis vectors are switched out if necessary. The algorithm complexity is of the same order as OMP, although it has one additional dictionary probe per step, and has costs associated with switching out of basis vectors.

Experimental results confirm that the Polytope Faces Pursuit algorithm produces good results on examples that are known to be challenging for MP, and that it is faster than the interior point method for Basis Pursuit (BP) on the experiments presented.

Acknowledgements

This work was partially supported by EPSRC grants GR/S82213/01, GR/S75802/01, EP/C005554/1 and EP/D000246/1. Some of the figures were generated using the Multi-Parametric Toolbox (MPT) for Matlab [15], WaveLab802 and Atomizer802 (`http://www-stat.stanford.edu/~wavelab/` and `http://www-stat.stanford.edu/~atomizer/`). Valuable comments from referees helped to improve the final paper.

References

1. Kreutz-Delgado, K., Murray, J.F., Rao, B.D., Engan, K., Lee, T.W., Sejnowski, T.J.: Dictionary learning algorithms for sparse representation. Neural Computation **15** (2003) 349–396
2. Chen, S.S., Donoho, D.L., Saunders, M.A.: Atomic decomposition by basis pursuit. SIAM Journal on Scientific Computing **20** (1998) 33–61
3. Mallat, S., Zhang, Z.: Matching pursuits with time-frequency dictionaries. IEEE Transactions on Signal Processing **41** (1993) 3397–3415
4. Pati, Y.C., Rezaiifar, R., Krishnaprasad, P.S.: Orthogonal matching pursuit: Recursive function approximation with applications to wavelet decomposition. In: Conference Record of The Twenty-Seventh Asilomar Conference on Signals, Systems and Computers, Pacific Grove, CA. (1993) 40–44
5. Gribonval, R., Nielsen, M.: Approximation with highly redundant dictionaries. In: Wavelets: Applications in Signal and Image Processing, Proc. SPIE'03, San Diego, USA. (2003) 216–227
6. Tropp, J.A.: Greed is good: Algorithmic results for sparse approximation. IEEE Transactions on Information Theory **50** (2004) 2231–2242
7. Fuchs, J.J.: On sparse representations in arbitrary redundant bases. IEEE Transactions on Information Theory **50** (2004) 1341–1344
8. Donoho, D.L., Elad, M.: Optimally sparse representation in general (nonorthogonal) dictionaries via ℓ^1 minimization. Proc. Nat. Aca. Sci. **100** (2003) 2197–2202
9. Donoho, D.L.: Neighborly polytopes and sparse solutions of underdetermined linear equations. Technical report, Statistics Department, Stanford University (2004)
10. Plumbley, M.D.: Polar polytopes and recovery of sparse representations (2005) Submitted for publication.
11. Schrijver, A.: Theory of Linear and Integer Programming. John Wiley & Sons Ltd, Chichester, UK (1998)
12. Grünbaum, B.: Convex Polytopes. Second edn. Graduate Texts in Mathematics 221. Springer-Verlag, New York (2003)
13. Natarajan, B.K.: Sparse approximate solutions to linear systems. SIAM J. Computing **25** (1995) 227–234
14. Andrle, M., Rebollo-Neira, L.: A swapping-based refinement of orthogonal matching pursuit strategies (2005) To appear in *Signal Processing*.
15. Kvasnica, M., Grieder, P., Baotić, M.: Multi-Parametric Toolbox (MPT) (2004)

ICA Based Semi-supervised Learning Algorithm for BCI Systems

Jianzhao Qin, Yuanqing Li, and Qijin Liu

Institute of Automation Science and Engineering,
South China University of Technology, Guangzhou 510640, China

Abstract. As an emerging technique, brain-computer interfaces (BCIs) bring us a new communication interface which can translate brain activities into control signals of devices like computers, robots etc. In this study, we introduce an independent component analysis (ICA) based semi-supervised learning algorithm for BCI systems. In this algorithm, we separate the raw electroencephalographic (EEG) signals into several independent components using ICA; then choose a best independent component for feature extraction and classification. To demonstrate the validity of our algorithm, we apply it to an data set from an EEG-based cursor control experiment implemented in Wadsworth Center. The data analysis results show that both ICA preprocessing and semi-supervised learning can improve prediction accuracy significantly.

1 Introduction

A brain-computer interface is a communication system that does not depend on brain's normal output pathways of peripheral nerves and muscles. It provides new augmentative communication technology to those who are paralyzed or have other severe movement deficits [1].

Electroencephalogram (EEG) is an electrical activity produced by the neurons and synapses of the central nervous system in the course of their operation and recorded from the scalp or from the cortical surface [2]. It can be taken as the input of BCI systems. Since noise including artifacts (e.g., eye movements, eye blinks and EMG) is inevitable in EEG signals, EEG based BCIs should contain a preprocessing procedure to separate the useful brain sources of EEG signals from noise. A good preprocessing method can improve the information transferring rate of BCIs significantly.

Independent component analysis (ICA), which finds a linear representation of nongaussian data such that the components are statistically independent [3], has been widely used in blind source separation [4,5,6], and EEG signal analysis [7], etc. We think ICA is a very promising preprocessing method to blindly separate the useful brain sources from noise (including artifacts) for BCI systems.

In this paper, we propose an algorithm for BCI systems. In this algorithm, ICA is introduced as a preprocessing method. That is, the raw EEG signals

J. Rosca et al. (Eds.): ICA 2006, LNCS 3889, pp. 214–221, 2006.

are first transformed into ICA components by an ICA algorithm, then features are extracted from the ICA component which is most relevant to the users' control intention and most suitable for feature extraction. Next, we apply a self-training based semi-supervised learning method for classification. The task of classification algorithm in BCI systems is to translate the features into control signals. Based on semi-supervised learning strategy, we train a classifier using small amount of labeled and large amount of unlabeled samples. The benefit of this method is in reducing training effort, which is often tedious and time-consuming, as stated in the description of Data Set IVa in [17].

We evaluated the ICA based semi-supervised learning method using the data set from an EEG-based cursor control experiment , which was implemented in Wadsworth Center. Three subjects' data were included in this data set. Data analysis results demonstrated the validity of our algorithm.

2 Methods

In this section, the presented methods include ICA preprocessing, feature extraction, and semi-supervised learning based classification.

2.1 ICA Preprocessing

In EEG-based BCI systems, the intention of the user is embedded in the EEG recordings. From the EEG recordings on the scalp, finding the hidden brain sources which present the users' intension is a very challenge task. Because, hidden brain sources are extremely weak, nonstationary signals and usually distorted by large noise, interference and on-going activity of the brain [11]. We think ICA, which can find statistically independent or as independent as possible components from the raw signal, is one of the promising methods that can be used to separate the hidden brain sources which present the users' intension from other noise including artifacts.

We assume that the multichannel EEG can be modelled by

$$\mathbf{X}(n) = \mathbf{AS}(n) \tag{1}$$

where $\mathbf{X}(n)$ is an EEG column vector at time n, the m unknown components $\mathbf{S} = [S_1, S_2, \ldots, S_m]^T$ are brain sources related to the intention of the users, artifacts etc., $\mathbf{A}$ is an unknown nonsingular square mixing matrix.

The task of blind source separation is to find a demixing matrix $\mathbf{C}$ such that the sources can be recovered as below

$$\mathbf{Y}(n) = \mathbf{CX}(n) \tag{2}$$

In this study, we apply a natural gradient-flexible ICA algorithm [12] (other ICA algorithms also can be used), which can blindly separate mixtures of sub- and super-gaussian sources, to find components.

2.2 Feature Extraction

After presenting ICA preprocessing, we will focus on feature extraction in this subsection. Note that the features will be extracted from ICA components instead of raw EEG signals. In the cursor control experiment mentioned before, the amplitude (i.e., the square root of power) in the mu and/or beta rhythm frequency band over sensorimotor cortex was used to control the cursor up and down [2]. Therefore, we still use the amplitude of mu or beta rhythm as the feature. However, in order to reflect the change of brain signals during a trial, we will extract a dynamic amplitude feature. That is, we separate the time interval of each trial into m overlapped time bins, and calculate the amplitude value of the mu or beta rhythm for each time bin. Consequently, a m dimensional amplitude feature vector is obtained for each trial.

Among the ICA components, only one will be chosen for amplitude feature extraction. How to select the component will be described in the next subsection. The dynamic amplitude feature of each trial is defined as,

$$\mathbf{PF} = [\sqrt{PF_1}, \sqrt{PF_2}, \sqrt{PF_3}, ..., \sqrt{PF_m}]^T, \tag{3}$$

$$PF_n = \sum_{f \in [11,14]} P_n(f), n = 1, ..., m, \tag{4}$$

where $P_n(f)$ is the power spectral density of the $n - th$ time bin, and 11–14Hz is the frequency band of mu rhythm in our off-line data analysis. While the providers reported to have used mu rhythm that is in 8–12Hz [9], our analysis and the authors of [14] found that roughly 11–14Hz (or 10–15Hz) is the discriminative frequency band.

The algorithm using to estimate the power spectral density is the Burg method [15]. The Burg method fits an AR model of the specified order (the order is 40 (the same as in [13]) in our power spectral density estimation) to the input signal by minimizing the arithmetic mean of the forward and backward prediction errors. The spectral density is then computed from the frequency response of the AR model.

2.3 Self-training Algorithm and ICA Component Selection

In this subsection, we first present a self-training algorithm for training our classifier, then discuss how to select an ICA component on which our feature extraction and classification are based.

For many EEG-based BCIs, a tedious and time consuming training process is needed to set system parameters. In BCI Competition 2005 [17], reducing the training process has been explicitly proposed as a task. In this paper, we resort to semi-supervised learning to train a classifier. Compared with the case of supervised learning, the training data set (labeled data) of semi-supervised learning is much smaller. Since unlabeled data are used for training, the performance of semi-supervised learning can still be satisfying.

A full Bayes classifier is used for classification. To train classifier parameters with a small training set, we use a self-training algorithm, which is a common semi-supervised learning technique. The outline of the self-training algorithm is as follows.

Self-training algorithm: Given two data sets: labeled data set D_l, and unlabeled data set D_u.

- Step 1: Train an initial Bayes classifier using D_l;
- Step 2: Estimate the labels of D_u using the current classifier;
- Step 3: Retrain the classifier using the data set D_u with estimated labels and the data set D_l;
- Step 4: If the change of the classifier's parameters are below a tolerance, terminate the iteration; otherwise, go back to Step 2.

Now we discuss the problem left in Section 2.2 on how to select an ICA component for feature extraction. Traditional feature selection methods used in BCI includes Fisher criterion, bi-serial correlation coefficients, cross-validation, weighting of sparse classifier, etc. A survey about these methods can be found in [16]. However, all these methods needs abundant labeled data. For selection of the best component from small amounts of labeled data and large amounts of unlabeled data, we introduce an extended entropy based criterion. This criterion is defined as,

$$J = -\sum_{n=1}^{M} P(n)\log(P(n)) + \frac{\left(\sum_{i=1}^{T}\sum_{n=1}^{M} P(n|\mathbf{F}_\mathbf{i})\log(P(n|\mathbf{F}_\mathbf{i}\mathbf{k}))\right)}{T}, \quad (5)$$

where M is the number of classes, T is the number of the training samples including the labeled and unlabeled samples, $P(n)$ is a prior of the nth class, $P(n|\mathbf{F}_\mathbf{i})$ is the posterior probability for the feature vector $\mathbf{F}_\mathbf{i}$ extracted from one independent component of the ith sample. Note that for both labeled data and unlabeled data, $P(n|\mathbf{F}_\mathbf{i})$ is from the bayes classifier trained by the above self-training algorithm. The second term of equation (5) is from a traditional entropy criterion [18], which can measure the separability of features if all the available data are labeled. The bigger the value, the better the separability of features. We extended the entropy criterion by adding the first term, which can impose a constraint for avoiding the case of serious unbalance of classes brought by classification.

According to this criterion, the ICA component corresponding to the highest J will be selected as the best component.

3 Experimental Data Analysis

In this section, we evaluate the ICA based semi-supervised learning method using the data set from an EEG-based cursor control experiment.

The EEG-based cursor control experiment was carried out in Wadsworth Center. In this experiment, the subjects sat in a reclining chair facing a video screen and was asked to remain motionless during performance. The subjects used mu or beta rhythm amplitude to control vertical position of a target located at the right edge of the video screen. The data set was recorded from three subjects (AA, BB, CC). Each subject's data included 10 sessions for each subject. The data set and the details of this experiment are available at `http://ida.first.fraunhofer.de/projects/bci/competition`.

For convenience, only the trials with the targets at the highest and lowest position of the right edge of the screen are used in our offline analysis (96*10 trials for each subject).

For evaluating our algorithm, we separate all the trials into three sets, i.e., labeled set, unlabeled set and independent test set. labeled set and unlabeled set are used for training. labeled set consists of 20 trials (10 trials for each target) from session 1; unlabeled set consists of 556 trials from the remaining trials of session 1 and all the trials of sessions 2–6; and the independent test set is composed of 384 trials of sessions 7–10.

In this study, we use the EEG signals of 18 channels with channel numbers 8–10, 12–17, 19–21, 48–50 and 52–54, which are just over the sensorimotor area, for the ICA preprocessing. To obtain a demixing matrix, we apply the natural gradient-flexible ICA algorithm [12] to the data matrix of 20 trials, which are randomly selected from the labeled and unlabeled data sets. Then using the demixing matrix to all the data set, we get the ICA components. Afterwards, we extract 5 time bins dynamic amplitude features of all the ICA components and perform ICA component selection. Finally, based on the selected ICA component and its amplitude feature, we perform classification to the unlabeled data set and independent test set.

In the following, we present our data analysis results. Figure 1 shows the first 6 ICA components of subject AA (left column) and their corresponding power spectral density calculated by burg algorithm (right column). Cn represents the nth independent component. $C6$ in this figure is obviously an artifact component. From the power spectral density of $C2$, we can see that an obvious frequency component in the frequency band 11–14hz; thus, $C2$ may be the component which can effectively control the cursor. In fact, by the criterion introduced in Section 2.3, $C2$ is indeed verified to be the best ICA components for classification.We apply our algorithm to the independent test set for test. By comparing the predicted target position for each trial with the true target position, the prediction accuracy rate is obtained. The accuracy rates for the three subjects are shown in the second row of Table 1.

For demonstrating the validity of ICA preprocessing, we perform similar analysis as above in the two cases: 1. Principal component analysis (PCA) is used to replace ICA for preprocessing; 2. No preprocessing is used.

Furthermore, we consider the second and third training settings to demonstrate the effectiveness of the self-training algorithm (the first setting):

1. self-training algorithm is used to train a bayes classifier;
2. only labeled data set is used to train a bayes classifier;
3. the true labels are assigned to the unlabeled data set, then use all the labeled data to train a bayes classifier.

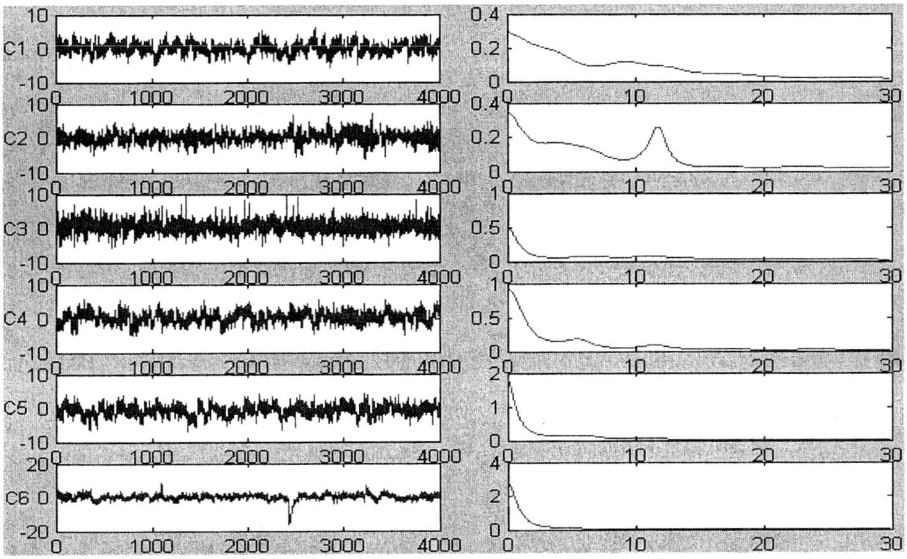

Fig. 1. Left column: 4000 samples of the first 6 ICA components of subject AA, right column: the corresponding power spectral density

Table 1. Accuracy rates (%) for the three subjects AA, BB and CC

Preprocessing method	training setting	AA Accuracy	BB Accuracy	CC Accuracy	Average Accuracy
ICA	1	94.52	89.74	89.52	**91.26**
PCA	1	87.99	85.53	89.52	87.68
No preprocessing	1	67.10	72.89	79.30	73.10
ICA	2	79.63	82.37	82.53	81.51
ICA	3	93.99	86.05	89.78	89.94

Table 1 lists the prediction accuracy rates obtained in all cases. First, the results show that the average accuracy rate obtained with ICA preprocessing is improved by 3.58% and 18.16% over those obtained with PCA preprocessing and without preprocessing, respectively. From Table 1, we can also find that semi-supervised learning improves the accuracy rate significantly (by 9.75%), compared with the accuracy rates obtained under the above training setting 2. Interestingly, even if we assign the true labels (i.e. the target positions of the unlabeled trials) to the unlabeled data set, and use all labeled data to train a classifier (training setting 3). The obtained accuracy rate is a little less than that obtained by semi-supervised learning. In most cases, the accuracy rate obtained in training setting 3 should be the best. But, in some cases, it is possible that small part of the samples are mislabeled (in the training procedure of cursor control BCI, it is possible that the user sometimes did not intend to hit the target), or exist some outliers. These wrong labeled samples or outliers will harm the performance of the classifier. Thus, in this case, the performance of training setting 3 may be poorer than the performance of training setting 1.

4 Conclusions

Through the offline analysis of the data set from a cursor control BCI experiment, we have shown that ICA based semi-supervised learning-based algorithm is an effective algorithm for BCI systems.

The ICA based semi-supervised learning algorithm offers several advantages. First, we think that the ICA preprocessing can separate useful source components from noise including artifacts to some extent. Thus we can choose one component which can best reflect brain's intention and most suitable for feature extraction. The results in Table 1 show that the prediction accuracy rate is improved significantly by ICA preprocessing, compared with the result of no preprocessing case. Second, since we choose only one ICA component for feature extraction, unlike using a small subset of channels of raw EEG signal [9], the computational burden of classification can be reduced significant . Furthermore, it follows from our data analysis results that ICA is superior than PCA, a similar preprocessing method, especially for some subjects. Finally, through ICA based semi-supervised learning, we use a small labeled data set and a large unlabeled training data set to train a classifier. The performance of the classifier is quite satisfying. Note that the labels of the unlabeled data set can also be predicted by this classifier. Therefore, the proposed ICA based semi-supervised learning algorithm can significantly reduce the time-consuming training effort for BCI systems.

Acknowledgements

This work was supported by National Natural Science Foundation of China under Grant 60475004, the Program for New Century Excellent Talents in China University under Grant NCET-04-0816, Guangdong Province Science Foundation for Research Team program under grant 04205783.

References

1. Wolpaw, J.R., Birbaumer, N., Heetderks, W.J., McFarland, D.J., Peckham, P.H., Schalk, G., Donchin, E., Quatrano, L.A., Robinson, C.J. and Vaughan, T.M.: Brain-Computer Interface Technology: A Review of the First International Meeting. IEEE Trans Rehab Eng **8** (2000) 164–173.
2. Fabiani, G.E., McFarland, D.J., Wolpaw, J.R., Pfurtscheller, G.: Conversion of EEG Activity Into Cursor Movement by a Brain-Computer Interface. IEEE Trans Rehab Eng **12** (2004) 331-338

3. Hyvärinen A., Oja, E.: Independent Component Analysis: Algorithms and Applications. Neural Networks **13** (2000) 411–430
4. Li, Y., Wang, J., Zurada, J. M.: Blind Extraction of Singularly Mixed Sources Signals. IEEE Trans. on Neural Networks, **11** (2000) 1413–1422
5. Li, Y., Wang, J.: Sequential Blind Extraction of Linearly Mixed Sources. IEEE Trans. on Signal Processing **50** (2002) 997–1006
6. Cichocki, A., Amari, S.: Adaptive Blind Signal and Image Processing (new revised and improved edition). John Wiley, New York (2003)
7. Makeig, S., Bell, A.J., Jung, T.-P., Sejnowski, T. J.: Independent Component Analysis of Electroencephalographic Data. Adv Neural Info Processing Systems **8** (1996) 145–151
8. Comon, P.: Independent Component Analysis - a New Concept? Signal Procesing **36** (1994) 287–314
9. BCI Competition 2003, Blankertz. B.: [Online]. Available: `http://ida.first.fhg.de/projects/bci/competition/`
10. Amari, S.: Natural Gradient Learning for Over- and Under-Complete Bases in ICA. Neural Computation **11** (1999) 1875–1883
11. Chichocki, A.: Blind Signal Processing Methods for Analyzing Multichannel Brain Signals. International Journal of Bioelectromagnetism **6** (2004)
12. Choi, S., Chichocki, A., Amari, S.: Flexible Independent Component Analysis. Proc. of the 1998 IEEE Workshop on NNSP (1998) 83–92
13. Cheng, M., Jia, W., Gao, X., Gao, S., Yang, F.: Mu rhythm-based Cursor Control: an offline analysis. Clinical Neurophysiology **115** (2004), 745–751
14. Blanchard, G. and Blankertz, B.: BCI Competition 2003–Data Set IIa: Spatial Patterns of Self-Controlled Brain Rhythm Modulations. IEEE Trans. on Biomedical Engineering **51** (2004) 1062–1066
15. Marple, S.L.: Digital Spectral Analysis. Prentice-Hall (1987) Chapter 7
16. Müller, K.R., Krauledat, M., Dornhege, G., Curio, G., Blankertz, B.: Machine learning techniques for brain-computer interfaces. Biomed. Tech. **49** (2004) 11–22
17. http://ida.first.fraunhofer.de/projects/bci/competition_iii
18. Godfried, T.T.: Comments on "A Modified Figure of Merit for Feature Selection in Pattern Recognition". IEEE Trans. on Information Theory **17** (1971) 618–620

State Inference in Variational Bayesian Nonlinear State-Space Models

Tapani Raiko, Matti Tornio, Antti Honkela, and Juha Karhunen

Helsinki University of Technology,
Neural Networks Research Centre,
P.O. Box 5400, FI-02015 HUT, Espoo, Finland
{tapani.raiko, matti.tornio, antti.honkela, juha.karhunen}@hut.fi

Abstract. Nonlinear source separation can be performed by inferring the state of a nonlinear state-space model. We study and improve the inference algorithm in the variational Bayesian blind source separation model introduced by Valpola and Karhunen in 2002. As comparison methods we use extensions of the Kalman filter that are widely used inference methods in tracking and control theory. The results in stability, speed, and accuracy favour our method especially in difficult inference problems.

1 Introduction

Many applications of source separation methods involve data with some kind of relations between consecutive observations. Examples include relations between neighbouring pixels in images and time series data. Using information on these relations improves the quality of separation results, especially in difficult nonlinear separation problems. Nonlinear modelling of relations may also be useful in linear mixing problems as the dynamics of the time series, for instance, may well be nonlinear.

A method for blind source separation using a nonlinear state-space model is described in [1]. In this paper we study and improve ways of estimating the sources or states in this framework. Efficient solution of the state estimation problem requires taking into account the nonlinear relations between consecutive samples, making it significantly more difficult than source separation in static models. Standard algorithms based on extensions of the Kalman smoother work rather well in general, but may fail to converge when estimating the states over a long gap or when used together with learning the model. We propose solving the problem by improving the variational Bayesian technique proposed in [1] by explicitly using the information on the relation between consecutive samples to speed up convergence.

To tackle just the state estimation (or source separation) part, we will simplify the blind problem by fixing the model weights and other parameters. In [2], linear and nonlinear state-space models are used for blind and semi-blind source separation. Also there the problem is simplified by fixing part of the model.

J. Rosca et al. (Eds.): ICA 2006, LNCS 3889, pp. 222–229, 2006.

2 Nonlinear State-Space Models

In nonlinear state-space models, the observation vectors $\mathbf{x}(t)$, $t = 1, 2, \ldots, T$, are assumed to have been generated from unobserved state (or source) vectors $\mathbf{s}(t)$. The model equations are

$$\mathbf{x}(t) = \mathbf{f}(\mathbf{s}(t)) + \mathbf{n}(t) \tag{1}$$

$$\mathbf{s}(t) = \mathbf{g}(\mathbf{s}(t-1)) + \mathbf{m}(t), \tag{2}$$

Both the mixing mapping $\mathbf{f}$ and the process mapping $\mathbf{g}$ are nonlinear. The noise model for both mixing and dynamical process is often assumed to be Gaussian

$$p(\mathbf{n}(t)) = \mathcal{N}\left[\mathbf{n}(t); \mathbf{0}; \mathbf{\Sigma}_x\right] \tag{3}$$

$$p(\mathbf{m}(t)) = \mathcal{N}\left[\mathbf{m}(t); \mathbf{0}; \mathbf{\Sigma}_s\right], \tag{4}$$

where $\mathbf{\Sigma}_x$ and $\mathbf{\Sigma}_s$ are the noise covariance matrices. In blind source separation, the mappings $\mathbf{f}$ and $\mathbf{g}$ are assumed to be unknown [1] but in this paper we concentrate on the case where they are known.

2.1 Inference Methods

The task of estimating a sequence of sources $\mathbf{s}(1), \ldots, \mathbf{s}(T)$ given a sequence of observations $\mathbf{x}(1), \ldots, \mathbf{x}(T)$ and the model is called inference. In case $\mathbf{f}$ and $\mathbf{g}$ in Eqs. (1) and (2) are linear, the state can be inferred analytically with an algorithm called the *Kalman filter* [3]. In a filter phase, evidence from the past is propagated forward, and in a smoothing phase, evidence from the future is propagated backwards. Only the most recent state can be inferred using the Kalman filter, otherwise the algorithm should be called the *Kalman smoother*. In [4], the Kalman filter is extended for blind source separation from time-varying noisy mixtures.

The idea behind *iterated extended Kalman smoother* [3] (IEKS) is to linearise the mappings $\mathbf{f}$ and $\mathbf{g}$ around the current state estimates using the first terms of the Taylor series expansion. The algorithm alternates between updating the state estimates by Kalman smoothing and renewing the linearisation. When the system is highly nonlinear or the initial estimate is poor, the IEKS may diverge.

The *iterative unscented Kalman smoother* [5,6] (IUKS) replaces the local linearisation of IEKS by a deterministic sampling technique. The sampled points are propagated through the nonlinearities, and a Gaussian distribution is fitted to them. The use of nonlocal information improves convergence and accuracy at the cost of doubling the computational complexity[1]. Still there is no guarantee of convergence.

A recent variant called *backward-smoothing extended Kalman filter* [8] searches for the maximum a posteriori solution to the filtering problem by a guarded Gauss-Newton method. It increases the accuracy further and guarantees convergence at the cost of about hundredfold increase in computational burden.

---

[1] An even better way of replacing the local linearisation when a multilayer perceptron network is used as a nonlinearity, is described in [7].

Particle filter [9] uses a set of particles or random samples to represent the state distribution. It is a Monte Carlo method developed especially for sequences. The particles are propagated through nonlinearities and there is no need for linearisation nor iterating. Given enough particles, the state estimate approaches the true distribution. Combining the filtering and smoothing directions is not straightforward but there are alternative methods for that. In [10], particle filters are used for non-stationary ICA.

2.2 Variational Bayesian Method

Nonlinear dynamical factor analysis (NDFA) [1] is a variational Bayesian method for learning nonlinear state-space models. The mappings $\mathbf{f}$ and $\mathbf{g}$ in Eqs. (1) and (2) are modelled with multilayer perceptron (MLP) networks whose parameters can be learned from the data. The parameter vector $\boldsymbol{\theta}$ include network weigths, noise levels, and hierarchical priors for them. The posterior distribution over the sources $\mathbf{S} = [\mathbf{s}(1), \ldots, \mathbf{s}(T)]$ and the parameters $\boldsymbol{\theta}$ is approximated by a Gaussian distribution $q(\mathbf{S}, \boldsymbol{\theta})$ with some further independency assumptions. Both learning and inference are based on minimising a cost function $\mathcal{C}_{\mathrm{KL}}$

$$\mathcal{C}_{\mathrm{KL}} = \int_{\boldsymbol{\theta}} \int_{\mathbf{S}} q(\mathbf{S}, \boldsymbol{\theta}) \ln \frac{q(\mathbf{S}, \boldsymbol{\theta})}{p(\mathbf{X}, \mathbf{S}, \boldsymbol{\theta})} d\mathbf{S} d\boldsymbol{\theta}, \tag{5}$$

where $p(\mathbf{X}, \mathbf{S}, \boldsymbol{\theta})$ is the joint probability density over the data $\mathbf{X} = [\mathbf{x}(1), \ldots, \mathbf{x}(T)]$, sources $\mathbf{S}$, and parameters $\boldsymbol{\theta}$. The cost function is based on Kullback-Leibler divergence between the approximation and the true posterior. It can be split into terms, which helps in studying only a part of the model at a time. The variational approach is less prone to overfitting compared to maximum a posteriori estimates and still fast compared to Monte Carlo methods. See [1] for details.

The variational Bayesian inference algorithm in [1] uses the gradient of the cost function w.r.t. state in a heuristic manner. We propose an algorithm that differs from it in three ways. Firstly, the heuristic updates are replaced by a standard conjugate gradient algorithm [11]. Secondly, the linearisation method from [7] is applied. Thirdly, the gradient is replaced by a vector of approximated total derivatives, as described in the following section.

2.3 Total Derivatives

When updates are done locally, information spreads around slowly because the states of different time slices affect each other only between updates. It is possible to predict this interaction by a suitable approximation. We get a novel update algorithm for the posterior mean of the states by replacing partial derivatives of the cost function w.r.t. state means $\overline{\mathbf{s}}(t)$ by (approximated) total derivatives

$$\frac{d\mathcal{C}_{\mathrm{KL}}}{d\overline{\mathbf{s}}(t)} = \sum_{\tau=1}^{T} \frac{\partial \mathcal{C}_{\mathrm{KL}}}{\partial \overline{\mathbf{s}}(\tau)} \frac{\partial \overline{\mathbf{s}}(\tau)}{\partial \overline{\mathbf{s}}(t)}. \tag{6}$$

They can be computed efficiently using the chain rule and dynamic programming, given that we can approximate the terms $\frac{\partial \overline{\mathbf{s}}(t)}{\partial \overline{\mathbf{s}}(t-1)}$ and $\frac{\partial \overline{\mathbf{s}}(t)}{\partial \overline{\mathbf{s}}(t+1)}$.

Before going into details, let us go through the idea. The posterior distribution of the state $\mathbf{s}(t)$ can be factored into three potentials, one from $\mathbf{s}(t-1)$ (the past), one from $\mathbf{s}(t+1)$ (the future), and one from $\mathbf{x}(t)$ (the observation). We will linearise the nonlinear mappings so that the three potentials become Gaussian. Then also the posterior of $\mathbf{s}(t)$ becomes Gaussian with a mean that is the weighted average of the means of the three potentials, where the weights are the inverse (co)variances of the potentials. A change in the mean of a potential results in a change of the mean of the posterior inversely proportional to their (co)variances.

The terms of the cost function (See Equation (5.6) in [1], although the notation is somewhat different) that relate to $\mathbf{s}(t)$ are

$$\begin{aligned}\mathcal{C}_{\mathrm{KL}}(\mathbf{s}(t)) = &\sum_{i=1}^{m}\left(-\frac{1}{2}\ln\widetilde{s}_{ii}(t)+\frac{1}{2}\Sigma_{sii}^{-1}\left\{\left[\overline{s}_i(t)-\overline{g}_i(\mathbf{s}(t-1))\right]^2+\widetilde{s}_i(t)\right\}\right)\\ &+\sum_{j=1}^{m}\frac{1}{2}\Sigma_{sjj}^{-1}\left\{\left[\overline{g}_j(\mathbf{s}(t))-\overline{s}_j(t+1)\right]^2+\widetilde{g}_j(\mathbf{s}(t))\right\}\\ &+\sum_{k=1}^{n}\frac{1}{2}\Sigma_{xkk}^{-1}\left\{\left[\overline{f}_k(\mathbf{s}(t))-\overline{x}_k(t)\right]^2+\widetilde{f}_k(\mathbf{s}(t))\right\},\end{aligned} \tag{7}$$

where $\overline{\alpha}$ and $\widetilde{\alpha}$ denote the mean and (co)variance of α over the posterior approximation q respectively and n and m are the dimensionalities of $\mathbf{x}$ and $\mathbf{s}$ respectively. Note that we assume diagonal noise covariances $\mathbf{\Sigma}$. Nonlinearities $\mathbf{f}$ and $\mathbf{g}$ are replaced by the linearisations

$$\widehat{\mathbf{f}}(\mathbf{s}(t)) = \overline{\mathbf{f}}(\mathbf{s}_{\mathrm{cur}}(t)) + \mathbf{J}_f(t)\left[\mathbf{s}(t)-\overline{\mathbf{s}}_{\mathrm{cur}}(t)\right] \tag{8}$$

$$\widehat{\mathbf{g}}(\mathbf{s}(t)) = \overline{\mathbf{g}}(\mathbf{s}_{\mathrm{cur}}(t)) + \mathbf{J}_g(t)\left[\mathbf{s}(t)-\overline{\mathbf{s}}_{\mathrm{cur}}(t)\right], \tag{9}$$

where the subscript cur denotes a current estimate that is constant w.r.t. further changes in $\mathbf{s}(t)$. The minimum of (7) with linearisations can be found at the zero of the gradient:

$$\widetilde{\mathbf{s}}_{\mathrm{opt}}(t) = \left[\mathbf{\Sigma}_s^{-1}+\mathbf{J}_g(t)^{\mathrm{T}}\mathbf{\Sigma}_s^{-1}\mathbf{J}_g(t)+\mathbf{J}_f(t)^{\mathrm{T}}\mathbf{\Sigma}_x^{-1}\mathbf{J}_f(t)\right]^{-1} \tag{10}$$

$$\begin{aligned}\overline{\mathbf{s}}_{\mathrm{opt}}(t) = \widetilde{\mathbf{s}}_{\mathrm{opt}}(t)\,\{&\mathbf{\Sigma}_s^{-1}\left[\overline{\mathbf{g}}(\mathbf{s}_{\mathrm{cur}}(t-1))+\mathbf{J}_g(t-1)(\overline{\mathbf{s}}(t-1)-\overline{\mathbf{s}}_{\mathrm{cur}}(t-1))\right]\\ &+\mathbf{J}_g(t)^{\mathrm{T}}\mathbf{\Sigma}_s^{-1}\left[\overline{\mathbf{s}}(t+1)-\overline{\mathbf{g}}(\mathbf{s}_{\mathrm{cur}}(t))\right]\\ &+\mathbf{J}_f(t)^{\mathrm{T}}\mathbf{\Sigma}_x^{-1}\left[\overline{\mathbf{x}}(t)-\overline{\mathbf{f}}(\mathbf{s}_{\mathrm{cur}}(t))\right]\}\,.\end{aligned} \tag{11}$$

The optimum mean reacts to changes in the past and in the future by

$$\frac{\partial\overline{\mathbf{s}}_{\mathrm{opt}}(t)}{\partial\overline{\mathbf{s}}(t-1)} = \widetilde{\mathbf{s}}_{\mathrm{opt}}(t)\mathbf{\Sigma}_s^{-1}\mathbf{J}_g(t-1) \tag{12}$$

$$\frac{\partial\overline{\mathbf{s}}_{\mathrm{opt}}(t)}{\partial\overline{\mathbf{s}}(t+1)} = \widetilde{\mathbf{s}}_{\mathrm{opt}}(t)\mathbf{J}_g(t)^{\mathrm{T}}\mathbf{\Sigma}_s^{-1}. \tag{13}$$

Finally, we assume that the Equations (12) and (13) apply approximately even in the nonlinear case when the subscripts opt are dropped out. The linearisation matrices $\mathbf{J}$ need to be computed anyway [7] so the computational overhead is rather small.

3 Experiments

To experimentally measure the performance of our proposed new method, we used two different data sets. The first data set was generated using a simulated double inverted pendulum system with known dynamics. As the second data set we used real-world speech data with unknown dynamics.

In all the experiments, IEKS and IUKS were run for 50 iterations and NDFA algorithm for 500 iterations. In most cases this was long enough for the algorithms to converge to a local minimum. For comparison purposes, the NDFA experiments were also repeated without using the total derivatives.

Even with a relatively high number of particles, particle smoother performed poorly compared to the iterative algorithms. The results for particle smoother are therefore omitted from the figures. They are however discussed where appropriate. Even though the particle smoother performed relatively poorly, it should be noted that many different schemes exists to improve the performance of particle filters [9], and therefore direct comparison between the iterative algorithms and the plain particle filter algorithm used in these experiments may be somewhat unjustified. The experiments were also repeated with the original NDFA algorithm presented in [1]. The results were quite poor, as was to be excepted, as the heuristic update rules are optimized for learning.

3.1 Double Inverted Pendulum

The double inverted pendulum system [6] (see Figure 1) is a standard benchmark in the field of control. The system consists of a cart and a two-part pole attached to the cart. The system has six states which are cart position on a track, cart velocity, and the angles and the angular velocities of the two attached pendulums. The single control signal is the lateral force applied to the cart. The dynamical equations for the double inverted pendulum system can be found e.g. in [6], in this experiment a discrete system with a time step of $\Delta t = 0.05$ s was simulated using the MATLAB ordinary differential equation solver `ode23`.

To make sure that the learning scheme did not favour the proposed algorithm, standard backpropagation algorithm was used to learn an MLP network to model the system dynamics using a relatively small sample of 2000 input-output pairs. To make this problem more challenging, only the velocity and position of the cart and the angle of the upper pendulum were available as observations, and the rest of the state had to be inferred from these. Experiments were run on ten different data sets with 50 samples each using 5 different initialisations. The final results can be seen in Figure 1.

IEKS suffered from quite serious convergence problems with this data set. These problems were especially bad during the early iterations, but several runs failed to converge to a meaningful result even after the iteration limit was reached. IUKS performed somewhat better, but suffered from some stability problems too. The proposed method was much more robust and did not suffer from stability issues and also performed better on average than the two

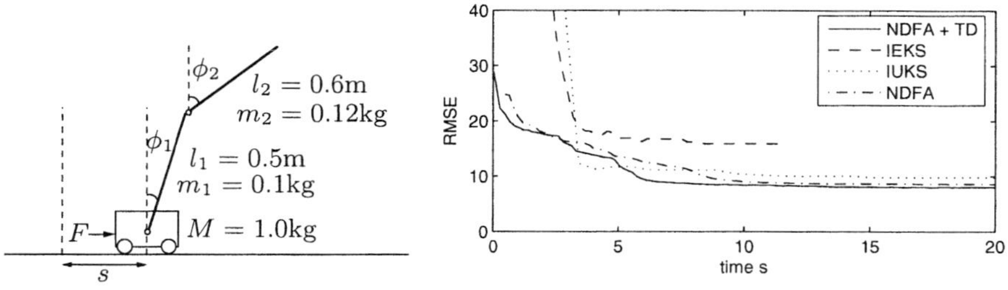

Fig. 1. Inference with the double inverted pendulum system. On the left the schematic of the system, on the right root mean square error plotted against computation time.

Kalman filter based algorithms. It should be noted, however, that in some experiments both IEKS and IUKS converged in only a few iterations, resulting in a superior performance compared to the proposed method. Therefore the problem with IEKS and IUKS may at least partially be related to poor choice of initialisations.

3.2 Speech Spectra

As a real world data set we used speech spectra. The data set consisted of 11200 21 dimensional samples which corresponds to 90 seconds of continuous human speech. The first 10000 samples were used to train a seven dimensional state-space model with the method from [1] and the rest of the data was used in the experiments. This data set poses a somewhat different problem from the double inverted pendulum system. The nonlinearities are not as strong as in the first experiment but the dimensionality of the observation and state spaces are higher, which emphasises the scalability of the methods.

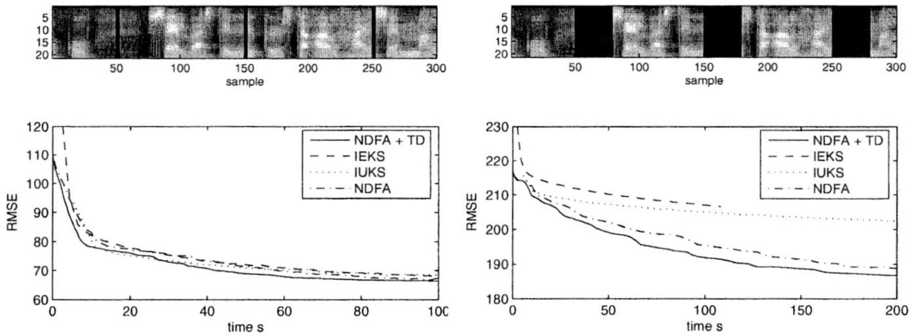

Fig. 2. Inference with the speech data and missing values. On the top one of the data sets used in the experiments (missing values marked in black), on the bottom root mean square error plotted against computation time. Left side figures use a small gap size, right side figures a large gap size.

The test data set was divided into three parts each consisting of 300 samples and all the algorithms were run for each data set with four random initialisations. The final results represent an average over both the different data sets and initialisations.

Since the true state is unknown in this experiment, the mean square error of the reconstruction of missing data was used to compare the different algorithms. Experiments were done with sets of both 3 and 30 consecutive missing samples. The ability to cope with missing values is very important when only partial observations are available or in the case of failures in the observation process. It also has interesting applications in the field of control as reported in [12].

Results can be seen in Figure 2. When missing values are present, especially in the case of the large gap size, the proposed algorithm performs clearly better than the rest of the compared algorithms. Compared to the double inverted pendulum data set, the stability issues with IEKS and IUKS were not as severe, but neither method could cope very well with long gaps of missing values.

4 Discussion and Conclusions

We proposed an algorithm for inference in nonlinear state-space models and compared it to some of the existing methods. The algorithm is based on minimising a variational Bayesian cost function and the novelty is in propagating the gradient through the state sequence. The results were slightly better than any of the comparison methods (IEKS and IUKS). The difference became large in a high-dimensional problem with long gaps in observations.

Our current implementation requires that the nonlinear mappings are modelled as multilayer perceptron networks. Part of the success of our method is due to a linearisation that is specialised to that case [7]. The idea presented in this paper applies in general.

When an algorithm is based on minimising a cost function, it is fairly easy to guarantee convergence. While the Kalman filter is clearly the best choice for inference in linear Gaussian models, the problem with many of the nonlinear generalisation (e.g. IEKS and IUKS) is that they cannot guarantee convergence. Even when the algorithms converge, convergence can be slow. A recent fix for convergence comes with a large computational cost [8] but this paper shows that stable inference can be fast, too.

While this paper concentrates on the case where nonlinear mappings and other model parameters are known, we aim at the case where they should be learned from the data [1]. Blind source separation involves a lot more iterations than the basic source separation. The requirements of a good inference algorithm change, too: There is always the previous estimate of the sources available and most of the time it is already quite accurate.

Acknowledgements

This work was supported in part by the Finnish Centre of Excellence Programme (2000-2005) under the project New Information Processing Principles and by the IST Programme of the European Community under the PASCAL Network of Excellence, IST-2002-506778. This publication only reflects the authors' views.

References

1. H. Valpola and J. Karhunen, "An unsupervised ensemble learning method for nonlinear dynamic state-space models," *Neural Computation*, vol. 14, no. 11, pp. 2647–2692, 2002.
2. A. Cichocki, L. Zhang, S. Choi, and S.-I. Amari, "Nonlinear dynamic independent component analysis using state-space and neural network models," in *Proc. of the 1st Int. Workshop on Independent Component Analysis and Signal Separation (ICA'99)*, (Aussois, France, January 11-15), pp. 99–104, 1999.
3. B. Anderson and J. Moore, *Optimal Filtering.* Englewood Cliffs, NJ: Prentice-Hall, 1979.
4. V. Koivunen, M. Enescu, and E. Oja, "Adaptive algorithm for blind separation from noisy time-varying mixtures," *Neural Computation*, vol. 13, pp. 2339–2357, 2001.
5. S. Julier and J. Uhlmann, "A new extension of the Kalman filter to nonlinear systems," in *Int. Symp. Aerospace/Defense Sensing, Simul. and Controls*, 1997.
6. E. A. Wan and R. van der Merwe, "The unscented Kalman filter," in *Kalman Filtering and Neural Networks* (S. Haykin, ed.), pp. 221–280, New York: Wiley, 2001.
7. A. Honkela and H. Valpola, "Unsupervised variational Bayesian learning of nonlinear models," in *Advances in Neural Information Processing Systems 17* (L. Saul, Y. Weiss, and L. Bottou, eds.), pp. 593–600, Cambridge, MA, USA: MIT Press, 2005.
8. M. Psiaki, "Backward-smoothing extended Kalman filter," *Journal of Guidance, Control, and Dynamics*, vol. 28, Sep–Oct 2005.
9. A. Doucet, N. de Freitas, and N. J. Gordon, *Sequential Monte Carlo Methods in Practice.* Springer Verlag, 2001.
10. R. Everson and S. Roberts, "Particle filters for non-stationary ICA," in *Advances in Independent Component Analysis* (M. Girolami, ed.), pp. 23–41, Springer-Verlag, 2000.
11. R. Fletcher and C. M. Reeves, "Function minimization by conjugate gradients," *The Computer Journal*, vol. 7, pp. 149–154, 1964.
12. T. Raiko and M. Tornio, "Learning nonlinear state-space models for control," in *Proc. Int. Joint Conf. on Neural Networks (IJCNN'05)*, (Montreal, Canada), pp. 815–820, 2005.

Quadratic MIMO Contrast Functions for Blind Source Separation in a Convolutive Context

Saloua Rhioui[1], Marc Castella[2], and Eric Moreau[1]

[1] STD, ISITV, av. G. Pompidou, BP56,
F-83162 La Valette du Var, Cedex, France
{rhioui, moreau}@univ-tln.fr
[2] GET/INT, Département CITI, UMR-CNRS 5157,
9 rue Charles Fourier, 91011 Évry Cedex, France
marc.castella@int-evry.fr

Abstract. This paper considers the problem of blind separation of sources mixed by a MIMO convolutive system. For both i.i.d. and non i.i.d. sources, quadratic separation criteria previously designed for the extraction of a single source are extended to parallel extraction in the MIMO case. These criteria are based on the use of so-called reference signals and a condition is given under which we obtain MIMO contrast functions. Simulations demonstrate that a particular choice of a set of reference signals ensures the contrast property. The performance offered by these criteria is investigated through simulations: it is shown that the proposed contrast functions avoid accumulation errors, contrary to deflation methods.

1 Introduction

We consider the problem of blind equalization of Linear Time Invariant (LTI) Multi-Input / Multi-Output (MIMO) systems. Such a problem is of interest e.g. in multi-user wireless communications where observed signals have to be equalized both in space and time in order to eliminate both intersymbol and cochannel interferences. These interferences are due to possible delays introduced by multi-path propagation and to possible multi-users. Examples are found in Space Division Multiple Access (SDMA) or Code Division Multiple Access (CDMA) communication systems.

Our approach is based on the use of a contrast function [6, 4]. In particular, this has the advantage to yield a sufficient condition for separation. In the context of MIMO systems and parallel extraction of all sources, classical contrast functions generally first require a pre-whitening stage on the observation signals in order to constrain the searched system to be para-unitary [4, 5, 8]. On the other hand, recent solutions have been shown to be very efficient when so-called reference signals are considered, either for equalization of a SISO or SIMO systems [3] or for extraction of one source [1] from a MIMO system. Our main goal in this paper is to propose a generalization of the latter results to the case of parallel extraction in convolutive MIMO systems. In particular notice that our results

J. Rosca et al. (Eds.): ICA 2006, LNCS 3889, pp. 230–237, 2006.

generalize the one in [2] and more importantly that we do not require any pre-whitening stage. The usefulness of such a wide family of criteria is illustrated by computer simulations.

2 Model and Problem Formulation

We consider a Q-dimensional ($Q \in \mathbb{N}, Q \geq 2$) discrete-time signal which is called vector of *observations* and denoted by $\mathbf{x}(n)$ (in the whole paper, n stands for any integer: $n \in \mathbb{Z}$). It results from a linear time invariant (LTI) multichannel system $\{\mathbf{M}\}$ described by the input-output relation:

$$\mathbf{x}(n) = \sum_{k \in \mathbb{Z}} \mathbf{M}(k)\mathbf{s}(n-k) \triangleq \{\mathbf{M}\}\mathbf{s}(n), \tag{1}$$

where $\mathbf{M}(n)$ is the sequence of (Q, N) impulse response matrices and $\mathbf{s}(n)$ is an N-dimensional ($N \in \mathbb{N}^*$) unknown and unobserved column vector, which is referred to as the vector of *sources.*

The multichannel blind deconvolution problem consists in estimating a multivariate LTI system $\{\mathbf{W}\}$ operating on the observations, such that the vector

$$\mathbf{y}(n) = \sum_{k \in \mathbb{Z}} \mathbf{W}(k)\mathbf{x}(n-k) \triangleq \{\mathbf{W}\}\mathbf{x}(n) \tag{2}$$

restores the N input sources. The problem is referred to as the blind source separation (BSS) problem, where blind means that no information is available on the mixing system and that the sources are unobservable. It is useful to define the (N, N) global LTI filter $\{\mathbf{G}\}$ by the following impulse response:

$$\mathbf{G}(n) = \sum_{k \in \mathbb{Z}} \mathbf{W}(k)\mathbf{M}(n-k). \tag{3}$$

We have then:

$$\mathbf{y}(n) = \sum_{k \in \mathbb{Z}} \mathbf{G}(n-k)\mathbf{s}(k) \triangleq \{\mathbf{G}\}\mathbf{s}(n). \tag{4}$$

In order to be able to solve the BSS problem, we have to introduce some assumptions on the source signals. The following one is known to play a key role:

A.1. The source vector components $s_i(n), i \in \{1, \ldots, N\}$ are *mutually independent*, stationary and zero-mean processes with unit variance. Their respective covariance functions are denoted by $\gamma_i(k), k \in \mathbb{Z}$ and are positive definite functions (i.e the corresponding spectrum density is positive).

Since the sources are assumed to be unobservable, some inherent indeterminations in their restitution remain: in the general case, their order cannot be restored and each of them is only recovered up to a permutation and a scalar filtering ambiguity. Consequently, the sources are said to be separated when the global transfer matrix $\mathbf{G}(z) \triangleq \sum_k \mathbf{G}(k)z^{-k}$ reads:

$$\mathbf{G}(z) = \mathbf{D}(z)\mathbf{P} \tag{5}$$

where $\mathbf{P}$ is permutation matrix and $\mathbf{D}(z) = \mathrm{Diag}(d_1(z), \dots, d_N(z))$ is a matrix with scalar filters on its main diagonal. Whenever the sources are assumed to be temporally i.i.d. (independent and identically distributed), the scalar filtering ambiguity reduces to a scaling factor and time delay. In this case, the sources are said to be separated when:

$$\mathbf{G}(z) = \mathbf{D}(z)\mathbf{\Lambda}\mathbf{P} \tag{6}$$

where $\mathbf{\Lambda}$ is a constant diagonal matrix and $\mathbf{D}(z) = \mathrm{Diag}(z^{-l_1}, \dots, z^{-l_N})$ with $(l_1, \dots, l_N) \in \mathbb{Z}^N$. Naturally, we assume that the mixing filter is invertible (which implies $Q \geq N$) in the sense that it is possible to obtain (6) or (5).

3 MIMO Separation Criteria

The concept of contrast function has been introduced in BSS so as to reduce the problem to an optimization one: by definition, a contrast function is a criterion which maximization leads to a separating solution. When a pre-whitening procedure has been applied, and under certain conditions, one of the simplest contrast [4, 5] in the context of a MIMO parallel extraction of all sources is given by

$$\mathcal{C}_R\{\mathbf{y}(n)\} \triangleq \sum_{i=1}^{N} |\mathrm{Cum}\{\underbrace{y_i(n), y_i(n), y_i(n), \dots, y_i(n)}_{R \text{ times},\ R \geq 3}\}| \tag{7}$$

where Cum denotes the cumulant. The main contribution of the paper consists in using criteria based on R-th order cross-cumulants, where $R-2$ variables are fixed. This choice yields a quadratic dependence with respect to the optimized parameter, which greatly simplifies the optimization task. We define the following R-th order cumulant, where $R \geq 3$:

$$\kappa_{R,z_i}\{y_i(n)\} = \mathrm{Cum}\{y_i(n), y_i(n), \underbrace{z_i(n), \dots, z_i(n)}_{R-2 \text{ times}}\} \tag{8}$$

where $z_i(n)$ are given signals to be precisely defined later. In previous works [1], they have been referred to as *reference* signals determined from prior information, but we will see that they may be chosen as observations whitened. We now define the following criterion:

$$\mathcal{C}_{R,\mathbf{z}}\{\mathbf{y}(n)\} \triangleq \sum_{i=1}^{N} |\kappa_{R,z_i}\{y_i(n)\}| \ . \tag{9}$$

This criterion is a MIMO extension of the results in [1]: it will allow the parallel extraction of all sources, contrary to [1] which allows the extraction of the sources one after the other. (9) also generalizes a result in [2].

3.1 Case of i.i.d. Sources

Main result. Since the sources have unit variance, one can restrict the multiplicative factors in (6) to $|\mathbf{\Lambda}| = \mathbf{I}$ by imposing the constraint $\mathrm{E}\{|y_i(n)|^2\} = 1$ for all $i \in \{1, \ldots, N\}$. For i.i.d. sources, this condition also reads:

$$\forall i \in \{1, \ldots, N\} \;, \quad \sum_{j=1}^{N} \sum_{k \in \mathbb{Z}} |G_{ij}(k)|^2 = 1 \tag{10}$$

We need to define the following supremum:

$$\kappa_{R,i}^{\max} \triangleq \max_{j=1}^{N} \sup_{k \in \mathbb{Z}} |\kappa_{R,z_i}\{s_j(n-k)\}| \tag{11}$$

The proof of Proposition 1 requires the following assumption:

A.2. $\forall i \in \{1, \ldots, N\}$, there exits (j_i, l_i) such that:

$$\kappa_{R,i}^{\max} = |\kappa_{R,z_i}\{s_{j_i}(n-l_i)\}| < +\infty \tag{12}$$

We can then state:

Proposition 1. *In the case of i.i.d. source signals and under the constraint* (10)*, the criterion* $\mathcal{C}_{R,\mathbf{z}}$ *is a contrast function if and only if each set*

$$\mathcal{I}_i \triangleq \{(j,k) \in \{1, \ldots, N\} \times \mathbb{Z} \mid |\kappa_{R,z_i}\{s_j(n-k)\}| = \kappa_{R,i}^{\max}\}, \tag{13}$$

where $i \in \{1, \ldots, N\}$*, contains a single element* (σ_i, k_i)*, where* σ *denotes a permutation in* $\{1, \ldots, N\}$*.*

Proof: We can write: $\kappa_{R,z_i}\{y_i(n)\} = \sum_{j=1}^{N} \sum_{k \in \mathbb{Z}} G_{ij}(k)^2 \kappa_{R,z_i}\{s_j(n-k)\}$ and, using (11) and (10), it follows

$$\mathcal{C}_{R,\mathbf{z}}\{\mathbf{y}(n)\} \leq \sum_{i=1}^{N} \sum_{j=1}^{N} \sum_{k \in \mathbb{Z}} |G_{ij}(k)|^2 |\kappa_{R,z_i}\{s_j(n-k)\}| \tag{14}$$

$$\leq \sum_{i=1}^{N} \kappa_{R,i}^{\max} \sum_{j=1}^{N} \sum_{k \in \mathbb{Z}} |G_{ij}(k)|^2 = \sum_{i=1}^{N} \kappa_{R,i}^{\max} \tag{15}$$

$$\leq \sum_{i=1}^{N} |\kappa_{R,z_i}\{s_{\sigma_i}(n-l_i)\}| = \mathcal{C}_{R,\mathbf{z}}\{\mathbf{s}(n-\mathbf{l})\}. \tag{16}$$

where $\mathbf{s}(n-\mathbf{l}) = (s_{\sigma_1}(n-l_1), \ldots, s_{\sigma_N}(n-l_N))^T$ is a vector of N source signals delayed. If the above upper-bound is reached (which is possible according to assumption A.2), then

$$\sum_{i=1}^{N} \sum_{j=1}^{N} \sum_{k \in \mathbb{Z}} |\mathbf{G}_{ij}(k)|^2 (\kappa_{R,i}^{\max} - |\kappa_{R,z_i}\{s_j(n-k)\}|) = 0. \tag{17}$$

All terms in the above sum being positive, if $\mathcal{I}_i$ contains a single element and σ is a permutation, one deduces, that $\{\mathbf{G}\}$ satisfies the equalization condition (6). Conversely, one can see that if $\mathcal{I}_i$ contains several elements, there exist non separating filters which maximize $\mathcal{C}_{R,\mathbf{z}}$. ■

Comments and alternative results. One should notice that no prewhitening has been required to prove the validity of the contrast $\mathcal{C}_{R,\mathbf{z}}$. This means that the global filter $\{\mathbf{G}\}$ need not be para-unitary for the result to hold but that (10) is sufficient. Incidently, one can notice that if (10) is replaced by: $\forall j \in \{1,\ldots,N\}$ $\sum_{i=1}^{N}\sum_{k\in\mathbb{Z}}|G_{ij}(k)|^2 = 1$, a result similar to the one given by Proposition 1 can be proved by changing the roles of i and j. Finally, one should notice also that, although no assumption has been explicitly made on the cumulants of the sources, no source should have vanishing R-th order cumulants in order to satisfy the conditions of Proposition 1.

3.2 Case of Non i.i.d. Sources

Non i.i.d. sources can only be recovered up to a scalar filtering. It is hence natural in this case to work on the scalar filters components which compose the MIMO system. Thanks to the definite positiveness assumed in A.1 we define the following j-norm:

$$\|h\|_j \triangleq \left(\sum_{(k_1,k_2)\in\mathbb{Z}^2} h(k_1)h^*(k_2)\gamma_j(k_2-k_1) \right)^{\frac{1}{2}} \tag{18}$$

Similarly to (10) we will impose a variance constraint on the output, which corresponds to a unit norm constraint on the row filters:

$$\forall i \in \{1,\ldots,N\} \quad \sum_{j=1}^{N} \|G_{ij}\|_j^2 = 1 \tag{19}$$

Corresponding to (11) we define $\tilde{\kappa}_{R,i}^{\max} \triangleq \max_{j=1}^{N} \sup_{\|\{\widetilde{G}_{ij}\}\|_j=1} |\kappa_{R,z_i}\{\{\widetilde{G}_{ij}\}s_j(n)\}|$ and corresponding to (12) we assume:

A.3. For all $i \in \{1,\ldots,N\}$ there exists $j_i \in \{1,\ldots,N\}$ and a filter $h_i^\sharp$ such that:

$$\tilde{\kappa}_{R,i}^{\max} = |\kappa_{R,z_i}\{\{h_i^\sharp\}s_{j_i}(n)\}| < +\infty \tag{20}$$

We can then state:

Proposition 2. *In the case of non i.i.d. sources and under the constraint* (19), *the criterion* $\mathcal{C}_{R,\mathbf{z}}$ *is a contrast if and only if each set*

$$\mathcal{I}_i \triangleq \{j \in \{1,\ldots,N\} \mid \sup_{\|h\|_j=1} |\kappa_{R,z_i}\{\{h\}s_j(n)\}| = \tilde{\kappa}_{R,i}^{\max}\}, \tag{21}$$

where $i \in \{1,\ldots,N\}$, *contains a single element* σ_i, *where* σ *denotes a permutation in* $\{1,\ldots,N\}$.

Proof: We have:$y_i(n) = \sum_{j=1}^{N}\{G_{ij}\}s_j(n) = \sum_{j=1}^{N}\|G_{ij}\|_j\{\widetilde{G}_{ij}\}s_j(n)$ where $\{\widetilde{G}_{ij}\}$ is defined by $\{\widetilde{G}_{ij}\} = \{\widetilde{G}_{ij}\}/\|\widetilde{G}_{ij}\|_j$ if $\|\widetilde{G}_{ij}\|_j \neq 0$ and $\{\widetilde{G}_{ij}\} = 0$ otherwise. Now we easily obtain:

$$\mathcal{C}_{R,\mathbf{z}}\{\mathbf{y}(n)\} \leq \sum_{i=1}^{N}\sum_{j=1}^{N} \|G_{ij}\|_j^2 |\kappa_{R,z_i}\{\{\widetilde{G}_{ij}\}s_j(n)\}| \leq \sum_{i=1}^{N}\sum_{j=1}^{N} \|G_{ij}\|_j^2 \bar{\kappa}_{R,i}^{\max} \quad (22)$$

and by arguments similar to the proof in the i.i.d. case, one obtains that the above upper-bound is reached if and only if the global filter is separating in the sense of Equation (5). ■

4 Simulations

4.1 Separation Procedure

Our MIMO contrast being quadratic, the optimization can be performed with a similar method to the one presented in [1] for MISO contrasts. The reference signals $z_i(n), i \in \{1, \ldots, n\}$ must be chosen according to A.1. Interestingly, the simulations have clearly demonstrated that it is practically a valid choice to choose them as the output of a prewhitening operation on the observations. This makes our method efficient and competitive compared to other methods.

By the optimization of $\mathcal{C}_{R,\mathbf{z}}$, one can estimate the sources. As has been done in [1], these estimation of the sources can in turn serve as reference signals: when this procedure is repeated iteratively, the number of iteration is denoted by N_I.

4.2 Results

Computer simulations are now presented to compare a deflation procedure to our proposed MIMO contrast $\mathcal{C}_{R,\mathbf{z}}$. We have used fourth order cumulants ($R = 4$). The separation criteria are the mean square estimation error (MSE) on each source for the PAM-4 i.i.d. source signals and $\tau_i \triangleq 1 - \frac{\max_j \|G_{ij}\|_j^2}{\sum_{j=1}^{N} \|G_{ij}\|_j^2}$ ($i \in \{1, \ldots, N\}$) for the CPM (Continuous Phase Modulation) non i.i.d source signals (modulation indices: 0.4, 0.7, 0.3 and 0.6). Note that $0 \leq \tau_i < 1$ and $\tau_i = 0$ if and only if $y_i(n)$ corresponds perfectly to one source. The mixing filter coefficients have systematically been randomly chosen according to a normal distribution.

Experiment 1. In Figure 1 (resp. Figure 2), we have plotted the cumulative distribution of the empirical values of the MSE (resp. values of τ_i) over 1000 Monte-Carlo runs. We have considered $N = 3$ source signals mixed on $Q = 4$ sensors with a filter of length $L = 3$, using $K = 10000$ samples and $N_I = 3$ iterations. We clearly notice that in all Monte-Carlo runs, the proposed method succeeded particularly well to separate the three different sources. This illustrates that choosing the output of a whitening filter for the references is a successful method. In addition, the proposed approach has an equal performance for the extraction of the three sources. As classically observed in deflation separation methods, the performance is worse for the last extracted source signals than for the first ones.

Experiment 2. We have considered $N = 3$ source signals mixed on $Q = 4$ sensors with a filter of length $L = 3$. The number of iterations for each source

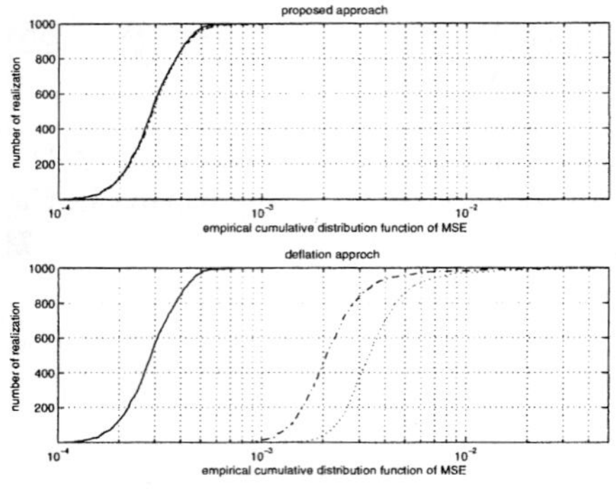

Fig. 1. Empirical cumulative distribution function of the MSE

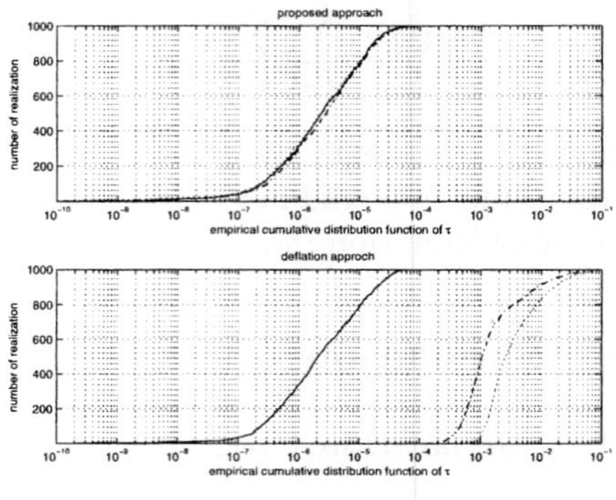

Fig. 2. Empirical cumulative distribution function of τ_i

Table 1. MSE for PAM-4 i.i.d sources and τ_i for CPM non i.i.d sources versus number of samples

	K	5000	10000	15000	20000	25000
PAM-4 sources	s_1	$6.20\ 10^{-4}$	$2.92\ 10^{-4}$	$1.92\ 10^{-4}$	$1.42\ 10^{-4}$	$1.17\ 10^{-4}$
proposed MIMO contrast	s_2	$6.08\ 10^{-4}$	$3.01\ 10^{-4}$	$1.97\ 10^{-4}$	$1.45\ 10^{-4}$	$1.17\ 10^{-4}$
	s_3	$6.07\ 10^{-4}$	$3.01\ 10^{-4}$	$1.94\ 10^{-4}$	$1.43\ 10^{-4}$	$1.17\ 10^{-4}$
PAM-4 sources	s_1	$6.22\ 10^{-4}$	$2.94\ 10^{-4}$	$1.97\ 10^{-4}$	$1.46\ 10^{-4}$	1.1710^{-4}
deflation appoach +	s_2	$6.87\ 10^{-3}$	$2.73\ 10^{-3}$	$2.21\ 10^{-3}$	$1.20\ 10^{-3}$	$9.68\ 10^{-4}$
quadratic MISO contrast	s_3	$1.17\ 10^{-2}$	$4.35\ 10^{-3}$	$4.35\ 10^{-3}$	$2.80\ 10^{-3}$	$1.71\ 10^{-3}$
CPM sources	τ_1	$5.76\ 10^{-6}$	$2.24\ 10^{-6}$	$1.21\ 10^{-6}$	$1.04\ 10^{-6}$	7.5110^{-7}
proposed MIMO contrast	τ_2	$4.82\ 10^{-6}$	$1.97\ 10^{-6}$	$1.44\ 10^{-6}$	$8.47\ 10^{-7}$	$7.9\ 10^{-7}$
	τ_3	$5.69\ 10^{-6}$	$2.29\ 10^{-6}$	$1.07\ 10^{-6}$	$8.72\ 10^{-7}$	$8.27\ 10^{-7}$
CPM sources	τ_1	$5.37\ 10^{-6}$	$2.67\ 10^{-6}$	$1.44\ 10^{-6}$	$1.08\ 10^{-6}$	5.2310^{-7}
deflation appoach +	τ_2	$5.03\ 10^{-3}$	$3.09\ 10^{-3}$	$4.06\ 10^{-3}$	$2.27\ 10^{-3}$	$2.16\ 10^{-3}$
quadratic MISO contrast	τ_3	$1.03\ 10^{-2}$	$8.51\ 10^{-3}$	$7.30\ 10^{-3}$	$6.75\ 10^{-3}$	$4.22\ 10^{-3}$

extraction was $N_I = 5$. In Table 1 we report both the average MSE of each source and τ_i for $i = 1, 2, 3$ on 100 Monte-Carlo runs for the three estimated sources versus the number of samples K. As intuitively expected, using the proposed MIMO contrast, a constant performance has been obtained for the three sources, contrary to the deflation procedure for which the performance is degraded for the extraction of s_2 and s_3.

Experiment 3. We now compare a deflation approach combined with the kurtosis based contrast $|\mathrm{Cum}\{y(n), y(n), y(n), y(n)\}|^2$ with our quadratic contrast. The kurtosis contrast has been optimized using a gradient ascent method. We have considered $N = 3$ source signals mixed on $Q = 4$ sensors with a filter of length $L = 3$. The number of samples is $K = 10000$. We plotted in Figure 3 the cumulative distribution of the empirical values of τ_i for the three

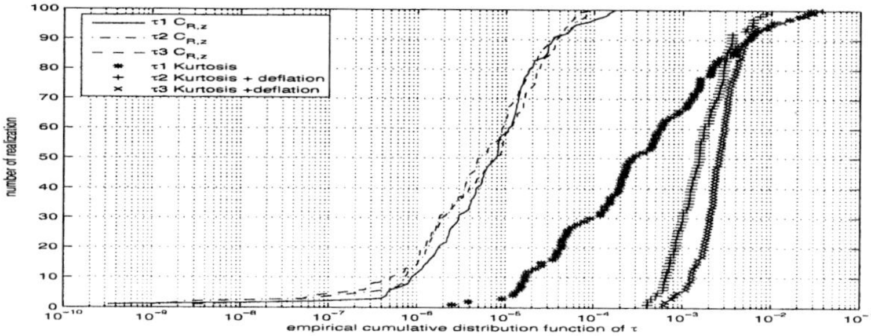

Fig. 3. Comparison of MIMO results by using the kurtosis based contrast and the proposed contrast $\mathcal{C}_{R,\mathbf{z}}$

CPM estimated sources on 100 Monte-Carlo runs for both contrasts. One can see that the quadratic approach outperforms the results obtained by the kurtosis contrast. Besides, the optimization of our contrast is much quicker than gradient optimization of the kurtosis.

References

1. M. Castella, S. Rhioui, E. Moreau and J.-C. Pesquet: "Source separation by quadratic contrast functions: a blind approach based on any higher-order statistics", Proc. of ICASSP 2005, pp. 569-572, Philadelphia, USA.
2. A. Adib, E. Moreau and D. Aboutajdine, "Source separation contrasts using a reference signal", *IEEE Signal Processing Letters*, Vol. 11, No. 3, pp 312-315, 2004.
3. B. Jelonnek, D. Boss and K.-D. Kammeyer, "Generalized Eigenvector Algorithm for Blind Equalization" *Signal Processing*, 61(3), pp. 237-264, September 1997.
4. P. Comon, "Contrasts for multichannel blind deconvolution", *IEEE Signal Processing Letters*, Vol. 3, No. 7, pp. 209-211, July 1996.
5. E. Moreau and J.-C. Pesquet, "Generalized contrasts for multichannel blind deconvolution of linear systems", *IEEE Signal Processing Letters*, Vol. 4, No. 6, pp. 182-183, June 1997.
6. C. Simon, P. Loubaton and C. Jutten, "Separation of a class of convolutive mixtures: a contrast function approach", *Signal Processing* Vol. 81, pp. 883-887, 2001
7. J.K. Tugnait, "Identification and deconvolution of multi-channel linear non-gaussian processes using higher order statistics and inverse filter criteria", *IEEE Trans. Signal Processing*, Vol. 45, No. 3, pp. 658-672, March 1997
8. R. Liu and Y. Inouye, "Blind equalization of MIMO-FIR channels driven by white but higher order colored source signals", *IEEE Trans. Information Theory*, Vol. 48, No. 5, pp. 1206-1214, May 2002.

A Canonical Genetic Algorithm for Blind Inversion of Linear Channels

Fernando Rojas[1], Jordi Solé-Casals[2], Enric Monte-Moreno[3], Carlos G. Puntonet[1], and Alberto Prieto[1]

[1] Computer Architecture and Technology Department, University of Granada 18071, Spain
{frojas, carlos}@atc.ugr.es, aprieto@ugr.es
http://www.atc.ugr.es/
[2] Signal Processing Group, University of Vic, Spain
jordi.sole@uvic.es
http://www.uvic.es/
[3] TALP Research Center, Polytechnic University of Catalonia, Spain
enric@gps.tcs.upc.es
http://gps-tsc.upc.es/veu/personal/enric/enric.html

Abstract. It is well known the relationship between source separation and blind deconvolution: If a filtered version of an unknown i.i.d. signal is observed, temporal independence between samples can be used to retrieve the original signal, in the same manner as spatial independence is used for source separation. In this paper we propose the use of a Genetic Algorithm (GA) to blindly invert linear channels. The use of GA may be more appropriate in the case of small number of samples, where other gradient-like methods fails because of poor estimation of statistics. The experimental results show that the presented method is able to invert unknown filters with good numerical results, even if only 100 samples or less are available.

1 Introduction

The problem of source separation may be formulated as the recovering of a set of unknown independent signals from the observations of mixtures of them without any prior information about either the sources or the mixture [1, 2]. The strategy used in this kind of problems is based on obtaining signals which maximize a certain independence criterion. In the bibliography multiple algorithms are proposed for solving the problem of source separation in instantaneous linear mixtures, from neural networks based methods [3], cumulants or moments methods [4, 5], geometric methods [6] or information theoretic methods [7]. In real world situations, however, the majority of mixtures can not be modeled as instantaneous and/or linear. This is the case of convolutive mixtures, where the effect of channel from source to sensor is modeled by a filter [8].

A particular case of blind separation is the case of blind deconvolution, which is presented in figure 1. Development of this framework is presented in the following section.

J. Rosca et al. (Eds.): ICA 2006, LNCS 3889, pp. 238 – 245, 2006.

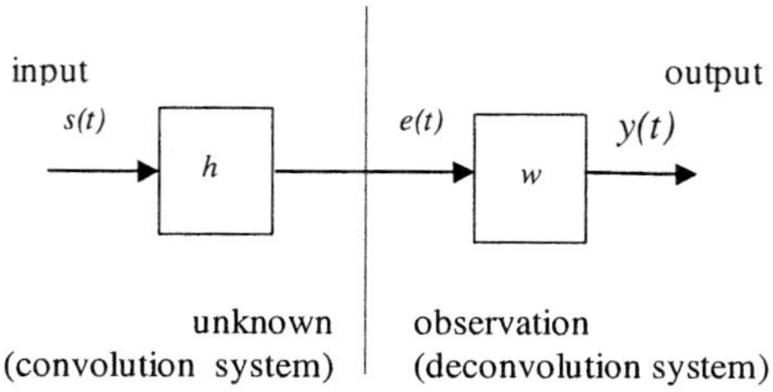

Fig. 1. Block diagram of the convolution system and blind deconvolution system. Both filter h and signal s(t) on the convolution process are unknown.

This paper proposes the use of Genetic Algorithms (GA) for the blind inversion of the (linear) channel. The theoretic framework for using source separation techniques in the case of blind deconvolution is presented in [9]. There, a quasi-nonparametric gradient approach is used, minimizing the mutual information of the output as a cost function to deal with the problem. A parametric approach can be found in [10]. The aim of the paper is to present a different optimization procedure to solve the problem, even if a small number of samples is available. In this case, gradient-like algorithms fail because of poor estimation of statistics. Our method is shown to be useful in this case, where other methods can not be used. This paper is organized as follows. Section 2 describes the linear model and presents the basic equations. Section 3 explains the Genetic Algorithm for blind deconvolution. Finally, section 4 presents the experiments showing the performance of the method.

2 Model and System Equations

2.1 Model

We suppose that the input of the system $S=\{s(t)\}$ is an unknown non-Gaussian independent and identically distributed (i.i.d.) process, and that subsystem h is a linear filter, unknown and invertible. We would like to estimate $s(t)$ by only observing the system output. This implies the blind estimation of the inverse structure composed of a linear filter w. Let **s** and **e** be the vectors of infinite dimension, whose t-th entries are $s(t)$ or $e(t)$, respectively. The unknown input-output transfer can be written as:

$$e = \mathrm{H}s \tag{1}$$

where:

$$\mathbf{H} = \begin{pmatrix} \cdots & \cdots & \cdots & \cdots & \cdots \\ \cdots & h(t+1) & h(t) & h(t-1) & \cdots \\ \cdots & h(t+2) & h(t+1) & h(t) & \cdots \\ \cdots & \cdots & \cdots & \cdots & \cdots \end{pmatrix} \tag{2}$$

is an infinite dimension Toeplitz matrix which represents the action of the filter h to the signal $s(t)$. The matrix **H** is non-singular provided that the filter h is invertible, i.e. satisfies $h^{-1}(t) * h(t) = h(t) * h^{-1}(t) = \delta(t)$, where $\delta(t)$ is the Dirac impulse. The infinite dimension of vectors and matrix is due to the lack of assumption on the filter order. If the filter h is a finite impulse response (FIR) filter of order N_h, the matrix dimension can be reduced to the size N_h. In practice, because infinite-dimension equations are not tractable, we have to choose a pertinent (finite) value for N_h.

2.2 Summary of Equations

From figure 1, we can write the mutual information of the output of the filter w using the notion of entropy rates of stochastic processes as:

$$I(Y) = \lim_{T \to \infty} \frac{1}{2T+1} \left\{ \sum_{t=-T}^{T} H(y(t)) - H(y_{-T}, \ldots, y_T) \right\} = H(y(\tau)) - H(Y) \tag{3}$$

where τ is arbitrary due to the stationary assumption. The input signal $S=\{s(t)\}$ is an unknown non-Gaussian i.i.d. process, $Y=\{y(t)\}$ is the output process and y denotes a vector of infinite dimension whose t-th entry is $y(t)$. We shall notice that $I(Y)$ is always positive and vanishes when Y is i.i.d.

After some algebra, Equation (2) can be rewritten as [10]:

$$I(Y) = H(y(\tau)) - \frac{1}{2\pi} \int_0^{2\pi} \log \left| \sum_{t=-\infty}^{+\infty} w(t) e^{-jt\theta} \right| d\theta - E[\mathcal{E}] \tag{4}$$

At this point we need to derive the optimization algorithm. One possibility is, for example, to use gradient-like algorithms, where the derivative of $I(Y)$ with respect to the coefficients of w filter is needed. In our system, a canonical genetic algorithm will be used, avoiding the calculus of hard statistics. The method is presented in next section.

3 Genetic Algorithm for Blind Deconvolution

3.1 Justifying the Use of a GA for Blind Deconvolution

A genetic algorithm (GA hereinafter) is a search technique used in computer science to find approximate solutions to combinatorial optimization problems. GAs are a particular class of evolutionary algorithms that use techniques inspired by evolutionary biology such as inheritance, mutation, natural selection, and recombination (or crossover) [11].

The process of blind deconvolution can be handled by a genetic algorithm which evolves individuals corresponding to different inverse filters and evaluate the estimated solutions according to a measure of statistical independence. This is a problem of global optimization: minimizing or maximizing a real valued function $f(\vec{x})$ in the parameter space $\vec{x} \in P$. This particular type of problems is suitable to be solved by a genetic algorithm. GAs are designed to move the population away from local minima that a traditional hill climbing algorithm might get stuck in. They are also easily parallelizable and their evaluation function can be any that assigns to each individual a real value into a partially ordered set (poset). GAs have already been successfully applied to linear and post-nonlinear blind source separation [12].

3.2 GA Characterization

The operation of the basic genetic algorithm needs the following features to be completely characterized:

- Encoding Scheme. The genes will represent the coefficients of the unknown deconvolution filter W (real coding). An initial decision must therefore be taken about the length of the inverse filter.
- Initialization Procedure. Coefficients of the deconvolution filter which form part of the chromosomes are randomly initialized.
- Fitness Function. The key point in the performance of a GA is the definition of the fitness function. In this case, we propose several fitness functions related with maximizing independence between the values of the estimated signal $y(t)$. Maximizing kurtosis absolute value is proposed as a first approach, followed by two cumulant-based expressions. Further details will be given in Section 3.2.1.
- Genetic Operators. Typical crossover and mutation operators will be used for the manipulation of the current population in each iteration of the GA. The crossover operator is "Simple One-point Crossover". The mutation operator is "Non-Uniform Mutation" [11]. This operator makes the exploration-exploitation trade-off be more favorable to exploration in the early stages of the algorithm, while exploitation takes more importance when the solution given by the GA is closer to the optimal.
- Parameter Set. Population size, number of generations, probability of mutation and crossover and other parameters relative to the genetic algorithm operation were chosen depending on the characteristics of the mixing problem. Generally a population of 80-100 individuals was used, stopping criteria was set between 60-100 iterations, crossover probability is 0.8 per chromosome and mutation probability is typically set between 0.05 and 0.08 per gene.

3.2.1 Evaluation Functions Proposed

One of the most remarkable advantages of genetic algorithms is its great flexibility for the application of new evaluation functions, being the only requirement that the evaluation function assigns a real value to each individual into a partially ordered set. Therefore, the evaluation function is extremely modular and independent from the rest of the GA. This ability will allow us to decide which evaluation function performs better in each situation. Generally, we will look for an evaluation function which gives higher scores for those chromosomes representing estimations which maximize statistical independence.

- Measuring nongaussianity by kurtosis. Absolute value of the kurtosis has been extensively used as a measure of nongaussianity in finding independent components [13]. Kurtosis is simple to compute. In this paper we propose using the absolute value of the normalized kurtosis as the first evaluation function:

$$|kurt(x)| = \left| \frac{E(x^4)}{E(x^2)^2} - 3 \right| \tag{5}$$

The evaluation function is directly derived from (5):

$$\text{eval}_{\text{Kurt}}(w) = |kurt(y)| \tag{6}$$

where y is the signal obtained after applying the filter w to the observation x.

- Measuring mutual information through approximation by cumulants. As it is well-known, kurtosis is very sensitive to outliers. Therefore we propose a different evaluation function. Using Edgewoth expansion, an approximation of mutual information can be reached using cumulants (higher-order statistics), as proposed in [14]:

$$\text{I}(y) = C - \left(\frac{1}{48} \sum_{i=1}^{n} \left[4\kappa_3(y_i)^2 + \kappa_4(y_i)^2 + 7\kappa_4(y_i)^4 - 6\kappa_3(y_i)^2 \cdot \kappa_4(y_i) \right] \right) \tag{7}$$

where $k_3(y_i) = m_3(y_i) = E\{y_i^3\}$, $k_4(y_i) = m_4(y_i) - 3 = E\{y_i^4\} - 3$ and C is a constant.

The proposed evaluation function splits the estimated signal y in a set of equal-size chunks. Subsequently, it approximates mutual information between the pieces of the signal according to equation (7). As mutual information must be minimal for the estimated source under the assumption of statistical independence, and the evaluation function is attempted to be maximized by the GA, the evaluation function for a given chromosome in this case will be:

$$\text{eval}_{\text{IM}}(W) = \frac{1}{48} \sum_{i=1}^{n} \left[4\kappa_3(y_i)^2 + \kappa_4(y_i)^2 + 7\kappa_4(y_i)^4 - 6\kappa_3(y_i)^2 \cdot \kappa_4(y_i) \right] \tag{8}$$

where y is the signal obtained after applying the filter W to the observation x and y_i is the i-th chunk from estimation y.

- Measuring negentropy through approximation by cumulants. Negentropy is a nonegative measure of nongaussianity. Finally, using again higher-order cumulants and the Gram-Charlier polynomial expansion, gives the approximation:

$$\text{J}(y) \approx \frac{1}{12} E\{y^3\}^2 + \frac{1}{48} \left[E\{y^4\} - 3 \right]^2 \tag{9}$$

The evaluation function is just equivalent to the approximation of negentropy, as the maximum values should give good estimations:

$$\text{eval}_{\text{NEG}}(W) = \text{J}(y) \tag{10}$$

where y is the signal obtained after applying the filter W to the observation x.

4 Experimental Results

Finally, in order to verify the effectiveness of the proposed algorithm, some experimental results using uniform random sources are presented. In all the experiments, the source signal is an uniform random source with zero mean and unit variance. As the performance criterion, we have used the output Signal to the Noise Ratio (SNR) measured in decibels.

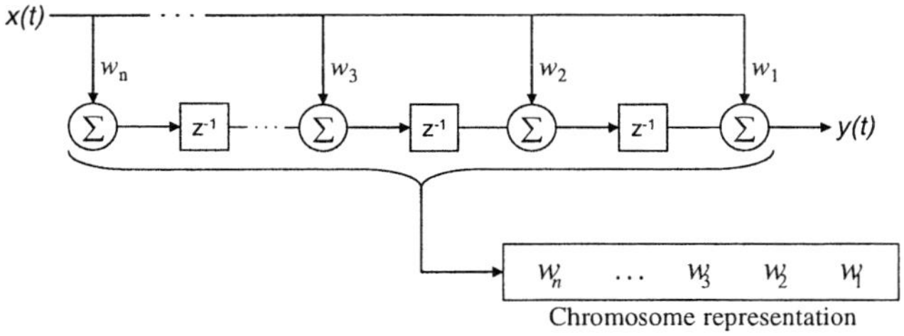

Fig. 2. Encoding scheme in a genetic algorithm for filter coefficients of linear blind deconvolution. The values of the variables stored in the chromosome are real numbers.

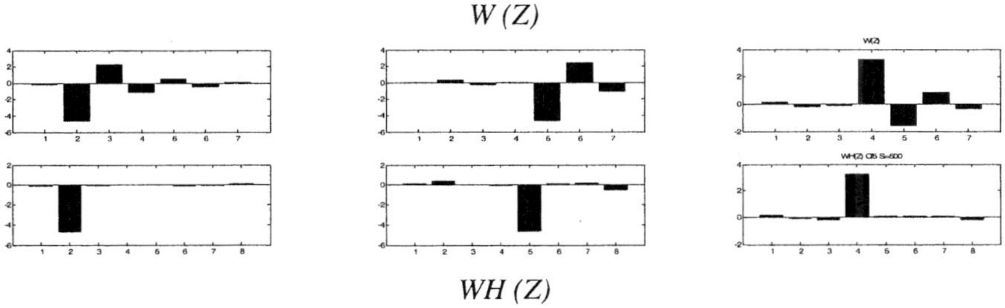

Fig. 3. On the top line, the inverse filter coefficients. Down figures represent coefficients of the convolution between filters *W(Z)* and *H(Z)*. From left to right: filter coefficients given by the GA using $\text{eval}_{\text{Kurt}}$ with an observed signal of 2000 samples, eval_{IM} with 1000 samples, and eval_{Neg} with 500 samples. Note that in some of the estimated filters, a delay due to the indeterminacies may appear.

In the first experiment, the (unknown) filter is the low-pass filter $H(Z) = 1 + 0.5z^{-1}$. Then, the proposed algorithm is used to obtain the inverse system. The parameters of the algorithm are: T = 5000 (number of observed samples), p = 7 (order of inverse filter), crossover probability was set to 0.8, mutation probability 0.075, population size is 80, and the stopping criterion is 100 generations.

In the second experiment, we diminish the number of available samples. Here, the problem is more difficult due to the lack of information. Parameters of the algorithm are set to: T = 1000 (number of observed samples), p = 5 (order of inverse filter), and we used the same parameters for the genetic part of the algorithm as in the first experiment. Figure 2 (center) shows the coefficients of the filters *W(Z)* and *WH(Z)* respectively when applying eval_{IM} .

In the last experiment, we reduced the number of available to T=500. The rest of the parameters remain the same as in the former simulations.

Figure 3 summarizes the results of the experiments in terms of crosstalk (left) and computational time (right). These experiments show that although the number of samples are low, the algorithm has been capable of estimating the inverse system. Crosstalk between the estimation and the source is situated between 20-30dB,

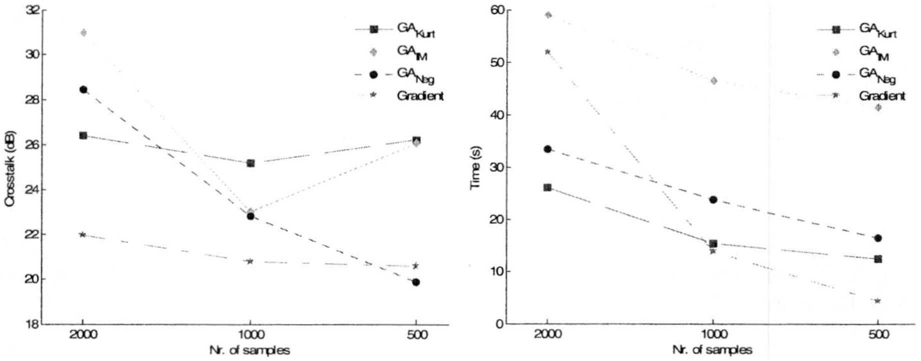

Fig. 4. Crosstalk and time comparison of the proposed GA using the three different evaluation functions and a gradient descent algorithm

depending on the length of the signal and the contrast function applied. When compared to a typical deconvolution gradient descent algorithm, the GA presents a better performance as the number of samples in the observed signal decreases.

5 Conclusion

A GA algorithm is presented for blind inversion of channels. The use of this technique is justified here because of the small number of samples. In this situation, gradient-like algorithm fails because it is very difficult to obtain a good estimation of statistics (score function, pdf, etc.). Optimization using GA avoids these calculations and gives us good results for inverting the unknown filter. Future research should extend the idea to Wiener systems (linear filter plus nonlinear function), where a Hammerstein structure can be used and all the parameters should be found by these optimization techniques.

References

1. Jutten, C., Hérault, J., Blind separation of sources, Part I: An adaptive algorithm based on neuromimetic architecture, Signal Processing 24, 1991.
2. Comon, P., Independent component analysis, a new concept?, Signal Processing 36, 1994.
3. Amari, S., Cichocki A., Yang, H.H., A new learning algorithm for blind signal separation, NIPS 95, MIT Press, 8, 1996
4. Comon, P., Separation of sources using higher order cumulants Advanced Algorithms and Architectures for Signal Processing, 1989.
5. Cardoso, J.P., Source separation using higher order moments, in Proc. ICASSP, 1989.
6. Puntonet, C., Mansour A., Jutten C., A geometrical algorithm for blind separation of sources, GRETSI, 1995
7. Bell, A.J., Sejnowski T.J., An information-maximization approach to blind separation and blind deconvolution, Neural Computation 7, 1995.

8. Nguyen Thi, H.L., Jutten, C., Blind source separation for convolutive mixtures, IEEE Transactions on Signal Processing, vol. 45, 2, 1995.
9. Taleb, A., Solé-Casals, J., Jutten, C., Quasi-Nonparametric Blind Inversion of Wiener Systems, IEEE Transactions on Signal Processing, Vol. 49, n°5, pp.917-924 (2001).
10. Solé-Casals, J., Jutten, C., Taleb A., Parametric approach to blind deconvolution of nonlinear channels. Neurocomputing, Vol. 48, pp.339-355 (2002)
11. Michalewicz, Z.: Genetic Algorithms + Data Structures = Evolution Programs. 3rd edn. Springer-Verlag, Berlin Heidelberg New York (1996)
12. F. Rojas, C.G. Puntonet, M. Rodríguez, I. Rojas, R. Martín-Clemente. Blind Source Separation in Post-Nonlinear Mixtures Using Competitive Learning, Simulated Annealing and a Genetic Algorithm. IEEE Trans. on SMC, Part C, Vol.34(4), pp.407-416, Nov. 2004.
13. Shalvi, O.; Weinstein, E. New criteria for blind deconvolution of non-minimum phase systems (channels). IEEE Trans. on Information Theory, Volume 36, Issue 2, March 1990 Page(s):312 – 321.
14. P. Comon, Independent Component Analysis, A new concept?. Signal Processing, vol.36, n°. 3, pp. 287-314. 1994.

Efficient Separation of Convolutive Image Mixtures*

Sarit Shwartz, Yoav Y. Schechner, and Michael Zibulevsky

Dept. Electrical Engineering, Technion - Israel Inst. Tech., Haifa 32000, Israel
psarit@tx.technion.ac.il, {yoav, mzib}@ee.technion.ac.il

Abstract. Convolutive mixtures of images are common in photography of semi-reflections. They also occur in microscopy and tomography. Their formation process involves focusing on an object layer, over which defocused layers are superimposed. Blind source separation (BSS) of convolutive image mixtures by direct optimization of mutual information is very complex and suffers from local minima. Thus, we devise an efficient approach to solve these problems. Our method is fast, while achieving high quality image separation. The convolutive BSS problem is converted into a set of instantaneous (pointwise) problems, using a short time Fourier transform (STFT). Standard BSS solutions for instantaneous problems suffer, however, from scale and permutation ambiguities. We overcome these ambiguities by exploiting a parametric model of the defocus point spread function. Moreover, we enhance the efficiency of the approach by exploiting the sparsity of the STFT representation as a prior.

1 Introduction

Typical blind source separation (BSS) methods seek separation when the mixing process is unknown. However, loose prior knowledge regarding the mixing process often exists, due to its physical origin. In particular, this process can be represented by a parametric form, rather than a trivial representation of raw numbers. For example, consider convolutive image mixtures caused by defocus blur. This blur can be parameterized, yet the parameters' values are unknown. Such mixtures occur in tomography and microscopy [1, 2]. They also occur in semi-reflections [1], e.g., from a glass window: a scene imaged behind the semi-reflector is superimposed on a reflected scene [3, 4]. Each scene is at a different distance from the camera, thus differently defocus blurred in the mixtures.

We claim that BSS can benefit from such a parametrization, as it makes the estimation more efficient while helping to alleviate ambiguities. In the case of semireflections, our goal is to decompose the mixed and blurred images into

* This research has been supported in parts by the "Dvorah" Fund of the Technion and by the HASSIP Research Network Program HPRN-CT-2002-00285, sponsored by the European Commission. The research was carried out in the Ollendorff Minerva Center. Minerva is funded through the BMBF. Yoav Schechner is a Landau Fellow-supported by the Taub Foundation, and an Alon Fellow.

J. Rosca et al. (Eds.): ICA 2006, LNCS 3889, pp. 246–253, 2006.

the separate scene layers, by minimizing the mutual information (MI) of the estimated objects. An attempt by Ref. [1] used exhaustive search, hence being computationally prohibitive. Ref. [5] attempted convolutive image separation by minimization of higher order cumulant. That method suffers from a scale ambiguity: the sources are reconstructed up to an unknown filter. Moreover, the method's complexity increases fast with the support of the separation kernel.

The complexity of convolutive source separation has been reduced in the domain of acoustic signals, by using frequency methods [6, 7]. There, BSS is decomposed into several small pointwise problems by applying a short-time-Fourier transform (STFT). Then, standard BSS tools are applied to each of the STFT channels. However, these tools suffer from fundamental ambiguities, which may reduce the overall separation quality. Ref. [8] suggested that these ambiguities can be overcome by nonlinear operations in the image domain. However, this method encountered performance problems when simulated over natural images.

We show that these problems can be efficiently solved by exploiting a parametric model for the unknown blur. Moreover, we use the sparsity of STFT coefficients to yield a practically unique solution, which is derived fast. The algorithm is demonstrated in simulations of semi-reflected natural scenes.

2 Problem Formulation

Let $\{s_1, \ldots, s_K\}$ be a set of K independent sources. Each source is of the form $s_k = s_k(\mathbf{x})$, $k = 1, \ldots, K$, where $\mathbf{x} = (x, y)$ is a two dimensional (2D) spatial coordinate vector in the case of images. Let $\{m_1, \ldots, m_K\}$ be a set of K measured signals, each of which is a linear mixture of a convolved version of the sources

$$m_i(\mathbf{x}) = a_{i1} * s_1(\mathbf{x}) + \ldots + a_{iK} * s_K(\mathbf{x}) \quad , i = 1, \ldots, K \ . \tag{1}$$

Here $*$ denotes convolution and $a_{ik}(\mathbf{x})$, $k = 1, \ldots, K$, are linear spatially invariant filters. Denote $\{\hat{s}_1, \ldots, \hat{s}_K\}$ as the set of the reconstructed sources. Reconstruction is done by applying a linear operator $\mathbf{W}$ on $\{m_1, \ldots, m_K\}$. Each of the reconstructed sources is of the form

$$\hat{s}_k(\mathbf{x}) = w_{k1} * m_1(\mathbf{x}) + \ldots + w_{kK} * m_K(\mathbf{x}) \quad , k = 1, \ldots, K \ , \tag{2}$$

where $w_{ik}(\mathbf{x})$ are linear spatially invariant filters. Our goal is: given only the measured signals $\{m_1, \ldots, m_K\}$, find a linear separation operator $\mathbf{W}$ that inverts the mixing process, thereby separating the sources. The mixing process is inverted by finding $\mathbf{W}$ that minimizes the MI of $\{\hat{s}_1, \ldots, \hat{s}_K\}$.

MI is expressed by using the marginal entropies $\mathcal{H}_{\hat{s}_k}$ and the joint entropy of the estimated sources $\mathcal{H}_{\hat{s}_1,\hat{s}_2}$ as $\mathcal{I}_{\hat{s}_1,\hat{s}_2} = \sum_{k=1}^{K} \mathcal{H}_{\hat{s}_k} - \mathcal{H}_{\hat{s}_1,\ldots,\hat{s}_K}$. However, estimation of the joint entropy may be unreliable. It can be avoided if the mixtures are pointwise, rather than convolutive. In pointwise mixtures, the separation operator $\mathbf{W}$ is a simple matrix, termed the separation matrix. In this case, the MI can be expressed as (see for example Ref. [9])

$$\mathcal{I}(\hat{s}_1, \hat{s}_2) = -\log|\det(\mathbf{W})| + \sum_{k=1}^{K} \mathcal{H}_{\hat{s}_k} \ . \tag{3}$$

It is desirable to do the same for convolutive mixtures. However, if $\mathbf{W}$ is a convolutive operator, Eq. (3) does not hold. We note that expressions similar to (3) have been developed for convolutive mixtures [10] assuming spatially white sources. Nevertheless, algorithms based on these expressions suffer from whitening of the separated sources, corrupting the estimation severely both in acoustic and in imaging applications.

3 Efficient Separation of Convolutive Image Mixtures

We may use Eq. (3) in convolutive mixtures, despite the fact that it is valid only in pointwise mixtures. This is achieved by decomposing the convolutive optimization problem into several smaller ones, which are apparently independent of each other. This approach is inspired by frequency domain algorithms developed for acoustic signals [6,7]. Nevertheless, this approach has its own fundamental limitations, which are discussed and solved in Secs. 4 and 5.

We apply STFT[1] to the data. Denote $\boldsymbol{\omega} = (\omega_x, \omega_y)$ as the index vector of the frequency variable of the 2D STFT. Assuming that the STFT window size is larger than the effective width[2] of the blur kernel [6], Eq. (1) becomes

$$m_i(\boldsymbol{\omega}, \mathbf{x}) \approx a_{i1}(\boldsymbol{\omega}) s_1(\boldsymbol{\omega}, \mathbf{x}) + \ldots + a_{iK}(\boldsymbol{\omega}) s_K(\boldsymbol{\omega}, \mathbf{x}), \quad i = 1, \ldots, K \ , \tag{4}$$

since convolution becomes a multiplication in this domain.

Eq. (4) exposes a fundamental problem in cases of energy-preserving convolution operators. In such operators $a_{ik}(\boldsymbol{\omega}) \to 1$ as $\boldsymbol{\omega} \to 0$ (the overall light energy over the image area is invariant to the convolution). This occurs in defocus blur, since change of focus does not cause light attenuation, only a different spread of the light energy across the sensor area [1,2]. As $a_{ik}(\boldsymbol{\omega}) \to 1$, Eq. (4) becomes

$$m_i(\boldsymbol{\omega}, \mathbf{x}) \approx s_1(\boldsymbol{\omega}, \mathbf{x}) + \ldots + s_K(\boldsymbol{\omega}, \mathbf{x}), \quad i = 1, \ldots, K \ . \tag{5}$$

This is a singular set of equations. Therefore, low spatial frequencies are not well reconstructed. Note that this has nothing to do with the ICA problem. Even if the blur kernels a_{ik} are *perfectly known*, the reconstruction is ill-conditioned in the low-frequency bands [1,2]. Keeping in mind this matter, we continue with the blind estimation process. Note that at each sub-band $\boldsymbol{\omega}$, Eq. (4) expresses a pointwise mixture of sub-band images. At each frequency channel, the mixed sources can be separated by simple ICA optimization. Then, all the separated sources from all the frequency channels may be combined by inverse STFT.

To describe the ICA optimization, denote $\mathbf{W}(\boldsymbol{\omega})$ as the separation matrix at channel $\boldsymbol{\omega}$. In addition, denote $\mathcal{I}^{\boldsymbol{\omega}}(\hat{s}_1, \hat{s}_2)$ and $\mathcal{H}^{\boldsymbol{\omega}}_{\hat{s}_k}$ as the MI and marginal entropies of the estimated sources at channel $\boldsymbol{\omega}$, respectively. Then, similarly to Eq. (3), the MI of the estimated sources at each channel is given by

$$\min_{\mathbf{W}(\boldsymbol{\omega})} \left\{ -\log |\det[\mathbf{W}(\boldsymbol{\omega})]| + \sum_{k=1}^{K} \hat{\mathcal{H}}^{\boldsymbol{\omega}}_{\hat{s}_k} \right\} \ , \tag{6}$$

[1] This operation is also termed as a *windowed Fourier transform*, which may be more appropriate for spatial coordinates as we use.

[2] A discussion regarding the STFT window width is given in Sec. 7.

where $\hat{\mathcal{H}}^{\omega}_{\hat{s}_k}$ is an estimator of the channel entropy of an estimated source. Hence, using this factorization, MI minimization of a convolutive mixture is expected to be both more accurate and more efficient to obtain.

Sparse Separation in the STFT Domain

Now, we exploit image statistics in order to achieve a computationally efficient solution for the sub-problems in each frequency channel. As shown in [11], sparsity of sources is a strong prior that can be exploited to achieve a very efficient separation. It is known from studies of image statistics (see for example [12]) that sub-band images are *sparse signals.* Motivated by [11,13], their quasi-maximum likelihood blind separation can be achieved via minimization of

$$\min_{\mathbf{W}(\omega)} \left\{ -\log\left|\det[\mathbf{W}(\omega)]\right| + (1/N)\sum\nolimits_{k=1}^{K}\sum\nolimits_{n=1}^{N}\left|\hat{s}_k(\omega,n)\right| \right\} . \tag{7}$$

Here n indexes the STFT shift (out of a total of N). This enables relative Newton optimization [14], which enhances the efficiency of sparse source separation.

4 Inherent Problems

The frequency representation brings efficiency of pointwise separation. With it, however, come fundamental ambiguities that are common in pointwise problems. The *permutation ambiguity* implies that the separated sub-band images appear at each channel in a random permutation. Some sub-band images corresponding to the "first" estimated source may actually belong to the "second" estimated source. When the channels are transformed back to the image domain using the inverse STFT, the reconstructed images can suffer from crosstalk. Even though source separation was achieved in each channel independently, distinct sub-band images from different sources are combined in the reconstruction.

In addition, the scale of different channels is unknown due to *scale ambiguity,* leading to imbalance between frequency channels. When the estimated channels of a source are transformed back to the image domain using the inverse STFT, the reconstructed image can appear unnatural and suffer from artifacts.

Moreover, the performance in each frequency channel is frequency dependent. Typically, there are a few frequency channels with good separation, a few channels with very poor separation and the rest of the channels have mediocre separation quality. There are several reasons for this phenomenon. One reason is related to the different sparsity of different frequency channels [15].

5 Inter-channel Knowledge Transfer

In this section we bypass the permutation and scale ambiguities by exploiting a prior about the unknown convolutive process. Blur caused by optical defocus can be parameterized [16]. As an example, consider a rough parametric model:

a simple 2D Gaussian kernel with different widths in the x and y directions [1]. Denote $\boldsymbol{\xi}_{i,k} = [\xi_{i,k,x}, \xi_{i,k,y}]$ as the vector of the unknown blur parameters of the blur kernel of source k at image i and

$$G_{\boldsymbol{\xi}_{i,k}}(\boldsymbol{\omega}) = \exp\left[-\omega_x^2/(2\xi_{i,k,x}^2)\right] \exp\left[-\omega_y^2/(2\xi_{i,k,y}^2)\right] \tag{8}$$

as the filter which preserves light energy. In addition to defocus, let us incorporate attenuation $g_{i,k}$ of each source k into any mixture i.

Assume that in each acquired image, one of the layers is focused,[3] i.e. $G_{\boldsymbol{\xi}_{k,k}} = 1$. Define $\mathbf{A}(\boldsymbol{\omega})$ as the mixing operator in frequency channel $\boldsymbol{\omega}$.

$$\mathbf{A}(\boldsymbol{\omega}) = \begin{bmatrix} 1 & g_{1,2}G_{\boldsymbol{\xi}_{1,2}}(\boldsymbol{\omega}) & \cdots & \cdots \\ g_{2,1}G_{\boldsymbol{\xi}_{2,1}}(\boldsymbol{\omega}) & 1 & \vdots & \vdots \\ \vdots & \vdots & \ddots & \vdots \\ \cdots & \cdots & g_{K,K-1}G_{\boldsymbol{\xi}_{K,K-1}}(\boldsymbol{\omega}) & 1 \end{bmatrix} . \tag{9}$$

Thus, the separation matrix in each channel is parameterized by $\boldsymbol{\xi}_{i,k}$ and $g_{i,k}$ and is of the form $\mathbf{W}(\boldsymbol{\omega}) = [\mathbf{A}(\boldsymbol{\omega})]^{-1}$. Note that the parameter $\boldsymbol{\xi}_{i,k}$ and $g_{i,k}$ are the *same for all frequency channels.* Hence, there is a small number of actual unknown blur variables. On the other hand, there is a large number of frequency channels upon which the estimation of these variables can be based.

As we explain is Sec. 5.2, we can automatically select three channels $\boldsymbol{\omega}^{\mathsf{a}}$, $\boldsymbol{\omega}^{\mathsf{b}}$ and $\boldsymbol{\omega}^{\mathsf{c}}$, that yield the best separation results according to a ranking criterion. Define $\tilde{\mathbf{A}}(\boldsymbol{\omega}^{\mathsf{a}}) = [\mathbf{W}(\boldsymbol{\omega}^{\mathsf{a}})]^{-1}$ and similarly $\tilde{\mathbf{A}}(\boldsymbol{\omega}^{\mathsf{b}})$ and $\tilde{\mathbf{A}}(\boldsymbol{\omega}^{\mathsf{c}})$. Let $\tilde{a}_{i,k}$ be the coefficients of $\tilde{\mathbf{A}}$. Then, for each blur kernel, we calculate the unknown blur parameters $\boldsymbol{\xi}_{i,k}$ and $g_{i,k}$ by solving the following set of equations:

$$\begin{cases} g_{i,k}G_{\boldsymbol{\xi}_{i,k}}(\boldsymbol{\omega}^{\mathsf{a}}) = \tilde{a}_{i,k}(\boldsymbol{\omega}^{\mathsf{a}})/\tilde{a}_{i,i}(\boldsymbol{\omega}^{\mathsf{a}}) \\ g_{i,k}G_{\boldsymbol{\xi}_{i,k}}(\boldsymbol{\omega}^{\mathsf{b}}) = \tilde{a}_{i,k}(\boldsymbol{\omega}^{\mathsf{b}})/\tilde{a}_{i,i}(\boldsymbol{\omega}^{\mathsf{b}}) \\ g_{i,k}G_{\boldsymbol{\xi}_{i,k}}(\boldsymbol{\omega}^{\mathsf{c}}) = \tilde{a}_{i,k}(\boldsymbol{\omega}^{\mathsf{c}})/\tilde{a}_{i,i}(\boldsymbol{\omega}^{\mathsf{c}}) \end{cases} , \tag{10}$$

We solve this set to find the parameters $\boldsymbol{\xi}_{i,k}$ and $g_{i,k}$, thus deriving the blur and attenuation parameters based on those few selected channels.[4]

Now, we can use these parameters and Eq. (8) to calculate $g_{i,k}G_{\boldsymbol{\xi}_{i,k}}(\boldsymbol{\omega})$ for all the frequency channels. This directly yields the separation operator $\mathbf{W}$ for *all the frequency channels.* We invert the mixing process by using this $\mathbf{W}$. It may be possible to achieve higher accuracy by representing each blur kernel using parametric models other than Gaussian, requiring more parameters. This would require selection of additional channels.

[3] We stress that we seek layer *separation* rather than *deblurring*. Therefore, if source k is defocused in all the images, we denote the least defocused version of source k as the effective source we aim to reconstruct. Then, we denote $G_{\boldsymbol{\xi}_{i,k}}(\boldsymbol{\omega})$ as the relative defocus filter between the effective source and the defocused source at image i .

[4] One might suggest optimizing the MI directly over the parameters $g_{i,k}$ and $\boldsymbol{\xi}_{i,j}$. However, this optimization scheme is not necessarily convex. A detailed discussion on this issue is given in [15].

5.1 Separation of Semi-reflections

Section 5 describes a parametric model for mixtures of blurred images. It consists of an attenuation factor $g_{i,k}$ and an energy preserving filter $G_{\boldsymbol{\xi}_{i,k}}$. However, in common applications such as semi-reflections [1] or widefield optical sectioning [2], no attenuation accompanies the change of focus. Hence, $g_{i,k} = 1$ for all i, k. For each signal, each source is affected only by two parameters in the Gaussian model. Thus, only two channels are needed to solve for the unknown $\boldsymbol{\xi}_{i,k}$. Moreover, in the special case of semi-reflections, we have only two sources. Therefore, the mixing operator and the separation operator are reduced to

$$\mathbf{A}(\boldsymbol{\omega}) = \begin{bmatrix} 1 & G_{\boldsymbol{\xi}_{1,2}} \\ G_{\boldsymbol{\xi}_{2,1}}(\boldsymbol{\omega}) & 1 \end{bmatrix}, \ \mathbf{W}(\boldsymbol{\omega}) = \begin{bmatrix} 1 & -G_{\boldsymbol{\xi}_{1,2}} \\ -G_{\boldsymbol{\xi}_{2,1}}(\boldsymbol{\omega}) & 1 \end{bmatrix} \{\det(|\mathbf{A}(\boldsymbol{\omega})|)\}^{-1} . \tag{11}$$

The equation system we need to solve in order to estimate $\boldsymbol{\xi}_{1,2}$ and $\boldsymbol{\xi}_{2,1}$ is

$$\begin{cases} -G_{\boldsymbol{\xi}_{1,2}}(\boldsymbol{\omega}^{\mathrm{a}}) = w_{1,2}(\boldsymbol{\omega}^{\mathrm{a}})/w_{1,1}(\boldsymbol{\omega}^{\mathrm{a}}) \\ -G_{\boldsymbol{\xi}_{1,2}}(\boldsymbol{\omega}^{\mathrm{b}}) = w_{1,2}(\boldsymbol{\omega}^{\mathrm{b}})/w_{1,1}(\boldsymbol{\omega}^{\mathrm{b}}) \\ -G_{\boldsymbol{\xi}_{2,1}}(\boldsymbol{\omega}^{\mathrm{a}}) = w_{2,1}(\boldsymbol{\omega}^{\mathrm{a}})/w_{2,2}(\boldsymbol{\omega}^{\mathrm{a}}) \\ -G_{\boldsymbol{\xi}_{2,1}}(\boldsymbol{\omega}^{\mathrm{b}}) = w_{2,1}(\boldsymbol{\omega}^{\mathrm{b}})/w_{2,2}(\boldsymbol{\omega}^{\mathrm{b}}) \end{cases} . \tag{12}$$

Here, $w_{i,k}$ are the coefficients of matrix $\mathbf{W}(\boldsymbol{\omega})$ and $\boldsymbol{\omega}^{\mathrm{a}}, \boldsymbol{\omega}^{\mathrm{b}}$ are the best and second best channels according to the ranking we describe next.[5]

We stress that thanks to this approach of parameter-based inter-channel knowledge transfer, the permutation, scale and sign ambiguities are solved: the sources are not derived in a random order or with inter-channel imbalance, but in a way that must be consistent with the blur model, hence with the image formation process. In addition, the problem of channel and data dependent performance is alleviated, since the separation operator is estimated based on selected channels performing well.

5.2 Selecting a Good Frequency Channels

The parameter estimation method requires *ranking* of the channels. The ranking relies on a quality criterion for the separation (i.e., independence) of $\hat{s}_1$ and $\hat{s}_2$ at each frequency channel $\boldsymbol{\omega}$, given the sparsity assumption.

The scatter plot of sparse independent signals has a cross shape aligned with the axes, in the $(\hat{s}_1, \hat{s}_2)$ plane, i.e., most of the samples should have small angles relative to the $\hat{s}_1$ and $\hat{s}_2$ axes. Define

$$\chi^{\boldsymbol{\omega}}_{\mathcal{L}_1} = \sum\nolimits_{k=1}^{2} \left(\left\{ \sum\nolimits_{n=1}^{N} |\hat{s}_k(\boldsymbol{\omega}, n)| \right\} / \left\{ \sum\nolimits_{n=1}^{N} [\hat{s}_k(\boldsymbol{\omega}, n)]^2 \right\} \right) . \tag{13}$$

This criterion increases as the samples in the scatter plot deviate from the $\hat{s}_1$ and $\hat{s}_2$ axes, and is reduced when each sample n has non-zero values exclusively in

[5] It might be possible to achieve better estimation by using more than two channels, for example, by solving a non-linear least squares problem.

Fig. 1. Simulation results: (a) Two original natural images. (b) The two convolved and mixed images. (c) Reconstructed layers.

$\hat{s}_1$ or $\hat{s}_2$. This closed form expression automatically determines which frequency channels yield the most separated sources, and are thus preferable.

Thus, in our algorithm, we first perform ICA in all the frequency channels. We then calculate $\chi^{\boldsymbol{\omega}}_{\mathcal{L}_1}$, thus ranking the channels. Then, we select the best channels as those that correspond to the smallest values of the penalty function $\chi^{\boldsymbol{\omega}}_{\mathcal{L}_1}$. These channels are used in Sec. 5.

6 Demonstration

The method was simulated using two natural images of size 122×162 pixels (Fig. 1a) as the two scene layers. The blur kernels we used are Gaussians with parameter vectors $\boldsymbol{\xi}_{1,2} = [1,\ 2]$ and $\boldsymbol{\xi}_{2,1} = [2,\ 1]$ pixels. We did not use attenuation coefficients because in photography of real semi-reflections, the image layers are only blurred but not attenuated by change of focus. We added i.i.d Gaussian noise with standard deviation of ~ 2.5 gray levels to the convolved and mixed images. The resulting mixed and noisy images are shown in Fig. 1b. Separation was performed using STFT having 13×13 frequency channels. The separation results are presented in Fig. 1c. The resulting images are indeed well separated. There is no visible crosstalk between the images. The contrast of the reconstructed images is reduced compared to the original images. This stems from inherent ill-conditioning of the mixing matrix at low frequencies (see Sec. 4), i.e., this is not associated with the blindness of the separation problem.

7 Discussion

The convolutive image separation algorithm has currently a single parameter to tweak: the width of the STFT window. It can affect the separation results, and the optimal size somewhat depends on the acquired images. As mentioned in Sec. 3, it must be larger than the effective width of the blur kernel. On the other hand, a very wide window can degrade the sparsity of the sub-band images. A detailed discussion is given in [15]. We determined the window width by trial and error, but we believe this can be automated. For example, multi-window STFT may be used, followed by selection of the the best window width using the criterion described in Sec. 5.2. This requires further research.

References

1. Schechner, Y.Y., Kiryati, N., Basri, R.: Separation of transparent layers using focus. Int. J. Computer Vision **89** (2000) 25–39
2. Macias-Garza, F., Bovik, A.C., Diller, K.R., Aggarwal, S.J., Aggarwal, J.K.: The missing cone problem and low-pass distortion in optical serial sectioning microscopy. In: Proc. ICASSP. Volume 2. (1988) 890–893
3. Schechner, Y.Y., Shamir, J., Kiryati, N.: Polarization and statistical analysis of scenes containing a semi-reflector. J. Opt. Soc. America A **17** (2000) 276–284
4. Bronstein, A.M., Bronstein, M.M., Zibulevsky, M., Zeevi, Y.Y.: Sparse ICA for blind separation of transmitted and reflected images. Intl. J. Imaging Science and Technology **15**(1) (2005) 84–91
5. Castella, M., Pesquet, J.C.: An iterative blind source separation method for convolutive mixtures of images. In: Proc. ICA2004. (2004) 922–929
6. Parra, L., Spence, C.: Convolutive blind separation of non-stationary sources. IEEE Trans. on Speech and Audio Processing **8** (2000) 320–327
7. Smaragdis, P.: Blind separation of convolved mixtures in the frequency domain. Neurocomputing **22** (1998) 21–34
8. Kasprzak, W., Okazaki, A.: Blind deconvolution of timely-correlated sources by homomorphic filtering in Fourier space. In: Proc. ICA2003. (2003) 1029–34
9. Hyvärinen, A., Karhunen, J., Oja, E.: Independent component analysis. John Wiley and Sons, NY (2001)
10. Pham, D.T.: Contrast functions for blind source separation and deconvolution of sources. In: Proc. ICA2001. (2001) 37–42
11. Zibulevsky, M., Pearlmutter, B.A.: Blind source separation by sparse decomposition in a signal dictionary. Neural Computations **13**(4) (2001) 863–882
12. Simoncelli, E.P.: Statistical models for images: Compression, restoration and synthesis. In: Proc. IEEE Asilomar Conf. Sig. Sys. and Computers. (1997) 673–678
13. Pham, D.T., Garrat, P.: Blind separation of a mixture of independent sources through a quasi-maximum likelihood approach. IEEE Trans. Sig. Proc. **45**(7) (1997) 1712–1725
14. Zibulevsky, M.: Blind source separation with relative newton method. In: Proc. ICA2003. (2003) 897–902
15. Shwartz, S., Schechner, Y.Y., Zibulevsky, M.: Efficient blind separation of convolutive image mixtures. Technical report, CCIT No. 553, Dep. Elec. Eng., Technion Israel Inst.Tech. (2005)
16. Born, M., Wolf, E.: Principles of optics. Pergamon, Oxford (1975)

Sparse Nonnegative Matrix Factorization Applied to Microarray Data Sets

K. Stadlthanner[1], F.J. Theis[1], E.W. Lang[1], A.M. Tomé[3], C.G. Puntonet[2], P. Gómez Vilda[4], T. Langmann[5], and G. Schmitz[5]

[1] Institute of Biophysics, University of Regensburg, 93040 Regensburg, Germany
kusta@web.de
[2] DATC, Universidad de Granada, E-18071 Granada, Spain
[3] DET / IEETA, Universidade de Aveiro, 3810-Aveiro, Portugal
[4] DATSI, Universidad Politécnica de Madrid, 28660-Madrid, Spain
[5] Institute for Clinical Chemistry and Laboratory Medicine, University Hospital, 93053 Regensburg, Germany

Abstract. Nonnegative Matrix Factorization (NMF) has proven to be a useful tool for the analysis of nonnegative multivariate data. However, it is known not to lead to unique results when applied to nonnegative Blind Source Separation (BSS) problems. In this paper we present first results of an extension to the NMF algorithm which solves the BSS problem when the underlying sources are sufficiently sparse. As the proposed target function has many local minima, we use a genetic algorithm for its minimization.

1 Matrix Factorization and Sparse Component Analysis

Since recently high throughput methods like microarrays allow to measure whole genom wide gene expression profiles. Intelligent data analysis tools are needed to unveil the information hidden in those microarray data sets. BSS might proof useful to go beyond simple clustering and decompose such data sets into component profiles which might be associated with underlying biological processes. Linear BSS can be considered a matrix factorization problem, where the $m \times T$ matrix of observations $\mathbf{X}$ is decomposed into an $m \times T$ matrix $\mathbf{S}$ of underlying sources and an $m \times m$ mixing matrix according to $\mathbf{X} = \mathbf{AS}$.

Thus the observations $\mathbf{X}$ represent weighted sums of m underlying sources which form the rows of the $m \times T$ matrix $\mathbf{S}$, and the element a_{ij} of the mixing matrix $\mathbf{A}$ forms the weight with which the j-th source contributes to the i-th observation. With BSS now, given only the matrix $\mathbf{X}$, a matrix factorization as in (1) is sought such that $\mathbf{A}$ and $\mathbf{S}$ are unique up to some scaling and permutation indeterminacies. Obviously, the problem is highly underdetermined and can only be solved uniquely if additional constraints are imposed onto the sources or the mixing matrix.

A variant of matrix factorization is nonnegative matrix factorization (NMF), where the matrices $\mathbf{S}$, $\mathbf{A}$ and $\mathbf{X}$ are assumed to be strictly nonnegative. But

J. Rosca et al. (Eds.): ICA 2006, LNCS 3889, pp. 254–261, 2006.

NMF cannot solve the BSS problem uniquely up to scaling and permutation indeterminacies, hence additional constraints are needed. Concerning microarrays, the assumption of sparsely represented sources, which have many zero entries, seems appropriate. Such sparseness constraints have already been exploited successfully in NMF based image analysis methods [2].

The basic idea is thus to determine two nonnegative matrices $\hat{\mathbf{A}}$ and $\hat{\mathbf{S}}$ such that a) the rows of the matrix $\hat{\mathbf{S}}$ are as sparse as possible and b) the reconstruction error of the mixtures $||\mathbf{X} - \hat{\mathbf{A}}\hat{\mathbf{S}}||^2$ is as small as possible. To solve this problem algorithmically we propose to minimize the following target function

$$E(\tilde{\mathbf{A}}, \tilde{\mathbf{S}}) = ||\mathbf{X} - \tilde{\mathbf{A}}\tilde{\mathbf{S}}||^2 - \lambda \sum_{i=1}^{N} \sigma(\tilde{\mathbf{s}}_i), \tag{1}$$

where σ is an appropriate sparseness measure defined as the fraction of its zero to its nonzero elements, λ is a positive weighting factor, and $\tilde{\mathbf{s}}_i$ denotes the i-th row of the matrix $\tilde{\mathbf{S}}$. To make this definition of sparseness practical we use a nonnegative threshold τ which defines the maximum value an entry of $\mathbf{s}$ may have in order to be regarded as a zero element. This leads to the following sparseness measure σ:

$$\sigma(\mathbf{s}) = \frac{\text{number of elements of } \mathbf{s} < \tau}{\text{number of elements of } \mathbf{s}}, \tag{2}$$

1.1 Genetic Algorithm Based Optimization

As the target function defined in Eq. 1 has many local minima we use a Genetic Algorithm (GA) for its minimization.

GAs are stochastic global search and optimization methods inspired by natural biological evolution. For the minimization of the target function in Eq. 1 the m^2 elements of the mixing matrix $\hat{\mathbf{A}}$ have to be determined. Because of the scaling indeterminacy we may assume that the columns of the original mixing matrix $\mathbf{A}$ are normalized such that its diagonal elements are all unity. Hence, only the $m(m-1)$ off-diagonal elements of the matrix $\hat{\mathbf{A}}$ have to be determined. Accordingly, each of the N_{ind} individuals of the GA algorithm consists of $m(m-1)$ parameters which are referred to as genes. As the original mixing matrix is assumed to have nonnegative entries only, all genes should be nonnegative, too. However, we allow the genes to become negative throughout the optimization procedure as we have observed in our experiments that otherwise the GA often fails to converge to the global minimum of the target function.

In every generation of the GA, the fitness of each individual for the optimization task has to be computed in order to determine the number of offsprings it will be allowed to produce. In order to limit the number of offsprings we use a linear scaling procedure to transform target function values to fitness values. To compute the target function values, for every individual, a matrix $\tilde{\mathbf{A}}_-$ is generated with off elements consisting of the genes as stored in the individual and with diagonal elements being unity. To avoid singular matrices $\tilde{\mathbf{A}}_-$ we replace matrices $\tilde{\mathbf{A}}_-$ with a conditional number higher than a user defined threshold τ_{sing} by a pseudo-random matrix with a conditional number lower than τ_{sing}, which also has ones on its diagonal. The genes of the corresponding individual are then adjusted accordingly.

Next, the inverse $\tilde{\mathbf{W}}_-$ of $\tilde{\mathbf{A}}_-$ is computed. The matrices $\tilde{\mathbf{S}}$ and $\tilde{\mathbf{A}}$ are then obtained by setting the negative elements of the matrices $\tilde{\mathbf{S}}_- = \tilde{\mathbf{W}}_-\mathbf{X}$ and $\tilde{\mathbf{A}}_-$, respectively, to zero. Inserting the matrices $\tilde{\mathbf{S}}$ and $\tilde{\mathbf{A}}$ into (1), the resulting target function value is assigned to the corresponding individual. The individuals are then arranged in ascending order according to their target function values, and their fitness values $F(p^{(i)})$, $i = 1, \ldots, N_{ind}$, are determined by

$$F(p^{(i)}) = 2 - \mu + 2(\mu - 1)\frac{p^{(i)} - 1}{N_{ind} - 1}, \tag{3}$$

where $p^{(i)}$ is the position of individual i in the ordered population. The scalar parameter μ, which is usually chosen to be between 1.1 and 2.0, denotes the selection pressure towards the fittest individuals.

We have used Stochastic Universal Sampling (SUS) [4] to determine the absolute number of offsprings an individual may produce. The offsprings are created in a two step procedure. In the first step, two individuals, which conform to the SUS criterion, are chosen at random and are used to create a new individual. Thereby, the genes of the new individual are generated by uniform crossover, i.e. each gene of the new individual is created by copying, each time with a probability of 50 %, the corresponding gene of the first or the second parent individuum. In the second step, called mutation, the new offsprings are obtained by altering a certain fraction r_{mut} of the genes of the new individuals. These genes are chosen at random and are increased or decreased by a random number in the range of $[0, m_{max}]$. The last action within each generation of a GA is the replacement of the parent individuals by their offsprings where we use an elitist reinsertion scheme. In order to prevent the algorithm from converging prematurely we make use of the concept of multiple populations.

To implement the algorithm we have used the functions provided by the *Genetic Algorithm Toolbox* [4] for all GA procedures except the mutation operator which we implemented by ourselves.

Despite using multiple populations, the algorithm failed in many experiments to recover the source and mixing matrix after its first run. Stable results could still be obtained by applying the algorithm repeatedly. The final estimates of the mixing matrix $\hat{\mathbf{A}}$ and the source matrix $\hat{\mathbf{S}}$ were obtained as $\hat{\mathbf{A}} = \prod_{j=1}^{K} \tilde{\mathbf{A}}^{(j)}$ and $\hat{\mathbf{S}} = \tilde{\mathbf{S}}^{(K)}$, respectively, as the matrix $\mathbf{X}$ can be factorized as $\mathbf{X} = \prod_{j=1}^{K} \tilde{\mathbf{A}}^{(j)}\tilde{\mathbf{S}}^{(K)}$.

2 Simulations

2.1 Reliability of the Proposed Algorithm

We generated 25 different observation matrices $\mathbf{X}^{(j)}$, $j = 1, \ldots, 25$, in order to evaluate the reliability of the proposed algorithm. Three nonnegative sources

$s_i^{(j)}$, $i = 1, \ldots, 3$, were created with each source consisting of 1000 nonnegative random elements uniformly distributed in the interval $(0, 1)$. We set $j = 900$ randomly selected elements of $s_1^{(j)}$, $j = 800$ of $s_2^{(29)}$ and $j = 700$ of $s_3^{(39)}$, respectively, to zero before adding random noise with a maximum amplitude of 0.001. Accordingly, source amplitudes smaller than $\tau = 0.001$ (cf. Eq. 2) were set to zero yielding sparseness values of the sources of $\sigma_1 = 0.9$, $\sigma_2 = 0.8$ and $\sigma_3 = 0.7$, respectively. These sources constituted the rows of 25 different source matrices $\mathbf{S}^{(j)}$. Next, 25 random nonnegative 3×3 mixing matrices $\mathbf{A}^{(j)}$ have been generated and the observation matrices $\mathbf{X}^{(j)}$ were computed as $\mathbf{X}^{(j)} = \mathbf{A}^{(j)}\mathbf{S}^{(j)}$. Based on these observation matrices only, the presented algorithm was used to recover the source and the mixing matrices, respectively. Multiple $N_{pop} = 8$ populations, each consisting of $N_{ind} = 50$ individuals, were indispensable to prevent premature convergence to suboptimal solutions. The populations were allowed to exchange $r_{mig} = 20\%$ of their fittest individuals after every $T_{ex} = 100$ generations. Thereby, each individual consisted of 6 genes corresponding to the 6 off-diagonal elements of the mixing matrix $\mathbf{A}$. Further matrices with condition numbers larger than $\tau_{sing} = 100$ have been replaced by random matrices with low condition numbers. In the simulations $\lambda = 0.01$ was used in (1), the selection pressure was set to $\mu = 1.5$, while $r_{mut} = 10\%$ of the genes of each individual were increased or decreased by maximally $m_{max} = 0.1$ during the mutation step. Furthermore, we used an elitist reinsertion scheme, where 98% of the individuals were replaced by their offsprings. Finally 2000 iterations and $K = 2 - 5$ repetitions of the algorithms were deemed sufficient.

Correlation coefficients between original and estimated sources and the cross-talking error (CTE) were determined in each case and are shown in Fig. 1. The algorithm lead in 76% of the simulations to correlation coefficients higher than 0.99 and to a cross-talking error between the estimated and the original mixing

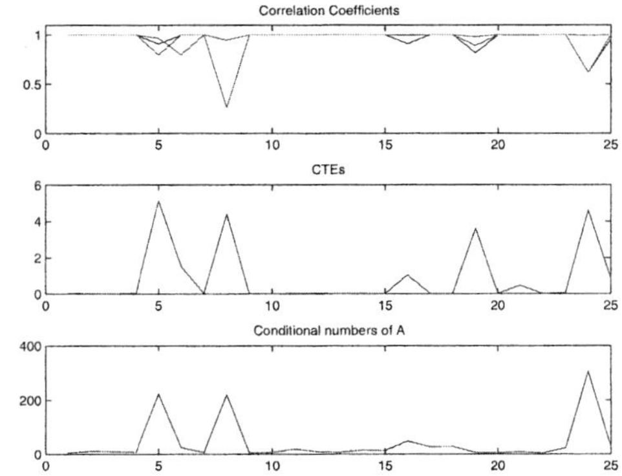

Fig. 1. Top: the correlation coefficients between the estimated and the original sources. Middle: crosstalking error between the estimated and the original mixing matrix. Bottom: the condition numbers of the original mixing matrices.

matrix below 1. Whenever the condition number of the mixing matrix was larger than $\tau_{sing} = 100$ the algorithm failed to converge. The reason is that in the case of poorly conditioned mixing matrices the global minimum is very narrow and therefore hard to locate during the optimization process. On the other hand, we have noticed that choosing τ_{sing} too large lead to worse results when the conditional number of the original mixing matrix was low. This happens because a low τ_{sing} narrows the search space and the global minimum is found easier. Hence, our algorithm is especially eligible for problems where the mixing matrix is not extremely poorly conditioned.

2.2 Recovery of Correlated Sources

In this section we show that the proposed method is capable of solving the BSS problem even if the underlying sources are correlated.

Three sources $\mathbf{s}_i$, $i = 1, \ldots, 3$, were generated as follows: $\mathbf{s}_1$ and $\mathbf{s}_2$ were nonnegative random vectors with 90% and 80% of their elements set to zero, $\mathbf{s}_3(n) = \mathbf{s}_2(n) + 0.001(n-1)$ where $\mathbf{s}_i(n)$ denotes the n-th element of the source $\mathbf{s}_i$ and $n = 1, \ldots, 1000$ (cf. Fig. 2). Note that $\mathbf{s}_2$ and $\mathbf{s}_3$ are correlated ($c = 0.65$), while $\mathbf{s}_1$ and $\mathbf{s}_2$ as well as $\mathbf{s}_1$ and $\mathbf{s}_3$ are uncorrelated. The observation matrix $\mathbf{X}$ (cf. Fig. 2) was generated by mixing the sources using a randomly generated nonnegative 3×3-matrix A which had a conditional number of 5.5.

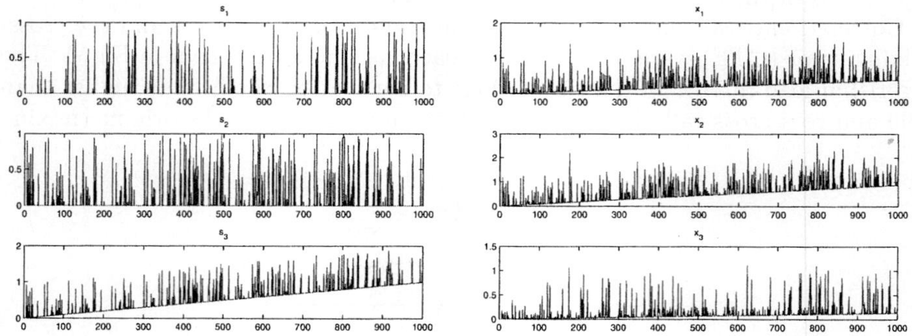

Fig. 2. *left*: The original sources. Note, that $\mathbf{s}_3$ was obtained from $\mathbf{s}_2$ by adding a linear function, *right*: The rows of the mixture matrix $\mathbf{X}$.

Table 1. Results obtained with the presented method (sparse NN BSS) and the fastICA algorithm. Displayed are the correlation coefficients c_i between the i-th original source and its corresponding estimate as well as the cross-talking error (CTE) between the estimated and the original mixing matrix.

	c_1	c_2	c_3	CTE
sparse NN BSS	1.00	1.00	1.00	0.39
fastICA	1.00	1.00	0.75	5.19

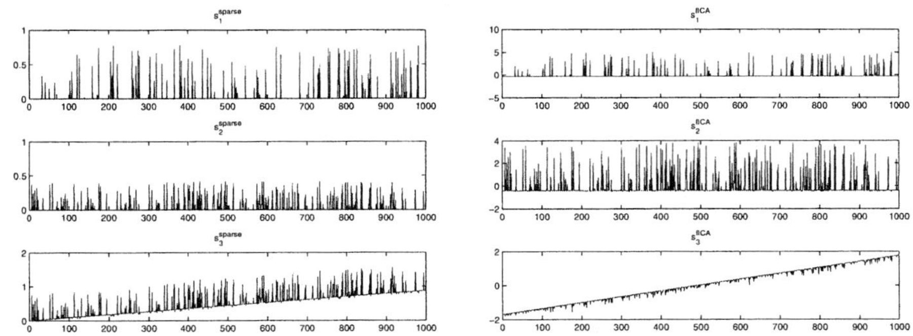

Fig. 3. *left*: Sources estimated by the nonnegative sparse BSS algorithm, *right*: Sources estimated by the fastICA algorithm

Using the proposed algorithm and the parameter set as in the last section, sources as well as the mixing matrix could be recovered almost perfectly (see Tab. 1 and Fig. 3) with only $K = 2$ repetitions of the algorithm. In contrast, ICA based BSS algorithms like fastICA [3] failed to recover the original correlated sources $\mathbf{s}_i$, $i = 1, \ldots, 3$, as well as the mixing matrix $\mathbf{A}$. Rather, fastICA tried to extract sources which are as independent as possible which, however, did not conform to the original sources sought after. Such poor performance of fastICA was to be expected as the independence assumption was violated deliberately.

3 Sparse NMF Analysis of PXE Microarray Data

Gene microarrays are the state of the art technique used to investigate cellular processes at the genetic level. Generally, the goal of a microarray experiment is to determine which genes are expressed to what extent by the biological cells under investigation. Current research, however, also focuses on clustering the detected genes according to their biological function. For this purpose, a huge group of identical cells is divided into several subgroups and each of the subgroups is then exposed to a specific ambient condition. The cells adapt themselves to these conditions by up- or down-regulating their biological processes as, for instance, their metabolism or the constitution of the ribosomes. If for each of the different ambient conditions a microarray experiment is carried out genes belonging to the same biological process should be jointly up or down-regulated and hence should be easy to detect by e.g. k-means clustering. However, many genes are known to participate simultaneously in several biological processes thus an assignment of those genes to only one cluster leads to incomplete information at best.

This problem can be overcome by linear BSS based clustering methods. Thereby, the microarray data collected constitute the rows of the data matrix $\mathbf{X}$, i.e. the element x_{mt}, $m = 1, ..., M$, and $t = 1, ..., T$, contains the expression level of the t-th gene on the microarray recorded under the m-th experimental condition. Here, M denotes the number of experiments and T is the total number of genes detected on the microarray chip. The rows of the matrix **S** are then considered to contain the genetic fingerprints of the individual biological processes occurring in the cell. This means that the element s_{mt}, $m = 1, ..., M$, $t = 1, ..., T$, contains the expression level of the t-th gene which would be observed if the gene expression of the m-th underlying biological process could be recorded in isolation in a microarray experiment. As mentioned above, the underlying biological processes lead to up- or down-regulated gene expressions depending on the ambient condition the cells were exposed to before the microarray experiment was carried out. These different levels of activity are encoded into the matrix **A**, where the element a_{mm} reflects the activity of the m-th biological process under the m-th experimental condition.

Obviously, the matrices **S** and **A**, respectively, are nonnegative as no negative expression of genes as well as no negative activity of biological processes exists. Furthermore, biological cells try to save energy by expressing as little genes as possible for each of their biological processes. Hence, the rows of the matrix **S** are supposed to contain only few active genes. Given these nonnegativity and sparseness constraints, the proposed sparse NMF algorithm seems appropriate to recover the genetic fingerprints of the biological processes in **S** as well as the activity matrix **A**.

We have applied the proposed sparse NMF algorithm to analyze microarray data which were recorded during an investigation of pseudoxanthoma elasticum (PXE), an inherited connective tissue disorder characterized by progressive calcification and fragmentation of elastic fibers in the skin, the retina, and the cardiovascular system. During the investigations M=8 microarray experiments have been carried out. In the first and the second experiment the PXE fibroblasts were incubated in Bovine Serum Albumin (BSA) whereas the incubation time was three hours in the first and 24 hours in the second experiment. In the third experiment the PXE fibroblasts were incubated for three hours in an environment with a high concentration of the Transcription Growth Factor beta and in the fourth experiment the cells were incubated for 24 hours in an environment which was rich in Interleukin 1 beta. The same experiments were then repeated with a control group of normal fibroblasts. The used Affymetrix HG-U133 plus 2.0 microarray chips are capable of detecting the expression levels of more than 54675 genes, of which, however, only T=10530 were expressed significantly in at least one of the experiments. Hence, only these 10530 genes were considered in the further data analysis.

This data set was used to constitute the 8×10530 observation matrix **X** which was then decomposed into the matrices $\hat{\mathbf{A}}$ and $\hat{\mathbf{S}}$ by the proposed sparse NMF algorithm. For the genetic algorithm we increased the number of sub-populations to $N_{pop} = 56$, the maximum number of iterations to 2500 and the number of algorithm repetitions to $K = 8$ while the remaining parameters were set as in section 2. Note, that after the fifth of the $K = 8$ repetitions of the algorithm the resulting matrices $\hat{\mathbf{A}}$ and $\hat{\mathbf{S}}$ did not change any further which indicated the overall convergence of the algorithm. After the sparse NMF analysis each row of the matrix S should ideally consist of the genetic fingerprint of one specific

Table 2. Number of genes related with calcium ion binding (#(cib)) for each of the eight estimated sources. Only genes which are rated exclusively as having a calcium ion binding molecular function in the Gene Ontology [5] database were considered. Most genes related with calcium ion binding are clustered into source 6.

	Source 1	Source 2	Source 3	Source 4	Source 5	**Source 6**	Source 7	Source 8
#(cib)	0	13	7	0	4	**35**	9	6

biological process only. It must be noted, however, that at least one hundred of such processes are occurring simultaneously in a biological cell while the number of available observations and hence the number of estimated sources was only eight. But despite these highly overcomplete settings the algorithm succeeded in grouping the majority of genes which are related with calcium ion binding (344 in total) and hence with the disease picture of PXE into the sixth estimated source (see Tab. 2). Furthermore, the calcium ion binding related genes in the sixth source seem to be specific for only one biological process as maximally 14 % of them could be found in any of the remaining sources.

For comparison, we have used Independent Component Analysis (ICA) to factorize the observation matrix $\mathbf{X}$ into the $m \times m$ mixing matrix $\mathbf{A}_{ICA}$ as well as into $m \times T$ source matrix $\mathbf{S}_{ICA}$. In ICA the nonnegativity and the sparseness constraints are replaced by the assumption that the rows of the matrix $\mathbf{S}_{ICA}$ are mutually independent. We have chosen the well-known fastICA algorithm [3] for the data analysis. In contrast to the results obtained with the sparse NMF algorithm, maximally 12 genes related with calcium ion binding could be found in one source. Hence, the proposed sparse NMF algorithm seems to be better suited for the analysis of PXE cells than the fastICA algorithm.

References

1. D.D. Lee and H.S. Seung. Learning the parts of objects by non-negative matrix factorization. Nature, 40:788-791, 1999.
2. P.O. Hoyer. Non-negative matrix factorization with sparseness constraints. Journal of Machine Learning Research, 5:1457-1469, 2004
3. A. Hyvärinen, Fast and robust fixed-point algorithms for independent component analysis, IEEE Transactions on Neuronal Networks, 10(3), 626-634, 1999
4. A. Chipperfield, P. Fleming, H. Pohlheim, C. Fonseca, Genetic Algorithm Toolbox, Evolutionary Computation Research Group, University fo Sheffield, www.shef.ac.uk/acse/research/ecrg/
5. Gene Ontology: tool for the unification of biology. The Gene Ontology Consortium (2000) Nature Genet. 25: 25-29.

Minimum Support ICA Using Order Statistics. Part I: Quasi-range Based Support Estimation*

Frédéric Vrins and Michel Verleysen**

Université catholique de Louvain, Machine Learning Group,
Place du Levant, 3, 1380 Louvain-la-Neuve, Belgium
{vrins, verleysen}@dice.ucl.ac.be

Abstract. The minimum support ICA algorithms currently use the extreme statistics difference (also called the *statistical range*) for support width estimation. In this paper, we extend this method by analyzing the use of (possibly averaged) differences between the $N-m+1$-th and m-th order statistics, where N is the sample size and m is a positive integer lower than $N/2$. Numerical results illustrate the expectation and variance of the estimators for various densities and sample sizes; theoretical results are provided for uniform densities. The estimators are analyzed from the specific viewpoint of ICA, i.e. considering that the support widths and the pdf shapes vary with demixing matrix updates.

1 Introduction

Recently, new contrasts for ICA have been developed for the separation of bounded sources, based on the fact that the output support width varies with the mixing coefficients; the independent components are recovered one by one, by finding directions in which the outputs have a minimum support convex hull measure [1, 3]. Such approach benefits from some advantages: on one hand the contrast is extremely simple and free of spurious maxima [1]; and on the other hand, its optimization can be easily handled, leading to interesting results in terms of speed and residual crosstalk.

The support estimation of a pdf f_X has been extensively studied in statistics and econometrics. Nevertheless, most methods require resampling techniques or tricky tuning parameters, and are thus not really appropriated to ICA algorithms. For instance, if the support $\Omega(X)$ of the output is (a, b), existing ICA methods currently use the *range* approximation to estimate the (Lebesgue) measure of the support $\mu[\Omega(X)] : b - a \simeq R(X) \triangleq \max_{i,j}[x_i - x_j]$, $1 \leq i, j \leq N$ where the x_j can either be considered as iid realizations of the common random variable (r.v.) X, or as a samples of a stationary stochastic process constituted of a sequence of N independent r.v. X_j, all sharing the same pdf f_X.

An extended form of this estimator will be considered here, using order statistics differences. The study is motivated by the idea that the extreme statistics

* The authors are grateful to D. Erdogmus for having inspirited this work by fruitful discussion on expectation and variance of cdf differences and m-spacings.

** Michel Verleysen is Research Director of the Belgian F.N.R.S.

J. Rosca et al. (Eds.): ICA 2006, LNCS 3889, pp. 262–269, 2006.

are not necessarily reliable. Let $x_{(1)} \leq \cdots \leq x_{(N)}$ be a rearranged version of the observed sample set $\mathcal{X}_N = \{x_1, ..., x_N\}$; each of the $x_{(j)}$ can be seen as a realization of a r.v. $X_{(j)}$. Obviously, the $X_{(j)}$ are not independent and do not share the same pdf. Both $x_{(j)}$ and $X_{(j)}$ are called the j-th order statistic of $\mathcal{X}_N$. This appellation is not related to the *(higher) order statistics*, frequently used in the ICA community. The *order statistics*, as defined in this paper, have already been used in the BSS context in [4] (see also [10], [9] and references therein). These ordered variates can be used to define the range $R_1(X) = X_{(N)} - X_{(1)}$, or the (symmetric) *quasi-ranges* (QR): $R_m(X) = X_{(N-m+1)} - X_{(m)}$, with $m < \lfloor N/2 \rfloor$. Such QR could be also used to estimate the quantity $b - a$. However, even if $R_m(X)$ is a generalization of $R(X) = R_1(X)$, both estimators only involve two sample points. In order to include more points in the estimation, we also compare $R_m(X)$ to $\langle R_m(X)\rangle \triangleq 1/m \sum_{i=1}^{m} R_i(X)$.

The QR-based support estimation is analyzed in Section 2, for various pdf and sample sizes. Specific phenomena are discussed in Section 3, keeping in mind that the pdf of X vary with time in ICA applications, due to the iterative updates of the demixing matrix. Note that the performance analysis of ICA algorithms using the above estimators is discussed in a separated paper [2].

2 Density, Bias and Variances of the QR

A large attention has been paid to order statistics and QRs in the statistic literature. For instance, the pdf $f_{R_m(X)}$ of $R_m(X)$ for $\Omega(X) = (a, b)$ has been established in [8]. If F_X denotes the cdf of X, the computation of $f_{X_{(j)}}$ yields

$$f_{R_m(X)}(r) = \frac{N!}{((m-1)!)^2(N-2m)!} \int_{-\infty}^{\infty} F_X^{m-1}(x) \left[F_X(x+r) - F_X(x)\right]^{N-2m} \times f_X(x) f_X(x+r) \left[1 - F_X(x+r)\right]^{m-1} dx \ .$$

It can be seen that the density $f_{R_m(X)}$ is a function of $f_X = F'_X$, as well as of N and m. Although, the above theoretical expression is of few use in practice; for most parent densities f_X, no analytical expression can be found for simple functions of $R_m(X)$, such as expectation and moments. Dealing with $f_{\langle R_m(X)\rangle}(r)$ is even worst, since $f_{\langle R_m(X)\rangle}$ depends on the joint density of $R_1(X), \cdots, R_m(X)$. A more reasonable way to compute the expectation and variances of $R_m(X)$ and $\langle R_m(X)\rangle$ is to prefer numerical simulations to theoretical developments that are valid for a single density only; this is done in Section 2.1. However, for comparison purposes, the exact expressions of $\mathrm{E}[R_m(X)]$, $\mathrm{VAR}[R_m(X)]$, $\mathrm{E}[\langle R_m(X)\rangle]$ and $\mathrm{VAR}[\langle R_m(X)\rangle]$ aer given in Section 2.2 in the case where f_X is the uniform pdf.

2.1 Empirical Expectation and Variance of QRs

Let us note U, L, T and V white r.v. having uniform, linear, triangular and 'V'-shape densities, respectively. We note the empirical expectations and variances

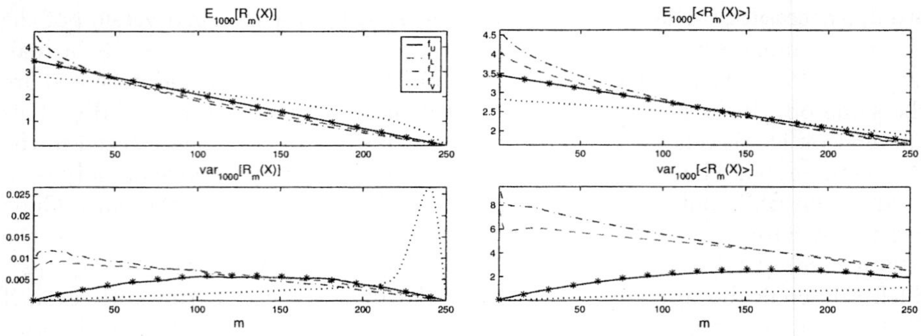

Fig. 1. Empirical expectations and variances of $R_m(X)$ (left) and $\langle R_m(X)\rangle$ (right) for $N = 500$ (1000 trials). The theoretical curves for the uniform case are labelled '$*$'.

of estimators taken over t trials as $\mathrm{E}_t[\cdot]$ and $\mathrm{VAR}_t[\cdot]$. The evolution of these quantities with respect to m is shown in Fig. 1. Three particular effects have to be emphasized.

- *Effect of m and f_X*: the estimation error increases with m for fixed N at a rate depending on f_X, and m has thus to be chosen small enough in comparison to N (the true support measures of the white r.v. are $\mu[\Omega(T)] = 2\sqrt{6} > \mu[\Omega(L)] = 3/2\sqrt{8} > \mu[\Omega(U)] = 2\sqrt{3} > \mu[\Omega(V)] = 2\sqrt{2}$). The support of V and U can be estimated with a low variance, even for a small m, contrarily to T and L. For instance, the variance of the estimators decreases with m for linear and triangular r.v., while this behavior cannot be observed for U or V; the variance of the estimators increases when unreliable points (i.e. corresponding to a low value of the pdf) are taken into account. The shape of $\mathrm{VAR}_t[R_m(U)]$ and $\mathrm{VAR}_t[\langle R_m(U)\rangle]$ are more surprising, but they have been confirmed by the analytical equations given in Section 2.2.
- *Effect of N*: it can be reasonably understood, though not visible on Fig. 1, that $R_m(X)$ and $\langle R_m(X)\rangle$ are asymptotically unbiaised estimators of $b - a$ if b and a are not isolated points, that is if the support $\Omega(X)$ includes some neighborhoods of b and a. Similarly, $\lim_{N\to\infty} \mathrm{VAR}[R_m(X)] = 0$ (for m fixed); We conjecture that the latter limit holds for $\langle R_m(X)\rangle$, with fixed m. Note that the convergence rate depends of f_X. These properties can be easily confirmed when X is uniform (see next section).
- *$R_m(X)$ vs $\langle R_m(X)\rangle$*: the error of $R_m(X)$ increases at a higher rate than the one of $\langle R_m(X)\rangle$ for increasing m and fixed N; this is a consequence of the regularization due to the average in $\langle R_m(X)\rangle$: $\Pr[\langle R_m(X)\rangle \geq R_m(X)] = 1$.

The above simulation results indicate that $\langle R_m(X)\rangle$ should be preferred to $R_m(X)$ for support estimation; for a small m compared to N, both the error and the variance are improved. The choice of m is difficult, though : it must be small enough to ensure a small error, but not too small if one desires to estimate the support of e.g. f_T or f_L or of noisy data; an optimal value of m depends of the unknown density.

2.2 Exact Expectation and Variance of QRs for Uniform Pdf

In this section, contrarily to the remaining of the paper, U denotes a normalized uniform r.v. with support equal to $(0,1)$, in order to simplify the mathematical developments, that are sketched in the Appendix.

Using the expression of $f_{R_m(X)}$ given in Section 2, it can be shown that $\mathrm{E}[R_m(U)] = (N-2m+1)/(N+1)$ and $\mathrm{VAR}[R_m(U)] = 2m(N-2m+1)/((N+2)(N+1)^2)$. Simple manipulations directly show that $R_m(U)$ is an asymptotically unbiaised estimator of the support measure, monotonously increasing with limit $b-a$, when m is kept fixed.

The expectation of $\langle R_m(U)\rangle$ can directly be derived from $\mathrm{E}[R_m(U)]$; if m is fixed, $\langle R_m(U)\rangle$ is asymptotically unbiased. On the contrary if we set $m = \lfloor N/k \rfloor$, $k \in \mathbb{Z}$, the asymptotic bias equals $1/k$. Such bias can be cancelled if $m(N)$ increases at a lower rate that N; this is e.g. the case if $m(N) = \lfloor \sqrt{N/k} \rfloor$. Regarding the variance of $\langle R_m(U)\rangle$, we have

$$\mathrm{VAR}[\langle R_m(U)\rangle] = \frac{-3m^3 + 2m^2(N-2) + 3mN + (N+1)}{3m(N+2)(N+1)^2} .$$

Using ad-hoc scaling coefficients, the related quantities can be obtained for *white* r.v. (no more confined in $(0,1)$). The theoretical curves (labelled using '$*$') are superimposed to the associated empirical ones if Fig. 1.

3 Estimating the Mixture Support

The above discussion gives general results regarding the estimation of the support convex hull measure of a one-dimensional r.v. Let us now focus on the support estimation of a single output (deflation scheme) of the 2-dimensional ICA application; the support varies with the associated row of transfer matrix. For the ease of understanding, we constrain the sources to share the same pdf. The instantaneous noise-free ICA mixture scheme, under whiteness constraint, leads to the following expression for an output:

$$Z_X(\phi+\varphi) = \cos(\phi+\varphi)S_1 + \sin(\phi+\varphi)S_2 , \tag{1}$$

where the S_i are the sources, and ϕ and φ are resp. the mixing-whitening and demixing angles. The subscript X means that the sources follow the pdf f_X. We define $\theta = \phi + \varphi$ as the input-output transfer angle.

The minimum support ICA approach updates the angle φ to minimize the objective function $\mu[\Omega(Z_X(\theta))]$. Since it has been shown that this cost function is concave in a given quadrant, a gradient-based method leads to $\theta = k\pi/2$, with $k \in \mathbb{Z}$. In practice however, $\mu[\Omega(Z_X(\theta))]$ has to be estimated, for example using the proposed form of estimators. The following subsections points out two phenomena that have to be considered.

3.1 The Mixing Effect

Fig. 2 shows the surface of the empirical expectation of the error ϵ, defined as

$$\epsilon(X, N, m, \theta) = \mu[\Omega(Z_X(\theta))] - \langle R_m(Z_X(\theta)) \rangle , \quad (2)$$

when f_X is a triangular or 'V'-shape density, θ ranges from 0 to $\pi/2$, and N from 2 to 500.

In addition to the bias dependency on f_X, we can observe what we call the 'mixing effect': the error increases for θ going from 0 or $\pi/2$ to $\pi/4$. This phenomenon can be explained as follows. The pdf of a sum of r.v. is obtained by convoluting the pdfs of the summed r.v. Therefore, the tails of resulting pdfs will be less sharp than the source pdfs. For instance, the pdf of a sum of two normalized uniform r.v. is triangular. The mixing effect phenomenon can now be understood, since for fixed N and m, the support measure of a pdf with sharp tails is better estimated than of a pdf with smoothly decreasing tails. The main consequence of this phenomenon is that the empirical contrast is more 'flat' than the true one with respect to the transfer angle.

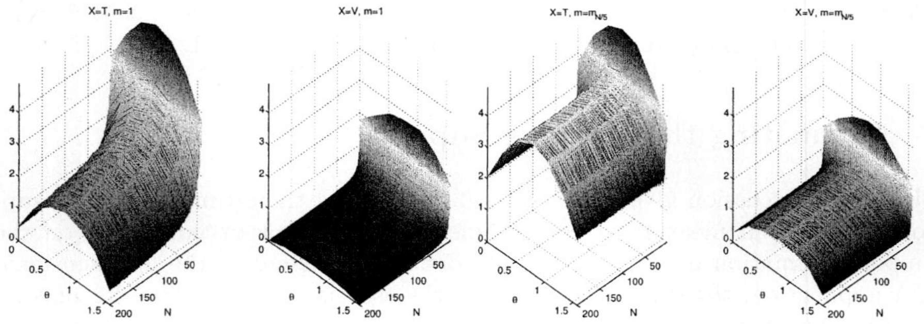

Fig. 2. Empirical error $\mathrm{E}_{100}\left[\epsilon(X, N, m, \theta)\right]$ for various source pdf f_X, with N ranging from 2 to 200, and $m_{N/5} \triangleq \max(1, \lfloor N/5 \rfloor)$

3.2 The Large-m Given N Effect

The mixing effect emphasizes that the support estimation quality depends of θ: the support of $Z_X(\pi/4)$ is always more under-estimated than the one of $Z_X(k\pi/2)$ by using QR or averaged QR estimators. This results from the fact that the output density depends of the transfer angle. In section 2.1, the effect of m on the expectation and variance of the estimators is shown to depend of the density f_X. In the ICA application it thus depends of θ: the bias increases with m, at a rate depending of $f_{Z_X(\theta)}$, i.e. of θ. This is a tricky point, since even if $\mu[\Omega(Z_X(\pi/4))] > \mu[\Omega(Z_X(k\pi/2))]$, this inequality evaluated using the support measure approximations can be violated. In this scenario, occurring for m greater than a threshold value $m^{\dagger}$, the contrast optimum will be obtained for

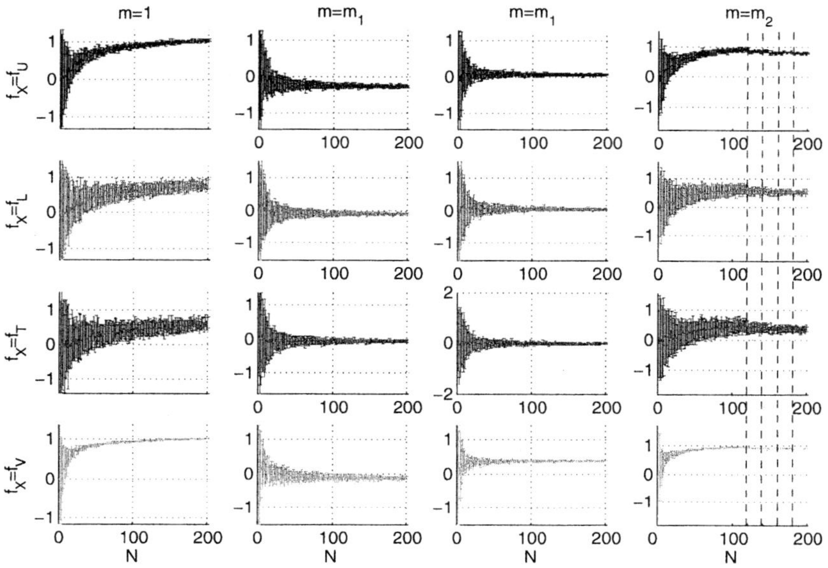

Fig. 3. Evolution of error means for various pdf f_X, N ranging from 2 to 200 and $m_1 = \max(1, \lfloor N/5 \rfloor)$,$m_2$ is given by eq. (3). $\mathrm{E}_{100}\left[\mathcal{E}_1(X)\right] = \mathrm{E}_{100}\left[\langle\mathcal{E}_1(X)\rangle\right]$ (first col.); $\mathrm{E}_{100}\left[\mathcal{E}_{m_1}(X)\right]$ (second col.); $\mathrm{E}_{100}\left[\langle\mathcal{E}_{m_1}(X)\rangle\right]$ (third col.); and $\mathrm{E}_{100}\left[\langle\mathcal{E}_{m_2}(X)\rangle\right]$ (last col.). The width of the curves reflects twice the empirical variance of $\mathcal{E}(X)$ and $\langle\mathcal{E}(X)\rangle$.

$\theta = \pi/4$ rather than $\theta \in \{0, \pi/2\}$, i.e. the algorithm will totally fail. For example, when dealing with $\langle R_m(Z_U(\theta))\rangle$ and two 500-sampled sources, $m^\dagger \simeq 100$. If $R_m(Z_U(\theta))$ is considered, $m^\dagger \simeq 40$. Indeed, the pdf of $Z_U(\pi/4)$ is triangular, and we see on Fig. 1 that the estimation of the support of a white triangular r.v. is lower than the estimated support of a white uniform r.v. for these values of N and $m > m^\dagger$. These values of $m^\dagger$ obviously decrease with decreasing N.

Fig. 3 illustrates the quantities $\mathcal{E}_m(X) \triangleq R_m(Z_X(\pi/4)) - R_m(Z_X(0))$ and $\langle\mathcal{E}_m(X)\rangle \triangleq \langle R_m(Z_X(\pi/4))\rangle - \langle R_m(Z_X(0))\rangle$: negative values of $\mathcal{E}_m(X)$ and $\langle\mathcal{E}_m(X)\rangle$ obtained for $m > m^\dagger$ clearly indicate that the optima of the empirical contrast, i.e. the corresponding estimators will lead to wrong source separation solutions. The last column shows the result obtained by using (3); the vertical dashed lines indicated that m has been incremented. This comment suggests to pay attention to the choice of m: it must be small enough by comparison to N to ensure a small error and $m < m^\dagger$, but greater than one for regularization purposes. Therefore, if $\overline{\alpha}$ denotes the nearest integer to α, we suggest to take m according to the rule

$$\max\left(1, \Re\left(\overline{\left[\left(\frac{N-18}{6.5}\right)^{0.65} - 4.5\right]}\right)\right) . \tag{3}$$

Though the above rule of the thumb will not be detailed here, we mention that rule (3) results from a distribution-free procedure for choosing m and valid for all θ and all source pdfs; the method is detailed in [2]. A positive point is that the critical value $m^\dagger$ does not seem to be sensitive to the number of sources, probably due to the compensation of two effects: even if the tails of $f_{Z_X(\theta)}$ tend to decrease exponentially when many sources are mixed since $f_{Z_X(\theta)}$ tend to be Gaussian-shape (inducing a large under-estimation of the support), large-value sample points can be observed due to the summed r.v. (so that the estimated mixture support should be larger than the estimated source support).

4 Conclusion and Future Work

In this paper, we have investigated the use of the quasi-ranges $R_m(X)$ and averaged symmetric quasi-ranges $\langle R_m(X)\rangle$ for support width minimization approaches to ICA. Note that the computation of the *true* QR requires the knowledge of the order statistics of X, that are unknown here; in this paper, the i-th order statistic $X_{(i)}$ was approximated by the i-th largest observed value $x_{(i)}$ of X. This work is motivated by the fact that extreme statistics can be unreliable. It is shown that m has to be chosen small in comparison to N, but greater than one to make the variance of the estimators decrease for several kinds of pdf. The main advantage of the averaged QR is that it takes $2m$ points into account. From both the expectation and variance points of view, the averaged QR has better performances than the simple QR. We have shown that an excessive value m with given N could lead the related ICA algorithms to totally fail; from this point of view too, the averaged QR has to be preferred to the QR. Future work should focus on a study involving specific noise, as well as a comparison with existing endpoint estimators.

References

1. Vrins, F., Jutten, C. & Verleysen, M. (2005) SWM : a Class of Convex Contrasts for Source Separation. In proc. ICASSP, *IEEE Int. Conf. on Acoustics, Speech and Sig. Process.*: V.161-V.164, Philadelphia (USA).
2. Vrins, F. and & Verleysen, M. (2006) Minimum Support ICA Using Order Statistics. Part II: Performance Analysis. In Proc. ICA, *Int. Conf. on Ind. Comp. Anal. and Blind Sig. Sep.*, Charleston SC (USA).
3. Cruces, S. & Duran, I. (2004) The Minimum Support Criterion for Blind Source Extraction: a Limiting Case of the Strengthened Young's Inequality. In proc. ICA, *Int. Conf. on Ind. Comp. Anal. and Blind Sig. Sep.*: 57–64, Granada (Spain).
4. Pham, D.-T. (2000) Blind Separation of Instantaneous Mixture of Sources Based on Order Statistics. *IEEE Trans. Sig. Process.*, **48**(2): 363–375.
5. Ebrahimi, N., Soofi, E.S, Zahedi, H. (2004) Information Properties of Order Statistics and Spacings. *IEEE Trans. Info. Theory*, **50**(1): 177–183.
6. Spiegel, M.R. (1974) "Mathematical Handbook of Formulas and Tables." McGraw-Hill, New York.
7. Papadatos, N. (1999) Upper Bound for the Covariance of Extreme Order Statistics From a Sample of Size Three. *Indian J. of Stat.*, Series A, Part 2 **61**: 229–240.
8. David, H.A. (1970). "Order Statistics." Wiley series in probability and mathematical statistics, Wiley, New York.

9. Even, J. (2003) "Contributions à la Séparation de Sources à l'Aide de Staitistiques d'Ordre." PhD Thesis, Univ. J. Fourier, Grenoble (France).
10. Blanco, Y. & Zazo, S. (2004) An Overview of BSS Techniques Based on Order Statistics: Formulation and Implementation Issues. In proc. ICA, *Int. Conf. on Indep. Comp. An. and Blind Sig. Sep.*: 73–80, Granada (Spain).

Appendix. Details of Some Results for Uniform Densities

• Expectation and Variance of $R_m(U)$

It is known e.g. from [5] that the pdf of the i-th order statistic a uniform r.v. is $f_{U_{(i)}}(u) = \frac{N!}{(i-1)!(N-i)!}[F_U(u)]^{i-1}[1 - F_U(u)]^{N-i} f_U(u)$. By using basic properties of expectation and $\Omega(U) = (0,1)$, we have $\mathrm{E}[U_{(i+1)} - U_{(i)}] = \frac{1}{N+1} \frac{N!}{i!(N-i-1)!} \int_0^1 x^i (1-x)^{N-i-1} dx$, where the integral equals $\frac{i!(N-i-1)!}{N!}$ [6]. It comes that $\sum_{i=m}^{N-m} \mathrm{E}[U_{(i+1)} - U_{(i)}] = \frac{N-2m+1}{N+1}$. Similar development as above on the i-th order statistic of a uniform variable on $(0,1)$ leads to $\mathrm{VAR}[U_{(i)}] = \frac{i(N-i+1)}{(N+2)(N+1)^2} = \mathrm{VAR}[U_{(N-i+1)}]$.

Since $\mathrm{CORR}[U_i, U_j]$ $(1 \le i < j \le N)$ is know from [7], we obtain

$$\mathrm{COV}[U_{(i+p)}, U_{(i)}] = \frac{i(N+1-i-p)}{(N+2)(N+1)^2} . \tag{4}$$

We find $\mathrm{VAR}[U_{(i+p)} - U_{(i)}] = \mathrm{VAR}[U_{(i+p)}] + \mathrm{VAR}[U_i] - 2\mathrm{COV}[U_{(i+p)}, U_{(i)}]$, which equals $\frac{p(N+1-p)}{(N+2)(N+1)^2}$. The results enounced in Section 2.2 comes when setting $i = m$ and $p = N - 2m + 1$.

• Expectation and Variance of $\langle R_m(U) \rangle$

We obviously have $\mathrm{E}[\langle R_m(X) \rangle] = \frac{1}{m} \sum_{p=1}^{m} \frac{N-2p+1}{N+1} = \frac{N-m}{N+1}$.

The computation of $\mathrm{VAR}[\langle R_m(U) \rangle]$ is more tricky. Observe first that:

$$m^2 \mathrm{VAR}[\langle R_m(U) \rangle] = \sum_{p=1}^{m} \mathrm{VAR}[R_p(U)] + 2 \sum_{1 \le i < j \le m} \mathrm{COV}\left[R_i(U), R_j(U)\right] . \tag{5}$$

Using eq. (4), we find: $\mathrm{COV}_{i<j}\left[R_i(U), R_j(U)\right] = 2i \frac{N+1-2j}{(N+2)(N+1)^2}$.

We have, using basic properties:

$$\sum_{p=1}^{m} \mathrm{VAR}[\langle R_p(U) \rangle] = \frac{(N+1)m(m+1) - 2/3m(m+1)(2m+1)}{(m+2)(m+1)^2} , \tag{6}$$

and

$$\sum_{1 \le i < j \le m} \mathrm{COV}\left[R_i(U), R_j(U)\right]] = \frac{m(m-1)}{6(N+2)(N+1)^2} \left\{-3m^2 + m(2N-3) + 2N\right\}.$$

which leads to the results presented in Section 2.2.

Minimum Support ICA Using Order Statistics. Part II: Performance Analysis

Frédéric Vrins and Michel Verleysen*

Université catholique de Louvain, Machine Learning Group,
Place du Levant, 3, 1380 Louvain-la-Neuve, Belgium
{vrins, verleysen}@dice.ucl.ac.be

Abstract. Linear instantaneous independent component analysis (ICA) is a well-known problem, for which efficient algorithms like FastICA and JADE have been developed. Nevertheless, the development of new contrasts and optimization procedures is still needed, e.g. to improve the separation performances in specific cases. For example, algorithms may exploit prior information, such as the sparseness or the non-negativity of the sources. In this paper, we show that support-width minimization-based ICA algorithms may outperform other well-known ICA methods when extracting bounded sources. The output supports are estimated using symmetric differences of order statistics.

1 Introduction and Motivation

Most of ICA researchers and practitioners agree with the idea that it does not exist a unique ICA algorithm outperforming all alternatives, and making the other methods useless. Obviously, certain approaches, like e.g. FastICA [11] or JADE [12] yield remarkable separation performances while simultaneously being fast. Nevertheless, at least three arguments for developing new ICA contrasts can be emphasized, even for the simplest (but also most widely used) linear, instantaneous and noise-free mixture scheme [10]. First, to extend the field of application of BSS techniques (specific procedures have been derived to deal with e.g. structured gaussian sources). Second, some contrasts can be handled easier than others; for example, the convexity property simplifies the optimization step. Third, the contrast performances may vary with the source densities, so that the separation performances depend on the cost function and on the application.

For example, we can cite BSS methods exploiting the non-negativity [9] or sparseness [8] of the sources, as well as their temporal dependency [13], etc. The minimum support approach has been independently suggested by Cruces & Duran [14] and Vrins et al. [1], to extract bounded sources in a deflation way. The theoretical framework has been well established; this approach has relationship to zero Renyi's entropy, and also with the Young and Brunn-Minkowski inequalities. On the other hand, this approach benefits from the discriminacy property, i.e. all the local optima of the theoretic criterion are relevant for ICA.

* Michel Verleysen is Research Director of the Belgian F.N.R.S.

J. Rosca et al. (Eds.): ICA 2006, LNCS 3889, pp. 270–277, 2006.

This property gives confidence in the solution obtained when the optimum is reached using gradient techniques. This is not the case for example when separating multimodal sources by minimizing the entropy or the mutual information [4]. It is interesting to note that the boundness prior on the sources have been used by Theis and Gruber to establish separability results in postnonlinear mixture schemes [7]. In addition, this approach can also be used to separate signals being correlated in some specific way, such as landscape images [3]. Finally, a related symmetric method with geometrical interpretation can be found in the nice paper of Pham [5].

However, the performances of the minimum-support ICA method on bounded source signals have not been detailed and compared to other methods. Similarly, the support estimation problem, which is a crucial issue though, is not discussed in this context. In this paper, we compare the performances of FastICA and maximum absolute kurtosis maximized using [6] (AKMICA) to minimum support algorithms called XSICA, OSICA and AVOSICA. In the three last algorithms, the support measure criterion are estimated in different ways, and is minimized using the optimization technique for non-differentiable criteria presented in [6]. We also analyze the performances of JADE, though it is a rather different method (highly limited by the number of sources, symmetric, algebraic and thus non-iterative).

We show that in the instantaneous noise-free and noisy cases, AVOSICA benefits from interesting signal interference ratio (SIR) performances results in comparison to other ICA algorithms, without added complexity.

2 The XSICA, OSICA and AVOSICA Algorithms

The recent minimum support approach to ICA requires support estimation; in [1,3] the statistical range is used, i.e. the output supports are estimated by the difference of the output extreme values. When this criterion is minimized using [6], we call this algorithm XSICA (extreme statistics ICA). Nevertheless, extreme values can be unreliable in the noisy case, so that alternative ways to estimate bounded support widths have to be derived. This can be easily done by using order statistics differences. The i-th order statistic of an observed sequence $\mathcal{X}_N = \{x_1, ..., x_N\}$ is noted $x_{(j)}$ and is the j-th largest observed sample, i.e. $\{x_{(1)} \leq \cdots \leq x_{(N)}\}$ [17]. The latter sequence is no other than an ordered version of the set $\mathcal{X}_N$. If we note by $R_m(X)$ ($1 \leq m < \lfloor N/2 \rfloor$, $m \in \mathbb{Z}$) the quasi-range defined by $x_{(N-m+1)} - x_{(m)}$, both the quantities $R_m(X)$ and $\langle R_m(X) \rangle \triangleq 1/m \sum_{i=1}^{m} R_m(X)$ can be seen as support width estimators, where m is a tuning parameter. Combining those criteria with the optimization procedure [6], the OSICA (order statistics ICA) and AVOSICA (average order statistics ICA) are obtained. Note that XSICA, OSICA and AVOSICA are equivalent when setting $m = 1$. In $\langle R_m(X) \rangle$, m equals twice the number of sample points used in the support estimation. The estimation of the support convex hull width by $\langle R_m(X) \rangle$ is analyzed in [2]; it is shown to be preferred to $R_m(X)$, but the performances of those practical criterion in terms of SIR are not discussed. In addition, no specific information about how to choose the tuning parameter m in $\langle R_m(X) \rangle$

is given. A small value of m cancels the regularization induced by the average, so that the criterion could be highly sensitive to noise; on the other hand, an excessive value of m may lead the algorithm to totally fail. Furthermore, even if only a small error is observed for large N and small m when estimating the support of a given random variable (r.v.), the shape of the function that links the variance of the estimator to m depends of the (unknown) pdf of the r.v.; the variance can either increases or decreases with m [2].

In the remaining of this paper, a meaningful procedure for choosing a satisfactory value for m given N is derived in Section 3. The performances of AVOSICA are then pointed out, in comparison to XSICA, OSICA, AKMICA, JADE and FastICA using the *gauss* non-linearity, for robustness purpose [10] (the *tanh* non-linearity gives similar results).

3 Towards a Meaningful Choice of m with Fixed N

In this section, we derive a procedure to set a default value for the tuning parameter m, for fixed N. We propose to find the maximum value m_0 of m given N, ensuring that the positive error $\mu[\Omega(X)] - \langle R_m(X)\rangle$ is lower than an error threshold $\mathcal{E}$ with a high probability, whatever is the density of X. In other words, we try to find m_0 such that for all $m \leq m_0$:

$$\Pr\left[\mu[\Omega(X)] - \langle R_m(X)\rangle \leq \mathcal{E}\right] \geq \mathcal{L}(m_0) \ , \tag{1}$$

where $\mathcal{L}(m_0)$ is a threshold ideally close to, but lower than one.

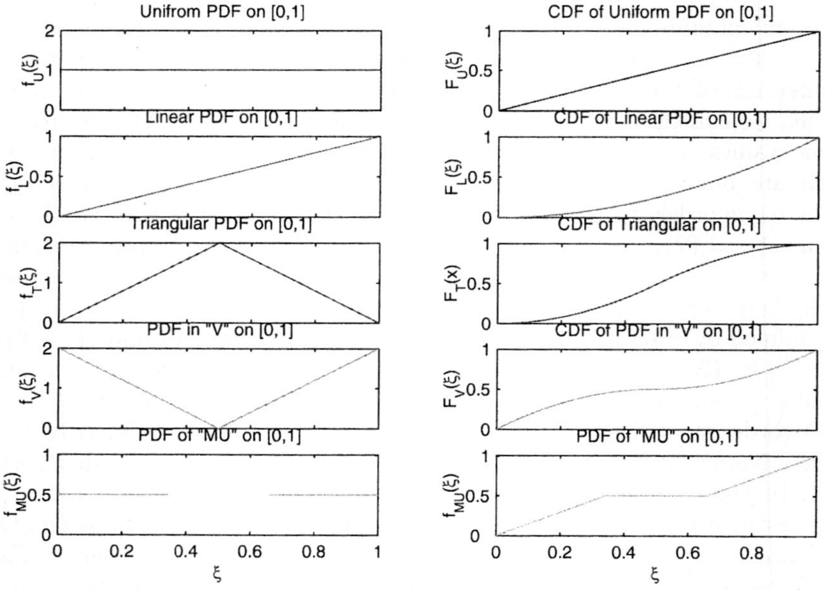

Fig. 1. Densities and (cumulative) distributions of the 5 sources

The main problem of this approach is that if $\mathcal{E}$ is a constant, we are not able to find an expression for $\mathcal{L}(m_0)$ that is i) useful, and ii) *distribution-free*, in the sense that it does not depends on f_X. For instance, the probability in (1) can be written as $1 - F_{\langle R_m(X)\rangle}(\mu[\Omega(X)] - \mathcal{E})$ where $F_{\langle R_m(X)\rangle}$ is the (cumulative) distribution of $\langle R_m(X)\rangle$, which depends on f_X through the order statistic densities $f_{X_{(i)}}$. The point is thus to include the density dependency into the error term $\mathcal{E}$. Let us approximate the support measure by using quantile differences, and define the error term as

$$\mathcal{E}(X) \triangleq \mu[\Omega(X)] - (\xi_q - \xi_p) \ , \tag{2}$$

where ξ_q and ξ_p ($0 \leq p < q \leq 1$) are the q-th and p-th quantiles of F_X, respectively. Note that $\mathcal{E}(X)$ is positive and tends to 0 for increasing q and decreasing p, whatever is the density of X, but at a various rate. For example, with $q = .95$ and $p = 1 - q$ we have $\mathcal{E}(T) = 31.6\%$ and $\mathcal{E}(V) = 5\%$ (see Fig. 1).

Observe that any lower bound of $\Pr\left[R_m(X) \geq \mu[\Omega(X)] - \mathcal{E}\right]$ can be used in the right hand side of eq. 1:

$$\begin{aligned}\Pr\left[\langle R_m(X)\rangle \geq \mu[\Omega(X)] - \mathcal{E}\right] = {} & \Pr\Big[\langle R_m(X)\rangle \geq \mu[\Omega(X)] - \mathcal{E} | R_m(X) \geq \mu[\Omega(X)] - \mathcal{E}\Big] \\ & \times \Pr\left[R_m(X) \geq \mu[\Omega(X)] - \mathcal{E}\right] \\ & + \Pr\Big[\langle R_m(X)\rangle \geq \mu[\Omega(X)] - \mathcal{E} | R_m(X) < \mu[\Omega(X)] - \mathcal{E}\Big] \\ & \times \Pr\left[R_m(X) < \mu[\Omega(X)] - \mathcal{E}\right] \\ \geq {} & \Pr\left[R_m(X) \geq \mu[\Omega(X)] - \mathcal{E}\right] \ , \end{aligned} \tag{3}$$

where the inequality result from the fact that $\langle R_m(X)\rangle \geq R_m(X)$ with probability one.

On the other hand, using the confidence interval for quantiles derived in [16], noting that $\Pr\Big[R_m(X) \geq R_{m_0}(X) | m \leq m_0\Big] = 1$ and setting $p = 1 - q$ in (2), the following inequality holds for for all $m \leq m_0$:

$$\Pr[R_m(X) \geq \xi_q - \xi_{1-q}] \geq \underbrace{\sum_{i=m_0}^{N} \binom{N}{i} q^{N-i}(1-q)^i - \sum_{i=N-m_0+1}^{N} \binom{N}{i} q^i(1-q)^{N-i}}_{\triangleq \mathcal{L}(q, m_0, N)}$$

and consequently, using inequality (3) and $\mathcal{E}(X)$ given by (2):

$$\begin{aligned}\Pr\Big[\mu[\Omega(X)] - \langle R_m(X)\rangle \leq \mathcal{E}(X)\Big] &= \Pr\left[\langle R_m(X)\rangle \geq \xi_q - \xi_{1-q}\right] \\ &\geq \mathcal{L}_+(q, m_0, N) \ , \end{aligned} \tag{4}$$

with $\mathcal{L}_+(q, m_0, N) \triangleq \max(0, \mathcal{L}(q, m_0, N))$. The latter inequality can be understood as follows: if q is chosen close enough to one, $\langle R_m(X)\rangle$ *nearly covers* the true support, with a probability higher than $\mathcal{L}_+(q, m_0, N)$. Note that q has to be chosen close enough to one, so that $\mathcal{E}(X)$ is small; otherwise the bound $\mathcal{L}_+$ in (4) is no more related to support estimation quality. The terms *close enough to one* depends of the cdf F_X. In practice however, if no information on the source

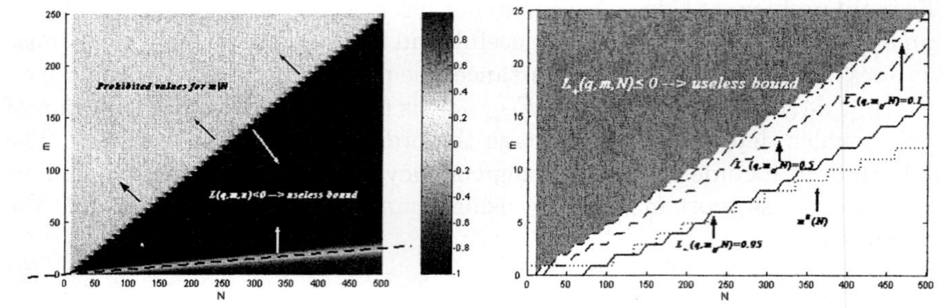

Fig. 2. Left: $\mathcal{L}(q,m,N)$, useful only for couples (m,N) below the dashed line; Right: selected iso-$\mathcal{L}_+$ curves for $q=0.95$. The curve $m^\sharp$ given by eq. (5) vs N has been also plotted.

densities is available, q can be a priori taken equal to e.g. 0.95. The value of m is thus fixed once the quantile number q and the probability threshold are fixed: we take m_0 as default value for m so that for a given quantile number q and $p = 1-q$, the probability lower bound $\mathcal{L}_+(q,m_0,N)$ is higher than or equal to a positive threshold lower than 1.

The single parameter m has thus been replaced by two parameters, but the proposed approach has two advantages, though. First, the new parameters have a concrete interpretation; q is related to the support estimation, and the bound $\mathcal{L}_+$ tells us the confidence that we can have in the support estimation. Second, in practice, q and $\mathcal{L}_+$ can be fixed, so that a direct relation between m and N is found, which can be used to set a default value for m.

Figure 2(a) shows $\mathcal{L}(q,m,N)$ versus m and N. The valid values of m for a given N are $m \le \lfloor N/2 \rfloor$. The bound is useful only for couples (m,N) below the dashed line illustrating $\mathcal{L}(q,m,N)=0$. In Figure 2(b) we plot the maximum value m_0 of m so that the quantity $\mathcal{L}_+(q,m_0,N)$ equals various fixed values (indicated on the related curve) with respect to N. Null values for m_0 indicate that it does not exist m_0 such that $\mathcal{L}_+(q,m,N) \ge 0.95$ for fixed N, $q=.95$ and all $m \le m_0$. In other words, each couple (m,N) located under these curves ensure that inequality (4) holds. Observe that for sufficiently large N, small m and for a given q, $\mathcal{L}_+(q,m,N)$ tends to one.

It must be stressed that some attention must be paid when evaluating $\mathcal{L}_+$ for large N; numerical problems may arise when dividing two factorial expressions of large numbers. Therefore, we suggest to use the logarithms when computing the binomial coefficients, i.e. $\binom{N}{i} = \exp\left[\sum_{j=1}^{N}\log j - \sum_{j=1}^{N-i}\log j - \sum_{j=1}^{i}\log j\right]$. If one desires to speed up the method, the following empirical law can be used for selecting a default value for m; we can take

$$m^\sharp(N) = \max\left(1, \Re\left(\overline{\left[\left(\frac{N-18}{6.5}\right)^{0.65} - 4.5\right]}\right)\right), \tag{5}$$

where $\overline{\alpha}$ denotes the nearest integer to α (see Fig. 2(b)).

4 Performances Comparison

In this section, we compare the extraction performances of 5 ICA algorithms: FastICA, JADE, AKMICA and three minimum-support approaches, OSICA (support estimated by $R_m(X)$), XSICA (support estimated by $R_1(X)$) and AVOSICA (support estimated by $\langle R_m(X) \rangle$). The default value for the parameter m was chosen equal to $m^\sharp$, given by (5). The algorithms have been tested on the extraction of 5 bounded and white sources from 5 mixtures. The pdf and cdf of the five sources (matched to (0,1)) are illustrated in Fig. 1. The mixing matrix is built from 25 random coefficients uniformly distributed in $(0, 1)$.

Figure 3 compares the histograms of the SIR for each extracted source in the noise-free case for $N = 2000$ and $m = m^\sharp(N)$. Remind that after having processed the permutation indetermination, the SIR criterion of the i-th source s_i reduces to $\mathrm{SIR}(s_i) = \sum_{j \neq i} |\mathbf{c}_i(j)| / |\mathbf{c}_i(i)|$. We can observe in Figure 3 that AVOSICA and XSICA give the most interesting results, in comparison to OSICA, AKMICA, JADE and FastICA (gauss), especially for the separation of sources with linear and triangular pdf. It must be stressed that even if AVOSICA and OSICA perform quite satisfactory for small values of N, the performances are improved for large N.

Table 1 summarizes the global SIR performance of ICA algorithms for various noise levels. Since we deal with SIR, the performance results are analyzed from the mixing matrix recovery point of view; the source denoising task is not

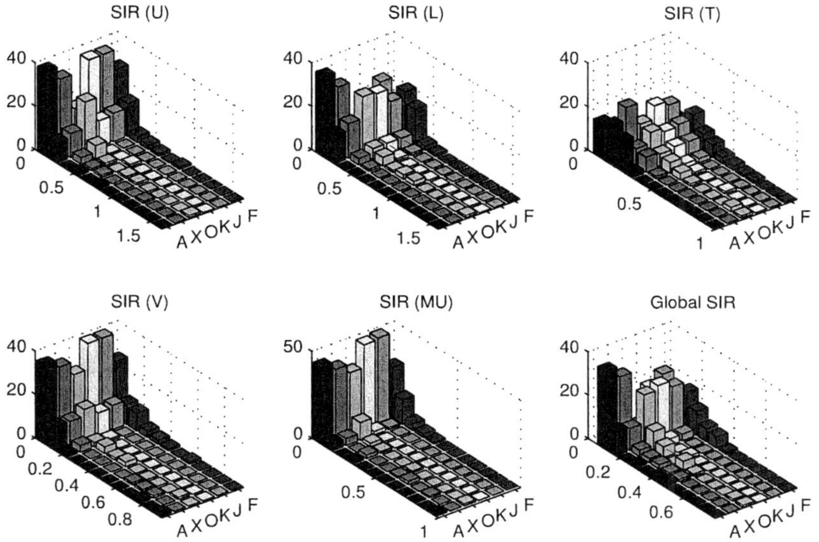

Fig. 3. 12-bins histograms of SIR for each extracted source, for 50 trials, $N = 2000$, and $m = m^\sharp(N) = 37$. The analyzed algorithms are AVOSICA ('A'), XSICA ('X'), OSICA ('O'), AKMICA ('K'), JADE ('J') and FastICA ('F'). The *global SIR* is the averaged SIR computed from the individual source SIRs for a given trial.

Table 1. 100-trials empirical means and variances of global SIR performances of several ICA algorithms (*global SIR* is the averaged SIR computed from the individual source SIRs for a given trial); $m = m^\sharp(N)$. Gaussian noise with standard deviation σ_n has been added to the whitened mixtures (so that for a given σ_n, the mixture SNRs equal $-10 \log \sigma_n^2$; they do not vary between trials, and do not depend of the mixing weights).

σ_n^2	N	AVOSICA	AKMICA	JADE	FastICA	XSICA	OSICA
0	500	**.106** (**.005**)	.135 (.006)	.127 (.006)	.194 (.023)	.139 (.025)	.208 (.03)
	2000	**.05** (.004)	.065 (.0012)	.06(**.0007**)	.097 (.004)	.082 (.02)	.087 (.003)
0.01	500	**.102** (**.003**)	.13 (.005)	.12 (.004)	.187 (.02)	.127 (.02)	.189 (.021)
	2000	**.047** (**.0006**)	.066 (.001)	.059 (.0008)	.105 (.005)	.08 (.02)	.085 (.0024)
0.05	500	**.105** (**.0027**)	.13 (.0039)	.122 (.0032)	.184 (.015)	.144 (.012)	.176 (.013)
	2000	**.051** (.0012)	.067 (.0012)	.06 (**.0006**)	.112 (.007)	.1 (.0225)	.09 (.006)

considered here. The global SIR, for a given trial, is obtained by computing the mean of the extracted sources SIR. The good results of AVOSICA can be observed, despite the fact that the value of m has not been chosen to optimize the results, i.e. we always have taken $m = m^\sharp(N)$. It must be stressed that the value of the parameter m is not critical when chosen around $m^\sharp(N)$.

JADE is a very good alternative when the dimensionality of the source space is low. The computational time of FastICA is its main advantage, contrarily to AKMICA.

5 Conclusion

In addition to existing results regarding the theoretical framework of minimum-support ICA and their specific advantages when separating sources correlated in a specific way, we have shown that these methods also yield competitive results in comparison to other ICA algorithms for extracting bounded sources in the noise-free and noisy cases. This is shown in the particular situation where the support measure is estimated using averaged quasi-ranges. We have further derived a rule to choose a default value for the tuning parameter m, for given sample size N. This choice is related to the confidence of support estimation quality. Numerical results illustrate that the proposed default value of m yield interesting SIR performances, that are comparable for m close to the suggested value.

References

1. Vrins, F., Jutten, C. & Verleysen, M. (2005) SWM : a Class of Convex Contrasts for Source Separation. In proc. ICASSP, *IEEE Int. Conf. on Acoustics, Speech and Sig. Process.*: V.161-V.164, Philadelphia (USA).
2. Vrins, F. & Verleysen, M. (2006) Minimum Support ICA Using Order Statistics. Part I: Quasi-Range Based Support Estimation. In proc. ICA, *Int. Conf. on Ind. Comp. Anal. and Blind Sig. Sep.*, Charleston SC (USA).

3. Vrins, F., Lee, J.A. & Verleysen, M. (2005) Filtering-free Blind Separation of Correlated Images. In proc. IWANN, *Int. Work-Conf. on Art. Neur. Net.*: 1091-1099, Barcelona (Spain).
4. Pham, D.T. & Vrins, F. (2005) Local Minima of Information-Theoretic Criteria in Blind Source Separation. *IEEE Sig. Process. Lett.*, **12**(11): 788-791.
5. Pham, D.-T. (2000) Blind Separation of Instantaneous Mixture of Sources Based on Order Statistics. *IEEE Trans. Sig. Process.*, **48**(2): 363-375.
6. Lee, J.A., Vrins, F. & Verleysen, M. (2005) A Simple ICA Algorithm for Non-Differentiable Contrasts. In proc. EUSIPCO, *Eur. Sig. Process. Conf.*: cr.1412.1-4, Antalya (Turkey).
7. Theis, F.J. & Gruber, P. (2004) Separability of Analytic Postnonlinear Blind Source Separation with Bounded Sources. In proc. ESANN, *Eur. Symp. Art. Neur. Net.*: 217-222, Bruges (Belgium).
8. Theis, F.J., Georgiev P. & Cichocki, A. (2004) Robust Overcomplete Matrix Recovery for Sparse Sources Using a Genralized Hough Transform. In proc. ESANN, *Eur. Symp. Art. Neur. Net.*: 343-348, Bruges (Belgium).
9. Plumbley, M.D. (2003) Algorithms for Nonnegative Independent Component Analysis. *IEEE Trans. Neur. Net.* **4**(3): 534-543.
10. Hyvarinen, A., Karhunen, J. & Oja, E. (2001). "Independent Component Analysis." Wiley series on adaptive and learning systems for signal processing, Wiley, New York.
11. Hyvarinen, A. & Oja, E. (1997) A Fast Fixed-point Algorithm Independent Component Analysis. *Neur. Comp.* **9**(7): 1483–1492.
12. Cardoso, J.F. & Souloumiac, A. (1993) Blind Beamforming for Non-Gaussian Signals. *IEE Proc.-F* **140**(6): 362–370.
13. Molgedey, J. & Schuster, H.G. (1994) Separation of a Mixture of Independent Signals Using Time Delayed Correlations. *Phys. Rev. Lett.* **72**: 3634–3636.
14. Cruces, S. & Duran, I. (2004) The Minimum Support Criterion for Blind Source Extraction: a Limiting Case of the Strengthened Young's Inequality. In proc. ICA, *Int. Conf. on Ind. Comp. Anal. and Blind Sig. Sep.*: 57–64, Granada (Spain).
15. Hoel, P.G. (1975) "Introduction to Mathematical Statistics." Wiley series in probability and mathematical statistics, Wiley, New York.
16. Chu, J.T. (1957). "Some Uses of Quasi-Ranges." Ann. Math. Statist. **28**: 173–180.
17. David, H.A. (1970). "Order Statistics." Wiley series in probability and mathematical statistics, Wiley, New York.

Separation of Periodically Time-Varying Mixtures Using Second-Order Statistics

Tzahi Weisman and Arie Yeredor

School of Electrical Engineering,
Tel-Aviv University, Israel
{itzhakwe, arie}@eng.tau.ac.il

Abstract. We address the problem of Blind Source Separation (BSS) in the context of instantaneous (memoryless) linear mixtures, where the unknown mixing coefficients are time varying, changing periodically in time. Such a mixing model is realistic, e.g., when considering a biological or physiological system where the mixing coefficients are affected by periodic processes like breathing, heart-beating etc. Assuming stationary sources with distinct spectra, we rely on second-order statistics (SOS) and offer an expansion of the classical Second Order Blind Identification (SOBI) algorithm, accommodating the periodic variation model. The proposed algorithm consists of estimating several types of correlation matrices related to the time-varying SOS of the observations, followed by applying generalized joint diagonalization, which leads to estimates of the parameters of the periodic mixing. These estimated parameters are used in turn to apply a time-varying unmixing operation, recovering the desired sources. In its basic form (as presented in here), the algorithm requires prior knowledge (or a good estimate) of the cyclic period. We demonstrate the performance improvement over SOBI in simulation.

1 Introduction

In quite a few Blind Source Separation (BSS) applications the classical assumption of static (i.e., time-invariant) mixing coefficients seems implausible. Often, the mixing medium undergoes some modifying processes during the observation interval, possibly causing slight variations in the mixing coefficients. Nevertheless, the case of time-varying mixtures has not been studied as extensively in the BSS community. It is common practice to rely of the adaptive nature of some sequential learning algorithms to track possible time-variations in the mixture - see, e.g., [1]. Yet another possible approach is to parameterize the time variation and try to estimate the associated variation parameters from the data, to be used, in turn, for constructing the time-varying separation. Such an approach was taken, e.g., in [2], where the time variations were modeled as linear in time.

In some bio-medical and other applications, such as in the case of multi-lead Electrocardiogram (EEG) (e.g., [3], [4]), it may be more reasonable to assume

J. Rosca et al. (Eds.): ICA 2006, LNCS 3889, pp. 278–285, 2006.

periodic variation of the mixing coefficients, as the medium (the human body) may be affected by the periodic breathing mechanism throughout the measurement interval.

The concept of periodic modeling of the mixtures' time-variations can be applied to various source models, thereby enhancing classical BSS algorithms based on respective properties of the sources. In this work we chose to concentrate on Second-Order Statistics (SOS), enhancing the well-known Second Order Blind Identification (SOBI) algorithm (Belouchrani et al., [5]). The resulting algorithm may thus be applied to any wide-sense stationary (WSS) source signals, as long as they are mutually uncorrelated and have non-similar spectra (i.e., no source is allowed to have a spectrum which is a scaled version of another source's spectrum).

2 Problem Formulation

We consider M zero-mean WSS mutually uncorrelated source signals with unknown (but distinct) spectra, denoted by $\boldsymbol{s}[n] = [s_1[n]s_2[n]...s_M[n]]^T$.

The most general representation of a square instantaneous time-varying (and noiseless) mixture model would be

$$\boldsymbol{x}[n] = \boldsymbol{A}[n]\boldsymbol{s}[n] \tag{1}$$

where $\boldsymbol{x}[n] = [x_1[n]x_2[n]...x_M[n]]^T$ are the M observations (at time-instant n).

Assuming that each element of $\boldsymbol{A}[n]$ varies periodically in time (with known[1] angular frequency ω), the most general expression for each element $A_{ij}[n]$ of $\boldsymbol{A}[n]$ would be (for $i, j = 1, 2, ..., M$):

$$A_{ij}[n] = A_{ij}^{(0)} + \sum_{k=1}^{\infty} A_{ij}^{(k)} \cos(k\omega n + \phi_{ij}^{(k)}). \tag{2}$$

where $\{A_{ij}^{(k)}\}_{k=0}^{\infty}$ and $\{\phi_{ij}^{(k)}\}_{k=1}^{\infty}$ are the unknown amplitudes and phases of the different harmonics. Note further, that using some trivial transformations, (2) can be written in matrix form as

$$\boldsymbol{A}[n] = (\boldsymbol{I} + \sum_{k=1}^{\infty}[\boldsymbol{\mathcal{E}}_c^{(k)} \cos(k\omega n) + \boldsymbol{\mathcal{E}}_s^{(k)} \sin(k\omega n)])\boldsymbol{A}_0 \tag{3}$$

where $\boldsymbol{A}_0$ is the "mean" constant matrix, $\boldsymbol{I}$ denotes the $M \times M$ identity matrix and $\{\boldsymbol{\mathcal{E}}_c^{(k)}\}_{k=1}^{\infty}$ and $\{\boldsymbol{\mathcal{E}}_s^{(k)}\}_{k=1}^{\infty}$ are the relative coefficients matrices of the quadrature components of the respective harmonics.

In order to simplify the discussion, we shall assume later that all the coefficients in the relative coefficients matrices have "small" absolute values (relative to 1). Further, we shall now assume that all relative coefficients pertaining to

[1] Often this is a reasonable assumption, as the mechanism which determines the frequency of the change in the mixing matrix is, in many cases, either known or can be conveniently measured externally.

second and higher harmonics are negligible, thereby reducing the discussion to a first-order Fourier approximation of the periodic variations. Hence, for notational convenience we shall, from now on, denote the remaining $\boldsymbol{\mathcal{E}}_c^{(1)}$ and $\boldsymbol{\mathcal{E}}_s^{(1)}$ simply as $\boldsymbol{\mathcal{E}}_c$ and $\boldsymbol{\mathcal{E}}_s$ (respectively).

To conclude, the mixing model is given by:

$$\boldsymbol{x}[n] = (\boldsymbol{I} + \boldsymbol{\mathcal{E}}_c \cos(\omega n) + \boldsymbol{\mathcal{E}}_s \sin(\omega n))\, \boldsymbol{A}_0 \boldsymbol{s}[n] \tag{4}$$

where $\boldsymbol{x}[n]$, $n = 1, 2, ..., N$ are the observed mixtures, from which it is desired to recover the sources $\boldsymbol{s}[n]$, possibly via estimation of $\boldsymbol{A}_0$, $\boldsymbol{\mathcal{E}}_c$ and $\boldsymbol{\mathcal{E}}_s$.

3 Derivation of the Algorithms

We begin by evaluating the SOS of $\boldsymbol{x}[n]$. It should be noted that despite the stationarity of the sources, the observations are obviously nonstationary, due to the time-varying nature of the mixture. First we establish the following:

$$\boldsymbol{R}_x[n,\ell] \triangleq E\left[\boldsymbol{x}[n+\ell]\boldsymbol{x}^T[n]\right] = E\left[\boldsymbol{A}[n]\boldsymbol{s}[n+\ell]\boldsymbol{s}^T[n]\boldsymbol{A}^T[n]\right] \triangleq \boldsymbol{A}[n]\boldsymbol{\Lambda}_\ell \boldsymbol{A}^T[n] \tag{5}$$

where $\boldsymbol{\Lambda}_\ell = E[\boldsymbol{s}[n+\ell]\boldsymbol{s}^T[n]]$ are the source signal's diagonal autocorrelation matrices at lag ℓ. For additional convenience we shall also denote $\boldsymbol{R}_x[n,\ell]$ as $\boldsymbol{R}_\ell[n]$. We assume that the correlation span of the sources is small relative to the variation period of the mixing, hence for all values of ℓ to be considered we may assume that $\cos(\omega\ell) \approx 1$ and $\sin(\omega\ell) \approx 0$. We therefore have $\cos(\omega(n+l)) \approx \cos(\omega n)$ and $\sin(\omega(n+l)) \approx \sin(\omega n)$, leading to

$$\begin{aligned}\boldsymbol{R}_\ell[n] \approx \boldsymbol{R}_\ell^{(0)} &+ (\boldsymbol{\mathcal{E}}_c\boldsymbol{R}_\ell^{(0)} + \boldsymbol{R}_\ell^{(0)}\boldsymbol{\mathcal{E}}_c^T)\cos(\omega n) + (\boldsymbol{\mathcal{E}}_s\boldsymbol{R}_\ell^{(0)} + \boldsymbol{R}_\ell^{(0)}\boldsymbol{\mathcal{E}}_s^T)\sin(\omega n)\\ &+ (\boldsymbol{\mathcal{E}}_c\boldsymbol{R}_\ell^{(0)}\boldsymbol{\mathcal{E}}_c^T)\cos^2(\omega n) + (\boldsymbol{\mathcal{E}}_s\boldsymbol{R}_\ell^{(0)}\boldsymbol{\mathcal{E}}_s^T)\sin^2(\omega n)\\ &+ (\boldsymbol{\mathcal{E}}_c\boldsymbol{R}_\ell^{(0)}\boldsymbol{\mathcal{E}}_c^T + \boldsymbol{\mathcal{E}}_s\boldsymbol{R}_\ell^{(0)}\boldsymbol{\mathcal{E}}_s^T)\cos(\omega n)\sin(\omega n).\end{aligned} \tag{6}$$

where

$$\boldsymbol{R}_\ell^{(0)} \triangleq \boldsymbol{A}_0\boldsymbol{\Lambda}_\ell\boldsymbol{A}_0^T. \tag{7}$$

Recalling that $\boldsymbol{\mathcal{E}}_c$ and $\boldsymbol{\mathcal{E}}_s$ are small (and both $\cos(\omega n)$ and $\sin(\omega n)$ are bounded), we now make some further approximation by neglecting all terms that are quadratic in either of these two matrices, namely we neglect the last three terms in (6), resulting in

$$\boldsymbol{R}_\ell[n] \approx \boldsymbol{R}_\ell^{(0)} + \boldsymbol{R}_\ell^{(c)}\cos(\omega n) + \boldsymbol{R}_\ell^{(s)}\sin(\omega n) \tag{8}$$

where

$$\boldsymbol{R}_\ell^{(c)} \triangleq \boldsymbol{\mathcal{E}}_c\boldsymbol{A}_0\boldsymbol{\Lambda}_\ell\boldsymbol{A}_0^T + \boldsymbol{A}_0\boldsymbol{\Lambda}_\ell\boldsymbol{A}_0^T\boldsymbol{\mathcal{E}}_c^T = \boldsymbol{\mathcal{E}}_c\boldsymbol{R}_\ell^{(0)} + \boldsymbol{R}_\ell^{(0)}\boldsymbol{\mathcal{E}}_c^T \tag{9a}$$

and

$$\boldsymbol{R}_\ell^{(s)} \triangleq \boldsymbol{\mathcal{E}}_s\boldsymbol{A}_0\boldsymbol{\Lambda}_\ell\boldsymbol{A}_0^T + \boldsymbol{A}_0\boldsymbol{\Lambda}_\ell\boldsymbol{A}_0^T\boldsymbol{\mathcal{E}}_s^T = \boldsymbol{\mathcal{E}}_s\boldsymbol{R}_\ell^{(0)} + \boldsymbol{R}_\ell^{(0)}\boldsymbol{\mathcal{E}}_s^T \tag{9b}$$

3.1 Estimating $\boldsymbol{R}_\ell^{(0)}$, $\boldsymbol{R}_\ell^{(c)}$ and $\boldsymbol{R}_\ell^{(s)}$

We now wish to estimate the unknown matrices $\boldsymbol{R}_\ell^{(0)}$, $\boldsymbol{R}_\ell^{(c)}$ and $\boldsymbol{R}_\ell^{(s)}$ from the available data $\boldsymbol{x}[1]$ to $\boldsymbol{x}[N]$. To this end, note that the (i,j)-th element of $\boldsymbol{R}_\ell[n]$, which is the expected value of the product $x_i[n+\ell]x_j[n]$, can be regarded as a "noisy" measurement thereof (with zero-mean noise). We can therefore arrange these samples in the following manner, applying the model specified by (8) casted as a linear least squares model in the unknown parameters:

$$\boldsymbol{y}(\ell,i,j) \triangleq \begin{bmatrix} x_i[1+\ell]x_j[1] \\ x_i[2+\ell]x_j[2] \\ \vdots \\ x_i[N+\ell]x_j[N] \end{bmatrix} \approx \begin{bmatrix} \boldsymbol{R}_\ell[1](i,j) \\ \boldsymbol{R}_\ell[2](i,j) \\ \vdots \\ \boldsymbol{R}_\ell[N](i,j) \end{bmatrix}$$

$$\approx \begin{bmatrix} 1 \ \cos(1\cdot\omega) \ \sin(1\cdot\omega) \\ 1 \ \cos(2\cdot\omega) \ \sin(2\cdot\omega) \\ \vdots \quad \vdots \quad \vdots \\ 1 \ \cos(N\cdot\omega) \ \sin(N\cdot\omega) \end{bmatrix} \begin{bmatrix} \boldsymbol{R}_\ell^{(0)}(i,j) \\ \boldsymbol{R}_\ell^{(c)}(i,j) \\ \boldsymbol{R}_\ell^{(s)}(i,j) \end{bmatrix} \triangleq \boldsymbol{H}\boldsymbol{\theta}(\ell,i,j) \quad (10)$$

where $\boldsymbol{\theta}(\ell,i,j)$ denotes the unknown parameters, namely the (i,j)-th elements of the three matrices $\boldsymbol{R}_\ell^{(0)}$, $\boldsymbol{R}_\ell^{(c)}$ and $\boldsymbol{R}_\ell^{(s)}$, and therefore needs to be estimated for each $1 \le i,j \le M$ and for all desired lags, say $0 \le \ell \le L$, where L is some maximum lag to be used. It is also assumed implicitly in (10), that $N+L$ samples are actually available, so that end effects are mitigated at the cost of not exploiting all the available samples for the shorter lags. Assuming $L << N$, the associated loss is quite negligible.

The Least Squares (LS) estimate is then given by

$$\hat{\boldsymbol{\theta}}(\ell,i,j) = (\boldsymbol{H}^T\boldsymbol{H})^{-1}\boldsymbol{H}^T\boldsymbol{y}(\ell,i,j)$$
$$= \begin{bmatrix} S_1 \ S_c \ S_s \\ S_c \ S_{cc} \ S_{cs} \\ S_s \ S_{cs} \ S_{ss} \end{bmatrix}^{-1} \cdot \sum_{n=1}^{N} \begin{bmatrix} x_i[n+\ell]x_j[n] \\ \cos(\omega n)x_i[n+\ell]x_j[n] \\ \sin(\omega n)x_i[n+\ell]x_j[n] \end{bmatrix}, \quad (11)$$

where the elements of the matrix $\boldsymbol{H}^T\boldsymbol{H}$ are given by the respective sums, and can be approximated by the associated integrals (assuming $1 \ll \frac{2\pi}{\omega}$) as indi-

Table 1. Explicit expressions for the elements of $\boldsymbol{H}^T\boldsymbol{H}$

Term	$\sum_{n=1}^{N}(\bullet)$	Approximating expression
S_1	1	N (exact)
S_c	$\cos(\omega n)$	$\frac{1}{\omega}\sin(\omega N)$
S_s	$\sin(\omega n)$	$\frac{1}{\omega}(1-\cos(\omega N))$
S_{cc}	$\cos^2(\omega n)$	$\frac{N}{2}+\frac{1}{4\omega}\sin(2\omega N)$
S_{ss}	$\sin^2(\omega n)$	$\frac{N}{2}-\frac{1}{4\omega}\sin(2\omega N)$
S_{cs}	$\cos(\omega n)\sin(\omega n)$	$\frac{1}{2\omega}\sin^2(\omega N)$

cated in Table 1. Once the parameters vector $\boldsymbol{\theta}(\ell,i,j)$ is estimated for each i, j and ℓ, the results can be plugged into the respective estimated matrices, obtaining $\hat{\boldsymbol{R}}_\ell^{(0)}$, $\hat{\boldsymbol{R}}_\ell^{(c)}$ and $\hat{\boldsymbol{R}}_\ell^{(s)}$. Note that while the true matrices $\boldsymbol{R}_\ell^{(0)}$, $\boldsymbol{R}_\ell^{(c)}$ and $\boldsymbol{R}_\ell^{(s)}$ are all symmetric, their estimated counterparts may be non-symmetric for $\ell \neq 0$. It is therefore proposed to "symmetrize" the estimates by replacing each $\hat{\boldsymbol{R}}_\ell^{(p)}$ with $\frac{1}{2}(\hat{\boldsymbol{R}}_\ell^{(p)} + \hat{\boldsymbol{R}}_\ell^{(p)T})$, $p = 0, c, s$. Using these estimated matrices, we now proceed to obtain estimates of the unknown mixing parameters $\boldsymbol{A}_0$, $\boldsymbol{\mathcal{E}}_c$ and $\boldsymbol{\mathcal{E}}_s$.

3.2 Estimating A_0, $\mathcal{E}_c$ and $\mathcal{E}_s$

Recalling the relations (7), (9a), (9b) between $\boldsymbol{A}$, $\boldsymbol{\mathcal{E}}_c$, $\boldsymbol{\mathcal{E}}_s$ and the matrices $\boldsymbol{R}_\ell^{(0)}$, $\boldsymbol{R}_\ell^{(c)}$ and $\boldsymbol{R}_\ell^{(s)}$, several approaches can be taken in extracting estimates of $\boldsymbol{A}$, $\boldsymbol{\mathcal{E}}_c$ and $\boldsymbol{\mathcal{E}}_s$ from the estimates of the correlation matrices, essentially by trying to attain the closest match when plugged into (7), (9a), (9b). To this end, we choose to employ the same sub-optimal (but computationally more simple) approach as in [2], i.e., attempting to match each term separately.

The first terms to be matched in the LS sense would be $\{\hat{\boldsymbol{R}}_\ell^{(0)}\}_{\ell=0}^L$, leading to the minimization of

$$\min_{\boldsymbol{A}_0,\boldsymbol{\Lambda}_0,\boldsymbol{\Lambda}_1,\ldots\boldsymbol{\Lambda}_L} \left\{ \sum_{\ell=0}^{L} ||\hat{\boldsymbol{R}}_\ell^{(0)} - \boldsymbol{A}_0\boldsymbol{\Lambda}_\ell\boldsymbol{A}_0^T||_F^2 \right\}, \tag{12}$$

which is exactly the static SOBI term, leading to a standard joint diagonalization approach. Note, however, that $\hat{\boldsymbol{R}}_\ell^{(0)}$ as obtained from (11) is (in general) not necessarily a positive definite matrix, which in turn implies that the standard whitening phase in SOBI may yield a complex whitening matrix. To mitigate this, we may use a non-orthogonal joint diagonalization algorithm such as [6], yielding $\hat{\boldsymbol{A}}_0$ and $\hat{\boldsymbol{\Lambda}}_\ell$, $\ell = 0, 1, 2, ..., L$.

LS matching of $\{\boldsymbol{R}_\ell^{(c)}\}_{\ell=0}^L$ and $\{\boldsymbol{R}_\ell^{(s)}\}_{\ell=0}^L$ requires the following minimization:

$$\min_{\boldsymbol{\mathcal{E}}_p} \left\{ \sum_{\ell=0}^{L} ||\hat{\boldsymbol{R}}_\ell^{(p)} - \boldsymbol{\mathcal{E}}_p\hat{\boldsymbol{R}}_\ell^{(0)} - \hat{\boldsymbol{R}}_\ell^{(0)}\boldsymbol{\mathcal{E}}_p^T||_F^2 \right\} \quad p = c, s. \tag{13}$$

Fortunately, the minimization criterion (13) is quadratic in $\boldsymbol{\mathcal{E}}_p$, and can therefore be minimized as follows. Denoting by $\boldsymbol{\varepsilon}_i$ the columns of $\boldsymbol{\mathcal{E}}_p^T$ such that $\boldsymbol{\mathcal{E}}_p^T = [\boldsymbol{\varepsilon}_1\boldsymbol{\varepsilon}_2...\boldsymbol{\varepsilon}_M]$ and by $\boldsymbol{r}_i^\ell$ the columns of $\hat{\boldsymbol{R}}_\ell^{(0)}$ such that $\hat{\boldsymbol{R}}_\ell^{(0)} = [\boldsymbol{r}_1^\ell\boldsymbol{r}_2^\ell...\boldsymbol{r}_M^\ell]$, we have (for $1 \leq i, j \leq M$, and exploiting the symmetry of $\hat{\boldsymbol{R}}_\ell^{(0)}$):

$$\left(\boldsymbol{\mathcal{E}}_p\hat{\boldsymbol{R}}_\ell^{(0)} + \hat{\boldsymbol{R}}_\ell^{(0)}\boldsymbol{\mathcal{E}}_p^T\right)_{(i,j)} = \boldsymbol{\varepsilon}_i^T\boldsymbol{r}_j^\ell + \boldsymbol{r}_i^{\ell T}\boldsymbol{\varepsilon}_j. \tag{14}$$

Denoting by vec($\bullet$) the operation of concatenating the columns of the argument $M \times M$ matrix into one $M^2 \times 1$ column, and denoting $\boldsymbol{\varepsilon}_p \triangleq \text{vec}(\boldsymbol{\mathcal{E}}_p^T)$, we obtain

$$\mathrm{vec}(\boldsymbol{\mathcal{E}}_p\hat{\boldsymbol{R}}_\ell^{(0)}) = \underbrace{\begin{bmatrix} \boldsymbol{I}\otimes\boldsymbol{r}_1^{\ell T} \\ \boldsymbol{I}\otimes\boldsymbol{r}_2^{\ell T} \\ \vdots \\ \boldsymbol{I}\otimes\boldsymbol{r}_M^{\ell T}\end{bmatrix}}_{\triangleq \boldsymbol{H}_1^\ell} \cdot \boldsymbol{\varepsilon}_p \quad , \quad \mathrm{vec}(\hat{\boldsymbol{R}}_\ell^{(0)}\boldsymbol{\mathcal{E}}_p^T) = \underbrace{\left[\boldsymbol{I}\otimes\hat{\boldsymbol{R}}_\ell^{(0)}\right]}_{\triangleq \boldsymbol{H}_2^\ell} \cdot \boldsymbol{\varepsilon}_p \tag{15}$$

where $\boldsymbol{I}$ denotes the $M \times M$ identity matrix and $\otimes$ denotes Kronecker's product. Consequently, the linear LS minimization problems (13) can be restated as

$$\min_{\boldsymbol{\varepsilon}_p} \left\{ \sum_{\ell=0}^{L} || \mathrm{vec}(\hat{\boldsymbol{R}}_\ell^{(p)}) - [\boldsymbol{H}_1^\ell + \boldsymbol{H}_2^\ell]\boldsymbol{\varepsilon}_p||^2 \right\} \tag{16}$$

(where $\boldsymbol{H}_1$ and $\boldsymbol{H}_2$ are defined in (15) above), whose solution is

$$\hat{\boldsymbol{\varepsilon}}_p = \left[\sum_{\ell=0}^{L}\left[\boldsymbol{H}_1^\ell + \boldsymbol{H}_2^\ell\right]^T \left[\boldsymbol{H}_1^\ell + \boldsymbol{H}_2^\ell\right]\right]^{-1} \cdot \sum_{\ell=0}^{L}\left[\boldsymbol{H}_1^\ell + \boldsymbol{H}_2^\ell\right]^T \cdot \mathrm{vec}(\hat{\boldsymbol{R}}_1^{(p)}). \tag{17}$$

Once the minimizing $\hat{\boldsymbol{\varepsilon}}_p = \mathrm{vec}(\hat{\boldsymbol{\mathcal{E}}}_p^T)$ is computed for $p = \{c, s\}$, $\hat{\boldsymbol{\mathcal{E}}}_c$ and $\hat{\boldsymbol{\mathcal{E}}}_s$ can be easily extracted by undoing the vec($\bullet$) operation, and we can construct an estimate of the periodically varying mixing matrix:

$$\hat{\boldsymbol{A}}[n] = (\boldsymbol{I} + \hat{\boldsymbol{\mathcal{E}}}_c\cos(\omega n) + \hat{\boldsymbol{\mathcal{E}}}_s\sin(\omega n))\hat{\boldsymbol{A}}_0 \quad , \quad n = 1, 2, ..., N \tag{18}$$

which, in turn, can be used for demixing the data: $\hat{\boldsymbol{s}}[n] = \hat{\boldsymbol{A}}[n]^{-1}\boldsymbol{x}[n]$. The signals $\hat{\boldsymbol{s}}[n]$ are the estimated source signals, up to the inherent scale and permutation ambiguities induces in any BSS problem. Note, however, that all scaling ambiguities are absorbed in the estimated $\boldsymbol{A}_0$ only, and there are no scaling ambiguities in the estimates of $\boldsymbol{\mathcal{E}}_c$ and $\boldsymbol{\mathcal{E}}_s$.

4 Regularization for Short Observation Intervals

When the observation length N is short relative to the variation period $\frac{2\pi}{\omega}$, the matrix $\boldsymbol{H}^T\boldsymbol{H}$ (of (11)) may become ill-conditioned, implying undetermined distribution of energy between the estimates of $\boldsymbol{R}_\ell^{(0)}$, $\boldsymbol{R}_\ell^{(c)}$ and $\boldsymbol{R}_\ell^{(s)}$. To mitigate this difficulty, we may introduce a Bayesian approach with respect to $\boldsymbol{R}_\ell^{(c)}$ and $\boldsymbol{R}_\ell^{(s)}$, conveying the information that they are supposed to be small relative to $\boldsymbol{R}_\ell^{(0)}$.

A convenient way for conveying that information is to augment each "measurement vector" $\boldsymbol{y}(\ell, i, j)$ in (10) (for all ℓ, i and j) with two zeros, similarly augmenting the matrices $\boldsymbol{H}$ with two rows: $\left[\begin{smallmatrix} 0 & 1 & 0 \\ 0 & 0 & 1\end{smallmatrix}\right]$. Then, a Weighted LS solution can be used, so as to attribute the proper weight to these fictitious measurements: assuming that the variation in all the true measurements are of the order of one[2], the additional measurements would be weighted by $\frac{1}{\sigma^2}$, where σ^2 reflects the allowable variation of the elements of $\boldsymbol{R}_\ell^{(c)}$ and $\boldsymbol{R}_\ell^{(s)}$, e.g., using $\sigma = 0.1$.

As a result, the only computational change that has to be applied in the algorithm is the substitution of $\boldsymbol{H}^T\boldsymbol{H}$ in (11) with $\boldsymbol{H}^T\boldsymbol{H} + \frac{1}{\sigma^2}\left[\begin{smallmatrix}0&0&0\\0&1&0\\0&0&1\end{smallmatrix}\right]$, where σ^2 is the regularization parameter. For longer observation intervals this operation is not necessary, but it may be left in anyway, since the associated weight would be overwhelmed by the weight of the data (or, in other words, the regularization matrix would be negligible with respect to $\boldsymbol{H}^T\boldsymbol{H}$).

5 Simulation Results

We demonstrate the performance with $M = 2$ sources, generated by filtering two uncorrelated zero-mean white Gaussian processes with the following digital filters: $H_1(z) = 0.8 - 0.2z^{-1} + 0.5z^{-2} - 0.3z^{-4}$ (for $s_1[n]$) and $H_2(z) = 0.8 + 0.3z^{-1} + 0.2z^{-2} - 0.5z^{-3}$ (for $s_2[n]$). The true mixing parameters were

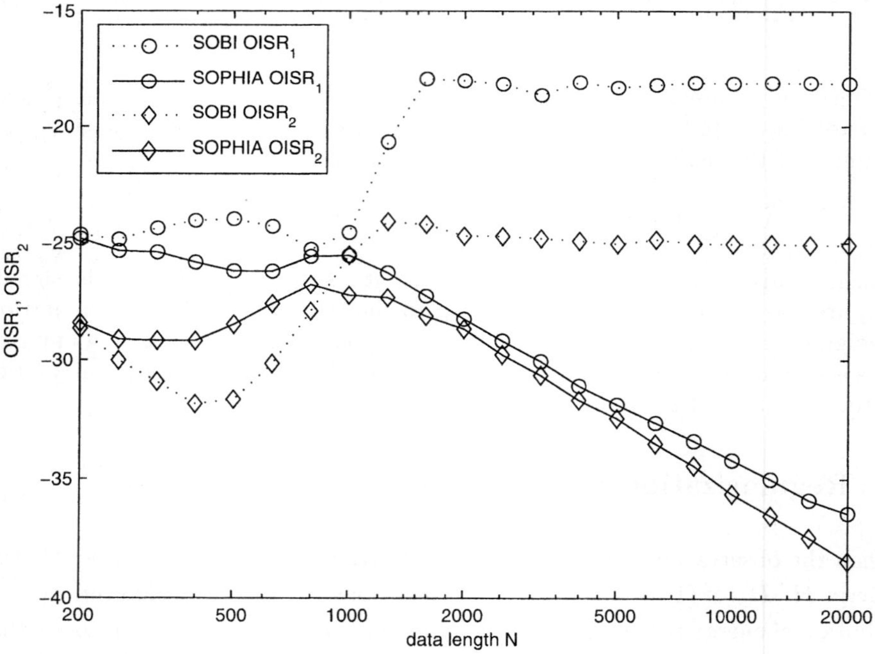

Fig. 1. OISR_1 and OISR_2 [dB] for SOPHIA, SOBI vs. N. Each point represents the average of 1000 trials.

[2] To justify this assumption, it is necessary to apply a pre-processing operation in which all the observations are normalized to unit energy. With Gaussian sources, this would guarantee that the variance of each element (of $\boldsymbol{y}(\ell, i, j)$) lies between 1 and 2 (depending on the true correlation value).

$\boldsymbol{A}_0 = \left[\begin{smallmatrix} 3 & -2 \\ -1 & 4 \end{smallmatrix}\right]$, $\boldsymbol{\mathcal{E}}_c = \left[\begin{smallmatrix} 0.01 & -0.02 \\ -0.07 & 0.03 \end{smallmatrix}\right]$ and $\boldsymbol{\mathcal{E}}_s = \left[\begin{smallmatrix} 0 & 0.08 \\ -0.02 & 0.05 \end{smallmatrix}\right]$. The periodic variation angular frequency was set to $\omega = \frac{2\pi}{1000}$.

We present the performance of the proposed algorithm (termed SOPHIA - Second-Order Periodic Hypothesis Identification Algorithm), as well as SOBI, vs. the observation length. Both algorithms were applied with five correlation lags ($L = 4$). The regularization parameter was set to $\sigma = 0.1$.

Our performance measure is the Overall Interference to Signal Ratio (OISR), defined as follows: Let $\boldsymbol{T}[n]$ denote the total (time-dependent) mixing-demixing effect, $\boldsymbol{T}[n] \triangleq \hat{\boldsymbol{A}}[n]^{-1}\boldsymbol{A}[n]$ for $n = 1, 2, ..., N$. Based on $\boldsymbol{T}[n]$, we compute the overall energy gains $\boldsymbol{T} \triangleq \sum_{n=1}^{N} \boldsymbol{T}[n] \odot \boldsymbol{T}[n]$, where $\odot$ denotes Hadamard's (element-wise) product (implying element-wise squaring in this case). Once $\boldsymbol{T}$ is calculated, the permutation ambiguity is resolved by selecting the row permutation which maximizes $\boldsymbol{T}$'s trace, and then the OISR for each source is the ratio of off-diagonal to diagonal energy in the respective row.

6 Conclusion

We presented an algorithm for blind separation of periodically time-varying mixtures, where the periodic variation introduces slight, smooth fluctuations about the mean values of the mixing parameters. The estimation algorithm consists of estimating the parameterized variations in the time-varying correlation matrices (at different lags), followed by a LS matching scheme which leads to the estimation of the time-varying mixing parameters. In a truly periodical mixture scenario, this approach offers considerable improvement over SOBI, especially with observation intervals considerably longer than the variation period. Future work would address the estimation of ω as well as the accommodation of more than a single frequency mode in the time-varying mixture.

References

1. Koivunen, V., Enescu, M., Oja, E.: Adaptive algorithm for blind separation from noisy time-varying mixtures. Neural Computations **13** (2001) 2339–2357
2. Yeredor, A.: TV-SOBI: An expansion of SOBI for linearly time-varying mixtures. Proceedings of The 4th International Symposium on Independent Component Analysis and Blind Source Separation (ICA2003) (2003)
3. Castells, F., Igual, J., Millet, J., Rieta, J.J.: Atrial activity extraction from atrial fibrillation episodes based on maximum likelihood source separation. Signal Processing **85** (2005) 523–535
4. Castells, F., Rieta, J.J., Millet, J., Zarzoso, V.: Spatiotemporal blind source separation approach to atrial activity estimation in atrial tachyarrhythmias. IEEE Trans. Biomedical Engineering **52** (2005) 258–267
5. Belouchrani, A., Abed-Meraim, K., Cardoso, J.F., Moulines, E.: A blind source separation technique using second-order statistics. IEEE Trans. Signal Processing **45** (1997) 434–444
6. Yeredor, A.: Non-orthogonal joint diagonalization in the least-squares sense with application in blind source separation. IEEE Trans. Signal Processing **50** (2002) 1545–1553

An Independent Component Ordering and Selection Procedure Based on the MSE Criterion

Edmond HaoCun Wu, Philip L.H. Yu, and W.K. Li

Department of Statistics & Actuarial Science,
The University of Hong Kong,
Pokfulam Road, Hong Kong
hcwu@hkusua.hku.hk, plhyu@hku.hk, hrntlwk@hku.hk

Abstract. Principal components (PCs) by construction have a natural ordering based on their cumulative proportion of variance explained. However, most ICA algorithms for finding independent components (ICs) are arbitrary, which limit the use of ICA in pattern discovery and dimension reduction. To solve this problem, we propose an efficient IC ordering approach and prove that this method guarantees to find the optimal ordering of ICs based on the MSE criterion. Furthermore, we employ the cross validation method to select the number of dominant ICs. Simulation experiments show that the proposed IC ordering and selection procedure is efficient and effective, which can be used to identify the dominant ICs as well as to reduce the number of ICs.

1 Introduction

Independent component analysis or blind source separation [4,11,12] is a statistical method which aims to express the observed data in terms of a linear combination of mutually independent latent variables called *independent components.* A typical ICA model is $\boldsymbol{x}(t) = \boldsymbol{A}\boldsymbol{s}(t)$, $t = 1, 2, \cdots, T$, where $\boldsymbol{x}(t) = [x_1(t), ..., x_p(t)]^T$ is the vector of p observed variables, $\boldsymbol{s}(t) = [s_1(t), ..., s_q(t)]^T$ is the vector of independent components, and $\boldsymbol{A}$ is an unknown mixing matrix. The task is to identify both the independent components and the matrix $\boldsymbol{A}$.

Many researchers proposed different methods to find ICs. For instance, a fast fixed point algorithm (FastICA) was presented by Hyvärinen and Oja [9,10]. The FastICA algorithm is a computationally efficient and robust fixed-point type algorithm for ICA. In [13], we found that ICA models demonstrate better forecasts of time-varying volatilities than PCA models. Although ICA is more powerful in modelling multivariate time series than PCA, PCA may be preferable in some cases (e.g., dimension reduction) because PCs have a natural ordering based on their cumulative proportion of variance explained. In contrast, the ICs found by most ICA algorithms lack of a proper ordering, which limits the use of ICA in many applications. For example, if we find 100 ICs from a high dimension dataset, how can we identify a modest number of dominant ICs that can well represent the data?

J. Rosca et al. (Eds.): ICA 2006, LNCS 3889, pp. 286–294, 2006.

In the literature, some methods have been proposed to determine a proper ordering of ICs. For example, Cheung *et. al* [1,2] suggested a so-called *Testing-and-Acceptance* (TnA) algorithm to determine locally optimal IC ordering and selection based on some error function. An error function they considered is Mean Square Error (MSE) which is defined as:

$$MSE(\hat{\boldsymbol{x}}) = \sum_{i=1}^{p} MSE(\hat{x}_i)$$

$$MSE(\hat{x}_i) = \frac{1}{T}\sum_{t=1}^{T}[x_i(t) - \hat{x}_i(t)]^2$$

where $\hat{\boldsymbol{x}}(t) = [\hat{x}_1(t), ..., \hat{x}_p(t)]^T$ is the vector of predicted $\boldsymbol{x}(t)$ obtained by ICA.

The basic idea of the TnA algorithm has similar spirit of backward elimination technique for variable selection in regression analysis. Starting from the component set consisting of all ICs, the TnA algorithm first picks an IC as the last one in the ordering, which minimizes the error function. Then, remove the IC from the component set and repeat the procedure on the remaining component set until all the ICs are ordered. As the exhaustive search of the optimal IC ordering is fairly time consuming, the TnA algorithm generally provides a fast but sub-optimal ordering of ICs because it does not guarantee that when m ICs are used, the first m ICs selected by TnA always minimize the error function.

In this paper, we show that if the error function is the MSE, the TnA algorithm always provides the optimal order of ICs. In addition, instead of searching sequentially as in the TnA algorithm, we consider a more efficient one-pass IC ordering procedure under the MSE criterion. The rest of the paper is organized as follows. In Section 2, we propose the IC ordering method and give the proof of the proposition. In Section 3, we present the cross validation method to determine the number of ICs to be selected. Some simulation studies are given in Section 4 and an application is illustrated in Section 5. Section 6 gives the conclusions.

2 Ordering Independent Components Under the MSE

Mean Square Error (MSE) is a widely accepted criterion in model assessment. For example, in dimension reduction, we can use MSE to measure the fitness of a low dimension representation to the original high dimension. In the following, we first prove a proposition and then present our IC ordering method.

Let $\boldsymbol{x}^{(m)}$ be the approximation of $\boldsymbol{x}$ based on a set of m independent components $\mathcal{S}_m = \{s_1, ..., s_m\}$. That is, $x_i^{(m)}(t) = a_{i1}s_1(t) + a_{i2}s_2(t) + ... + a_{im}s_m(t)$. Let $MinMSE(m)$ be the minimum MSE of $\boldsymbol{x}^{(m)}$ among all subsets $\{\mathcal{S}_m\}$ of ICs of size m. Denote the optimal subset of ICs by $\mathcal{S}_{opt}(m)$.

Proposition 1. *Given a set of ICs, $\mathcal{S}_q = \{s_1, ..., s_q\}$, where the ICs are placed in an order such that*

$$\sum_{t=1}^{T}\sum_{i=1}^{p} a_{i1}^2 Var(s_1(t)) \geq \sum_{t=1}^{T}\sum_{i=1}^{p} a_{i2}^2 Var(s_2(t)) \geq \cdots \geq \sum_{t=1}^{T}\sum_{i=1}^{p} a_{iq}^2 Var(s_q(t)). \quad (1)$$

Then we have $\mathcal{S}_{opt}(m) = \{s_1, ..., s_m\}$, $m = 1, ..., q$.

Proof. Without loss of generality, we assume that every IC has zero mean. Suppose the ICs do not necessarily satisfy (1). Note that

$$MSE(\boldsymbol{x}_i^{(m)}) = \frac{1}{T}\sum_{t=1}^{T}[x_i(t) - x_i^{(m)}(t)]^2 = \frac{1}{T}\sum_{t=1}^{T}\left[\sum_{j=m+1}^{q} a_{ij}s_j(t)\right]^2$$
$$= \frac{1}{T}\sum_{t=1}^{T}\left[\sum_{j=m+1}^{q} a_{ij}^2 s_j^2(t)\right] = \frac{1}{T}\sum_{t=1}^{T}\left[\sum_{j=m+1}^{q} a_{ij}^2 Var(s_j(t))\right]$$

The second-last equality holds because any two ICs are independent and hence have zero covariance and the last equality holds because of zero mean ICs. The mean square error MSE of $\boldsymbol{x}^{(m)}$ is

$$MSE(\boldsymbol{x}^{(m)}) = \frac{1}{T}\sum_{j=m+1}^{q}\sum_{t=1}^{T}\sum_{i=1}^{p} a_{ij}^2 Var(s_j(t)).$$

It is now easy to see that the $MinMSE(m)$ can be attained by selecting the m largest ICs in the order in (1) and thus $\mathcal{S}_{opt}(m) = \{s_1, ..., s_m\}$.

This proposition implies that when increasing the number of independent components, the larger optimal IC sets are the supersets of the smaller optimal IC sets under the MSE criterion, i.e., $\mathcal{S}_{opt}(m) \subset \mathcal{S}_{opt}(m+1)$, $m=1, ..., q-1$. □

By using this proposition, we can rank order the ICs in a way such that (1) is satisfied. Given the ordering of ICs, the remaining issue is to identify the dominant ICs. This is particularly useful when the number of ICs is large.

3 Selecting Independent Components by Cross Validation

In [1], Cheung *et. al* proposed a cost function $J(m)$ as the IC selection criterion which is defined as:

$$J(m) = Q(\boldsymbol{x}_L^{(m)}) - Q(\boldsymbol{x}_L^{(m-1)}), \quad 2 \leq m \leq q$$

where $\boldsymbol{x}_L^{(m)}$ is the predicted $\boldsymbol{x}$ based on the first m ICs ordered according to a list L, and Q is an error function such as MSE. Cheung *et. al* commented that there may exist a big drop in the error function when one dominant IC is added and adding non-dominant ICs should make little change in the error function.

Therefore, they recommended identifying the global minimal point $m = m^*$ of the curve of $J(m)$ and select the first m^* ICs in the list L as the dominant ICs.

As found in our empirical application in Section 5, this method may recommend a very large m which may not be too sensible in practice. Therefore in the following, we propose an alternative criterion for IC selection. The criterion used here is the cost-complexity function:

$$C(m) = Q(m) + \alpha m$$

where m is the number of ICs selected, $Q(m)$ is an error function based on m ICs (we use Mean Square Error MSE in this paper), and α is a non-negative constant. For a fixed α, we can choose the optimal m such as $C(m)$ is minimized. Notice that when $\alpha = 0$, the minimum is attained at $m = q$. Therefore, the α in $C(m)$ controls the penalty factor to m.

However, how to specify the value of α? To do this, we can adopt a cross validation approach [5], the one often used to evaluate the prediction performance of a model. The basic idea is to remove some of the data before training begins, after training is done (i.e., the optimal m is determined), the data that was removed can be used to test the prediction performance of the m chosen ICs on the removed data.

In a k-fold cross-validation (we use $k = 10$ here), we first divide the data into k subsets of equal size. Then we train the dataset k times, each time leaving out one of the subsets from training, and using only the leaving-out subset (or testing subset) to compute the error function Q. Consider a set of plausible values of α, the best α is the one having the smallest total error of the k testing subsets.

For a given α and a given training dataset, we can find the optimal m by minimizing $C(m)$. But what range of the values of α do you consider? Here, we choose 0 as the lower bound of α because when $\alpha = 0$, $C(q) = Q(q) \leq Q(m) = C(m)$ for all m and hence the optimal m must be q. When α is getting larger and larger, the $C(q)$ is more likely to exceed other $C(m)$. Therefore, we define the upper bound of α as $\alpha_{upper} = \frac{Q(1)-Q(q)}{q-1}$, where q is the total number of ICs. This is because when $\alpha = \frac{Q(1)-Q(q)}{q-1}$, $C(1) = C(q)$. In short, the searching space of α is set to be the interval $(0, \frac{Q(1)-Q(q)}{q-1}]$. The step size of α depends on the value of α_{upper}. When α_{upper} is large, more steps may be required. The default is 1,000 steps. We summarize the major steps of IC selection as follows:

1. Construct the cost-complexity function $C(m) = Q(m) + \alpha m$.
2. Divide the multivariate time series dataset into k subsets of equal size, with each subset having T/k observations. Use any $k-1$ subsets to form a training set and the remaining subset as a testing set.
3. For each training set, use ICA model to find the mixing matrix $\boldsymbol{A}_{train}$.
4. For each testing set, the ICs is obtained by $\boldsymbol{s}_{test}(t) = \boldsymbol{A}_{train}^{-1}\boldsymbol{x}_{test}(t)$.
5. Then, $\boldsymbol{x}^{(m)}(t)$ is predicted by a linear combination of the m ICs in $\boldsymbol{s}_{test}(t)$, that is, $\boldsymbol{x}^{(m)}(t) = \boldsymbol{A}_{train}\boldsymbol{s}_{test}^{(m)}(t)$.
6. Determine the range of α: $(0, \frac{Q(1)-Q(q)}{q-1}]$ and set the step size of α.

7. For each α (starting from 0),
   (a) find the minimal $Q(m_1^*), ..., Q(m_k^*)$ for the k testing sets, respectively Here, $Q(m) = MSE(\boldsymbol{x}^{(m)})$.
   (b) sum up $Q(m_1^*), ..., Q(m_k^*)$ and denote the sum by Q_α
8. Determine the optimal α that minimizes Q_α and denote this α by α^*.
9. Determine the optimal m that minimizes $C^*(m)$, where $C^*(m)$ is the cost-complexity function based on the whole data set and $\alpha = \alpha^*$.

4 Simulation

Experiment 1. We design an experiment to validate the effectiveness of our proposed model for IC ordering and selection. We select four time series sources which are normal distributed noise (S_1), uniform distributed noise (S_2), Mod (S_3), and Sin (S_4), respectively. They are shown in the first column of Fig 1. We also simulate four time series which are the mixtures of the sources. $X_1 = 0.5 \times S_3 + 0.5 \times S_4$, $X_2 = 0.5 \times S_1 + 0.5 \times S_4$, $X_3 = 0.5 \times S_2 + 0.5 \times S_3$, $X_4 = 0.5 \times S_1 + 0.5 \times S_3$. Then, we use our model to determine how many dominant ICs are needed to model the time series and the order of the dominant ICs. The ordering result is shown in the third column of Fig 1 (From top to bottom, the order is 1st IC, 2nd IC, 3rd IC, and 4th IC).

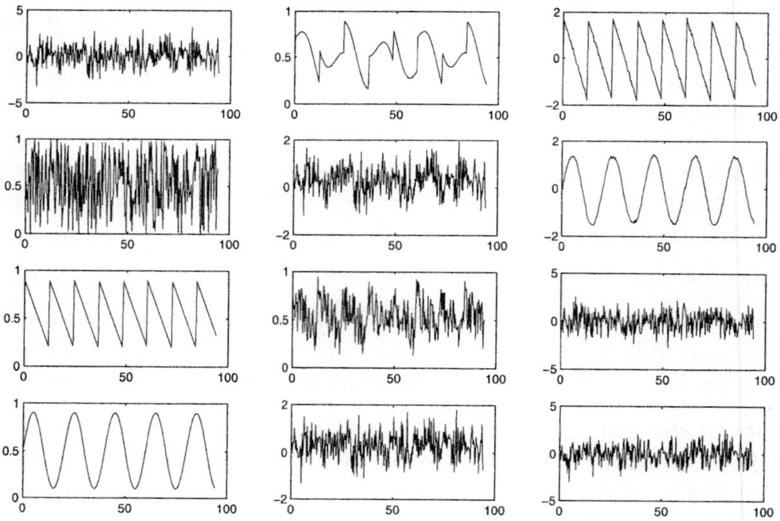

(1st column: Original Sources, 2nd column: Mixtures, 3rd column: Ordered ICs)

Fig. 1. A Simulation of IC Ordering and Selection

By using our IC ordering and selection method, we get the optimal $m = 2$. The result also suggest that the first two dominant ICs are approximate the Mod

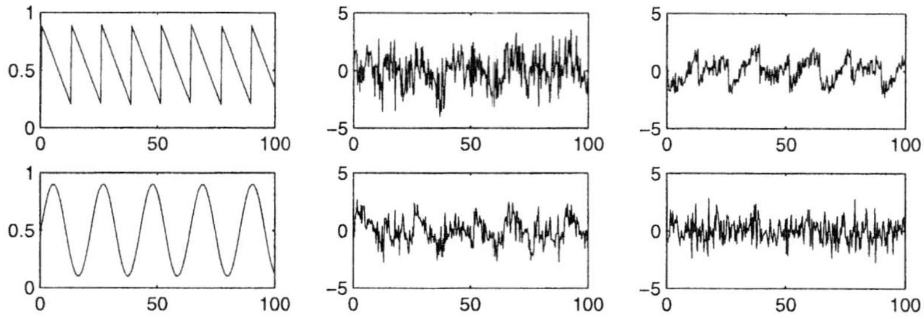

(1st column: first two ordered ICs, 2nd column: first two PCs, 3rd column: the two ICs obtained from first two PCs)

Fig. 2. Comparison of the first 2 ICs and PCs

and Sin sources. The result is consistent with our expectation. Because we think the noise sources are not very useful to identify underlying patterns, so a good IC selection model should be able to filter out such noises. As comparison, when we plot the first two PCs (2nd column in Fig 2), we cannot clearly identify the original sources. The two ICs (3rd column in Fig 2) found by the two PCs also lack of satisfactory recovery of the two most important patterns in the original multivariate time series. In contrast, the first two ICs found by our method demonstrate good recovery of the original two patterns (1st column in Fig 2).

From this experiment, we validate that our IC selection model can correctly identify the most important sources from a given time series dataset. Therefore, the cross validation method for IC selection is feasible. In addition, from the results of this experiment, we suggest that in IC selection, it's better to order all the ICs first, and then select the ICs. This method is superior to using PCA for dimension reduction first, e.g., finding 2 ICs from first 2 PCs, because the dimension reduction will distort the ICs discovered, as we have shown in the example.

Experiment 2. We also compare the performance of ICA and PCA in identifying dominant ICs. The four sources are unchanged while the mixtures are randomly generated for each simulation. Here, $X_1 = (0.5 \pm \epsilon) \times S_3 + (0.5 \pm \epsilon) \times S_4$, $X_2 = (0.5 \pm \epsilon) \times S_1 + (0.5 \pm \epsilon) \times S_4$, $X_3 = (0.5 \pm \epsilon) \times S_2 + (0.5 \pm \epsilon) \times S_3$, $X_4 = (0.5 \pm \epsilon) \times S_1 + (0.5 \pm \epsilon) \times S_3$, where ϵ is iid r.v. which follows uniform distribution $U(0, 0.25)$. Then, we use PCA to find the fours PCs for the mixture. Because the PCs are ordered based on the cumulative proportion of variance explained, we can regard this ordering as the relatively importance of the PCs. Also, we employ our proposed method to order the four ICs found. To get a more reliable performance analysis, we run the simulation 100 times to avoid bias results by chance.

Among the 100 simulation trials, ICA can correctly identify the two non-noisy source patterns as the first two dominant sources for 84 trials while PCA only succeed in 41 trials. This experiment strongly suggests that the ICA ordering and selection approach is feasible in finding the underlying and meaningful patterns and it is more robust than PCA in filtering out noise sources.

5 A Real Application

In this case study, we try to validate our proposed IC ordering method and test the performance by employing real financial data. We select daily stock prices of 27 HSI Constituent Stocks from 26/11/2001 to 26/11/2004 as the multivariate time series. The daily returns $r_i(t)$ are calculated by $r_i(t) = \ln(p_i(t)) - \ln(p_i(t-1))$, where $p_i(t)$ is the closing price of stock i on the trading day t.

By using FastICA, we found 27 ICs from the return series. Using our IC selection procedure, the optimal α is found to be 0.0005. Fig 3. shows the plot of the cost function values (with $\alpha = 0.0005$) against the number of ordered ICs m. In this case, the first 4 ICs is the optimal choice since when $m = 4$, the cost function is minimized.

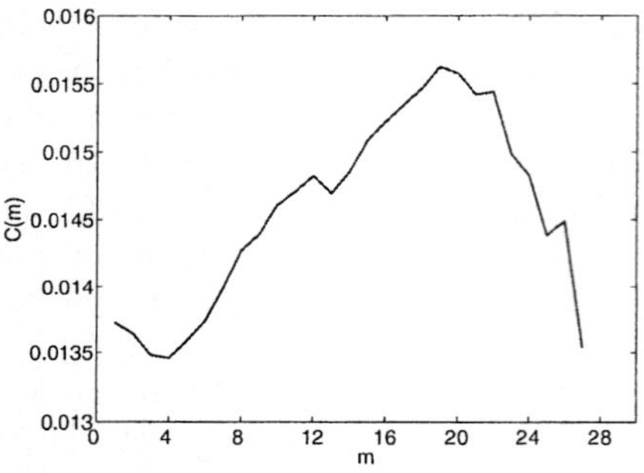

Fig. 3. C(m) function for IC Selection

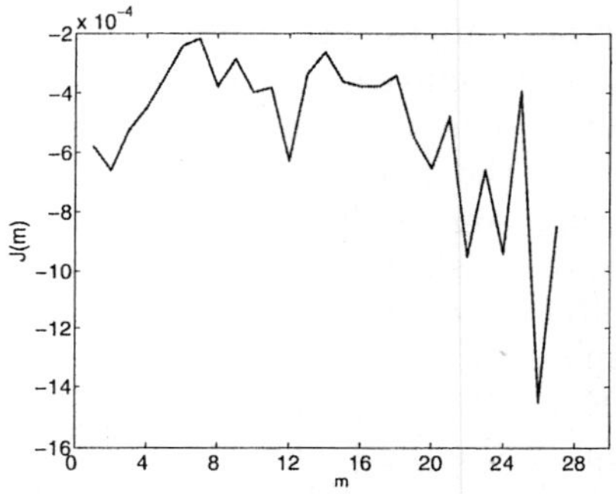

Fig. 4. J(m) function for IC Selection

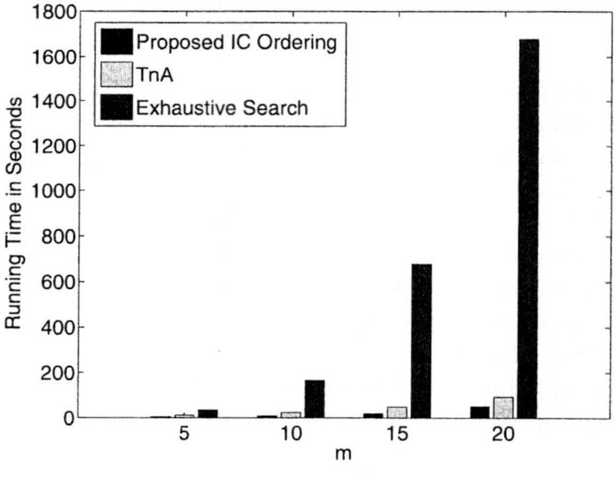

Fig. 5. Comparison of Running time

In contrast, using the IC selection method proposed in [1], the optimal number of ICs found should be 26 under the MSE criterion (See Fig. 4). Obviously, a smaller number of ICs is more reasonable as most empirical studies found in the financial literature suggest at most 10 components [6].

The optimal sets of ICs found by exhaustive search are consistent with our ordering method. This result validates our finding in Proposition 1. In Fig 5, we note that the running time of exhaustive search is extremely huge, and even impossible to achieve when the number of ICs is large. Therefore, by using our proposed IC ordering approach, we can efficiently find the sets of dominant ICs for effective time series analysis. The TnA algorithm also achieves this result, however, its computational cost is greater than our proposed method because the TnA is a sequential search method.

6 Conclusion

In this article, we propose an independent component ordering criterion based on the ICs' loadings and variances. We prove that the sets of ICs selected according the ordering method can maximally reduce the total MSE. We also find that the optimal sets of ICs under the MSE criterion are nested, this property provides us an efficient way to find the optimal ICs. Based on this, we suggest the cross validation method for selecting the optimal number of ICs. The simulation experiments demonstrate that it is more effective to use our proposed method to reduce the number of ICs instead of using PCs to restrict the number of ICs. In the future, we intend to find similar IC ordering properties under other reconstruction error measures.

References

1. Yiu-ming Cheung and Lei Xu, An Empirical Method to Select Dominant Independent Components in ICA for Time Series Analysis, *Proceedings of 1999 International Joint Conference on Neural Networks* (IJCNN'1999), Vol. 6, pp. 3883-3887, Washington, DC, July 10-16, 1999.
2. Yiu-ming Cheung and Lei Xu, Independent Component Ordering in ICA Time Series Analysis, *Neurocomputing*, Vol. 41, No. 1-4, pp145-152, 2001.
3. Carol Alexander, *Market Models: A Guide to Financial Data Analysis*, UK, J. Wiley, 2001.
4. P. Comon, Independent component analysis: a new concept?, *Signal Processing*, 36, 287-314, 1994.
5. Craven, P. and Wahba, G., Smoothing noisy data with spline functions. *Numer. Math.*, 31: 377-403, 1979.
6. Fama, E.F., French, K.R., 1993. Common risk factors in the returns on stocks and bonds. *Journal of Financial Economics* 33, 3-56.
7. Golub, G., Heath, M. and Wahba, G., Generalized cross validation as a method for choosing a good ridge parameter. *Technometrics*, 21: 215-224, 1979.
8. Heath, D., R. Jarrow, and A. Morton, Bond Pricing and the Term Structure of Interest Rates: A New Methodology, *Econometrica*, 60(1): 77-105, 1992.

9. A. Hyvärinen, Fast and robust fixed-point algorithms for independent component analysis, *IEEE Transactions on Neural Networks*, 10(3), 626-634, 1999.
10. A. Hyvärinen and E. Oja, A fast fixed-point algorithm for independent component analysis, *Neural Computation*, 9, 1483-1492, 1997.
11. A. Hyvärinen, J. Karhunen, E. Oja, *Independent Component Analysis*, New York, J. Wiley, 2001.
12. C. Jutten and J. Herault, Blind separation of sources, part I: An adaptive algorithm based on neuromimetic architecture, *Signal Processing*, 24, 1-10, 1991.
13. Edmond H. C. Wu, Philip L. H. Yu, Volatility modelling of multivariate financial time series by using ICA-GARCH models, In *Intelligent Data Engineering and Automated Learning - IDEAL 2005, Lecture Notes in Computer Science*, Volume 3578, (M. Gallagher, J. Hogan, and F. Maire (Eds)), 571-579. Springer-Verlag: Berlin, 2005.

Riemannian Optimization Method on the Flag Manifold for Independent Subspace Analysis

Yasunori Nishimori[1], Shotaro Akaho[1], and Mark D. Plumbley[2]

[1] Neuroscience Research Institute,
National Institute of Advanced Industrial Science and Technology (AIST),
AIST Central2, 1-1-1, Umezono, Tsukuba, Ibaraki 305-8568, Japan
y.nishimori@aist.go.jp, s.akaho@aist.go.jp
[2] Department of Electronic Engineering, Queen Mary University of London,
Mile End Road, London E1 4NS, UK
mark.plumbley@elec.qmul.ac.uk

Abstract. Recent authors have investigated the use of manifolds and Lie group methods for independent component analysis (ICA), including the Stiefel and the Grassmann manifolds and the orthogonal group $O(n)$. In this paper we introduce a new class of manifold, the *generalized flag manifold*, which is suitable for independent subspace analysis. The generalized flag manifold is a set of subspaces which are orthogonal to each other, and includes the Stiefel and the Grassmann manifolds as special cases. We describe how the independent subspace analysis problem can be tackled as an optimization on the generalized flag manifold. We propose a Riemannian optimization method on the generalized flag manifold by adapting an existing geodesic formula for the Stiefel manifold, and present a new learning algorithm for independent subspace analysis based on this approach. Experiments confirm the effectiveness of our method.

1 Introduction

Independent component analysis (ICA) assumes a statistical generative model of the form $x = As$ where $s = (s_1, \ldots, s_n)^\top \in \mathbb{R}^n$, the components s_i are generated from statistically independent sources, and $A \in \mathbb{R}^{n\times n}$ is an invertible matrix. Most algorithms for ICA use whitening as a preprocessing step, giving $z = Bx$ such that $E[zz^\top] = I_n$. After the whitening operation, solving the ICA task reduces to an optimization on the orthogonal group [8, 9], i.e. over the set of orthogonal demixing matrices $\{W \in \mathbb{R}^{n\times n} | W^\top W = I_n\}$ where W is a demixing matrix in $y = (y_1, \ldots, y_n)^\top = W^\top z$ which attempts to recover the original signals up to a scaling and permutation ambiguity.

Optimization on a special class of manifolds related to the orthogonal group such as the Stiefel and the Grassmann manifolds frequently appear in the context of neural networks, signal processing, pattern recognition, computer vision, numerics and so on [3]. Principal, minor, and independent component analysis are formalized as optimization on the Stiefel manifold, subspace tracking and application-driven dimension reduction can be solved by optimization

J. Rosca et al. (Eds.): ICA 2006, LNCS 3889, pp. 295–302, 2006.

on the Grassmann manifold. Generally, optimization on manifolds raises more intricate problems than optimization on Euclidean space, however, optimization on the Stiefel and the Grassmann manifolds can be tackled in a geometrically natural way based on Riemannian geometry. Because those manifolds are homogeneous, we can explicitly compute various Riemannian quantities relative to a Riemannian metric g, including Riemannian gradient, Hessian, geodesic, parallel translation vector, and curvature. Therefore, by replacing, for instance, an ordinary gradient vector of a cost function by a Riemannian gradient vector, and updating a point along the geodesic in direction to the Riemannian gradient, we get a Riemannian gradient descent method on that manifold:

$$W_{k+1} = W_k - \eta \nabla f(W_k) : \text{Euclidean}$$

$$W_{k+1} = \varphi_M(W_k, -\operatorname{grad}_W f(W_k), \eta) : \text{Riemannian},$$

where $\varphi_M(W, V, t)$ denotes the geodesic equation on a manifold M starting from $W \in M$ in direction $V \in T_W M$ relative to a Riemannian metric g on M. As such, other iterative optimization methods on Euclidean space are directly modified for the Stiefel and the Grassmann manifolds just replacing everything by Riemannian counterparts. Riemannian optimization methods are performed along piecewise geodesic curve on the manifold, so updated points always stays on the manifold and guarantees the stability against the deviation from the manifold. For more information about this approach, see e.g. [4].

In this paper we introduce a new class of manifold: the *generalized flag manifold.* This new manifold naturally arises when we slightly relax the assumption of ICA and consider independent subspace analysis (ISA), allowing dependence between signals within to different subspaces. So far researchers in neural networks, signal processing have mainly concentrated on the Stiefel and Grassmann manifolds for optimization: the generalized flag manifold is a generalization of both the Stiefel and the Grassmann manifolds. We extend the Riemannian optimization method to this new manifold using our previous geodesic formula for the Stiefel manifold [9], and based on it propose a new algorithm for ISA. Finally, we present computer experiments comparing the ordinary gradient method for ISA with the new Riemannian gradient geodesic method based on the flag manifold.

2 Independent Subspace Analysis

Hyvärinen and Hoyer introduced independent subspace analysis (ISA) [7], by relaxing the usual statistically independent condition of each source signal in ICA. The source signal s is decomposed into d_i-tuples ($i = 1, \ldots, r$) where signals within a particular tuple are allowed to be dependent on each other, while signals belonging to different tuples are statistically independent. Since the ISA algorithm uses pre-whitening, we have $W^\top W = I_n$ as for normal ICA. However, because of the statistical dependence of signals within tuples, the manifold of candidate matrices for ISA is no longer simply the Stiefel manifold, but rather it is the Stiefel manifold with an additional symmetry.

Therefore solving ISA task can be regarded as an optimization on this new manifold, which is known as the *generalized flag manifold.*

3 What Is the Generalized Flag Manifold?

Let us introduce the generalized flag manifold. A *flag* in $\mathbb{R}^n$ is an increasing sequence of subspaces $V_1 \subset \cdots \subset V_r \subset \mathbb{R}^n$ of $\mathbb{R}^n$, $1 \leq r \leq n$. Given any sequence $(n_1, \ldots, n_r)$ of nonnegative integers with $n_1 + \cdots + n_r \leq n$, the *generalized flag manifold* $\mathrm{Fl}(n_1, n_2, \ldots, n_r)$ is defined as the set of all flags $(V_1, \ldots, V_r)$ of vector spaces with $V_1 \subset \cdots \subset V_r \subset \mathbb{R}^n$ and $\dim V_i = n_1 + \cdots + n_i, i = 1, 2, \ldots, r$. $\mathrm{Fl}(n_1, n_2, \cdots, n_r)$ is a smooth, connected, compact manifold. We can also consider a modified version of the generalized flag manifold, where we instead consider the set of the vector spaces V

$$V = V_1 \oplus V_2 \oplus \cdots \oplus V_r \subset \mathbb{R}^n, \tag{1}$$

where $\dim V_i = d_i,\ d_1 \leq d_2 \leq \cdots \leq d_r, \dim V = \sum_{i=1}^r d_i = p \leq n$. With the mapping $V_i \mapsto \bigoplus_{j=1}^i V_j$ we can see that the set of all V forms a manifold isomorphic to the original definition so this is also a generalized flag manifold, which we denote by $\mathrm{Fl}(n, d_1, d_2, \cdots, d_r)$. We represent a point on this manifold by a n by p orthogonal matrix W, i.e. $W^\top W = I_p$, which can be decomposed as

$$W = [W_1, W_2, \cdots, W_r],$$
$$W_i = (w_1^i, w_2^i, \cdots, w_{d_i}^i),$$

where $w_k^i \in \mathbb{R}^n$, $k = 1, \cdots, d_i$ for some i, form the orthogonal basis of V_i. Note that we are not so concerned with the individual frame vectors w_k^i themselves, rather with the subspace in $\mathbb{R}^n$ spanned by that set of vectors for some i. In other words any two matrices W_1, W_2 related by

$$W_2 = W_1 \begin{pmatrix} R_1 & & & \\ & R_2 & & \\ & & \ddots & \\ & & & R_r \end{pmatrix} \equiv W_1 \mathrm{diag}(R_1, R_2, \cdots, R_r) \tag{2}$$

where $R_i\ (1 \leq i \leq r) \in O(d_i)$, i.e. $R_i R_i^T = R_i^T R_i = I_{d_i}$, correspond to the same point on the manifold: we say we *identify* these two matrices. The generalized flag manifold is a generalization of both the Stiefel and the Grassmann manifolds. If all $d_i\ (1 \leq i \leq r) = 1$, it reduces to the Stiefel manifold, if $r = 1$, it reduces to the Grassmann manifold [3].

4 Geometry of the Flag Manifold

4.1 Tangent Space Structure

A tangent space of a manifold is an analogue of a tangent plane of a hypersurface in Euclidean space. For W to represent points in $\mathrm{Fl}(n, d_1, d_2, \cdots, d_r)$ we have

$$W^\top W = I_p \tag{3}$$

$$W\mathrm{diag}(R_1, R_2, \ldots, R_r), \quad R_i \in O(d_i), \quad 1 \le i \le r \quad \text{are identified.} \tag{4}$$

A tangent vector $V = [V_1, V_2, \ldots, V_r]$ of $\mathrm{Fl}(n, d_1, d_2, \ldots, d_r)$ at $W = [W_1, W_2, \ldots, W_r]$ must satisfy the equation obtained by differentiating (3)

$$W^\top V + V^\top W = O, \tag{5}$$

Now let us consider the following curve on the flag manifold passing through W.

$$W\mathrm{diag}(R_1(t), R_2(t), \ldots, R_r(t))$$

where $R_i(t) \in O(d_i)$, $R_i(0) = I_{d_i}$, $(1 \le i \le r)$. Since we neglect the effect of each rotation $R_i(t)$, V must be orthogonal to

$$W\mathrm{diag}(R_1'(0), R_2'(0), \ldots, R_r'(0)) = W\mathrm{diag}(X_1, \ldots, X_r),$$

where X_i $(1 \le i \le r)$ is a $d_i \times d_i$ skew symmetric matrix. Thus

$$0 = \mathrm{tr}\left\{\mathrm{diag}(X_1^\top, \ldots, X_r^\top) W^\top V\right\} = -\sum_{i=1}^{n} \mathrm{tr} X_i W_i^\top V_i \text{ for all } X_i$$

so we can show that $W_i^\top V_i$ is symmetric. Now the i-i block of (5) yields

$$W_i^\top V_i + V_i^\top W_i = O.$$

and therefore $W_i^\top V_i = O$, $(1 \le i \le r)$.

So to summarize, a tangent vector $V = [V_1, V_2, \ldots, V_r]$ of $\mathrm{Fl}(n, d_1, d_2, \ldots, d_r)$ at $W = [W_1, W_2, \ldots, W_r]$ is characterized by

$$W^\top V + V^\top W = O \quad \text{and} \quad W_i^\top V_i = O, \quad i = 1, \ldots, r. \tag{6}$$

4.2 Natural Gradient

In this section we derive the natural gradient of a function over the generalized flag manifold.

We use the following notations.

$$G = I - \frac{1}{2} W W^\top \quad \text{with} \quad G^{-1} = I + W W^T \tag{7}$$

$$X = \nabla_W f = \left(\frac{\partial f}{\partial w_{ij}}\right) = [X_1, \ldots, X_r] \tag{8}$$

$$Y = G^{-1} \nabla_W f = (I + W W^\top) \nabla_W f, \qquad Y_i = G^{-1} X_i \tag{9}$$

Now the natural gradient V of a function f on $\mathrm{Fl}(n, d_1, d_2, \ldots, d_r)$ is equal to the orthogonal projection of Y to $T_W\,\mathrm{Fl}(n, d_1, d_2, \ldots, d_r)$. In other words, $T_W\,\mathrm{Fl}(n, d_1, d_2, \ldots, d_r)$ is obtained through the minimization of $(V - Y)^t$

$G(V-Y)$ under the constraints (6). This can be solved by the Lagrange multiplier method. Let us introduce

$$L = \mathrm{tr}\{(V-Y)^\top G(V-Y)\} - \sum_i \mathrm{tr}(A_i^\top W_i^\top V_i) - \sum_i \sum_{j\neq i} \mathrm{tr}\{B_{ij}^\top (W_i^\top V_j + V_i^\top W_j)\}$$
$$= \mathrm{tr}\{(V-Y)^\top G(V-Y)\} - \sum_i \mathrm{tr}(A_i^\top W_i^\top V_i) - \sum_i \sum_{j\neq i} \mathrm{tr}(B_{ij}^\top W_i^\top V_j + B_{ji} W_i^\top V_j).$$

Differentiating L with respect to V_i and equating to zero leads to

$$\frac{\partial L}{\partial V_i} = 2G(V_i - Y_i) - W_i A_i - \sum_{j\neq i} W_j (B_{ij} + B_{ji}^\top) = O. \tag{10}$$

$$V_i = Y_i + \frac{G^{-1}}{2} W_i A_i + \frac{G^{-1}}{2} \sum_{j\neq i} W_j (B_{ij} + B_{ji}^\top) \tag{11}$$

Substituting $G^{-1} = I + WW^\top = I + \sum_i W_i W_i^\top$ into (11), we get

$$V_i = Y_i + W_i A_i + \sum_{j\neq i} W_j (B_{ij} + B_{ji}^\top).$$

Therefore, the condition of the tangent vector yields

$$W_i^\top V_i = W_i^\top Y_i + A_i = O \Leftrightarrow A_i = -W_i^\top Y_i, \tag{12}$$
$$W_i^\top V_j + V_i^\top W_j = W_i^\top Y_j + (B_{ji} + B_{ij}^\top) + Y_i^\top W_j + B_{ij}^\top + B_{ji} = O. \tag{13}$$

Thus,

$$B_{ij} + B_{ji}^\top = -\frac{1}{2}(W_j^\top Y_i + Y_j^\top W_i) \quad (i \neq j),$$

$$V_i = X_i - (W_i W_i^\top X_i + \sum_{j\neq i} W_j X_j^\top W_i). \tag{14}$$

It is easy to check that this formula includes the natural gradient formula for the Stiefel and the Grassmann manifolds [3] as special cases.

4.3 Geodesic of the Flag Manifold

We use our geodesic formula of the Stiefel manifold relative to the normal homogeneous metric for the Stiefel manifold: $G = I - \frac{1}{2}WW^\top$ obtained in [9].

$$\varphi_{\mathrm{St}(n,p)}(W, -\mathrm{grad}_W f, t) = \exp(-t(\nabla f(W) W^\top - W \nabla f(W)^\top)) W \tag{15}$$
$$\varphi_{\mathrm{St}(n,p)}(W, V, t) = \exp(t(DW^\top - WD^\top)) W, \tag{16}$$

where $D = (I - \frac{1}{2}WW^\top)V$.

We decompose the tangent space of $\mathrm{St}(n,p)$ at W into the vertical space V_W and the horizontal space H_W with respect to G. $T_W\,\mathrm{St}(n,p) = H_W \oplus V_W$. The vertical space does not depend on the metric; it is determined from the quotient space structure of $T_W\,\mathrm{Fl}(n,d_1,d_2,\ldots,d_r)$:

$$V_W = W\mathrm{diag}(X_1^\top,\ldots,X_r^\top), \text{where } X_i \in TO(d_i) \text{ (skew symmetric matrices)}.$$

We need to lift a tangent vector V at $T_W\,\mathrm{Fl}(n,d_1,d_2,\ldots,d_r)$ to $\tilde{V} \in H_W$. It turns out that $\tilde{V} = V$, because

$$g_W^{\mathrm{St}(n,p)}(X,V) = \mathrm{tr}(X^\top GV) \tag{17}$$

$$= -\mathrm{tr}\{\mathrm{diag}(X_1^\top,\ldots,X_r^\top)W^\top(I - \frac{1}{2}WW^\top)V\} \tag{18}$$

$$= -\frac{1}{2}\mathrm{tr}\{(\mathrm{diag}(X_1^\top,\ldots,X_r^\top))\,W^\top V\} = 0, \text{ for all } X \in V_W. \tag{19}$$

Because the projection $\pi : \mathrm{St}(n,p) \to \mathrm{Fl}(n,d_1,\ldots,d_r)$ $(\pi(W) = W)$ is a Riemannian submersion, the following theorem guarantees that the geodesic on the Stiefel manifold starting from W in direction V yields the geodesic on the flag manifold emanating from W with the initial velocity V. Both geodesics are based on the same Riemannian metric G.

Let $p : \tilde{M} \to M$ be a Riemannian submersion, $\tilde{c}(t)$ be a geodesic of $(\tilde{M}, g^{\tilde{M}})$. If the vector $\tilde{c}'(0)$ is horizontal, then $\tilde{c}'(t)$ is horizontal for any t, and the curve $p(\tilde{c}(t))$ is a geodesic of (M,g) of the same length as $\tilde{c}(t)$ [5, 9].

5 Application to Independent Subspace Analysis

In order to validate the effectiveness of the proposed algorithm, we performed independent subspace analysis experiments of natural image data [7]. In this experiment, we attempt to decompose a gray-scale image $I(x,y)$ into linear combination of basis images $a_i(x,y)$ as

$$I(x,y) = \sum_{i=1}^{n} s_i a_i(x,y), \tag{20}$$

where s_i is a coefficient. Let the inverse filter of this model be

$$s_i = \langle w_i, I\rangle = \sum_{x,y} w_i(x,y)I(x,y). \tag{21}$$

The problem is to estimate s_i (or equivalently $w_i(x,y)$) when a set of images are given. For this purpose, we apply the independent subspace criterion: proposed by Hyvärinen et al. [7] Components are partitioned into disjoint subspaces $S_1,\ldots,S_r$, and s_i and s_j are statistically independent if i and j belong to different subspaces. As a cost function, we take a negative log-likelihood:

$$f(\{w_i\}) = -\sum_{k=1}^{K} \log L(I_k;\{w_i\}) = -\sum_{k=1}^{K}\sum_{j=1}^{r} \log p\left(\sum_{i\in S_j}\langle w_i, I_k\rangle^2\right) \tag{22}$$

where the suffix k denotes the index of samples and p is the exponential distribution $p(x) = \alpha \exp(-\alpha x)$, where the parameter α does not appear in learning rule and hence can be ignored. The cost function is invariant under rotation within the subspace.

We applied the above model to small image patches of size 16×16 pixels. We prepared 10000 image patches at random locations extracted from monochrome photographs of natural images. (The dataset and ISA code is distributed by Hyvärinen http://www.cis.hut.fi/projects/ica/data/images).

In the preprocessing phase, the mean gray-scale value of each image patch was subtracted, and then the dimension of the image was reduced from 256 to 160 by principal component analysis ($n = 160$), and finally the data were whitened. Independent subspace analysis was performed for this dataset, in which each subspace is 4-dimensional ($d_i = 4$), and accordingly the 160 dimensional subspaces were separated into 40 subspaces ($r = 40$).

We compared two methods to extract independent subspaces from the dataset. One is the ordinary gradient method used in [7] ($\eta = 1$) and the other is the proposed method based on geodesic of the flag manifold ($\eta = 1.1$). The best learning constant was chosen for each algorithm. In the ordinary gradient method, the extraction matrix was projected to the orthogonal group by singular value decomposition in each step. Both algorithms were implemented in MATLAB on 3.8GHz, 2.00GB machine.

The behavior of the cost function along iterations is shown in Fig. 1(a). In early stages of learning, the cost of the geodesic approach decreased much faster than the ordinary gradient method. The average run time per iteration was 9.06 seconds for the ordinary gradient method; 8.95 seconds for the geodesic method. The recovered inverse filters $w_i(x, y)$ are shown in Fig. 1(b). The filters were clearly separated into groups. We found no significant difference between the points of convergence of the two methods, and neither method appeared to get 'stuck' in a local optimum.

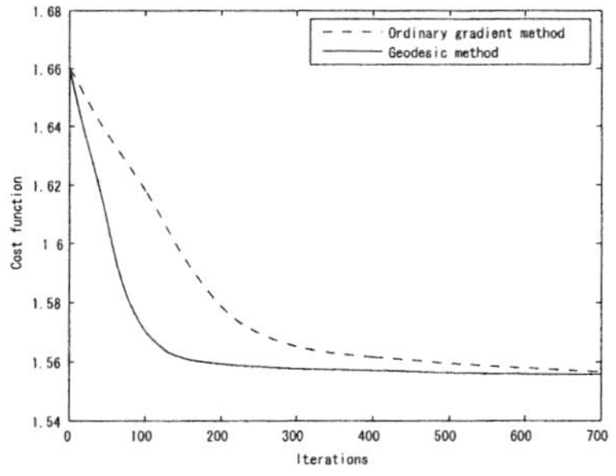

(a)

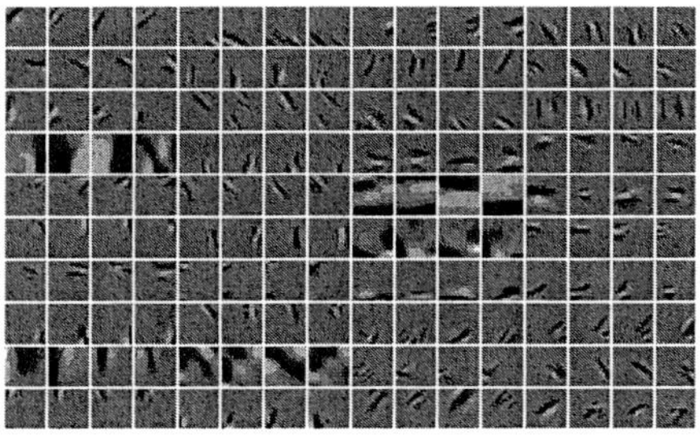

(b)

Fig. 1. Results, showing (a) learning curves; (b) recovered inverse filters

6 Conclusion

We have introduced a new manifold, the generalized flag manifold, for solving the independent subspace analysis (ISA) problem, and have developed a Riemannian gradient descent geodesic method on the generalized flag manifold. Computer experiments confirm that our algorithm gives good performance compared with the ordinary gradient descent method.

While we have concentrated on the gradient descent method in this paper, conjugated gradient and the Newton methods could also be used for searching over manifold using geodesics. Also, while we used orthogonal matrices to represent points on the flag manifold, the algorithm could be described using non-orthogonal matrices, as Absil et al have done for the Grassmannian [1].

Acknowledgements

This work is supported in part by JSPS Grant-in-Aid for Exploratory Research 16650050, MEXT Grant-in-Aid for Scientific Research on Priority Areas 17022033, and EPSRC grants GR/S75802/01, GR/S82213/01, EP/D000246/1 and EP/C005554/1.

References

1. P-A. Absil, R. Mahony, and R. Sepulchre, Riemannian geometry of Grassmann manifolds with a view on algorithmic computation, *Acta Applicandae Mathematicae*, **80**(2), pp.199-220, 2004.
2. S. Amari, Natural gradient works efficiently in Learning, *Neural Computation*, **10**, pp.251-276, 1998.
3. A. Edelman, T.A. Arias, and S.T. Smith, The Geometry of algorithms with orthogonality constraints, *SIAM Journal on Matrix Analysis and Applications*, **20**(2), pp.303-353, 1998.
4. S. Fiori, Quasi-Geodesic Neural Learning Algorithms over the Orthogonal Group: A Tutorial, *Journal of Machine Learning Research*, **6**, pp.743-781, 2005.
5. S. Gallot, D. Hulin, and J. Lafontaine, *Riemannian Geometry*, Springer, 1990.
6. U. Helmke, J. B. Moore, *Optimization and dynamical systems*, Springer, 1994.
7. A. Hyvärinen and P.O. Hoyer, Emergence of phase and shift invariant features by decomposition of natural images into independent feature subspaces. *Neural Computation*, **12**(7), pp.1705-1720, 2000.
8. Y. Nishimori, Learning Algorithm for Independent Component Analysis by Geodesic Flows on Orthogonal Group, *Proceedings of International Joint Conference on Neural Networks (IJCNN1999)*, **2**, pp.1625-1647, 1999.
9. Y. Nishimori and S. Akaho, Learning Algorithms Utilizing Quasi-Geodesic Flows on the Stiefel Manifold, *Neurocomputing*, **67** pp.106-135, 2005.
10. M. D. Plumbley, Algorithms for non-negative independent component analysis. *IEEE Transactions on Neural Networks*, **14**(3), pp.534-543, 2003.

A Two-Stage Based Approach for Extracting Periodic Signals

Zhi-Lin Zhang[1,2] and Liqing Zhang[1]

[1] Department of Computer Science,
Shanghai Jiao Tong University, Shanghai 200030, China
[2] School of Computer Science and Engineering,
University of Electronic Science and Technology of China,
Chengdu 610054, China
zlzhang@uestc.edu.cn, zhang-lq@cs.sjtu.edu.cn

Abstract. In many applications, such as biomedical engineering, it is often required to obtain specific periodic source signals. In this paper, we propose a two-stage based approach for extracting periodic signals. At the first stage, the autocorrelation property of the desired source signal is exploited to roughly extract the desired source signal. At the second stage, the extracted signal is further processed as cleanly as possible, based on the higher-order statistics. Simulations on artificially generated data and real-world ECG data have showed its better performance, compared with many existing extraction algorithms.

1 Introduction

Blind source extraction (BSE) [1] is a powerful technique that is closely related to blind source separation (BSS) [2]. The basic task of BSE is to estimate some components of source signals that are linearly combined in observations.

Compared with BSS, BSE has many advantages [1]. One advantage is that it can extract only the "interesting" signals from noisy mixed signals by exploiting their desired properties. That is to say, it requires certain additional *a priori* information of the desired source signal. Thus it generally is implemented in a semi-blind way. In many applications [3–6], such as the fetal ECG extraction [7, 8], the desired source signal is periodic or quasi-periodic. Therefore the period property can be used to extract the desired source signals.

Barros and Cichocki [3] first proposed an algorithm that can quickly extract the desired source signal with a specific period. But the algorithm's performance strongly depends on the precise estimation of an optimal time delay. In addition, the literature did not provide methods to find the optimal time delay, and only used the fundamental or multiple period of the desired source signal as the optimal time delay. In fact, the fundamental or multiple period is not necessarily the optimal time delay [7].

To overcome the drawbacks, several approaches recently have been proposed. Jafari *et al.* [6] proposed a fast algorithm that can instantaneously extract all the periodic source signals from the mixtures. It only needs to know the period

J. Rosca et al. (Eds.): ICA 2006, LNCS 3889, pp. 303–310, 2006.

of one of the source signals to extract, without knowing the optimal time delay. Another advantage is its tolerance of large estimate errors of the period. But its performance degrades if the period is not small enough, or if the periods of the source signals are close to each other.

On the other hand, we proposed an extraction algorithm [8], whose performance is not affected by the value of the period of the desired source signal. But the algorithm also needs to know the optimal time delay. In practice, it also takes the fundamental period of the desired source signal as the optimal time delay. However, compared with the one in [3], the algorithm can achieve better extraction performance due to exploitation of higher-order statistics information, and to some extent it is tolerant of estimate errors of the period. But the algorithm may fail in some complicated situations (see simulations in Section 4).

Recently, Lu *et al.* [9] proposed the so-called constrained ICA algorithm, which needs to elaborately design a reference signal that is closely related to the desired source signal. However, to design such reference signal, one should obtain lots of *a priori* information, which is not available in many cases.

In this paper, we propose an extraction algorithm, which only needs to estimate the period of the desired signal, and it is non-sensitive to the estimate errors of the period. Simulations on the artificially generated data and real-world data have showed its validity and good performance.

2 Problem Statement

Suppose one observes an n-dimentional stochastic signal vector $\mathbf{x}$ that is regarded as the linear transformation of an m-dimensional *mutually independent* zero-mean and unit-variance source vector $\mathbf{s}$, i.e., $\mathbf{x} = \mathbf{A}\mathbf{s}$, where $\mathbf{A}$ is an unknown mixing matrix. The goal of source extraction is to find a vector $\mathbf{w}$ such that $y = \mathbf{w}^T\mathbf{x} = \mathbf{w}^T\mathbf{A}\mathbf{s}$ is an estimated source signal up to a scalar. To cope with ill-conditioned cases and to make algorithms simpler and faster, whitening is often used to transform the observed signals $\mathbf{x}$ to $\tilde{\mathbf{x}} = \mathbf{V}\mathbf{x}$ such that $E\{\tilde{\mathbf{x}}\tilde{\mathbf{x}}^T\} = \mathbf{I}$, where $\mathbf{V}$ is a whitening matrix. For convenience, in the following discussion we assume that $\mathbf{x}$ are the whitened observed signals and $n = m$.

Since our goal is to extract the periodic source signal, we further assume the desired periodic signal s_i satisfies the following relations for a specific integer τ^*:

$$\begin{cases} E\left\{s_i(k)s_i(k-\tau^*)\right\} > 0 \\ E\left\{s_j(k)s_j(k-\tau^*)\right\} = 0 \quad \forall j \neq i \end{cases} \tag{1}$$

where s_j are other source signals, k is the time index, and τ^* is the so-called optimal time delay [3].

Ideally, under the constraint $\|\mathbf{w}\| = 1$, maximizing the objective function

$$J(\mathbf{w}) = E\{y(k)y(k-\tau^*)\} = \mathbf{w}^T E\{\mathbf{x}(k)\mathbf{x}(k-\tau^*)^T\}\mathbf{w} \tag{2}$$

leads to perfect recovery of the desired periodic source signal s_i. The reason for this formulation is that for the desired signal s_i, this delayed autocorrelation has a large positive value, while for other source signals this value is zero.

Using the standard gradient approach and neglecting the small difference between $R_x(\tau^*)$ and $R_x(\tau^*)^T$, from the objective function (2) we can derive the Barros's algorithm [3]:

$$\begin{cases} \mathbf{w}^+ = R_x(\tau^*)\mathbf{w} \\ \mathbf{w} = \mathbf{w}^+/\|\mathbf{w}^+\| \end{cases} \tag{3}$$

where $R_x(\tau^*) = E\{\mathbf{x}(k)\mathbf{x}(k-\tau^*)^T\}$. Note here the Barros's algorithm is derived from a different aspect. The algorithm is simple and fast. However, some important practical issues should be considered.

One important issue is the optimal time delay τ^*. In most cases such optimal time delay does not exist. In other words, although the desired signal s_i is strongly autocorrelated at the time delay τ^*, several other source signals may be also autocorrelated at the delay τ^* (i.e., $E\{s_j(k)s_j(k-\tau^*)\} \neq 0, j \neq i$).

Another issue is the effect of finite samples [7, 10]. Even if the source signals are strictly mutually uncorrelated, in practice the calculated correlations of source signals using limited samples are generally non-zero, due to the fact that the expectation operator is replaced by the mathematical average. That is to say, even if $E\{s_i(k)s_j(k-\tau^*)\} = 0$ and $E\{s_j(k)s_j(k-\tau^*)\} = 0, j \neq i$, it is very possible that $\sum_{k=\tau^*}^{N-1} s_i(k)s_j(k-\tau^*)/(N-\tau^*) \neq 0$ and $\sum_{k=\tau^*}^{N-1} s_j(k)s_j(k-\tau^*)/(N-\tau^*) \neq 0$.

According to the results in [7], the performance of the Barros's algorithm (3) greatly degrades due to the joint effect of the above two issues.

The third issue is the exploitation of the higher-order statistics. In many applications the source signals are physically mutually independent. Therefore, suitable use of the higher-order statistics, rather than only using the second-order statistics, is expected to improve the extraction performance.

Considering the above issues, in the next section we will propose an efficient two-stage algorithm, which achieves better performance than many existing algorithms.

3 Proposed Algorithm

3.1 Framework of the Proposed Algorithm

First, we estimate the fundamental period τ of the desired source signal. For estimating τ, there are many methods, such as the autocorrelation method [3], and the instantaneous frequency estimation technique [4]. In addition, in some applications, e.g., biomedical signal processing, this type of information is often readily available [3, 5, 9]. Note that τ is not necessarily the optimal time delay.

The following procedure is roughly divided into two stages. The first stage is called the capture stage. In this stage, the algorithm coarsely extracts the desired source signal by using the estimated period. After the algorithm converges, we obtain the weight vector $\hat{\mathbf{w}}$. But due to some reasons (discussed below), $\hat{\mathbf{w}}$ is only close to the ideally optimal weight vector $\mathbf{w}_*$ (in the sense that the extracted desired source signal $y_* = \mathbf{w}_*^T\mathbf{x}$ is not mixed by any crosstalk noise). Therefore the captured source signal $\hat{y} = \hat{\mathbf{w}}^T\mathbf{x}$ is mixed by some noise and interference.

Next, in the second stage, we run the fixed-point algorithm [1, 11] on the original mixtures, using $\hat{\mathbf{w}}$ as its initial weight vector. The initial weight vector $\hat{\mathbf{w}}$ can ensure the fixed-point algorithm converges to the sub-optimal solution $\bar{\mathbf{w}}$, which is much closer to $\mathbf{w}_*$ than $\hat{\mathbf{w}}$ is. Then we finally obtain the estimated desired source signal $\bar{y} = \bar{\mathbf{w}}^T\mathbf{x}$, which is almost not mixed by any interference.

3.2 The First Stage: Coarse Capture

In the first stage, the goal is to roughly extract the desired source signal, exploiting its autocorrelation structure. First, consider the objective function (2):

$$\begin{aligned} J(\mathbf{w}) &= \frac{1}{2}J(\mathbf{w}) + \frac{1}{2}J(\mathbf{w})^T \\ &= \frac{1}{2}\mathbf{w}^T E\{\mathbf{x}(k)\mathbf{x}(k-\tau)^T\}\mathbf{w} + \frac{1}{2}\mathbf{w}^T E\{\mathbf{x}(k-\tau)\mathbf{x}(k)^T\}\mathbf{w} \\ &= \frac{1}{2}\mathbf{w}^T\Big(\mathbf{R}_x(\tau) + \mathbf{R}_x(\tau)^T\Big)\mathbf{w} \\ &= \frac{1}{2}\mathbf{w}^T\overline{\mathbf{R}}\mathbf{w}, \end{aligned} \tag{4}$$

which implies that under the constraint $\|\mathbf{w}\| = 1$ maximizing (2) is equivalent to finding the eigenvector corresponding to the maximal eigenvalue of the real symmetric matrix $\overline{\mathbf{R}}$. Thus we directly obtain the following algorithm:

$$\mathbf{w} = EIG(\overline{\mathbf{R}}), \tag{5}$$

where $EIG(\overline{\mathbf{R}})$ is the operator that calculates the normalized eigenvector corresponding to the maximal eigenvalue of the real symmetric matrix $\overline{\mathbf{R}}$. After convergence, the algorithm gives the solution $\hat{\mathbf{w}}$.

If τ is the optimal time delay, then the solution $\hat{\mathbf{w}}$ is just the optimal solution $\mathbf{w}_*$, and the captured signal $\hat{y} = \hat{\mathbf{w}}^T\mathbf{x}$ is just the desired source signal, without any noise or distortion. However, the optimal time delay generally does not exist, then the optimal solution $\mathbf{w}_*$ is ideal. Therefore, in fact the solution $\hat{\mathbf{w}}$ is only near $\mathbf{w}_*$, and the captured desired source signal is noisy.

Note that if $\hat{\mathbf{w}}$ is not close enough to $\mathbf{w}_*$, the algorithms in the second stage may converge to other local maxima, and thus cannot obtain the desired source signal. Therefore we should make $\hat{\mathbf{w}}$ approximate $\mathbf{w}_*$ as closely as possible. In [7] we have showed that the larger the autocorrelations of undesired source signals at the delay τ are, and/or the larger the absolute value of the cross-correlation between any two source signals at the delay τ is, the farther $\hat{\mathbf{w}}$ deviates from $\mathbf{w}_*$. To ensure $\hat{\mathbf{w}}$ is close enough to $\mathbf{w}_*$, we modify the former objective function (2) as follows:

$$J(\mathbf{w}) = \mathbf{w}^T\Big\{\sum_{l=1}^{P}\Big(\mathbf{R}_x(l\tau) + \mathbf{R}_x(l\tau)^T\Big)\Big\}\mathbf{w}, \tag{6}$$

and its corresponding algorithm is given by

$$\mathbf{w} = EIG\Big(\sum_{l=1}^{P}\Big(\mathbf{R}_x(l\tau) + \mathbf{R}_x(l\tau)^T\Big)\Big), \tag{7}$$

where P is a positive integer, and τ is the fundamental period of the desired source signal. The algorithm is based on the averaged eigen-structure of correlation matrices of source signals over multi-delays. In [7] we have showed that both the averaged auto-correlations of undesired source signals and the averaged absolute value of cross-correlation between any two source signals at the delay τ tend to zero with P increasing. Thus the converged solution $\hat{\mathbf{w}}$ is closer to $\mathbf{w}_*$, ensuring the successful fine extraction of the second stage.

3.3 The Second Stage: Fine Extraction

In this stage we use the one-unit fixed-point algorithm [11] to make the solution $\hat{\mathbf{w}}$ from the previous stage further close to the optimal solution $\mathbf{w}_*$, which implies the extracted source signal is cleaner, with less noise and interference. $\hat{\mathbf{w}}$ is taken as the initial weight value of the fixed-point algorithm:

$$\begin{cases} \mathbf{w}^+ = E\{\mathbf{x}(\mathbf{w}^T\mathbf{x})^3\} - 3\mathbf{w} \\ \mathbf{w} = \mathbf{w}^+/\|\mathbf{w}^+\|. \end{cases} \tag{8}$$

To improve the robustness to outliers and spiky noise, we can adopt the modified fixed-point algorithm [1]:

$$\begin{cases} \mathbf{w}^+ = \frac{E\{y^3\mathbf{x}\}}{E\{y^4\}} - \mathbf{w} \\ \mathbf{w} = \mathbf{w}^+/\|\mathbf{w}^+\|. \end{cases} \tag{9}$$

When the algorithm (8) or (9) has converged, we obtain the solution $\bar{\mathbf{w}}$, which is much closer to the optimal solution $\mathbf{w}_*$ than $\hat{\mathbf{w}}$ is. Therefore, we finally get the estimated desired source signal $\bar{y} = \bar{\mathbf{w}}^T\mathbf{x}$.

4 Simulations

In the first simulation, we generated five zero-mean and unit-variance source signals, shown in Fig.1(a). Each signal has 3000 samples. Three signals are periodic (or quasi-periodic), respectively given by $s_1(k) = \sin(2\pi f_1 k + 6\cos(2\pi 200k))$, $s_2(k) = \cos(2\pi f_2 k)$, and $s_3(k) = \cos(2\pi f_3 k+2)$, where $t_s = 1\times 10^{-4}$, $f_1 = 0.061$, $f_2 = 0.054$, and $f_3 = 0.028$. t_s is the sampling period, and $f_i(i = 1, 2, 3)$ are normalized frequencies. The other two signals are Gaussian noise. Note that s_1 is the desired source signal, and its fundamental period is assumed to be known.

The source signals were randomly mixed (Fig.1(b)). After whitening the mixed signals, we ran the Barros's algorithm [3], the algorithm in [8], the one in [7] and the proposed one in this paper. The results are shown in Fig.1(c), from which it is clear to see that the algorithms in [3] and the one in [8] extracted the wrong source signal, while the other two algorithms correctly extracted the desired source signals. To evaluate the extraction performance of the two algorithms that obtained the correct signal, we adopted the following measure:

$$PI = -10E\{lg(s(k) - \tilde{s}(k))^2\}, \quad (dB) \tag{10}$$

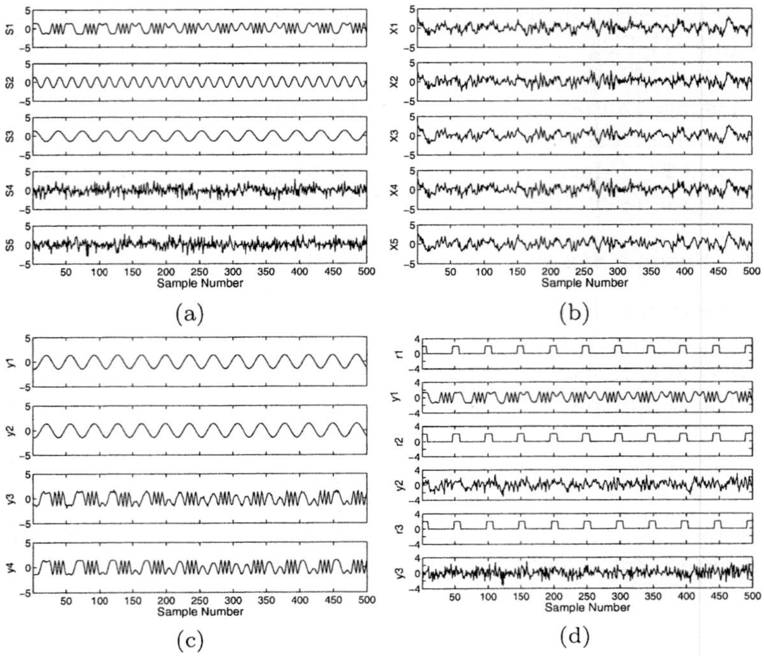

Fig. 1. Simulation on artificially generated data. (a) The five source signals. (b) The randomly mixed signals. (c) The extracted signals, respectively by the algorithm in [3] (y_1), the one in [8] (y_2), the one in [7] (y_3) and the proposed one in this paper (y_4). (d) The reference signals (r_1, r_2, r_3) and the corresponding extracted signals (y_1, y_2, y_3) by the constrained ICA [9].

where $s(k)$ was the desired source signal, and $\tilde{s}(k)$ was the extracted signal (both of them were normalized to be zero-mean and unit-variance). The higher PI was, the better the performance was. The averaged performance over 400 independent running of the algorithm in [7] and of the proposed algorithm was, respectively, 23.4 dB and 48.9 dB. The results showed the proposed algorithm has better extraction performance.

Next, we used the constrained ICA [9] to extract the desired source signal. Fig.1(d) shows the results, where r_1, r_2, r_3 are the very similar reference signals, having the same fundamental period. The rectangular pulse width of r_2 is larger than that of r_1 by only a sampling period, while keeping the same pulse occurrence time. r_3 has the same rectangular pulse width as r_1, but is delayed by a sampling period. y_1, y_2 and y_3 are the extracted signals, respectively by using the reference signal r_1, r_2 and r_3. Clearly, only y_1 was well recovered. This means that the algorithm's performance is greatly affected by the reference signal. To achieve good performance, the elaborately designed reference signal is necessary, which cannot be obtained in many cases. In addition, the selection of some parameters of the algorithm is crucial to the algorithm.

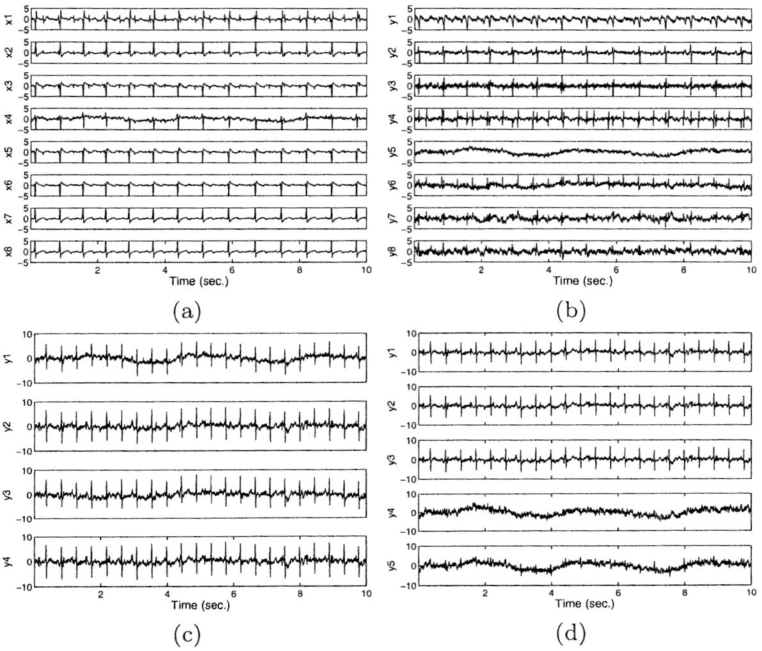

Fig. 2. Simulation on real-world ECG data. (a) The ECG data. (b) The separated signals by [6]. (c) The extracted signals, respectively by the algorithm in [3], the one in [8], the one in [7] and the proposed one by this paper, from the top down. (d)The extracted signals when the estimate errors were introduced. y_1-y_3 were extracted by the proposed algorithm, while y_4 and y_5 by the Barros's algorithm [3].

Next we used the real-world ECG data [12] (Fig.2(a)). Our goal was to extract the fetal ECG, which was very weak and almost only visible in x_1. Using the method in [8] we estimated that the period of the fetal ECG was 112 sampling periods. After whitening the sensor signals, we respectively ran the algorithm in [6], the one in [3], the one in [8], the one in [7] and the proposed one. The results are shown in Fig.2(b) and (c). We can see that the algorithm in [6] had poor performance. The reason is that the period of the fetal ECG was not small, violating the basic assumption of the algorithm. In addition, the extracted fetal ECG by the Barros's algorithm [3] also showed the poor performance.

In practice, the estimate errors of the period of the desired source signal are inevitable. Suppose the estimated period of the Fetal ECG deviates from its true value. The extraction results are shown in Fig.2(d), where y_1, y_2, y_3 were the corresponding extracted signals by the proposed algorithm when the estimated period was 108, 114, and 118 sampling periods, respectively. y_4, y_5 were the extracted signals by the Barros's algorithm [3] when the estimated period was 110 and 113 sampling periods, respectively. Clearly, the proposed algorithm is not sensitive to the estimate errors.

5 Conclusions

In this paper we present a two-stage algorithm for extracting periodic signals. It converges quickly, due to the use of efficient eigenvalue decomposition methods at the first stage and the fast fixed-point algorithm in the second stage. And it has good extraction performance. Furthermore, the algorithm is tolerant of estimate errors of the period of the desired source signal. In addition, the algorithm does not need to design the so-called reference signals.

Acknowledgements

The work was supported by the National Basic Research Program of China (Grant No. 2005CB724301) and National Natural Science Foundation of China (Grant No.60375015).

References

1. Cichocki, A., Amari, S.: Adaptive Blind Signal and Image Processing: Learning Algorithms and Applications. John Wiley & Sons, New York (2002)
2. Cruces-Alvarez, S.A., Cichocki, A., Amari, S.: From Blind Signal Extraction to Blind Instantaneous Signal Separation: Criteria, Algorithm, and Stability. IEEE Trans. Neural Networks **15 (4)** (2004) 859-873
3. Barros, A.K., Cichocki, A.: Extraction of Specific Signals with Temporal Structure. Neural Computation **13 (9)** (2001) 1995-2003
4. Barros, A.K., Ohnishi, N.: Heart Instantaneous Frequency (HIF): An Alternative Approach to Extract Heart Rate Variability. IEEE Trans. Biomedical Engineering **48 (8)** (2001) 850-855
5. Hansen, L.K., Nielsen, F.Å., Larsen, J.: Exploring FMRI Data for Periodic Signal Components. Artificial Intelligence in Medicine **25 (1)** (2002) 35-44
6. Jafari, M.G., *et al.*: Sequential Blind Source Separation Based Exclusively on Second Order Statistics Developed for a Class of Periodic Signals. IEEE Trans. Signal Processing (in press)
7. Zhang, Z.-L., Yi, Z.: Robust Extraction of Specific Signals with Temporal Structure. Neurocomputing (in press)
8. Zhang, Z.-L., Yi, Z.: Extraction of Temporally Correlated Sources with Its Application to Non-invasive Fetal Electrocardiogram Extraction. Neurocomputing (in press)
9. Lu, W., Rajapakse, J.C.: Approach and Applications of Constrained ICA. IEEE Trans. Neural Networks **16 (1)** (2005) 203-212
10. Bermejo, S.: Finite Sample Effects in Higher Order Statistics Contrast Functions for Sequential Blind Source Separation. IEEE Signal Processing Letters **12 (6)** (2005) 481-484
11. Hyvärinen, A., Oja, E.: A Fast Fixed-Point Algorithm for Independent Component Analysis. Neural Computation **9 (7)** (1997) 1483-1492
12. De Moor, D.(eds.): Daisy: Database for the Identification of Systems. Available online at: http://www.esat.kuleuven.ac.be/sista/daisy

ICA by PCA Approach: Relating Higher-Order Statistics to Second-Order Moments

Kun Zhang and Lai-Wan Chan*

Department of Computer Science and Engineering,
The Chinese University of Hong Kong, Shatin, Hong Kong
{kzhang, lwchan}@cse.cuhk.edu.hk

Abstract. It is well known that principal component analysis (PCA) only considers the second-order statistics and that independent component analysis (ICA) exploits higher-order statistics of the data. In this paper, for whitened data, we give an elegant way to incorporate higher-order statistics implicitly in the form of second-order moments, and show that ICA can be performed by PCA following a simple transformation. This method is termed P-ICA. Kurtosis-based P-ICA is equivalent to the fourth-order blind identification (FOBI) algorithm [2]. Analysis of the transformation form enables us to give the robust version of P-ICA, which exploits the trade-off of all even order statistics of sources. Experimental comparisons of P-ICA with the prevailing ICA methods are presented. The main advantage of P-ICA is that it enables any PCA system, especially the dedicated hardware, to perform ICA after slight modification.

1 Introduction

In independent component analysis (ICA), we have an observable random vector $\mathbf{x} = [x_1, x_2, ..., x_n]^T$. Variables x_i are assumed to be linear mixtures of some mutually statistically independent variables s_i, which form the random vector $\mathbf{s} = [s_1, s_2, ..., s_n]^T$ (here we assume that the number of observations is equal to that of sources, and that s_i are zero-mean). Mathematically, $\mathbf{x}$ is generated according to $\mathbf{x} = \mathbf{A}\mathbf{s}$. By using some linear transform, the task of ICA is to find the random vector $\mathbf{y} = [y_1, y_2, ..., y_n]^T$:

$$\mathbf{y} = \mathbf{W}\mathbf{x} \tag{1}$$

whose components are as mutually independent as possible, such that it provides an estimate of $\mathbf{s}$.

Since they involve no iterative optimization and the computation is extremely simple, closed-form solutions to ICA were developed by using high-order cumulants [4, 6, 7, 10] or the derivatives of mixtures [9]. However, the applicability of these results is limited since they only apply for the two-source case. Generally

* This work was partially supported by a grant from the Research rants Council of the Hong Kong Special Administration Region, China.

J. Rosca et al. (Eds.): ICA 2006, LNCS 3889, pp. 311–318, 2006.

speaking, ICA does not have closed-form solutions, and an ordinary ICA algorithm consists of two parts—an objective function (contrast function) and an optimization method used to optimize it.

In this paper, by investigating the relationship between ICA and PCA, we present a way to solve the ICA problem by explicitly applying principal component analysis (PCA). We consider the case where sources have different distributions. It is shown that for whitened data, with a tricky transformation, the linear transformation for PCA of the transformed data is exactly that for ICA of the original data. Although this method, namely P-ICA, may not give optimal results, it is very simple and efficient, as PCA can be easily done by eigenvalue decomposition (EVD) of the data covariance matrix or singular value decomposition (SVD) of the data. In addition, it just explicitly involves second-order statistics and has no local optima. As a consequence, second-order statistics-based systems, especially the hardware ones, can also perform ICA with P-ICA.

2 PCA and ICA

Given the data $\mathbf{x}$, PCA and ICA both aim to find the linear transformation given in Eq. 1. However, they are based on different criteria and exploit data information of different aspects. PCA aims at finding an orthogonal transformation $\mathbf{W}$ which gives uncorrelated outputs, or equivalently, in PCA, each principal component defines a projection that encapsules the maximum amount of variation in the data and is uncorrelated to the previous principal components. It only considers the second order statistical characteristics of the data. In other words, PCA just uses the joint Gaussian distribution to fit the data and finds an orthogonal transformation which makes the joint Gaussian distribution factorable, regardless of the true distribution of the data. While in ICA, we find the linear transformation which makes the true joint distribution of the transformed data factorable, such that the outputs are mutually independent.

Statistically speaking, mutual independence is much stronger than uncorrelatedness between the variables, i.e. independence guarantees uncorrelatedness. In ICA, the independent components y_i are uncorrelated, at least approximately, since they are as independent as possible.Some ICA algorithms, such as FastICA [8] and JADE [3], require whitening the observed data $\mathbf{x}$ as a preprocessing step. The method proposed in this paper also requires this step.

Whitening of the data $\mathbf{x}$ can be performed by PCA or EVD of the covariance matrix of $\mathbf{x}$. Denote the covariance matrix of $\mathbf{x}$ by $\mathbf{\Sigma_x}$, i.e. $\mathbf{\Sigma_x} = E\{\mathbf{x}\mathbf{x}^T\}$. Let the EVD of $\mathbf{\Sigma_x}$ be $\mathbf{\Sigma_x} = \mathbf{E}\mathbf{D}\mathbf{E}^T$, where $\mathbf{E}$ is the orthogonal matrix consisting of eigenvectors of $\mathbf{\Sigma_x}$ as its columns and $\mathbf{D}$ the diagonal matrix of the corresponding eigenvalues. Whitening of $\mathbf{x}$ can be done using the matrix $\mathbf{V} = \mathbf{E}\mathbf{D}^{-1/2}\mathbf{E}^T$. Denote the whitened data by $\mathbf{v}$, i.e.

$$\mathbf{v} = \mathbf{V}\mathbf{x} \tag{2}$$

After whitening, we just need to find an orthogonal transformation $\mathbf{U}$ to make components of $\mathbf{y} = \mathbf{Uv}$ mutually independent. The transformation matrix $\mathbf{W}$ in Eq. 1 can then be constructed as

$$\mathbf{W} = \mathbf{UV} \tag{3}$$

3 ICA by PCA

Without loss of generality, we assume the sources s_i are zero-mean and of unit variance, i.e. $E\{s_i\} = 0$ and $E\{s_i^2\} = 1$. Let $\mathbf{B} = \mathbf{VA}$. One can easily see that $\mathbf{B}$ is orthogonal. Now let us consider the case where s_i have different kurtosis.

3.1 Based on Kurtosis

Under the condition that $s_1, ..., s_n$ have different kurtosis, we have the following theorem.

Theorem 1. *Let* $\mathbf{s}$, $\mathbf{v}$, *and* $\mathbf{z}$ *be random vectors such that* $\mathbf{v} = \mathbf{Bs}$, *where* B *is an orthogonal matrix, and*

$$\mathbf{z} = ||\mathbf{v}|| \cdot \mathbf{v} \tag{4}$$

Suppose additionally $\mathbf{s}$ *has zero-mean independent components and these components have different kurtosis. Then the orthogonal matrix* $\mathbf{U}$ *which gives principal components of* $\mathbf{z}$ *(without centering) performs ICA on* $\mathbf{v}$.

Proof. Let $\mathbf{s} = [s_1, ..., s_n]^T$, $\mathbf{U} = [\mathbf{u}_1, ..., \mathbf{u}_n]$, and $\mathbf{C} = [\mathbf{c}_1, ..., \mathbf{c}_n] = [c_{ij}]_{n\times n} = \mathbf{UB}$. Since $\mathbf{B}$ is orthogonal, we have $||\mathbf{v}|| = ||\mathbf{Bs}|| = ||\mathbf{s}||$. The second-order origin moment of the projection of $\mathbf{z}$ on $\mathbf{u}_i$ is

$$\begin{aligned} E(\mathbf{u}_i^T\mathbf{z})^2 &= E(\mathbf{u}_i^T \cdot ||\mathbf{v}|| \cdot \mathbf{v})^2 = E\{||\mathbf{s}||^2 \cdot (\mathbf{u}_i^T\mathbf{Bs})^2\} = E\Big\{\sum_{k=1}^{n}(\mathbf{c}_i^T\mathbf{s})^2 \cdot s_k^2\Big\} \\ &= \sum_{k=1}^{n} c_{ik}^2 E(s_k^4) + \sum_{k=1}^{n}\sum_{\substack{p=1\\p\neq k}}^{n} c_{ip}^2 E(s_p^2 s_k^2) + \sum_{k=1}^{n}\sum_{\substack{p=1,\\p\neq k}}^{n}\sum_{\substack{q=1,\\q\neq p}}^{n} c_{ip}c_{iq}E(s_p s_q s_k^2) \end{aligned} \tag{5}$$

When $q \neq p$, obviously at least one of q and p is different from k. Suppose $q \neq k$. We then have $E(s_p s_q s_k^2) = E(s_q)E(s_p s_k^2) = 0$ since s_i are independent and zero-mean. We also have $\sum_{k=1}^{n} c_{ik}^2 = 1$ since $\mathbf{C}$ is orthogonal. Equation 5 then becomes

$$\begin{aligned} E(\mathbf{u}_i^T\mathbf{z})^2 &= \sum_{k=1}^{n} c_{ik}^2 E(s_k^4) + \sum_{k=1}^{n}\sum_{\substack{p=1\\p\neq k}}^{n} c_{ip}^2 E(s_p^2)E(s_k^2) \\ &= \sum_{k=1}^{n} c_{ik}^2 E(s_k^4) + \sum_{k=1}^{n}\sum_{\substack{p=1\\p\neq k}}^{n} c_{ip}^2 = \sum_{k=1}^{n} c_{ik}^2 \mathrm{kurt}(s_k) + (n+2) \end{aligned} \tag{6}$$

Therefore $E(\mathbf{u}_i^T\mathbf{z})^2$ is the weighted average of $\mathrm{kurt}(s_i)$ plus a constant. As s_i are assumed to have different kurtosis, without loss of generality, we assume

$\text{kurt}(s_1) > \text{kurt}(s_2) > \cdots > \text{kurt}(s_n)$. From Eq. 6 we can see that maximization of $E(\mathbf{u}_1^T\mathbf{z})^2$ gives $\mathbf{c}_1 = [\pm 1, 0, ..., 0]^T$, which means that $y_1 = \mathbf{u}_1^T\mathbf{v} = \mathbf{u}_1^T\mathbf{B}\mathbf{s} = \mathbf{c}_1^T\mathbf{s} = \pm s_1$. After finding $\mathbf{u}_1$, under the constraint that $\mathbf{u}_2$ is orthogonal to $\mathbf{u}_1$, the maximum of $E(\mathbf{u}_2^T\mathbf{z})^2$ is obtained at $\mathbf{c}_2 = [0, \pm 1, 0, ..., 0]^T$. Consequently $y_2 = \mathbf{u}_2^T\mathbf{v} = \mathbf{c}_2^T\mathbf{s} = \pm s_2$. Repeating this procedure, finally all independent components can be estimated as $y_i = \mathbf{u}_i^T\mathbf{v} = \pm s_i$ where $\mathbf{u}_i$ maximizes $E(\mathbf{u}_i^T\mathbf{z})^2$. In other words, the orthogonal matrix $\mathbf{U}$ performing PCA on $\mathbf{z}$ (without centering of $\mathbf{z}$) actually performs ICA on $\mathbf{v}$. □

Note that $\mathbf{z}$ may not be zero-mean since $\mathbf{v}$ may be nonsymmetrical. And we should not do centering of $\mathbf{z}$ when performing PCA on $\mathbf{z}$—in fact, here PCA is used to maximize the origin moment of $\mathbf{u}_i^T\mathbf{z}$, rather than its variance.

We term this method as P-ICA. Kurtosis-based P-ICA, which is actually equivalent to the fourth-order blind identification (FOBI) algorithm [2], consists of three steps: 1. do whitening of $\mathbf{x}$ (Eq. 2); 2. do the transformation given in Eq. 4; 3. find $\mathbf{U}$ by PCA on $\mathbf{z}$. After estimating the orthogonal matrix $\mathbf{U}$, the de-mixing matrix for the original data, $\mathbf{W}$, can be constructed according to Eq. 3. With PCA done by EVD, below we give the Matlab code implementing this algorithm:

```
% whitening
[E,D]=eig(cov(x)); v=E*D∧(-.5)*E'*x;
% Data transformation and ICA by PCA
z=sqrt(sum(v.∧2)).*v; [EE,DD]=eig(z*z'); y=EE'*v;
```

3.2 An Illustration

For clarity, let us use a simple example to illustrate how P-ICA works. We use a sinusoid waveform and a Gaussian white signal as the independent sources, shown in Fig. 1 (a). The mixing matrix $\mathbf{A}$ is randomly chosen. The observed data $\mathbf{x} = \mathbf{A}\mathbf{s}$ are shown in Fig. 1 (b).

The whitened version of the observations, $\mathbf{v} = \mathbf{V}\mathbf{x}$, is shown in Fig. 1 (c). Fig. 1 (d) shows the waveforms of the transformed data $\mathbf{z} = ||\mathbf{v}|| \cdot \mathbf{v}$ together with the scatterplot. From this figure we can see that the axes corresponding to the independent sources s_i are almost the same as those giving principal components of $\mathbf{z}$. The orthogonal matrix for PCA of $\mathbf{z}$ can be obtained by applying EVD on the covariance matrix of $\mathbf{z}$:

$$\mathbf{U} = \begin{bmatrix} -0.2878 & -0.9577 \\ -0.9577 & 0.2878 \end{bmatrix}$$

According to Theorem 1, the independent components of $\mathbf{x}$ can then be obtained as $\mathbf{y} = \mathbf{U}\mathbf{v}$ (or equivalently, $\mathbf{y} = \mathbf{U}\mathbf{V}\mathbf{x}$), as shown in Fig. 1 (e). Clearly the independent sources s_i have been successfully recovered.

3.3 For Robustness

From Eq. 6 we can see that the vector $\mathbf{u}_i$ found by maximizing $E(\mathbf{u}_i^T\mathbf{z})$ actually depends on the kurtosis of s_i. It is well known that kurtosis can be very sensitive to outliers, so it is useful to develop a robust version for P-ICA.

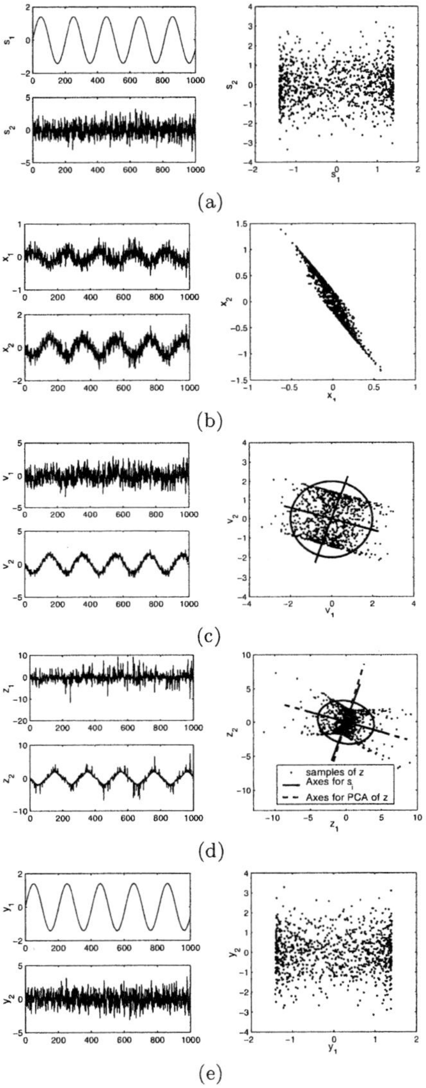

Fig. 1. An illustrative example. (a) The independent sources s_i (left) and their joint scatterplot (right). (b) The observations x_i (left) and their joint scatterplot (right). (c) The whitened data v_i (left) and their joint scatterplot (right). The straight lines in the figure on the right show the axes of the independent sources s_i. The ellipse, which is actually a circle, is a contour of the joint Gaussian distribution fitting $\mathbf{v}$. (d) The transformed data $\mathbf{z} = ||\mathbf{v}|| \cdot \mathbf{v}$ (left) and their joint scatterplot (right). The ellipse shows a contour of the joint Gaussian distribution fitting $\mathbf{z}$. (e) The estimated independent component $\mathbf{y} = \mathbf{U}\mathbf{v} = \mathbf{U}\mathbf{V}\mathbf{x}$ (left) and their joint scatterplot (right). Clearly y_i provide a good estimate for the independent sources s_i.

Now let us replace $||\mathbf{v}||$ in Eq. 4 by $f^{\frac{1}{2}}(||\mathbf{v}||^2)$, i.e. $\mathbf{z} = f^{\frac{1}{2}}(||\mathbf{v}||^2) \cdot \mathbf{v}$, where f is a sufficiently smooth and monotonically increasing function and $f(0) = 0$. We have the Taylor expansion about the origin for $f(||\mathbf{v}||^2)$:

$$f(||\mathbf{v}||^2) = f(||\mathbf{s}||^2) = f(s_1^2 + \cdots + s_n^2)$$
$$= \sum_{\substack{e_1,\ldots,e_n=0 \\ \Pi e_i \neq 0}}^{\infty} \alpha_{(e_1,e_2,\ldots,e_n)} \cdot s_1^{2e_1} s_2^{2e_2} \cdots s_n^{2e_n} \quad (7)$$

where $\alpha_{(e_1,e_2,\ldots,e_n)}$ denotes the coefficients of the Taylor expansion.

Suppose s_i have finite moments. We then have

$$E(\mathbf{u}_i^T\mathbf{z})^2 = E\{f(||\mathbf{v}||^2) \cdot (\mathbf{u}_i^T\mathbf{B}\mathbf{s})^2\} = E\Big\{f(||\mathbf{s}||^2) \cdot \Big(\sum_{p=1}^{n}\sum_{q=1}^{n} c_{ip}c_{iq}s_p s_q\Big)\Big\}$$
$$= \sum_{p=1}^{n}\sum_{\substack{q=1 \\ q\neq p}}^{n} \sum_{\substack{e_1,\ldots,e_n=0 \\ \Pi e_i \neq 0}}^{\infty} c_{ip}c_{iq}\alpha_{(e_1,e_2,\ldots,e_n)} \cdot E(s_1^{2e_1} \cdots s_p^{2e_2+1} \cdots s_q^{2e_q+1} \cdots s_n^{2e_n})$$
$$+ \sum_{p=1}^{n} c_{ip}^2 E\{s_p^2 \cdot f(||\mathbf{s}||^2)\} \quad (8)$$

Provided that at most one of the sources s_i is nonsymmetrical, also taking into account that s_i are mutually independent, the first term in Eq. 8 vanishes. Define $G_p \equiv E\{s_p^2 \cdot f(||\mathbf{s}||^2)\}$. As the expectation of s_p^2 weighted by $f(||\mathbf{s}||^2)$, G_p is actually a function of the moments of s_i of all even orders according to Eq. 7:

$$G_p = \sum_{\substack{e_1,\ldots,e_n=0 \\ \Pi e_i \neq 0}}^{\infty} \alpha_{(e_1,e_2,\ldots,e_n)} \cdot E(s_1^{2e_1}) \cdots E(s_p^{2e_p+2}) \cdots E(s_n^{2e_n})$$

Suppose G_p vary according to p. Roughly speaking, this condition could be enforced by assuming s_i have different distributions. According to Eq. 8, $E(\mathbf{u}_i^T\mathbf{z})^2$ is the weighted average of G_p: $E(\mathbf{u}_i^T\mathbf{z})^2 = \sum_{p=1}^n c_{ip}^2 G_p$. This equation is similar to Eq. 6, and we can see that the orthogonal matrix $\mathbf{U}$ which gives principal components of $\mathbf{z}$ (without centering) performs ICA on $\mathbf{v}$.

In order to improve the robustness, $f(\cdot)$ should be a function increasing slower than the linear one. Many functions can be chosen as f for such a purpose. In our experiments, we choose

$$f(||\mathbf{v}||^2) = \log(1 + ||\mathbf{v}||^2) \quad (9)$$

which behaves very well. Note that we have made the assumption that at most one of s_i is nonsymmetrical when using the robust version of P-ICA, while in kurtosis-based P-ICA, this assumption is unnecessary.

4 Experiments

The illustrative example in Subsection 3.2 has demonstrated the validity of the P-ICA method. Now we give two additional experiments. The Amari performance index P_{err} [1] is used to assess the quality of $\mathbf{W}$ for separating observations generated by the mixing matrix $\mathbf{A}$. The smaller P_{err}, the better $\mathbf{W}$.

The P-ICA method assumes that sources have different distributions. The first experiment demonstrates the necessity of this assumption and how the separation performance is affected by violation of this assumption. s_1 is generated by the Cornish-Fisher (C-F) expansion [5] with skewness 0 and kurtosis 1. s_2 is also generated by the C-F expansion with skewness 0, but the kurtosis varies from -1 and 6. Each source has 1000 samples. We compare five methods, which are kurtosis-based P-ICA, robust P-ICA (with f given in Eq. 9), FastICA [8] with pow3 nonlinearity, FastICA with tanh nonlinearity, and JADE [3], for separating s_1 and s_2 from their mixtures. For each value of $\text{kurt}(s_2)$, we randomly generate s_2 and the mixing matrix $\mathbf{A}$, and repeat the algorithms 40 runs. The average performance index is shown in Fig. 2. As expected, when s_1 and s_2 have the same distribution (here their distribution only depends on the skewness and kurtosis as they are both generated by the C-F expansion), P-ICA fails to separate the sources. When $\text{kurt}(s_2)$ is far from $\text{kurt}(s_1)$, the performance of the robust P-ICA is very close to the best, and the other three methods produce similar results.

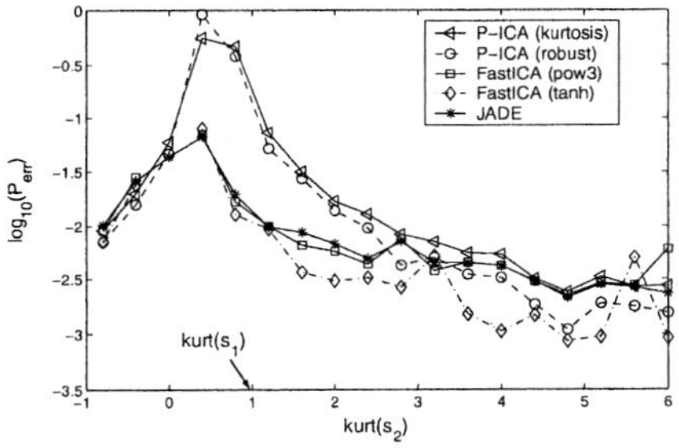

Fig. 2. Logarithm of the average performance index as function of $\text{kurt}(s_2)$. Note that $\text{kurt}(s_1) = 1$, and s_1 and s_2 are both generated by the Cornish-Fisher expansion.

The second experiment tests how the source number affects the separation performance. Four sources are involved, which are a non-central t distributed white signal (with $v = 5$ and $\delta = 1$, s_1), a uniformly distributed white signal (s_2), a Gaussian white signal (s_3), and a square waveform (s_4). Each source has 1000 samples. The source number n varies from 2 to 4. For each source number, we randomly generate the sources and the mixing matrix, and repeat the methods 50 runs. The average performance index, together with its standard deviation, is given in Table 1. Compared to other methods, the performance of P-ICA becomes worse when the source number increases. There may be two reasons for this phenomenon. First, as the source number increases, the difference between source distributions becomes more and more insignificant. Second, in developing P-ICA, we treat the last term in Eq. 5 (and the first term in Eq. 8) as zero, which theoretically holds. However, in practice, due to the finite-sample effect and the fact that sources may not be completely independent, this term may

Table 1. The average P_{err} of the five algorithms with different source number for 50 runs. The values in parentheses are the standard deviations.

Source number (n)	P-ICA (kurtosis)	P-ICA (robust)	FastICA (pow3)	FastICA (tanh)	JADE
2 (s_1&s_2)	0.1137(0.0663)	**0.0786**(0.0502)	0.1137(0.0663)	**0.0619**(0.0410)	0.1288(0.0773)
3 ($s_1 \sim s_3$)	0.1381(0.0591)	0.1104(0.0454)	0.1234(0.0632)	**0.0901**(0.0516)	**0.1069**(0.0535)
4 ($s_1 \sim s_4$)	0.1109(0.0366)	0.0851(0.0284)	0.0852(0.0284)	**0.0568**(0.0171)	**0.0735**(0.0236)

not vanish and the error accumulates as the source number increases. However, we should remark that the computational cost of P-ICA is very light since it is just a PCA procedure after a simple transformation.

5 Conclusion

We investigated the relationship between PCA and ICA, and consequently showed that ICA with sources having different distributions can be solved by PCA following a simple transformation. Such a method, termed P-ICA, does not explicitly involve higher-order statistics and has low computational cost. Kurtosis-based P-ICA, which just exploits the kurtoses of sources, is the same as FOBI [2]. We further developed the robust version of P-ICA, which is insensitive to outliers. P-ICA enables PCA (hardware) systems to perform ICA after sligh modification. Experimental results showed the validity and restriction of P-ICA.

References

1. S. Amari, A. Cichocki, and H. H. Yang. A new learning algorithm for blind signal separation. In *Advances in Neural Information Processing Systems*, 1996.
2. J. F. Cardoso. Source separation using higher order moments. In *ICASSP'89*, pages 2109–2112, 1989.
3. J. F. Cardoso and A. Souloumiac. Blind beamforming for non-Gaussian signals. *IEE Proceeding-F*, 140(6):362–370, 1993.
4. P. Comon. Separatin of stochastic processes. In *Proceedings of the Workshop on Higher-Order Spectral Analysis*, pages 174–179, Vail, Colorado, USA, June 1989.
5. E. A. Cornish and R. A. Fisher. Moments and cumulants in the specification of distributions. *Review of the International Statistical Institute*, 5:307–320, 1937.
6. F. Harroy and J. L. Lacoume. Maximum likelihood estimators and Cramer-Rao bounds in source separation. *Signal Processing*, 55(2):167–177, 1996.
7. F. Herrmann and A. K. Nandi. Blind separation of linear instantaneous mixtures using closed-form estimators. *Signal Processing*, 81:1537–1556, 2001.
8. A. Hyvärinen. Fast and robust fixed-point algorithms for independent component analysis. *IEEE Transactions on Neural Networks*, 10(3):626–634, 1999a.
9. S. Lagrange, L. Jaulin, V. Vigneron, and C. Jutten. Analytical solution of the blind source separation problem using derivatives.
10. V. Zarzoso and A. K. Nandi. Closed-form estimators for blind separation of sources—part I: Real mixtures. *Wireless Personal Communications*, 21:5–28, 2002.

Separability of Convolutive Mixtures: Application to the Separation of Combustion Noise and Piston-Slap in Diesel Engine

Moussa Akil and Christine Servière

Laboratoire des Images et des Signaux, ENSIEG BP 46,
38402 Saint-Martin d'Hères, France
christine.serviere@lis.inpg.fr

Abstract. We focus on convolutive mixtures, expressed in time-domain. Separation is known to be obtained by testing the independence between delayed outputs. This criterion can be much simplified and we prove in this paper that testing the independence between the contributions of all sources on the same sensor at same time index also leads to separability. We recover the contribution by using Wiener filtering (or Minimal Distorsion Principal) which is included in the separation filters. The independence is tested here with the mutual information. It is minimized only for non-delayed outputs of the Wiener filters. The test is easier and shows good results on simulation and experimental signals for the separation of piston slap and combustion in diesel engine.

1 Introduction

Blind source separation (BSS) is a method for recovering a set of unknown source signals from the observation of their mixtures. Among open issues, recovering the sources from their linear convolutive mixtures remains a challenging problem. Many solutions have been addressed in the frequency-domain, particularly for the separation of non-stationary audio signals. In the BSS of stationary signals, two problems remain open in time domain. It has been proved [1] that convolutive mixtures are separable, that is, the independence of the outputs insures the separation of the sources, up to a few indeterminacies. However, the meaning of the independence is not the same in convolutive and instantaneous contexts. In the convolutive context, the outputs have to be independent in the sense of stochastic processes [2] which requires the independence of the random variables $y_i(n)$ and $y_j(n-m)$ for all discrete times n and m. The independence criteria are therefore more complicated and computationally expensive. Several ideas are given in [3,4] to test the independence in function of time delays m, using the mutual information criterion. The second problem is coming from the inherent indeterminacy of the definition of a source in the BSS model. Indeed, any linear transform of a source can also be considered as a source and there is an infinity of separators that can extract sources. Some constraints can be added either on the source signals (they are usually supposed to be normalized) or on the separator system (Minimal Distorsion Principal [5]). In [5], one proposition is to choose the separator

J. Rosca et al. (Eds.): ICA 2006, LNCS 3889, pp. 319 – 326, 2006.

which minimizes the quadratic error between sensors and outputs, also known as Wiener filter. In this paper, we deal with convolutive mixtures and express the model in time-domain. The aim is to quantify the proportions of mechanical noise coming from piston slap or thermal noise and received on accelerometers, placed on one cylinder of a diesel engine. We are only interested in the contribution of these two sources recorded on each sensor. These signals are uniquely defined, which removes the filter indeterminacy. It can also help to simplify the independence criterion and we prove in this paper that testing the independence between the contributions of all sources on the same sensor at same time index n also leads to separability. We recover these contributions $z_i(n)$ by using Wiener filters which are included in the separation filters. The independence criterion is therefore less complicated as it requires only the independence between the outputs $z_i(n)$ and $z_j(n)$ (and no more $y_i(n)$ and $y_j(n-m)$). The mutual information is used here and shows good results on simulation and experimental signals for the separation of piston slap and combustion noise in diesel engine.

2 Modelization of the Observations

Let us consider the standard convolutive mixing model with M inputs and M outputs. Each sensor $x_j(n)$ ($j=1, .., M$) receives a linear convolution (noted *) of each source $s_i(n)$ ($i=1,\ldots,M$) at discrete time n:

$$x_j(n) = \sum_{i=1}^{M} h_{ji} * s_i(n) \tag{1}$$

where h_{ij} represents the impulse response from source i to sensor j. The inverse of mixing filters are not necessarily causal, so the aim of BSS is to recover non-causal filters with impulse responses g_{ij} between sensor i and output j, such that the output vector $\mathbf{y}(n)$ estimates the sources, up to a linear filter :

$$y_j(n) = \sum_{i=1}^{M} \sum_{k=-L}^{L} g_{ji}(k) x_i(n-k) \tag{2}$$

Any linear transform of a source can also be considered as a source and there is an infinity of separators g_{ij} that can extract sources. We focus here on the estimation of the signals $h_{ij} * s_i(n)$, coming from source i on sensor j. These signals are uniquely defined, which removes the filter indeterminacy. Let be a 2 sources 2 sensors scheme. For sake of simplicity, we call here sources the two contributions on the first sensor. So, $x_1(n)$ is equal to : $x_1(n) = s_1(n) + s_2(n)$. Let be $y_1(n)$ and $y_2(n)$, two outputs :

$$y_j(n) = \sum_{i=1}^{2} \sum_{k=-L}^{L} g_{ji}(k) x_i(n-k) \tag{3}$$

If $y_j(n)$ is any linear filtering of one source, than the contribution of this source on the first sensor is calculated by an (eventually non causal) Wiener filter $W_j(z)$ such that the quadratic error between $x_1(n)$ and $y_j(n)$ is minimized. The two contributions on the first sensor are so given by:

$$z_j(n) = \sum_{i=1}^{2} \sum_{k=-L}^{L} w_j(k) y_i(n-k) \tag{4}$$

where the discrete Fourier Transforms (DFT) of the Wiener filters $w_j(k)$ are computed in function of the cross-spectra of $x_1(n)$ and $y_j(n)$:

$$W_1(f) = \frac{\gamma_{Y1X1}(f)}{\gamma_{Y1}(f)} \quad ; \quad W_2(f) = \frac{\gamma_{Y2X1}(f)}{\gamma_{Y2}(f)} \tag{5}$$

3 Separability of the Source Contributions on One Sensor

In specific cases, testing the independence between $y_1(n)$ and $y_2(n)$ is sufficient [6] to ensure the separation. For example, for i.i.d. normalized sources, the sum of fourth-order cumulants of the outputs is a contrast function [7] under a condition on separating filters [6]. For linear filtering of i.i.d. signals, the same result is obtained after a first step of whitening of the data. However, in a general case, delays must be introduced in the contrast function and the separability of convolutive mixtures is obtained only when the components of the output vector $\mathbf{y}(n)$ are independent in the sense of stochastic variables : $y_1(n)$ and $y_2(n-m)$ have to be independent for all discrete time delays m. For example, a solution is to minimize the criterion J :

$$J = \sum_m I\left(y_1(n), y_2(n-m)\right) \tag{6}$$

where I represents the mutual information (7). I is nonnegative and equal to zero if and only if the components are statistically independent.

$$I(y) = \int_R p_y(y) \ln\left(\frac{p_y(y)}{\prod_{i=1}^{M} p_{yi}(y_i)}\right) dy \tag{7}$$

The delays m can be taken in an *a priori* set [-K, ..., K], which depends on the degree of the filters corresponding to the whole mixing-separating system. The criterion (6) is computationally expensive. In [3], a gradient-based algorithm minimizes (6): at each time iteration, a random value of delay m is chosen and $I(y_1(n), y_2(n-m))$ is used as the current separation criterion.

We propose to study here the separability of $z_1(n)$ and $z_2(n)$ (4) versus $y_1(n)$ and $y_2(n)$. We show that it is simpler and that no time delay $(n-m)$ is needed. Suppose now any outputs $y_1(n)$ and $y_2(n)$. To ensure the separation, it is necessary (but not sufficient) that the mutual information $I(y_1(n), y_2(n))$ is zero. Two cases can happen. If each output $y_j(n)$ only depends on one source, the outputs are also independent in the sense of stochastic processes (the separation has been effected) and it will be also verified for $z_1(n)$ and $z_2(n)$. So $I(z_1(n), z_2(n))=0$. In the second case, the outputs $y_j(n)$ can be independent ($I(y_1(n), y_2(n))=0$ at time delay 0) but remain mixtures of sources. For example, in the case of i.i.d sources, the two following outputs $y_j(n)$ are independent (8):

$$y_1(n) = s_1(n) + s_2(n)$$
$$y_2(n) = s_1(n-1) + s_2(n-1) \tag{8}$$

It occurs (typically for i.i.d. sources) when one source is common in the two outputs but with two different time index *(n-n0)* and *(n-n1)*. In that case, $y_j(n)$ are independent but surely not the components of $z_j(n)=W_i(z)\, y_j(n)$, as common time index can appear after linear filtering. It can be seen intuitively, since Wiener filtering aims at the maximization of the correlation between $z_1(n)$ and $x_1(n)$ (respectively $z_2(n)$ and $x_1(n)$). We will prove theoretically that indeed $I(z_1(n), z_2(n))$ is not equal to zero. As a consequence, testing the cancellation of $I(y_1(n), y_2(n))$ and $I(z_1(n), z_2(n))$ will ensure the separability.

Suppose that $y_1(n)$ and $y_2(n)$ are mixtures of the sources (even if $I(y_1(n), y_2(n))=0$). So are $z_1(n)$ and $z_2(n)$ after Wiener filtering. Let be $Z_1(f)$ and $Z_2(f)$, their DFT's. They are of the form (10). The transfer functions $W_1(f)$ and $W_2(f)$ (5) of the Wiener filters are expressed in function of the DFT of filters $g_{ij}(k)$, $G_{ji}(f)$, and the source spectra:

$$W_1(f) = \frac{\overline{G}_{11}(f)\gamma_{S1}(f) + \overline{G}_{12}(f)\gamma_{S2}(f)}{\gamma_{Y1}(f)} \quad ; \quad W_2(f) = \frac{\overline{G}_{21}(f)\gamma_{S1}(f) + \overline{G}_{22}(f)\gamma_{S2}(f)}{\gamma_{Y2}(f)} \tag{9}$$

$$Z_1(f) = \frac{|G_{11}(f)|^2\gamma_{S1}(f) + G_{11}(f)\overline{G}_{12}(f)\gamma_{S2}(f)}{\gamma_{Y1}(f)} S_1(f) + \frac{\overline{G}_{11}(f)G_{12}(f)\gamma_{S1}(f) + |G_{12}(f)|^2\gamma_{S2}(f)}{\gamma_{Y1}(f)} S_2(f)$$
$$Z_2(f) = \frac{|G_{21}(f)|^2\gamma_{S1}(f) + G_{21}(f)\overline{G}_{22}(f)\gamma_{S2}(f)}{\gamma_{Y2}(f)} S_1(f) + \frac{\overline{G}_{21}(f)G_{22}(f)\gamma_{S1}(f) + |G_{22}(f)|^2\gamma_{S2}(f)}{\gamma_{Y2}(f)} S_2(f) \tag{10}$$

$z_j(n)$ are linear filtering of $s_1(n)$ and $s_2(n)$ as $y_1(n)$ and $y_2(n)$. Call $u_{ij}(k)$, the new mixing filters: $u_{ij}(k)=[w_j * g_{ij}](k)$ where * stands for the linear convolution. $z_j(n)$ are expressed as :

$$z_j(n) = \sum_{i=1}^{2}\sum_{k=-L}^{L} u_{ij}(k)s_i(n-k) \tag{11}$$

The two signals $z_1(n)$ and $z_2(n)$ cannot be independent ($I(z_1(n), z_2(n))$ is not zero) if some coefficients $u_{11}(k)$ and $u_{12}(k)$ are non zero for common time delays k. And, at least, we prove that one coefficient, $u_{ij}(k)(0)$, is non zero. Suppose that the DFT is computed on N time samples :

$$u_{11}(0) = \sum_{f=0}^{N-1} \frac{|G_{11}(f)|^2\gamma_{S1}(f) + G_{11}(f)\overline{G}_{12}(f)\gamma_{S2}(f)}{\gamma_{Y1}(f)} \tag{12}$$

$$\begin{aligned} |u_{11}(0)|^2 &= \left(\sum_f |G_{11}(f)|^2 \frac{\gamma_{S1}(f)}{\gamma_{Y1}(f)}\right)^2 + \left|\sum_f G_{11}(f)\overline{G}_{12}(f)\frac{\gamma_{S2}(f)}{\gamma_{Y1}(f)}\right|^2 \\ &+ \sum_f |G_{11}(f)|^2 \frac{\gamma_{S1}(f)}{\gamma_{Y1}(f)}\left(\sum_f \frac{\gamma_{S2}(f)}{\gamma_{Y1}(f)}(G_{11}(f)\overline{G}_{12}(f) + \overline{G}_{11}(f)G_{12}(f))\right) \end{aligned} \tag{13}$$

If the third term of the sum is positive or null, then $u_{11}(k)(0)$ cannot be null. If it is negative, $(u_{11}(k)(0))^2$ is always superior to a strictly positive value (14). Similar computations can be done with $u_{12}(k)(0)$, $u_{21}(k)(0)$ and $u_{22}(k)(0)$.

$$\begin{aligned} \left|u_{11}(0)\right|^2 &> \left(\sum_f |G_{11}(f)|^2 \frac{\gamma_{S1}(f)}{\gamma_{Y1}(f)}\right)^2 + \left|\sum_f G_{11}(f)\overline{G}_{12}(f)\frac{\gamma_{S2}(f)}{\gamma_{Y1}(f)}\right|^2 \\ &-\sum_f |G_{11}(f)|^2 \frac{\gamma_{S1}(f)}{\gamma_{Y1}(f)}\left(\left|\sum_f \frac{\gamma_{S2}(f)}{\gamma_{Y1}(f)}(G_{11}(f)\overline{G}_{12}(f)\right| + \left|\sum_f \frac{\gamma_{S2}(f)}{\gamma_{Y1}(f)}\overline{G}_{11}(f)G_{12}(f))\right|\right) \\ &\geq \left|\sum_f |G_{11}(f)|^2 \frac{\gamma_{S1}(f)}{\gamma_{Y1}(f)} - \left|\frac{\gamma_{S2}(f)}{\gamma_{Y1}(f)}(G_{11}(f)\overline{G}_{12}(f)\right|\right|^2 \end{aligned} \tag{14}$$

So, for any outputs $y_j(n)$ which verify $I(y_1(n), y_2(n))=0$, then after Wiener filtering projected on the same sensor (here the first one) $I(z_1(n), z_2(n))$ is non zero. The only exception concerns the outputs $y_j(n)$ which depend on one source and it means that the separation has been achieved. Same results can also be obtained with M sources.

As a consequence, testing $I(y_1(n), y_2(n))=0$ and $I(z_1(n), z_2(n))=0$, ensures the separability. The criterion is much more easier to test than the mutual information of delayed outputs as it can be verified in an iterative way. Moreover the outputs are directly the contribution of the sources on the processed sensor.

4 Separating Algorithm and Simulations

The final separating algorithm for convolutive mixtures is based here on the minization of the mutual information as in [3] but the previous proof of separability could be exploited with another independence test.

Initialization : $\mathbf{y}(n)= \mathbf{x}(n)$
Repeat until convergence :

- Estimate the score function difference between $y_1(n)$ and $y_2(n)$: $\beta(y_1(n),y_2(n))$
- Update : $\mathbf{y}(n) \leftarrow \mathbf{y}(n)- \mu\, \beta(y_1(n),y_2(n))$
- Compute the Wiener filters $W_i(z)$, and the contributions : $z_j(n)=W_i(z)\, y_j(n)$
- Replace : $\mathbf{y}(n) \leftarrow \mathbf{z}(n)$

The performances are shown in figures 1 and 2 with simulations results. Each source (of 1500 samples) is constituted of the sum of a uniform random signal and a sinusoid. They are mixed with filters :

$$H(z) = \begin{bmatrix} 1+0.2z^{-1}+0.1z^{-2} & 0.5+0.3z^{-1}+0.1z^{-2} \\ 0.5+0.3z^{-1}+0.1z^{-2} & 1+0.2z^{-1}+0.1z^{-2} \end{bmatrix} \tag{15}$$

The mutual information (between $z_1(n)$ and $z_2(n)$) and the quadratic error between $z_1(n)$ and the exact contribution are plotted in fig.1 and 2 with marks, for each iteration. They are averaged on 50 realizations of the sources. It shows good results for the convergence speed and the residual quadratic error. The results can still be improved by adding some constraints. Indeed, four contributions must be computed in this scheme by projecting $y_1(n)$ (respectively $y_2(n)$) on the two sensors: $z_{11}(n)$, $z_{21}(n)$ (respectively $z_{12}(n)$, $z_{22}(n)$). The convergence speed is increasing by adding the mutual information between the projections on the second sensor $I(z_{21}(n), z_{22}(n))$ to $I(z_{11}(n), z_{12}(n))$ (as previously) in the minimization. The results are displayed in figures 1 and 2 in solid line and show the increasing of the convergence. So, the new algorithm is:

Initialization : $y(n)= x(n)$
Repeat until convergence :

- Estimate the score function differences: $\beta(z_{11}(n), z_{12}(n))$, $\beta(z_{21}(n), z_{22}(n))$
- Update : $y(n) \leftarrow y(n)- \mu\, [\beta(z_{21}(n), z_{22}(n))+\beta(z_{11}(n), z_{12}(n))]$
- Compute the Wiener filters $W_{ij}(z)$, and the contributions : $z_{ij}(n)=W_{ij}(z)\, y_j(n)$
- Replace : $y(n) \leftarrow [z_{11}(n), z_{12}(n))]$

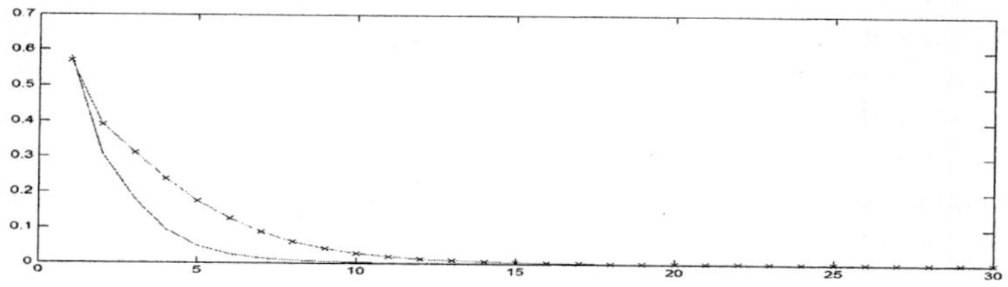

Fig. 1. Mutual information versus iterations

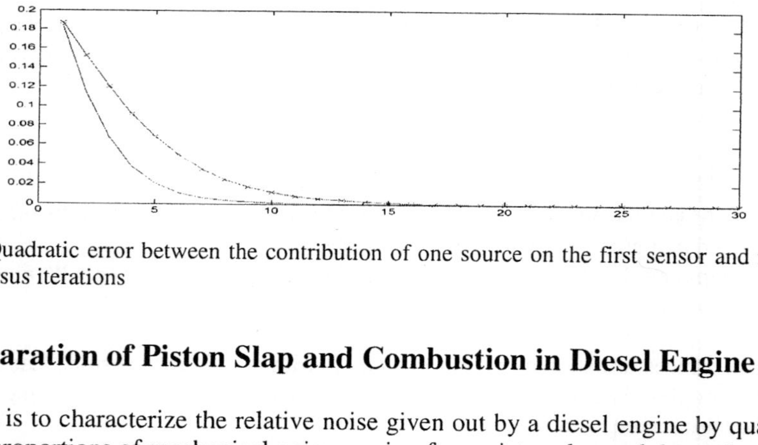

Fig. 2. Quadratic error between the contribution of one source on the first sensor and its estimate, versus iterations

5 Separation of Piston Slap and Combustion in Diesel Engine

The aim is to characterize the relative noise given out by a diesel engine by quantifying the proportions of mechanical noise coming from piston slap and thermal noise or combustion. Signals are issued of ten accelerometers, placed on a four-stroke and four cylinder diesel engine. They record thermal and mechanical phenomena that are temporally superposed around the TDC, as well as spectrally overlapping. Some sensors respond to vertical moves or horizontal ones, according to their positions. Therefore, some accelerometers are more sensitive to combustion noise whereas the other ones receive more mechanical noise as piston-slap. Nevertheless, all accelerometer signals are convolutive mixtures of thermal and mechanical sources. Signals have been sampled at 25600Hz. In figure 3, we show the power measured by one sensor in the angular window [-40°, 80°] of the crankshaft (in green). This sensor responded to horizontal moves as it was placed on one side of the liner and received more piston-slap. The figure 3 includes the measured pressure and the injection control pulses. The contributions of the two sources on the sensor have been estimated by the algorithm proposed in section 4 and their powers are shown in figures 4 and 5.

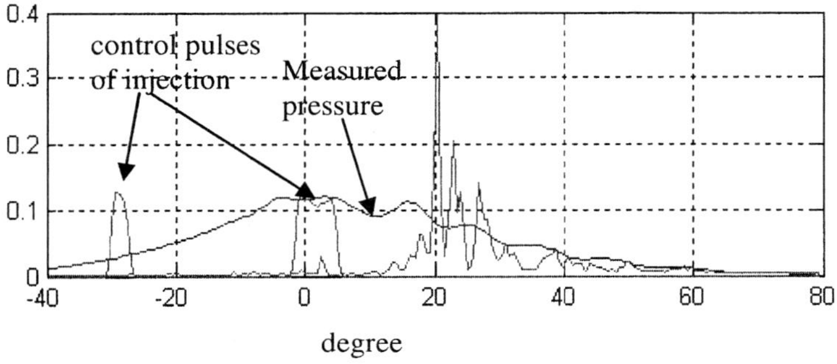

Fig. 3. Power of one sensor versus the crankshaft angle in degree

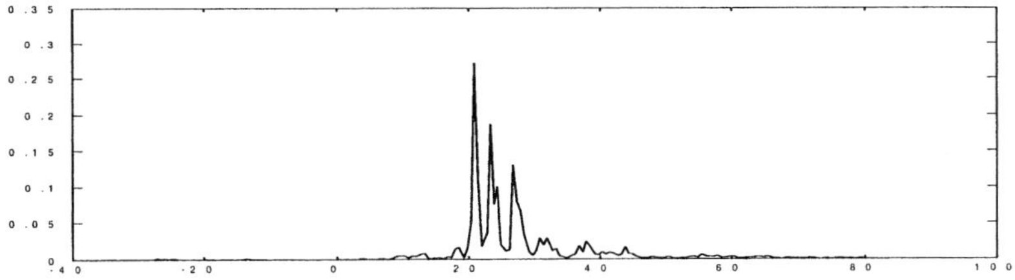

Fig. 4. Power of the first separated source versus the crankshaft angle

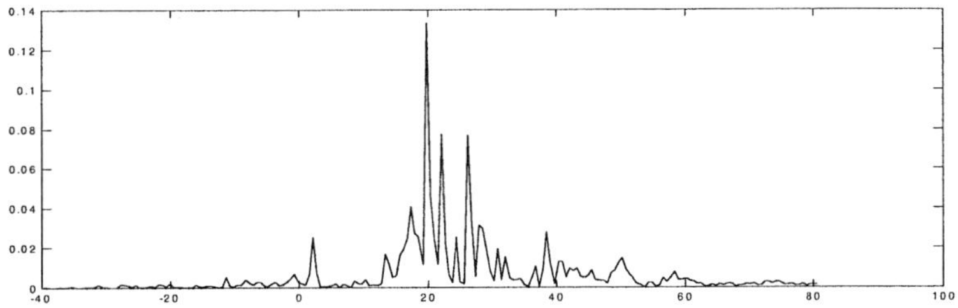

Fig. 5. Power of the second separated source versus the crankshaft angle

A first experiment has been done without injection and therefore no combustion noise is present. It helps to know the exact localization of the mechanical and thermal phenomena. This experiment (not presented here) shows that the sensor registers three mechanical shocks, at 20°5, 23° and 27°. By difference, we can conclude that the combustion is present between 10° and 20°, including the position of the main shock and around 0° for the pre-combustion. The two phenomena are well noticeable in figure 3. After separation, we can see that the first contribution is really the most important. It can be correctly attributed to mechanical shocks as they take place at 21°, 23°8 and 28°8. Besides no pre-combustion is seen around 0° and the main combustion (between 10° and 20°) is well separated. The second contribution is a good estimation of the thermal noise as we recover the pre-combustion and the main combustion. Moreover, the position of the pre-combustion is validated by the localization of the control pulses of injection (seen in figure 3).

6 Conclusion

We focus on the separability of convolutive mixtures, expressed in time-domain. In the convolutive context, the outputs $y_i(n)$ have to be independent in the sense of stochastic processes which requires the independence of $y_i(n)$ and $y_i(n-m)$ for all discrete times n and m. The independence criteria are therefore complicated and computationally expensive. The criterion has been simplified as we recover only the contribution of all sources on all sensors, by using Wiener filtering (or Minimal Distorsion Principal). It has been proved that testing the independence between these contributions on the same sensor also leads to separability, without testing an independence test of delayed outputs. The criterion is easier to test and is implemented here by minimizing the mutual information of the outputs after Wiener filtering. It shows good results on simulation and experimental signals for the separation of piston slap and combustion in diesel engine.

References

1. D. Yellin, E. Weinstein: Criteria for multichannel signal separation, IEEE Trans. On Signal Processing, pp 2158-2168, August 1994
2. A. Papoulis: Probability, random variables and stochastic processes, Mc Graw-Hill, 1991
3. M. Babaie-Zadeh, C. Jutten: Mutual information minimization: application to blind source separation, Signal Processing, 85(5):975-995, May 2005
4. M. El Rahbi, G. Gelle, H. Fenniri, G. Delaunay: A penalized mutual information criterion for blind separation of convolutive mixtures, Signal Processing 84(4), pp 1979-1984
5. K. Matsuoka, Y.Ohba, Y.Toyota, S. Nakashima: Blind separation for convolutive mixtures of many voices, Int. workshop on acoustic echo and noise control, Sep 2003, Kyoto, Japan
6. P. Comon: Contrasts for multichannel blind deconvolution, IEEE Signal Processing Letters, vol 3, n°7, pp 209-211, 1996
7. P. Comon: Independent component analysis, a new concept, IEEE Trans. On Signal Processing, vol 36, n°3, pp 287-314, 1994

Blind Signal Separation on Real Data: Tracking and Implementation

Paul Baxter, Geoff Spence, and John McWhirter

QinetiQ, St Andrews Road, Malvern, Worcs, WR14 3PS, United Kingdom
p.baxter@signal.qinetiq.com

Abstract. There are numerous algorithms available for blind signal separation (BSS) of multiple signals, but most of these are optimised for short blocks of data, stationary signals and time invariant mixing matrices. As such, they are unsuitable for real-world applications, which often require tracking BSS carried out in real time with as small a lag as possible. This paper looks at the problems encountered in applying BSS to real data sets and addresses the issue of computationally efficient tracking BSS based on well-understood two-stage block-based approaches. An example is included where the technique is applied to a five-minute section of twin foetal electrocardiogram (ECG) data.

1 Introduction

A commonly desired objective of signal processing is to recover a signal of interest from sensor recordings in which it may be masked by noise and by other, interfering, signals. Often there will be a large number of signals present, and any individual sensor can only receive a mixture of these signals: in general, only limited information about the signals of interest can be recovered from such a mixture. Blind signal separation (BSS) aims to separate signals by utilizing multiple sensors, commonly using the assumption that the signals are independent (Independent Component Analysis). BSS has been successfully used in many different application areas, e.g. artefact suppression in electroencephlogram (EEG) recordings [8], foetal electrocardiogram (ECG) analysis [10] and image enhancement [5].

The term 'blind' is used to indicate that no prior information, either concerning the individual signals (other than the independence assumption), or the manner in which they combine at the sensors is available. Unknown factors generally include the number of signals, the locations of the signal sources and the sensor locations.

Many different techniques have been developed for carrying out BSS. Some of the best performing, or best known, are JADE [3], FastICA [6], BLISS [7], EASI [4], InfoMax [2] and kernelICA [1]. All of these have been developed to solve the BSS problem in the theoretical case, so without modification they are not necessarily suitable for processing real data sets, especially over long periods of time. Although the difficulties arising in processing real data sets are often unique to the type of data, there are many common problems. In this paper we describe some of these common problems.

Most real data sets contain non-stationary signals and time-varying mixing, especially if they are recorded over long periods of time. We describe how to extend

J. Rosca et al. (Eds.): ICA 2006, LNCS 3889, pp. 327 – 334, 2006.

single block-based BSS algorithms to produce an efficient tracking BSS approach. We provide a demonstration of this technique, applied to a twin foetal ECG data set over five minutes. This shows the utility of the technique.

In section 2 we introduce the basic BSS model, investigate the two-stage approach to solving it and observe some of the difficulties encountered when using it on real data. Our tracking BSS approach is developed in section 3, and is demonstrated on foetal twin ECG data in section 4. Conclusions are drawn in section 5.

2 Basic BSS

2.1 Data Model

The basic linear BSS data model, assuming m sensors and T samples is:

$$\mathbf{X} = \mathbf{AS} + \mathbf{N} \tag{1}$$

The (m x T) matrix **X** denotes the observed sensor data, so the rows of **X** contain the sensor outputs. The (m x m) matrix **A** denotes the time-invariant mixing matrix and the rows of the (m x T) matrix **S** contain the independent signals, assumed to be stationary. The (m x T) matrix **N** contains the sensor noise, usually assumed to be white Gaussian noise, uncorrelated between the sensors. For the sake of simplicity we assume that the number of signals is equal to the number of sensors (m), although for most techniques it is only necessary for there to be at least as many sensors as signals. The following conditions also apply to the data model:

- The mixing process is assumed to be linear and instantaneous (time delays between sensors can be represented as phase shifts);
- The mixing process is assumed to be time invariant;
- At most one of the signals has a Gaussian distribution (only required for complete signal separation).

2.2 Algorithms

Many of the basic BSS algorithms operate on the whole data block at once, using a two-stage approach to achieve signal separation. Firstly, in the *second-order stage*, the sensor outputs are decorrelated and normalised using a method such as the singular value decomposition (SVD) as shown in equation (2). Here, the columns of the (m x m) matrix **U** contain the orthonormal steering (spatial) vectors. The estimated orthonormal signals are contained in the rows of the (m x T) matrix $\mathbf{V}^T$. The (m x m) diagonal matrix α contains the singular values. The orthonormal signals are related to the independent signals by a (m x m) 'hidden' rotation matrix **R**. The *higher-order stage* of separation determines **R**, as shown in equation 3. The matrix $\mathbf{RV}^T = \hat{\mathbf{S}}$ contains the estimates of the signals and $\mathbf{U}\alpha\mathbf{R}^T$ denotes the estimated mixing matrix.

$$\mathbf{X}=\mathbf{U}\alpha\mathbf{V}^T \tag{2}$$

$$\mathbf{X}=\mathbf{U}\alpha\mathbf{R}^T\mathbf{R}\mathbf{V}^T \tag{3}$$

Many of the block-based BSS algorithms differ only in the method used for computing **R**. JADE, BLISS, FastICA and KernelICA all use this two-stage approach, with the same second-order stage but different methods for the higher-order stage.

Whichever BSS algorithm is used, certain difficulties tend to arise when they are applied to real data sets. Some of the commonest of these are:

- **Computational Cost:** In general the computational cost associated with applying BSS to a data block is high, at least $O(m^2)$ and much higher in the case of some algorithms such as KernelICA. This leads to two, more specific, problems:
  - **Real Time Processing:** In many cases real time processing is required, so computationally expensive procedures require high processing power, which is expensive and possibly unobtainable;
  - **Small Processing Lag:** Even if processing power is available to process data in real time, computationally expensive procedures lead to a large lag between the data arriving at the sensors and the processed data being available;
- **Time variation:** Although short sections of many real data sets are sufficiently time-invariant for the basic BSS algorithms to run, longer sections of such data sets are time-varying, e.g. foetal ECG recordings are often time-invariant over 10 second blocks, but the mixing is time-varying and the signals are non-stationary over 1 hour blocks. Three types of time variation are commonly seen:
  - **Non-Stationary Signals:** Usually the signal power varies, this includes the onset of interfering (jamming) signals and signal births and deaths;
  - **Non-Stationary Noise Levels:** Can lead to portions of the data where the signals are swamped by noise and signal separation may not be possible. Such events should not be allowed to bias the overall tracking process;
  - **Time-Varying Steering Vectors:** The relative locations of the signals and sensors can change during the data collection.

In this paper we develop a tracking BSS algorithm for data with significant time variation, and wherever possible try to reduce the computational costs involved.

3 Tracking BSS

We present a method for extending two-stage BSS techniques to time-varying data sets. The basic principle is to use a moving data window, where the signals are separated in each window using a block-based approach. However, the use of a moving window technique alone is not sufficient for many real applications; here we address the following issues:

- **Computational Load:** Processing the individual windows in isolation is inefficient;
- **Signal Swapping:** In each window the signals may be separated in a different order. This is due to both the inherent signal ordering indeterminacy in the higher-order stage, and to the second-order stage ordering the signals by their powers.

Our tracking BSS technique uses two techniques based on second and higher-order statistics, but for the purposes of tracking, these are first **initialised** using past information and then **updated using small rotations**. This is similar in concept to the EASI algorithm [4], but unlike EASI this approach seeks to utilize the fast convergence of block-based approaches, albeit in a tracking context.

The tracking BSS technique we present here can be implemented using either overlapping or contiguous data windows. The first window defines the acquisition phase; in this window the data is processed by a normal two-stage BSS technique. It is inefficient to process the remaining windows in isolation. The use of the SVD, equation (2), in the remaining windows can be problematic. It successfully decorrelates the signals, but also orders the signals according to their power. This can lead to signals being ordered differently in adjacent windows if their powers vary (signal swapping). Similarly, even if initialised near one solution, most higher-order techniques are not guaranteed to converge to this solution, but instead will find a permuted version of this solution.

The tracking BSS technique presented here can overcome these problems by using information estimated from the previous window to initialise the current one and then applying small updates for both the second and higher-order stages.

3.1 Second-Order Stage

Consider the second-order stage in a given tracking window. The signals can be made orthonormal by combining an initialisation process (using the second-order information from the previous window), a decorrelation method (via a Jacobi diagonalisation) for updating the orthogonality of the signals and a new normalization step.

The Jacobi method for diagonalisation can be used as the decorrelation method, where each pairwise rotation is constrained to choose the smallest of the possible angles to rotate the signals by [9]. For example, if **x** and **y** represent the (1 x T) i'th and j'th vectors the pairwise orthogonalisation step of Jacobi diagonalisation can be done by diagonalising their symmetric correlation matrix, i.e. by finding a rotation matrix **Q,** parameterised by θ, that zeros the off-diagonal elements of the matrix **B**:

$$\mathbf{B} = \mathbf{Q}^T \begin{bmatrix} \mathbf{x} \\ \mathbf{y} \end{bmatrix} \begin{bmatrix} \mathbf{x}^T & \mathbf{y}^T \end{bmatrix} \mathbf{Q} = \begin{bmatrix} c & s \\ -s & c \end{bmatrix}^T \begin{bmatrix} a_{ii} & a_{ij} \\ a_{ji} & a_{jj} \end{bmatrix} \begin{bmatrix} c & s \\ -s & c \end{bmatrix}, \tag{4}$$

where c=cos(θ) and s=sin(θ). The correct θ therefore satisfies:

$$\tan(2\theta) = 2a_{ij} / (a_{jj} - a_{ii}) \tag{5}$$

The LHS of equation (5) can be expressed as $2t/(1-t^2)$, where t=tan(θ). Thus, there are two solutions for θ in the range [−π/2, π/2]. The orthogonality of two vectors has been initialised using information from the last window, so **x** and **y** are nearly orthogonal to begin with. Thus the two solutions for θ are close to 0 and π/2. A normal SVD chooses between these by ordering the outputs according to their power; this can cause the vectors to be rotated by approximately $\pi/2$ and hence introduce signal swapping. In our tracking BSS technique we avoid this by insisting on the solution closest to 0 being used.

A mathematical summary of the second-order update stage is shown in equations (7) to (10), where subscript k denotes values belonging to the k'th data window, e.g. $\mathbf{X}_k$ denotes the data in the k'th window. We first note that the symmetric (unnormalised) correlation matrix of the data has as its eigenvalue decomposition

$$\mathbf{X}_k\mathbf{X}_k^T = \mathbf{U}_k\boldsymbol{\alpha}_k^2\mathbf{U}_k^T = \mathbf{U}_k\boldsymbol{\lambda}_k\mathbf{U}_k^T, \tag{6}$$

where $\mathbf{U}_k$ contains the eigenvectors and $\boldsymbol{\lambda}_k$ contains the eigenvalues.

In equation (7) the eigenvectors for the last window $\mathbf{U}_{k-1}$ are used to initialise the eigenvalue decomposition of the current covariance matrix; for a slowly changing mixing matrix, $\boldsymbol{\lambda}_k'$ will be nearly diagonal. Thus $\boldsymbol{\lambda}_k'$ can be simply diagonalised by $\mathbf{U}_z$, equation (8), found by the Jacobi method with the small rotation constraint described above. The updated eigenvectors are found via equation (9) and the estimated orthonormal vectors are given by equation (10).

Equation (9) shows how $\mathbf{U}_k$ is calculated as the product of an **initialisation process**, $\mathbf{U}_{k-1}$, and a **small update**, $\mathbf{U}_z$.

$$\boldsymbol{\lambda}_k' = \mathbf{U}_{k-1}^T\mathbf{X}_k\mathbf{X}_k^T\mathbf{U}_{k-1} \tag{7}$$

$$\boldsymbol{\lambda}_k = \mathbf{U}_z^T\boldsymbol{\lambda}_k'\mathbf{U}_z = \mathbf{U}_z^T\mathbf{U}_{k-1}^T\left(\mathbf{X}_k\mathbf{X}_k^T\right)\mathbf{U}_{k-1}\mathbf{U}_z \tag{8}$$

$$\mathbf{U}_k = \mathbf{U}_{k-1}\mathbf{U}_z \tag{9}$$

$$\mathbf{V}_k^T = \boldsymbol{\lambda}_k^{-0.5}\mathbf{U}_k^T\mathbf{X}_k \tag{10}$$

3.2 Higher-Order Stage

The higher-order stage is carried out in a similar way to the second-order stage. The independence of the orthonormal vectors $\mathbf{V}_k^T$ can be initialised by the rotation matrix $\mathbf{R}_{k-1}$ derived from the last window, equation (11). For a slowly changing mixing matrix, and where signal swapping has been avoided in the second-order stage, then $\mathbf{R}_{k-1}\mathbf{R}_k \approx \mathbf{I}$. Then the initialized signal estimates, $\hat{\mathbf{S}}_k'$, are nearly separated and need to be updated using small angle rotations to avoid introducing signal swapping. Most two-stage BSS algorithms can be easily modified so they find a rotation matrix, $\mathbf{R}_z$, only using small angles. Equations (12) and (13) show how $\mathbf{R}_z$ is used to find the independent signal estimates and to update the rotation matrix.

$$\hat{\mathbf{S}}_k' = \mathbf{R}_{k-1}\mathbf{V}_k^T = \mathbf{R}_{k-1}\mathbf{R}_k\hat{\mathbf{S}}_k \tag{11}$$

$$\hat{\mathbf{S}}_k = \mathbf{R}_z\left(\mathbf{R}_{k-1}\mathbf{V}_k^T\right) \tag{12}$$

$$\mathbf{R}_k = \mathbf{R}_z\mathbf{R}_{k-1} \tag{13}$$

4 Demonstration of Concept

In this section the results of applying the tracking BSS technique using the BLISS algorithm [7] for the higher-order stage to foetal ECG recordings are presented. It is possible to monitor single or multiple foetuses by placing ECG sensors on the mother's abdomen and analysing the signals. It is hard to observe the weak foetal signals in the outputs from a single sensor due to maternal signals and other electrical interference. BSS offers a way to separate out the weak foetal signals, but this analysis needs to be carried out over long periods of time where the stationarity hypothesis is not valid.

Figure 1 shows the first 10 seconds of the 12 sensor recordings, sampled at 512Hz, demonstrating the small magnitude of the foetal signals in the sensor outputs. This data set is relatively time-invariant has very few interfering signals e.g. muscle noise; this means that the signals are clear and good separation should be achievable.

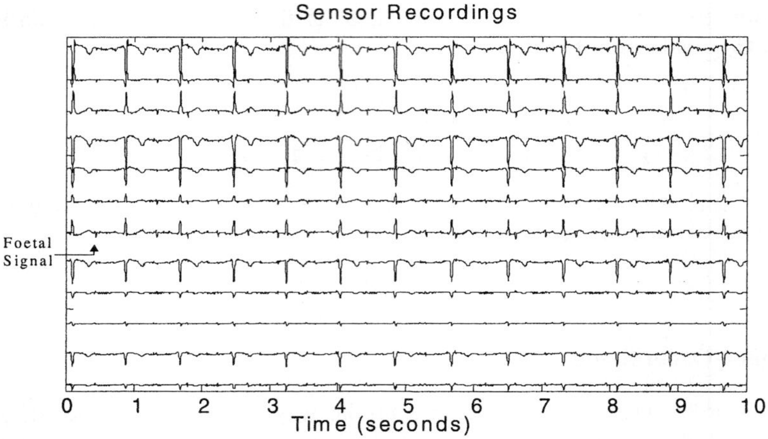

Fig. 1. Section of the sensor recordings

As in many real data analysis problems, qualitative performance measures are hard to find, so qualitative assessments on the quality of the separation must be made.

In figure 2, the first ten seconds of the signals separated by the EASI algorithm (learning rate 0.001) are shown. Note a larger learning rate caused the EASI algorithm to introduce signal swapping. The convergence problems of the EASI algorithm can be seen; signal breakthrough occurs up to 4000 samples into the data set. This effect will follow any sudden change in signal powers or steering vectors. The EASI algorithm did provide good separation on the remaining 290 seconds of this data set.

Figure 3 shows the first ten seconds of the signals separated by the tracking BSS technique (block size 5120 samples, 50% block-to-block overlap); the separation is clearer as no breakthrough is visible, and the F2 is more clearly separated. The time taken to process 1 second of input data by the tracking BSS algorithm was 0.04 seconds – demonstrating that the algorithm can lead to real-time processing. The algorithm was coded in C++ and run on a Dell Latitude 1.2GHz computer, and the figure quoted is for the average of 30 trials. Other experiments have shown that the tracking BSS technique can work on heavily artifact-corrupted data sets.

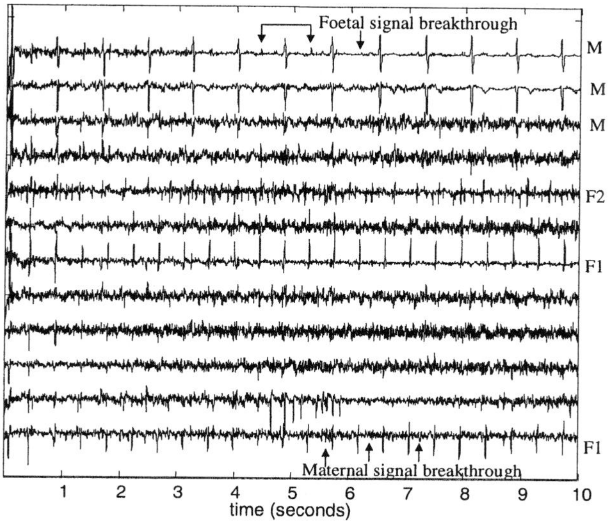

Fig. 2. First ten seconds extracts of the signals separated by EASI. The signals are denoted by M – maternal, F1 – foetus 1 and F2 – foetus 2.

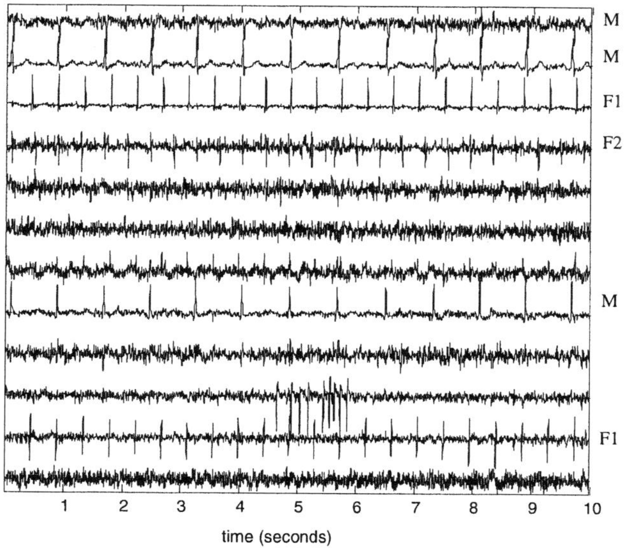

Fig. 3. First ten seconds extracts of the signals separated by the tracking BSS technique developed at QinetiQ. The signals are denoted by M - maternal, F1 - foetus 1 and F2 - foetus 2.

5 Conclusions

In this paper it has been shown that block-based blind signal separation (BSS) methods, combined with a moving window approach, can in principle be used for tracking real, non-stationary signals. However, the use of the moving window principle alone is not sufficient due to the introduction of signal swapping. We overcome this problem and show how past estimates can be efficiently used in the tracking process, to reduce the overall computational cost.

A demonstration of this tracking BSS approach is shown, where it is applied to a five-minute recording of twin foetal ECG data.

References

1. Bach, F., Jordan, M.: Kernel Independent Analysis. Report No UCB/CSD-01-1166, UCSE Berkerley, Presented at ICA2001, November 2001.
2. Bell, A., Sejnowski, T.: An Information-maximisation approach to blind separation and blind deconvolution, Neural Computation 7, 6, 1004-1034, 1995.
3. Cardoso, J., Souloumiac, A.: Blind beamforming for non-Gaussian signals, IEE proc-F, Vol. 140, no. 6, pp. 362-370, Dec 1993.
4. Cardoso, J., Laheld, B.: Equivariant adaptive source separation, IEEE Trans. on Signal Processing, vol. 44, no 12, pp. 3017-3030, Dec. 1996.
5. Haykin, S., Unsupervised adaptive filtering, John Wiley and sons, 2000.
6. Hyvarinen, A., Karhunen, J., Oja, E.: Independent Component Analysis, Wiley, 2001.
7. McWhirter, J., Clarke, I., Spence, G.: Multi-linear algebra for independent component analysis, SPIE's 44th Annual meeting, The international symposium on Optical Science, Engineering and Instrumentation, Denver USA, 18-23rd July, 1999.
8. Romero, S., Mananas, M., Clos, S., Gimenez, S., Barbanoj, M.: Reduction of EEG artifacts by ICA in different sleep stages, 25th annual conference of the IEEE EMBS, September 2003.
9. Spence, G., Clarke, I., McWhirter, J.: Dynamic Blind Signal Separation, UK Patent Application No. 0326539.4, Patent application filed 14th November 2003.
10. Taylor, M., Smith, M., Thomas, M., Green, A., Chenga, F., Oseku-Affula, S., Wee, L.: Non-invasive fetal electrocardiography in singleton and multiple pregnancies, BJOG: an International Journal of Obstetrics and Gynaecology, Vol. 110, pp. 668–678, July 2003.

Compression of Multicomponent Satellite Images Using Independent Components Analysis

Isidore Paul Akam Bita[1], Michel Barret[2], and Dinh-Tuan Antoine Pham[1]

[1] Laboratoire de Modélisation et Calcul, BP 53, 38041 Grenoble, France
{Isidore-Paul.Akambita, Dinh-Tuan.Pham}@imag.fr
[2] Supélec, 2 Rue Edouard Belin, 57070 Metz, France
Michel.Barret@supelec.fr

Abstract. In this paper we propose some compression schemes for multicomponent satellite images. These compression schemes use a classical bi-dimensional discrete wavelet transform (DWT) for spatial redundancy reduction, associated with linear transforms that reduce the spectral redundancy in an optimal way, for uniform scalar quantizers. These transforms are returned by Independent Component Analysis (ICA) algorithms which have been modified in order to maximize the compression gain under the assumption of high rate quantization and entropy coding. One algorithm, called `ICA_opt`, returns an optimal asymptotical linear transform and the other, called `ICA_orth`, returns an optimal asymptotical orthogonal transform. We compare the performance in high and medium rate coding of the Karhunen Loeve Transform (KLT) with the transforms returned by the modified ICA algorithms. These last transforms perform better than the KLT in term of compression gain in all cases, and in some cases the gain becomes significant.

1 Introduction

Multicomponent satellite images represent a very large amount of data that have to be transmitted and stored for different applications. The different components of the image generally represent the same scene with different views depending on the wavelength. That means that there is a high degree of dependence (or redundancy) between components. The aim of transform coding is to minimize these redundancy as far as possible in order to optimize the compression. In the last years, a lot of compression schemes have been proposed, which are generally characterized by the transformations used for redundancy reduction. The problem with the multicomponent images is that there is two kinds of redundancy: the spatial redundancy between the different pixels of a component and the spectral redundancy between the different components. It is important for compression to find a transformation which minimizes these two redundancies as far as possible. The DWT is known to be a good candidate for spatial redundancy reduction in each component, it is used in the new standard JPEG2000 and in other coding systems associated with zero tree coders like SPIHT (cf. [8] and its references). It has been shown in [5] that after the DWT a few redundancy remains between coefficients in a same subband or between coefficients in different

J. Rosca et al. (Eds.): ICA 2006, LNCS 3889, pp. 335–342, 2006.

subbands of a single component. However they are generally not very significant and can be partially captured by quite complex entropy coders like EBCOT [8]. The KLT is recommended in JPEG2000 [8] for spectral decorrelation[1] and is known to be optimal under the assumption of Gaussian sources and scale invariant quantizers [3] like uniform scalar quantizers. Narozny *et al.* [6] proposed two modified ICA algorithms which provide optimal linear transforms in transform coding, under the single assumptions of high rate entropy coding and scalar uniform quantizers (*without the Gaussian assumption of the data*). It was shown in this paper that if a multicomponent signal $\mathbf{X}(t) = [\mathbf{X}_1(t), \mathbf{X}_2(t), ..., \mathbf{X}_N(t)]$, $t = 1, \ldots, T$, is linearly transformed into $\mathbf{Y}(t) = \mathbf{AX}(t)$, with $\mathbf{A}$ an invertible matrix, and if the components $(Y_i(t))_{1\leq i\leq N}$ of $\mathbf{Y}(t)$ are quantized uniformly with an optimal allocation between quantizers and entropy coded, then the transform $\mathbf{A}$ that, for any (high) given bit rate, minimizes the distortion (mean square error) between $\mathbf{X}(t)$ and the reconstructed signal $\mathbf{A}^{-1}\mathbf{Y}^q(t)$ (with $\mathbf{Y}^q$ the vector of the quantized components of $\mathbf{Y}$), for any (high) given bit rate, is the one that minimizes the criterion $C(\mathbf{A}) = I(Y_1; \cdots; Y_N) + C_O(\mathbf{A})$, where $I(Y_1; \cdots; Y_N)$ is the classical mutual information criterion of ICA and

$$C_O(\mathbf{A}) = \frac{1}{2}\log_2 \frac{\det \operatorname{diag}(\mathbf{A}^{-T}\mathbf{A}^{-1})}{\det(\mathbf{A}^{-T}\mathbf{A}^{-1})}$$

diag denoting the operator which builds a diagonal matrix from the diagonal of its argument. The term $C_O(\mathbf{A})$ is non negative and can be zero if and only if $\mathbf{A}$ has orthogonal rows, hence can be viewed as a penalization term for non orthogonality. Based on the algorithm for minimizing the mutual information criterion of [7], Narozny *et al.* [6] have developd two modified versions called `Ica_opt` and `Ica_orth`, which minimize $C(\mathbf{A})$ without constraint and under the constraint that $\mathbf{A}$ has orthogonal rows.

In this paper, we extended the notion of generalized coding gain introduced in [6] to multicomponent images subjected to both bi-dimensional DWT for intra-component redundancy reduction and linear transforms for inter-component redundancy reduction. We say that those algorithms are optimal if they asymptotically minimize the mean rate coding subjected to a given distortion, which is the main goal in compression. Moreover, it can be shown that the transform provided with `ICA_orth` remains optimal when it is associated with any DWT and that the transform returned by `ICA_opt` remains optimal when it is associated with any orthogonal DWT.

The main idea of this paper is to associate a transform that accomplishes spatial redundancy reduction in each component with another one that accomplishes spectral redundancy reduction. The latter is accomplished thanks to the linear transforms returned by the modified ICA algorithms. Such a kind of structure has been studied in [9] on multispectral images using the KLT for spectral redundancy reduction and a vector quantization before coding. Another close study has been done by Dragotti *et al.* [2], where many cases have been compared including different transforms for the spatial and spectral redundancy reductions.

[1] It renders the spectral components uncorrelated but not necessary independent.

In section 2, we present the different compression structures studied in this paper. In section 3, we present two figures of merit for comparing transformations in transform coding, under the assumption of high rate entropy coding, one is called the generalized coding gain and the other the generalized reducible bits. In section 4, we show and discuss the results that have been obtained with the different schemes and various transforms.

2 The Different Compression Schemes

The block diagram of the compression schemes is represented in Fig. 1. The transform is divided into two distinct parts: the spatial redundancy reduction (accomplished by the DWT) and the spectral redundancy reduction which can be accomplished by the KLT, or the transform returned by either `ICA_opt` or `ICA_orth`.

For a given multispectral image $\mathbf{X}$ with N spectral components denoted $\mathbf{X}_i$ ($1 \leq i \leq N$), we first apply the same DWT to each component with the same level of decomposition (we use the 9/7 Daubechies DWT). Then we apply the spectral redundancy reduction transform in two different ways. One way consists of computing as many as optimal linear transforms as subbands, i.e., the spectral reduction of redundancy is accomplished subband by subband, and the other consists of computing only a same optimal linear transform for all the subbands. In the following we refer to them as the *first compression scheme* and the *second compression scheme*. For both compression schemes, each subband of each component is quantized with an infinite uniform scalar quantizer before entropy coding. We use the first order entropy in order to estimate the rate of quantized coefficients. The quantization step for each quantizer is chosen in such a way that the optimal bit allocation between quantizers is achieved under the assumption of high rate entropy coding [4]. The structures we study *differ* from the one recommended in JPEG2000 for multicomponent images, since in JPEG2000, the spectral redundancy reduction is performed *before* the reduction of spatial redundancy. As in [2], we observed that the best scores are obtained when the spatial redundancy reduction is accomplished first, like in our schemes. In order to reconstruct a decoded image, the reverse coding process is applied to the quantized coefficients, using the inverse transforms of the coding process. As a measure of distortion, we use the mean square error between the original image and the decoded image. It is well known that the Huffman coding or the arithmetic coding achieve performance very close to the first order entropy [1]. We can then evaluate the asymptotic rate-distortion curve of the image associated with the specific spectral transform used.

For hyperspectral images, the number N of spectral components is very large: a few thousands in general. In order to avoid the computation and memory usage needed for spectral redundancy reduction transform, we divide the image (along the spectral axis) into blocks of K components. We have thus $N_b = N/K$ blocks, assuming for simplicity that N is proportional to K so that N_b is an integer. We then consider each block as a multispectral image and we apply the compression

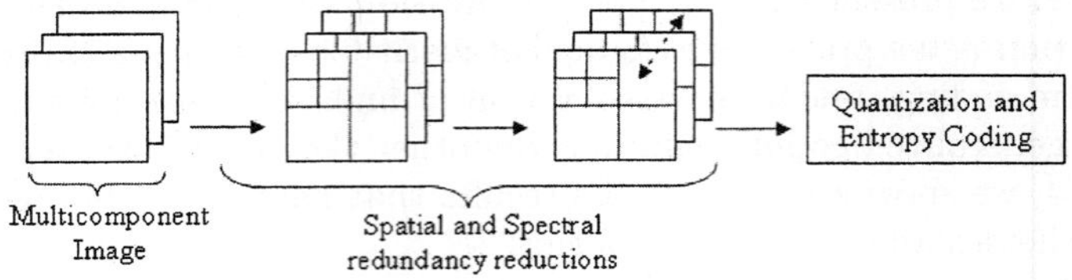

Fig. 1. The block diagram of the compression schemes

schemes described above to it. The final asymptotic rate-distortion curve of the hyperspectral image is taken as the average of the rate-distortion curves of the N_b different blocks.

3 Expression of the Asymptotic Distortion

In this section we give an approximation of the end-to-end distortion under some assumptions for multispectral images. First we give simple expressions of the generalized coding gain and the generalized reducible bits.

3.1 Generalized Coding Gain and Reducible Bits

In order to compare spectral transforms for high rate entropy coding, we define two figures of merit that are related one to the other. The first one is called the *generalized reducible bits*, expressed in bits per pixel (bpp) and the second one is called the *generalized coding gain*, expressed in decibels (dB). For both compression schemes (the first and the second), and for any spectral transformation $\mathcal{A}$, the generalized reducible bits associated with the distortion D is defined by

$$G_R(\mathcal{A}; D) = R_D(\mathcal{A}_0) - R_D(\mathcal{A}), \tag{1}$$

and the generalized coding gain associated with the rate R is given by

$$G_D(\mathcal{A}; R) = \text{PSNR}_R(\mathcal{A}) - \text{PSNR}_R(\mathcal{A}_0), \tag{2}$$

where $\mathcal{A}_0$ is the reference transform for spectral redundancy reduction (we will use either the identity transform or the KLT). The identity transform means that no spectral redundancy reduction transform is applied after the spatial one. The quantity $R_D(\mathcal{A})$ is the average rate per pixel and per component obtained with any of the two above described compression schemes, when the spectral redundancy reduction transform $\mathcal{A}$ is used with an optimal bit allocation between quantizers and an end-to-end distortion equals to D. The Peak Signal to Noise Ratio (PSNR) is a measure of distortion and is more often used in image compression than the mean square error. It is related to D and is defined by $\text{PSNR} = 10 \log_{10} \frac{(2^c-1)^2}{D}$, where c is the precision (i.e., the number of bits per pixel) of the original image; generally $c \in \{8, 12, 16\}$. The quantity $\text{PSNR}_R(\mathcal{A})$

is the PSNR associated with the spectral transformation $\mathcal{A}$ with an optimal bit allocation between quantizers and an average rate per pixel equals to R.

We can see that both definitions of generalized coding gain and generalized reducible bits are dual. Indeed, they depend on the asymptotic rate-distortion curve assuming that the optimal bit allocation between quantizers is achieved.

3.2 Expression of the Distortion in High Rate Entropy Coding

Under the assumption that the quantization noises are zero mean and independent, it can be shown that the end-to-end distortion D between the original image and the reconstructed one is well approximated by

$$D \approx \frac{1}{N} \sum_{m=1}^{M} \pi_m w_m \left[\sum_{i=1}^{N} w_i^{(m)} D_i^{(m)} \right] \tag{3}$$

where $D_i^{(m)}$ is the mean square error between the transformed coefficients of the mth subband of the ith component and their quantized values. Let M denote the number of subbands. Since we choose high rate scalar uniform quantizers, the mean square error $D_i^{(m)}$ depends only on the quantization step $q_i^{(m)}$: $D_i^{(m)} \approx [q_i^{(m)}]^2/12$. The factor π_m is the ratio between the number of wavelet coefficients of the subband m and the number of pixels of a single component. The weighting factor w_m depends only on the filters used in the inverse DWT. In [10], a method for evaluating these weighting factors in the case of monodimensional wavelets is clearly explained, some explanations for extending the method to bi-dimensional wavelets are also given. In the first compression scheme, the coefficient $w_i^{(m)}$ depends only on the spectral redundancy reduction transformation associated with the subband m and the component i. It can be computed from the relation

$$w_i^{(m)} = \sum_{j=1}^{N} [\mathcal{A}_{ji}^{(m)-1}]^2 \tag{4}$$

where $(\mathcal{A})_{ji}^{(m)-1}$ is the matrix element on the jth row and the ith column of the inverse spectral transform associated with the mth subband. The expression between the brackets in (3) can be interpreted as the mean square error of quantization for all the transformed coefficients of the mth subband belonging to the N components. For an orthogonal transform, like KLT or the one returned by `ICA_orth`, all the coefficients $w_i^{(m)}$ are equal to one. In the second compression scheme, the factors $w_i^{(m)}$ do not depend on m. Indeed, we have $w_i^{(m)} = \sum_{j=1}^{N} (\mathcal{A}_{ji}^{-1})^2$, where $\mathcal{A}$ is the linear transform applied in each subband to reduce spectral redundancy.

3.3 Optimal Bit Allocation

Under high rate assumption, it is well known [4] that the first order entropy $H(Q[X])$ of a uniform scalar quantized random variable X, is well approximated by the Bennet's approximation

$$H(Q[X]) \approx h(X) - \frac{1}{2}\log_2(12D_q) \tag{5}$$

where D_q is the mean square error between X and its quantized value and $h(X) = -\int f(x)\log_2 f(x)\,dx$ is the differential entropy of X. The average rate per pixel and per component of the first compression scheme is given by $R = \frac{1}{N}\sum_{i=1}^{N}\sum_{m=1}^{M}\pi_m R_i^{(m)}$, where $R_i^{(m)}$ denotes the first order entropy of the quantized coefficients of the mth subband of the ith component. The optimal bit allocation problem can be stated as follows: for a given rate R_c, how to allocate the scalar quantization step of each uniform scalar quantizer in such a way that the end-to-end distortion is minimized with the constraint that the resulting global average R is not greater than R_c. Due to the duality of the rate-distortion and the distortion-rate curves, this problem can also be stated as: given an end-to-end distortion D_c, how to find the optimal quantization step of each uniform scalar quantizer in order to minimize the global average rate with the constraint that the resulting end-to-end distortion D is not greater than D_c. This constrained problem can easily be solved using the Lagrange multiplier method [1].

Using the Bennet's approximation for evaluating the first order entropy, combined with the approximation (3) of the end-to-end distortion and the global mean rate expression, we find that the optimal bit allocation is obtained when the mean square error $D_i^{(m)}$ satisfies the condition

$$D_i^{(m)} = \frac{D_c}{w_m w_i^{(m)}} \tag{6}$$

Hence, since $D_i^{(m)} \approx [q_i^{(m)}]^2/12$, we can find out the quantization step $q_i^{(m)}$ of each subband and each component.

4 Evaluation of the Generalized Coding Gain

We tested the both compression schemes on satellite images, some are multispectral images of towns[2] in France and others are AVIRIS[3] hyperspectral images. The multispectral images are sights of Vannes, Moissac and Strasbourg. All of them have four spectral components and an original resolution of 12 bpp. The hyperspectral images have each 224 spectral components (the wavelength varying from the visible to the infra-red) and an original resolution of 16 bpp. We evaluated the generalized coding gain and the generalized reducible bits of three specific spectral redundancy reduction transforms: KLT, ICA_orth (returned by `ICA_orth`) and ICA_opt (returned by `ICA_opt`). Since the KLT is commonly used (it cancels statistical correlation and hence is optimal under the assumption of Gaussian data), we compare the performance of the modified ICA transforms with the KLT.

[2] These images have been given as a favor by the French National Spatial Studies Center, CNES.

[3] These images have been given as a favor by the NASA.

Table 1. Generalized coding gain in dB for the two compression schemes

	First scheme			Second scheme		
	KLT	ICA_orth	ICA_opt	KLT	ICA_orth	ICA_opt
Moissac	6.1	6.16	6.2	5.95	6.01	6.03
Strasbourg	5.84	5.91	5.93	5.78	5.88	5.90
AVIRIS1	10.7	13.8	14	10.6	12.3	12.8
AVIRIS2	12	14.4	14.5	11	13.1	13.6

First, we verified that the approximation (3) is a good one for the end-to-end distortion. Indeed, for different images and distortions chosen at random, the relative approximation error using (3) is less than 1% for high and medium rates, which comforts us in using the relation (6) for optimal bit allocation. The generalized coding gain are plotted for the multispectral image Vannes in Fig. 2. The results obtained with the others multispectral images are summarized in Table 1. Looking at the results we can see that for the two compression

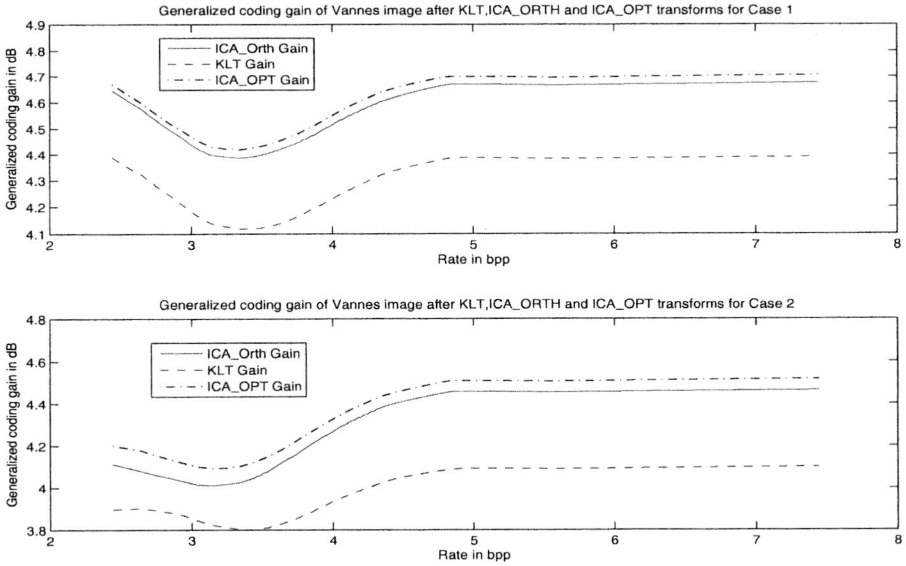

Fig. 2. Generalized coding gain compared with the identity transform of Vannes. Top: for the first scheme, bottom: for the second scheme.

schemes, the new transforms ICA_orth and ICA_opt, obtained from modified ICA algorithms, perform very well and give significant compression gain compare to the identity spectral transform. They also perform better than the KLT, with more than 1 dB of compression gain in some cases. Also, we can see that the performance of the two compression schemes are quite close, and the second has the advantage of less memory and computing complexities.

The generalized coding gain of Table 1 are obtained by averaging the generalized coding gain for rates between 2.5 and 5 bpp per component, in which it varies slightly. One inconvenience of the proposed schemes is that the spectral transforms need a lot of memory ressources and computing time. That is due to the fact that the modified ICA algorithms work iteratively and need to perform some complex statistic functions for an important quantity of data.

5 Conclusion

We proposed two compression schemes for multicomponent satellite images using both spatial and spectral redundancy reductions. The latter is achieved by using new transforms returned by modified ICA algorithms. The results of our simulations show a good performance of the methods and in many cases, the modified ICA algorithms perform better than the KLT. The second compression scheme suffers little loss of performance with respect to the first one, but it has the advantage of being much simpler in term of complexity of computations.

References

1. Thomas M. Cover and Joy A. Thomas, *Elements of Information Theory*, John Wiley and Sons (1991).
2. P. L. Dragotti, G. Poggi and A. R P. Ragozini, Compression of Multispectral Images by Three-Dimensional SPIHT Algorith, IEEE Transaction on Geoscience and Remote Sensing, Vol. 38 No. 1 (2000) 416–428.
3. K. Goyal, J. Zhuang and M. Vetterli, Transform Coding With Backward Adaptive Updates, IEEE Transactions Information Theory, Vol. 46 No. 3 (1997) 1623–1633.
4. A. Gersho and R. M. Gray, *Vector quantization and signal compression*, Kluwer Academic Publisher (1992).
5. J. Liu and P. Moulin, Information-Theoretic Analysis of Interscale and Intrascale Dependencies Between Image Wavelet Coefficients, IEEE Transactions on Image Processing, Vol. 10 No. 11 (2001) 1647–1658.
6. M. Narozny, D.-T. Pham, M. Barret and I. P. Akam Bita, Modified ICA Algorithm for Finding Optimal Transforms in Transform coding, IEEE 4th Int. Symp. on Image and Signal Processing and Analysis (2005) 111–116.
7. D.-T. Pham, Fast Algorithms for Mutual Information Based Independent Component Analysis, IEEE Transactions on Signal Processing, Vol. 52 No. 10 (2004).
8. D. S. Taubman and M. W. Marcellin, *JPEG2000, Image compression fundamentals, standards and practice.* Kluwer Academic Publishers (2004).
9. J. Vaisey, M. Barlaud and M. Antonini, Multispectral Image Coding Using Lattice VQ and the Wavelet Transform, IEEE International Conference on Image Processing, Chicago, USA (1998).
10. J. W. Woods and T. Naveen, A Filter Based Bit Allocation Scheme for Subband Compression HDTV, IEEE Transactions on Image Processing, Vol. 1 No. 3 (1992) 436–440.

Fixed-Point Complex ICA Algorithms for the Blind Separation of Sources Using Their Real or Imaginary Components

Scott C. Douglas[1], Jan Eriksson[2], and Visa Koivunen[2]

[1] Department of Electrical Engineering,
Southern Methodist University, Dallas, Texas 75275, USA
[2] Signal Processing Laboratory, SMARAD CoE,
Helsinki University of Technology,
Espoo 02015, Finland

Abstract. The complex-valued signal model is useful for several practical applications, yet few algorithms for separating complex linear mixtures exist. This paper develops two algorithms for separating mixtures of independent complex-valued signals in which statistical independence of the real and imaginary components is assumed. The procedures extract sources assuming that the kurtoses of either the real or imaginary components are non-zero. Simulations indicate the efficacy of the methods in performing source separation for wireless communications models.

1 Introduction

The goal of blind source separation is to find an $(m \times m)$ matrix $\mathbf{B}$ such that

$$\mathbf{y}(k) = \mathbf{B}\mathbf{x}(k) \quad \text{and} \quad \mathbf{x}(k) = \mathbf{A}\mathbf{s}(k), \tag{1}$$

where $\mathbf{y}(k)$ contains estimates of the m sources in $\mathbf{s}(k)$, $\mathbf{A}$ is full rank, and $\mathbf{s}(k)$ is typically assumed to contain independent signals. This paper focuses on complex-valued source separation, in which all quantities in (1) are complex-valued. Few algorithms have been developed for complex ICA [1, 2, 3, 4, 5]. The complex FastICA procedure in [2] uses a circular contrast and may not perform well with mixtures containing non-circular sources such as real-valued BPSK signals.

In this paper, we consider algorithms for separating complex-valued signal mixtures using fourth-moment contrasts, in which the sources in $\mathbf{s}(k)$ are assumed to have independent real- and imaginary components. Such an assumption is quite reasonable in some applications, particularly in multiple-input, multiple-output (MIMO) wireless communications systems where higher-order modulation schemes are used. We develop two procedures that employ modified versions of the FastICA algorithm to extract each of the m complex sources based on the statistics of either their real or imaginary component. Simulations show the efficacy of the proposed methods for complex-valued source separation.

J. Rosca et al. (Eds.): ICA 2006, LNCS 3889, pp. 343–351, 2006.

2 On Mixtures of Complex-Valued Signals

Without loss of generality, assume that the sources in $\mathbf{s}(k) = \mathbf{s}_R(k) + j\mathbf{s}_I(k)$ are zero-mean and strong-uncorrelated [6, 7], such that the source covariance and pseudo-covariance matrices are $E\{\mathbf{s}(k)\mathbf{s}^H(k)\} = \mathbf{I}$ and $E\{\mathbf{s}(k)\mathbf{s}^T(k)\} = \boldsymbol{\Lambda}$, and $\boldsymbol{\Lambda}$ is a diagonal matrix of real-valued circularity coefficients λ_i with $0 \leq \lambda_i \leq 1$.

Consider an algorithm that adjusts a single row $\mathbf{b} = [b_1 \ \cdots \ b_m]^T$ of $\mathbf{B}$ in (1) to extract a source $s_i(k)$ based on the statistics of its real or imaginary component $s_{R,i}(k)$ or $s_{I,i}(k)$. The output signal is $y(k) = y_R(k) + jy_I(k) = \mathbf{b}^T\mathbf{x}(k) = \mathbf{c}^T\mathbf{s}(k)$, where $\mathbf{c} = \mathbf{A}^T\mathbf{b} = \mathbf{c}_R + j\mathbf{c}_I$. Thus, $y_R(k) = \mathbf{c}_R^T\mathbf{s}_R(k) - \mathbf{c}_I^T\mathbf{s}_I(k)$ and $y_I(k) = \mathbf{c}_I^T\mathbf{s}_R(k) + \mathbf{c}_R^T\mathbf{s}_I(k)$. The normalized kurtoses of $s_{R,i}(k)$ and $s_{I,i}(k)$ are

$$\kappa_{R,i} = \frac{2E\{s_{R,i}^4(k)\}}{(1+\lambda_i)^2} - 3 \ \text{ and } \ \kappa_{I,i} = \begin{cases} \dfrac{2E\{s_{I,i}^4(k)\}}{(1-\lambda_i)^2} - 3, & \text{if } 0 \leq \lambda_i < 1 \\ 0, & \text{if } \lambda_i = 1 \end{cases} . \tag{2}$$

The quantities $\kappa_{R,i}$ and $\kappa_{I,i}$ are related to the symmetric kurtosis of $s_i(k)$ as

$$\kappa_i = \frac{1}{4}\left[(1+\lambda_i)^2\kappa_{R,i} + (1-\lambda_i)^2\kappa_{I,i}\right]. \tag{3}$$

Theorem 1. *Under the above conditions, the real and imaginary components $y_R(k)$ and $y_I(k)$ of $y(k)$ have the following second moments and kurtoses:*

$$E\{y_R^2(k)\} = \frac{1}{2}\sum_{i=1}^{m}(1+\lambda_i)c_{R,i}^2 + (1-\lambda_i)c_{I,i}^2 \tag{4}$$

$$E\{y_I^2(k)\} = \frac{1}{2}\sum_{i=1}^{m}(1-\lambda_i)c_{R,i}^2 + (1+\lambda_i)c_{I,i}^2, \quad E\{y_R(k)y_I(k)\} = \sum_{i=1}^{m}\lambda_i c_{R,i}c_{I,i} \tag{5}$$

$$\begin{aligned}\kappa[y_R(k)] = &\sum_{i=1}^{m}\left[\frac{\kappa_{R,i}}{4}(1+\lambda_i)^2c_{R,i}^4 + \frac{\kappa_{I,i}}{4}(1-\lambda_i)^2c_{I,i}^4\right] \\ &+\frac{3}{2}\sum_{i=1}^{m}\left[E\{s_{R,i}^2(k)s_{I,i}^2(k)\} - \frac{1}{4}(1-\lambda_i^2)\right]c_{R,i}^2c_{I,i}^2 \\ &-4\sum_{i=1}^{m}c_{R,i}^3c_{I,i}E\{s_R^3(k)s_I(k)\} + c_{R,i}c_{I,i}^3E\{s_R(k)s_I^3(k)\}\end{aligned} \tag{6}$$

$$\begin{aligned}\kappa[y_I(k)] = &\sum_{i=1}^{m}\left[\frac{\kappa_{R,i}}{4}(1+\lambda_i)^2c_{I,i}^4 + \frac{\kappa_{I,i}}{4}(1-\lambda_i)^2c_{R,i}^4\right] \\ &+\frac{3}{2}\sum_{i=1}^{m}\left[E\{s_{R,i}^2(k)s_{I,i}^2(k)\} - \frac{1}{4}(1-\lambda_i^2)\right]c_{R,i}^2c_{I,i}^2 \\ &+4\sum_{i=1}^{m}c_{I,i}^3c_{R,i}E\{s_R^3(k)s_I(k)\} + c_{I,i}c_{R,i}^3E\{s_R(k)s_I^3(k)\}\end{aligned} \tag{7}$$

Corollary 1.1: Under the additional assumption that $s_{R,i}(k)$ and $s_{I,i}(k)$ are independent for all $1 \leq i \leq m$,

$$\kappa[y_R(k)] = \sum_{i=1}^{m} \left[\frac{\kappa_{R,i}}{4}(1+\lambda_i)^2 c_{R,i}^4 + \frac{\kappa_{I,i}}{4}(1-\lambda_i)^2 c_{I,i}^4\right] \tag{8}$$

$$\kappa[y_I(k)] = \sum_{i=1}^{m} \left[\frac{\kappa_{R,i}}{4}(1+\lambda_i)^2 c_{I,i}^4 + \frac{\kappa_{I,i}}{4}(1-\lambda_i)^2 c_{R,i}^4\right] \tag{9}$$

The fourth-order statistical structures of the real and imaginary components of linearly-mixed, statistically-independent strong-uncorrelated, and possibly non-circular complex sources is not as simple as in the real-valued case. If all $s_{R,i}(k)$ and $s_{I,i}(k)$ are jointly statistically-independent, however, (8) and (9) are similar in structure to the real-valued case, leading to the following theorem.

Theorem 2. *Consider the single-unit extraction criterion*

$$\mathcal{J}(\mathbf{b}) = \left|\frac{\kappa[y_R(k)]}{(E\{|y_R(k)|^2\})^2}\right| \tag{10}$$

where $y(k) = \mathbf{b}^T\mathbf{x}(k) = \mathbf{c}^T\mathbf{s}(k) = y_R(k) + jy_I(k)$. Assume that all *of the sources are statistically-independent with statistically-independent real and imaginary parts, and at least one of the sources has a real and/or imaginary part with $\kappa_{R,i} \neq 0$ and/or $\kappa_{I,i} \neq 0$. Then, maximization of $\mathcal{J}(\mathbf{b})$ over $\mathbf{b}$ under the constraint $E\{y_R^2(k)\} = 1$ yields one of the columns of $\mathbf{A}^{-1}$ for which $\kappa_{R,i} \neq 0$ or $\kappa_{I,i} \neq 0$, up to a complex scaling factor $e^{j\pi p/2}$, where p is an integer.*

Proof: Define the $(2m)$-dimensional real-valued vector $\overline{\mathbf{c}}_{\mathcal{R}} = \sqrt{2}[\mathbf{c}_R^T[\mathbf{I}+\mathbf{\Lambda}]^{-1/2}\ \mathbf{c}_I^T[\mathbf{I}-\mathbf{\Lambda}]^{+/2}]^T$, with entries $\{\overline{c}_i\}$, where $\mathbf{N}^+$ denotes the pseudo-inverse of a square matrix $\mathbf{N}$. Let $\overline{\kappa}_i$ denote the $(2m)$-element sequence $\{\kappa_{R,1},\ldots,\kappa_{R,m}, \kappa_{I,1},\ldots,\kappa_{I,m}\}$. Substituting these relations into (8) and and (4) yields

$$\kappa_{\mathcal{R}}[y_R(k)] = \sum_{i=1}^{2m} \overline{\kappa}_i \overline{c}_i^4 \quad \text{and} \quad E\{y_R^2(k)\} = \sum_{i=1}^{2m} \overline{c}_i^2. \tag{11}$$

The relations in (11) are identical to those in the $2m$-dimensional real-valued separation case. Thus, constrained maximization of $\mathcal{J}(\mathbf{b})$ results in an extracted source with a non-zero-kurtosis real or imaginary component. The one non-zero coefficient of $\mathbf{b}^T\mathbf{A}$ equals $e^{j\pi p/2}$ because (i) absolute signs of the $\{\overline{c}_i\}$ do not matter, and (ii) the real or imaginary component of a source could be extracted.

3 FastICA Algorithms for Extracting a Single Source with Independent Real and Imaginary Components

We now develop fast-converging single-unit procedures to extract one source from mixtures of sources having independent real and imaginary components. Two methods are considered. The first algorithm relies on the strong-uncorrelating transform $\mathbf{G}$ that diagonalizes both the sample covariance and pseudo-covariance matrices $\mathbf{R}_{XX} = E\{\mathbf{x}(k)\mathbf{x}^H(k)\}$ and $\mathbf{P}_{XX} = E\{\mathbf{x}(k)\mathbf{x}^T(k)\}$, respectively [7]. Let $\boldsymbol{\Gamma} = \mathbf{GA} = \boldsymbol{\Gamma}_R + j\boldsymbol{\Gamma}_I$ and $\mathbf{v}(k) = \mathbf{Gx}(k) = \mathbf{v}_R(k) + j\mathbf{v}_I(k)$. Then, $\mathbf{s}(k)$ and $\mathbf{v}(k)$ are related as

$$\begin{bmatrix} \mathbf{v}_R(k) \\ \mathbf{v}_I(k) \end{bmatrix} = \begin{bmatrix} \boldsymbol{\Gamma}_R & -\boldsymbol{\Gamma}_I \\ \boldsymbol{\Gamma}_I & \boldsymbol{\Gamma}_R \end{bmatrix} \begin{bmatrix} \mathbf{s}_R(k) \\ \mathbf{s}_I(k) \end{bmatrix}. \tag{12}$$

The matrix premultiplying $[\mathbf{s}_R^T(k)\ \mathbf{s}_I^T(k)]^T$ on the right-hand side of (12) is real-valued and orthogonal. Moreover, under strong-uncorrelation, $E\{\mathbf{v}_R(k)\mathbf{v}_R^T(k)\} = \frac{1}{2}(\mathbf{I} + \widehat{\boldsymbol{\Lambda}})$, $E\{\mathbf{v}_I(k)\mathbf{v}_I^T(k)\} = \frac{1}{2}(\mathbf{I} - \widehat{\boldsymbol{\Lambda}})$, and $E\{\mathbf{v}_R(n)\mathbf{v}_I^T(n)\} = \mathbf{0}$, where $\widehat{\boldsymbol{\Lambda}} = \mathbf{GP}_{XX}\mathbf{G}^T$ is diagonal. Since the elements of $\mathbf{v}_R(k)$ and $\mathbf{v}_I(k)$ are not unit variance as required by the real-valued FastICA algorithm, define

$$\underline{\mathbf{v}}(k) = \begin{bmatrix} \overline{\mathbf{v}}_R(k) \\ -\overline{\mathbf{v}}_I(k) \end{bmatrix} = \begin{bmatrix} \sqrt{2}(\mathbf{I} + \widehat{\boldsymbol{\Lambda}})^{-1/2}\mathbf{v}_R(k) \\ -\sqrt{2}(\mathbf{I} - \widehat{\boldsymbol{\Lambda}})^{+1/2}\mathbf{v}_I(k) \end{bmatrix}. \tag{13}$$

$$\underline{\mathbf{s}}(k) = \begin{bmatrix} \overline{\mathbf{s}}_R(k) \\ -\overline{\mathbf{s}}_I(k) \end{bmatrix} = \begin{bmatrix} \sqrt{2}(\mathbf{I} + \widehat{\boldsymbol{\Lambda}})^{-1/2}\mathbf{s}_R(k) \\ -\sqrt{2}(\mathbf{I} - \widehat{\boldsymbol{\Lambda}})^{+1/2}\mathbf{s}_I(k) \end{bmatrix}. \tag{14}$$

The scaling operations in (13)–(14) are not valid in the space of complex matrices. Despite this fact, the orthonormal mixing properties between the sources in $\mathbf{s}(k)$ and the prewhitened mixture $\mathbf{v}(k)$ are maintained with this scaling.

Theorem 3. *Let* $\overline{\mathbf{v}}(k) = \overline{\mathbf{v}}_R(k) + j\overline{\mathbf{v}}_I(k)$ *and* $\overline{\mathbf{s}}(k) = \overline{\mathbf{s}}_R(k) + j\overline{\mathbf{s}}_I(k)$. *Then, under strong-uncorrelation, the relationship between* $\overline{\mathbf{v}}(k)$ *and* $\overline{\mathbf{s}}(k)$ *is identical to that between* $\mathbf{v}(k)$ *and* $\mathbf{s}(k)$, *i.e.* $\overline{\mathbf{v}}(k) = \boldsymbol{\Gamma}\overline{\mathbf{s}}(k)$ with $\boldsymbol{\Gamma}\boldsymbol{\Gamma}^H = \boldsymbol{\Gamma}^T\boldsymbol{\Gamma}^* = \mathbf{I}$.

Proof: The proof is obtained by considering the structure of linearly-mixed strong-uncorrelated random variables as described in [5, 7] and is omitted for brevity.

The above theorem allows us to proceed with the specification of the FastICA algorithm in this case, as $E\{\underline{\mathbf{s}}(k)\underline{\mathbf{s}}^T(k)\} = \mathbf{I}$. All of the identifiability, uniqueness, and separability results for complex-valued ICA are preserved [7].

Given the relationship $\mathbf{w}_t = \mathbf{w}_{R,t} + j\mathbf{w}_{I,t}$, let $\underline{\mathbf{w}}_t = [\mathbf{w}_{R,t}^T\ \ \mathbf{w}_{I,t}^T]^T$, and define the output of the single-unit extraction system as $\overline{y}_t(k) = \underline{\mathbf{w}}_t^T\underline{\mathbf{v}}(k)$. It can be easily shown that $\overline{y}_t(k) = \Re e[\mathbf{w}_t^T\overline{\mathbf{v}}(k)]$. Since $\underline{\mathbf{v}}(k)$ contains an orthogonally-mixed set of $(2m)$ independent, real-valued sources with zero means and unit variances, we can use the standard real-valued FastICA procedure with kurtosis contrast to adjust the coefficients in $\underline{\mathbf{w}}_t$ as

$$\widetilde{\underline{\mathbf{w}}}_t = \left(\frac{1}{N}\sum_{n=1}^{N} \overline{y}_t^3(n)\underline{\mathbf{v}}(n)\right) - 3\underline{\mathbf{w}}_t, \qquad \underline{\mathbf{w}}_{t+1} = \frac{\widetilde{\underline{\mathbf{w}}}_t}{\sqrt{\widetilde{\underline{\mathbf{w}}}_t^T\widetilde{\underline{\mathbf{w}}}_t}}. \tag{15}$$

As has been shown in [8] for real-valued mixtures, this algorithm is guaranteed to converge to an extracting solution, which for our data structure means that one of the real or imaginary components of $\mathbf{s}(k)$ is obtained in $\overline{y}_t(k)$ with unit-variance scaling and (possibly) a sign change.

The algorithm in (15) requires the strong-uncorrelating transform, which requires specialized code to compute in the general case. It is possible to design a single-unit FastICA procedure to separate mixtures of complex-valued sources using only ordinary prewhitening. In this version, find any prewhitening matrix $\widehat{\mathbf{G}}$ satisfying $\widehat{\mathbf{G}}\mathbf{R}_{XX}\widehat{\mathbf{G}}^H = \mathbf{I}$, and set $\mathbf{v}(k) = \widehat{\mathbf{G}}\mathbf{x}(k)$. The pseudo-covariance matrix, which is not needed here, is not diagonal. The relationship between $\mathbf{v}(k)$ and $\mathbf{s}(k)$ is given in complex form by $\mathbf{v}(k) = \widehat{\boldsymbol{\Gamma}}\mathbf{s}(k)$, or in real form as in (12) with $\boldsymbol{\Gamma} = \widehat{\boldsymbol{\Gamma}}_R + j\widehat{\boldsymbol{\Gamma}}_I$. Define the $(2m)$-dimensional vectors of real-valued elements $\underline{\mathbf{v}}(k)$ and $\underline{\mathbf{s}}(k)$ as

$$\underline{\mathbf{v}}(k) = \begin{bmatrix} \mathbf{v}_R(k) \\ -\mathbf{v}_I(k) \end{bmatrix} \text{ and } \underline{\mathbf{s}}(k) = \begin{bmatrix} \mathbf{s}_R(k) \\ -\mathbf{s}_I(k) \end{bmatrix} \tag{16}$$

Then, the sample autocorrelation matrix of $\mathbf{v}_R(k)$ is

$$\widehat{\mathbf{R}}_{VV} = \frac{1}{N}\sum_{n=1}^{N} \underline{\mathbf{v}}(n)\underline{\mathbf{v}}^T(n) \tag{17}$$

$$= \begin{bmatrix} \widehat{\boldsymbol{\Gamma}}_R & \widehat{\boldsymbol{\Gamma}}_I \\ -\widehat{\boldsymbol{\Gamma}}_I & \widehat{\boldsymbol{\Gamma}}_R \end{bmatrix} \left(\frac{1}{N}\sum_{n=1}^{N} \begin{bmatrix} \mathbf{s}_R(n)\mathbf{s}_R^T(n) & \mathbf{s}_R(n)\mathbf{s}_I^T(n) \\ \mathbf{s}_I(n)\mathbf{s}_R^T(n) & \mathbf{s}_I(n)\mathbf{s}_I^T(n) \end{bmatrix}\right) \begin{bmatrix} \widehat{\boldsymbol{\Gamma}}_R^T & -\widehat{\boldsymbol{\Gamma}}_I^T \\ \widehat{\boldsymbol{\Gamma}}_I^T & \widehat{\boldsymbol{\Gamma}}_R^T \end{bmatrix} \tag{18}$$

In the limit at $N \to \infty$, $\widehat{\mathbf{R}}_{VV}$ is not diagonal. Moreover, the powers in the $(2m)$ real-valued sources in $\underline{\mathbf{s}}(k)$ are not unity. Thus, a pair of fundamental assumptions about the FastICA procedure do not hold. Even so, we can derive a modified FastICA procedure to obtain one of the non-zero-kurtosis sources in $\underline{\mathbf{s}}(k)$; see [9] for a similar derivation of a different algorithm. Define $\underline{\mathbf{w}}_t$ as the system vector, and let $\underline{y}_t(k) = \underline{\mathbf{w}}_t^T\underline{\mathbf{v}}(k)$. Define

$$\underline{\mathbf{c}}_t = \begin{bmatrix} \mathbf{c}_{R,t} \\ -\mathbf{c}_{I,t} \end{bmatrix}, \; \underline{\widehat{\boldsymbol{\Gamma}}} = \begin{bmatrix} \widehat{\boldsymbol{\Gamma}}_R & \widehat{\boldsymbol{\Gamma}}_I \\ -\widehat{\boldsymbol{\Gamma}}_I & \widehat{\boldsymbol{\Gamma}}_R \end{bmatrix}, \quad \text{and} \quad \underline{\boldsymbol{\Lambda}}_S = \begin{bmatrix} \mathbf{I}+\boldsymbol{\Lambda} & \mathbf{0} \\ \mathbf{0} & \mathbf{I}-\boldsymbol{\Lambda} \end{bmatrix}. \tag{19}$$

Then, we set $\underline{y}_t(k) = \underline{\mathbf{c}}_t^T\underline{\mathbf{s}}(k)$ and $\underline{\mathbf{c}}_t = \underline{\widehat{\boldsymbol{\Gamma}}}^T\underline{\mathbf{w}}_t$. Consider the fourth moment term

$$E\{|\underline{y}_t(k)|^4\} = \underline{\mathbf{c}}_t^T E\{\underline{\mathbf{s}}(k)\underline{\mathbf{s}}^T(k)\underline{\mathbf{c}}_t\underline{\mathbf{c}}_t^T\underline{\mathbf{s}}(k)\underline{\mathbf{s}}^T(k)\}\underline{\mathbf{c}}_t. \tag{20}$$

It is straightforward to show that

$$E\{\underline{\mathbf{s}}(k)\underline{\mathbf{s}}^T(k)\underline{\mathbf{c}}_t\underline{\mathbf{c}}_t^T\underline{\mathbf{s}}(k)\underline{\mathbf{s}}^T(k)\} = 2\underline{\boldsymbol{\Lambda}}_S\underline{\mathbf{c}}_t\underline{\mathbf{c}}_t^T\underline{\boldsymbol{\Lambda}}_S + \underline{\boldsymbol{\Lambda}}\underline{\mathbf{c}}_t^T\underline{\boldsymbol{\Lambda}}_S\underline{\mathbf{c}}_t + \underline{\mathbf{K}}\text{diag}[\underline{\mathbf{c}}_t\underline{\mathbf{c}}_t^T], \tag{21}$$

such that the desired update is

$$\widetilde{\underline{\mathbf{c}}}_t = E\{\underline{\mathbf{s}}(k)\underline{\mathbf{s}}^T(k)\underline{\mathbf{c}}_t\underline{\mathbf{c}}_t^T\underline{\mathbf{s}}(k)\underline{\mathbf{s}}^T(k)\}\underline{\mathbf{c}}_t - 3\underline{\boldsymbol{\Lambda}}_S\underline{\mathbf{c}}_t\underline{\mathbf{c}}_t^T\underline{\boldsymbol{\Lambda}}_S\underline{\mathbf{c}}_t. \tag{22}$$

The power constraint changes to $E\{\underline{y}_t^2(k)\} = \underline{\mathbf{c}}_t^T \boldsymbol{\Lambda}_S \underline{\mathbf{c}}_t = (1 \pm \lambda_i)$, and it is met when $\underline{\mathbf{c}}_t$ has only one non-zero unity-valued element, which is equivalent to $\underline{\mathbf{w}}_t^T \underline{\mathbf{w}}_t = 1$. Transforming back to the coordinates $\underline{\mathbf{w}}_t$, we obtain the update

$$\widetilde{\underline{\mathbf{w}}}_t = \left(\frac{1}{N}\sum_{n=1}^{N} \underline{y}_t^3(n)\underline{\mathbf{v}}(n)\right) - 3\widehat{\mathbf{R}}_{VV}\underline{\mathbf{w}}_t\left(\frac{1}{N}\sum_{n=1}^{N}\underline{y}^2(n)\right), \quad \underline{\mathbf{w}}_{t+1} = \frac{\widetilde{\underline{\mathbf{w}}}_t}{\sqrt{\widetilde{\underline{\mathbf{w}}}_t^T \widetilde{\underline{\mathbf{w}}}_t}} \tag{23}$$

4 Designing Multiple-Component Extraction Procedures

For either of our single-source extraction procedures in (15) or (23) , we now develop extensions that employ multiple parallel systems to extract each of the source components within the mixture. We exploit the structure of the complex prewhitened mixing system as indicated in (12) to make this task easier. Suppose that $\mathbf{w}_t = \mathbf{w}_{R,t} + j\mathbf{w}_{it}$ of either algorithm has converged such that $\Re e[\mathbf{w}_t^T \mathbf{v}(k)] = \underline{d}_{R,i} s_{R,i}(k)$ for some real-valued scalar $\underline{d}_{R,i}$. Then, $\Im m[\mathbf{w}_t^T \mathbf{v}(k)] = \underline{d}_{I,i} s_{I,i}(k)$ for some real-valued scalar $\underline{d}_{I,i}$. Similarly, if $\Re e[\mathbf{w}_t^T \mathbf{v}(k)] = \underline{d}_{I,i} s_{I,i}(k)$, then $\Im m[\mathbf{w}_t^T \mathbf{v}(k)] = \underline{d}_{R,i} s_{R,i}(k)$. In other words, extracting any real (or imaginary) component of a source in the mixture gives the corresponding imaginary (or real) component of that source via the complex conjugate of the complex-valued system output. We only need to run m single-unit real-valued extraction procedures and employ each extracted coefficient vector twice to deflate the signal space as sources are extracted. If Gram-Schmidt deflation is employed, the multi-source extension of the algorithm in (23) for the ith separation stage is

$$\underline{y}_{it}(k) = \underline{\mathbf{w}}_{it}^T \underline{\mathbf{v}}(k) \tag{24}$$

$$\widetilde{\underline{\mathbf{w}}}_{it} = \left(\frac{1}{N}\sum_{n=1}^{N} \underline{y}_{it}^3(n)\underline{\mathbf{v}}(n)\right) - 3\left(\frac{1}{N}\sum_{n=1}^{N}\underline{y}_{it}^2(n)\right)\widehat{\mathbf{R}}_{VV}\underline{\mathbf{w}}_{it} \tag{25}$$

for $n = 1$ to $i-1$ do

$$\overline{\underline{\mathbf{w}}}_{it} = \widetilde{\underline{\mathbf{w}}}_{it} - \underline{\mathbf{w}}_n \underline{\mathbf{w}}_n^T \widetilde{\underline{\mathbf{w}}}_{it} \tag{26}$$

$$\overline{\overline{\underline{\mathbf{w}}}}_{it} = \overline{\underline{\mathbf{w}}}_{it} - \underline{\mathbf{m}}_n \underline{\mathbf{m}}_n^T \overline{\underline{\mathbf{w}}}_{it} \tag{27}$$

end

$$\underline{\mathbf{w}}_{i(t+1)} = \frac{\overline{\overline{\underline{\mathbf{w}}}}_{it}}{\sqrt{\overline{\overline{\underline{\mathbf{w}}}}_{it}^T \overline{\overline{\underline{\mathbf{w}}}}_{it}}} \tag{28}$$

where the vectors $\underline{\mathbf{w}}_n = [\mathbf{w}_{R,n}^T \quad \mathbf{w}_{I,n}^T]^T$ and $\underline{\mathbf{m}}_n = [-\mathbf{w}_{I,n}^T \quad \mathbf{w}_{R,n}^T]^T$ are the coefficient vectors from the previous extraction steps. After convergence of all units,

$$\mathbf{W} = \begin{bmatrix} \mathbf{w}_{R,1}^T + j\mathbf{w}_{I,1}^T \\ \vdots \\ \mathbf{w}_{R,m}^T + j\mathbf{w}_{I,m}^T \end{bmatrix}, \qquad \mathbf{y}(k) = \mathbf{W}\mathbf{v}(k). \tag{29}$$

The following theorem describes the separating capabilities of this algorithm.

Theorem 4. *Suppose* $\mathbf{x}(k)$ *contains a mixture of* m *complex-valued statistically-independent sources that all have statistically-independent real and imaginary parts in which all but one of the sources has either a real or an imaginary component with a non-zero kurtosis. Then, either of the algorithms in (15) or (23) combined with (26)–(29) extracts all complex-valued sources in* $\mathbf{s}(k)$.

Remarks: The above theorem allows for each source to have a zero-kurtosis real or imaginary part that could be Gaussian-distributed. Thus, our algorithms can extract several BPSK sources measured in Gaussian noise in a complex baseband representation of an array processing system in wireless communications. In addition, note that our techniques are more powerful than a general $(2m)$-dimensional FastICA procedure applied to a set of prewhitened signal mixtures generated from the real and imaginary parts of $\mathbf{x}(k)$. The latter procedure would require all but one of the $2m$ *total real and imaginary parts* of the complex sources to have a non-zero kurtosis.

5 Simulations

We now explore the numerical performances of the proposed algorithms. All evaluations are performed on synthetic data using the MATLAB technical com-

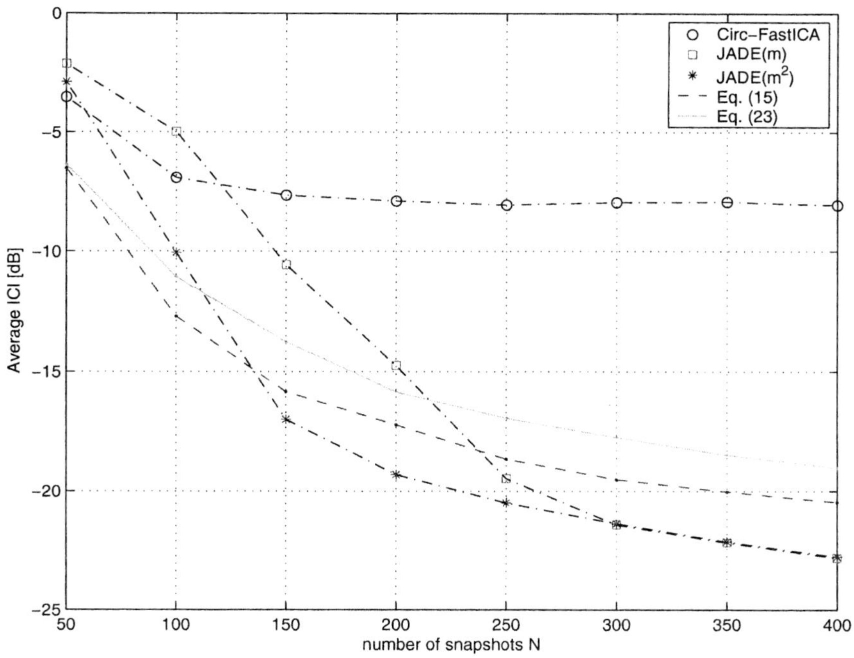

Fig. 1. Convergence of $E\{ICI_t\}$ for the various algorithms in a noiseless six-source separation task

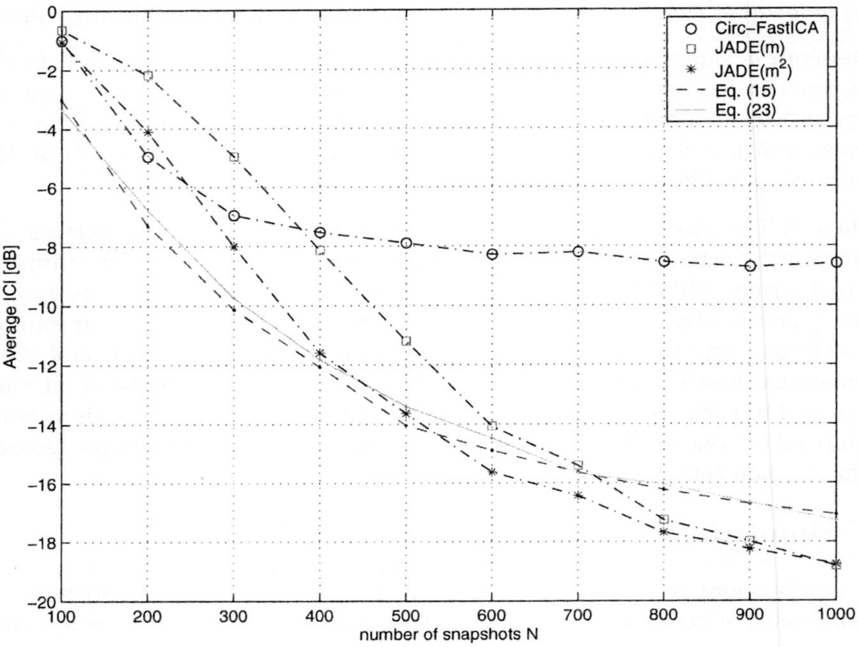

Fig. 2. Convergence of $E\{ICI_t\}$ for the various algorithms in a noisy seven-source separation task

puting environment. Three BPSK and three 16-QAM sources were mixed using $\mathbf{A} = \mathbf{U}\boldsymbol{\Sigma}\mathbf{V}^H$, where $\mathbf{U}$ and $\mathbf{V}$ are random complex orthogonal and the complex diagonal elements of $\boldsymbol{\Sigma}$ have amplitudes in the range $[0.2, 1]$. The average inter-channel interference (ICI) is used to measure performance, as given by

$$ICI = \frac{1}{m}\sum_{i=1}^{m}\left(\frac{\sum_{l=1}^{m}|c_{il}|^2 - \max_{1\leq k\leq m}|c_{ik}|^2}{\max_{1\leq k\leq m}|c_{ik}|^2}\right), \tag{30}$$

where $c_{il} = [\mathbf{W}\widehat{\mathbf{G}}\mathbf{A}]_{il}$. Shown in Figure 1 are the performances of the multi-unit versions of the algorithms in (15) and (23) along with those of two different versions of JADE using m and m^2 cumulant matrices [1], and the circular complex FastICA algorithm [2] with asymmetric deflation and $G(|y|^2) = 0.5|y|^2$. The proposed methods outperform the algorthm in [2], and they also perform better than JADE(m) for $N \leq 200$. Figure 2 shows a more-realistic situation in which an additional Gaussian was included in the $m = 7$-source mixtures and additive circular uncorrelated Gaussian noises with variances $\sigma_\nu^2 = 0.001$ was used as measurement interference. In this case, the proposed methods perform as well as or better than both JADE versions for $N \leq 500$ snapshots.

Per-unit convergence of the proposed algorithms is as fast as the original real-valued FastICA algorithm; only a few iterations of (15) and (23) are required at each stage.

6 Conclusions

In this paper, we have derived two novel algorithms for extracting independent sources from complex-valued mixtures using the fourth-moment properties of their real or imaginary components. The algorithms are computationally-simple and converge quickly. Simulations on mixtures of complex-valued signals typically found in wireless communications applications show the methods' efficacies.

References

1. J.-F. Cardoso and A. Soloumiac, "Blind beamforming for non-Gaussian signals," *IEE Proc. F.,* vol. 140, pp. 362-370, Dec. 1993.
2. E. Bingham and A. Hyvarinen, "A fast fixed-point algorithm for independent component analysis of complex-valued signals," *Int. J. Neural Syst.,* vol. 10, no. 1, pp 1-8, Feb. 2000.
3. S. Amari, S.C. Douglas, A. Cichocki, and H.H. Yang, "Multichannel blind deconvolution using the natural gradient," *Proc. 1st IEEE Workshop Signal Proc. Adv. Wireless Commun.,* Paris, France, pp. 101-104, Apr. 1997.
4. V. D. Calhoun, T. Adali, G. D. Pearlson, P. C. van Zijl, and J. J. Pekar, "Independent component analysis of fMRI data in the complex domain," *Magn Reson. Med.*, vol. 48, pp. 180-192, 2002.
5. J. Eriksson, A.-M. Seppola, and V. Koivunen, "Complex ICA for circular and non-circular sources," *Proc. EUSIPCO 2005,* Antalya, Turkey, Sept. 2005.
6. J. Eriksson and V. Koivunen, "Complex-valued ICA using second order statistics," *Proc. IEEE Workshop Machine Learning Signal Processing,* Sao Luis, Brazil, pp. 183-191, Oct. 2004.
7. J. Eriksson and V. Koivunen, "Complex random vectors and ICA models: Identifiability, uniqueness, and separability," *IEEE Trans. Inform. Theory,* vol. 52, no. 3, Mar. 2006 (in press).
8. A. Hyvarinen, J. Karhunen, and E. Oja, *Independent Component Analysis* (New York: Wiley, 2001).
9. S.C. Douglas, "Fixed-point FastICA algorithms for the blind separation of complex-valued signal mixtures," *Proc. 39th Asilomar Conf. Signals, Syst., Comput.*, Pacific Grove, CA, Oct. 2005.

Blind Estimation of Row Relative Degree Via Constrained Mutual Information Minimization

Jani Even* and Kenji Sugimoto

Graduate School of Information Science,
Nara Institute of Science and Technology,
Nara, 630-0192, Japan
even@is.naist.jp, kenji@is.naist.jp

Abstract. This paper studies a method for blind (input signals being unknown) estimation of the row relative degrees of a system non invertible at infinity. The proposed method uses a blind signal deconvolution scheme: A system, called demixer, is applied to the observed signals and is updated in order to minimize the mutual information. A key point is that the demixer is constrained to be biproper whereas the system is not invertible at infinity, consequently deconvolution is not achievable. But, the row relative degrees can be obtained in two steps: i) minimizing the mutual information at the output of the demixer. ii) using second order statistics of the obtained outputs. Although convergence has not yet been proved, extensive numerical simulation shows the effectiveness of this method.

1 Introduction

Blind signal separation has recently attracted much attention, and many efficient statistical methods have appeared in the last decade. In blind signal separation, the focus is on the recovery of unknown signals using only observed mixtures of these signals. During recovery, the dynamical system corresponding to the mixture is also partially identified (see book [1] for details). This approach is hence promising in control engineering too, because we often encounter the case where some of the input signals are unavailable due to noise, saturation, or failure; see, for example, [4].

Since control systems are often strictly proper, i.e. non invertible at infinity, many theoretical developments assume that the degrees of the rational transfer matrix representing the system are partially known. Some of the most important parameters are the row relative degrees (see Sect.2.2 and [2, 5]). However few methods enable the determination of these parameters, thus trial and error scheme is often used in practical applications.

This paper studies a method that enables the blind estimation of the row relative degrees of a system. However, the blind treatment has a cost in term of

* This work is supported in part by Scientific Research Grant-In-Aid from the Japan Society for the Promotion of Science under "JSPS Postdoctoral Fellowship for Foreign Researchers" program.

J. Rosca et al. (Eds.): ICA 2006, LNCS 3889, pp. 352–359, 2006.

indeterminacy: Only the difference of relative degree among the rows is obtained. The blind estimation is achieved in two steps:

i) Minimizing the mutual information with some structural constraints on the demixer: It does not result in a blind deconvolution because of the constraints,
ii) exploiting the second order statistics of the output signals obtained after i).

Although convergence has not yet been proved, extensive numerical simulation shows the effectiveness of this method.

The proposed method provides a good insight in the system's structure because the approximate inverse of the system is factorized in two terms, one of those corresponding to the row relative degrees information. Traditional blind signal separation methods do not perform such a factorization because they focus on the recovery of sources. Since the use of blind deconvolution techniques based on mutual information is not widespread yet in control community, this paper is also an attempt to show their potential in this field.

Some notations below will be used in this paper: For a matrix A: $A^{(i,:)}$ denotes the ith row of A and $A^{(i,j)}$ is the element of A in the ith row and the jth column.

$\mathcal{O}_{m\times p}$ is a null matrix of size $m \times p$ and $\mathcal{I}_m$ is the identity matrix of size m. $\delta_{i,j}$ is the Kronecker's delta equal to 1 if $i = j$ and 0 if $i \neq j$. All transfer matrices are assumed to be square of size $m \times m$ with $m > 1$.

2 Preliminaries

2.1 Blind Deconvolution

Throughout the paper we treat discrete-time signals. The goal of blind deconvolution is to recover the unknown input signals, called source signals, $\mathbf{s}(t) = [s_1(t), \ldots, s_m(t)]^T$ applied to an unknown transfer matrix $H(z)$, called "mixer", when only the observed signals $\mathbf{v}(t) = [v_1(t), \ldots, v_m(t)]^T = H(z)\mathbf{s}(t)$ are available. Throughout the paper, $H(z)$ is assumed to be stable rational, proper, and of minimal phase with full normal rank.

A transfer matrix $W(z)$, called "demixer", is applied to the observations in order to obtain the estimates $\mathbf{y}(t) = [y_1(t), \ldots, y_m(t)]^T$ of the sources as illustrated in Fig. 1. A common hypothesis used in blind signal deconvolution is to assume that each source $s_i(t)$ is an independent identically distributed ($i.i.d.$) process and that all sources are mutually statistically independent. It is also necessary to assume that at most one of the sources has a Gaussian distribution. With these hypotheses, blind signal deconvolution can be achieved by adapting the demixer $W(z)$ in order to obtain signals $\mathbf{y}(t)$ whose components are mutually statistically independent [3].

Even under the above conditions the blind identification of the transfer matrix has indeterminacies: We can detect neither a permutation of the outputs, the delay, nor the scale of each output. This is formulated by the relation:

$$W(z)H(z) = P\,\Lambda(z), \tag{1}$$

where P is a permutation matrix and $\Lambda(z)$ is a diagonal transfer matrix with entries of the form $\alpha_i\, z^{-\lambda_i}$.

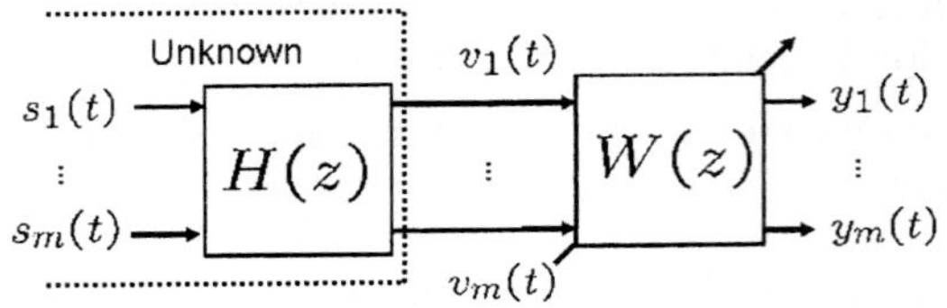

Fig. 1. Blind signal deconvolution scheme

2.2 Row Relative Degree

A transfer matrix $H(z)$ is said to be biproper, i.e. invertible at infinity, if the first matrix that appears in its power series expansion $H(z) = H_0 + H_1 z^{-1} + \ldots$ is invertible. A transfer matrix $H(z)$ of size $m \times m$ is said to be non invertible at infinity if $\lim_{|z| \to \infty} H(z) = M$ is not an invertible matrix.

The relative degree of a rational polynomial fraction $N(z)/D(z)$ is $\deg D(z) - \deg N(z)$. The matrix of relative degrees of a polynomial transfer matrix $H(z) = \{H_{ij}(z)\}_{i,j \in [1,m]}$ is $\{d_{ij}\}_{i,j \in [1,m]}$ with d_{ij} relative degree of $H_{ij}(z)$. The ith row relative degree of $H(z)$ is $d_i = \min_j d_{ij}$.

In the remainder of the paper, we assume that $H(z)$ is non invertible at infinity. We further assume that $H(z)$ is such that

$$\mathrm{diag}\left(z^{d_1} \ldots z^{d_m}\right) H(z) \tag{2}$$

is invertible at infinity. Namely, shifting each output signal by a number of samples equal to the row relative degree results in a biproper transfer matrix.

Considering the two transfer matrices $H(z)$ and $\tilde{H}(z) = H(z)\Lambda(z)$, where $\Lambda(z) = \mathrm{diag}(z^{-\alpha_j})_{j \in [1,m]}$, their row relative degrees are $d_i = \min_j d_{ij}$ and $\tilde{d_i} = \min_j (d_{ij} + \alpha_j)$, respectively, and are different. However, the row relative degree differences $r_{pq} = d_p - d_q$ and $\tilde{r}_{pq} = \tilde{d_p} - \tilde{d_q}$ are same if either one of the conditions below is fulfilled

i) $\alpha_j = \alpha$ for all j: $\tilde{r}_{pq} = \min_j (d_{pj}) + \alpha - \min_j (d_{qj}) - \alpha = r_{pq}$,
ii) $d_{ij} = d_i$ for all j: $\tilde{r}_{pq} = d_p + \min_j \left(\alpha_{\sigma_p(j)}\right) - d_q - \min_j \left(\alpha_{\sigma_q(j)}\right) = r_{pq}$.

In the case ii), the indeterminacies of blind deconvolution in Eq.(1) do not prevent to estimate the row relative degree differences (permutation of column has no effect on the row relative degrees).

3 Main Results

3.1 Adaptation of $W(z)$

The proposed method exploits a classical blind deconvolution scheme. However structures of the mixer and demixer are incompatible: The demixer is constrained to be biproper and thus cannot be the inverse of the mixer which is non invertible at infinity.

The demixer $W(z)$ is a finite impulse response (FIR) system

$$W(z) = W_0 + W_1 z^{-1} + \ldots + W_l z^{-l}.$$

The matrices W_j are adapted with a batch algorithm based on the on-line method proposed in [1]. The method minimizes the mutual information of the outputs

$$\mathrm{MI}[\mathbf{y}(t)] = -H(\mathbf{y}(t)) + \sum_{i=1}^{m} H(y_i(t)),$$

where $H(X) = -\int P_X(X) \ln P_X(X) dX$ is the entropy. Adaptation rule is derived from the relative gradient of $\mathrm{MI}[\mathbf{y}(t)]$.

Let $W_i(k)$ denote the matrix W_i at iteration k and $\mu(k)$ be the positive adaptation step used at iteration k. The adaptation law is

$$W_i(k+1) = W_i(k) - \mu(k) \Delta W_i(k),$$

$$\text{with} \quad \Delta W_i(k) = \sum_{j=0}^{i} (\delta_{j,0} \mathcal{I}_m - < \psi(\mathbf{y}(t)) \mathbf{y}^T(t-j) >_t) W_{i-j}(k)$$

$$\Delta W_0(k) = (\mathcal{I}_m - < \psi(\mathbf{y}(t)) \mathbf{y}^T(t) >_t)\ W_0(k)$$

where $< . >_t$ denotes time average on the data block. $\psi(.) = [\psi_1(.) \ldots \psi_m(.)]^T$ is a vector containing approximations of the score functions associated with the source signals: $\psi_{\text{real}}(s) = -\partial ln[P_s(s)]/\partial s$ (see [1] for derivation and discussion on approximation of score). In a single iteration k, the whole block of signal $\mathbf{y}(t)$ has to be computed (hence this is a batch algorithm).

At initialization $W(z) = \mathcal{I}_m$. An important property of the adaptation law is that when $W(z)$ is initialized to a biproper filter, it remains biproper during adaptation [1]. Hence blind deconvolution cannot be attained because $H(z)$ is not biproper and $\mathrm{MI}[\mathbf{y}(t)]$ reaches a local minimum by the above adaptation.

For simplicity, we illustrate our discussion with 2×2 transfer matrices. Assume that first and second rows of $H(z)$ have relative degrees $d_1 = 0$ and $d_2 = d > 0$, respectively. In this case, the power series expansion of $H(z)$ is

$$H(z) = \left[\frac{H_0^{(1,:)}}{\mathcal{O}_{1\times 2}} \right] + \ldots + \left[\frac{H_{d-1}^{(1,:)}}{\mathcal{O}_{1\times 2}} \right] z^{-(d-1)} + H_d z^{-d} + H_T z^{-(d+1)} + \ldots$$

Conjecture. *If the demixer $W(z)$ is initialized to identity and adapted with the above adaptation law, then the cascade $G(z) = W(z)H(z)$ (for an even d) converges to*

$$\begin{aligned} G(z) \approx & \left[\frac{G_0^{(1,:)}}{\mathcal{O}_{1\times 2}} \right] + \left[\frac{G_1^{(1,:)}}{\mathcal{O}_{1\times 2}} \right] z^{-1} + \ldots + \left[\frac{G_{r-1}^{(1,:)}}{\mathcal{O}_{1\times 2}} \right] z^{-(r-1)} \\ & + \left[\frac{G_r^{(1,:)}}{G_r^{(2,:)}} \right] z^{-r} + \left[\frac{\mathcal{O}_{1\times 2}}{G_{r+1}^{(2,:)}} \right] z^{-r-1} + \ldots + \left[\frac{\mathcal{O}_{1\times 2}}{G_d^{(2,:)}} \right] z^{-d} \end{aligned} \qquad (3)$$

with the integer $r = d/2$ and the two rows of G_r being orthogonal.

Let us explain a ground of this conjecture. First note that minimizing $\mathrm{MI}[\mathbf{y}(t)]$ is to force that $G(z) = W(z)H(z)$ has statistically orthogonal outputs that have non Gaussian distributions [3]. Namely, both of the following conditions are to be fulfilled:

i) The two rows $G_j^{(1,:)}$ and $G_j^{(2,:)}$ of each matrix G_j are orthogonal.
ii) The vectors $[G_1^{(i,:)}, \ldots, G_l^{(i,:)}]$ for $i = 1, 2$ have only one non null element.

Conditions i) and ii) cannot be satisfied completely because the demixer is biproper, but the adaptation tries to attain them. In particular, the first matrix W_0 is solution of:

$$W_0 \left[\frac{H_0^{(1,:)}}{\mathcal{O}_{1\times m}}\right] = \left[\frac{W_0^{(1,1)} H_0^{(1,:)}}{W_0^{(2,1)} H_0^{(1,:)}}\right] = G_0.$$

At initialization $W_0^{(1,1)} = 1$ and $W_0^{(2,1)} = 0$, consequently i) implies that during adaptation the second row of G_0 remains null and the first row is proportional to $H_0^{(1,:)}$.

The second row of G_d is initialized to $G_d^{(2,:)} = W_0^{(2,2)} H_d^{(2,:)}$. During adaptation, W_0 is constrained to be invertible and $W_0^{(2,1)}$ is null consequently $W_0^{(2,2)}$ cannot be null. Therefore the second row of G_d remains non null during adaptation but i) implies that the first row $G_d^{(1,:)}$ converges to zero.

Ideally all the other matrices G_j should be set to zero during adaptation in order to fulfill "as well as possible" the condition ii). For $j > d$ it is possible to do so. But due to the constraints imposed by the non invertibility of $H(z)$ and biproperness of $W(z)$, all the coefficients cannot be simultaneously set to zero for $j \in [1, d-1]$. However, because of the structure of $H(z)$, most of the G_j have one null row. Consequently i) is fulfilled: All G_j have orthogonal rows. But ii) is not achieved and the algorithm obtains Eq.(3). Extensive numerical simulation shows that the repartition of these non null coefficients is balanced between the two rows. (*Note: If d is odd with $r = (d-1)/2$ then second row of G_j for $j \in [0, r]$ and first row for $j \in [r+1, d]$ are null.*)

3.2 Row Relative Degree Difference Estimation

After minimizing the mutual information $\mathrm{MI}[\mathbf{y}(t)]$, the row relative degree difference between the rows of $H(z)$ are determined by using the off-diagonal terms of the covariance $\Gamma(\mathbf{y}, \tau)$. Considering the 2×2 case, the off-diagonal term is:

$$\mathcal{C}_{12}(\mathbf{y}, \tau) = \mathcal{E}\{y_1(t) y_2(t+\tau)^T\}. \tag{4}$$

By hypothesis the source signals are statistically independent and have unit variance, as a result their covariance is: $\mathcal{E}\{s_1(p)s_2(q)^T\} = \delta_{p,q}$. Since the transfer from sources to output signals is of the form Eq.(3), thus Eq. (4) gives:

$$\begin{aligned}\mathcal{C}_{12}(\mathbf{y}, \tau) = {} & G_0^{(1,:)} G_d^{(2,:)T} \delta_{\tau,d} + \left[G_0^{(1,:)} G_{d-1}^{(2,:)T} + G_1^{(1,:)} G_d^{(2,:)T}\right] \delta_{\tau,d-1} + \ldots \\ & + \left[G_{r-1}^{(1,:)} G_r^{(2,:)T} + G_r^{(1,:)} G_{r+1}^{(2,:)T}\right] \delta_{\tau,1}\end{aligned} \tag{5}$$

The covariance is null if τ is not in $[1, d]$. Therefore after minimizing $\mathrm{MI}[\mathbf{y}(t)]$, the row relative degree difference d can be estimated by inspecting the covariance of the output signals: d is equal to the largest delay τ for which $C_{12}(\mathbf{y}, \tau)$ is not null. (*Note: In the case of a negative relative degree difference d, the covariance is null if τ is not in $[d, -1]$.*)

3.3 Proposed Method

In practice, a threshold β, function of estimation variance, is chosen and the largest delay τ_0 such that $C_{12}(\mathbf{y}, \tau) > \beta$ is the row relative degree difference estimation $\widehat{r_{12}} = \tau_0$. But finding such a threshold β is not an easy task.

However, a nice property appears when a relatively hight threshold β is chosen in order to avoid selecting delay out of $[1, d]$. The estimated row relative degree difference is $\widehat{r_{12}} = r_{12} - \epsilon$ with $0 \leq \epsilon < r_{12}$ an integer representing the error. Then consider the shifted observations:

$$\begin{bmatrix} v_1(t + \widehat{r_{12}}) \\ v_2(t) \end{bmatrix} = \begin{bmatrix} z^{d_1 - d_2 - \epsilon} & 0 \\ 0 & 1 \end{bmatrix} H(z)s(t) = \begin{bmatrix} z^{d_1 - \epsilon} & 0 \\ 0 & z^{d_2} \end{bmatrix} H(z)s(t - d_2)$$

- When $\epsilon = 0$, $\mathrm{diag}(z^{d_1}\ z^{d_2})H(z)$ is biproper. Consequently, using the same adaptation rule to adapt $W(z)$ after shifting the observation results in a blind deconvolution. Thus $C_{12}(\mathbf{y}, \tau)$ is null for all τ because output signals are statistically independent.
- When $\epsilon \neq 0$ the row relative degree difference of the system whose outputs are the shifted observations is $\epsilon \in [1, r_{12} - 1]$. Thus using the same adaptation rule to adapt $W(z)$ after shifting the observation leads again to a cascade of the form Eq.(3) and $C_{12}(\mathbf{y}, \tau)$ still presents non null values.

Thus iterating the same procedure ensures that $\epsilon \rightarrow 0$. In order to exploit this property, the proposed method is iterative:

1. initialization $\widehat{r_{12}} := 0$,
2. adapt the biproper demixer to minimize the mutual information,
3. compute $C_{12}(\mathbf{y}, \tau)$ for delay in $[-\tau_{\max}, \tau_{\max}]$,
4. if $C_{12}(\mathbf{y}, \tau) < \beta$ for all $\tau \in [-\tau_{\max}, \tau_{\max}]$ then stop iteration,
5. otherwise: Select τ_0 the largest delay such that $C_{12}(\mathbf{y}, \tau) > \beta$, update the estimation $\widehat{r_{12}} := \widehat{r_{12}} + \tau_0$, shift the first observation $v_1(t) := v_1(t + \tau_0)$ and go to step 2.,

4 Numerical Simulation

Consider the system $H(z) = \begin{bmatrix} \frac{1}{z+0.7} & \frac{0.9}{z-0.6} \\ \frac{-0.5}{(z-0.3)(z+0.7)(z+0.4)} & \frac{1}{(z+0.2)(z-0.4)(z+0.6)} \end{bmatrix}$, whose impulse response is given in Fig.2-(a). The row relative degrees are $d_1 = 1$ and $d_2 = 3$. Unknown source signals are *i.i.d.* processes uniformly distributed with zero mean and unit variance. The number of samples used in this example was $T = 10000$. 100 experiments were performed. The FIR filter has $l = 20$ coefficients.

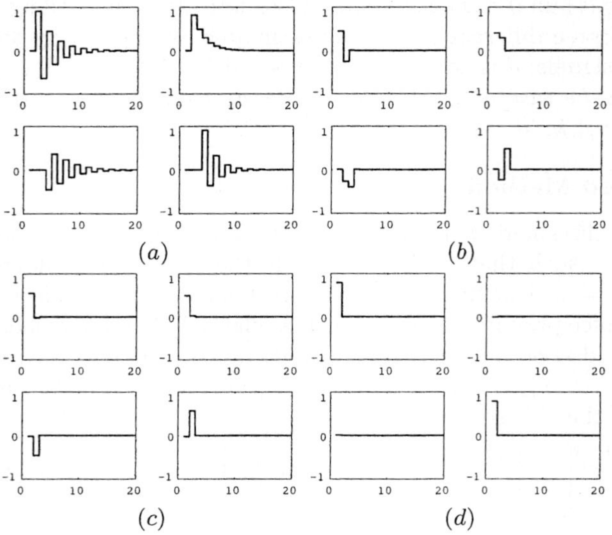

Fig. 2. Impulse response of $H(z)$ (a) and mean (variances are all less than $2.5e^{-4}$) of the impulse response of $W(z)H(z)$ after: First, second and third iterations respectively in (b), (c) and (d) (the subplot ij is the transfer from input j to output i)

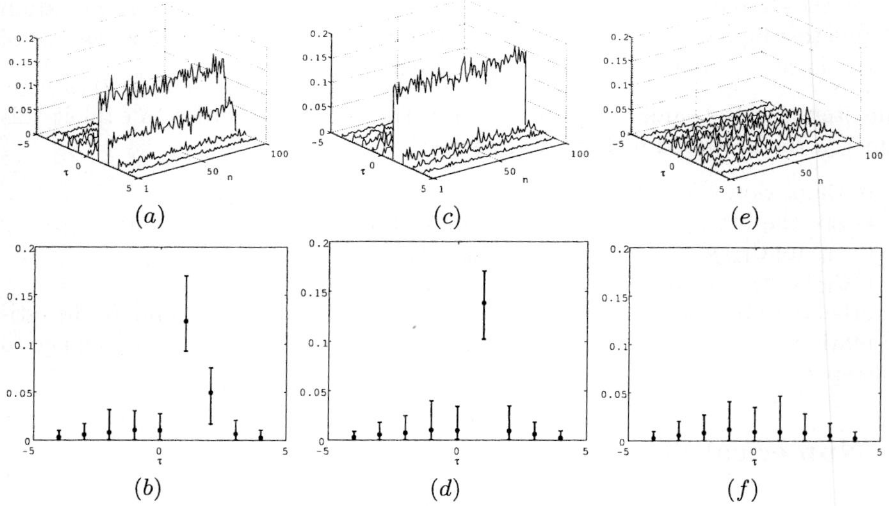

Fig. 3. $\mathcal{C}_{12}(\mathbf{y}, \tau)$ for: First (a)(b), second (c)(d) and third (e)(f) iterations. The upper row shows results for all experiments (index n) and the bottom row shows the mean, the minimum and the maximum of the covariance computed on all experiments.

The evolution of the impulse response of the cascade is presented in Fig.2-(b), (c) and (d). After adapting $W(z)$, the impulse response has the form of Eq.(3) in Fig.2-(b) and (c) but finally after the row relative degree difference was estimated the cascade is equal to identity because the blind deconvolution is achieved, Fig.2-(d).

The evolution of the covariance is depicted in Fig.3: First, second and third iterations (in left, middle and right column respectively). Fig.3-(a), (c) and (e) show $\mathcal{C}_{12}(\mathbf{y}, \tau)$ versus the delay τ for all the 100 experiments. The mean, minimum and maximum value of $\mathcal{C}_{12}(\mathbf{y}, \tau)$ are also plotted in Fig.3-(b), (d) and (f). During first iteration, the largest values are obtained for $\tau = 1, 2$ as expected from Eq.(5). But a threshold β such that $\tau = 2$ is selected for all experiments does not exist, see Fig.3-(b). For second iteration (after shifting first observation by one sample for all experiments: $\widehat{r_{12}} = 1$), the value $\tau = 1$ is selected in all experiments. Thus the true row relative degree difference of two, i.e. $\widehat{r_{12}} + 1 = 2$, is correctly estimated for all experiments. Consequently, after shifting again of one sample the first observation for all experiments and minimize the mutual information, the covariance has only very small values and the algorithm stops.

5 Conclusion

In this paper we show how to blindly estimate the row relative degree difference of a class of transfer matrices non-invertible at infinity by means of an iterative method based on a blind signal deconvolution setting.

References

1. Cichocki, A., Amari, S.: Adaptive Blind Signal and Image Processing : Learning Algorithms and Applications. John Wiley & Sons. (2002)
2. Kailath, T.: Linear systems. Prentice-Hall. (1980)
3. Shalvi, O., Weinstein, E.: New criteria for blind deconvolution of non-minimum phase systems. IEEE Trans. Inf. Theory. **36(2)** (1990) 312–321
4. Sugimoto, K., Suzuki, A., Nitta, M., Adachi, N.: Polynomial Matrix Approach to Independent Component Analysis: (Part II) Application. IFAC World Congress, Prague, Czech Republic (2005)
5. Wolovich, W.A., Falb, P.L.: Invariants and Canonical Forms under Dynamical Compensation. SIAM J. Control and Optimization. **14** (1976) 996–1008

Undoing the Affine Transformation Using Blind Source Separation

Nazli Guney and Aysin Ertuzun

Department of Electrical and Electronics Engineering,
Bogazici University, Istanbul, 34342 Turkey
{naz, ertuz}@boun.edu.tr

Abstract. Recognition of planar objects from their images taken from different viewpoints requires affine invariants calculated from the object boundaries. The equivalence between affine transformation and the source mixture model simplifies the recognition problem. The rotation, scaling, skewing, and translation effects of the affine transformation can be undone by using blind source separation (BSS) techniques to go back to a canonical object view. Then, the problem is reduced from a more involved affine invariant search to a simple shape matching task. Point correspondence between two curves initially related by an affine transformation is obtained with further processing.

1 Introduction

Object recognition is one of the major goals in machine vision. Typically, the images of an object taken from different viewpoints should be registered as belonging to the same object. For this purpose, invariants, which are shape descriptors that remain unchanged even when the viewing point changes are needed. The transformation that the object has gone through in an image compared to another view is approximated with the affine transformation if the object is viewed from a distance that is an order or more greater than the maximum object diameter along the direction of view [1]. Since this assumption always holds for planar objects, recognition problem becomes a search for affine invariants [2]. The affine transformation includes rotation, scaling, skewing and translation, and it is known to preserve parallel lines and equispaced points along a line [3].

A large number of affine invariants for planar object recognition have been proposed in the literature [1], [3]-[5]. These are classified as either boundary or region based techniques. Yet another classification for invariants is whether they are local or global. Local invariants, which use higher order derivatives, are susceptible to noise [4]. Global invariants, on the other hand, suffer from the occlusion of the object. The invariants proposed in [3]-[5] are all based on the wavelet transform. These methods produce global invariants, since the whole set of boundary points are used when calculating the wavelet coefficients. The wavelet transform decomposes the object boundary into a number of scales and the invariants use a subset of these scales. Although the wavelet transform based

J. Rosca et al. (Eds.): ICA 2006, LNCS 3889, pp. 360–367, 2006.

methods are easy to compute, their performance is very much dependent on whether point correspondence between the object boundaries related by an affine transformation exists or not.

In this paper, we show that planar object recognition problem can be simplified by employing blind source separation (BSS) techniques based on methods that use time structure. Here, the separation is achieved by the second order blind identification method in [6], where a set of covariance matrices are jointly diagonalized. As a result; the rotation, scaling, skewing and translation effects of the affine transformation are eliminated and the affine transformed object is brought to its canonical view. With further processing, point correspondence between the object boundaries extracted from two different images of the same object is obtained. Then, it is possible either to use the wavelet transform based shape descriptors [3]-[5] or any other shape descriptor to match objects, where the shape descriptor does not need to have the affine invariant property. In [7], Herault and Jutten have developed an adaptive algorithm, which uses non-linear decorrelation concepts, to estimate the unknown independent sources. They have applied this algorithm to a similar image processing problem.

The organization of the paper is as follows. Section II is about the mathematics of the object recognition problem. The BSS techniques using time structure and the equivalence between the source mixture model and the affine transformation are the subjects of Section III. The algorithm for obtaining point correspondence is introduced in Section IV. Some experimental results are given in Section V. Conclusions are made in Section VI.

2 Object Recognition

A typical recognition task involves the identification of an object from its image taken from an arbitrary viewpoint. The boundaries of the objects in both the reference images in the database and the query image have to be analyzed. The coordinates of the boundaries of the objects in two images, which portray the same object, are related by an affine transformation

$$\begin{bmatrix} \tilde{x} \\ \tilde{y} \end{bmatrix} = \begin{bmatrix} a_1 & a_2 \\ a_3 & a_4 \end{bmatrix} \begin{bmatrix} x \\ y \end{bmatrix} + \begin{bmatrix} b_1 \\ b_2 \end{bmatrix} . \tag{1}$$

(1) can be expressed in vector-matrix notation as

$$\tilde{\mathbf{x}} = \mathbf{A}\mathbf{x} + \mathbf{b}, \tag{2}$$

where $\mathbf{x}$ and $\tilde{\mathbf{x}}$ are the vectors that contain the coordinates of the original and the affine transformed image boundaries, respectively. Here, the nonsingular matrix $\mathbf{A}$ represents the scaling, skewing and rotation, and $\mathbf{b}$ is the translation in the affine transformation. Thus, affine invariants are calculated from the boundaries, which are expected to be identical whatever the particular value of the matrix $\mathbf{A}$ is if the boundaries belong to the images of the same object taken from different viewpoints. The invariants should be distinct for different objects. The degree of match between two invariant functions, I_1 and I_2 is given by their normalized correlation value [3]

$$\frac{\sum_{n=0}^{N-1} I_1[n] I_2[n]}{\sqrt{\sum_{n=0}^{N-1} I_1^2[n] \sum_{n=0}^{N-1} I_2^2[n]}}, \tag{3}$$

where N is the length of the invariant function. The invariants calculated from two curves related by an affine transformation should be identically one if there is perfect point correspondence and no noise. Deviation from this assumption decreases the amount of correlation.

3 Blind Source Separation Techniques

The mixture model is given by

$$\begin{bmatrix} m_1(t) \\ m_2(t) \end{bmatrix} = \begin{bmatrix} w_{11} & w_{12} \\ w_{21} & w_{22} \end{bmatrix} \begin{bmatrix} s_1(t) \\ s_2(t) \end{bmatrix} \tag{4}$$

$$\mathbf{m} = \mathbf{W}\mathbf{s}, \tag{5}$$

where $m_i(t)$ $i \in \{1,2\}$ and $s_i(t)$ $i \in \{1,2\}$ are the mixtures and the sources, respectively. The signals in the mixture model are assumed to have zero mean. Both $\mathbf{W}$ and $\mathbf{s}$ are unknown and should be estimated. Various techniques have been developed by assuming that the mixtures and the sources are random variables instead of time signals. While some of those increase the nongaussianity of the sources with respect to the mixtures, some others are based on nonlinear decorrelation to increase the statistical independence of the sources [8]. In the end, the sources are estimated up to a sign and amplitude ambiguity. Moreover, the order of the sources cannot be determined.

If the sources are time signals, on the other hand, they carry a lot more structure than when they are random variables [8]. The ordering of the time series data can be exploited in many ways. Methods using time structure have been developed based on covariances. The covariance matrix for lag τ of the zero mean mixture signals is

$$\begin{aligned} \mathbf{C}_\tau^{\mathbf{m}} &= E\{\mathbf{m}(t)\mathbf{m}(t-\tau)^T\} \\ &= \begin{bmatrix} E\{m_1(t)m_1(t-\tau)\} & E\{m_1(t)m_2(t-\tau)\} \\ E\{m_2(t)m_1(t-\tau)\} & E\{m_2(t)m_2(t-\tau)\} \end{bmatrix}, \end{aligned} \tag{6}$$

where $(\cdot)^T$ denotes transposition. The whitening of the mixtures is one of the most important pre-processing operations. Specifically, the matrix $\mathbf{V}$ is applied to the mixtures so that the resultant signals given by the vector $\mathbf{z}$ are white with uncorrelated components having variances equal to unity: $\mathbf{z} = \mathbf{V}\mathbf{m}$. Information from the covariance matrices corresponding to lags $\tau \neq 0$ is needed for source separation. Sources are estimated with the transformation $\tilde{\mathbf{s}} = \mathbf{B}\mathbf{z}$.

The simplest method using time structure is the **A**lgorithm for **M**ultiple **U**nknown **S**ignals **E**xtraction (AMUSE) algorithm in [9], [10]. It uses only one

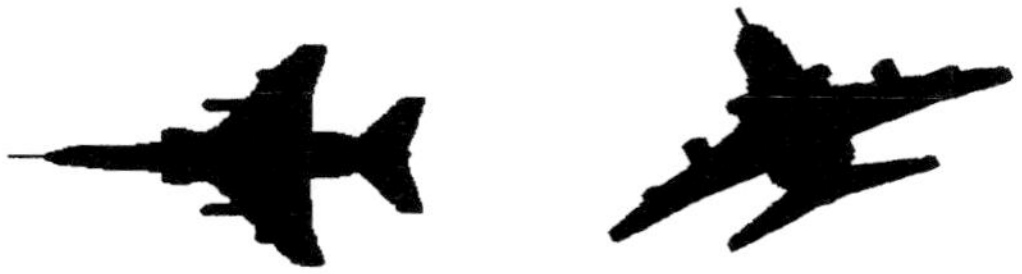

(a) original (b) affine transformed

Fig. 1. The original object and its affine transformed version

lag, τ. The separating matrix $\mathbf{B}$ is estimated by whitening the zero mean data $\mathbf{m}$ to obtain $\mathbf{z}$, computing the eigendecomposition for $\bar{\mathbf{C}}_\tau^{\mathbf{z}}$, and finally taking the eigenvectors as the rows of the separating matrix. A modified covariance matrix is used so that the matrix is symmetric and the eigendecomposition can be done: $\bar{\mathbf{C}}_\tau^{\mathbf{z}} = \frac{1}{2}\left[\mathbf{C}_\tau^{\mathbf{z}} + (\mathbf{C}_\tau^{\mathbf{z}})^T\right]$. The covariance matrix of the mixtures is $\bar{\mathbf{C}}_\tau^{\mathbf{z}} = \mathbf{B}^T\bar{\mathbf{C}}_\tau^{\mathbf{s}}\mathbf{B}$, which is, in fact, $\bar{\mathbf{C}}_\tau^{\mathbf{z}} = \mathbf{B}^T\mathbf{D}\mathbf{B}$, where $\mathbf{D}$ is a diagonal matrix and this equality follows due to the uncorrelatedness of the sources. Thus, $\mathbf{B}$ is part of the eigendecomposition of the matrix $\bar{\mathbf{C}}_\tau^{\mathbf{z}}$. The algorithm fails if for the time lag chosen, the eigenvectors of the covariance matrix are not distinct. The AMUSE algorithm has been extended to several time lags by the **S**econd **O**rder **B**lind **I**dentification (SOBI) algorithm in [6]. For successful demixing of sources, it is enough that only one of the covariance matrices has distinct eigenvectors. The covariance matrices are diagonalized jointly by a matrix $\mathbf{U}$.

The affine transformation (2) and the mixture model (5) are seen to be equivalent if the object centroid given by the means of the x and y coordinates of the boundary is moved to the origin (i.e., so that they have zero mean). The coordinates of an affine transformed object boundary may be visualized as a mixture of the coordinates in some canonical view. Fig. 1 shows an object from the airplane object database in [3] and its affine transformed version, where the affine transformation matrix has $a_1 = 1$, $a_2 = 2$, $a_3 = 7$ and $a_4 = -3$ in (1). The vector, $\mathbf{b}$, is assumed to be an all-zeros vector for simplicity. The x and y coordinates of the object boundaries are shown in Fig. 2. The s_i's found by using the SOBI algorithm are demonstrated in Fig. 3. From the figure, it is seen that the SOBI algorithm has introduced a sign and order ambiguity to the coordinates of the original object. Moreover, the s_i's of the affine transformed object are related to those of the original object by a sign and

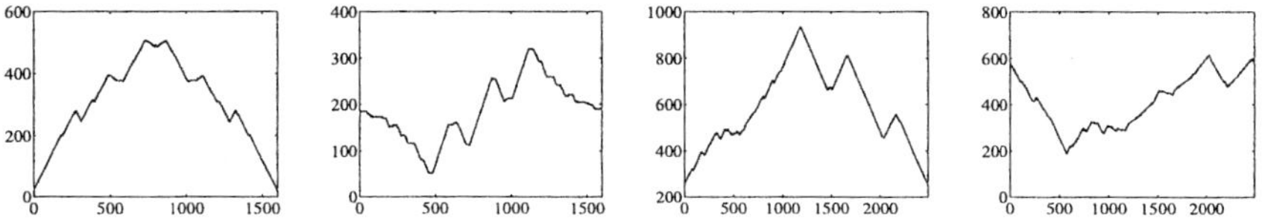

(a) $x_{original}$ (b) $y_{original}$ (c) x_{affine} (d) y_{affine}

Fig. 2. The x and y coordinates of the original and transformed object boundaries, horizontal axis - sample number, vertical axis - x or y coordinates

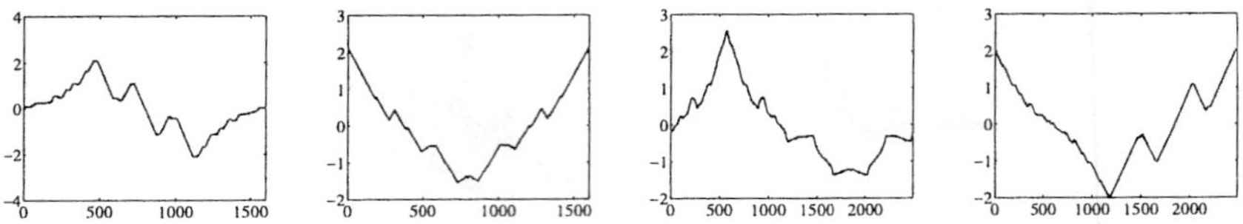

(a) $s_{1,original}$ (b) $s_{2,original}$ (c) $s_{1,affine}$ (d) $s_{2,affine}$

Fig. 3. The s_i's of the original and transformed object boundaries, horizontal axis - sample number, vertical axis - x or y coordinates

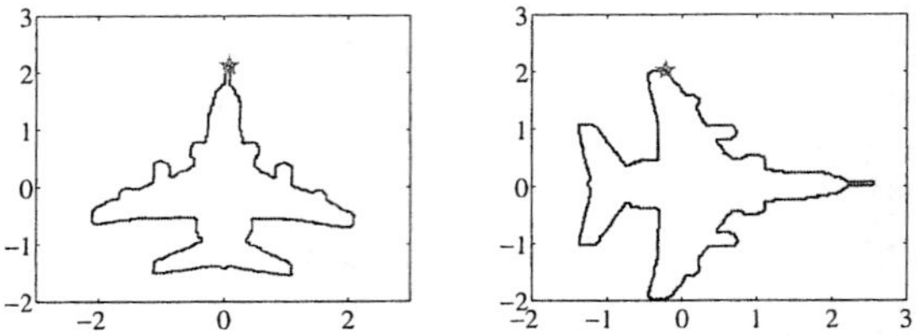

(a) original (b) affine transformed

Fig. 4. Boundaries of the original and transformed objects processed with SOBI, stars indicate the starting points of the boundaries

order ambiguity. The starting points of the boundaries, which are shown with stars in Fig. 4 are different, as the boundary tracking algorithms can start from any point.

4 Post Processing

Although applying BSS techniques to the object boundaries brings them to a canonical view, there are still significant problems that need to be addressed. Specifically, these techniques employ linear transformations of the mixtures to arrive at the sources. According to affine invariants, this is yet another affine transformation. Those that suffer from a lack of point correspondence between the object boundaries still exhibit low correlation values.

First of all, the order and sign ambiguities are removed. The aim of SOBI, is to make s_1 uncorrelated with s_2 for all possible time lags. The correlations between the s_i's of the original and transformed object boundaries for all possible shifts of the transformed boundary show the correct order of the transformed s_i's and their signs with respect to the original s_i's. The calculated correlations should be maximized in magnitude when the corresponding s_i's are used. For instance, if there is an order ambiguity, then s_1 (s_2) of the original boundary is related to the s_2 (s_1) of the transformed boundary. The correlation between original s_1 and affine s_2 is maximized in magnitude when they are exactly aligned (i.e. they have the same starting point). Note that the correlation between original s_2 and affine s_1 should be maximized for the same shift in theory. Hence, the starting points for the boundaries can be made the same. Before the correla-

Table 1. Maximum correlation values between s_i's

	Correlation	Shift
(i) $s_{1,\text{original}}$-$s_{1,\text{affine}}$	-0.9500	529
(ii) $s_{1,\text{original}}$-$s_{2,\text{affine}}$	-0.9993	264
(iii) $s_{2,\text{original}}$-$s_{1,\text{affine}}$	0.9672	265
(iv) $s_{2,\text{original}}$-$s_{2,\text{affine}}$	-0.9445	530

tions are calculated, the boundaries are parameterized with arc length, which means that the boundary samples are at approximately equal distances from each other.

At this stage, we go back to the example in the previous section and give the maximizing correlation values in Table 1. For this particular example, none of the correlations is maximized for the same shift, but they are close enough to be considered as equal. Here, both of the boundaries have a length of 2^{10}. The decision as to whether there is a sign or order ambiguity and the proper alignment of the boundaries is done looking at the magnitudes mag((i)×(iv)) and mag((ii)×(iii)) from Table 1. This tells us that the sign of affine s_2 should be reversed, and the order of the first and second s_i's of the affine transformed boundary should be switched. They should both be shifted in time by 265. For this shift, the correlations (i) and (iv) are 0.0133 and 0.0093, respectively. The algorithm for obtaining point correspondence may be summarized as follows:

1. Extract object boundaries from the images related by an affine transformation and perform SOBI algorithm to obtain s_i's.
2. Parameterize the boundaries with arc length and calculate the correlations (i)-(iv) in Table 1 for all possible shifts of the transformed s_i's.
3. Choose maximum of mag((i)×(iv)) and mag((ii)×(iii)). If mag((i)×(iv)) is chosen, then s_i's of the affine object are in correct order. Reverse the sign of affine s_1 or s_2 if (i) or (iv) is negative, respectively. Shift the starting point of the boundary to the average of shift (i) and (iv). If mag((ii)×(iii)) is chosen, reverse the sign of affine s_1 or s_2 if (iii) or (ii) is negative, respectively. Switch the orders of the s_i's. Shift the starting point of the boundary to the average of shift (ii) and (iii). Finally, do a parametrization of the boundaries with arc length.

The importance of the steps above and that the algorithm works is shown via the affine invariant wavelet function in [3]. This invariant uses six wavelet scales. When the boundary is sampled to a length of 2^7, seven scales of coefficients are calculated and the finest scale, which is very much affected by noise is avoided. The wavelet in [11] is being used. When the boundaries extracted from images are used as they are, the correlation in (3) between the invariants is 0.5163. The algorithm outlined above increases the correlation value to 0.9703. The final form of the affine transformed object boundary normalized to a length of 2^7 and the invariants are displayed in Fig. 5.

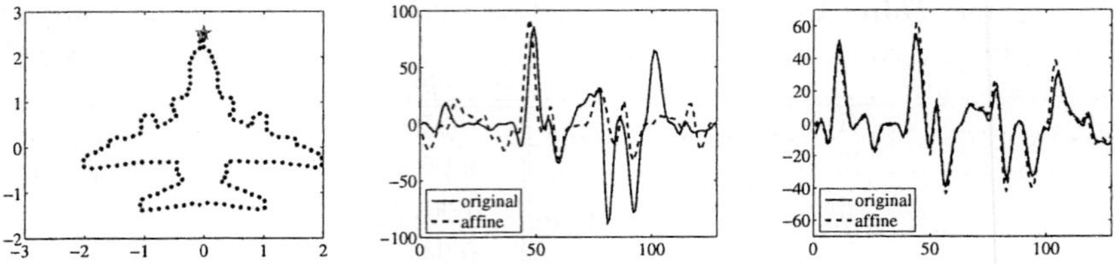

(a) the final boundary (b) no processing (c) processed

Fig. 5. The final form of the transformed boundary and the invariant functions

5 Experimental Results

Two experiments have been carried out to test the proposed algorithm. The airplane object database in [3] with 20 objects is used. In the first experiment, uniformly distributed white noise at different signal-to-noise ratio (SNR) values has been added to the boundaries of the affine transformed objects before the normalized correlation in (3) is calculated. SNR is defined as the ratio of the average squared distance of the boundary points from the centroid to the variance of the uniformly distributed noise. Each object in the database is transformed using $a_1 = 1$, $a_2 = 2$, $a_3 = 7$ and $a_4 = -3$. For each SNR value, fifty realizations of noise are added to the boundary of each affine transformed object and the correlation values for the 1000 realizations are averaged. Fig. 6a shows the result of the experiment. When SNR = 20 dB, it is not possible to distinguish the boundaries belonging to different, but similar objects, since the details are lost. Uniformly distributed noise moves the boundary points, which is an ordered set initially, in random directions. Then, the object boundary is not a collection of ordered points like it should be. Until SNR increases so much that the order of the boundary points is not altered, the algorithm does not have so high correlation values. SOBI algorithm exploits the order of the boundary points to arrive at the canonical object view.

In the second experiment, the transformation $a_1 = 20(1-\gamma), a_2 = 10, a_3 = -10$ and $a_4 = 20(1 + \gamma)$ with $\gamma = \{0, 0.1, \ldots, 0.9\}$ is used to test how the algorithm

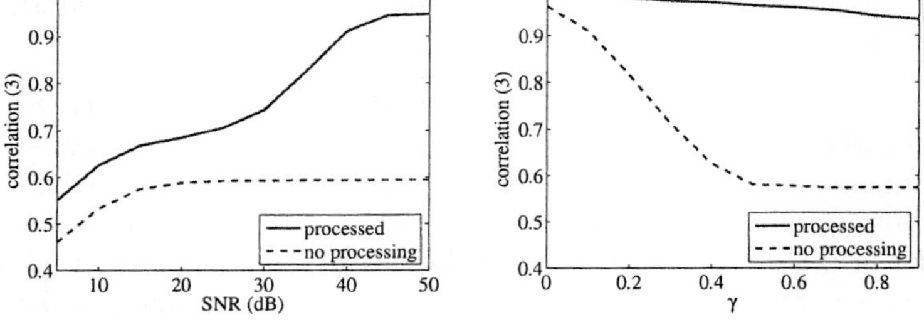

(a) varying SNR (b) varying γ

Fig. 6. Experimental results

works under increasing amount of affine distortion. The case $\gamma = 0$ corresponds to a similarity transformation, which consists of scaling and rotation. Again, for each transformation 20 airplane objects are used and the normalized correlation value in (3) is averaged. The results are given in Fig. 6b. $\gamma \geq 0.5$ implies a large affine distortion of the object. The algorithm proposed still has very high correlation for these values of γ as it undoes the affine transformation.

6 Conclusion

In this paper, we have shown that the recognition of planar objects problem can be simplified from an affine invariant search to a simple shape matching task. The rotation, scaling, skewing and translation introduced by the affine transformation can be eliminated by using BSS techniques. Approximate point correspondence between two curves related by an affine transformation is obtained with the algorithm proposed. The proposed algorithm is shown to be successful under practical levels of noise and large affine distortions.

References

1. D. Forsyth, J. L. Mundy, C. Coelho, A. Heller, and C. Rothwell, "Invariant descriptors for 3-D object recognition and pose,", *IEEE Trans. PAMI,* vol. 13, pp. 971-991, Oct. 1991.
2. I. Weiss, "Geometric invariants and object recognition," *International Journal of Computer Vision,* vol. 10, no. 3, pp. 207-231, 1993.
3. M. I. Khalil and M. M. Bayoumi, "A dyadic wavelet affine invariant function for 2-D shape recognition," *IEEE Trans. PAMI,* vol. 23, pp. 1152-1164, Oct. 2001.
4. R. Alferez and Y. F. Wang, "Geometric and illumination invariants for object recognition," *IEEE Trans. PAMI,* vol. 21, pp. 505-536, June 1999.
5. Q. M. Tieng and W. W. Boles, "Wavelet-based affine invariant representation: a tool for recognizing planar objects in 3D space," *IEEE Trans. PAMI,* vol. 19, pp. 846-857, August 1997.
6. A. Belouchrani, K. Abed-Meraim, J. F. Cardoso, and E. Moulines, "A blind source separation technique using second-order statistics," *IEEE Trans. Signal Process.,* vol. 45, pp. 434-444, Feb. 1997.
7. C. Jutten and J. Herault, "Blind separation of sources, Part I: An Adaptive algorithm based on neuromimetic architecture," *Signal Processing,* vol. 24, pp. 1-10, 1991.
8. A. Hyvärinen, J. Karhunen, and E. Oja, "Independent component analysis," John Wiley and Sons, Inc., 2001.
9. L. Tong, V. C. Soon, Y. F. Huang, and R. Liu, "AMUSE: A new blind identification algorithm," in *Proc. 1990 IEEE ISCAS,* New Orleans, L.A., May 1990.
10. L. Tong, R. Liu, V. C. Soon, and Y. F. Huang, "Indeterminacy and identifiability of blind identification," *IEEE Trans. CAS,* vol. 38, pp. 499-509, May 1991.
11. S. Mallat and S. Zhong, "Characterization of signals from multiscale edges," *IEEE Trans. PAMI,* vol. 14, pp. 710-732, July 1992.

Source Separation of Astrophysical Ice Mixtures

Jorge Igual, Raul Llinares, and Addisson Salazar

Universidad Politecnica de Valencia, Departamento de Comunicaciones,
Pza. Ferrandiz y Carbonell 1, CP 03801, Alcoy, Spain
{jigual, rllinares, asalazar}@dcom.upv.es

Abstract. Infrared spectroscopy provides direct information on the composition of interstellar dust grains, which play a fundamental role in the evolution of interstellar medium ISM, from cold, quiet and low density molecular clouds to warm, active and dense protostellar ones. The determination of these components is fundamental to predict under the appropriate environmental conditions their evolution, including the appearance of new molecules, radicals and complex organics. The absorption spectrum of the astrophysical ice can be considered the additive linear absorption spectra of the multiple molecules present in the ice, so a linear instantaneous ICA mixture model is appropriate. We present the ICA statement of the problem, discussing the convenience of the model and its advantages in front of supervised methods. We obtain the MAP estimate of the mixing matrix, including its non-negative entries as a prior. We present the results carried out with an ice analogs database, confirming the suitability of the ICA approach.

1 Introduction

The origin of the biogenic elements in Earth is an intriguing question [1]. It is thought that prebiotic molecules were present on primitive Earth. Different processes could be responsible for their evolution into more complex ones. One of the most famous experiments was carried out by Miller [2]. In this experiment, glycine was formed by energetic processes such as the electric discharge of a mixture of HCN and aldehydes acids. Other experiments [3] demonstrate that other molecules of interest such as carbonic acid (H_2CO_3) could be formed from simple molecules such as CO by protons implantation. The understanding of chemical evolution of elementary molecules as water, carbon monoxide, methane or carbon dioxide is one of the key questions for understanding the origin of life on Earth.

These molecules are combined in different quantities, called concentrations, in other compounds and found usually in the frozen state, named from now on ice mixtures.

One way to study this chemical evolution consists of the simulation of these processes in the laboratory. In the laboratory, we are interested in the analysis of the composition of different ices and their behavior in specific conditions that reproduce situations in which these substances are found outside of the laboratory. For example: to raise the temperature gradually or to radiate the samples with ions that are similar to the ones originated outside of the planet Earth.

J. Rosca et al. (Eds.): ICA 2006, LNCS 3889, pp. 368 – 375, 2006.

The signal that identifies each molecule is the infrared absorption spectrum, where the absorption bands correspond to specific vibrational mode of the molecules, each one with different atoms and bonds. We know that the frequencies corresponding to the middle infrared spectrum (4000-400 cm^{-1}; 2.5-25 μm) span the same range as the vibrational frequencies of the adjacent atoms in molecules associated with the cosmogenically most abundant species. The components are called endmembers and the spectra signatures or digital tracks.

ICA performs a "multi-ice" unsupervised decomposition of the data in different spectra instead of a classical matched filter detection approach. The supervised approach raises other difficulties in our application; not only the requirement of an expert with the risk of misinterpretations, but furthermore, the library of endmembers could be very large because considering that the spectrum vary depending on many variables such as temperature or irradiation, we would need a description of every compound for every case. In fact, even under very controlled conditions, like in a laboratory, the spectra registered in different experiments for the same mixed ice can sometimes change. In Figure 1 we represent the CO_2 spectrum band around 2350 cm^{-1} obtained in Leiden and Polytechnic University of Valencia UPV laboratories. The mixed ice production consists of defining the mixture, preparing the gases necessary in the corresponding abundance, introducing them in the gas container using the gas law to convert pressures to abundances, and depositing on a cold substrate. The final ice can vary due to different problems, e.g., the purity of the gases, small changes in the temperature, sequential mixing of gases, or the influence of the molecular mass in the deposition rate [4].

2 Infrared Absorption Features of Astrophysical Ices

Life on Earth originated in stars which formed out of interstellar gas and dust [5]. The lifecycle of an interstellar ice mantle consists of the following stages: a.) formation and growth by surface reactions on grains, b.) the ices are subject to energetic phenomena associated to the origin of stars that lead to c.) thermal evaporation or photolysis into more complex organic solids and d.) the material is returned to the ISM or accreted to providing the necessary conditions for the formation of planetary bodies. To understand all this process, knowledge of the different components of the ices is fundamental and it is achieved through the infrared absorption signatures corresponding to each molecule. The spectrum is formed by absorption bands around some specific wavelengths determined by their atomic composition and bond structure, e.g., the 4.67 μm (2140 cm^{-1}) C≡O stretching band; their peak position and width depend on the presence or absence of some molecules that can affect their dipolar moment, e.g.., when the break-up of the hydrogen bond weakens the O-H stretching feature of fully H bonded water at 3 μm (3280 cm^{-1}) [6], in addition to temperature and particle shape. Besides, these bands usually have an area and a width related to the compound concentration in the ice.

There are different magnitudes related to the absorption. One of them is the optical depth τ , which is defined as the integral of the absorption coefficient times the density along the path. It can be calculated for every frequency υ as:

$$I(\upsilon) = I_0(\upsilon)e^{-\tau_\upsilon} \tag{1}$$

where I is the intensity which passes through the ice and I_0 is the initial infrared beam intensity. The optical depth ranges between zero, when the radiation impinging goes through the material and arrives to the detector, therefore $I = I_0$, i.e., there is no absorption at all, and infinite if the ice is able to absorb all the arriving radiation $I = 0$. As an example, in Fig. 2 we show the infrared spectra of the methane and water registered at the UPV laboratory. Obviously a background spectrum is first registered before the measurements of the ice under test.

Some artifacts can appear due to preprocessing tasks, complex physics and noise sources, e.g., changes with time and temperature, calibrations or sensor noise. For example, in Fig. 2 we can see wavelengths where the optical depth takes negative values, such as around the CH_4 methane peak around 10 μm.

Of all the preprocessing tasks, the most important one is the baseline removal, because in very attenuate absorption bands a bad approximation of the baseline can mask some compounds or produce negative values of the optical depth.

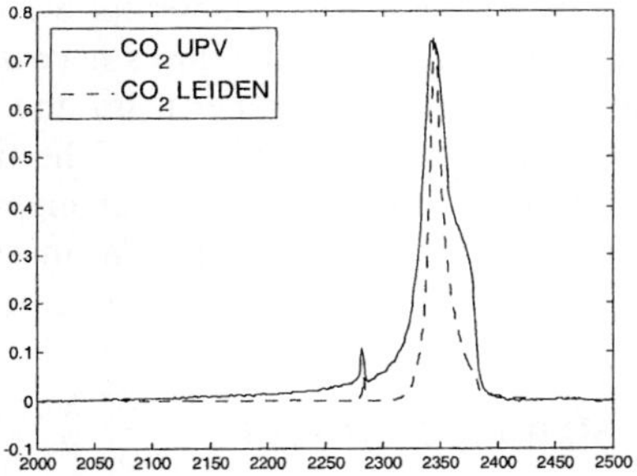

Fig. 1. CO_2 spectrum. Absorbance vs. wavenumber.

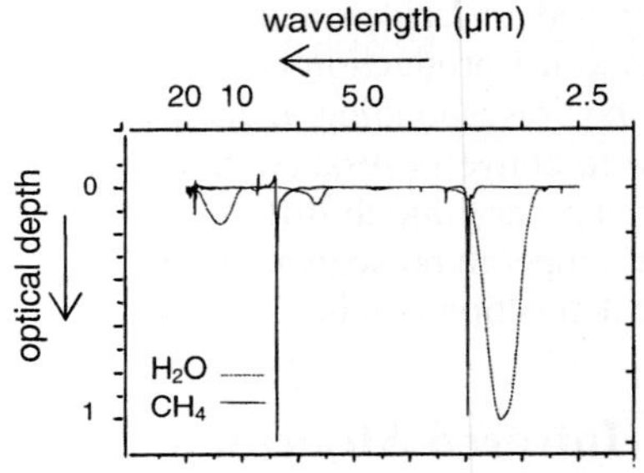

Fig. 2. Water and methane spectrum. Optical depth vs. wavelength.

3 ICA Framework

ICA has been applied in different astrophysical applications, e.g., to study celestial sources [7], to separate the cosmic microwave background, the galactic dust radiation, the synchrotron radiation and the free-free radiation components of the microwave sky radiation [8] or the mineralogical identification with the OMEGA spectra on board Mars Express [9]. The infrared spectrum has also been used in source separation problems [10], [11].

Only the sign factor indeterminacy of ICA must be considered, taking into account that it is really straightforward to detect the correct one, since the optical depth of the sharp absorption bands must be also positive. The restriction of at most one Gaussian source is not either a problem; the histogram analysis of pure molecules is far from corresponding to a Gaussian random variable.

3.1 ICA Model

The ICA formulation of the problem is: the infrared absorption spectrum x_{ij} measured for $i = 1,...,M$ ice mixtures in the spectral band $j = 1,...,P$, typically corresponding to 4000 up to 400 cm^{-1} with resolutions 1 or 2 cm^{-1}, is the linear combination of the independent absorption spectra s_{kj} of the molecules (sources) $k = 1,...,K$ present in those ices. The concentration of molecule k in ice i is the mixing matrix entry a_{ik}. The concentrations a_{ik} and absorption spectra s_{kj} are non-negative, although some preprocessing tasks such as baseline removal, noise and complex physics in the measurement process can produce a negative x_{ij}. All these processes will be resumed in a noise term n_{ij} for ice i in wavelength j. In matrix notation, for a given wavelength, we have:

$$\mathbf{x} = \mathbf{A}\mathbf{s} + \mathbf{n} \tag{2}$$

3.2 Discussion of the Model

We will review some physical considerations about the absorption spectra and endmembers in order to explain and discuss the conditions when the ICA model **(2)** is suitable for the problem.

Independency of the sources. The sources correspond to the spectral signatures of the endmembers. At a first glance, the hypothesis sounds rather intuitive that the absorption of different molecules has nothing to do due to the fact that vibrational frequencies of different components are not correlated. This is true, in general, but not always. As an example, at 10 K, the band width of the CO-stretch in CO/O_2 mixtures increases with O_2 concentration from 2.2 cm^{-1} up to 5.5 cm^{-1} when the amounts are equal, and then decreases again when O_2 is more abundant [6]. To model these relationships with conditional densities for a general case is very complicated. The same problem arises when the spectrum changes with temperature and radiation, so the only thing we can do in such ill-conditioned situations is for practical purposes to suppose that there are different kinds of CO, i.e., the number of sources must be increased.

Statistics of the sources. Considering the histogram of the spectra of the endmembers, most of them correspond to supergaussian signals, due to there being no absorption in most wavelengths.

Instantaneous mixture. The instantaneous model is clearly appropriate since absorption bands are not related. Nevertheless, a convolutive model could be suggested for the case aforementioned in the discussion about independency.

Linear mixture. The linear mixing model is also a usual supposition, i.e., the total absorption spectrum is the linear addition of the absorption features of every ice compound. But it is known that this is a simplified model in environments such as the planetary atmospheres, where the mixing is exponential [12].

Noise. It is basically due to measurement noise, although the preprocessing can also distort the actual signal. It is usually considered a white or coloured Gaussian noise. In our case, as the sensor noise level is relatively low (high quality measurements in the laboratory), it will not be considered in our model, so the relation between sources and mixtures becomes deterministic and equation (2) simplifies.

Number of mixtures. Usually $M > K$, i.e., more ices than simple molecules are available. In fact, it will be usual in the whitening step to reduce the dimension of the problem. The list of sources usually include CO, CO_2, H_2O, NH_3 and CH_4.

Priors. Absorption spectra are non-negative. However, as we explained before, the input samples can include some few low magnitude negative values, so this prior will not be included. The concentrations are also non-negative. This positive constraint is introduced in the model with a proper simple constraint. If both of them were considered, decompositions such as non negative matrix factorization [13] could be appropriate. Finally, the supergaussianity of the sources is used in the approximation of the densities of the independent components.

3.3 A MAP Estimate

Ignoring the noise term in (2) the problem becomes:

$$\mathbf{x} = \mathbf{A}\mathbf{s} \tag{3}$$

For a given observation $\mathbf{x}$ of the optical depth in the ice mixtures at a given wavelength, the posterior probability of $\mathbf{A}$ using Bayes rule is:

$$p(\mathbf{A}/\mathbf{x}) = \frac{p(\mathbf{x}/\mathbf{A})\,p(\mathbf{A})}{p(\mathbf{x})} \tag{4}$$

where $p(\mathbf{A})$ is the prior probability of the mixing matrix and $p(\mathbf{x}/\mathbf{A})$ is:

$$p(\mathbf{x}/\mathbf{A}) = \left|\det \mathbf{A}^{-1}\right| \prod_{k=1}^{K} p_k(s_k) \tag{5}$$

due to (3) and to the independence of the K sources. To maximize (4), the term $p(\mathbf{x})$ is not relevant. Considering the posterior probability for the whole set of wavelengths $\mathbf{x}(j),\ \ j = 1,\ldots,P$, normalizing and taking logarithms, the maximum a posteriori estimate is obtained:

$$\mathbf{A}_{map} = \arg\max_{\mathbf{A}} \frac{1}{P}\sum_{j=1}^{P}\sum_{k=1}^{K} \log p_k(s_k(j)) - \log\left|\det \mathbf{A}\right| + \log p(\mathbf{A}) \tag{6}$$

This is the maximum likelihood estimate plus a term $\log p(\mathbf{A})$ that tries to penalize $\mathbf{A}$ entries with low probability in the prior knowledge. The demixing problem is:

$$\mathbf{y} = \mathbf{B}\mathbf{x} \tag{7}$$

where $\mathbf{B}$ is the recovering matrix and $\mathbf{y}$ the recovered signals. For the square case, $\mathbf{B} = \mathbf{A}^{-1}$, the equivalent function to (6) to be maximized is:

$$\mathbf{B}_{map} = \arg\max_{\mathbf{B}} \frac{1}{P}\sum_{j=1}^{P}\sum_{k=1}^{K} \log p_k(\mathbf{b}_k^T\mathbf{x}(j)) + \log\left|\det \mathbf{B}\right| + \log p(\mathbf{B}^{-1}) \tag{8}$$

At this point we introduce the prior of the mixing matrix $p(\mathbf{A})$. Because its entries must be non-negative, a simple constraint imposing that $p(a_{ij})=0$ for $a_{ij}<0$ is enough. To maximize **(8)** we need to know the factorized pdf of the independent sources, but due to the supergaussianity of the sources, the derivative of the logarithm of the unknown pdf's can be substituted by other non linear functions. The stochastic version of the algorithm that maximizes **(8)** is:

$$\mathbf{B}_{new} = \mathbf{B}_{old} + \alpha(\mathbf{I} + \mathbf{f}(\mathbf{y})\mathbf{y}^T)\mathbf{B}_{old} \tag{9}$$

subject to $\mathbf{B}^{-1}$ does not have negative entries; α is the usual learning rate and $\mathbf{f}$ the non-linear component-wise function; i.e., the same updating rule as in the infomax algorithm [14] with the prior restriction. Note that **(9)** corresponds to the natural gradient version of the algorithm. The algorithm can be understood as finding $\mathbf{B}$ such that the distribution of $\mathbf{y}$ in **(7)** is as close as possible in a Kullback-Leibler divergence sense to the supposed distribution of the sources [15] without violating the constraint; e.g., if $f(y) = -2\tanh(y)$, then $\log p_i(s_i) = \beta - 2\log\cosh(s_i)$ with β a parameter fixed to make this the logarithm of a pdf. As we mentioned before, usually there will be more ices than molecules, so a first step of prewhitening will be carried out to reduce the dimension of the problem. This step also accelerates the convergence of the algorithm and obviously the same transformation must be applied to the prior.

4 Results

We tested the algorithm with the ice analogs database of Leiden [16]. This database contains the infrared spectra of laboratory analogs of interstellar ices. Different mixtures of molecules (from one up to three components selected from H_2, H_2O, NH_3, CH_4, CO, H_2CO, CH_3OH, O_2, N_2 and CO_2) at different temperatures and UV doses were produced, being the final spectrum calculated ratioing the measured and the background spectrum. The units of the data are cm^{-1} and absorbance, which is defined as optical

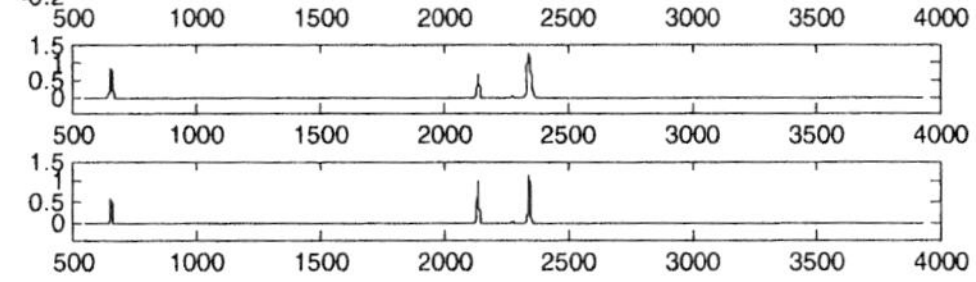

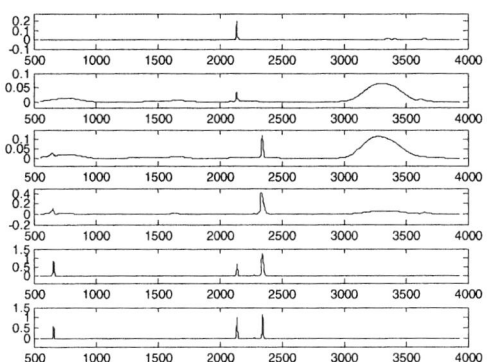

Fig. 3. Ice mixtures. Top to bottom, H_2O+CO (10:100), H_2O+CO (100:33), H_2O+CO_2 (100:14), H_2O+CO_2 (100:125), CO+CO_2 (100:70), CO+CO_2 (100:23).

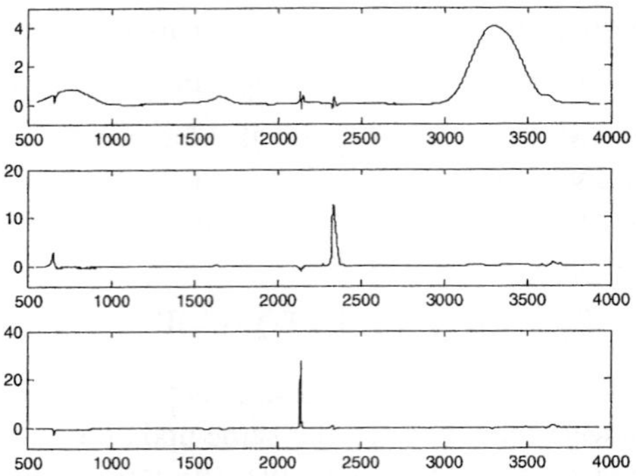

Fig. 4. Recovered Sources. Top to bottom, H2O, CO2 and CO.

length/ln10. The baseline was removed with Origin software and the useful wavelengths intervals were selected.

In Fig. 3 we show the ice mixtures, in this case ices containing H_2O, CO and CO_2. The three estimated components are shown in Fig. 4, where we can observe a CO and a CO_2 residual in the H_2O component due to the sharp peak value identifying the two components around 2150 cm^{-1} and 2350 cm^{-1} respectively.

5 Conclusions

In this paper we have presented the application of ICA to astrophysical ices. Mixture signals are the infrared absorption spectra of ices, being the sources the spectra of the different compounds present in the ices and the entries their concentrations. The study and decomposition of the components of these astrophysical ices is necessary in future studies where, depending on the composition and environmental conditions, each ice can produce a different complex compound. An additional restriction is imposed in order to obtain a constrained MAP estimate: the entries of the mixing matrix are non-negative. The effectiveness of ICA to extract the molecules present in a collection of laboratory ices has been proved.

The ICA approach has other advantages with respect to classical supervised methods: we do not need a complete library of endmembers for different situations, it is a multichannel method, concentration and endmembers can be estimated at the same time, and more challenging, it can extract signals corresponding to artifacts and unknown compounds not included in the library. Future work will try to model the dependencies between components for some molecules and to exploit the wavelength structure of the spectra.

Acknowledgements

We thank to M.A. Satorre, head of the Astrophysical Ices Laboratory of the UPV, for his helpful comments about the understanding of the physics and contribution in Fig.1 and 2. This work has been supported by Spanish Administration under grant TEC 2005-01820, GVA under grant GV05/250 and UPV under 2003-0554 project.

References

1. Van Dishoeck, E.F., Blake, G.A., Draine, B.T., Lunine, J.I.: Protostars and Planets III, University of Arizona Press. Tucson, (1993) 163
2. Miller, S.L., Ann. N.Y. Acad. Sci., 69, (1957) 123
3. Strazzulla, G., Brucato, J.R., Palumbo, M.E. Satorre, M.A.: Ion implantation in ices, NIMB, 116, (1996) 289-293
4. Gerakines P., Schutte W., Greenberg J., van Dishoeck E.: The infrared band strengths of H_2O, CO and CO_2 in laboratory simulations of astrophysical ice mixtures. Astronomy and astrophysics, (1997)
5. Pendleton Y. J.: The nature and evolution of interstellar organics. ASP Conference Series, From stardust to planetesimals, Vol. 122, (1997)
6. Ehrenfreund, P., Boogert, A., Gerakines, P., Tielens, A., van Dishoeck, E.: infrared spectroscopy of interstellar apolar ice analogs. Astronomy and astrophysics, (1997)
7. Nuzillard D., Bijaoui A.:Blind source separation and analysis of multispectral astronomical images. Astronomy and astrophysics suplement series (2000) 129-138
8. Maino D. Farusi, A. Baccigalupi, C. Perrotta, F. Banday, A. J. Bedini, L. Burigana, C. De Zotti, G. Górski, K. M. and Salerno, E.. All-sky astrophysical component separation with (FastICA). Monthly Notices of the Royal Astronomical Society, (2001)
9. Forni O., Poulet F., Bibring J.P., Erard S. Component separation of OMEGA spectra with ICA. Lunar and planetary Science XXXVI, (2005)
10. Parra L., Mueller K.R., Spence C., Ziehe A., Sajda P.: Unmixing Hyperspectral Data, Advances in Neural Information Processing Systems 12, MIT Press, , (2000) 942-948
11. Bayliss J., Gualtieri J., Cromp R.: Analysing hyperspectral data with ICA. Proc. SPIE Applied Image and Pattern Recognition Wrkshp. (1997)
12. Bandfeld, J., Christensen P.: spectral data set factor analysis and end-member recovery: application to analysis of Martian atmospheric particulates. Journal of Geophysical Research, Vol. 105, No. E4, (2000) 9573-9587
13. Lee D.D., Seung, S.: Learning the parts of objects by non-negative matrix factorization. Nature, Vol. 401, (1999) 788-791
14. Bell A.J., Sejnowsky T.J.: An information-maximization approach to blind separation and blind deconvolution. Neural Computation 7 (1995) 1129-1159
15. Cardoso, J.F.: Higher-order contrasts for ICA, Neural Computation 11(1999) 157-192
16. http://www.strw.leidenuniv.nl/~schutte/database.html

Improvement on Multivariate Statistical Process Monitoring Using Multi-scale ICA

Fei Liu and Chang-Ying Wu

Institute of Automation, Southern Yangtze University,
Wuxi, 214122, P.R. China
fliu@sytu.edu.cn

Abstract. A multi-scale independent component analysis (ICA) approach is investigated for industrial process monitoring. By integrating the ability of wavelet on multi-scale analysis and that of ICA on extracting independent components for non-Gaussian process variables, the multivariate statistical monitoring techniques can obtain improved performance. Contrastive tests have been carried out on the famous benchmark chemical plant among ICA-like and PCA-like methods, which reveals that multi-scale ICA approach has lower missed detection rate of faults.

1 Introduction

Modern chemical processes, which are equipped with instrument and data collector, contain thousands of measured variables, such as temperatures, pressures and flow rates. The correlation among the process variables exists as a result of either association or causation. Instead of univariable statistical process control (SPC), it is necessary and possible to apply multivariate statistical process control (MSPC) to extract relevant information in the redundant process data, and to detect if statistically significant abnormalities occur. As a methodology, MSPC monitors whether the process is in control through the analysis of the various control charts, such as T2 and SPE. While numerous procedures of univariable SPC are available in manufacturing processes and are likely to be part of a basic industrial training program, MSPC procedures are being used to monitor chemical processes that are inherently multivariate [1].

Basically, as a mathematical tool, principal component analysis (PCA) can essentially identifies important characteristics in multivariate redundant data and has successfully been applied to performance monitoring and fault diagnosis for industrial process [2], [3]. PCA makes variables de-correlated by means of maximizing the variance within the process data, which follows a Gaussian probability distribution or independent identical distribution. Unfortunately, the process variables and its statistical information are very complex in actual industrial production, it is difficult to make certain about process variable's probability distribution [4].

As a blind source separation technique, independent component analysis (ICA) has been founded wide applications in processing of medical signals, compressing of images [5], and machine fault detection [6], etc. Compared with PCA,

J. Rosca et al. (Eds.): ICA 2006, LNCS 3889, pp. 376–383, 2006.

ICA represents a set of random variables as linear combination of statistically independent component variables, and is less sensitive to process variable's probability. Benefited from this, recently, ICA has been introduced to process monitoring [7], [8], which can be more efficacious in a non-Gaussian context. In the real-world stochastic processes, the energy or power spectrum of variables often changes with time or frequency, as a result, almost all the industrial processes are multi-scale in nature. Accordingly, while the existing ICA process monitoring has been adopted at a single scale, it may be more significant to improve the process monitoring by means of multi-scale ICA. In fact, the concept of multi-scale analysis already exists in the project of image processing [9], multi-scale PCA monitoring [10] and blind source separation [11].

In this paper, a multi-scale independent component analysis (MSICA) approach is investigated for process monitoring, which integrates the ability of wavelet on multi-scale analysis and that of ICA on extracting independent components for non-Gaussian variables. While the process is in control, the genuine model is decided by the reconstructed signals from the selected wavelet coefficients, which violate the threshold of the ICA model at the significant scales. By means of the presented MSICA, the statistical monitoring method is discussed on the famous benchmark plant, Tennessee-Eastman (TE) chemical process. Compared with traditional MSPC methods (including existing PCA and MSPCA), MSICA reveals lower missed detection rate of faults.

2 Multi-scale ICA Monitoring

2.1 Wavelet-Based MSICA Model

For an original data set $X = (x(1), x(2), \cdots, x(m)) \in R^{n \times m}$ with m measure variables and n samples, the standardization is firstly performed to make measure variables have zero mean and unit variance. By applying l steps wavelet decomposes to every measure variable $x(i)$, $i = 1, 2, \cdots, m$, there are l detail coefficient vectors $a_{k,i}$ on scale $k = 1, 2, \cdots, l$, and an approximate coefficient vector $b_{l,i}$. Assuming p is the sample length of $a_{k,i}$ and $b_{l,i}$, on the kth scale, there exist a detail coefficient matrix $A_k = (a_{k,1}, a_{k,2}, \cdots, a_{k,m})^T \in R^{m \times p}$ and an approximate coefficient matrix $B_l = (b_{l,1}, b_{l,2}, \cdots, b_{l,m})^T \in R^{m \times p}$. Moreover, let $A_k = C\widetilde{S}$ or $\widetilde{S} = FA_k$, where $\widetilde{S} \in R^{p \times m}$ is source signal matrix, $C \in R^{p \times p}$ is a nonsingular constant matrix, and $F = C^{-1}$. In general, A_k may be whitened by the transformation QA_k, where $Q = \Lambda^{-1/2}U^T$, and Λ and U come from the eigen-decomposition of the covariance, i.e. $E(A_k A_k^T) = U \Lambda U^T$. To separate statistical independent source signals from A_k, the notion of entropy is used, which is a measure of uncertainty of a continuous stochastic variable. Under the condition of the same variance, the smaller the differential entropy of a random variable is, the greater non-Gaussianity it has. It is known that the Gaussian random variable has maximal differential entropy and non-Gaussian implies indepedence [12]. This gives a way to judge the independent of stochastic variable by comparing its differential entropy.

Consider a continuous stochastic variable y with probability density function $f(y)$, its differential entropy is given by $H(y) = -\int f(y)logf(y)dy$. To avoid estimating the probability density function $f(y)$, the negentropy of $H(y)$ is defined as $J(y) = H(y_{Gauss}) - H(y)$, where $H(y_{Gauss})$ is a Gaussian stochastic variable with the same variance as y. As a fast algorithm,$J(y) \approx [E\{G(y)\} - E\{G(y_{Gauss})\}]^2$ is adopted here instead of negentropy definition [13], where $G(\cdot)$ is a non-quadratic function.

For $\widetilde{S} = FA_k$, single source signal can be expressed as $\widetilde{s}(i) = f \cdot A_k$, where f is a row vector in matrix F. In the setting of entropy, the modeling of independent component is translated to following optimization problem: find an optimal f satisfying $E\{fA_k \cdot (fA_k)^T\} = \|f\|^2 = 1$ to make $J(f \cdot A_k)$ maximum. Keep in mind negentropy definition, the optimization problem is equivalent to following:

$$\min_f E\{G(f \cdot A_k)\}$$
$$s.t. E\{fA_k \cdot (fA_k)^T\} = \|f\|^2 = 1 \tag{1}$$

Based on the Kuhn-Tucker conditions, that $E\{G(f \cdot A_k)\}$ is optimal while $E\{A_k g(f \cdot A_k)\} - \beta f = 0$ where $\beta = E\{f_0 A_k g(f_0 \cdot A_k)\}$ with optimum f_0, Newton's method is borrowed to solve above optimization problem (1), and here is omitted. Obviously, all the row vectors of F can be estimated in the same way. And then constant matrix C and source signal matrix $\widetilde{S}$ can be computed via $\widetilde{S} = FA_k$.

Retaining d independent source signals, $\overline{A}_k = Q^{-1} F_d \widetilde{S}_d$ is the reconstruction of detail coefficients matrix A_k, where $F_d \in R^{m \times d}$, $\widetilde{S}_d \in R^{d \times p}$ are the corresponding retained de-mixing matrix and source signal matrix, respectively [8]. Simultaneously, a threshold is set based on the median of serial signal $\overline{a}_{k,i}$ for removing residual of signals and acquiring information of significant events [14]. Similarly, B_l is reconstructed as A_k is done.

Applying reverse wavelet transformation on the retained detail coefficient at every scale and approximate coefficient at the coarsest scale [9], the reconstructed process data $Y \in R^{n \times m}$ is obtained and then modeled in ICA form, $Y^T = TS + E$, where $S = (s(1), s(2), \cdots, s(d))^T \in R^{d \times n}$ is independent components matrix of process, $T \in R^{m \times d}$ is coefficient matrix, and $E \in R^{m \times n}$ is residual. For mathematical convenience, assume $E = 0$, $d = m$, then $S = T^{-1} \cdot Y^T = W \cdot Y^T$. Based on above multi-scale ICA model, in next subsection, the process monitoring technique is investigated.

2.2 Statistical Monitoring

Two universal statistics, I^2 or T^2 (I^2 used in ICA-based monitoring method, while T^2 in PCA-based method) and Q (i.e. Square Prediction Error, SPE), have been employed in real-time process monitoring. At the tth sample, $I^2(t) = S_d^T(t) S_d(t)$, where $S_d(t)$ is the tth column vector of the projection of the original data in the directions of independent components, and $SPE = e^T(t) \cdot e(t)$, where $e(t) = x(t) - \widehat{x}(t)$ is the residual between sample $x(t)$ and prediction of model $\widehat{x}(t)$.

Because independent components may not follow normal distribution, the control limit of statistic can not be decided by a special probability function, as done in PCA. Instead, kernel density estimation is introduced, in which the univariate kernel estimator is defined as $f(x) = \frac{1}{nh}\sum_{i=1}^{n} K\{(x - x_i)/h\}$, where $K(\cdot)$ is the kernel function, x is the data point under consideration, x_i is the sample data, $i = 1, 2, \cdots, n$, and h is the smoothing parameter. In practice, Gaussian function is usually chosen as the kernel function. To get the point z, which occupies the 99% area of density function $f(x)$, the following equation is used,

$$\int_{-\infty}^{z} f(x)dx = \int_{-\infty}^{z} \frac{1}{nh}\sum_{i=1}^{n} K\{(x - x_i)/h\}dx$$
$$= \int_{-\infty}^{z} \frac{1}{nh}\sum_{i=1}^{n}\{exp[(x - x_i)^2/(2h^2)]/\sqrt{2\pi}\}dx = 0.99 \quad (2)$$

The detailed selection of h may be found in [15]. The control limits of process normal operation are then easily obtained.

3 Cases Study

3.1 Missed Detection Rate Comparison

The TE process is a realistic industrial plant for evaluating process control and monitoring methods, which consists of five major units: reactor, condenser, compressor, separator, stripper. It contains eight components: reactants A, C, D, E, inert B fed to the reactor, products G, H and by-product F formed in the reactor. The process contains 41 measured variables and 11 manipulated variables, which are sampled per 3 minutes. All the process measurements involved faults are introduced at sample 301. The reader may refer to [16] for details.

For some typical statistical control methods and statistics, the missed detection rates for all 20 faults are shown in Table 1. Because the variables in Fault 3, 9 and 15 have no remarkable mean and standard deviation changing, their missed detection rates are high for all the statistics. It is conjectured that any statistic based on data-driven methods will result in high missed detection rates for these faults [2], thus in Table 1 they are marked by asterisk. Besides the three faults, the minimum missed detection rates of all faults are denoted by bold style. As a whole, it maybe true that MSICA has superiority in process monitoring and detection.

3.2 Monitoring and Detection Tests

Introduce fault 4 to TE process, which is a step change in the reactor cooling water inlet temperature. This leads a step change of the cooling water flow rate as shown in Fig. 1(Left), and a sudden jump of the reactor temperature as shown in Fig. 1(Right). The other 50 measure and manipulated variables retain steady, the variance of the mean and the standard deviation of each variable is inapparent.

Table 1. Missed detection rates for the testing set

Fault	$T^2(PCA)$	Q	$T^2(MSPCA)$	Q	$I^2(MSICA)$	Q
1	0.0076	0.0045	0.0106	0.0061	**0.0030**	**0.0030**
2	0.1515	0.0136	0.0258	0.0152	**0.0061**	**0.0045**
3*	0.9682	0.9394	0.9652	0.9788	**0.7242**	**0.8833**
4	0.7833	**0.0091**	0.9061	0.2485	**0.0515**	0.0318
5	0.7015	0.6485	0.7242	0.6227	**0**	**0**
6	0.0106	**0**	**0**	**0**	**0**	**0**
7	0.4364	**0**	**0**	**0**	**0**	**0**
8	**0.0227**	0.0258	0.0545	0.0242	0.0303	**0.0167**
9*	0.9680	0.9409	0.9621	0.9758	**0.8318**	**0.8788**
10	0.5240	0.4273	0.6277	0.4318	**0.1985**	**0.1621**
11	0.6680	0.2440	0.8076	**0.0403**	**0.1682**	0.2289
12	0.0260	0.0227	0.0394	0.0167	**0.0030**	**0.0136**
13	0.0772	0.0651	0.0924	0.6515	**0.0606**	**0.0470**
14	0.1773	0.0015	0.0015	0.0030	**0.0014**	**0.0015**
15*	0.9970	0.9030	0.9545	0.9610	**0.7575**	**0.9000**
16	0.7591	0.6455	0.7515	0.5954	**0.0742**	**0.1379**
17	0.3197	**0.0727**	0.1682	0.1848	**0.0379**	0.1000
18	0.1318	0.0969	0.0985	0.1091	**0.0818**	**0.0758**
19	0.9878	0.6091	0.7864	0.9136	**0.1303**	**0.5424**
20	0.7727	0.4379	0.7848	0.5530	**0.0970**	**0.1970**

All these make the detection and diagnosis of fault 4 more challenging than other faults. The T^2 statistic charts based on PCA, multi-scale PCA (MSPCA) are shown in the Fig. 2. It is obvious that T^2 statistics are not beyond the threshold after sample 301. However, I^2 statistic chart of MSICA gives an exciting result as shown in the Fig. 3. After sample 300, it can be seen that I^2 statistic goes beyond the control limit distinctly, its ability to detect fault 4 is better than that of other methods.

Based on the process monitoring, furthermore, the fault detection and diagnosis is carried out to identify the observation variables most closely related to the

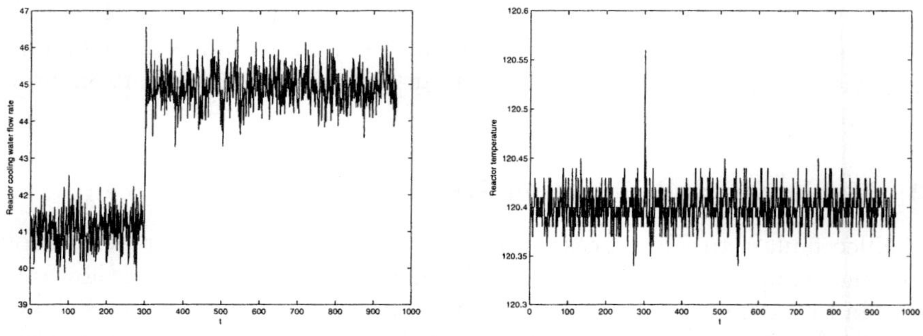

Fig. 1. Cooling water flow rate (Left) and reactor temperature (Right)

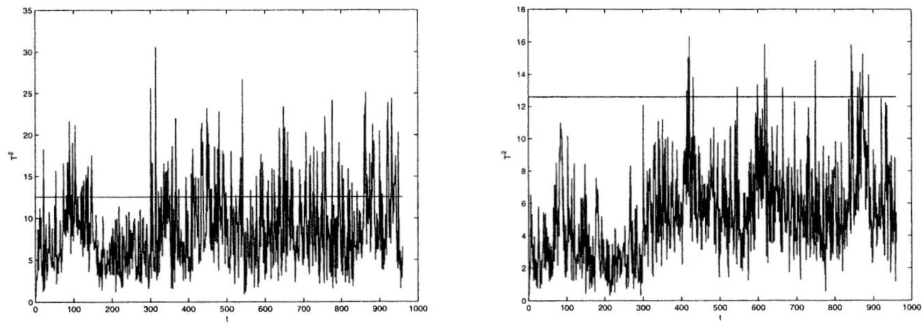

Fig. 2. The T^2 statistics of PCA (Left) and MSPCA (Right) powered by fault 4

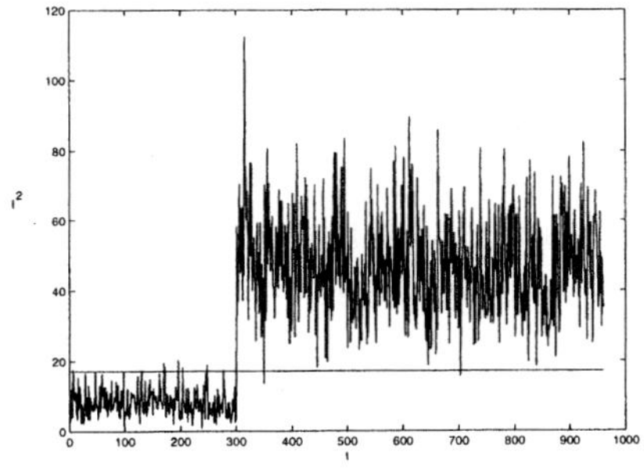

Fig. 3. The I^2 statistic of MSICA powered by fault 4

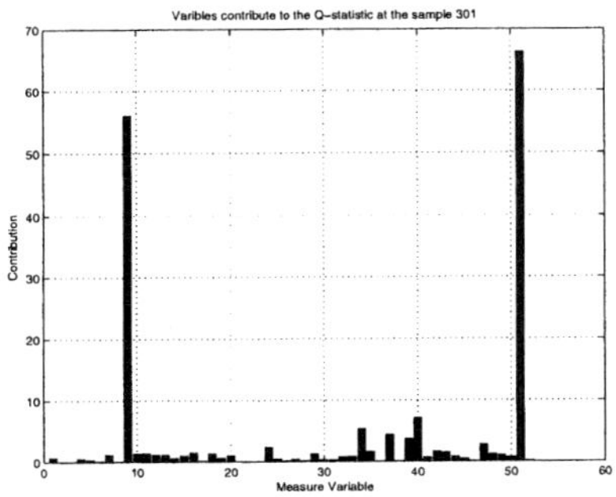

Fig. 4. The average contribution of fault 4 for the MSICA-based SPE

faults. As a basic tool, the average contribution of fault 4 at sample 301 is shown in Fig. 4. Obviously, variable 9 (Reactor temperature) and variable 51(Reactor cooling water flow rate) correlate with the fault 4 deeply. In fact, the two control loops of reactor cooling water flow rate and reactor temperature are cascade. This is consistent with the fault description.

Another interesting test is carried out by introducing fault 10, which involves a random change in the temperature of stream 4 (C feed). Even though this

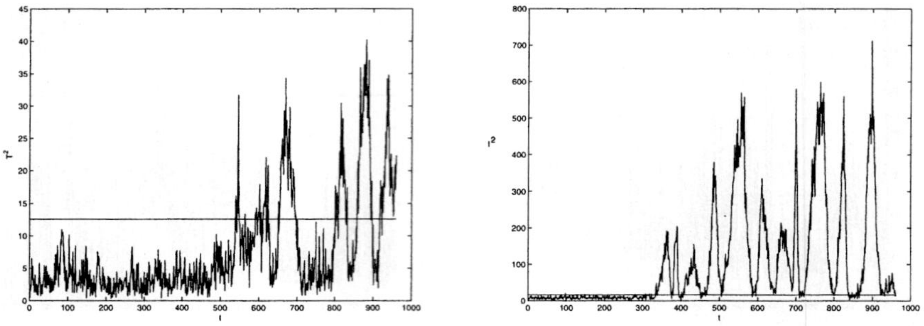

Fig. 5. The T^2 statistic of MSPCA (Left) and the I^2 statistic MSICA (Right) powered by fault 10

fault has no some effect on product quality by the titer of product in the stream 11, it is a hidden trouble on the safety of the process because of its effect on the pressure of the reactor. The T^2 statistic chart based on MSPCA and the I^2 statistic chart based on MSICA are shown in Fig. 5. It is obvious that PCA-like methods do not give an alarm in time, in which T^2 statistic is beyond the threshold after sample 500. Contrastively, the monitoring response of MSICA is acceptable.

4 Conclusion

By integrating the advantage of wavelet transform and independent component analysis, a MSICA approach is introduced to the process monitoring. The application results on TE plant illuminate some advantage over traditional MSPC methods. The ICA-based monitoring methods can give some compellent outcome in actual process operation.

References

1. Mason, R. L., Young, J. C.: Multivariate Statistical Process Control with Industrial Application. SIAM, USA (2002)
2. Chiang, L.H., Russell, E. L., Braatz, R. D.: Fault Detection and Diagnosis in Industrial Systems. Springer-Verlag, London (2001)
3. Kano, M., Nagao, K., Hasebe, S., Hashimoto, I., Bakshi, B.: Comparison of Statistical Process Monitoring Methods: Application to the Eastman Challenge Problem. Computers and Chemical Engineering 24(2000) 175-181
4. Johnson, R.A, Wichern, D.W.: Applied Multivariate Statistical Analysis. 4th edn. Englewood Cliff, Prentice Hall (1998)
5. Roberts, S., Everson, R.: Independent Component Analysis: Principles and Practice. Cambridge University (2001)
6. Ypma, A., Tax, D.M.J., Duin, R.P.W.: Robust Machine Fault Detection with Independent Component Analysis and Support Vector Data Description. Proc. IEEE Neural Networks for Signal Processing (1999) 67-76

7. Hyvarinen, A., Oja, E.: Independent Component Analysis: Algorithms and Applications. Neural Networks 13(2000) 411-430
8. Lee, J.-M., Yoo, C., Lee, I.-B.: Statistical Process Monitoring with Independent Component Analysis. Journal of Process Control 14(2004) 467-485
9. Mallat, S. G.: A Theory for Multiresolution Signal Decomposition: The Wavelet Representation. IEEE Transaction of Pattern Analysis and Machine Intelligence 11(7)(1989) 674-693
10. Bakshi, B. R.: Multiscale PCA with Application to Multivariate Statistical Process Monitoring. AICHE J. 44(7)(1998) 1596-1610
11. Kisilev, P., Zibulevsky, M., Zeevi, Y.Y.: A Multiscale Framework for Blind Source Separation of Linearly Mixed Signals. Journal of Machine Learning Research 4(2003) 1339-1364
12. Cover, T.M, Thomas, J.A.: Elements of Information theory. Wiley, New York (1991)
13. Hyvarinen, A.: Fast and Robust Fixed-Point Algorithms for Independent Component Analysis. IEEE Trans. Neural Networks 10(3)(1999) 626-634
14. Daubechies, I.: Ten Lectures on Wavelet. Capital City Press (1992)
15. Wang, M. P., Jones, M. C.: Kernel Smoothing. Chapman & Hall, London, UK (1995)
16. Downs J.J, Vogel, E. F.: A Plant-Wide Industrial Process Control Problem. Computers and Chemical Engineering 17(3)(1993) 245-255

Global Noise Elimination from ELF Band Electromagnetic Signals by Independent Component Analysis

Motoaki Mouri[1], Arao Funase[1,2], Andrzej Cichocki[2], Ichi Takumi[1], Hiroshi Yasukawa[3], and Masayasu Hata[4]

[1] Nagoya Institute of Technology, Graduate School of Engineering, Gokiso-cho, Showa-ku, Nagoya 466-8555, Japan
[2] RIKEN BSI, Laboratory for Advanced Brain Signal Processing 2-1 Hirosawa, Wako, Saitama 351-0198, Japan
[3] Aichi Prefectural University, Department of Applied Information Technology, Nagakute-cho, Aichi-gun, Aichi, 480-1198, Japan
[4] Chubu University, Matsumoto-cho, Kasugai-city, Aichi, 487-8501, Japan

Abstract. Anomalous radiations of environmental electromagnetic (EM) waves are reported as the portents of earthquakes. We have been measuring the Extremely Low Frequency (ELF) range all over Japan in order to predict earthquakes. The observed signals contain global noise which has stronger power than local signals. Therefore, global noise distorts the results of earthquake-prediction. In order to overcome this distortion, it is necessary to eliminate global noises from the observed signals. In this paper, we propose a method of global noise elimination by Independent Component Analysis (ICA) and evaluate the effectiveness of this method.

1 Introduction

Japan is a country where earthquakes occur frequently, and has received extensive damage from huge earthquakes. The occurrence of giant earthquakes will be worried about in the near future, too. The Earthquake Research Committee of Japan reported in 2001 that the occurrence possibility of very giant earthquakes of Nankai and Tohnankai earthquakes (magnitude over 8) within 30 years reached between 40% and 50% [1].

It is urgent task to achieve an accurate earthquake prediction to help minimize the damages caused by earthquakes. Anomalous radiations of environmental electromagnetic (EM) waves have been reported as the portent to be the earth's crustal motion including earthquakes [2],[3]. We have been measuring Extremely Low Frequency (ELF) magnetic fields all over Japan since 1989, in order to try to predict earthquakes.

However, the EM radiation data contains signals other than the earth's crustal motion. The recorded data are distorted by noise. It is important to remove noise as a preprocessing step for earthquake prediction.

J. Rosca et al. (Eds.): ICA 2006, LNCS 3889, pp. 384–391, 2006.

In the present paper, we propose the method for noise elimination by Independent Component Analysis (ICA) to extract the anomalous EM radiation data more accurately, and evaluate effectiveness of this method.

2 Outline of ELF Band Observation of EM Radiation Data

We have been observing power of 223Hz in EM radiation in about 40 places around the country (Fig.1). This frequency band has been a little influenced by solar activity and the global environment (Fig.2). Observation systems have three axial loop antennas with east-west, north-south, and vertical ranges. Observation devices sample EM levels and average the signals over 6-second periods. These data are transported to our institute on the Public Telephone Network.

The observed signals are composed of the local signals and the global noise. The local signals are caused by regional EM radiation; for example, the earth's

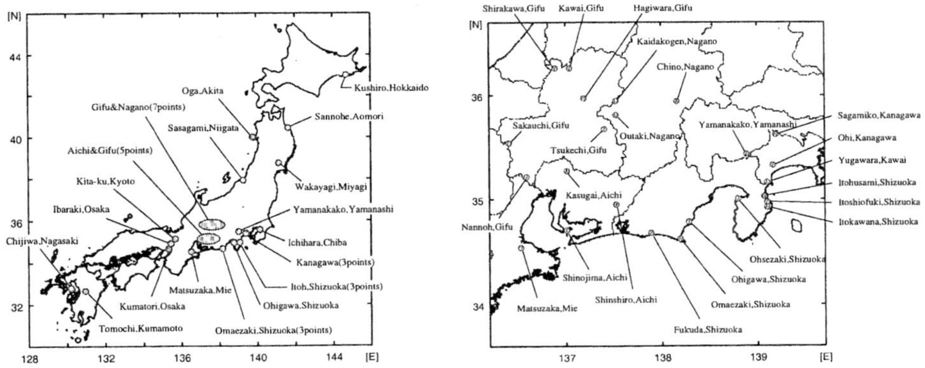

Fig. 1. Arrangement of observation points

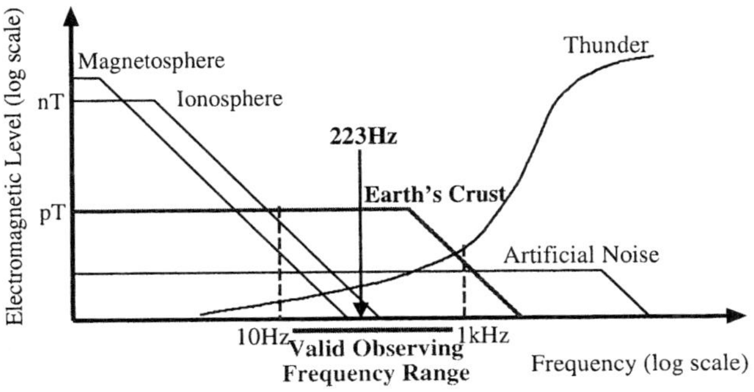

Fig. 2. EM radiation levels of each source

crustal motion, thunderclouds, or other interferences. The local signals in each observation point have different values. The global noise is caused by global EM radiation; for example, heat thunderclouds in lower latitudes, the solar activity, and many others. The radiation from southern heat thunderclouds is influenced by the ionosphere in the spread process. The global noise has a circadian rhythm because the ionosphere changes in a day by the effect of the sun. The global noises in each observation point have almost same values because the global noise is recorded all over the observation point.

The influence of the global noise is especially stronger than many other noises. Eliminating global noise (or extracting local signals) is important as preprocessing of earthquake prediction. The sources of each EM radiations are mutually independent. We estimate the global noise by separating observed signals using ICA next to remove it from the observed data.

3 Method of Global Noise Elimination Using ICA

To estimate the global noise and the local signals accurately, we must analyze the data from all the observation points. However, it is unrealistic that all the observed signals are processed because the number of all source signals is over the number of separable signals. It is necessary to decrease the number of source signals contained in the observed signals. Therefore, we approximately estimate the components from the good signals recorded several observation points. In that case, note that it is difficult to directly separate the global noise and the local signals from all observed signals. We solve this problem by the idea of subtracting global noise from the observed signals.

Procedures of global noise elimination by NG-FICA are as follow.

1. Selecting several good observation points from all observation points.
2. Estimating independent components by ICA from the observed signals recorded in the selected observation points.
3. Selecting a global noise component from among the estimated signals.
4. Calculating the amplitudes of the global noises corresponding to each observed signals.
5. Estimating the enhanced local signals by subtracting each global noises from the observed signals.

3.1 Selecting Observation Points

Since the global noise is large, we must select signals similar to global noise. Global noise has a strong correlation with all the observed signals because global noise is the main component of the observed signals. Therefore, we establish the observed signals priority list in the descending order of following expression.

$$|r_{xx_j}| = \left| \sum_i \frac{E\left[(x_i - \overline{x_i})(x_j - \overline{x_j})\right]}{\sqrt{E\left[(x_i - \overline{x_i})^2\right]}\sqrt{E\left[(x_j - \overline{x_j})^2\right]}} \right| \tag{1}$$

where $|r_{xx_j}|$ is the absolute of sum correlation coefficient between x_j of all the observed signals. We select observation points by hand based on this priority. Experience shows that selecting around 6 observed signals give the best results.

3.2 Source Separation

In ICA, various algorithms are proposed with a focus on assumption independence. The algorithm adopted by this paper is NG-FICA (Natural Gradient - Flexible ICA) [4]. NG-FICA uses kurtosis as independency criterion and uses natural gradient for the learning algorithm. This algorithm is implemented as a part of the package, "ICALAB for signal processing" [5].

In NG-FICA, input data of vector $\boldsymbol{x}$ is applied sphering (prewhitening) as a linear transformation.

$$\boldsymbol{z} = \boldsymbol{Q} \cdot (\boldsymbol{x} - \overline{\boldsymbol{x}}), \quad \boldsymbol{Q} = \left\{ E\left[(\boldsymbol{x} - \overline{\boldsymbol{x}})(\boldsymbol{x} - \overline{\boldsymbol{x}})^T \right] \right\}^{-1/2} \tag{2}$$

where vector $\overline{\boldsymbol{x}}$ is the mean of $\boldsymbol{x}$. $\boldsymbol{Q}$ is calculated by Principal Component Analysis (PCA). The presumption method uses vector $\boldsymbol{z}$ as the input data.

The update nonlinear functions are based on the following expressions.

$$\Delta \boldsymbol{W} = \eta \left(\boldsymbol{I} - E\left[\boldsymbol{y}\boldsymbol{y}^{\boldsymbol{T}} - (\boldsymbol{\varphi}\boldsymbol{y}^{\boldsymbol{T}} + \boldsymbol{y}\boldsymbol{\varphi}^{\boldsymbol{T}}) \right] \right) \boldsymbol{W} \tag{3}$$

$$\varphi_i = |y_i|^{\alpha_i - 1} \mathrm{sgn}(y_i) \quad (i = 1, 2, \cdots, n) \tag{4}$$

where η is the appropriate learning rate (constant number), $\boldsymbol{y}$ is the temporary estimated signal ($= \boldsymbol{W}\boldsymbol{z}$), and sgn($y_i$) is the signum function of y_i. Gaussian exponent α_i is decided based on the kurtosis κ_i $\left(= \frac{E[{y_i}^4]}{\{E[{y_i}^2]\}^2} - 3 \right)$ of y_i; α_i is decided near 0 if κ_i is big, but α_i is decided 4 if κ_i is small. Finally, the independent components $\boldsymbol{y}$ are estimated as:

$$\boldsymbol{y} = \boldsymbol{W}\boldsymbol{Q}\boldsymbol{x}. \tag{5}$$

3.3 Selection of Global Noise Component

By the ICA, a global noise is extracted as one of the estimated signals. However, the estimated signals come out in a random fashion due the permutation ambiguity. Therefore, it is necessary to identify the global noise component from the estimated signals. We select one component y_g which has a maximal value by the following expression.

$$|r_{xy_g}| = \left| \sum_i \frac{E\left[(x_i - \overline{x_i})(y_g - \overline{y_g}) \right]}{\sqrt{E\left[(x_i - \overline{x_i})^2 \right]} \sqrt{E\left[(y_g - \overline{y_g})^2 \right]}} \right| \tag{6}$$

3.4 Calculation of Local Signals

The amplitudes of the estimated global noise component and the actual global noise are not the same, because the estimated components may have arbitrary scale factors. Therefore, it is necessary to adjust the amplitude of the global noise component for each observed signals. When the amplitude of global noise component was appropriately weighted, the mean square error (MSE) between the observed signal and global noise component would be minimized. The MSE between the observed signal x_j and the weighted global noise component $b_j y_g$ is calculated as $E[((x_j - \overline{x_j}) - b_j(y_g - \overline{y_g}))^2]$. The appropriately weight b_j which gives the least MSE, is obtained by the following expression:

$$b_j = \frac{E\left[(x_j - \overline{x_j})(y_g - \overline{y_g})\right]}{E\left[(y_g - \overline{y_g})^2\right]} \tag{7}$$

Using vector $\boldsymbol{b}$ constructed from b_j, local signals are calculated as:

$$\boldsymbol{x_L} = \boldsymbol{x} - \boldsymbol{b} y_g. \tag{8}$$

4 Sample Case of Global Noise Elimination

4.1 Processing Data

An anomalous signal was observed for two days starting from January 4, 2001, at the observation point in Nannoh, Gifu Prefecture (call Nannoh after this). We tried to obtain local signals about this day by eliminating the global noise by the proposed method. The recorded signals from Nannoh might have anomalous signals related to the earthquake, because an earthquake (magnitude 4.8) occurred in Tohnoh, Gifu Prefecture on January 6.

4.2 Results

Fig.3 and Fig.4 show the signals that were observed in Kawai, Gifu Prefecture (call Kawai after this) and Nannoh on January 4, 2001. The vertical axes show the EM levels (pT$\sqrt{\text{Hz}}$) and the horizontal axes indicate the time courses (hours). Both of these signals have high amplitudes in nighttime and have low amplitudes in daytime. Changes like these are mostly observed for all observation points throughout the year. In other words, these signals have global noises like a circadian rhythm.

Fig.5 shows one of the separated components from the observed signals by the ICA algorithm [4],[5]. The vertical axis shows the amplitude of estimated signal and the horizon axis indicates the time course. This component has global noise features, because it has a strong correlation to each observed signal (Table1). and this signal has a high amplitude in nighttime and low in daytime. Therefore, this component is selected as global noise.

The local signals of Kawai and Nannoh are shown in Fig.6 and Fig.7. Their axes are the same as those in Fig.3. These signals do not have circadian rhythm like the raw observed signals. In addition, the local signal in Nannoh has clearly anomalous signals since about 6 a.m. From these results, it is evident that the proposed method can eliminate global noise from observed signals.

Table 1. Coefficients of observed signals and global noise (from Fig.5)

Kushiro, Hokkaido	—
Sannohe, Aomori	-0.82223442
Oga, Akita	-0.83796101
Wakayagi, Miyagi	-0.94404159
Ichihara, Chiba	0.05078791
Sasagami, Niigata	-0.87207873
Yugawara, Kanagawa	-0.81599893
Sagamiko, Kanagawa	-0.64286322
Yamanakako, Yamanashi	-0.80240557
ItohUsami, Shizuoka	-0.95962168
ItohShiofuki, Shizuoka	-0.75264114
ItohKawana, Shizuoka	-0.81597344
Oosezaki, Shizuoka	-0.80150511
Omaezaki, Shizuoka	-0.64272118
Ohtaki, Nagano	-0.94814968
KaidaKougen, Nagano	-0.89521541
Kawai, Gifu	-0.79121470
Shirakawa, Gifu	-0.84631972
Hagiwara, Gifu	-0.88124143
Tsukechi, Gifu	-0.80214358
Sakauchi, Gifu	-0.94820738
Nannoh, Gifu	-0.62218944
Kasugai, Aichi	—
Shinojima, Aichi	—
Matsuzaka, Mie	-0.53462919
Kitaku, Kyoto	-0.92216432
Kumatori, Osaka	-0.71976842
Ibaraki, Osaka	-0.90787270
Chijiwa, Nagasaki	-0.88037934
Tomochi, Kumamoto	-0.81322911
Average	-0.784147501

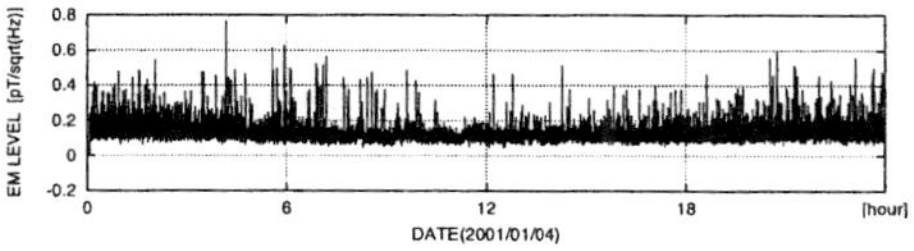

Fig. 3. Observed signal (January 4, 2001 at Kawai, Gifu)

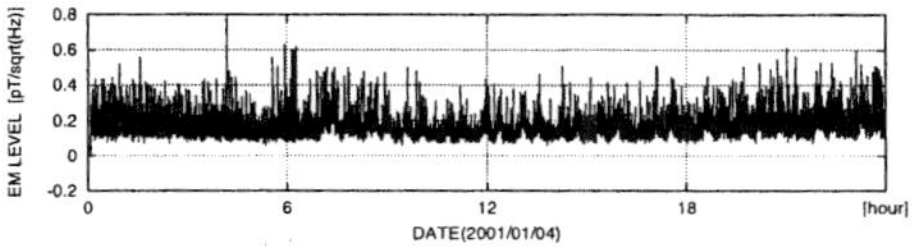

Fig. 4. Observed signal (January 4, 2001 at Nannoh, Gifu)

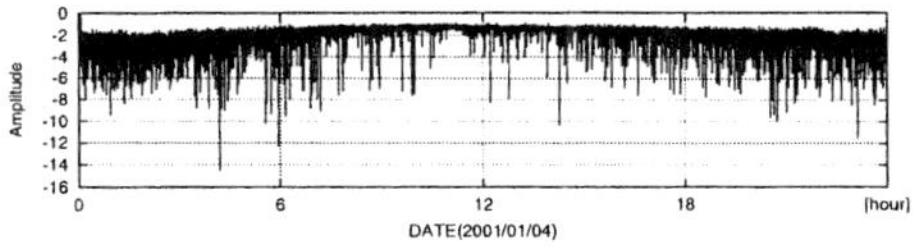

Fig. 5. Global noise component

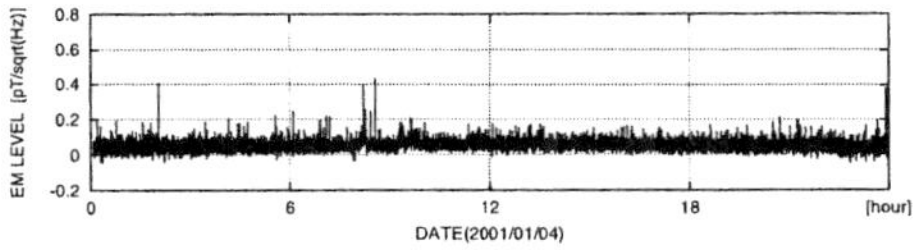

Fig. 6. Local signal (January 4, 2001 at Kawai, Gifu)

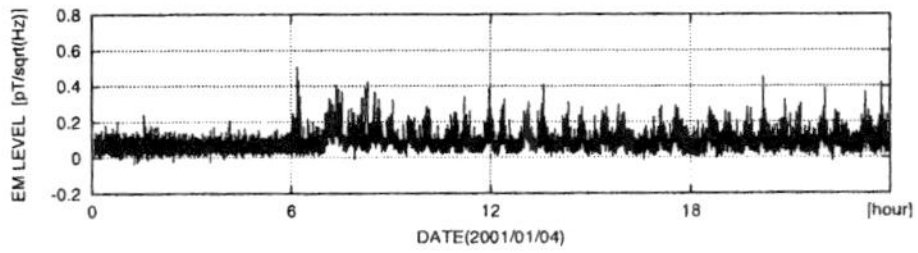

Fig. 7. Local signal (January 4, 2001 at Nannoh, Gifu)

5 Effectiveness of ICA in Global Noise Elimination

In above results, we confirmed that the proposed method can remove the global noise from observed signals. In this section, we confirm how well this method can remove the global noise compared with an observed signal and a signal processed by the conventional method. In order to compare results, we focus on frequency of processed signals because the period of global noises is 1 day. The conventional method is similar to the proposed method, but uses PCA instead of ICA.

5.1 Procedure and Processing Data

1. k = 1.
2. Extracting observed signals from k day to (k+3) day.
3. Computing local signals from the extracted observed signals by the proposed and conventional methods.
4. Normalizing the observed and local signals.
5. Applying the Blackman window to each signal.
6. Processing the DFT of each signal.
7. k = k+1, and go to step 2.

The processing data are observed from Kawai, Gifu for January 2001. These data did not have any anomalous signals.

5.2 Results

Fig.8 shows the amplitude spectrums of the observed signals from Kawai. The vertical axis shows the amplitude spectrums and the horizon axis indicates the

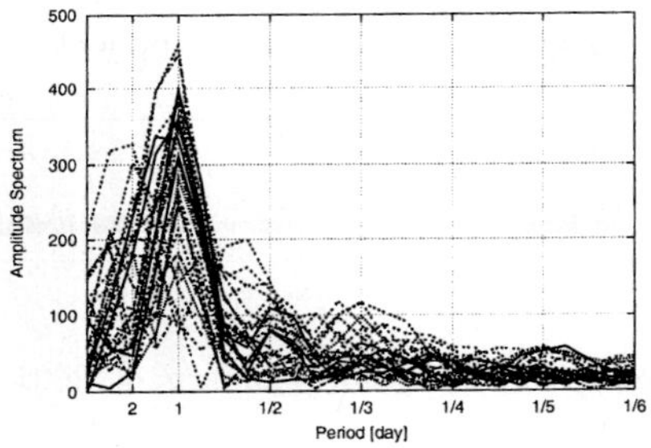

Fig. 8. Amplitude spectrum of observed signal

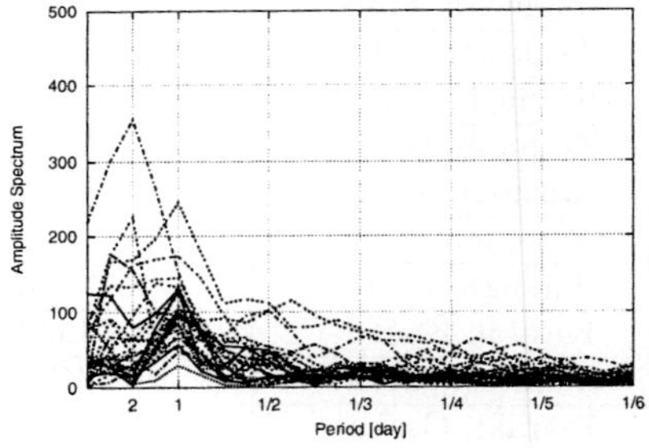

Fig. 9. Amplitude spectrum of local signal using conventional method

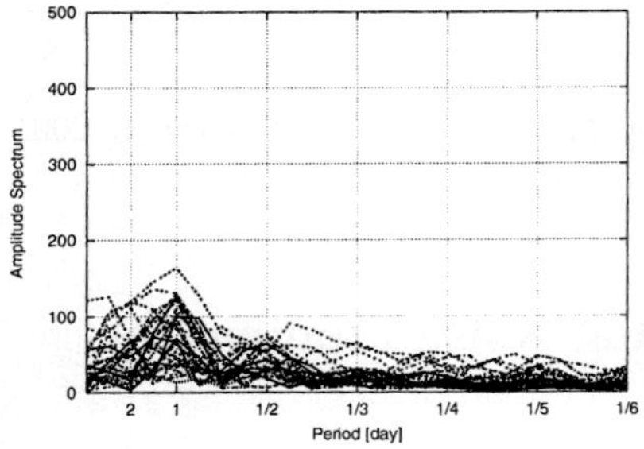

Fig. 10. Amplitude spectrum of local signal using proposed method

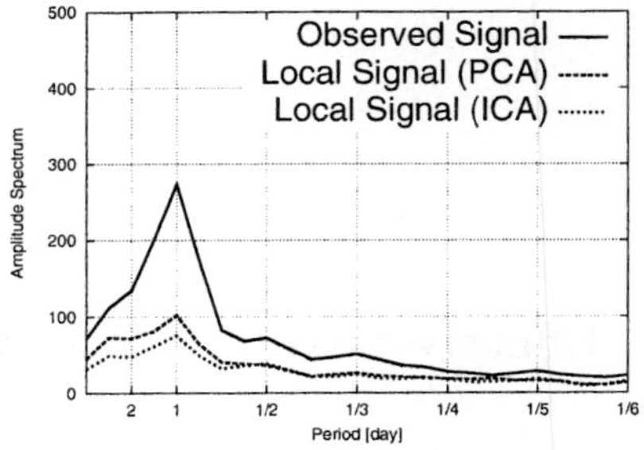

Fig. 11. Average amplitude spectrum of each signal

period. The plotted processing results during one month (28 lines) are overlapping. From this figure, all observed signals have the peak at 1 day. The main factor of these peaks is the circadian rhythm of the global noise.

Fig.9 and 10 show the amplitude spectrums of the local signals calculated by the conventional method (Fig.9) and the proposed method (Fig.10). By the conventional method, a few results have a large value at 1 day. Global noise cannot be perfectly eliminated. On the other hand, proposed method succeeds at elimination global noises from all data.

Fig.11 shows the average amplitude spectrums of each signals shown in Fig.8,9, and 10. The shrinkage of the circadian rhythm in the case of the proposed method is smaller than in the case of conventional method.

Thus, The proposed method is more effective in eliminating global noise than the conventional method.

6 Conclusion

In this paper, we proposed a global noise elimination method by ICA. The proposed method actually calculated local signals in sample case. Compared with the conventional method by PCA, the proposed method can eliminate global noise more effectively.

In future works, we plan to automatically select observation points. The current selection method involves a heavy workload because it needs trial and error. Investigation of applying alternative ICA algorithms is also important, because NG-FICA sometimes does not provide good results for extracting global noise. It is necessary to modify a preprocessing and/or apply more robust ICA algorithms. Moreover, we will verify the effectiveness of the proposed method by anomalous detection and source estimation.

Acknowledgment

This research was supported in part by JSPS for the Grant-in-Aid for Scientific Research (A)17206042.

References

1. The Headquarters for Earthquake Research Promotion Earthquake Research Committee, “Long-term evaluations of Earthquake on Nankai Trough (in Japanese)”, 8th Conference of Subcommittee for Instituting Results in Society (2001).
2. Gokhberg,M.B., V.A.Morgunov, T.Yoshino and I.Tomizawa, “Experimental measurements of EM emissions possibly related to earthquakes in Japan”, J.Geophys.Res., 87, pp.7824-7829, (1982).
3. Edit by M. Hayakawa and Y. Fujisawa “EM phenomena related to earthquake prediction”, Terra Scientific(TERAPUB), Tokyo, (1994).
4. S. Choi, A. Cichocki, and S. Amari, “Flexible independent component analysis”, Journal of VLSI Signal Processing, Vol.26, No.1/2, pp.25-38, (2000).
5. A. Cichocki, S. Amari, K. Siwek, T. Tanaka, et al., ICALAB toolboxes [available online at http://www.bsp.brain.riken.jp/ICALAB/]

BLUES from Music: BLind Underdetermined Extraction of Sources from Music*

Michael Syskind Pedersen, Tue Lehn-Schiøler, and Jan Larsen

Intelligent Signal Processing, IMM, Technical University of Denmark
{msp, tls, jl}@imm.dtu.dk

Abstract. In this paper we propose to use an instantaneous ICA method (BLUES) to separate the instruments in a real music stereo recording. We combine two strong separation techniques to segregate instruments from a mixture: ICA and binary time-frequency masking. By combining the methods, we are able to make use of the fact that the sources are differently distributed in both space, time and frequency. Our method is able to segregate an arbitrary number of instruments and the segregated sources are maintained as stereo signals. We have evaluated our method on real stereo recordings, and we can segregate instruments which are spatially different from other instruments.

1 Introduction

Finding and separating the individual instruments from a song is of interest to the music community. Among the possible applications is a system where e.g. the guitar is removed from a song. The guitar can then be heard by a person trying to learn how to play. At a later stage the student can play the guitar track with the original recording. Also when transcribing music to get the written note sheets it is a great benefit to have the individual instruments in separate channels. Transcription can be of value both for musicians and for people wishing to compare (search in) music. On a less ambitious level identifying the instruments and finding the identity of the vocalist may aid in classifying the music and again make search in music possible. For all these applications, separation of music into its basic components is interesting. We find that the most important application of music separation is as a preprocessing step.

Examples can be found where music consists of a single instrument only, and much of the literature on signal processing of music deals with these examples. However, in the vast majority of music several instruments are played together, each instrument has its own unique sound and it is these sounds in unison that produce the final piece. Some of the instruments are playing at a high pitch and

* This work is supported by the Danish Technical Research Council (STVF), through the framework project "Intelligent Sound", STVF no. 26-04-0092, the PASCAL network, contract no. 506778. and the Oticon Foundation.

J. Rosca et al. (Eds.): ICA 2006, LNCS 3889, pp. 392–399, 2006.

some at a low, some with many overtones some with few, some with sharp onset and so on. The individual instruments furthermore each play their own part in the final piece. Sometimes the instruments are played together and sometimes they are played alone. Common for all music is that the instruments are not all playing at the same time. This means that the instruments to some extent are separated in time and frequency. In most modern productions the instruments are recorded separately in a controlled studio environment. Afterwards the different sources are mixed into a stereo signal. The mixing typically puts the most important signal in the center of the sound picture hence often the vocal part is located here perhaps along with some of the drums. The other instruments are placed spatially away from the center. The information gained from the fact that the instruments are distributed in both space, frequency and time can be used to separate them.

Independent component analysis (ICA) is a well-known technique to separate mixtures consisting of several signals into independent components [1]. The most simple ICA model is the instantaneous ICA model. Here the vector $\mathbf{x}(n)$ of recorded signals at the discrete time index n is assumed to be a linear superposition of each of the sources $\mathbf{s}(n)$ as

$$\mathbf{x}(n) = \mathbf{A}\mathbf{s}(n) + \boldsymbol{\nu}(n), \tag{1}$$

where $\mathbf{A}$ is the mixing matrix and $\boldsymbol{\nu}(n)$ is additional noise. If reverberations and delays between the microphones are taken into account, each recording is a mixture of different filtered versions of the source signals. This model is termed the *convolutive* mixing model.

The separation of music pieces by ICA and similar methods has so far not received much attention. In the first attempts ICA was applied to separation of mixed audio sources [2]. A standard (non-convolutive) ICA algorithm is applied to the time-frequency distribution (spectrogram) of different music pieces. The resulting model has a large number of basis functions and corresponding source signals. Many of these arise from the same signal and thus a postprocessing step tries to cluster the components. The system is evaluated by listening tests by the author and by displaying the separated waveforms. Plumbley et al. [3] presents a range of methods for music separation, among these are an ICA approach. Their objective is to transcribe a polyphonic single instrument piece. The convolutive ICA model is trained on a midi synthesized piece of piano music. Mostly, only a single note is played making it possible for the model to identify the notes as a basis. The evaluation by comparing the transcription to the original note sheets showed good although not perfect performance. Smaragdis et al. has presented both an ICA approach [4] and a Non-negative Matrix Factorization (NMF) approach [5] to music separation. The NMF works on the power spectrogram assuming that the sources are additive. In [6] the idea is extended to use convolutive NMF. The NMF approach is also pursued in [7] where an artificial mixture of a flute and piano is separated and in [8] where the drums are separated from polyphonic music. In [9] ICA/NMF is used along with a vocal discriminant to extract the vocal.

Time-Frequency (T-F) masking is another method used to segregate sounds from a mixture (see e.g. [10]). In *computational auditory scene analysis*, the technique of T-F masking has been commonly used for years. Here, source separation is based on organizational cues from auditory scene analysis [11]. When the source signals do not overlap in the time-frequency domain, high-quality reconstruction can be obtained [12]. However, when there are overlaps between the source signals good separation can still be obtained by applying a binary time-frequency mask to the mixture [12, 13]. Binary masking is also consistent with constraints from auditory scene analysis such as people's ability to hear and segregate sounds [14]. More recently the technique has also become popular in blind source separation, where separation is based on non-overlapping sources in the T-F domain [15]. T-F masking is applicable to source separation/ segregation using one microphone [10, 16] or more than one microphone [12, 13]. In order to segregate stereo music into independent components, we propose a method to combine ICA with T-F masking in order to iterative separate music into spatially independent components. ICA and T-F masking has previously been combined. In [17], ICA has been applied to separate two signals from two mixtures. Based on the ICA outputs, T-F masks are estimated and a mask is applied to each of the ICA outputs in order to improve the signal to noise ratio.

Section 2 provides a review of ICA on stereo signals. In section 3 it is described how to combine ICA with masking in the time frequency domain. In section 4 the algorithm is tested on real music. The result is evaluated by comparing the separated signals to the true recordings given by the master tape containing the individual instruments.

2 ICA on Stereo Signals

In stereo music, different music sources (song and instruments) are mixed so that the sources are located at spatially different positions. Often the sounds are recorded separately and mixed afterwards. A simple way to create a stereo mixture is to select different amplitudes for the two signals in the mixture. Therefore, we assume that the stereo mixture $\mathbf{x}$ at the discrete time index n can be modeled as an instantaneous mixture as in eqn. (1), i.e.

$$\begin{bmatrix} x_1(n) \\ x_2(n) \end{bmatrix} = \begin{bmatrix} a_{11} & \cdots & a_{1N} \\ a_{21} & \cdots & a_{2N} \end{bmatrix} \begin{bmatrix} s_1(n) \\ \vdots \\ s_N(n) \end{bmatrix} + \begin{bmatrix} \nu_1(n) \\ \nu_2(n) \end{bmatrix}. \quad (2)$$

Each row in the mixing matrix $[a_{1i}\ a_{2i}]^T$ contains the gain of the i'th source in the stereo channels. The additional noise could e.g. be music signals which do not origin from a certain direction. If the gain ratio a_{1i}/a_{2i} of the i'th source is different from the gain ratio from any other source, we can segregate this source from the mixture. A piece of music often consists of several instruments as well as singing voice. Therefore, it is likely that the number of sources is greater than two. Hereby we have an *underdetermined* mixture. In [18] it was shown

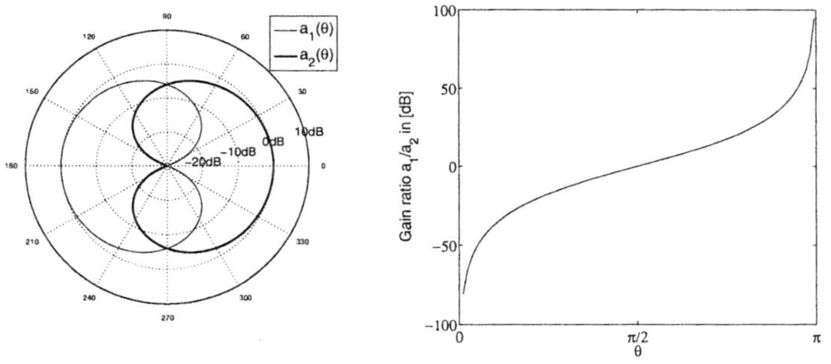

Fig. 1. The two stereo responses $a_1(\theta)$ and $a_2(\theta)$ are shown as function of the direction θ. The monotonic gain ratio is shown as function of the direction θ.

how to extract speech signals iteratively from an underdetermined instantaneous mixture of speech signals. In [18] it was assumed that a particular gain ratio a_{1i}/a_{2i} corresponded to a particular spatial source location. An example of such a location-dependant gain ratio is shown in Fig 1. This gain ratio is obtained by selecting the two gains as $a_1(\theta) = 0.5\big(1 - \cos(\theta)\big)$ and $a_2(\theta) = 0.5\big(1 + \cos(\theta)\big)$.

2.1 ICA Solution as an Adaptive Beamformer

When there are no more sources than sensors, an estimate $\tilde{\mathbf{s}}(n)$ of the original sources can be found by applying a (pseudo) inverse linear system, to eqn. (1).

$$\mathbf{y}(n) = \mathbf{W}\mathbf{x}(n) = \mathbf{W}\mathbf{A}\mathbf{s}(n) \tag{3}$$

where $\mathbf{W}$ is a 2×2 separation matrix. From eqn. (3) we see that the output $\mathbf{y}$ is a function of $\mathbf{s}$ multiplied by $\mathbf{WA}$. Hereby we see that $\mathbf{y}$ is just a different weighting of $\mathbf{s}$ than $\mathbf{x}$ is. If the number of sources is greater than the number of mixtures, not all the sources can be segregated. Instead, an ICA algorithm will estimate $\mathbf{y}$ as two subsets of the mixtures which are as independent as possible, and these subsets are weighted functions of $\mathbf{s}$. The ICA solution can be regarded as an adaptive beamformer which in the case of underdetermined mixtures places the zero gain directions towards different groups of sources. By comparing the two outputs, two binary masks can be found in the T-F domain. Each mask is able to remove the group of sources towards which one of the ICA solutions places a zero gain direction.

3 Extraction with ICA and Binary Masking

A flowchart of the algorithm is presented in Fig. 2. As described in the previous section, a two-input-two-output ICA algorithm is applied to the input mixtures, disregarding the number of source signals that actually exist in the mixture. As shown below the binary mask is estimated by comparing the amplitudes of the two ICA outputs and hence it is necessary to deal with the arbitrary scaling

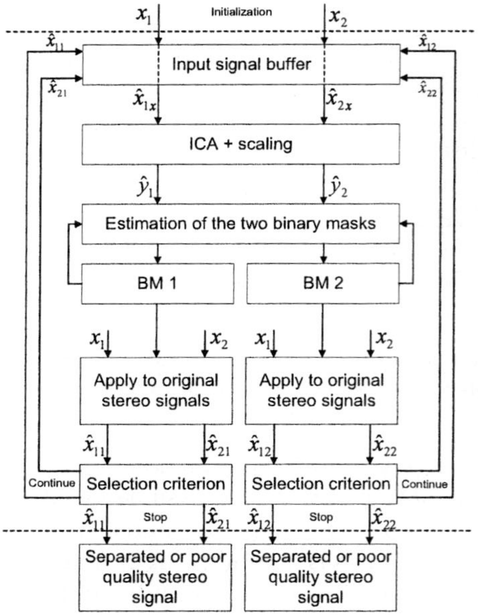

Fig. 2. Flowchart showing the main steps of the algorithm. From the output of the ICA algorithm, binary masks are estimated. The binary masks are applied to the original signals which again are processed through the ICA step. Every time the output from one of the binary masks is detected as a single signal, the signal is stored. The iterative procedure stops when all outputs only consist of a single signal. The flowchart has been adopted from [18].

of the ICA algorithm. As proposed in [1], we assume that all source signals have the same variance and the outputs are therefore scaled to have the same variance. From the two re-scaled output signals, $\hat{y}_1(n)$ and $\hat{y}_2(n)$, spectrograms are obtained by use of the Short-Time Fourier Transform (STFT):

$$y_1 \rightarrow \hat{y}_1 \rightarrow Y_1(\omega, t) \tag{4}$$
$$y_2 \rightarrow \hat{y}_2 \rightarrow Y_2(\omega, t), \tag{5}$$

where ω is the frequency and t is the time index. The binary masks are then found by a bitwise amplitude comparison between the two spectrograms:

$$\text{BM1}(\omega, t) = \tau|Y_1(\omega, t)| > |Y_2(\omega, t)| \tag{6}$$
$$\text{BM2}(\omega, t) = \tau|Y_2(\omega, t)| > |Y_1(\omega, t)|, \tag{7}$$

where τ is a threshold that determines the sparseness of the mask. As τ is increased, the mask is sparser. We have chosen $\tau = 1.5$. Next, each of the two binary masks is applied to the original mixtures x_1 and x_2 in the T-F domain, and by this non-linear processing, some of the music signal are *removed* by one of the masks while other parts of music are removed by the other mask. After the masks have been applied to the signals, they are reconstructed in the time domain by the inverse STFT and and two sets of masked output signals $(\hat{x}_{11}, \hat{x}_{21})$ and $(\hat{x}_{12}, \hat{x}_{22})$ are obtained.

In the next step, it is considered whether the masked output signals consists of more than one signal. The masked output signals are divided into three group defined by the selection criterion in section 3.1. It is decided whether there is one signal in the segregated output signal, more than one signal in the segregated output, or if the segregated signal contains too little energy, so that the signal is expected to be of too poor quality.

There is no guarantee that two different outputs are not different parts of the same separated source signal. By considering the correlation between the segregated signals in the time domain, it is decided whether two outputs contains the same signal. If so, their corresponding two masks are merged. Also the correlation between the segregated signals and the signals with too poor quality is considered. From the correlation coefficient, it is decided whether the mask of the segregated signal is extended by merging the mask of the signal of poor quality. Hereby the overall quality of the new mask is higher.

When no more signal consist of more than one signal, the separation procedure stops. After the correlation between the output signals have been found, some masks still have not been assigned to any of the source signal estimates. All these masks are then combined in order to create a *background mask*. The background mask is then applied to the original two mixtures, and possible sounds that remain in the background mask are found. The separation procedure is then applied to the remaining signal to ensure that there is no further signal hidden. This procedure is continued until the remaining mask does not change any more. Note that the final output signals are maintained as stereo signals.

3.1 Selection Criterion

It is important to decide whether the algorithm should stop or whether the processing should proceed. The algorithm should stop separating when the signal consists of only one source or when the mask is too sparse so that the quality of the resulting signal is unsatisfactory. Otherwise, the separation procedure should proceed. We consider the covariance matrix between the output signals to which the binary mask has been applied, i.e. $\mathbf{R}_{xx} = \langle \mathbf{x}\mathbf{x}^H \rangle$. If the covariance matrix is close to singular, it indicates that there is only one source signal. To measure the singularity, we find the condition number of $\mathbf{R}_{xx}$. If the condition number is below a threshold, it is decided that $\mathbf{x}$ contains more than one signal and the separation procedure continues. Otherwise, it is assumed that $\mathbf{x}$ consists of a single source and the separation procedure stops.

4 Results

The method has been applied to different pieces of music. The used window length was 512, the FFT length was 2048. The overlap between time frames was 75%. The sampling frequency is 10 kHz. Listening tests confirm that the

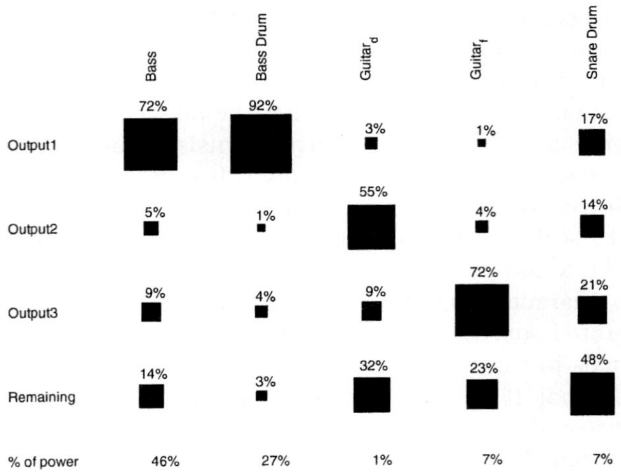

Fig. 3. Correlation coefficients between the extracted channels and the original stereo channels. The coefficients has been normalized such that the columns sum to one. The last row shows the percentage of power of the tracks in the mixture.

method is able to segregate individual instruments from the stereo mixture. We do not observe that correlations can be heard. However, musical artifacts are audible. Examples are available on-line for subjective evaluation [19]. In order to evaluate the method objectively, the method has been applied to 5 seconds of stereo music, where each of the different instruments has been recorded separately, processed from a mono signal into a stereo signal, and then mixed. We evaluate the performance by calculating the correlation between the segregated channels and the original tracks. The results are shown in Fig. 3 As it can be seen from the figure, the correlation between the estimated channels and the original channels is quite high. The best segregation has been obtained for those channels, where the two channels are made different by a gain difference. Among those channels is the guitars, which are well segregated from the mixture. The more omnidirectional (same gain from all directions) stereo channels cannot be segregated by our method. However, those channels are mainly captured in the remaining signal, which contains what is left when the other sources has been segregated. Some of the tracks have the same gain difference. Therefore, it is hard to segregate the 'bass' from the 'bass drum'.

5 Conclusion

We have presented an approach to segregate single sound tracks from a stereo mixture of different tracks while keeping the extracted signals as stereo signals. The method utilizes that music is sparse in the time, space and frequency domain by combining ICA and binary time-frequency masking. It is designed to separate tracks from mixtures where the stereo effect is based on a gain difference. Experiments verify that real music can be separated by this algorithm and results on an artificial mixture reveals that the separated channel is highly correlated with the original recordings.

We believe that this algorithm can be a useful preprocessing tool for annotation of music or for detecting instrumentation.

References

1. Hyvärinen, A., Karhunen, J., Oja, E.: Independent Component Analysis. Wiley (2001)
2. Casey, M., Westner, A.: Separation of mixed audio sources by independent subspace analysis. In: Proc. ICMC. (2000)
3. Plumbley, M.D., Abdallah, S.A., Bello, J.P., Davies, M.E., Monti, G., Sandler, M.B.: Automatic music transcription and audio source separation. Cybernetics and Systems **33** (2002) 603–627
4. Smaragdis, P., Casey, M.: Audio/visual independent components. Proc. ICA'2003 (2003) 709–712
5. Smaragdis, P., Brown, J.C.: Non-negative matrix factorization for polyphonic music transcription. Proc. WASPAA 2003 (2003) 177–180
6. Smaragdis, P.: Non-negative matrix factor deconvolution; extraction of multiple sound sourses from monophonic inputs. Proc. ICA'2004 (2004) 494–499
7. Wang, B., Plumbley, M.D.: Musical audio stream separation by non-negative matrix factorization. In: Proc. DMRN Summer Conf. (2005)
8. Helén, M., Virtanen, T.: Separation of drums from polyphonic music using non-negative matrix factorization and support vector machine. In: Proc. EUSIPCO'2005. (2005)
9. Vembu, S., Baumann, S.: Separation of vocals from polyphonic audio recordings. In: Proc. ISMIR2005. (2005) 337–344
10. Wang, D.L., Brown, G.J.: Separation of speech from interfering sounds based on oscillatory correlation. IEEE Trans. Neural Networks **10** (1999) 684–697
11. Bregman, A.S.: Auditory Scene Analysis. 2 edn. MIT Press (1990)
12. Yilmaz, O., Rickard, S.: Blind separation of speech mixtures via time-frequency masking. IEEE Trans. Signal Processing **52** (2004) 1830–1847
13. Roman, N., Wang, D.L., Brown, G.J.: Speech segregation based on sound localization. J. Acoust. Soc. Amer. **114** (2003) 2236–2252
14. Wang, D.L.: On ideal binary mask as the computational goal of auditory scene analysis. In Divenyi, P., ed.: Speech Separation by Humans and Machines. Kluwer, Norwell, MA (2005) 181–197
15. Jourjine, A., Rickard, S., Yilmaz, O.: Blind separation of disjoint orthogonal signals: Demixing n sources from 2 mixtures. In: Proc. ICASSP. (2000) 2985–2988
16. Hu, G., Wang, D.L.: Monaural speech segregation based on pitch tracking and amplitude modulation. IEEE Trans. Neural Networks **15** (2004) 1135–1150
17. Kolossa, D., Orglmeister, R.: Nonlinear postprocessing for blind speech separation. In: Proc. ICA'2004, Granada, Spain (2004) 832–839
18. Pedersen, M.S., Wang, D.L., Larsen, J., Kjems, U.: Overcomplete blind source separation by combining ICA and binary time-frequency masking. In: Proc. MLSP. (2005) 15–20
19. http://www.intelligentsound.org/demos/demos.htm.

On the Performance of a HOS-Based ICA Algorithm in BSS of Acoustic Emission Signals

Carlos G. Puntonet[1], Juan-José González de-la-Rosa[2], Isidro Lloret, and Juan-Manuel Górriz

[1] University of Granada, Department of Architecture and Computers Technology, ESII, C/Periodista Daniel Saucedo. 18071, Granada, Spain
carlos@atc.ugr.es

[2] University of Cádiz, Electronics Area, Research Group TIC168 - Computational Electronics Instrumentation, EPSA, Av. Ramón Puyol S/N. 11202, Algeciras-Cádiz, Spain
juanjose.delarosa@uca.es

Abstract. A cumulant-based independent component analysis (Cum-ICA) is applied for blind source separation (BSS) in a synthetic, multi-sensor scenario, within a non-destructive pipeline test. Acoustic Emission (AE) sequences were acquired by a wide frequency range transducer (100-800 kHz) and digitalized by a 2.5 MHz, 8-bit ADC. Four common sources in AE testing are linearly mixed, involving real AE sequences, impulses and parasitic signals from human activity. A digital high-pass filter achieves a SNR up to -40 dB.

1 Introduction

AE signal processing usually deals with the problem of separation multiple events which sequentially occur in several measurement points during a non-destructive test. In most situations, the test involves the study of the behavior of secondary events, or reflections, resulting from an excitation (the main event). These echoes carry information related with the medium through which they propagate, as well as reflecting surfaces [1].

But, in almost every measurement scenario, an acquired sequence contains information regarding not only the AE under study, but also additive noise processes (mainly from the measurement equipment) and other parasitic signals, e.g. originated by human activity or machinery vibrations. As a consequence, in non-favorable SNR cases, BSS should be accomplished before characterization [2], in order to obtain the most reliable *fingerprint* of the AE event.

The main goal of this paper is to show how an ICA algorithm (based in cumulants) can separate signals from a multi-sensor array, which comprises synthetics of AE events and additive signals, widespread used in non-destructive vibration tests. The algorithm have proven success for a SNR=-40 dB situation, and uses the cross-cumulants of the measured time-series to maximize the goal function. These higher-order statistics take advantage from their noise rejection capabilities to extract sources. A high-pass filter completes the post-processing in the cases of low-frequency couplings.

J. Rosca et al. (Eds.): ICA 2006, LNCS 3889, pp. 400–405, 2006.

The paper is structured as follows: in Section 2 we make a brief progress report on AE characterization. Section 3 summarizes the ICA model and outlines its properties. Results are displayed in section 4. Finally, conclusions and achievements are drawn in section 5.

2 Analysis of Acoustic Emission Signals

AE is defined as the class of phenomena whereby transient elastic waves are generated by the rapid (and spontaneous) release of energy from localized sources within a material, or the transient elastic wave(s) so generated. Elastic energy travels through the material as a stress or strain wave and is typically detected using a piezoelectric transducer, which converts the surface displacement (vibrations) to an electrical signal.

AE signal processing is used for the detection and characterization of failures in non-destructive testing and identification of low-level biological signals [2]. Most AE signals are non-stationary and they consist of overlapping bursts with unknown amplitude and arrival time. These characteristics can be described by modelling the signal [1], by means of neural networks, and using wavelet transforms.

The above second-order techniques have been also applied in an automatic analysis context of the estimation of the time of occurrence and amplitude of the bursts. Multiresolution has provided good performance in de-noising (up to SNR=-30 dB) and estimation of time instances, due to the selectivity of the filters banks implemented in the wavelets [3].

Higher order statistics (HOS) have enhanced characterization in analyzing biological signals due to the capability for rejecting noise [4].

3 The ICA Model and Its Properties

3.1 Outline of ICA

BSS by ICA is receiving attention because of its applications in many fields such as speech recognition, medicine and telecommunications [5]. Statistical methods in BSS are based in the probability distributions and the cumulants of the mixtures. The recovered signals (the source estimators) have to satisfy a condition which is modelled by a contrast function. The underlying assumptions are the mutual independence among sources and the non-singularity of the mixing matrix [6],[7],[8].

Let $\mathbf{s}(t) = [s_1(t), s_2(t), \ldots, s_m(t)]^T$ be the transposed vector of sources (statistically independent). The mixture of the sources is modelled by

$$\mathbf{x}(t) = \mathbf{A} \cdot \mathbf{s}(t) \tag{1}$$

where $\mathbf{x}(t) = [x_1(t), x_2(t), \ldots, x_m(t)]^T$ is the available vector of observations and $\mathbf{A} = [a_{ij}] \in \Re^{m \times n}$ is the unknown mixing matrix, modelling the environment in which signals are mixed, transmitted and measured [9]. We assume that $\mathbf{A}$ is a non-singular n×n square matrix. The goal of ICA is to find a non-singular n×m separating matrix **B** such that extracts sources via

$$\hat{\mathbf{s}}(t) = \mathbf{y}(t) = \mathbf{B} \cdot \mathbf{x}(t) = \mathbf{B} \cdot \mathbf{A} \cdot \mathbf{s}(t) \tag{2}$$

where vector $\mathbf{y}(t) = [y_1(t), y_2(t), \ldots, y_m(t)]^T$ is an estimator of the sources. The separating matrix has a scaling freedom on each of its rows because the relative amplitudes of sources in $\mathbf{s}(t)$ and columns of **A** are unknown [7]. The transfer matrix $\mathbf{G} \equiv \mathbf{BA}$ relates the vector of independent original signals to its estimators.

3.2 CumICA

High order statistics, known as cumulants, are used to infer new properties about the data of non-Gaussian processes. Before cumulants, such processes had to be treated as if they were Gaussian. Cumulants and polyspectra reveal information about amplitude and phase, whereas second order statistics are phase-blind. The relationship among the cumulant of r stochastic signals and their moments of order $p, p \leq r$, can be calculated by using the *Leonov-Shiryayev* formula [10]:

$$\begin{aligned} Cum(x_1, ..., x_r) = \sum (-1)^k \cdot (k-1)! \cdot E\{\prod_{i \in v_1} x_i\} \\ \cdot E\{\prod_{j \in v_2} x_j\} \cdots E\{\prod_{k \in v_p} x_k\} \end{aligned} \tag{3}$$

where the addition operator is extended over all the set of v_i $(1 \leq i \leq p \leq r)$ and v_i compose a partition of *1,. . . ,r*.

It has been proved that a set of random variables are statistically independent if their cross-cumulants are zero. This property is used to define a contrast function, by minimizing the distance between the cumulants of the sources **s**(t) and the outputs **y**(t). As sources are unknown, it is necessary to involve the observed signals. Separation can be developed using the following contrast function based on the entropy of the outputs [6]:

$$H(\mathbf{z}) = H(\mathbf{s}) + log[det(\mathbf{G})] - \sum \frac{\mathbf{C}_{1+\beta,y_i}}{1+\beta} \tag{4}$$

where $\mathbf{C}_{1+\beta,y_i}$ is the $1+\beta$th-order cumulant of the ith output, **z** is a non-linear function of the outputs y_i, **s** is the source vector, **G** is the global transfer matrix of the ICA model and $\beta > 1$ is an integer verifying that $\beta + 1$-order cumulants are non-zero.

Using equation 4, the separating matrix can be obtained by means of the following recurrent equation [9]

$$\mathbf{B}^{(h+1)} = [\mathbf{I} + \mu^{(h)}(\mathbf{C}^{1,\beta}_{y,y}\mathbf{S}^{\beta}_{y} - I)]\mathbf{B}^{(h)} \tag{5}$$

where $\mathbf{S}^{\beta}_{y}$ is the matrix of the signs of the output cumulants. Equation 5 can be interpreted as a quasi-Newton algorithm of the cumulant matrix $\mathbf{C}^{1,\beta}_{y,y}$. The learning rate parameters $\mu^{(h)}$ and η are related by

$$\mu^{(h)} = \min(\frac{2\eta}{1+\eta\beta}, \frac{\eta}{1+\eta\|\mathbf{C}^{1,\beta}_{y,y}\|_p}) \tag{6}$$

with $\eta < 1$ to avoid $\mathbf{B}^{(h+1)}$ being singular; $\|.\|_p$ denotes de p-norm of a matrix. The adaptative equation 5 converges, if the matrix $\mathbf{C}^{1,\beta}_{y,y}\mathbf{S}^{\beta}_{y}$ tends to the identity.

Provided with the mathematical foundations the experimental results are outlined.

4 Experimental Results

The sensor is attached to the outer surface of the pipeline, which is under mechanical excitation. Each sequence comprises 2502 points (sampling frequency of 2.5 MHz and 8 bits of resolution), and assembles the main AE event and the subsequent reflections (echoes).

Four sources have been considered and linearly mixed. The real AE event, an uniform white noise ($SNR = -40\ dB$), a damped sine wave and an impulse-like event. The damping sine wave models a mechanical vibration which may occur, i.e. as a consequence of a maintenance action. It has a damping factor of 2000 and a frequency of 8000 Hz. Finally, the impulse is included as a very common signal registered in vibration monitoring.

The results of the algorithm are depicted in figure 1. The damping sinusoid is considered as a frequency component of the impulse-like event because IC3 and IC4 are almost the same. The final independent components are obtained filter-

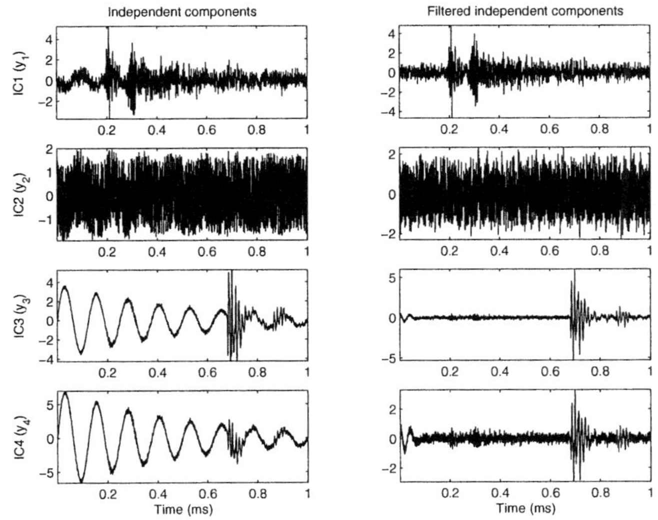

Fig. 1. Estimated sources (ICs; Independent Components) and filtered estimated sources

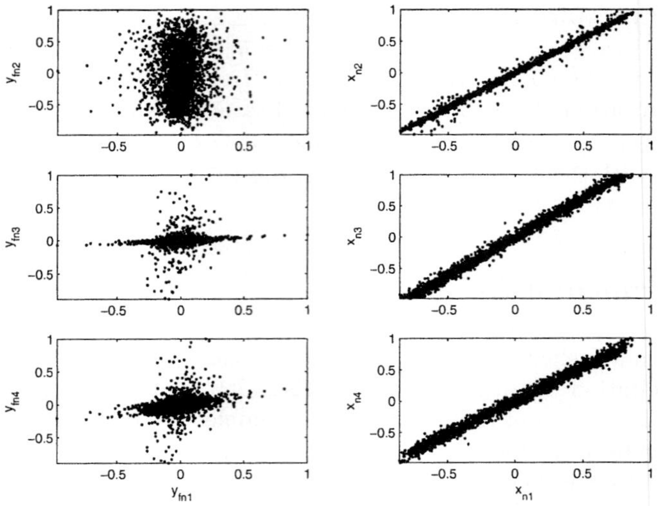

Fig. 2. Joint distributions of the mixtures and the independent components

ing the independent components by a 5th-order *Butterworth* high-pass digital filter (20000 kHz).

Finally, to test the independence of the independent components, some relevant joint distributions have been included in figure 2. The left column shows how for any IC, the values are quite random. This means that for a value (point) of an IC, almost all the values of the another IC are allowed. On the other hand, the joint distributions of the mixtures are linearly shaped, which leads us to infer a dependency before separating by ICA.

The above results lead us to some conclusions on the use of HOS as a characterizing and separating tool to be considered in a non-destructive measurement system.

5 Conclusions and Future Work

ICA is far different from traditional methods, as power spectrum, which obtain an energy diagram of the different frequency components, with the risk that low-level sounds could be masked. This experience shows that the algorithm is able to separate the sources with small energy levels in comparison to the background noise. This is explained away by statistical independence basis of ICA, regardless of the energy associated to each frequency component. The post filtering action let us work with very low SNR signals. The next step is oriented in a double direction. First, a stage involving four real mixtures will be developed. Second, and simultaneously, the computational complexity of the algorithms have to be reduced to perform an implementation.

Acknowledgement

The authors would like to thank the *Spanish Ministry of Education and Science* for funding the projects DPI2003-00878, TEC2004-06096 and PTR1995-0824-OP.

References

1. Piotrkowski, R., Gallego, A., Castro, E., García-Hernández, M., Ruzzante, J.: Ti and Cr nitride coating/steel adherence assessed by acoustic emission wavelet analysis. Non Destructive Testing and Evaluation (NDT and E) International (Ed. Elsevier) **8** (2005) 260–267
2. de la Rosa, J.J.G., Puntonet, C.G., Lloret, I.: An application of the independent component analysis to monitor acoustic emission signals generated by termite activity in wood. Measurement (Ed. Elsevier) **37** (2005) 63–76 Available online 12 October 2004.
3. de la Rosa, J.J.G., Puntonet, C.G., Lloret, I., Górriz, J.M.: Wavelets and wavelet packets applied to termite detection. Lecture Notes in Computer Science (LNCS) **3514** (2005) 900–907 Computational Science - ICCS 2005: 5th International Conference, GA Atlanta, USA, May 22-25, 2005, Proceedings, Part I.
4. Puntonet, C.G., de la Rosa, J.J.G., Lloret, I., Górriz, J.M.: Recognition of insect emissions applying the discrete wavelet transform. Lecture Notes in Computer Science (LNCS) **3686** (2005) 505–513 Third International Conference on Advances in Pattern Recognition, ICAPR 2005 Bath, UK, August 22-25, 2005, Proceedings, Part I.
5. Mansour, A., Barros, A.K., Onishi, N.: Comparison among three estimators for higher-order statistics. In: The Fifth International Conference on Neural Information Processing, Kitakyushu, Japan (1998)
6. Puntonet, C.G., Mansour, A.: Blind separation of sources using density estimation and simulated annealing. IEICE Transactions on Fundamental of Electronics Communications and Computer Sciences **E84-A** (2001)
7. Hyvärinen, A., Oja, E.: Independent Components Analysis: A Tutorial. Helsinki University of Technology, Laboratory of Computer and Information Science (1999)
8. Lee, T.W., Girolami, M., Bell, A.J.: A unifying information-theoretic framework for independent component analysis. Computers and Mathematics with Applications **39** (2000) 1–21
9. de la Rosa, J.J.G., Puntonet, C.G., Górriz, J.M., Lloret, I.: An application of ICA to identify vibratory low-level signals generated by termites. Lecture Notes in Computer Science (LNCS) **3195** (2004) 1126–1133 Proceedings of the Fifth International Conference, ICA 2004, Granada, Spain.
10. Swami, A., Mendel, J.M., Nikias, C.L.: Higher-Order Spectral Analysis Toolbox User's Guide. (2001)

Two Applications of Independent Component Analysis for Non-destructive Evaluation by Ultrasounds

Addisson Salazar, Jorge Gosálbez, Jorge Igual, Raul Llinares, and Luis Vergara

Universidad Politécnica de Valencia, Departamento de Comunicaciones, Camino de Vera s/n, 46022 Valencia, Spain
asalazar@dcom.upv.es, jorgocas@dcom.upv.es, jigual@dcom.upv.es, lvergara@dcom.upv.es, rllinares@dcom.upv.es

Abstract. This paper presents two novel applications of ICA in Non Destructive Evaluation by ultrasounds applied to diagnosis of the material consolidation status and to determination of the thickness material profiles in restoration of historical buildings. In those applications the injected ultrasonic pulse is buried in backscattering grain noise plus sinusoidal phenomena; this latter is analyzed by ICA. The mixture matrix is used to extract useful information concerning to resonance phenomenon of multiple reflections of the ultrasonic pulse at non consolidated zones and to improve the signals by detecting interferences in ultrasonic signals. Results are shown by real experiments at a wall of a Basilica's restored cupola. ICA is used as pre-processor to obtain enhanced power signal B-Scans of the wall.

1 Introduction

Non-destructive evaluation (NDE) by ultrasounds is a very useful technique applied in fields such as construction, food, and biomedicine. The technique has basically two operation modes: pulse-echo (one sensor as emitter and receiver) and transmission (one emitter and one receiver). An ultrasound pulse is injected in the inspected material and a response of the material microstructure is received [1,2]. The measured signal can contain echoes produced from discontinuities, inhomogeneities, borders of the material, plus material grain noise (superimposition of many small echoes due to the material microstructure). All of this information can be used for quality control and characterization of materials [3,4]. The present study used the pulse-echo technique, given the inspected material consisted of a wall with no possible access from opposite sides. This wall was a zone at the cupola of the *Basilica de la Vírgen de los Desamparados in Valencia*, Spain.

This paper includes two novel applications of ICA [5,6] as pre-processor in ultrasound NDE applied to historical building restoration. The first application consists in using the mixture matrix to discern on useful material information of the consolidation process that consists in injecting a product to fill cracks at the wall. The second application is detecting interferences in the recorded signals to cancel them improving

J. Rosca et al. (Eds.): ICA 2006, LNCS 3889, pp. 406 – 413, 2006.

the quality of the signals. This procedure was applied to recorded signals for estimating thickness layer profile at the wall.

Interferences can be due to the internal clocks of the measurement equipment, interferences with other equipments, and so on. In many applications, the recording of a high quality raw data is a difficult task, especially in situations where the conditions can not be controlled during the experiment. One difficulty to have the measurements at the cupola was the use of a plastic for covering the transducer to avoid direct contact of the ultrasonic coupling gel with artistic paintings on the walls. This kind of measurement produced attenuated signals

B-Scans diagrams were used to visualize consolidated and non consolidated material zones to check the quality of restoration and to detect interfaces between different materials in the wall. B-Scan is a 2D representation of a signal set. The evolution in time windows of a parameter such as power or spectrum was calculated for each one of the signals. Then all of the calculated information was put together in a representation of the measurement point versus the signal parameter evolution. Figure 1 shows different points of a B-Scan measured by ultrasound at the cupola.

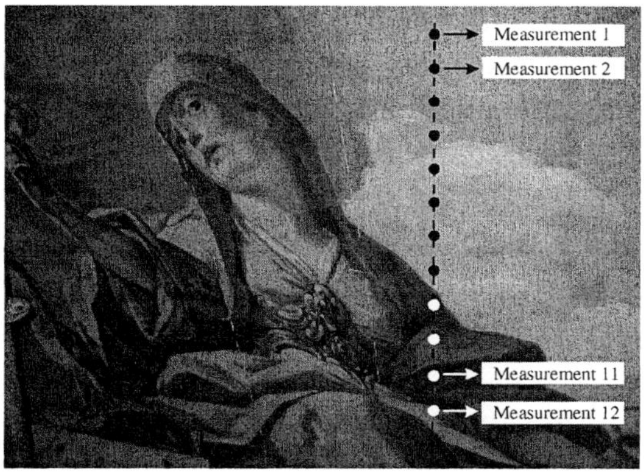

Fig. 1. Ultrasound inspection at the cupola

Following sections describe the ICA model of the problem; the performed experiments, including the equipment setup, a comparison between the B-Scans obtained using or not using ICA as a pre-processor and the sensitivity to detect interferences. Finally the conclusions and future work.

2 ICA Statement of the Problem

The recorded signals are modelled as the superposition of the backscattered signal plus sinusoidal phenomena. This latter sinusoidal contribution should be determined to know if it is due to useful information on the material structure, such as material resonances, or interferences due to the instrumentation during measurement. ICA statement of the problem is:

$$x_k(t) = s_k(t) + \sum_{i=1}^{N-1} \alpha_{ik} e^{j(\omega_i t + \theta_{ik})} \qquad k = 1...M \tag{1}$$

where M is the number of measurements, $x_k(t)$ is the received signal from the material at the position k of the B-Scan, $s_k(t)$ is the backscattering signal that depends on the material microstructure, and $\alpha_{ik} e^{j(\omega_i t + \theta_{ik})}$ $i = 1 \ldots N-1, k = 1 \ldots M$ are the sinusoidal sources to be analyzed.

The backscattering signal, under certain assumptions related to the wavelength of the ultrasonic wave and the scattering size, can be modeled as a stochastic process given by:

$$\{\tilde{Z}(\mathbf{x},t)\} = \sum_{n=1}^{N(\mathbf{x})} \tilde{A}_n(\mathbf{x}) f(t - \tilde{\tau}_n(\mathbf{x})) \tag{2}$$

where $\mathbf{x}$ is the transducer location (we obtain different backscattering registers for different transducer locations). The random variable, $\tilde{A}_n$, is the scattering cross-section of the nth scatter The r.v. $\tilde{\tau}_n$ is the delay of the signal backscattered by the nth scatter and $N(\mathbf{x})$ is the number of scatters contributing from this position. The function $f(t)$ is a complex envelope of the ultrasonic frequency pulse, that is

$$f(t) = p(t) e^{j\omega_0 t} \tag{3}$$

where $p(t)$ is the pulse envelope and ω_0 the transducer central frequency.

Backscattering model of equation (2) is composed of a homogeneous non-dispersive media, and randomly distributed punctual scatters depicting the composite nature of the received grain noise signal instead of a rigorous description of the material microstructure [7].

In the simplest case consisting of a homogeneous material and only one harmonic of the sinusoidal components, the ICA model of equation (1) is

$$x_k(t) = s(t) + \alpha_k e^{j(\omega_i t + \theta_k)} \qquad k = 1...M \tag{4}$$

As we know, in usual ICA (no prior information ICA model included) we need as many mixtures as sources. In the case of equation (1), a B-Scan of 2 points would be enough. In the proposed applications M = 12 and 10, therefore 12 and 10 mixtures were used in order to include the anomalies of the material and allow a relative high number of interferences. Even more; if we think that there is not enough with the M points registered, the number of sensors can be virtually increased if we record responses to different pulses, considering that the echo is the same and the pulse repetition period is not a multiple of the sinusoid period [8].

Obviously the sinusoidal components have the same frequencies along the B-Scan, with possibly changing amplitude and phase. From a statistical point of view, considering the interference or resonance as a sinusoid with deterministic but unknown amplitude and uniform random phase, it is clearly guaranteed that the backscattering signal and it are statistically independent.

3 Experiments and Results

The objectives of the experiments were visualizing non consolidated zones and to calculate layer thickness at the wall of the cupola. Ultrasound transducers have a working transmission frequency; the higher transducer frequency the higher capacity to detect small details, but also lower capacity of material penetration. Therefore using high frequencies is possible to detect smaller details but they have to be closer to material surface. The transducer used for consolidation analysis (application 1) was 1 MHz and transducer used for thickness layer profile was 5 MHz (application 2). This latter was selected because we were interested in obtaining information of the superficial layers.

3.1 Equipment Setup

The equipment setup used for NDE of the historical building was the following:

Table 1. Equipment setup

Ultrasound equipment setup		Acquisition equipment setup	
Ultrasound Equipment	*Matec PR5000*	Acquisition equipment	*Oscilloscope TDS3012 Tektronix*
Transducers	*1 MHz (application 1)* *5 MHz (application 2)*	Sampling frequency	*10 MHz and 250 MHz*
Pulse width	*0.9 μs*	Sample number	*10000*
Pulse amplitude	*80 %*	Observation time	*1 ms and 40 μs*
Analog filter	*200 kHz – 2.25 MHz (ap. 1)* *1 MHz – 7 MHz (ap. 2)*	Vertical resolution	*16 bits*
Excitation signal	*Tone burst 1 MHz and 5 MHz*	Dynamic range	*± 2.5V*
Operation mode	*Pulse-echo*	Average	*64 acquisitions*
Amplifier gain	*65 dB*	PC connection	*GPIB*

Due to the temporal structure of recorded signals we selected the Temporal Decorrelation Separation TDSEP algorithm based on simultaneous diagonalisation of time-lagged correlation matrices. This algorithm exploits the temporal structure of the signals and can separate more than one Gaussian [9]. The mixture matrix obtained by ICA was used to separate the information concerning to the sinusoidal phenomena.

3.2 Diagnosis of the Material Consolidation Status

BSS by ICA was selected for this application because, on the contrary of the classic spectral analysis techniques [10], BSS is an unsupervised method that does not require any estimation of noise autocorrelation matrix in data corrupted by the sinusoidal interference, considerations on sort of noise or model assumptions such as filter order in model based methods.

Figure 2 shows the B-Scan estimated by signal power using a conventional non-stationary analysis applying a moving window over the 12 ultrasonic recorded signals. Figure 2a shows two zones clearly differentiated; the first corresponds to consolidated zone (low level of signal) and the second corresponds to non consolidated zone (high level of signal). The signal penetrates well into the wall at the consolidated zone and it is attenuated before reflecting any kind of signal. Conversely, the signal level is increased in a non consolidated zone due to multiple reflections of the ultrasonic pulse, see Figure 2b.

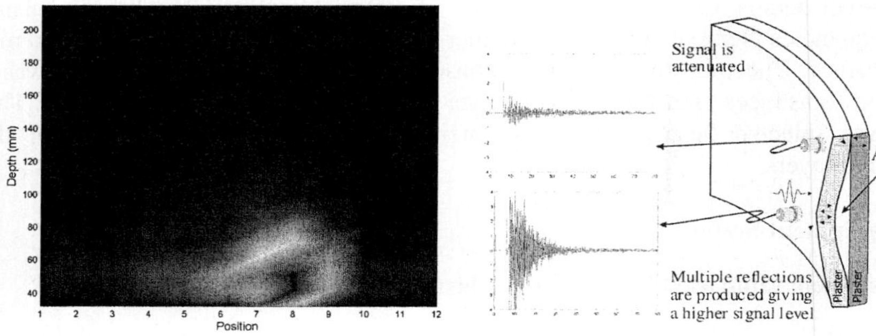

2a. Power B-Scan **2b.** Scheme of the wall

Fig. 2. Power signal B-Scan by non-stationary analysis

From the spectral analysis, two frequencies (181 and 356 kHz) were found in all the recorded signals. After estimating B-Scan of Figure 2 was not clear enough the origin of those frequencies, they could be interferences or material resonances. Then we applied ICA to obtain more information from the mixture matrix and recovered sources. Figure 3 shows the recorded signals and the recovered sources by ICA; the sample numbers processed were from 600 to 6000.

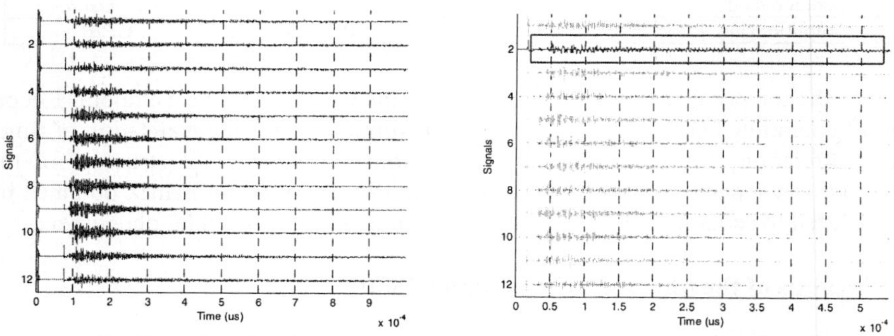

3a. Recorded signals **3b.** Recovered sources

Fig. 3. Recorded signals and recovered sources (the supposed "interference" is highlighted)

Figure 4a and 4b show two B-Scans obtained from the mixture matrix corresponding to $\mathbf{x} = \sum_{i=1}^{12} \mathbf{a}_i s_i,\ s_i = 0\ (i \neq 2)$ and $\mathbf{x} = \sum_{i=1}^{12} \mathbf{a}_i s_i,\ s_i = 0\ (i = 2)$ respectively. The first B-Scan represents the sinusoidal phenomenon depicting the non consolidated zone.

Thus this phenomenon can be associated with the shape of the material non consolidated zone. The second B-Scan is the complementary information concerning to the consolidated zone. The diagrams obtained from ICA information depict more precisely the two different zones of the material than the one obtained by non-stationary analysis.

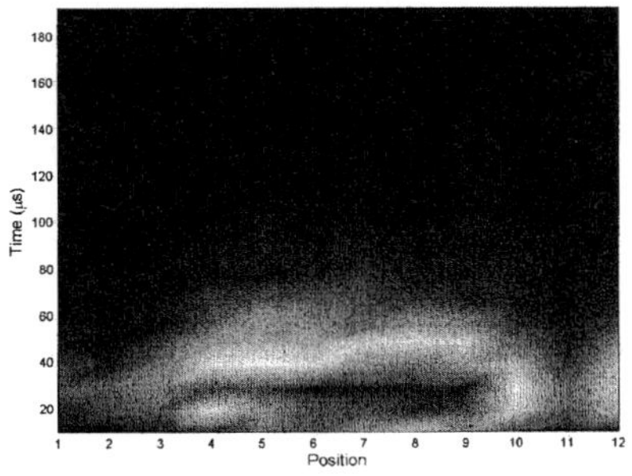

4a. B-Scan from sinusoidal components

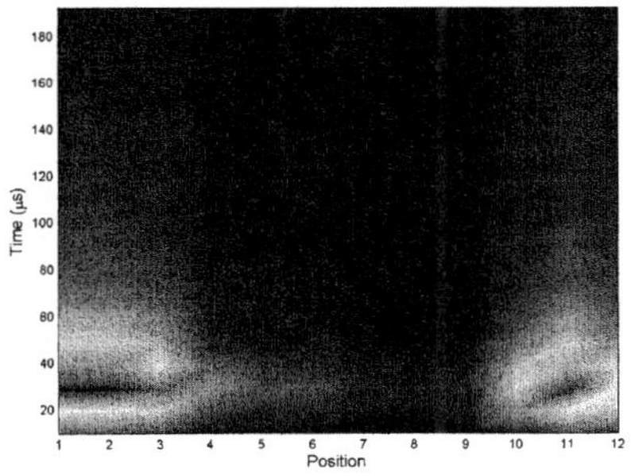

4b. B-Scan from backscattered components

Fig. 4. Power B-Scan after ICA preprocessing

3.3 Thickness Material Layer Profile

Figure 5a and 5b show the recorded signals plus 1 MHz artificial interference added and the corresponding B-Scan calculated by the evolution of the centroid frequency [11]. Following information is represented in the diagram: i.) axis x: transducer position; from position 0 to 10, ii.) axis y: depth axis, and iii.) axis z: depicted by colours that denotes the parameter level at a given position in a given depth.

The depth is obtained by $depth = velocity * time / 2$ where factor 2 is due to the round trip travel of the ultrasound pulse between the material surface and the layer. The first two layers of the cupola wall were composed of mortar and plaster respectively. For calculation of depth, an average ultrasound propagation velocity of 1600 m/s was used calculated from lab probes. Due to the 1 MHz interference, the B-Scan is not clear enough to represent a profile of a layer.

Figure 6a and 6b show the results obtained by applying ICA on the ultrasonic signals. To assess the sensitivity of the ICA in detecting the interference, a controlled interference was added to the signals, trying with different frequencies and amplitudes of the interference. Figure 6a shows the error in the extraction of the interference versus the ratio power interference to power signal ($P_{interference}/P_{signal}$) for different interference frequencies. The higher interference amplitude the better extraction of the interference and the higher interference frequency the worst extraction of the interference. Figure 6b depicts an enhanced centroid frequency B-Scan (cancelled interference) with a layer clearly defined at 4 mm. corresponding to the mortar layer of the cupola wall.

Other alternatives for sinusoid extraction are based on the use of the so-called notch filtering [12]. They can be designed assuming prior knowledge of the pulsations to be cancelled. In this sense, BSS could be used as a prior step to notch filtering, but implying transient effects, possible instability problems and some distortion of the so obtained interference-free records (because of the finite notch bandwidth).

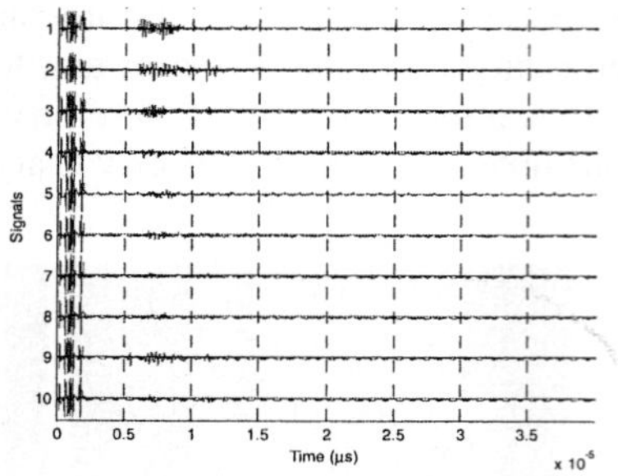

5a. Recorded signals

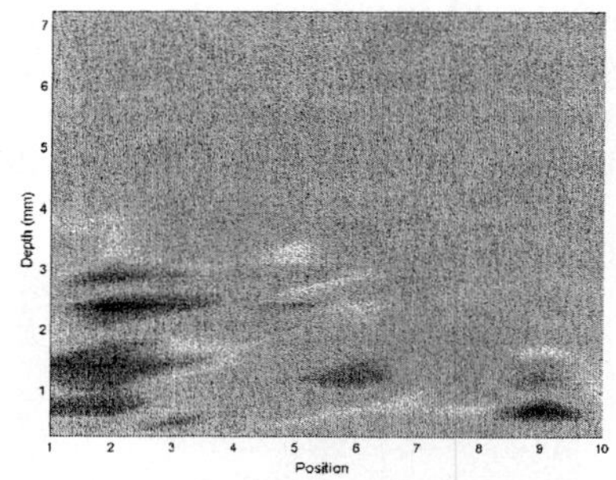

5b. Centroid frequency B-Scan

Fig. 5. Recorded signals and centroid frequency B-Scan by non-stationary analysis

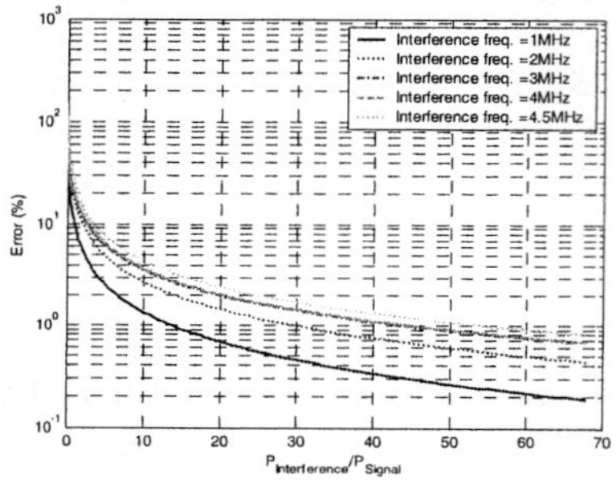

6a. Error percentage vs. $(P_{interference}/P_{signal})$

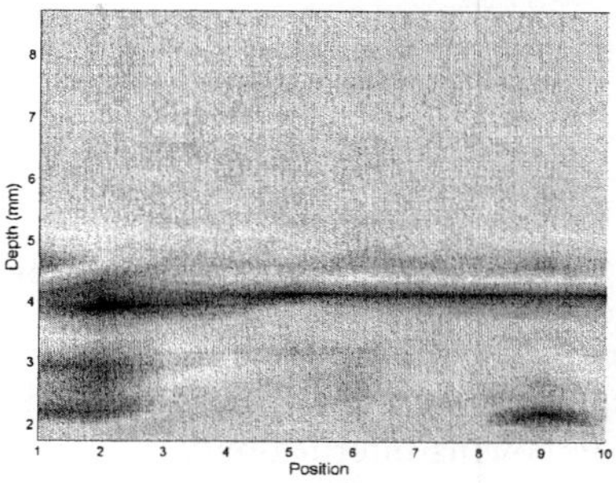

6b. Centroid frequency B-Scan with cancelled interferences

Fig. 6. Recorded signals and centroid frequency B-Scan by non-stationary analysis

Results obtained from the Basilica's cupola inspection were validated in testing an architectonic scale model replica. In addition some material samples were extracted from the replica and measured at lab to obtain an accurate calculation of material wave propagation.

4 Conclusions

The ICA model for ultrasound evaluation as the superposition of the backscattered signal plus sinusoidal phenomena has been tested by means of two novel applications.

The application of ICA to NDE by ultrasounds has enabled the diagnosis of the consolidation status in restoration of historical buildings. The proposed procedure allowed separating the sources corresponding to the contributions of consolidated and non consolidated zones to the backscattered recorded signals.

The application of ICA to NDE by ultrasounds made possible the determination of the thickness layer profile allowed cancelling interferences from the recorded signals.

The enhanced B-Scan enabled determining the first thickness layer of mortar for the analyzed wall. ICA works well in the case of a relatively high interference level with respect to the ultrasonic signal.

Enhanced power and centroid frequency B-Scans were obtained using ICA as preprocessor of the non-stationary analysis. Future work is being addressed to apply the ICA for classification and characterization of materials.

Acknowledgements

This work has been supported by Spanish Administration under grant TEC 2005-01820, *Universidad Politécnica de Valencia* under interdisciplinary grant 2004-0900, and *Generalitat Valenciana* under grant GVA 05/250.

References

1. Krautkrämer J., Ultrasonic Testing of Materials, Springer, 4th edition, Berlin, 1990.
2. Cheeke J.D., Fundamentals and Applications of Ultrasonic Waves, CRC Press LLC, USA, 2002.
3. Salazar A., Vergara L., Igual J., Gosalbez J., Blind source separation for classification and detection of flaws in impact-echo testing, Mechanical Systems and Signal Processing, Elsevier, v. 19 n. 6, pp. 1312-1325, 2005.
4. Vergara L., et al., "NDE Ultrasonic Methods to Characterize the Porosity of Mortar", NDT&E International, Elsevier, v. 34 n. 8, pp. 557-562, 2001.
5. Hyvärinen A.: Independent Component Analysis. John Wiley & Sons, 2001.
6. Cichocki A. and Amari S.: Adaptive Blind Signal and Image Processing: Learning algorithms and applications. Wiley, John & Sons, 2001.
7. Miralles R., Vergara L., and Gosalbez J., Material Grain Noise Analysis by Using Higher-Order Statistics. Signal Processing, Elsevier, 84(1):197-205, January 2004.
8. Igual J., Camacho A., Vergara L., Cancelling sinusoidal interferences in ultrasonic applications with a BSS algorithm for more sources than sensors, Proceedings of the Independent Componente Analysis Workshop , ICA, San Diego, 2001.
9. Ziehe A. and Müller K.R., TDSEP- An efficient algorithm for blind separation using time structure, in Proc. Int. Conf. Artificial Neural Networks, L. Niklasson, M. Bodén, and T. Ziemke, Eds., Skövde, pp. 675-680, Sweden, 1998.
10. Kay S., Spectral estimation: Theory and application. Prentice-Hall, 1988.
11. Vergara L., Gosalbez J., Fuente J.V., Miralles R., and Bosch I., "Measurement of cement porosity by centroid frequency profiles of ultrasonic grain noise", Signal Processing, Elsevier, v. 84, n. 12, pp. 2315-2324, 2004.
12. Regalia P. A., Adaptive IIR filtering in signal processing and control. Marcel Dekker, 1994.

Blind Spatial Multiplexing Using Order Statistics for Time-Varying Channels*

Santiago Sazo[1], Yolanda Blanco-Archilla[2], and Lino García[2]

[1] Centro de Domótica Integral – CeDInt. Universidad Politécnica de Madrid,
ETSIT - Ciudad Universitaria s/n,
28040 Madrid, Spain
santiago@gaps.ssr.upm.es
[2] Escuela Superior Politécnica, Universidad Europea de Madrid,
28670 Villaviciosa de Odón, Madrid, Spain
{myolanda.blanco, lino.garcia}@uem.es

Abstract. Spatial multiplexing is currently one of the most promising techniques exploiting the spatial dimension to increase data rates. Most of existing methods are based on coherent detection techniques that imply multichannel estimation. This procedure, especially for time-varying channels, increases the overhead rate due to the periodical training requirement. A suitable approach dealing with this scenario proposes the use of Blind Source Separation (BSS) principles to minimize the mentioned overhead still offering the increased data rate. The authors have developed in previous publications a new BSS technique based on Order Statistics (OS) labeled as ICA-OS with very satisfactory performance in static scenarios. In these studies it was already realized that the amount of data required for convergence was significantly less than other well known methods. Therefore, in this current contribution we present some results showing the capability of our procedure to deal with time-varying channels typical of mobile applications without training requirement.

1 Introduction

Modern communications require increased data rates without extra bandwidths in order to provide a suitable service for the incoming applications. Exploitance of the spatial dimension has been traditionally based on standard beamforming or spatial diversity. Additionally, few years ago it was pointed out the possibility of spatial multiplexing to increase data rate without involving other resources but just extra complexity at the RF parts and also more complex reception techniques. [1] and references therein showed that capacity in MIMO (Multiple Input – Multiple Output) systems maybe even be multiplied by the minimum number of antennas at any side. This amazing result has driven the attention of researchers all over the world in the recent years.

* This work has been partly supported by National Spanish Projects PCT-350100-2004-1, TEC2004-06915-C03-02/TCM and the European Project AST-CT-2003-502910.

J. Rosca et al. (Eds.): ICA 2006, LNCS 3889, pp. 414 – 421, 2006.

Most of the practical schemes trying to get this promised benefit are based on matrix eigendecompositions (if channel is known at both sides) or in a more attractive approach, BLAST-like implementations just require channel knowledge at the receiver side. Training, especially for time varying channels, reduces significantly the desired data rate due to the periodic interleaving of known data. This overhead is much more significant in MIMO applications because several antennas require specific training reducing the promised data rate gain.

In this contribution we deal with this problem proposing a BSS approach suitable for time-varying channels. The main handicap of existing BSS techniques applied to this scenario is the very large amount of data required to get satisfactory performance, especially those based on Higher Order Statistics (HOS) estimation. This characteristic is acceptable for delay non sensitive applications or static mixtures but seems to be less useful or impractical in MIMO communications.

The authors proposed in several previous publications a new BSS method based on OS. In these proposals the Cummulative Density Function (cdf) estimation using several order statistics was shown to provide suitable means for signals separation. This procedure provided very satisfactory results for both subGaussian and super-Gaussian distributions but specially promising for digital modulations as QAM. More indeed, the amount of data required for appropriate separation was much less than other well known procedures. In those previous publications this remarkable result was just mentioned as a procedure saving computational burden.

In this contribution we are focused on MIMO communications using QAM communications through time-varying channels, showing that ICA-OS is more suitable than other BSS as JADE [10] as one of the most representative schemes. Basically, JADE is a cumulant batch algorithm which optimizes a 4th order measure of independence among the whole set of outputs.

2 Our Model for ICA Procedures for Time-Varying Channels

The standard linear mixture can be easily generalized to a time-varying MIMO communication channel.

$$\mathbf{y}[k] = \mathbf{H}[k]\mathbf{x}[k] + \mathbf{n}[k] \tag{1}$$

Where $\mathbf{y}$ is the received vector whose dimension is the number of receiving antennas N_r and $\mathbf{x}$ is the transmitted vector of size $N_t \times 1$ where N_t is the number of transmit antennas. $\mathbf{n}$ is the additive Gaussian noise and $\mathbf{H}$ is the MIMO matrix sized $N_t \times N_r$. We have remarked the time dependence using the variable k. This model assumes flat fading transmission through a linear mixture with additive noise.

MIMO channel model assumed considers uncorrelated spatial fading. This hypothesis is realistic if the scatterers are located around the antennas and the antennas are enough separated. For terminals, typically this is fulfilled even for less than a lambda separation while for base stations with antennas under the roofs, a few lambdas may be enough. Thus, this model is especially suitable for urban or indoor communications which is the case where more data rate is required and spatial multiplexing is more demanded.

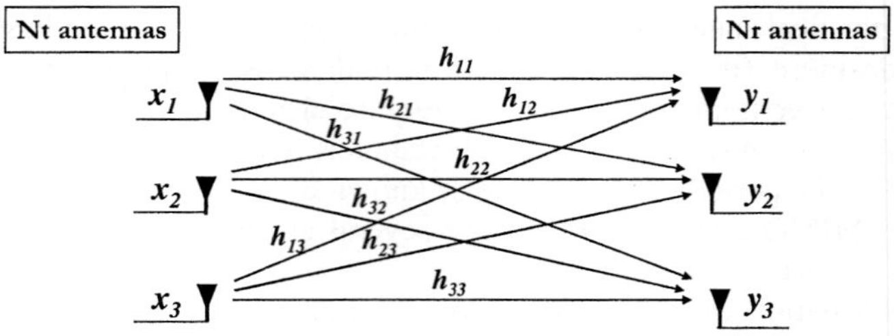

Fig. 1. MIMO channel Mode

Time evolution of the spatial uncorrelated coefficients of matrix **H** is implemented using the procedure described by Hoeher in [2] for the flat fading case. This model is in fact based on an intuitive interpretation of the Bello's model [3] as an incoherent superposition of N echoes, where each echo is characterized by a random phase, a random delay and a random Doppler shift. For the flat fading case, each element mn of matrix **H** collapses to

$$h_{mn}[k] = \lim_{N \to \infty} \frac{1}{\sqrt{N}} \sum_{l=1}^{N} e^{j\theta_l} e^{jk2\pi f_{Dl}/f_s} \tag{2}$$

where θ_l are the random phases, f_{Dl} are the Doppler random variables following the standard Jake's spectrum characterized by f_{Dmax}, and f_s is the sampling frequency. In practical cases, $N = 25$ seems to provide satisfactory accuracy.

Most of the BSS methods would collect a certain amount of **y** vectors along some time interval in order to estimate the needed statistics from the measured data to perform different algorithms. The degradation of different proposals will depend on the amount of data required for satisfactory separation. This behavior requires static mixtures that can not be guaranteed in time varying scenarios.

3 The ICA-OS Method Applied to Communications

Previously, let us remark that the separation scheme in the communication scenery presented at previous section is:

$$\mathbf{w} = \mathbf{BDy} \tag{3}$$

where **w** ($N_t \times n$) are the output channels to be updated towards the N_t estimated sources $\boldsymbol{x}$, **B** is the ($N_r \times N_t$) unknown orthogonal separation matrix, **D** is the decorrelation matrix obtained through the well known whitening preprocessing scheme [11] and **y** ($N_r \times n$) are the set of n symbols collected at every N_r antenna.

ICA-OS has been proposed and explained by the authors in several papers [6-9], therefore let us just expose the main ideas:

ICA-OS is a deflation ICA procedure where at each stage, the separation vector $\mathbf{b_i}$ is updated by means of the maximization of a certain non-Gaussianity measure; consequently, one non-gaussian original source is obtained at the $\mathbf{w_i}$ output channel. In other words, the non-Gaussianity measure $J(\mathbf{b_i})$ is used like ICA cost function. In the practical implementation J is maximized through a gradient rule plus some restriction which forces orthogonality between separation vectors:

$$\mathbf{b}_i[t+1]=\mathbf{b}_i[t]+\mu\nabla\mathbf{J}_i|_{\mathbf{b}_i[t]} \quad (4)$$
$$\mathbf{b}_i\perp\{\mathbf{b}_1\mathbf{b}_2\cdots\mathbf{b}_{i-1}\}$$

In this sense, our main contributions have been a new family of Gaussianity measures based on Order Statistics [7, 8] as well as a new multistage deflation algorithm which decreases the dimension of the problem with each stage [9].

Our recommended Gaussianity distance (among the quoted family) for communication signals is the infinite norm applied to the difference between the implied inverse cdf's -denoted as Q -:

$$J(w_i(b_i))=\max_u|Q_{w_i}(u)-Q_g(u)| \quad (5)$$

Q_{wi} corresponds to the analyzed signal w_i and Q_g to the equivalent Gaussian distribution g.

The practical way to calculate previous distance is based on the estimation of Q´s through the extreme Order Statistics (OS). It was proved in [6] after some mathematical development that Eq. (5) can be estimated through the following expression:

$$\hat{J}(w_i(b_i))=|w_{i(k)}-w_{i(l)}+2Q_g(\frac{l}{n})| \quad (6)$$
$$where(k,l)\cong(n,1)$$

Where $w_{i(k)}$ is the k order statistic obtained in a simple way ordering a set large enough of n samples:

$$w_{i(1)}<w_{i(2)}<....<w_{i(l)}<...<w_{i(k)}<w_{i(n)} \quad (7)$$

and the known value $Q_g(l/n)$ is the l order statistic of the Gaussian distribution. The advantages of using previous infinite norm instead of others *p-norms* [6, 7] either Gaussian distances are:

- It just needs a couple of order statistics to be estimated in front of the whole set of OS used by other norms.
- The OS are estimated easily just ordering the samples, instead of the complexity involved in HOS [10] and non-linearities used by other Gaussianity measures [11].
- Besides, it works more efficiently and robustly with a few samples (around n=100) compared with other ICA methods, especially when the sources are subGaussians which usually are the kind of communication distributions, (see compared performance indexes in [7, 8]). This behavior is the main reason to use this measure like ICA cost Function in slowly variant MIMO channels (see next section).

At this point, it is necessary to expose the gradient expression (for more details see [8, 9]):

$$\left|\nabla \mathbf{J}\right|_{t\, \mathbf{b}_j[t]} = S\mathbf{z}(\mathbf{d}_k - \mathbf{d}_l) \tag{8}$$

$$where\ \ S = sign(w_{i(k)}(\mathbf{b}_t) - w_{i(l)}(\mathbf{b}_t) + 2Q_g(\frac{l}{n}))_{\mathbf{b}_j[t]}$$

In previous equation **z** is the vector ($N_r \times n$) obtained as **z** = **Dy** (from Eq. (3)) after the well known sphering preprocessing [6, 11]. Other wise vector **d** is calculated by means of:

$$\mathrm{d}_r[m] = \left\{ \begin{matrix} 1 & if\ w_i[m] = w_{i(r)} \\ 0 & otherwise \end{matrix} \right\}_{m=1\ldots n} \tag{9}$$

Interested reader may review [6] in order to clarify the whole multistage procedure. We would like to remark that this scheme is also very efficient because in consecutive stages, the dimension of the vector space is consequently reduced.

4 Simulations

In order to show the ability of our method to cope with time variant channels we have run a set of simulations for different cases related to the ratio between the Doppler and the sampling frequencies. The maximum Doppler frequency is considered for pedestrian speed, 3 Km/h, and carrier frequency 2.4 GHz. These values are motivated by the fact that the most realistic scenario related to spatial multiplexing is probably the wireless Local Area Networks (WLAN). Assuming isotropic distribution, the Doppler density was derived by Clarke [4] and sometimes dubbed Jakes distribution. The main parameter to control the Doppler rate in terms of the symbol rate is defined in our simulations considering

$$f_s = Mf_{D\max} \tag{10}$$

where factor M controls the relationship between both frequencies. When M increases, that is the symbol rate is much larger than the fading rate, the channel is nearly constant along the processed data. If M is not so large, the batch size must be reduced decreasing the expected performance.

In order to get a wide view about the performance of our method we have also implemented one of the most representative BSS methods, JADE [10], and also we have implemented the V-BLAST (see for instance [5] for an exhaustive and detailed review of BLAST and related techniques) as a non blind spatial multiplexing procedure to evaluate the loss of blind methods in front of trained procedures. Comparison with trained schemes includes the effect of some estimation noise modelled as AWGN added to every component of **H** whose power is the effective noise power.

We have to remark that the implemented method requires some minimum training at the beginning of the transmission in order to solve the inherent ambiguity in the order of the recovered sources and also the phase ambiguity (although this point maybe overcome using noncoherent modulations). After ICA-OS a ZF demodulator is used to obtain the set of estimated bits.

Performance will be evaluated in terms of BER for a constellations 4QAM and different values of the parameter M in Eq. (11).

a) Scenario 1. The following simulation shows the degradation of the separation scheme according to the MIMO channel variability increases (M decreases). Communication sources are 4-QAM with 4 antennas at both ends. Fig. 2 shows the effect of the window size for all the cases under consideration. Noise is added at reception antennas with $Eb/N0$ = 30 dBs. It can be observed that there should be a trade off between the window size and the channel variability. Clearly, for static channels, the longer the window size, the better performance. However, for dynamic scenarios there is an optimum window size where the channel remains nearly static and the separation is performed. Some of these values are summarized in the following table:

Table 1. Optimal window size

Factor M	Windows Size
500	50
5000	100
15000	250

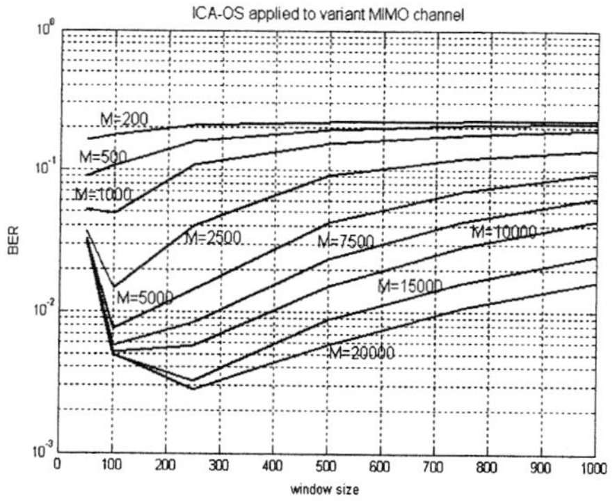

Fig. 2. Performance in terms of window and M

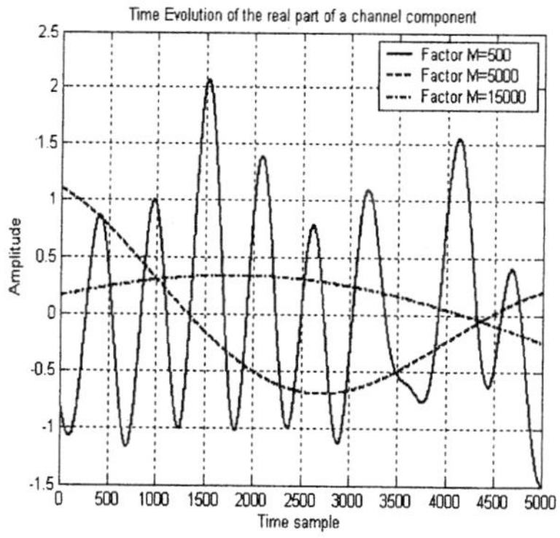

Fig. 3. Time evolution of the channel

Fig. 3 complements this view with the time evolution of an arbitrary channel component (the real part for simplicity). It can be observed how fast it is the channel evolution depending on the aforementioned factor M.

b) Scenario 2. Communication sources 4QAM; 4 antennas at both sides and different sampling frequencies where the window size is fixed according to the optimum value provided in Table 1. It can be observed in Fig. 4 that if the channel is highly variant (M=500) none of the schemes (ICA-OS and JADE) is able to separate the involved signals. However, as the channel becomes more static, the better performance of the ICA-OS is shown in front of JADE. It has to be remarked that these schemes are able to perform with satisfactory performance for high SNR scenarios, while for values below 10 dB are not suitable. We are currently working towards the improvement of our approach combining the ICA-OS source separation with more complex MultiUser Detectors (MUDs) in a second stage.

c) Additionally, in a stationary environment we have compared ICA-OS with a second stage using ZF or BLAST criteria in front of trained ZF and BLAST in Fig. 5. Main ideas are the following:

1. ICA-OS method estimates the unknown MIMO channel **H**, afterwards ZF demodulators either is used to recover source bits.
2. On the other side, ICA-OS estimates the unknown channel **H** and afterwards BLAST demodulates the separated symbols using the estimation of **H**.
3. When **H** is known, but with channel estimation errors using instantaneous estimators, BLAST algorithm and ZF schemes obtains the demodulated bit sources

Results shown in Fig. 5 remark that V-BLAST is better than ZF in the trained mode, as expected. Logically blind techniques are worse than the previous ones that know the channel; anyway degradation could be acceptable in many applications. It can be also observed that ICA-OS with blast performs worse than ICA-OS with ZF due probably to the error propagation effect related to BLAST approaches, this last point remarks that BLAST techniques seems to be very sensitive to channel errors associated to estimation procedures. Although these are preliminary results, this approach combining BSS with more complex MUDs seems to be very promising for further work.

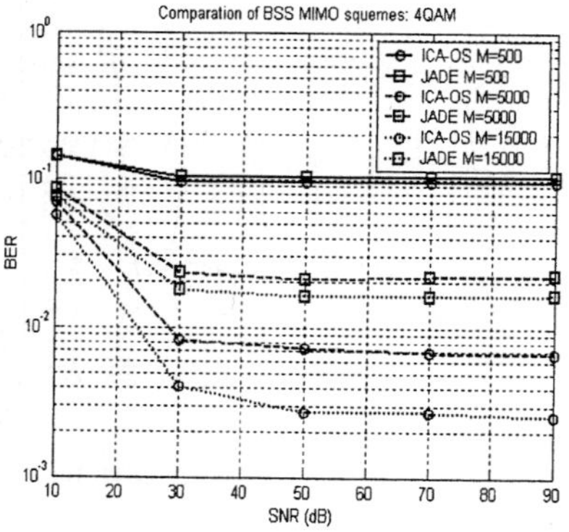

Fig. 4. Comparison of BSS MIMO schemes for 4QAM

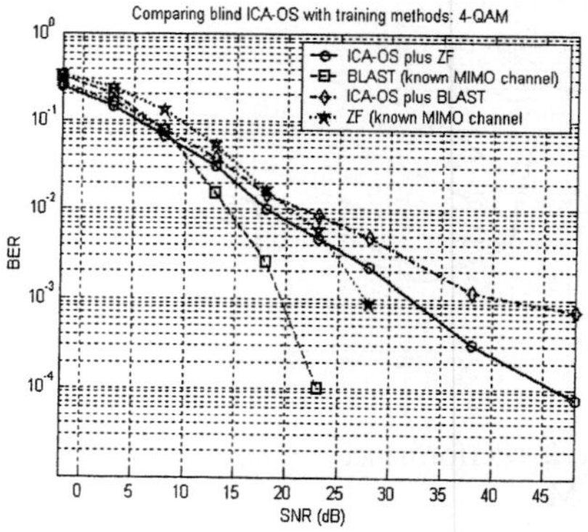

Fig. 5. Comparison of different approaches for blind and trained MIMO processing

5 Conclusions

This paper addresses the implementation and assessment of the ICA-OS procedure for spatial multiplexing in MIMO time varying scenarios. This procedure is shown to perform very satisfactorily in front of other well known BSS method where the amount of data available is shortened in order to deal with significant Doppler scenarios. These results may envision the possibility of spatial multiplexing without the overhead related to training the MIMO structure. The loss of performance of trained

ZF and trained BLAST is not very significant. Future work must be done in order to improve present results with the combination of BSS and more complex MUDs as BLAST.

References

1. I.E. Telatar. Capacity of Multi-Antenna Gaussian Channels. European Trans. On Telecommunications, Vol.10, November - December 1999.
2. P. Hoeher. An statistical discrete-time model for the WSSUS multipath channel. IEEE Transactions on Vehicular Technology, Vol. 41, pp. 461-468, Nov.1992.
3. P.A. Bello. Characterization of randomly time-variant linear channels. IEEE Transactions on Communication Systems, Vol. CS-11, pp.360-393, December 1963.
4. R.H. Clarke. A statistical theory of mobile radio reception. Bell Syst. Tech. J., Vol. 47, pp.957-1000, July-August 1968.
5. B. Vucetic, J. Yuan. Space time coding. John Wiley 2003.
6. Y. Blanco, S. Zazo, J.C. Principe. Alternative Statistical Gaussianity Measure using the Cumulative Density Function. Proceedings of the Second Workshop on Independent Component Analysis and Blind Signal ICA'00. 537-542. June 2000. Helsinky (Finland).
7. Y. Blanco, S. Zazo. New Gaussianity Measures based on Order Statistics. Application to ICA. Neurocomputing (Elseviver Science) Vol: 51C, pp 303-320.
8. Y. Blanco, S. Zazo. An overview of BSS techniques based on Order Statistics: Formulation and Implementation Issues. Fifth Workshop on BSS-ICA. ICA'04. LNCS, vol 3195. pages 73-88.
9. Y. Blanco, S. Zazo, J.M. Páez-Borrallo. Adaptive ICA with Order statistics in Multidimensional Scenarios. Proceedings of the 6th International Work-conference on Artificial and Natural Neural Networks IWANN'01. (Vol II- 770-778) June 2001. Granada (Spain).
10. J.F. Cardoso. High – order contrasts for independent component analysis, Neural Computation 11(1):157-192, 1999.
11. A. Hyvarien, J,Karhunen, E. Oja. Independent Component Analysis. Ed. J. Wiley and Sons, 2001.

Semi-blind Equalization of Wireless MIMO Frequency Selective Communication Channels

Oomke Weikert, Christian Klünder, and Udo Zölzer

Department of Signal Processing and Communications,
Helmut-Schmidt-University - University of Federal Armed Forces Hamburg, Germany
Holstenhofweg 85, 22043 Hamburg, Germany
{oomke.weikert, udo.zoelzer}@hsu-hamburg.de, c.kluender@gmx.de

Abstract. The semi-blind equalization for a wireless multiple-input multiple-output (MIMO) system with frequency selective Ricean channels is addressed. A reformulation of the problem accepting higher complexity allows to apply ordinary complex Independent Component Analysis (ICA). An algorithm presented here resolves the increased number of permutations due to the reformulation of the convolutive blind signal separation problem. The remaining ambiguities are as many as in the non-frequency selective case and solved by a short preamble. The efficiency of the proposed method is illustrated by numerical simulations. According to the Bit Error Rate the semi-blind equalization shows a good performance in comparison to training based channel estimation & equalization.

1 Introduction

Multiple-input multiple-output (MIMO) systems provide higher data rates that data-demanding applications require. Training based channel estimation of frequency selective MIMO channels reduces the effective transmission rate. The semi-blind equalization approaches can reduce the amount of training to a minimum. A well-known challenge is to resolve the remaining ambiguities left by blind signal separation.

The blind equalization of frequency selective MIMO channels is also known as convolutive blind signal separation (BSS) or blind separation of convolutive mixtures [1]. Methods for blind separation of convolutive mixtures can be subdivided into direct and indirect approaches. In a direct approach the separated signals are extracted without explicit identification of the mixing matrix, while indirect methods identify the unknown channels before separation and equalization. A variety of two-step algorithms using the indirect approach are proposed. The linear prediction (LP) based approaches [2] and the subspace method [3] are most popular among them. A further approach is proposed in [4].

A different approach [5] uses Orthogonal Frequency Division Multiplexing (OFDM). The convolutive mixture is transformed associating each subcarrier with an instantaneous ICA problem. This special case of performing ICA in the frequency domain requires to solve the frequency dependent permutation problem.

J. Rosca et al. (Eds.): ICA 2006, LNCS 3889, pp. 422–429, 2006.

In this paper the direct approach is considered. A reformulation of the convolutive blind signal separation problem is used to apply ordinary complex ICA. The increased number of ambiguities due to the reformulation of the convolutive blind signal separation problem are resolved by a three step algorithm. In the first step the special structure of the reordered mixing matrix, a generalized Sylvester matrix, is used to reduce the unknown permutations. The remaining ambiguities are as many as in the non-frequency selective case. Considering a communication system with M-QAM modulation, in a second step the phase ambiguity can be reduced up to a multiple of $\pi/2$. Using differential modulation the remaining phase ambiguity can be solved while the permutations remain. As the permutations can not be solved with a differential modulation here a non-differential M-QAM modulation scheme is utilized. To resolve the permutations preamble symbols are transmitted. It will be shown that the required number of preamble symbols is equal to the number of transmit antennas.

The paper is organized as follows. The MIMO system and the frequency selective MIMO channel model is described in Section 2. Section 3 covers the semi-blind equalization of the frequency selective wireless MIMO system. Simulation results are presented in Section 4. As a performance measure the blind equalization method is compared with a training based channel estimation & equalization. A summary and conclusion marks can be found in Section 5.

2 System Description

2.1 Transmitter

We consider a frequency selective MIMO wireless system with n_T transmit and n_R receive antennas (see Fig. 1). The serial data stream is split into n_T parallel substreams, one for every transmit antenna. The substreams are mapped into M-QAM symbols and organized in frames. Each frame of length N_F consists of N_I information symbols and N_P preamble symbols. The preamble sequences are orthogonal to each other. The symbol substreams are subsequently transmitted over the n_T antennas at the same time. The symbol transmitted by antenna m

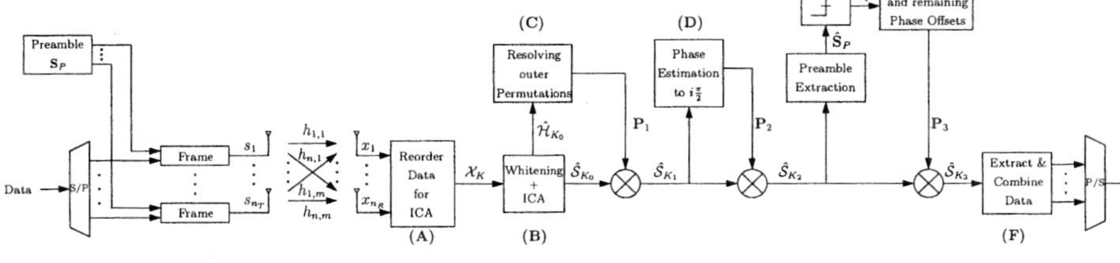

Fig. 1. Wireless MIMO system with semi-blind equalization of frequency selective channels and ambiguity resolution

at time instant k is denoted by $s_m(k)$. The transmitted symbols are arranged in vector $\mathbf{s}(k) = [s_1(k), \ldots, s_{n_T}(k)]^T$ of length n_T, where $(\cdot)^T$ denotes the transpose operation.

2.2 Channel

Between every transmit antenna m and every receive antenna n there is a frequency selective channel impulse response (CIR) of length $L+1$, described by the vector $\mathbf{h}_{n,m} = [h_{n,m}(0), \ldots, h_{n,m}(L)]^T$. Assuming the same channel order L for all channels, the frequency selective MIMO channel can be described by $L+1$ complex channel matrices $\mathbf{H}(k)$, $k = 0, ..., L$, with the dimension $n_R \times n_T$.

$$\mathbf{H}(k) = \begin{bmatrix} h_{1,1}(k) & \cdots & h_{1,n_T}(k) \\ \vdots & \ddots & \vdots \\ h_{n_R,1}(k) & \cdots & h_{n_R,n_T}(k) \end{bmatrix} \tag{1}$$

We suppose that the channels remain constant over the transmission of a frame and vary independently from frame to frame (*block fading channel*).

The elements $h_{n,m}(k), k = 0, \ldots, L$, of the channel impulse responses are complex random variables with a Gaussian distributed real and imaginary part, zero mean and variance $\sigma^2(k)$. The first element $h_{n,m}(0)$ includes also the direct component with the amplitude $p_{n,m}$ and the power $p_{n,m}^2 = c_R \cdot \sigma^2(0)$, where c_R is the Rice factor. The power delay profile describes by the variances $\sigma^2(k), k = 0, \ldots, L$ how the power is distributed over the taps of the channel impulse response. Here the variances decrease exponentially with k.

The channel energy is normalized by the condition

$$\sum_{k=0}^{L} E\left\{|h_{n,m}(k)|^2\right\} = p_{n,m}^2 + \sum_{k=0}^{L} \sigma^2(k) := 1. \tag{2}$$

We assume additive white Gaussian noise (AWGN) with zero mean and variance σ_n^2 per receive antenna.

2.3 Receiver

The symbol received by antenna n at time instant k is denoted by $x_n(k)$. The symbols received by the n_R antennas are arranged in a vector $\mathbf{x}(k) = [x_1(k), \ldots, x_{n_R}(k)]^T$ of length n_R, which can be expressed with (1) and $\mathbf{n}(k)$ as noise vector of length n_R as

$$\mathbf{x}(k) = \sum_{i=0}^{L} \mathbf{H}(i)\, \mathbf{s}(k-i) + \mathbf{n}(k). \tag{3}$$

3 Semi-blind Equalization

3.1 Blind Signal Separation

A. Reorder Data for ICA

The convolutive mixture in (3) is reordered to apply ordinary ICA [1]. Considering the receive vector $\mathbf{x}(k)$ of length n_R at K successive time instances the receive symbols are arranged in a vector $\mathbf{x}_K(k) = [\,\mathbf{x}(k),\ \ldots,\ \mathbf{x}(k+K-1)\,]^T$ of length $K \cdot n_R$. The transmitted symbols that are the Independent Components (IC) are arranged in a vector $\mathbf{s}_K(k) = [\,\mathbf{s}(k-L),\ \ldots,\ \mathbf{s}(k+K-1)\,]^T$ of length $(K+L)\cdot n_T$. As the channel is assumed to be a block fading channel the received data is processed in frames, which is represented by the receive signal matrix $\mathcal{X}_K = [\mathbf{x}_K(0),\ \mathbf{x}_K(1),\ \ldots,\ \mathbf{x}_K(N_F-K)]$ and the associated transmit signal matrix $\mathcal{S}_K = [\mathbf{s}_K(0),\ \mathbf{s}_K(1),\ \ldots,\ \mathbf{s}_K(N_F-K)]$. With $\mathbf{x}_K(k)$ and $\mathbf{s}_K(k)$ the mixing matrix $\mathcal{H}_K$ becomes a Toeplitz matrix (called generalized Sylvester matrix) of size $K \cdot n_R \times (K+L)\cdot n_T$:

$$\mathcal{H}_K = \begin{bmatrix} \mathbf{H}(L) & \cdots & \mathbf{H}(0) & \cdots & \mathbf{0} \\ \vdots & \ddots & \ddots & \ddots & \vdots \\ \mathbf{0} & \cdots & \mathbf{H}(L) & \cdots & \mathbf{H}(0) \end{bmatrix}, \tag{4}$$

containing the channel matrices $\mathbf{H}(k)$, $k = 0, ..., L$. The convolutive mixture in (3) simplifies to a multiplicative one, that is the reformulated ICA problem, given by

$$\mathcal{X}_K = \mathcal{H}_K \cdot \mathcal{S}_K + \mathcal{N}. \tag{5}$$

where $\mathcal{N}$ denotes the noise.

As ICA requires the number of observed mixtures to be at least equal to the number of ICs, the matrix $\mathcal{H}_K$ should have at least as many rows as columns, which is expressed as $K \cdot n_R \geq (K+L)\cdot n_T$. For K the following relation holds

$$K \geq \left\lceil \frac{n_T \cdot L}{n_R - n_T} \right\rceil. \tag{6}$$

As a consequence, applying ICA to frequency selective MIMO systems requires more receive than transmit antennas.

B. Whitening + ICA

Ordinary complex ICA algorithms like complex FastICA [6] with symmetric orthogonalization or JADE [7] are applied to obtain estimates of the separated signals $\hat{\mathcal{S}}_{K_0}$ and an estimate $\hat{\mathcal{H}}_{K_0}$ for the generalized Sylvester matrix. Recalling the restrictions of ICA [1], every estimated IC shows a different unknown phase offset. In addition the estimated ICs are arbitrary permuted. These permutations are divided into outer permutations in $\hat{\mathcal{S}}_{K_0}$, i.e. permutations of the vectors $\mathbf{s}(k)$ in $\mathbf{s}_K(k)$, and inner permutations, i.e. permutations of $s_m(k)$ in $\mathbf{s}(k)$.

3.2 Ambiguity Resolution

C. Resolving Outer Permutations
To resolve the outer permutations, the special structure of the reordered mixing matrix $\mathcal{H}_K$ is used. Due to permutations of the rows in $\hat{\mathcal{S}}_{K_0}$, only the columns of $\hat{\mathcal{H}}_{K_0}$ are permuted. The same applies to the unknown phase offsets.

An algorithm to rearrange the matrix $\hat{\mathcal{H}}_{K_0}$ to a Toeplitz matrix is presented for $n_T = 2$, which can easily be extended to more transmit antennas. The channel matrix $\mathbf{H}(k)$ for $n_T = 2$ is split into subcolumns $\mathbf{H}(k) = [\mathbf{h}_1(k)\, \mathbf{h}_2(k)]$. The algorithm is described in two major steps:

1. The n_T rightmost columns in $\hat{\mathcal{H}}_{K_0}$ can be found by two criteria:
   (a) Assuming an exponential decreasing power delay profile channel, the sum of the absolute values of subcolumn $\mathbf{h}_m(0)$ in (7) is maximal compared to other subcolumns in the same row.
   (b) Considering the part of the column above subcolumn $\mathbf{h}_m(0)$, the sum of the absolutes values should be nearly zero.

   These criteria require $K \geq 2$, elsewise no zero submatrices could be found in $\hat{\mathcal{H}}_{K_0}$. The found columns, in (7) depicted by the boxed columns, will be moved to the right side of the matrix as shown in (8). The columns including $\mathbf{h}_1(0)$ and $\mathbf{h}_2(0)$ can be permuted (inner permutations).

$$\begin{bmatrix} \ddots & \vdots & & \vdots & \cdots \\ \cdots & \mathbf{0}\, e^{j\Delta\varphi_1} & \cdots & \mathbf{0}\, e^{j\Delta\varphi_2} & \cdots \\ \cdots & \mathbf{h}_1(0)\, e^{j\Delta\varphi_1} & \cdots & \mathbf{h}_2(0)\, e^{j\Delta\varphi_2} & \cdots \end{bmatrix} \tag{7}$$

$$\left[\begin{array}{ccc|cc} \ddots & \vdots & & \vdots & \vdots \\ \cdots & \boxed{\mathbf{h}_2(0)\, e^{j\Delta\varphi_3}} & \cdots & \mathbf{0}\, e^{j\Delta\varphi_1} & \mathbf{0}\, e^{j\Delta\varphi_2} \\ \cdots & \mathbf{h}_2(1)\, e^{j\Delta\varphi_3} & \cdots & \mathbf{h}_1(0)\, e^{j\Delta\varphi_1} & \boxed{\mathbf{h}_2(0)\, e^{j\Delta\varphi_2}} \end{array}\right] \tag{8}$$

2. The row one subcolumn above $\mathbf{h}_2(0)$ is searched for a subcolumn having the same absolute values as $\mathbf{h}_2(0)$, depicted by the boxed subcolumns in (8). The matching column is moved left to the already sorted ones as depicted in (9).

$$\left[\begin{array}{c|ccc} \ddots & \vdots & \vdots & \vdots \\ \cdots & \mathbf{h}_2(0)\, e^{j\Delta\varphi_2} & \mathbf{0}\, e^{j\Delta\varphi_1} & \mathbf{0}\, e^{j\Delta\varphi_2} \\ \cdots & \mathbf{h}_2(1)\, e^{j\Delta\varphi_2} & \mathbf{h}_1(0)\, e^{j\Delta\varphi_1} & \mathbf{h}_2(0)\, e^{j\Delta\varphi_2} \end{array}\right] \tag{9}$$

The boxed submatrices in (8) have the same phase angle but different phase offsets. To relate the phase offsets the differential angle $\Delta\psi$ is determined by

$$\Delta\psi = \angle(\mathbf{h}_2(0)\, e^{j\Delta\varphi_3}) - \angle(\mathbf{h}_2(0)\, e^{j\Delta\varphi_2}) = \Delta\varphi_3 - \Delta\varphi_2. \tag{10}$$

By multiplying the found column by $e^{-j\Delta\psi}$ we get

$$\mathbf{h}_2(i)\, e^{j\Delta\varphi_3} \cdot e^{-j\Delta\psi} = \mathbf{h}_2(i)\, e^{j\Delta\varphi_2}\ , \ i = 0, 1 \tag{11}$$

and the same phase offset $\Delta\varphi_2$ is achieved. This processing step is repeated until $\hat{\mathcal{H}}_{K_0}$ is sorted.

The Toeplitz matrix is resorted up to the inner permutations in $\mathbf{H}(k)$. As the rows in $\hat{\mathcal{S}}_{K_0}$ are permuted in the same way as the columns in $\hat{\mathcal{H}}_{K_0}$, during the execution of the algorithm a permutation matrix $\mathbf{P}_1$ is arranged to obtain the corrected signal matrix $\hat{\mathcal{S}}_{K_1} = \mathbf{P}_1 \cdot \hat{\mathcal{S}}_{K_0}$.

D. Phase Estimation to $i\frac{\pi}{2}$
By using M-QAM modulated signals the phase can be estimated by a set of symbols and fourth order cumulants [8] up to a multiple of $\frac{\pi}{2}$. The estimated phase of source m is denoted with φ_{o_m}. With the unitary matrix $\mathbf{P}_2$

$$\mathbf{P}_2 = \mathbf{I}_{K+L} \otimes \mathrm{diag}(e^{-j\hat{\varphi}_{o_1}}, \ldots, e^{-j\hat{\varphi}_{o_{n_T}}}), \tag{12}$$

where $\otimes$ denotes the Kronecker tensor product and $\mathbf{I}$ is the identity matrix, the corrected signal matrix $\hat{\mathcal{S}}_{K_2} = \mathbf{P}_2 \cdot \hat{\mathcal{S}}_{K_1}$ is obtained.

E. Resolving Inner Permutations and the Remaining Phase Offsets
The remaining ambiguities are solved by orthogonal preamble sequences, designed using Hadamard matrices, with a preamble length $N_P \geq n_T$. The symbols of the extracted preamble $\hat{\mathbf{S}}_P$ are hard decided, that is denoted by $\tilde{\mathbf{S}}_P$. The unitary cross correlation matrix $\mathbf{R}_{\tilde{\mathbf{S}}\mathbf{S}}$, given by

$$\mathbf{R}_{\tilde{\mathbf{S}}\mathbf{S}} = \frac{1}{N_P}\tilde{\mathbf{S}}_P\mathbf{S}_P^H, \tag{13}$$

where $(\cdot)^H$ denotes the complex-conjugate (Hermitian) transpose, will be an identity matrix if all phase offsets are zero and no inner permutations exist. As the outer permutations are solved, the matrix

$$\mathbf{P}_3 = \mathbf{I}_{K+L} \otimes \mathbf{R}_{\tilde{\mathbf{S}}\mathbf{S}}^H \tag{14}$$

is used to obtain the corrected signal matrix $\hat{\mathcal{S}}_{K_3} = \mathbf{P}_3 \cdot \hat{\mathcal{S}}_{K_2}$.

F. Extract & Combine Data
By solving the ICA problem in (5) the transmitted symbols $s_m(k)$ are estimated $(K+L)$ times. Prior to a demodulation the multiple estimated transmit symbols can be combined to achieve a lower Bit Error Rate.

4 Simulation Results

For a performance comparison a training based channel estimation and equalization is considered. A training sequence is used to estimate the channel matrices by the Minimum Mean Square Error (MMSE) - estimator [9, 10]. We use random training sequences [10] of length $N_{P,T} = 2 \cdot n_T \cdot (L+1) + L$ per transmit antenna, which is slightly above the optimal training sequence length [10] of $N_{P,T} = 1.8 \cdot n_T \cdot (L+1) + L$. The channel estimate is used to equalize the received data. For the training based equalization the successive FS-Blast [11] is used.

In Fig. 2(a) one can see a comparison of semi-blind and training based channel estimation & equalization based on the Bit Error Rate (BER) versus the Signal-to-Noise-Ratio (SNR). Also shown is the BER for the case of perfect knowledge of the channel impulse response (CIR) at the receiver. As expected, the result with training based channel estimation is closer to the result with perfect knowledge than that with semi-blind equalization. The presented semi-blind approach requires a 3 dB increased Signal-to-Noise-Ratio (SNR) to achieve a Bit Error Rate of $4 \cdot 10^{-3}$ compared to training based channel estimation & equalization. Comparing the two ICA algorithms JADE shows a better performance than FastICA. One can also see, that an increased number of preamble symbols N_P does not significantly change the BER. It is sufficient to use the minimal desired number, which is $N_P = 2$ for $n_T = 2$. The mean square error (MSE) versus the

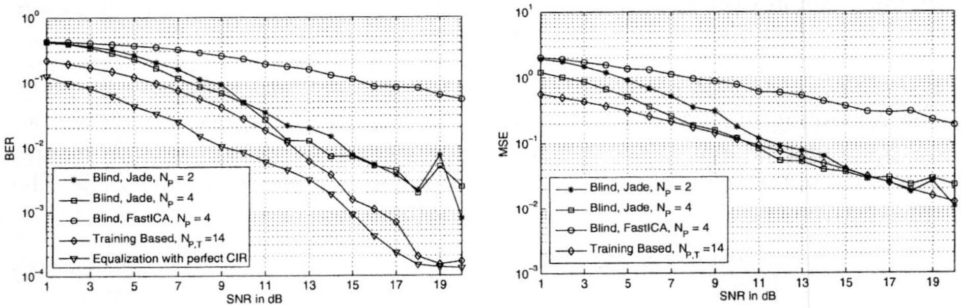

(a) BER versus SNR (b) MSE versus SNR

Fig. 2. Comparison of semi-blind and training based channel estimation & equalization, Simulation with $n_T = 2$ and $n_R = 4$, frequency selective MIMO channel order $L = 2$, $c_R = 1$, 4-QAM modulation, Semi-blind equalization with 500 symbols per frame with combining multiple estimations

SNR for semi-blind and training based channel estimation is shown in Fig. 2(b). The MSE of the channel estimation is calculated as

$$\text{MSE} = E\left\{\sum_{k=0}^{L} \frac{1}{n_T \cdot n_R} \sum_{m=1}^{n_T} \sum_{n=1}^{n_R} \left| h_{n,m}(k) - \rho_m \hat{h}_{n,m}(k) \right|^2 \right\}, \tag{15}$$

in which $\hat{h}_{n,m}(k)$ is the estimated channel impulse response and ρ_m is used to compensate the scalar ambiguity associated with the estimation results for the m-th antenna. This calculation of the MSE is equal to the one used in the literature since (2) holds. For high SNR greater than 10 dB the MSE of the training based approach does not differ from the MSE achieved with semi-blind equalization. The BER of the training based approach in contrast is less than the one obtained by semi-blind equalization. It is important to note that the separation of the independent components do not depend on the estimated mixing matrix and their accuracy.

5 Conclusions

We showed that ordinary complex ICA can be used for semi-blind equalization of MIMO frequency selective channels. An algorithm was presented that resolve the increased number of ambiguities due to the reformulation of the convolutive blind signal separation problem. Compared with training based channel estimation & equalization the presented semi-blind approach requires a 3 dB increased Signal-to-Noise-Ratio (SNR) to achieve a Bit Error Rate of $4 \cdot 10^{-3}$.

References

1. Hyvärinen, A., Karhunen, J., Oja, E.: Independent Component Analysis. John Wiley & Sons, New York (2001) ISBN 0-471-40540-X.
2. Papadias, C.: Unsupervised receiver processing techniques for linear space-time equalization of wideband multiple input / multiple output channels. IEEE Transactions on Signal Processing **52**(2) (2004) 472 – 482
3. Gorokhov, A., Loubaton, P.: Subspace-based techniques for blind separation of convolutive mixtures with temporally correlated sources. IEEE Transactions on Circuits and Systems—Part I: Fundamental Theory and Applications **44**(9) (1997) 813 – 820
4. Zeng, Y., Ng, T.S., Ma, S.: Blind MIMO channel estimation with an upper bound for channel orders. In: Proc. IEEE International Conference on Communications. (2005) 1996 – 2000
5. Obradovic, D., Madhu, N., Szabo, A., Wong, C.S.: Independent component analysis for semi-blind signal separation in MIMO mobile frequency selective communication channels. In: Proc. IEEE International Joint Conference on Neural Networks. (2004) 53–58
6. Bingham, E., Hyvärinen, A.: A Fast Fixed-Point Algorithm for Independent Component Analysis of Complex Valued Signals. International Journal of Neural Systems **10**(1) (2000) 1–8
7. Cardoso, J.F., Souloumiac, A.: Blind Beamforming for Non Gaussian Signals. In: IEE-Proceedings-F, vol. 140, no. 6. (1993) 362–370
8. Cartwright, K.V.: Blind Phase Recovery in General QAM Communication Systems Using Higher Order Statistics. In: IEEE Signal Processing Letters, Vol. 6. (1999) 327–329
9. Fragouli, C., Al-Dhahir, N., Turin, W.: Training-Based Channel Estimation for Multiple-Antenna Broadband Transmission. IEEE Transactions on Wireless Communications **2**(2) (2003) 384–391
10. Weikert, O., Hagebölling, F., Zimmermann, C., Zölzer, U.: On the Influence of Power Delay Profiles of MIMO Frequency Selective Channels on Equalization and Channel Estimation. (submitted to ICC 2006)
11. Wübben, D., Kühn, V., Kammeyer, K.D.: Successive Detection Algorithm for Frequency Selective Layered Space-Time Receiver. In: Proc. IEEE ICC'03. Volume 4. (2003) 2291 – 2295

Comparison of BSS Methods for the Detection of α-Activity Components in EEG

Sergey Borisov[1,2], Alexander Ilin[1], Ricardo Vigário[1], and Erkki Oja[1]

[1] Laboratory of Computer and Information Science, Helsinki University of Technology, P.O. Box 5400, FIN-02015 HUT, Espoo, Finland
[2] Department of Human and Animal Physiology, Faculty of Biology, Moscow State University, Leninskie Gori, Moscow, 119992, Russia
basirov@mail.ru

Abstract. In this paper, we tested the efficiency of a two-step blind source separation (BSS) approach for the extraction of independent sources of α-activity from ongoing electroencephalograms (EEG). The method starts with a denoising source separation (DSS) of the recordings, and is followed by either an independent component analysis (ICA) or a temporal decorrelation algorithm (FastICA and TDSEP, respectively). This two-step method was compared with DSS, ICA and TDSEP alone. The tests were performed with simulated data based on real EEG signal, to guarantee the existence of a "ground truth". The most efficient algorithm, for proper component extraction (regardless of the amount of α-activity in their spectra) is a combination of DSS and ICA. It provided also more stable results than ICA alone. TDSEP, in combination with DSS, was efficient only for the extraction of the components with prominent α-activity.

1 Introduction

There is a considerable amount of work devoted to studying brain dynamics reflected in EEG/MEG signals using independent component analysis (ICA) (e.g., see [10, 4, 9]). One particular application of this signal processing method is analysis of EEG α-activity [6, 2]. At the same time, the study of dynamics of ongoing EEG α-activity taking into account the nonstationary nature of the EEG signal is very promising [5]. Using both approaches, it would be interesting to study the independent sources for different quasi-stationary segments of EEG α-activity differentiated, for example, by their mean amplitude.

It has been shown that, in noisy environments and in presence of insufficiently informative data set, as compared to the measuring space dimension, ICA tends to overfit into local minima (see [8]). Indeed, when we applied ICA for analysis of ongoing EEG α-activity, we faced local minima problems, which resulted in the instability of the obtained results. This problem has motivated us to use in [1] a new BSS framework called denoising source separation (DSS) which takes into account a prior information about separated signals [7].

The independence criterion assumed in this method was the uncorrelatedness of α-content in different sources. Provided that this assumption is valid, the method is able to separate alpha sources containing different amount of α-activity. In practice,

J. Rosca et al. (Eds.): ICA 2006, LNCS 3889, pp. 430 – 437, 2006.

however, this difference was not large enough in all cases and therefore the reliable separation was not always achieved.

Trying to solve both mentioned problems we suggest in this paper a combination of the α-frequency based DSS and ICA in a two-step algorithm. In the first stage, DSS identifies the α-subspace, while ICA separates statistically independent α-components in the second stage. We also investigate the use of temporal decorrelation methods such as TDSEP, as a possible second step in the proposed approach (for details of TDSEP, see [12]).

The goal of the present work is to test the efficiency of the proposed approach for the separation of the α-components and to compare the results of the tests with those obtained by other single-step BSS algorithms such as ICA, DSS and TDSEP alone. We tested these approaches using simulated data, maximally resembling the properties of real EEG signals.

2 Methods

2.1 Simulated EEG Signal

The "reference" 16-channel signal (i.e., targets to the BSS decomposition) was compiled from independent components extracted by ICA from EEG recordings registered from 16 healthy adolescents in rest condition with eyes closed (1 min, 7680 points totally). From every subject, only one component was taken. The choice of the components was based on the amount of certain frequencies in their spectra. Two types of "reference" signals were compiled. The first dataset contained seven components (of 16) with prominent α-activity (7-13 Hz) in their spectra (see Fig. 1), and the second dataset contained three such components. The rest of the components in both datasets could also contain some amount of alpha, but it was not the main frequency in their spectra. All components were normalized to unit variance.

The simulated signal was produced with three different mixing matrices to reach some sort of statistical measure of result consistency. All matrices were obtained using aforementioned real EEG data, as result of ICA decomposition. Each matrix

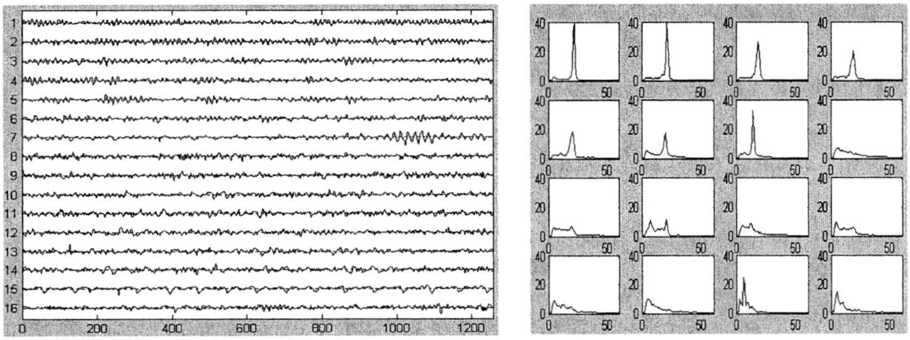

Fig. 1. Reference 16-channel signal containing seven components with prominent α-activity (10 s fragment, 1280 points; left) and spectral power of each component of the signal (right)

corresponds to a spatial distribution of a set of sources. They were chosen to reflect different weightings from the brain regions, hence various alpha contributions. Such contribution C_j of each source, say, the jth source, to the total variance of the mixed signal is given by the squared norm of the corresponding mixing vector a_j, normalized by the total variance of the mixing matrix:

$$C_j = \frac{\sum_i a_{ij}^2}{\sum_i \sum_j a_{ij}^2}, \quad (1)$$

where a_{ij} is the ij-th element of mixing matrix A. This characteristic can be interpreted as a power of certain source in terms of forming the mixed signal. The first mixing matrix, (a), emphasized sources 4, 5, 6, 9 and 12; the second matrix, (b), highlighted sources 1-4, 9, 11, and 13, whereas the third, (c), gave clear relevance to source 1 and, in essentially lesser degree, - to source 2 (see Fig. 2).

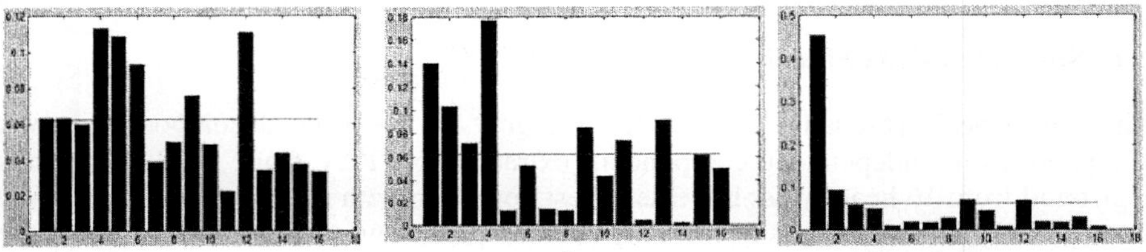

a b c

Fig. 2. Contributions of each component of reference signal into forming mixed signal as determined by mixing matrixes (*a, b, c*). Each column corresponds to C_j for each mixing vector of certain mixing matrix (see description in the text).

To simulate additional noise artifacts in the simulated data we added to each of the mixed signals white noise with variance equal to 36% of the variance of the reference signal.

The generation of the simulated signal is summarized in Fig. 3. With two types of reference signal and three different mixing matrices, we generated six different simulated datasets to be analyzed with a set of algorithms.

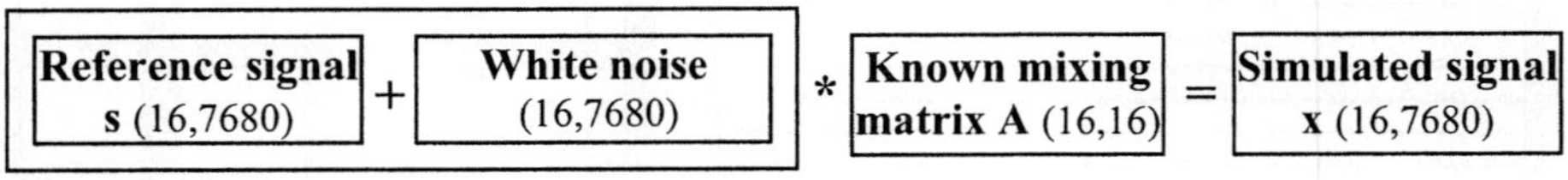

Fig. 3. Scheme of generation of the simulated signal

2.2 Algorithms

The main approach we propose and test here was a two-step algorithm containing both DSS and ICA. The first step in each combined algorithm consisted of a frequency-based DSS, targeting the α-band content in the recordings. It can be summarized in a series of operations starting by spatial whitening, followed by filtering the white data in the frequency of interest. Then a new projection from the original recordings is sought which maximizes the power of the filtered data.

Only eight components obtained by DSS, containing most pronounced alpha in their spectra, were selected for further analysis in a second step. In the second step, we used two well-known BSS algorithms. One was FastICA [13], a fast and robust fixed-point algorithm, which maximizes the non-Gaussianity of the projected components (for details, see [3]). Another algorithm tested was TDSEP [14] - an algorithm for temporal decorrelating the output signals, based on the joint diagonalization of delayed correlation matrices [12]. This method should be suitable for extracting time-structured sources, in particular, α-activity.

In both two-steps algorithms (with ICA or TDSEP), we also studied the influence of adding an intermediate filtering of the components obtained in the first stage (result of DSS). This filtering targeted the α-band (7-13 Hz). Although the DSS procedure aims at boosting this frequency band, we have often observed that other frequencies were also included in the estimates. The additional filtering stage aims at reducing the effect of such undesired frequencies, and hence results in an improvement in the detection of the α-components.

As a control for described testing, we processed the same datasets using DSS, ICA and TDSEP separately. Thus, seven algorithms were tested in total; see Fig. 4.

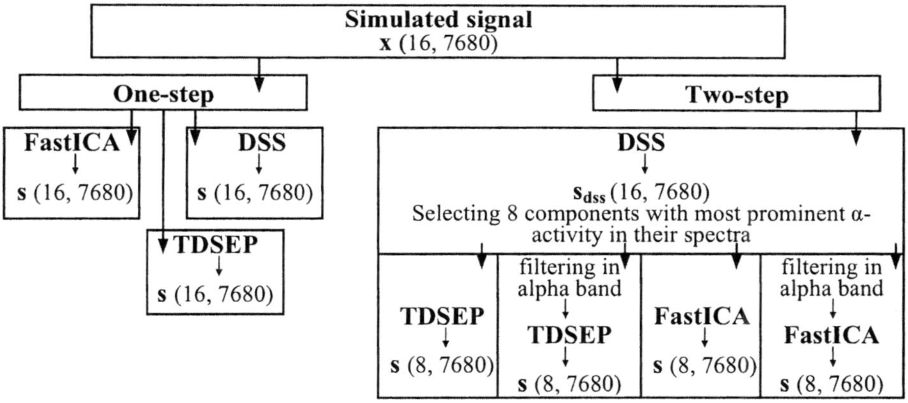

Fig. 4. Scheme of processing the simulated signal with different algorithms

2.3 Estimation of Algorithm Accuracy

The source separation efficiency, for the compared algorithms, was estimated using cross-correlation calculated between separated sources and the "reference" signals. The obtained cross-correlation matrices were then analyzed: we selected the maximum value of the cross-correlation coefficient (CC) for each of eight rows of the matrices. Because we are only interested in α-activity, only the eight rows of CC corresponding to the components with most pronounced alpha were analyzed. The obtained values had to correspond to the cases when the given channel of the obtained signal correlates well with some channel of the reference signal. We averaged these eight values, for each type of simulated signal, thus getting an integral estimation of the algorithm effectiveness in extracting those eight components from the six 16-channel simulated signals.

It has been observed that different runs of ICA on finite data sets may result in somewhat different estimates of the independent sources [11]. To test the stability of the results obtained by the algorithms containing ICA, we run those algorithms 50 times changing initial conditions. Then we calculated the CC between the obtained components and the reference signal and estimated the mean and the standard deviation of mean (STD) for every entry of the 50 CC matrices. We calculated the coefficient of variation (CV) of CC as a ratio of mean and STD values for every entry, and than averaged CV across the rows of the matrix. This value reflected the average level (for a certain row) of CC variations across the repeated calculations, hence was an estimate of the stability for each of eight extracted components.

3 Results

3.1 Processing the Signals Containing Seven Alpha-Components

The results of processing simulated signal containing seven alpha components show quite high effectiveness for four algorithms: DSS, ICA in combination with DSS, both with and without intermediate filtering (DSS_f_ICA; DSS_ICA), and TDSEP in combination with DSS, with α-filtering (DSS_f_TDSEP). The average maximum CC were always greater than 0.8, for all three types of the simulated signal (Table 1). At the same time, the STD values were seldomly above 12% of the mean values of CC. This fact suggests that all eight components were extracted by the mentioned methods quite accurately. ICA alone could also extract the components well, but only from dataset mixed with mixing matrix (a). TDSEP and TDSEP in combination with DSS (without intermediate α-filtering) show the same, rather poor, results. Since the results were very close for both approaches, we present only the results for TDSEP alone.

Table 1. The average maximum CC (± STD) between the components extracted from the simulated signal containing seven alpha sources, and reference signal. *a*, *b* and *c* – results for the simulated signals mixed with mixing matrixes (a), (b) or (c) respectively. The coefficients were averaged across all eight extracted components. Every column represents the result for one of the applied BSS methods. The values above 0.8 are marked with bold font.

	DSS	**ICA**	**DSS_ICA**	**DSS_f_ICA**	**TDSEP**	**DSS_f_TDSEP**
a	**0.85±0.05**	**0.87±0.04**	**0.84±0.12**	**0.82±0.09**	0.67±0.11	**0.87±0.05**
b	**0.87±0.05**	0.74±0.17	**0.82±0.10**	**0.83±0.08**	0.71±0.08	**0.86±0.09**
c	**0.84±0.04**	0.73±0.12	**0.89±0.04**	**0.89±0.04**	0.67±0.11	**0.88±0.05**

3.2 Processing the Signals Containing Three Alpha-Components

Results obtained for the signals containing only three components with prominent α-activity differ from the ones described above. The maximum CC were found between the reference signal and the components extracted by ICA and particularly by DSS_ICA (see Table 2). According to the STD values for average CC, CC for data obtained by these algorithms were high for all eight extracted components.

Table 2. The average maximum CC (± STD) between components extracted from the simulated signal containing three alpha sources and reference signal. For the legend see Table 1.

	DSS	ICA	DSS_ICA	DSS_f_ICA	TDSEP	DSS_f_TDSEP
a	0.72±0.11	**0.87±0.04**	**0.89±0.02**	0.77±0.12	0.76±0.08	0.74±0.17
b	0.74±0.15	**0.80±0.12**	**0.90±0.03**	0.67±0.11	0.62±0.12	0.79±0.14
c	0.75±0.15	**0.85±0.08**	**0.84±0.09**	0.71±0.16	0.65±0.03	0.74±0.17

The components extracted by DSS, DSS_f_ICA and DSS_f_TDSEP had lower values of CC compared with the two algorithms mentioned above. At the same time, the STD values for these average CC were essentially higher (Table 2). The last fact should be interpreted so that the CC between the reference signal and components obtained by DSS, DSS_f_ICA and DSS_f_TDSEP are low for some components and are high for others. Indeed, more detailed analysis of the CC obtained for the simulated signal mixed with mixing matrix (b) confirms that suggestion (see Fig. 4). The high correlation with reference signal (about 0.8 and higher) was found for three and four components extracted by DSS and DSS_f_TDSEP respectively, and only for two components extracted by DSS_f_ICA. At the same time, there are five such components extracted by ICA alone and all eight – extracted by DSS_ICA.

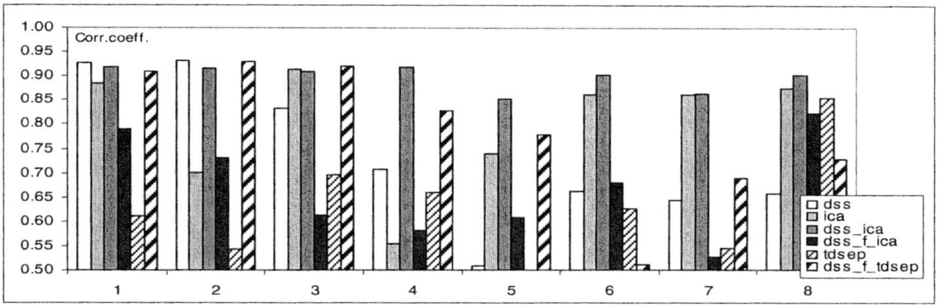

Fig. 4. The maximum correlation coefficients between eight components extracted from the simulated signal (mixed with mixing matrix (b)) containing 3 α-sources and reference signal. Every bar represents the result for one of the applied BSS methods.

It is interesting to note that the highest values of CC for components extracted by DSS and DSS_f_TDSEP were observed for the components containing high amount of alpha (see the first three sets of bars in Fig. 4). That means that in case when there are only few α-components in the mixed signal, these methods will be able to extract these components accurately, but they would not extract properly other components. Yet, ICA in combination with DSS without alpha-filtering could extract well both types of components with and without α-activity. This is to be expected, because the ICA block will be able to use additional information present in the other components. Remember that, in real applications, we do not have access to the number of the alpha components, and are, therefore, not able to restrict our search only to that frequency.

3.3 Stability of Results Obtained by ICA-Based Algorithms

From the results presented above, the DSS_f_ICA algorithm was the least effective ICA-based method. Hence, we exclude it from testing the stability results. The average CV of CC for components extracted with multiple runs of ICA and DSS_ICA are presented in Table 3.

Table 3. The CV values averaged across the rows of the average cross-correlation matrix for 50 CC, for components obtained by ICA-based algorithms. The calculations were performed for both types (*7 alpha, 3 alpha*) of simulated signal mixed with mixing matrix (b). The lowest values of CV comparing between the methods are marked in bold.

7 alpha	ICA	0.94	1.39	1.13	0.88	0.85	1.22	0.89	0.93
	DSS_ICA	**0.26**	**0.64**	**0.74**	**0.64**	**0.39**	**0.27**	**0.25**	**0.33**
3 alpha	ICA	0.91	1.40	1.15	0.92	0.78	1.04	0.94	0.87
	DSS_ICA	**0.70**	**0.59**	**1.05**	**0.76**	**0.47**	**0.57**	**0.20**	**0.17**

As follows from Table 3, the CV values were higher (in most cases more than twice higher) for both types of the signal, for the results obtained by ICA. That allows us to conclude that results obtained by DSS_ICA are more stable than the ones obtained by ICA alone.

4 Discussion

In the present work we tested a set of BSS methods for the extraction of independent components of EEG α-activity. To estimate the methods' efficiency, we generated simulated recordings with known parameters. These recordings were based on real EEG signals taken from different subjects. The signals were linearly mixed with matrices obtained from real EEG data. Hence, we insured to have clear target signals for the separation, while preserving their independence and properties close to real EEG. Yet, we need to notice that the present results were obtained using modeled data and are therefore not guaranteed to be completely reproducible on real data.

Testing the suggested algorithms using two simulated dataset types demonstrates that the most effective approach is ICA following DSS, without an intermediate α-filtering. This algorithm shows good results of component extraction for the set containing seven α-sources (see Table 1), and the best results for the set with three α-sources (see Table 2). In the second data set, ICA alone was also quite efficient, but for the set with seven α-components its efficiency was rather weak. Besides, in both cases the stability of the results obtained by ICA alone was noticeably lower.

The approach of DSS followed by TDSEP, with intermediate filtering, was much more efficient in terms of α-component extraction than TDSEP alone or DSS_TDSEP. In fact, DSS_f_TDSEP algorithm could extract the α-components very accurately, in some cases even better than DSS_ICA (see Table 1). It means that this algorithm can be useful for the extraction of prominent α-components from real data. Yet, since we can not know what kind of sources a real signal contains, the validation of the obtained components is required. One such verification is possible, for instance, by its discriminative power in the corresponding frequency band, *i.e.*, how prominent the peak is as compared to the background spectrum. Another method is to run the DSS step using white Gaussian data as inputs, and see the power of the α-components that will be estimated. The highest power of these components will be some sort of lower bound to true estimates found on real data.

At the same time, as follows from our results, it is not clear that DSS_f_TDSEP will accurately estimate "not-prominent" α-components containing other frequencies in addition to α-activity. On the contrary, using DSS_ICA we can expect that most of components will be extracted properly, regardless of the amount of α-activity they contain.

References

1. Borisov S., Ilin A., Vigário R., Kaplan A.: Source localization of low- and high-amplitude alpha activity: A segmental and DSS analysis. 11th Annual Meeting of Organization for Human Brain Mapping (OHBM, June, 2005). Neuroimage (2005) 26. Suppl. 1. 38
2. Delorme A., Makeig S.: EEG changes accompanying learned regulation of 12-Hz EEG activity. IEEE Trans Neural Syst Rehabil Eng. 11. (2003) 133-137
3. Hyvarinen A., Karhunen J., Oja E.: Independent Component Analysis. J. Wiley. (2001)
4. Jung T.-P., Makeig S. McKeown M.J. Bell A.J. Lee T.-W. Sejnowski T.J. Imaging Brain Dynamics Using Independent Component Analysis. Proceedings of the IEEE. 7. Vol. 89. (2001) 1107-1122
5. Kaplan A.Ya., Fingelkurts An.A., Fingelkurts Al.A., Borisov S.V., Darkhovsky B.S.: Nonstationary nature of the brain activity as revealed by EEG/MEG: methodological, practical and conceptual challenges. Signal Processing. 85. (2005) 2190-2212
6. Makeig S., Enghoff S., Jung T.P., Sejnowski T.J.: A natural basis for efficient brain-actuated control. IEEE Trans Rehabil Eng. 8. (2000) 208-211
7. Särelä J., Valpola H.: Denoising Source Separation. Journal of machine learning research. 6. (2005) 233–272
8. Särelä J., Vigário R.: Overlearning in marginal distribution-based ICA: analysis and solutions. Journal of machine learning research. 4. (2003) 1447-1469
9. Tang A., Pearlmutter B., Malaszenko N., Phung D. Reeb B.: Independent Components of Magnetoencephalography: Localization. Neural Computation. 8. Vol.14. (2002) 1827-1858
10. Vigário R, Särelä J, Jousmäki V, Hämäläinen M, Oja E.: Independent component approach to the analysis of EEG and MEG recordings. IEEE transactions on biomedical engineering. 47. (2000) 589-593
11. Ylipaavalniemi J., Vigário R.: Analysis of Auditory fMRI Recordings via ICA: A Study on Consistency, Proceedings of the 2004 International Joint Conference on Neural Networks (IJCNN, July, 2004) V.1. (2004) P. 249-254
12. Ziehe A., Müller K.-R.: TDSEP - An Effective Algorithm for Blind Separation Using Time Structure. Proceedings of the 8th International Conference on Artificial Neural Networks (ICANN, September, 1998) V. 8. (1998) 675-680
13. FastICA Package online at http://www.cis.hut.fi/research/ica/fastica
14. TDSEP Package online at http://wwwold.first.fhg.de/~ziehe/download.html

Analysis on EEG Signals in Visually and Auditorily Guided Saccade Task by FICAR

Arao Funase[1,2], Yagi Tohru[3,4], Motoaki Mouri[1], Allan K. Barros[5], Andrzej Cichocki[2], and Ichi Takumi[1]

[1] Graduate School of Engineering, Nagoya Institute of Technology, Gokiso-cho, Showa-ku, Nagoya, 466-8555, Japan
[2] Brain Science Institute, RIKEN, 2-1, Hirosawa, Wako, 351-0198, Japan
[3] Graduate School of Information Science and Engineering, Tokyo Institute of Technology, 1-12-1, Ookayama, Meguro-ku, Tokyo, 152-8555, Japan
[4] Bio-Mimetic Control Research Center, RIKEN 2271-130, Anagahora, Shimoshidami, Moriyama-ku, Nagoya, 463-0003, Japan
[5] Technological Center, Universidade Federal do Maranhão, Rua dos Guriatans, Qd. 5, casa 22, Renascenca II, Sao Luis, MA, 65075000, Brazil

Abstract. Recently an independent component analysis (ICA) becomes powerful tools to processing bio-signals. In our studies, the ICA is applied to processing on saccade-related EEG signals in order to predict saccadic eye movements because an ensemble averaging, which is a conventional processing method of EEG signals, is not suitable for real-time processing. We have already detected saccade-related independent components (ICs) by ICA. However, features of saccade-related EEG signals and saccade-related ICs were not compared. In this paper, saccade-related EEG signals and saccade-related ICs in visually and auditorily guided saccade task are compared in the point of the latency between starting time of a saccade and time when a saccade-related EEG signal or an IC has maximum value and in the point of the peak scale where a saccade-related EEG signal or an IC has maximum value.

1 Background

Nowadays, many researchers have been researching "Brain Computer Interfaces (BCIs)". BCIs connect computers and human by EEG signals. BCIs have some advantages compared with conventional interfaces. First is that users do not use inputs with body movements but use inputs with thinking, emotion, and motivation. Second is that computers work before user's movements because EEG signals include information for predicting beginning time of user's movement.

The BCI, which is introduced by our group, predict eye movements by EEG signal before eye movements and move a mouse cursor by the EEG signal before eye movements. This BCI can improve the latency between beginning time of eye movements and beginning time of working system. Therefore, this advantage of our BCI is attractive for the alarm of inattentive driving and the high-speed targeting system. In conventional research, a saccade-related EEG was detected

J. Rosca et al. (Eds.): ICA 2006, LNCS 3889, pp. 438–445, 2006.

before eye movements by ensemble averaging method [1]. However, ensemble averaging method has a disadvantage because ensemble averaging method needs many repetitive trials. Therefore independent component analysis (ICA) method was applied to analysis on saccade-related EEG signals because the ICA method can process raw EEG signals and find indepedent components (ICs) related to various EEG activities.

In our previous research, the ICA method can extract saccade-related independent components [2]. However, saccade-related ICs were extracted in case of visually guided saccade task. Therefore, in this paper, we confirm whether saccade-related ICs can be extracted in case of auditorily guided saccade task and compare saccade-related ICs in case of auditorily guided saccade task with saccade-related ICs in case of visually guided saccade task.

2 Fast ICA with Reference Signal [3]

Recently, the ICA method has been introduced in the field of bio-signal processing as a promising technique for separating independent sources [4]. The ICA is based on the following principle. Assuming that the original (or source) signals are mutual independent and have been linearly mixed, and that these mixed signals are available, the ICA finds in a blind manner a linear combination of the mixed signals which recovers the original source signals, possibly re-scaled and randomly arranged in the outputs.

The $\mathbf{s} = [s_1, s_2, \cdots, s_n]^{\mathrm{T}}$ means n independent signals from EEG sources in the brain, for example. The mixed signals (or the recorded signals) $\mathbf{x}$ are thus given by $\mathbf{x} = \mathbf{As}$, where $\mathbf{A}$ is an $n \times n$ invertible matrix. $\mathbf{A}$ is the matrix for mixing independent signals. In the ICA, only $\mathbf{x}$ is observed. The value for $\mathbf{s}$ is calculate by $\mathbf{s} = \mathbf{Wx}$ ($\mathbf{W} = \mathbf{A}^{-1}$). However, it is impossible to calculate $\mathbf{A}^{-1}$ algebraically because information for $\mathbf{A}$ and $\mathbf{s}$ is not already known. Therefore, in the ICA algorithm, $\mathbf{W}$ is estimated non-algebraically. In order to calculate $\mathbf{W}$, the assumption of the ICA algorithm that $\mathbf{s}$ is mutually independent is used. Different cost functions, which are defined from the assumption, are used in the literature, usually involving a non-linearity that shapes the probability destiny function of the source signals. However, high-order statistics, such as the kurtosis, are widely used as well. The kurtosis shows how independent a signal is because the kurtosis is the classical measure of non-gaussianity. The Fast ICA which is one of the ICA algorithms, is based on a cost function minimization or maximization that is a function of the kurtosis $\kappa(\mathbf{w}^{\mathrm{T}}\mathbf{x}) = \boldsymbol{E}(\mathbf{w}^{\mathrm{T}}\mathbf{x})^4 - 3[\boldsymbol{E}\{\mathbf{w}^{\mathrm{T}}\mathbf{x}\}^2]^2 = \boldsymbol{E}\{(\mathbf{w}^{\mathrm{T}}\mathbf{x})^4\} - 3||\mathbf{w}||^4$; $\mathbf{w}$ is one of the raw of $\mathbf{W}$. Then Fast ICA changes the weight $\mathbf{w}$ to extract an IC with the fixed-point algorithm.

From among several ICA algorithms, we selected the "Modified Fast ICA with Reference signal (FICAR)" algorithm to use in this study [3]. This algorithm can extract only the desired components by initializing the algorithm with prior information on the signal of interest. The main advantage of this approach is users can give instructions to extract a desired signal more strongly.

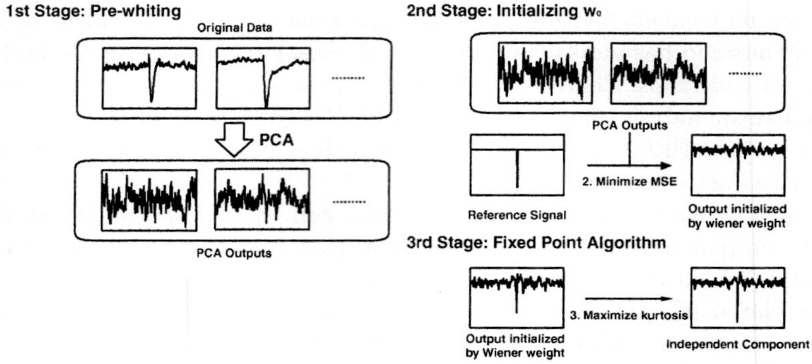

Fig. 1. Conceptual three stage for extraction of desired ICs

Fig.1 shows an overview of the procedures of FICAR algorithm. First, principal component analysis (PCA) outputs are calculated from original recorded signals to speed up the convergence of the algorithm. Second, this algorithm initialized $\mathbf{w}_k$ ($k = 0$; k is the iteration number.) using some priori information included in a reference signal, $\mathbf{d}$, correlated with $\mathbf{s}_i$, i.e. $\boldsymbol{E}[\mathbf{d}s_i] \neq 0$. This algorithm estimates a weight vector $\mathbf{w}$. Therefore, we calculate the error ε between $\mathbf{d}$, which is a reference signal, and $\mathbf{u} = \mathbf{w}^{\mathrm{T}}\mathbf{x}$; $\varepsilon = \mathbf{d} - \mathbf{u}$. The initial weight $\mathbf{w}_0$ are calculate by the minimization of the mean-squared error (MSE) given by $\boldsymbol{E}[\varepsilon^2]$. To calculate the MSE, the least mean square (LMS) is used in order to calculate the MSE. After some calculations, the optimum weight (also called the Wiener weight) to minimize the MSE was found to be $\mathbf{w}^* = \boldsymbol{E}[\mathbf{dx}]$. This algorithm initialized $\mathbf{w}_0 = \boldsymbol{E}[\mathbf{dx}]/||\boldsymbol{E}[\mathbf{dx}]||$. Third, this algorithm calculates $\mathbf{w}_{k+1}$ by $\mathbf{w}_{k+1} = \boldsymbol{E}[\mathbf{x}(\mathbf{w}_k^{\mathrm{T}}\mathbf{x})^3] - 3\mathbf{w}$ to maximize kurtosis. Then this algorithm can extract an IC closest to a reference signal or strictly speaking an IC which is correlated with the reference signal.

3 Experimental Settings

There are four tasks in this study. The first task is to record EEG signals during a saccade to a visual target that is on his right side or left side. The second task is to record EEG signals as a control condition when a subject dose not perform a saccade even though a visual stimulus has been displayed. First task and second task are called visual experiments. On the other hand, the third task is to record EEG signals during a saccade to a auditory target that is on his right side or left side. The fourth task is to record EEG signals as a control condition against the third task when a subject dose not perform a saccade even if a auditory stimulus has been turned on. The third task and fourth task are called auditory experiments. Each experiment is comprised of 50 trials in total: 25 trials on the right side and 25 trials on the left side.

The EEG signals are recorded through 19 electrodes (Ag-AgCl), which are placed on a subject's head in accord with the international 10-20 electrode position system. The electrooculogram (EOG) signals are simultaneously recorded through two pairs of electrodes (Ag-AgCl) attached to the top-bottom side and right-left side of the right eye.The number of subjects is five (Subject A, B, C, D, E). All subjects are men and have normal vision. All subjects are right-handed. All data are smpled at 1000[Hz], and stored on a hard disk for off-line data processing after post-amplification.

In this paper, the shape of the reference signal is that of an impulse signal having one peak. This shape is caused for two reasons. First, the saccade-related EEG had a sharp change like am impulse [1]. Second, the main components of an EEG signal are the neural responses, and the waveform of the neural responses is resembled to impulse.

4 Experimental Results and Discussion

4.1 Results of Ensemble Averaging

Fig.2 shows the experimental results obtained for "Subject A" when the visual stimulus on the right side is illuminated. This EEG data is processed with ensemble averaging and high-pass filter (cut-off 4 [Hz]). Fig.2-a and 2-b show the data with and without eye movement to right side, respectively. Black lines indicate results in visual experiments and gray lines represent results in auditory experiments. The top boxes show the voltage generated in response to the LED becoming illuminated. The middle boxes indicate the potential of EOG signals. The increase of EOG signals means an eye movement to right side. The bottom boxes represent EEG potential recorded at the right occipital lobe (at O2 in the international 10-20 electrode position system). The horizontal axes indicate the time span, where 0 [ms] indicates the start point of the eye movement. The

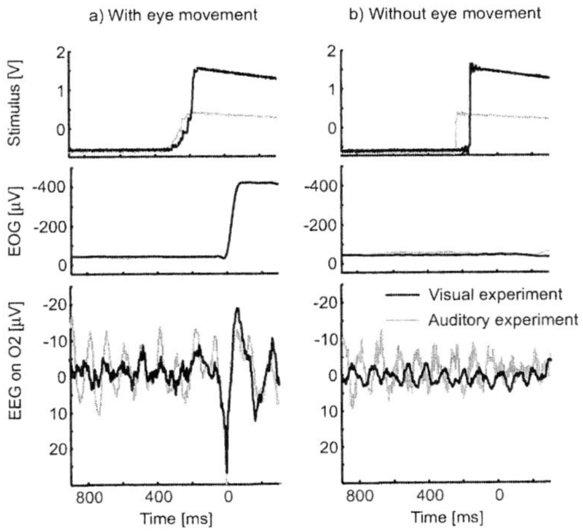

Fig. 2. Saccade-related EEG recorded on O2

Table 1. Peak time and amplitude on sharp change of EEG

	In visual experiments		In auditory experiments	
	Peak time [ms]	$n = \frac{\bar{x}-\mu}{s}$	Peak time [ms]	$n = \frac{\bar{x}-\mu}{s}$
	Right/Left	Right/Left	Right/Leftt	Right/Left
A	-3 / -2	8.6 / 9.3	-4 / -4	6.3 / 7.0
B	-5 / -3	6.3 / 7.8	-6 / -2	4.4 / 4.7
C	-3 / -4	7.0 / 6.9	-4 / -4	6.3 / 7.0
D	-3 / -2	8.2 / 8.2	-3 / -3	8.0 / 6.7
E	-3 / -3	7.8 / 7.9	-4 / -5	6.5 / 7.3
Ave.	-3.4 / -2.8	7.6 / 8.0	-4.2 / -3.6	6.3 / 6.5
STD	0.9 / 0.8	0.9 / 0.9	1.1 / 1.1	1.3 / 1.1

amplitude of the EEG signal is sharply changed just before eye movement in the case of an eye movement. However, there was no change for the case of no eye movement. The same tendency was observed for all five subjects in the case of both visual and auditory experiments.

In order to focus on features of saccade-related EEG signal, a time when saccade-related EEG signals have maximum amplitude and maximum amplitude is checked in Table 1. Amplitude was defined as n which is how many times standard deviation during 1000 [ms] before saccade is difference between mean of EEG potential during 1000 [ms] before saccade and maximum amplitude.

$n = \frac{\bar{x}-\mu}{s}$; where $\bar{x}$ is mean of EEG potential during 1000 [ms] before saccade, μ is maximum amplitude, and s is standard deviation during 1000 [ms] before saccade.

Peak time when saccade-related EEG signal is from -6 [ms] to -2[ms] (Ave. = -3.5, STD = 1.1) and n is from 4.4 to 9.3 (Ave. = 7.1, STD = 1.1) in Table 1. From Table 1, features of saccade-related EEG signals were observed before saccade and these features were observed remarkably.

4.2 Results of FICAR

We prepared about 500 reference signals for use in this experiment. As describe above, a reference signal has one peak point because waveform of a reference signal is a impulse wave. The signals differ in the time it took each to peak. The first reference signal has a peak when the stimulus is illuminated, and the time when the second reference signal has a peak is (*the time when the first reference signal has a peak*)+1 [ms]. The time when each reference signal has a peak is (*the time when the previous reference signal has a peak*)+1 [ms]. The final reference signal peaked in 300 [ms] after an eye movement.

Fig.3 shows the experimental results obtained when a subject move his eyes toward a visual and auditory target on the right side. These data are processed using the FICAR against the raw EEG data. The left figures indicate results in visual experiments and the right figures show results in auditory experiments. Top boxes represent the shapes of reference signals and bottom boxes indicate the amplitude

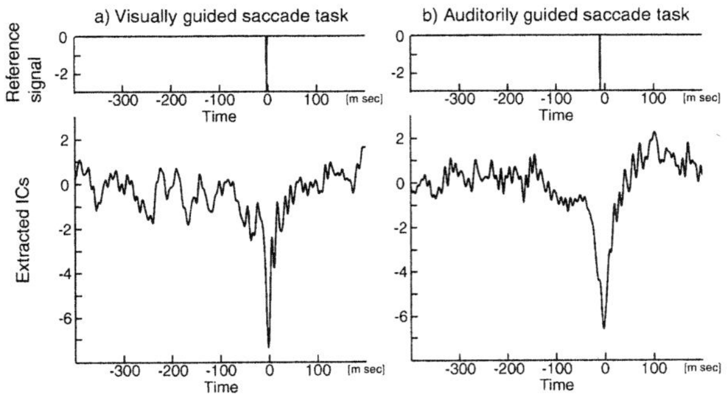

Fig. 3. Extracted signals for FICAR in visual and auditory experiments

of the ICs obtained by using the FICAR. The horizontal axes in these graphs represent the time course, where 0 [ms] indicates the start point of eye movement.

The results show that the amplitude of the signal obtained by the FICAR is sharply changed when a reference signal is set just before eye movements. The shape of the IC that is obtained when the peak of the reference signal occurred prior to an eye movements resembles the shape obtained with the ensemble averaging method (See Fig.2 and 3). The IC which has a peak just before eye movements bears a resemblance to the features of ensemble averaging in respect to the time when the potential incurs a sharp change. In the case of all subject and trials, this component is extracted. Therefore, we conclude that this pre-movement component is related to the saccade-related IC.

4.3 Extraction Rate

Next, we will determine how many of the saccade-related ICs obtained by using the FICAR. Table 2-(a) and 2-(b) represents the rate for extracting saccade-related ICs from the raw EEG data. The extraction rate is defined by ratio:
(the number of trials in which saccade-related IC are extracted)
/ (The total number oftrials).

Table 2. Extraction rate for extracting saccade-related ICs in visual and auditory experiments

(a) In visual experiments

Subject	Subject B	Subject C	Subject D	Subject E
Right / Left	Right / Left	Right / Left	Right / Left	Right / Left
60%/88%	60%/64%	52%/64%	80%/80%	88%/80%

(b) In auditory experiments

Subject A	Subject B	Subject C	Subject D	Subject E
Right / Left	Right / Left	Right / Left	Right / Left	Right / Left
92%/84%	68%/72%	76%/68%	60%/80%	52%/88%

The lowest rate was 52%. However, the rate for most of the subjects was over 60% and the highest rate was 92%. The average rate was 72.8%.

In the ensemble averaging results, a sharp change of the EEG signal is recorded each time; however, a subject had to perform the task over 20 trials. On the other hand, in the case of the FICAR, the rate for extracting saccade-related IC is below 100%. However, the saccade-related IC was extract in only two trials, and the ICA method extracted the same feature as the ensemble averaging results in a shorter time than ensemble averaging. Therefore, from the results, we find that the ICA method is more suitable for extracting saccade-related components than the ensemble averaging method. In other words, we have confirmed that ICA is potentially useful for developing BCI.

4.4 Comparison Between Saccade-Related EEG Signal and IC

In order to compare the saccade-related EEG with saccade-related IC, we focus on a time when saccade-related ICs have a maximum amplitude and maximum amplitude. Table 3 shows a time when saccade-related ICs have maximum amplitude and maximum amplitude in results of FICAR. Definition of n was the same as results of ensemble averaging. Value of each cell was calculated by averaging.

Table 3. Peak time and amplitude on sharp change of ICs

	In visual experiments		In auditory experiments	
	Peak time [ms]	$n = \frac{x-\mu}{s}$	Peak time [ms]	$n = \frac{x-\mu}{s}$
	Right/Left	Right/Left	Right/Leftt	Right/Left
A	-12.7 / -12.7	5.6 / 5.2	-13.0 / -16.0	4.8 / 4.6
B	-8.9 / -11.9	3.3 / 5.6	-19.1 / -13.1	3.7 / 3.8
C	-7.8 / -12.5	3.5 / 4.8	-13.4 / -18.6	3.7 / 3.6
D	-12.4 / -16.1	5.9 / 6.1	-7.8 / -13.8	4.5 / 5.0
E	-13.8 / -15.1	6.8 / 6.4	-7.8 / -9.7	4.5 / 5.7
Ave.	-11.5 / -13.7	5.0 / 5.6	-12.2 / -14.2	4.3 / 4.5
STD	3.0 / 1.8	1.5 / 0.6	4.7 / 3.3	0.5 / 0.9

Peak time when saccade-related ICs have maximum amplitude is from -19.1 [ms] to -7.8[ms] (Ave = -12.9, STD = 3.3) and n is from 3.3 to 6.8 (Ave. = 4.9, STD = 1.0) in Table 3. From Table 3, features of saccade-related ICs were observed before saccade and these features were observed remarkably.

Comparing results of saccade-related EEG signal with results of saccade-related ICs, Peak time when saccade-related ICs have maximum amplitude is earlier than peak time when saccade-related EEG signals have maximum amplitude. This is big advantage in the case of developing proposed BCI, the alarm of inattentive driving, and the high-speed targeting system. However amplitude calculated as n in the case of saccade-related ICs is not larger than in the case of saccade-related EEG signal. Therefore, in the point of S/N ratio, results of ensemble averaging are better than results of FICAR. However, if pre-processing is used before EEG signals are processed by ICA, S/N ratio become better in the case of ICA results.

5 Conclusion

This paper present extraction of saccade-related ICs and compared features of saccade-related EEG signals and saccade-related ICs in the point of a time when saccade-related ICs have a maximum amplitude and maximum amplitude in visual experiments and auditory experiments. Our study shows that EEG signals related to saccade can be extracted by the ICA method. The extraction rate for the saccade-relate IC was 72.8%. This rate is not high enough to apply the ICA method to signal processing for BCIs. Therefore, EEG signals must be used with pre-processing. Comparing results of saccade-related EEG signals with results of saccade-related ICs, peak time when saccade-related ICs have maximum amplitude is earlier than peak time when saccade-related EEG signals have maximum amplitude. This is very important advantage for developing our BCI. However, S/N ratio in being processed by FICAR is not improved comparing S/N ratio in being processed by ensemble averaging. In the future, we will try to obtain a higher extraction rate for extracting the saccade-related ICs and to improve S/N ratio in being processed by FICAR using by advanced ICA algorithms and pre-processing.

References

1. A. Funase, T. Yagi, Y. Kuno, Y. Uchikawa, gA Study on electro-encephalo-gram (EEG) in Eye Movementsh, Studies in Applied Electromagnetics and Mechanics,@Vol. 18, pp.709-712, 2000.
2. A. Funase, T. Yagi, A. K. Barros, Y. Kuno, Y. Uchikawa, gAnalysis on saccade-related EEG with independent component analysish, International Journal of Applied Electromagnetics and Machanics, Vol. 14, pp. 353-358, 2002.
3. A. K. Barros, R. Vigario, V. Jousmaki, N. Ohnishi, "Extraction of Event-Related Signals from Multichannel Bioelectrical Measurements", IEEE Transactions on Biomedical Engineering, Vol. 47, No. 5, pp. 583-588, 2000.
4. A. Hyvarinen, E. Oja, "Independent Component Analysis: Algorithms and Applications", Neural Networks, Vol. 13, pp. 411-430, 2000.

Cogito Componentiter Ergo Sum

Lars Kai Hansen and Ling Feng

Informatics and Mathematical Modelling,
Technical University of Denmark, DK-2800 Kgs. Lyngby, Denmark
{lkh, lf}@imm.dtu.dk
www.imm.dtu.dk

Abstract. Cognitive component analysis (COCA) is defined as the process of unsupervised grouping of data such that the ensuing group structure is well-aligned with that resulting from human cognitive activity. We present evidence that independent component analysis of abstract data such as text, social interactions, music, and speech leads to low level cognitive components.

1 Introduction

During evolution human and animal visual, auditory, and other primary sensory systems have adapted to a broad ecological ensemble of natural stimuli. This long-time on-going adaption process has resulted in representations in human and animal perceptual systems which closely resemble the information theoretically optimal representations obtained by independent component analysis (ICA), see e.g., [1] on visual contrast representation, [2] on visual features involved in color and stereo processing, and [3] on representations of sound features. For a general discussion consult also the textbook [4]. The human perceptional system can model complex multi-agent scenery. Human cognition uses a broad spectrum of cues for analyzing perceptual input and separate individual signal producing agents, such as speakers, gestures, affections etc. Humans seem to be able to readily adapt strategies from one perceptual domain to another and furthermore to apply these information processing strategies, such as, object grouping, to both more abstract and more complex environments, than have been present during evolution. Given our present, and rather detailed, understanding of the ICA-like representations in primary sensory systems, it seems natural to pose the question: *Are such information optimal representations rooted in independence also relevant for modeling higher cognitive functions?* We are currently pursuing a research programme, trying to understand the limitations of the ecological hypothesis for higher level cognitive processes, such as grouping abstract objects, navigating social networks, understanding multi-speaker environments, and understanding the representational differences between self and environment.

Wagensberg has pointed to the importance of independence for successful 'life forms' [5]

> A living individual is part of the world with some identity that tends to become independent of the uncertainty of the rest of the world

J. Rosca et al. (Eds.): ICA 2006, LNCS 3889, pp. 446–453, 2006.

Thus natural selection favors innovations that increase independence of the agent in the face of environmental uncertainty, while maximizing the gain from the predictable aspects of the niche. This view represents a precision of the classical Darwinian formulation that natural selection simply favors adaptation to given conditions. Wagensberg points out that recent biological innovations, such as nervous systems and brains are means to decrease the sensitivity to un-predictable fluctuations. An important aspect of environmental analysis is to be able to recognize event induced by the self and other agents. Wagensberg also points out that by creating alliances agents can give up independence for the benefit of a group, which in turns may increase independence for the group as an entity. Both in its simple one-agent form and in the more tentative analysis of the group model, Wagensberg's theory emphasizes the crucial importance of *statistical independence* for evolution of perception, semantics and indeed cognition. While cognition may be hard to quantify, its direct consequence, human behavior, has a rich phenomenology which is becoming increasingly accessible to modeling. The digitalization of everyday life as reflected, say, in telecommunication, commerce, and media usage allows quantification and modeling of human patterns of activity, often at the level of individuals. Grouping of events or objects in categories is fundamental to human cognition. In machine learning, classification is a rather well-understood task when based on *labelled* examples [6]. In this case classification belongs to the class of *supervised* learning problems. Clustering is a closely related *unsupervised* learning problem, in which we use general statistical rules to group objects, without a priori providing a set of labelled examples. It is a fascinating finding in many real world data sets that the label structure discovered by unsupervised learning closely coincides with labels obtained by letting a human or a group of humans perform classification, labels derived from human cognition. *We thus define cognitive component analysis (COCA) as unsupervised grouping of data such that the ensuing group structure is well-aligned with that resulting from human cognitive activity* [7]. This presentation is based on our earlier results using ICA for abstract data such as text, dynamic text (chat), web pages including text and images, see e.g., [8,9,10,11,12].

2 Where Have We Found Cognitive Components?

Text Analysis. Symbol manipulation as in text is a hallmark of human cognition. Salton proposed the so-called vector space representation for statistical modeling of text data, for a review see [13]. A term set is chosen and a document is represented by the vector of term frequencies. A document database then forms a so-called term-document matrix. The vector space representation can be used for classification and retrieval by noting that similar documents are somehow expected to be 'close' in the vector space. A metric can be based on the simple Euclidean distance if document vectors are properly normalized, otherwise angular distance may be useful. This approach is principled, fast, and language independent. Deerwester and co-workers developed the concept of latent semantics based on principal component analysis of the term-document

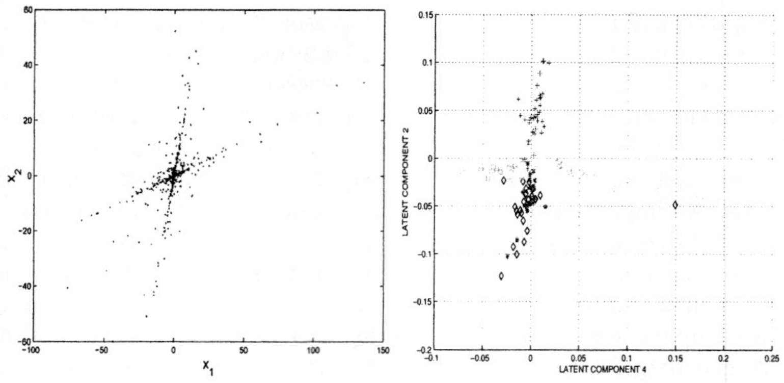

Fig. 1. Generic feature distribution produced by a linear mixture of sparse sources (left) and a typical 'latent semantic analysis' scatter plot of principal component projections of a text database (right). The characteristics of a sparse signal is that it consists of relatively few large magnitude samples on a background of small signals. Latent semantic analysis of the so-called MED text database reveals that the semantic components are indeed very sparse and does follow the laten directions (principal components). Topics are indicated by the different markers. In [16] an ICA analysis of this data set post-processed with simple heuristic classifier showed that manually defined topics were very well aligned with the independent components. Hence, constituting an example of cognitive component analysis: Unsupervised learning leads to a label structure corresponding to that of human cognitive activity.

matrix [14]. The fundamental observation behind the latent semantic indexing (LSI) approach is that similar documents are using similar vocabularies, hence, the vectors of a given topic could appear as produced by a stochastic process with highly correlated term-entries. By projecting the term-frequency vectors on a relatively low dimensional subspace, say determined by the maximal amount of variance one would be able to filter out the inevitable 'noise'. Noise should here be thought of as individual document differences in term usage within a specific context. For well-defined topics, one could simply hope that a given context would have a stable core term set that would come out as a eigen 'direction' in the term vector space. The orthogonality constraint of co-variance matrix eigenvectors, however, often limits the interpretability of the LSI representation, and LSI is therefore more often used as a dimensional reduction tool. The representation can be post-processed to reveal cognitive components, e.g., by interactive visualization schemes [15]. In Figure 1 (right) we indicate the scatter plot of a small text database. The database consists of documents with overlapping vocabulary but five different (high level cognitive) labels. The 'ray'-structure signaling a sparse linear mixture is evident.

Social Networks. The ability to understand social networks is critical to humans. Is it possible that the simple unsupervised scheme for identification of independent components could play a role in this human capacity? To investigate this issue we have initiated an analysis of a well-known social network of

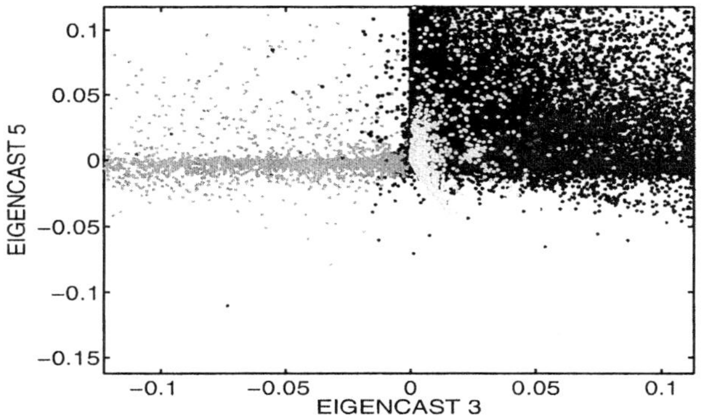

Fig. 2. The so-called actor network quantifies the collaborative pattern of 382.000 actors participating in almost 128.000 movies. For visualization we have projected the data onto principal components (LSI) of the actor-actor co-variance matrix. The eigenvectors of this matrix are called 'eigencasts' and they represent characteristic communities of actors that tend to co-appear in movies. The network is extremely sparse, so the most prominent variance components are related to near-disjunct sub-communities of actors with many common movies. However, a close up of the coupling between two latent semantic components (the region $\sim (0,0)$) reveals the ubiquitous signature of a sparse linear mixture: A pronounced 'ray' structure emanating from (0,0). The ICA components are color coded. We speculate that the cognitive machinery developed for handling of independent events can also be used to locate independent sub-communities, hence, navigate complex social networks.

some practical importance. The so-called *actor network* is a quantitative representation of the co-participation of actors in movies, for a discussion of this network, see e.g., [17]. The observation model for the network is not too different from that of text. Each movie is represented by the *cast*, i.e., the list of actors. We have converted the table of the about $T = 128.000$ movies with a total of $J = 382.000$ individual actors, to a sparse $J \times T$ matrix. For visualization we have projected the data onto principal components (LSI) of the actor-actor co-variance matrix. The eigenvectors of this matrix are called 'eigencasts' and represent characteristic communities of actors that tend to co-appear in movies. The sparsity and magnitude of the network means that the components are dominated by communities with very small intersections, however, a closer look at such scatter plots reveals detail suggesting that a simple linear mixture model indeed provides a reasonable representation of the (small) coupling between these relative trivial disjunct subsets, see Figure 2. Such insight may be used for computer assisted navigation of collaborative, peer-to-peer networks, for example in the context of search and retrieval.

Musical Genre. The growing market for digital music and intelligent music services creates an increasing interest in modeling of music data. It is now feasible to estimate consensus musical genre by *supervised* learning from rather short music segments, say 5-10 seconds, see e.g., [18], thus enabling computerized handling of music request at a high cognitive complexity level. To understand the possibilities and limitations for unsupervised modeling of music data we here visualize a small music sample using the latent semantic analysis framework. The intended use is for a music search engine function, hence, we envision that a largely text based query has resulted in a few music entries, and the algorithm is going to find the group structure inherent in the retrieval for the user. We represent three tunes (with human genre labels: `heavy, jazz, classical`) by their spectral content in overlapping small time frames (w = 30msec, with an overlap of 10msec, see [18], for details). To make the visualization relatively independent of 'pitch', we use the so-called mel-cepstral representation (MFCC, K = 13 coefficients pr. frame). To reduce noise in the visualization we have further 'sparsified' the amplitudes. PCA provided unsupervised latent semantic dimensions and a scatter plot of the data on the subspace spanned by two such dimensions is shown in Figure 3. For interpretation we have coded the data points with signatures of the three genres involved. The ICA ray structure is striking, however, we note that the situation is not one-to-one as in the small text

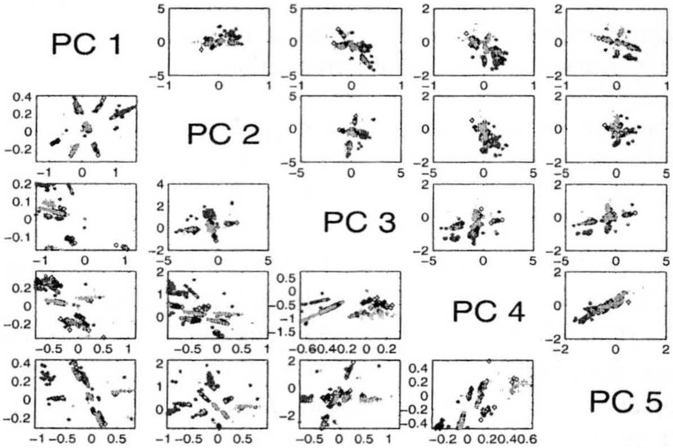

Fig. 3. We represent three music tunes (genre labels: `heavy metal, jazz, classical`) by their spectral content in overlapping small time frames (w = 30msec, with an overlap of 10msec, see [18], for details). To make the visualization relatively independent of 'pitch', we use the so-called mel-cepstral representation (MFCC, K = 13 coefficients pr. frame). To reduce noise in the visualization we have 'sparsified' the amplitudes. This was achieved simply by keeping coefficients that belonged to the upper 5% magnitude percentile. The total number of frames in the analysis was $F = 10^5$. Latent semantic analysis provided unsupervised subspaces with maximal variance for a given dimension. We show the scatter plots of the data of the first 1-5 latent dimensions. The scatter plots below the diagonal have been 'zoomed' to reveal more details of the ICA 'ray' structure. For interpretation we have coded the data points with signatures of the three genres involved: classical (∗), heavy metal (diamond), jazz (+). The ICA ray structure is striking, however, note that the situation is not one-to-one (ray to genre) as in the small text databases. A component (ray) quantifies a characteristic musical 'theme' at the temporal level of a frame (30msec), i.e., an entity similar to the 'phoneme' in speech.

databases. A component quantifies a characteristic 'theme' at the temporal scale of a frame (30msec), it is an issue for further research whether genre *recognition* can be done from the salient themes, or we need to combine more than one theme to reach the classification performance obtained in [18].

Phonemes as Cognitive Components of Speech. There is a strong recent interest in representations and methods for computational auditory scene analysis, see e.g., Haykin and Chen's review on the cocktail party problem [19]. Low level cognitive components of speech encompass language specific features such as phonemes and speaker's voice prints. Such features can be considered 'pre-semantic' and would be recognized by human cognition without comprehension of the spoken message. We have recently investigated such low-level features and found generalizable features using ICA representations [20,21], here we give an example of such analysis based on four simple utterances s, o, f, a. We analysed 40 msec windows of length (95% overlap). The windows were represented by 16 Mel-cepstrum coefficients. After variance normalization the features were sparsified based on energy zeroing windows of normalized magnitudes with a statistical $z < 1.7$. This threshold process retains 55from original features. LSI/PCA was performed on the sparsified feature coefficients to get the most variant PCA components. The results in figure 4 seem to indicate that cognitive components corresponding to the phoneme /ae/ which opens the utterances s and f, can be identified using linear component analysis. For more details on such analysis see [20,21].

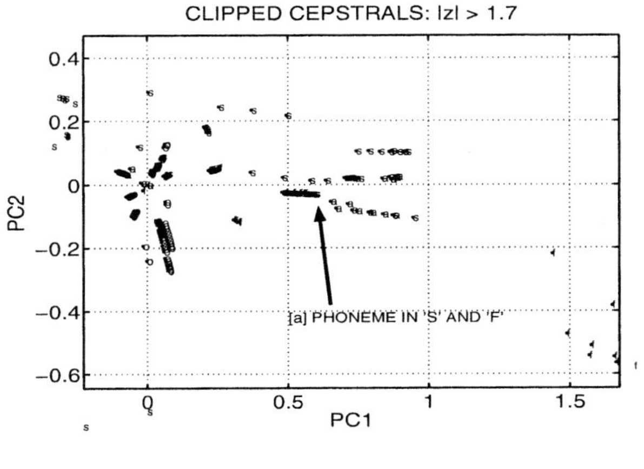

Fig. 4. Four simple utterances s, o, f, a were analysed. We analysed 40 msec windows of length (95% overlap). The windows were represented by 16 Mel-cepstrum coefficients. After variance normalization the features were sparsified based on energy zeroing windows of normalized magnitudes with a statistical $z \leq 1.7$. This threshold process retains 55% of the power in the original features. LSI/PCA was then performed on the sparsified feature coefficients for visualization. The results seem to indicate that generalizable cognitive components corresponding to the phoneme /ae/ opening the utterances s and f, can be identified using linear component analysis.

3 Conclusion

Cognitive component analysis (COCA) has been defined as the process of unsupervised grouping of data such that the ensuing group structure is well-aligned with that resulting from human cognitive activity. It is well-established that information theoretically optimal representations, similar to those found by ICA, are in use in several information processing tasks in human and animal perception. By visualization of data using latent semantic analysis-like plots, we have shown that independent components analysis is also relevant for representing semantic structure, in text and other abstract data such as social networks, musical features, and speech. We therefore speculate that the cognitive machinery developed for analyzing complex perceptual signals from multi-agent environments may also be used in higher brain function. Hence, we hypothesize that independent component analysis –given the right representation– may be a quite generic tool for COCA.

Acknowledgments

This work is supported by the Danish Technical Research Council, through the framework project 'Intelligent Sound', www.intelligentsound.org (STVF No. 26-04-0092). We thank our coworkers in the project for providing data for this presentation.

References

1. Bell, A.J., Sejnowski, T.J.: The 'independent components' of natural scenes are edge filters. Vision Research **37** (1997) 3327–3338
2. Hoyer, P., Hyvrinen, A.: Independent component analysis applied to feature extraction from colour and stereo images. Network: Comput. Neural Syst. **11** (2000) 191–210
3. Lewicki, M.: Efficient coding of natural sounds. Nature Neuroscience **5** (2002) 356–363
4. Hyvarinen, A., Karhunen, J., Oja, E.: Independent Component Analysis. John Wiley & Sons (2001)
5. Wagensberg, J.: Complexity versus uncertainty: The question of staying alife. Biology and philosophy **15** (2000) 493–508
6. Bishop, C.: Neural Networks for Pattern Recognition. Oxford University Press, Oxford (1995)
7. Hansen, L.K., Ahrendt, P., Larsen, J.: Towards cognitive component analysis. In: AKRR'05 -International and Interdisciplinary Conference on Adaptive Knowledge Representation and Reasoning, Pattern Recognition Society of Finland, Finnish Artificial Intelligence Society, Finnish Cognitive Linguistics Society (2005) Best paper award AKRR'05 in the category of Cognitive Models.
8. Hansen, L.K., Larsen, J., Kolenda, T.: On independent component analysis for multimedia signals. In: Multimedia Image and Video Processing. CRC Press (2000) 175–199

9. Hansen, L.K., Larsen, J., Kolenda, T.: Blind detection of independent dynamic components. In: IEEE International Conference on Acoustics, Speech, and Signal Processing 2001. Volume 5. (2001) 3197–3200
10. Kolenda, T., Hansen, L.K., Larsen, J.: Signal detection using ICA: Application to chat room topic spotting. In: Third International Conference on Independent Component Analysis and Blind Source Separation. (2001) 540–545
11. Kolenda, T., Hansen, L.K., Larsen, J., Winther, O.: Independent component analysis for understanding multimedia content. In et al. H.B., ed.: Proceedings of IEEE Workshop on Neural Networks for Signal Processing XII, Piscataway, New Jersey, IEEE Press (2002) 757–766 Martigny, Valais, Switzerland, Sept. 4-6, 2002.
12. Larsen, J., Hansen, L., Kolenda, T., Nielsen, F.: Independent component analysis in multimedia modeling. In ichi Amari et al. S., ed.: Fourth International Symposion on Independent Component Analysis and Blind Source Separation, Nara, Japan (2003) 687–696 Invited Paper.
13. Salton, G.: Automatic Text Processing: The Transformation, Analysis, and Retrieval of Information by Computer. Addison-Wesley (1989)
14. Deerwester, S.C., Dumais, S.T., Landauer, T.K., Furnas, G.W., Harshman, R.A.: Indexing by latent semantic analysis. JASIS **41** (1990) 391–407
15. Landauer, T.K., Laham, D., Derr, M.: From paragraph to graph: latent semantic analysis for information visualization. Proc Natl Acad Sci **101** (2004) 5214–5219
16. Kolenda, T., Hansen, L.K., Sigurdsson, S.: Independent components in text. In: Advances in Independent Component Analysis. Springer-Verlag (2000) 229–250
17. Barabasi, A.L., Albert, R.: Emergence of scaling in random networks. Science **286** (1999) 509–512
18. Ahrendt, P., Meng, A., Larsen, J.: Decision Time Horizon For Music Genre Classification Using Short Time Features. In: EUSIPCO, Vienna, Austria (2004) 1293–1296
19. Haykin, S., Chen, Z.: The cocktail party problem. Neural Computation **17** (2005) 1875–1902
20. Feng, L., Hansen, L.K.: Phonemes as short time cognitive components. In: Submitted for ICASSP'06. (2005)
21. Feng, L., Hansen, L.K.: On low level cognitive components of speech. In: International Conference on Computational Intelligence for Modelling (CIMCA'05). (2005)

Kernel Independent Component Analysis for Gene Expression Data Clustering

Xin Jin, Anbang Xu, Rongfang Bie, and Ping Guo

Department of Computer Science, Beijing Normal University,
Beijing 100875, China
xinjin796@126.com, anbangxu@mail.bnu.edu.cn
rfbie@bnu.edu.cn, pguo@bnu.edu.cn

Abstract. We present the use of KICA to perform clustering of gene expression data. Comparison experiments between KICA and two other methods, PCA and ICA, are performed. Three clustering algorithms, including weighted graph partitioning, *k*-means and agglomerative hierarchical clustering, and two similarity measures, including Euclidean and Pearson correlation, are also evaluated. The results indicate that KICA is an efficient feature extraction approach for gene expression data clustering. Our empirical study showed that clustering with the components instead of the original variables does improve cluster quality. In particular, the first few components by KICA capture most of the cluster structure. We also showed that clustering with components has different impact on different algorithms and different similarity metrics. Overall, we would recommend KICA before clustering gene expression data.

1 Introduction

Monitoring tens of thousands of genes in parallel under different experimental environments or across different tissue types provides a systematic genome-wide approach to help in understanding a wide range of problems, such as gene functions in various cellular processes, gene regulations in different cellular signaling pathways, the diagnose of disease conditions, and the effects of medical treatments. There are two major methods of measurement for gene expression data at a certain point in time: the Microarray method and the SAGE method [1,3,17].

A key step in the analysis of gene expression data is the clustering of biologically relevant groups of genes or tissue samples that have similar expression patterns [2, 9]. However, one main challenge in this task is the high dimensional gene data. Feature extraction methods, whether in linear or a nonlinear form, produce preprocessing transformations of high dimensional input data, which may increase the overall performance of clustering.

Principal Component Analysis (PCA), also known as Karhunen-Loeve expansion, is a classical feature extraction and data representation technique widely used in the areas of pattern recognition. Independent Component Analysis (ICA) [11] is a relatively recent method that can be considered as an extension of PCA. ICA is a general-purpose statistical method that originally arose from the study of blind source separation (BSS). Another application of ICA is unsupervised feature extraction,

J. Rosca et al. (Eds.): ICA 2006, LNCS 3889, pp. 454 – 461, 2006.

where the aim is to linearly transform the input data into uncorrelated components, along which the distribution of the sample set is the least Gaussian. However ICA is a linear method in nature, so it is inadequate to describe the complex nonlinear variations. The kernel trick is one of the crucial tricks for machine learning. Its basic idea is to project the input data into a high-dimensional implicit feature space F with a nonlinear mapping at first, and then the data are analyzed in F space so that nonlinear relations of the input data can be described.

In this paper, we will illustrate the potential of Kernel-ICA for gene expression data clustering. This paper is organized as follows: In section 2, three feature extraction methods, PCA, ICA and KICA, are described. Section 3 focuses on three clustering algorithms and two similarity measures. In section 4, we describe the gene expression dataset, performance measures and the results. Finally, the main conclusions are presented in section 5.

2 Components Based Feature Extraction

PCA: Principal Component Analysis (PCA) is a classical multivariate data analysis method that is useful in linear feature extraction [8,12]. One simple approach to PCA is to use singular value decomposition (SVD). Let us denote the data covariance matrix by $R_x(0) = E\{x(t)x^T(t)\}$. Then the SVD of $R_x(0)$ gives $R_x(0) = UDU^T$, where $U = [U_s, U_n]$ is the eigenvector matrix (i.e. modal matrix) and D is the diagonal matrix whose diagonal elements correspond to the eigenvalues of $R_x(0)$ (in descending order). Then the PCA transformation from m-dimensional data to n-dimensional subspace is given by choosing the first n column vectors, i.e., n principal component vector y is given by $y = U_s^T x$.

ICA: Independent Component Analysis (ICA) first performs the dimensionality reduction by data sphering (whitening) which project the data onto its subspace as well as normalizing its variance. In other words, the data sphering transformation Q is given by $Q = D_s^{-1/2} U_s^T$, The whitened vector $z \in R^n$ is given by $z = Qx$. The orthogonal factor V in ICA can be found by minimizing the mutual information in z. The natural gradient in orthogonality constraint [13] or relative gradient (EASI algorithm) [14] leads to the learning algorithm that has the form:

$$\Delta V = \eta_t \{I - y y^T - \varphi(y) y^T + y \varphi^T(y)\} V \tag{1}$$

where $y = Vz$ and $\varphi(y)$ is an elementwise non-linear function whose i_{th} element is given by:

$$\varphi_i(y_i) = -\frac{d \log p_i(y_i)}{dy_i} \tag{2}$$

where $\{p_i(\cdot)\}$ are the probability density functions of sources. Then the ICA transformation W' is given by $y = W'x$, where $W' = VQ$. Since we do not know the probability density functions of sources in advance, we have to rely on the hypothesized density functions. The ICA [15] adopts a generalized Gaussian density which is able to approximate all kinds of unimodal distributions.

KICA: KICA is the kernel counterpart of ICA [10]. Let the inner product be implicitly defined by the kernel function k in F with associated transformation Φ. Now we will extend nonlinearly the centering and whitening of the data.

Centering in F: We shift the data $\Phi(x_i)$ (i=1,...,k) with its mean $E(\Phi(x))$, to obtain data $\Phi'(x_i) = \Phi(x_i) - E(\Phi(x))$ with a mean of 0.

Whitening in F: The goal of this step is to find a transformation matrix $\hat{\mathbf{Q}}$ such that the covariance matrix of the samples $\hat{\Phi}(x_i) = Q\Phi'(x_i)$ (i=1,...,k) is a unit matrix.

Transformation of test vectors: For an arbitrary test vector $z \in X$ the Kernel-ICA transformation can be made using $z^* = \hat{\mathbf{W}}\hat{\mathbf{Q}}\Phi(z) = \hat{\mathbf{W}}\mathbf{A}_k(X, z)$. Here $\hat{\mathbf{W}}$ denotes the orthogonal transformation matrix we obtained as the output from the iterative section of ICA, while $\hat{\mathbf{Q}}$ is the matrix obtained from kernel centering and whitening. Practically speaking, Kernel-ICA = Kernel-Centering + Kernel-Whitening + ICA. Selecting an appropriate kernel function for a particular application area is very important. In this paper, we adopt cosine kernel for gene expression data.

3 Clustering Algorithms

We used three clustering algorithms in our empirical study: the hierarchical average-link algorithm, the k-means algorithm (with random initialization, 20 times for each experiment) and weighted graph partitioning.

Hierarchical Clustering: Hierarchical clustering techniques produce a nested sequence of partitions, with a single, all-inclusive cluster at the top and singleton clusters of individual points at the bottom [16]. Each intermediate level can be viewed as combining two clusters from the next lower level (or splitting a cluster from the next higher level). There are two basic approaches to generating a hierarchical clustering: Agglomerative and Divisive. Agglomerative algorithm is more common, and this is the technique that we will use. We summarize the traditional agglomerative hierarchical clustering procedure as follows.

1. Compute the similarity between all pairs of clusters, i.e., calculate a similarity matrix whose ij^{th} entry gives the similarity between the ith and j^{th} clusters.
2. Merge the most similar (closest) two clusters.
3. Update the similarity matrix to reflect the pairwise similarity between the new cluster and the original clusters.
4. Repeat steps 2 and 3 until only a single cluster remains.

Hard Partitional Clustering: In contrast to hierarchical techniques, hard partitional clustering techniques create a one-level (unnested) partitioning of the data points. If k is the desired number of clusters, then partitional approaches typically find all k clusters at once. There are a number of partitional techniques, among which the k-means algorithm is mostly widely used [1]. K-means is based on the idea that a center point can represent a cluster. In particular, for k-means we use the notion of a centroid, which is the mean or median point of a group of points.

The basic k-means clustering technique is presented below.

1. Select k points as the initial centroids.
2. Assign all points to the closest centroid.
3. Recompute the centroid of each cluster.
4. Repeat steps 2 and 3 until the centroids don't change or change little.

Weighted Graph Partitioning: Weighted graph partitioning is a graph based clustering method that has the characteristics of both hierarchical and partitional techniques. The objects to be clustered can be viewed as a set of vertices V, two points X and Y (or vertices V_1 and V_2) are connected with an undirected edge of positive weight $w(X, Y)$, or $(X, Y, w(X,Y)) \in E$. The cardinality of the set of edges $|E|$ equals the number of non-zero similarities between all pairs of samples. A set of edges whose removal partitions a graph $G=(V, E)$ into k pairwise disjoint sub-graphs $G_i=(V_i, E_i)$, is called an edge separator. Our objective is to find such a separator with a minimum sum of edge weights. While striving for the minimum cut objective, the number of objects in each cluster has to be kept approximately equal. We use OPOSSUM [4], which produces balanced (equal sized) clusters from the similarity matrix using multilevel multi-constraint graph partitioning [5]. Balanced clusters are desirable because each cluster represents an equally important share of the data. However, some natural classes may not be equal size. By using a higher number of clusters we can account for multi-modal classes (e.g., XOR-problem) and clusters can be merges at a latter stage. In gene expression data clustering, sparsity can be induced by looking only at the v strongest edges or at the sub-graph induced by pruning all edges except the v nearest neighbors for each vertex. Sparsity makes this approach feasible for large datasets.

We used two similarity measures for each of the three clustering algorithms described above.

Euclidean: The Minkowski distances $L_p(X,Y)=(\Sigma_i|X_i-Y_i|^p)^{1/p}$ are commonly used metrics for clustering problems. For $p=2$ we obtain the Euclidean distance. For Euclidean space, we chose to relate distances d and similarities S using $S=pow(e^{-d},2)$. Consequently, we define Euclidean [0,1] normalized similarity as

$$S^{(E)}(X,Y) = e^{-\|X-Y\|_2^2} \tag{3}$$

which has important properties that the commonly adopted $S(X,Y)=1/(1+\| X-Y\|_2)$ lacks.

Pearson Correlation: In collaborative filtering, correlation is often used to predict a feature from a highly similar mentor group of objects whose features are known. The [0,1] normalized Pearson correlation is defined as

$$S^{(P)}(X,Y) = \frac{1}{2}\left(\frac{(X-\overline{X})\cdot(Y-\overline{Y})}{\|X-\overline{X}\|_2 \times \|Y-\overline{Y}\|_2}+1\right) \tag{4}$$

where $\overline{X}$ denotes the average feature value of X over all dimensions.

4 Case Studies

We used two standard data sets to evaluate the effectiveness of PCA, ICA and KICA for clustering gene expression.

The Leukemia dataset: The Microarray dataset, which is described in [3] and available at [18], comes from a study of gene expression in two types of acute leukemia: acute lymphoblastic leukemia (ALL) and acute myeloid leukemia (AML). Gene expression levels were measured using Affymetrix high-density oligonucleotide arrays containing 6,817 human genes. The dataset comprise 47 cases of ALL (38 ALL B-cell and 9 ALL T-cell) and 25 cases of AML.

NCBI SAGE data: The SAGE gene expression data we used is based on 52 Hs (human sapiens) SAGE brain libraries which are publicly available on the NCBI SAGE website [6]. These libraries are made of samples from human brain and fall in to four categories: Astrocytoma (11 libraries), Ependymoma (9 libraries), Glioblastoma (8 libraries) and Medulloblastoma (24 libraries). There are 64558 genes in the dataset.

4.1 Performance Measures

Mutual Information: While entropy and purity are suitable for measuring a single cluster's quality, they are biased to favor smaller clusters. Consequently, for the overall (not cluster-wise) performance evaluation, we use a symmetric measure called mutual information. Given g categories (classes) S_h ($h \in \{1,\ldots,g\}$, $X_i \in S_h \Leftrightarrow \kappa_i = h$), we use the "true" classification labels κ to evaluate the performance. Let $n_j^{(h)}$ denote the number of objects in cluster C_j that are classified to be h as given by κ. Then mutual information is defined as:

$$\Lambda^{(M)}(\kappa,\lambda) = \frac{1}{n}\sum_{j=1}^{k}\sum_{h=1}^{g} n_j^{(h)} \frac{\log\left(\frac{n_j^{(h)} n}{\sum_{i=1}^{k} n_i^{(h)} \sum_{i=1}^{g} n_j^{(i)}}\right)}{\log(k \cdot g)} \tag{5}$$

We use the mutual information criterion because it successfully captures how related the labeling and categorizations are without a bias towards smaller clusters.

Precision: Since mutual information is not intuitive, we choose another measure to evaluate the clustering performance, i.e. precision. We choose the class label that shares with most samples in a cluster as the class label. Then, the precision for each cluster is defined as:

$$\mathrm{P(A)} = \frac{1}{|\mathrm{A}|}\max(|\{\mathrm{X_i} \mid \mathrm{label}(\mathrm{X_i}){=}\mathrm{C_j}\}|) \tag{6}$$

In order to avoid the possible bias from small clusters which have very high precision, the final precision is defined by the weighted sum of the precision for all clusters, as shown in the following equation:

$$\mathrm{P} = \sum_{k=1}^{G} \frac{|\mathrm{A_k}|}{\mathrm{N}} \mathrm{P(A_k)} \tag{7}$$

4.2 Results and Discussion

Here are the overall results from our empirical study

1. Feature extraction does improve the performance of high dimensional clustering.
2. The performance of clustering results (i.e., mutual information and precision) on the KICA transformed data is higher than PCA, ICA and than on the original data.
3. In most cases, the first few IC's do give the highest performance.

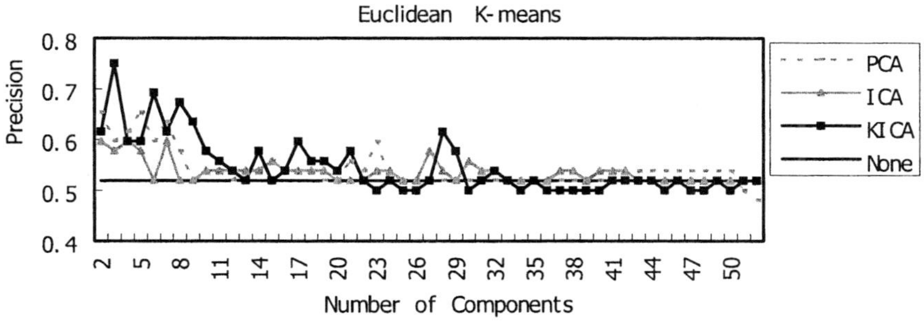

Fig. 1. Feature extraction results on the SAGE data using K-means as the clustering algorithm and Euclidean as the similarity metric

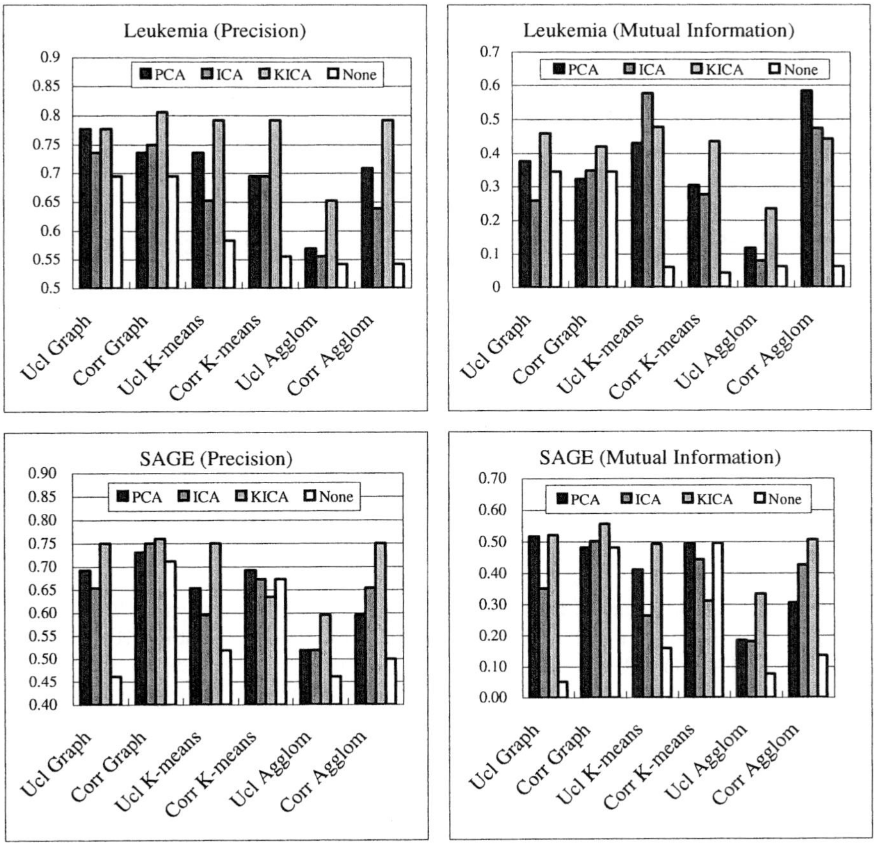

Fig. 2. Feature extraction results on the Leukemia dataset and SAGE dataset using three clustering algorithms: Graph (weighted graph partitioning), *K*-mean and Agglom (agglomerative hierarchical clustering) with two similarity measures: Ucl (Euclidean) and Corr (Pearson Correlation)

Fig.1 shows the results on the SAGE data using *k*-means as the clustering algorithm and Euclidean as the similarity metric. The best clustering performance is achieved by KICA when the first three components are used. The figure shows that clustering with the first few components instead of the original data can help extract the clustering in the data. Other clustering algorithms and similarity metrics have similar results on both SAGE and Microarray datasets

Fig. 2 shows the best performance of PCA, ICA, KICA and None-feature-extraction on both data sets using three clustering algorithms with two similarity measures. Clustering based on extracted features is better than clustering on the original data. Among the three feature extraction methods, KICA is the best.

5 Conclusions

This paper presents the use of KICA to perform clustering of gene expression data obtained from two major techniques: Microarray and SAGE. A number of different clustering algorithms (hierarchical average-link clustering, *k*-means and weighted graph partitioning) and similarity measures (Euclidean and Pearson Correlation) are compared. The results (using Mutual Information and Precision as performance measures) show that KICA based clustering performs better than those clustered by the PCA, ICA and the raw data.

Acknowledgement

The research work described in this paper was supported by grants from the National Natural Science Foundation of China (Project No. 60275002, No. 60273015 and No. 10001006).

References

1. Cai Li, Huang Haiyan, Blackshaw Seth, Liu Jun, Cepko Connie, Wong Wing: Clustering Analysis of SAGE Data Using a Poisson Approach. Genome Biology, 5:R51 (2004)
2. Eisen, M.B., et al.: Cluster Analysis and Display of Genome-wide Expression Patterns. Proc. Natl. Acad. Sci. USA 95: 14863-14868 (1998)
3. Golub TR, et al.: Molecular Classification of Cancer: Class Discovery and Class Prediction by Gene Expression Monitoring. Science 286:531–537. doi: 10.1016/S0378-4371(00)00404-0 (1999)
4. Alexander Strehl and Joydeep Ghosh: Value-based Customer Grouping from Large Retail Data-sets. In Proc. SPIE Conference on Data Mining and Knowledge Discovery, Orlando, volume 4057, pp. 33-42. SPIE, April (2000)
5. George Karypis and Vipin Kumar: A Fast and High Quality Multilevel Scheme for Partitioning Irregular Graphs. SIAM J. Sci. Comput., 20(1):359-392 (1998)
6. NCBI SAGE website: http://www.ncbi.nlm.nih.gov/SAGE (2005)
7. I. S. Dhillon, D. S. Modha: Concept Decomposition for Large Sparse Text Data using Clustering, Machine Learning, 42(1):143-175 (2001)

8. Xiang-Yan Zeng, Yen-Wei Chen, Zensho Nakao and Hanqing Lu: A New Texture Feature based on PCA Maps and Its Application to Image Retrieval. IEICE Trans. Inf. and Syst., Vol.E86-D, No.5, pp.929-936 (2003)
9. J. Sander, R.T. Ng, et al.: A Methodology for Analyzing SAGE Libraries for Cancer Profiling. ACM Transactions on Information Systems, 23(1) 35-60 (2005)
10. A. Kocsor and J. Csirik: Fast Independent Component Analysis in Kernel Feature Spaces. In Proc. of SOFSEM, Springer Verlag LNCS Series, Vol. 2234, pp. 271-281 (2001)
11. A. Hyvarinen: New Approximations of Differential Entropy for Independent Component Analysis and Projection Pursuit. In Advances in Neural Information Processing Systems, Vol. 10, pp. 273-279, MIT Press (1998)
12. K.I. Diamantaras and S.Y. Kung: Principal Component Neural Networks: Theory and Applications. John Wiley & Sons, INC (1996)
13. S. Amari: Natural Gradient for Over- and Under-complete Bases in ICA. Neural Computation, 11(8), 1875-1883 (1999)
14. J.F. Cardoso and B.H. Laheld: Equivariant Adaptive Source Separation. IEEE Trans. Signal Processing, 44(12), 3017-3030, Dec (1996)
15. S. Choi, A. Cichocki and S. Amari: Flexible Independent Component Analysis. Journal of VLSI Signal processing, 26, 25-38, Aug (2000)
16. A. El-Hamdouchi and P. Willet: Comparison of Hierarchic Agglomerative Clustering Methods for Document Retrieval, the Computer Journal, Vol. 32, No. 3 (1989)
17. V. E. Velculescu, L. Zhang, B. Vogelstein, and K. W. Kinzler: Serial Analysis of Gene Expression. Science, 270:484 (1995)
18. Cancer Genomics Publications Datasets: http://www-genome.wi.mit.edu/cgi-bin/cancer/datasets.cgi (2005)

Topographic Independent Component Analysis of Gene Expression Time Series Data

Sookjeong Kim and Seungjin Choi

Department of Computer Science,
Pohang University of Science and Technology,
San 31 Hyoja-dong, Nam-gu, Pohang 790-784, Korea
{koko, seungjin}@postech.ac.kr

Abstract. Topographic independent component analysis (TICA) is an interesting extension of the conventional ICA, which aims at finding a linear decomposition into approximately independent components with the dependence between two components is approximated by their proximity in the topographic representation. In this paper we apply the topographic ICA to gene expression time series data and compare it with the conventional ICA as well as the independent subspace analysis (ISA). Empirical study with yeast cell cycle-related data and yeast sporulation data, shows that TICA is more suitable for gene clustering.

1 Introduction

Microarray technology allows us to measure expression levels of thousands of genes simultaneously, producing gene expression profiles that are useful in discriminating cancer tissues from healthy ones or in revealing biological functions of certain genes. Successive microarray experiments over time, produces gene expression time series data. Main issues in these experiments (over time), are to detect cellular processes underlying regulatory effects, to infer regulatory networks, and ultimately to match genes with associated biological functions.

Linear model-based methods explicitly describe expression levels of genes as linear functions of common hidden variables which are expected to be related to distinct biological causes of variations such as regulators of gene expression, cellular functions, or responses to experimental treatments. Such linear model-based methods include principal component analysis (PCA) [1], factor analysis [2] independent component analysis (ICA) [3,4], and independent subspace analysis (ISA) [5,6]. Standard clustering methods (such as k-means and hierarchical clustering) assign a gene (involving various biological functions) to one of clusters, however linear model-based methods allow the assignment of such a gene to null, single, or multiple clusters.

In the context of bioinformatics, Liebermeister [4] showed that expression modes and their influences, extracted by ICA, could be used to visualize the samples and genes in lower-dimensional space and a projection to expression

J. Rosca et al. (Eds.): ICA 2006, LNCS 3889, pp. 462–469, 2006.

modes could highlight particular biological functions. In addition, ICA was successfully applied to gene clustering [7,8]. ISA [9] is a generalization of ICA where invariant feature subspace is incorporated with multidimensional ICA, allowing components in the same subspace to be dependent but requiring independence between feature subspace. It was shown in [5,6] that ISA is more useful in gene clustering and gene-gene interaction analysis, compared to ICA.

Topographic independent component analysis (TICA) is a further generalization of ISA, which aims at finding a linear decomposition into approximately independent components with the dependence between two components is approximated by their proximity in the topographic representation [10]. In other words, TICA incorporates some nonlinear dependency into a linear model, which is more suitable for gene expression time series data where there might exist some dependency between expression modes. In this paper we apply TICA to gene expression time series data and compare it with the conventional ICA as well as the independent subspace analysis (ISA). Empirical study with yeast cell cycle-related data and yeast sporulation data, shows that TICA is more suitable for gene clustering.

2 Methods: ICA, ISA, TICA

2.1 ICA

ICA is a statistical method that decomposes a multivariate data into a linear sum of non-orthogonal basis vectors with basis coefficients being statistically independent. The simplest form of ICA consider the linear generative model where the data matrix $\boldsymbol{X} = [X_{ij}]$ (where the element X_{ij} represents the expression level of gene i associated with the jth sample, $i = 1, \ldots, m$, $j = 1, \ldots, N$) is assumed to be generated by

$$\boldsymbol{X} = \boldsymbol{S}\boldsymbol{A}, \tag{1}$$

where $\boldsymbol{S} \in \mathbb{R}^{m \times n}$ is a matrix consisting of latent variables (or encoding variables) and the row vectors of $\boldsymbol{A} \in \mathbb{R}^{n \times N}$ are basis vectors corresponding to *linear modes* in [4].

Given a matrix $\boldsymbol{X} \in \mathbb{R}^{m \times N}$, its row and column vectors are denoted by $\boldsymbol{x}_i$, $i = 1, \ldots, m$ and by $\boldsymbol{x}^j$, $j = 1, \ldots, N$. Throughout this paper, we assume that the data matrix $\boldsymbol{X}$ is already whitened. In other words, the row vectors of $\boldsymbol{A}$ are confined to be orthogonal each other and to be normalized to have unit norm. Non-orthogonal factor is reflected in a whitening transform. In order to avoid an abuse of notations, we use the notation $\boldsymbol{X}$ for the whitened data matrix and $n \leq N$ represents an intrinsic dimension estimated by PCA.

ICA searches for a parameter matrix $\boldsymbol{W} \in \mathbb{R}^{n \times n}$ which maximizes the normalized log-likelihood $\mathcal{L}_{ica}$ given by

$$\mathcal{L}_{ica} = \frac{1}{m} \sum_{t=1}^{m} \sum_{i=1}^{n} \log p\left(\langle \boldsymbol{w}_i, \boldsymbol{x}_t \rangle\right) + \log \left|\det \boldsymbol{W}\right|, \tag{2}$$

where $\langle \cdot, \cdot \rangle$ denotes the inner product between two arguments. The estimated parameter matrix $\boldsymbol{W}$ leads us to calculate the latent variable matrix by $\boldsymbol{S} = \boldsymbol{X}\boldsymbol{W}^T$ ($\boldsymbol{W}^T = \boldsymbol{A}^{-1}$). For an orthogonal matrix $\boldsymbol{A}$, the row vectors of $\boldsymbol{W}$ coincide with the row vectors of $\boldsymbol{A}$.

2.2 ISA

In contrast to ICA, multidimensional ICA [11] assumes that latent variables $s_i = \langle \boldsymbol{w}_i, \boldsymbol{x} \rangle$ are divided into J number of κ-tuples (where κ represents the dimension of subspace) and find a linear decomposition such that J κ-tuples are independent with allowing the components in the same tuple to be dependent. For the sake of simplicity, we assume identical dimension, κ for every feature subspace. ISA [9] incorporates the invariant feature subspace into the multidimensional ICA. To this end, the pooled energy $E_j(\boldsymbol{x})$ for the jth feature subspace $\mathcal{F}_j$ is defined by

$$E_j(\boldsymbol{x}) = \sum_{i \in \mathcal{F}_j} \langle \boldsymbol{w}_i, \boldsymbol{x} \rangle^2 . \tag{3}$$

With these definitions, the normalized log-likelihood $\mathcal{L}_{isa}$ of the data given the ISA model, is given by

$$\mathcal{L}_{isa} = \frac{1}{m} \sum_{t=1}^{m} \sum_{j=1}^{J} \log p \left(\sum_{i \in \mathcal{F}_j} \langle \boldsymbol{w}_i, \boldsymbol{x}_t \rangle^2 \right) + \log |\det \boldsymbol{W}| , \tag{4}$$

where $p\left(\sum_{i \in \mathcal{F}_j} s_i^2\right) = p_j(s_i, i \in \mathcal{F}_j)$ represents the probability density inside the jth κ-tuple of s_i.

The parameter matrix $\boldsymbol{W}$ which maximizes the log-likelihood (4), finds a linear decomposition such that pooled energies $E_j(\boldsymbol{x})$ are independent but the components $s_i \in \mathcal{F}_j$ are allowed to be dependent. Learning $\boldsymbol{W}$ can be carried out by a gradient-ascent method. More details on ISA can be found in [9].

2.3 TICA

TICA is a further generalization of ISA, which aims at finding a linear decomposition into approximately independent components with the dependence between two components is approximated by their proximity in the topographic representation [10].

The following normalized log-likelihood $\mathcal{L}_{tica}$ was considered for TICA,

$$\mathcal{L}_{tica} = \frac{1}{m} \sum_{t=1}^{m} \sum_{j=1}^{n} \Psi \left(\sum_{i=1}^{n} h(i,j) \langle \boldsymbol{w}_i, \boldsymbol{x}_t \rangle^2 \right) + \log |\det \boldsymbol{W}| , \tag{5}$$

where $\Psi(\cdot)$ is a function of local energies that plays a similar role to the log-density in the conventional ICA and $h(i,j)$ is a neighborhood function. See [10] for more details.

3 Experiments and Results

3.1 Datasets

Our experiments were conducted with publicly available yeast cell cycle datasets [12,13] and yeast sporulation dataset [14] (see Table 1).

Table 1. Four datasets used in our experiments, are summarized. First three datasets are yeast cell cycle-related data that was also used in [12,13] and the last dataset is yeast sporulation data [14]. Experiments in yeast cell cycle-related data, are named by the method used to synchronize yeast cells. The number of open read frames (ORFs) is the number of time-series that have no missing values in them, time interval is the interval between measurements, # time points is the number of measurements, and # of eigenvectors indicated the number of eigenvectors chosen by PCA-L [15].

no	experiment	# of ORFs	time interval	# time points	# of eigenvectors
1	alpha	4579	7 min	18	6
2	cdc15	5490	10-20 min	24	8
3	cdc28	3167	10 min	17	6
4	sporulation	6118	0.5-3 hr	7	4

3.2 Procedures

Procedures that we took from a preprocessing till statistical significance test, are summarized below.

1) *Preprocessing:* The gene expression data matrix X was preprocessed such that each element is associated with $X_{ij} = \log R_{ij} - \log G_{ij}$ where R_{ij} and G_{ij} represent red and green light intensity, respectively. In practice, gene expression data usually contain missing values. We removed genes whose profiles have missing values more than 10%. Then we applied the *KNNimput* method [16], in order to fill in missing values. The data matrix was doubly centered such that each row and each column have zero mean.

2) *Data whitening:* Given the gene expression data matrix $X \in \mathbb{R}^{m \times N}$ where m is the number of genes and N is the number of arrays (time points), we chose the dimension n using the *PCA-L* method [15]. Data whitening was carried out through PCA with n principal eigenvectors.

3) *Decomposition by ICA, ISA, and TICA:* We applied ICA, ISA, and TICA algorithms, to whitened data matrix, in order to estimate the parameter matrix $W \in \mathbb{R}^{n \times n}$.

4) *Gene clustering:* For each column vector $\boldsymbol{s}^i$, genes with strong positive and negative values are grouped, which leads to two clusters related to induced and repressed genes. We considered standard deviation σ for each column vector as a threshold. Genes with expression levels higher than $c \times \sigma$ and with expression levels lower than $-c \times \sigma$, are grouped as two significant clusters. In our experiments, we chose $c = 1.5$.

5) *Statistical significance test:* To determine statistical significance of functional category enrichment for each cluster, we used the Gene Ontology (GO) annotation database [17] where genes were assigned to an associated set of functional categories. We calculated p-values for statistical significance test, using the hypergeometric distribution that is used to obtain the chance probability of observing the number of genes from a particular GO category within each cluster. The p-value is the probability to find at least k genes from a functional category within a cluster of size c:

$$p = 1 - \sum_{i=0}^{k-1} \frac{\binom{f}{i}\binom{g-f}{c-i}}{\binom{g}{c}}, \tag{6}$$

where f is the total number of genes within a functional category and g is the total number of genes within the genome [18].

3.3 Results

It was shown in [7] that ICA-based gene clustering method outperformed several existing methods such as PCA, k-means, and hierarchical clustering. However, the conventional ICA model do not take into account the temporal dependence of gene expression time series data. The inherent time dependencies in the data suggest that clustering techniques which reflect those dependencies yield improved performance. ISA and TICA-based gene clustering methods consider somewhat dependencies of gene expression patterns. Clustering results with 4 different data sets (described in Table 1), confirm that TICA and ISA indeed yield better clustering, compared to ICA (see Fig. 1).

For TICA, a square neighborhood function of size 3×3, was used. The intrinsic dimension n determined by the PCA-L for each data set is summarized in Table 1. For ISA, the number of feature subspace, was chosen as $J = 2$ for Dataset 1, 3, 4 and $J = 4$ for Dataset 2. Thus, the dimension of the feature subspace is $\kappa = 3$ for Dataset 1, 3, 4 and $\kappa = 2$ for Dataset 2.

For each data set, we determined n latent variables by ICA, ISA, and TICA and investigate the biological coherence of $2n$ clusters consisting of genes with significantly high and low expression levels within independent components. For each cluster, we calculated p-values and considered only p-values less than 10^{-5}. Scatter plots of the negative logarithm of p-value, are shown in Fig. 1.

The TICA decomposition of the gene expression data matrix of rank n, leads to n temporal modes (corresponding to n basis vectors). Each temporal mode defines two gene clusters that show a strong positive or negative response. These clusters contain subgroups related to particular biological functions, mostly consistent with the temporal modes. Fig. 2 depicts 3 temporal modes during sporulation. The temporal modes mainly reflect the sporulation behavior (see also Table 2).

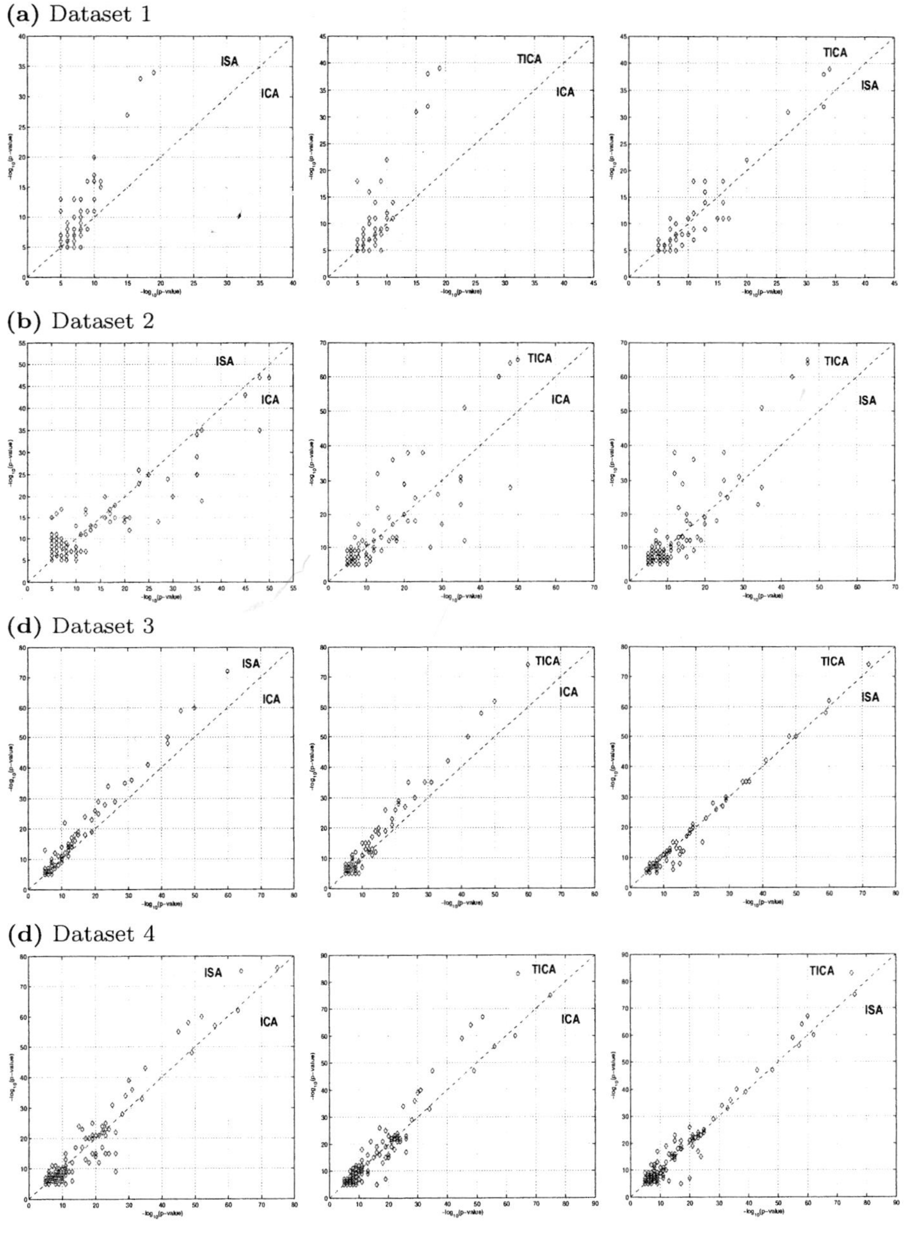

Fig. 1. Performance comparison of three different ICA methods (ICA, ISA, and TICA) with yeast cell cycle-related data and yeast sporulation data (see Table 1). Through Dataset 1-4, TICA has more points above the (anti-diagonal) line representing equal performance, compared to ICA and ISA, which indicates the enrichment of the TICA-based clustering.

Table 2. Temporal modes extracted by TICA from Dataset 4 (sporulation). The modes were characterized according to functionally related clusters.

Mode	Induced functions	Repressed functions
1	sporulation, spore wall assembly,	alcohol metabolism, carbohydrate metabolism, oxidoreductase activity
2	ribosome biogenesis and assembly, rRNA processing, rRNA metabolism, cytosolic ribosome, ribosome, structural constituent of ribosome	organic acid metabolism, carboxylic acid metabolism, amine metabolism
3	sulfur metabolism, cytosolic ribosome (sensu Eukarya), ribosome	cell cycle, cell proliferation, nuclear division, chromosome

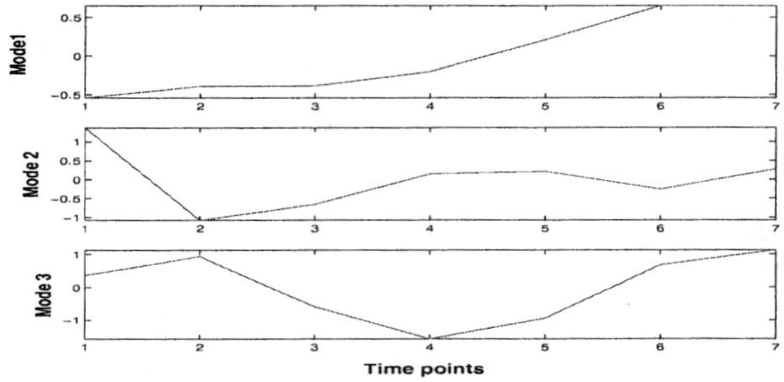

Fig. 2. Temporal modes of clusters computed by TICA are shown for Dataset 4 (sporulation)

4 Conclusions

In this paper we have applied the method of topographic ICA to gene expression time series data, in order to evaluate its performance in the task of gene clustering. Empirical comparison to the conventional ICA and the independent subspace analysis, have shown that the topographic ICA is more suitable in grouping genes into clusters containing genes associated with similar functions.

Acknowledgments. This work was supported by National Core Research Center for Systems Bio-Dynamics and POSTECH Basic Research Fund.

References

1. Raychaudhuri, S., Stuart, J.M., Altman, R.B.: Principal components analysis to summarize microarray experiments: Application to sporulation time series. In: Proc. Pacific Symp. Biocomputing. (2000) 452–463
2. Girolami, M., Breitling, R.: Biologically valid linear factor models of gene expression. Bioinformatics **20** (2004) 3021–3033

3. Hori, G., Inoue, M., Nishimura, S., Nakahara, H.: Blind gene classification based on ICA of microarray data. In: Proc. ICA, San Diego, California (2001)
4. Liebermeister, W.: Linear modes of gene expression determined by independent component analysis. Bioinformatics **18** (2002) 51–60
5. Kim, H., Choi, S., Bang, S.Y.: Membership scoring via independent feature subspace analysis for grouping co-expressed genes. In: Proc. Int'l Joint Conf. Neural Networks, Portland, Oregon (2003)
6. Kim, H., Choi, S.: Independent subspaces of gene expression data. In: Proc. IASTED Int'l Conf. Artificial Intelligence and Applications, Innsbruck, Austria (2005)
7. Lee, S., Batzoglou, S.: ICA-based clustering of genes from microarray expression data. In: Advances in Neural Information Processing Systems. Volume 16., MIT Press (2004)
8. Kim, S., Choi, S.: Independent arrays or independent time courese for gene expression data. In: Proc. IEEE Int'l Symp. Circuits and Systems, Kobe, Japan (2005)
9. Hyvärinen, A., Hoyer, P.O.: Emergence of phase and shift invarianct features by decomposition of natural images into independent feature subspaces. Neural Computation **12** (2000) 1705–1720
10. Hyvärinen, A., Hoyer, P., Inki, M.: Topographic independent component analysis. Neural Computation **13** (2001) 1525–1558
11. Cardoso, J.F.: Multidimensional independent component analysis. In: Proc. ICASSP, Seattle, WA (1998)
12. Cho, R.J., Campbell, M.J., Winzeler, E.A., Steinmetz, L., Conway, A., Wodicka, L., Wolfsberg, T.G., Gabrielian, A.E., Landsman, D., Lcokhart, D.J., Davis, R.W.: A genome-wide transcriptional analysis of the mitotic cell cycle. Mol. Cell **2** (1998) 65–73
13. Spellman, P.T., Sherlock, G., Zhang, M.Q., Iyer, V.R., Anders, K., Eisen, M.B., Brown, P.O., Botstein, D., Futcher, B.: Comprehensive identification of cell cycle-regulated genes of the yeast *saccharomyces cerevisiae* by microarray hybridization. Molecular Biology of the Cell **9** (1998) 3273–3297
14. Chu, S., DeRisi, J., Eisen, M., Mulholland, J., Botstein, D., Brown, P.O., Herskowitz, I.: The transcriptional program of sporulation in budding yeast. Science **282** (1998) 699–705
15. Minka, T.P.: Automatic choice of dimensionality for PCA. In: Advances in Neural Information Processing Systems. Volume 13., MIT Press (2001)
16. Troyanskaya, O., Cantor, M., Sherlock, G., Brown, P., Hastie, T., Tibshirani, R., Botstein, D., Altman, R.B.: Missing value estimation methods for DNA microarrays. Bioinformatics **17** (2001) 520–525
17. Gene Ontology Consortium: Creating the gene ontology resource: Design and implementation. Genome Research **11** (2001) 1425–1433
18. Tavazoie, S., Hughes, J.D., Campbell, M.J., Cho, R.J., Church, G.M.: Systematic determination of genetic network architecture. Nature Genetics **22** (1999) 281–285

Blind Source Separation of Cardiac Murmurs from Heart Recordings

Astrid Pietilä[1], Milad El-Segaier[2], Ricardo Vigário[1], and Erkki Pesonen[2]

[1] Neural Networks Research Centre, Helsinki University of Technology, Finland
Astrid.Pietila@tkk.fi
[2] Department of Paediatrics, Division of Paediatric Cardiology, Lund University Hospital, Sweden

Abstract. A significant percentage of young children present cardiac murmurs. However, only one percent of them are caused by a congenital heart defect; others are physiological. Auscultation of the heart is still the primary diagnostic tool for judging the type of cardiac murmur. An automated system for an initial recording and analysis of the cardiac sounds could enable the primary care physicians to make the initial diagnosis and thus decrease the workload of the specialised health care system.

The first step in any automated murmur classifier is the identification of different components of cardiac cycle and separation of the murmurs. Here we propose a new methodological framework to address this issue from a machine learning perspective, combining Independent Component Analysis and Denoising Source Separation. We show that such a method is rather efficient in the separation of cardiac murmurs. The framework is equally capable of separating heart sounds S1 and S2 and artifacts such as voices recorded during the measurements.

1 Introduction

Recent advances in data recording technology and digital signal processing have enabled systematic collection and analysis of heart recordings [1,2,3,4]. In computer analyses, it is crucial that different components of the heart cycle are identified and some of their most prominent features are accurately timed [5,6]. The computer analysis of acoustic heart signals has appeared to be particularly sensitive, specific and cost effective in the diagnosing and evaluating congenital cardiac defects [7,8,9].

The fact that cardiac specialists are able to distinguish between the pathological and physiological murmurs indicates that there are physiological and hemodynamical principles behind the generation of the different murmur types. There are special characteristics in the cardiac murmurs that can be used by the specialists for classifying them into pathological and physiological groups.

The computer analysis has some advantages over the auscultation by specialists, though: the findings will be automatically documented, and the computer has a higher frequency sensitivity than the human ear [10].

J. Rosca et al. (Eds.): ICA 2006, LNCS 3889, pp. 470–477, 2006.

Because of the overlap between heart sounds and systolic murmur, it has been difficult to identify the beginning and the end of systolic murmurs, let alone some further details in the overlapping periods. In earlier reports, the early and late parts of systole have been removed [6,7]. Removing the early parts of systole may potentially wholly remove some murmurs. This may increase the risk of totally missing the diagnosis of certain cardiac defects. The murmurs caused by small muscular ventricular septal defect or mild semilunar valve stenosis occur early in the systole and are thereby usually overlapped by the first heart sound.

In this study, we suggest a blind source separation (BSS) method for the identification of cardiac murmurs. This solution is sought as a combination of Singular Spectrum Analysis (SSA, [11,12]), Independent Component Analysis (ICA, [13]) and Denoising Source Separation (DSS, [14]). The first two steps take into account the convolutive nature of the recordings and the intrinsic statistical characteristics of the sources, while producing a good set of filtering templates for DSS. In this way, we isolate murmurs from the main cardiac sounds, conventionally labelled as S1 and S2. This is achieved even in presence of some overlap between the murmur and S1 or S2, without the need for truncation of potentially relevant information. We also show that we can extract measuring artifacts such as doctor or patient voices, captured by the recording apparatus.

2 Data and Methods

2.1 The Data

The cardiac sounds were recorded from nine patients with known congenital cardiac defects, aged from 3 to 17 years, the median being six years. The length of the recordings varied between 45 and 50 seconds. A PC-based, six-channel recording system (cf. [15]) developed at the Helsinki University of Technology was used. The cardiac sounds were recorded at right parasternal intercostal spaces 2 and 3, left parasternal intercostal spaces 2, 3 and 4, and at the cardiac apex, all in a supine position. The signal was amplified by 40 dB in order to compensate for the signal intensity attenuation, due to the acoustic properties of the chest wall, such as subcutaneous fat. The amplified signal was digitised with 16 bits resolution and 11.025 kHz sampling frequency. Customised software was written for recording and monitoring of the sound signals. The recordings were supplied to the first author without the information about the type of cardiac defects they contained. Thus, the processing of the data sets was done in a completely blind manner.

2.2 Independent Component Analysis and Deconvolution

In BSS, we assume that a set of observations is generated via an instantaneous linear mixing of underlying source signals, following the standard model: $\mathbf{x} = \mathbf{A}\mathbf{s}$. There, $\mathbf{x}$ is a vector with the observations, $\mathbf{s}$ are the underlying source signals, and $\mathbf{A}$ expresses the mixing process. In order to solve the BSS problem, a set of general assumptions needs to be made, either on the sources or on the mixing.

ICA, one of the most widely used tools to estimate the BSS solutions, assumes the sources to be statistically independent. An additional assumption is the non-Gaussian distribution of those sources. Algorithms performing ICA can be based on concepts such as negentropy, maximum likelihood, or mutual information (for further details on the theory, as well as historical background of ICA, cf. [13]).

There is a considerable amount of algorithms capable of performing ICA. Without a great loss of generality, we have used the FastICA algorithm in this study, because of its robustness and convergence speed (a Matlab™ package for FastICA can be found online in [16]). It is a fixed-point algorithm that estimates the sources by maximizing an approximation of their negentropy.

Although instantaneous ICA has been successfully applied to many biomedical signal processing problems, its model does not fully apply to the current data. In fact, because of finite speed of sound and the multiple paths that cardiac sounds take to reach the microphone array, the mixture process can not be considered instantaneous, but rather convolutive. To deconvolve the recordings, we follow the suggestion in [13], and perform instantaneous ICA on the embedded recordings. With the embedding of the recordings, the number of parameters to be estimated also increases, and a dimension reduction may be needed. If the data is not sufficiently informative and the reduction is not performed, the ICA algorithm may tend to overfit the data [17].

2.3 Singular Spectrum Analysis

Singular Spectrum Analysis (SSA) [11,12] consists of the Principal Component Analysis (PCA) of an augmented data set, containing the original data and copies of it lagged by a range of time steps. Often this method is proposed for single channel analysis, but multichannel versions exist.

The eigenvectors resulting from SSA decomposition contain information of the frequency content of the data. Hence, it is often used to isolate the dynamics of a given signal into a trend, oscillatory components and noise. Another use is to build data-driven filters for the signals.

As stated earlier, embedded ICA requires a certain degree of dimension reduction in the whitening stage. Aside from the scaling, this whitening is simply SSA. Reducing the dimension of the whitened data corresponds to the removal of certain undesired frequencies.

2.4 Denoising Source Separation

Denoising Source Separation [14] is a recently proposed algorithmic framework, where source separation algorithms are constructed around denoising principles. In this way, different kinds of prior information about the sources can easily be added to solve the BSS problem. DSS can be seen as a generalization of ICA: in ICA, the additional prior information is the non-Gaussianity assumption of the sources.

The general DSS framework assumes that a pre-whitening stage has been performed, and comprises of the additional four steps:

1. Estimate one source:
   $\mathbf{s} = \mathbf{w}^T\mathbf{X}$.
   $\mathbf{X}$ holds information from all channels and samples. $\mathbf{s}$ is the source estimate and $\mathbf{w}$ is the estimate of the demixing vector.
2. Denoise the estimate:
   $\mathbf{s}^+ = \mathbf{f}(\mathbf{s})$.
   $\mathbf{f}(\mathbf{s})$ is selected according to the available information about the sources.
3. Optimize the linear mapping to the filtered estimate:
   $\mathbf{w}^+ = \mathbf{X}\mathbf{s}^{+T}$.
4. Normalization step:
   $\mathbf{w}_{new} = \frac{\mathbf{w}^+}{\|\mathbf{w}^+\|}$.

If not fully converged, or further information can be retrieved from the new estimates, one can return to the first step, and apply further filtering strategies. More sources can also be estimated by imposing orthogonalization between the estimated components. Step 3 can also be performed by a simple PCA projection of the filtered estimates.

The denoising function $\mathbf{f}(\mathbf{s})$ reflects the knowledge one has of the sought sources. When there is clear information about them, one way to perform the filtering step consists of a simple masking of the sections of interest in the data. This can be done either in time, frequency or time-frequency domains. If possible, sections in the data that are completely uninteresting are then removed from the estimate. In practice, often the best we can do is to pick sections dominated by the characteristics of interest (i.e. energy or frequency contents). The denoising procedure will enhance these characteristics, possibly revealing them even in sections previously dominated by noise. One can then design an even more refined mask for further search.

2.5 The Methodological Framework

In this study, we have used a two-dimensional masking approach for denoising in DSS. The mask was designed to be applied in spectrograms, hence aiming at both time and frequency detection of the signals of interest.

The proposed approach can be described by the following steps:

1. Preprocessing with SSA
The original signals are first embedded. The degree of embedding is 11. It is chosen based on the maximum time delay between the heart sounds and the various microphones used. The distances in children's chests are usually quite short, hence the low degree of embedding. PCA then projects the signals onto components that account for different frequency ranges in the data. In order to partially reduce noise, while rendering the ICA analysis more reliable, the lowest and highest frequency bins are pruned away with SSA.

2. Mask creation and development
FastICA with a deflation approach and third power nonlinearity is used for extracting independent components from the SSA-preprosessed data. Due to

embedding there are now 66 channels in the data. After a dimension reduction, the amount of independent components will be 24. One of the found components is chosen to be a model for the S1 mask. It often also presents strong S2 peaks. The mask is defined by lowpass-filtering the selected component in dct domain. In mask-based DSS, the function $\mathbf{f}(\mathbf{s})$ (see section 2.4) equals element-wise product of the mask and the signals.

Before going to the next step, the mask model found by ICA can be further tuned with a time-based DSS procedure.

3. Mask-based DSS in the time and frequency domains
A mask-based two-dimensional denoising will separate components that have broad frequency content, and short time span. Hence, S1 and S2 are expected to be isolated from other components. The other components are then projected back to the original signal space, now without the contamination from S1 and S2.

Naturally, as we isolate particular components, the remaining signals are free to show their characteristics and possibly dictate the design of further masks.

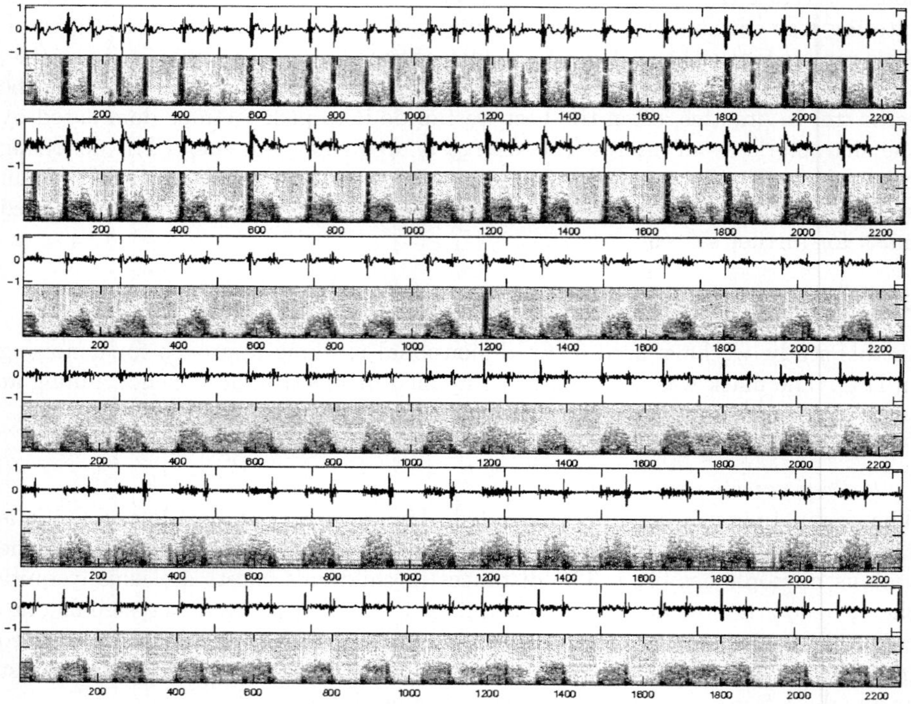

Fig. 1. Recording from one patient. All six channels are shown together with their spectrograms. The S1 and S2 are clearly visible in the first spectrogram as periodic pairs of vertical bars covering all the frequencies. Murmur is visible in the systole, between the S1 and S2, present on all six recordings.

3 Results

We can see in Fig. 1 one example of the recordings. All six channels recorded on the chest of the patient are depicted together with their spectrograms. Only the first 9 seconds of the signals are shown so the features are more clearly seen. It is easy to see the different time and frequency characteristics of the heart sounds S1 and S2, and of the murmurs. While the former are burst-like, with broad frequency content, the latter have frequencies typically over 35 Hz, and occupy a much longer time interval, typically between S1 and S2. In addition, one can see also occasional overlap between the murmur and either S1 or S2.

Figure 2 shows the results of our signal separation, targeting both S1 and S2 (Fig. 2a) or the murmurs (Fig. 2b). Note that we are capable of isolating

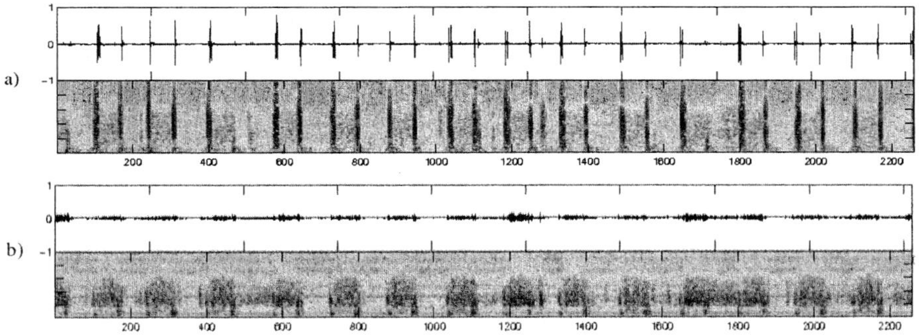

Fig. 2. Examples of results obtained from the recordings shown in Fig. 1. In a) are some estimates of heart sounds S1 and S2. In b) are extracted heart murmurs.

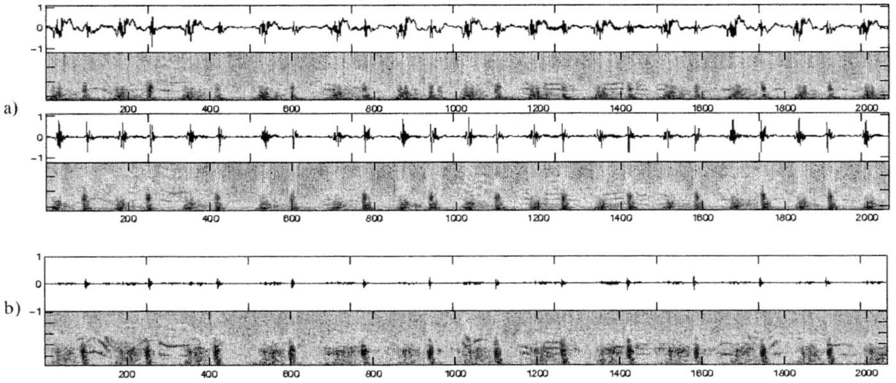

Fig. 3. In a) we show two signals from another patient together with their spectrograms. In b), the speech is separated from the recordings. The heart sounds S1 and S2 are still somewhat visible, but their relative power was decreased significantly.

each of these signals with minimal interference from the other. These signals are better evaluated acoustically, hence we recommend the visit to the internet-based additional information [18].

Finally, we show in Fig. 3 that we can also identify some artifacts present in the recordings. In Fig. 3a we show a couple of signals collected from another patient: in those recordings, clear speech contamination is present. In Fig. 3b we show the results of our signal separation. Note the clear formant structure present in the estimate, characteristic of any vowel utterance.

4 Discussion

Cardiac auscultation and murmur characterization is a highly sensitive and cost effective method in diagnosing congenital heart diseases. An automated analysis of cardiac recordings and the primary classification of murmurs to physiological and pathological would lead to a significant saving of specialized health care resources. We have tackled the first step in the separation of recorded signals into sounds of interest.

In addition to the ability to isolate sound artifacts such as the doctor and patient's voices, we have shown that DSS, with a suitable pre-processing by SSA and ICA, gives clear estimations of S1, S2 and the murmurs. Our statistical signal processing method also avoids the need for excluding parts of the signals from the analysis, in particular when murmurs partially overlap the S1 and S2. Therefore, we expect to be able to decrease the risk of missing the diagnoses of small muscular ventricular septal defect and mild semilunar valve stenosis.

Still, our methodological framework includes a manual step in which a researcher chooses the most appropriate S1 model among all components. In the future, this selection could be done automatically by using statistical measures for selecting the best S1 among the components that have first been normalized.

To conclude, many of the important murmur characteristics, such as timing and frequency content, can now be calculated and analyzed. Additional classification of the murmurs into pathological vs. physiological or septal vs. valvular groups should now be addressed. By identifying the features that discriminate between these sub-classes, one may in the future be able to also use the DSS framework in this context.

References

1. Balster DA, Chan DP, Rowland DG, Allen HD. Digital acoustic analysis of precordial innocent versus ventricular septal defect murmurs in children. Am J Cardiol 1997;79(11):1552-5.
2. Donnerstein RL. Continuous spectral analysis of heart murmurs for evaluating stenotic cardiac lesions. Am J Cardiol 1989;64(10):625-30.
3. Nygaard H, Thuesen L, Hasenkam JM, Pedersen EM, Paulsen PK. Assessing the severity of aortic valve stenosis by spectral analysis of cardiac murmurs (spectral vibrocardiography). Part I: Technical aspects. J Heart Valve Dis 1993;2(4):454-67.

4. Nygaard H, Thuesen L, Terp K, Hasenkam JM, Paulsens PK. Assessing the severity of aortic valve stenosis by spectral analysis of cardiac murmurs (spectral vibrocardiography). Part II: Clinical aspects. J Heart Valve Dis 1993;2(4):468-75.
5. Iwata A, Ishii N, Suzumura N and Ikegaya K. Algorithm for detecting the first and the second heart sounds by spectral tracking. Med Biol Eng Comput 1980;18(1):19-26.
6. El-Segaier M, Lilja O, Lukkarinen S, Sörnmo L, Sepponen R and Pesonen E. Computer-based detection and analysis of heart sound and murmur. Ann Biomed Eng 2005;33(7):937-42.
7. Thompson WR, Hayek CS, Tuchinda C, Telford JK and Lombardo JS. Automated cardiac auscultation for detection of pathologic heart murmurs. Pediatr Cardiol 2001;22(5):373-9.
8. Tavel ME and Katz H. Usefulness of a new sound spectral averaging technique to distinguish an innocent systolic murmur from that of aortic stenosis. Am J Cardiol 2005;95(7):902-4.
9. Kim D and Tavel ME. Assessment of severity of aortic stenosis through time-frequency analysis of murmur. Chest 2003;124(5):1638-44.
10. Rangayyan, RM and Lehner RJ. Phonocardiogram signal analysis: A review. Crit. Rev. Biomed. Eng. 1987;15(3):211236.
11. Vautard R and Ghil M. Singular spectrum analysis in nonlinear dynamics, with applications to paleoclimatic time series. Physica D 1989;35:395424
12. Yiou P, Sornette D and Ghil M. Data-adaptive wavelets and multi-scale singular-spectrum analysis. Physica D 2000;142(3-4):254-290.
13. Hyvärinen A, Karhunen J and Oja E. Independent Component Analysis. 1st edn. Wiley-Interscience, New York, NY.2001.
14. Särelä J, and Valpola H. Denoising Source Separation. Journal of Machine Learning Research 2005;6:233-272.
15. Lukkarinen S, Noponen A-L, Sikiö K and Angerla A. A New Phonocardiographic Recording System. Computers in Cardiology 1997;24:117-120.
16. www.cis.hut.fi/projects/ica/fastica. A FastICA Matlab package;2005
17. Särelä J, and Vigário. Overlearning in Marginal Distribution-Based ICA: Analysis and Solutions. Journal of Machine Learning Research 2003;4:1447-1469.
18. www.cis.hut.fi/astrid/publications/ICA06. More comprehensive results on the study are presented in the internet.

Derivation of Atrial Surface Reentries Applying ICA to the Standard Electrocardiogram of Patients in Postoperative Atrial Fibrillation

José Joaquín Rieta[1], Fernando Hornero[2], César Sánchez[3], Carlos Vayá[1], David Moratal[1], and Juan Manuel Sanchis[1]

[1] Bioengineering Electronic & Telemedicine. Valencia University of Technology, Carretera Nazaret–Oliva s/n, 46730, Gandía (Valencia), Spain
{jjrieta, carvasa, dmoratal, jmsanch}@eln.upv.es
[2] Cardiac Surgery Service, General University Hospital of Valencia Foundation, Avenida Tres Cruces s/n, 46014, Valencia, Spain
hornero_fer@gva.es
[3] Innovation in Bioengineering, Castilla–La Mancha University, Camino del Pozuelo s/n, 16071, Cuenca, Spain
cesar.sanchez@uclm.es

Abstract. In this study a set of patients undergoing cardiac surgery, that developed postoperative atrial fibrillation, were selected to verify if the information available on the atrial surface can be derived with the only use of body surface recordings. Standard electrocardiograms were obtained and processed by independent component analysis (ICA) to extract a unified atrial activity (AA) that takes into account the atrial contribution from each surface lead. Next, this AA has been compared with internal recordings. Main atrial frequency, cross-correlation between power spectral densities and spectral coherence have been obtained in this study. Results show that information provided by surface ICA-estimated AA allows to derive atrial surface reentries in AF patients, thus improving the noninvasive knowledge of atrial arrhythmias when internal atrial recordings are unavailable.

1 Introduction

Atrial fibrillation (AF) is the most commonly diagnosed sustained arrhythmia in clinical practice and affects up to 1% of the general population. Considering its prevalence with age, this arrhythmia affects up to 15% of the population older than 70 and has an incidence that doubles with each advancing decade [1]. There exist evidence that AF is one of the main causes of embolic events that, in 75% of the cases, develop complications associated with cerebrovascular accidents, provoking that a patient with AF has twice the risk of death than a healthy person [2]. On the other hand, AF is one of the most common complication of cardiothoracic surgery, affecting up to 60% of the patients undergoing this procedure, especially in the first days after the intervention, thus increasing morbidity, hospital stay and associated costs [3].

J. Rosca et al. (Eds.): ICA 2006, LNCS 3889, pp. 478–485, 2006.

AF occurs when the electrical impulses in the atria degenerate into a chaotic pattern, resulting in an irregular and rapid heartbeat due to the unpredictable conduction of these impulses from the atria to the ventricles [1]. Reentries are the main consequence of this chaotic wavefront propagation within the atrial tissue, involving one or more circuits caused by the continuous spreading of multiple wavelets wandering throughout the atria [1]. The fractionation of the wavefronts as they propagate results in self-perpetuating independent wavelets. On the ECG, AF is described by the replacement of P waves by fibrillatory waves that vary in size, shape and timing [2].

Independent Component Analysis (ICA) and methods related to blind signal separation (BSS) have been applied successfully to different biomedical challenges during last years. Regarding the electrocardiogram (ECG) and AF, principal component analysis (PCA) has been used to extract the atrial activity (AA) from the 12–lead surface ECG in patients with AF in order to study the effects of drug administration [4], to measure the degree of local organization of this arrhythmia [5] and to study linear ablation procedures [6]. With respect to ICA, it has also been applied for the extraction of AA in AF episodes from the surface ECG [7,8], the suppression of artifacts from internal epicardial recordings [9] and the discrimination among supraventricular arrhythmias [10].

To study AF from surface ECG, AA and ventricular activity (VA) have to be separated. One solution consists of applying ICA-based methods, which are able to use the multi-lead information provided by the ECG to obtain a unified AA [7, 8]. On the other hand, the relation between body surface and epicardial (heart's surface) atrial waveforms is of clinical interest for non-invasive assessment of electrical remodeling in AF [11]. This phenomenon is related with the progressive shortening of effective refractory periods, which involves the perpetuation of the desease [1]. Considering the aforementioned interest, it has to be demonstrated clinically if the result obtained via the ICA-based AA estimation methodology is able to offer the same (or similar) information than that provided by atrial epicardial recordings. The present work is focused on clinically assess the similarity and equivalence of both informations and, as a consequence, validate the ICA-based estimation methodology to derive the main epicardial atrial activation patterns. Obviously, this way to derive the atrial reentries from body surface recordings has the great advantage of being a noninvasive technique. In addition, this procedure may serve as the starting point of ECG and ICA-based studies where internal recordings are unavailable.

2 Selection of Patients and Recording Procedure

In this work 15 patients undergoing cardiac surgery that developed postoperative AF were selected. For these type of patients it is usual to attach a set of electrodes on the heart's surface (epicardial electrodes) during the surgery intervention. This will allow a very precise and comfortable postoperative monitoring. Two epicardial electrodes were placed on the right atrium free wall in order to follow those patients that developed AF after the surgery.

For each patient, Standard 12 lead ECGs were recorded during several minutes. Next, one segment of 8 seconds in length was selected for each patient trying to obtain the most significant AF characteristics and avoid patient movements and other disturbances. The sampling frequency was 1kHz and the resolution of the digital recording system was 16 bits, thus providing a sensibility better than 0.4μV. The 12 leads were recorded following standard procedures except for monopolar leads V3 and V4, that were connected to the atrial epicardial electrodes. In this way it was possible to record simultaneously body surface and epicardial activity with the only use of a standard 12-lead ECG system, which is the most popular equipment in clinical cardiology.

The fact of discarding V3 and V4 from surface recordings has been motivated because AA information is mainly present in leads II, aVF and V1 [1] and, in addition, it is preferred to preserve leads V2 and V5 to record 3-dimensional information of the ECG [12].

3 Methods

3.1 ICA Estimation of Reentries

Before the application of ICA, the selected segments from each ECG recording have been normalized in amplitude, notch filtered (f_n = 50Hz), to cancel out powerline interference, and high-pass filtered (f_h = 0.7Hz) to suppress base line wandering. Next, after the ICA stage, the recordings have been low-pass filtered (f_l = 70Hz) to reduce high frequency noise. The sequence of these operations is justified because preprocessing can have a significant impact on the separation performance of ICA-based AA estimation methods. Specifically, it has been proved that low-pass filtering, though itself is a linear operation, involves a data reduction that decreases the quality of the ICA-based AA estimation [13]. To solve this problem, the ICA approach has to be applied before the low-pass filtering and then, any other post-ICA processing could be performed over the data. In addition, the impact of notch and baseline wander filtering is not relevant and, therefore, can be applied before ICA in order to reduce noise before the blind separation stage. All the filtering operations have been performed using a forward–backward procedure to avoid phase distortion of the signals.

Each ECG recording has been processed with the FastICA algorithm, which is a robust and fast way of solving blind signal separation problems [14]. In addition, FastICA can operate in a deflation mode, in which the independent components are estimated one by one. Hence, the algorithm can be stopped as soon as the AA sources have been extracted, with the consequent benefit in computational complexity. There was not performed any dimensionality reduction in the whitening process before ICA computation. The use of ICA to estimate the AA from ECG surface recordings allows to take into account the atrial contribution in every lead to generate a unified signal estimate condensing the AA information [7]. This procedure of extracting the AA globally can be considered as an improvement in those situations where 12-lead ECGs are available [4,6,7].

After the ICA stage, the subgaussian character of AA as opposed to the supergaussian behaviour of VA allows its identification using a kurtosis-based source reordering [7]. In general VA presents high values within the heart beat (QRS complex) and low values in the rest of the cardiac cycle. Hence the histogram analysis of VA reveals an impulsive (supergaussian) behaviour with typical kurtosis values above 15. On the other hand, the AA of an AF episode has been modeled as a saw-tooth signal, consisting of a sinusoid with several harmonics, which behaves as a subgaussian random process [15]. As a consequence, the kurtosis-based reordering arranges first the subgaussian sources, associated to AA, then the Gaussian ones, associated to noise and artifacts, and finally the supergaussian sources, corresponding to VA. Therefore, the signals with lower kurtosis are considered as the AA, taking as the main AA source the signal with largest spectral power around the main atrial frequency [7].

As an example, Fig. 1 plots surface lead V1 (widely accepted as the lead with larger atrial contribution [1]), atrial epicardial leads, AA_{E1} and AA_{E2}, and the estimated AA via ICA from surface recordings AA_S for patient #14 in the database. As can be observed, the ICA-based estimation is able to extract the main reentry patterns observed in epicardial recordings.

After the ICA stage it was always possible to identify the AA source among the whole set of separated sources. The identification was carried out following the aforementioned steps based on reordering the sources from lower to higher kurtosis, obtaining and analyzing the power spectral density of the sources with subgaussian kurtosis and, finally, visually inspecting the fibrillatory waves in the original ECG against the estimated AA source obtained by the ICA separation.

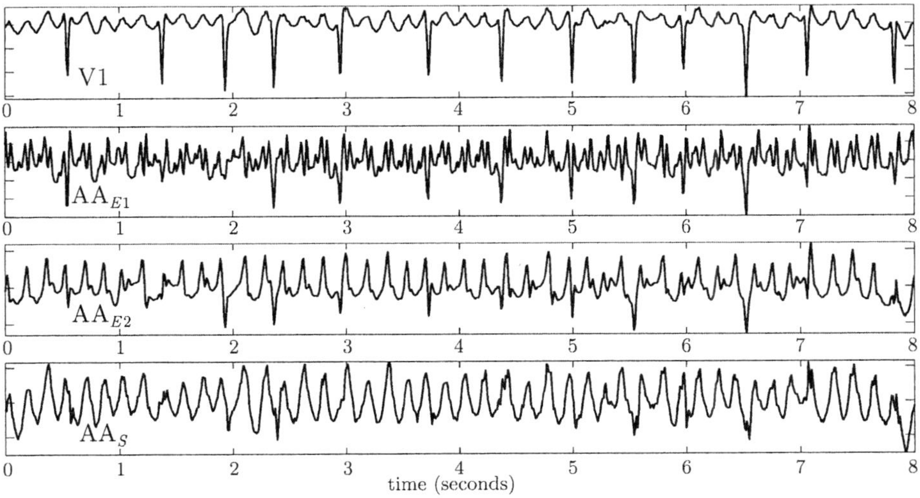

Fig. 1. Comparison between surface, epicardial and estimated atrial activities. Recording V1 comes from surface lead V1, AA_{E1} and AA_{E2} are from epicardial electrodes and AA_S is the ICA-based AA estimation from body surface electrodes.

3.2 Frequency Domain Assessment

During last years it has been usual to perform AF studies making use of the AA power spectral density to analyze the main atrial frequency and its variations with time or with drug administration [2,4,15]. This frequency is identified with the fundamental harmonic of the arial spectrum.

Through the study of the fibrillatory frequency it has been extracted relevant information about the success of electric cardioversion for AF treatment [16]. Moreover, atrial fibrillatory frequency correlates well with intraatrial cycle length, a parameter of primary importance in AF domestication and response to therapy [11]. Because of that, the comparison and estimation parameters defined in this study are focused in the frequency domain. Therefore, after the ICA stage in which the full signal bandwidth is employed, the AA frequency band is limited from 0 to 20Hz, which is a wide enough margin to include all the atrial spectral information [2,4,15].

In order to compare surface and epicardial signals, the power spectral density (PSD) has been computed using Welch–WOSA modified periodogram, with 4096-point Hamming windowing, 50% overlapping and 8192-point FFTs. As a comparison reinforcement, the atrial frequency associated to the main atrial peak has been calculated. Fig. 2 shows an example, obtained for patient #14, of epicardial (P_{EE1} and P_{EE2}) and ICA-estimated surface spectra (P_{SS}). Taking as a starting point the PSD, additional spectral parameters have been defined. The spectral cross-correlation, $|R_{P_{SS}P_{EE}}|$, is computed as

$$|R_{P_{SS}P_{EE}}| = \left| \frac{C_{P_{SS}P_{EE}}}{\sigma_{P_{SS}} \sigma_{P_{SS}}} \right| \tag{1}$$

where $C_{P_{SS}P_{EE}}$ is the covariance of P_{SS} and P_{EE}, $\sigma_{P_{SS}}$ and $\sigma_{P_{EE}}$ being the standard deviation of the PSDs. On the other hand, the spectral coherence between epicardial and surface recordings $S_{P_{SS}P_{EE}}$ has been obtained as

$$S_{P_{SS}P_{EE}} = \frac{|P_{SE}|^2}{P_{SS}P_{EE}} \tag{2}$$

where P_{SE} is the cross PSD, in other words, the averaged product between surface (P_S) and epicardial (P_E) spectra, respectively. Because the result of

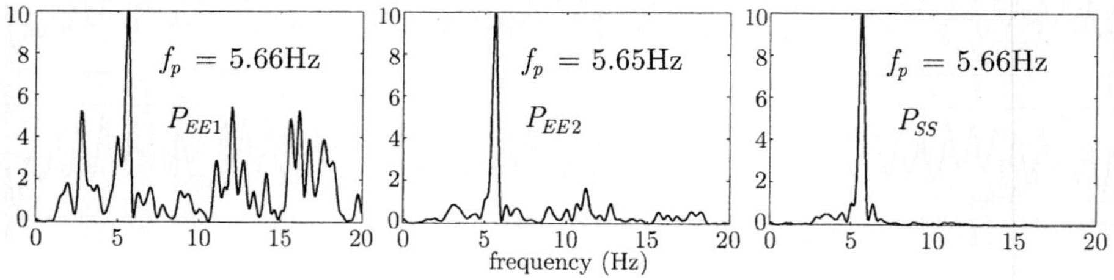

Fig. 2. Normalized power spectral density (patient #14) for atrial epicardial activities, P_{EE1} and P_{EE2}, and the body surface ICA-estimated AA, P_{SS}. It is also indicated the main atrial frequency for each case.

spectral coherence is a vector, with the same number of points than the original PSDs involved in its computation, it has been defined an index to evaluate the spectral similarity. This index has been obtained as the average value of the spectral coherence vector in the AA frequency band of interest (0–20Hz). For simplicity, this index will also be called spectral coherence from now on.

4 Results and Discussion

The exposed methodology has been applied to the patients' database. After the ICA stage, atrial and ventricular components were separated and identified satisfactory. For each patient, main atrial frequencies have been obtained both from epicardial and ICA-based recordings. In all the cases, a high degree of similarity has been observed between pairs of frequencies. The obtained set of atrial frequencies had normal values, 5.45±1.27Hz (range 3.41–7.69Hz) and the difference between epicardial and surface frequencies is negligible, the average result being 0.026±0.039Hz (range 0–0.125Hz).

To condense comfortably all the information computed for the analyzed cases, Fig. 3 plots a normalized diagram illustrating the spectral cross-correlation results ($R_{P_{SS}P_{EE}}$) and mean spectral coherence ($\bar{S}_{P_{SS}P_{EE}}$) between epicardial and surface recordings. The average value of the spectral cross-correlation has been 85.34±11.08% (range 60.57–97.31%) and for the spectral coherence 70.10±9.46% (range 54.92–83.95%). As can be observed in both indexes, results indicate a great similarity between epicardial and ICA-surface information.

With respect to the ICA-estimated AA waveform, it has been observed its regularity and the maintenance of the fundamental period associated with the dominant atrial reentry. Of course, this affirmation can be made thanks to the parallel observation of the epicardial recordings. However, some specifical comments have to be made with respect to a concrete epicardial recording.

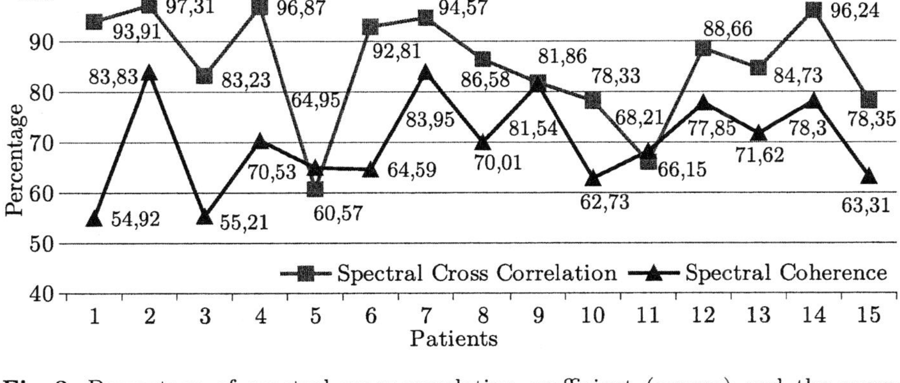

Fig. 3. Percentage of spectral cross-correlation coefficient (square) and the average spectral coherence (triangle) for the 15 patients under study

Firstly, an epicardial recording comes from an electrode attached directly to the atrial tissue, and will provide the AA present in that point. Therefore, if a concrete reentry is unable to reach the point, that AA could be hidden in the recording provided by the epicardial lead. In other cases, that reentry could be masked by other large amplitude reentries present at the same place and time instant. On the other hand, it is also possible the existence of a local reentry very close to the place where the electrode has been attached to. In this case, local reentry will be registered by the electrode but, at the same time, other epicardial electrodes attached to different places could miss or mask this activity because its locality. This fact can also be applicable to the surface ECG.

As a consequence, the signal provided by an epicardial electrode is strongly dependent both on its placement and on the concrete atrial depolarization pattern of that patient and recording instant. Therefore, epicardial signal variability could be significative, as has been shown in Fig. 1. Anyway, the most usual situation will provoke the dominant reentries to be manifested in all epicardial leads. In addition, each lead will present its local singularities. This fact is the explanation of the differences between epicardial waveforms in Fig. 1 and spectra in Fig. 2. Moreover, in this latter case, the signals (AA_{E1} and AA_{E2}) present some ventricular contamination and, hence, it is normal to observe additional frequency components than that present in the surface ICA-based AA estimation (AA_S). Finally, in all the spectra of Fig. 2 it is possible to observe the existence of a main reentry, with an atrial frequency of 5.66Hz, clearly showing the maintenance of the fundamental regularity associated to that reentry.

5 Conclusions

The present study has assessed that atrial activity estimation, using body surface recordings and ICA techniques, is a valid and useful tool to obtain information very close to that provided by epicardial recordings, thus allowing to study atrial activation patterns in AF patients. In addition, the great advantage of ICA to estimate AA is the consideration of multi-lead information and its unification to provide a single AA waveform. This information can be cardiologically relevant in pharmacologic and surgical treatment of AF. Considering the satisfactory results obtained, these techniques could become a useful and valuable tool for atrial activation pattern analysis and reentry study in patients with AF where epicardial recordings are unavailable or where their obtention needs an expensive and uncomfortable procedure.

Acknowledgements

The authors would like to thank the Cardiac Surgery Service, from the General University Hospital of Valencia (Spain), for the clinical advice provided and kind help in obtaining the signals. This work was partly funded by the research incentive program of the Valencia University of Technology and the action IIARC0/2004/249 from the Consellería de Empresa, Universidad y Ciencia.

References

[1] Fuster, V., Ryden, L.E., Asinger, R.W., Cannom, D.S., et al.: ACC/AHA/ESC guidelines for the management of patients with atrial fibrillation. Journal of the American College of Cardiology **38** (2001) 1266/I–1266/LXX

[2] Bollmann, A., Kanuru, N.K., McTeague, K.K., Walter, P.F., DeLurgio, D.B., Langberg, J.J.: Frequency analysis of human atrial fibrillation using the surface electrocardiogram and its response to ibutilide. American Journal of Cardiology **81** (1998) 1439–1445

[3] Sedrakyan, A., Treasure, T., Browne, J., Krumholz, H., et al: Pharmacologic prophylaxis for postoperative atrial tachyarrhythmia in general thoracic surgery: Evidence from randomized clinical trials. The Journal of Thoracic and Cardiovascular Surgery **129** (2005) 997–1005

[4] Raine, D., Langley, P., Murray, A., Dunuwille, A., Bourke, J.P.: Surface atrial frequency analysis in patients with atrial fibrillation: A tool for evaluating the effects of intervention. Jrnl. Card. Electrophysiology **15** (2004) 1021–1026

[5] Faes, L., Nollo, G., Kirchner, M., Olivetti, E., et al.: Principal component analysis and cluster analysis for measuring the local organisation of human atrial fibrillation. Med. Biol. Eng Comput. **39** (2001) 656–663

[6] Raine, D., Langley, P., Murray, A., Furniss, S.S., Bourke, J.P.: Surface atrial frequency analysis in patients with atrial fibrillation: Assessing the effects of linear left atrial ablation. Jrnl. of Cardiovascular Electrophysiology **16** (2005) 838–844

[7] Rieta, J.J., Castells, F., Sanchez, C., Zarzoso, V., Millet, J.: Atrial activity extraction for atrial fibrillation analysis using blind source separation. IEEE Transactions on Biomedical Engineering **51** (2004) 1176–1186

[8] Lemay, M., Vesin, J.M., Ihara, Z., Kappenberger, L.: Suppression of ventricular activity in the surface electrocardiogram of atrial fibrillation. Lecture Notes in Computer Science **3195** (2004) 1095–1102

[9] Liu, J.H., Kao, T.: Removing artifacts from atrial epicardial signals during atrial fibrillation. International Conference on Independent Component Analysis and Blind Signal Separation (ICA) **4** (2003) 179–183, Nara, Japan

[10] Rieta, J.J., Millet, J., Zarzoso, V., Castells, F., Sanchez, C., Garcia, R., Morell, S.: Atrial fibrillation, atrial flutter and normal sinus rhythm discrimination by means of blind source separation and spectral parameters extraction. IEEE Computers in Cardiology **29** (2002) 25–28, Memphis, TN

[11] Husser, D., Stridh, M., Sornmo, L., Olsson, S.B., Bollmann, A.: Frequency analysis of atrial fibrillation from the surface electrocardiogram. Indian Pacing and Electrophysiology Journal **4** (2004) 122–136

[12] Malmivuo, J., Plonsey, R.: Bioelectromagnetism: Principles and Applications of Bioelectric and Biomagnetic Fields. Oxford University Press (1995)

[13] Rieta, J.J., Sanchez, C., Sanchis, J.M., Castells, F., Millet, J.: Mixing matrix pseudostationarity and ECG preprocessing impact on ICA-based atrial fibrillation analysis. Lecture Notes in Computer Science **3195** (2004) 1079–1086

[14] Hyvarinen, A., Karhunen, J., Oja, E.: Independent Component Analysis. John Wiley & Sons, Inc. (2001)

[15] Stridh, M., Sörnmo, L., Meurling, C.J., Olsson, S.B.: Characterization of atrial fibrillation using the surface ECG: time-dependent spectral properties. IEEE Transactions on Biomedical Engineering **48** (2001) 19–27

[16] Tai, C., Chen, S., Liu, A.S., Yu, W., Ding, Y., Chang, M., Kao, T.: Spectral analysis of chronic atrial fibrillation and its relation to minimal defibrillation energy. Pace-Pacing and Clinical Electrophysiology **25** (2002) 1747–1751

Wavelet Denoising as Preprocessing Stage to Improve ICA Performance in Atrial Fibrillation Analysis

César Sánchez[1], José Joaquín Rieta[2], Carlos Vayá[2], David Moratal Perez[2], Roberto Zangróniz[1], and José Millet[2]

[1] Innovation in Bioengineering, Castilla-La Mancha University, Camino del Pozuelo s/n 16071, Cuenca, Spain
[2] Bioengineering, Electronics and Telemedicine, Valencia University of Technology, Carretera Nazaret-Oliva s/n 46730, Gandía, Spain

Abstract. Blind Source Separation (BSS) has been probed as one of the most effective techniques for atrial activity (AA) extraction in supraventricular tachyarrhythmia episodes like atrial fibrillation (AF). In these situations, a wavelet transform denoising stage can improve the extraction quality with low computational cost. Each ECG lead is processed to obtain its representation in the wavelet domain where the BSS systems improve their performance. The comparison of spectral parameters (main peak and power spectral density concentration) and statistics values (kurtosis) proves that the sparse decomposition in the wavelet domain of the observed mixtures reduces Gaussian contamination of these signals, speeds up the convergence and increase the quality of the extracted signal. The easy and fast implementation, robustness and efficiency are some of the main advantages of this technique making possible the application in real time systems as a support tool to clinical diagnostics.

1 Introduction

Atrial fibrillation (AF) is one of the most common arrhythmias and causes the highest number of admissions in casualty department of hospitals.

The prevalence of this supraventricular tachyarrhythmia is estimated at 0.4% of the general population and the median age of the AF patients is 75 years. The incidence increases with the age [1] and is slightly more common in women. About the 1% of the people above 60 years suffer from AF. These numbers raise by 6% in people above 80. AF is relatively rare in the people below 20 years. Symptoms associated with AF depend on several factors but most patients experience palpitations, chest pain, lightheadedness, presyncopes, dizziness, fatigue and dyspnea. Clearly, this arrhythmia may directly impact the quality of life. Nowadays, the treatment and analysis of AF is not completely satisfactory, and the high levels of morbidity, mortality and associated costs give rise to many scientific works and publications about this theme [2].

J. Rosca et al. (Eds.): ICA 2006, LNCS 3889, pp. 486–494, 2006.

Blind Source Separation (BSS) has been probed as one of the most effective techniques for the atrial activity (AA) extraction in supraventricular tachyarrhythmia episodes like atrial fibrillation (AF) [3], [4]. Nevertheless, a reduced number of available reference signals (leads) and the presence of noise can decrease the efficiency of these methods. Low-pass filtering as a possibility to remove noise can reduce the information in the data, since high-frequency features of the data are lost. Hence, this information reduction may involve a reduction of independence. In addition, low-pass filtering performs some kind of averaging over the data, and sums tend to increase Gaussianity, thus, decreasing performance [5]. In recent works, a denoising wavelet stage has been proved as a very efficient preprocessing technique which improves the performance of certain BSS applications [6], [7], [8]. The definition of the observed mixtures in the wavelet domain reduces the Gaussian contamination of these signals and speeds up the convergence. This paper shows how the use of the wavelet domain can give higher quality AA extraction indexes than the isolated BSS systems when 12 leads ECG registers are used. The comparison of the spectral concentration, main peak and kurtosis values justify the convenience of the proposed algorithm.

1.1 Wavelet Transform Principles

Wavelet analysis is used to transform the signal under investigation into another representation that presents the signal information in a more useful form, joining spectral and temporal analysis [9].

Mathematically speaking, the Wavelet Transform (WT) is a convolution of the wavelet function $\psi_{a,b}$ (a dilated and displaced version of the "mother wavelet" ψ, where the parameters a and b indicate scale and translation, respectively) with the signal $x(t)$.

The discrete wavelet transform (DWT) results from discretizing scale and translation parameters. The definition of parameters as $a = 2^j$ and $b = k \cdot 2^j$ leads to the dyadic DWT (DyWT), expressed as (1).

$$c(j,k) = \sum_{n \in Z} \left[x[n] \psi_{j,k}[n] \right]$$

$$\psi_{j,k}(n) = 2^{-j/2} \psi \left[2^{-j} n - k \right] \tag{1}$$

$$a = 2^j \qquad b = k 2^j \qquad (j,k) \in Z^2$$

The implementation of the DyWT is very easy and can be done with a hierarchical structure decomposition. It consists of a filter cascade banks which provide the detail and approximation coefficients in several steps of filtering and downsampling. The decomposition is obtained by iterating an M-channel filter bank on its lowpass output. The cut-off and pass bands of the resulting filters in every family of wavelet sets the final resolution quality of the system. The possibility of reconstructing the original signal from some of the obtained basic blocks without loss of information is other important advantage of this discrete transforms. This mathematical synthesis is called the Inverse Discrete Wavelet Transform (IDWT) and has the next representation in the discrete domain:

$$x[n] = \sum_{j \in Z} \sum_{k \in Z} c(j,k)\psi_{j,k}[n] \tag{2}$$

2 Database

The signals are created from recordings of an own database of 12 leads ECG, with signals obtained at the Cardiac Electrophysiology Laboratory of the University Clinical Hospital in Valencia and diagnosed by cardiologists. 35 real AF registers of 22 patients have been selected. All the registers have been pre-processed and normalized to remove possible fluctuations of the base line, interferences, etc. The sampling frequency was 1kHz and the ECG selected segments were 6 seconds in length.

3 Method

The simplest model of BSS takes on the presence of n statistically independent signals and n observed linear and instantaneous mixtures. In this work, the independence and nongaussianity of the atria and ventricles as signal sources is taken on. Recent works have studied the propagation mechanisms and uncoordinated atrial activation in AF to demonstrate this assumption [11].

The BSS model in its more compact form is given by

$$\mathbf{x}(t) = \mathbf{A} \cdot \mathbf{s}(t) \tag{3}$$

$$\hat{\mathbf{s}}(t) = \mathbf{y}(t) = \mathbf{B} \cdot \mathbf{x}(t) \tag{4}$$

where $\hat{\mathbf{s}}(t)$ is the vector of estimated sources and $\mathbf{B}$ is the separation matrix that recovers the independent sources. Taking the classic definition of the BSS problem as a starting point, Equation (3), the wavelet coefficients of each signal can be expressed as:

$$\mathbf{c}_x(j,k) = \sum \mathbf{x}(t)\psi_{j,k}(t) \tag{5}$$

$$\mathbf{c}_s(j,k) = \sum \mathbf{s}(t)\psi_{j,k}(t) \tag{6}$$

where $\mathbf{c}_x$ and $\mathbf{c}_s$ represent the wavelet coefficient vectors of the mixtures and the original sources respectively. Using a de-noising stage, some of these coefficients become zero and the decomposition can be considered as optimal if the sparse representability have to be exploited [12], [13], [14]. After that, a new formulation of the blind separation problem can be done:

$$\mathbf{c}_x(j,k) = \mathbf{A} \cdot \mathbf{c}_s(j,k) \tag{7}$$

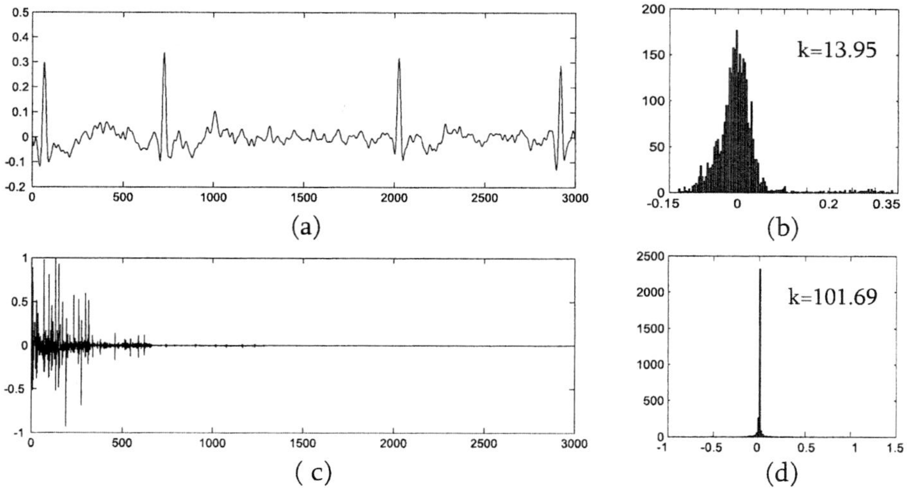

Fig. 1. Example of AF recording, histogram and kurtosis value in time domain (a-b) and wavelet domain (c-d). The wavelet domain decomposition presents statistics of the source signals less Gaussian than in the time domain.

These new mixtures present higher kurtosis values (see Figure 1) becoming less Gaussian [6]. According with (4), in the wavelet domain, the estimated sources are expressed by:

$$\hat{\mathbf{c}}_s(j,k) = \mathbf{y}_s(j,k) = \mathbf{B} \cdot \mathbf{c}_x(j,k) \qquad (8)$$

Finally, the original sources are obtained with the inverse wavelet transform of the vectors defined by (8):

$$\mathbf{s}(t) = \sum_j \sum_k \mathbf{c}_s(j,k)\psi_{j,k}(t) \qquad (9)$$

Several wavelet functions (daubechies, symlets, splines) have been tested but the results are similar in all cases. The function *biorthogonal 4.4* and five levels of decomposition are the wavelet family and the configuration with the best performance respectively. The FastICA algorithm was preferred to perform the BSS process, due to its fast convergence and robust performance, previously demonstrated in a variety of different applications [15], [16].

Spectral analysis has been used to identify the AA between the obtained signals. The signal with a main frequency peak in the band of 5-8 Hz and higher spectral concentration in this range -as it is usual in an AF episode- is identified as AA.

4 Results

Both methods have been applied to the selected registers and the final results with their mean value and standard deviation have been presented in Table 1.

Spectral parameters have been considered, for example, principal peak (*fp*) and spectral energy concentration in the band of 5-8 Hz (named as *SCBP*). The high energy concentration in this band in respect of the total energy is typical in AF episodes that present a dominant peak in this range. An increase of this concentration involves a greater purity of the extracted AA and can be considered as a quality extraction. The Welch's averaged, modified periodogram method (50% overlap, Hamming window, NFFT=8192) has been used to estimate the power spectral density (PSD). On the other hand, kurtosis is a measure of non-Gaussianity of a signal and can be used to identify the convenience of the transformation to the wavelet domain in this work.

As can be observed in Table 1, there is a clare improvement in the quality of the extracted AA. The obtained main peak values denote the high remanent presence of ventricular activity in the case of the isolated ICA method. The increase of the spectral concentration levels suggests the greater efficiency of the Wavelet-ICA algorithm.

The calculated kurtosis values from the wavelet coefficients in each lead explain the better convergence and prove the advisability of the proposed method, see Table 2. In this case, the statistics of the sources become clearly less Gaussian (greater kurtosis) than they are in the time domain. In 9 out of the 35 considered registers the isolated ICA method has presented convergence

Table 1. Spectral Parameters; main peak and spectral energy concentration in the band 5-8 Hz (mean value and standard deviation)

	FastICA	Wavelet+FastICA
Fp(Hz)	4.89±0.13	6.52±0.06
SCBP	0.21±0.08	0.36±0.10

Table 2. Kurtosis values of the mixtures in the wavelet and time domain (mean value and standard deviation)

	I	II	III
Wavelet Domain	214.3±12.3	406.1±15.1	379.2±9.3
Time Domain	19.9±0.92	12.7±1.10	12.7±1.32
	aVR	aVL	AVF
Wavelet Domain	421.8±19.3	344.2±11.2	393.2±16.4
Time Domain	14.3±0.76	12.4±0.10	7.6±2.52
	V1	V2	V3
Wavelet Domain	178.7±9.93	182.8±10.2	339.6±9.3
Time Domain	7.6±1.00	8.9±2.23	13.7±0.62
	V4	V5	V6
Wavelet Domain	429.4±11.2	500.8±10.9	400.4±12.5
Time Domain	17.7±1.09	18.1±1.99	14.9±2.65

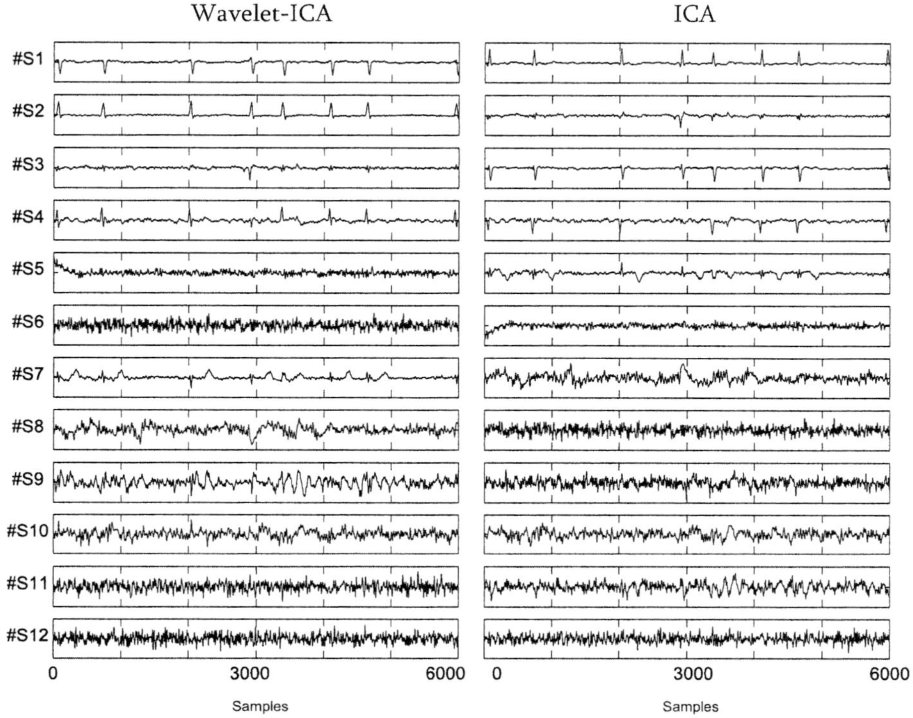

Fig. 2. Obtained sources with Wavelet-ICA and ICA method

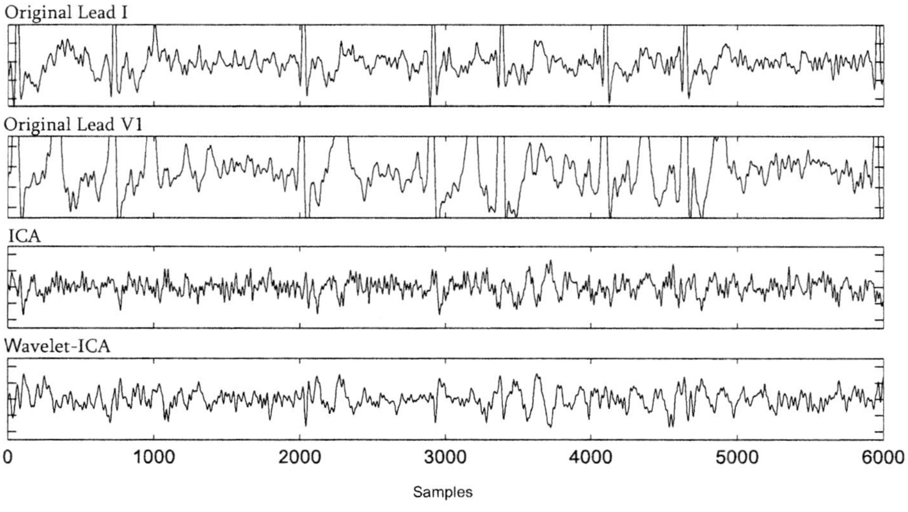

Fig. 3. Wave form comparison between extracted AA and original f waves

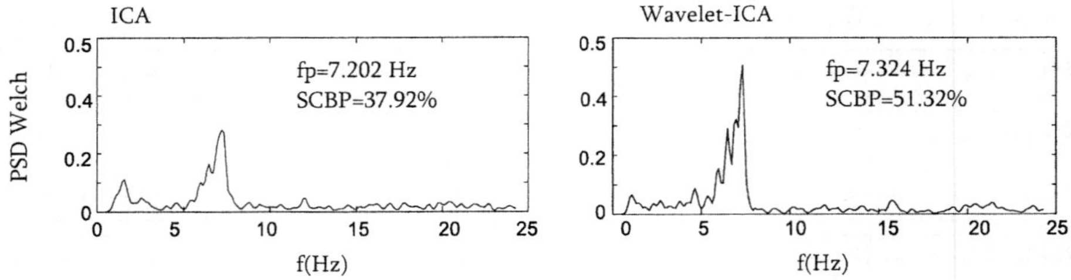

Fig. 4. Spectral comparison of extracted AA PSD

problems and high computational load (up to two minutes with a PC Pentium IV 3 GHz, 512 Mb RAM), whereas the Wavelet-ICA algorithm has presented these problems in only one register with computational time lower than 12 seconds in all cases.

Figure 2 shows the obtained signals using the isolated ICA method and the Wavelet-ICA method, respectively. Strictly speaking, movements of the heart, such as contraction of the atria and ventricles, could violate the ICA assumption of spatial stationarity of the physical sources but, in general, the authors consider that these possible variations do not significantly affect the BSS instantaneous linear mixing model for AF episodes [3]. This consideration is reinforced by the fact that results providing the estimation of the main atrial frequency of AA using this ICA-based BSS technique are the same as those obtained through the application of other accepted AA extraction techniques [11].

In Figure 3, the comparison between the wave forms in leads I and V1 and the extracted AA with the analysed methods is presented. The correllation in the f waves segments is clearly high. Figure 4 demonstrates the increase of the quality extraction using Wavelet-ICA according with the obtained PSD of each extracted AA in a selected register.

5 Conclusions

As the low pass filter used a preprocessing stage can impact the performance of ICA methods in the case of AA extraction from AF episodes, the Wavelet preprocessing method is presented as an alternative to these traditional de-noising systems. Some additional examples of the applicability of this method can be the fetal ECG extraction [6], analysis of supraventricular arrhythmias in Holter registers [7], etc.

The Gaussian feature reduction of the mixtures in the wavelet domain improves the convergence of the ICA algorithm. This increase of the computational speed makes the proposed method suitable for real time applications.

The positive results reported in this paper show that the Wavelet-ICA process presents evident advantages in contrast with the isolated ICA systems; less processing time and convergence problems and increase of quality AA extraction.

Therefore, the proposed method could be a powerful clinical diagnostics tool as a previous stage for early detection and AF classification algorithms.

Acknowledgements

This work was partly funded by several research grants from *Junta de Comunidades de Castilla-La Mancha* (Ref: PAC-05-008-1) and *Consellería de Empresa Universidad y Ciencia de la Generalitat Valenciana* (Ref: IIARC0/2004/249).

References

1. Falk R.H.: Medical progress: Atrial Fibrillation. New England Journal of Medicine. **344 (14)** (2001) 1067–1078.
2. Igbal M.B., Taneja A.K., Lip G.Y. and Flather M.: Recent developments in atrial fibrillation. BMJ. **330** (2005) 238–243
3. Rieta J.J., Castells F., Sánchez C., Zarzoso V., Millet J.: Atrial activity extraction for atrial fibrillation analysis using blind source separation. IEEE Trans Biomed Eng. **51 (7)** (2004) 1176–1186.
4. Castells F., Rieta J.J., Millet J., Zarzoso V.: Spatiotemporal blind source separation approach to atrial activity estimation in atril tachyarrhythmias. IEEE Trans Biomed Eng. **52(2)** (2005) 258–267
5. Rieta J.J., Sánchez C., Sanchis J.M., Castells F. and Millet J.: Mixing Matrix Pseudostationarity and ECG Preprocessing Impact on ICA-Based Atrial Fibrillation Analysis. Lecture Notes in Computer Science. Independent Component Analysis and Blind Signal Separation. Springer. **3195** (2004) 1079-1086.
6. Jafari M. and Chambers J.: Fetal electrocardiogram extraction by sequential source separation in the wavelet domain. IEEE Trans Biomed Eng. **52 (3)** (2005) 390–400.
7. Sánchez C., Rieta J. J., Castells F., Alcaraz R. and Millet J.: Wavelet Domain Blind Signal Separation to Analyze Supraventricular Arrhtymias from Holter Registers. Lecture Notes in Computer Science. Independent Component Analysis and Blind Signal Separation. Springer. **3195** (2004) 1111-1117.
8. Sarela J., Valpola H.: Denoising source separation. Journal of Machine Learning Research. **6** (2005) 233-272.
9. Addison P.S.: The Illustrated Wavelet Transform Handbook. Introductory Theory and Applications in Science, Engineering, Medicine and Finance. Institute of Physics Publishing. (2002).
10. Cichocki A., Amari S.: Adaptive Blind Signal and Image Processing: Learning Algorthms and Applications. John Wiley and Sons. (2003).
11. Rieta J.J., Castells F., Sánchez C., Moratal-Pérez D. and Millet J.: Bioelectric model of atrial fibrillation: Applicability of blind source separation techniques for atrial activity estimation in atrial fibrillation episodes. Computers in Cardiology. Los Alamitos, CA: IEEE. **30** (2003) 525–528
12. Zibulesvsky M., Zeevi Y.Y.: Extraction of a source from a multichannel data using sparse decomposition. Neurocomputing. **49 (1)** (2002) 163-173.
13. Zibulevsky M., Pearlmutter B. A. , Bofill P. , and Kisilev P.: Blind source separation by sparse decomposition. Independent Components Analysis: Principles and Practice (S. J. Roberts and R. M. Everson, eds.). Cambridge University Press (2001).

14. Pham D. and Cardoso J.F.: Blind separation of instantaneous mixtures of non stationary sources. IEEE Transactions on Signal Processing. **49 (9)** (2001) 18371848.
15. Sánchez C., Rieta J.J., Vayá C., Moratal D., Cervigón R., Blas J.M. and Millet J.: Atrial Activity Enhancement by Blind Sparse Sequential Separation. Computers in Cardiology. Los Alamitos, CA: IEEE. **32** (2005) In press.
16. Hyvarinen A.: Fast and robust fixed-point algorithms for independent component analysis. IEEE Trans. Neural Networks **10** (1999) 626-634.

Performance Study of Convolutive BSS Algorithms Applied to the Electrocardiogram of Atrial Fibrillation

Carlos Vayá[1], José Joaquín Rieta[1], César Sánchez[2], and David Moratal[1]

[1] Bioengineering, Electronics, Telemedicine and Medical Computer Science Research Group, Valencia University of Technology, Carretera Nazaret-Oliva s/n, 46730, Gandía (Valencia), Spain
{carvasa, jjrieta, dmoratal}@eln.upv.es

[2] Innovation in Bioengineering, Castilla-La Mancha University, Camino del Pozuelo, s/n, 16071, Cuenca, Spain
cesar.sanchez@uclm.es

Abstract. Atrial Fibrillation (AF) is one of the atrial cardiac arrythmias with highest prevalence in the elderly. In order to use the electrocardiogram (ECG) as a noninvasive tool for AF analysis, we need to separate the atrial activity (AA) from other cardioelectric signals. In this matter, Blind Source Separation (BSS) techniques are able to perform a multi-lead analysis of the ECG with the aim to obtain a set of independent sources where the AA is included. Two different assumptions on the mixing model in the human body can be done. Firstly, the instantaneous mixing model can be assumed in spite of the inaccuracy of this approximation. Secondly, the convolutive model is a more realistic model where weighted and delayed contributions in the generation of the electrocardiogram signals are considered. In this paper, a comparison between the performance of both models in the extraction of the AA in AF episodes is developed by analyzing the reults of five distinct BSS algorithms.

1 Introduction

Atrial fibrillation (AF) is one of the most commonly encountered atrial arrhythmias in routine clinical practice [1]. The analysis of the electrocardiogram (ECG) is the most extended noninvasive technique in medical treatment of AF. The exhaustive analysis of the AF requires previously to separate the atrial activity (AA) component from other cardioelectric signals like the ventricular activity (VA). The early extraction techniques worked in time domain and obtained the atrial activity by the substraction of the average QRS complex and the average T wave.[1] This family of techniques, widely used in medical applications, are only applied to the ECG lead where atrial fibrillation is more easily distinguishable,

[1] The tracing recorded from the electrical activity of the heart forms a series of waves and complexes that are labelled in alphabetical order. The depolarization of the ventricles produces the QRS complex and their repolarization causes the T wave.

J. Rosca et al. (Eds.): ICA 2006, LNCS 3889, pp. 495–502, 2006.

e.g. V1, and they do not make use of the information included in every lead. On the contrary, Blind Source Separation (BSS) techniques make a multi-lead statistical analysis with the aim to obtain a set of independent sources that include a unified AA signal [2].

The applicability of BSS techniques for AA estimation in AF episodes is justified in [3]. Firstly, in AF episodes the bioelectric sources of the heart generating AA and VA are proved to be uncoupled and statistically independent, given the uncoordinated operation of AA and VA during AF episodes. Secondly, it is also proved that both activities present a non-Gaussian behavior (AA has a subgaussian probability distribution whereas the VA is clearly supergaussian). Finally, it is demonstrated that ECG recordings are linear mixtures of bioelectric signals that depend on the position of the ECG electrodes. Therefore, the extraction of the AF from the ECG can be tackled as a BSS problem where the mixture of AA, VA, noise and some other bioelectric signals produces the registered ECG signals of every lead [3], as depicted in Fig. 1. The standard 12-leads ECG consists of three limb leads (I, II and III), three augmented limb leads (aV_R, aV_L and aV_F) and six precordial leads (from V1 to V6).

Two different mixing models of the bioelectric signals can be considered under the assumption of linearity. On the one hand, in some BSS algorithms the instantaneous mixture of the cardioelectric sources in the human body is assumed. For instance, the FastICA algorithm [4], which has already been applied to the extraction of the AA from ECGs of AF episodes, is based on the instantaneous mixing model. Therefore, FastICA assumes that no propagation delay of the cardioelectric signals exists in the human body. The error introduced by FastICA as a consequence of assuming instantaneous mixtures resulted to be negligible and quite good results in the extraction of the AA have been obtained [3]. On the other hand, in convolutive BSS (CBBS) algorithms the more realistic case of weighted and, besides, delayed contributions in the generation of the observed signals is considered. CBSS algorithms have not been applied yet to the extraction of the AA. The main objective of this essay is to test the performance of convolutive BSS algorithms in the extraction of atrial activity. Therefore, a comparison between the performance of FastICA and CBSS algortihms in the extraction of the AA in AF episodes is developed in this work.

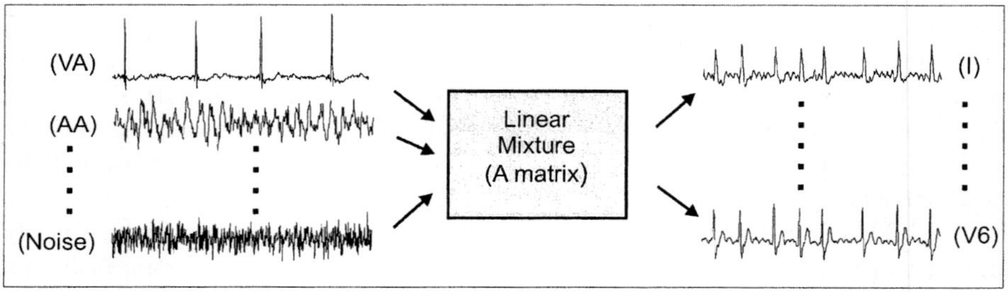

Fig. 1. Generation of the standard 12-leads ECG as the linear mixture of atrial activity (AA), ventricular activity (VA), and other independent cardioelectric sources

2 Tested Algorithms

Four CBSS algorithms have been selected as those that optimize the separation of audio sources in reverberant spaces (i.e. convolutive mixtures) [5], and they are briefly described in the following subsections.

2.1 MBLMS

Lambert [6] has obtained a cost function for the LMS (Least-Mean Square) algorithm that can be extended to the multiple inputs case:

$$\Psi = trace \quad E\left\{(\mathbf{y} - g((\mathbf{s}))(\mathbf{y} - g(\mathbf{s})^H\right\} \tag{1}$$

The algorithm that use the previous cost function is called the Multi-channel Blind Least-Mean Square (MBLMS) algorithm. With the cost function described above, the update of the adaptative filter weights can be expressed as a function that only depends on the observed signals and the pdf of the original sources [6].

2.2 TDD

Ikeda and Murata[7] proposed a BSS algorithm that extends the application of the TDD (Time-Delayed Decorrelation) algorithm to the case of convolutive mixtures of signals. This algorithm uses the windowed-Fourier transform (i.e. the spectrogram) to transform mixed source signals into the time-frequency domain. Then, the TDD algorithm is applied to the signals of each frequency independently.

One main difficulty of BSS algorithms that work in time-frequency domain is the ambiguities of amplitude and permutation. In [7] this is solved by using the inverse of the decorrelating matrices and the develop of signals. On one the hand, the problem of amplitude ambiguity is solved by putting back the separated independent components to the sensor input with the inverse matrices $\mathbf{B}(w)$. On the other hand, the problem of permutation is solved by the similarity among envelopes. As a result, the algorithm proposed by Ikeda and Murata obtains separated spectrograms $S_i(f,t)$ which inverse Fourier transforms yield the estimated original time-domain signal sources $s_i(t)$[7].

2.3 Infomax

Asano et al [8] proposed to combine the Infomax principle of maximizing the output entropy with two array signal processing techniques to enhance the performance of blind separation. The first technique is based on the subspace method for reducing the effect of reflections and noise. The second technique is a new method for solving the permutation in the frequency domain. In this method the coherency of the mixing matrix in adjacent frequencies is utilized, which is termed Inter-Frequency Coherency (IFC). We will henceforth refer this combination of techniques as the Infomax algorithm [8].

2.4 CoBliSS

Schobben and Sommen worked in the BSS problem applied to the separation of multiple speakers in a room using multiple microphones [9]. They presented a new BSS algorithm that was entirely based on Second Order Statistics (SOS), which was entitled 'Convolutive Blind Signal Separation (CoBliSS)' algorithm. In this algorithm, the optimization is done by minimizing the cross-correlation among the outputs of the multichannel separating filter. This criterion is transformed to the frequency domain in order to achieve a computationally inexpensive algorithm with fast convergence. The filter coefficients are calculated in the frequency domain such that the cross-correlation become equal to zero. An iterative method is proposed in which the weights are adjusted iteratively in alternately one and the other domain [9].

3 Notation

The performance of the algorithms is measured by using two parameters. On the one hand, SIR_{AA} measures the performance as an improvement of a signal to interference ratio. Considering x_i as the observation with the highest contribution of AA, the signal to interference ratio of x_i is defined as [5]:

$$SIR_{AA}^{o} = 10 \log \frac{\mathbf{E}\{(h_{ij} * s_j)^2\}}{\mathbf{E}\{(\sum_{\substack{k=1 \\ k \neq j}}^{N} h_{ik} * s_k)^2\}} \tag{2}$$

where h_{ij} are the FIR filters of the $\mathbf{A}$ mixing matrix. In the same way, considering that $\hat{s_p}$ is the estimated source with the highest contribution of AA, the signal to interference ratio of $\hat{s_p}$ is [5]:

$$SIR_{AA}^{e} = 10 \log \frac{\mathbf{E}\{(g_{pj} * s_j)^2\}}{\mathbf{E}\{(\sum_{\substack{k=1 \\ k \neq p}}^{N} g_{pk} * s_k)^2\}} \tag{3}$$

where g_{pk} are the FIR filters of the $\mathbf{G}$ global system matrix so that $\mathbf{G} = \mathbf{W} * \mathbf{A}$, and $\mathbf{G}$ is the estimated separation matrix. Finally, by using logarithmic units the SIR_{AA} is defined as:

$$SIR_{AA} = SIR_{AA}^{e} - SIR_{AA}^{o} \tag{4}$$

On the other hand, we also measure the performance of the extraction as a cross-correlation between the original AA and the estimated AA [4]:

$$R_{AA} = \frac{\mathbf{E}\{s_{AA} \cdot \hat{s}_{AA}\}}{\sqrt{\mathbf{E}\{s_{AA}^2\}\mathbf{E}\{\hat{s}_{AA}^2\}}} \tag{5}$$

where s_{AA} and $\hat{s}_{AA}$ are the original and the estimated AA respectively.

4 ECG Database

The calculation of the parameters defined in the previous section needs the original sources and the mixing matrix to be known. Given that all of them are unknown in the case of real ECGs, we have established two environments of synthesized AF ECGs, so that the measure parameters can be calculated.

In the first environment, 15 pairs of separated AA and VA recordings of AF ECG episodes are convolutively mixed by aleatory $\mathbf{A}$ mixing matrices in which FIR filters length has been changed from 1 to 8. All recordings are 12 seconds long and were obtained at a sampling rate (f_s) of 1 kHz. The length of the filters of the $\mathbf{W}$ separation matrix is an adjustable parameter in the CBSS algorithms. It has been changed in the tests from 2 to 32. The value of two is the lowest value allowed by all the tested CBSS algorithms. The value of 32 has been chosen so that the controlled maximum length of the g_{ij} filters is lower than 40 (40 ms in duration), accordingly to the aforementioned maximum value of N_m[6].

In the second environment, AF 12-leads ECG are synthesized by adding AA and VA of every lead, previously separated from real AF ECGs by using QRST cancellation:

$$\mathbf{x} = \mathbf{x}_{AA} + \mathbf{x}_{AV} \tag{6}$$

where $\mathbf{x}_{AA}$ is a matrix that contains the 12 atrial signals, $\mathbf{x}_{AV}$ is a matrix that contains the 12 corresponding ventricular signals and $\mathbf{x}$ is the 12 leads synthesized ECG. All resulting ECG recordings last for 8 seconds and are sampled at 1 KHz. This second environment comprises 20 synthesized ECGs. The maximum value of N_s has been fixed to 128 by considering 128 ms as a reasonable maximum propagation delay for all the bioelectric signals in the human body [10].

5 Results

In Fig. 2 mean values of SIR_{AA} obtained in the first environment are presented for different lengths of the mixing matrix filters (N_m). Four different values of N_m (1, 2, 4 and 8) have been tested. This figure shows the results of the four aforementioned CBSS algorithms. Also the results obtained by FastICA are included . This was made to compare ICA and CBSS methods. We can appreciate that FastICA SIR_{AA} mean values are higher than MBLMS SIR_{AA} mean values for any value of N_m. More specifically, maximum FastICA SIR_{AA} mean values are around 40 dB whereas MBLMS SIR_{AA} mean values are always lower than 5 dB. In other words, the application of MBLMS to mixtures of AA an VA does not yield any source signal separation. On the contrary, the values obtained by TDD are much better than values obtained by MBLMS. In fact, SIR_{AA} mean values of FastICA are only around 10 dB higher than TDD SIR_{AA} mean values for instantaneous mixtures ($N_m = 1$). Furthermore, both values decrease and tend to be equal when N_m increases. Indeed, TDD SIR_{AA} exceeds FastICA SIR_{AA} when N_m equals to 8. Infomax algorithm presents a similar behavior to TDD, that is, both SIR_{AA} and R_{AA} are quite lower than the respective FastICA values for instantaneous mixtures and much more similar when N_m

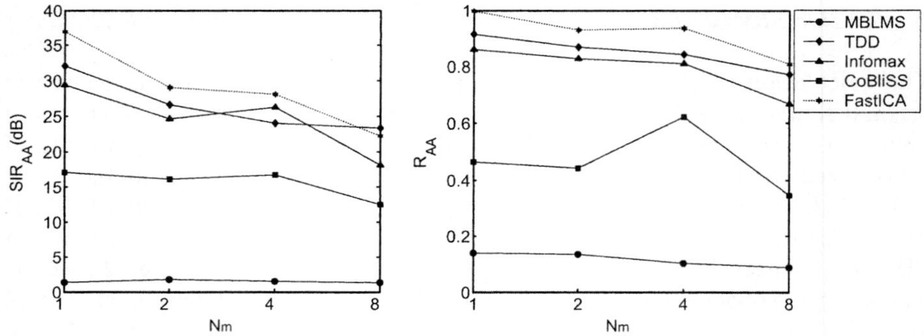

Fig. 2. First environment SIR_{AA} and R_{AA} mean values for four lengths of the mixing matrix FIR filters (N_m). Four CBSS algorithms and FastICA results are included.

increases. Finally, CoBliSS SIR_{AA} mean values are around two decades lower than FastICA SIR_{AA} mean values, that is, the performance in the extraction of the AA is much better than the performance obtained by MBLMS. However, it does not reach the performance obtained by TDD and Infomax.

Also in Fig. 2 we can observe the same tendencies of the quality of the extraction in terms of the R_{AA} parameter. We can see that FastICA R_{AA} values are always near to one (i.e. the original and estimated AA are very similar). R_{AA} of CBSS algorithms is always lower than R_{AA} of FastICA. TDD and Ifomax are the CBSS algorithms which R_{AA} mean values are nearest to one, and they tend to the FastICA R_{AA} values when N_m increases.

Fig. 3 illustrates the influence of the length of the separation matrix filters (N_s) in the quality of the extraction. Five different cases of N_s have been tested (2, 4, 8, 16 and 32). The mean values of FastICA are included only as a reference constant value, given that FastICA does not match the convolutive model and, therefore, the filters length parameter cannot be chosen. In the four considered CBSS algorithms, SIR_{AA} and R_{AA} decreases when N_s increases and they are always lower than the respective FastICA parameters. TDD and Infomax are

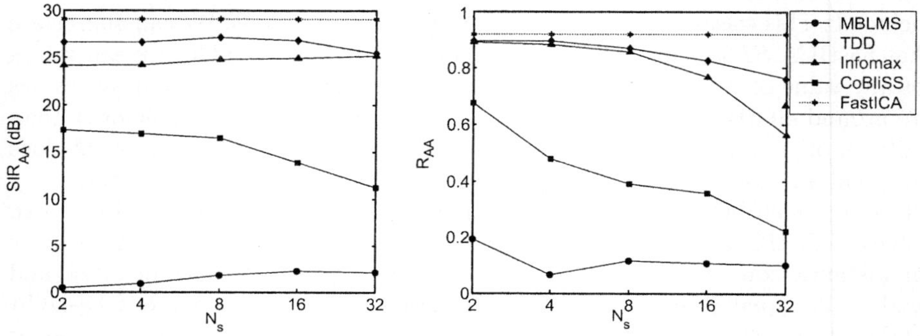

Fig. 3. First environment SIR_{AA} and R_{AA} mean values for five lengths of the separation matrix FIR filters (N_s). Four CBSS algorithms and FastICA are included.

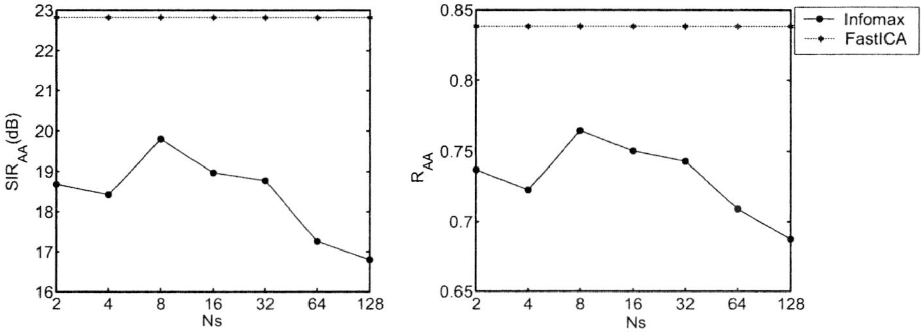

Fig. 4. Second environment mean values of SIR_{AA} and R_{AA} for seven lengths of the separation matrix FIR filters (N_s). Infomax algorithm and FastICA results are included.

the CBSS algorithms that obtain the highest values of SIR_{AA} and R_{AA} for any value of N_s.

Second environment performance was only tested for the Infomax algorithm, given that this is the only analyzed CBSS algorithm that simultaneously offers good AA extraction quality in the first environment and the possibility of being successfully adapted to the 12 leads ECG case. We summarize the results of the second environment in Fig. 4. We consider seven different lengths of the separation matrix FIR filters (N_s = 2, 4, 8, 16, 32, 64 and 128). Also here FastICA mean values are included only as a reference constant value. It can be seen that FastICA SIRAA mean value is several decibels greater than Infomax SIRAA mean values. Furthermore, FastICA correlation is nearer to one than Infomax correlation. In other words, the AA estimated by FastICA is more similar to the original AA than the AA estimated by Infomax.

6 Conclusions

The differences between performances of the tested CBSS algorithms reveal that not all of them are useful to extract AA from ECGs of AF episodes. Only TDD and Infomax obtained acceptable results in comparison with FastICA. With regard to the N_m parameter in the first environment, FastICA always exceeds the results obtained by the analyzed CBSS algorithms in the case of instantaneous mixtures ($N_m = 1$). Hence the instantaneous linear mixing model can be considered as an adequate model for the bioelectric mixtures in the human body. Obviously, the instantaneous linear mixing model is the particular case of the convolutive mixing model when N_m is equal to one. Consequently, CBSS algorithms need an improvement to reach at least the performance of ICA algorithms in the instantaneous case. By contrast, TDD and Infomax tend to reach FastICA results when N_m increases. Therefore, in the case of convolutive mixtures, these two CBSS algorithms are feasible to be used in the extraction of the AA in AF episodes.

In relation to N_s in the first environment, the optimal values of SIR_{AA} and R_{AA} are obtained for the shortest length of the separation matrix FIR filters

($N_s = 2$) in all the CBSS algorithms. TDD and Infomax are the CBSS algorithms that are least influenced by the value of N_s. This means that these are the algorithms that best adapt their filters coefficients in any case of convolutive mixtures.

To sum up, Infomax and TDD are the CBSS algorithms that obtain the best quality in the AA extraction. Furthermore, the Infomax algorithm has been easily adapted to the standard 12-leads ECG of the second environment. An in-depth analysis of the Infomax algorithm and its adjustment to the special features of ECG signals are possible subjects for future research.

Acknowledgements

This work was partly funded by the research incentive program of the Valencia University of Technology and the action IIARC0/2004/249.

References

1. Fuster, V., Ryden, L., Asinger, R.W., et al: CC/AHA/ESC guidelines for the management of patients with atrial fibrillation. Journal of the American College of Cardiology **38**(4) (2001) 1266/I–1266/LXX
2. Rieta, J.J., Castells, F., Sánchez, C., Zarzoso, V.: Atrial activity extraction for atrial fibrillation analysis using blind source separation. IEEE Transations of Biomedical Engineeringy **51**(7) (2004) 1176–1186
3. Rieta, J.J., Castells, F., Sánchez, C., Moratal-Pérez, D., Millet, J.: Bioelectric model of atrial fibrillation: applicability of blind source separation techniques for atrial activity estimation in atrial fibrillation episodes. IEEE Computers in Cardiology **30** (2003) 525–528
4. Hyvärinen, A., Karhunen, J., Oja, E.: Independent Component Analysis. John Willey & Sons, Inc (2001)
5. Sanchis, J.M., Castells, F., Rieta, J.J.: Convolutive acoustic mixtures approximation to an instantaneous model using a stereo boundary microphone configuration. Lecture Notes in Computer Science **3195** (2004) 816–823
6. Lambert, R.H.: Multichanel Blind Deconvolution: FIR matrix algebra and separation of multipath mixtures. PH.D, University of Southern California (1996)
7. Ikeda, S., Murata, N.: A method of blind separation on temporal structure of signals. In: Proceedings of The Fifth International Conference on Neural Information Processing (ICONIP'98), Kitakyushu, Japan (1998) 737–742
8. Asano, F., Ikeda, S., Ogawa, M., Asoh, H., Kitawaki, N.: A combined approach of array processing and independent component analysis for blind separation of acoustic signals. In: Proceedings of the IEEE Conference on Acoustics, Speech and Signal Processing, (Salt Lake City, USA) 2001
9. Schobben, D., P, S.: A new convolutive blind signal separation algortithm based on second order statistics. In: Proc.Int.Conf on Signal and Image Processing. (1998) 564–569
10. Plonsey, R., Heppner, D.B.: Considerations of quasi-stationarity in electrophysiological systems. Bulletin of Mathematical Biophysics **29**(4) (1967) 657–664

Brains and Phantoms: An ICA Study of fMRI

Jarkko Ylipaavalniemi[1], Seppo Mattila[2,3], Antti Tarkiainen[3], and Ricardo Vigário[1]

[1] Neural Networks Research Centre,
Laboratory of Computer and Information Science,
Helsinki University of Technology, P.O. Box 5400, FI-02015 TKK, Finland
jarkko.ylipaavalniemi@tkk.fi
[2] Brain Research Unit, Low Temperature Laboratory,
Helsinki University of Technology, P.O. Box 2200, FI-02015 TKK, Finland
[3] Advanced Magnetic Imaging Centre,
Helsinki University of Technology, P.O. Box 3000, FI-02015 TKK, Finland

Abstract. Biomedical signal processing is arguably the most successful application of independent component analysis (ICA) to real world data. For almost a decade, its use in connection with functional magnetic resonance imaging (fMRI) has allowed for data-driven analysis, partly removing the constraints for stringent experimental setups, which are often required by traditional methods based on the use of temporal references. Recent studies on the consistency of independent components have resulted in a series of tools enabling a more reliable use of ICA. In particular, it is now rather easy to detect algorithmic overfitting and isolate subspaces of related activation. Yet, often the nature of the components may not be determined unambiguously. Focal fMRI signals, seemingly originating from within a subject's brain and showing physiologically plausible temporal behavior, are typically considered relevant. This paper presents a study, which makes use of a standard homogeneous spherical phantom and shows evidence for artifacts caused by the measuring device or environment, with characteristics that could easily be misinterpreted as physiological. Our results suggest that reliable analysis of fMRI data using ICA may be far more difficult than previously thought. At least, artificial behavior revealed by phantom analysis should be considered when conclusions are drawn from real subject measurements.

1 Introduction

Searching for a set of generative source signals from their linear mixing, with little to no knowledge on the sources or the mixing process, is referred to as blind source separation (BSS). Independent component analysis (ICA) is possibly the most widely used data-driven method to solve such problems (a good introduction to ICA, including its historical debuts and theoretical frameworks can be found in the textbook [1]; further reading and applications can also be found in [2, 3]). Biomedical signal processing is arguably the most successful application of ICA to real world data (representative examples can be found in, *e.g.*, the following review papers [4, 5, 6, 7]).

J. Rosca et al. (Eds.): ICA 2006, LNCS 3889, pp. 503–510, 2006.

On the other hand, functional magnetic resonance imaging (fMRI) has secured a strong position in non-invasive studies of the living human brain. It provides indirect information on neural activity, by measuring the blood oxygenation level dependent (BOLD) signal (*cf.*, [8]). When analyzing fMRI data under the statistical parametric mapping (SPM) framework [9], researchers validate an active brain region through the match of its temporal activation pattern with a carefully predetermined experimental setup. Therefore, they can only validate predictions. When performing a data-driven analysis, such as ICA, the researcher is given greater freedom and is thus capable of detecting unforeseen activity. This allows the study of a whole new set of more complex research questions. Yet, the problem of interpreting the nature of the detected components remains, since not all components have a physiological origin. The key rationales often used in identifying components of interest include the focal nature and potential symmetry of the spatial patterns; whether the activation is located inside the brain and if it falls on expected regions for given stimuli; and how plausible the corresponding time-courses are.

Recent studies on the consistency of independent components resulted in a series of tools enabling a more reliable use of ICA (*cf.*, [10, 11, 12]). In particular, it is now rather easy to detect algorithmic overfitting and isolate subspaces of related activation. However, some results (*cf.*, [13, 11, 14]) suggest that identifying the relevant components may, in fact, not be as straightforward as previously thought, *e.g.*, in the presence of artifacts with characteristics matching the aforementioned rationales on activation volumes and temporal patterns.

In this paper we confront the analyses of brain responses to auditory stimuli (presented earlier in [14]) and of recently collected data from a standard MRI phantom. Our results suggest that the analysis of fMRI data using ICA may be more difficult than previously thought.

2 Data and Methods

2.1 Functional Magnetic Resonance Imaging

Varying concentrations of oxygen in the blood result in changes of its magnetic properties. Because active brain areas produce a local increase in the blood flow, measuring the MR-signal during rest periods and during task conditions, *e.g.* when attending to stimulus presentation, results in detectable differences in the measured images. This is the general basis of fMRI.

When using controlled stimuli, it is common to look for voxels in the brain with a temporal activation pattern that matches the time-courses of the stimuli (*cf.*, [9]). However, the use of general data-driven methods, such as ICA, have been suggested when attempting to observe epiphenomena that are hard to tie to the stimuli or tasks, or when searching for brain reaction to unlabeled stimuli.

2.2 Phantom fMRI Measurements

Artifacts caused by the measuring device or environment can corrupt the fMRI signal. Hence, their characteristics have to be assessed in order to reliably separate them from genuine BOLD signals. One example of a known artifact in fMRI signal is the low frequency drift [15], but also more complex artifacts do exist.

Test objects, *i.e.*, phantoms, are often used for assessing the quality of data collected by MRI equipments (*cf.*, [16, 17]), in terms of image properties, such as, signal-to-noise ratio and image uniformity. They can also be used for testing and guiding further analysis applied to the data (*cf.*, [13]).

Here, we use a homogeneous spherical phantom, provided by GE Healthcare. It has a diameter of about 15 cm, and is filled with silicone gel, in order to produce images with intrinsically uniform brightness over the whole phantom. Half an hour before starting the measurements, the phantom was placed within the head coil and the patient bed moved into measurement position to avoid any movement induced artifacts. We used the Gradient Echo (GRE) Echo Planar Imaging (EPI) technique, commonly used for fMRI. All the data were acquired using a 3.0 Tesla MRI scanner (Signa EXCITE 3.0T; GE Healthcare, Chalfont St. Giles, UK), with a quadrature birdcage head coil at the Advanced Magnetic Imaging Centre of Helsinki University of Technology.

The data consisted of 300 fMRI time points with TR = 3000 ms (*i.e.*, a 3 second time resolution). The first four scans were excluded from further processing. For each time point, we acquired 37 axial 3.0 mm slices (spacing = 0 mm) with a 96×96 acquisition matrix and a field of view of 200 mm. We used a flip angle of 90 deg and TE = 32 ms, all typical values in fMRI studies. Since we were interested in seeing possible artifacts arising from the measurement equipment or the environment, the data was not preprocessed before the analysis.

2.3 Real Auditory Measurements

To compare the components found from the phantom data to the ones found from real measurements, we also used fMRI data of 14 human subjects attending auditory word stimuli. The stimuli consisted of repetitions of resting and listening periods. The data consisted of 80 fMRI time points and were acquired with the same imaging parameters, head coil, and MRI scanner as the phantom data (further information about the experimental paradigm, and data analysis with a reliable ICA procedure can be found in [14]). However, in contrast to the phantom data, these measurements were acquired prior to the EXCITE upgrade for the imaging equipment.

2.4 Reliable ICA

In BSS, the measured data is an instantaneous linear mixture of generative source signals, *i.e.*, $\mathbf{X} = \mathbf{AS}$, for $\mathbf{X}$, $\mathbf{A}$ and $\mathbf{S}$, respectively, the observed data, the mixing matrix and the underlying sources. The goal is to identify both the sources and the mixing process with as few assumptions as possible. ICA solves the BSS problem by assuming only that the generative sources are statistically independent from each other. Hence, when applied to fMRI data, we often look for spatially independent neuronal activity, with the columns of the mixing matrix giving the temporal activation of such components.

Theoretically, statistical independence means that the joint probability density of the sources is factorisable on its marginal densities. In practice, several estimation algorithms have been proposed to perform ICA, mainly based on concepts such as negentropy, mutual information or maximum likelihood (for further information, *cf.*, [1, 2, 3]). The experiments in this study use the FastICA algorithm [18], an iterative method with fixed-point optimization. The considerations based on the FastICA algorithm should also be valid for other implementations.

When analyzing a finite data set, the estimated components may change slightly each time the analysis is performed. This behavior can be caused by many factors. For example, the theoretical assumption of statistical independence may not hold for the analyzed data [19, 4]. In this case, the somewhat less restrictive sparse constraint for the underlying sources may still hold, as suggested in [5]. Also, the algorithmic implementation of ICA may be inherently stochastic. Furthermore, additive noise or other data features can cause variations in the solutions. When the degrees of freedom is high, there is also a tendency for overfitting the data [20]. For ICA, this corresponds to bumps or spikes, which occur quite randomly each time the algorithm is run.

In this paper, the consistency of the estimated signals is tested by running the algorithm with many different initial conditions, and bootstrapping of the data. FastICA was used in *symmetric* mode with *tanh* nonlinearity, other parameters where left at default values. The solutions found are combined according to their similarities. Estimates that differ greatly from run to run are less likely to correspond to true components, whereas the ones with small variances are considered reliable (further details, including the combination strategy, can be found in [14]). Similar approaches for consistency analysis and visualization can be found in, *e.g.*, [10, 11, 12].

3 Results

A set of independent components showing the simplest or the most structured time-courses found by analyzing the phantom measurement is shown in Fig. 1. The disks on the left hand side show the spread of the estimates from the different runs. The upper disk depicts the intra-group and the lower disk the inter-group distances. The circles represent mean distances. In ideal estimate discrimination, the upper disk should fit within the hole in the lower disk. The slices represent, clockwise from top-left, the sum of the whole volume; the slice containing the highest power; the slice with the maximum voxel value; and the slice with the minimum voxel value. The mean time-course is shown as a line, superimposed on the spread from all the runs, shown as quantile bands with different intensities. More details on the reliability measures can be found in [14].

All the shown components were found to be very reliable, although some exhibit a small amount of variability. It appears that none of the components results from the signal processing related to the analysis itself, or random noise in the data. Furthermore, the time-courses of the components show clear and

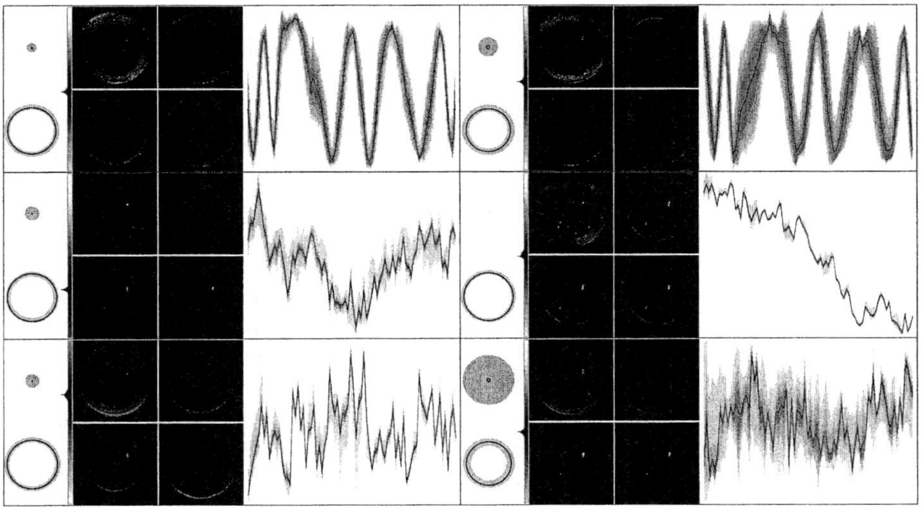

Fig. 1. A set of independent components with simple or structured time-courses, found from the phantom measurement. A time period of 240 seconds is shown.

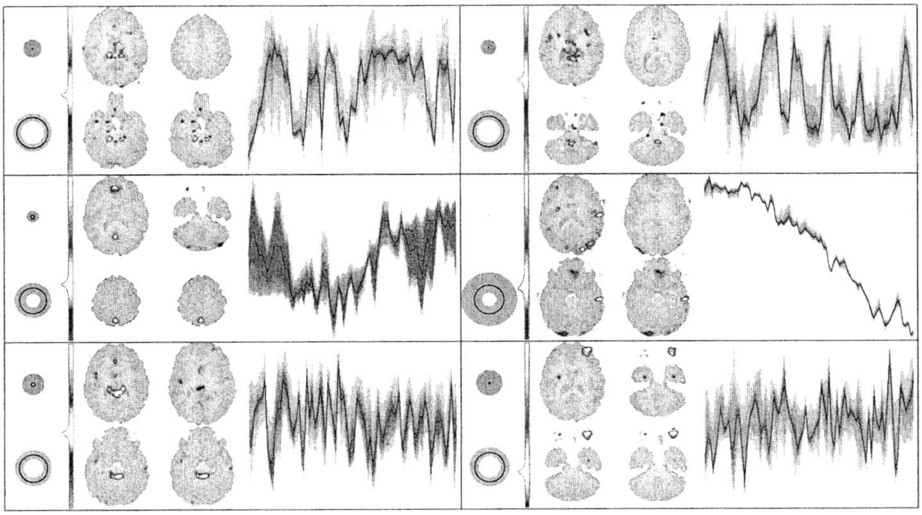

Fig. 2. A set of independent components, found from the measurement of a real subject, which resemble the ones found from the phantom data. A time period of 240 seconds is shown.

systematic structures on time scales of several seconds. In other words, components like these could be considered as relevant activity, in real brain data.

Another set of components found from the real brain measurements with time-courses resembling the ones found from the phantom data are shown in Fig. 2. The results are from a single subject, but similar components were found in the data of all the 14 subjects studied. Here the slices are superimposed on a structural MRI of the subject. The first component has a well structured time-course, which could easily be misinterpreted as being related to the on-off type stimuli. In addition, the volume contains a very focal and symmetric activation pattern in the mid-brain area. However, the time-course, *i.e.*, mixing vector, shares so much of the characteristics of the first component of the phantom data, that a more probable explanation is that both components are manifestations of the same scanner or environment induced artifact. This, however, would be impossible to notice without having performed the phantom analysis.

Similarly, the other components shown in Fig. 2 exhibit focal activation patterns within the brain and/or structured time-courses, *e.g.*, periodic or slowly varying. Therefore fitting the aforementioned rationales for identifying interesting components. A close match for many of the time-courses can be found from the phantom data, with correlations reaching as high as 90%. However, simple mathematical measures, such as cross-correlation do not show the complete picture. For example, temporal delays may need to be taken into account. Also, at times, the phantom components show temporal behavior close to the one used in the block design of the experiment, causing very strong correlations with several components. The crucial question is whether in human evaluation the characteristics of the phantom and brain components are confusingly similar, *e.g.* the time-course associated with the primary auditory cortex is remarkably close to the first two phantom components in Fig. 1 (mean correlations 43% and 38%, respectively). However, in this study, the auditory component can unquestionably be labeled as physiological. Yet, it may still contain contribution from an artificial signal. Furthermore, artificial signals can be much harder to rule out in less controlled experimental setups.

The spatial differences between the phantom and brain components may be attributed to the homogeneity of the phantom, in contrast to the highly non-uniform MR signal of the human brain. The structural differences may also affect the magnitude of the measured components. Some of the components could be related to, *e.g.*, heating of the gradient coils during the imaging or time-dependent changes in the magnetic fields. Other hardware instabilities and the imaging environment can also produce artifacts (*cf.*, [15] and references therein). However, a detailed discussion on the origin of the components is beyond the scope of this paper.

With a real subject, artifacts can also be caused by, *e.g.*, cardiac pulsation or head movements. For instance, some of the components in Fig. 2 show characteristics typical for head movements that have not been completely compensated in the preprocessing of the data. Interestingly, similar behavior in the phantom data suggests that they may in fact be caused by other phenomena.

4 Discussion

Phantom measurements are routinely used for verifying and calibrating the quality of MRI machinery. However, data-driven analysis of phantom fMRI data has been largely overlooked, possibly due to the lack of a method for assessing the reliability of the solutions. Although some earlier work (*cf.*, [13]) has shown that consistent independent components can indeed be found from phantom measurements, to our knowledge, such components have never before been shown publicly and compared with components found from fMRI studies with real subjects. However, the results presented here strongly suggest that such comparisons may be very valuable for the whole research field.

The presented results from analyzing phantom data using ICA reveal evidence for possible misinterpretations in ICA studies with real subjects. The evidence suggests that analyzing fMRI data using ICA may actually be far more difficult than previously thought. It is possible that other methods than ICA are also affected. For example, the reference time-course of the stimuli could have points of coincidence with artificial signals. The analysis would thus mix real brain activations with artifacts.

Although not shown, the results also suggest that the imaging parameters affect the scanner induced components. Therefore, it is important that the phantom measurements are made with the same parameters as those used with the real subjects. It is also expectable that the artifacts can differ with, *e.g.*, time, scanner and measurement coils used. This suggests that data-driven analysis, such as ICA, of phantom data may be useful for quality control of fMRI machinery. The possible effects of different preprocessing steps, typical in fMRI analysis, could also be tested with a similar approach.

The purpose of this paper is to be a word of warning for the ICA community involved in analyzing fMRI data. Clearly, we need a better understanding of the artificial, scanner or environment induced, signals, and of the way they are manifested in phantom and real brain measurements. Possible methods for automatic exclusion of such artifacts should also be considered. If artifacts with systematic characteristics are observed, they could be used for designing real brain measurements such that the stimulus timing does not coincide with the known artifacts. However, the present results strongly suggest that if a researcher wants to base conclusions on components with a purely physiological origin, the ICA results should be compared with phantom measurements.

References

1. Hyvärinen, A., Karhunen, J., Oja, E.: Independent Component Analysis. 1st edn. Wiley-Interscience, New York, NY (2001)
2. Cichocki, A., Amari, S.I.: Adaptive Blind Signal and Image Processing: Learning Algorithms and Applications. 1st edn. Wiley-Interscience, New York, NY (2002)
3. Stone, J.V.: Independent Component Analysis : A Tutorial Introduction. 1st edn. MIT Press/Bradford Books, Cambridge, MA (2004)

4. Jung, T.P., Makeig, S., McKeown, M.J., Bell, A.J., Lee, T.W., Sejnowski, T.J.: Imaging Brain Dynamics Using Independent Component Analysis. Proceedings of the IEEE **89** (2001) 1107–1122
5. Vigário, R., Särelä, J., Jousmäki, V., Hämäläinen, M., Oja, E.: Independent Component Approach to the Analysis of EEG and MEG Recordings. IEEE Transactions on Neural Networks **47** (2000) 589–593
6. Tang, A., Pearlmutter, B., Malaszenko, N., Phung, D., Reeb, B.: Independent Components of Magnetoencephalography: Localization. Neural Computation **14** (2002) 1827–1858
7. Calhoun, V.D., Adali, T., Hansen, L.K., Larsen, J., Pekar, J.J.: ICA of functional MRI Data: An Overview. In: Proc. 4th Int. Symp. on Independent Component Analysis and Blind Signal Separation (ICA2003), Nara, Japan (2003) 281–288
8. Haacke, E.M., Brown, R.W., Thompson, M.R., Venkatesan, R.: Magnetic Resonance Imaging: Physical Principles and Sequence Design. 1st edn. Wiley-Interscience, New York, NY (1999)
9. Worsley, K.J., Friston, K.J.: Analysis of fMRI Time-Series Revisited — Again. NeuroImage **2** (1995) 173–235
10. Meinecke, F., Ziehe, A., Kawanabe, M., Müller, K.R.: A Resampling Approach to Estimate the Stability of One-Dimensional or Multidimensional Independent Components. IEEE Transactions on Biomedical Engineering **49** (2002) 1514–1525
11. McKeown, M.J., Hansen, L.K., Sejnowski, T.J.: Independent Component Analysis of functional MRI: What Is Signal and What Is Noise? Current Opinion in Neurobiology **13** (2003) 620–629
12. Himberg, J., Hyvärinen, A., Esposito, F.: Validating the independent components of neuroimaging time series via clustering and visualization. NeuroImage **22** (2004) 1214–1222
13. McKeown, M.J., Varadarajan, V., Huettel, S., McCarthy, G.: Deterministic and stochastic features of fMRI data: implications for analysis of event-related experiments. Journal of Neuroscience Methods **118** (2002) 103–113
14. Ylipaavalniemi, J., Vigário, R.: Analysis of Auditory fMRI Recordings via ICA: A Study on Consistency. In: Proceedings of the 2004 International Joint Conference on Neural Networks (IJCNN 2004). Volume 1., Budapest, Hungary (2004) 249–254
15. Smith, A.M., Lewis, B.K., Ruttimann, U.E., Ye, F.Q., Sinnwell, T.M., Yang, Y., Duyn, J.H., Frank, J.A.: Investigation of Low Frequency Drift in fMRI Signal. NeuroImage **9** (1999) 526–533
16. Lerski, R.A., McRobbie, D.W., Straughan, K., Walker, P.M., de Certaines, J.D., Bernard, A.M.: Multi-center trial with protocols and prototype test objects for the assessment of MRI equipment. Magnetic Resonance Imaging **6** (1988) 201–214
17. Price, R.R., Axel, L., Morgan, T., Newman, R., Perman, W., Schneiders, N., Selikson, M., Wood, M., Thomas, S.R.: Quality assurance methods and phantoms for magnetic resonance imaging: Report of AAPM nuclear magnetic resonance Task Group No. 1. Medical Physics **17** (1990) 287–289
18. FastICA: MATLAB Package. In: http://www.cis.hut.fi/research/ica/fastica. (1998)
19. McKeown, M.J., Sejnowski, T.J.: Independent Component Analysis of fMRI Data: Examining the Assumptions. Human Brain Mapping **6** (1998) 368–372
20. Särelä, J., Vigário, R.: Overlearning in Marginal Distribution-Based ICA: Analysis and Solutions. Journal of Machine Learning Research **4** (2003) 1447–1469

Comparison of ICA Algorithms for the Isolation of Biological Artifacts in Magnetoencephalography

Heriberto Zavala-Fernández[1], Tilmann H. Sander[2], Martin Burghoff[2], Reinhold Orglmeister[1], and Lutz Trahms[2]

[1] Technische Universität Berlin, Institut für Elektronik, Einsteinufer 17, 10587 Berlin, Germany
hzavimed@mailbox.tu-berlin.de
[2] Physikalisch-Technische Bundesanstalt, Abbestr. 2-10, 10587 Berlin, Germany

Abstract. The application of Independent Component Analysis (ICA) to achieve blind source separation is now an accepted technique in the field of biosignal processing. The reduction of biological artifacts in magneto- and electroencephalographic recordings is a frequent goal. Four of the most common ICA methods, extended Infomax, FastICA, JADE, and SOBI are compared here with respect to their ability to isolate magnetoencephalographic (MEG) artifacts. The four algorithms are applied to the same data set containing heart beat and eye movement artifacts. For a quantification of the result simple spatial and temporal correlation measures are suggested and the usage of reference signals. Of the four algorithms only JADE was marginally less successful.

1 Introduction

For the analysis of magnetoencephalographic (MEG) recordings the suppression of unwanted signal components is an important preprocessing step. Independent Component Analysis (see recent introductory texts [1-3] for references) is widely used for this purpose. From the multitude of algorithms described theoretically it is very difficult for the practitioner to choose a suitable algorithm for the task at hand. Only very few studies address this for experimental data important issue. The power line interference artifact in MEG was isolated in [4] using three ICA algorithms. In [5] respiratory and eye movement artifacts were removed from MEG data using a second-order algorithm followed by a higher order algorithm. Four algorithms were applied to remove eye movement and blinking artifacts from EEG data in [6].

These studies differ considerably in their methodology. In the present study the focus is on the two most common biological artifacts in MEG data: heart beat and eye movements. Investigating these artifacts we suggest a testing framework using temporal and spatial correlation. To start, results for four frequently used algorithms are presented: extended Infomax, FastICA, JADE, and SOBI.

Success of ICA can only be proven on simulated data. As a simulation is not possible in all situations (ICA is often applied because the signal content is unknown) the more practical route of validation using separately recorded reference signals is applied here.

J. Rosca et al. (Eds.): ICA 2006, LNCS 3889, pp. 511 – 518, 2006.

2 Algorithms Used

These algorithms are derived from the ICA concept, which supposes that each observed signal x_i of a multi-channel recording with m channels can be described by a linear superposition

$$x_i(t) = \sum_{j=1}^{n} a_{ij} s_j(t) \tag{1}$$

of n source signals s_j, i.e. number of components n equals number of sensors m.

Assuming that the sources are statistically independent the joint probability density function of the signals s_j factorizes. Then the sources can be separated theoretically by estimating a demixing matrix $\mathbf{W}$. Estimates y_i of the original sources s_j are found by applying the demixing matrix to the measured variables: $\mathbf{y}(t) = \mathbf{W}\mathbf{x}(t)$.

Before the ICA algorithms are applied the observed signals are pre-processed in two steps: Centering[1-3] and whitening[1-3].

FastICA. The FastICA[7] algorithm estimates the non-Gaussianity of the signal distributions using higher-order statistics and the negentropy, which is a non-negative function of the differential entropy. FastICA is based on an iteration scheme for finding a projection $\mathbf{u}=\mathbf{W}^T\mathbf{x}$ maximizing non-Gaussianity. It can be summarized by the update rule

$$\mathbf{W} \propto \left\langle \mathbf{x} g(\mathbf{W}^T\mathbf{x}) \right\rangle - \left\langle g'(\mathbf{W}^T\mathbf{x}) \right\rangle \mathbf{W} , \tag{2}$$

and the subsequent normalization of the updated $\mathbf{W}$ until convergence is reached. There exist several possible choices for the non-linear function $g(\mathbf{u})$.

Extended Infomax. This extended Infomax[9] is an extension of the Infomax algorithm, which is based on the information maximization principle[8] with the ability of separate mixed signals with sub- and super-Gaussian distributions. This is achieved by introducing a learning rule able to switch between both distributions.

The switching criterion between the sub- and super-Gaussian distributions for $\mathbf{y}(t)$ is contained in the following learning rule

$$\Delta\mathbf{W} \propto \left[\mathbf{I} - \mathbf{K}\tanh(\mathbf{y})\mathbf{y}^T - \mathbf{y}\mathbf{y}^T\right]\mathbf{W} \quad \begin{cases} k_i = 1 & \text{:supergaussian} \\ k_i = -1 & \text{:subgaussian.} \end{cases} \tag{3}$$

The elements of the diagonal matrix $\mathbf{K}$ are obtained according to

$$k_i = \operatorname{sign}\left[\left\langle \operatorname{sech}^2(y_i) \right\rangle \left\langle y_i^2 \right\rangle - \left\langle [\tanh(y_i)] y_i \right\rangle\right]. \tag{4}$$

The Infomax algorithm maximizes the entropy of the outputs $\mathbf{H}(\mathbf{y})$. The maximization of this joint entropy consists of maximizing the individual entropies while minimizing the mutual information $\mathbf{I}(\mathbf{x})$ shared between them. When this latter quantity is zero, the variables are statistically independent.

JADE. The Joint Approximate Diagonalization of Eigenmatrices[10,11] (JADE) algorithm uses fourth-order cumulants $Q = Cum(x_p, x_j, x_k, x_l)$, which present in a clear form the additional information provided for the high-order statistics.

The JADE algorithm aims to reduce the mutual information contained in the cumulant matrices by looking for a rotation matrix such that the cumulant matrices are as diagonal as possible. The joint diagonalization is found by the Jacobi technique.

SOBI. In contrast to the three algorithms sketched so far the Second Order Blind Identification algorithm(SOBI[12], TDSEP[13,14]) takes advantage of the temporal structure in the observed data.

The basis of the SOBI algorithm is a set of time-lagged covariance matrices

$$\mathbf{R_x}(\tau)=<x(t+\tau)x^T(t)> \qquad \tau \neq 0 \,. \tag{5}$$

For independent sources these matrices have to be diagonal. To estimate the sources a joint diagonalization of the time-lagged covariance matrices is performed similarly to the JADE algorithm. The approach to use a set of τ values is intended to avoid an inferior source separation as there is no theoretically proven choice of τ values.

3 MEG Measurement and ICA Calculations

The MEG recordings were performed in a shielded room, where the level of technical interference signals was reduced. The data were measured using a helmet shaped MEG sensor (www.eagle-tek.com) with 93 channels. For the purpose of generating typical biological artifacts the subject was instructed along the following protocol during the measurement: Rest for 30 s, horizontal eye movements for 30 s, rest for 30 s, vertical eye movements for 30 s, rest for 30 s, and eye blinking for 30 s. To simplify the artifact identification in the MEG the Electrooculogram (EOG: single lead below the eye relative to the forehead) and the Electrocardiogram (ECG: single lead on sternum relative to forehead) signals were simultaneously recorded.

The raw data sampled at 2 kHz were downsampled offline by a factor of 8 to 250 samples/s to reduce the computational load for the ICA algorithms. In total 62500 data points (250 s * 250 samples/s) were input into the ICA calculations for each of the 93 channels without dimension reduction of the signal space.

This study was realized using the software package EEGLAB[15]. EEGLAB is a freely available MATLAB® package for the analysis of single-trial EEG dynamics including various ICA algorithms. The following parameters were chosen: FastICA with $g(\mathbf{u}) = \mathbf{u}^3$, SOBI with vector $\tau = \{1,2,\ldots,100\}$ (no specific parameters for JADE and extended Infomax).

4 Results

After the ICA calculation the component due to the heart beat was identified manually by comparison with the ECG signal and the result is shown in Fig. 1. All algorithms have successfully isolated a Cardiac Artifact (CA) component. On the left side the time series for the CA components (FastICA = FICA, extended Infomax = eINFO) are shown together with the ECG reference signal. The algorithm associated with each time series is indicated and the time series were rescaled for ease of comparison. The time series shown in Fig. 1 a) is only a short section of the time series input into the calculation. On the right side the CA maps for the four different algorithms are shown. The maps are interpolations using a projection of the three dimensional MEG sensor coordinates onto a plane showing the magnetic field as level curves. The maps shown are views onto the top of the head with nose and ears indicated.

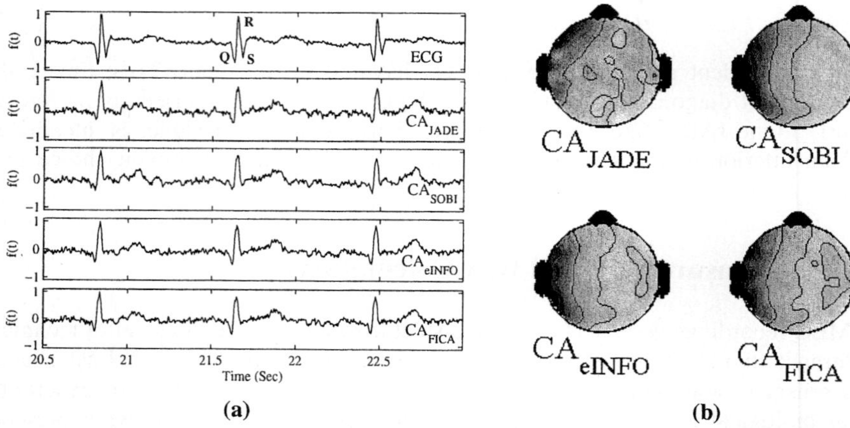

Fig. 1. a) Typical ECG signal (*top trace*) and the time series of the CAs isolated from MEG by the ICA algorithms, **b**) associated CA maps (*nose and ears indicated*)

The four ICA time series in Fig. 1a) agree with each other, but they are different from the ECG (S-peak not visible as in the ECG trace). Such differences are well known[17] and reflect the complementary nature of electrical and magnetic measurements. Comparing the maps in Fig. 1b) visually it can be seen that the CA_{JADE} map is different from the others and the CA_{eINFO} and CA_{SOBI} maps are most similar. The CA map is expected[16] to exhibit a homogeneous (smooth) field distribution due to the relatively large sensor to source distance. The small extrema in the CA_{JADE} map indicate a suboptimal identification of the CA. The strongest field in all maps is on the left side in agreement with the position of the heart on the left side of the body.

For a quantitative analysis of the ICA result two types of comparison were made. Firstly the appropriate reference time series (ECG, EOG) was correlated using Eq. 7 with the full length time series $y_i(t)$ resulting from the ICA after demixing, i.e. inverting Eq. 1, and the ICA time series were correlated with each other. Secondly the vector angle between the ICA field maps, i.e. the ICA base vectors, was calculated using Eq. 6, where $\vec{v}_i$ with $\mathbf{V}=\mathbf{W}^{-1}$ denotes the base vector.

$$\alpha = \cos^{-1}\left(\frac{\vec{v_i}\cdot\vec{v_j}}{\left|\vec{v_i}\right|\left|\vec{v_j}\right|}\right) \tag{6}$$

$$\rho_{fy} = \frac{\langle (f(t)-\bar{f})(y_i(t)-\bar{y}_i)\rangle}{\sigma(f)\sigma(y_i)} \qquad f(t) = ECG, EOG, y_i. \tag{7}$$

The angles between the CA components maps resulting from the different ICA calculations are shown in matrix format in Fig. 2a), where the labels are FastICA = F, extended Infomax = eI, J = JADE, S = SOBI and the grey scale for the angles is indicated on the right side. It can be seen immediately that the JADE results is different from the other results. The angle between CA_{JADE} and the others CAs has a minimum value of 29° and a maximum of 41°. In comparison the angles between the others CAs are less than 12° (Figure 2a).

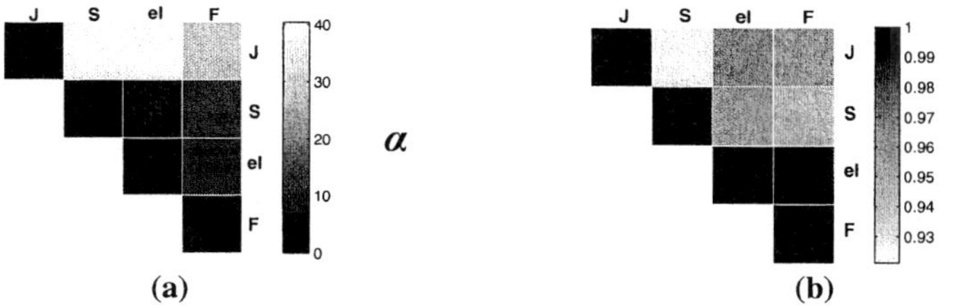

Fig. 2. a) Matrix representation of angles between the CA component maps calculated using Eq. 6, **b**) the correlation values between the CA time series calculated using Eq. 7. The labels are F = FastICA, eI = extended Infomax, J = JADE, S = SOBI.

The correlations between the CA time series are shown in Fig. 2b). The correlations have generally high values between 0.92 and 0.99 and the lowest correlation occurs between the time series of SOBI and JADE. In contrast to the high correlations between the results from the ICA algorithms the correlation between ECG and the ICA result was 0.42 to 0.45 (values not shown in Fig. 2b). This is a consequence of the differences in signal morphology as mentioned above.

The ICA results for the horizontal eye movement artifact (hEMA) are shown in Fig. 3, which displays time series on the left and maps on the right as in Fig. 1. All algorithms identified the hEMA as can be seen from the time series in Fig. 3 a), although the JADE time series appears noisier. The maps on the right side have basically the same structure with the strongest signals at the front. This is expected for the signal due to eye movements. Similar to the behavior observed for the CA the JADE map of the hEMA appears less regular compared to the other maps.

A quantification of these observations was made using Eqns. 6 and 7 on the hEMA maps and time series and additionally the correlation between EOG and hEMA time series was calculated. The smallest angle between ICA maps is 6° occurring between extended Infomax and FastICA. The angle between the JADE hEMA map and the other maps is 13° to 20° indicating that the JADE result is slightly different from the others. For the correlation values the result is similar: The JADE time series has always the lowest correlations in the range from 0.94 to 0.96. The correlations between the other algorithms and the EOG signal always exceed 0.98.

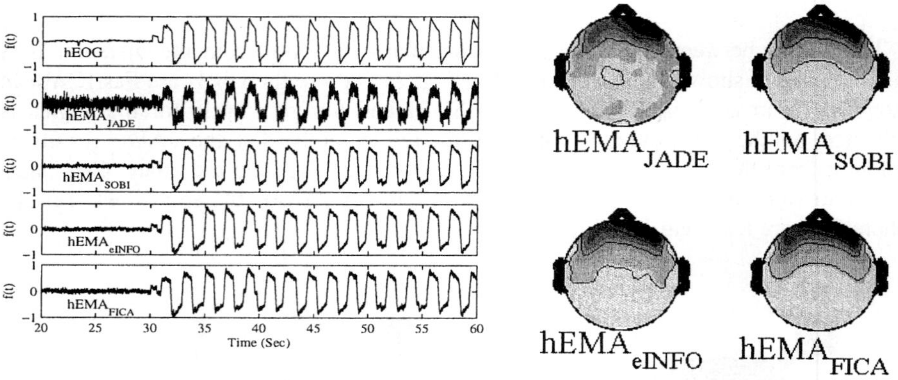

(a) (b)

Fig. 3. **a**) EOG signal (*top trace*) and the time series of the horizontal EMAs isolated by the ICA algorithms, **b**) associated EMA maps (*nose and ears indicated*)

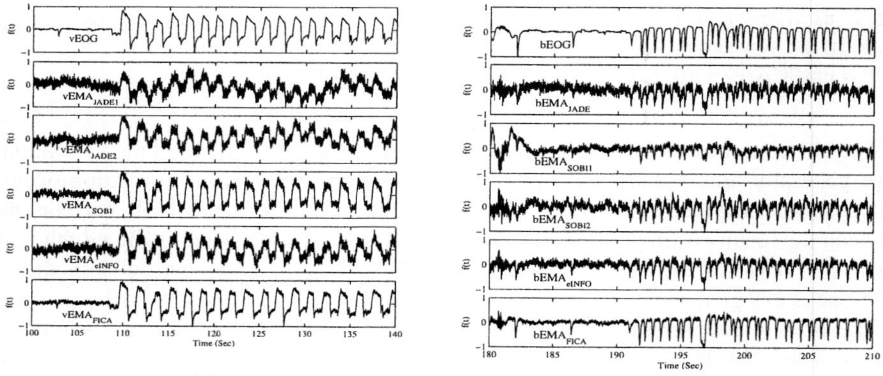

(a) (b)

Fig. 4. EOG signal (*top trace*) and the time series of: **a**) the vertical EMAs and **b**) the blinking EMAs isolated by the ICA algorithms

The ICA results for the vertical and blinking eye movement artifact (vEMA and bEMA respectively) are shown in Fig. 4 displaying the time series only. In case of vEMA in Fig. 4a), JADE found two separate components related to the vertical EOG, while ext. Infomax, SOBI and FastICA extracted only a single component appearing less noisier than the others. The JADE and ext. Infomax time series have relatively low correlations (Eq. 7) to the vertical EOG in the range from 0.6 to 0.88, while the corresponding correlations for the other algorithms always exceed 0.95.

The separation of the bEMA in Fig. 4b) shows that SOBI found two components related to the blinking EOG, while the other algorithms found only one. The correlations between bEOG and bEMA of SOBI are in the range from 0.5 to 0.83. The corresponding JADE and ext. Infomax correlations are in the range from 0.8 to 0.9. The best correlation was 0.95 for the FastICA algorithm. The angles calculated using Eq. 6 do not contradict the correlation result (omitted due to lack of space).

Principal Component Analysis (PCA) was also applied to the full data set. The five strongest PCA components are artifact related. The CA is well identified in one component, but in contrast to the ICA algorithms PCA was not able to separate the EMAs into single PCA components. Furthermore their time series show temporal overlap between CA and the EMAs. As it is well known decorrelation is not sufficient to achieve independence.

5 Conclusions

Four different ICA algorithms were applied to a 93 channel MEG data set containing heart beat and eye movement artifacts signals. Using the simultaneously measured ECG and EOG and prior knowledge artifact related independent components could be identified in the result from all four algorithms. Comparing the results from the four algorithms using a correlation between the time series and the angle between the maps it was found that JADE performs slightly inferior to the other algorithms for the single data MEG data set used here. The computation time needed to obtain the independent components using JADE was ten times longer compared to the other ICA algorithms.

All algorithms were capable of isolating the super-Gaussian probability distribution of the heart beat and most algorithms succeeded for the essentially bimodal distributions of the eye movements.

In contrast to the continuous presence of the heart beat eye movements can be controlled to a certain degree during an experiment and the data set used seems to be unrealistic in this respect. There were two reasons for this choice: Firstly, patients or elderly people often cannot control their eye movements, and secondly, infrequently occurring artifacts violate the stationarity assumption of ICA. Therefore a comparative study using a data set with infrequent artifacts might mainly test the ability of algorithms to cope with non-stationarity.

To summarize it seems that the suggested comparative framework is of high practical value as demonstrated on a limited data set. Future work will assess the separability between cortical activity and artifact signals.

Acknowledgments

Help with the measurements by Alf Pretzell, the DAAD (German Academic Exchange Service) scholarship for H.Z.F. (PKZ: A/04/21558), and the Berlin Neuroimaging Centre (BMBF 01GO0208 BNIC) are gratefully acknowledged.

References

1. Hyvärinen, A., Karhunen, J., Oja, E.: ICA, John Wiley and Sons, New York, 2001.
2. Roberts, S., Everson, R., Eds.: ICA - Principles and Practice, Cambridge University Press, Cambridge, 2001.
3. Cichocki, A., Amari, S.: Adaptive Blind Signal and Image Processing. John Wiley and Sons, New York, 2002.
4. Ziehe, A., Nolte, G., Sander, T.H., Mueller, K.-R., Curio, G.: A comparison of ICA-based artifact reduction methods for MEG, in *Proc of Biomag2001*, J. Nenonen, R.J. Ilmoniemi, and T. Katila Eds., Helsinki Univ. of Technology, pp. 895–899, 2001.
5. Moran, J.E., Drake, C.L., Tepley, N.: ICA Methods for MEG Imaging, *Neurol. And Clin. Neurophysiol.*, vol. 72, pp. 1-4, 2004.
6. Joyce, C.A., Gorodnitsky, I.F., Kutas, M.: Automatic Removal of Eye Movement and Blink Artifacts from EEG Data Using Blind Component Separation, *Psychophysiol.*, vol. 72, pp. 313-325, 2005.
7. Hyvärinen, A., Oja, E.: A Fast Fixed-Point Algorithm for Independent Component Analysis, *Neural Computation*, vol. 9, p. 1483–1492, 1997.
8. Bell, A., Sejnowski, T.: An Information Approach to Blind Separation and Blind Deconvolution, *Neural Comput.*, 7, pp. 1129-1159, 1995.
9. Lee, T.-W., Girolami, M., Sejnowski, T.-J.: ICA Using an Extended Infomax Algorithm for Mixed Sub- and Supergaussian Sources. *Neural Comp.*, vol. 11, pp. 417–441, 1999.
10. Cardoso, J.-F., Souloumiac, A.: Blind Beamforming for Non Gaussian Signals. *IEE-Proceedings-F*, vol. 140, no 6, pp. 362–370, 1993.
11. Cardoso, J.-F.: High-Order Contrasts for Independent Component Analysis. *Neural Computation*, vol. 11, pp. 157–192, 1999.
12. Belouchrani, A., Abed-Meraim, K., Cardoso, J.-F., Moulines, E.: A Blind Source Separation Technique Based on Second-Order Statistics, *IEEE Trans. on Sig. Proc.*, vol. 45, pp. 434–444, 1997.
13. Ziehe, A., Müller, K.-R.: TDSEP – An Efficient Algorithm for Blind Separation Using Time Structure, in *Proceedings of the 8th ICANN*, L. Niklasson, M. Bodén, and T. Ziemke Eds., pp. 675–680, Springer Verlag, 1998.
14. Köhler, B.-U., Orglmeister, R.: A Blind Source Separation Algorithm Using Weighted Time Delays, in *Proc. of the 2nd Intern. Workshop on ICA and BSS*, P. Pajunen and J. Karhunen Eds., Helsinki, pp. 471–475, 2000.
15. Delorme, A., Makeig, S.: EEGLAB: An Open Source Toolbox for Analysis of Single-Trial EEG Dynamics Including Independent Component Analysis, *Journal of Neuroscience Methods.*, vol. 134, pp. 9-21, 2003.
16. Sander, T.H., Wuebbeler, G., Lueschow, A., Curio, G., Trahms, L.: Cardiac Artifact Subspace Identification and Elimination in Cognitive MEG-Data Using Time-Delayed Decorrelation, *IEEE Trans. Biomed. Eng.*, vol. 49, p. 345–354, 2002.
17. Brockmeier, K., Schmitz, L., Bobadilla-Chavez, J.-J., Burghoff, M., Koch, H., Zimmermann, R., Trahms, L.: Magnetocardiography and 32-Lead Potential Mapping: Repolarization in Normal Subjects During Pharmacologically Induced Stress, *J. Cardiovasc. Electrophys.*, vol. 8, p. 615-626, 1997.

Automatic De-noising of Doppler Ultrasound Signals Using Matching Pursuit Method

Yufeng Zhang, Le Wang, Yali Gao, Jianhua Chen, and Xinling Shi

Department of Electronic Engineering, Yunnan University, Kunming, Yunnan Province, 650091, The People's Republic of China
yfengzhang@yahoo.com,
{wangle, gaoyl, chenjh, shixl}@ynu.edu.cn

Abstract. A novel de-noising method, called matching pursuit method, for improving the signal-to-noise ratio (SNR) of Doppler ultrasound blood flow signals is proposed. Using this method, the Doppler ultrasound signal is first decomposed into a linear expansion of waveforms, called time-frequency atoms, which are selected from a redundant dictionary named Gabor functions. Then a decay parameter-based algorithm is employed to determine the decomposition times. Finally, the de-noised Doppler signal is reconstructed using the selected components. The SNR improvements and the maximum frequency estimation precision with simulated Doppler blood flow signals have been used to evaluate a performance comparison based on the wavelet, the wavelet packets and the matching pursuit de-noising algorithms. From the simulation and clinical experiment results, it is concluded that the performance of the matching pursuit approach is the best for the Doppler ultrasound signal de-noising.

1 Introduction

Doppler ultrasound provides a noninvasive assessment of the hemodynamic flow condition within blood vessels and cardiac cavities [1, 2]. Diagnostic information is extracted from spectrograms of the Doppler blood flow signal which results from the backscattering of the ultrasound beam by moving red blood cells. The spectral estimation of Doppler ultrasound signals is normally performed using the short-time Fourier transform (STFT). Then the mean and maximum frequency waveforms, which are used to evaluate the mean flow volume and determine the flow statues, are extracted from the estimated spectrogram. Meanwhile, indices estimated from the maximum frequency waveform, such as SBI (spectral broadening index), S/D (the ratio of systolic maximum velocity to diastolic velocity), PI (pulsatility index) and RI (resistive index), are used for quantification of vascular diseases' severity [1, 2]. Any extra frequency component in the Doppler signal coming from noise may reduce the spectral estimation resolution, which harms extraction of maximum frequency waveform and further processing. Therefore, it is a preliminary and important step to de-noise the Doppler ultrasound signal, especially when the signal-to-noise ratio (SNR) is low.

The wavelet shrinkage method with the standard discrete wavelet transform (DWT), first proposed by Donoho [3], has been employed to de-noise Doppler ultrasound signals [4]. Here the DWT using real-valued wavelet filter coefficients was applied to the data, and the resulting wavelet coefficients were soft-thresholded. The inverse

J. Rosca et al. (Eds.): ICA 2006, LNCS 3889, pp. 519 – 526, 2006.

DWT then returned the de-noised data and the signal's spectrogram was found through the STFT. An alternative method, called discrete wavelet frame (DWF)[5, 6], was also used to de-noise the Doppler ultrasound signals. Unlike the standard DWT, the output of the filter bank in this method is not sub-sampled, leading to the wavelet coefficients from the DWF decomposition invariant with respect to the translations of the input signal. This method has proved to be superior to methods based on the DWT.

The matching pursuit (MP) algorithm, introduced by Mallat and Zhang [7], relies on adaptive decomposition of a signal into waveforms called atoms from a large and redundant dictionary of functions. Unlike the DWT and the wavelet packets (WPs) algorithms, which select the basis that is best adapted to the global signal properties, the MP is a greedy algorithm that chooses at each iteration a waveform that is best adapted to approximate part of the signal. The global optimization does not perform well for decomposition of non-stationary signal because the information is diluted across the whole basis, as opposed the greedy approach of a matching pursuit. Thus the MP algorithm is a locally adaptive method, which is efficient to represent non-stationary signals, such as the Doppler blood flow signals, which have a relative wide bandwidth changing rapidly with time. The present paper is the first attempt to de-noise Doppler ultrasound signals using the MP algorithm. Commonly used methods, the DWT and the WPs, are compared with the adaptive decomposition method based on the MP technique with a Gabor dictionary.

In this paper we shall first briefly describe the mathematical background of the used de-noising methods. The simulation of the Doppler blood flow signals, experiments on simulated Doppler signals and clinical cases based on three methods will be presented in Section III, followed by the results, discussion, and conclusion.

2 Methods

2.1 Wavelet and Wavelet Packets De-noising Algorithms

The mathematical description of the continuous wavelet transform (CWT) of $f \in \mathbf{L}^2(\mathbb{R})$ was described by Mallat [8] as

$$WT_f(u,s) = \int_{-\infty}^{+\infty} f(t)\psi_{u,s}^*(t)dt, \quad \psi_{u,s}(t) = \frac{1}{\sqrt{|s|}}\psi\left(\frac{t-u}{s}\right) \tag{1}$$

where $\psi_{u,s}(t)$ is a family of orthogonal wavelets. s and u denote the dilation and translation parameters respectively. The DWT calculates the wavelet coefficients at discrete intervals of time and scale instead of at all scales. In this case, a vector space is generated by scaling and translating two basic functions, $\phi(t)$ and $\varphi(t)$, defined as

$$\phi_{j,k}(t) = 2^{j/2}\phi(2^j t - k), \quad \varphi_{j,k}(t) = 2^{j/2}\varphi(2^j t - k) \tag{2}$$

By combining the two, the function $f(t)$ is precisely obtained as

$$f(t) = \sum_k c_{j_0}(k)\varphi_{j_0,k}(t) + \sum_k \sum_{j=j_0}^{\infty} d_j(k)\phi_{j,k}(t) \tag{3}$$

In Eq. (3), two kinds of coefficients, called discrete wavelet transform, are used as projections in the vector space: the scaling coefficients $c_{j_0}(k)$ which are rough details, and the wavelet coefficients $d_j(k)$ which are finer details.

Mallat developed an efficient way to implement this algorithm, which is known as a two-channel sub-band coder[8]. For a single level of decomposition, this algorithm passes the signal through two complementary (high-pass and low-pass) filters resulting in approximations which are high-scale, low-frequency components of the signal, and details which are low-scale, high-frequency components of the signal. For further levels of decomposition, successive approximations may be iteratively broken down into details and approximations. WPs also spread the DWT, decomposing the high pass filter output, that is, the finer details. This results in a binary tree filter bank with a number of levels depending on the desired scale of resolution. The binary tree can be considered as a library of bases called WPs [9]. The objective is to select the best base to represent the signal by adequately pruning the tree. This is done by using a certain criterion of measuring the information cost of each node. In this paper, Shannon entropy has been used [9]. When reconstructing the signal, coefficients below a certain level are regarded as noise and thresholded out. Thresholding may be soft or hard. The soft thresholding-based de-noising algorithm has been proven to have the advantages of optimizing mean square error and keeping the smoothness of the de-noised signal [3]. A Gaussian white noise with a noise level σ is presumed in Doppler signals. A soft-thresholding nonlinearity is applied to all wavelet detail coefficients with a threshold defined as

$$T = r_1\sigma\sqrt{2\log(L)} \tag{4}$$

where the constant r_1 can be set to 1 while choosing the orthogonal wavelet, L is the length of the signal. The thresholded wavelet coefficients $\hat{d}_j(k)$ can be obtained by:

$$\hat{d}_j(k) = \begin{cases} \operatorname{sgn}(d_j(k))(|d_j(k)| - T) & |w_m| \geq T \\ 0 & |w_m| < T \end{cases} \tag{5}$$

It is easy to find that the threshold T is proportional to the noise level σ, which can be estimated from the standard deviation of the detail coefficients at the first resolution level because the Doppler ultrasound signal is band-limited[4, 6]. Then the noise-free signals based on the DWT and WP algorithms can be estimated from the reconstructions based on the thresholded coefficients, respectively.

2.2 Matching Pursuit De-noising Algorithm

MP decomposition of a signal $f(t)$ is carried out by first approximating the signal with its orthogonal projection onto an atom that can best match the local structure of the signal [7]. Then, the same procedure is repeated on the residual vector until the signal is decomposed into a series of time-frequency atoms in decreasing energy order

$$f(t) = \sum_{n=0}^{m-1} \left\langle R^n f, g_{\gamma_n} \right\rangle g_{\gamma_n} + R^m f \tag{6}$$

where $R^m f$ is the residual vector after $f(t)$ is approximated by m atoms, and $\left\langle R^n f, g_{\gamma_n} \right\rangle g_{\gamma_n}$ represents the projection of $R^n f$ onto an atom g_{γ_n} (note $R^0 f = f$).

The discrete implementation of a matching pursuit is explained for a dictionary of Gabor time-frequency atoms. A real Gabor function can be expressed as

$$g_\gamma(t) = K(\gamma) e^{-\pi((t-u)/s)^2} \sin(2\pi \frac{\omega}{N}(t-u) + \phi) \tag{7}$$

where N is the size of the signal for which the dictionary is constructed, and $K(\gamma)$ is such that $\|g_\gamma\| = 1$. $\gamma = \{u, \omega, s, \phi\}$ denotes parameters of the dictionary's functions (time-frequency atoms). In the dictionary implemented originally by Mallat [7], the parameters of the atoms are chosen from dyadic sequences of integers. Parameters u and ω are sampled for each octave j with interval $s = 2^j$.,$0 \le j \le L$, (signal size $N = 2^L$).

MP decomposition is an iterative algorithm. In this work, decomposition is stopped after extracting the first M coherent structures of the signal. The first M coherent structures are determined using a decay parameter, denoted by

$$\lambda(M) = \sqrt{1 - \frac{\|R^M f\|^2}{\|R^{M-1} f\|^2}} \tag{8}$$

where $R^M f$ is the residual energy level at the Mth iteration. The decomposition is continued until the decay parameter does not reduce any further. At this stage, the selected components represent the coherent structures, and the residue represents the incoherent structures in the signal with respect to the dictionary. The residue can be assumed to be due to random noise, as it does not show any time frequency localization. The signal reconstructed using M coherent structures, i.e.

$$f = \sum_{n=0}^{M} \left\langle R^n f, g_{\gamma_n} \right\rangle g_{\gamma_n} . \tag{9}$$

3 Experiments

In the experimental study, the Symmlet wavelets are selected to implement the DWT and WPs algorithms. The eighth-order wavelet is used in the standard DWT, and sixth-order wavelets are tested in the WPs. The computational complexity of the DWT and the WPs increase with the analyzing resolutions, and the low frequency part of the Doppler ultrasound signal is usually filtered away by the wall filter. Concerning these two aspects, the coarsest resolution I is chosen as 6 in this paper. For the MP method, the decomposition is continued until the decay parameter is less than or equal to 0.01, which approximately means that it does not reduce any further.

In simulation study, the Doppler blood flow signal model proposed by Wang and Fish [10] has been used. Normal carotid artery signals are simulated based on this model. The sample frequency is 20 kHz, above the Nyquist rate; the cardiac cycle period is 1000 ms. The prespecified SNRs (SNR = 0 dB and SNR = 10 dB) are obtained by adding white Gaussian noise to the simulated Doppler signals. The quantitative indices, SNR of the Doppler signal is defined as:

$$SNR = 10 \times \log_{10} \frac{S_0}{N_0}, \quad S_0 = \frac{1}{L}\sum_{l=0}^{L-1}(S(l)-\mu_S)^2, \quad N_0 = \frac{1}{L}\sum_{l=0}^{L-1}(N(l)-\mu_N)^2 \tag{10}$$

where S and N are the simulated signal and the noise of the length L, μ_S and μ_N are the mean values of the S and N, respectively. The Doppler spectrograms are calculated by using the STFT with a 10 ms Gaussian window ($\alpha = 3$). The theoretical maximum frequency of the simulated Doppler signal can be determined by $f_{\max}(t) = f_m(t) + 3\sigma_f(t)$. The maximum frequency waveforms of the de-noised Doppler signals are extracted from the estimated spectrograms using the percentile method, whose percentile factor is chosen as 0.1 in this study. The SNR and the RMS error of the maximum frequency waveform are calculated before and after de-noising to compare the performance improvements of the three de-noising algorithms.

In the clinical study, the measurement system employed consists of the pulsed Doppler unit of a HP SONOS 5500 ultrasound imaging system, analog/digital interface card (sound Blaster Pro-16 bit), one PC and one printer. The Doppler unit is used in pulse mode, and the applied frequency of the ultrasound is 2.7 MHz. The clinical Doppler signals are recorded from a child's aorta by placing sample volume just near the center of the aortic arch. The audio Doppler signals are sampled using the Sound Blaster Card in the PC. The sample rate is set to 22.05 kHz. The three de-noising algorithms with the same parameters described in simulation study are used to de-noise the clinical signals. The overall enhancement of the spectrogram and the smoothness of the maximum frequency waveform are used to compare the de-noising performance.

4 Results and Discussion

Fig.1 shows the SNR values of the results of the three de-noising methods applied to the simulated noisy Doppler blood flow signals. From this figure, it can be found that the MP technique has provided the best de-noising result (the highest SNR value) of the three methods studied.

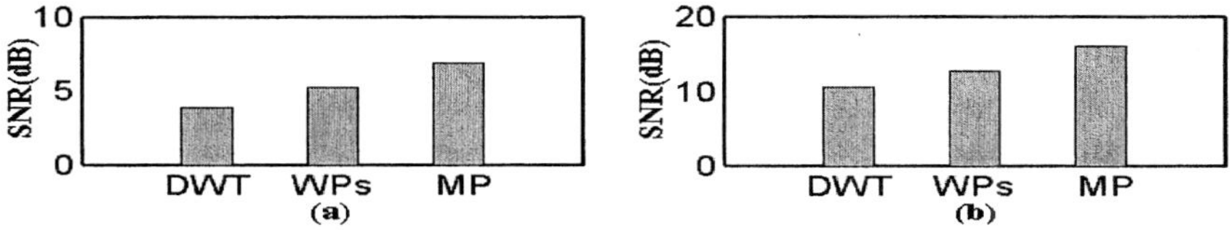

Fig. 1. The SNR values of the results of the three de-noising methods applied to the simulated Doppler blood flow signal with SNRs of 0 dB (a) and 10dB (b)

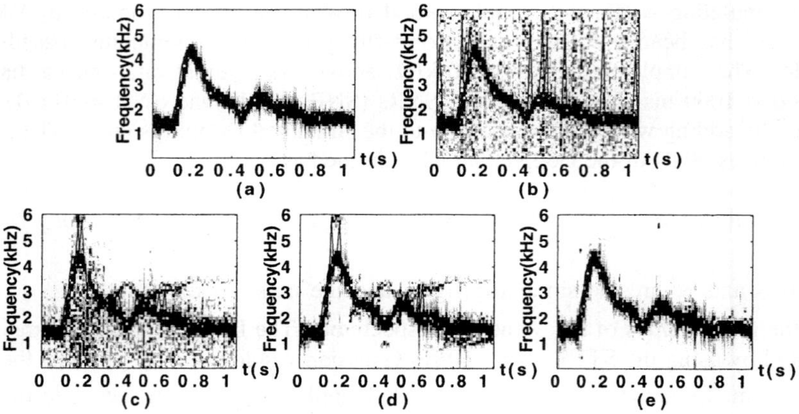

Fig. 2. The spectrogram and the maximum frequency waveform, which is superimposed on the spectrogram with a solid curve, of the simulated original signal (a), the signal added noise (SNR=5 dB) (b). The de-noised versions using the DWT (c), the WPs (d), and the MP (e).

Fig. 2 presents the estimated Doppler spectrograms of one original simulated signal, the noisy signal with SNR=5 dB and the de-noised signals. In Fig. 2(b), it is observed that the Doppler spectrograms include a mass of disturbance distributed in whole frequency band. The maximum frequency waveform extracted from the estimated spectrogram includes considerable distortion, which indicates the difficulty in finding correct indices used for quantification of vascular diseases' severity from the spectrograms under the lower SNR situation. Fig. 2 (c)-(e) show the enhancements of the spectrograms and the improvements of the maximum frequency waveforms after de-noising using the DWT, the WPs and the MP, respectively. It is obvious that there are the least extra frequency components in the spectrogram and the least distortion in maximum frequency waveform shown in Fig. 2 (e), de-noised using the MP algorithm.

Table1. The mean $\mu_{f\max}$ and standard deviation $\sigma_{f\max}$ of the RMS errors of the maximum frequency waveforms extracted from the original noisy signal (SNR=5 dB), the de-noised signals based on three different algorithms

Signal	Noisy signal	De-noised by DWT	De-noised by WPs	De-noised by MP
$\mu_{f\max}$ (Hz)	1461	308	211	104
$\sigma_{f\max}$ (Hz)	191	66	31	17

Table 1 lists the RMS errors of the maximum frequency waveforms extracted from the de-noised spectrograms over 30 independent realizations of noisy Doppler signal with SNR=5 dB. It is found that all RMS errors of the maximum frequency waveform of the de-noised signals have decreased much more than those before de-noising. The maximum frequency waveform of the signal de-noised using the WPs has a lower RMS error and deviation than that associated with the DWT. However, the mean and standard deviation of the RMS error in the maximum frequency are the lowest when the MP method is used to de-noise the Doppler signal.

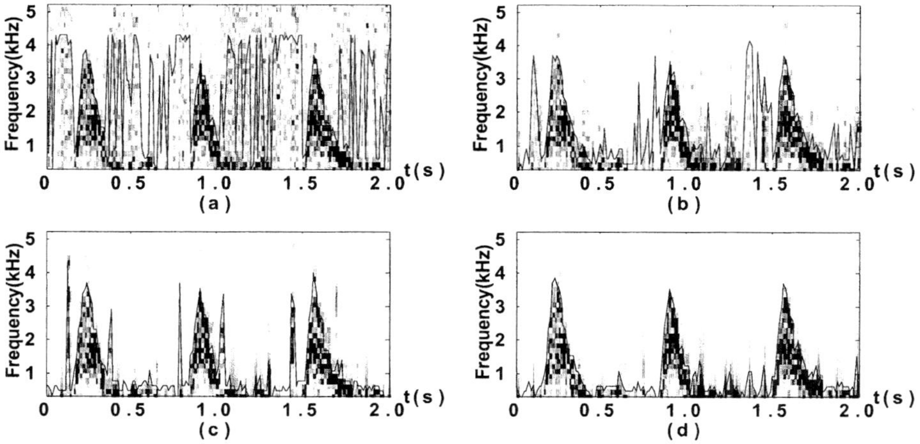

Fig. 3. The spectrogram and the maximum frequency waveform, which is superimposed on the spectrogram with a solid curve, of the clinical original noisy signal (a), The de-noised versions using the DWT (b), the WPs (c), and the MP (d)

The three different algorithms are applied to de-noise the clinical Doppler ultrasound signals. The spectrograms and maximum frequency waveforms are estimated for the comparison. As an illustration, a segment of Doppler ultrasound signal recorded from a child's aortic arch is shown in Fig. 3. (a). The de-noised versions based on the DWT, the WPs and the MP are shown in Fig.3 (b)-(d). It can be found that all three algorithms have removed noises and distortions in the spectrograms and maximum frequency waveforms. From Fig. 3. (c), we can also observe that the spectrogram of the signal after de-noising using the WPs has fewer additional noises than that using the DWT shown in Fig. 3. (b). As observed in Fig. 3(d), it is obvious that the most significant amount of random noise has been removed from the original signal by the MP de-noising method. The superimposed maximum frequency waveforms also confirm that the algorithm using the MP achieves the best performance.

It is worthwhile to mention that the de-noising results with wavelets and the WPs are highly dependent on the selection of the threshold value for the coefficients. In the case of the MP method, the decay parameter is a more objective measure.

5 Conclusion

A novel approach to de-noise Doppler ultrasound signals for enhanced spectrogram estimation using the MP is proposed. The MP algorithm isolates signal structures that are coherent with respect to a given dictionary. After removing the coherent structures from a signal, the residue behaves like a realization of the noise. Thus, the MP method is based on local optimization and very suitable for decomposition of nonstationary signals, whereas the wavelet techniques are well adapted to global signal properties. From the simulation and clinical experiment results, it is concluded that the performance of the decay parameter-based approach using the matching pursuit algorithm is the best for the Doppler ultrasound signal de-noising.

References

1. Kalman, P.G., Johnston, K.W., Zuech, P., Kassam, M., Poots, K.: In vitro comparison of alternative methods for quantifying the severity of Doppler spectral broadening for the diagnosis of carotid arterial occlusive disease, Ultrasound Med. Biol. 11 (1985) 435–440.
2. Johnston, K.W., Taraschuk, I.: Validation of the role of pulsatility index in quantitation of the severity of peripheral arterial occlusive disease, Am. J. Surg. 131 (1976) 295–297
3. Donoho, D.L.: De-noising by soft-thresholding, IEEE Trans. Inform. Theory. 41 (1995) 613–627.
4. Liu, B., Wang, Y., Wang, W.: Spectrogram enhancement algorithm: a soft thresholding-based approach, Ultrasound Med. Biol. 25 (1999) 839–846.
5. Lang, M., Guo, H., Odegard, J. E., Burrus, C. S., Wells, R. O.: Noise reduction using an undecimated discrete wavelet transform, IEEE Signal Processing Letters. 3 (1996) 10–12.
6. Zhang, Y., Wang, Y., Wang, W., Liu, B.: Doppler ultrasound signal de-noising based on wavelet frames, IEEE Ultrason. Ferroelect. Freq. Contr. 48 (2001) 709–716.
7. Mallat, S.G., Zhang, Z.: Matching pursuits with time-frequency dictionaries, IEEE Trans. Signal Processing. 41 (1993) 3397–3415.
8. Mallat, S.G.: A wavelet tour of signal processing, Academic Press, 1999.
9. Coifman, R.R., Winckerhauser, M.V.: Entropy-based algorithms for best basis selection. IEEE Trans. Inform. Theory. 38 (1992) 713–718.
10. Wang, Y., Fish, P.J.: Arterial Doppler signal simulation by time domain processing, Eur J Ultrasound. 3 (1996) 71–81.

A Novel Normalization and Regularization Scheme for Broadband Convolutive Blind Source Separation

Robert Aichner, Herbert Buchner, and Walter Kellermann

Multimedia Communications and Signal Processing,
University of Erlangen-Nuremberg,
Cauerstr. 7, D-91058 Erlangen, Germany
{aichner, buchner, wk}@LNT.de

Abstract. In this paper we propose a novel blind source separation (BSS) algorithm for convolutive mixtures combining advantages of broadband algorithms with the computational efficiency of narrowband techniques. It is based on a recently presented generic broadband algorithm. By selective application of the Szegö theorem which relates properties of Toeplitz and circulant matrices, a new normalization is derived which approximates well the exact normalization of the generic broadband algorithm presented in [2]. The new scheme thus results in a computationally efficient and fast converging algorithm while still avoiding typical narrowband problems such as the internal permutation problem or circularity effects. Moreover, a novel regularization method for the generic broadband algorithm is proposed and subsequently also derived for the proposed algorithm. Experimental results in realistic acoustic environments show improved performance of the novel algorithm compared to previous approximations.

1 Introduction

The problem of recovering signals from multiple observed linear mixtures is commonly refered to Blind source separation (BSS) [4]. To deal with the convolutive mixing case as encountered, e.g., in acoustic environments, we are interested in finding a corresponding demixing system, where the output signals $y_q(n)$, $q = 1, \dots, P$ are described by $y_q(n) = \sum_{p=1}^{P} \sum_{\kappa=0}^{L-1} w_{pq,\kappa} x_p(n-\kappa)$. Here $w_{pq,\kappa}$, $\kappa = 0, \dots, L-1$ denote the current weights of the MIMO filter taps from the p-th sensor channel $x_p(n)$ to the q-th output channel. In this paper the number of *active source* signals Q is less or equal to the number of microphones P. BSS algorithms are solely based on the fundamental assumption of mutual statistical independence of the different source signals. The separation is achieved by forcing the output signals y_q to be mutually statistically decoupled up to joint moments of a certain order.

In [2] a general derivation of broadband convolutive BSS algorithms based on second-order statistics has been presented for both, the time and frequency

J. Rosca et al. (Eds.): ICA 2006, LNCS 3889, pp. 527–535, 2006.

domain. This broadband derivation yields an algorithm which possesses an inherent normalization of the coefficient update leading to fast convergence also for colored signals such as speech. However, for realistic acoustic environments large correlation matrices have to be inverted for every output channel. An approximation of this matrix by a diagonal matrix led to a very efficient algorithm which allows real-time implementation using a block-online update structure [1]. In this paper we first briefly summarize the generic second-order broadband algorithm and then present a novel normalization strategy leading to an improved algorithm combining broadband advantages with narrowband efficiency. Moreover, a new regularization method for the obtained algorithm is presented.

2 Generic Broadband Algorithm

A block processing broadband algorithm simultaneously exploiting nonwhiteness and nonstationarity of the source signals was derived in [2] based on the natural gradient of the cost function $\mathcal{J}$:

$$\nabla_{\mathbf{W}}^{\mathrm{NG}} \mathcal{J}(m) = 2\sum_{i=0}^{\infty} \beta(i,m)\mathbf{W}\left\{\mathbf{R}_{\mathbf{yy}}(i) - \mathrm{bdiag}\,\mathbf{R}_{\mathbf{yy}}(i)\right\} \mathrm{bdiag}^{-1}\,\mathbf{R}_{\mathbf{yy}}(i). \quad (1)$$

The variables m and i denote the block time index. $\mathbf{W}$ is the demixing matrix containing the filter weights and exhibits a Sylvester structure, i.e., each column of the submatrices $\mathbf{W}_{pq}$ is shifted by one sample containing the current weights of the MIMO sub-filter of length L from the p-th sensor channel to the q-th output channel [2]. The short-time correlation matrix $\mathbf{R}_{\mathbf{yy}}(i)$ of size $PL \times PL$ is composed of channel-wise $L \times L$ submatrices $\mathbf{R}_{\mathbf{y}_p\mathbf{y}_q}(i)$ each containing L time-lags and thus exploiting nonwhiteness of the source signals. The bdiag operation on a partitioned block matrix consisting of several submatrices sets all submatrices on the off-diagonals to zero. Therefore, the update becomes zero if and only if $\mathbf{R}_{\mathbf{y}_p\mathbf{y}_q}$, $p \neq q$, i.e., all *output cross-correlations over all time-lags* become zero. Thus, the algorithm explicitly exploits the nonwhiteness property of the output signals. The variable β denotes a weighting function with finite support that is normalized according to $\sum_{i=0}^{m} \beta(i,m) = 1$ allowing for offline, online or block-online realizations of the algorithm. In [1] a block-online update rule was derived for the coefficient update (1) by specifying $\beta(i,m)$ such that it leads to a combination of an online update and an offline update exploiting the nonstationarity of the source signals (in addition to their nonwhiteness).

In principle, there are two basic methods for the block-based estimation of the short-time output correlation matrices $\mathbf{R}_{\mathbf{y}_p\mathbf{y}_q}(i)$ for nonstationary signals: the so-called *covariance method* and the *correlation method*, as they are known from linear prediction problems [5]. In contrast to [2] the computationally less complex correlation method was used in [1] which is obtained by assuming stationarity within each block i. This method is also considered here as it leads to a *Toeplitz structure* of the $L \times L$ matrices $\mathbf{R}_{\mathbf{y}_p\mathbf{y}_q}(i)$. Using the correlation method, the Toeplitz matrix $\mathbf{R}_{\mathbf{y}_p\mathbf{y}_q}$ can be written as a matrix product

$$\mathbf{R}_{\mathbf{y}_p\mathbf{y}_q}(i) = \tilde{\mathbf{Y}}_p^H(i)\tilde{\mathbf{Y}}_q(i) \tag{2}$$

where $\tilde{\mathbf{Y}}_q$ is a $N + L - 1 \times L$ Sylvester matrix (N denotes the block length for estimation of the correlations), i.e., for each subsequent column the first N output signal values are shifted by one sample.

$$\tilde{\mathbf{Y}}_q(m) = \begin{bmatrix} y_q(mL) & \cdots & 0 \\ y_q(mL+1) & \ddots & \vdots \\ \vdots & \ddots & 0 \\ y_q(mL+N-1) & \ddots & y_q(mL) \\ 0 & \ddots & y_q(mL+1) \\ \vdots & & \vdots \\ 0 & \cdots & y_q(mL+N-1) \end{bmatrix} \tag{3}$$

3 Normalization Strategies

3.1 Exact Normalization Based on Matrix Inverse

The update of the generic algorithm based on the natural gradient (1) exhibits a normalization by the inverse of a block-diagonal matrix. This means that for the correlation method the $L \times L$ Toeplitz matrices $\mathbf{R}_{\mathbf{y}_q\mathbf{y}_q}$, $q = 1, \ldots, P$, given by (2), have to be inverted in (1). The complexity of a Toeplitz matrix inversion is $\mathcal{O}(L^2)$. For realistic acoustic environments, large values for L (e.g., 1024) are required which are prohibitive for a real-time implementation of the exact normalization on most current hardware platforms.

3.2 Normalization Based on Diagonal Matrices in the Time-Domain

In [1] an approximation of the matrix inverse has been used to obtain an efficient algorithm suitable for real-time implementation. There, the off-diagonals of the auto-correlation submatrices have been neglected, so that for the correlation method it can be approximated by a diagonal matrix with the output signal powers, i.e., $\mathbf{R}_{\mathbf{y}_q\mathbf{y}_q}(i) \approx \text{diag}\left\{\mathbf{R}_{\mathbf{y}_q\mathbf{y}_q}(i)\right\} = \sigma_{y_q}^2(i)\mathbf{I}$ for $q = 1, \ldots, P$, where the diag operator applied to a matrix sets all off-diagonal elements to zero. Thus, the matrix inversion is replaced by an element-wise division.

3.3 Novel Approximation of Exact Normalization Based on the Szegö Theorem

The broadband algorithm given by (1) can also be formulated equivalently in the frequency domain as has been presented in [2]. Additionally it has been shown that by certain approximations to this frequency-domain formulation a purely narrowband version of the broadband algorithm can be obtained. In this section we will derive a novel algorithm combining broadband and narrowband techniques by using two steps. First, the exact normalization is formulated equivalently in the frequency domain. In a second step the Szegö theorem [3] is applied to the normalization to obtain an efficient version of the exact normalization. The Szegö theorem allows a *selective* introduction of narrowband approximations to specific parts of the algorithm. This approach allows to combine both, the advantages of the broadband algorithm (e.g. avoiding internal, i.e., discrete Fourier transform (DFT) bin-wise permutation ambiguity and circularity problem) and the low complexity of a narrowband approach.

Exact Normalization Expressed in the Frequency Domain. In [3] it was shown that any Toeplitz matrix can be expressed equivalently in the frequency domain by first generating a circulant matrix by proper extension of the Toeplitz matrix. Then the circulant matrix is diagonalized by using the DFT matrix $\mathbf{F}_R$ of size $R \times R$ where $R \geq N + L$ denotes the transformation length. These two steps are given for the Toeplitz output signal matrix $\tilde{\mathbf{Y}}_q$ as

$$\tilde{\mathbf{Y}}_q = \mathbf{W}^{01_{N+L}}_{N+L\times R}\mathbf{C}_{\tilde{\mathbf{Y}}_q}\mathbf{W}^{1_L 0}_{R\times L} \tag{4}$$

$$= \mathbf{W}^{01_{N+L}}_{N+L\times R}\mathbf{F}_R^{-1}\underline{\tilde{\mathbf{Y}}}_q\mathbf{F}_R\mathbf{W}^{1_L 0}_{R\times L}, \tag{5}$$

where $\mathbf{C}_{\tilde{\mathbf{Y}}_q}$ is an $R \times R$ circulant matrix, and the window matrices are given as

$$\mathbf{W}^{01_{N+L}}_{N+L\times R} = [\mathbf{0}_{N+L\times R-N-L}, \mathbf{I}_{N+L\times N+L}] \tag{6}$$

$$\mathbf{W}^{1_L 0}_{R\times L} = [\mathbf{I}_{L\times L}, \mathbf{0}_{R-L\times L}]. \tag{7}$$

Here the convention is used that the lower index of a window matrix denotes its dimensions and the upper index describes the positions of ones and zeros. The size of the unity submatrices is indicated in subscript (e.g., "01_L"). The matrix $\underline{\tilde{\mathbf{Y}}}_q$ exhibits a diagonal structure containing the eigenvalues of the circulant matrix $\mathbf{C}_{\tilde{\mathbf{Y}}_q}$ on the main diagonal. The eigenvalues are calculated by the DFT of the first column of $\mathbf{C}_{\tilde{\mathbf{Y}}_q}$ and thus $\underline{\tilde{\mathbf{Y}}}_q$ can be interpreted as the frequency-domain counterpart of $\tilde{\mathbf{Y}}_q$:

$$\underline{\tilde{\mathbf{Y}}}_q = \mathrm{Diag}\{\mathbf{F}_R[0,\ldots,0,y_q(iL),\ldots,y_q(iL+N-1),0,\ldots,0]^T\}. \tag{8}$$

The operator Diag $\{\mathbf{a}\}$ denotes a square matrix with the elements of vector $\mathbf{a}$ on its main diagonal. As an illustration of (4), the circulant matrix $\mathbf{C}_{\tilde{\mathbf{Y}}_q}$ and the window matrices, which constrain the circular matrix to the original matrix $\tilde{\mathbf{Y}}_q$, are shown in Fig. 1. With (5) we can now write $\mathbf{R}_{\mathbf{y}_p\mathbf{y}_q}$ as

$$\mathbf{R}_{\mathbf{y}_p\mathbf{y}_q} = \mathbf{W}^{1_L 0}_{L\times R}\mathbf{F}_R^{-1}\underline{\tilde{\mathbf{Y}}}_p^H\mathbf{F}_R\mathbf{W}^{01_{N+L}}_{R\times N+L}\mathbf{W}^{01_{N+L}}_{N+L\times R}\mathbf{F}_R^{-1}\underline{\tilde{\mathbf{Y}}}_q\mathbf{F}_R\mathbf{W}^{1_L 0}_{R\times L}. \tag{9}$$

It can be seen in the upper left corner of the illustration in Fig. 1 that by extending the window matrix $\mathbf{W}^{01_{N+D}}_{N+D\times R}$ to $\mathbf{W}^{01_R}_{R\times R} = \mathbf{I}_{R\times R}$ only rows of zeros are introduced at the beginning of the matrix $\tilde{\mathbf{Y}}_q$. These appended zeros have

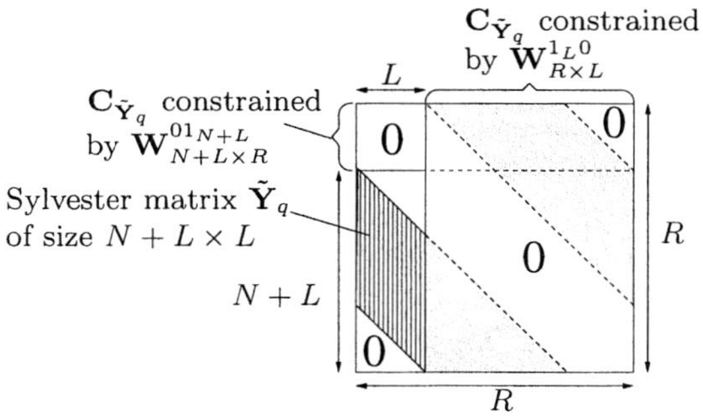

Fig. 1. Illustration of (4) showing the relation between circulant matrix $\mathbf{C}_{\tilde{\mathbf{Y}}_q}$ and Toeplitz matrix $\tilde{\mathbf{Y}}_q$

no effect on the calculation of the correlation matrix $\mathbf{R}_{\mathbf{y}_p\mathbf{y}_q}$ and thus we can replace the multiplication of the window matrices in (9) by

$$\mathbf{W}^{01_R}_{R\times R}\mathbf{W}^{01_R}_{R\times R} = \mathbf{I}_{R\times R}. \tag{10}$$

This leads to

$$\mathbf{R}_{\mathbf{y}_p\mathbf{y}_q} = \mathbf{W}^{1_L 0}_{L\times R}\mathbf{F}_R^{-1}\underline{\tilde{\mathbf{Y}}}_p^H\underline{\tilde{\mathbf{Y}}}_q\mathbf{F}_R\mathbf{W}^{1_L 0}_{R\times L} \tag{11}$$
$$= \mathbf{W}^{1_L 0}_{L\times R}\mathbf{C}_{\tilde{\mathbf{Y}}_p\tilde{\mathbf{Y}}_q}\mathbf{W}^{1_L 0}_{R\times L} \tag{12}$$

The correlation matrix in (11) is an equivalent expression to (2) in the frequency domain. Thus, the normalization based on the inversion of (11) or (12) for $p = q = 1, \ldots, P$ still corresponds to the exact normalization based on the matrix inverse of a Toeplitz matrix as described in Sect. 3.1. In the following it is shown how the inverse of (12) can be approximated to obtain an efficient implementation.

Application of the Szegö Theorem. In the tutorial paper [3] the Szegö theorem is formulated and proven for finite-order Toeplitz matrices. A finite-order Toeplitz matrix is defined as an $R \times R$ Toeplitz matrix where a finite L exists such that all elements of the matrix with the row or column index greater than L are equal to zero. It was shown in [3] that the $R \times R$ Toeplitz matrix of order L is asymptotically equivalent to the $R \times R$ circulant matrix generated from an appropriately complemented $L \times L$ Toeplitz matrix. Moreover, if the two matrices are of hermitian structure, then the Szegö theorem on the asymptotic eigenvalue distribution states:

1. The eigenvalues of both matrices lie between a lower and an upper bound.
2. The arithmetic means of the eigenvalues of both matrices are equal if the size R of both matrices approaches infinity.

Then, the eigenvalues of both matrices are said to be asymptotically equally distributed.

It can be seen in (12) that the autocorrelation matrix necessary for the normalization can either expressed as an $L \times L$ Toeplitz matrix $\mathbf{R}_{\mathbf{y}_q\mathbf{y}_q}$ or an $R \times R$ circulant matrix $\mathbf{C}_{\tilde{\mathbf{Y}}_q\tilde{\mathbf{Y}}_q}$ generated from the Toeplitz matrix by extending it appropriately and multiplying it with some window matrices. According to [3] both matrices are asymptotically equivalent. As both, the Toeplitz and the circulant matrices are hermitian, it is possible to apply the Szegö theorem. The eigenvalues of $\mathbf{C}_{\tilde{\mathbf{Y}}_q\tilde{\mathbf{Y}}_q}$ are given in (11) as the elements on the main diagonal of the diagonal matrix $\underline{\tilde{\mathbf{Y}}}_q^H\underline{\tilde{\mathbf{Y}}}_q$. The Szegö theorem states that the eigenvalues of the $R \times R$ Toeplitz matrix generated by appending zeros to $\mathbf{R}_{\mathbf{y}_q\mathbf{y}_q}$ can be asymptotically approximated by $\underline{\tilde{\mathbf{Y}}}_q^H\underline{\tilde{\mathbf{Y}}}_q$ for $R \to \infty$. The benefit of this approximation becomes clear if we take a look at the inverse of a circulant matrix. The inverse of a circulant matrix can be easily calculated by inverting its eigenvalues

$$\mathbf{C}^{-1}_{\tilde{\mathbf{Y}}_q\tilde{\mathbf{Y}}_q} = \mathbf{F}_R^{-1}\left(\underline{\tilde{\mathbf{Y}}}_q^H\underline{\tilde{\mathbf{Y}}}_q\right)^{-1}\mathbf{F}_R. \tag{13}$$

By using the Szegö theorem we can now approximate the inverse of the Toeplitz matrix $\mathbf{R}_{\mathbf{y}_q\mathbf{y}_q}$ by the inverse of the circulant matrix (13) for $R \to \infty$

$$\mathbf{R}^{-1}_{\mathbf{y}_q\mathbf{y}_q} \approx \mathbf{W}^{1_L0}_{L\times R}\mathbf{F}_R^{-1}\left(\underline{\tilde{\mathbf{Y}}}_q^H\underline{\tilde{\mathbf{Y}}}_q\right)^{-1}\mathbf{F}_R\mathbf{W}^{1_L0}_{R\times L}. \tag{14}$$

This can also be denoted as narrowband approximation because the eigenvalues $\underline{\tilde{\mathbf{Y}}}_q^H\underline{\tilde{\mathbf{Y}}}_q$ can easily be determined as the DFT of the first column of the circulant matrix $\mathbf{C}_{\tilde{\mathbf{Y}}_q\tilde{\mathbf{Y}}_q}$. The inverse in (14) can now be efficiently implemented as a scalar inversion because $\underline{\tilde{\mathbf{Y}}}_q^H\underline{\tilde{\mathbf{Y}}}_q$ denotes a diagonal matrix. Moreover, it is important to note that the inverse of a circulant matrix is also circulant. Thus, after the windowing by $\mathbf{W}^{1_L0}_{\cdots}$ the resulting matrix $\mathbf{R}^{-1}_{\mathbf{y}_q\mathbf{y}_q}$ exhibits again a Toeplitz structure.

In summary, (14) can be efficiently implemented as a DFT of the first column of $\mathbf{C}_{\tilde{\mathbf{Y}}_q\tilde{\mathbf{Y}}_q}$ followed by a scalar inversion of the frequency-domain values and then applying the inverse DFT. After the windowing operation these values are then replicated to generate the Toeplitz structure of $\mathbf{R}^{-1}_{\mathbf{y}_q\mathbf{y}_q}$. This approach reduces the complexity from $\mathcal{O}(L^2)$ to $\mathcal{O}(\log R)$ (e.g., experiments in Sect. 5: $R = 4L$) which resulted in a real-time implementation on a regular laptop. Obtaining a Toeplitz matrix after the inversion has the advantage that in the update equation (1) again a product of Toeplitz matrices has to be calculated which can be efficiently implemented using fast convolutions. For more details see [1].

4 Regularization of the Matrix Inverse

Prior to inversion of the autocorrelation Toeplitz matrices according to (2) a regularization is necessary as these matrices may be ill-conditioned. Here we propose to attenuate the off-diagonals of $\mathbf{R}_{\mathbf{y}_q\mathbf{y}_q}$ by multiplying them with the factor ρ

$$\breve{\mathbf{R}}_{\mathbf{y}_q\mathbf{y}_q} = \rho\mathbf{R}_{\mathbf{y}_q\mathbf{y}_q} + (1-\rho)\text{diag}\left\{\mathbf{R}_{\mathbf{y}_q\mathbf{y}_q}\right\}, \tag{15}$$

where $\mathrm{diag}\left\{\mathbf{R}_{\mathbf{y}_q\mathbf{y}_q}\right\} = \sigma_{y_q}^2\mathbf{I}$. The weighting factor ρ is chosen such that $0 \leq \rho \leq 1$. Using this regularization the algorithm performs well even if there is just one active source. It should be noted that for $\rho = 0$ the previous approach in [1] summarized in Sect. 3.2 can be seen as a special case of the regularized version of the novel normalization presented in Sect. 3.3.

As outlined in Section 3.3 the selective narrowband approximation leads to an inversion of circulant matrices $\mathbf{C}_{\tilde{\mathbf{Y}}_p\tilde{\mathbf{Y}}_q}$ instead of Toeplitz matrices $\mathbf{R}_{\mathbf{y}_q\mathbf{y}_q}$. Thus, analogously to (15) it is desirable for the proposed algorithm to also regularize $\mathbf{C}_{\tilde{\mathbf{Y}}_p\tilde{\mathbf{Y}}_q}$ prior to inversion:

$$\breve{\mathbf{C}}_{\tilde{\mathbf{Y}}_q\tilde{\mathbf{Y}}_q} = \rho\mathbf{C}_{\tilde{\mathbf{Y}}_q\tilde{\mathbf{Y}}_q} + (1-\rho)\mathrm{diag}\left\{\mathbf{C}_{\tilde{\mathbf{Y}}_q\tilde{\mathbf{Y}}_q}\right\}. \tag{16}$$

In Section 3.3 it was pointed out that every circulant matrix can be expressed using the DFT and inverse DFT matrix and a diagonal matrix

$$\mathbf{C}_{\tilde{\mathbf{Y}}_q\tilde{\mathbf{Y}}_q} = \mathbf{F}_R^{-1}\underline{\tilde{\mathbf{Y}}}_q^H\underline{\tilde{\mathbf{Y}}}_q\mathbf{F}_R. \tag{17}$$

As shown in [3] the diagonal matrix $\underline{\tilde{\mathbf{Y}}}_q^H\underline{\tilde{\mathbf{Y}}}_q$ contains the DFT elements of the first column of the circulant matrix on its diagonal. Thus, by applying the diag operator on $\mathbf{C}_{\tilde{\mathbf{Y}}_q\tilde{\mathbf{Y}}_q}$ we can write

$$\begin{aligned}\mathrm{diag}\left\{\mathbf{C}_{\tilde{\mathbf{Y}}_q\tilde{\mathbf{Y}}_q}\right\} &= r_{y_qy_q}(0)\cdot\mathbf{I} = \sigma_{y_q}^2\cdot\mathbf{I}\\ &= \mathbf{F}_R^{-1}\,\sigma_{y_q}^2\cdot\mathbf{I}\cdot\mathbf{F}_R. \end{aligned}\tag{18}$$

Thus, (16) can be simplified to a narrowband regularization in each frequency bin as

$$\breve{\mathbf{C}}_{\tilde{\mathbf{Y}}_p\tilde{\mathbf{Y}}_q} = \rho\mathbf{F}_R^{-1}\underline{\tilde{\mathbf{Y}}}_q^H\underline{\tilde{\mathbf{Y}}}_q\mathbf{F}_R + (1-\rho)\sigma_{y_q}^2\mathbf{I} \tag{19}$$

$$= \mathbf{F}_R^{-1}\left(\rho\underline{\tilde{\mathbf{Y}}}_q^H\underline{\tilde{\mathbf{Y}}}_q + (1-\rho)\sigma_{y_q}^2\mathbf{I}\right)\mathbf{F}_R. \tag{20}$$

It should be noted that the regularization in (19) can also be applied to purely narrowband algorithms (e.g., Sect. IV-C in [2]). There, considerable separation performance improvements compared to a regularization which adds a constant to the diagonal have been observed, too.

5 Experiments

The experiments were conducted using speech data convolved with measured impulse responses of (a) speakers in a real room with reverberation time $T_{60} = 250$ms at $\pm 45°$ and 2m distance of the sources to the array and (b) impulse responses of a driver and co-driver in a car ($T_{60} = 50$ms) with the array mounted to the rear mirror. In the car scenario recorded background noise with 0dB SNR was added. The sampling frequency was $f_s = 16$kHz. A two-element microphone array with a spacing of 20cm was used for both recordings. To evaluate the performance, the signal-to-interference ratio (SIR) was calculated which is defined as the ratio of the signal power of the target signal to the signal power from the jammer signal. The demixing filter length L was chosen to 1024 taps, the block length $N = 2L$ and the DFT length $R = 4L$. The frameshift was L samples, $K = 8$ blocks have been used to exploit nonstationarity and $j_{\max} = 5$ iterations have been used as number of iterations for the offline update. The adaptive stepsize proposed in [1] has been used with the minimum and maximum values $\mu_{\min} = 0.0001$, $\mu_{\max} = 0.01$, respectively and the forgetting factor $\lambda = 0.2$. The factor ρ for the novel regularization was set to $\rho = 0.5$. The demixing filters were initialized with a shifted unit impulse where $w_{qq,20} = 1$ for $q = 1, \ldots, P$ and zeros elsewhere.

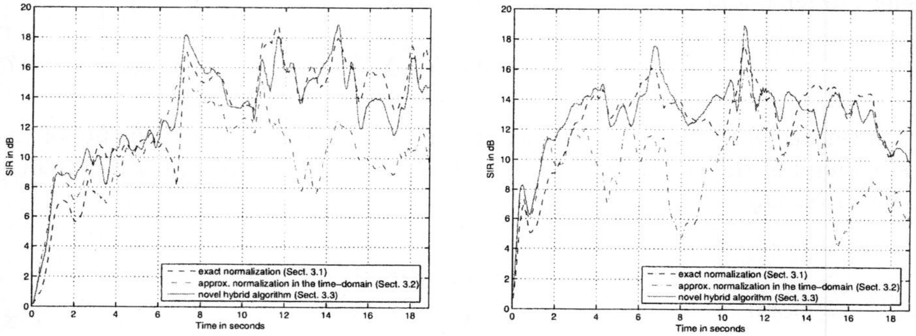

Fig. 2. SIR results for reverberant room (left) and car environment (right)

In Fig. 2 the results of the broadband algorithm with the three different normalization schemes presented in Sect. 3 are shown. It can be seen that the novel normalization scheme (solid) approximates the exact normalization (dashed) very well and yields improved performance compared to the time-domain approximation (dash-dotted). Sometimes the novel algorithm even seems to slightly outperform the exact normalization. This can be explained by the usage of an adaptive stepsize [1] which may result in slightly different convergence speeds for all three algorithms.

6 Conclusion

In this paper a novel efficient normalization scheme was presented resulting in a novel algorithm combining advantages of broadband algorithms with the efficiency of narrowband techniques. Moreover a regularization method was proposed leading to improved convergence behaviour. Experimental results in realistic acoustic environments confirm the efficiency of the proposed approach.

References

1. R. Aichner, H. Buchner, F. Yan, and W. Kellermann. A real-time blind source separation scheme and its application to reverberant and noisy acoustic environments. *Signal Processing*, 2005. to appear.
2. H. Buchner, R. Aichner, and W. Kellermann. A generalization of blind source separation algorithms for convolutive mixtures based on second-order statistics. *IEEE Trans. Speech Audio Processing*, 13(1):120–134, Jan. 2005.
3. R.M. Gray. On the asymptotic eigenvalue distribution of Toeplitz matrices. *IEEE Trans. on Information Theory*, 18(6):725–730, 1972.
4. A. Hyvaerinen, J. Karhunen, and E. Oja. *Independent Component Analysis*. John Wiley & Sons, 2001.
5. J.D. Markel and A.H. Gray. *Linear Prediction of Speech*. Springer Verlag, Berlin, 1976.

A Robust Method to Count and Locate Audio Sources in a Stereophonic Linear Instantaneous Mixture

Simon Arberet, Rémi Gribonval, and Frédéric Bimbot

IRISA, France

Abstract. We propose a robust method to estimate the number of audio sources and the mixing matrix in a linear instantaneous mixture, even with more sources than sensors. Our method is based on a multiscale Short Time Fourier Transform (STFT), and relies on the assumption that in the neighborhood of some (unknown) scales and time-frequency points, only one source contributes to the mixture. Such time-frequency regions provide local estimates of the corresponding columns of the mixing matrix. Our main contribution is a new clustering algorithm called DEMIX to estimate the number of sources and the mixing matrix based on such local estimates. In contrast to DUET or other similar sparsity-based algorithms, which rely on a global scatter plot, our algorithm exploits a local confidence measure to weight the influence of each time-frequency point in the estimated matrix. Inspired by the work of Deville, the confidence measure relies on the time-frequency local persistence of the activity/inactivity of each source. Experiments are provided with stereophonic mixtures and show the improved performance of DEMIX compared to K-means or ELBG clustering algorithms.

1 Introduction

The problem of estimating the number of audio sources and the mixing matrix is considered in a possibly degenerate noisy linear instantaneous mixture $x_m(\tau) = \sum_{n=1}^{N} a_{mn}s_n(\tau) + e_m(\tau)$, $1 \leq m \leq M$, more conveniently written in matrix form $\mathbf{x}(\tau) = \mathbf{A}\mathbf{s}(\tau) + \mathbf{e}(\tau)$. While the M signals $x_m(\tau)$ are observed, the number N of sources as well as the $M \times N$ mixing matrix $\mathbf{A}$, the N source signals $s_n(\tau)$ and the noise signals $e_m(\tau)$ are unknown.

Our approach relies on assumptions similar to those of DUET [1] and TIFROM [2,3]. It exploits the fact that for each source, there is at least one time-frequency region where it is the only source contributing to the mixture. This assumption is related to sparsity of the time-frequency representation of the sources, which is a well-known property of a variety of audio sources. In many sparsity-based source separation approaches [4,5,1] this property is exploited globally by drawing a scatter plot of the time-frequency values $\mathbf{X}(t,f)\}_{t,f}$ – which more or less displays lines directed by the columns $\mathbf{a}_n$ of the mixing matrix – and cluster them into N clusters. Such a global clustering approach is sensitive to the parameters of the clustering algorithm, and to the fact that the direction of some sources of weak energy might not appear clearly in the global scatter plot. Rather than using a full scatter plot, our approach is to exploit the local time-frequency persistence [2,3] of the activity/inactivity of each source to get a robust estimation of the number N of sources and the mixing matrix $\mathbf{A}$. This is similar to the TIFROM [2,3] method,

J. Rosca et al. (Eds.): ICA 2006, LNCS 3889, pp. 536–543, 2006.

which –in the stereophonic case– uses the variance of the ratio $\frac{X_2(t,f)}{X_1(t,f)}$ within a time-frequency region to determine whether the region contains a single active source or more. Our main contributions are to:

1. use a multi-resolution framework (multiple window STFT) to account for the different possible durations of audio structures in each source.
2. rely on a local confidence measure to determine how valid is the assumption that only one source contributes to the mixture in a given time-frequency region;
3. propose a new clustering algorithm called DEMIX, based on the confidence measure, that counts the sources and locates them.

In Section 2, after some reminders on related approaches to estimate the mixing matrix, we give the outline of our approach and describe the confidence measure. In Section 3 we describe the new clustering algorithm DEMIX, and Section 4 is devoted to experiments that compare several methods on audio mixtures.

2 Exploiting Sparsity and Persistence

Let us analyze briefly the most simple sparse source model: assume that at each time τ, only one source $n := n(\tau)$ is active ($s_n(\tau) \neq 0$ and $s_k(\tau) = 0\ \forall k \neq n$). In such a case, the noiseless mixture at time τ is $\mathbf{x}(\tau) = \mathbf{a}_n s_n(\tau)$. In other word each point $\mathbf{x}(\tau) \in \mathbb{R}^M$ is aligned on one of the columns $\mathbf{a}_n$ of the mixing matrix $\mathbf{A}$. In fact this simple model is not very sparse, but (the real and imaginary parts of) STFT values $\mathbf{X}(t,f)$ approximately displays such a behaviour, since the linear mixture model $\mathbf{X}(t,f) = \mathbf{AS}(t,f) + \mathbf{E}(t,f)$ holds and in *many* time-frequency points (t,f), only one source is dominant compared to the others. However, there are points where several sources are similarly active, which can make it difficult to estimate the mixing matrix by simply clustering the global scatter plot.

2.1 Related Work

Many source separation methods for the stereophonic case ($M = 2$) use the idea of sparsity in order to find mixing directions. In Bofill and Zibulevsky's algorithm [4] and DUET [1], the global (time-frequency) scatter plot is transformed into angular values $\theta(t,f) = \tan^{-1}(X_2(t,f)/X_1(t,f))$, and the columns of the mixing matrix are estimated by finding maxima in an energy weighted smoothed histogram of these values. One of the difficulties with this approach is that it seems difficult to adjust how much smoothing must be performed on the histogram to resolve close directions without introducing spurious peaks.

Another approach is the TIFROM method [2,3] which consists in selecting only time-frequency points that have a great chance of being generated by only one source. In TIFROM, for each time-frequency point (t,f), the mean $\bar{\alpha}_{t,f}$ and variance $\sigma^2_{t,f}$ of Time-Frequency Ratios Of Mixtures $\alpha(t',f') = \widehat{x_2}(t',f')/\widehat{x_1}(t',f')$ are computed using all times t' within a neighborhood of t and $f' = f$. By searching for the lowest value of the variance, a time-frequency domain is located where essentially one source is present, and the corresponding column of $\mathbf{A}$ is identified as being proportional to $(1, \bar{\alpha}_{t,f})^T$.

However, it seems quite difficult to exploit TIFROM to actually determine *how many* sources are present in the mixture and find their directions. In addition, the *asymmetric* roles given by $\alpha(t',f')$ to the left and right channels of a stereophonic mixture is not fully satisfying as for sources located almost on the first channel (i.e., with mixing column close to $(0,1)^T$), the corresponding variance are likely to remain high, even at good time-frequency points.

2.2 Proposed Approach

We propose to overcome these limitations of TIFROM by replacing the local variance and mean of the ratios $\frac{\widehat{x_2}(t,f)}{\widehat{x_1}(t,f)}$ with the principal direction of the local scatter plot $(\widehat{x_1}(t,f),\widehat{x_2}(t,f))$, together with a measure of how strongly it points in its principal direction. For this, we first define time-frequency neighborhoods $\Omega_{t,f}$ around each time-frequency point (t,f). A discrete STFT with a window of size L computed with half overlapping windows and no zero padding provides values on the discrete time-frequency grid $t = kL/2$, $k \in \mathbb{Z}$ and $f = l/L$, $0 \leq l \leq L/2$. A possible shape of time-frequency neighborhood of a time-frequency point (t,f) is $\Omega_{t,f} = \{(t+kL/2, f+k'/L), |k| \leq S_T, |k'| \leq S_F\}$ but the approach is amenable to using or combining several shapes and size of neighborhoods. Each neighborhood provides a local scatter plot corresponding to a $M \times \mathrm{card}(\Omega_{t,f})$ matrix $\mathbf{X}_{\Omega_{t,f}}$ with entries $\mathrm{Re}[\mathbf{X}(t',f')]$ and $\mathrm{Im}[\mathbf{X}(t',f')]$ for $(t',f') \in \Omega_{t,f}$. Performing a Principal Component Analysis (PCA) on $\mathbf{X}_{\Omega_{t,f}}$ we obtain a principal direction as a unit vector $\hat{\mathbf{u}}(t,f) \in \mathbb{R}^M$. In the stereophonic case $M = 2$, the direction of the estimated principal unit vector $\hat{\mathbf{u}}(t,f) \in \mathbb{R}^2$ is equivalently translated into an angle $\hat{\theta}(t,f)$.

2.3 A Confidence Measure

To have an idea of how likely it is that the unit principal vector $\hat{\mathbf{u}}(t,f)$ corresponds to a direction of the mixing matrix, we need to know with what confidence we can trust the fact that a single source is active in the corresponding local scatter plot. We propose to rely again on PCA to define the confidence measure

$$\widehat{\mathcal{T}}(t,f) := \hat{\lambda}_1(t,f) / \sum_{i=2}^{M} \hat{\lambda}_i(t,f) \tag{1}$$

where $\hat{\lambda}_1(t,f) \geq \ldots \geq \hat{\lambda}_M(t,f)$ are the eigenvalues of the $M \times M$ matrix $\mathbf{X}_{\Omega_{t,f}} \mathbf{X}_{\Omega_{t,f}}^T$. As explained in Appendix A, this measure can be viewed as a local signal to noise ratio between the dominant source and the contribution of the other ones together with the noise, so we will often express it in deciBels, that is to say $20\log_{10}\widehat{\mathcal{T}}$.

Figure 1(a)-(b) shows the local scatter plot in two time-frequency regions: one where many sources are simultaneously active, and another one where essentially one source is active. It illustrates the good correlation of the value of the confidence measure with the validity of the tested hypothesis.

Figure 2(a) displays the collection of pairs $(\hat{\theta}(t,f), 20\log_{10}\hat{\mathcal{T}}(t,f))$, or *direction-confidence scatter plot* (DCSP), obtained by PCA for all time-frequency regions of the

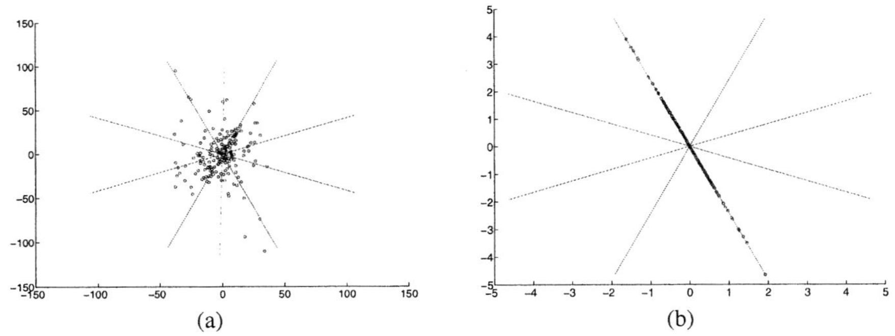

Fig. 1. Two local scatter plots for a stereophonic noiseless mixture of four audio sources. Solid lines indicate all possible true directions, the dashed line indicates the direction estimated by PCA. (a) Local scatter plot in a region where multiple sources contribute to the mixture. The measured confidence value is low (9.4 dB) (b) Region where essentially only one source contributes to the mixture. The measured confidence value is high (101.4 dB) and the dashed line coincides with one of the solid lines.

signal, together with four lines indicating the angles corresponding to the true underlying directions. One can observe that the higher the confidence, the smaller the average distance between the point and one of the true directions. We discuss in Appendix A a statistical analysis of the Significance of the confidence measure in the stereophonic case, which is used to build the DEMIX clustering algorithm described in the next section.

3 The DEMIX Algorithm

We propose a clustering algorithm called DEMIX (Direction Estimation of Mixing matrIX) which estimates both the number of sources and the directions of the columns of the mixing matrix. The algorithm is deterministic and does not rely on a prior knowledge on the number N of columns of $\mathbf{A}$. However, in the case where this number is known the algorithm can be adapted to incorporate this information. The algorithm is described in the stereophonic case $M = 2$ using angles $\hat{\theta}$ to denote mixing directions, but the approach extends to $M > 2$ mixtures by clustering the directions $\hat{\mathbf{u}}(t, f)$ instead.

The first step of the algorithm consists in iteratively creating K clusters by selecting points $(\widehat{\theta}_k, \widehat{\mathcal{T}_k})$ with highest confidence and aggregating sufficiently close points around them. The second step is to estimate the direction $\widehat{\theta}_k^c$ of each cluster. Finally, we use a statistical test to eliminate non significant clusters and keep $\widehat{N} \leq K$ clusters which centroids provide the estimated directions of the mixing matrix.

3.1 Step 1: Cluster Creation

DEMIX iteratively create K clusters $C_k \subset P$ –where P is the DCSP– starting from $K = 0$, $P_K = P_0 = P$:

1. find the point $(\widehat{\theta}_K, \widehat{\mathcal{T}}_K) \in P_K$ with the highest confidence;
2. create a cluster C_K with all points $(\widehat{\theta}, \widehat{\mathcal{T}}) \in P$ "sufficiently close" to $(\widehat{\theta}_K, \widehat{\mathcal{T}}_K)$;
3. if $P_{K+1} := P_K \setminus C_K = \emptyset$, stop; otherwise increment $K \leftarrow K+1$ and go back to 1.

Note that in step 2 the newly created cluster might interesect previous clusters. To give a precise meaning to the notion of being "sufficiently close" to $(\widehat{\theta}_K, \widehat{\mathcal{T}}_K)$, we rely on the statistical model developped in Appendix A and include in C_K all points $(\hat{\theta}, \widehat{\mathcal{T}})$ such that $|\hat{\theta} - \widehat{\theta}_K| \leq \sigma(\widehat{\mathcal{T}}, \widehat{\mathcal{T}}_K)$ where the expression of $\sigma(\widehat{\mathcal{T}}, \widehat{\mathcal{T}}_K)$ is given in Equation (8).

3.2 Step 2: Direction Estimation

Since the clusters might intersect, the estimation of the centroid $\widehat{\theta}_k^c$ of a cluster C_k is based on a subset $C_k'' \subset C_k$ of "unbiased" points that belong *exclusively* to C_k. Due to lack of space we skip the description of how these subsets are selected. In light of the statistical model developped in Appendix A, the points $(\hat{\theta}, \hat{\mathcal{T}}) \in C_k''$ are assumed independent and distributed as $\widehat{\theta} \sim \mathcal{N}(\theta_k^{true}, \sigma_\theta^2(\hat{\mathcal{T}}))$ where θ_k^{true} is the unknown underlying direction and $\sigma_\theta^2(\widehat{\mathcal{T}})$ is defined in equation (6). The centroid of the cluster if therefore defined as the minimum variance unbiased estimator of θ_k^{true}

$$\widehat{\theta}_k^c := \sum_{(\widehat{\theta}, \widehat{\mathcal{T}}) \in C_k''} \sigma_\theta^{-2}(\widehat{\mathcal{T}})\widehat{\theta} / \sum_{(\widehat{\theta}, \widehat{\mathcal{T}}) \in C_k''} \sigma_\theta^{-2}(\widehat{\mathcal{T}}). \tag{2}$$

3.3 Step 3: Cluster Elimination

The last step aims at removing possibly spurious clusters among the K that have been built. We propose to use the variance $1/\sum_{(\widehat{\theta}, \hat{\mathcal{T}}) \in C_k''} \sigma_\theta^{-2}(\hat{\mathcal{T}})$ of the centroid estimator $\widehat{\theta}_k^c$ to help decide which clusters should be kept. We define two strategies: (DEMIXN) if we know the true number N of true directions, we keep the directions of the N clusters with the smallest centroid variance; (DEMIX) otherwise, we remove the directions of a clusters C_j whenever there is another cluster $C_o \neq C_j$ with

$$|\widehat{\theta}_j^c - \widehat{\theta}_o^c| \leq q_2 / \sum_{(\widehat{\theta}, \hat{\mathcal{T}}) \in C_j''} \sigma_\theta^{-2}(\hat{\mathcal{T}}) \tag{3}$$

where the quantile q_2 defines a confidence interval (see the Appendix). It is also possible to replace σ_θ with a slightly modified version $\hat{\sigma}_\theta$ relying on a quantile q_1 to define a confidence interval, see Eq. (7). To finish, we recompute the centroids of the clusters defined by the remaining directions, as described in Sections 3.1 and 3.2.

4 Experiments

We compared on several test mixtures the proposed algorithms (DEMIX and DEMIXN) and the classical K-means [6] and ELBG [7] clustering algorithms. Two variants of

K-means and ELBG were considered, one on the scatter plot of $tan^{-1}(X_2/X_1)(t,f)$, the other one on that of the angles $\hat{\theta}(t,f)$ obtained after the proposed local PCA. The mixtures were based on signals taken from a set of 200 Polish voice excerpts of 5 seconds sampled at 4kHz[1]. Noiseless linear instantaneous mixtures were performed with mixing matrices in the most favorable shape where all directions are equally spaced (as in [4]), with a number of directions ranging from $N = 2$ to $N = 15$. For each N, we chose $T = 20$ differents configurations of signals sources among the 200 available. A first measure of performance was the rate of success in the estimation of the number of sources (for DEMX and DEMXN only, because K-means and ELBG have a fix number of clusters). We observed that up to $N = 8$ sources, DEMIX estimates correctly the number of directions in more than four cases out of five, but when $N > 10$ it always fails to count the number of sources. DEMIXN is similarly successful up to $N = 10$ sources and always fails for $N > 12$. The reason why DEMIXN can fail in finding the right number of sources while it is known is that the cluster creation stage might result in $K < N$ clusters. In case success, we could also measure the *angular mean error* (AME) which is the mean distance in degrees between true directions and estimated ones. Distances are computed in the best way to pair estimated directions with the true ones. For each tested algorithm, we computed the *average* AME among test mixtures where $\widehat{N} = N$. Since K-means and ELBG are randomly initialized, we ran them $I = 10$ times for each test mixture and focussed on the smallest AME over these 10 runs, which gives an optimistic estimate of their performance.

As can be seen on Figure 2(b), DEMIX and DEMIXN algorithms obtain the best performance. Since the AME for DEMIX and DEMIXN can only be measured when a correct number of sources is estimated, it is not computed when $N > 10$ (resp. $N > 12$) for DEMIX (resp. DEMIXN).

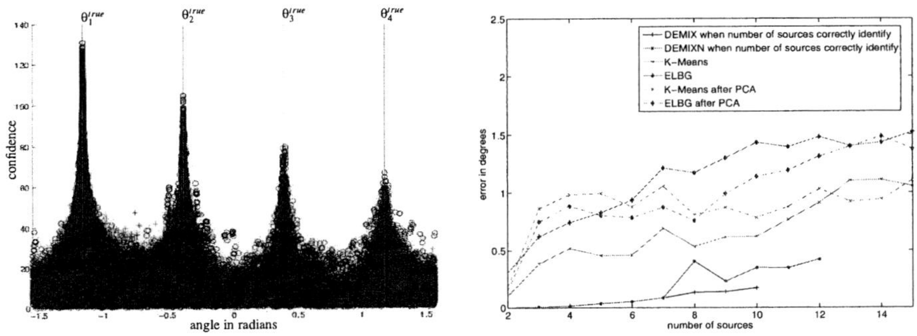

(a) Direction-confidence scatter plot (DCSP) (b) Average AME as a function of the number of sources

Fig. 2. (a) Direction-confidence scatter plot of points $(\hat{\theta}, 20\log_{10}\hat{\mathcal{T}})$ obtained by PCA on time-frequency regions based on a single STFT with window size is $L = 4096$ and neighborhoods of size $|\Omega_{t,f}| = 10$. (see section 2.3). (b) Experimental results of section 4.

[1] The signals are available at http://mlsp2005.conwiz.dk/index.php?id=30

5 Conclusion

We designed,developped, and evaluated a new algorithm to estimate the source directions of the mixing matrix in the instantaneous underdetermined two-sensor case. The proposed DEMIX algorithm yields better experimental results than those obtained by K-means and ELBG clustering algorithms on the same multiscale STFT data. Furthermore DEMIX estimates itself the number of mixing sources. This algorithm was designed using a confidence measure which is one of the main contribution of the article. The confidence measure allows to well detect regions of time-frequency points where essentially one source is active. This confidence measure could also be used in the source separation process, in addition with the estimated mixing matrix, to determine which source should be estimated in which time-frequency region, possibly providing a fully adaptive local (pseudo) Wiener filter. Further works include the extension of the DEMIX algorithm to delayed and convolved mixtures. We are also looking into the practical aspects and validation of the algorithm for source separation with more than two sensors.

References

1. Yilmaz, O., Rickard, S.: Blind separation of speech mixtures via time-frequency masking. In: IEEE Transactions on Signal Processing. Volume 52. (2002) 1830–1847
2. F. Abrard, Y. Deville, P.W.: From blind source separation to blind source cancellation in the underdetermined case: a new approach based on time-frequency analysis. In: ICA. (2001)
3. F.Abrard, Y.: Blind separation of dependent sources using the "time-frequency ratio of mixtures" approach. In: ISSPA 2003, Paris, France, IEEE (2003)
4. P. Bofill, M.Z.: Underdetermined blind source separation using sparse representations. In: Signal Processing. Volume 81. (2001) 2353–2362
5. Paul D.O'Grady, B.A., T.Rickard, S.: Survey of sparse and non-sparse methods in source separation. IJIST (International Journal of Imaging Systems and Technology) (2005)
6. MacQueen, J.B.: Some methods for classification and analysis of multivariate observations. In: 5-th Berkeley Symposium on Mathematical Statistics and Probability. (1967)
7. Patanè, G., Russo, M.: The enhanced LBG algorithm. Neural Networks **14**(9) (2001) 1219–1237
8. Härdel, W., Simar, L., eds.: Applied multivariate statistical analysis. Spinger-Verlag (2003)

A Statistical Analysis in the Stereophonic Case

In this appendix we make a statistical model in the stereophonic case ($M = 2$) to better understand the significance of the confidence measure $\widehat{\mathcal{T}}(t,f)$ as a measure of how robustly $\widehat{\theta}(t,f)$ estimates the "true" underlying direction of the dominant source. For that, we model the STFT coefficients of the most active source in the time-frequency region $\Omega_{t,f}$ with a centered normal distribution of (large) variance σ^2, and the contribution of all other sources, plus possibly noise, as 2-dimensional centered normal distribution with covariance matrix $\widetilde{\sigma}^2\mathbf{Id}_2$. Letting $\mathbf{a}$ be the normalized ($\|\mathbf{a}\|^2 = 1$) column of the mixing matrix $\mathbf{A}$ which corresponds to the most active source, then the model is that for $(t',f') \in \Omega_{t,f}$ we have:

$$\mathbf{x}(t',f') = s(t',f')\mathbf{a} + \mathbf{n}(t',f') \tag{4}$$

where

$$s(t',f') \sim \mathcal{N}\left(0,\sigma^2\right),\ \mathbf{n}(t',f') \sim \mathcal{N}\left(0,\tilde{\sigma}^2\mathbf{Id}_2\right) \tag{5}$$

therefore $\mathbf{x}(t',f') \sim \mathcal{N}\left(0,\tilde{\sigma}^2\mathbf{Id}_2+\sigma^2\mathbf{aa}^T\right)$. Let $\lambda_1 \geq \lambda_2$ be the eigenvalues of the covariance matrix $\Sigma := \tilde{\sigma}^2\mathbf{Id}_2+\sigma^2\mathbf{aa}^T$ and $\mathbf{u}=(u_1,u_2)^T$ be a unit eigenvector corresponding with λ_1. By elementary linear algebra we have $\frac{\lambda_1}{\lambda_2} = \frac{\tilde{\sigma}^2+\sigma^2}{\tilde{\sigma}^2} = 1+\frac{\sigma^2}{\tilde{\sigma}^2}$ and, if $\lambda_1 > \lambda_2$ (i.e., $\sigma > 0$), $\mathbf{u}$ is colinear to $\mathbf{a}$. Therefore, the true direction $\theta^{true} = \tan^{-1}(\frac{a_2}{a_1})$ is given by the direction of the principal component. Note that in this model λ_1/λ_2 is related to the "local signal to noise ratio" $\sigma^2/\tilde{\sigma}^2$ between the most active source and the others.

A.1 Precision of PCA

Since the values $\widehat{\theta}(t,f)$ and $\widehat{\mathcal{T}}(t,f) = \hat{\lambda}_1/\hat{\lambda}_2$ are computed by PCA on sample of $m := \mathrm{card}(\Omega_{t,f})$ points, they only provide estimates of the true direction and of the "true" confidence λ_1/λ_2 with a finite precision which we want to estimate as a function of the sample size m. For that, we use the following result which is an immediate application of [8, Theorems 4.11, 5.7, 9.4] : for large sample size, $\widehat{\mathcal{T}}/(\lambda_1/\lambda_2)$ converges in law to $\mathcal{N}\left(1,\sigma_{\mathcal{T}}^2\right)$ with $\sigma_{\mathcal{T}}^2 = 4/(m-1)$, and $\widehat{\theta}$ converges in law to $\mathcal{N}(\theta^{true},\sigma_\theta^2(\lambda_1/\lambda_2))$ with

$$\sigma_\theta^2(\mathcal{T}) := \frac{1}{m-1}\frac{\mathcal{T}}{(\mathcal{T}-1)^2}. \tag{6}$$

A.2 Confidence Intervals

If λ_1/λ_2 is known, then we know the standard deviation of the estimated angle $\hat{\theta}$ with respect to the true one. Since we know the distribution of the confidence measure $\hat{\mathcal{T}}$ which is close, but not equal to λ_1/λ_2, we can only predict the deviation of $\hat{\theta}$ with respect to a "true" direction" using confidence intervals. With probability exceeding $1-\alpha(q_1)/2$, we have $\lambda_1/\lambda_2 \geq \hat{\mathcal{T}}/(1+q_1\sigma_{\mathcal{T}})$. Therefore, instead of $\sigma_\theta^2(\hat{\mathcal{T}})$ we can use

$$\hat{\sigma}_\theta^2(\hat{\mathcal{T}}) := \sigma_\theta^2\left(\hat{\mathcal{T}}/(1+q_1\sigma_{\mathcal{T}})\right) \tag{7}$$

and model $\hat{\theta}$ as $\hat{\theta} \sim \mathcal{N}\left(\theta^{true},\hat{\sigma}_\theta^2(\hat{\mathcal{T}})\right)$ instead of $\hat{\theta} \sim \mathcal{N}\left(\theta^{true},\sigma_\theta^2(\hat{\mathcal{T}})\right)$.

Neglecting the possible dependencies between $\hat{\theta}$ and $\hat{\mathcal{T}}$ and following the same path, we get a statistical *upper bound* $|\hat{\theta}-\theta^{true}| \leq q_2\hat{\sigma}_\theta(\hat{\mathcal{T}})$ with confidence level $1-\alpha(q_2)/2$. We use it to determine whether two points belong to the same cluster in the cluster creation step. This leads to the definition

$$\sigma(\widehat{\mathcal{T}},\widehat{\mathcal{T}}^c) = q_2\left(\hat{\sigma}_\theta(\widehat{\mathcal{T}})+\hat{\sigma}_\theta(\widehat{\mathcal{T}}^c)\right) \tag{8}$$

We use quantil values $q_1 = q_2 = 2.33$ to provide confidence levels of 99 percent.

Convolutive Demixing with Sparse Discrete Prior Models for Markov Sources

Radu Balan and Justinian Rosca

Siemens Corporate Research,
755 College Road East,
Princeton, NJ 08540
{radu.balan, justinian.rosca}@siemens.com

Abstract. In this paper we present a new source separation method based on dynamic sparse source signal models. Source signals are modeled in frequency domain as a product of a Bernoulli selection variable with a deterministic but unknown spectral amplitude. The Bernoulli variables are modeled in turn by first order Markov processes with transition probabilities learned from a training database. We consider a scenario where the mixing parameters are estimated by calibration. We obtain the MAP signal estimators and show they are implemented by a Viterbi decoding scheme. We validate this approach by simulations using TIMIT database, and compare the separation performance of this algorithm with our previous extended DUET method.

1 Introduction

Signal Separation is a well studied topic in signal processing. Many studies were published during the past 10 years, each of them considering the separation problem from different points of view. Once can use model complexity to classify these studies into four categories:

1. Simple models for both sources and mixing. Typical signals are modeled as independent random variables, in their original domain, or transformed domain (e.g. frequency domain). The mixing model is either instantaneous, or anechoic. The ICA problem [1], DUET algorithm ([2]), or [3] belong to this category;
2. Complex source models, but simple mixing models. An example of this type is separation of two speech signals from one recording using one microphone. In this case, source signals are modeled using complex stochastic models, e.g. AR processes in [4], HMMs in [5], or generalized exponentials in [6];
3. Complex mixing models, but simple source models. This is the case of standard convolutive ICA. For instance source signals are i.i.d. but the mixing operator is composed of unknown transfer functions. Thus the problem turns into a blind channel estimation as in e.g. [7-9];
4. Complex mixing and source models. For instance [10] uses AR to model source signals, and FIR transfer functions for mixing.

J. Rosca et al. (Eds.): ICA 2006, LNCS 3889, pp. 544–551, 2006.

We chose the complexity criterion in order to point out the basic trade-off of signal separation algorithms. A more complex mixing or source model may yield a better performance provided it fits well the data. However more complex models are less robust to mismatches than a simpler model, and may perform unexpectedly worse on real world data. In our prior experiments [11] we found that simple signal and mixing models yield surprisingly good results on real world data. Robustness to model uncertainties explains these good results. Indeed this is the case with DUET. The basic idea of the DUET approach is the assumption that for any time-frequency point, only one signal from the ensemble of source signals would use that time-frequency point. In [12] we extended this assumption in a system with D sensors to what we called *generalized W-disjoint orthogonality hypothesis* by allowing up to $D-1$ source signals to use simultaneously any time-frequency point. In both cases source signals were assumed mutually independent across both time and frequency. In other words, any two different time-frequency coefficients of the same source are assumed independent. However we would like to increase the power of source separation particularly when there exists prior knowledge about the sources (see also [5,6,13]). In this paper we propose an incremental increase in source model complexity combined with simple mixing model that conforms to our basic belief that models should not be more complicated than what is really needed in order to solve the problem. For this we allow for statistical dependencies of source signals across time. More precisely [14] postulates a signal model that states that the time-frequency coefficient $S(k,\omega)$ of a (speech) signal $s(t)$ factors as a product of a continuous random variable, say $G(k,\omega)$, and a 0/1 Bernoulli $b(k,\omega)$, $S(k,\omega) = b(k,\omega)G(k,\omega)$. This formula models sparse signals. See also [15] for a similar signal model. Denoting by q the probability of b to be 1, and by $p(\cdot)$ the p.d.f. of G, the p.d.f. of S turns into $p_S(S) = qp(S) + (1-q)\delta(S)$, with δ, the Dirac distribution. For L independent signals $S_1, \ldots, S_L$, the joint p.d.f. is obtained by conditioning with respect to the Bernoulli random variables. The rank k term, $0 \le k \le N$, is associated to a case when exactly k sources are active, and the rest are zero. In [12] we showed that by truncating to the first N+1 terms the approximated joint p.d.f. corresponds to the case when *at most N sources are active simultaneously*, which constitutes the *generalized W-disjoint hypothesis*. This paper extends the signal model introduced before by assuming the Bernoulli variables are generated by a Markov process, while the complex amplitudes $G(k,\omega)$ are modeled as unknown deterministic variables. The application we target is a meeting transcription system (see Figure 1) where an array of microphones records the meeting, and the convolutive mixing parameters are learned during an initial calibration phase. Section 3 describes the statistical signal estimators. We show that signal estimation is similar to a Viterbi decoding scheme. Section 4 presents the methods for learning the transition probabilities of source models, and of the mixing parameters. Section 5 contains numerical results, and is followed by the conclusion section.

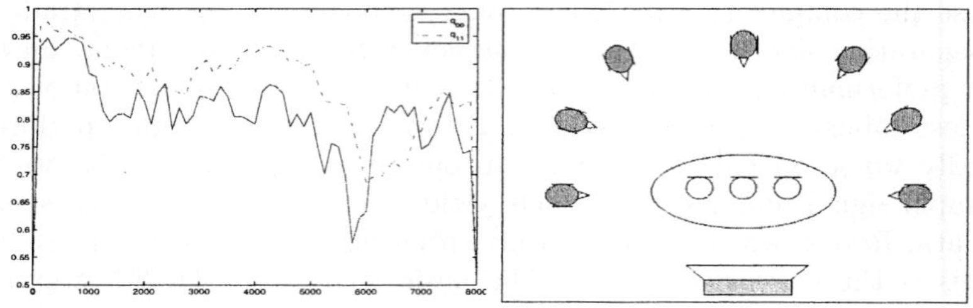

Fig. 1. Transition probabilitites of one signal for $\tau = 0.1$ (left plot), and the experimental setup (right plot)

2 Signal and Mixing Models

2.1 Convolutive Mixing Model

Consider the measurements of L source signals by an array of D sensors. In time domain the mixing model is $x_d(t) = \sum_{l=1}^{L} h_{d,l} \star s_l(t) + n_d(t)$, $1 \leq d \leq D$ where $n_1, \ldots, n_D$ are sensor noises, and $h_{d,l}$ are the impulse responses from source l to sensor d. We renormalize the sources by absorbing $h_{1,l}$ into the definition of source s_l.

We denote by $X_d(k,\omega)$, $S_l(k,\omega)$, $N_d(k,\omega)$ the short-time Fourier transform of signals $x_d(t), s_l(t)$, and $n_d(t)$, respectively, with respect to a window $W(t)$, where k is the frame index, and ω the frequency index. Then the convolutive mixing model turns into $X_d(k,\omega) = \sum_{l=1}^{L} A_{d,l}(\omega) S_l(k,\omega) + N_d(k,\omega)$. When no danger of confusion arises, we drop the arguments k, ω in X_d, S_l and N_d.

2.2 Signal Model

Consider a source signal $s(t)$, $1 \leq t \leq T$, and its associated short-time Fourier transform $S(k,\omega)$, $1 \leq k \leq K_{max}$, $0 \leq \omega \leq \Omega$. Each time-frequency coefficient $S(k,\omega)$ is modeled by the product $b(k,\omega)G(k,\omega)$ as before, where b is a Bernoulli (0/1) random variable, and G is an unknown deterministic complex amplitude. In previous work we assumed $\{b(k,\omega)\ ;\ k,\omega\}$ is a set of independent random variables. In this paper we preserve independence along the frequency index, but we introduce a Markov dependence along the time index. The independence in frequency is supported by the remark that local stationarity in time domain implies decorrelation of frequency components. Along the time index, our assumption amounts to $P(b(k,\omega)|b(k-1,\omega), b(k-2,\omega), \ldots, b(1,\omega)) = P(b(k,\omega)|b(k-1,\omega)) = \pi_\omega(b(k,\omega), b(k-1,\omega))$ where $\{\pi_\omega\}$ is the set of 2×2 matrices of probabilities of transition. By successive conditioning we obtain that: $P(\{b(k,\omega)\ ;\ k,\omega\}) = \prod_\omega P(b(1,\omega)) \prod_{k=2}^{K_{max}} \pi_\omega(b(k,\omega), b(k-1,\omega))$. For each source in the mixture we assume we have a database of training signals where we learn the matrices of transition probabilities and the set of initial probabilities (see Section 5).

For a collection of L source signals, we assume that only N Bernoulli variables are nonzero; the rest are zero. We denote by $\{(b_l(k,\omega))_{1\leq l\leq L}; k,\omega\}$ the collection of Bernoulli random variables, $\sigma(k,\omega) = \{l \; ; \; b_l(k,\omega) = 1\}$ the N-set of nonzero components of $S(k,\omega)$, $(\pi^l_\omega)_{1\leq l\leq L, 0\leq\omega\leq\Omega}$ the collection of transition probability matrices, $(P^l_\omega)_{1\leq l\leq L,0\leq\omega\leq\Omega}$ the collection of initial probabilities. Then the joint pdf becomes:

$$P(\{b_l(k,\omega) \; ; \; l,k,\omega\}) = \prod_\omega Q^0_\omega(\sigma(1,\omega)) \prod_{k\geq 2} Q_\omega(\sigma(k,\omega),\sigma(k-1,\omega))$$

where $Q_\omega(\sigma(k,\omega),\sigma(k-1,\omega)) = \prod_{l=1}^L \pi^l_\omega(b_l(k,\omega), b_l(k-1,\omega))$, $Q^0_\omega(\sigma(1,\omega)) = \prod_{l=1}^L P^l_\omega(b_l(1,\omega))$. The collection of all subsets $\sigma(k,\omega)$ defines a trajectory through the selection space S_L^N, the set of N-subsets of $\{1,2,\ldots,L\}$. Thus for each frequency ω we associate $\Sigma_\omega = \{\sigma(k,\omega) \; ; \; 1\leq k\leq K_{max}\}$ the selection space trajectory. Source estimation is then equivalent to estimating both the selection space trajectories $(\Sigma_\omega)_\omega$ and the complex amplitudes $\{G_l(k,\omega) \; ; \; l\in \sigma(k,\omega)\}$.

In this paper we assume that the mixing model is given by a convolutive mixture, signals $S_l(k,\omega)$ satisfy the signal model above, and noise components $N_d(k,\omega)$ are Gaussian i.i.d. with zero mean and spectral variance σ^2.

Our problem is: Estimate the source signals $(s_1(t), \ldots, s_L(t))_{1\leq t\leq T}$ given measurements $(x_1(t), \ldots, x_D(t))_{1\leq t\leq T}$ of the linear convolutive mixing model, and assuming the following:

1. The mixing matrix $A = (A_{d,l}(\omega))_{1\leq d\leq D, 1\leq l\leq L}$ is known;
2. The noise $\{n(t)\}$ is i.i.d Gaussian with zero mean and known spectral power σ^2;
3. The components of signal S are independent and satisfy the stochastic model presented before, with known probabilities of transition $(\pi^l_\omega)_{l,\omega}$ and initial probabilities P^l_ω;
4. At every time-frequency point (k,ω) at most N components of $S(k,\omega)$ are non-zero, and N is known.

3 MAP Signal Estimation

In this paper we estimate the signals $(s_l(t))_{l,t}$ by maximizing the posterior distribution of the Bernoulli variables, and the likelihood of the complex amplitudes. Alternatively, using a uniform prior model on the amplitudes, our solution is a MAP estimator of both the selection variables and the complex amplitudes. The criterion to maximize is:

$$I = \prod_\omega P(\{X(k,\omega); 1\leq k\leq K_{max}\}|\{b_l(k,\omega), G(k,\omega); l, 1\leq k\leq K_{max}\}) \times P(\{b_l(k,\omega); l, 1\leq k\leq K_{max}\}) \quad (1)$$

We replace the Bernoulli variables by the set-valued variables $\Sigma_\omega = (\sigma(k,\omega))_{k,\omega}$, and we consider the reduced complex amplitude N-vector $\mathbf{G}_r(k,\omega)$ corresponding to nonzero components of S (in turn selected by $\sigma(k,\omega)$). We let $A_r(k,\omega)$ denote the $D \times N$ mixing matrix whose columns corresponds to the nonzero components of $S(k,\omega)$: $(A_r(k,\omega))_{d,m} = A_{d,l(m)}(\omega)$, where $l(m)$ is the mth element of $\sigma(k,\omega)$. The first term decouples into a product of likelihoods at each time k; the second term is estimated before. Putting these two expressions together, the criterion to maximize becomes (up to a multiplicative constant term):

$$I(\Sigma, \mathbf{G}_r) = \prod_{\omega} \left[\prod_{k} exp\{-\frac{1}{\sigma^2}\|X - A_r\mathbf{G}_r\|^2\} \right]$$

$$\times \left[\prod_{k\geq 2} Q_\omega(\sigma(k,\omega), \sigma(k-1,\omega)) \right] Q^0_\omega(\sigma(1,\omega))$$

Given $\sigma(k,\omega)$, at every (k,ω) we can solve for $\mathbf{G}_r(k,\omega)$ and obtain $\mathbf{G}_r(k,\omega) = (A_r^* A_r)^{-1} A_r^* X$. Taking the logarithm, flipping the sign, ignoring some constants, and replacing $\mathbf{G}_r$ by the above estimate, we obtain the following optimization problem

$$min_\Sigma \sum_\omega \sum_k [X^*(1 - A_r(A_r^*A_r)^{-1}A_r^*)X - \sigma^2 \log Q_\omega(\sigma(k,\omega), \sigma(k-1,\omega))] - \sigma^2 \log Q^0_\omega(\sigma(1,\omega))$$

Let us denote by

$$C(\sigma(k,\omega)) = X(k,\omega)^*(1 - A_r(k,\omega)(A_r^*(k,\omega)A_r(k,\omega))^{-1}A_r^*(k,\omega))X(k,\omega)$$

and

$$D(\sigma(k,\omega), \sigma(k-1,\omega)) = -\sigma^2 \log Q(\sigma(k,\omega), \sigma(k-1,\omega))$$

for $k \geq 2$. Then the optimization becomes

$$\min_{\Sigma_\omega} \sum_{k\geq 2} C(\sigma(k,\omega)) + D(\sigma(k,\omega), \sigma(k-1,\omega)) + C(\sigma(1,\omega) - Q^0_\omega(\sigma(1,\omega))$$

at every frequency ω. The solution represents a trajectory Σ_ω in the selection space $(S_L^N)^{K_{max}}$. The optimization can be efficiently implemented using a backward-forward best path propagation algorithm (Viterbi) widely used in channel decoding problems. The algorithm is as follows:

Algorithm

Step 1. (Initialization) Set $k = K_{max}$, and $J_k^*(s) = 0$ for all $s \in S_L^N$.

Step 2. (Backward propagation) For all $s \in S_L^N$ N-subsets of $\{1, 2, \ldots, L\}$ repeat

- For all $s' \in S_L^N$ compute $J(s,s') = J_k^*(s') + C(s') + D(s',s)$
- Find the minimum over s', and set $J_{k-1}^*(s) = min_{s'} J(s,s')$

Step 3. Decrement $k = k - 1$, and if $k > 1$ go Step 2.

Step 4. At $k = 1$, replace $C(s')$ by $C(s') - \sigma^2 \log Q^0_\omega(\sigma(1,\omega))$ and perform Step 2. Denote $\sigma^*(1,\omega) = argmin_s J_1^*(s)$.

Step 5. (Forward iteration) Set $k = 2$ and repeat until $k = K_{max}$:

- For all $s \in S_L^N$ compute $J(s) = C(s) + D(s, \sigma^*(k-1,\omega))$
- Find the minimum and set $\sigma^*(k,\omega) = argmin_s J(s)$
- Increment $k = k + 1$.

4 Model Training

4.1 Transition and Intial Probabities Estimation

For training we used a fixed sentence uttered by the corresponding speaker. We assumed the recorded voice is made of two components: one part which is critical to understanding, and a second component which can be removed losslessly from an information point of view. Thus $s = s_{critic} + s_{extra}$. Assuming the first component has a Laplace (or even peackier) distribution in frequency domain whereas the second component is Gaussian, the estimation of s_{critic} is done by (soft, or hard) thresholding of the measured signal. We chose a threshold proportional to square root of signal spectral power. Thus, in case of hard thresholding $S_{critical}(k,\omega) = S(k,\omega)$ if $|S(k,\omega)| \geq \tau\sqrt{R_s(\omega)}$, and is zero otherwise. The factor τ is chosen so that the thresholded signal sounds almost identical to the original signal s. Subjective experimentation showed that a factor $\tau = 0.1$ satisfies this requirement. Once $\{S_{critical}(k,\omega); k,\omega\}$ has been obtained, we estimate the binary sequence $\{b(k,\omega); k,\omega\}$ simply by setting $b(k,\omega) = 1$ for $S_{critical}(k,\omega) \neq 0$, and 0 otherwise. From the binary sequence $\{b(k,\omega); k,\omega\}$ we estimate the transition probability matrices π_ω and initial probabilitites P_ω by maximum likelihood estimators: $\pi_\omega(0,0) = \frac{N_{00}}{N_{00}+N_{01}}$, $\pi_\omega(1,0) = 1 - \pi_\omega(0,0)$, $\pi_\omega(1,1) = \frac{N_{11}}{N_{10}+N_{11}}$, $\pi_\omega(0,1) = 1 - \pi_\omega(1,1)$, $P_\omega(1) = \frac{N_1}{N_0+N_1}$, $P_\omega(0) = 1 - P_\omega(1)$, where N_0, N_1, N_{00}, N_{01}, N_{10}, N_{11} are, respectively, the number of 0's, 1's, 00's, 01's, 10's, 11's in the binary training sequence $(b(k,\omega))_k$. Figure 1 plots an example of the distributions $\pi_\omega(0,0)$ and $\pi_\omega(1,1)$.

4.2 Mixing Parameters Estimation

Consider the case one source only is active. Then the frequency representation of the recorded signal turns into $X(k,\omega) = a(\omega)S(k,\omega) + N(k,\omega)$, where $a(\omega)$ is the "steering vector" associated to source S. We use the maximum likelihood estimation to estimate a. Assuming Gaussian i.i.d. noise, the resulting maximum likelihood estimator yields $\hat{a}(\omega)$ the eigenvector corresponding to the largest eigenvalue of the sampled covariance matrix, normalized so that $\hat{a}_1 = 1$, $R\hat{a} = \lambda\hat{a}$, $R(\omega) = \sum_k X(k,\omega)X^*(k,\omega)$.

5 Experimental Evaluation

Consider the setup of a meeting recording system as depicted in Figure 1: $L = 7$ speakers placed around a conference table are recorded by a video camera (for eventual postprocessing) and an array of microphones. During the calibration phase both the source model parameters and the mixing parameters were learned. In our simulations we used a linear array with inter-microphone distance $d_a = 5$ cm and sampling frequency $f_s = 16$ KHz. The simulated mixing environment was weakly echoic with a reverberation time below 10ms. We used 4 female and 3 male speakers from the TIMIT database at positions located at multiple of 30 degrees. Testing was done on wavefiles of around 10 seconds of normal speech. We added Gaussian noise with $\sigma = 0.1$ (note σ is an absolute value rather than relative to signals). We tested for $N = 1$ and $N = 2$ (the number of simultaneous speakers), even though all $L = 7$ speakers were active most of the time. We estimated each source using the MAP-based Estimation Algorithm presented in Section 4 for four choices of priors: 1) use the initial distribution and transition probabilities learned from the training database as presented before; 2) use uniform initial distribution probabilities but the transition probabilities learned from the training database; 3) use uniform transition probabilities, but initial probabilities learned from the training database; 4) use uniform distributions for both the initial distribution and for the transition probabilities. This last combination of priors turns our MAP algorithm into the extended DUET presented in [12].

We compared these algorithms with respect to the Signal To Interference Plus Noise Ratio (SINR) Gain. The SINR gain for component l is defined by:

$$SINRg_l = oSINR - iSINR = 10 \log_{10} \frac{E(x_1 - s_l)}{E(\hat{s}_l - s_l)}$$

where $E(z)$ is the energy of signal z, and $x_1, s_l, \hat{s}_l$ are respectively, the microphone 1 measured signal, input signal l at microphone 1, and the lth estimated signal. The larger the $SINRg$ the better. We experimentally verified that the choice for initial distribution probabilities does not have almost any effect on the outputs. In Figure 2 we plot the SINR gain as function of number of microphones D, for our setup with $L = 7$ sources, and a variable number of microphones ranging from 2 to 6, for two hypotheses: $N = 1$ and $N = 2$, respectively. We notice the gain is an increasing function of number of microphones, and our MAP algorithm (called Markov, in Figure) outperforms DUET by about 1 dB in average.

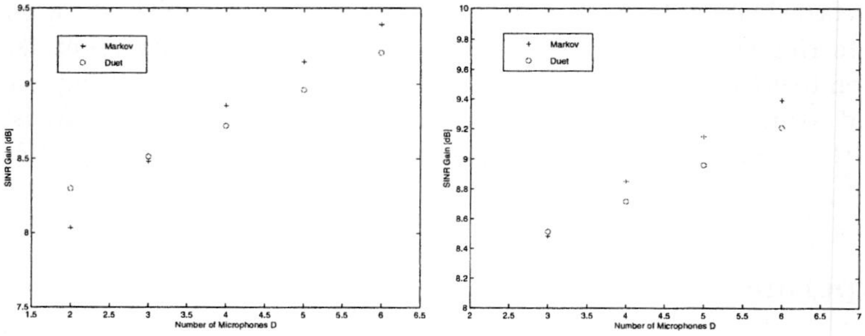

Fig. 2. SINR Gain for $N = 1$ (left plot) and $N = 2$ (right plot), for $L = 7$ sources and a variable array ranging from 2 to 6 microphones

6 Conclusions

In this paper we presented a novel signal separation algorithm that extends our past DUET algorithm. The algorithms works for underdetermined cases, when there are fewer sensors than sources, and in the presence of noise. The main assumptions are: (i) source signals have sparse time-frequency representations (although another representation, such as time-scale, would work as well); (ii) each frequency is independent from one another; (iii) the binary selection variables obey a homogeneous Markov process model, with transition and initial probabilities learned from a training database. We derived the MAP estimator of binary selection variables and ML of the complex signal TF coefficients, and show it can be efficiently implemented using a Viterbi decoding scheme. Next we validated our solution in a 7-voice, and 2 to 6 calibrated microphone array setup. We obtained an improvement of about 1 dB compared with the previous DUET algorithm, and no noticeable distortions.

References

1. Pierre Comon, "Independent component analysis, a new concept?," *Signal Processing*, vol. 36, no. 3, pp. 287–314, 1994.
2. A. Jourjine, S. Rickard, and O. Yilmaz, "Blind separation of disjoint orthogonal signals: Demixing N sources from 2 mixtures," in *Proc. ICASSP*, 2000.
3. M. Aoki, M. Okamoto, S. Aoki, and H. Matsui, "Sound source segregation based on estimating incident angle of each frequency component of input signals acquired by multiple microphones," *Acoust. Sci. & Tech.*, vol. 22, no. 2, pp. 149–157, 2001.
4. R. Balan, A. Jourjine, and J Rosca, "Ar processes and sources can be reconstructed from degenerate mixtures," in *Proc. ICA*, 1999, pp. 467–472.
5. S. T. Roweis, "One microphone source separation," in *Neural Information Processing Systems 13 (NIPS)*, 2000, pp. 793–799.
6. G.J. Jang and T-W Lee, "A probabilistic approach to single channel blind signal separation," in *Proc. of NIPS*, 2002.
7. H. Sawada, R. Mukai, S. Araki, and S. Makino, "A robust and precise method for solving the permutation problem of frequency-domain blind source separation," *IEEE Trans. SAP*, vol. 12, no. 5, pp. 530–538, 2004.
8. A. Hyvarinen, J. Karhunen, and E. Oja, *Independent component analysis*, John Wiley and Sons, 2001.
9. J.F. Cardoso, "Infomax and maximum likelihood for blind source separation," *IEEE Signal Processing Letters*, vol. 4, no. 4, pp. 112–114, April 1997.
10. E. Weinstein, A.V. Oppenheim, M. Feder, and J.R. Buck, "Iterative and sequential algorithms for multisensor signal enhancement," *IEEE Trans. on SP*, vol. 42, no. 4, pp. 846–859, 1994.
11. R. Balan, J. Rosca, and S. Rickard, "Robustness of parametric source demixing in echoic environments," in *Proc. ICA*, December 2001.
12. J. Rosca, C. Borss, and R. Balan, "Generalized sparse signal mixing model and application to noisy blind source separation," in *Proc. ICASSP*, 2004.
13. S. Hosseini, C. Jutten, and D. Pham, "Markovian source separation," *IEEE Trans. on Sig. Proc.*, vol. 51, pp. 3009–3019, 2003.
14. R. Balan and J. Rosca, "Statistical properties of STFT ratios for two channel systems and applications to blind source separation," in *Proc. ICA-BSS*, 2000.
15. P.J. Wolfe, S.J. Godsill, and W.J. Ng, "Bayesian variable selection and regularization for time-frequency surface estimation," *J.R.Statist.Soc.B*, vol. 66, no. Part 3, pp. 575–589, 2004.

Independent Component Analysis for Speech Enhancement with Missing TF Content

Doru-Cristian Balcan[1] and Justinian Rosca[2]

[1] Carnegie Mellon University, Computer Science Department,
5000 Forbes Ave., Pittsburgh, PA 15123, USA
dbalcan@cs.cmu.edu

[2] Siemens Corporate Research, Princeton NJ 08540, USA
justinian.rosca@scr.siemens.com

Abstract. We address the problem of Speech Enhancement in a setting where parts of the time-frequency content of the speech signal are missing. In telephony, speech is band-limited and the goal is to reconstruct a wide-band version of the observed data. Quite differently, in Blind Source Separation scenarios, information about a source can be masked by noise or other sources. These masked components are "gaps" or missing source values to be "filled in". We propose a framework for unitary treatment of these problems, which is based on a relatively simple "spectrum restoration" procedure. The main idea is to use Independent Component Analysis as an adaptive, data-driven, linear representation of the signal in the speech frame space, and then apply a vector-quantization-based matching procedure to reconstruct each frame. We analyze the performance of the reconstruction with objective quality measures such as log-spectral distortion and Itakura-Saito distance.

1 Introduction

Speech enhancement for real-world environments is still a signal processing challenge in spite of steady progress over the last forty years [1]. The work reported here addresses this problem in a setting where signal corruption consists of "hiding", zeroing out, various parts of the time-frequency (TF) content. Alternatively, we refer to this TF information as "missing" from the observed signal spectrogram. Our objective is to reconstruct the signal, or at least to infer the missing content, using the observable TF content and clean speech training data.

Two very different speech enhancement applications motivate this approach. On one side, present digital telephony operates at the minimum bandwidth requirements (e.g. 300Hz $-$ $4k$Hz) although speech is much richer. Wide-band speech is desirable. This is the bandwidth extension problem (BWE) [2] and is typically approached using clean speech models. Linear predictive coding analysis decomposes the estimation problem into the extension of the excitation signal and the extension of the spectral envelope. The excitation can be extended e.g. with a spectral copy of the low-frequency excitation or by bandpass modulated Gaussian noise [3,4]. The spectral envelope can be extended using pattern recognition techniques relying on a Hidden Markov Model (HMM) of clean speech.

J. Rosca et al. (Eds.): ICA 2006, LNCS 3889, pp. 552–560, 2006.

On the other side, source extraction using blind source separation (BSS) techniques results in a limited fidelity reconstruction of the speech sounds in complex auditory scenarios. One reason is that many of the BSS techniques make simplifying assumptions about the underlying signal or mixing models, e.g. anechoic or single-path propagation models or sparsity and disjointness of TF representations, in order to be able to deal with more difficult realistic cases [5]. Such techniques do not presently take advantage of prior statistical speech models [6]. Statistical models of clean speech (e.g. [7]) capture the principal spectral shapes, or speech units, and their dynamics in order to be able to distinguish and recognize the speech units. This is the case of Hidden Markov Models in present Automatic Speech Recognition technology.

In our approach, the pattern of missing data could be either time invariant or variable in time, however the conceptual approach for restoring the signal does not depend on it and is rather general. We organize the description of the method as follows. The formal description of the problem in Section 2 is followed by the training-based procedure for the inference of missing data, and by the analysis of alternative choice of transforms. Section 5 presents experimental results. Section 6 highlights the main contributions and future work.

2 Formal Description and Notations

Given an observed signal $\tilde{x}$ as an altered version of a "clean" signal x_c, we want to reconstruct x_c. Let $\tilde{X}$ and respectively X_c be the Short-Time Fourier Transform (STFT) versions of $\tilde{x}$ and x_c, where the frame length is denoted by L, the overlap by δ, and the length of $\tilde{x}$ is N. We can therefore consider $\tilde{X}$ and X_c as $L \times \lfloor (N-\delta) / (L-\delta) \rfloor$ complex matrices.

To intuitively describe the missing data locations, we introduce the concept of "mask" associated with an observed signal. A mask M is simply a binary (0-1) matrix over TF domain, of the same size as $\tilde{X}$, with the obvious meaning: $M(\omega, t) = 0$, if and only if the TF point at "location" (ω, t) in the spectrogram $\tilde{X}$ is missing. In this case, we regard $\tilde{X}$ as being a "masked" version of X_c:

$$\tilde{X} = M \odot X_c \tag{1}$$

where $\odot$ is point-wise multiplication. The equation above immediately implies, for each time frame t:

$$\tilde{X}(:,t) = M(:,t) \odot X_c(:,t) = diag(M(:,t)) \cdot X_c(:,t) \tag{2}$$

Let us denote $\mathcal{F}$ and $\mathcal{F}^{-1}$ to represent the direct and the inverse Fourier Transform operators [1] respectively, and define $\mathbf{x}(t) = \mathcal{F}^{-1}\left(\tilde{X}(:,t)\right)$, and analogously $\mathbf{x}_c(t) = \mathcal{F}^{-1}(X_c(:,t))$. It follows that:

[1] Here, they are considered $L \times L$ complex matrices, normalized such that $\mathcal{F}^{-1} = \bar{\mathcal{F}}$.

$$\mathbf{x}(t) = \mathcal{F}^{-1} \cdot diag\left(M\left(:,t\right)\right) \cdot \mathcal{F} \cdot \mathbf{x}_c\left(t\right) \tag{3}$$

From here it follows that the observed speech frame is the filtered version of some clean speech frame. However, $\mathbf{x}_c$ can not be uniquely recovered just from equation 3. To reconstruct the clean frame $\mathbf{x}_c$, we will exploit the valuable information given by mask M about how the original clean data was transformed.

We can identify various particular cases of the problem, by inspecting the distribution of 1-entries in the mask. For example, if all the columns of M are identical, we denote this as the "constant missing data pattern" case, otherwise we refer to M as the description of a "variable missing data pattern". If all the entries of M corresponding to low-frequency entries in $\tilde{X}$ (up to some cut-off frequency F_{cut}) are equal to 1, we have a particular sub-case of the "constant pattern" case, namely BWE. Alternatively, we may obtain the mask by applying the BSS algorithm in [5]. Namely, when a mixture of 3 or more speech signals is given, the TF contents of these sources are assumed mutually disjoint, and a maximum-likelihood map is generated for each of them, to indicate the most informative positions of the sources in the TF plane. In this case, the masks we work with will have a variable, random-looking aspect, and the source restoration will fall under the "variable pattern" case [2].

In the following, we show how clean speech can be used to aid in the speech enhancement problem, present a unified methodology to treat these different situations, and identify the virtues and the limitations of our approach.

3 Inference of Missing Data

To restore the spectrogram of the corrupted speech signal, we need to design a procedure for the reconstruction of each data frame. As we noted in the previous section, we have good knowledge about how the clean data is processed to produce the observed data. Unfortunately, the information we have is not sufficient, since the frame filtering involved here is not an invertible process except in the trivial case when the entries in the corresponding mask column are all 1. Equation 3 has in general infinitely many solutions for $\mathbf{x}_c(t)$.

Fortunately, not all these possible choices are likely to correspond to real speech data frames. Furthermore, two (or more) consecutive observed frames should have originated in consecutive clean speech frames, and the smoothness of clean speech TF representation induces strong additional constraints on the reconstructed speech signal. Clean speech training data will help us restrict our search, by giving valuable hints of optimal (or plausible) solutions.

[2] In the constant pattern case, it is possible to give a very intuitive meaning to $\mathbf{x}(t)$, namely $\mathbf{x}(t) = \mathcal{W} \odot \left(\tilde{x}\left(\left(t-1\right)\left(L-\delta\right)+1, ..., \left(t-1\right)\left(L-\delta\right)+L\right)^T\right)$ is a windowed version of the tth frame extracted from $\tilde{x}$. In the variable pattern case, approximating consecutive frames in $\tilde{x}$ with filtered versions of consecutive frames in x_c is generally not accurate.

Now, we introduce several notations that are useful in describing the proposed reconstruction procedure. Let $\mathbf{X}_{train}$ denote a $L \times T$ matrix whose columns are speech frames, extracted at random positions from clean speech example sentences. We will denote by $\mathbf{x}_{train}^{m}(t)$ the result of filtering $\mathbf{x}_{train}(t)$, (the tth column of $\mathbf{X}_{train}$) using mask vector m, as in Eqn. 3:

$$\mathbf{x}_{train}^{m}(t) = \mathcal{F}^{-1} \cdot diag(m) \cdot \mathcal{F} \cdot \mathbf{x}_{train}(t) \tag{4}$$

The procedure we propose to infer the missing data is relatively simple. For a certain observed frame $\mathbf{x}(t)$, and known $M(:,t)$ we filter the training frames accordingly, thus obtaining $\mathbf{x}_{train}^{M(:,t)}(t_1), t_1 = 1, ..., T$. Then, for a distance criterion d of our choice, we find the index:

$$t_* = \arg\min_{t_1} d\left(\mathbf{x}(t), \mathbf{x}_{train}^{M(:,t)}(t_1)\right) \tag{5}$$

To obtain the reconstructed frame $\hat{\mathbf{x}}(t)$, we simply add the "missing" part of $\mathbf{x}_{train}(t_*)$ to the observed frame $\mathbf{x}(t)$:

$$\hat{\mathbf{x}}(t) = \mathbf{x}(t) + \left(\mathbf{x}_{train}(t_*) - \mathbf{x}_{train}^{M(:,t)}(t_*)\right) = \mathbf{x}(t) + \mathbf{x}_{train}^{1-M(:,t)}(t_*) \tag{6}$$

The matching step (that is, finding t_* in Eqn. 5) is solvable by exhaustive search for each test frame in the training database[3]. However, matching can be efficiently defined and performed by vector quantization [8].

The quality of the reconstruction is strongly dependent on the distance criterion we choose. To address this problem, let us emphasize first that the signal representation (set of vectors in L-dimensional real or complex space) lends itself to very natural distance measures, among which the most intuitive is the Euclidean distance. It should be very clear though that this distance measure is only useful if the distribution of the data is suitably described in the "frame space" (for instance, if it is axis-aligned). Otherwise, a small distance might not necessarily mean good perceptual similarity and the best match might actually be completely erroneous from reconstruction point of view. There are two ways (not necessarily orthogonal) to avoid this obstacle.

One way is to choose several best matches (e.g. best ten) and combine them to produce a better reconstruction of the current frame. The second one refers to finding and using the best description of the data as linear combination of vectors in the space.

Independent Component Analysis (ICA) is an adaptive, linear representation method. When applied to speech frames, ICA provides a linear representation that maximizes the statistical independence of its coefficients, and therefore finds the directions with respect to which the coefficients are as sparsely distributed as possible. It is therefore a good idea to employ an ICA algorithm such as FastICA [9], which learns the ICA representation matrices from the clean training data. We denote the ICA basis function matrix by $\mathbf{A}$, and the filter matrix by $\mathbf{W}$.

[3] From now on, we refer to the observed data frames as "test data", to fit into the "testing vs. training" Machine Learning terminology.

The next step is to transform both the testing data and the filtered training data into the ICA domain (by multiplying each frame vector by matrix $\mathbf{W}$) and obtain the new vectors:

$$\mathbf{s}(t) = \mathbf{W} \cdot \mathbf{x}(t) \ ; \qquad \mathbf{s}_{train}^{M(:,t)} t) = \mathbf{W} \cdot \mathbf{x}_{train}^{M(:,t)}(t) \tag{7}$$

Next, find the best match by using the Euclidean distance in this space rather, than in the frame space (further referred to as "time domain"). Further justification and analysis of the inference in the ICA domain is given in [10].

Using ICA to represent speech has already been investigated, and its characteristics have been reported in several papers (e.g. [11], etc.). The ICA filters (rows of $\mathbf{W}$) tend to have high sensitivity to relatively narrow frequency bands, and therefore the ICA basis functions (columns of A) will tend to represent primarily frequency information within these bands. We can easily demonstrate this behavior in Figure 1.

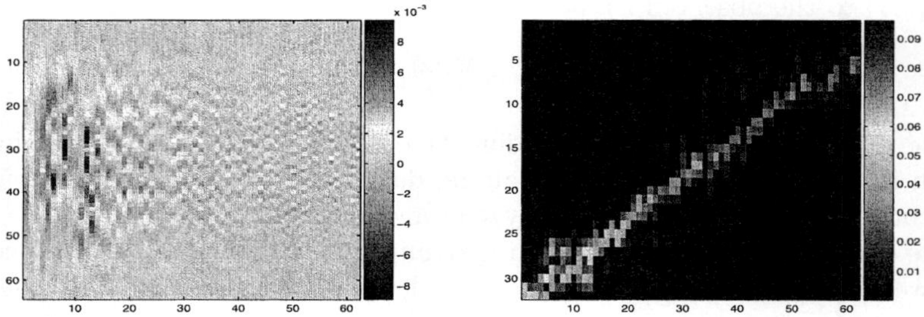

Fig. 1. Frequency selectivity of ICA basis functions: representation of both $\mathbf{A}$ and the Fourier domain representation of the columns of $\mathbf{A}$; Columns are sorted according to the peak frequency response

As a general practice, we note that we can obtain completely new distance criteria simply by mapping the test data and the filtered training data into some new space (either by a linear, or by a nonlinear mapping) and compute the Euclidean distance between the mapped vectors. In the following, we investigate the performance of computing these distances in various domains.

4 Alternative Domain Transforms

Let us examine particular cases of distance matching criteria, as well as their performance in speech signal reconstruction. Alternatively, we can regard a particular distance choice as the Euclidean distance in a certain transformed space, and we can talk about the choice of transform, instead of the choice of distance.

4.1 Euclidean Distance in Time-Domain

Let $\Lambda = diag\left(M\left(:,t\right)\right)$. The expression of this distance is:

$$\begin{aligned}
\left\|\mathbf{x}\left(t\right)-\mathbf{x}_{train}^{M(:,t)}\left(t_1\right)\right\|_2^2 &= \left(\mathbf{x}\left(t\right)-\mathbf{x}_{train}^{M(:,t)}\left(t_1\right)\right)^T \overline{\left(\mathbf{x}\left(t\right)-\mathbf{x}_{train}^{M(:,t)}\left(t_1\right)\right)}\\
&= \left(\mathbf{x}\left(t\right)-\mathcal{F}^{-1}\cdot\Lambda\cdot\mathcal{F}\mathbf{x}_{train}\left(t_1\right)\right)^T\cdot\overline{\left(\mathbf{x}\left(t\right)-\mathcal{F}^{-1}\cdot\Lambda\cdot\mathcal{F}\mathbf{x}_{train}\left(t_1\right)\right)}\\
&= \left(\mathcal{F}\mathbf{x}\left(t\right)-\Lambda\cdot\mathcal{F}\mathbf{x}_{train}\left(t_1\right)\right)^T\cdot\left(\mathcal{F}^{-1}\right)^T\cdot\overline{\mathcal{F}^{-1}}\cdot\overline{\left(\mathcal{F}\mathbf{x}\left(t\right)-\Lambda\cdot\mathcal{F}\mathbf{x}_{train}\left(t_1\right)\right)}\\
&= \left(\Lambda\cdot\mathcal{F}\mathbf{x}_c\left(t\right)-\Lambda\cdot\mathcal{F}\mathbf{x}_{train}\left(t_1\right)\right)^T\cdot\overline{\left(\Lambda\cdot\mathcal{F}\mathbf{x}_c\left(t\right)-\Lambda\cdot\mathcal{F}\mathbf{x}_{train}\left(t_1\right)\right)}\\
&= \left(\mathcal{F}\mathbf{x}_c\left(t\right)-\mathcal{F}\mathbf{x}_{train}\left(t_1\right)\right)^T\cdot\Lambda\cdot\overline{\left(\mathcal{F}\mathbf{x}_c\left(t\right)-\mathcal{F}\mathbf{x}_{train}\left(t_1\right)\right)}\\
&= \left(\mathbf{x}_c\left(t\right)-\mathbf{x}_{train}\left(t_1\right)\right)^T\cdot\left(\mathcal{F}\Lambda\mathcal{F}^{-1}\right)\cdot\overline{\left(\mathbf{x}_c\left(t\right)-\mathbf{x}_{train}\left(t_1\right)\right)}
\end{aligned}$$

As we can see, this distance reflects somehow the "closeness" between the clean frame (from which the observed data supposedly originated) and the clean training data, but the weighting matrix $\mathcal{F}\cdot\Lambda\cdot\mathcal{F}^{-1}$ alters this natural closeness according to the missing data pattern $M\left(:,t\right)$ in a rather non-intuitive way. Speech signals do not have a too informative description in the (frame-based) time domain representation and our setting is apparently quite sensitive to this issue, especially when training data is scarce.

4.2 Euclidean Distance in Fourier Domain

Instead of $\left\|\mathbf{x}\left(t\right)-\mathbf{x}_{train}^{M(:,t)}\left(t_1\right)\right\|_2$, we use $\left\|\mathcal{F}\mathbf{x}\left(t\right)-\mathcal{F}\mathbf{x}_{train}^{M(:,t)}\left(t_1\right)\right\|_2$:

$$\begin{aligned}
\left\|\mathcal{F}\mathbf{x}\left(t\right)-\mathcal{F}\mathbf{x}_{train}^{M(:,t)}\left(t_1\right)\right\|_2^2 &= \left(\mathcal{F}\mathbf{x}\left(t\right)-\mathcal{F}\mathbf{x}_{train}^{M(:,t)}\left(t_1\right)\right)^T\cdot\overline{\left(\mathcal{F}\mathbf{x}\left(t\right)-\mathcal{F}\mathbf{x}_{train}^{M(:,t)}\left(t_1\right)\right)}\\
&= \left(\mathbf{x}\left(t\right)-\mathbf{x}_{train}^{M(:,t)}\left(t_1\right)\right)^T\overline{\left(\mathbf{x}\left(t\right)-\mathbf{x}_{train}^{M(:,t)}\left(t_1\right)\right)} = \left\|\mathbf{x}\left(t\right)-\mathbf{x}_{train}^{M(:,t)}\left(t_1\right)\right\|_2^2
\end{aligned}$$

Under our working assumptions, using Euclidean distance in time or Fourier domain will produce exactly the same results.

4.3 Euclidean Distance in ICA Domain

One problem with the previous two distance measures is that neither representation is the most appropriate one for describing the distribution of speech frames. One reason for their limited power is that they are both fixed, data independent representations. Since the distribution of speech frames is different from one speaker to another it is meaningful to consider adaptive representations of the data. We mentioned earlier ICA as a data-driven linear representation, whose objective is to maximize the statistical independence of the coefficients. Under this assumption, the basis functions will point into directions that determine maximally sparse (sharply peaked, with heavy tails) marginal distributions of the coefficients. Let us inspect what the Euclidean distance is in the ICA domain:

$$\begin{aligned}\left\|\mathbf{s}(t)-\mathbf{s}_{train}^{M(:,t)}(t_1)\right\|_2^2 &= \left(\mathbf{s}(t)-\mathbf{s}_{train}^{M(:,t)}(t_1)\right)^T\cdot\overline{\left(\mathbf{s}(t)-\mathbf{s}_{train}^{M(:,t)}(t_1)\right)}\\ &= \left(\mathbf{Wx}(t)-\mathbf{Wx}_{train}^{M(:,t)}(t_1)\right)^T\overline{\left(\mathbf{Wx}(t)-\mathbf{Wx}_{train}^{M(:,t)}(t_1)\right)}\\ &= \left(\mathbf{x}(t)-\mathbf{x}_{train}^{M(:,t)}(t_1)\right)^T\cdot\mathbf{W}^T\mathbf{W}\cdot\overline{\left(\mathbf{x}(t)-\mathbf{x}_{train}^{M(:,t)}(t_1)\right)}\end{aligned}$$

Since matrix $\mathbf{W}$ is in general not orthogonal, the weighted distance it induces will be different from the previous two distance measures. Furthermore, we can use the Mahalanobis distance (in either time or ICA domains) to normalize the data covariance, and thus to reduce the dependency of the distance on the relative scaling along the principal axes of the corresponding space.

5 Experimental Results

To demonstrate the performance of the proposed method, we tested it in several special settings, on data taken from the NTT database. The signals represent sentences, several seconds long, sampled at $F_s = 16k\text{Hz}$, and read by either male or female native English language speakers. We illustrate the reconstruction performance both qualitatively, by displaying the spectrogram of corrupted and reconstructed speech signals, and quantitatively, by computing objective reconstruction quality measures, such as log-spectral distortion (LSD) [2] and Itakura-Saito distance (IS) [1].

Table 1. (a) Left - Numerical values of mean values for LSD and IS, in the case of BWE ($F_{cut} = 4k\text{Hz}$). (b) Right - Mean LSD and IS values, in the case of alternative constant pattern setting. Amount of training data is either 20000 or 100000 frames of clean speech.

(a)	20,000		100,000		(b)	20,000		100,000	
Domain	LSD	IS	LSD	IS		LSD	IS	LSD	IS
Time	5.728	0.166	5.518	0.0756		5.935	0.164	5.827	0.132
ICA	5.574	0.157	5.462	0.0670		5.829	0.147	5.670	0.106

First, we applied the proposed method for BWE, where the observed signal was obtained by low-pass filtering a wideband speech signal. As the value for the cut-off frequency, we used $F_{cut} = 4k\text{Hz}$. We display the results obtained (see Table 1 (a)) when using Euclidean distance in time domain and in ICA domain. Although the difference is not big, the experiment shows that the performance in the latter case was consistently better. Also, by increasing the number of training speech frames, the performance is improved. Overall, results are comparable in terms of LSD with the state-of-the-art [2].

In the second experiment, we chose a fixed, multi-band pattern and again we tried to reconstruct a wideband signal (see example in Figure 2). The missing frequency bands were $1k\text{Hz}-2k\text{Hz}$, $3k\text{Hz}-4k\text{Hz}$, $5k\text{Hz}-6k\text{Hz}$, and $7k\text{Hz}-8k\text{Hz}$.

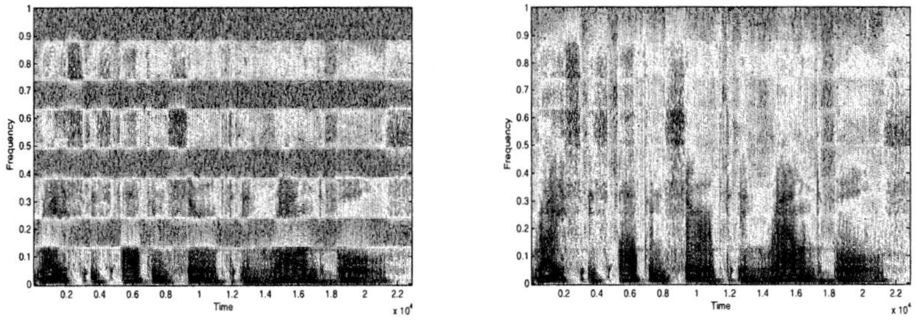

Fig. 2. Example of observed male speech data and the reconstructed signal

The numerical values obtained for LSD and IS resulted after the bandwidth extension are shown in 1 (b). The results display the same behavior we noticed in the previous experiment.

6 Conclusions

This paper addresses the problem of speech enhancement in a setting where parts of the time-frequency content of the speech signal are missing. We analyze a simple but general algorithm for inferring the missing spectral data, and demonstrate the wide applicability of the method to bandwidth expansion, to a multi spectral band inference problem, and show applicability to enhancing speech resulting from TF BSS methods. Interesting areas of further work are coding for VQ-based training data compression and algorithm speed-up, applications to BSS, and extensions of the algorithm to exploit models of temporal relationships between successive frames.

References

1. Deng, L., O'Shaughnessy, D.: Speech Processing. M. Dekker (2003)
2. Jax, P.: Ch. 6. In: Audio Bandwidth Extension. J. Wiley and Sons (2004)
3. Qian, Y., Kabal, P.: Combining equalization and estimation for bandwidth extension of narrowband speech. ICASSP (2004) I–713–I–716
4. Unno, T., McCree, A.: A robust narrowband to wideband extension system featuring enhanced codebook mapping. In: ICASSP 2005, (DSPS R&D Center, Texas Instruments, Dallas, TX) I–805 – I–808
5. Rickard, S., Balan, R., Rosca, J.: Real-time time-frequency based blind source separation. (In: ICA 2001, San Diego, CA) 651–656
6. Attias, H., Platt, J., Acero, A., Deng, L.: Speech denoising and dereverberation using probabilistic models. In: Neural Information Processing Systems 13. (2001) 758–764
7. Roweis, S.T.: One microphone source separation. In: Neural Information Processing Systems 13 (NIPS). (2000) 793–799
8. Gray, R., Neuhoff, D.: Quantization. IEEE Trans. on Information Theory **44** (1998) 2325–2383

9. Hyvarinen, A., Karhunen, J., Oja, E.: Independent component analysis. John Wiley and Sons (2001)
10. Rosca, J., Gerkmann, T., Balcan, D.: Statistical inference with missing data in the ica domain. (In: Submitted to ICASSP 2006)
11. Lewicki, M.: Efficient coding of natural sounds. Nature Neurosci. **5** (2002) 356–363

Harmonic Source Separation Using Prestored Spectra

Mert Bay and James W. Beauchamp

School of Music and Dept. of Electrical and Computer Engineering,
University of Illinois at Urbana-Champaign,
Urbana, IL 61801
mertbay@uiuc.edu, jwbeauch@uiuc.edu

Abstract. Detecting multiple pitches (F0s) and segregating musical instrument lines from monaural recordings of contrapuntal polyphonic music into separate tracks is a difficult problem in music signal processing. Applications include audio-to-MIDI conversion, automatic music transcription, and audio enhancement and transformation. Past attempts at separation have been limited to separating two harmonic signals in a contrapuntal duet (Maher, 1990) or several harmonic signals in a single chord (Virtanen and Klapuri, 2001, 2002). Several researchers have attempted polyphonic pitch detection (Klapuri, 2001; Eggink and Brown, 2004a), predominant melody extraction (Goto, 2001; Marolt, 2004; Eggink and Brown, 2004b), and instrument recognition (Eggink and Brown, 2003). Our solution assumes that each instrument is represented as a time-varying harmonic series and that errors can be corrected using prior knowledge of instrument spectra. Fundamental frequencies (F0s) for each time frame are estimated from input spectral data using an Expectation-Maximization (EM) based algorithm with Gaussian distributions used to represent the harmonic series. Collisions (i.e., overlaps) between instrument harmonics, which frequently occur, are predicted from the estimated F0s. The uncollided harmonics are matched to ones contained in a pre-stored spectrum library in order that each F0`s harmonic series is assigned to the appropriate instrument. Corrupted harmonics are restored using data taken from the library. Finally, each voice is additively resynthesized to a separate track. This algorithm is demonstrated for a monaural signal containing three contrapuntal musical instrument voices with distinct timbres.

1 Introduction

Ordinarily, before separating individual instrument voices into separate tracks, polyphonic pitch detection must be performed on a monaural file instrument mixture. However, we considered two cases: 1) Obtaining F0 data and spectrum analysis from solo recordings before mixing them to monaural. 2) Obtaining F0 data directly from the monaural polyphonic mixture. While our ultimate objective is to solve the more general second case, because of the difficulty of polyphonic pitch detection, we have decided, for now, to focus on the first method. Moreover, starting with solo signals is necessary for evaluating the performance of our separation algorithm by comparing the original solo and separated signals, via listening and computing rms spectral errors. See Fig. 1 for an overview of the pitch detection/separation method.

J. Rosca et al. (Eds.): ICA 2006, LNCS 3889, pp. 561 – 568, 2006.

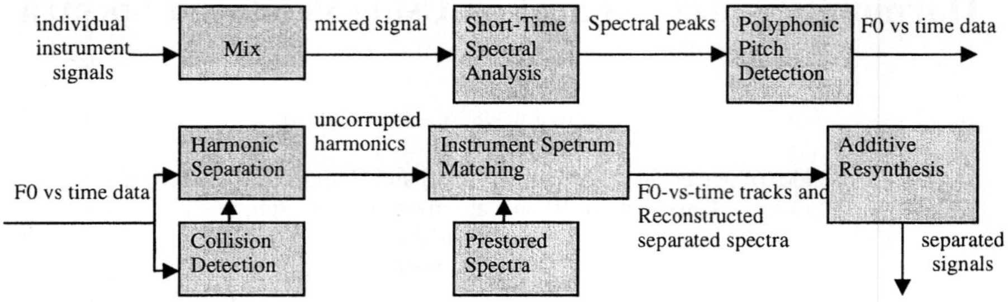

Fig. 1. Flow diagram of the pitch detection/separation algorithm

2 Method

2.1 Spectral Analysis

The first stage of the method performs short-time peak-tracking spectral analysis of the test signal to find a set of spectral peaks for each frame (McAulay and Quatieri, 1986; Smith and Serra, 1987; Beauchamp, 1993). Fig. 2 shows the spectral peaks for a single frame corresponding to 1.4 s from the start of a three instrumental voice mixture (Bb clarinet, trombone, and alto saxophone). The 5 s clarinet and saxophone solo passages were clipped from a jazz CD (Art Pepper, 1996) and from a Mozart's *Requiem* trombone solo recorded by Jay Bulen at the University of Iowa. Obviously, the solos were not intended to harmonize or synchronize in any way.

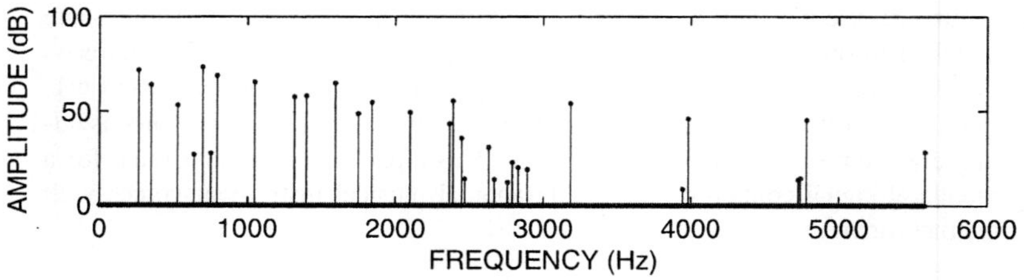

Fig. 2. Spectrum of three instrument mixture at t=1.4 s

2.2 Polyphonic Pitch Detection

For each frame, each fundamental frequency (F0) candidate is represented as mixture of 10 Gaussian PDFs whose means are located at integer multiples of F0 and whose STD bandwidths are 30 Hz. Then the expectation of this candidate F0 is calculated by integrating the product of the mixture of Gaussian distruburions with the input spectrum (see Fig. 3). This is in essence is the correlation of the input with the GMM in the frequency domain.

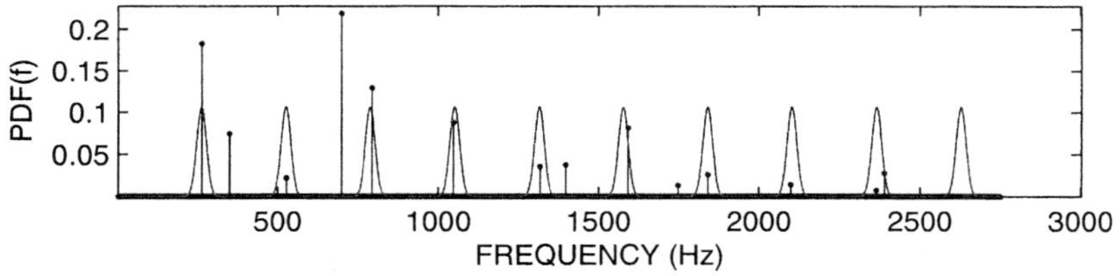

Fig. 3. Using a mixture of Gaussian distributions to calculate the expectation of an F0 candidate

Assuming that the input signal contains N simultaneous instrumental voices, the expectations of all possible combinations of N F0s are calculated in a specified F0 search range. The optimum combination which yields the highest expectation is chosen. However, as mentioned above, this method has so far only proved robust for N=1, so at this point we are using F0s based on the original individual tracks.

2.3 Harmonic Collision Detection and Initial Separation

For each frame, the frequencies of collided harmonics are calculated theoretically according to the location of the harmonics of the estimated F0s, within the resolution of the spectral analysis. These harmonics are ignored in the spectrum matching step. (see Fig. 4). So at this point, three spectra with missing harmonics for the current frame have been resolved, but they haven't been identified as paricular instruments yet.

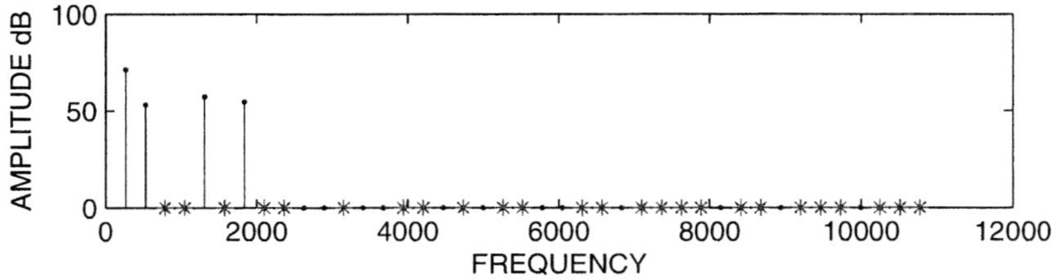

Fig. 4. Initially separated uncorrupted harmonics (denoted by .) for one of the estimated F0s from the spectrum of Fig. 2. * denotes the positions of estimated collisions where harmonic amplitudes are set to zero.

2.4 Instrument Spectrum Library

An instrument spectrum library (the training set) was created using University of Iowa musical instrument samples (Fritts, 1997-). This database includes individual tones performed at three different dynamics (*pp*, *mf*, and *ff*) in semitone F0 increments for clarinet, saxophone, and trombone. For each F0, the tones were analyzed (Beauchamp, 1993) and a spectrum space created consisting of the harmonic spectra of all of the frames for the three tones performed at that F0. The number of harmonics for each F0 is given by floor(.5f_s /F0), where f_s is the sampling frequency. Then a K-means clustering algorithm (Rabiner and Juang, 1993) partitioned the space into 10 different clusters, and each cluster's centroid was calculated. (Fig. 5 shows an example K-means "cluster spectrum".) 10 spectra, which form a "sublibrary", were chosen as a compromise between providing adequate spectral diversity while having a sufficient number of candidates to average within each cluster. We have also experimented with clustering according to spectral centroid ranges of the training data and calculating the average spectrum for each spectral centroid cluster (Beauchamp and Horner, 1995). Both methods yield similar results, but K-means avoids the problem of sparsity of data for some clusters.

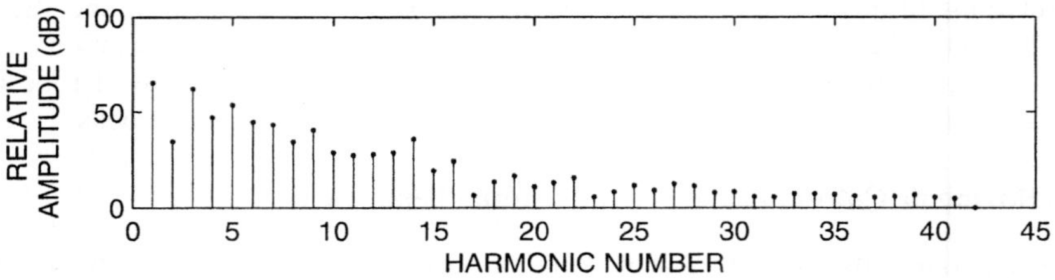

Fig. 5. One of the K-means cluster spectra from the clarinet library for F0=261.6Hz

2.5 Instrument Spectrum Matching

To replace the corrupted harmonics in an initially separated harmonic spectrum, the corresponding F0 sublibraries of the entire spectrum library are searched to find the best match to the uncorrupted input harmonics (see Fig. 4). A least squares (LMS) algorithm is used to obtain the optimum scaling factor between the input and prestored spectra. Basically, the instrument matching part is a nearest-neighbor classifier where the distance measure is a (possibly frequency-weighted) Euclidian distance between the corresponding harmonics of the initially separated and the sublibrary spectra. However, we have found that even 10 cluster spectra are insufficient to avoid artifacts that occur when switching between spectra. Therefore, after choosing the instrument library for the initially separated spectra, LMS is applied again to find an optimum interpolation between the best two spectra out of the 10. This improves matching for individual frames while smoothing transitions as the spectrum changes from one frame to the next. For synthesis we can either replace only the corrupted harmonics (see Fig. 6) or replace all of the spectrum components from the interpolated library spectra. While the former method may yield better fidelity to the test spectra, the latter method can yield a result with fewer audible artifacts.

Finally the reconstituted spectra are resynthesized to the individual instrument tracks using sinusoidal additive synthesis (Beauchamp, 1993). Frequencies and amplitudes of the corresponding harmonics are linearly interpolated and phases accumulated between frames, with initial phases set to random values.

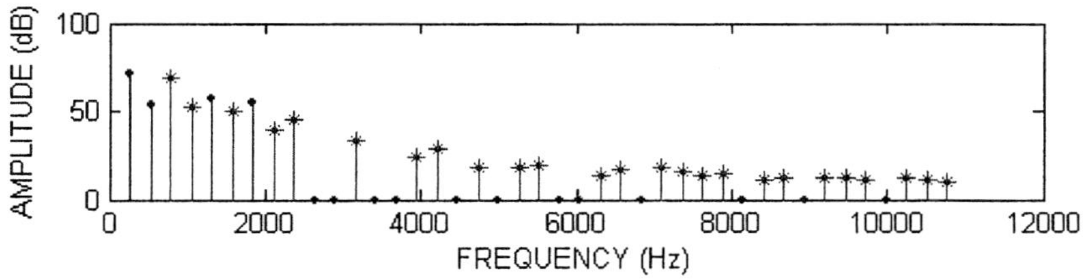

Fig. 6. The uncorrupted harmonics of an initially separated spectrum (denoted by .) are classified as clarinet and the collisions (denoted by *) are replaced from the best-match prestored spectrum. Note that zero values in the initially separated spectrum are due to the test clarinet's spectrum above 2500 Hz being weaker than the training clarinet's spectrum in the same frequency range.

3 Results

Since at the current stage of our research our pitch detection algorithm does not perform well enough for subsequent instrument separation, we used F0s obtained from the solo tracks (see Fig. 7). Nevertheless, instrument matching was blind with respect to the source of each F0. For each frame and each F0, instrument classification resulted from matching the three corrupted harmonic spectra across all three instrument libraries. As it turned out, with the correct F0 contours, the correct instruments were chosen with 100% accuracy.

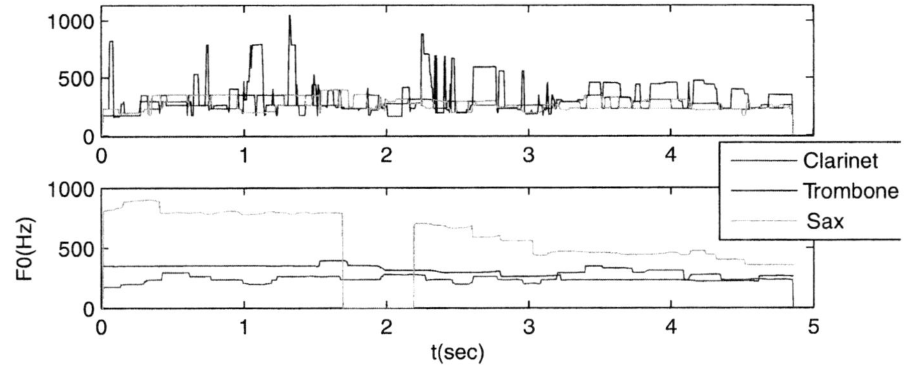

Fig. 7. Pitch contours estimated from the mixture signal (upper) and individual solos (lower)

Original and the separated tracks were compared by listening and by measuring spectral rms error. Most audible artifacts in the separated tracks seemed to be due to unison and octave collisions. Nearly all harmonics of two instruments played in unison are corrupted, while in the octave case every second harmonic of the lower voice and nearly all harmonics of the higher voice are corrupted. However, in practice, two voices are usually not pitched exactly an octave apart, so we could retrieve some upper harmonics of the higher tone in order to estimate its lower harmonics. Figs. 8, 9, and 10 each show spectrograms of the original instrument solo tracks and the corresponding tracks separated from the mixture with collided harmonics replaced.

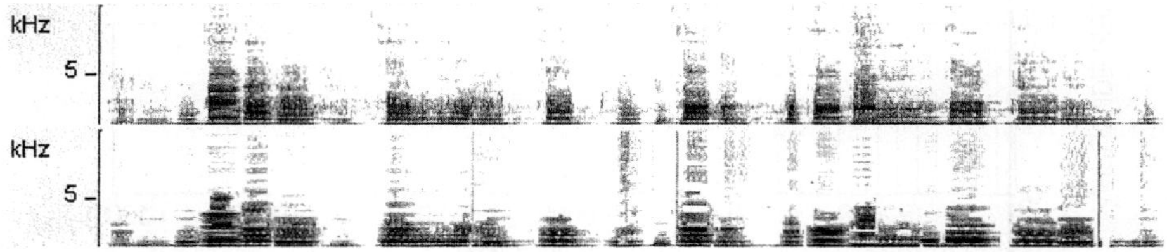

Fig. 8. Original (upper one) and separated (lower one) clarinet spectra

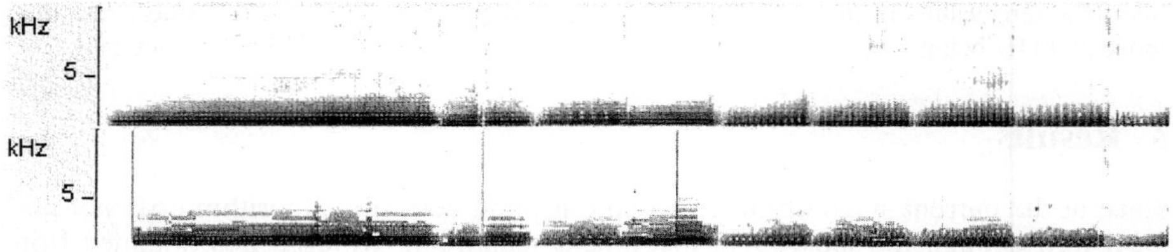

Fig. 9. Original (upper one) and separated (lower one) trombone spectra

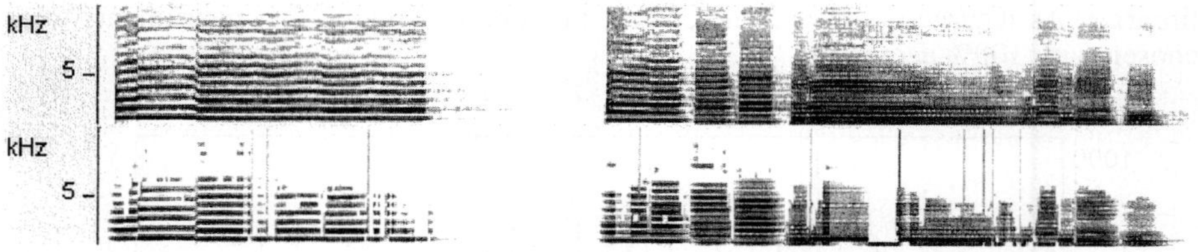

Fig. 10. Original (upper one) and separated (lower one) saxophone spectra

Audible differences between the original and synthetic tracks include a) inherent differences between original and best-match library spectra, b) loss of reverberation and other noise, c) occasional sound "bobbling" due to high occurrence of harmonic collisions, resulting in insufficient data to estimate the corrupted harmonics correctly, thus resulting in sharp discontinuities, and d) upper harmonic "chattering" due to switching between different library spectra. The latter effect is alleviated by LMS interpolation between the two best library spectra matches.

Since the resynthesized tracks are not phase-locked with the originals, we cannot compute an accurate time-domain difference residual. However, we can compute the rms difference between the time-varying harmonic amplitudes of the separated tracks and the originals. Fig. 11 shows graphs of relative-amplitude spectral rms error vs. time for the three instruments. The rms error was calculated using the equation

$$error_{rms}(j) = 20\log10\left(\frac{\sqrt{\sum_{k=1}^{K}(x_{jk} - \alpha\hat{x}_{jk})^2}}{\frac{1}{J}\sum_{j=1}^{J}\sqrt{\sum_{k}^{K} x_{jk}^2}}\right)$$

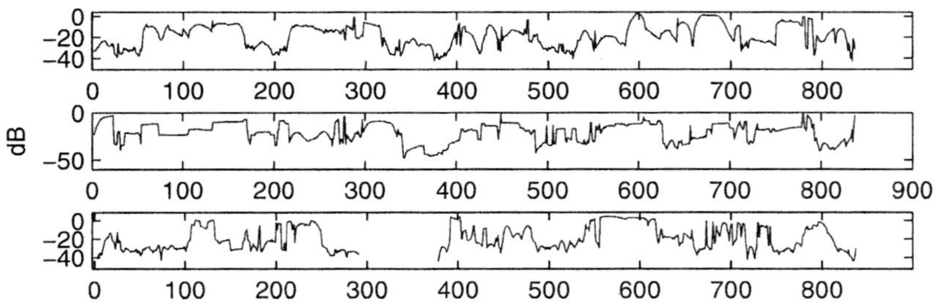

Fig. 11. Spectral rms error for clarinet, trombone, and saxophone

where j = frame number, J = number of frames, k = harmonic number, K = number of harmonics, x_{jk} = original track harmonic amplitude, $\hat{x}_{jk}$ = separated track harmonic amplitude, and α is a constant scale factor which minimizes the error for the entire signal. The rms error is normalized by the average rms amplitude of the original signal instead of rms value for that frame because in the latter case when the amplitude is very small the error gets extremely high although it is not really audible.

Averaging the rms errors over time yielded 21.57% for the clarinet, 14.65% for the trombone, and 27.27% for the alto saxophone. Evaluating the performance by the spectral rms difference might be misleading because although the separated trombone has the lowest rms error, in the authors' opinion the clarinet sounds better. The original, mixture, and separated samples can be found at http://ems.music.uiuc.edu/beaucham/sounds/separation/.

4 Summary

Using pitch-vs.-time tracks derived from three non-harmonizing instrument solos with distinctive timbres and prestored independent instrument spectra to correct collided harmonics, we were able to separate the solos from their monaural mixture with reasonable preservation of quality. rms spectral accuracy varied from about 14% to 27%. An objective evaluation of separation quality would be highly desireable but is non-trivial because of the necessity of comparing to a standard level of degradation (Thiede *et al.*, 2000). We attempted to derive the pitch-vs.-time tracks directly from the monaural mixture, but the results were not accurate enough for reasonable quality separation. However, our method has demonstrated that it is not necessary to have prior knowledge of each initially separated spectrum's instrument identity, because this is sorted out in the spectral matching process.

5 Future Work

First, we plan to try out different polyphonic pitch detection algorithms in an effort to improve this important, and ultimately necessary, analysis step. Transition probabilities between notes may be utilized. Second, we plan to increase the size of the spectral data base to handle a wider variety of instruments. Third, we will attempt to use estimates of corrupting spectra in order to estimate the true amplitudes of corrupted harmonics so as to obviate their replacement. Fourth, we will attempt to find note boundaries and optimize spectral choices over notes. Fifth, we will attempt to utilize time behavior over notes (vibrato, beating) to more intelligently separate partials.

References

1. J. Beauchamp. (1993) "Unix Workstation Software for Analysis, Graphics, Modification, and Synthesis of Musical Sounds", Audio Eng. Soc. Preprint No. 3479, pp. 1-17.
2. J.W. Beauchamp and A. Horner (1995). "Wavetable Interpolation Synthesis Based on Time-Variant Spectral Analysis of Musical Sounds", Audio Eng. Soc. Preprint No. 3960, pp. 1-17.
3. J. Eggink and G. J. Brown (2003). "A missing feature approach to instrument identification in polyphonic music", Proc. Int. Conf. on Acoustics, Speech, and Signal Processing (ICASSP-03), pp. 553-556.
4. J. Eggink and G. J. Brown (2004a). "Instrument recognition in accompanied sonatas and concertos", Proc. Int. Conf. on Acoustics, Speech, and Signal Processing (ICASSP-04), pp. IV217-220.
5. J. Eggink and G. J. Brown (2004b). "Extracting melody lines from complex audio", Proc. 5th Int. Conf. on Music Information Retrieval (ISMIR-04), pp. 84-91.
6. L. Fritts (1997-). "University of Iowa Musical Instrument Samples", on-line at http://theremin.music.uiowa.edu/MIS.html.
7. M. Goto (2001). "A predominant-F0 estimation method for CD recordings: MAP estimation using EM algorithm for adaptive tone models". Proc. Int. Conf. on Acoustics, Speech, and Signal Processing (ICASSP-01), pp. 3365-3368.
8. A. Klapuri (2001). "Multipitch estimation and sound separation by the spectral smoothness principle", Proc. ICASSP'01, pp. 3381-3384.
9. R. Maher (1990). "Evaluation of a method for separating digitized duet signals", J. Audio Eng. Soc. 38(12), pp. 957-979.
10. M. Marolt (2004). "Gaussian mixture models for extraction of melodic lines from audio recordings", Proc. 5th Int. Conf. on Music Information Retrieval (ISMIR'04), pp. 80-83.
11. R. J. McAulay and T. F. Quatieri (1986). "Speech analysis/synthesis based on a sinusoidal representation", IEEE Trans. Acoust. Speech, Signal Processing, ASSP-34, pp. 744-754.
12. A. Pepper (1996). *The Intimate Art Pepper* (music CD), tracks 5 & 7.
13. L. Rabiner and B.-H. Juang (1993). *Fundamentals of Speech Recognition*, Prentice Hall, pp. 125-128.
14. J. O. Smith and X. Serra (1987). "PARSHL: An analysis/synthesis program for non-harmonic sounds based on a sinusoidal representation", Proc. 1987 Int. Computer Music Conf., pp. 290-297.
15. T. Thiede, W. C. Treurniet, R. Bitto, C. Schmidmer, T. Sporer, J. G. Beerends, Ca. Colomes, M.l Keyhl, G. Stoll, K. Brandenburg, and B. Feiten (2000). "PEAQ-The ITU Standard for Objective Measurement of Perceived Audio Quality", J. Audio Eng. Soc. 48(1/2), pp. 3-29.
16. T. Virtanen and A. Klapuri (2001). "Separation of harmonic sounds using multipitch analysis and iterative parameter estimation", IEEE Workshop on Applicatioins of Signal Processing to Audio and Acoustics (WASPAA-01), pp. 83-86.
17. T. Virtanen, A. Klapuri (2002), "Separation of Harmonic Sounds Using Linear Models for the Overtone Series", IEEE Int. Conf. on Acoustics, Speech and Signal Processing, (ICASSP-02).

Underdetermined Convoluted Source Reconstruction Using LP and SOCP, and a Neural Approximator of the Optimizer

Pau Bofill[1] and Enric Monte[2]

[1] Departament d'Arquitectura de Computadors–UPC, Campus Nord, Mòdul D6, Jordi Girona 1-3, 08034 Barcelona, Spain
pau@ac.upc.edu
[2] Departament de Teoria del Senyal i Comunicacions–UPC, Campus Nord, Mòdul D5, Jordi Girona 1-3, 08034 Barcelona, Spain
enric@gps.tsc.upc.edu

Abstract. The middle-term goal of this research project is to be able to recover several sound sources from a binaural life recording, by previously measuring the acoustic response of the room. As a previous step, this paper focuses on the reconstruction of n sources from m *convolutive* mixtures when $m < n$ (*underdetermined* case), assuming the mixing matrix is *known*.

The reconstruction is done in the frequency domain by assuming that the source components are Laplacian in their real and imaginary parts. By posterior likelihood optimization, this leads to norm 1 minimization subject to the mixing equations, which is an instance of *linear programming* (LP). Alternatively, the assumption of Laplacianity imposed on the magnitudes leads to *second order cone programming* (SOCP).

Performance experiments are run from synthetic mixtures based on realistic simulations of each source-microphone impulse response. Two sets of sources are used as benchmarks: four speech utterances and six short violin melodies. Results show S/N reconstruction ratios around 10dB. If any, SOCP performs slightly better.

SOCP is probably too slow for real-time processing. In the last part of this paper we train a neural network to predict the response of the optimizer. Preliminary results show that the approach is feasible but yet inmature.

1 Introduction

Our long-term goal is to be able to separate each of the sources from a binaural life recording of several instruments or voices playing simultaneously in a room. This is an instance of the *blind source separation* problem, in a convolutive underdetermined setting (room reverberation and more sources than microphones):

$$\begin{aligned} x_1(t) &= h_{11}(t) * s_1(t) + \ldots + h_{1n}(t) * s_n(t) \\ &\cdots \\ x_m(t) &= h_{m1}(t) * s_1(t) + \ldots + h_{mn}(t) * s_n(t) \,, \end{aligned} \tag{1}$$

J. Rosca et al. (Eds.): ICA 2006, LNCS 3889, pp. 569–576, 2006.

where $x_i(t)$ is the signal recorded at microphone i, $s_j(t)$ is the signal from source j, h_{ij} is the room's impulse response from source j to microphone i, and $*$ is the convolution operator. The general problem is to blindly determine the sources $s_j(t)$, given only the mixtures $x_i(t)$.

In the discrete frequency domain, for a given K-sample short-time window or *frame*, the mixing equations are rewritten as

$$\begin{aligned} X_1^k &= H_{11}^k S_1^k + \ldots + H_{1n}^k S_n^k \\ &\ldots \\ X_m^k &= H_{m1}^k S_1^k + \ldots + H_{mn}^k S_n^k , \end{aligned} \tag{2}$$

for $k = 0 \ldots K/2$, where k stands for the frequency bin (since all signals are real, only $K/2+1$ bins are necessary). In matrix notation,

$$\mathbf{X}^k = \mathbf{Z}^k \mathbf{S}^k, \quad k = 0, \ldots, K/2 . \tag{3}$$

In this model, for every frequency bin k, the data is the set of spectral mixture vectors corresponding to every time frame.

The problem of *blind source separation* was first formulated for $m = n$ and the *instantaneous mixture* case (i.e., $h_{ij}(t) = \beta_{ij}$ and $\mathbf{Z}^k = \mathbf{B}$, an attenuation matrix of real coefficients). The usual approach to the separation has been *indepent component analysis*, which solves the system by assuming only the statistical independence of the source components. Of particular interest to our work is the so called *sparse* case [1,2], which is usually modelled by further assuming that the source components have a Laplacian pdf.

A harder problem is the *delayed mixture* case, corresponding to the anechoic room recording. That is, when signal delays are taken into account, without echoes or reverberation (i.e., $h_{ij}(t) = \beta_{ij}\delta(t - \tau_{ij})$ and $\mathbf{Z}^k = \mathbf{B} \cdot e^{j2\pi k\mathbf{T}/K}$ with $\mathbf{T}$ a delay matrix, and $\cdot$ the component to component product). The delayed case is considered, for instance, in [3].

Equation 3 describes the most general and hardest situation, the *convolutive mixture* case. For $n = m$ the convolutive case has been tackled either in the time domain [4] or in the frequency domain [5,6], among others. In the time domain the difficulty is the number of echoes that have to be considered for a realistic characterization of the impulse response, whereas in the frequency domain the difficulty lies in the gain and ordering undeterminacies.

When the number m of sensors is smaller than the number n of sources, inferring the mixing matrix, on the one side, and inferring the sources, on the other, can be formulated as two separate problems. Even when the mixing matrix is known, the mixing equations are still *underdetermined* and some additional assumption must be imposed on the sources to solve the system unambiguously.

In our previous work [7,8], the assumption of sparsity turns out to be of help. The reason why sparse representations lend themselves to good separability is because, since only a few coefficients are significantly different from zero, the likelihood that a particular data point belongs mainly to a single source is high. In other words, sparse representations are separable because sparse sources are, to some extent, disjoint. Sparsity depends on the representation domain. For speech and music signals, the time-domain representation is definetly not sparse. Yet, the frequency-domain representation, being pseudo-harmonic, is significantly sparse. Therefore, we use the frequency-domain.

Results for the delayed underdetermined case and for the convolutive underdetermined case can be found in [9,10] and [11], respectively.

In this paper we tackle the recovery of the sources *given the convolution matrix*. After all, the approach is feasible since the impulse response of the room can be measured [12,13].

In a maximum posterior likelihood framework, this paper presents an experimental comparison between LP and SOCP [14] for the reconstruction of the sources, and suggests the use of a neural network approximator to predict the behaviour of the optimizer. SOCP was proposed for the delayed underdetermined case in [9]. A different analysis of the performance of SOCP in a setting very similar to this paper can be found in [15].

2 A Maximum Posterior Likelihood Formulation

Reverting to Eqn. 3, our goal is to solve

$$\mathbf{X}^k = \mathbf{Z}^k \mathbf{S}^k, \quad k = 0, \ldots, K/2,$$

given $\mathbf{X}^k$ and $\mathbf{Z}^k$ when $m < n$, with m the dimension of $\mathbf{X}^k$ and n the dimension of $\mathbf{S}^k$. Since the system is underdetermined, some additional restriction must be imposed to get an unambiguous solution.

As in our previous work [7,9] the assumption we are going to use is the Laplacianity of the sources. But, since we are working in the frequency domain, such assumption can be imposed either on the real and imaginary parts of $\mathbf{S}^k$ (Section 2.1) or on the magnitude of $\mathbf{S}^k$ (Section 2.2).

2.1 Laplacianity of the Real and Imaginary Parts: An Instance of Linear Programming

The separation in this case is modeled by rewritting Eqn. 3 in terms of the real and imaginary parts. Let

$$\mathcal{X} = \begin{pmatrix} \mathrm{Re}(\mathbf{X}) \\ \mathrm{Im}(\mathbf{X}) \end{pmatrix}$$

be the aggregate vector of real and imaginary components, and let

$$\mathcal{S} = \begin{pmatrix} \mathrm{Re}(\mathbf{S}) \\ \mathrm{Im}(\mathbf{S}) \end{pmatrix}$$

and

$$\mathcal{Z} = \begin{pmatrix} \mathrm{Re}(\mathbf{Z}) & -\mathrm{Im}(\mathbf{Z}) \\ \mathrm{Im}(\mathbf{Z}) & \mathrm{Re}(\mathbf{Z}) \end{pmatrix} .$$

Then,

$$\mathcal{X}^k = \mathcal{Z}^k \mathcal{S}^k, \quad k = 0, \ldots, K/2\,, \tag{4}$$

which, for every k, is a linear underdetermined mixture model.

The assumptions we make in this case are the following: (a) The source components are statistically independent, (b) their real and imaginary parts follow a Laplacian distribution with (c) equal variances. These assumptions can be modeled by

$$P(\mathcal{S}_l^k) = e^{-\mu|\mathcal{S}_l^k|}\,, \tag{5}$$

with $|\cdot|$ the absolute value. Maximizing the posterior probability of the source components ($\mathcal{S}$) given the data ($\mathcal{X}$),

$$max_{\mathcal{S}} P(\mathcal{S}^k|\mathcal{X}^k) \propto max_{\mathcal{S}} P(\mathcal{X}^k|\mathcal{S}^k) P(\mathcal{S}^k)\,. \tag{6}$$

In the absense of noise, $\mathcal{X}^k = \mathcal{Z}^k \mathcal{S}^k$, and $P(\mathcal{X}^k|\mathcal{S}^k) = 1$. Therefore, under the assumptions above,

$$max_{\mathcal{S}} P(\mathcal{S}^k|\mathcal{X}^k) \propto \Pi_l e^{-\mu|\mathcal{S}_l^k|}\,. \tag{7}$$

Taking logarithms, the above is equivalent to

$$\begin{array}{c} min_{\mathcal{S}} \sum_l |\mathcal{S}_l^k| \\ \text{subject to} \quad \mathcal{X}^k = \mathcal{Z}^k \mathcal{S}^k\,, \end{array} \tag{8}$$

which can easily be expressed as a linear programming problem [16].

2.2 Laplacianity of the Magnitudes: An Instance of Second Order Cone Programming

Going back to the mixing model of Eqn. 3, and like we did for the delayed case in [9], an alternative formulation comes from the following assumptions: (a) the source components are statistically independent, (b) their phases are uniformly distributed, and (d) their magnitudes follow a Laplacian distribution with (d) equal variances. That is,

$$P(\mathcal{S}_l^k) \propto e^{-\mu \text{Mag}(\mathcal{S}_l^k)}\,. \tag{9}$$

Proceeding like in the previous section, the posterior likelihood formulation leads to

$$\begin{array}{c} min_{\mathbf{S}} \sum_l \text{Mag}(\mathcal{S}_l^k) \\ \text{subject to} \quad \mathbf{X}^k = \mathbf{Z}^k \mathbf{S}^k\,. \end{array} \tag{10}$$

This problem is an instance of second-order cone programming.

Notice that, even though the objective function depends only on the magnitudes, the phase information is preserved by the constraints.

3 Simulations in a Synthetic Room

Experiments were conducted on two data sets: *FourVoices*, a set of 4 speech utterances of 2.9 s of duration, and *SixFluteMelodies*, a set of 6 flute excerpts of 5.7 s of duration. Details of the signal analysis and resynthesis can be found in [7].

Synthetic mixtures were obtained in the frequency domain by previously simulating the impulse response of the room for each source-microphone pair. The reflections of the sound on the room were computed by speculation using the simulator in [17], using 0.6 as the reflection coefficient of the room. The impulse response was truncated to fit the length of the frames (100ms and 400ms for each data set, respectively). The room settings were those of Fig. 1.

Reconstruction of the sources was done by SOCP using the software package in [18].

After resynthesis back to the time domain, the S/N reconstruction error for a given source l is defined as

$$S/N = 10 \log \frac{||\hat{s}_l(t)||^2}{||\hat{s}_l(t) - s_l(t)||^2}, \tag{11}$$

with $\hat{s}_l(t)$ the reconstructed signal and $||\cdot||^2$ the sum of squares, and t stands for discrete time.

Results are shown in Table 1. As can be seen, in all the experiments the separation is around 10dB which, for speech signals, corresponds roughly to a 97% intelligibility at the sentence level [19,13]. SOCP performs sligthly better on the *FourVoices* data, and *SixFluteMelodies* is the hardest of the two. Audio files will be played at the conference.

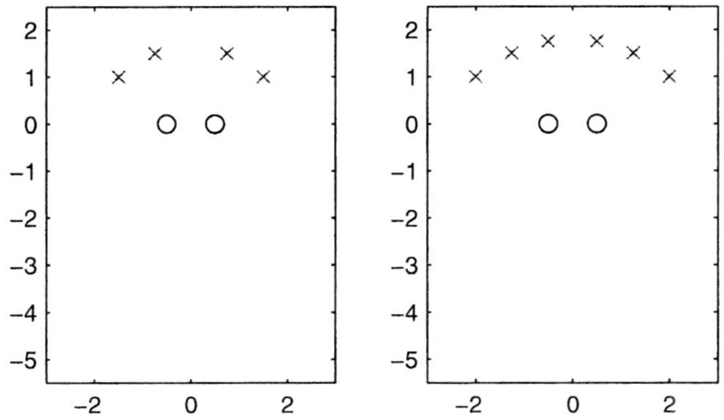

Fig. 1. Spatial arrangement of the sources for the *FourVoices* and the *SixFluteMelodies* experiments. Crosses represent the sources and circles represent the microphones. Distances are represented in meters. Both sources and microphones are 1.6m above the floor, and the room is 2.5m high.

Table 1. S/N ratios (dB) of the recovered sources with LP and SOCP. Average and worst case values.

	LP	SOCP
FourVoices	12.5 9.3	12.0 9.0
SixFluteMelodies	9.9 6.9	11.4 7.4

For the sake of comparison, in [7] S/N ratios around 19dB were found on the same data for the instantaneous mixture case, and results presented here are of the same order of magnitude as those in [11], even though the actual experiments are completely different.

4 A Neural Approximator of the Optimizer

In an attempt to eventually replace the SOCP optimizer by a feed forward network, we trained a multilayer perceptron with the sensor data as input and the reconstructed sources as target. In fact, since this was done in the frequency domain, a neural network was used for every frequency bin k. The input was $\mathcal{X}^k$, the aggregate vector of real and imaginary parts of the mixtures as defined in Section 2.1, and the target was $\hat{\mathcal{S}}^k$ the aggregate vector of real and imaginary parts of the reconstructed sources. Three layers were used with $2m$,3 and $2n$ units, respectively.

The data was divided in a training set with 50% of the frames, a validation set with 25% of the frames, and a test set with the remaining 25% frames (for every 4 consecutive frames, the first two were assigned to the training set, the next one was assigned to the validation set, and the last one was assigned to the test set).

The training was done with the Matlab/network toolbox [20], using an hyperbolic tangent transfer function for the hidden layer, a linear transfer function for the output layer, an the Levenberg-Marquardt algorithm with regularization as the learning rule (in the toolbox terminology, the network was initialized using the *newff* function and the 'tansig', 'purelin' and 'trainbr' parameter settings, respectively). The stop condition was based on the validation set. Each network was restarted seven times, and the best configuration on the validation test was chosen.

After training, the networks were applied to the test set yielding $\mathcal{R}^k$, the approximated sources for the subset of frames in the test set. A S/N prediction ratio was defined for every signal l as

$$S/N = 10 \log \frac{||\mathcal{R}_l^k||^2}{||\mathcal{R}_l^k - \hat{\mathcal{S}}_l^k{}_{\text{test}}||^2} \,. \tag{12}$$

Preliminary results are shown in Table 2 for the SOCP data. The *SixFluteMelodies* experiment results are poor, but still, the shape of the data is predicted for most of the signals. Further experiments will be reported in the conference.

Table 2. S/N ratios (dB) of the output approximated by the neural network, with respect to the sources recosntructed by the SOCP optimizer, for the subset of frames of the training set. Average and worst case values.

	SOCP
FourVoices	6.5 4.9
SixFluteMelodies	4.1 0.6

5 Conclusions and Further Work

In this paper we presented a comparison between LP and SOCP for the reconstruction of audio sources using the simulated impulse response of a synthetic recording room over two signal sets, *FourVoices* and *SixFluteMelodies*, using two microphones. SOCP performed slightly better. A neural network was then used to predict the output of the SOCP optimizer. Preliminary results were encouraging.

As future work, we plan to actually record the mixed sources in a physical room, using an actual measurement of its transfer function. In relation to the neural approximator, the approach needs further evaluation, both in terms of performance and in terms of computational cost.

References

1. Olshausen B.A. and Field D.J.: Sparse coding with an overcomplete basis set: A strategy employed by V1?. In: Vision Research, 37: 3311-3325, 1997.
2. Zibulevsky M. and Pearlmutter B.A.: Blind Source Separation by Sparse Decomposition. TR No. CS99-1, University of New Mexico, Albuquerque, July 1999. http://www.cs.unm.edu/~bap/papers/sparse-ica-99a.ps.gz
3. Yeredor A.: Blind Source Separation with Pure Delay Mixtures. In: Proc. of ICA'01, pp 522-527, 2001.
4. Parra L. and Spence C.: Blind Source Separation based on Multiple Decorrelations. In: IEEE Trans on Speech and Audio Processing, pp 320-327, May 2000.
5. Smaragdis P.: Blind Separation of Convolved Mixtures in the Frequency Domain. In: Int. Worshop on Independence and ANN, Tenerife, Spain, Feb. 1998.
6. Araki S., Makino S., Mukai R., Nishikawa T. and Saruwatari H..: Fundamental Limitation of Frequency Domain Blind Source Separation for Convolved Mixture of Speech. In: Proc. of ICA'01, pp 132-137, 2001. Available at http://www.kecl.ntt.co.jp/icl/signal/makino/
7. Bofill P. and Zibulevsky M.: Underdetermined Blind Source Separation using Sparse Representations. In: Signal Processing, 81 (2001), pp 2353-2362, 2001. Preprint available at http://www.ac.upc.es/homes/pau/
8. Zibulevsky M., Pearlmutter B.A., Bofill P. and Kisiliev P.: Blind Source Separation by Sparse Decomposition in a Signal Dictionary. In: Roberts S.J and Everson R.M., (eds.): Independent Component Analysis: Principles and Practice, Cambridge University Press, Chapter 7, pp 181-208, 2000.

9. Bofill P.: Underdetermined Blind Separation of Delayed Sound Sources in the Frequency Domain. In: Neurocomputing, Special issue: Evolving Solution with Neural Networks, Fanni A. and Uncini A. (Eds), Vol 55, Issues 3-4, pp 627-641, October 2003. Preprint available at http://www.ac.upc.es/homes/pau/
10. Rickard S., Balan R., Rosca J.: Real-Time Time-Frequency Based Blind Source Separation. In: Proc. of ICA'01, pp 651-656, 2001.
11. S. Araki, H. Sawada, R. Mukai, and S. Makino: A novel blind source separation method with observation vector clustering. In: Proc. IWAENC2005, pp. 117-120, Sept. 2005. Available at http://www.kecl.ntt.co.jp/icl/signal/makino/
12. Schroeder M.R.: New Method of Measuring Reverberation Time. In: J. Acoust.Soc. Am. 37.3, pp 402-. 419, 1965.
13. Bonavida A. "Acustica", Course notes, Edited by Cpet, ETSETB-UPC, 1979.
14. Lobo M.S., Vandenberghe L., Boyd S. and Lebret H.: Applications of Second-order Cone Programming. In Linear Algebra and its Applications. 284, pp 193-228, 1998.
15. S. Winter, H. Sawada, and S. Makino: On real and complex valued L1-norm minimization for overcomplete blind source separation. In: Proc. WASPAA2005, pp. 86-89, Oct. 2005. Available at http://www.kecl.ntt.co.jp/icl/signal/makino/
16. M.S. Lewicki and T.J. Sejnowski: Learning overcomplete representations. In: Neural Computation, vol 12, num 2, pp. 337-365, 2000. Available at citeseer.nj.nec.com/lewicki98learning.html.
17. McGovern S.G., http://www.steve-m.us/code/fconv.m, paper available at http://www.steve-m.us/rir.html
18. Sturm J.F.: Using SeDuMi 1.0x, a Matlab Toolbox for Optimization over Symmetric Cones. 1999. http://www.unimaas.nl/ sturm
19. L. Kinsler, A. Frey, A. Coppens, J. Sanders: Fundamentals of Acoustics. John Wiley & Sons,1989
20. The Mathworks, Matlab/network toolbox.

Utilization of Blind Source Separation Algorithms for MIMO Linear Precoding*

Paula Maria Castro, Héctor J. Pérez-Iglesias, Adriana Dapena, and Luis Castedo

Departamento de Electrónica e Sistemas, Universidade da Coruña,
Campus de Elviña s/n, 15.071. A Coruña, Spain
Tel.: ++34-981-167000; Fax.: ++34-981-167160
{pcastro, hperez, adriana, luis}@udc.es

Abstract. In this paper we investigate the application of Blind Source Separation (BSS) algorithms for the decoding of linearly precoded MIMO communication systems and for the design of limited feedback channels that send the Channel Status Information (CSI) from the receiver to the transmitter. The advantage of using BSS is that the MIMO channel can be continuously tracked without the need of pilot symbols. CSI is only sent through the feedback channel when the BSS algorithm indicates the presence of a strong channel variation.

Keywords: Adaptive blind source separation, MIMO time–varying channels, linear precoding.

1 Introduction

The continuous development of the wireless communication industry creates an enormous demand of high bit rate radio interfaces. Recently, it has been demonstrated that it is possible to achieve higher spectral efficiencies when using multiple antennas at both transmission and reception [1]. Transmitting over these Multiple–Input/Multiple–Output (MIMO) channels requires sophisticated signal processing methods in order to compensate the channel impairments. In particular, the receiver has to perform a Space–Time (ST) equalization to separate the streams transmitted through the multiple antennas. ST equalization is a difficult task that is traditionally carried out at the receiving side thus increasing complexity and cost of receivers. The cost of ST equalization at reception can be considerably reduced if an important part of the channel compensation is performed at the transmitter by means of precoding techniques. Besides, jointly optimal ST precoder and decoder designs provide better performance when compared to ST optimization only in the receiver side [2, 3].

Contrary to non linear approaches [4, 5], this paper focuses in linear precoding schemes that end up with the simplest possible receivers. In linearly precoded

* This work has been supported by Xunta de Galicia, Ministerio de Educación y Ciencia of Spain and FEDER funds of the European Union under grants number PGIDT05PXIC10502PN and TEC2004-06451-C05-01.

J. Rosca et al. (Eds.): ICA 2006, LNCS 3889, pp. 577–584, 2006.

systems, the signals received by the antennas (observations) are instantaneous mixtures of the original signals (sources). The mixing system results from the joint consideration of the precoding matrix and the channel matrix. As a consequence, Blind Source Separation (BSS) algorithms [6] can be employed in order to decode the observations.

An important issue to consider in precoding schemes is that the encoding matrix must be adapted to changes in the channel. Towards this aim, we will apply Adaptive BSS (ABSS) algorithms [7, 8] to update the separating matrix in each data slot in accordance with channel time variations. In order to obtain a good tracking performance, a one–bit flag will be transmitted to the encoder over a limited–rate feedback channel [9]. By means of that bit the transmitter knows if the channel has changed or not and, therefore, if it has to request a training sequence for estimating it. In addition, we will propose a simple way to initialize ABSS algorithms in precoding systems in order to obtain faster convergence.

This paper is organized as follows. Section 2, describes the signal model corresponding to a MIMO communication system with linear precoding. Section 3 presents three novel strategies to decode the received signals using adaptive BSS algorithms. Illustrative computer simulations are presented in Section 4 and some concluding remarks are made in Section 5.

2 Signal Model

Let us consider the precoded MIMO communication system shown in Figure 1. The input bit–stream is modulated to generate the output stream of (possibly complex) symbols $\mathbf{s}(n)$. Let $\mathbf{s}(n) = [s_1(n), ..., s_{N_T}(n)]^T$ be the vector formed by N_T original signals. We assume that they are zero-mean, stationary, temporally-white, non-Gaussian distributed and statistically independent. The signals $\mathbf{s}(n)$ are then filtered using a linear precoder system represented by an $N_T \times N_T$ complex-valued matrix $\mathbf{F}$. As a consequence, the coded symbols, $\mathbf{s}_t(n)$, and the original ones, $\mathbf{s}(n)$, are related by the following expression

$$\mathbf{s}_t(n) = \mathbf{F}\mathbf{s}(n) \quad (1)$$

These signals arrive at an array of N_R antennas whose output at time n, denoted by $\mathbf{x}(n) = [x_1(n), ..., x_{N_R}(n)]^T$, is given by

$$\mathbf{x}(n) = \mathbf{H}(n)\mathbf{F}\mathbf{s}(n) + \mathbf{v}(n) \quad (2)$$

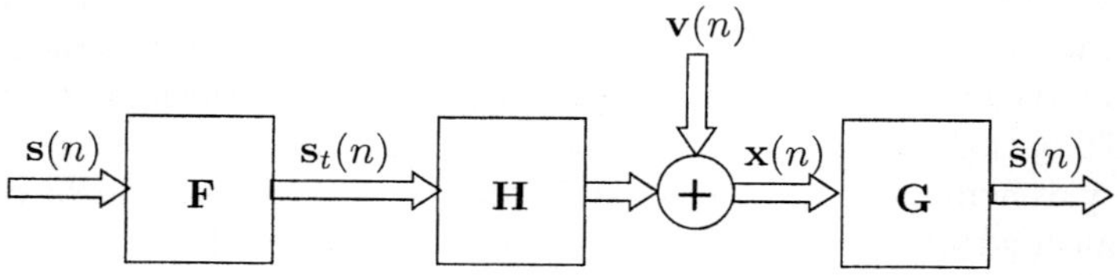

Fig. 1. Precoding communication scheme

where $\mathbf{H}(n)$ is a $N_R \times N_T$ matrix representing the MIMO channel and $\mathbf{v}(n)$ is the white Gaussian noise.

Throughout this paper we will assume that all entries into $\mathbf{H}(n)$ are complex Gaussian with i.i.d. real and imaginary parts with zero mean and unit variance. Denoting the gain from transmit antenna j and the receiver antenna i by $h_{ij}(n)$, the magnitudes of the channel gains $|h_{ij}(n)|$ will have a Rayleigh distribution. We also consider a block fading channel in which the channel matrix response remains constant for blocks of L symbols and changes according to an autoregressive model of order $p = 1$ from one block to another in the following form,

$$\mathbf{H}(n) = \mathbf{A}\mathbf{H}(n-1) + (\mathbf{I} - \mathbf{A})\mathbf{w}(n) \tag{3}$$

where $\mathbf{A}$ is a diagonal matrix whose entries a_{kk} are given by $J_o(2\pi f_D T\tau)$, being f_D the Doppler frequency, T the duration of a data frame, J_o the zero-order Bessel function of the first kind and $\tau = 1$. Vector $\mathbf{w}(n)$ is a zero–mean, i.i.d. and complex Gaussian vector process. The speed in channel changes is decided by means of the parameter L. For low values of L we will have fast fading channels whereas higher values of L lead to more static channels. A flat fading channel is assumed in which the symbol time–period is much larger than the channel delay–spread.

Finally, we assume that a linear decoder is employed to produce estimates of the transmitted symbols. The decoder is represented by a $N_T \times N_R$ matrix $\mathbf{G}$ and the symbol estimates are given by

$$\hat{\mathbf{s}}(n) = \mathbf{G}\mathbf{x}(n) \tag{4}$$

2.1 Linear Transmission/Reception Optimization

The goal in precoding schemes is to design the matrices $\mathbf{F}$ and $\mathbf{G}$ that provide the best performance with respect to some optimization criterion. In this paper, we employ a Zero–Forcing (ZF) approach in order to minimize the symbol estimation errors under a transmit power constraint (see, for instance [3, 10]). The optimization problem can be formulated as

$$\{\mathbf{F}_{opt}^{ZF}, \mathbf{G}_{opt}^{ZF}\} = arg \min_{\mathbf{F},\mathbf{G}} E\{||\mathbf{s}(n) - \hat{\mathbf{s}}(n)||_2^2\} \tag{5}$$

subject to $\mathbf{GHF} = \mathbf{I}$ and $tr(\mathbf{F}\mathbf{R}_s\mathbf{F}^H) = P_t$. In order to obtain the optimal linear precoding/decoding matrices $\mathbf{F}$ and $\mathbf{G}$ we define the following channel eigenvalue decomposition

$$\mathbf{H}^H(n)\mathbf{R}_n^{-1}(n)\mathbf{H}(n) = \mathbf{V}(n)\mathbf{\Delta}(n)\mathbf{V}^H(n) \tag{6}$$

Applying the Lagrangian method, it is possible to demonstrate that the joint ZF solution for the design of the linear precoder/decoder is given by [10]

$$\mathbf{F}_{opt}^{ZF} = \sqrt{\frac{P_t}{tr(\Delta^{-1/2})}}\mathbf{V}\Delta^{-1/4} \tag{7}$$

$$\mathbf{G}_{opt}^{ZF} = \sqrt{\frac{tr(\Delta^{-1/2})}{P_t}}\Delta^{-3/4}\mathbf{V}^H\mathbf{H}^H\mathbf{R}_n^{-1} \tag{8}$$

3 Proposed Decoding Schemes

It is interesting to note that the received signals (observations) $\mathbf{x}(n)$ given by equation (2) are instantaneous mixtures of the original signals $\mathbf{s}(n)$, where $\mathbf{H}(n)\mathbf{F}$ represents the mixing system. Consequently the decoding matrix $\mathbf{G}$ can be interpreted as the separating system needed to recover the original signals from the observations and it can be estimated using many BSS algorithms. Based on this idea, we propose in the sequel several decoding strategies for time–varying MIMO channels.

3.1 Approach I: Basic Decoding Scheme

Figure 2 shows a block diagram of a general linearly precoded MIMO system with limited feedback channel. We propose to use an adaptive BSS algorithm to find the separating (decoding) matrix $\mathbf{G}_{ABSS}$ from the received symbols. The decoded (separated) signals are thus obtained according to

$$\hat{\mathbf{s}}(n) = \mathbf{G}_{ABSS}\mathbf{x}(n) \tag{9}$$

In order to detect variations in the channel we make use the following performance index $LI(\mathbf{W}^{(l)})$ that measures the likeliness between the decoding matrices obtained for consecutive symbols

$$LI(\mathbf{W}^{(i)}) = \sum_{i=1}^{N}\left(\sum_{j=1}^{N}\frac{|w_{ij}|^2}{max_l(|w_{il}|^2)} - 1\right) + \sum_{j=1}^{N}\left(\sum_{i=1}^{N}\frac{|w_{ij}|^2}{max_l(|w_{lj}[k]|^2)} - 1\right) \tag{10}$$

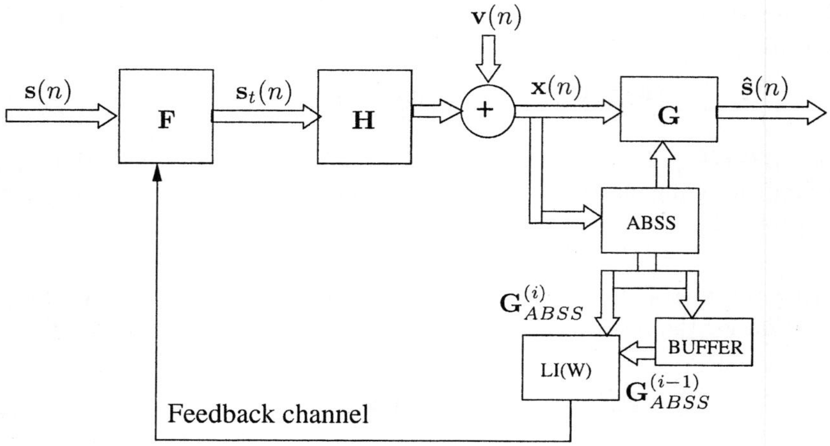

Fig. 2. Basic Decoding Scheme

where $\mathbf{W}^{(i)} = \mathbf{G}_{ABSS}^{(i-1)}(\mathbf{G}_{ABSS}^{(i)})^{-1}$. The superscript (i) denotes the symbol for which the decoding matrix has been calculated[1]. We consider that a channel change has occurred when this performance criterion exceeds a threshold value ϵ, i.e., when

$$LI(\mathbf{W}^{(i)}) - \frac{1}{i - i_c} \sum_{l=i_c}^{i-1} LI(\mathbf{W}^{(l)}) > \epsilon \tag{11}$$

where i_c denotes the last symbol for which a channel change has been detected.

When a change in the channel is detected, a bit is transmitted through the feedback channel to indicate that a training sequence must be transmitted. From this training sequence, the receiver estimates the channel matrix $\mathbf{H}(n)$ using a supervised algorithm. Finally, the feedback channel is also used to send from the receiver to the transmitter the channel estimate needed to adapt $\mathbf{F}$ in the transmit side by evaluating (7).

Note also that since $\mathbf{G}_{ABSS}$ is an estimation of the mixing system inverse, the channel matrix $\mathbf{H}(n)$ can be easily estimated without using training sequences by using a totally blind method such that $\hat{\mathbf{H}}(n) = (\mathbf{F}\mathbf{G}_{ABSS})^{-1}$. However, we have observed that this approach produces a poor performance in terms of bit rate error.

3.2 Approach II: Decoding Using a Stored Matrix G

The performance of the previous approach can be substantially improved for slow fading channels (see Section 4) by evaluating equation (8) at the receiver to obtain matrix $\mathbf{G}$ each time a change in the channel is detected. Thus, this matrix will be used in the receiver, instead of $\mathbf{G}_{ABSS}^{(i)}$, to decode the received signals.

3.3 Adaptive Algorithms: Initial Conditions

The initial conditions of the separating system is a crucial issue to consider when adaptive algorithms are used to obtain the separating coefficients. In the previously proposed approaches, the algorithm is used to obtain an estimation of the decoding matrix $\mathbf{G}$ given in equation (8). For this reason it is sensitive to think that equation (8) is a good starting point. Recall that this matrix can be obtained for the channel matrix $\mathbf{H}(n)$ each time a change in the channel is detected. This initialization has two important advantages: the convergence speed is increased and the permutation indeterminacy inherent to BSS algorithms is avoided.

4 Computer Simulations

In this section we present the results of several computer simulations that we carried out to validate the proposed systems. It is assumed that the sources have

[1] The index $LI(\mathbf{W})$ has been used in previous BSS work to measure the performance of BSS algorithms [11].

been passed through a mixing system with $N_T = 3$ and $N_R = 4$ antennas. We considered 10,000 QPSK symbols transmitted over normalized Rayleigh channels such that $E[||\mathbf{H}||_F^2] = 1$, where $||\cdot||_F$ denotes the Frobenius norm. The SNR plotted is given by $input\ SNR = 10\, log P_T\, / \sigma_v^2$, where P_T is the transmit power. Note that possible channel attenuation/gain is not considered. We set the transmit power to $P_t = 20$. In order to reduce the computational cost associated to track the channel variations, we have evaluated the decision criterion (11) after processing 10 received symbols instead of symbol by symbol.

Many conventional BSS algorithms can be used in the proposed schemes to estimate the mixing system. Among of all of them, we have selected the adaptive EASI algorithm proposed by [7, 8]. As regards EASI parameters, we have considered $\epsilon = 0.05$ and a constant adaptation step equal to $\lambda = 0.1$, with non–linear functions given by $g_i = (diag(\hat{\mathbf{s}}\hat{\mathbf{s}}^H))_i \hat{\mathbf{s}}_i$, for $1 \leq i \leq N_T$ (see [7, 8]). We will track the MIMO channel with the matrix given by the adaptive BSS algorithm according to the system described in Section 3. The channel is constant along $L = 1$ or $L = 1000$ data symbols. Obviously, the size of L determine of speed in channel time variations. You can see in the Figure 3 how for larger blocks (i.e., more slow fading channels) a better performance could be obtained. Finally, you can see in this figure how employing the optimum decoding matrix $\mathbf{G}_{opt}^{ZF}$ according to the approach II described in Section 3 produces too better results. This is because the ABSS algorithm provides a suboptimum decoding matrix. In the figure can be also seen the curve when perfect channel information is available, i.e., when $\epsilon = 0$, at the transmitter side. Smaller values of ϵ will yield to better BER performances due to the larger number of channel re–estimations and precoder updates at cost of greater number of initializations and therefore, greater

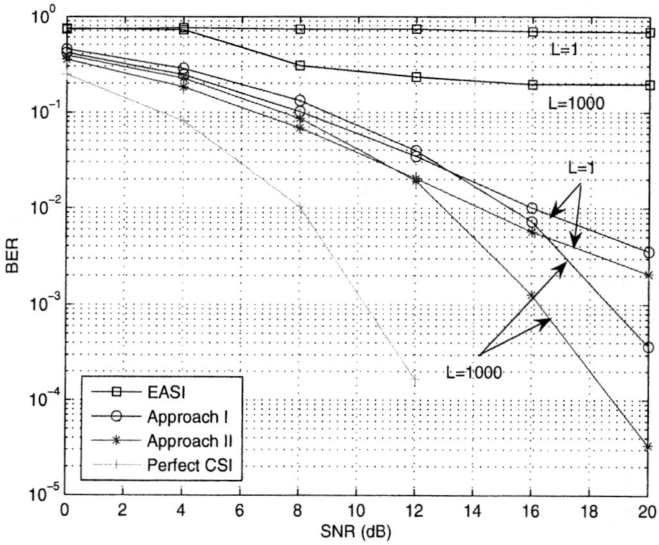

Fig. 3. BER performance vs. SNR

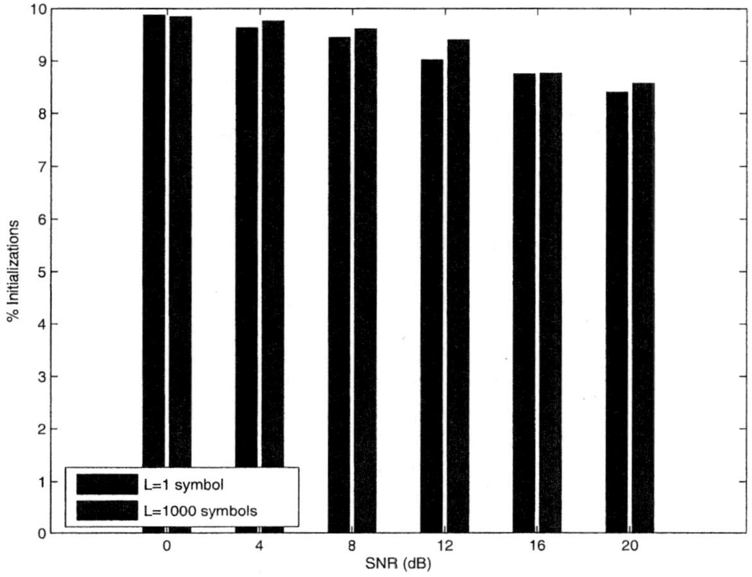

Fig. 4. Percentage of number of initializations vs SNR

overhead in the feedback channel. This means that the optimal value of ϵ must be modified depending on the fading speed to ensure a good performance.

Figure 4 plots the number of initializations of the matrix $\mathbf{G}_{ABSS}$ (and updates of $\mathbf{F}$ and $\mathbf{G}$) for different values of SNR as a function of L when approach II is employed. This value will depend weakly on the block size L and on the signal to noise ratio employing the proposed approach. Obviously, for lower SNR the precoder matrix $\mathbf{F}$ will be adapted more times in order to get a correct channel tracking. Note that if we evaluate the proposed decision criterion each received symbol the algorithm complexity is not only increased but the number of initializations would be extend greatly, specially for low SNR.

5 Conclusions

In this work we have studied the utilization of Blind Source Separation algorithms for decoding linearly precoded MIMO communication systems. The basic idea is to consider that the received precoded signals are instantaneous mixtures of the sources and that they can be decoded using adaptive BSS algorithms. This simple strategy has been combined together with the low rate characteristic of limited feedback channels available in wireless communications to track channel variations. Simulation results show that the performance of this scheme can be improved for slow varying channels by including a buffer in the receiver that contains previous decoding matrices if an adequate start matrix for the

BSS algorithm is selected. In this approach, we examine the likeness between the separating matrix and the matrices stored in the receiver side in order to track the channel variations.

References

1. G. Foschini and M. Gans, "On limits of wireless communication in a fading environment when using multiple antennas", *Wireless Personal Communications*, pp. 311-335, March 1998.
2. L. Hanzo, C. H. Wong and R. M. S. Yee, *Adaptive Wireless Transceivers*, John Wiley & Sons Ltd, 2002.
3. R. F. H. Fischer, *Precoding and Signal Shaping for Digital Transmission*, John Wiley & Sons, Inc., Publication, New York, 2002.
4. M. Tomlinson, "New Automatic Equalizer Employing Modulo Arithmetic", *Electronic Letters*, pp. 138-139, March 1971.
5. H. Harashima, Miyakawa, "Matched–Transmission Technique for Channels with Intersymbol Interference", *IEEE Journal on Communications*, pp. 774-780, August 1972.
6. T-W. Lee, *Independent Component Analysis: Theory and Applications*, Kluwer Academic Publishers, Boston, 1998.
7. J. F. Cardoso and B. H. Laheld, "Equivariant adaptive source separation", *IEEE Transactions on Signal Processing*, vol. 44, pp. 3017-3030, December 1996.
8. B. H. Laheld and J. F. Cardoso, "Adaptive source separation with uniform performance", *EUSIPCO*, 2004.
9. David J. Love, Robert W. Heath Jr., Wiroonsak Santipach and Michael L. Honig, "What is the value of the limited feedback for MIMO channels?", *IEEE Communications Magazine*, pp. 54-59, October 2004.
10. T. P. Kurpjuhn, M. Joham, and J. A. Nossek, "Optimization Criteria for Linear Precoding in Flat Fading TDD–MIMO Downlink Channels with Matched Filter Receivers", in *Proc. VTC 2004 Spring*, vol. 2, pp. 809-813, May 2004.
11. E. Moreau and O. Macchi, "High-Order Contrasts for Self-Adaptive Source Separation", *International Journal of Adaptive Control and Signal Processing, pp. 19–46, 1996.*

Speech Enhancement Based on the Response Features of Facilitated EI Neurons

André B. Cavalcante[1], Danilo P. Mandic[2], Tomasz M. Rutkowski[3], and Allan Kardec Barros[1]

[1] Laboratory for Biological Information Processing, Universidade Federal do Maranhão, Brazil
abcborges@hotmail.com, allan@ufma.br
http://pib.dee.ufma.br/

[2] Department of Electrical and Electronic Engineering, Imperial College London, United Kingdom
d.mandic@imperial.ac.uk
http://www.commsp.ee.ic.ac.uk/

[3] Laboratory for Advanced Brain Signal Processing, Brain Science Institute Riken, Japan
tomek@brain.riken.jp
http://www.bsp.brain.riken.jp/

Abstract. A real-time approach for the enhancement of speech at zero degree azimuth is proposed. This is achieved inspired by the response features of the "Facilitated EI neurons". This way, frequency segregation through a bandpass filter bank is followed by "supression analysis" which inhibits sources that are not at "facilitated" positions. Unlike with the existing approaches for the solution of cocktail party problem, where the performance under low SNR (signal-to-noise ratio) reverberation conditions is severely limited, the proposed approach has the capability to circumvent these problems. This is quantified through both objective and subjective performance measures and supported by real world simulation examples.

1 Introduction

The cocktail party effect is an important and comprehensively addressed phenomenon in both psychology and machine learning [1,2]. Indeed, the area of blind source separation (BSS) focuses on recovering a desired source or a group of unobservable sources from their observed mixtures. Despite the relative success and excitement associated with the early BSS techniques, in real world situations several critical issues need to be tackled prior to achieving this goal, these include reverberation effects together to the ambiguities coming from different power levels of the interference sources.

In terms of human physiology, it is now well understood that the peripheral and central auditory system employ sets of highly specialized structures dedicated to the psycho-acoustic separation, these aim specifically at tackling the generic cocktail party problem [3]. Consequently, research on speech enhancement has been influenced by results related to the response features of both the

J. Rosca et al. (Eds.): ICA 2006, LNCS 3889, pp. 585–592, 2006.

inner ear and the neural auditory complex. Thus, for instance, results from [4] show that based on a subspace of basis functions and Wiener coefficients, it is possible to recover speech discourse from a single corrupted signal. The basis functions employed were generated based on the independent component analysis (ICA) class of algorithms [2]. The success of this approach was also due to the underlying use of "efficient coding", a concept borrowed from the research on cochlear nerves [5]. In [6], binaural cues, that is interaural time difference (ITD) and interaural intensity difference (IID), were used to aid source separation, and were extracted by means of second order time and frequency features. This has also enabled features to create a training set for the associated learning algorithms. Despite the success and excellent performance of these methods, an open problem remains the need for training, which is not guaranteed or feasible in a real-world context.

The real-time models proposed are heavily based on either the properties of human auditory system [7] or harmonic characteristics of speech [8]. The former approaches employ subtraction methods in order to mimic the auditory masking phenomenon. This however assumes unrealistic assumtions, such as uncorrelated noise sources, which is not realistic in reverberant environments. The latter approaches are based on the extraction of the fundamental frequency and its harmonics from a corrupted speech signal. Notice that harmonic tracking implies constraints on the power ratio of the target and interference signal, in addition to the smearing effect caused by reverberation (which degrades the harmonic content).

The purpose of this work is therefore to propose a real-time binaural approach for the enhancement of the unknown speech the source at zero degree azimuth. This is achieved based on the response features of "Facilitated EI neurons" reported in [9]. These cells are located in an important nucleus for localization tasks called the inferior colliculus, and they possess high spatial selectivity, resulting in high signal enhancement in even very noisy and reverberant environments. Our proposed approach aims at mimicing the behavior of these cells, based upon building of suppression curves through an evaluation of IID cue.

2 Methods

Let $\mathbf{S} = [s_1(t), s_2(t), \ldots, s_n(t)]$ be a set of source signals at time t, and let us assume a set of two receivers $\mathbf{X} = [x_1(t), x_2(t)]$. To simulate a reverberant environment, let model a binaural receiver signal as

$$x_i(t) = \sum_{j=1}^{n} [h_{ij}(t) * s_j(t)] \quad i = 1, 2 \tag{1}$$

where symbol "$*$" denotes the convolution operator and $h_{ij}(t)$ represents the room impulse response between the jth sound source and i^{th} receiver. We employ the time required for the signal energy to decay down to $60dB$ below the initial level as a measure to characterise the room impulse response (the reverberation time RT_{60}).

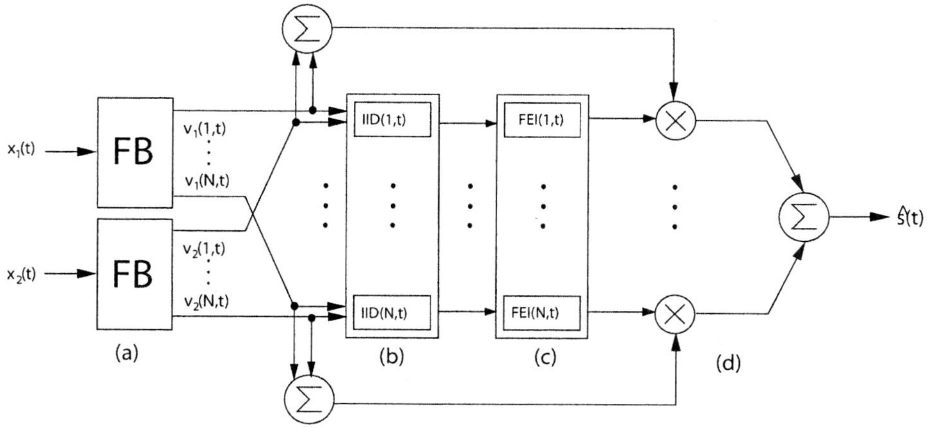

Fig. 1. Proposed speech enhancement structure. (a) Frequency segregation by a bandpass filter bank. (b) IID functions used to create position cue over the time. (c) Facilitated EI neurons bank to perform the selection of the target source. (d) Synthesis.

Notice that a room impulse response is a non-minimum phase function, this makes the enhancement of the target source from (1) based on $h_{ij}(t)$ to be a difficult and ill–posed inverse statistical problem. To that cause, we propose a model, shown in Figure 1, which does not require the solution of this inverse problem. Instead, it is based on the knowledge coming from the human psychology and the physics of biological cells. The following subsections provide a full explanation of the structure from Figure 1.

2.1 Frequency Segregation

The mathematical description of the inner ear is usually based on a bank of bandpass filters. Following this idea, the first step in our proposed model performs frequency segregation of the binaural input into narrow subbands. In techical terms, we explore spectral sparseness of the sound sources. To decompose such spectrum into N subbands, we employ a bank of N non–overlapping and equidistantly distributed Butterworth bandpass filters. These filters are designed so as to have the same bandwidth $\Delta\omega$; this way the kth subband can be expressed as

$$v_i(k,t) = \int_t f(k,\tau)x_i(t+\tau)d\tau \quad i = 1,2 \tag{2}$$

where $f(k,t)$ denotes the kth filter within the filter bank.

2.2 Evaluation of IID Functions

Notice that the central auditory nuclei employ IID cues for psycho-acoustical tasks [9], whereas the majority of work in speech enhancement research employ time windowed power differences of subbands as IID cues [6]. However, we used here a different interpretation. To conform with the biological mechanism behind speech enhancement, we propose a simple measure of instantanaous intensity differences for every subband, given by

$$IID(k,t) = v_1(k,t) - v_2(k,t) \quad . \tag{3}$$

2.3 Facilitated EI Neurons (FEI)

The central auditory system comprises an intricate circuitry of nuclei [10], where the acoustic information is systematically processed and encoded. This circuitry consists of monaural and binaural nuclei, where monaural nuclei receive information from only one ear, whereas binaural nuclei receive information from both the left and right ear.

The binaural nuclei are formed by excitatory/inhibitory (EI) cells. These cells are excited by one ear and inhibited by the other ear through excitatory/inhibitory projections provided by other nuclei. The inferior colliculus (IC) is an example of nucleus formed by EI cells. Within the IC, the EI cells are excited by the contralateral ear and inhibited by the ipsilateral ear. To explain this, assume that a positive IID points out that the acoustical stimuli has more intensity at the contralateral plane. A negative IID points out the opposite. Thus, the EI cells in IC are excited by acoustical stimuli with positive IID and inhibited by those with negative IID. Furthermore, the higher the IID, the more pronounced the activity of these cells.

On the other hand, there is a group of EI cells in the inferior colliculus of some animals that does not follow that behavior: the Facilitated EI neurons (FEI). These cells are only excited over a small range of IID, in contrast to common EI cells, hence the stimulus must be at a specific position which is "facilitated" by FEI cells.

In this work, we create a mechanism inspired by the response features of FEI neurons. This mechanism roughly represents these features and is based on "suppress curves" which supress acoustic stimuli that do not correspond to the "facilitated" positions. This phenomenon can be expressed as have

$$g(k,t) = \begin{cases} 1 \text{ if } |IID(k,t)| \leq \phi \\ 0 \text{ otherwise} \end{cases} \quad , \tag{4}$$

where symbol ϕ represents a directional bias. To enhance the source at zero degree azimuth, ϕ must be a very small number. Observe that $g(k,t)$ is a binary function which does not correspond to the activity pattern of nervous cells, and does not mimic the humans who have a limited sensitivity to brief changes in binaural cues such as IID [11]. To remedy this, we represent the real behavior of FEI neurons as a soft version of $g(k,t)$ where only low frequencies are preserved in order to exclude the influence of brief changes in $IID(k,t)$. We thus consider a FEI neuron function given by

$$FEI(k,t) = \int_t l(\tau) g(k,t+\tau) d\tau \quad , \tag{5}$$

where $l(t)$ is lowpass filter with a normalized cutoff frequency ω_N.

2.4 Synthesis

At this step, our model performs the recovering process using the subbands extracted from the binaural input signal and the FEI functions. The enhanced signal is given by

$$\hat{s}(t) = \sum_{k=1}^{N} [v(k,t)FEI(k,t)] \quad , \tag{6}$$

where $v(k,t) = v_1(k,t) + v_2(k,t)$.

We summarize the proposed algorithm as it is implemented in the experiments with the following procedure:

Step 1: Repeat the following steps for k=1,2,...,N.
Step 2: Extract the kth set of subbands $[v_1(k,t), v_2(k,t)]$ from the inputs $[x_1(t), x_2(t)]$.
Step 3: Let $IID(k,t) = v_1(k,t) - v_2(k,t)$.
Step 4: If $|IID(k,t)| \leq \phi$ at time instant t, then $g(k,t) = 1$. Otherwise, $g(k,t) = 0$.
Step 5: Compute $FEI(k,t)$ functions by removing high frequencies from g(k,t).
Step 6: Let the recovered signal $s(t) = s(t) + [v_1(k,t) + v_2(k,t)]FEI(k,t)$.

3 Experimental Results and Discussions

We carried out simulations based on three speech signals, two male and one female. Each utterance was reverberated according to three conditions: $RT_{60} = \{0.1 \text{ s}, 0.2 \text{ s}, 0.3 \text{ s}\}$. The mixtures were obtained according to (1) and adjusted over a range of low SNR(signal-to-noise ratio) levels: -15 dB$\leq SNR \leq$0 dB. The room impulse responses were simulated for a small room (5 m x 5 m x 3 m) using the approach from [12]. Each source was placed at the same distance to the receivers. The female source was fixed at zero degree azimuth so that it was the target source. Each utterance was sampled at f_s = 16kHz. The spectrum was segregated into subbands with $\Delta\omega$ = 10Hz. The directional bias in (4) and the normalized cutoff frequency of $l(t)$ in (5) were respectively $\phi = 0.0001$ and $\omega_N = 0.05$.

3.1 SNR Evaluation

Robust criteria to measure the quality of recovered signals have been subject of much research [13]; we here, for rigour, employ both objective and subjective measurements.

As the objective measurement, we employ the standard signal to noise ratio (SNR), given by

$$SNR = 10\log_{10}\frac{\sum_t s^2(t)}{\sum_t (\hat{s}(t) - s(t))^2} \quad , \tag{7}$$

where $s(t)$ is a clean speech signal (target source) and $\hat{s}(t)$ denotes the enhanced signal.

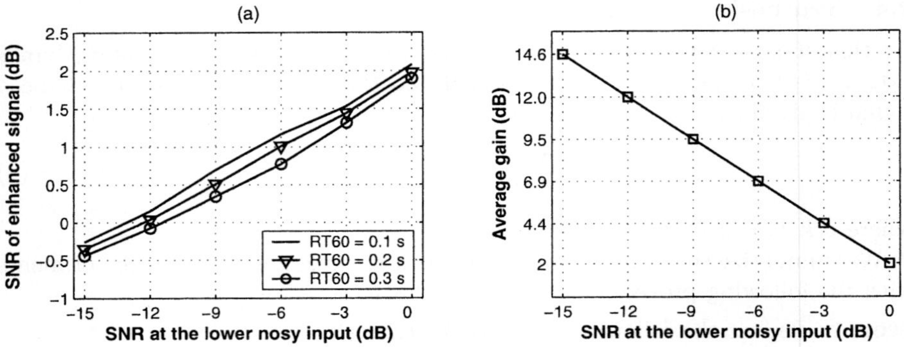

Fig. 2. SNR improvements. (a) Performance over the ranges of SNR and RT_{60}. (b) Average gain of SNR over the RT_{60} range.

Figure 2.(a) showns the SNR values of the enhanced signal for the three analysed reverberation times. It is clear that our approach is quite robust to the changes in reverberation time once the three curves are very close. We attribute this behavior to the fact that our IID evaluation is based on an instant intensity difference instead of power differences over a time window. Consequently, our architecture is less influenced by the smearing effect of reverberation. The average gain in SNR value for the three reverberant conditions is shown in Figure 2.(b). Observe that the system acts by gradually adjusting the interference power level to the clean speech signal power level, which for negative SNRs means (according to (7)) an increase in the SNR value. This is highly desirable since fornegative SNRs the interference power is greater than the power of the clean speech signal.

Next, since the SNR measure, despite revealing the presence of noise, is not a criteriong of intelligibility, we employ a hearing quality test as a subjective measurement.

3.2 Hearing Quality Test

The human auditory system is highly qualified to tackle the cocktail party problem. With this in mind, five listeners were asked to quantify the speech intelligibility of our model. They were requested to consider various criteria, such as the presence of interference sources and speech distortion. Listeners were instructed to quantify the intelligibility of each sentence according to: 1$\leq$ *poor* $<$2, 2$\leq$ *understandable* $<$3, 3$\leq$ *good* $<$4, 4$\leq$ *excellent* $<$5.

Figure 3 illustrates the average score of the hearing quality test before and after the enhancement, over a range of SNR levels. The reverberation time was $RT_{60} = 0.3$ s.

The hearing quality tests showed that the proposed model also improves the intelligibility of the speech discourse of the target source.

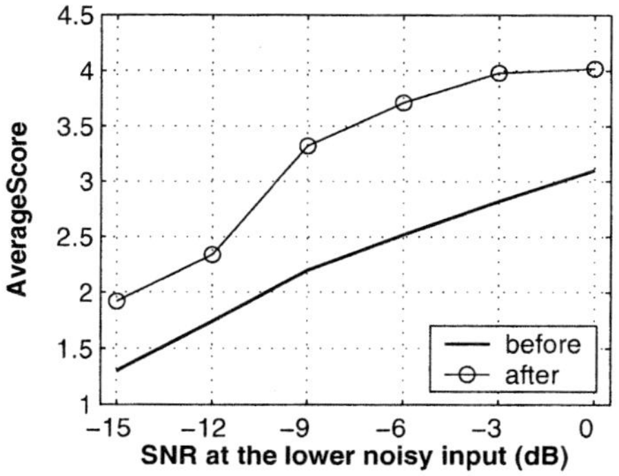

Fig. 3. Comparison of score of hearing quality test before and after the enhancement for $RT_{60} = 0.3$ s

3.3 Spectral Analysis

In Figure 4 the spectral analysis of our results is presented; notice the smearing effect caused by reverberation, indicating the spreading of energy around the fundamental frequency of the clean signal as seen in Figure 4.(a)(b). However, our model showed the capability to recover the fundamental frequency of the target signal as illustrated in Figure 4.(c).

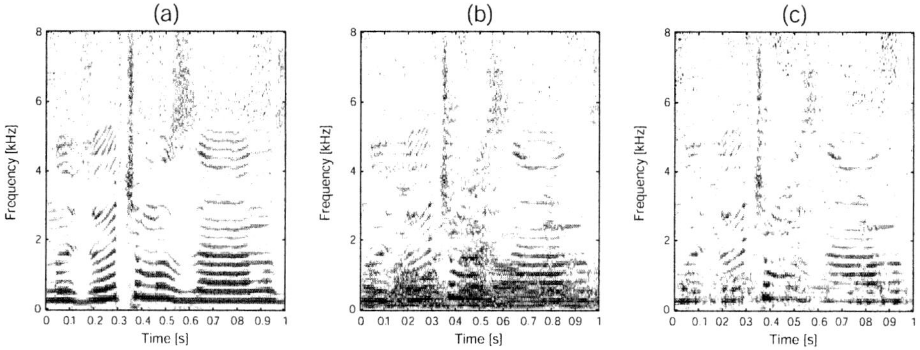

Fig. 4. Spectral analysis. (a) Clean speech signal. (b) Mixed and reverberated input at 0 dB and $RT_{60} = 0.1$ s. (c) Enhanced signal.

4 Conclusions

We have proposed a real-time approach for the enhancement of sound sources at zero degree azimuth. This is achieved based on the response features of the "Facilitated EI neurons". This enabled us to circumvent most constraints associated with other models of this kind. The high improvement in speech quality has been achieved under very noisy and reverberant conditions, this was quantified through both objective and subjective performance criteria.

References

1. Cherry, E.C.: Some experiments on the recognition of speech, with one and two ears. Journal of the Acoustic Society of America **25** (1953) 975–979
2. Hyvarinen, A., Oja, E.: A fast fixed-point algorithm for independent component analysis. Neural Computation **9** (1997) 1483–1492
3. de Cheveigne, A.: The auditory system as a separation machine. In: Proceedings of International Symposium of Hearing. (2000)
4. Barros, A.K., Ohnishi, N.: Single channel speech enhancement by efficient coding. Signal Processing **85** (2005) 1805–1812
5. Lewicki, M.: Efficient coding of natural sounds. Nature Neuroscience **5**(4) (2002) 356–363
6. Roman, N., Wang, D., Brown, G.: Speech segregation based on sound localization. Journal of Acoustical Society of America **114**(1) (2003) 2236–2252
7. Virgag, N.: Single channel speech enhancement based on masking properties of the human auditory system. IEEE Transactions on Signal Processing **7** (1999) 126–137
8. Barros, A.K., Rutkowski, T.M., Itakura, F., Ohnishi, N.: Estimation of speech embedded in a reverberant and noisy environment by independent component analysis and wavelets. IEEE Transactions on Neural Networks **13**(4) (2002)
9. Pollak, G., Burger, R., Park, T., Klug, A., Bauer, E.: Roles of inhibition for transforming binaural properties in the brainstem auditory system. Hearing Research **168**(1–2) (2002) 60–78
10. Pollak, G., Burger, R., Klug, A.: Dissecting the circuitry of the auditory system. Trends in Neuroscience **26** (2004) 33–39
11. Grantham, G.: Discrimination of dynamic interaural intensity differences. Journal of Acoustical Society of America **76**(1) (1984) 71–76
12. Allen, J., Berkley, D.: Image method for efficiently simulating small-room acoustics. Journal of Acoustical Society of America **65** (1979) 943–950
13. Hansen, J., Pellom, B.L.: An effective qualiy evaluation protocol for speech enhancement algorithms. In: Proceedings of ICSLP 1998. Volume 7. (1998) 2819–2822

Blind Separation of Sparse Sources Using Jeffrey's Inverse Prior and the EM Algorithm

Cédric Févotte and Simon J. Godsill

Engineering Dept., University of Cambridge, Cambridge, CB2 1PZ, UK
{cf269, sjg}@eng.cam.ac.uk
http://www-sigproc.eng.cam.ac.uk/~cf269/

Abstract. In this paper we study the properties of the Jeffrey's inverse prior for blind separation of sparse sources. This very sparse prior was previously used for Wavelet-based image denoising. In this paper we consider separation of 3×3 and 2×3 noisy mixtures of audio signals, decomposed on a MDCT basis. The hierarchical formulation of the inverse prior allows for EM-based computation of MAP estimates. This procedure happens to be fast when compared to a standard more complex Markov chain Monte Carlo method using the flexible Student t prior, with competitive results obtained.

Blind Source Separation (BSS) consists of estimating n signals (the sources) from the sole observation of m mixtures of them (the observations). If many efficient approaches exist for (over)determined ($m \geq n$) non-noisy linear instantaneous, in particular within the field of Independent Component Analysis, the general linear instantaneous case, with mixtures possibly noisy and/or underdetermined ($m < n$) is still a very challenging problem.

A now common approach for BSS, in particular for underdetermined mixtures, consists of exploiting source sparsity assumptions. Sparsity means that only "few" expansion coefficients of the sources on a given basis are significantly different from zero and its use to handle source separation problem (possibly underdetermined) was introduced in the seminal papers [1,2].

In [3,4] we modeled the expansion coefficients of the sources by identically and independently distributed (i.i.d) Student t processes and a Gibbs sampler (a standard MCMC simulation method) was proposed to sample from the posterior distribution of the mixing matrix, the input noise variance, the source coefficients and hyperparameters of the Student t distributions. The method was successfully applied to determined and underdetermined noisy audio mixtures, decomposed on a MDCT basis (a local cosine basis). In this paper, we give the source coefficients the Jeffrey's inverse prior $p(x) \propto 1/|x|$. This prior was used for image denoising and sparse regression in [5,6]. It provides very sparse signal estimates and, as shown in [5] in the context of denoising, is good compromise between soft and hard thresholding. Though Jeffrey's prior corresponds to an improper probability density function, it admits a hierarchical formulation which leads to proper posterior densities, and allows for efficient EM-based computa-

J. Rosca et al. (Eds.): ICA 2006, LNCS 3889, pp. 593–600, 2006.

tion of Maximum A Posteriori (MAP) estimates of the source coefficients, the mixing matrix and the noise variance.

The paper is organized as follows: Section 1 introduces notations and assumptions. Section 2 presents the different priors used for the source coefficients, the mixing matrix, and the input noise variance. Section 3 gives the EM updates of each of the latter parameters. Section 4 provides separation results for determined and underdetermined mixtures of audio sources. The Jeffrey's prior is shown to have good denoising properties, and the proposed method happens to be fast when compared to the more complex MCMC approach using the flexible Student t prior proposed in [4], with competitive results obtained. Section 5 draws some conclusions.

1 Notations

1.1 Mixture and Aim

We consider the following standard linear instantaneous model, $\forall t = 1, \ldots, N$:

$$\mathbf{x}(t) = \mathbf{A}\,\mathbf{s}(t) + \mathbf{n}(t) \tag{1}$$

where $\mathbf{x}(t) = [x_1(t), \ldots, x_m(t)]^T$ is a vector of size m containing the observations, $\mathbf{s}(t) = [s_1(t), \ldots, s_n(t)]^T$ is a vector of size n containing the sources and $\mathbf{n}(t) = [n_1(t), \ldots, n_m(t)]^T$ is a vector of size m containing additive noise. Variables without time index t denote whole sequences of samples, *e.g*, $\mathbf{x} = [\mathbf{x}(1), \ldots, \mathbf{x}(N)]$ and $x_1 = [x_1(1), \ldots, x_1(N)]$.

The aim of the following work is to estimate the sources $\mathbf{s}$ and the mixing matrix $\mathbf{A}$ up to the standard BSS indeterminacies on gain and order, that is, compute $\hat{\mathbf{s}}$ and $\hat{\mathbf{A}}$ such that ideally $\hat{\mathbf{A}} = \mathbf{A}\,\mathbf{D}\,\mathbf{P}$ and $\hat{\mathbf{s}} = \mathbf{P}^T\,\mathbf{D}^{-1}\,\mathbf{s}$, where $\mathbf{D}$ is a diagonal matrix and $\mathbf{P}$ is a permutation matrix.

1.2 Time Domain / Transform Domain

Let $x \in \mathbb{R}^{1 \times N} \to \tilde{x} \in \mathbb{R}^{1 \times N}$ denote a bijective linear transform, preferably orthonormal. Denoting for $k = 1, \ldots, N$, $\tilde{\mathbf{x}}_k = [\tilde{x}_{1,k}, \ldots, \tilde{x}_{1,m}]^T$ and $\tilde{\mathbf{n}}_k$, $\tilde{\mathbf{s}}_k$ similarly, by linearity of the t-f transform we have

$$\tilde{\mathbf{x}}_k = \mathbf{A}\,\tilde{\mathbf{s}}_k + \tilde{\mathbf{n}}_k \tag{2}$$

Furthermore, the t-f transform being bijective, solving the problem defined by Eq. (1) in the time domain is equivalent to solving Eq. (2) in the transform domain.

1.3 Some Assumptions

We make the following assumptions:

- $\mathbf{A}$ is full-column rank,
- The source coefficients $\{\tilde{s}_{i,k}\}$ are assumed to be *sparse* and mutually independent,

- The noise $\mathbf{n}_t$ is assumed to be i.i.d Gaussian with covariance $\sigma^2 \mathbf{I}_n$. When the transform $x \to \tilde{x}$ is orthonormal, $\tilde{\mathbf{n}}_k$ is equivalently i.i.d Gaussian with covariance $\sigma^2 \mathbf{I}_n$.

2 Priors

2.1 Source Coefficients

The source coefficients are given Jeffrey's inverse prior:

$$p(\tilde{s}_{i,k}) \propto \frac{1}{|\tilde{s}_{i,k}|} \tag{3}$$

This is a very heavy-tailed prior, symmetrical and centered at 0, and thus a relevant model for sparsity. This prior is *scale-invariant*: if $p(x) \propto 1/x$ and if $y = a\,x$ where a is a constant, then, by applying the rule for the change of variable in a pdf, $p(y) \propto 1/y$. Thus, oppositely to Student t or Laplace priors, the inverse prior does not require to update any scale parameter for the sources. As noted in [5], this prior is so heavy-tailed that it is actually *improper* (its integral is not finite). Improper priors are common in Bayesian inference, and can be used as long as they lead to proper posterior distributions. In fact the prior $p(\tilde{s}_{i,k}) \propto 1/|\tilde{s}_{i,k}|$ used with a Gaussian likelihood leads to an improper posterior distribution for the sources. Thus, as in [5,6], we rather use the following hierarchical formulation of the inverse prior:

$$p(\tilde{s}_{i,k}|v_{i,k}) = \mathcal{N}(s_{i,k}|0, v_{i,k}) \quad \text{with} \quad p(v_{i,k}) \propto 1/v_{i,k} \tag{4}$$

where $\mathcal{N}(x|\mu, \sigma^2)$ denotes the density of the normal distribution. With these assumptions, elementary integration yields

$$\int_0^\infty p(\tilde{s}_{i,k}|v_{i,k})\, p(v_{i,k})\; dv_{i,k} \propto 1/|\tilde{s}_{i,k}| \tag{5}$$

The inverse prior $p(v_{i,k}) \propto 1/v_{i,k}$ being itself a limiting case of the inverted-Gamma distribution, the prior $p(\tilde{s}_{i,k}) \propto 1/|\tilde{s}_{i,k}|$ may be regarded as a special case of the Student t prior which was used for source separation purposes in [4]. In the following we will denote $\mathbf{v}_k = [v_{1,k}, \ldots, v_{n,k}]^T$ and $\mathbf{v} = [\mathbf{v}_1, \ldots, \mathbf{v}_N]$.

2.2 Noise Variance

σ^2 is given a inverted-Gamma conjugate prior:

$$p(\sigma^2|\alpha_\sigma, \beta_\sigma) = \mathcal{IG}(\sigma^2|\alpha_\sigma, \beta_\sigma) \tag{6}$$

where $\mathcal{IG}(x|\alpha, \beta) = \frac{\beta^\alpha}{\Gamma(\alpha)}\, x^{-(\alpha+1)} \exp(-\frac{\beta}{x})$, $x \in [0, +\infty)$. The inverted-Gamma distribution has a unique mode, which is found at $x = \beta/(\alpha + 1)$.

2.3 Mixing Matrix

Let $\mathbf{r}_1, \dots, \mathbf{r}_m$ be the $n \times 1$ vectors denoting the transposed rows of $\mathbf{A}$, such that $\mathbf{A}^T = [\mathbf{r}_1 \ \dots \ \mathbf{r}_m]$. We give each row $\mathbf{r}_i$ a Gaussian conjugate prior with zero mean:

$$p(\mathbf{r}_j|\sigma^2) = \mathcal{N}(\mathbf{r}_j|0, \sigma_r^2 \, \mathbf{I}_n) \tag{7}$$

3 EM Updates

We now describe how to find a MAP estimates of the parameters $\boldsymbol{\theta} = \{\tilde{\mathbf{s}}, \mathbf{A}, \sigma^2\}$, using the Expectation Maximization algorithm (EM) [7]. With the variances $\mathbf{v}$ treated as missing data, the EM algorithm is based on the alternate evaluation and maximization of the following function:

$$\begin{aligned} Q(\boldsymbol{\theta}|\boldsymbol{\theta}') &= \int_{\mathbf{v}} \log p(\boldsymbol{\theta}|\mathbf{v}, \tilde{\mathbf{x}}) \; p(\mathbf{v}|\boldsymbol{\theta}', \tilde{\mathbf{x}}) \; d\mathbf{v} \\ &= \int_{\mathbf{v}} \log p(\tilde{\mathbf{s}}, \mathbf{A}, \sigma|\mathbf{v}, \tilde{\mathbf{x}}) \; p(\mathbf{v}|\tilde{\mathbf{s}}') \; d\mathbf{v} \\ &= \int_{\mathbf{v}} \log p(\tilde{\mathbf{s}}|\mathbf{A}, \sigma, \mathbf{v}, \tilde{\mathbf{x}}) \; p(\mathbf{v}|\tilde{\mathbf{s}}') \; d\mathbf{v} + \log p(\mathbf{A}|\tilde{\mathbf{s}}, \sigma^2, \tilde{\mathbf{x}}) + \log p(\sigma^2|\tilde{\mathbf{s}}, \tilde{\mathbf{x}}) \end{aligned} \tag{8}$$

One iteration of the EM algorithm is as follows:

E-step	Evaluate $Q(\boldsymbol{\theta}|\boldsymbol{\theta}^{(l)})$
M-step	$\boldsymbol{\theta}^{(l+1)} = \text{Argmax } Q(\boldsymbol{\theta}|\boldsymbol{\theta}^{(l)})$

The derivation of the update steps for every parameter is mainly a matter of finding the modes of the posterior distributions involved in Eq. (8). In the following we skip derivations, details of the calculations of the posterior distributions can be found in [4].

3.1 Missing Data Posterior Distribution

The E-step requires integration over the posterior distribution of the missing data $\mathbf{v}$:

$$p(\mathbf{v}|\tilde{\mathbf{s}}) = \prod_{i,k} p(v_{i,k}|\tilde{s}_{i,k}) \quad \text{with} \quad p(v_{i,k}|\tilde{s}_{i,k}) = \mathcal{IG}\left(v_{i,k}|\frac{1}{2}, \frac{\tilde{s}_{i,k}^2}{2}\right) \tag{9}$$

3.2 Update of $\tilde{\mathbf{s}}$

The posterior distribution of $\tilde{\mathbf{s}}$ is $p(\tilde{\mathbf{s}}|\mathbf{A}, \sigma^2, \mathbf{v}, \tilde{\mathbf{x}}) = \prod_k p(\tilde{\mathbf{s}}_k|\mathbf{A}, \sigma^2, \mathbf{v}_k, \tilde{\mathbf{x}}_k)$ with

$$p(\tilde{\mathbf{s}}_k|\mathbf{A}, \sigma^2, \mathbf{v}_k, \tilde{\mathbf{x}}_k) = \mathcal{N}(\tilde{\mathbf{s}}_k|\boldsymbol{\mu}_{\tilde{\mathbf{s}}_k}, \boldsymbol{\Sigma}_{\tilde{\mathbf{s}}_k}) \tag{10}$$

where $\boldsymbol{\Sigma}_{\tilde{\mathbf{s}}_k} = \left(\frac{1}{\sigma^2}\,\mathbf{A}^T\,\mathbf{A} + \mathrm{diag}\,(\mathbf{v}_k)^{-1}\right)^{-1}$ and $\boldsymbol{\mu}_{\tilde{\mathbf{s}}_k} = \frac{1}{\sigma^2}\,\boldsymbol{\Sigma}_{\tilde{\mathbf{s}}_k}\,\mathbf{A}^T\,\tilde{\mathbf{x}}_k$. The value of $\tilde{\mathbf{s}}_k$ is simply updated to the mode $\boldsymbol{\mu}_{\tilde{\mathbf{s}}_k}$ of the posterior distribution, with $\mathrm{diag}\,(\mathbf{v}_k)^{-1}$ being integrated over the missing data posterior distribution:

$$\tilde{\mathbf{s}}_k = \left(\mathbf{A}^T\,\mathbf{A} + \sigma^2\,\mathrm{diag}\left(\mathbf{c}_k^{'}\right)\right)^{-1}\mathbf{A}^T\,\tilde{\mathbf{x}}_k \tag{11}$$

where

$$\mathbf{c}_k = [c_{1,k},\ldots,c_{n,k}]^T \quad \text{and} \quad c_{i,k} = \mathrm{E}\left\{\frac{1}{v_{i,k}}|\tilde{s}_{i,k}\right\} = \frac{1}{\tilde{s}_{i,k}^2} \tag{12}$$

3.3 Update of A

The rows of $\mathbf{A}$ are a posteriori mutually independent with

$$p(\mathbf{r}_j|\tilde{\mathbf{s}},\sigma^2,\tilde{\mathbf{x}}) = \mathcal{N}(\mathbf{r}_j|\boldsymbol{\mu}_{\mathbf{r}_j},\boldsymbol{\Sigma}_{\mathbf{r}}) \tag{13}$$

where $\boldsymbol{\Sigma}_{\mathbf{r}} = \left(\frac{1}{\sigma^2}\sum_k \tilde{\mathbf{s}}_k\,\tilde{\mathbf{s}}_k^T + \frac{1}{\sigma_r^2}\,\mathbf{I}_n\right)^{-1}$ and $\boldsymbol{\mu}_{\mathbf{r}_j} = \frac{1}{\sigma^2}\,\boldsymbol{\Sigma}_{\mathbf{r}}\sum_k \tilde{x}_{j,k}\,\tilde{\mathbf{s}}_k$. Row j is updated to the mode $\boldsymbol{\mu}_{\mathbf{r}_j}$ of the posterior distribution: [1]

$$\mathbf{r}_j = \left(\sum_k \tilde{\mathbf{s}}_k\,\tilde{\mathbf{s}}_k^T + \frac{\sigma^2}{\sigma_r^2}\,\mathbf{I}_n\right)^{-1}\sum_k \tilde{x}_{j,k}\,\tilde{\mathbf{s}}_k \tag{14}$$

3.4 Update of σ^2

The posterior distribution $p(\sigma^2|\tilde{\mathbf{s}},\tilde{\mathbf{x}})$ is written

$$p(\sigma^2|\tilde{\mathbf{s}},\tilde{\mathbf{x}}) = \mathcal{IG}(\sigma^2|\alpha,\beta) \tag{15}$$

with $\alpha = \frac{(N-n)m}{2} + \alpha_\sigma$ and

$$\beta = \sum_{j=1}^{m}\left(\left(\sum_k \tilde{x}_{j,k}^2\right) - \left(\sum_k \tilde{x}_{j,k}\tilde{\mathbf{s}}_k^T\right)\left(\sum_k \tilde{\mathbf{s}}_k\tilde{\mathbf{s}}_k^T\right)^{-1}\left(\sum_k \tilde{x}_{j,k}\tilde{\mathbf{s}}_k\right)\right) + \beta_\sigma \tag{16}$$

σ^2 is updated to the mode of the distribution:

$$\sigma^2 = \frac{\beta}{\alpha+1} \tag{17}$$

4 Results

We present results of separation of 3×3 and 2×3 mixtures of audio sources. Results are discussed in Section 5.

[1] In practice the columns of $\mathbf{A}$ are normalized to 1 after each iteration to solve the BSS indeterminacy on gain.

4.1 Determined Mixture

We first study a mixture of $n = 3$ audio sources (speech, piano, guitar) with $m = 3$ observations. The mixing matrix is given in Table 1. We set $\sigma = 0.1$, which corresponds to approximatively 9.5dB noise on each observation. The signals are sampled at 8kHz with length $N = 65356$ ($\approx 8s$). We used a MDCT basis to decompose the observations, using a sine bell and 50% overlap, yielding a time resolution of 64ms (half the window length). The MDCT is a local cosine transform known to provide sparse representations of audio signals [8]. Time resolutions of 32ms and 128ms were also tried but led to slightly poorer results. The proposed method was compared with the method in [4], which uses a Student t prior on the sources and computes MMSE estimate of the parameters using a Gibbs sampler. We also show the results provided by the standard ICA algorithm JADE [9] applied to $\tilde{\mathbf{x}}$, which estimates a separating matrix via joint-diagonalization of a set of cumulant matrices, and apply the obtained matrix to the data, without denoising of the sources estimates. In our EM algorithm $\mathbf{A}$ is initialized with the JADE estimate and σ^2 is initialized to 1. We used noninformative priors $\sigma_r = +\infty$ and $\alpha_\sigma = \beta_\sigma = 0$ for $\mathbf{A}$ and σ^2.

Table 1 shows the mixing matrices estimated by the two methods. Table 2 provides separation quality criteria for the sources estimates. The criteria used are described in [10]. Basically, the SDR (Source to Distortion Ratio) provides an overall separation performance criterion, the SIR (Source to Interferences Ratio) measures the level of interferences from the other sources in each source estimate, SNR (Source to Noise Ratio) measures the error due to the additive noise on the sensors and the SAR (Source to Artifacts Ratio) measures the level of artifacts in the source estimates. The higher are the ratios, the better is quality of estimation. We point out that the performance criteria are invariant to a change of basis, so that figures can be computed either on the time sequences ($\hat{\mathbf{s}}$ compared to $\mathbf{s}$) or the MDCT coefficients ($\hat{\tilde{\mathbf{s}}}$ compared to $\tilde{\mathbf{s}}$). The estimated sources can be listened to at `http://www-sigproc.eng.cam.ac.uk/~cf269/ica06/sound_files.html`, which is perhaps the best way to assess the audio quality of the results.

Table 1. Estimates of **A** for the determined mixture

Original matrix	Jeffrey + EM
$\mathbf{A} = \begin{bmatrix} 1 & 1 & 1 \\ 0.8 & 1.3 & -0.9 \\ 1.2 & -0.7 & 1.1 \end{bmatrix}$	$\hat{\mathbf{A}} = \begin{bmatrix} 1 & 1 & 1 \\ 0.8090 & 1.3097 & -0.8922 \\ 1.1921 & -0.7341 & 1.0593 \end{bmatrix}$
t + MCMC	JADE
$\hat{\mathbf{A}} = \begin{bmatrix} 1 & 1 & 1 \\ 0.7914 & 1.3063 & -0.9004 \\ \pm(0.0049) & (\pm 0.0049) & (\pm 0.0047) \\ 1.1922 & -0.6980 & 1.1079 \\ \pm(0.0063) & \pm(0.0050) & \pm(0.0045) \end{bmatrix}$	$\hat{\mathbf{A}} = \begin{bmatrix} 1 & 1 & 1 \\ 0.8403 & 1.3430 & -0.9085 \\ 1.1543 & -0.9408 & 0.8823 \end{bmatrix}$

Table 2. Performance criteria for the determined mixture

	$\hat{s}_1$				$\hat{s}_2$				$\hat{s}_3$			
	SDR	SIR	SAR	SNR	SDR	SIR	SAR	SNR	SDR	SIR	SAR	SNR
Jeffrey + EM	10.6	34.5	10.6	29.2	14.0	36.3	14.0	32.1	11.7	43.4	11.8	29.9
t + MCMC	11.8	40.3	12.5	20.4	15.1	47.0	15.9	22.9	13.1	43.4	13.9	20.7
JADE	5.4	32.7	-	5.4	7.5	17.1	-	8.1	5.9	14.9	-	6.6

4.2 Undetermined Mixture

We now consider the more difficult case consisting of discarding one observation of the previous mixture, thus yielding an underdetermined problem. Results are found in Tables 3 and 4. In the EM algorithm the mixing matrix was initialized with the result of the simple clustering method described in [1] [2] which yielded $\mathbf{A}_{init} = [1\ 1\ 1; 0.6348\ 1.4405\ \ -0.9216]$.

Table 3. Estimates of **A** for the underdetermined mixture

Jeffrey + EM	t + MCMC
$\hat{\mathbf{A}} = \begin{bmatrix} 1 & 1 & 1 \\ 0.7083 & 1.5630 & -0.9109 \end{bmatrix}$	$\hat{\mathbf{A}} = \begin{bmatrix} 1 & 1 & 1 \\ 0.7715 & 1.3072 & -0.9089 \\ (\pm 0.0059) & (\pm 0.0060) & (\pm 0.0052) \end{bmatrix}$

Table 4. Performance criteria for the underdetermined mixture

	$\hat{s}_1$				$\hat{s}_2$				$\hat{s}_3$			
	SDR	SIR	SAR	SNR	SDR	SIR	SAR	SNR	SDR	SIR	SAR	SNR
Jeffrey + EM	1.0	19.5	1.1	26.6	5.9	16.2	6.4	29.8	9.6	21.0	10.0	29.9
t + MCMC	0.7	14.1	1.4	14.6	6.4	23.3	6.7	19.8	11.4	28.5	13.9	15.3

5 Conclusions

Not surprisingly the performances of both method are better in the determined case than in the underdetermined one, notably in terms of SIRs. This is because when **A** is square, $\tilde{\mathbf{s}}_k$ given by Eq. (11) is simply the application of the weighted pseudo-inverse of **A** to $\tilde{\mathbf{x}}_k$. Tables 2 and 4 also show that Jeffrey's inverse prior leads to higher SNRs than the Student t + MCMC method. This is because the inverse prior leads to much sparser representations and actually sets many coefficients to zero (see [5]). This was confirmed by computing sparsity indexes on the obtained source coefficients estimates. Though the SARs obtained with both methods are quite similar in each case, when listening to the source estimates

[2] This method simply consists in projecting the observations on the sphere and run a K-Means algorithm to identify the slope of the clusters generated by the mixing matrix columns.

those obtained with the Jeffrey's + EM approach tend to suffer from stronger musical noise. We believe this is because the residual error originating from the additive noise on the observations masks part of the artifacts in the Student t + MCMC estimates.

One major advantage of the Jeffrey's + EM approach is computational cost: it takes 9 min to achieve convergence (50 iterations) in the underdetermined case, when it takes 6h to run the 2000 iterations necessary to obtain convergence of the Student t + MCMC method (and the MMSE estimates were computed from the next 500 samples). But on the other hand, as in many cases, the EM approach happened to be very sensitive to initialization and could lead to local maxima, while the MCMC approach scans all the posterior distribution of the parameters and thus provides reliable estimates, independently of the initializations.

Acknowledgement. This work was supported by the European Commission funded Research Training Network HASSIP (HPRN-CT-2002-00285).

References

1. Zibulevsky, M., Pearlmutter, B.A., Bofill, P., Kisilev, P.: Blind source separation by sparse decomposition. In Roberts, S.J., Everson, R.M., eds.: Independent Component Analysis: Principles and Practice. Cambridge University Press (2001)
2. Jourjine, A., Rickard, S., Yilmaz, O.: Blind separation of disjoint orthogonal signals: Demixing n sources from 2 mixtures. In: Proc. ICASSP. Volume 5., Istanbul, Turkey (2000) 2985–2988
3. Févotte, C., Godsill, S.J., Wolfe, P.J.: Bayesian approach for blind separation of underdetermined mixtures of sparse sources. In: Proc. 5th International Conference on Independent Component Analysis and Blind Source Separation (ICA 2004), Granada, Spain (2004) 398–405
4. Févotte, C., Godsill, S.J.: A Bayesian approach for blind separation of sparse sources. IEEE Trans. Speech and Audio Processing (In press) Preprint available at `http://www-sigproc.eng.cam.ac.uk/~cf269/`.
5. Figueiredo, M.A.T., Nowak, R.D.: Wavelet-based image estimation: An empirical Bayes approach using Jeffrey's noninformative prior. IEEE Trans. Image Processing **10** (2001) 1322–1331
6. Figueiredo, M.A.T.: Adaptive sparseness for supervised learning. IEEE Trans. Pattern Analysis and Machine Intelligence **25** (2003) 1150–1159
7. Dempster, A.P., Laird, N.M., Rubin, D.B.: Maximum likelihood from incomplete data via the em algorithm. Journal of the Royal Statistical Society, Series B **39** (1977) 1–38
8. Mallat, S.: A wavelet tour of signal processing. Academic Press (1998)
9. Cardoso, J.F., Souloumiac, A.: Blind beamforming for non Gaussian signals. IEE Proceedings-F **140** (1993) 362–370
10. Gribonval, R., Benaroya, L., Vincent, E., Févotte, C.: Proposals for performance measurement in source separation. In: Proc. 4th Symposium on Independent Component Analysis and Blind Source Separation (ICA'03), Nara, Japan (2003)

Solution of Permutation Problem in Frequency Domain ICA, Using Multivariate Probability Density Functions

Atsuo Hiroe

Intelligent Systems Research Laboratory, Information Technologies Laboratories, Sony Corporation, 6-7-35 Kitashinagawa, Shinagawa-ku, Tokyo 141-0001, Japan
Atsuo.Hiroe@jp.sony.com

Abstract. Conventional Independent Component Analysis (ICA) in frequency domain inherently causes the permutation problem. To solve the problem fundamentally, we propose a new framework for separation of the whole spectrograms instead of the conventional binwise separation. Under our framework, a measure of independence is calculated from the whole spectrograms, not individual frequency bins. For the calculation, we introduce some multivariate probability density functions (PDFs) which take a spectrum as arguments. To seek the unmixing matrix that makes spectrograms independent, we demonstrate a gradient-based algorithm using multivariate activation functions derived from the PDFs. Through experiments using real sound data, we have confirmed that our framework is effective to generate permutation-free unmixed results.

Keywords: independent component analysis, frequency domain, permutation problem, multivariate probability density function, multivariate activation function, spherical distribution.

1 Introduction

As a method to separate convolutive mixtures, Independent Component Analysis (ICA) in frequency domain has been often used. The frequency domain ICA has an advantage in terms of faster convergence than time domain deconvolution, however it also has a large problem called permutation, inconsistency of output channels among frequency bins [1].

Recently to treat the permutation problem, two major approaches have been done: 1) Postprocesses to correct permutation [2], and 2) Time-frequency modeling [3]. Both, however, are still challenging.

As for the postprocesses, the envelope similarity method is likely to misjudge in some bins with low power, and the direction-of-arrival estimation method is sensitive to microphones' configuration. Additionally, the postprocesses themselves expense computational power.

The other, time-frequency modeling, means to build some relations among bins into the learning algorithm. It however doesn't guarantee permutation-free results, or its algorithm is limited to the case of two microphones [3].

J. Rosca et al. (Eds.): ICA 2006, LNCS 3889, pp. 601–608, 2006.

What is more desirable in the frequency domain ICA is that the ICA directly generates permutation-free spectrograms. To realize such unmixing, we will present a new framework in Sect. 3.

2 Overview of Conventional Frequency Domain ICA

Before explaining our framework, we briefly review the conventional methods. The unmixing process in conventional ICA is formulated as:

$$\begin{bmatrix} Y_1(\omega,t) \\ \vdots \\ Y_n(\omega,t) \end{bmatrix} = \begin{bmatrix} w_{11}(\omega) & \dots & w_{1n}(\omega) \\ \vdots & \ddots & \vdots \\ w_{n1}(\omega) & \dots & w_{nn}(\omega) \end{bmatrix} \begin{bmatrix} X_1(\omega,t) \\ \vdots \\ X_n(\omega,t) \end{bmatrix} \tag{1}$$

$$\Leftrightarrow \quad \boldsymbol{Y}(\omega,t) = \boldsymbol{W}(\omega)\boldsymbol{X}(\omega,t), \tag{2}$$

where ω and t are the index of frequency bin and frame respectively, and $X_k(\omega,t)$ and $Y_k(\omega,t)$ are observation and unmixed signal in ωth bin of kth channel, and $w_{ij}(\omega)$ denotes a weight from $X_j(\omega,t)$ to $Y_i(\omega,t)$. To seek an unmixing matrix $\boldsymbol{W}(\omega)$, Smaragdis [1] applied the natural gradient algorithm [5], such as:

$$\Delta\boldsymbol{W}(\omega) = \left\{\boldsymbol{I} + E_t\left[\varphi(\boldsymbol{Y}(\omega,t))\boldsymbol{Y}(\omega,t)^H\right]\right\}\boldsymbol{W}(\omega) \quad (\boldsymbol{I} : \text{identity matrix}) \tag{3}$$

$$\boldsymbol{W}(\omega) \leftarrow \boldsymbol{W}(\omega) + \eta\Delta\boldsymbol{W}(\omega) \tag{4}$$

$$\varphi(\boldsymbol{Y}(\omega,t)) = \left[\varphi(Y_1(\omega,t)), \cdots, \varphi(Y_n(\omega,t))\right]^T \tag{5}$$

$$\varphi(Y_k(\omega,t)) = \frac{\partial}{\partial Y_k(\omega,t)} \log P_{Y_k(\omega)}(Y_k(\omega,t)), \tag{6}$$

where $P_{Y_k(\omega)}(Y_k(\omega,t))$ denotes the probability density function (PDF) corresponding to $Y_k(\omega,t)$, and $\varphi(Y_k(\omega,t))$ is the derivative of $\log P_{Y_k(\omega)}(Y_k(\omega,t))$, called activation function or score function. These functions do not have to exactly correspond to the distribution of $Y_k(\omega,t)$.

In the above framework, however, unmixing is done just in individual bins and thus no relations over bins are considered (Fig. **1**). This is a fundamental reason why the conventional frequency domain ICA causes the permutation problem.

To reflect some relations among frequency bins, Mitianoudis and Davies [3,4] applied a time-frequency model, such as:

$$\varphi(Y_k(\omega,t)) = \frac{1}{\beta_k(t)} \frac{|Y_k(\omega,t)|}{Y_k(\omega,t)} \tag{7}$$

$$\beta_k(t) = \operatorname*{mean}_{\omega}\left[\,|Y_k(\omega,t)|\,\right] . \tag{8}$$

They proposed that this $\beta_k(t)$ term imposes frequency coupling between bins. However they also reported that its effect is still limited. Hence to prevent permutation inconsistencies, another framework is needed.

3 New Framework to Solve the Permutation Problem

Consider the unmixing in the whole spectrograms, like Fig. **2**. Such process should directly generate unmixed results without permutation inconsistencies. In order

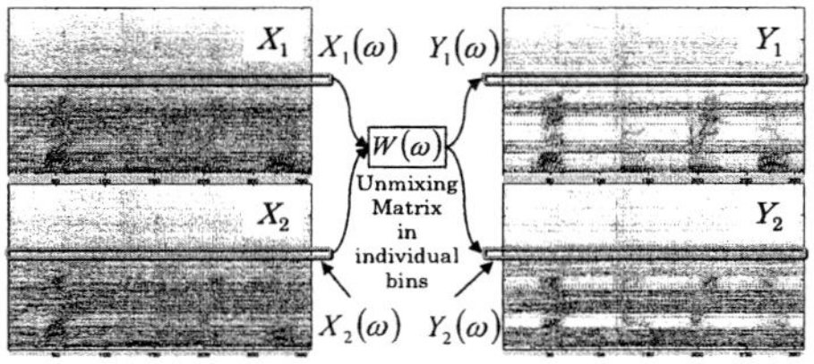

Fig. 1. Unmixing in individual bins (conventional)

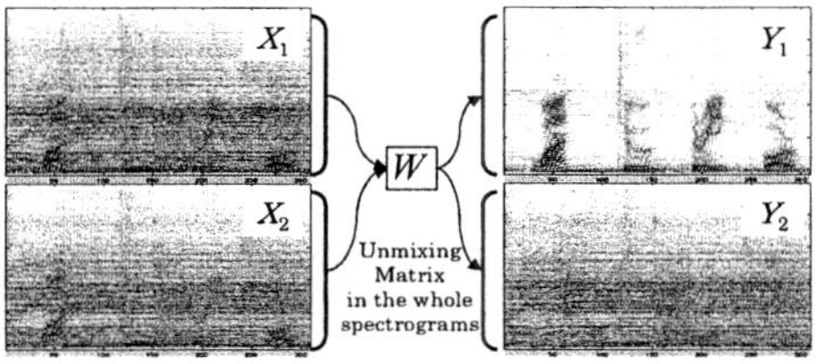

Fig. 2. Unmixing at the whole spectrograms (proposed)

to realize it, we suggest to use following two features, and will explain them in the below subsections.

1. Formula for the whole unmixing.
2. Measure of independence in the whole spectrograms.

3.1 Formula for the Whole Unmixing

Let us construct a formula corresponding to Fig.**2**, unmixing over all frequency bins. Such formula can be made through developing (1) to all bins. As a result, we obtain a formula with a matrix of diagonal matrices:

$$\left[\begin{array}{c} Y_1(1,t) \\ \vdots \\ Y_1(M,t) \\ \hline \vdots \\ \hline Y_n(1,t) \\ \vdots \\ Y_n(M,t) \end{array}\right] = \left[\begin{array}{cc|c|cc} w_{11}(1) & \mathbf{0} & & w_{1n}(1) & \mathbf{0} \\ & \ddots & \cdots & & \ddots \\ \mathbf{0} & w_{11}(M) & & \mathbf{0} & w_{1n}(M) \\ \hline & \vdots & \ddots & & \vdots \\ \hline w_{n1}(1) & \mathbf{0} & & w_{nn}(1) & \mathbf{0} \\ & \ddots & \cdots & & \ddots \\ \mathbf{0} & w_{n1}(M) & & \mathbf{0} & w_{nn}(M) \end{array}\right] \left[\begin{array}{c} X_1(1,t) \\ \vdots \\ X_1(M,t) \\ \hline \vdots \\ \hline X_n(1,t) \\ \vdots \\ X_n(M,t) \end{array}\right] \tag{9}$$

$$\Leftrightarrow \boldsymbol{Y}(t) = \boldsymbol{W}\boldsymbol{X}(t) \tag{10}$$

$$\Leftrightarrow \begin{bmatrix} \boldsymbol{Y_1}(t) \\ \vdots \\ \boldsymbol{Y_n}(t) \end{bmatrix} = \begin{bmatrix} \boldsymbol{W_{11}} & \cdots & \boldsymbol{W_{1n}} \\ \vdots & \ddots & \vdots \\ \boldsymbol{W_{n1}} & \cdots & \boldsymbol{W_{nn}} \end{bmatrix} \begin{bmatrix} \boldsymbol{X_1}(t) \\ \vdots \\ \boldsymbol{X_n}(t) \end{bmatrix}, \tag{11}$$

where n and M are number of microphones and frequency bins respectively (assuming that number of outputs is same as that of microphones). Using $\boldsymbol{Y_k}(t)$ and $\boldsymbol{X_k}(t)$, a spectrum of the kth channel, we can represent (9) also as (11). This channel-wise notation is important in introducing multivariate probability density functions in the next subsection.

3.2 Measure of Independence from the Whole Spectrograms

Let $P_Y(\boldsymbol{Y}(t))$ be the probability density function (PDF) of a vector $\boldsymbol{Y}(t)$, and $P_{Y_k}(\boldsymbol{Y_k}(t))$ be the PDF of a vector $\boldsymbol{Y_k}(t)$. Note that $P_{Y_k}(\boldsymbol{Y_k}(t))$ is also a multivariate PDF unlike the conventional $P_{Y_k(\omega)}(\boldsymbol{Y_k}(\omega,t))$.

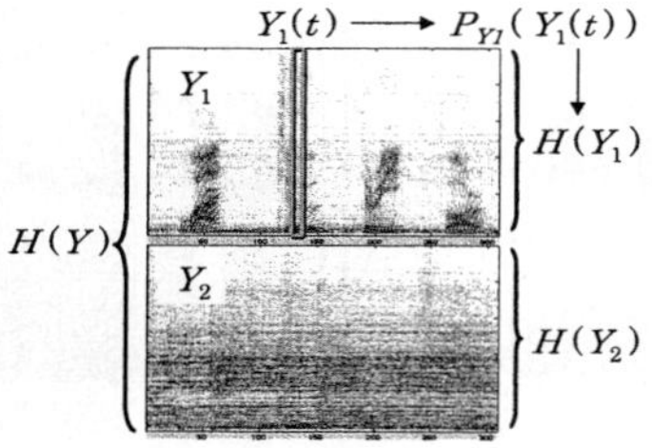

Fig. 3. Entropy of the whole spectrograms

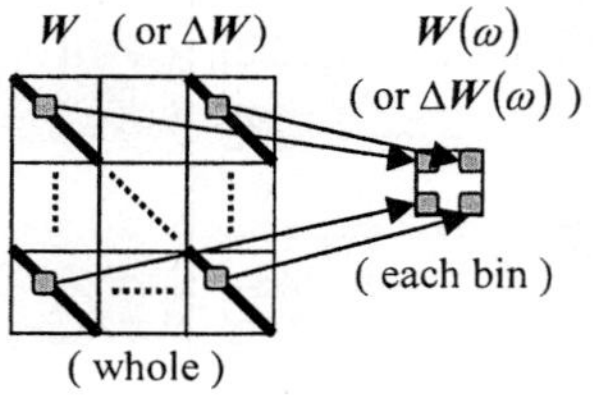

Fig. 4. Unmix matrices

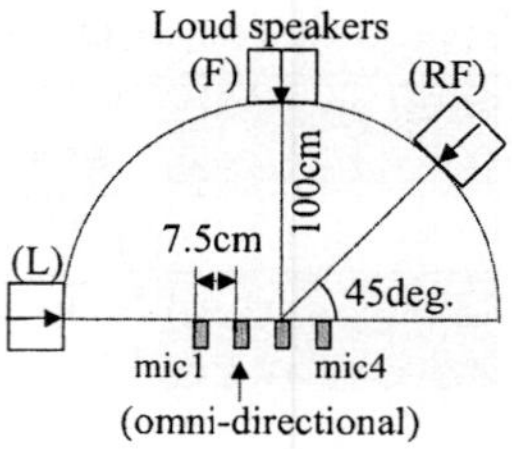

Fig. 5. Recording environment

When $\boldsymbol{Y_1}(t)$, ..., $\boldsymbol{Y_n}(t)$ are mutually independent, $P_Y(\boldsymbol{Y}(t))$ should be decomposed to $\prod_k P_{Y_k}(\boldsymbol{Y_k}(t))$. It means that Kullback-Leibler divergence (KLD) between $P_Y(\boldsymbol{Y}(t))$ and $\prod_k P_{Y_k}(\boldsymbol{Y_k}(t))$ should work as a measure of independence in the whole spectrograms. The Kullback-Leibler divergence $\mathrm{KLD}(\boldsymbol{Y})$ is calculated as:

$$\mathrm{KLD}(\boldsymbol{Y}) = \sum_{k=1}^{n} H(\boldsymbol{Y_k}) - H(\boldsymbol{Y}) \tag{12}$$

$$= \sum_{k=1}^{n} E_t\left[-\log P_{Y_k}(\boldsymbol{Y_k}(t))\right] - \log\left|\det(\boldsymbol{W})\right| - H(\boldsymbol{X}), \tag{13}$$

where $H(\boldsymbol{Y_k})$ is differential entropy of $\boldsymbol{Y_k}(t)$, $H(\boldsymbol{Y})$ is joint entropy of $\boldsymbol{Y}(t)$, and $E_t[]$ means expectation over frames (Fig.**3**). Equation (13) is derived with both (10) and the definition of $H(\boldsymbol{Y_k})$, indicating $P_Y(\boldsymbol{Y}(t))$ is unnecessary any longer.

As preparatory experiments, we calculated the KLD from artificially permuted spectrograms, and found that less permuted spectrograms lead to lower value of KLD.

3.3 Derivation of Learning Rule

The unmixing matrix $\boldsymbol{W}$ that makes $\boldsymbol{Y_1},\ldots,\boldsymbol{Y_n}$ independent should minimize the KLD in (13). To seek such $\boldsymbol{W}$, we apply the natural gradient rule [5] to (13) then obtain the following learning rule:

$$\Delta\boldsymbol{W} = -\frac{\partial\,\mathrm{KLD}(\boldsymbol{Y})}{\partial\boldsymbol{W}}\boldsymbol{W}^H\boldsymbol{W} = \left\{\boldsymbol{I} - \frac{\partial}{\partial\boldsymbol{W}}\left[\sum_{k=1}^{n} H(\boldsymbol{Y_k})\right]\boldsymbol{W}^H\right\}\boldsymbol{W}. \tag{14}$$

In (14), however, derivation of $\partial\sum H(\boldsymbol{Y_k})/\partial\boldsymbol{W}$ is difficult to express in a simple formula. Hence instead of $\Delta\boldsymbol{W}$, we derive a rule of $\Delta\boldsymbol{W}(\omega)$, submatrix of $\Delta\boldsymbol{W}$ corresponding to ωth frequency bin (Fig.**4**). Then we obtain the learning rule:

$$\Delta\boldsymbol{W}(\omega) = \left\{\boldsymbol{I} + E_t\left[\varphi_\omega(\boldsymbol{Y}(t))\boldsymbol{Y}(\omega,t)^H\right]\right\}\boldsymbol{W}(\omega) \tag{15}$$

$$\varphi_\omega(\boldsymbol{Y}(t)) = \left[\varphi_{1\omega}(\boldsymbol{Y_1}(t)),\cdots,\varphi_{n\omega}(\boldsymbol{Y_n}(t))\right]^T \tag{16}$$

$$\varphi_{k\omega}(\boldsymbol{Y_k}(t)) = \frac{\partial}{\partial Y_k(\omega,t)}\log P_{Y_k}(\boldsymbol{Y_k}(t)). \tag{17}$$

Note that the learning rule (15) updates just non-zero elements in (9), thus it doesn't increase total number of parameters to be estimated. Comparing (15) with the conventional rule (3), the difference is in arguments of the activation function. In (15), the value of $\varphi_{k\omega}(\boldsymbol{Y_k}(t))$ is determined from a spectrum of channel k, namely whole frequency bins, while the value of the conventional activation functions is determined from only $Y_k(\omega, t)$, ωth bin's data.

3.4 Multivariate Probability Density Functions

In our framework, what is the most crucial is what multivariate PDF to use. As a first challenge, we are employing a class of PDFs called 'spherical distribution' [6], which is represented as an assignment of vector's norm into a proper scalar function $f(x)$, such as:

$$P_{Y_k}(\boldsymbol{Y_k}(t)) = hf\left(\|\boldsymbol{Y_k}(t)\|_2\right) \quad (h\text{: Normalization factor}) \tag{18}$$

$$\|\boldsymbol{Y_k}(t)\|_2 = \left(\sum_{\omega} |Y_k(\omega, t)|^2\right)^{1/2} . \tag{19}$$

Specifying $f(x)$ generates various PDFs and corresponding activation functions as in the following table, where m and K are proper positive constants. Compared with the conventional activation functions, the uniquely increased calculation is only on the norm $||\boldsymbol{Y_k}(t)||_2$.

$f(x)$	$P_{Y_k}(\boldsymbol{Y_k}(t))$	$\varphi_{k\omega}(\boldsymbol{Y_k}(t))$
$\frac{1}{\cosh^m(Kx)}$	$\frac{h}{\cosh^m(K\|\boldsymbol{Y_k}(t)\|_2)}$	$-mK\tanh(K\|\boldsymbol{Y_k}(t)\|_2)Y_k(\omega,t)/\|\boldsymbol{Y_k}(t)\|_2$
$\exp\{-K\|x\|\}$	$h\exp\{-K\|\boldsymbol{Y_k}(t)\|_2\}$	$-KY_k(\omega,t)/\|\boldsymbol{Y_k}(t)\|_2$

In the research for conventional scalar PDFs, distinction between super- and sub-gaussianity of the PDF is significant for stable and correct separation. In our framework, however, it is open question how the distinction should be applied.

3.5 Preprocesses and Postprocesses

Before the learning, observations $X_k(\omega, t)$ are normalized to unit variance over frames, as in the conventional binwise method. The normalization can also suppress permutation inconsistencies, since a frequency bin with low power (variance) has few contribution to the KLD in (13) and thus it is easily permuted.

After the learning, unmixed spectrograms $\boldsymbol{Y}$ have scaling ambiguities among bins, thus rescaling each bin is still necessary. In our framework, rescaling can simply be represented with the whole unmixing matrix $\boldsymbol{W}$, although its operation is equivalent to binwise one. For example, the rescaling based on the minimal distortion principle [7] is represented as $\boldsymbol{W} \leftarrow \mathrm{diag}\left(\boldsymbol{W}^{-1}\right)\boldsymbol{W}$.

4 Experimental Results

We demonstrate results of the separation of a set of data which were recorded separately and mixed on a computer. Recording was done through playing each

Table 1. Signal-to-interference ratios

Test No.	Sources (from)	Signals	SIR [db] to src1	src2	src3	KLD ($\times 10^4$)
1	src1: beet.wav (RF)	Observations	0.04	0.11	—	13.60
	src2: mike.wav (F)	Unmixed with (20)	16.95	19.69	—	5.82
		Unmixed with (21)	3.93	1.14	—	6.87
2	src1: beet.wav (RF)	Observations	-2.46	-2.12	-2.77	16.82
	src2: mike.wav (F)	Unmixed with (20)	13.46	11.24	9.20	8.13
	src3: beet9.wav (L)	Unmixed with (21)	-2.02	0.76	4.98	8.62

wave file which is put in "ICA '99 Synthetic Benchmarks" [8] from each loud speaker designated in Fig.**5**. The format of recorded data is in 16kHz sampling rate and 8 seconds length. As observations, two different mixtures were generated: mixture of two sources and that of three sources (see Table **1**). The observations were transformed to spectrograms of 257 bins ($M = 257$) and 1000 frames, using 512 point Fast Fourier Transform (Hanning window, 128 shift).

In the learning process, we employed the following two activation functions:

$$\varphi_{k\omega}(\boldsymbol{Y_k}(t)) = -K\frac{Y_k(\omega,t)}{\|\boldsymbol{Y_k}(t)\|_2} \qquad (K = \sqrt{M} \simeq 16.03) \tag{20}$$

$$\varphi(Y_k(\omega,t)) = -\frac{Y_k(\omega,t)}{|Y_k(\omega,t)|} . \tag{21}$$

Equation (20) is a proposed multivariate activation function, while (21) is conventional scalar one. As common parameters, we used the learning rate $\eta = 0.3$ and 300 iterations. No postprocesses to correct permutation were done.

After obtaining frequency domain results $\{\boldsymbol{Y_1}, \cdots, \boldsymbol{Y_n}\}$ and the corresponding time-domain signals $\{\boldsymbol{y_1}, \cdots, \boldsymbol{y_n}\}$, we calculated signal-to-interference ratio (SIR) as following steps:

1. Let $\boldsymbol{x_{k,s_j}}$ be the contribution from the source $\boldsymbol{s_j}$ to the kth microphone. Decompose each $\boldsymbol{y_k}$ to the target signal, component along with $\boldsymbol{x_{k,s_j}}$, and the interference.
2. Calculate signal-to-interference ratio $\mathrm{SIR}(\boldsymbol{y_k}, \boldsymbol{x_{k,s_j}})$ as $10\log_{10} \|\mathrm{target}\|^2 / \|\mathrm{interference}\|^2$. (In step 1 and 2, we used BSS EVAL Toolbox [9].)
3. Define $\mathrm{SIR}(\boldsymbol{s_j}) = \max_k \mathrm{SIR}(\boldsymbol{y_k}, \boldsymbol{x_{k,s_j}})$ as total SIR to the source $\boldsymbol{s_j}$. (Operation 'max' means specifying which $\boldsymbol{y_k}$ corresponds to $\boldsymbol{s_j}$)

We show the results in Table **1**, where Test 1 is on the mixture of two sources and Test 2 is on three sources. In both tests, we found that any SIR with (20) is more improved due to the permutation-free results.

Fig.**6**–Fig.**8** are spectrograms on Test 2. Fig.**6** shows the contribution from src1 (voice) to the microphone2, as a close-up of $32 \le \omega \le 128$ (1k–4k[Hz]) and $0 \le t \le 188$ (0–1.5[sec]). Fig.**7** and Fig.**8** show the corresponding unmixed result with (20) and (21) respectively. Fig.**7**, with the proposed, doesn't have the permutation inconsistency, while Fig.**8**, with the conventional, has many permutation inconsistencies.

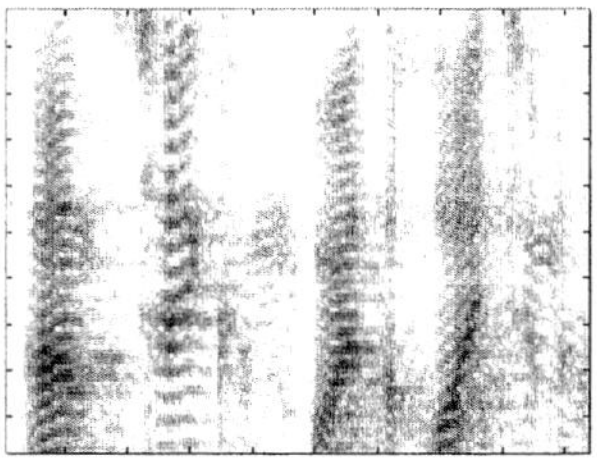

Fig. 6. Contribution of src2 to mic2

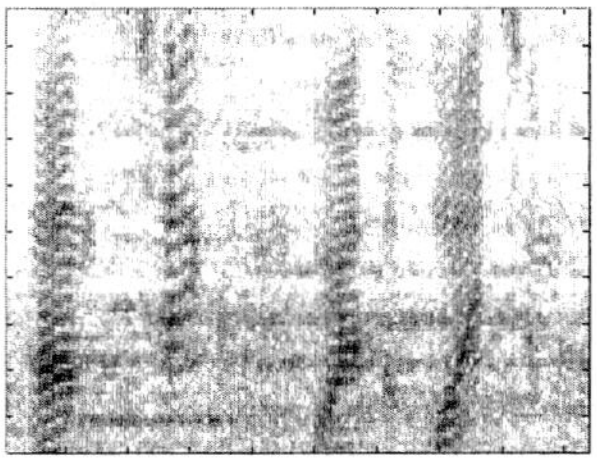

Fig. 7. $\boldsymbol{Y_2}$, Unmixed with (20) (proposed)

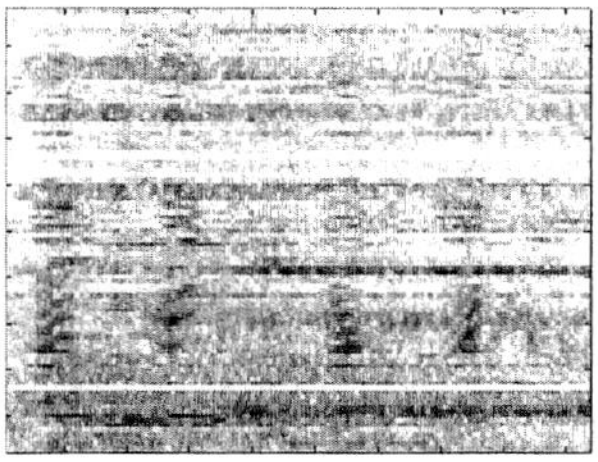

Fig. 8. $\boldsymbol{Y_2}$, Unmixed with (21) (conventional)

The rightmost column in Table **1** denotes values of Kullback-Leibler divergence calculated with (13) and $P_{Y_k}(\boldsymbol{Y_k}(t)) = h \exp\{-K\|\boldsymbol{Y_k}(t)\|_2\}$, assuming that $H(\boldsymbol{X}) = 0$ and $h = 1$. It also indicates the fact that in our framework less permuted spectrograms lead to lower value of KLD as well as more unmixed spectrograms do.

Fig.**1** and Fig.**2** indicate another set of unmixed results, using `rsm2_m[AB].wav` which are put in Te-won Lee's web page [10]. Fig.**2** shows the results with (20) and no permutation inconsistencies occur, while Fig.**1** shows the results with (21) and band-like permutations occur.

5 Discussion

From the view of multivariate probabilistic density functions, the conventional methods can be reinterpreted as follows. In (15)–(17), applying (22) leads to the conventional binwise rule (3). Similarly, (23) leads to the rule by Mitianoudis and Davies (7). These facts mean they correspond to each particular case.

$$P_{Y_k}(\boldsymbol{Y_k}(t)) = \prod_{\omega} P_{Y_k(\omega)}(Y_k(\omega, t)) \tag{22}$$

$$P_{Y_k}(\boldsymbol{Y_k}(t)) = \frac{h}{\left(\sum_{\omega} |Y_k(\omega, t)|\right)^M} \tag{23}$$

Equation (22) does enforce excessive assumption that all frequency bins even within the identical channel (spectrogram) are independent, thus it causes permutation inconsistencies. Another case (23) doesn't meet a requirement of PDF that all summation of probability must be 1, rather it diverges.

On the other hand, our framework just enforce an assumption that all spectrograms are mutually independent. This should be suitable to separate observation spectrograms to permutation-free outputs.

6 Conclusion

We presented a new framework to separate the whole spectrograms, instead of individual frequency bins, where a measure of independence is calculated from the whole spectrograms using multivariate probability density functions. We also demonstrated the learning rule using multivariate activation functions, and specific instances of these functions. Experimental results indicate that our framework generates permutation-free outputs and improves SIRs.

At present, however, following issues are still open: How robust our framework is to actual various observations in terms of convergence or unmixing performance; What kind of activation functions are suitable in terms of stability or permutation-free separation; How ours can be applied to algorithms except the gradient. Of course, research of multivariate PDFs itself is also challenging, including their super- or sub-gaussianity.

We hope that our proposal can be a new step in frequency domain ICA.

Acknowledgment

We specially thank Dr. Helmute Lucke, a former member of our group, for discussing this issue with us during early phase of our research.

References

1. P. Smaragdis, "Blind separation of convolved mixtures in the frequency domain," *Neurocomputating*, 10(2):251–276, 1998.
2. H. Sawada and R. Mukai and S. Araki and S. Makino, "A Robust and Precise Method for Solving the Permutation Problem of Frequency-Domain Blind Source Separation," *Proc. ICA 2003*, pp.505–510, April, 2003.
3. N. Mitianoudis and M. Davies, "Audio source separation of convolutive mixtures," *Trans. Audio and Speech Processing*, vol.11, no.5, pp.489–497, 2003.
4. M. Davies, "Audio Source Separation," Chapter in *Mathematics in Signal Processing V*, Oxford University Press, 2002.
5. S. Amari and A. Cichocki and HH. Yang, "A new learning algorithm for blind signal separation," In: *Advances in Neural Information Processing Systems*, 8, MIT Press., 1996
6. `http://www.quantlet.com/mdstat/scripts/mva/htmlbook/mvahtmlnode42.html`
7. K. Matsuoka and S. Nakashima, "Minimal distortion principle for blind source separation," *Proc. ICA 2001*, pp.722–727, Dec. 2001.
8. `http://sound.media.mit.edu/ica-bench/sources/`
9. C. Févotte, R. Gribonval and E. Vincent, *BSS EVAL Toolbox User Guide*, IRISA Technical Report 1706, Rennes, France, April 2005. `http://www.irisa.fr/metiss/bss eval/`
10. `http://www.cnl.salk.edu/~tewon/Blind/blind_audio.html`

ICA-Based Speech Features in the Frequency Domain

Włodzimierz Kasprzak, Adam F. Okazaki, and Adam B. Kowalski

Institute of Control and Computation Engineering,
Warsaw University of Technology,
ul. Nowowiejska 15-19, PL - 00-665 Warsaw, Poland
W.Kasprzak@ia.pw.edu.pl
http://www.ia.pw.edu.pl/

Abstract. We apply the technique of independent component analysis to Fourier power coefficients of speech signal frames for a blind detection of basic vectors (sources). A subset of sources corresponding to the noisy influence of basic frequency is identified and its corresponding features could be eliminated. The mixing coefficients for such sources are then determined for every speech sample. We compare our features with the Mel Frequency Cepstrum Coefficient (MFCC) features, widely used today for phoneme-based speech recognition.

1 Introduction

It is common in automatic speech recognition systems to apply a frame-based segmentation of the signal, i.e. to use short-time frames [1], [2] in which a windowed Fourier transform is performed. Although specific features of a single frame can be detected already in the time-domain (like LPC features), there are widely used Mel Frequency Cepstrum Coefficients (MFCC) [2], [3] which are computed in the "cepstral" space (this needs a homomorphic filtering via the Fourier space back to the time domain and a post-processing step called "liftering").

It was observed, that statistical cues could offer increased power to speaker recognition systems [4], [5]. In this context the two techniques - PCA and ICA - can be considered [6], [7]. Different authors derive the principal component analysis (PCA) or ICA [4] of the power spectra vectors, which are also smoothed using Mel-scale triangular filters. The authors of [5] assume that the spectra of sounds generated by a given speaker can be synthesized using a set of speaker specific basis functions - the unknown source in the ICA model.

In this paper we follow this idea and we expect the Fourier power coefficients of a single frame to be mixtures of a set of basic, statistically independent vectors. In section 2 the problem of speech feature detection is introduced. The proposed approach is described in section 3 and simulation results follow in section 4.

J. Rosca et al. (Eds.): ICA 2006, LNCS 3889, pp. 609–616, 2006.

2 The MFCC Features for Speech

The task of ICA is to find the waveforms $s_i(t)$ of the sources, knowing only the mixtures $x_j(t)$ and the number m of sources [6]. A well-known iterative optimization method the stochastic gradient (or gradient descent) search [8] can be applied in this context. Especially for the ICA problem different gradient approaches were developed (e.g. the *natural gradient descent* [9]). An efficient "batch" approach is the method "FastICA" [7]. The batch processing allows a preliminary "whitening" step for the zero-mean mixture signals, which improves the convergence speed of the ICA procedure.

2.1 Energy of Speech Samples

As illustrated in Fig. 1 and 2 the energy distribution in time of the same spoken word significantly differs from sample to sample and from speaker to speaker. Hence, we need a feature scheme which is rather interested in the "waveform" or relative (normalized) energy pattern than in the global energy distribution. In some applications the possibility to achieve ICA demixing results with respect to a scaling factor only is a disadvantage of the ICA approach. In case of speech features no such drawback should appear.

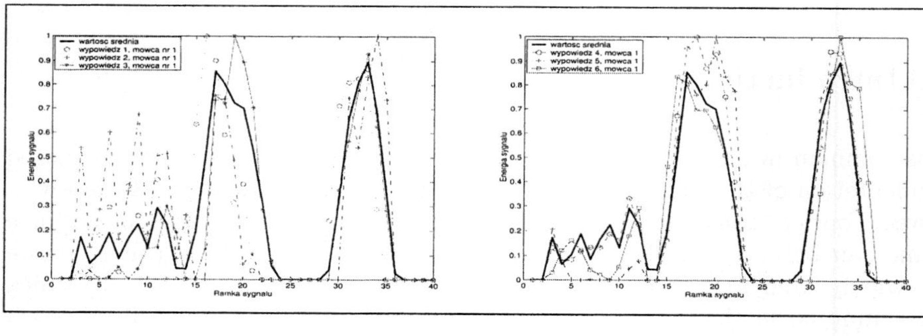

Fig. 1. Energy distribution in time of some polish word "pusc" (release): 3 (left) and 3 (right drawing) samples with their averages (bold lines) for one speaker

In order to limit the variability of energy distribution among speakers and due to different emotional attitude of the speaker we make an energy normalization step before feature detection. As our goal is to extract ICA-based features and to compare them with the MFCC features only, without performing general word recognition, we can deal with the necessary time stretching by performing an interpolation-based resampling in the time domain in advance of the feature detection step. In this way we assure that the current utterance and the pattern utterance have both the same number of samples (for every word a different number of samples is usually required).

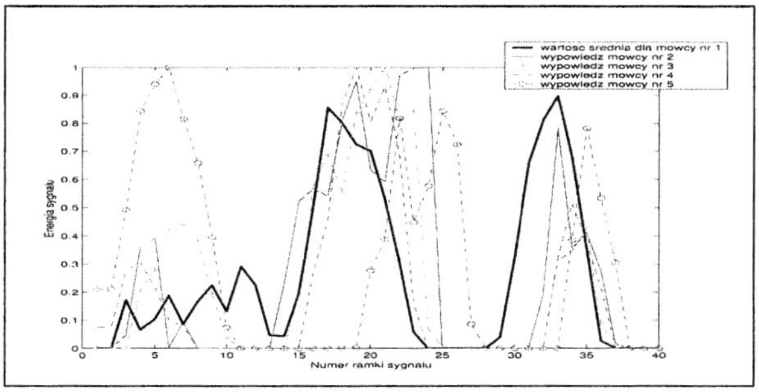

Fig. 2. Energy distribution in time of polish word "pusc" (release) - the averages for 5 different speakers

2.2 The Standard MFCC Features

The Mel-cepstrum features are the result of the characteristic (homomorphic) transformation $MFCC(h) = FT^{-1}\{MFC\{FT\{h\}\}\}$ for $\mathbf{h} = \mathbf{x} \otimes \mathbf{w}$ (a convolution of $\mathbf{x}$ with $\mathbf{w}$).

The short-term power spectrum is computed by applying the discrete Fourier Transform (DFT) (in fact the FFT) to each windowed signal and taking directly the magnitudes of Fourier coefficients raised to the power of two. The power spectrum is usually represented on a log scale.

A MEL scale (empirical result) adopts the frequency bandwidths to the bandwidths recognized by the human auditory system. The set of Fourier features is reduced by considering bandwidths, centered around some MEL scale frequencies. Usually one uses a set of l triangle filters $D(l,t)$ to compute l so called Mel-spectral coefficients $MFC(k,t)$ for every signal frame t.

A disadvantage of Fourier coefficients, even after consolidation by triangle filters, is the joint correlation of neighbor frequency coefficients. Since the vocal tract is smooth, energy levels in adjacent bands tend to be correlated. To compensate this smoothing of features the inverse DFT (in fact only the cosine transform as the transformed MFC's are real-valued) is applied, which converts the set of logarithm-scaled energies to a set of cepstrum coefficients (for example, m = 12), which are largely un-correlated:

$$MFCC(k,t) = \sum_{l=0}^{M-1} log[MFC(l,t)] \cos\left[\frac{k(2l+1)\pi}{2M}\right], \quad k = 1, ..., 12. \tag{1}$$

Another disadvantage of this scheme is that noisy oscillations of the human larynx are overlayed onto the energy of basic frequency and some of its first harmonic frequencies. To reduce it a so called *liftering* of the MFCC features is finally performed [2], [3]. Let c_n be the n-th MFCC. Then its liftering is as follows:

$$c_n^{lift} = \left[1 + \frac{L}{2}\sin\left(\frac{\pi n}{L}\right)\right] c_n, \; n = 1, 2, ..., K < L, \tag{2}$$

where L is related to the feature index for the basic frequency. Usually the final number of features L is set by default to the number of triangle filters, as the on-line computation of this parameter for every consecutive frame is not feasible: (a) a variable number of features could appear for different frames, (b) although the basic frequency is related to the individual speaker, it is variable even for the same speaker, as it depends on the accentuation and emotional standing.

3 The Approach

3.1 Applying ICA for Source Separation

The Fourier coefficients obtained for given frame $FC(a, t)$ constitute a vector $\mathbf{x}(t)$ - a single (mixture) input to the ICA learning procedure (the size of this vector is N). This vector is expected to be a particular mixture of $m < N$ independent sources. For every spoken word, that we can detect in the speech sample, we get a learning set of frames: $x_i(t)(i = 1, ..., n; t = 1, ..., N)$.

The basic mixing model in ICA (without noise) is assumed. $\mathbf{x}(t)$ is a matrix of n time-varying vector signals, each of size N. $\mathbf{a}_i$ is a set of n mixing vectors (each of size m) combined to a mixing matrix $\mathbf{A}$ (every $\mathbf{a}_i$ is a single row of matrix $\mathbf{A}$). $\{\mathbf{s}_i(t)\}$ is a set of m sources - each one consists of N time samples.

After running the ICA method both unknown sources and unknown mixing coefficients are determined - on base of given sequence of observations (frames) $x_i(t)$ the vector $\mathbf{s}$ and weight matrix $\mathbf{W}$ are estimated. The sources need to be normalized and reordered, while the weights are of no importance during learning.

3.2 Matching of Source Sets

During learning we need to establish a correspondence between existing source set (the reference components) and the newly created source set for current signal frame. During this comparison a proper permutation index, the scaling and even the sign of amplitude must be adjusted [10]:

(1) The amplitudes of all components are re-scaled to the interval of $< -1, 1 >$.
(2) FOR all tested components $y_i, (i = 1, ..., n)$ DO:
FOR all reference components $s_j, (j = 1., ..., n)$ DO:
compute the mean square error of approximating s_j by y_i or by $-y_i$:

$$MSE[y_i, s_j] \text{ and } MSE[-y_i, s_j]$$

and select the better one, i.e. with lower value;
(3) All selected MSE-s are transformed into elements of a new created matrix

$$\mathbf{P} = [a_{i,j}]_{n \times n}, \text{ where } a_i = \frac{1}{\sqrt{MSE[y_i, s_j]}}.$$

(4) The error index $EI(\mathbf{P})$ is computed as:

$$\frac{1}{n}\left[\sum_{i=1}^{n}\sum_{j=1}^{n}\frac{a_{ij}}{max_i(a_{ij})} - n\right] + \frac{1}{n}\left[\sum_{j=1}^{n}\sum_{i=1}^{n}\frac{a_{ij}}{max_j(a_{ik})} - n\right].$$

The first part of above sum expresses the average error for matching a tested ICA component with one reference component, whereas the second part is equivalent

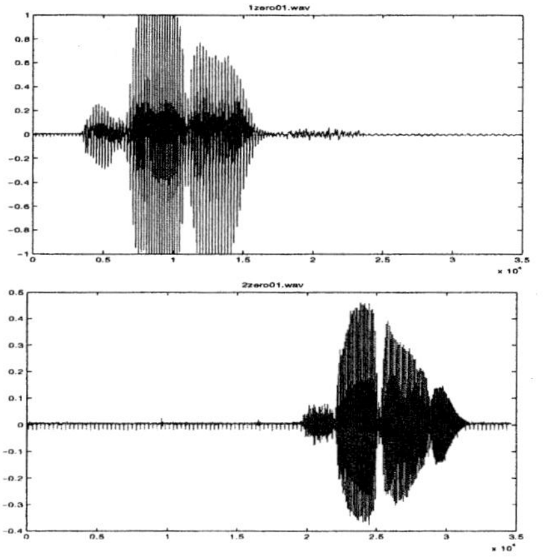

Fig. 3. Waveforms of the word "zero" pronounced by two speakers (male and female)

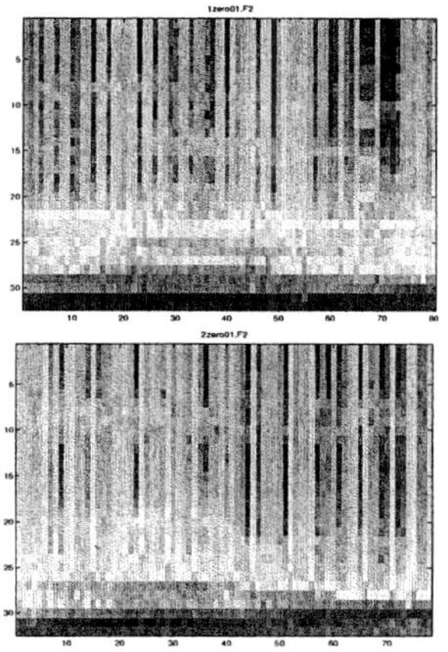

Fig. 4. The spectrograms (selected frames with sufficient energy only) for above words "zero"

to a penalty score, if a single reference component is matched with more than one tested component.

3.3 Larynx Noise Detection

Some of the sources correspond to the noisy influence of basic oscillations of the larynx. In MFCC scheme they are tried to be eliminated by the "liftering" processing. In case of our ICA scheme these "noisy" sources are detected by their continuously decreasing waveform, with its highest value at the index of 0. The remaining sources are equipped with one or several local maxima at particular frequency indices (see Fig. 3 and 5).

3.4 Feature Extraction

For every signal frame we need to determine a feature vector in the previously established ICA space determined by the selected source set. These features are equivalent to the unknown mixing coefficients of ICA sources that lead to the power spectrum vector for current frame. Hence, let us assume the matrix $\mathbf{S}$, with rows representing the reference ICA sources in frequency space $s_i(\omega)$, was established during the learning phase. One part of these sources forms the feature-relevant base $\mathbf{S_F}$ and the other part - the larynx-related part $\mathbf{S_L}$ of matrix $\mathbf{S}$. Then we estimate the unknown mixing coefficients for current window k of the speech signal as: $a_k^T = x_k(\omega)\mathbf{S}^{-1}$, where $x_k(\omega)$ is the vector of power spectra for the k-th window of speech. The final feature vector is a sub-vector of a_k corresponding to the subspace determined by $\mathbf{S_F}$.

An illustration of ICA features detected for the source set in Fig. 5 is specified in Fig.7 and 8. We observe that the coefficients $\mathbf{W}$ for different words "jeden" and "dwa" with the same ICA components are quite different, but for the same word and even different speaker - these coefficients are similar.

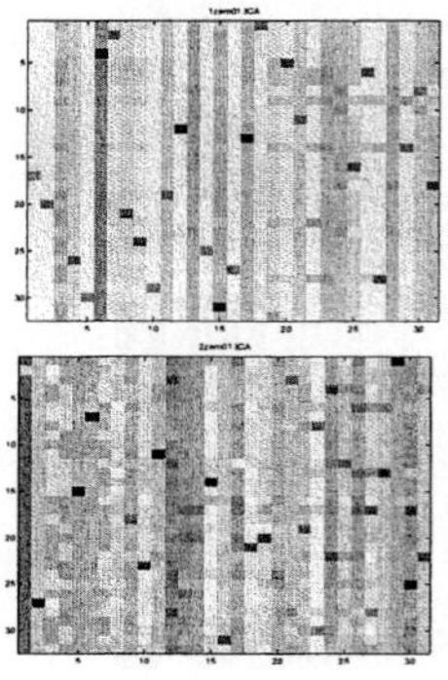

Fig. 5. The detected 31 basic vectors (one column represents one vector with 32 elements) after ICA was applied to above two spectral images

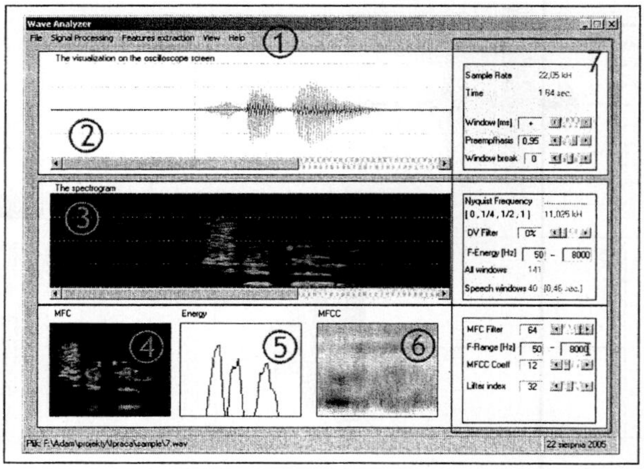

Fig. 6. The main window of our test application: (1) menu, (2) oscillogram, (3) spectrogram, (4) MFC, (5) energy, (6) MFCC or ICA, (7) analysis parameters

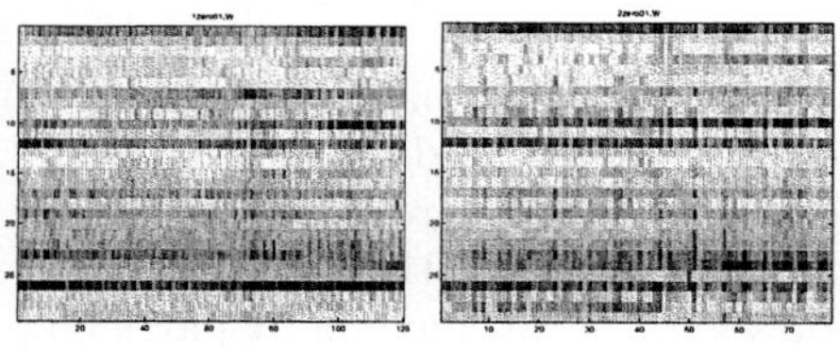

Fig. 7. The coefficients **W** (one column represents one vector of coefficients for one signal frame) for two speech samples of word "zero" from different speakers. Great similarities appear.

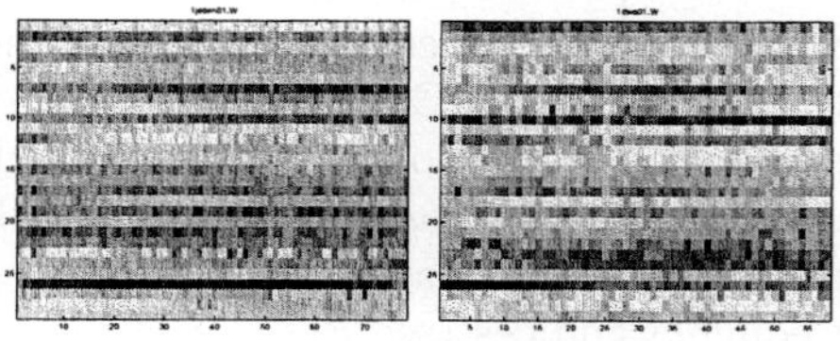

Fig. 8. The coefficients **W** for different words "jeden" (one) and "dwa" (two) from the same speaker. Large differences appear.

4 Experimental Results

Both the MFCC- and ICA-based approaches for speech frame feature detection were implemented and tested on speech signal examples, acquired with the sampling frequency of 22 kHz. Speech samples from 18 persons (both male and female) were available for testing (Fig. 6).

The MFCC and ICA features are quite stable for different samples of the same word and speaker (see Fig. 9). For different speakers a larger variability appears (Fig. 10).

Some experiments of both approaches are summarized in tables 1-2, where the EI index values were computed while matching the tested sample components

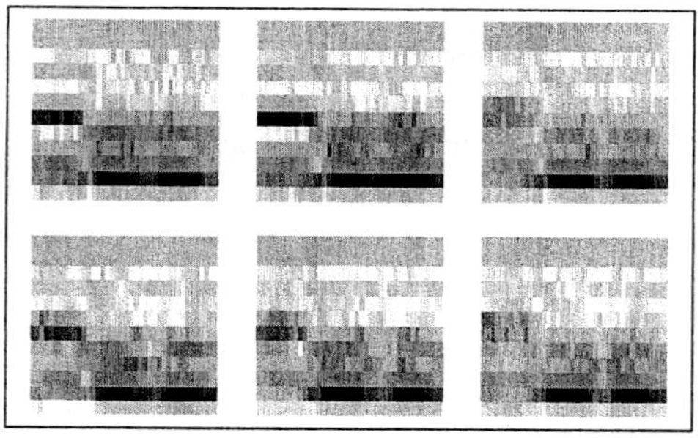

Fig. 9. MFCC features for speaker 1 and word "pusc" (release) - 6 different samples

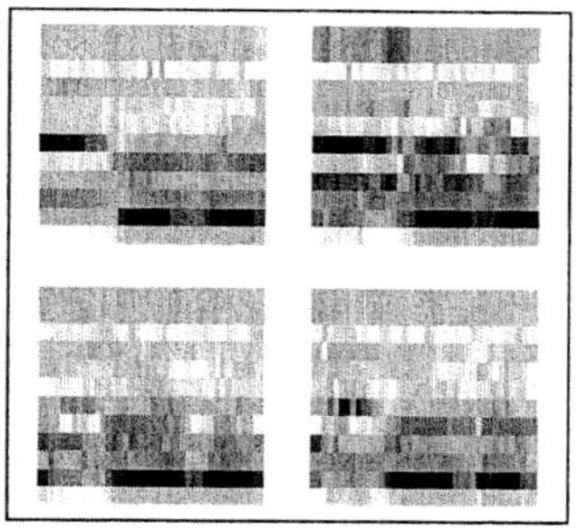

Fig. 10. MFCC features for speakers 2-4 and word "pusc" (release)

Table 1. Comparison of the error index $EI(\mathbf{P})$ for components - sources - of the same word ("zero") but for 4 different speakers (31 sources with 32 elements)

Reference / *Tested*	M1	F1	M2	F2
Male 1	6.04	4.46	5.15	3.90
Female 1	6.15	4.62	5.85	5.56
Male 2	6.21	4.47	5.13	4.70
Female 2	9.03	8.47	7.45	7.92

Table 2. Comparison of the error index for components - sources - of different words ("zero", "jeden", "dwa") but the same speaker (31 sources with 32 elements)

Reference / *Tested*	"zero"	"jeden"	"dwa"
"zero"	3.46	2.50	1.98
"jeden"	2.33	2.82	1.20
"dwa"	2.66	2.94	1.85

Table 3. The classification success rate for the MFCC- and ICA-based feature sets (20 classes with 26 learning and 12 verification samples for each class - different speakers)

Feature set / *Word*	MFCC	ICA
"zero"	66%	100%
"jeden" (one)	58%	100%
"dwa" (two)	63%	66%
"trzy" (three)	58%	100%
"cztery" (four)	58%	100%
...		
"dziewiec" (nine)	83%	100%
"start"	66%	66%
"stop"	92%	83%
"lewo" (left)	66%	91%
"prawo" (right)	66%	66%
"gora" (up)	89%	100%
"doł" (down)	75%	83%
"pusc" (release)	91%	91%
"złap" (catch)	91%	91%
"os" (axis)	83%	91%
"chwytak" (grab)	66%	100%
average	75.6 %	90.55%

with the proper reference components. From Table 1 it is evident, that the components are quite independent from the speaker. From Table 2 we conclude that ICA sources, obtained for different words of one speaker, are also similar. Hence, ICA produces a general base for speech features.

The last table 3 summarizes a comparison between MFCC features and ICA-based features. A word reference (class) was represented by an average map of all the learned feature maps for given word. We applied a simple minimum-distance classifier for the classification of feature sets, computed in both schemas - MFCC and ICA. A success was noted if the minimum distance was achieved for the proper reference word and the distance between current feature map and reference map was below half of the standard deviation of samples for given class.

5 Conclusion

We have proposed an ICA-based method for speech feature detection in a frame-based speech recognition system. A subset of sources detected by ICA provides base vectors of the feature space in the frequency domain, whereas the mixing coefficients in ICA mixing model constitute the feature vectors. The experiments show a better quality (in terms of the recognition success rate) of such features if compared to standard MFCC features.

Acknowledgment. The work reported in this paper was supported by the grant MNiI - 4T11A 003 25.

References

1. Junqua, J.-C., Haton, J.-P.: Robustness in automatic speech recognition. Kluwer Academic Publications, Boston etc. (1996)
2. Rabiner, L., Juang, B.: Fundamentals of Speech Recognition. Prentice Hall, New York (1993)
3. Grocholewski, S.: Statystyczne podstawy systemu ARM dla jezyka polskiego. Vol. 362 of "Rozprawy", Poznan University of Technology Press (2001)
4. Ding, P., Kang, X., Zhang, L.: Personal recognition using ICA. Proceedings ICONIP (2001)
5. Rosca, J., Kopfmehl, A.: Cepstrum-like ICA representations for text independent speaker recognition. Procceedings of ICA'2003, (Nara, Japan, April 2003), Publ. by NTT Kyoto (2003) 999–1004.
6. Cichocki, A., Amari, S.: Adaptive Blind Signal and Image Processing. John Wiley, Chichester, UK (2002)
7. Hyvarinen, A., Karhunen, J., Oja, E.: Independent Component Analysis. John Wiley & Sons, New York etc. (2001)
8. Duda, R.O., Hart, P.E., Stork, D.: Pattern classification. 2nd edition. John Wiley, New York (2001)
9. Amari, S., Douglas, S.C., Cichocki, A., Yang, H.Y.: Novel on-line adaptive learning algorithms for blind deconvolution using the natural gradient approach. IEEE Signal Proc. Workshop on Signal Processing Advances in Wireless Communications, (Paris, April 1977), 107–112
10. Kasprzak, W.: Adaptive computation methods in digital image sequence analysis. Vol. 127 of "Prace Naukowe - Elektronika", Warsaw University of Technology Press, Warszawa, Poland (2000)

Monaural Music Source Separation: Nonnegativity, Sparseness, and Shift-Invariance

Minje Kim and Seungjin Choi

Department of Computer Science,
Pohang University of Science and Technology,
San 31 Hyoja-dong, Nam-gu Pohang 790-784, Korea
{minjekim, seungjin}@postech.ac.kr

Abstract. In this paper we present a method for polyphonic music source separation from their monaural mixture, where the underlying assumption is that the harmonic structure of a musical instrument remains roughly the same even if it is played at various pitches and is recorded in various mixing environments. We incorporate with *nonnegativity, shift-invariance*, and *sparseness* to select representative spectral basis vectors that are used to restore music sources from their monaural mixture. Experimental results with monaural instantaneous mixture of voice/cello and monaural convolutive mixture of saxophone/viola, are shown to confirm the validity of our proposed method.

1 Introduction

The nonnegative matrix factorization (NMF) [1] or its extension such as nonnegative matrix deconvolution (NMD) [2] and sparse coding [3], was shown to be useful in polyphonic music description [4, 5], in the extraction of multiple music sound sources [2, 6], and in general sound classification [7]. Some of these methods regard each note as a source, which might be appropriate for music transcription and work for source separation in a very limited case.

In this paper we present a method for monaural polyphonic music separation, the goal of which is to restore the whole melody generated by each musical instrument from a single channel mixture of several polyphonic musical sounds. We assume that the harmonic structure of a musical instrument approximately remains the same, even if it is played at different pitches and is recorded in different environments. Different musical instruments are assumed to have different spectral characteristics (harmonic structure).

The main idea is to select a few representative spectral basis vectors in the auditory spectrogram of measurement data, assuming that there are some sections in the auditory spectrogram where only a single note from a single source appears. Rather than learning basis vectors, we select a few appropriate nonnegative basis vectors using the sparseness of spectral coefficients. These shift-invariant nonnegative basis vectors are fixed and associated encoding variables are learned by the overlapping NMF [8] which incorporates with the shift-invariant representation, in order to restore music sources. The method is related to our earlier work [9] and the generalized prior subspace analysis [10].

J. Rosca et al. (Eds.): ICA 2006, LNCS 3889, pp. 617–624, 2006.

However, the key distinction lies in a way of selecting shift-invariant basis vectors. Promising results with monaural instantaneous mixture of voice/cello and convolutive mixture of saxophone/viola, are presented to confirm the validity of our proposed method.

2 Overlapping NMF: Nonnegativity and Shift-Invariance

Nonnegative matrix factorization (NMF) is a simple but efficient factorization method for decomposing multivariate data into a linear combination of basis vectors with nonnegativity constraints for both basis and encoding matrix [1].

Given a nonnegative data matrix $\boldsymbol{V} \in \mathbb{R}^{m \times N}$ (where $V_{ij} \geq 0$), NMF seeks a factorization

$$\boldsymbol{V} \approx \boldsymbol{W}\boldsymbol{H}, \tag{1}$$

where $\boldsymbol{W} \in \mathbb{R}^{m \times n}$ $(n \leq m)$ contains nonnegative basis vectors in its columns and $\boldsymbol{H} \in \mathbb{R}^{n \times N}$ represents the nonnegative encoding variable matrix. Appropriate objective functions and associated multiplicative updating algorithms for NMF can be found in [1].

The overlapping NMF is an interesting extension of the original NMF, where transform-invariant representation and a sparseness constraint are incorporated with NMF [8]. Some of basis vectors computed by NMF could correspond to the transformed versions of a single representative basis vector. The basic idea of the overlapping NMF is to find transformation-invariant basis vectors such that fewer number of basis vectors could reconstruct observed data. Given a set of transformation matrices, $\mathcal{T} = \left\{\boldsymbol{T}^{(1)}, \boldsymbol{T}^{(2)}, \ldots, \boldsymbol{T}^{(K)}\right\}$, the overlapping NMF finds a nonnegative basis matrix $\boldsymbol{W}$ and a set of nonnegative encoding matrix $\left\{\boldsymbol{H}^{(k)}\right\}$ (for $k = 1, \ldots, K$) which minimizes

$$\mathcal{J}(\boldsymbol{W}, \boldsymbol{H}) = \frac{1}{2} \left\| \boldsymbol{V} - \sum_{k=1}^{K} \boldsymbol{T}^{(k)} \boldsymbol{W} \boldsymbol{H}^{(k)} \right\|_F^2, \tag{2}$$

where $\|\cdot\|_F$ represents Frobenious norm. The multiplicative updating rules for the overlapping NMF were derived in [8], which are summarized below.

Algorithm Outline: Overlapping NMF [8]

Step 1. Calculate the reconstruction: $\boldsymbol{R} = \sum_{k=1}^{K} \boldsymbol{T}^{(k)} \boldsymbol{W} \boldsymbol{H}^{(k)}$.
Step 2. Update the encoding matrix by

$$\boldsymbol{H}^{(k)} \leftarrow \boldsymbol{H}^{(k)} \odot \frac{\boldsymbol{W}^{\top} \left[\boldsymbol{T}^{(k)}\right]^{\top} \boldsymbol{V}}{\boldsymbol{W}^{\top} \left[\boldsymbol{T}^{(k)}\right]^{\top} \boldsymbol{R}}, \quad k = 1, \ldots, K, \tag{3}$$

where $\odot$ denotes the Hadamard product and the division is carried out in an element-wise fashion.

Step 3. Calculate the reconstruction $\boldsymbol{R}$ again using the encoding matrix $\boldsymbol{H}^{(k)}$ updated in Step 2, as in Step 1.

Step 4. Update the basis matrix by

$$\boldsymbol{W} \leftarrow \boldsymbol{W} \odot \frac{\sum_{k=1}^{K} \left[\boldsymbol{T}^{(k)}\right]^{\top} \boldsymbol{V} \left[\boldsymbol{H}^{(k)}\right]^{\top}}{\sum_{k=1}^{K} \left[\boldsymbol{T}^{(k)}\right]^{\top} \boldsymbol{R} \left[\boldsymbol{H}^{(k)}\right]^{\top}}. \tag{4}$$

3 Spectral Basis Selection: Sparseness

The goal of spectral basis selection is to choose R representative vectors $\boldsymbol{V}_r = [\boldsymbol{v}_{r_1} \cdots \boldsymbol{v}_{r_R}]$ (R is the number of music sources) from $\boldsymbol{V} = [\boldsymbol{v}_1 \cdots \boldsymbol{v}_N]$ where $\boldsymbol{V}$ is the data matrix associated with the spectrogram of mixed sound. Each column vector $\boldsymbol{v}_t$ corresponds to the power spectrum of the mixed sound at time $t = 1, \ldots, N$. Selected representative vectors are fixed as basis vectors that are used to learn an associated encoding matrix set through the overlapping NMF with sparseness constraint, in order to restore unmixed musical sound.

Our spectral basis selection method is based on the assumption that there are some sections where only a single note from a single source appears. In the spectrogram of mixed sound, solo sections are searched partly through the sparseness value of $\boldsymbol{v}_t$ over time. Our earlier work can be found in [9].

Fig. 1 shows the schematic diagram of the spectral basis selection method, consisting of two parts. The first part is to select several candidate vectors $\boldsymbol{V}_c = [\boldsymbol{v}_{c_1}\, \boldsymbol{v}_{c_2} \cdots \boldsymbol{v}_{c_K}]$ from $\boldsymbol{V}$ using a sparseness measure and a clustering-elimination method. The second part involves determining representative basis vectors from candidate vectors, through the overlapping NMF. More detailed description is summarized below.

Part 1

1. **Sparseness calculation.** We calculate the sparseness value for input vectors $\boldsymbol{v}_t$ for $t = 1, \ldots, N$, using the measure in [11],

$$\xi_t = \text{sparseness}(\boldsymbol{v}_t) = \frac{\sqrt{m} - (\sum_i |v_{it}|)/\sqrt{\sum_i v_{it}^2}}{\sqrt{m} - 1}, \tag{5}$$

where v_{it} is the ith element of the m-dimensional vector $\boldsymbol{v}_t$.

2. **Normalization.** We normalize input vectors $\boldsymbol{v}_t$ for $t = 1, \ldots, N$ such that each vector has unit Euclidean norm,

$$\boldsymbol{v}_t \leftarrow \frac{\boldsymbol{v}_t}{\|\boldsymbol{v}_t\|}. \tag{6}$$

3. **Alignment.** We calculate the index $f_i = t^*$ which involves the largest sparseness value among $\{\xi_t\}_{t=1}^{N}$, i.e,

$$t^* = \arg\max_{1 \le t \le N} \xi_t. \tag{7}$$

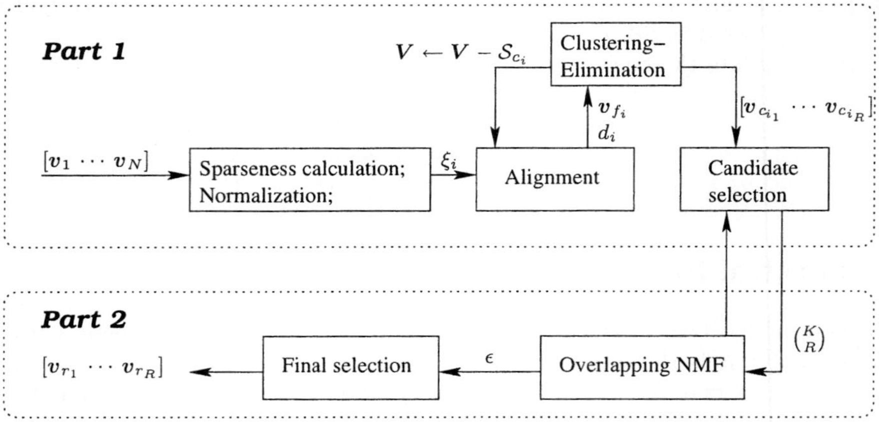

Fig. 1. Schematic diagram of our spectral basis selection method, is shown, where 'Part 1' involves the selection of candidate vectors and 'Part 2' determines a few representative spectral basis vectors from candidate vectors found in 'Part 1'

The vector v_{f_i} associated with the index $f_i = t^*$, is referred to as a *foundation vector* that has the largest sparseness value among $\{v_t\}$. Then we align each vector v_j in L remaining input vectors (initially $L = N$ but L represents the number of remaining vectors after the clustering-elimination procedure in step 4) with respect to the current foundation vector v_{f_i} such that the Euclidean distance between v_{f_i} and vertically shift-up or -down version of v_j, is minimized. In other words, vectors v_j are vertically shifted-up or -down such that their shifted version provides the minimal Euclidean distance from the foundation vector v_{f_i}.

4. **Clustering-Elimination.** The goal of the clustering-elimination step is to eliminate vectors belonging to the cluster where the foundation vector is contained, since those vectors are regarded as redundant vectors. To this end, we first apply the k-means clustering method to dichotomize the aligned vectors (including the foundation vector), leading to two groups $\mathcal{S}_{c_i}$ and $\bar{\mathcal{S}}_{c_i}$. The cluster containing the foundation vectors, $\mathcal{S}_{c_i}$, is further grouped into R sub-clusters, producing $\{v_{c_{i_1}}, \ldots, v_{c_{i_R}}\}$ that is a collection of mean vectors of R sub-clusters.
5. **Candidate selection.** Add the mean vector of the cluster $\mathcal{S}_{c_i}$ to the candidate set.
6. **Repeat.** Repeat steps 3-5 with data excluding vectors in $\mathcal{S}_{c_i}$, i.e, $V - \mathcal{S}_{c_i}$, until we choose a pre-specified number of candidate vectors or there is no remaining input vector.

Part 2

This second part involves determining the final representative spectral basis vectors $\{v_{r_1}, \ldots, v_{r_R}\}$ from $K \geq R$ candidate vectors $\{v_{c_1}, \ldots, v_{c_K}\}$ (where K

is the integral multiples of R, depending on the number of loops in the clustering-elimination) found in the first part.

1. **Overlapping NMF.** Repeat the following step for all possible $\binom{K}{R}$ combination. Construct a small set of input vectors $\widetilde{\boldsymbol{V}}$ by random sampling and treat them at input vectors for the overlapping NMF. Choose R candidate vectors from $\{\boldsymbol{v}_{c_1}, \ldots, \boldsymbol{v}_{c_K}\}$ and fix them (denoted by $\widetilde{\boldsymbol{W}}$) as basis vectors. Run the overlapping NMF with these $\widetilde{\boldsymbol{V}}$ and $\widetilde{\boldsymbol{W}}$ to calculate the reconstruction error.
2. **Final selection.** Choose spectral basis vectors that give the lowest reconstruction error.

4 Numerical Experiments

We present two simulation results for monaural instantaneous mixtures of voice and cello and monaural convolutive mixtures of saxophone and viola. We apply our spectral basis selection method with the overlapping NMF to these two data sets transformed to auditory spectrograms using the NSL toolbox [12]. Experimental results are shown in Fig. 2 and 3 where figure captions describe detailed results. Note that the mixture in Fig. 3 (c) is a convolutive mixture and

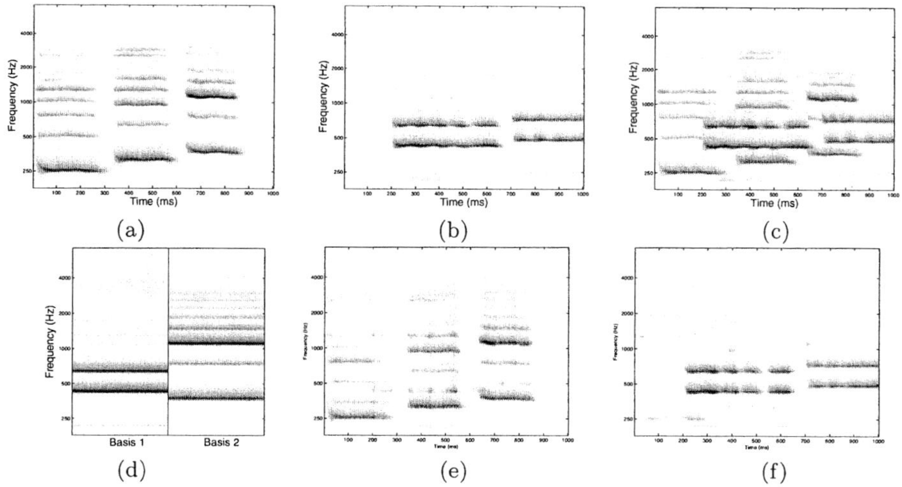

Fig. 2. Auditory spectrograms of original sound of /*ah*/ voice and a single string of a cello are shown in (a) and (b), respectively. Horizontal bars reflect the harmonic structure. One can see that every note is the vertically-shifted version of each other if their musical instrument sources are the same. Monaural mixture of voice and cello is shown in (c) and final two representative spectral basis vectors in (d) which give the smallest reconstruction error in the overlapping NMF are selected by our algorithm in Fig. 1. Each of these two basis vectors is a representative one for voice and a string of cello. Unmixed sound is shown in (e) and (f) for voice and cello, respectively.

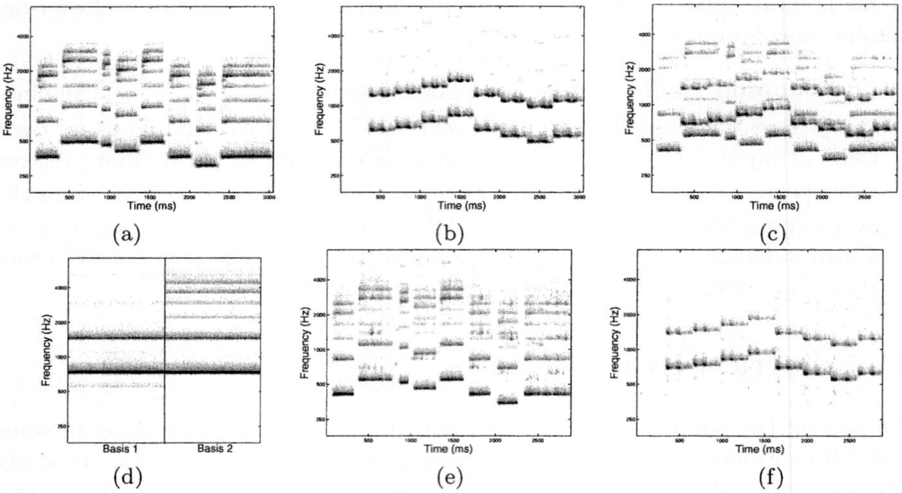

Fig. 3. Auditory spectrograms of original sound of saxophone and viola are shown in (a) and (b), respectively. Every note is artificially generated by changing the frequency of a real sample sound, so that the spectral character of each instrument is constant in all the variations of notes. We mixed these two signals by convolving them with two impulse response signals measured in a studio environment (reverberation time is about 150ms and the frequency response makes a peak at around 27Hz). The monaural convolutive mixture is shown in (c) and finally selected two representative spectral basis vectors are in (d). Unmixed sound is shown in (e) and (f) for saxophone and viola, respectively.

we can apply our framework even in that case without any modification if the reverberation time is not too long.

Fig. 4 shows the reusability of our obtained spectral basis vectors. The mixture in Fig. 4 (c) is another part of the same song used in Fig. 3. In this example, we do not have to find out the spectral basis vectors of saxophone and viola again, but can simply reuse the previous results of Fig. 3. Note that if some input data do not satisfy the horizontal sparseness, which means that there is no section occupied by only one instrument, our spectral basis selection method will fail in this case. However we can attack this problem by reusing the previously obtained spectral basis vectors of the same source instruments. Audio demo can be found in `http://home.postech.ac.kr/~minjekim/demo.php`.

The set of transformation matrices, $\mathcal{T}$, that we used, is

$$\mathcal{T} = \left\{ \boldsymbol{T}^{(k)} \,\middle|\, \boldsymbol{T}^{(k)} = \overset{k-m}{\overrightarrow{\boldsymbol{I}}}, \quad 1 \leq k \leq 2m-1 \right\}, \tag{8}$$

where $\boldsymbol{I} \in \mathbb{R}^{m \times m}$ is the identity matrix and $\overset{j}{\overrightarrow{\boldsymbol{I}}}$ leads to the shift-up or shift-down of row vectors of $\boldsymbol{I}$ by j, if j is positive or negative, respectively. After shift-up or -down, empty elements are zero-padded.

For the case where $m = 3$, $\boldsymbol{T}^{(2)}$ and $\boldsymbol{T}^{(5)}$ (they means that $k = 2$ and $k = 5$) are defined as

$$\boldsymbol{T}^{(2)} = \overset{\overset{2-3}{\longmapsto}}{\boldsymbol{I}} = \begin{bmatrix} 0 & 0 & 0 \\ 1 & 0 & 0 \\ 0 & 1 & 0 \end{bmatrix}, \quad \boldsymbol{T}^{(5)} = \overset{\overset{5-3}{\longmapsto}}{\boldsymbol{I}} = \begin{bmatrix} 0 & 0 & 1 \\ 0 & 0 & 0 \\ 0 & 0 & 0 \end{bmatrix}. \tag{9}$$

Multiplying a vector by these transformation matrices, leads to a set of vertically-shifted vectors.

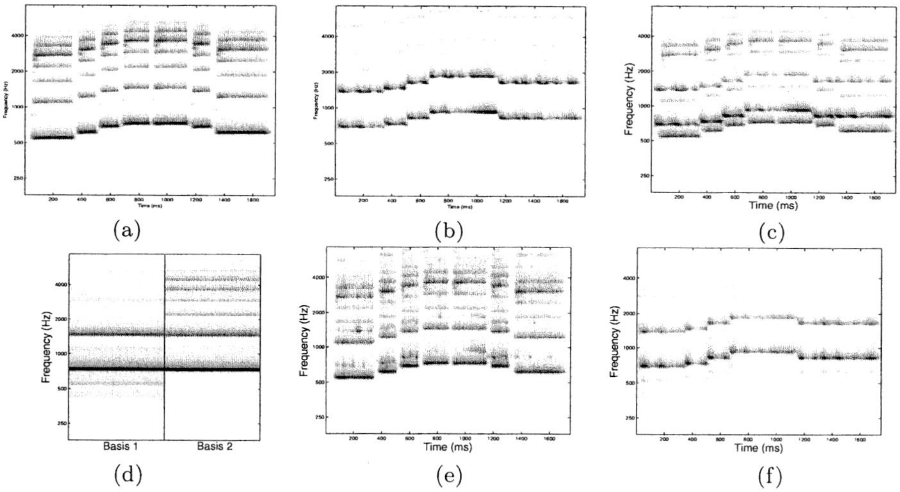

Fig. 4. These figures show the reusability of spectral basis vectors. Auditory spectrograms of original sound of saxophone and viola are shown in (a) and (b), respectively. Every note is generated in the same manner of Fig. 3 but the melody is totally different from it since this is another part of the same song. The mixing process is also the same with the previous experiment. The monaural convolutive mixture is shown in (c). Instead of finding out representative basis vectors, we reused the basis vectors (d) found in previous example. Unmixed sound is shown in (e) and (f) for saxophone and viola, respectively.

5 Discussions

We have presented a method of spectral basis selection for monaural music source separation, where we incorporated with the harmonics, sparseness, clustering, and the overlapping NMF. Rather than learning spectral basis vectors from the data, our approach is to select a few representative spectral vectors among given data and fix them as basis vectors to learn associated encoding variables through the overlapping NMF, in order to restore unmixed sound. The success of our approach lies in the two assumptions. The one is that the distinguished timbre of a given musical instrument can be expressed by a transform-invariant time-frequency representation, even though their pitches are varying. The other is that there is solo sections in a musical sound where the contribution of each source instrument appears. Our experimental results showed that the proposed methods are reasonable in both instantaneous and convolutive mixture cases.

Acknowledgments. This work was supported by ITEP Brain Neuroinformatics Program and ITRC CMEST.

References

1. Lee, D.D., Seung, H.S.: Learning the parts of objects by non-negative matrix factorization. Nature **401** (1999) 788–791
2. Smaragdis, P.: Non-negative matrix factor deconvolution: Extraction of multiple sound sources from monophonic inputs. In: Proc. Int'l Conf. Independent Component Analysis and Blind Signal Separation, Granada, Spain (2004) 494–499
3. Plumbley, M.D., Abdallah, S.A., Bello, J.P., Davies, M.E., Monti, G., Sandler, M.B.: Automatic transcription and audio source separation. Cybernetics and Systems (2002) 603–627
4. Smaragdis, P., Brown, J.C.: Non-negative matrix factorization for polyphonic music transcription. In: Proc. IEEE Workshop on Applications of Signal Processing to Audio and Acoustics, New Paltz, NY (2003) 177–180
5. Abdallah, S.A., Plumbley, M.D.: Polyphonic music transcription by non-negative sparse coding of power spectra. In: Proc. Int'l Conf. Music Information Retrieval, Barcelona, Spain (2004) 318–325
6. Helén, M., Virtanin, T.: Separation of drums from polyphonic music using non-negative matrix factorization and support vector machine. In: Proc. European Signal Processing Conference, Antalaya, Turkey (2005)
7. Cho, Y.C., Choi, S.: Nonnegative features of spectro-temporal sounds for classfication. Pattern Recognition Letters **26** (2005) 1327–1336
8. Eggert, J., Wersing, H., Körner, E.: Transformation-invariant representation and NMF. In: Proc. Int'l Joint Conf. Neural Networks. (2004)
9. Kim, M., Choi, S.: On spectral basis selection for single channel polyphonic music separation. In: Proc. Int'l Conf. Artificial Neural Networks. Volume 2., Warsaw, Poland, Springer (2005) 157–162
10. FitzGerald, D., Cranitch, M., Coyle, E.: Generalised prior subspace analysis for polyphonic pitch transcription. In: Proc. Int'l Conf. Digital Audio Effects. (2005)
11. Hoyer, P.O.: Non-negative matrix factorization with sparseness constraints. Journal of Machine Learning Research **5** (2004) 1457–1469
12. Ru, P., Chi, T., Shamma, S.: NSL Toolbox (1997)

Complex FastIVA: A Robust Maximum Likelihood Approach of MICA for Convolutive BSS

Intae Lee, Taesu Kim, and Te-Won Lee

Institute for Neural Computation, University of California,
San Diego 9500 Gilman Drive, La Jolla, CA 92093
{intelli, taesu, tewon}@ucsd.edu

Abstract. We tackle the frequency-domain blind source separation problem in a way to avoid permutation correction. By exploiting the facts that the frequency components of a signal have some dependency and that the mixing of sources is restricted to each frequency bin, we apply the concept of multidimensional independent component analysis to the problem and propose a new algorithm that separates independent groups of dependent source components. We introduce general entropic contrast functions for this analysis and a corresponding likelihood function with a multidimensional prior that models the dependent frequency components. We assume circularity for the complex variables and derive a fast algorithm by applying Newton's method learning rule. The algorithm separates mixed sources even in very challenging acoustic settings.

1 Introduction

In order to deal with the problem of separating acoustic signals, researchers have proposed algorithms that extend to convolutive mixtures in both the time domain and the frequency domain to handle reverberation and time delayed mixing of the sources. Dealing with the signals in the frequency domain has the advantage of increased performance mostly due to the fact that it can better handle longer filter lengths and that the convolution problem is reduced to a complex instantaneous mixing problem for each frequency bin. However, a permutation correction problem arises. Previous solutions match the permutation by direction of arrival estimation [1], inter-frequency correlations of signal envelopes [2], or the combination of the two [3]. Although these methods provide a good intuitive solution, they show limited robustness.

We take a fundamentally different approach to this problem by assuming that the source signal of interest has certain dependency in the frequency domain that can be modeled in a multidimensional prior. Here, instead of running independent component analysis (ICA) algorithms in each frequency bin and correcting the permutation, we extract the originally dependent sources together as a group using the multidimensional prior. This model is an extension of maximum likelihood approach to the multidimensional ICA (MICA) [4] and we will call it

J. Rosca et al. (Eds.): ICA 2006, LNCS 3889, pp. 625–632, 2006.

independent vector analysis (IVA). Although, in the analysis, there is no consideration of how to separate the mixture, if any, of dependent sources, the analysis can still be applied to the frequency domain blind source separation (BSS) problem since mixing is restricted to each frequency bin and hence there is no mixing between the dependent inter-frequency components. Experiments with real recordings demonstrate the effectiveness of the new algorithm that uses a proper multidimensional prior.

In this paper, because of the limited number of pages that is allowed, we concentrate more on the analysis, algorithm derivation, and discussion than on the simulation results.

2 Independent Vector Analysis

As ICA can be represented by the mutual information of the outputs $\mathbf{y}$, $I(\mathbf{y})$, as its contrast function, IVA or MICA can be represented by

$$D(Pr(\mathbf{y})||\prod_i Pr(\mathbf{y}_i)) \tag{1}$$

as its contrast function where $D(\cdot||\cdot)$ is the KL divergence and $Pr(\mathbf{y}_i)$ is the marginal distribution of $\mathbf{y}_i$ such that each $\mathbf{y}_i$ becomes the group of dependent components of interest when the contrast is minimized to 0. Note that this is a valid contrast function since the contrast is minimized to 0 if and only if $\mathbf{y}_i$'s are independent of each other $\left(Pr(\mathbf{y}) = \prod_i Pr(\mathbf{y}_i)\right)$.

2.1 Contrasts of IVA for White Data

For the sake of simplicity, we assume zero-mean data $\mathbf{x}$ and process $\mathbf{y} = \mathbf{Wx}$ to be zero-mean and white. This is a common task in ICA algorithms for increasing the learning speed, and is done by prewhitening the zero-mean data $\mathbf{x}$ and by constraining the rows of the unmixing matrix $\mathbf{W}$ to be orthonormalized ($\mathbf{WW}^H = \mathbf{I}$). The prewhitening should not be always done for IVA algorithms since there is no guarantee that the dependent source components in each group are uncorrelated unless known to be so in advance. However, the uncorrelatedness can mostly be assumed for frequency-domain components of a natural signal, and here we restrict our analysis to the assumption that the dependent sources are uncorrelated.

By analysis taken with care, we show that the contrasts for IVA closely resemble the classic entropic contrasts for ICA. The given contrast (1) can be looked at and analyzed in information geometry, as was done for ICA (Cardoso, Ch.4 of [5]). We define the independent vector manifold as the exponential family distribution with a constant base measure and all possible features except for any cross-term between any component of $\mathbf{y}_i$ and any component of $\mathbf{y}_j$ ($i \neq j$) such that the vectors $\mathbf{y}_i$ and $\mathbf{y}_j$ are independent of each other. We also introduce the Gaussian manifold which is the exponential family distribution with all first and second order features and a constant base measure. We denote the Gaussian manifold and independent vector manifold as $\mathcal{G}$ and $\mathcal{P}$ respectively and denote $Pr(\mathbf{y}^N)$ as the information projection of $Pr(\mathbf{y})$ onto $\mathcal{G}$ such that $Pr(\mathbf{y}^N) \triangleq \arg\min_{p\in\mathcal{G}} D(Pr(\mathbf{y})||p)$. Note that $\prod_i Pr(\mathbf{y}_i)$ and $\prod_i Pr(\mathbf{y}_i^N)$ are the information projections of $Pr(\mathbf{y})$ and $\prod_i Pr(\mathbf{y}_i)$ onto $\mathcal{P}$ and $\mathcal{G}$ respectively. Since we constrain $\mathbf{y}$ to be zero-mean and white, we will consider the subspace of zero-mean white distributions. In this subspace, the Gaussian manifold shrinks such that not only it becomes a single point because the zero-mean whiteness constraint fixes all parameters of the first and second order features right away, but also it gets included in the independent vector manifold ($\mathcal{G} \subset \mathcal{P}$) since uncorrelatedness of a Gaussian distribution implies independence. Hence all information projections of probability distributions to the Gaussian manifold are the same distribution and thus $Pr(\mathbf{y}^N) = \prod_i Pr(\mathbf{y}_i^N)$. Now, by applying the Pythagorean relation in information geometry and introducing negentropy $N(\mathbf{y}) \triangleq D(Pr(\mathbf{y})||Pr(\mathbf{y}^N))$, we acquire

$$N(\mathbf{y}) = D\big(Pr(\mathbf{y})||\prod_i Pr(\mathbf{y}_i)\big) + \sum_i N(\mathbf{y}_i). \tag{2}$$

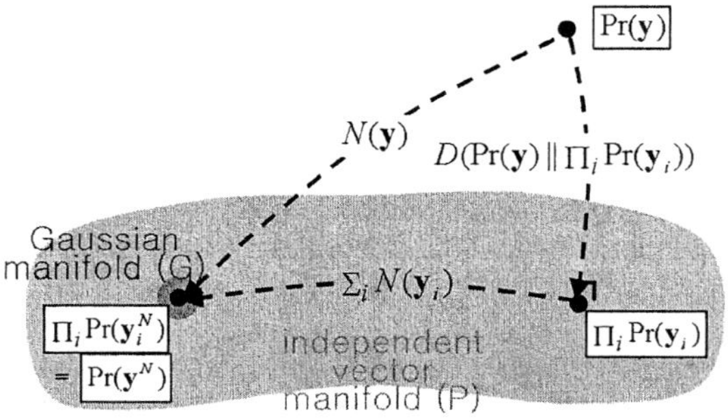

Fig. 1. Information geometry: Entropic contrasts of IVA for zero-mean white data

These relations are drawn in Fig.1. Since $N(\mathbf{y})$ remains constant for any invertible transformation, minimizing $D\big(Pr(\mathbf{y})||\prod_i Pr(\mathbf{y}_i)\big)$ is equivalent to maximizing $\sum_i N(\mathbf{y}_i)$.

Furthermore, from the following equations,

$$\sum_i N(\mathbf{y}_i) = \sum_i \big(\underbrace{H(\mathbf{y}_i^N)}_{\text{const.}} - H(\mathbf{y}_i)\big) = \sum_i E_{\mathbf{y}_i}\big[\log\big(Pr(\mathbf{y}_i)\big)\big] + \text{const.}, \tag{3}$$

since $\mathbf{y}$ is zero-mean and white, we know that $Pr(\mathbf{y}_i^N)$ is always an independent multivariate zero-mean Gaussian distribution such that $H(\mathbf{y}_i^N)$ is a constant. Thus maximizing the sum of negentropies is equivalent to minimizing the sum of entropies.

In spite of the validity of these entropic contrasts, there are some problems in using them directly such as the difficulty in obtaining the true distribution $Pr(\mathbf{y}_i)$'s from finite data size and the heavy computational load. One way to tackle these problems is to exploit prior information of the source and to use empirical distribution, such as using the (normalized) log-likelihood function,

$$\sum_i \hat{E}\big[\log\big(\hat{P}_{\mathbf{s}_i}(\mathbf{y}_i)\big)\big]\Big(= \hat{E}\big[\log\big(\hat{P}_{\mathbf{s}}(\mathbf{y})\big)\big] = \hat{E}\big[\log\big(\hat{P}_{\mathbf{W}^{-1}\mathbf{s}}(\mathbf{x})\big)\big]\Big), \quad (4)$$

where $\hat{E}[\cdots] = \frac{1}{N}\sum_n \cdots$, and $\hat{P}_{\mathbf{s}_i}(\cdot)$ denotes the estimated source prior. Note that this is an equivalent form to the entropic contrast in the rightmost expression of (3) given the source prior in (4) is exact and there are large enough number of data samples. It is known that proper target distributions, even if not exact, are available for good separation results (See Ch.4 in [5]).

3 A Fast Learning Algorithm for Frequency-Domain BSS

In order to apply the discussed model to the frequency-domain BSS problem, we have to clarify the management of complex-valued variables and have to find a valid source target for the frequency-domain components.

3.1 Complex Variables: Optimization and Assumptions

In the discussion of probability distributions, contrasts, and contrast optimization, the question of how to deal with complex-valued variables arises because real-valued contrast functions of complex variables are not analytic. There are standard ways to handle the problem of optimizing such functions. The first is to write everything in terms of the real quantities using $s = u + iv$ and then use the standard tools. An alternative is to follow the lines of [6], which is equivalent to the first method described, but notationally cleaner. Basically, these methods are regarding the functions as real valued functions of real variables which means that a complex variable is manipulated as separated two dimensional real variables and hence, the function gets optimized according to the real variables separately. As the relation between the separated real variables, we assume circularity in the source variables such that $E[\mathbf{ss}^T] = \mathbf{O}$. Hence for zero-mean white $\mathbf{y}$ we have, as in [7], $\hat{E}[\mathbf{yy}^H] = \mathbf{I}$, and $\hat{E}[\mathbf{yy}^T] = \mathbf{O}$.

3.2 Source Prior: Symmetric Exponential Norm Distribution

In roughly modeling an acoustic signal in the frequency domain, whiteness, circularity aussumption, and sparseness of the norm were taken into account. Here, we introduce a symmetric exponential norm distribution (SEND), which is meant to be a joint distribution of real-valued variables and can be regarded as an extension of double exponential distirbution, a.k.a. Laplace distribution,

$$\hat{P}_{\mathbf{s}_i}(\mathbf{s}_i) \propto \frac{e^{-\sqrt{\frac{2}{F}}\sqrt{\sum_f |s_i^f|^2}}}{\sqrt{\sum_f |s_i^f|^2}^{\,2F-1}} \quad (5)$$

where the superscript f corresponds to each frequency component and F denotes the total number of f's. This distribution was derived such that it has spherical contour line and the norm of the variables follows an exponential distribution. This distribution shrinks to the Laplace Distribution if the variables are real-valued and the dimension is 1. The variance was adjusted for the variables to be white. Note that this distribution satisfies whiteness, sparseness, circularity, and also variance dependency.

After replacing $\hat{P}_{\mathbf{s}_i}(\cdot)$ in the contrast $\hat{E}\left[\log\left(\hat{P}_{\mathbf{s}_i}(\mathbf{y}_i)\right)\right]$ with (5), the contrast becomes

$$\sum_i \hat{E}[G(\sum_f |y_i^f|^2)] = \sum_i \hat{E}[G(\sum_f |\mathbf{w}_i^{fH}\mathbf{x}^f|^2)] \tag{6}$$

where $G(x) = \sqrt{\frac{2x}{F}} + (F - \frac{1}{2})\log x$ with the constraint that $\mathbf{w}_i^f$'s are normalized. Note that the nonlinear function $G(\cdot)$ corresponds to the source prior with the relation of

$$G(\sum_f |y_i^f|^2) \equiv -\log \hat{P}_{\mathbf{s}_i}(\mathbf{y}_i), \tag{7}$$

and also the contrast has changed its sign to be minus likelihood. By using the Lagrange multiplier λ_i^f's, we can have the normalization constraint together in the contrast function, which results in

$$\sum_i \left[\hat{E}[G(\sum_f |\mathbf{w}_i^{fH}\mathbf{x}^f|^2)] - \sum_f \lambda_i^f(\mathbf{w}_i^{fH}\mathbf{w}_i^f - 1)\right]. \tag{8}$$

3.3 Contrast Optimization: Newton's Method

Once the contrast function is selected, we can derive our separation algorithm by choosing the optimization method. ICA algorithms using Newton's method for contrast optimization are called FastICA algorithms. They can avoid choosing proper learning rate and have the advantage of fast convergence speed when compared to other gradient-descent type methods. While applying Newton's method to our contrast function, as discussed earlier, we follow the standard way of dealing with the optimization of real-valued functions of complex variables. However, we avoid introducing new variables such as $[\mathrm{Re}(\mathbf{w}); \mathrm{Im}(\mathbf{w})]$ or $[\mathbf{w}; \mathbf{w}^*]$ such that the derivation is easier to follow.

For this, we start from the Taylor series expansion of a real-valued function $f(\mathbf{w})$, where $\mathbf{w}$ is complex. Using the definitions for complex derivatives and complex gradients in [6], it can be shown that the Taylor series expansion of $f(\mathbf{w})$ up to the second order is

$$\begin{aligned} f(\mathbf{w}) \approx\ & f(\mathbf{w}_o) + \frac{\partial f(\mathbf{w}_o)}{\partial \mathbf{w}^T}(\mathbf{w} - \mathbf{w}_o) + \frac{\partial f(\mathbf{w}_o)}{\partial \mathbf{w}^H}(\mathbf{w} - \mathbf{w}_o)^* \\ & + \frac{1}{2}(\mathbf{w} - \mathbf{w}_o)^T \frac{\partial^2 f(\mathbf{w}_o)}{\partial \mathbf{w} \partial \mathbf{w}^T}(\mathbf{w} - \mathbf{w}_o) + \frac{1}{2}(\mathbf{w} - \mathbf{w}_o)^H \frac{\partial^2 f(\mathbf{w}_o)}{\partial \mathbf{w}^* \partial \mathbf{w}^H}(\mathbf{w} - \mathbf{w}_o)^* \\ & + (\mathbf{w} - \mathbf{w}_o)^H \frac{\partial^2 f(\mathbf{w}_o)}{\partial \mathbf{w}^* \partial \mathbf{w}^T}(\mathbf{w} - \mathbf{w}_o). \end{aligned} \tag{9}$$

This can be verified by defining a vector $[\mathbf{w};\mathbf{w}^*]$ and applying the Taylor series expansion form as in [6]. The $\mathbf{w}$ that optimizes the function $f(\mathbf{w})$ will set the gradient of it, $\frac{\partial f(\mathbf{w})}{\partial \mathbf{w}^*}$, to zero and hence from (9),

$$\frac{\partial f(\mathbf{w})}{\partial \mathbf{w}^*} \approx \frac{\partial f(\mathbf{w}_o)}{\partial \mathbf{w}^*} + \frac{\partial^2 f(\mathbf{w}_o)}{\partial \mathbf{w}^* \partial \mathbf{w}^T}(\mathbf{w}-\mathbf{w}_o) + \frac{\partial^2 f(\mathbf{w}_o)}{\partial \mathbf{w}^* \partial \mathbf{w}^H}(\mathbf{w}-\mathbf{w}_o)^* \equiv \mathbf{O}. \quad (10)$$

Note that this equation is equivalent to the Newton step equation (37) in [6]. Using this, we can derive a fast algorithm with the given contrast function in (8). Letting $\mathbf{w} \equiv \mathbf{w}_i^f$ and setting $f(\mathbf{w}_i^f) \equiv \hat{E}\Big[G(\sum_f |\mathbf{w}_i^{fH}\mathbf{x}^f|^2)\Big] - \sum_f \lambda_i^f(\mathbf{w}_i^{fH}\mathbf{w}_i^f - 1)$, the derivative terms in (10) become

$$\frac{\partial f(\mathbf{w}_{i,o}^f)}{\partial \mathbf{w}_i^{f*}} = \hat{E}\Big[y_{i,o}^{f*}\, G'(\sum_f |y_{i,o}^f|^2)\mathbf{x}^f\Big] - \lambda_i^f \mathbf{w}_{i,o}^f \quad (11)$$

$$\frac{\partial^2 f(\mathbf{w}_{i,o}^f)}{\partial \mathbf{w}_i^{f*}\partial \mathbf{w}_i^{fT}} = \hat{E}\Big[\Big(G'(\sum_f |y_{i,o}^f|^2) + |y_{i,o}^f|^2\, G''(\sum_f |y_{i,o}^f|^2)\Big)\mathbf{x}^f\mathbf{x}^{fH}\Big] - \lambda_i^f \mathbf{I} \quad (12)$$

$$\approx \hat{E}\Big[G'(\sum_f |y_{i,o}^f|^2) + |y_{i,o}^f|^2\, G''(\sum_f |y_{i,o}^f|^2)\Big]\hat{E}\big[\mathbf{x}^f\mathbf{x}^{fH}\big] - \lambda_i^f \mathbf{I} \quad (13)$$

$$\approx \Big(\hat{E}\Big[G'(\sum_f |y_{i,o}^f|^2) + |y_{i,o}^f|^2\, G''(\sum_f |y_{i,o}^f|^2)\Big] - \lambda_i^f\Big)\mathbf{I}, \quad (14)$$

$$\frac{\partial^2 f(\mathbf{w}_{i,o}^f)}{\partial \mathbf{w}_i^{f*}\partial \mathbf{w}_i^{fH}} = \hat{E}\Big[(y_{i,o}^{f*})^2\, G''(\sum_f |y_{i,o}^f|^2)\mathbf{x}^f\mathbf{x}^{fT}\Big] \quad (15)$$

$$\approx \hat{E}\Big[(y_{i,o}^{f*})^2\, G''(\sum_f |y_{i,o}^f|^2)\Big]\,\hat{E}\big[\mathbf{x}^f\mathbf{x}^{fT}\big] \quad (16)$$

$$\approx \mathbf{O}, \quad (17)$$

where $y_{i,o}^f$ denotes $\mathbf{w}_{i,o}^{fH}\mathbf{x}^f$ and some approximations were done in (13, 16) and (14, 17) by separation of expectations as in [7] and the previous circularity and whiteness assumptions respectively.

By (17), the Newton step equation (10) reduces to

$$\mathbf{w}_i^f - \mathbf{w}_{i,o}^f = -\Big(\frac{\partial^2 f(\mathbf{w}_{i,o}^f)}{\partial \mathbf{w}_i^{f*}\partial \mathbf{w}_i^{fT}}\Big)^{-1}\frac{\partial f(\mathbf{w}_{i,o}^f)}{\partial \mathbf{w}_i^{f*}}, \quad (18)$$

and by (14), the inverse of a matrix term in (18) becomes a constant multiplication term. This shows that the fast algorithm for complex variables we are deriving is a variant of gradient descent algorithm as it is true for FastICA algorithms with real-valued variables. By substitution, our corresponding iterative algorithm becomes as follows,

$$\mathbf{w}_i^f = \mathbf{w}_{i,o}^f - \frac{\hat{E}\Big[(y_{i,o}^{f*})\, G'(\sum_f |y_{i,o}^f|^2)\mathbf{x}^f\Big] - \lambda_i^f \mathbf{w}_{i,o}^f}{\hat{E}\Big[G'(\sum_f |y_{i,o}^f|^2) + |y_{i,o}^f|^2\, G''(\sum_f |y_{i,o}^f|^2)\Big] - \lambda_i^f}, \quad (19)$$

where it can be easily evaluated that the Lagrange multiplier should be $\lambda_i^f = \hat{E}\left[|y_{i,o}^f|^2 G'\left(\sum_f |y_{i,o}^f|^2\right)\right]$. Also, instead of evaluating λ_i^f, we can remove it by multiplying the numerator in (19) on both sides of the equation. Hence, with the need of normalization, the learning rule becomes

$$\mathbf{w}_i^f = \hat{E}\left[G'\left(\sum_f |y_{i,o}^f|^2\right) + |y_{i,o}^f|^2 G''\left(\sum_f |y_{i,o}^f|^2\right)\right]\mathbf{w}_{i,o}^f - \hat{E}\left[(y_{i,o}^{f*})\, G'\left(\sum_f |y_{i,o}^f|^2\right)\mathbf{x}^f\right]. \tag{20}$$

In addition to normalization, the rows of the unmixing matrix $\mathbf{W}$ need to be decorrelated. For the maximum likelihood approach, symmetric decorrelation $\left(\mathbf{W} \leftarrow (\mathbf{W}\mathbf{W}^H)^{-\frac{1}{2}}\mathbf{W}\right)$ should be used instead of the deflationary decorrelation scheme since the source prior is not SEND itself but the product of SENDs. It should be noted that decorrelation is actually done in each frequency bin $\left(\mathbf{W}^f \leftarrow (\mathbf{W}^f\mathbf{W}^{fH})^{-\frac{1}{2}}\mathbf{W}^f\right)$.

4 Experiments

Our BSS algorithm was applied to real recordings and the performance was compared with the perviously mentioned permutation correction algorithms. However, because of the limit of pages in this paper, we will just show the comparison result with Sawada's algorithm [3] which is regarded as the most improved version of them. Three real spoken speeches and one hip-hop music played by a fixed speaker were recorded in a real office environment. For the separation, 2048 FFT points and 8 seconds of speech length were used. The results are shown in Fig.2. While Sawada's algorithm extracted only one speech source, our algorithm was able to separate all sources.

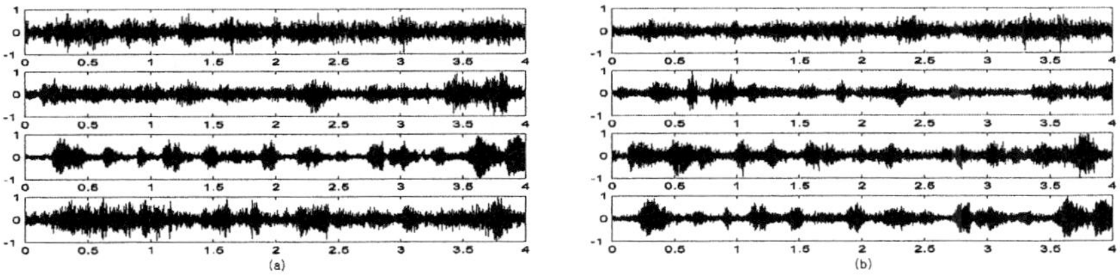

Fig. 2. Separation results of (a) Sawada's algorithm (b) our algorithm

5 Discussion

We showed the effectiveness of applying MICA to the frequency-domain BSS problem by using a valid multidimensional prior. This analysis adds an inter-frequency source-dependency constraint to the learning of each component such that the permutation problem is avoided, or solved robustly. Interestingly and similarly to ICA, we had certain flexibilities in defining or modeling the source prior. This can be seen in [8] where a different source prior and a gradient-descent type algorithm show good performance.

Although independent groups of sources or independent groups of features can be separated by several algorithms [4,9], the task to separate the mixture of dependent sources or features is still difficult. This problem is also inherent in our proposed IVA algorithm which can be easily seen from its entropic contrasts. While the contrast (1) is minimized to 0, the dependent sources can still remain unseparated since the mixture between dependent sources does not change the value of the contrast. However, a significant advantage of applying such source constraint to the frequency-domain BSS problem can be seen from the following mixing structure, which is a simple 2 × 2, 2 frequency case;

$$\begin{bmatrix} x_1^1[n] \\ x_2^1[n] \\ \hline x_1^2[n] \\ x_2^2[n] \end{bmatrix} = \left[\begin{array}{cc|cc} a_{11}^{11} & a_{12}^{11} & 0 & 0 \\ a_{21}^{11} & a_{22}^{11} & 0 & 0 \\ \hline 0 & 0 & a_{11}^{22} & a_{12}^{22} \\ 0 & 0 & a_{21}^{22} & a_{22}^{22} \end{array}\right] \begin{bmatrix} s_1^1[n] \\ s_2^1[n] \\ \hline s_1^2[n] \\ s_2^2[n] \end{bmatrix}. \tag{21}$$

The 0 terms restrict the mixing and the unmixing to each frequency bin, and thus, no mixture between components of different frequency bins can exist.

References

1. Kurita, S., Saruwatari, H., Kajita, S., Takeda, K., Itakura, F.: Evaluation of blind signal separation method using directivity pattern under reverberant conditions. In: Proc. IEEE Int. Conf. on Acoustics, Speech, and Signal Processing. (2000) 3140–3143
2. Murata, N., Ikeda, S., Ziehe, A.: An approach to blind source separation based on temporal structure of speech signals. Neurocomputing **41** (2001) 1–24
3. Sawada, H., Mukai, R., Araki, S., Makino, S.: A robust and precise method for solving the permutation problem of frequency-domain blind source separation. In: Proc. Int. Conf. on Independent Component Analysis and Blind Source Separation. (2003) 505–510
4. Cardoso, J.F.: Multidimensional independent component analysis. In: Proc. IEEE Int. Conf. on Acoustics, Speech, and Signal Processing. (1998)
5. Haykin, S.: Unsupervised Adaptive Filtering. Volume 1. John Wiley and Sons (2000)
6. van den Bos, A.: Complex gradient and hessian. In: IEE Proceedings on Vision, Image and Signal Processing. Volume 141. (1994) 380–382
7. Bingham, E., Hyvärinen, A.: A fast fixed-point algorithm for independent componenet analysis of complex-valued signals. Int. J. of neural systems **10**(1) (2000) 1–8
8. Kim, T., Attias, H.T., Lee, S.Y., Lee, T.W.: Blind source separation exploiting higher-order frequency dependencies. To appear in IEEE Trans. Speech Audio Processing (2006)
9. Hyvärinen, A., Hoyer, P.O.: Emergence of phase and shift invariant features by decomposition of natural images into independent feature subspaces. Neural Computation **12**(7) (2000) 1705–1720

Under-Determined Source Separation: Comparison of Two Approaches Based on Sparse Decompositions

Sylvain Lesage, Sacha Krstulović, and Rémi Gribonval

METISS project, IRISA-INRIA,
Campus de Beaulieu, 35042 Rennes Cedex, France
firstname.lastname@irisa.fr

Abstract. This paper focuses on under-determined source separation when the mixing parameters are known. The approach is based on a sparse decomposition of the mixture. In the proposed method, the mixture is decomposed with Matching Pursuit by introducing a new class of multi-channel dictionaries, where the atoms are given by a spatial direction and a waveform. The knowledge of the mixing matrix is directly integrated in the decomposition. Compared to the separation by multi-channel Matching Pursuit followed by a clustering, the new algorithm introduces less artifacts whereas the level of residual interferences is about the same. These two methods are compared to Bofill & Zibulevsky's separation algorithm and DUET method. We also study the effect of smoothing the decompositions and the importance of the quality of the estimation of the mixing matrix.

1 Introduction

The source separation problem [1] consists in retrieving unknown signals (the sources) from the only knowledge of mixtures of these signals (the channels). Each channel x_n is the linear combination of the sources, $x_n(t) = \sum_{i=1}^{I} a_{n,i} \cdot s_i(t)$, where $a_{n,i}$ is a constant setting the level of the source s_i in the mixture x_n. Thus the mixture can be written in linear algebra as $\mathbf{x} = \mathbf{A}\mathbf{s}$, where $\mathbf{A}$ is the mixing matrix, and the rows of the matrices $\mathbf{x}$ and $\mathbf{s}$ are respectively the signals x_n and s_i. In the determined (resp. over-determined) case, where the number of observed channels is equal to (resp. greater than) the number of sources, estimating the mixing matrix and estimating the sources are equivalent problems. Conversely, in the under-determined case, the knowledge of the mixing matrix or its estimate is not sufficient to recover the sources, and a model of the sources is generally needed to estimate them [2]. Generally, it is a difficult task to distinguish, in the performances of a given algorithm, the effect of the quality of the matrix estimation from the effect of the mismatch to the model.

In this article, we focus on the under-determined case. Our approach uses models based on the existence of sparse representations of the sources [3], and assumes the perfect knowledge of the mixing matrix. We compare two separation algorithms based on variants of Matching Pursuit (MP) [4]. The first variant consists in decomposing the multi-channel mixture without knowing the mixing matrix, and then using the mixing matrix to classify the coefficients of the decomposition and affecting them to the sources to estimate [5,6]. The second variant consists in using the mixing matrix in the sparse

J. Rosca et al. (Eds.): ICA 2006, LNCS 3889, pp. 633–640, 2006.

decomposition step itself, and no additional classification step is needed. The performance of these two algorithms are compared to the best linear separator (BLS) [7], to the Bofill & Zibulevski's algorithm (BZ) [8] and to the DUET algorithm [6].

This article is organized as follows : in section 2, we recall the general definition of Matching Pursuit. Multi-channel MP and its various separation algorithms are described in section 3 and we detail the experimental conditions and the results in section 4.

2 Matching Pursuit

A signal x (considered as a vector of the Hilbert space $\mathcal{H}$ of finite-energy signals) admits a sparse decomposition over the dictionary $\mathcal{D} = \{\phi_k\}$ of atoms ϕ_k – or elementary signals ϕ_k – if it can be written as a linear combination $x = \sum_k c_k \phi_k$ where few coefficients $\{c_k\}$ are non-negligible. In this framework, MP iteratively computes sparse approximations of the form $x = \sum_{m=1}^{M} c_{k_m} \phi_{k_m} + R^M$ where R^M is a residual that tends to zero as the number of iterations M tends to infinity. The principle of the algorithm is to select, at each step, the atom that is the most correlated to the residual, then to update the residual by removing the contribution of this atom.

The most current stopping criteria are based on the absolute or relative level of energy of the residual or/and on a fixed number of iterations to run. In addition, the Gabor dictionary is classically used to sparsely decompose audio signals. It is composed of a collection of time-frequency Gabor atoms $\phi_{s,u,\xi}(t) = w\left(\frac{t-u}{s}\right) \cdot \exp\left(2j\pi\xi(t-u)\right)$. These atoms are defined by the choice of a window w of unit energy (Hanning, Gaussian, ...), a scale factor s, a time localization u, and a frequency ξ. Such a dictionary allows a fast computation of the inner products between the signal and the atoms by applying some windowed-FFTs.

3 Source Separation with Matching Pursuit

Source separation techniques based on sparse approximations of multi-channel signals on a dictionary have been proposed in the multi-channel case [3,6]. More specifically, in the MP framework, the method proposed in [5,9] uses multi-channel MP, followed by a clustering (note that the base idea of this method could be developed for other multi-channel sparse decomposition algorithms, e.g. [10].) After recalling the principle of the method based on MP plus clustering, we propose a variant where the definition of the dictionary includes knowledge of the mixing matrix $\mathbf{A}$.

3.1 Multi-channel Matching Pursuit

For the sparse decomposition of multi-channel signals, we use a dictionary $\boldsymbol{\mathcal{D}}$ composed of multi-channel atoms $\boldsymbol{\phi}$. These atoms are defined by $\boldsymbol{\phi} = (c_1\phi, c_2\phi, \ldots, c_N\phi)$, where $\phi \in \mathcal{D}$ is a mono-channel atom from a dictionary $\mathcal{D}$ and where the coefficients $c_1, \ldots, c_N$ satisfy $\sum_{n=1}^{N} c_n^2 = 1$. After M iterations, multi-channel MP leads to a decomposition of the form $(x_1, \ldots, x_N) = \widehat{\mathbf{x}}^M + (R_1^M, \ldots, R_N^M)$, with $\widehat{\mathbf{x}}^M := \sum_{m=1}^{M} (c_{1,k_m}\phi_{k_m}, \ldots, c_{N,k_m}\phi_{k_m})$. The algorithm is composed of the following steps :

1. Initialization : $M = 1, \quad R_n^0 = x_n, \quad c_{n,k} = 0, \quad \forall n, \forall k;$
2. Computation of the inner product between each channel of the residual R_n^{M-1} and each atom ϕ_k of the mono-channel dictionary.
3. Selection of $k_M = \arg\max_k \sum_{n=1}^{N} |\langle R_n^{M-1}, \phi_k\rangle|^2$
4. For each channel n, update of the residual : $R_n^M = R_n^{M-1} - \langle R_n^{M-1}, \phi_{k_M}\rangle \phi_{k_M}$ and of the coefficients : $c_{n,k_M}^M = c_{n,k_M}^{M-1} + \langle R_n^{M-1}, \phi_{k_M}\rangle$
5. If the stopping criterion has not been reached, $M \leftarrow M+1$, then go back to 2.

The multi-channel signal $\widehat{\mathbf{x}}^M$, approximated by multi-channel Matching Pursuit, allows to estimate each mono-channel source signal s_i using the atoms of the decomposition that are allocated to it, in the following manner : assuming the mixing matrix $\mathbf{A}$ is known with unit columns $\|\mathbf{a}_i\|_2 = \sum_n a_{n,i}^2 = 1$, the atom k_M is attributed to the source of index $\hat{i}_M = \arg\max_i |\langle \mathbf{c}_{k_M}, \mathbf{a}_i\rangle|$. This corresponds to partitioning the multi-channel coefficient space $\{\mathbf{c} = (c_n)_{1\le n\le N} \in \mathbb{C}^N\}$ into I subsets corresponding to the columns $\mathbf{a}_i$ of $\mathbf{A}$ (I being the number of sources). The source s_i is reconstructed by :

$$\hat{s}_i = \sum_{M|\hat{i}_M = i} \langle \mathbf{c}_{k_M}, \mathbf{a}_i\rangle \phi_{k_M}. \tag{1}$$

We call this separation algorithm MP1. Alternately, MP2 is a variant consisting in attributing each atom to the N closest sources. This second selection, also used in Bofill & Zibulevsky's algorithm [8] in the stereophonic case ($N = 2$), corresponds to the minimization of the l_1 norm of the projection of the coefficients $\mathbf{c}_{k_M}$ on N directions of the mixing matrix :

$$\hat{J}_M = \arg\min_{J\subset[1,I]} \|\mathbf{A}_J^{-1}\mathbf{c}_{k_M}\|_1 \text{ , with } \mathbf{A}_J = [\mathbf{a}_i]_{i\in J} \tag{2}$$

3.2 Demixing Pursuit

Combining the expression of the linear instantaneous mixtures $x_n = \sum_{i=1}^{I} a_{n,i} \cdot s_i$ and that of a candidate sparse decomposition $s_i = \sum_{k=1}^{K} c_{i,k}\phi_k$ of each source s_i on the mono-channel dictionary $\mathcal{D}$, we can write $x_n = \sum_{i,k} a_{n,i}\, c_{i,k}\, \phi_k$. This is translated in linear algebra as $\mathbf{x} = \mathbf{AC\Phi}^T$, with $\mathbf{\Phi}^T$ the matrix which rows are the mono-channel atoms ϕ_k, and $\mathbf{C} = \{c_{i,k}\}_{i,k}$ a matrix of sparse components. This decomposition can also be written $\mathbf{x} = \sum_{i,k} c_{i,k}\, \mathbf{a}_i\, \phi_k$, that is to say that $\mathbf{x}$ admits a sparse decomposition on the "directional" multi-channel dictionary constituted of the atoms $\mathbf{a}_i\, \phi_k = (a_{1,i}\, \phi_k, \ldots, a_{N,i}\, \phi_k)$. One can therefore get a decomposition of this type by applying MP on the latter dictionary. The inner products are then computed as $\langle \mathbf{R}^M, \mathbf{a}_i\phi_k\rangle = \mathbf{a}_i^T \mathbf{R}^M \phi_k^T$ and the source s_i is reconstructed by :

$$\hat{s}_i = \sum_k c_{i,k}\phi_k. \tag{3}$$

This new algorithm is called demixing pursuit (DP) and its theoretical properties have been studied in [11]. Using a directional dictionary is equivalent to applying multi-channel MP with the constraint that the components $\mathbf{c}_{k_M}$ of section 3.1 shall be proportional to a column $\mathbf{a}_i$ of $\mathbf{A}$.

4 Experiments

We compare the algorithms MP1, MP2 and DP described previously to three reference algorithms. The experiments are performed on a stereophonic linear instantaneous mixture of three musical sources (a cello, some drums and a piano). The sampling frequency of the signals is 8kHz, and their length is 2.4s (19200 samples). The mixing matrix is the following :

$$\begin{bmatrix} \cos(\pi/8) & \cos(\pi/4) & \cos(3\pi/8) \\ \sin(\pi/8) & \sin(\pi/4) & \sin(3\pi/8) \end{bmatrix}$$

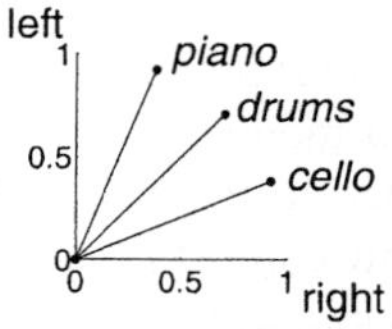

The energy of the drums, located in the middle, is about twice weaker than the energies of the piano and cello, which are quite similar.

We use the measures of separation performance proposed in [7], that allow to finely analyze the origin of the distortions between the estimated source and the original one. These measures, expressed in decibels, are based on the decomposition of an estimated source signal into parts due to original source, interferences and algorithmic artifacts. The relative ratios between the energies of these three parts define the Source to Distortion Ratio (SDR, global distortion), Source to Interference Ratio (SIR) and Source to Artifacts Ratio (SAR). For these three measures, of the same nature as the classical Signal to Noise Ratio, higher ratio mean better performances.

4.1 Reference Algorithms

The performances of MP1, MP2 and DP are compared to those of three reference algorithms : the best linear separator (BLS) [7], DUET [6] and the Bofill & Zibulevski's algorithm (BZ) [8].

The first one only consists in the application of a matrix $\mathbf{B}$ to the signal. $\mathbf{B}$ is such that the estimated sources $\hat{\mathbf{s}} = \mathbf{B}\mathbf{x}$ minimize the distortion due to the interferences [7]. If the sources are assumed to be mutually orthogonal, if the mixing matrix $\mathbf{A}$ is known, and if we denote $\mathbf{D}$ the diagonal matrix of the norms of the sources, then, with $\hat{\mathbf{A}} = \mathbf{AD}$, the matrix $\mathbf{B}$ is given by : $\mathbf{B} = \mathbf{D}\hat{\mathbf{A}}^H(\hat{\mathbf{A}}\hat{\mathbf{A}}^H)^{-1}$.

The algorithm DUET [6] applies a short-time Fourier transform (STFT) to each channel of the signal, then applies a mask that assumes only one source to be active for each time-frequency "box", and finally inverts the STFT to build the estimated source.

The Bofill & Zibulevski's algorithm [8] relies on the same principle as DUET, the only difference being that each time-frequency box is attributed to the two nearest sources. This attribution is determined by an l_1 norm minimization (see Eq.2.)

In all the experiments, DUET and BZ are applied with a Hanning window of 4096 samples, with an overlap of 2048 samples (50% of the size of the window). Their performances strongly depend on the size of the window, and we have observed that a greater, or more critically, smaller window size strongly decreases the performances in the studied cases. Therefore, the results shown below employ an *a posteriori* optimal window size. Note that in practice it might be hard to choose the optimal window size, since the performances can't be not known.

In the experiments, the mixing matrix $\mathbf{A}$ is fixed *a priori* when using BLS, DUET and BZ. Thus, we did not use the mixing matrix estimations described in [6,8].

4.2 Different Versions of MP Algorithms

In this experiment, we study the influence of the number of iterations, of the composition of the dictionary, of the exploitation of the residual, and of a smoothing post-treatment using the DP algorithm. Two dictionaries may be used for the decomposition :

- a "small" dictionary made of Gabor atoms of length $s = 4096$ with an overlap of half the length ($u = ns/2$, $n \in \mathbb{N}$). This corresponds to the STFT used by the DUET and BZ algorithms.
- a "large" dictionary made of Gabor atoms which length goes from $s = 64$ to 16384 (by powers of two). The overlap between two successive atoms is also 50% of the length of an atom.

The time needed for computing the MP-based algorithms is largely higher than for computing DUET and BZ. DUET and BZ were computed on the small dictionary in 0.2 second ; 20000 iterations of DP were computed on the small dictionary in 5 minutes and were computed on the large dictionary in 20 minutes. Note that the computation of the MP-based algorithms were made tractable by a fast implementation available at [13].

Figure 1 represents the SDR, SIR and SAR of the "piano" source estimated by the different algorithms, against the number of iterations (the results are similar for the two other sources).

Firstly, we can remark that for any number of iterations, using the large dictionary leads to a better separation than using the small dictionary. Indeed, in the case of the large dictionary, MP chooses the optimal window size automatically. The need to optimize *a priori* the window size is removed, contrarily to the BZ and DUET algorithms.

In addition, we can notice that the performance improvement is monotonic when the number of iterations increases. More precisely, artifacts, which dominate the distortion,

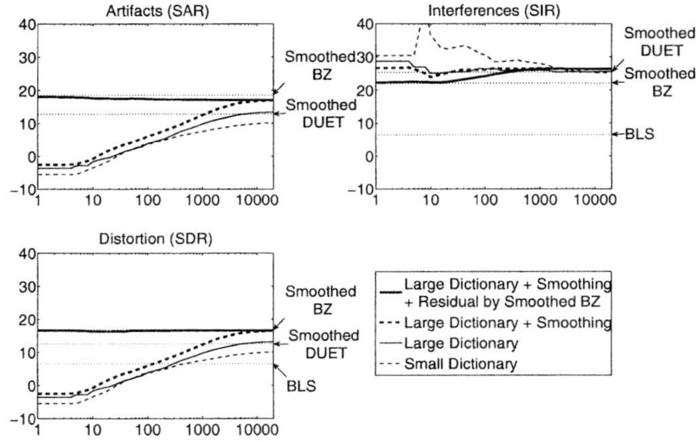

Fig. 1. Distortions (dB), "piano" source estimated by DP

are important when the sources are reconstructed with few atoms, and decrease when more iterations are performed, thanks to the contribution of new atoms. After a sufficient number of iterations, DP becomes better than DUET in terms of artifacts (SAR) and global distortion (SDR).

Using the hypothesis that the smoothing introduced by the overlap of the windows of the STFT in the BZ algorithm plays a role in its good performance [12], we tried to smooth the sources estimated by DUET, BZ and MP-based algorithms. This smoothing consists in performing several estimations of the sources from shifted versions of the dictionary and producing the mean of these estimations. The effect is to transform the binary time-frequency masking in a smoother masking. The amelioration brought by the smoothing is very clear for the artifacts (SAR improved by $\sim$ 4dB for DP, and $\sim$ 1dB for DUET and BZ), but not systematic for the interferences (SIR). Note that for clarity, only the smoothed versions of DUET and BZ are shown.

In order to compensate for the distortion due to the small number of atoms, the residual of the decomposition $\mathbf{R}^M$ can be separated using the linear separator $\mathbf{A}^H(\mathbf{A}\mathbf{A}^H)^{-1}\mathbf{R}^M$, or DUET, or BZ and then added to the estimated sources. The former linear separator assumes that all the residuals of the sources have the same energy. Asymptotically, the hypothesis is verified, the more energetic sources having their atoms selected in the first place. In figure 1, the upper curve shows the performance when the residual of the smoothed DP is separated by the smoothed BZ. This method is a trade-off between the two algorithms, tuned by the number of iterations. Notably, the artifacts are lower than with simple smoothed DP for a small number of iterations, as the smoothed BZ produces less artifacts. The same kind of trade-off is obtained when separating with the linear separator or the smoothed DUET.

For MP1 and MP2, changing the dictionary, adding the residual and the smoothing produces the same type of effects than for DP.

4.3 What If the Mixing Matrix Is Imprecisely Known ?

The following experiment evaluates the capacity of the different algorithms to maintain a good separation when the mixing matrix is no longer known, but only estimated. A voluntary imprecision is introduced by a rotation of the true matrix. The directions of the three sources are shifted by the same angle, which varies between $-\pi/16$ and $\pi/16$ (half the distance between two sources). The experiments are done with the "large" dictionary. They include the separation of the residual with smoothed BZ and the smoothing, and use 5000 iterations. Smoothing is also applied to DUET and BZ. The performances are given on Figure 2, depending on the perturbation angle, for the piano.

Evolution of the SAR – The studied methods keep an approximately constant level of artifacts for any angle of perturbation. BZ and DP introduce the least artifacts, followed by MP2, MP1 and DUET that present equivalent performances. The levels of artifacts are intrinsic to the underlying models of each method.

Evolution of the SIR – For the methods MP1 and DUET, the time-frequency atoms are only attributed to one source. Therefore, these methods produce the least interferences and stay robust to a perturbation of the mixing matrix. The large decrease in the level of interferences for negative angles is due to the location of the piano: it has no neighbor

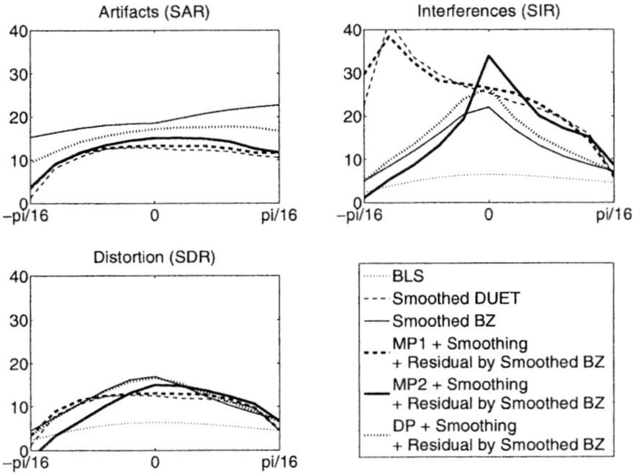

Fig. 2. Distortions (dB), source "piano" depending on the perturbation angle

instrument in this direction, thus most of the atoms in this direction actually belong to the piano source. In the case of MP2, DP and BZ, allotting time-frequency atoms to several sources introduces a larger sensibility to the perturbation on the mixing matrix. For a well-estimated mixing matrix, MP2 produces the least interferences.

Evolution of the SDR – By definition, global distortion (SDR) is dominated by the minimum of SAR and SIR. For a well-estimated mixing matrix, by decreasing order of performance, the methods are scaled as : BZ, DP, MP2, MP1, DUET, and BLS. On the other hand, when a perturbation is introduced on the mixing matrix, the methods MP1 and DUET (attribution to one direction) prove to be more robust than DP (selection of the atoms by Matching Pursuit only on the estimated directions of the sources) and than the methods MP2, BZ (attribution to two directions, that lead to a larger sensibility to interferences).

5 Conclusions

We have compared several methods for under-determined source separation by sparse decomposition, assuming that the mixing matrix is known. In the algorithms MP1 and MP2, the mixing matrix is used *a posteriori* to classify and gather the atoms resulting from the decomposition by Matching Pursuit. In the Demixing Pursuit, the knowledge of the mixing matrix is included *a priori* in the definition of the dictionary. The version of DP with the addition of the smoothing gives better performances, for global distortion and artifacts, than the method DUET but worse than BZ. A trade-off between reference algorithms and MP-based algorithms is obtained when the residual from Matching Pursuit is separated by a reference algorithm. When the mixing matrix is well estimated, BZ, DP and MP2 give the best results. On the other hand, MP1 and DUET seem to be more robust to an error on the estimation of the mixing matrix.

The proposed formalism allows to perform separation in the case of under-determined convolutive mixtures, provided that the mixing filters are known. In that case, the atoms of the multi-channel dictionary represent on each channel what is obtained at the sensor when each mono-channel atom is passed through the mixing filters. The algorithm is then just the application of Matching Pursuit on these normalized multi-channel atoms and the sources are reconstructed as in DP. The related experiments are currently being developed.

Another perspective is to consider the joint estimation of the mixing matrix and the sources in the linear instantaneous case, or of the filters and the sources in the convolutive case. Alternately, we are investigating possible improvements of the sparse decomposition by learning dictionaries adapted to the mixture, notably directional multichannel dictionaries for demixing pursuit.

References

1. J.-F. Cardoso, "Blind signal separation: statistical principles," *Proc. IEEE, Special issue on blind identification and estimation*, vol. 9, no. 10, pp. 2009–2025, Oct. 1998.
2. O. Bermond and J.-F. Cardoso, "Mthodes de sparation de sources dans le cas sous-dtermin," in *Proc. GRETSI, Vannes, France*, 1999, pp. 749–752.
3. M. Zibulevsky and B. Pearlmutter, "Blind source separation by sparse decomposition in a signal dictionary," *Neural Computations*, vol. 13, no. 4, pp. 863–882, 2001.
4. S. Mallat and Z. Zhang, "Matching pursuit with time-frequency dictionaries," *IEEE Trans. on Signal Processing*, vol. 41, pp. 3397–3415, Dec. 1993.
5. R. Gribonval, "Sparse decomposition of stereo signals with matching pursuit and application to blind separation of more than two sources from a stereo mixture," in *Proc. Int. Conf. Acoust. Speech Signal Process. (ICASSP'02), Orlando, Florida, USA, May 2002*, 2002.
6. O. Yilmaz and S. Rickard, "Blind separation of speech mixtures via time-frequency masking," *IEEE Transactions on Signal Processing*, vol. 52, no. 7, pp. 1830–1847, July 2004.
7. R. Gribonval, L. Benaroya, E. Vincent, and C. Févotte, "Proposals for performance measurement in source separation," in *Proc. 4th Int. Symp. on Independent Component Analysis and Blind Signal Separation (ICA2003)*, Nara, Japan, Apr. 2003, pp. 763–768, see "BSS EVAL Toolbox", `http://bass--db.gforge.inria.fr`.
8. P. Bofill and M. Zibulevsky, "Blind separation of more sources than mixtures using sparsity of their short-time fourier transform," in *Proc. ICA2000*, Helsinki, june 2000, pp. 87–92.
9. R. Gribonval, "Piecewise linear source separation," in *Proc. SPIE'03 – "Wavelets: Applications in Signal and Image Processing", San Diego, California, USA*, vol. 5207, August 2003, pp. 297–310.
10. B. D. Rao, S. Cotter, and K. Engan, "Diversity measure minimization based method for computing sparse solutions to linear inverse problems with multiple measurement vectors," in *Proceedings in Acoustics, Speech, and Signal Processing (ICASSP' 04)*, may 2004.
11. R. Gribonval and M. Nielsen, "Beyond sparsity : recovering structured representations by l1-minimization and greedy algorithms. – application to the analysis of sparse underdetermined ICA–," IRISA, Tech. Rep. 1684, Jan. 2005, http://www.irisa.fr/metiss/gribonval/.
12. S. Araki, S. Makino, H. Sawada, and R. Mukai, "Reducing musical noise by a fine-shift overlap-add method applied to source separation using a time-frequency mask," in *Proceedings in Acoustics, Speech, and Signal Processing (ICASSP' 05)*, vol. 3, march 2005, pp. 81–84.
13. R. Gribonval and S. Krstulović, "The Matching Pursuit ToolKit", see http://mptk.gforge.inria.fr.

Separation of Mixed Audio Signals by Source Localization and Binary Masking with Hilbert Spectrum

Md. Khademul Islam Molla[1], Keikichi Hirose[2], and Nabuaki Minematsu[1]

[1] Graduate School of Frontier Sciences
[2] Graduate School of Information Science and Technology,
The University of Tokyo, 7-3-1 Hongo, Bunkyo-ku, Tokyo 113-0033, Japan
{molla, hirose, mine}@gavo.t.u-tokyo.ac.jp

Abstract. The Hilbert transformation together with empirical mode decomposition (EMD) produces Hilbert spectrum (HS) which is a fine-resolution time-frequency (TF) representation of any nonlinear and non-stationary signal. A method of audio signal separation from stereo mixtures based on the spatial location of the sources is presented in this paper. The TF representation of the audio signal is obtained by HS. The sources are localized in the space of time and intensity differences between two microphones' signals. The separation is performed by masking the target signal in TF domain considering that the sources are disjoint orthogonal. The experimental results of the proposed method show a noticeable improvement of separation efficiency.

1 Introduction

When the audio recording is performed using two microphones in an adverse acoustical environment, time difference (TD) and intensity difference (ID) are introduced between the mixed signals. Those are termed as interaural differences and used to localize the audio sources in spatial domain. Such source localization method is used in [1], [2], [3] to separate the individual audio source from two mixtures. In [2], the mixtures are produced by convoluting monaural signal with measured head related transfer functions (HRTFs) [4], whereas two microphones are used for recording in [1], [3]. The use HRTF introduces TD and ID in the mixture signals. The TD is the main localization cue at low frequencies and ID dominates the high frequency range. The partition between these two ranges of frequency depends on the spacing between the microphones [2]. To cover the entire frequency range, the TD and ID are jointly used in localization. The TF masks are used to segregate the individual sources in TF domain. The principal assumption of the masking based separation in TF domain is that, the audio sources are disjoint orthogonal i.e. not more than one source is active at any TF point [1]. The short-time Fourier tranform (STFT) is an usual approch to represent the time domain signal in TF domain [1], [2], [3]. The STFT based TF representation includes a remarkable amount of cross-spectral energy due to the harmonic assumption and window overlapping. The both time and frequency resolution can not be extended independently. Those two limitations of STFT degrades the disjoint orthogonality of the audio sources and hence the separation efficiency by using masking method in TF domain.

J. Rosca et al. (Eds.): ICA 2006, LNCS 3889, pp. 641 – 648, 2006.

In this paper a novel technique to separate audio sources stereo mixtures based on spatial localization is described. The separation efficiency can be improved by maximizing the resolution and minimizing the cross-spectral energy terms in TF space. The proposed separation method employs HS as the TF representation. HS does not include noticeable amount of cross-spectral energy terms. The empirical mode decomposition (EMD), a new technique for nonlinear and non-stationary time series analysis [5] and Hilbert transformation are employed together to derive HS. Based on the TD and ID between two mixtures, the TF spaces (HSs of two mixtures) are clustered in TD-ID space to localize the audio sources. The TF space of each source is segregated by binary masking method [3], and the time domain signals are recostructed by applying the inverse transformations. The HS has better TF resolutions as well as less cross-spectral energy than STFT and hence more suitable for source disjoint orthogonality consideration.

Regarding the arrangement of this paper, the EMD and HS are illustrated in section 2, the source localization and separation methods are described in sections 3 and 4 respectively. The concept of disjoint orthogonality is presented in section 5. The experimental results are shown in section 6 and finally some concluding remarks are included in section 7.

2 The Modification of EMD and Hilbert Spectrum

The EMD represents the mixture signal as a collection of oscillatory basis components $C_m(t)$ termed as intrinsic mode functions (IMFs) containing some basic properties [5, 6]. The decomposition process can also be considered as dyadic filter-bank as proved by analysis of white noise [6], [7]. Each IMF should satisfy two basic conditions: (i) in the whole data set, the number of extrema and the number of zero crossing must be the same or differ at most by one, (ii) the mean value of the envelope defined by the local maxima and the envelope defined by the local minima is always zero. There exist many approaches of computing EMD [6]. The following algorithm is adopted here to decompose the signal $s(t)$ into a set of IMF components.

a) Initialize the residue $r_0(t)=s(t)$ and index of IMF $m=1$
b) (i) set $g_0(t)=r_{m-1}(t)$ and $i=1$
  (ii) Find the extrema (minima and maxima) of $g_{i-1}(t)$
  (iii) Compute upper and lower envelopes $h_{i-1}(t)$ and $l_{i-1}(t)$
  (iv) Find mean envelope $\mu_{i-1}(t)=[h_{i-1}(t)+l_{i-1}(t)]/2$
  (v) Update $g_i(t)=g_{i-1}(t)- \mu_{i-1}(t)$ and $i=i+1$
  (vi) Repeat steps (ii)-(v) until $g_i(t)$ being an IMF satisfying the above mentioned two basic conditions. If so, the mth IMF $C_m(t)=g_i(t)$and update residue $r_m(t)=r_{m-1}(t)-C_m(t)$
c) Repeat step (b) with the index of IMF $m=m+1$

At the end of the decomposition the signal $s(t)$ is represented as:

$$s(t) = \sum_{m=1}^{M} C_m + r_M \tag{1}$$

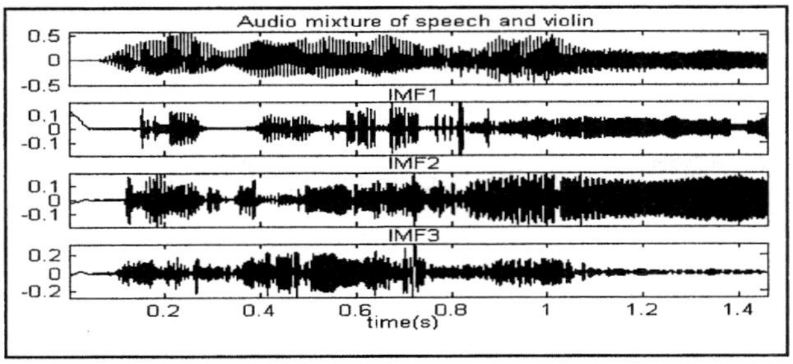

Fig. 1. The *EMD* of an audio mixture (speech and flute sound) showing first three IMFs out of 14

where M is the number of IMF components and r_M is the final residue. The r_M monotonously converges to a constant or takes a function with only one maxima and minima such that no more IMF can be derived. A band-limited (80Hz-4kHz) audio mixture signal and the decomposed IMF components are shown in Figure 1.

The IMFs computed by the basic EMD include energy at frequencies that cannot be associated with the original data. This phenomenon obviously includes unwanted signal energy in HS and hence degrades the separation performance. To eliminate such unwanted signals, a band-pass filtering method is proposed to be included in the original EMD algorithm. This attempt ensures to run every IMF inside the given frequency band. The proposed modification also increases the number of IMF components that improves the frequency resolution of the decomposition. The analyzing signal $s(t)$ is first passed through a zero phase band-pass filter (BPF). The same filter is included in step (vi) of the original algorithm. The procedure is as follows: first generate the IMF $C_m(t)$, filter it to yield the filtered IMF $\hat{C}_m(t)$ and compute the residue $\hat{r}_m(t)=r_{m-1}(t)-\hat{C}_m(t)$ to generate $\hat{C}_{m+1}(t)$. After completing the decomposition, the modified EMD can be represented by the same way as in Eq. (1). Experimentally it is found that the modified EMD generates 23 IMFs whereas, original one produces 14 IMFs from the same signal of Fig. 1. All the subsequent operations (computing instantaneous frequency, constructing Hilbert spectrum) are performed on the modified EMD.

2.1 Instantaneous Frequency

Instantaneous frequency (IF) represents signal's frequency at an instance, and is defined as the rate of change of the phase angle at the instant of the "analytic" version of the signal. Every IMF is a real valued signal. The discrete Hilbert transform (HT) denoted by $\hbar_d[.]$ is used to compute the analytic signal for an IMF. HT provides a phase-shift of $\pm\pi/2$ to all frequency components, whilst leaving the magnitudes unchanged [5]. Then the analytic version of the mth IMF $\hat{C}_m(t)$ is defined as:

$$z_m(t) = \hat{C}_m(t) + j\hbar_d[\hat{C}_m(t)] = a_m(t)e^{j\theta_m(t)} \tag{2}$$

where $a_m(t)$ and $\theta_m(t)$ are instantaneous amplitude and phase respectively of the mth IMF. The IF of mth IMF is then computed by the derivative of the phase $\theta_m(t)$ as: $f_m(t) = \frac{d\tilde{\theta}(t)}{dt}$, where $\tilde{\theta}_m(t)$ represents the unwrapped version of $\theta_m(t)$. The median smoothing filter is used to tackle the discontinuities of IF computed by discrete time derivative of the phase vector.

2.2 Hilbert Spectrum

Hilbert Spectrum represents the distribution of the signal energy as a function of time and frequency. It is also designated as Hilbert amplitude spectrum $H(\omega,t)$ or simply Hilbert spectrum (HS). This process first normalizes the IF vectors of all IMFs between 0 to 0.5. Each IF vector is multiplied by the scaling factor η=0.5/(IF_{max}-IF_{min}), where IF_{max}=Max(f_1, f_2,...,f_m,...,f_M) and IF_{min}=Min(f_1, f_2,...,f_m,...,f_M). The bin spacing of the HS is 0.5/B, where B is the number of desired frequency bins selected arbitrarily. Each element H(ω,t) is defined as the weighted sum of the instantaneous amplitudes of all the IMFs at ω^{th} frequency bin,

$$H(\omega, t) = \sum_{m=1}^{M} a_m(t) w_m^{(\omega)}(t) \tag{3}$$

where the weight factor $w_m^{(\omega)}(t)$ takes 1 if $\eta \times f_m(t)$ falls within ω^{th} band, otherwise is 0. After computing the elements over the frequency bins, H represents the instantaneous signal spectrum in TF space as a 2D table. The time resolution of H is equal to the sampling rate and the frequency resolution can be chosen up to Nyquest limit. Fig. 2 represents the Hilbert spectrum of the audio signal shown in Fig. 1 using 256 frequency bins (with sapling rate 16kHz).

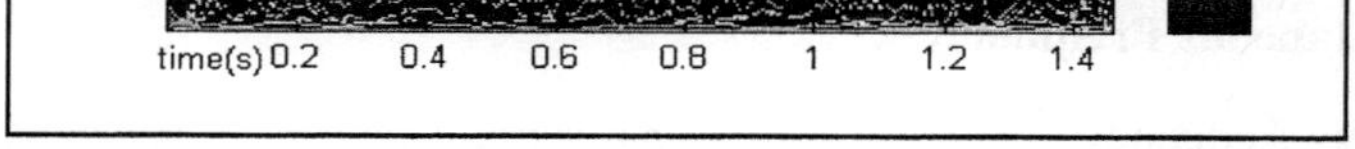

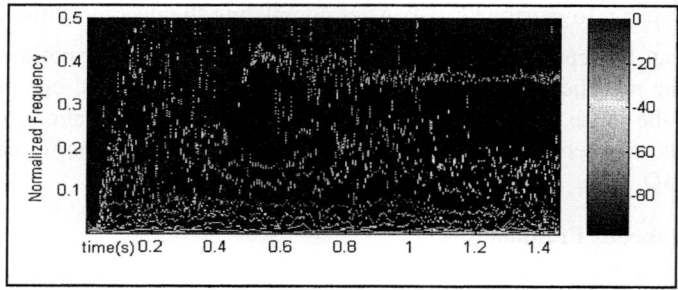

Fig. 2. Hilbert spectrum with 256 frequency bins. The amplitude is in dB.

3 Source Localization

The audio sources are localized in TD-ID space. There is a one-to-one mapping between the azimuth location and a region in TD-ID space. The TD and ID are computed from the relative phase and energy differences of the TF spaces of two mixtures. If $H_L(\omega,t)$ and $H_R(\omega,t)$ are the Hilbert spectrum of binaural mixtures $x_l(t)$ and $x_r(t)$ respectively, the TD and ID can easily be computed as [1], [3]:

$$TD(\omega,t) = \frac{1}{\omega}[\tilde{\phi}_L(\omega,t) - \tilde{\phi}_R(\omega,t)]$$
$$ID(\omega,t) = 20\log\left(\frac{|H_R(\omega,t)|}{|H_L(\omega,t)|}\right) \tag{4}$$

where $\tilde{\phi}_L(\omega,t)$ and $\tilde{\phi}_R(\omega,t)$ are the unwrapped phases corresponding to H_L and H_R respectively. The difference between the phase terms remains within $(-\pi,\pi)$. The intensity (energy) and phase information are smoothed in TF space by average filtering with the time frame of length 1ms. It improves the stability of the instantaneous energy and phase response computed by the discrete derivative of the analytic signals.

The values of TD and ID computed by Eq. (4) are quantized into discrete levels (50 levels). Then the histogram ψ(TD, ID) is constructed by mapping each TF point into quantized TD-ID space. Fig. 3 shows the TD-ID space localization of three sources placed at 50°, 90° and 110° azimuths. The three peaks (with some degree of spreading) correspond to distinct active sources. The histogram is weighted by the energy function in the TF space of the mixture.

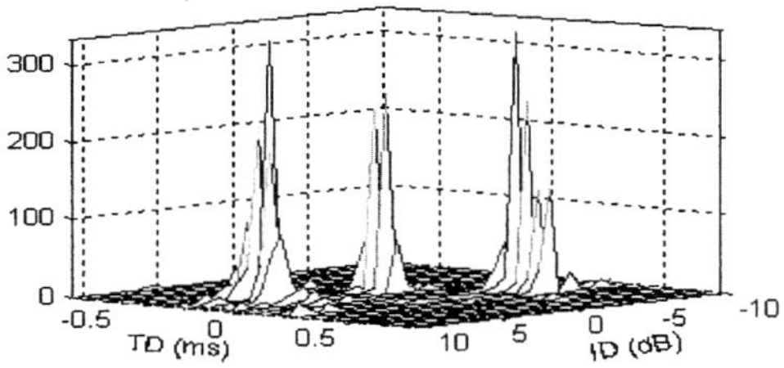

Fig. 3. TD-ID space localization of three sources

4 Source Separation

The individual source placed at different azimuth locations has the unique regions in the histogram ψ (TD, ID). Such a mapping allows to construct the TF mask corresponding to each region and it is used to mask H_L or H_R to yield the TF representation of the original source. If δ_n and κ_n are the set of TD and ID respectively representing the rectangle of the peak region of the nth source in ψ (TD, ID), its TF mask can be computed as:

$$M^{(n)}(\omega,t) = \begin{cases} 1 : [(TD(\omega,t) \in \delta_n) and\ (ID(\omega,t) \in \kappa_n)] \\ 0 : otherwise \end{cases} ; \forall\, \omega, t \tag{5}$$

The binary mask nullifies TF points of interfering sources. The HS of the nth source can be computed as: $H^{(n)}(\omega,t) = M^{(n)}(\omega,t)H_L(\omega,t)$ or $H^{(n)}(\omega,t) = M^{(n)}(\omega,t)H_R(\omega,t)$.

During the Hilbert transform the real part of the signal remains unchanged. The time domain signal of nth source is reconstructed by filtering out the imaginary part from the

HS and summing over frequency bins as: $s^{(n)}(t)=\sum_{\omega}H^{(n)}(\omega,t)\cdot\cos[\phi(\omega,t)]$. where $\phi\,(\omega,t)$ is the phase matrix of H_L (or H_R). The phase matrix is saved during the construction of HS to be used in re-synthesis.

5 Disjoint Orthogonality in TF Space

If $Y_1(\omega,t)$ and $Y_2(\omega,t)$ are the TF representation of the signals $y_1(t)$ and $y_2(t)$ respectively, the disjoint orthogonality assumption can be stated as: $Y_1(\omega,t)Y_2(\omega,t)=0;\forall\omega,t$. In order to better measure of a signal at a particular time and frequency (ω, t), it is natural to desire that Δ_t and Δ_ω be as narrow as possible. In STFT based TF representation Δ_t and Δ_ω has to satisfy an uncertainty inequality $\Delta_t\Delta_\omega\geq 0.5$ which is the trade-off of the selection of TF resolution. The Hilbert spectrum has better time-frequency resolution and improved disjoint orthogonality (DO) of audio sources in TF space. The signal to interference ratio (*SIR*) is used as basis to measure the DO. The *SIR* for the nth source signal is,

$$SIR_n=\sum_{\omega}\sum_{t}\frac{X_n(\omega,t)}{Y_n(\omega,t)};Y_n(\omega,t)\neq 0 \qquad (6)$$
$$Y_n(\omega,t)=\sum_{\substack{i=1\\ i\neq n}}^{N}X_i(\omega,t)$$

where N is the number of audio signal considered to be disjoint orthogonal, $X_n(\omega,t)$ is the TF representation (using STFT or HS) of the nth signal. The dimension of *TF* representation using STFT and HS may be different, hence the DO is defined in percentage computed over the entire TF space. Finally the average disjoint orthogonality (*ADO*) is the average of all *SIR*s of individual signal as: $ADO=\frac{1}{N}\sum_{n=1}^{N}SIR_n$. The same process is applied to measure $ADO\in(0, 1)$ for STFT and HS based TF representation of the audio signals. Some experimental results are presented to compare STFT and HS as the TF representation tools of audio signals in terms of disjoint orthogonality.

6 Experimental Results

The separation efficiency of the proposed algorithm is evaluated by separating the signals from two mixtures of three audio sources: speech of two male persons (sm1 and sm2) and speech of a female (sf1). The recording is performed in an anechoic room. The spacing between two microphones is 10cm placed at 1.5m distance from each source. The sources are placed at different azimuth locations (0° to180°). The sampling rate of all the recording was set to 16kHz with 16-bit amplitude resolution.

Three binaural mixtures (m1, m2 and m3) are produced by arranging the sources at different azimuth locations as: m1{sm1(70°), sm2(100°), sf1(140°)}, m2{sm1(50°), sm2(80°), sf1(90°)}, m3{sm1(130°), sm2(90°), sf1(150°)}. The average value of short time energy ratio between original and separated signal is proposed as the criterion to measure the separation efficiency. It is termed as OSSR (original to separated signal ratio) and defined as:

$$OSSR = \left| \frac{1}{T} \sum_{t=1}^{T} \log 10 \left(\frac{\sum_{i=1}^{w} s_{original}^2 (t+i)}{\sum_{i=1}^{w} s_{separated}^2 (t+i)} \right) \right| \quad (7)$$

where $s_{original}$ and $s_{separated}$ are the original and separated signal respectively, w is frame length (10 ms) and T is the data length. If the two signals are same, OSSR=0 and any other value is a measure of their dissimilarity. Smaller value of OSSR indicates better separation. Table 1 shows the average OSSR of each signal for every mixture. It is observed that the separation efficiency is degraded when the sources are placed closely. The separation accuracy is better for larger apart angle between the sources. The separation efficiency is compared for two types of TF representations: HS and STFT. Also the efficiency is compared for two types of stereo mixtures: using HRTF and recorded by two microphones (Mic2). It is noticed that the HS based TF representation improves the separation performance than STFT. Although, HRTF based mixing system has better separation efficiency, it is less applicable in real world applications.

The individual audio signal is projected to TF space using HS and STFT separately to produce some experimental results of DO. Fig. 4(a) shows the comparison between HS and STFT (using Hamming and Hanning window with 60% overlapping) in terms of ADO as a function of the number of frequency bins, whereas Fig. 4(b) presents the comparison as a function of window overlapping. The ADO of the audio signals of HS is better than that of STFT based TF representation. It is obvious to produce better source separation by the proposed method with HS as the TF representation.

Table 1. Experimental results of proposed separation algorithm

Mixture	TF	OSSR of sm1		OSSR of sm2		OSSR of sf1	
		Mic2	HRTF	Mic2	HRTF	Mic2	HRTF
m1	HS	0.0472	0.0401	0.0532	0.0491	0.0642	0.0586
	STFT	0.0781	0.0617	0.0743	0.0687	0.0831	0.0817
m2	HS	0.0519	0.0485	0.0882	0.0803	0.0817	0.0783
	STFT	0.0827	0.0758	0.1035	0.0975	0.1106	0.1047
m3	HS	0.0817	0.0737	0.0534	0.0478	0.0784	0.0711
	STFT	0.1073	0.0902	0.0903	0.0817	0.1012	0.0983

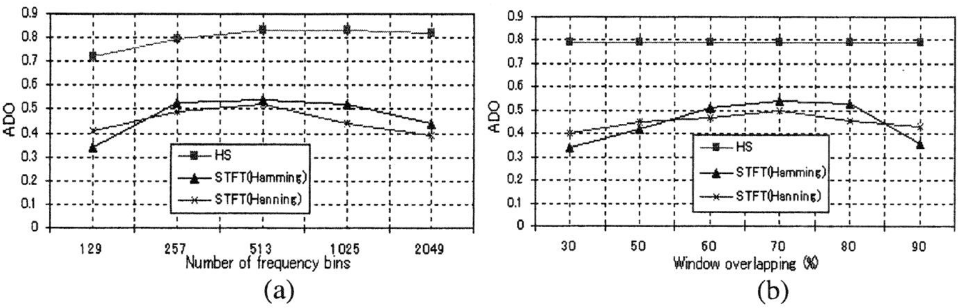

Fig. 4. ADO of HS and STFT as a function of (a)frequency bins, (b) window overlapping

6 Conclusions

We have presented a method of separating mixed audio signals by localizing the sources in TD-ID space. It is assumed that the sources are disjoint orthogonal and the separation is obtained by estimating the binary masks for individual source signal in TF space. The use of HS as the TF representation improves the separation efficiency for both of the mixtures using HRTF as well as microphone pair. The specialty of HS is that the time resolution can be as precise as the sampling period and the frequency resolution depends on the choice up to Nyquist frequency. Hence it serves as the potential TF representation for the consideration of disjoint orthogonality of audio sources. The robust localization and separation of moving sources are the main concern as the future works.

References

1. Yilmaz, O., Rickard, S.:Blind Separation of Speech Mixtures via Time-Frequency Masking, IEEE Transactions on Signal Processing, Vol. 52, No. 7, pages 1830-1847, July 2004.
2. Roman, N., Wang, D., Brown, G. J.: Speech segregation based on sound localization. Acost. Soc. of America, 114(4): 2236-2252, 2003.
3. Baeck, M., Zolzer, U.: Real-Time Implementation of Source Separation Algorithm. DAFx-03, London, UK, 2003.
4. http://sound.media.mit.edu/KEMAR.html
5. Huang, N.E, et. al.: The empirical mode decomposition and Hilbert spectrum for nonlinear and non-stationary time series analysis. Proc. Roy. Soc. London A, Vol. 454: 903-995, 1998.
6. Flandrin, P., Rilling, G., Goncalves, P.: Emperical Mode Decomposition as a filter bank. IEEE Sig. Proc. Letter, 2003.
7. Wu, B. Z., Huang, N. E.: A study of the characteristics of white noise using the empirical mode decomposition method. Proc. R. Soc. Lond. A (460), pp: 1597-1611, 2004.

ICA and Binary-Mask-Based Blind Source Separation with Small Directional Microphones

Yoshimitsu Mori[1], Hiroshi Saruwatari[1], Tomoya Takatani[1], Kiyohiro Shikano[1], Takashi Hiekata[2], and Takashi Morita[2]

[1] Nara Institute of Science and Technology, Nara 630-0192, Japan
yoshim-m@is.naist.jp
[2] Kobe Steel,Ltd., Kobe, 651-2271, Japan
t-hiekata@kobelco.jp

Abstract. A new two-stage blind source separation (BSS) for convolutive mixtures of speech is proposed, in which a Single-Input Multiple-Output (SIMO)-model-based ICA and binary mask processing are combined. SIMO-model-based ICA can separate the mixed signals, not into monaural source signals but into SIMO-model-based signals from independent sources as they are at the microphones. Thus, the separated signals of SIMO-model-based ICA can maintain the spatial qualities of each sound source. Owing to the attractive property, binary mask processing can be applied to efficiently remove the residual interference components after SIMO-model-based ICA. The experimental results using small directional microphone array reveal that the separation performance can be considerably improved by using the proposed method in comparison to the conventional source separation methods.

1 Introduction

Blind source separation (BSS) is the approach taken to estimate original source signals using only the information of the mixed signals observed in each input channel. This technique is based on *unsupervised* filtering, and much attention has been paid to the BSS technique in many fields of signal processing.

In recent works of BSS based on independent component analysis (ICA), various methods have been proposed for acoustic-sound separation [1,2,3]. In this paper, we mainly address the BSS problem under highly reverberant conditions which often arise in many practical audio applications. The separation performance of the conventional ICA is far from being sufficient in such a case because too long separation filters is required but the unsupervised learning of the filter is not so easy. Therefore, one possible improvement is to partly combine ICA with another supervised signal enhancement technique, e.g., spectral subtraction. However, in the conventional ICA framework, each of the separated outputs is a *monaural* signal, and this leads to the drawback that many kinds of superior *multichannel* techniques cannot be applied.

To solve the problem, we propose a novel two-stage BSS algorithm which is applicable to an array of directional microphones. This approach resolves the

J. Rosca et al. (Eds.): ICA 2006, LNCS 3889, pp. 649–657, 2006.

BSS problem into two stages: (a) a Single-Input Multiple-Output (SIMO)-model-based ICA [4] and (b) binary mask processing [5,6] in the time-frequency domain for the SIMO signals obtained from the preceding SIMO-model-based ICA. Here the term "SIMO" represents the specific transmission system in which the input is a single source signal and the outputs are its transmitted signals observed at multiple microphones. SIMO-model-based ICA can separate the mixed signals, not into monaural source signals but into SIMO-model-based signals from independent sources as they are at the microphones. Thus, the separated signals of SIMO-model-based ICA can maintain the spatial qualities of each sound source. After the SIMO-model-based ICA, the residual components of the interference, which are often staying in the output of SIMO-model-based ICA as well as the conventional ICA, can be efficiently removed by the following binary mask processing. The experimental results using small directional microphone array reveal that the proposed method can successfully achieve the BSS for speech mixtures even under a realistic reverberant condition.

2 Mixing Process and Conventional BSS

2.1 Mixing Process

In this study, the number of microphones is K and the number of multiple sound sources is L, where we deal with the case of $K = L$. In the frequency domain, the observed signals in which multiple source signals are mixed are given by

$$\boldsymbol{X}(f) = \boldsymbol{A}(f)\boldsymbol{S}(f), \tag{1}$$

where $\boldsymbol{X}(f) = [X_1(f), \cdots, X_K(f)]^{\mathrm{T}}$ is the observed signal vector, and $\boldsymbol{S}(f) = [S_1(f), \cdots, S_L(f)]^{\mathrm{T}}$ is the source signal vector. Also, $\boldsymbol{A}(f) = [A_{kl}(f)]_{kl}$ is the mixing matrix, where $[X]_{ij}$ denotes the matrix which includes the element X in the i-th row and the j-th column. The mixing matrix $\boldsymbol{A}(f)$ is assumed to be complex-valued because we introduce a model to deal with the arrival lags among microphones and room reverberations.

2.2 Conventional ICA-Based BSS

In the frequency-domain ICA (FDICA), first, the short-time analysis of observed signals is conducted by frame-by-frame discrete Fourier transform (DFT). By plotting the spectral values in a frequency bin for each microphone input frame by frame, we consider them as a time series. Hereafter, we designate the time series as $\boldsymbol{X}(f,t) =[X_1(f,t), \cdots, X_K(f,t)]^{\mathrm{T}}$.

Next, we perform signal separation using the complex-valued unmixing matrix, $\boldsymbol{W}(f) = [W_{lk}(f)]_{lk}$, so that the L time-series output $\boldsymbol{Y}(f,t)=[Y_1(f,t), \cdots, Y_L(f,t)]^{\mathrm{T}}$ becomes mutually independent; this procedure can be given as

$$\boldsymbol{Y}(f,t) = \boldsymbol{W}(f)\boldsymbol{X}(f,t). \tag{2}$$

We perform this procedure with respect to all frequency bins. The optimal $\boldsymbol{W}(f)$ is obtained by, for example, the following iterative updating equation [7]:

$$\boldsymbol{W}^{[i+1]}(f) = \eta\left[\boldsymbol{I} - \left\langle \boldsymbol{\Phi}(\boldsymbol{Y}(f,t))\boldsymbol{Y}^{\mathrm{H}}(f,t)\right\rangle_t\right]\boldsymbol{W}^{[i]}(f) + \boldsymbol{W}^{[i]}(f), \tag{3}$$

where $\boldsymbol{I}$ is the identity matrix, $\langle\cdot\rangle_t$ denotes the time-averaging operator, $[i]$ means the value of the i th step in the iterations, η is the step-size parameter, and $\boldsymbol{\Phi}(\cdot)$ is the appropriate nonlinear function.

2.3 Conventional Binary-Mask-Based BSS

Binary mask processing [5,6] is one of the alternative approach which is aimed to solve the BSS problem, but is not based on ICA. We estimate a binary mask by comparing the amplitudes of the observed signals, and pick up the target sound component which arrives at the *better microphone* closer to the target speech. This procedure is performed in time-frequency regions, and is to pass the specific regions where target speech is dominant and mask the other regions. Under the assumption that the l-th sound source is close to the l-th microphone and $L = 2$, the l-th separated signal is given by

$$\hat{Y}_l(f,t) = m_l(f,t)X_l(f,t), \tag{4}$$

where $m_l(f,t)$ is the binary mask operation which is defined as $m_l(f,t) = 1$ if $X_l(f,t) > X_k(f,t)$ $(k \neq l)$; otherwise $m_l(f,t) = 0$.

This method requires very few computational complexities, and this property is well applicable to real-time processing. The method assumes the sparseness in the spectral components of the sound sources, i.e., there are no overlaps in time-frequency components of the sources. However the assumption does not hold in an usual audio application (indeed, e.g., a mixture of speech and common broadband stationary noise has many overlaps).

3 Proposed Two-Stage BSS Algorithm

3.1 Motivation and Strategy

In the previous research, SIMO-model-based ICA was proposed by, e.g., Takatani et al. [4], and they showed that SIMO-model-based ICA can separate the mixed signals into SIMO-model-based signals at the microphone points. This finding has motivated us to combine the SIMO-model-based ICA and binary masking. That is, the binary mask technique can be applied to the SIMO components of each source obtained from SIMO-model-based ICA. The configuration of the proposed method is depicted in Fig. 1(a). Binary masking which follows SIMO-model-based ICA can remove the residual component of the interference effectively without adding huge computational complexities.

It is worth mentioning that the novelty of this strategy mainly lies in the two-stage idea of the unique combination of SIMO-mode-based ICA and the SIMO-model-based binary mask. To illustrate the novelty of the proposed method, we

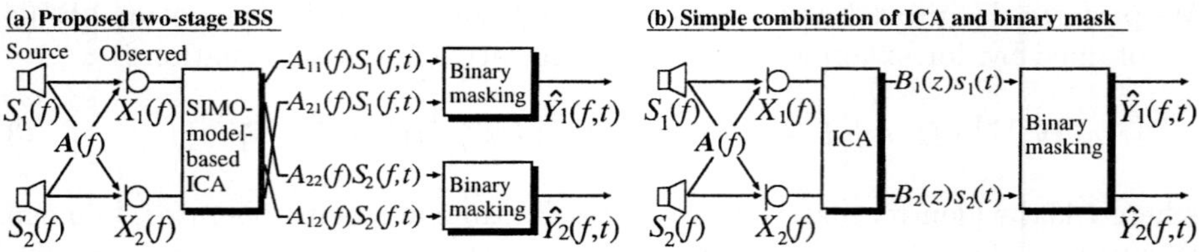

Fig. 1. Input and output relations in (a) proposed two-stage BSS and (b) simple combination of conventional ICA and binary masking, where $K = L = 2$

compare the proposed combination with a simple two-stage combination of a conventional monaural-output ICA and binary masking (see Fig. 1(b)).

In general, the conventional ICAs can only supply the source signals $Y_l(f,t) = B_l(f)S_l(f,t) + E_l(f,t)$ $(l = 1, \cdots, L)$, where $B_l(f)$ is an unknown arbitrary distortion filter and $E_l(f,t)$ is a residual separation error which is mainly caused by an insufficient convergence in ICA. The residual error $E_l(f,t)$ should be removed by binary masking in the next post-processing stage. However, the combination is very problematic and cannot function well because of the existence of the spectral overlaps in the time-frequency domain. For instance, if all sources have nonzero spectral components (i.e., sparseness assumption does not hold) in the specific frequency subband and these are comparable, the decision in binary masking for $Y_1(f,t)$ and $Y_2(f,t)$ is vague and the output results in a ravaged signal. Thus the simple combination of the conventional ICA and binary masking is not valid for solving the BSS problem.

On the other hand, our proposed combination contains the special SIMO-model-based ICA in the first stage. The aim of the SIMO-model-based ICA is to supply the specific SIMO signals with respect to each of sources, $A_{kl}(f)S_l(f,t)$, up to the possible delay of the filters and the residual error. Needless to say, the obtained SIMO components is well applicable to binary masking because of the spatial properties that the separated SIMO component at the specific microphone closer to the target sound still maintains the large gain. Thus, after having the SIMO components, we can introduce the binary mask for the efficient reduction of the remaining error in ICA, even when the sparseness assumption does not hold.

To illustrate the theory with examples, we performed a preliminary experiment in that the binary mask is applied to the ideal solutions of the two types of the ICAs (SIMO-ICA and the simple conventional ICA) under a real acoustic condition described in Sect. 4. A large distortion of 4.6 dB was observed if we directly use binary masking after straight-pass components of each source (binary mask is applied to $A_{11}(f)S_1(f,t)$ and $A_{22}(f)S_2(f,t)$); this means that the simple combination of ICA and binary masking is likely to involve the sound distortion. On the other hand, a small distortion of 0.1 dB was measured in the use of binary masking after SIMO components of each source (e.g., binary mask is applied to $A_{11}(f)S_1(f,t)$ and $A_{21}(f)S_1(f,t)$ for picking up the source 1).

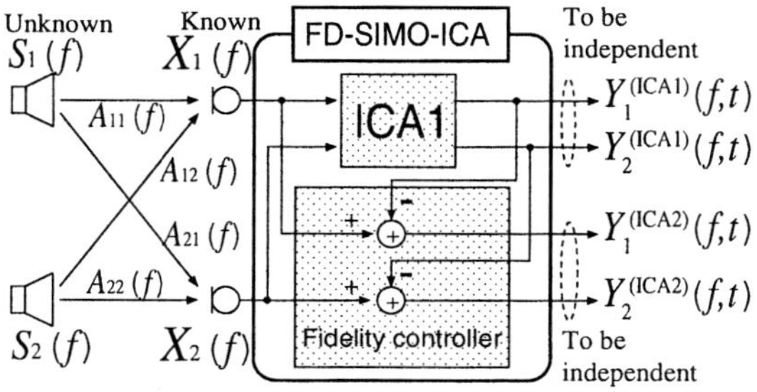

Fig. 2. Input and output relations in the proposed FD-SIMO-ICA, where $K = L = 2$

3.2 Algorithm

Time-domain *SIMO-ICA* [4] has recently been proposed by one of the authors as a means of obtaining SIMO-model-based signals directly in the ICA updating. In this paper, we extend the time-domain SIMO-ICA to frequency-domain SIMO-ICA (FD-SIMO-ICA). FD-SIMO-ICA is conducted for extracting the SIMO-model-based signals corresponding to each of sources. The FD-SIMO-ICA consists of $(L-1)$ FDICA parts and a *fidelity controller*, and each ICA runs in parallel under the fidelity control of the entire separation system (see Fig. 2). The separated signals of the l-th ICA $(l = 1, \cdots L-1)$ are defined by

$$\boldsymbol{Y}_{(\mathrm{ICA}l)}(f,t) = [Y_k^{(\mathrm{ICA}l)}(f,t)]_{k1} = \boldsymbol{W}_{(\mathrm{ICA}l)}(f)\boldsymbol{X}(f,t), \tag{5}$$

where $\boldsymbol{W}_{(\mathrm{ICA}l)}(f) = [W_{ij}^{(\mathrm{ICA}l)}(f)]_{ij}$ is the separation matrix in the l-th ICA.

Regarding the fidelity controller, we calculate the following signal vector $\boldsymbol{Y}_{(\mathrm{ICA}L)}(f,t)$, in which the all elements are to be mutually independent,

$$\boldsymbol{Y}_{(\mathrm{ICA}L)}(f,t) = \boldsymbol{X}(f,t) - \sum_{l=1}^{L-1} \boldsymbol{Y}_{(\mathrm{ICA}l)}(f,t). \tag{6}$$

Hereafter, we regard $\boldsymbol{Y}_{(\mathrm{ICA}L)}(f,t)$ as an output of a *virtual* "L-th" ICA. The reason we use the word "*virtual*" here is that the L-th ICA does not have own separation filters unlike the other ICAs, and $\boldsymbol{Y}_{(\mathrm{ICA}L)}(f,t)$ is subject to $\boldsymbol{W}_{(\mathrm{ICA}l)}(f)$ $(l=1,\cdots,L-1)$.

If the independent sound sources are separated by (5), and simultaneously the signals obtained by (6) are also mutually independent, then the output signals converge on unique solutions, up to the permutation, as

$$\boldsymbol{Y}_{(\mathrm{ICA}l)}(f,t) = \mathrm{diag}\big[\boldsymbol{A}(f)\boldsymbol{P}_l^{\mathrm{T}}\big]\boldsymbol{P}_l\boldsymbol{S}(f,t), \tag{7}$$

where $\boldsymbol{P}_l$ $(l = 1,\cdots,L)$ are exclusively-selected permutation matrices which satisfy $\sum_{l=1}^{L}\boldsymbol{P}_l = [1]_{ij}$. Regarding a proof of this, see [4] with an appropriate modification into the frequency-domain representation. Obviously the solutions given by (7) provide necessary and sufficient SIMO components, $A_{kl}(f)S_l(f,t)$, for each l-th source. Thus, the separated signals of SIMO-ICA can maintain the spatial qualities of each sound source. For example in the case of $L = K = 2$, one possibility is given by

$$[Y_1^{(\mathrm{ICA1})}(f,t),\ Y_2^{(\mathrm{ICA1})}(f,t)]^{\mathrm{T}} = [A_{11}(f)S_1(f,t),\ A_{22}(f)S_2(f,t)]^{\mathrm{T}}, \quad (8)$$

$$[Y_1^{(\mathrm{ICA2})}(f,t),\ Y_2^{(\mathrm{ICA2})}(f,t)]^{\mathrm{T}} = [A_{12}(f)S_2(f,t),\ A_{21}(f)S_1(f,t)]^{\mathrm{T}}, \quad (9)$$

where $\boldsymbol{P}_1 = \boldsymbol{I}$ and $\boldsymbol{P}_2 = [1]_{ij} - \boldsymbol{I}$.

In order to obtain (7), the natural gradient of Kullback-Leibler divergence of (6) with respect to $\boldsymbol{W}_{(\mathrm{ICA}l)}(f)$ should be added to the existing nonholonomic iterative learning rule [1] of the separation filter in the l-th ICA ($l = 1, \cdots, L-1$). The new iterative algorithm of the l-th ICA ($l = 1, \cdots, L-1$) is given as

$$\begin{aligned}
&\boldsymbol{W}^{[j+1]}_{(\mathrm{ICA}l)}(f) \\
&= \boldsymbol{W}^{[j]}_{(\mathrm{ICA}l)}(f) - \alpha \cdot \mathrm{offdiag}\left\langle \boldsymbol{\Phi}(\boldsymbol{Y}^{[j]}_{(\mathrm{ICA}l)}(f,t))\boldsymbol{Y}^{[j]}_{(\mathrm{ICA}l)}(f,t)^{\mathrm{H}}\right\rangle_t \boldsymbol{W}^{[j]}_{(\mathrm{ICA}l)}(f) \\
&\quad + \alpha \cdot \mathrm{offdiag}\left\langle \boldsymbol{\Phi}\left(\boldsymbol{X}(f,t) - \sum_{l=1}^{L-1}\boldsymbol{Y}^{[j]}_{(\mathrm{ICA}l)}(f,t)\right)\left(\boldsymbol{X}(f,t) - \sum_{l=1}^{L-1}\boldsymbol{Y}^{[j]}_{(\mathrm{ICA}l)}(f,t)\right)^{\mathrm{H}}\right\rangle_t \\
&\quad \left(\boldsymbol{I} - \sum_{l=1}^{L-1}\boldsymbol{W}^{[j]}_{(\mathrm{ICA}l)}(f)\right),
\end{aligned} \quad (10)$$

where α is a step-size parameter, and $\boldsymbol{\Phi}(\cdot)$ is a nonlinear function as [7]:

$$\boldsymbol{\Phi}(\boldsymbol{Y}(f,t)) \equiv \left[\tanh(|Y_l(f,t)|)e^{\jmath \cdot \arg(Y_l(f,t))}\right]_{l1}. \quad (11)$$

Also, the initial values of $\boldsymbol{W}_{(\mathrm{ICA}l)}(f)$ for all l should be different.

After FD-SIMO-ICA, binary masking processing is applied. For example in the case of (8) and (9), the resultant output signal corresponding to the source 1 is obtained as follows:

$$\hat{Y}_1(f,t) = m_1(f,t)Y_1^{(\mathrm{ICA1})}(f,t), \quad (12)$$

where $m_1(f,t)$ is the binary mask operation which is defined as $m_1(f,t) = 1$ if $Y_1^{(\mathrm{ICA1})}(f,t)$ is greater than $Y_2^{(\mathrm{ICA2})}(f,t)$; otherwise $m_1(f,t) = 0$. Also, the resultant output signal corresponding to the source 2 is given by

$$\hat{Y}_2(f,t) = m_2(f,t)Y_2^{(\mathrm{ICA1})}(f,t), \quad (13)$$

where $m_2(f,t)$ is the binary mask operation which is defined as $m_2(f,t) = 1$ if $Y_2^{(\mathrm{ICA1})}(f,t)$ is greater than $Y_1^{(\mathrm{ICA2})}(f,t)$; otherwise $m_2(f,t) = 0$. The extension to the general case of $L = K > 2$ can be easily implemented in the same manner.

4 Experiments in Real Acoustic Room

4.1 Conditions for Experiments

We carried out sound-separation experiments using acoustical source signals recorded in the real room illustrated in Fig. 3, where two sources and two directional microphones are set. The reverberation time in this room is 200 ms.

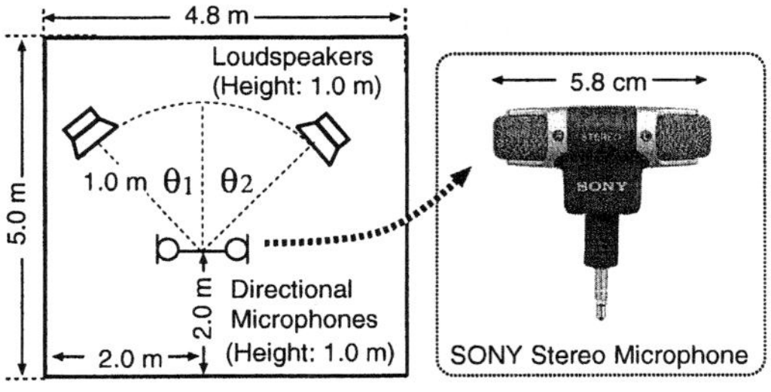

Fig. 3. Layout of reverberant room used in experiments

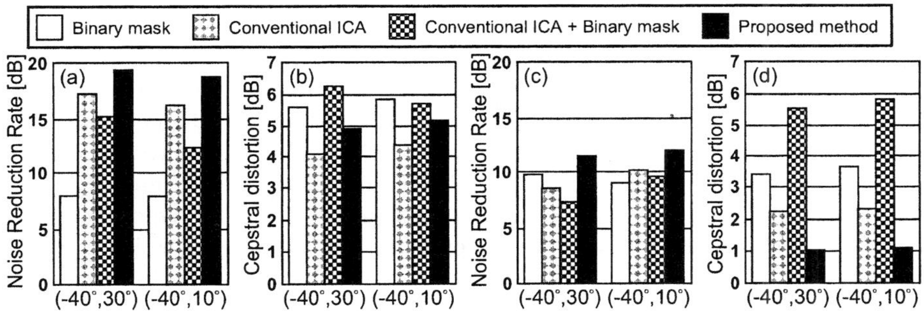

Fig. 4. (a) Results of NRR for speech-speech mixing, (b) CD for speech-speech mixing, (c) NRR for speech-noise mixing, and (d) CD for speech-noise mixing

Two speech signals are assumed to arrive from different directions, θ_1 and θ_2, where we prepare two kinds of source direction patterns as follows; $(\theta_1, \theta_2) = (-40°, 30°)$ or $(-40°, 10°)$. We used the speech signals spoken by two male and two female speakers, and colored stationary noise as the source samples. The sampling frequency is 8 kHz and the length of each sample is limited to 3 s. The DFT size of $\boldsymbol{W}(f)$ is 1024. We use an initial value which is given by null beamformers [3] whose directions of sources are $(-60°, 60°)$. We compare four methods as follows: (A) the conventional binary-mask-based BSS given in Sect. 2.3, (B) the conventional ICA-based BSS given in Sect. 2.2, where the scaling ambiguity can be properly solved by [1], (C) simple combination of the conventional ICA and binary masking, and (D) the proposed two-stage BSS.

4.2 Experimental Evaluation on Separation Performance

Noise reduction rate (NRR) [3], defined as the output signal-to-noise ratio (SNR) in dB minus the input SNR in dB, is used as the objective indication of separation performance. The SNRs are calculated under the assumption that the speech signal of the undesired speaker is regarded as noise.

Figure 4(a) shows the results of NRR for speech-speech mixing under different speaker allocations. These scores are the averages of 12 speaker combinations. Also, Fig. 4(c) shows the results of NRR for the mixing of speech and stationary noise. From the results, we can confirm that the proposed two-stage BSS can improve the separation performance regardless the speaker directions and the noise condition, and the proposed BSS outperforms all of the conventional methods. On the contrary the simple combination of the conventional ICA and binary masking shows deteriorations, and this result is well consistent with the discussion provided in Sect. 3.1.

4.3 Experimental Evaluation on Sound Distortion

Since NRR score indicates only the degree of interference reduction, we could not evaluate the sound quality, i.e., degree of sound distortion in the previous section. In order to assess the distortion of the separated signals, we introduce a measure of *Cepstral Distortion* (CD) which indicates the distance between the spectral envelope of the original source signal and the target component in the separated output. Note that the CD cannot take into account the degree of interference reduction unlike NRR, and thus the CD and NRR are complementary scores. The 20th-order cepstrum based on the smoothed FFT spectrum is used. The CD will be decreased to zero if the processing gives no distortion.

Figure 4(b) depicts the CDs for speech-speech mixing, and Fig. 4(d) for speech-noise mixing. As can be confirmed, the CDs of both the conventional ICA and the proposed method are relatively small in comparison to those of the binary masking and its simple combination with ICA. This means that (1) the conventional binary-mask-based methods involve a heavy distortion due to the improper time variant masking arising in the non-sparse frequency subband, (2) but the proposed method cannot be affected by such an improperness. These facts are promising evidences on the feasibility of the proposed combination technique of SIMO-model-based ICA and binary masking.

5 Conclusion

We proposed a new BSS framework in which the SIMO-model-based ICA and binary mask processing are efficiently combined. In order to evaluate its effectiveness, a separation experiment was carried out under a reverberant condition. The experimental results revealed that the proposed method outperforms the combination of the conventional ICA and binary mask processing as well as the simple ICA and binary mask processing.

Acknowledgment. The authors thank Dr. H. Sawada, Mr. R. Mukai, and Mrs. S. Araki of NTT CS-lab. for their kind discussion on this work. This work was partly supported by CREST "Advanced Media Technology for Everyday Living" of JST in Japan.

References

1. N. Murata and S. Ikeda, "An on-line algorithm for blind source separation on speech signals," *Proc. NOLTA98*, vol.3, pp.923–926, 1998.
2. L. Parra and C. Spence, "Convolutive blind separation of non-stationary sources," *IEEE Trans. Speech & Audio Processing*, vol.8, pp.320–327, 2000.
3. H. Saruwatari, S. Kurita, K. Takeda, F. Itakura, T. Nishikawa and K. Shikano, "Blind source separation combining independent component analysis and beamforming," *EURASIP Journal on Applied Sig. Process.*, vol.2003, pp.1135–1146, 2003.
4. T. Takatani, T. Nishikawa, H. Saruwatari and K. Shikano, "High-fidelity blind separation of acoustic signals using SIMO-model-based ICA with information-geometric learning," *Proc. IWAENC2003*, pp.251–254, 2003.
5. R. Lyon, "A computational model of binaural localization and separation," *Proc. ICASSP83*, pp.1148–1151, 1983.
6. N. Roman, D. Wang and G. Brown, "Speech segregation based on sound localization," *Proc. IJCNN01*, pp.2861–2866, 2001.
7. H. Sawada, R. Mukai, S. Araki and S. Makino, "Polar coordinate based nonlinear function for frequency domain blind source separation," *IEICE Trans. Fundamentals*, vol.E86-A, no.3, pp.590–596, 2003.

Blind Deconvolution with Sparse Priors on the Deconvolution Filters

Hyung-Min Park[1], Jong-Hwan Lee[1],
Sang-Hoon Oh[2], and Soo-Young Lee[1]

[1] Brain Science Research Center and Dept. Biosystems,
Korea Advanced Institute of Science and Technology,
Daejeon, 305-701, Republic of Korea
hmpark@kaist.ac.kr
[2] Dept. Information Communication Eng.,
Mokwon University, Daejeon, 302-729, Republic of Korea

Abstract. In performing blind deconvolution to remove reverberation from speech signal, most acoustic deconvolution filters need a great many number of taps, and acoustic environments are often time-varying. Therefore, deconvolution filter coefficients should find their desired values with limited data, but conventional methods need lots of data to converge the coefficients. In this paper, we use sparse priors on the acoustic deconvolution filters to speed up the convergence and obtain better performance. In order to derive a learning algorithm which includes priors on the deconvolution filters, we discuss that a deconvolution algorithm can be obtained by the joint probability density of observed signal and the algorithm includes prior information through the posterior probability density. Simulation results show that sparseness on the acoustic deconvolution filters can be successfully used for adaptation of the filters by improving convergence and performance.

1 Introduction

Blind deconvolution has become an important topic for research and development in digital signal processing because it has high potential for broad applications in speech enhancement as well as communications. Especially, blind deconvolution in acoustic environments is a very challenging problem because natural acoustic signals are time-correlated and deconvolution for acoustic environments are very complex.

For example, let us consider the teleconferencing problem, in which people talk into a microphone located not at their mouth as in usual telephone conversation, but located some distance away. The speech is reverberated and can have interference among phonemes. In that case, the speech intelligibility is degraded. We can model the situation as a single-input-single-output (SISO) discrete-time linear system, in which the relationship between the input and the output signal is given by

$$x(n) = \sum_{k=0}^{L_m-1} h(k)s(n-k) + v(n). \tag{1}$$

J. Rosca et al. (Eds.): ICA 2006, LNCS 3889, pp. 658–665, 2006.

The goal of blind deconvolution is to recover the input signal $s(n)$ from the output $x(n)$ when the channel $h(k)$ is unknown. Typically, the noise sequence $v(n)$ is modeled by a zero-mean white Gaussian noise process.

Many researchers have studied on the problem and proposed a number of blind deconvolution algorithms [1, 2]. In most of the blind deconvolution methods, a causal finite-impulse-response (FIR) filter as a linear deconvolutive system is used to recover the input signal $s(n)$. Hence, the deconvolutive model can be formulated by

$$u(n) = \sum_{k=0}^{L_a-1} w(k)x(n-k), \tag{2}$$

where $w(k)$ is a filter coefficient of the deconvolution filter. The overall system is shown in Fig. 1. Since the blind deconvolution methods do not have a training sequence, adaptation of $w(k)$ usually makes use of some *a priori* statistical knowledge of the recovered output signal $u(n)$.

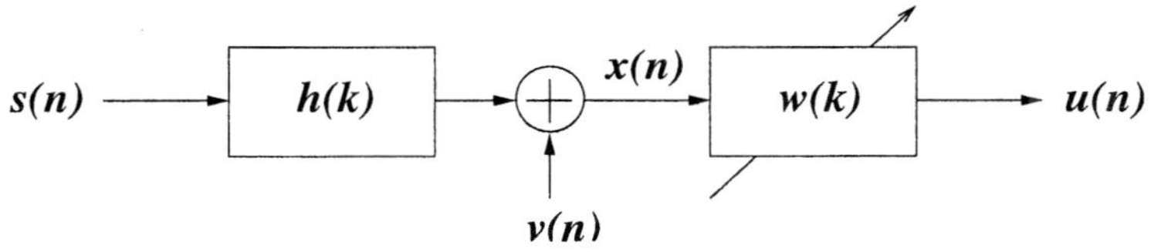

Fig. 1. Overall system for convolution and deconvolution

Among various signals, speech is so time-correlated and non-stationary that many algorithms do not work well to deconvolve it. In order to derive a robust algorithm, entropy might be a good candidate [3]. By forming the cost function as negative entropy of the output probability density function (pdf) $p(u)$ and minimizing the cost using its gradient with respect to coefficients of the deconvolution filter, the learning rule can be obtained in the frequency domain as

$$\Delta W \propto (1/W^* - \text{fft}\{g(\underline{u})\}X^*)|W|^2, \tag{3}$$

where W and X are the discrete Fourier transform of the deconvolution filter and the observed signal $x(n)$, respectively [3]. For the nonlinear function $g(u)$, we can use $-p'(u)/p(u)$.

Although some prior knowledge on the distribution of the recovered output signal is used to adapt the deconvolution filter as we mentioned in the previous paragraph, we do not usually have any assumption on the deconvolution filter $w(k)$. From a point of view, this is advantageous because the learning permits that the filter may have any kind of types. In some application fields, however, we can obtain some knowledge on the deconvolution filter or assume statistics on the filter. In those cases, the filter may be estimated more exactly or easily if we make use of the knowledge or statistics. Especially, when the deconvolution filter is too complex or the number of observed data is too limited to adapt the filter, prior information on the deconvolution filter can play an important role.

Even though assumption on a form of the mixing matrix in independent component analysis gives successful estimation of the image features [4], most of the image feature extraction problems do not suffer from lack of data by obtaining sufficient number of image patches. Thus, the assumption hardly affect the results as critical information. In this paper, we try to set priors on acoustic deconvolution filters for blind deconvolution of speech signals. It is known that most of the acoustic deconvolution filters require lots of taps. In addition, acoustic environments are often time-varying, and deconvolution algorithm should adapt the deconvolution filter in a short time period (i.e., with limited number of data). Therefore, prior information on the deconvolution filter is likely to give better estimates of the filter. The deconvolution filters actually estimate the inverse of acoustic reverberation, and the filters require very different numbers of taps according to the acoustic reverberation. In order to deconvolve various types of acoustic reverberation, we need to use a sufficient number of taps for the deconvolution filter, and a large number of taps are almost zero in most cases. Therefore, we can impose sparse priors on the acoustic deconvolution filters, and the plausible priors help us estimate more exact filters.

2 Another Derivation on the Blind Deconvolution Algorithm

For simple derivation, the deconvolutive model, Eq. (2), can be represented in the form using time-delay operator z^{-1} as

$$u(n) = W(z)x(n), \tag{4}$$

where

$$W(z) = \sum_{k=0}^{L_a-1} w(k)z^{-k}. \tag{5}$$

In order to derive the blind deconvolution algorithm, Eq. (3), let us consider the input and the output signal of the deconvolutive model over a N sample block, defined by the following vectors:

$$\begin{aligned} \mathbf{x} &= [x(0),\ x(1),\ \cdots,\ x(N-1)]^T, \\ \mathbf{u} &= [u(0),\ u(1),\ \cdots,\ u(N-1)]^T. \end{aligned} \tag{6}$$

Both the input and the output signal, $x(n)$ and $u(n)$ are zeros for $n < 0$.

Then, we can write the output signal vector $\mathbf{u}$ as

$$\mathbf{u} = \begin{bmatrix} w(0) & 0 & \cdots & 0 \\ w(1) & w(0) & \cdots & 0 \\ \vdots & \vdots & \ddots & \vdots \\ w(N-1) & w(N-2) & \cdots & w(0) \end{bmatrix} \mathbf{x}. \tag{7}$$

Here, $w(L_a+1) = w(L_a+2) = \cdots = w(N-1) = 0$ by assuming that the length of the channel L_a is much smaller than N.

The joint probability density of the observed signal vector $\mathbf{x}$ can be given by

$$p(\mathbf{x}|W(z)) = |w(0)^N| p(\mathbf{u}), \tag{8}$$

and $p(\mathbf{u}) = p^N(u(n))$ for an i.i.d. signal. Therefore, the log-likelihood of Eq. (8) is

$$L(W(z)) = N \log |w(0)| + N \log p(u(n)). \tag{9}$$

By maximizing the log-likelihood with respect to $w(k)$, the natural gradient algorithm [5, 6] for updating $w(k)$ is given by

$$\Delta w(k) \propto w(k) - \varphi(u(n)) r_k(n), \tag{10}$$

where $\varphi(u(n))$ denotes the score function given by $-p'(u(n))/p(u(n))$, and

$$r_k(n) = \sum_{l=0}^{L_a - 1} w(l) u(n - k + l). \tag{11}$$

The algorithm has the almost same form as in [7].

As an efficient way to implement the algorithm, a frequency-domain processing using the short-time Fourier transform for sample blocks can be considered. The resulting algorithm is the same as Eq. (3).

3 Imposing Sparse Priors on the Deconvolution Filters

In order to make use of priors on the deconvolution filters during adaptation of the filter coefficients, let us reconsider the joint probability density of the observed signal vector, $p(\mathbf{x}|W(z))$. In addition, we assume that the joint probability density of the deconvolution filter $p(W(z))$ is known as prior information. In that case, the logarithm of the posterior probability density can be expressed as

$$\log p(W(z)|\mathbf{x}) = \log p(\mathbf{x}|W(z)) + \log p(W(z)) - \log p(\mathbf{x}). \tag{12}$$

Maximizing $\log p(W(z)|\mathbf{x})$ with respect to $W(z)$ provides a learning algorithm for adapting $W(z)$ with priors on the deconvolution filter. Since the third term of the right side in Eq. (12) does not depend on the deconvolution filter, it does not affect the learning algorithm. Note that the first term is the same as the log-likelihood of Eq. (9). Therefore, Eq. (3) can be used for updating $W(z)$. Adding to Eq. (3), we have to maximize the second term $\log p(W(z))$ with respect to $W(z)$ as

$$\frac{d \log p(W(z))}{dW(z)} = \frac{\frac{dp(W(z))}{dW(z)}}{p(W(z))}. \tag{13}$$

As we mentioned in Section 1, a sufficient number of taps are used for the deconvolution filter, and a large number of taps are almost zero in most cases. Therefore, sparseness can be imposed on the acoustic deconvolution filters. As a simple and general pdf for sparse distribution, Laplacian distribution can be considered. For simple formulation, we also assume that coefficients of the deconvolution filter are i.i.d. Thus, equation for updating $w(k)$ from the second term in Eq. (12) is

$$\begin{aligned}\Delta w(k) &\propto \frac{d \log p(W(z))}{dw(k)} = \frac{dL_a \log p(w(k))}{dw(k)} \\ &\propto -\mathrm{sgn}(w(k)).\end{aligned} \tag{14}$$

In maximizing the posterior probability density, we obtained two equations for learning the deconvolution filter. One of them is processed in the frequency domain whereas the other is performed in the time domain. In real implementation even without prior information on the deconvolution filter, the inverse Fourier transform of the filter should be computed whenever the filter is adapted in the frequency domain. This is because one has to take the part corresponding to length of the deconvolution filter and pad zeros to the remaining part of the block. Therefore, we can easily apply Eq. (15) after the deconvolution filter in the time domain is computed.

Overall procedure for updating the deconvolution filter is as follows:

- Begin
  1. Transform the deconvolution filter into the frequency domain.
  2. Make a sample block.
  3. Transform the block into the frequency domain.
  4. Update the deconvolution filter using Eq. (3).
  5. Transform the deconvolution filter into the time domain.
  6. Update the filter using Eq. (15) and set the outside of the deconvolution filter to zero.
  7. Go to step 1.
- End

4 Experimental Results

We have performed experiments on blind deconvolution to show the effect of sparse priors on the deconvolution filters. Experimental results were compared in terms of the intersymbol interference (ISI) [3, 8], which is computed by

$$\mathrm{ISI}(dB) = 10 \log \left(\frac{\sum_k |t(k)|^2 - \max_k |t(k)|^2}{\max_k |t(k)|^2} \right), \tag{15}$$

where $t(k) = w(k) * h(k)$.

As input data to the SISO linear system of Eq. (1), some speech files from a male speaker were selected in the TIMIT database [9]. The total signal had about 27 second length, and the sampling rate was converted into $8kHz$. It is known that speech signal approximately follows Laplacian distribution. Therefore, sgn(·) was used as $g(\cdot)$ in Eq. (3). In addition, note that speech signal is not i.i.d. When one performs blind deconvolution with speech signal, the deconvolution filter learns from the signal in order to remove not only reverberation in acoustic environments but also correlation or dependence of the speech signal. To avoid the side-effect that deconvolution algorithm removes dependence of speech, a pre-whitening filter has been learned from speech and then used for whitening the signal which was reverberated by a convolutive channel.

The convolutive channel to generate the output signal $x(n)$ of the SISO linear system of Eq. (1) was a 32 tap non-minimum phase filter as shown in Fig. 2. The channel was a part of the impulse response measured in a normal office room. In order to deconvolve the channel, we have employed a 512 tap filter for $w(k)$ with tap-centering initialization. The block size was 1024 to apply the fast Fourier transform. As shown in the SISO linear system, the observed signal $x(n)$ is generally corrupted by noise which may come from various noise sources. In this paper, additive white Gaussian noise was used to corrupt the observed signal.

Fig. 3 displays the ISI for the deconvolution algorithm with sparse priors on the deconvolution filter. In this experiment, the signal-to-noise ratio (SNR) of the observed signal was 15dB. For comparison, the simulation on the algorithm

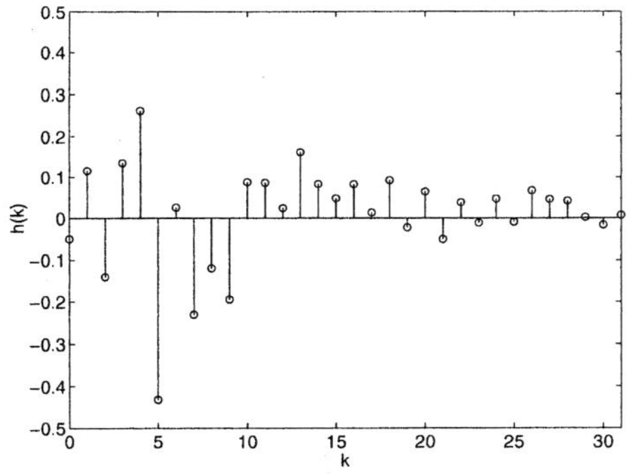

Fig. 2. A 32 tap non-minimum phase convolutive channel

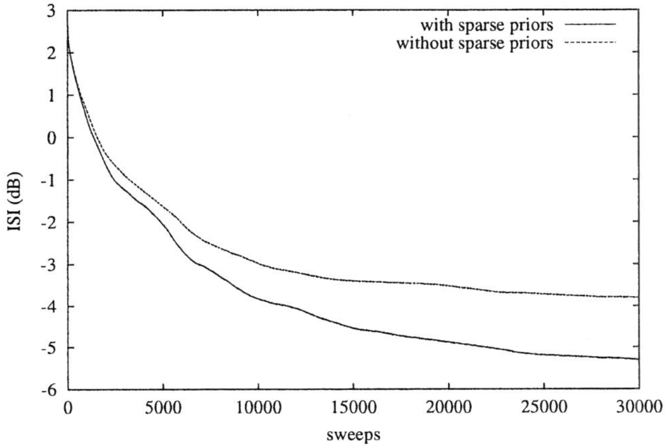

Fig. 3. The ISI for the deconvolution algorithm with the signal whose SNR is 15dB

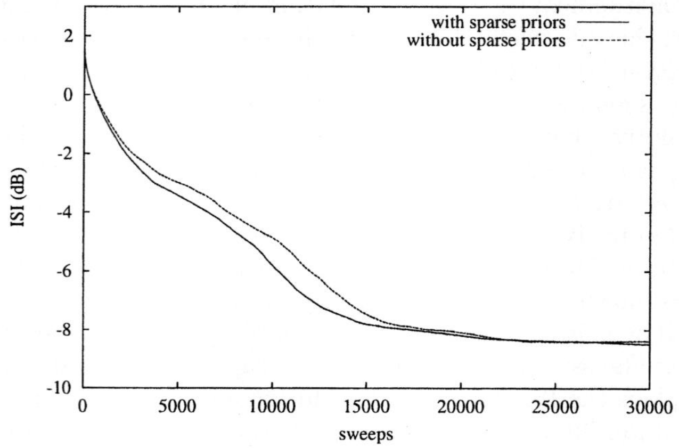

Fig. 4. The ISI for the deconvolution algorithm with the signal whose SNR is 20dB

without sparse priors has been performed with the same parameters, and the result was included. Although additional computation requirement to impose sparse priors was negligible, the deconvolution algorithm with sparse priors on the filter showed faster convergence and better performance than that without the priors. The result indicated that the deconvolution algorithm could not converge to a desired solution without sparse priors and the priors on the filter provided useful information to learn it.

In order to consider the effect of noise, we repeated the simulation for the signal whose SNR was 20dB, and Fig. 4 shows the result. The difference between the ISIs for the deconvolution algorithm with and without the priors was smaller than that in the previous experiment. In this simulation, the observed signal was contaminated with less noise. Even without resort to the priors, hence, the deconvolution algorithm could easily adapt the filter by using the observed data which contained more clear information on the convolutive channel. However, note that convolutive channels in real-world situation are not fixed as in this experiment but time-varying and the noise comes from very various sources such as measurement error and distributed noise.

Strengthening sparse priors excessively by increasing the step-size of Eq. (15) might accelerate the convergence speed in the early stage but disturb the frequency domain update algorithm of Eq. (3). Therefore, the convergence speed might slow down. In order to avoid the disturbance, we need to choose a moderate step-size, and using time-decaying step-sizes can be an appropriate strategy to give fast convergence in the beginning part and not to disturb the frequency domain update algorithm in the ending part.

5 Conclusion

In this paper, we imposed sparse priors on acoustic deconvolution filters for blind deconvolution to remove reverberation from speech signal. In order to include the sparse priors in the deconvolution algorithm, we maximized the posterior probability density of convolved signal with respect to the filters. The resulting algorithm needed negligible additional computation. Simulations indicated that sparseness imposed on the filters could provide useful information to accelerate convergence speed of the filters and provide better performance comparing with completely blind prior information.

Acknowledgments

This work was supported by the Brain Neuroinformatics Research Program sponsored by Korean Ministry of Science and Technology.

References

1. Haykin, S., ed.: Blind Deconvolution. Prentice-Hall, Englewood Cliffs, NJ (1994)
2. Ding, Z., Li, Y., eds.: Blind Equalization and Identification. Marcel Dekker, New York, NY (2001)
3. Lambert, R.H.: Multichannel Blind Deconvolution: FIR Matrix Algebra and Separation of Multipath Mixtures. PhD thesis, University of Southern California, Los Angeles (1996)
4. Hyvärinen, A., Karthikesh, R.: Imposing sparsity on the mixing matrix in independent component analysis. Neurocomputing **49**(1-4) (2002) 151–162
5. Amari, S.: Natural gradient works efficiently in learning. Neural Computation **10**(2) (1998) 251–276
6. Cardoso, J.F., Laheld, B.: Equivariant adaptive source separation. IEEE Transactions on Signal Processing **44**(12) (1996) 3017–3030
7. Amari, S., Douglas, S.C., Cichocki, A., Yang, H.H.: Novel on-line algorithms for blind deconvolution using natural gradient approach. In: IFAC Symposium on System Identification, Kitakyushu, Japan (1997) 1057–1062
8. Sklar, B., ed.: Digital Communications (Fundamentals and Applications). Prentice-Hall, Englewood Cliffs, NJ (1988)
9. DARPA TIMIT acoustic-phonetic continuous speech corpus cd-rom documentation. NIST Speech Disc 1-1.1 (1993)

Estimating the Spatial Position of Spectral Components in Audio

R. Mitchell Parry and Irfan Essa

Georgia Institute of Technology,
College of Computing / GVU Center,
85 5th Street, NW, Atlanta, Georgia, USA
{parry, irfan}@cc.gatech.edu

Abstract. One way of separating sources from a single mixture recording is by extracting spectral components and then combining them to form estimates of the sources. The grouping process remains a difficult problem. We propose, for instances when multiple mixture signals are available, clustering the components based on their relative contribution to each mixture (*i.e.*, their spatial position). We introduce novel factorizations of magnitude spectrograms from multiple recordings and derive update rules that extend independent subspace analysis and non-negative matrix factorization to concurrently estimate the spectral shape, time envelope and spatial position of each component. We show that estimated component positions are near the position of their corresponding source, and that multichannel non-negative matrix factorization can distinguish three pianos by their position in the mixture.

1 Introduction

One way of separating sources from a single mixture recording is by extracting a number of spectral components. These components are then grouped to form the original sources. The difficulty is deciding how to group these components. We propose, for instances when multiple mixture signals are available, grouping components by their relative contribution to each mixture (*i.e.*, their spatial position). A source's spatial position is an important element of the classic problem of blind source separation. One way to represent this information is the spatial mixing matrix $\mathbf{A}$:

$$\mathbf{x}(t) = \mathbf{A}\mathbf{s}(t) \tag{1}$$

where $\mathbf{x}(t)$ is the $M \times 1$ time-varying vector mixture, $\mathbf{s}(t)$ is the $N \times 1$ time-varying vector source, and $\mathbf{A}$ is the $M \times N$ mixing matrix. We use independent component analysis (ICA) to estimate $\mathbf{A}$, $\mathbf{s}$, or the unmixing matrix $\mathbf{W}$ [1]:

$$\hat{\mathbf{s}}(t) = \mathbf{W}\mathbf{x}(t) \tag{2}$$

where $\hat{\mathbf{s}}(t)$ is an estimate of the original sources $\mathbf{s}(t)$.

The fundamental limitation of basic ICA algorithms is that they can only separate as many sources as mixtures (*i.e.*, $N \leq M$). If $N > M$, the system is

J. Rosca et al. (Eds.): ICA 2006, LNCS 3889, pp. 666–673, 2006.

underdetermined and information is lost during the mixing process that cannot be recovered without strong assumptions during unmixing. To address this, independent subspace analysis (ISA) applies the assumption that each source is the sum of one or more spectral components [2-4]. When a classic ICA algorithm operates on the magnitude spectrogram of a *single* mixture signal, it extracts spectral shapes and amplitude envelopes. The difficulty is then grouping these spectral components into source streams. Even though magnitude spectrogram data is always non-negative, ISA decomposes it into components that can be negative. Because of this mismatch, non-negative matrix factorization (NMF) is an alternative way to decompose a magnitude spectrogram into spectral components [5-7]. The main difficulty of both ISA and NMF is how to group spectral components into source streams.

Because multiple mixture signals are often available, it seems reasonable to try to group spectral components according to their spatial positions. Recently, FitzGerald et al. extended non-negative matrix factorization of a single mixture to non-negative tensor factorization of multiple mixtures [8]. We present a different matrix factorization for NMF and ISA that leverages the additional spatial information in multiple mixtures.

2 Fundamental Technologies

We briefly review some of the fundamental technologies that we use for this research: non-negative matrix factorization, independent subspace analysis, and undercomplete independent component analysis.

Non-negative Matrix Factorization. Non-negative matrix factorization (NMF) decomposes a $K \times T$ matrix, $\mathbf{X}$, into the product of a $K \times P$ matrix $\mathbf{B}$ and a $P \times T$ matrix $\mathbf{S}$ [9]. If $\mathbf{X}$ is a spectrogram, K is the number of frequency bins, T is the number of time frames, and P is the desired number of components. Each component comprises a spectral shape in $\mathbf{B}$ and an amplitude envelope in $\mathbf{S}$. NMF minimizes the squared Euclidian error using the following update rules [9]:

$$\mathbf{B}_{kn}^{(i+1)} \leftarrow \mathbf{B}_{kn}^{(i)} \frac{(\mathbf{X}\mathbf{S}^{(i)^{\mathrm{T}}})_{kn}}{(\mathbf{Z}^{(i)}\mathbf{S}^{(i)^{\mathrm{T}}})_{kn}} \tag{3}$$

$$\mathbf{S}_{nt}^{(i+1)} \leftarrow \mathbf{S}_{nt}^{(i)} \frac{(\mathbf{B}^{(i)^{\mathrm{T}}}\mathbf{X})_{nt}}{(\mathbf{B}^{(i)^{\mathrm{T}}}\mathbf{Z}^{(i)})_{nt}} \tag{4}$$

where $\mathbf{Z}^{(i)} = \mathbf{B}^{(i)}\mathbf{S}^{(i)}$ is the current estimate of $\mathbf{X}$. We extend NMF to learn the spectral shape, amplitude envelope, and spatial position of each component. The additional spatial information can be used to cluster notes into instrument streams.

Independent Subspace Analysis. Casey and Westner [2] introduce independent subspace analysis (ISA) in order to separate multiple sources from a single mixture. They use ICA to extract spectral components from the magnitude spectra

$\mathbf{x}(t)$ of a mixture signal, and then group components according to spectral similarity. ICA is performed using a classic algorithm such as Bell and Sejnowski's information maximization algorithm that maximizes the entropy of a nonlinear function of the estimated sources [10]. This leads to an update rule for unmixing matrix $\mathbf{W}$ proportional to the following:

$$\Delta\mathbf{W} \propto \mathbf{W}^{-\mathrm{T}} - 2\mathbf{y}\mathbf{x}^{\mathrm{T}} \tag{5}$$

During an initial whitening stage, they remove less important principal components to achieve a $K : P$ dimensionality reduction. Alternatively, we consider undercomplete independent component analysis for this reduction.

Undercomplete Independent Component Analysis. Undercomplete independent component analysis performs dimensionality reduction during ICA without throwing away information during whitening [11]. This leads to an update rule that does not require a square unmixing matrix as in (5):

$$\Delta\mathbf{W} \propto \left(\mathbf{W}\mathbf{C}\mathbf{W}^{\mathrm{T}}\right)^{-1}(\mathbf{W}\mathbf{C}) - 2\mathbf{y}\mathbf{x}^{\mathrm{T}} \tag{6}$$

We extend ISA to additionally estimate the spatial position of each component. Because we have K frequency bins for each of the M mixtures, we use undercomplete ICA to accomplish the $KM : P$ dimensionality reduction.

3 Multichannel Non-negative Matrix Factorization

We extend non-negative matrix factorization to handle multiple mixtures by concurrently estimating the spatial positions of spectral components. Our underlying assumption is that sources maintain their spatial position and the components maintain their spectral shape across channels. Therefore, a single component may be modeled as a single spectral shape, spatial position, and amplitude envelope.

To accommodate multiple mixtures we introduce an additional $M \times P$ matrix $\mathbf{Q}$. Each column of $\mathbf{Q}$ contains the spatial position of the spectral component represented by the corresponding column in $\mathbf{B}$ and row in $\mathbf{S}$. In order to apply a factorization on magnitude spectra from multiple recordings, $\mathbf{X}_m$ $(1 \le m \le M)$, we construct $\hat{\mathbf{X}} \approx \hat{\mathbf{B}}\hat{\mathbf{Q}}\mathbf{S}$, where $\hat{\mathbf{B}}$ is the multichannel spectral mixing matrix and $\hat{\mathbf{Q}}$ is the multichannel spatial mixing matrix. For $M = 2$,

$$\hat{\mathbf{X}} = \begin{bmatrix}\mathbf{X}_1\\ \mathbf{X}_2\end{bmatrix} \approx \begin{bmatrix}\mathbf{B} & \mathbf{0}\\ \mathbf{0} & \mathbf{B}\end{bmatrix}\begin{bmatrix}\mathbf{Q}_1\\ \mathbf{Q}_2\end{bmatrix}\mathbf{S}, \tag{7}$$

where $\mathbf{Q}_m$ is a diagonal matrix containing the m-th row of $\mathbf{Q}$ on the diagonal. Figure 1 illustrates this factorization highlighting one component with $K = 5$, $M = 2$, $P = 3$,and $T = 7$. We minimize the squared Euclidian error between $\hat{\mathbf{X}}$ and $\hat{\mathbf{B}}\hat{\mathbf{Q}}\mathbf{S}$ and derive the following updates:

$$\mathbf{B}_{kn}^{(i+1)} \leftarrow \mathbf{B}_{kn}^{(i)} \frac{\sum_{m=1}^{M} (\mathbf{X}_m {\mathbf{S}^{(i)}}^{\mathrm{T}} {\mathbf{Q}_m^{(i)}}^{\mathrm{T}})_{kn}}{\sum_{m=1}^{M} (\mathbf{Z}_m^{(i)} {\mathbf{S}^{(i)}}^{\mathrm{T}} {\mathbf{Q}_m^{(i)}}^{\mathrm{T}})_{kn}} \tag{8}$$

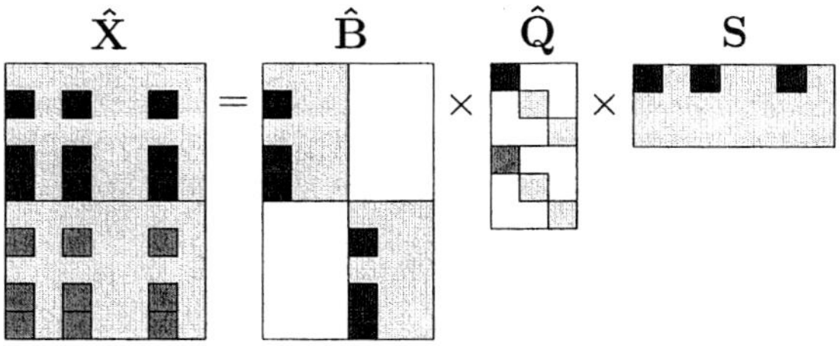

Fig. 1. Multichannel formulation for non-negative matrix factorization

$$\mathbf{Q}_{mn}^{(i+1)} \leftarrow \mathbf{Q}_{mn}^{(i)} \frac{(\mathbf{B}^{(i)^{\mathrm{T}}} \mathbf{X}_m \mathbf{S}^{(i)^{\mathrm{T}}})_{nn}}{(\mathbf{B}^{(i)^{\mathrm{T}}} \mathbf{Z}_m^{(i)} \mathbf{S}^{(i)^{\mathrm{T}}})_{nn}} \tag{9}$$

$$\mathbf{S}_{nt}^{(i+1)} \leftarrow \mathbf{S}_{nt}^{(i)} \frac{\sum_{m=1}^{M} (\mathbf{Q}_m^{(i)^{\mathrm{T}}} \mathbf{B}^{(i)^{\mathrm{T}}} \mathbf{X}_m)_{nt}}{\sum_{m=1}^{M} (\mathbf{Q}_m^{(i)^{\mathrm{T}}} \mathbf{B}^{(i)^{\mathrm{T}}} \mathbf{Z}_m^{(i)})_{nt}} \tag{10}$$

where $\mathbf{Z}_m^{(i)} = \mathbf{B}^{(i)} \mathbf{Q}_m^{(i)} \mathbf{S}^{(i)}$ is the current estimate of $\mathbf{X}_m$, and subscript pairs indicate matrix indexing.

4 Multichannel Independent Subspace Analysis

For multichannel independent subspace analysis, we introduce an $M \times P$ matrix $\mathbf{U}$ containing the spatial unmixing parameters for each component in its columns. We factorize the unmixing system as $\mathbf{S} = \hat{\mathbf{U}}\hat{\mathbf{V}}\hat{\mathbf{X}}$, where $\hat{\mathbf{U}}$ is the multichannel spatial unmixing matrix and $\hat{\mathbf{V}}$ is the multichannel spectral unmixing matrix:

$$\mathbf{S} = \begin{bmatrix} \mathbf{U}_1 \ \mathbf{U}_2 \end{bmatrix} \begin{bmatrix} \mathbf{V} & \mathbf{0} \\ \mathbf{0} & \mathbf{V} \end{bmatrix} \begin{bmatrix} \mathbf{X}_1 \\ \mathbf{X}_2 \end{bmatrix} \tag{11}$$

where $\mathbf{U}_m$ is a diagonal matrix containing the m-th row of $\mathbf{U}$, $\hat{\mathbf{U}} = \hat{\mathbf{Q}}^{\#}$ and $\mathbf{V} = \mathbf{B}^{\#}$, where $^{\#}$ is the Moore-Penrose pseudoinverse [12]. Figure 2 shows the multichannel ISA factorization using the same dimensions as Figure 1. We derive update rules proportional to the following:

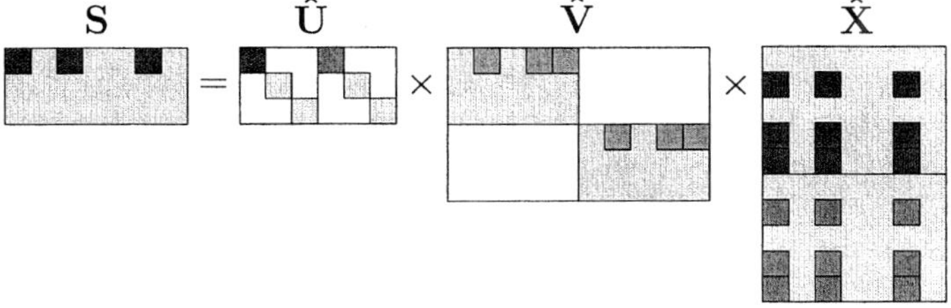

Fig. 2. Multichannel factorization for independent subspace analysis

$$\Delta \mathbf{U}_{mn} \propto \left[(\mathbf{R}_m \mathbf{C} \mathbf{R}_m^{\mathrm{T}})^{-1} \mathbf{R}_m \mathbf{C} \mathbf{V}^{\mathrm{T}} - 2 \mathbf{Y}_m \mathbf{X}_m^{\mathrm{T}} \mathbf{V}^{\mathrm{T}} \right]_{nn} \tag{12}$$

$$\Delta \mathbf{V}_{nk} \propto \sum_{m=1}^{M} \left[\mathbf{U}_m^{\mathrm{T}} (\mathbf{R}_m \mathbf{C} \mathbf{R}_m^{\mathrm{T}})^{-1} \mathbf{R}_m \mathbf{C} - 2 \mathbf{U}_m \mathbf{Y}_m \mathbf{X}_m^{\mathrm{T}} \right]_{nk} \tag{13}$$

where $\mathbf{R}_m = \mathbf{U}_m \mathbf{V}$ and $\mathbf{Y}_m = \tanh(\mathbf{U}_m \mathbf{V} \mathbf{X}_m)$. In order to decrease computation we initially whiten the data in $\hat{\mathbf{X}}$ using a block diagonal "whitening" matrix:

$$\hat{\mathbf{D}} = \begin{bmatrix} \mathbf{D} & \mathbf{0} \\ \mathbf{0} & \mathbf{D} \end{bmatrix} \tag{14}$$

where $\mathbf{D}$ is the whitening matrix for the average magnitude spectra across all channels.

5 Results

We demonstrate our multichannel extensions to NMF and ISA on mixtures of drum and piano music sampled at 11025 Hz. We mix the tracks in the time domain via (1). Then, we compute the magnitude spectrogram of the mixture signals using a Hanning window of 512 samples with 50% overlap and a fast Fourier transform of 1024 samples. We generate the mixtures panning the piano to the right and drum to the left with the following mixing matrix (*i.e.*, $M = N = 2$):

$$\mathbf{A} = \begin{bmatrix} 0.2 & 0.8 \\ 0.8 & 0.2 \end{bmatrix} \tag{15}$$

where each column of $\mathbf{A}$ is the spatial position of a source.

For multichannel NMF, we apply a gradient descent algorithm to the drum and piano mixture. To initialize $\mathbf{B}$ and $\mathbf{S}$, we apply successive updates of (3) and (4) on the average magnitude spectrogram of the mixtures. After convergence, we set the minimum value in $\mathbf{B}$ and $\mathbf{S}$ to a small factor to avoid clamping at zero with the multiplicative updates. Finally, we alternately apply (8), (9), and (10) to extract $P = 7$ components. Throughout the estimation, we maintain unit norm columns of $\mathbf{B}$ and $\mathbf{Q}$. For multichannel ISA, we apply a block whitening matrix $\hat{\mathbf{D}}$ that reduces the dimensionality from KM to $50M$ before alternate updates using (12) and (13) to extract $P = 7$ components.

The left and right side of Figure 3 shows the extracted components using multichannel NMF and multichannel ISA, respectively. Figure 3(a) and 3(b) plots the time envelope of the components. The envelopes show that components 2, 6, and 7 from NMF and components 1 and 2 from ISA represent the short spiked attacks of the drums. The other components are from the piano. Because the NMF components contain only non-negative values, they are generally easier to interpret than the ISA components. For example, the piano components in Figure 3(a) have sharp attacks and smooth decay illustrated by roughly right-triangular onsets. This detail is less prevalent in the ISA components especially at lower energy levels.

The component spectra in Figure 3(c) show the harmonic content of the piano and the noisy or low-frequency content of the drums. The larger peaks in the piano components occur at roughly linearly spaced frequencies indicating a harmonic relationship between them. This structure is more apparent in NMF components 3, 4, and 5. The noisy frequency content in component 2, and low-frequency concentration in components 6 and 7 are characteristic of the drums. This structure is difficult to see in the ISA components in Figure 3(d). Figure 3(e) and 3(f) show the component positions. These positions verify what we can see in the temporal envelopes and frequency content of the components. The drum components cluster on the left and the piano components cluster on the right.

When applied to more difficult examples, multichannel ISA was less predictable and generally less informative than multichannel NMF. For example, sources that contain highly similar spectra are difficult for ISA to handle. When

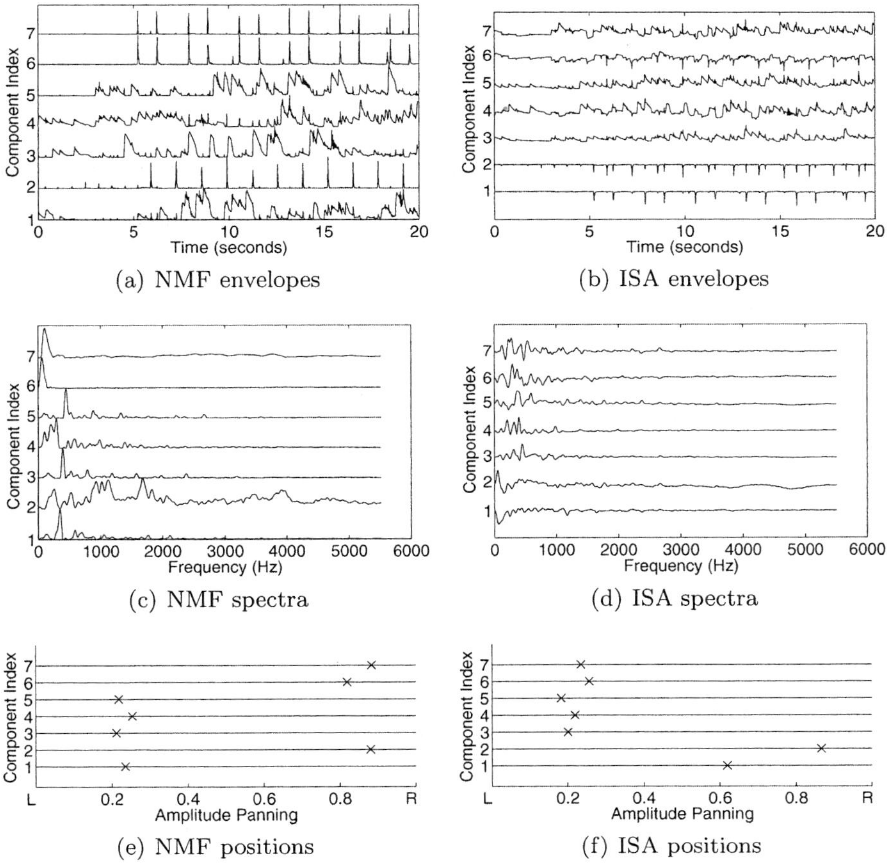

Fig. 3. Extracted components from drums and piano using multichannel NMF (*left*) and multichannel ISA (*right*)

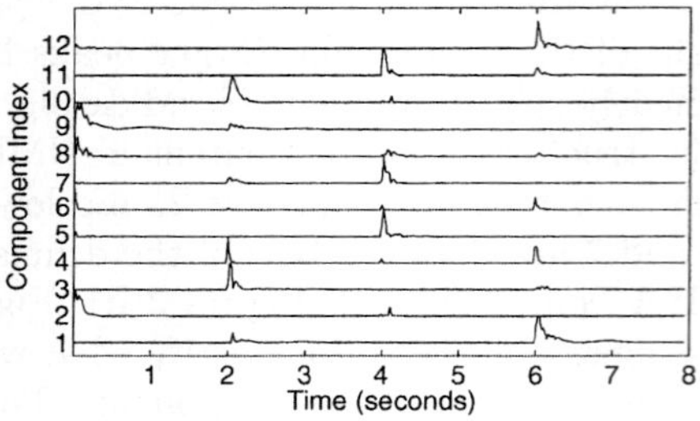

(a) NMF envelopes

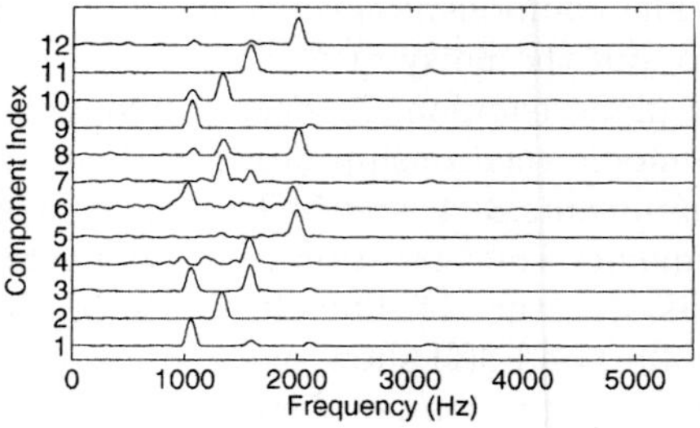

(b) NMF spectra

Fig. 4. Extracted component envelopes and spectra for multichannel NMF and three piano sources

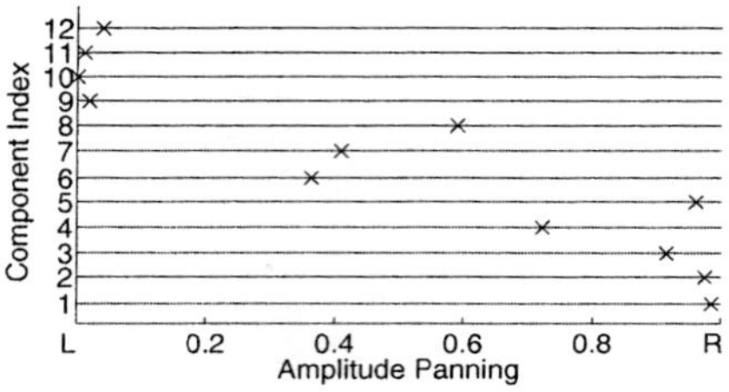

Fig. 5. Extracted component positions for multichannel NMF and three piano sources

applied to magnitude spectrograms, ICA generates linearly independent spectral shapes. Therefore, it is impossible for two components to represent the same spectra. In contrast, multichannel NMF only requires the non-negativity of source components.

We apply multichannel NMF to three pianos playing the same four notes in different orders. Piano 1, 2 and 3, are positioned to the left, center, and right in the stereo mixture, respectively. Figure 4 shows $P = 12$ extracted components. Components 9–12 clearly represent piano 1 playing the notes in order from low to high. Each component is roughly one note represented by a temporal spike in Figure 4(a), one dominating frequency in Figure 4(b), and cluster together on the left side of Figure 5. In a similar way, components 1, 2, 3, and 5 represent piano 3, except component 3 contains two frequency peaks instead of one. The remaining components capture parts of piano 2. However, each contain multiple frequency concentrations and are generally less distinct. In spite of this, each source can be distinguished by its stereo position in Figure 5.

6 Conclusion

We propose leveraging the additional spatial information in multichannel mixtures to cluster components spatially. We introduce novel factorizations of magnitude spectrograms from multiple mixture signals. We derive update rules that extend independent subspace analysis and non-negative matrix factorization to concurrently estimate the spectral shape, temporal envelope, and spatial position of each component. Finally, we show that estimated component positions are near the position of their corresponding source. On a mixture of three pianos playing the same notes in different orders, multichannel NMF extracts components that can be distinguished by their spatial positions.

References

1. Hyvärinen, A.: Independent Component Analysis. New York: Wiley (2001)
2. Casey, M., Westner, W.: Separation of mixed audio sources by independent subspace analysis. In: Proceedings of the International Computer Music Conference, Berlin (2000)
3. Smaragdis, P.: Redundancy Reduction for Computational Audition, a Unifying Approach. PhD thesis, MAS Department, Massachusetts Institute of Technology (2001)
4. FitzGerald, D., Coyle, E., Laylor, B.: Sub-band independent subspace analysis for drum transcription. In: Proceedings of International Conference on Digital Audio Effects, Hamburg, Germany (2002) 65–69
5. Brown, J.C., Smaragdis, P.: Independent component analysis for automatic note extraction from musical trills. Journal of the Acoustical Society of America **115**(5) (2004) 2295–2306
6. Abdallah, S.A., Plumbley, M.D.: Polyphonic transcription by non-negative sparse coding of power spectra. In: Proceedings of the International Conference on Music Information Retrieval, Barcelona, Spain (2004) 318–325
7. Smaragdis, P., Brown, J.C.: Non-negative matrix factorization for polyphonic music transcription. In: Workshop on Applications of Signal Processing to Audio and Acoustics, New Paltz, NY (2003) 177–180
8. FitzGerald, D., Cranitch, M., Coyle, E.: Non-negative tensor factorisation for sound source separation. In: Proceedings of Irish Signals and Systems Conference, Dublin, Ireland (2005)
9. Lee, D.D., Seung, H.S.: Algorithms for non-negative matrix factorization. In: Advances in Neural Information Processing Systems 13. MIT Press (2001) 556–562
10. Bell, A., Sejnowski, T.J.: An information-maximization approach to blind separation and blind deconvolution. Neural Computation **7** (1995) 1129–1159
11. Stone, J.V., Porrill, J.: Undercomplete independent component analysis for signal separation and dimension reduction. Technical report, Department of Psychology, University of Sheffield, Sheffield, England (1997)
12. Trefethen, L.N., Bau, D.B.: Numerical Linear Algebra. Philadelphia: SIAM (1997)

Separating Underdetermined Convolutive Speech Mixtures

Michael Syskind Pedersen[1,2], DeLiang Wang[3], Jan Larsen[1], and Ulrik Kjems[2]

[1] Informatics and Mathematical Modelling, Technical University of Denmark, Richard Petersens Plads, Building 321, DK-2800 Kgs. Lyngby, Denmark
[2] Oticon A/S, Kongebakken 9, DK-2765 Smørum, Denmark
[3] Department of Computer Science and Engineering & Center for Cognitive Science, The Ohio State University, Columbus, OH 43210-1277, USA
{msp, jl}@imm.dtu.dk, uk@oticon.dk, dwang@cse.ohio-state.edu

Abstract. A limitation in many source separation tasks is that the number of source signals has to be known in advance. Further, in order to achieve good performance, the number of sources cannot exceed the number of sensors. In many real-world applications these limitations are too restrictive. We propose a method for underdetermined blind source separation of convolutive mixtures. The proposed framework is applicable for separation of instantaneous as well as convolutive speech mixtures. It is possible to iteratively extract each speech signal from the mixture by combining *blind source separation* techniques with *binary time-frequency masking*. In the proposed method, the number of source signals is not assumed to be known in advance and the number of sources is not limited to the number of microphones. Our approach needs only two microphones and the separated sounds are maintained as stereo signals.

1 Introduction

Blind source separation (BSS) addresses the problem of recovering N unknown source signals $\mathbf{s}(n) = [s_1(n), \ldots, s_N(n)]^T$ from M recorded mixtures $\mathbf{x}(n) = [x_1(n), \ldots, x_M(n)]^T$ of the source signals. The term 'blind' refers to that only the recorded mixtures are known. An important application for BSS is separation of speech signals. The recorded mixtures are assumed to be linear superpositions of the source signals. Such a linear mixture can either be instantaneous or convolutive. The instantaneous mixture is given as

$$\mathbf{x}(n) = \mathbf{A}\mathbf{s}(n) + \boldsymbol{\nu}(n), \quad (1)$$

where $\mathbf{A}$ is an $M \times N$ mixing matrix and n denotes the discrete time index. $\boldsymbol{\nu}(n)$ is additional noise. A method to retrieve the original signals up to an arbitrary permutation and scaling is independent component analysis (ICA) [1]. In ICA, the main assumption is that the source signals are independent. By applying ICA, an estimate $\mathbf{y}(n)$ of the source signals can be obtained by finding a (pseudo)inverse $\mathbf{W}$ of the mixing matrix so that

$$\mathbf{y}(n) = \mathbf{W}\mathbf{x}(n). \quad (2)$$

J. Rosca et al. (Eds.): ICA 2006, LNCS 3889, pp. 674–681, 2006.

Notice, this inversion is not exact when noise is included in the mixing model. When noise is included as in (1), $\mathbf{x}(n)$ is a nonlinear function of $\mathbf{s}(n)$. Still, the inverse system is assumed to be approximated by a linear system.

The convolutive mixture is given as

$$\mathbf{x}(n) = \sum_{k=0}^{K-1} \mathbf{A}_k \mathbf{s}(n-k) + \boldsymbol{\nu}(n) \tag{3}$$

Here, the source signals are mixtures of filtered versions of the original source signals. The filters are assumed to be causal and of finite length K. The convolutive mixture is more applicable for separation of speech signals because the convolutive model takes reverberations into account. The separation of convolutive mixtures can either be performed in the time or in the frequency domain. The separation system for each discrete frequency ω is given by

$$\mathbf{Y}(\omega, t) = \mathbf{W}(\omega)\mathbf{X}(\omega, t), \tag{4}$$

where t is the time frame index. Most methods, both instantaneous and convolutive, require that the number of source signals is known in advance. Another drawback of most of these methods is that the number of source signals is assumed not to exceed the number of microphones, i.e. $M \geq N$.

If $N > M$, even if the mixing process is known, it may not be invertible, and the independent components cannot be recovered exactly [1]. In the case of more sources than sensors, the *underdetermined/overcomplete* case, successful separation often relies on the assumption that the source signals are sparsely distributed in the time-frequency domain [2], [3]. If the source signals do not overlap in the time-frequency domain, high-quality reconstruction could be obtained [3].

However, there is overlap between the source signals. In this case, good separation can still be obtained by applying a binary time-frequency (T-F) mask to the mixture [2], [3]. In *computational auditory scene analysis*, the technique of T-F masking has been commonly used for years (see e.g. [4]). Here, source separation is based on organizational cues from auditory scene analysis [5]. More recently the technique has also become popular in blind source separation, where separation is based on non-overlapping sources in the T-F domain [6]. T-F masking is applicable to source separation/ segregation using one microphone [4],[7],[8] or more than one microphone [2], [3]. T-F masking is typically applied as a binary mask. For a binary mask, each T-F unit is either weighted by one or zero. An advantage of using a binary mask is that only a binary decision has to be made [9]. Such a decision can be based on, e.g., clustering [2], [3], [6], or direction-of-arrival [10]. ICA has been used in different combinations with the binary mask. In [10], separation is performed by first removing $N-M$ signals via masking and afterwards applying ICA in order to separate the remaining M signals. ICA has also been used in the other way around. In [11], it has been applied to separate two signals by using two microphones. Based on the ICA outputs, T-F masks are estimated and a mask is applied to each of the ICA outputs in order to improve the signal to noise ratio (SNR).

In this paper, we propose a method to segregate an arbitrary number of speech signals in a reverberant environment. We extend a previously proposed method for separation of instantaneous mixtures [12] to separation of convolutive mixtures. Based on the output of a square (2×2) blind source separation algorithm and binary T-F masks, our method segregates speech signals iteratively from the mixtures until an estimate of each signal is obtained.

2 Blind Extraction by Combining BSS and Binary Masking

With only two microphones, it is not possible to separate more than two signals from each other because only one null direction can be placed for each output. This fact does not mean that the blind source separation solution is useless in the case of $N > M$. In [12] we examined what happened if an ICA algorithm was applied to an underdetermined 2-by-N mixture. When the two outputs were considered, we found that the ICA algorithm separates the mixtures into subspaces, which are as independent as possible. Some of the source signals are mainly in one output while other sources mainly are present in the other output.

A flowchart for the algorithm is given in Fig. 1. As described in the previous section, a two-input-two-output blind source separation algorithm has been applied to the input mixtures, regardless the number of source signals that actually exist in the mixture. The two output signals are arbitrarily scaled. Different methods have been proposed in order to solve the scaling ambiguity. Here, we assume that all source signals have the same variance as proposed in [1] and the outputs are therefore scaled to have the same variance.

The two re-scaled output signals, $\hat{y}_1(n)$ and $\hat{y}_2(n)$, are transformed into the frequency domain e.g. using the Short-Time Fourier Transform STFT so that two spectrograms are obtained:

$$\hat{y}_1 \rightarrow Y_1(\omega, t) \tag{5}$$
$$\hat{y}_2 \rightarrow Y_2(\omega, t), \tag{6}$$

where ω denotes the frequency and t is the time frame index. The binary masks are then determined for each T-F unit by comparing the amplitudes of the two spectrograms:

$$\text{BM1}(\omega, t) = \tau |Y_1(\omega, t)| > |Y_2(\omega, t)| \tag{7}$$
$$\text{BM2}(\omega, t) = \tau |Y_2(\omega, t)| > |Y_1(\omega, t)|, \tag{8}$$

where τ is a threshold. Next, each of the two binary masks is applied to the original mixtures in the T-F domain, and by this non-linear processing, some of the speech signals are *removed* by one of the masks while other speakers are removed by the other mask. After the masks have been applied to the signals, they are reconstructed in the time domain by the inverse STFT. If there is only a single signal left in the masked output, defined by the selection criteria in Section 2.3, i.e. all but one speech signal have been masked, this signal is considered extracted from the mixture and it is saved. If there are more than one signal left in the masked outputs, the procedure is applied to the two masked signals again and a new set of masks are created based on (7), (8) and the previous masks. The use of the previous mask ensures that T-F units that have been removed from the mixture are not reintroduced by the next mask. This is done by an element-wise multiplication between the previous mask and the new mask. This iterative procedure is followed until all masked outputs consist of only a single speech signal. When the procedure stops, the correlation between the segregated sources are found in order to determine whether a source signal has been segregated more than once. If so, the source is re-estimated by merging the two correlated masks. It is important to notice that the iteratively updated mask always is applied to the original mixtures and not to the previously masked signal. Hereby a deterioration of the signal due to multiple iterations is avoided.

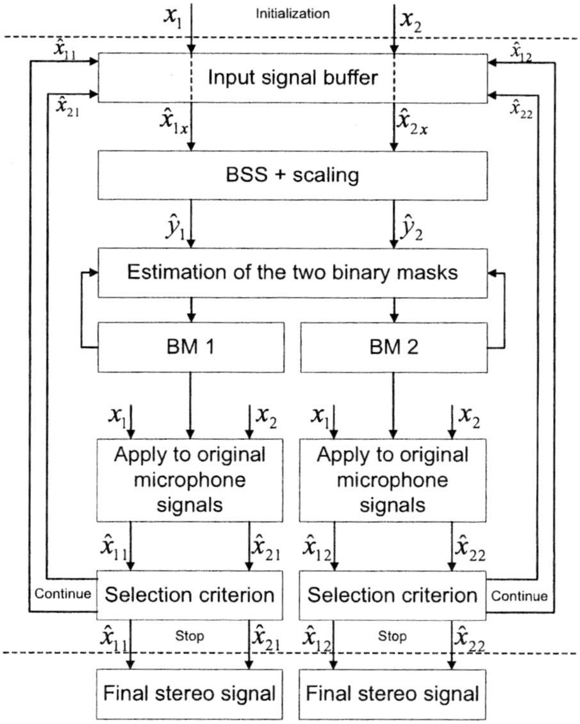

Fig. 1. Flowchart showing the main steps of the proposed algorithm. From the output of the BSS algorithm, binary masks are estimated. The binary masks are applied to the original signals which again are processed through the BSS step. Every time the output from one of the binary masks is detected as a single signal, the signal is stored. The iterative procedure stops when all outputs only consist of a single signal. The flowchart has been adopted from [12].

2.1 Finding the Background Signals

Since some signals may have been removed by both masks, all T-F units that have not been assigned the value '1' are used to create a *background mask*, and the procedure is applied to the mixture signal after the remaining mask is applied, to ensure that all signals are estimated. Notice that this step has been omitted from Fig. 1.

2.2 Extension to Convolutive Mixtures

Each convolutive mixture is given by a linear superposition of filtered versions of each of the source signals. The filters are given by the impulse responses from each of the sources to each of the microphones. An algorithm capable of separating convolutive mixtures is used in the BSS step. Separation still relies on the fact that the source signals can be grouped such that one output mainly contains one part of the source signals and the other output mainly contains the other part of the signals. In order to avoid arbitrary filtering, only the cross channels of the separation filters have been estimated. The direct channel is constrained to be an impulse. Specifically, we employ the frequency domain convolutive BSS algorithm by Parra and Spence [13][1].

2.3 Selection Criterion

In order to decide if all but one signal have been removed, we consider the envelope statistics of the signal. By considering the envelope histogram, it can be determined whether one or more than one signal is present in the mixture. If only one speech signal is present, many of the amplitude values are close to zero. If more speech signals are present, less amplitude values are close to zero. In order to discriminate between one and more than one speech signals in the mixture, we measure the width of the histogram as proposed in [14] as the distance between the 90% and the 10% percentile normalized to the 50% percentile, i.e.

$$\text{width} = \frac{P_{90} - P_{10}}{P_{50}}. \quad (9)$$

Further processing on a pair of masked signals should be avoided if there is one or zero speech signals in the mixture. If the calculated width is smaller than two, we assume that the masked signal consists of more than one speech signal. We discriminate between zero and one signal by considering the energy of the segregated signal. This selection criterion is more robust to reverberations than the correlation-based criterion used in [12].

3 Evaluation

The algorithm described above has been implemented and evaluated with instantaneous and convolutive mixtures. For the STFT, an FFT length of 2048

[1] Matlab code is available from `http://ida.first.gmd.de/~harmeli/download/download_convbss.html`

has been used. A Hanning window with a length of 512 samples has been applied to the FFT signal and the frame shift is 256 samples. A high frequency resolution is found to be necessary in order to obtain good performance. The sampling frequency of the speech signals is 10 kHz, and the duration of each signal is 5 s. The thresholds have been found from initial experiments. In the ICA step, the separation matrix is initialized by the identity matrix, i.e. $\mathbf{W} = \mathbf{I}$. When using a binary mask, it is not possible to reconstruct the speech signal as if it was recorded in the absence of the interfering signals, because the signals partly overlap. Therefore, as a computational goal for source separation, we employ the *ideal binary mask*[9]. The ideal binary mask for a signal is found for each T-F unit by comparing the energy of the desired signal to the energy of all the interfering signals. Whenever the signal energy is higher, the T-F unit is assigned the value '1' and whenever the interfering signals have more energy, the T-F unit is assigned the value '0'. As in [8], for each of the separated signals, the percentage of energy loss P_{EL} and the percentage of noise residue P_{NR} are calculated as well as the signal to noise ratio (SNR) using the resynthesized speech from the ideal binary mask as the ground truth:

$$P_{EL} = \frac{\sum_n e_1^2(n)}{\sum_n I^2(n)}, \quad P_{NR} = \frac{\sum_n e_2^2(n)}{\sum_n O^2(n)}, \quad \text{SNR} = 10\log_{10}\left[\frac{\sum_n I^2(n)}{\sum_n (I(n)-O(n))^2}\right],$$

where $O(n)$ is the estimated signal, and $I(n)$ is the recorded mixture resynthesized after applying the ideal binary mask. $e_1(n)$ denotes the signal present in $I(n)$ but absent in $O(n)$ and $e_2(n)$ denotes the signal present in $O(n)$ but absent in $I(n)$. The input signal to noise ratio, SNR_i, is found too, which is the ratio between the desired signal and the noise in the recorded mixtures.

Convolutive mixtures consisting of four speech signals have also been separated. The signals are uniformly distributed in the interval $0° \leq \theta \leq 180°$. The mixtures have been obtained with room impulse responses synthesized using the image model [15]. The estimated room reverberation time is $T_{60} \approx 160$ ms. The distance between the microphones is 20 cm. The method has been evaluated with and without the proposed selection criterion described in Section 2.3. When the selection criterion was not used, it has been decided when a source signal has been separated by listening to the signals. The separation results are shown in Table 1 and Table 2. The average input SNR is −4.91 dB. When the selection criterion was applied manually, the average SNR after separation is 1.91 dB

Table 1. Separation results for four convolutively mixed speech mixtures. A manual selection criterion was used.

Signal No.	P_{EL}(%)	P_{NR}(%)	SNR_i (dB)	SNR (dB)
1	66.78	20.41	-4.50	1.35
2	32.29	41.20	-4.50	1.24
3	52.86	19.08	-3.97	2.12
4	15.78	30.39	-6.67	2.91
Average	41.93	27.77	-4.91	1.91

Table 2. Separation results for four convolutively mixed speech mixtures. The selection criterion as proposed in Section 2.3 was used.

Signal No.	P_{EL}(%)	P_{NR}(%)	SNR$_i$ (dB)	SNR (dB)
1	39.12	46.70	-4.50	0.63
2	64.18	18.62	-4.50	1.45
3	26.88	33.73	-3.97	2.40
4	45.27	32.49	-6.67	1.69
Average	43.86	32.88	-4.91	1.54

with an average SNR gain of 6.8 dB. When selection criterion was applied as proposed, the average SNR after separation is 1.45 dB with an average SNR gain of 6.4 dB, which is about half a dB worse than selecting the segregated signals manually. It is not always that all the sources are extracted from the mixture. Therefore the selection criterion could be further improved. For separation of instantaneous mixtures an SNR gain of 14 dB can be obtained, which is significantly higher than that for the reverberant case. This may be explained by several factors. Errors such as misaligned permutations are introduced from the BSS algorithm. Also, convolutive mixtures are not as sparse in the T-F domain as instantaneous mixtures. Further, the assumption that the same signals group into the same groups for all frequencies may not hold. Some artifacts (musical noise) exist in the segregated signals. Especially in the cases, where the values of P_{EL} and P_{NR} are high. Separation results are available for listening at `www.imm.dtu.dk/~msp`.

As mentioned earlier, several approaches have been recently proposed to separate more than two sources using two microphones by employing binary T-F masking [2], [3], [10]. These methods use clustering of amplitude and time differences between the microphones. In contrast, our method separates speech mixtures by iteratively extracting individual source signals. Our results are quite competitive although rigorous statements about comparison are difficult because the test conditions are different.

4 Concluding Remarks

A novel method of blind source separation of underdetermined mixtures has been described. Based on sparseness and independence, the method iteratively extracts all the speech signals. The linear processing from BSS methods alone cannot separate more sources than the number of recordings, but with the additional nonlinear processing introduced by the binary mask, it is possible to separate more sources than the number of sensors. Our method is applicable to separation of instantaneous as well as convolutive mixtures and the output signals are maintained as stereo signals. An important part of the method is the detection of when a single signal exists at the output. Future work will include better selection criteria to detect a single speech signal, especially in a reverberant environment. More systematic evaluation and comparison will also be given in the future. The assumption of two microphones may be relaxed and the method may also be applicable to other signals than speech which also have significant redundancy.

Acknowledgements

The work was performed while M.S.P. was a visiting scholar at The Ohio State University Department of Computer Science and Engineering. M.S.P was supported by the Oticon Foundation. M.S.P and J.L are partly also supported by the European Commission through the sixth framework IST Network of Excellence: PASCAL. D.L.W was supported in part by an AFOSR grant and an AFRL grant.

References

1. Hyvärinen, A., Karhunen, J., Oja, E.: Independent Component Analysis. Wiley (2001)
2. Roman, N., Wang, D.L., Brown, G.J.: Speech segregation based on sound localization. J. Acoust. Soc. Amer. **114** (2003) 2236–2252
3. Yilmaz, O., Rickard, S.: Blind separation of speech mixtures via time-frequency masking. IEEE Trans. Signal Processing **52** (2004) 1830–1847
4. Wang, D.L., Brown, G.J.: Separation of speech from interfering sounds based on oscillatory correlation. IEEE Trans. Neural Networks **10** (1999) 684–697
5. Bregman, A.S.: Auditory Scene Analysis. 2 edn. MIT Press (1990)
6. Jourjine, A., Rickard, S., Yilmaz, O.: Blind separation of disjoint orthogonal signals: Demixing N sources from 2 mixtures. In: Proc. ICASSP. (2000) 2985–2988
7. Roweis, S.: One microphone source separation. In: NIPS'00. (2000) 793–799
8. Hu, G., Wang, D.L.: Monaural speech segregation based on pitch tracking and amplitude modulation. IEEE Trans. Neural Networks **15** (2004) 1135–1150
9. Wang, D.L.: On ideal binary mask as the computational goal of auditory scene analysis. In Divenyi, P., ed.: Speech Separation by Humans and Machines. Kluwer, Norwell, MA (2005) 181–197
10. Araki, S., Makino, S., Sawada, H., Mukai, R.: Underdetermined blind separation of convolutive mixtures of speech with directivity pattern based mask and ICA. In: Proc. ICA'2004. (2004) 898–905
11. Kolossa, D., Orglmeister, R.: Nonlinear postprocessing for blind speech separation. In: Proc. ICA'2004, Granada, Spain (2004) 832–839
12. Pedersen, M.S., Wang, D.L., Larsen, J., Kjems, U.: Overcomplete blind source separation by combining ICA and binary time-frequency masking. In: Proceedings of the MLSP workshop, Mystic, CT, USA (2005)
13. Parra, L., Spence, C.: Convolutive blind separation of non-stationary sources. IEEE Trans. Speech and Audio Processing **8** (2000) 320–327
14. Büchler, M.C.: Algorithms for Sound Classification in Hearing Instruments. PhD thesis, Swiss Federal Institute of Technology, Zurich (2002)
15. Allen, J.B., Berkley, D.A.: Image method for efficiently simulating small-room acoustics. J. Acoust. Soc. Amer. **65** (1979) 943–950

Two Time-Frequency Ratio-Based Blind Source Separation Methods for Time-Delayed Mixtures

Matthieu Puigt and Yannick Deville

Laboratoire d'Astrophysique de Toulouse-Tarbes,
Observatoire Midi-Pyrénées - Université Paul Sabatier Toulouse 3,
14 Av. Edouard Belin, 31400 Toulouse, France
mpuigt@ast.obs-mip.fr, ydeville@ast.obs-mip.fr

Abstract. We propose two time-frequency (TF) blind source separation (BSS) methods suited to attenuated and delayed (AD) mixtures. They consist in identifying the columns of the (filtered permuted) mixing matrix in Constant-Time TF zones where they detect that a single source occurs, using TIme-Frequency Ratios Of Mixtures (hence their name AD-TIFROM-CT). We thus identify columns of scale coefficients and time shifts. Unlike various previously reported TF-BSS approaches, these methods set very limited constraints on the source sparsity and overlap. They are especially suited to non-stationary sources.

1 Introduction

Blind source separation (BSS) consists in estimating a set of N unknown sources from a set of P observations resulting from mixtures of these sources through unknown propagation channels. Most of the approaches that have been developed to this end are based on Independent Component Analysis [1]. More recently, several methods based on ratios of time-frequency (TF) transforms of the observed signals have been reported. Some of these methods, i.e. DUET and its modified versions, are based on an anechoic mixing model, involving attenuations and delays (AD) (this is not the general convolutive model). However, they require the sources to have no overlap in the TF domain [2], which is quite restrictive. On the contrary, only slight differences in the TF representations of the sources are requested by our Linear Instantaneous (LI) TIFROM method [3]. We here propose two novel TF-BSS methods, inspired by this LI-TIFROM approach, but suited to more general mixtures involving time shifts. We thus avoid the restriction[1] of the DUET method concerning the sparsity of the sources in the TF domain, while addressing the same class of mixtures.

2 Problem Statement

In this paper, we assume that N unknown source signals $s_j(n)$ are transferred through AD channels and added, thus providing a set of N mixed observed

[1] Note however that DUET also applies to underdetermined mixtures, which is not, at this stage, the case of the methods that we propose in this paper.

J. Rosca et al. (Eds.): ICA 2006, LNCS 3889, pp. 682–690, 2006.

signals $x_i(n)$. This reads

$$x_i(n) = \sum_{j=1}^{N} a_{ij}\, s_j(n - n_{ij}) \qquad i = 1 \ldots N, \tag{1}$$

where a_{ij} are real-valued strictly positive constant scale coefficients and n_{ij} are integer-valued time shifts. We here handle the scale/filter indeterminacies inherent in the BSS problem by extending to AD mixtures an approach that we introduced in another type of LI-BSS method, i.e. LI-TIFCORR [4]. This approach may be defined as follows. We consider an arbitrary permutation function $\sigma(.)$, applied to the indices j of the source signals, which yields the permuted source signals $s_{\sigma(j)}(n)$. We then introduce scaled and time-shifted versions of the latter signals, equal to their contributions in the first mixed signal, i.e.

$$s'_j(n) = a_{1,\sigma(j)}\, s_{\sigma(j)}\left(n - n_{1,\sigma(j)})\right). \tag{2}$$

The mixing equation (1) may then be rewritten as

$$x_i(n) = \sum_{j=1}^{N} a_{i,\sigma(j)}\, s_{\sigma(j)}\left(n - n_{i,\sigma(j)})\right) = \sum_{j=1}^{N} b_{ij}\, s'_j\left(n - \mu_{ij})\right) \tag{3}$$

with

$$b_{ij} = \frac{a_{i,\sigma(j)}}{a_{1,\sigma(j)}} \qquad \text{and} \qquad \mu_{ij} = n_{i,\sigma(j)} - n_{1,\sigma(j)}. \tag{4}$$

The Fourier transform of Eq. (3) reads

$$X_i(\omega) = \sum_{j=1}^{N} b_{ij}\, e^{-j\omega\mu_{ij}}\, S'_j(\omega) \qquad i = 1 \ldots N. \tag{5}$$

This yields in matrix form

$$\underline{X}(\omega) = B(\omega)\, \underline{S'}(\omega) \tag{6}$$

where $\underline{S'}(\omega) = [S'_1(\omega) \cdots S'_N(\omega)]^T$ and

$$B(\omega) = \left[b_{ij} e^{-j\omega\mu_{ij}}\right] \qquad i, j = 1 \ldots N. \tag{7}$$

In this paper, we aim at introducing methods for estimating $B(\omega)$.

3 Proposed Basic TIFROM Method for AD Mixtures

3.1 Time-Frequency Tool and Assumptions

We recently proposed [3] a LI-BSS method based on TIme-Frequency Ratios Of Mixtures, that we therefore called "LI-TIFROM". Starting from this method, we here develop extensions intended for AD mixtures. These approaches are called AD-TIFROM-CT, since they are shown below to only use "Constant-Time analysis zones". The TF transform of the signals considered in these approaches is the Short-Time Fourier Transform (STFT) defined as:

$$U(n,\omega) = \sum_{n'=-\infty}^{+\infty} u(n')h(n'-n)e^{-j\omega n'} \tag{8}$$

where $h(n'-n)$ is a shifted windowing function, centered on time n. $U(n,\omega)$ is the contribution of the signal $u(n)$ in the TF window corresponding to the short time window centered on n and to the angular frequency ω.

The AD-TIFROM-CT approach uses the following definitions and assumptions.

Definition 1. A source is said to "occur alone" in a TF area (which is composed of several adjacent above-defined TF windows) if only this source has a TF transform which is not equal to zero everywhere in this TF area.

Definition 2. A source is said to be "visible" in the TF domain if there exist at least one TF area where it occurs alone.

Assumption 1. Each source is visible in the TF domain.

Note that this is a very limited sparsity constraint !

Assumption 2. There exist no TF areas where the TF transforms of all sources are equal to zero everywhere[2].

Assumption 3. When several sources occur in a given set of adjacent TF windows, they should vary so that at least one of the moduli of ratios of STFTs of observations, $|X_i(n,\omega)/X_1(n,\omega)|$, with $i = 2\dots N$, does not take the same value in all these windows . Especially, i) at least one of the sources must take significantly different TF values in these windows and ii) the sources should not vary proportionally.

3.2 Overall Structure of the Basic AD-TIFROM-CT Method

The AD-TIFROM-CT method aims at estimating the mixing matrix $B(\omega)$ defined in Eq. (7), i.e. the parameters b_{im} and μ_{im}, with $i = 2\dots N$ and $m = 1\dots N$ ($i = 1$ yields $b_{ij} = 1$ and $\mu_{ij} = 0$: see Eq. (4)). The basic version of this method is composed of a pre-processing stage and 3 main stages:

1. The pre-processing stage consists in deriving the STFTs $X_i(n,\omega)$ of the mixed signals, according to Eq. (8).
2. The detection stage aims at finding "constant-time TF analysis zones" where a source occurs alone, using the method introduced in Section 3.3.

[2] This assumption is only made for the sake of simplicity: it may be removed in practice, thanks to the noise contained by real recordings, as explained in [3].

3. The identification stage aims at estimating the columns of $B(\omega)$ in the above single-source analysis zones, using the method proposed in Section 3.4.
4. In the combination stage, we eventually derive the output signals. They may be obtained in the frequency domain by computing $\underline{Y}(\omega) = B^{-1}(\omega)\,\underline{X}(\omega)$ where $\underline{Y}(\omega) = [Y_1(\omega) \cdots Y_N(\omega)]^T$ is the vector of Fourier transforms of the output signals. The time-domain versions of these signals are then obtained by applying an inverse Fourier transform to $\underline{Y}(\omega)$.

3.3 Detection of Single-Source Constant-Time TF Analysis Zones

As stated above, the BSS method that we here introduce first includes a detection stage for finding single-source TF zones. The frequency-domain mixture equations corresponding to Eq. (1) read

$$X_i(\omega) = \sum_{j=1}^{N} a_{ij}\, e^{-j\omega n_{ij}}\, S_j(\omega) \qquad i = 1 \dots N. \tag{9}$$

This relationship between the observations and sources remains almost exact when expressed in the TF domain if the time shifts n_{ij} are small enough as compared to the temporal width of the windowing function $h(.)$ used in the STFT transform. We here assume that this condition is met and thus that the STFTs of the observations can be expressed wrt. the STFTs of the sources as

$$X_i(n,\omega) = \sum_{j=1}^{N} a_{ij}\, e^{-j\,\omega n_{ij}}\, S_j(n,\omega) \qquad i = 1 \dots N. \tag{10}$$

Let us consider the ratio of STFTs of mixtures

$$\alpha_i(n,\omega) = \frac{X_i(n,\omega)}{X_1(n,\omega)} = \frac{\sum_{j=1}^{N} a_{ij}\, e^{-j\,\omega n_{ij}}\, S_j(n,\omega)}{\sum_{j=1}^{N} a_{1j}\, e^{-j\,\omega n_{1j}}\, S_j(n,\omega)}. \tag{11}$$

If a source $S_k(n,\omega)$ occurs alone in the considered TF window (n_p, ω_l) then

$$\alpha_i(n_p,\omega_l) = \frac{a_{ik}}{a_{1k}} e^{-j\omega(n_{ik}-n_{1k})} = b_{im} e^{-j\omega\mu_{im}} \tag{12}$$

with b_{im} and μ_{im} defined by Eq. (4) and $k = \sigma(m)$. Since we assumed all mixing coefficients a_{ik} to be real and positive, all resulting scale coefficients b_{im} are also real and positive. The modulus of the parameter value $\alpha_i(n_p,\omega_l)$ provided in Eq. (12) is therefore equal to b_{im}. If only source $S_k(n,\omega)$ occurs in several frequency-adjacent windows (n_p,ω_l), then $|\alpha_i(n_p,\omega_l)|$ is constant over these adjacent windows. On the contrary, it takes different values over these windows for at least one index i if several sources are present, due to Assumption 3. To exploit this phenomenon, we compute the sample variance of $|\alpha_i(n,\omega)|$ on "constant-time analysis zones" that we define as series of M frequency windows corresponding to adjacent ω_l, applying this approach independently to each time index n_p. This set of frequency points ω_l is denoted Ω hereafter and the corresponding TF zone is therefore denoted (n_p, Ω). We respectively define the sample mean and variance of $|\alpha_i(n_p, \omega_l)|$ on (n_p, Ω) as

$$\overline{|\alpha_i|}(n_p, \Omega) = \frac{1}{M} \sum_{l=1}^{M} |\alpha_i(n_p, \omega_l)|, \quad (13)$$

$$var\left[|\alpha_i|\right](n_p, \Omega) = \frac{1}{M} \sum_{l=1}^{M} \left| |\alpha_i(n_p, \omega_l)| - \overline{|\alpha_i|}(n_p, \Omega) \right|^2. \quad (14)$$

We first compute these parameters independently for each i, with $i = 2 \ldots N$. We then derive the mean over i of these variances $var\left[|\alpha_i|\right](n_p, \Omega)$, i.e.

$$MVAR\left[|\alpha|\right](n_p, \Omega) = \frac{1}{N-1} \sum_{i=2}^{N} var\left[|\alpha_i|\right](n_p, \Omega). \quad (15)$$

Similarly, we compute the inverse ratios and their means and variances on each considered analysis zone, i.e.

$$\beta_i(n, \omega) = \frac{1}{\alpha_i(n, \omega)} = \frac{X_1(n, \omega)}{X_i(n, \omega)} \quad (16)$$

$$\overline{|\beta_i|}(n_p, \Omega) = \frac{1}{M} \sum_{l=1}^{M} |\beta_i(n_p, \omega_l)| \quad (17)$$

$$var\left[|\beta_i|\right](n_p, \Omega) = \frac{1}{M} \sum_{l=1}^{M} \left| |\beta_i(n_p, \omega_l)| - \overline{|\beta_i|}(n_p, \Omega) \right|^2. \quad (18)$$

The mean over i of these variances $var\left[|\beta_i|\right](n_p, \Omega)$ then reads

$$MVAR\left[|\beta|\right](n_p, \Omega) = \frac{1}{N-1} \sum_{i=2}^{N} var\left[|\beta_i|\right](n_p, \Omega). \quad (19)$$

This mean $MVAR\left[|\beta|\right](n_p, \Omega)$ has lower or higher values than the above mean $MVAR\left[|\alpha|\right](n_p, \Omega)$, depending on mixing scale coefficients. The best single-source TF zones are those where $min\left\{MVAR\left[|\alpha|\right](n_p, \Omega), MVAR\left[|\beta|\right](n_p, \Omega)\right\}$ takes the lowest values.

3.4 Identification Stage

Thanks to expression (12) of the parameters $\alpha_i(n, \omega)$ in single-source analysis zones, a natural idea for estimating the time shifts μ_{im} consists in taking advantage of the phase of $\alpha_i(n, \omega)$. We consider independently each time position n_p associated to TF windows and for each such position, we unwrap the phase of $\alpha_i(n_p, \omega)$ over all associated frequency-adjacent TF points. If we assume that

$S_k(n, \omega)$ occurs alone in an analysis zone (n_p, Ω) and we consider the unwraped phase $\phi_i(n_p, \omega_l)$ of $\alpha_i(n_p, \omega_l)$ in this zone, due to Eq. (12) we have

$$-\omega_l \mu_{im} = \phi_i(n_p, \omega_l) + 2q_{im}(n_p)\pi, \tag{20}$$

where $q_{im}(n_p)$ is an unknown integer. Eq (20) shows that the curve associated to the variations of the phase $\phi_i(n_p, \omega_l)$ wrt. ω_l in a single-source zone (n_p, Ω) is a line and that its slope, equal to $-\mu_{im}$, does not depend on the value of $q_{im}(n_p)$. This slope therefore allows us to identify μ_{im}, with no phase indeterminacy. Our method for identifying the set of parameters μ_{im} associated to a column of $B(\omega)$ therefore operates as follows. In the selected constant-time single-source analysis zone, for each observed signal with index i, we consider the M points which have two coordinates, resp. defined as the frequencies ω_l and the corresponding values $\phi_i(n_p, \omega_l)$ of the unwraped phase of the identification parameter. We determine the least-mean square regression line associated to these points. The estimate of the parameter μ_{im} is then set to the integer which is the closest to the opposite of the slope of this regression line.

The overall identification stage consists in successively considering the analysis zones ordered according to increasing values of $min \{MVAR [|\alpha|] (n_p, \Omega)$, $MVAR [|\beta|] (n_p, \Omega)\}$. For each such zone, the estimates of b_{im} associated to a column of $B(\omega)$ are set to the values of $\overline{|\alpha_i|}(n_p, \Omega)$ or $1/\overline{|\beta_i|}(n_p, \Omega)$, depending whether respectively the parameter $MVAR [|\alpha|]$ or $MVAR [|\beta|]$ takes the lowest value in this zone. A new column of b_{im} is kept if its distance wrt. each previously found column of b_{im} is above a user-defined threshold ϵ_1. If a column of b_{im} is identified and kept, the corresponding column of μ_{im} is simultaneously identified, by using regression lines in the same analysis zone as explained above. The identification procedure ends when the number of columns of $B(\omega)$ thus kept becomes equal to the specified number N of sources to be separated.

4 Proposed Improved TIFROM Method for AD Mixtures

For $N > 2$ or when the time shifts μ_{im} are non-negligible wrt. to the length of STFT windows in the case $N = 2$, the above basic method turned out to yield false results in a significant number of experimental tests: on the one hand, we obtained columns of scale coefficients which did not correspond to actual columns of $B(\omega)$. On the other hand, we only achieved a coarse identification of the associated time shifts. Both problems can be solved thanks to clustering techniques. We now detail such an approach. In this approach, we form clusters of "points" where each point consists of a tentative column of parameters b_{im}. To this end, we first compute the parameters $MVAR [|\alpha|] (n_p, \Omega)$ and $MVAR [|\beta|] (n_p, \Omega)$ for all analysis zones and we then only consider the zones which are such that

$$min \{MVAR [|\alpha|] (n_p, \Omega), MVAR [|\beta|] (n_p, \Omega)\} \leq \epsilon_2, \tag{21}$$

where ϵ_2 is a small positive user-defined threshold. We thus only keep single-source zones, which correspond to the beginning of the ordered list created in the detection stage. We successively consider each of the first and subsequent analysis zones in this beginning of the ordered list and we use them in a slightly different way than in the basic identification procedure that we described above. Here again, for each considered analysis zone, the estimates of the parameters b_{im} are set to the values of $\overline{|\alpha_i|}(n_p, \Omega)$ or $1/\overline{|\beta_i|}(n_p, \Omega)$, depending on which of the parameters $MVAR\,[|\alpha|]$ and $MVAR\,[|\beta|]$ takes the lowest value in this zone. The estimated column associated to the first zone in the ordered list is kept as the first point in the first cluster. Each subsequently estimated column is then used as follows. We compute its distances wrt. all clusters created up to this stage, where the distance wrt. a cluster is defined as the distance wrt. the first point which was included in it. If such a distance is below a user-defined threshold ϵ_1, this new column is inserted as a new point in the corresponding cluster. Otherwise, this new column is kept as the first point of a new cluster. This is repeated for all analysis zones which fulfill condition (21). If the threshold ϵ_1 is low enough, the number of clusters thus created is at least equal to the specified number N of sources to be extracted. We then keep the N clusters which contain the highest numbers of points. For each cluster, we eventually derive a representative, by selecting its point which corresponds to the lowest value of $min\,\{MVAR\,[|\alpha|]\,(n_p, \Omega), MVAR\,[|\beta|]\,(n_p, \Omega)\}$ and thus presumably to the best single-source zone. This yields the N columns of estimates of b_{im}.

We estimate the parameters μ_{im} as follows. Independently, for each of the above N clusters of columns of b_{im}, we first compute the parameters μ_{im} in the same TF zones as these scale coefficients b_{im}. We then derive the histograms of these parameters μ_{im}, independently for each index i. We eventually keep the peak value in each histogram as the estimate of μ_{im}.

5 Experimental Results

We now present tests performed with $N = 2$ sources of English speech signals sampled at 20 kHz. These signals consist of 2.5 s of continuous speech from different male speakers. The performance achieved in each test is measured by the overall signal-to-interference-ratio (SIR) Improvement achieved by this system, denoted $SIRI$ below, and defined as the ratio of the output and input SIRs of our BSS system. The mixing matrix is set to

$$A(\omega) = \begin{bmatrix} 1 & 0.9\,e^{-j\omega 75} \\ 0.9\,e^{-j\omega 75} & 1 \end{bmatrix}. \tag{22}$$

The input SIR is thus equal to 0.9 dB. The number d of samples per STFT window is varied geometrically from 1024 to 16384. The number M of windows per analysis zone is set to 8 when $d = 1024$. This value of M is then increased geometrically with d. Thus, the absolute width of the frequency bands associated to the frequency domain Ω of the analysis zones (n_p, Ω) takes the same value

Table 1. Performance (*SIRI* in dB) for both methods vs STFT window size d.

Method	STFT window size d				
	1024	2048	4096	8192	16384
Basic	-2.8	14.8	6.2	26.8	invisible
Improved	20.2	14.8	23.9	26.8	invisible

whatever d. This value is 156.25 Hz. In each test, the temporal overlap between STFT windows is fixed to 75%. The resulting $SIRI$s are given in Table 1.

The cluster-based method yields better or same results as the basic one. The mean $SIRI$s are resp. equal to 11.2 and 24.3 dB with the basic and cluster-based approaches. When $d = 16384$, one source is not visible in the TF plane. Two results illustrate the usefulness of clustering techniques in our approaches: when $d = 1024$, with the basic method, the Frobenius norm of the difference between the actual and theoretical matrices of scales coefficients b_{im} is equal to 1.3, while this norm is only equal to 5.4e-2 with the cluster-based approach. We explain this phenomenon as follows: with the basic method, the parameters b_{im} were identified in analysis zones which were selected because they were at the beginning of the ordered list created in the detection stage, but these identified columns did not correspond to the actual (filtered permuted) mixing matrix, so that the outputs of our BSS system did not provide well separated sources. As only a few occurrences are obtained for each false column value, clustering techniques solved this problem. The case when $d = 4096$ is interesting too: we obtain the same matrix of scale coefficients with both methods (the above-defined Frobenius norm is equal to 3.6e-2). The estimated values of time shifts are equal to the theoretical ones with the cluster-based method, while we have slight differences with the basic approach: the estimated time shifts are equal to 74 and -76 while theoretical ones are ± 75. This clearly demonstrates the usefulness of clustering techniques.

6 Conclusion and Extensions

In this paper, we proposed two TF BSS methods for AD mixtures. They avoid the restrictions of the DUET method, which needs the source to be (approximately) W-disjoint orthogonal. Our methods consist in first finding the TF zones where a source occurs alone and then, identifying in these zones the parameters of the (filtered permuted) mixing matrix. Thanks to this principle, these approaches apply to non-stationary sources, but also to stationary and/or dependent sources (we could extend the discussion in [3] to AD mixtures), provided there exists at least a tiny TF zone per source where this source occurs alone. We experimentally showed the usefulness of clustering techniques in our methods. Our future investigations will consist in a more detailed characterization of the experimental performance of the proposed approaches. We will also aim at extending these methods to general convolutive mixtures.

References

1. A. Hyvärinen, J. Karhunen, E. Oja: Independent Component Analysis, Wiley-Interscience, New York, 2001.
2. A. Jourjine, S. Rickard, O. Yilmaz : Blind separation of disjoint orthogonal signals: Demixing N sources from 2 mixtures, Proceedings of ICASSP 2000, IEEE Press, Istanbul, Turkey, June 5-9, 2000, vol. 5, pp. 2985-2988.
3. F. Abrard, Y. Deville: A time-frequency blind signal separation method applicable to underdetermined mixtures of dependent sources, Signal Processing, Vol. 85, Issue 7, pp. 1389-1403, July 2005.
4. Y. Deville, Temporal and time-frequency correlation-based blind source separation methods, Proceedings of ICA 2003, pp. 1059-1064, Nara, Japan, April 1-4, 2003.

On Calculating the Inverse of Separation Matrix in Frequency-Domain Blind Source Separation

Hiroshi Sawada, Shoko Araki, Ryo Mukai, and Shoji Makino

NTT Communication Science Laboratories, NTT Corporation,
2-4 Hikaridai, Seika-cho, Soraku-gun, Kyoto 619-0237, Japan
{sawada, shoko, ryo, maki}@cslab.kecl.ntt.co.jp

Abstract. For blind source separation (BSS) of convolutive mixtures, the frequency-domain approach is efficient and practical, because the convolutive mixtures are modeled with instantaneous mixtures at each frequency bin and simple instantaneous independent component analysis (ICA) can be employed to separate the mixtures. However, the permutation and scaling ambiguities of ICA solutions need to be aligned to obtain proper time-domain separated signals. This paper discusses the idea that calculating the inverses of separation matrices obtained by ICA is very important as regards aligning these ambiguities. This paper also shows the relationship between the ICA-based method and the time-frequency masking method for BSS, which becomes clear by calculating the inverses.

1 Introduction

With acoustical applications of blind source separation (BSS), such as solving a cocktail party problem, signals are generally mixed in a convolutive manner with reverberations. Let $\mathrm{s}_1, \ldots, \mathrm{s}_N$ be N source signals and $\mathrm{x}_1, \ldots, \mathrm{x}_M$ be M sensor observations. Then, the convolutive mixture model is formulated as

$$\mathrm{x}_j(t) = \sum_{k=1}^{N} \sum_{l} \mathrm{h}_{jk}(l)\, \mathrm{s}_k(t-l), \quad j=1, \ldots, M, \tag{1}$$

where t represents time and $\mathrm{h}_{jk}(l)$ represents the impulse response from source k to sensor j. If we consider sounds mixed in a practical room situation, impulse responses $\mathrm{h}_{jk}(l)$ tend to have hundreds or thousands of taps even with an 8 kHz sampling rate. This makes the convolutive BSS problem very difficult compared with the BSS of simple instantaneous mixtures.

A practical approach for such convolutive mixtures is frequency-domain BSS [1-10], where we apply a short-time Fourier transform (STFT) to the sensor observations $\mathrm{x}_j(t)$. Then, the convolutive model (1) can be approximated as an instantaneous mixture model at each frequency:

$$x_j(f,t) = \sum_{k=1}^{N} h_{jk}(f) s_k(f,t), \quad j=1, \ldots, M, \tag{2}$$

J. Rosca et al. (Eds.): ICA 2006, LNCS 3889, pp. 691–699, 2006.

where f represents frequency, t is now down-sampled with the distance of the frame shift, $h_{jk}(f)$ is the frequency response from source k to sensor j, and $s_k(f,t)$ is a frequency-domain time-series signal of $s_k(t)$ obtained with an STFT. The vector notation of (2) is

$$\mathbf{x}(f,t) = \sum_{k=1}^{N} \mathbf{h}_k(f) s_k(f,t), \tag{3}$$

where $\mathbf{x} = [x_1, \ldots, x_M]^T$ and $\mathbf{h}_k = [h_{1k}, \ldots, h_{Mk}]^T$.

Once we assume instantaneous mixtures at each frequency, and also if the number of sources N does not exceed the number of sensors M, we can apply standard instantaneous independent component analysis (ICA) [11] to the mixtures $\mathbf{x}(f,t)$ to obtain separated frequency components:

$$\mathbf{y}(f,t) = \mathbf{W}(f)\,\mathbf{x}(f,t), \tag{4}$$

where $\mathbf{y} = [y_1, \ldots, y_N]^T$ is the vector of separated frequency components and $\mathbf{W} = [\mathbf{w}_1, \ldots, \mathbf{w}_N]^H$ is an $N \times M$ separation matrix [1-6]. However, the ICA solution has the permutation and scaling ambiguities. We need to align these ambiguities to obtain proper time-domain separated signals.

Various studies have tried to solve these permutation and scaling problems because they constitute a critical issue. Some of these studies have attempted to solve the problems by using information obtained from (4), i.e the separation matrix $\mathbf{W}$ and/or the separated signals $\mathbf{y}$, or by imposing some constraints on $\mathbf{W}$. By contrast, we believe that the inverses of separation matrices $\mathbf{W}$ provide useful information for solving these problems. The main topic of this paper is to discuss the procedures for solving these problems by calculating the inverses.

There is also a frequency-domain BSS method that is based on time-frequency (T-F) masking [7-10]. It does not employ a standard ICA to separate the mixtures, and can be applied even if the number of sources N exceeds the number of sensors M. The method relies on the sparseness of source signals. It classifies the mixtures $\mathbf{x}(f,t)$ based on spatial information extracted from them. As the second topic of this paper, we show a link between the ICA-based method and the T-F masking method. The link becomes clear once we have the decomposition (6) of mixtures by calculating the inverse of $\mathbf{W}$. Based on the link, we see that some of the techniques used in solving the permutation problem can also be used for classifying the mixtures in the T-F masking method, and vice versa.

2 Calculating the Inverses of Separation Matrices

Figure 1 shows the flow of ICA-based frequency-domain BSS that we consider in this paper. The inverse of separation matrix $\mathbf{W}$ is represented as

$$[\mathbf{a}_1, \cdots, \mathbf{a}_N] = \mathbf{W}^{-1}, \ \mathbf{a}_i = [a_{1i}, \ldots, a_{Mi}]^T, \tag{5}$$

which we call basis vectors obtained by ICA, because the mixture $\mathbf{x}(f,t)$ is represented by their linear combination by multiplying $\mathbf{W}^{-1}$ and (4):

$$\mathbf{x}(f,t) = \sum_{i=1}^{N} \mathbf{a}_i(f) y_i(f,t). \tag{6}$$

The basis vectors provide the key information with which to solve the permutation and scaling problems as shown in the following sections. If $\mathbf{W}$ is not square, we use the Moore-Penrose pseudoinverse instead of the inverse. It is not difficult to make $\mathbf{W}$ invertible by using an appropriate ICA procedure, such as whitening followed by unitary transformation (e.g. FastICA [11]).

3 Solving the Permutation Problem

Various methods have been proposed for solving the permutation problem:

1. making the separation matrices $\mathbf{W}(f)$ smooth along frequencies f [1, 2],
2. maximizing the correlation of separated signal envelopes $|y_i|$ [3],
3. analyzing the directivity patterns calculated from $\mathbf{W}(f)$ [4],
4. manipulating basis vectors $\mathbf{a}_i(f)$ [5, 6].

The third and fourth methods utilize the same information because $\mathbf{W}$ and $\mathbf{a}_i$ are related to each other through the inversion (5). However, the fourth method is easier to apply when there are more than two sources [5, 6], because basis vectors $\mathbf{a}_i$ directly represent estimated mixing system information (6). This section describes how to utilize this information for solving the permutation problem.

3.1 Assumption and Basic Idea

If ICA works well, we obtain separated signals $y_i(f,t)$ that should be close to source signals $s_k(f,t)$ up to the permutation and scaling ambiguities. If we compare (3) and (6), we see that the basis vectors $\mathbf{a}_i(f)$, which are obtained by ICA and the subsequent inversion of $\mathbf{W}$, should be close to the vectors $\mathbf{h}_k(f)$, again, up to the permutation and scaling ambiguities. The use of different subscripts, i and k, indicates the permutation ambiguity.

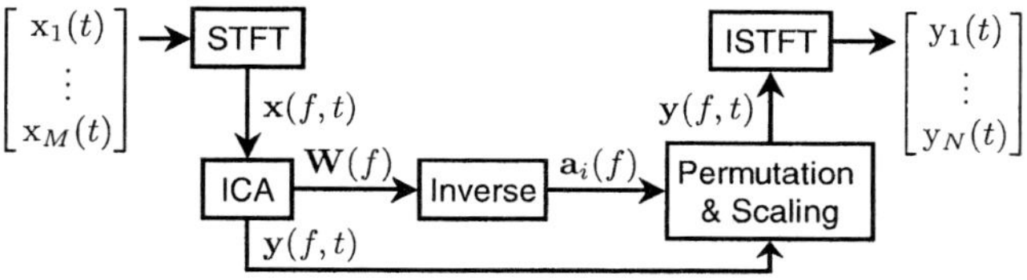

Fig. 1. Flow of ICA-based frequency-domain BSS

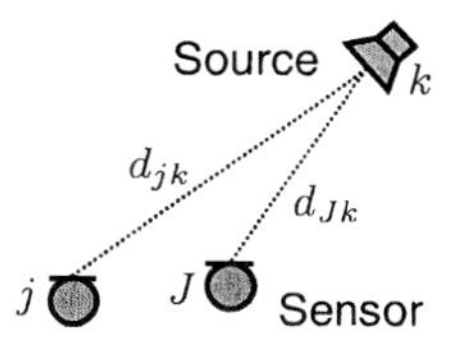

Fig. 2. Direct-path model

The method presented here assumes a direct-path model (Fig. 2) for the vector $\mathbf{h}_k = [h_{1k}, \ldots, h_{Mk}]^T$, even though in reality signals are mixed in a multi-path model (1). This simplified model is expressed in the frequency domain:

$$h_{jk}(f) \approx \lambda_{jk} \cdot e^{-\jmath 2\pi f \tau_{jk}}, \tag{7}$$

where τ_{jk} and $\lambda_{jk} \geq 0$ are the time delay and attenuation from source k to sensor j, respectively. Since we cannot distinguish the phase (or amplitude) of $s_k(f,t)$ and $h_{jk}(f)$, these two parameters can be considered to be relative (this fact causes the scaling ambiguity). Thus, without loss of generality, we normalize them and align the scaling ambiguity by

$$\tau_{jk} = (d_{jk} - d_{Jk})/c, \tag{8}$$

$$\textstyle\sum_{j=1}^{M} \lambda_{jk}^2 = 1, \tag{9}$$

where d_{jk} is the distance from source k to sensor j (Fig. 2), and c is the propagation velocity of the signal. Normalization (8) makes $\tau_{Jk} = 0$, i.e. the relative time delay is zero at a reference sensor J.

By following the normalizations (8) and (9), the scaling ambiguity of basis vectors $\mathbf{a}_i$ is aligned by the operation

$$\mathbf{a}_i \leftarrow \frac{\mathbf{a}_i}{||\mathbf{a}_i||} e^{-\jmath \arg(a_{Ji})} \tag{10}$$

which makes $\arg(a_{Ji}) = 0$ and $||\mathbf{a}_i|| = 1$. Now, the task as regards the permutation problem is to determine a permutation Π_f that relates the subscript i and k with $i = \Pi_f(k)$, and to estimate parameters τ_{jk}, λ_{jk} that make the model (7) match the $a_{ji}(f)$ element of a basis vector. This can be formulated so as to find Π_f, τ_{jk} and λ_{jk} that minimize the cost function:

$$\mathcal{J} = \sum_{f \in \mathcal{F}} \sum_{k=1}^{N} \sum_{j=1}^{M} |a_{ji}(f) - \lambda_{jk} \cdot e^{-\jmath 2\pi f \tau_{jk}}|^2, \quad i = \Pi_f(k), \tag{11}$$

where $\mathcal{F}$ is the set of frequencies that we have to consider.

3.2 Clustering Frequency-Normalized Basis Vectors

If we consider the frequency range where spatial aliasing does not occur:

$$\mathcal{F} = \{f : -\pi < 2\pi f \tau_{jk} < \pi, {}^\forall j, k\} \tag{12}$$

we can introduce the frequency normalization technique [6] to minimize the cost function (11). Let $d_{\max}$ be the maximum distance between the reference sensor J and any other sensor. Then the relative time delay is bounded by

$$\max_{jk} |\tau_{jk}| \leq d_{\max}/c \tag{13}$$

and therefore the frequency range $\mathcal{F}$ can be expressed with

$$\mathcal{F} = \{f : 0 < f < \frac{c}{2d_{\max}}\}. \tag{14}$$

The frequency normalization technique [6] removes frequency dependence from the elements of scale-normalized basis vectors (10):

$$\bar{a}_{ji}(f) \leftarrow |a_{ji}(f)| \exp\left[\jmath \frac{\arg[a_{ji}(f)]}{4fc^{-1}d_{\max}}\right]. \tag{15}$$

The rationale of dividing the argument by $4fc^{-1}d_{\max}$ is discussed in [6]. With this operation, the cost function (11) is converted to

$$\bar{\mathcal{J}} = \sum_{f\in\mathcal{F}} \sum_{k=1}^{N} \sum_{j=1}^{M} |\bar{a}_{ji}(f) - \bar{h}_{jk}|^2, \quad i = \Pi_f(k) \tag{16}$$

where

$$\bar{h}_{jk} = \lambda_{jk} \cdot \exp[-\jmath \frac{\pi}{2} \cdot \frac{c \cdot \tau_{jk}}{d_{\max}}] \tag{17}$$

is a frequency-normalized model. In a vector notation

$$\bar{\mathcal{J}} = \sum_{f\in\mathcal{F}} \sum_{k=1}^{N} ||\bar{\mathbf{a}}_i(f) - \bar{\mathbf{h}}_k||^2, \quad i = \Pi_f(k), \tag{18}$$

where $\bar{\mathbf{a}}_i = [\bar{a}_{1i}, \ldots, \bar{a}_{Mi}]^T$ and $\bar{\mathbf{h}}_k = [\bar{h}_{1k}, \ldots, \bar{h}_{Mk}]^T$. Because the model $\bar{\mathbf{h}}_k$ do not depend on frequency, $\bar{\mathcal{J}}$ can be minimized efficiently by a clustering algorithm that iterates the following two updates until convergence:

$$\Pi_f \leftarrow \text{argmin}_{\Pi} \sum_{k=1}^{N} ||\bar{\mathbf{a}}_{\Pi(k)}(f) - \bar{\mathbf{h}}_k||^2, \quad \text{for each } f \in \mathcal{F}, \tag{19}$$

$$\bar{\mathbf{h}}_k \leftarrow \sum_{f\in\mathcal{F}} \bar{\mathbf{a}}_{\Pi_f(k)}(f), \;\; \bar{\mathbf{h}}_k \leftarrow \bar{\mathbf{h}}_k / ||\bar{\mathbf{h}}_k||, \quad \text{for each } k = 1, \ldots, N. \tag{20}$$

The first update (19) optimizes the permutation Π_f for each frequency f with the current model $\bar{\mathbf{h}}_k$. The second update (20) calculates the most probable model $\bar{\mathbf{h}}_k$ with the current permutations. This set of updates is very similar to that of the k-means algorithm [12].

After the algorithm has converged, the permutation ambiguity in each frequency bin is aligned by

$$\mathbf{a}_k(f) \leftarrow \mathbf{a}_{\Pi_f(k)}(f), \;\; y_k(f,t) \leftarrow y_{\Pi_f(k)}(f,t), \;\; k = 1, \ldots, N. \tag{21}$$

In addition to aligning the permutations, the method estimates the model parameters by

$$\tau_{jk} = -\frac{2}{\pi} \cdot \frac{d_{\max}}{c} \arg(\bar{h}_{jk}), \quad \lambda_{jk} = |\bar{h}_{jk}|. \tag{22}$$

From these parameters and sensor array geometry, we can perform source localization, such as direction-of-arrival (DOA) estimation.

4 Solving the Scaling Problem

The ultimate goal as regards the scaling problem is to recover each source $\mathrm{s}_k(t)$, i.e. multichannel blind deconvolution. However, this is very difficult with colored source signals, such as speech. A feasible goal [13, 14] is simply to recover the observation of each source k at a reference sensor J

$$\textstyle\sum_l \mathrm{h}_{Jk}(l)\,\mathrm{s}_k(t-l)\,. \tag{23}$$

If we consider the frequency-domain counterpart of the above discussion, there is no practical way to recover the amplitude and phase of $s_k(f,t)$ blindly, but there is a feasible way to recover those of

$$h_{Jk}(f)s_k(f,t) \tag{24}$$

instead [3, 13]. We use this criterion for the scaling problem.

Calculating the inverse (5) and obtaining the linear combination form (6) of $\mathbf{x}(f,t)$ provides an instant solution to the scaling problem. If ICA works well and the permutation ambiguity is solved, we obtain separated signals $y_k(f,t)$ that should be close to source signals $s_k(f,t)$, now only up to the scaling ambiguity. If we compare (3) and (6), we see that $\mathbf{a}_k(f)y_k(f,t)$ should be close to $\mathbf{h}_k(f)s_k(f,t)$ and therefore $a_{Jk}(f)y_k(f,t)$ should be close to $h_{Jk}(f)s_k(f,t)$. Thus the scaling alignment can be performed simply by

$$y_k(f,t) \leftarrow a_{Jk}(f)y_k(f,t)\,. \tag{25}$$

In other words, there is no scaling ambiguity to be considered in (6) if we do not discriminate between $\mathbf{a}_i(f)$ and $y_i(f,t)$.

5 A Link to the Time-Frequency Masking Method

This section reveals a link between the ICA-based method and the time-frequency (T-F) masking method. The link becomes clear by the linear combination form (6) of $\mathbf{x}(f,t)$ obtained by the inverse (5) of the ICA separation matrix $\mathbf{W}(f)$.

Let us explain the T-F masking method, in which we assume the sparseness of source signals, i.e. at most only one source is active for each time-frequency slot (f,t). Based on this assumption, the mixture model (3) can be simplified as

$$\mathbf{x}(f,t) = \mathbf{h}_k(f)s_k(f,t),\;\; k \in \{1,\ldots,N\}. \tag{26}$$

where k depends on each time-frequency slot (f,t). Then, the method classifies the observation vectors $\mathbf{x}(f,t), {}^\forall f,t$ into N clusters $C_1,\ldots,C_N$ so that the k-th cluster contains observation vectors in which the k-th source is the only active source. After the classification, time domain separated signals $\mathrm{y}_k(t)$ are obtained by applying an inverse STFT (ISTFT) to the following classified frequency components

$$y_k(f,t) = \begin{cases} x_J(f,t) & \mathbf{x}(f,t) \in C_k, \\ 0 & \text{otherwise.} \end{cases} \tag{27}$$

In the classification, the spatial information expressed in $\mathbf{x}(f,t)$ is extracted and used. Typically, the phase difference normalized with frequency and/or the amplitude difference between two sensors:

$$\frac{\arg[x_2(f,t)/x_1(f,t)]}{2\pi f} \quad \text{and/or} \quad \left|\frac{x_2(f,t)}{x_1(f,t)}\right| \tag{28}$$

are calculated for the classification [7-9]. However, these papers presented only cases with two sensors. Recently, we proposed a new technique for using all the information of more than two sensors [10]. The technique used there is similar to that presented in Sec. 3. Thus, we consider that various techniques for classifying observation vectors $\mathbf{x}(f,t)$ in the T-F masking method can be used to classify basis vectors $\mathbf{a}_i(f,t)$ for solving the permutation problem in the ICA-based method, and vice versa.

Let us discuss this relationship in the following. If the sparseness assumption is satisfied, the linear combination form (6) obtained by ICA is reduced to

$$\mathbf{x}(f,t) = \mathbf{a}_i(f) y_i(f,t), \quad i \in \{1, \ldots, N\}. \tag{29}$$

where i depends on each time-frequency slot (f,t). If we compare (26) and (29), we see that $\mathbf{h}_k(f)s_k(f,t)$ should be close to $\mathbf{a}_i(f)y_i(f,t)$ for each time-frequency slot (f,t). Thus, the spatial information expressed in observation vectors $\mathbf{x}(f,t)$ with the sparseness assumption (26) is the same as that of basis vectors $\mathbf{a}_i(f,t)$ up to the scaling ambiguity. Therefore, we can use the same techniques for extracting spatial information from observation vectors $\mathbf{x}(f,t)$ and basis vectors $\mathbf{a}_i(f,t)$.

The normalization formulas (10) and (15) and the clustering procedure of (19) and (20) can be used not only for the ICA-based method but also the T-F masking method. Of course, we need to replace $\mathbf{a}_i(f)$ with $\mathbf{x}(f,t)$ and modify (19) for a standard clustering algorithm such as the k-means algorithm [12]. Figure 3 shows the flows of both methods in accordance with this idea.

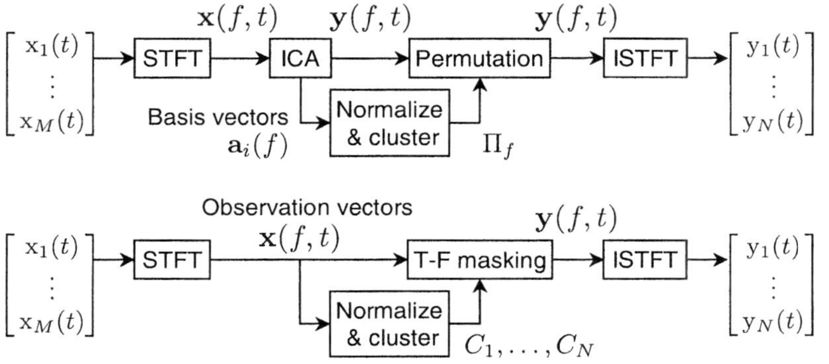

Fig. 3. Flows of ICA-based method (above) and time-frequency (T-F) masking method (below)

6 Experimental Results

We have performed experiments with the conditions shown in Fig. 4. We used 16 combinations of three speeches to evaluate the separation performance. The system did not have to know the sensor geometry for solving the permutation problem, but just the maximum distance $d_{\max} = 4$ cm between the reference sensor and any other sensor. The computational time was less than 3 seconds for 3-second speech mixtures, meaning that real-time processing was possible.

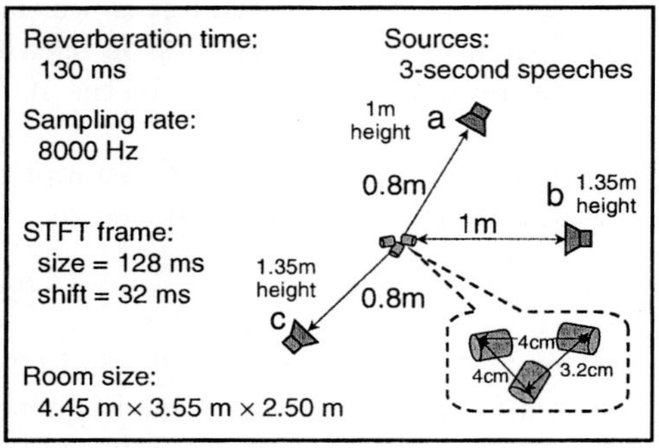

Fig. 4. Experimental conditions

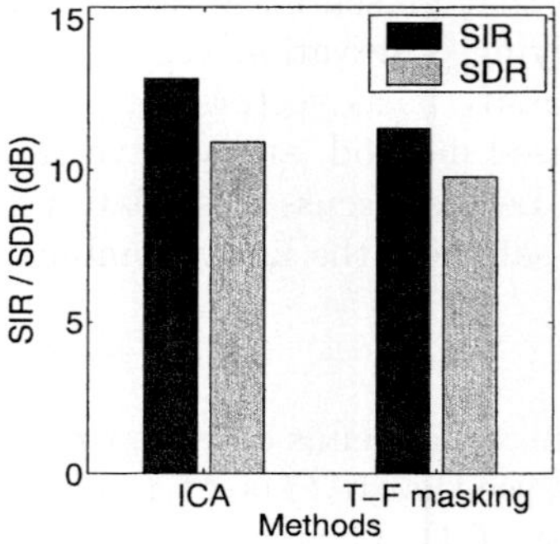

Fig. 5. Separation performance

Figure 5 shows the average signal-to-interference ratio (SIR) and signal-to-distortion ratio (SDR), whose detailed definitions can be found in [10]. Basically, the SIR indicates how well the mixtures are separated into the sources, and the SDR indicates how close each separated signal is to the observation of the corresponding source at the reference sensor. Since the number of sensors was sufficient for the number of sources in this case, ICA-based method worked better than T-F masking as shown in Fig. 5. We have also already obtained results with another setup where the number of sensors was insufficient ($N = 4, M = 3$) and the T-F masking method still worked [10].

7 Conclusion

In the ICA-based frequency-domain BSS, the permutation and scaling ambiguity of the ICA solution should be aligned. Once we have the form (6) by calculating the inverses of ICA separation matrices $\mathbf{W}$, the scaling ambiguity does not have to be considered. To align the permutation ambiguity, we can exploit the mixing system information represented in basis vectors $\mathbf{a}_i$, as Sec. 3 presents an efficient method. The form (6) clarifies the relationship between the ICA-based method and the T-F masking method. The same technique as that presented in Sec. 3 can be used in the T-F masking method for clustering observations $\mathbf{x}(f, t)$ and extracting the mixing system information.

References

1. Smaragdis, P.: Blind separation of convolved mixtures in the frequency domain. Neurocomputing **22** (1998) 21–34
2. Parra, L., Spence, C.: Convolutive blind separation of non-stationary sources. IEEE Trans. Speech Audio Processing **8** (2000) 320–327
3. Murata, N., Ikeda, S., Ziehe, A.: An approach to blind source separation based on temporal structure of speech signals. Neurocomputing **41** (2001) 1–24
4. Saruwatari, H., Kurita, S., Takeda, K., Itakura, F., Nishikawa, T., Shikano, K.: Blind source separation combining independent component analysis and beamforming. EURASIP Journal on Applied Signal Processing **2003** (2003) 1135–1146
5. Mukai, R., Sawada, H., Araki, S., Makino, S.: Frequency domain blind source separation for many speech signals. In: Proc. ICA 2004 (LNCS 3195). Springer-Verlag (2004) 461–469
6. Sawada, H., Araki, S., Mukai, R., Makino, S.: Blind extraction of a dominant source signal from mixtures of many sources. In: Proc. ICASSP 2005. Volume III. (2005) 61–64
7. Aoki, M., Okamoto, M., Aoki, S., Matsui, H., Sakurai, T., Kaneda, Y.: Sound source segregation based on estimating incident angle of each frequency component of input signals acquired by multiple microphones. Acoustical Science and Technology **22** (2001) 149–157
8. Rickard, S., Balan, R., Rosca, J.: Real-time time-frequency based blind source separation. In: Proc. ICA2001. (2001) 651–656
9. Yilmaz, O., Rickard, S.: Blind separation of speech mixtures via time-frequency masking. IEEE Trans. Signal Processing **52** (2004) 1830–1847
10. Araki, S., Sawada, H., Mukai, R., Makino, S.: A novel blind source separation method with observation vector clustering. In: Proc. 2005 International Workshop on Acoustic Echo and Noise Control (IWAENC 2005). (2005)
11. Hyvärinen, A., Karhunen, J., Oja, E.: Independent Component Analysis. John Wiley & Sons (2001)
12. Duda, R.O., Hart, P.E., Stork, D.G.: Pattern Classification. 2nd edn. Wiley Interscience (2000)
13. Matsuoka, K., Nakashima, S.: Minimal distortion principle for blind source separation. In: Proc. ICA 2001. (2001) 722–727
14. Takatani, T., Nishikawa, T., Saruwatari, H., Shikano, K.: Blind separation of binaural sound mixtures using SIMO-model-based independent component analysis. In: Proc. ICASSP 2004. Volume IV. (2004) 113–116

Nonnegative Matrix Factor 2-D Deconvolution for Blind Single Channel Source Separation

Mikkel N. Schmidt and Morten Mørup

Technical University of Denmark,
Informatics and Mathematical Modelling,
Richard Petersens Plads, Building 321,
DK-2800 Kgs. Lyngby, Denmark
{mns, mm}@imm.dtu.dk

Abstract. We present a novel method for blind separation of instruments in single channel polyphonic music based on a non-negative matrix factor 2-D deconvolution algorithm. The method is an extention of NMFD recently introduced by Smaragdis [1]. Using a model which is convolutive in both time and frequency we factorize a spectrogram representation of music into components corresponding to individual instruments. Based on this factorization we separate the instruments using spectrogram masking. The proposed algorithm has applications in computational auditory scene analysis, music information retrieval, and automatic music transcription.

1 Introduction

The separation of multiple sound sources from a single channel recording is a difficult problem which has been extensively addressed in the literature. Many of the proposed methods are based on matrix decompositions of a spectrogram representation of the sound. The basic idea is to represent the sources by different frequency signatures which vary in intensity over time.

Non-negative matrix factorization (NMF) [2, 3] has proven a very useful tool in a variety of signal processing fields. NMF gives a sparse (or parts-based) decomposition [3] and under certain conditions the decomposition is unique [4] making it unnecessary to impose constraints in the form of orthogonality or independence. Efficient algorithms for computing the NMF have been introduced by Lee and Seung [5]. NMF has a variety of applications in music signal processing; recently, Helén and Virtanen [6] described a method for separating drums from polyphonic music using NMF and Smaragdis and Brown [7] used NMF for automatic transcription of polyphonic music.

When polyphonic music is modelled by factorizing a magnitude spectrogram with NMF, each instrument note is modelled by an instantaneous frequency signature which can vary over time. Smaragdis [1] introduced an extension to NMF, namely the non-negative matrix factor deconvolution (NMFD) algorithm, in which each instrument note is modelled by a time-frequency signature which varies in intensity over time. Thus, the model can represent components with

J. Rosca et al. (Eds.): ICA 2006, LNCS 3889, pp. 700–707, 2006.

temporal structure. Smaragdis showed how this can be used to separate individual drums from a real recording of drum sounds. This approach was further pursued by Wang and Plumbley [8] who separated mixtures of different musical instruments. Virtanen [9] presented an algorithm based on similar ideas and evaluated its performance by separating mixtures of harmonic sounds.

In this paper, we propose a new method to factorize a log-frequency spectrogram using a model which can represent both temporal structure and the pitch change which occurs when an instrument plays different notes. We denote this the non-negative matrix factor 2-D deconvolution (NMF2D). We use a log-frequency spectrogram because a pitch change in a log-frequency domain simply corresponds to a displacement on the frequency axis, whereas in a linear-frequency domain a pitch change would also change the distance between the harmonics. Where previous methods need one component to model each note for each instrument, the proposed model represents each instrument compactly by a single time-frequency profile convolved in both time and frequency by a time-pitch weight matrix. This model impressively decreases the number of components needed to model various instruments and effectively solves the blind single channel source separation problem for certain classes of musical signals. In section 2 we introduce the NMF2D model and derive the update equations for recursively computing the factorization based on two different cost functions. In section 3 we show how the algorithm can be used to analyze and separate polyphonic music and we compare the algorithm to the NMFD method of Smaragdis [1]. This is followed by a discussion of the results.

2 Method

We start by recalling the non-negative matrix factorization (NMF) problem: $\mathbf{V} \approx \mathbf{WH}$, where $\mathbf{V}$, $\mathbf{W}$, and $\mathbf{H}$ are non-negative matrices. The task is find factors $\mathbf{W}$ and $\mathbf{H}$ which approximate $\mathbf{V}$ as well as possible according to some cost function. Lee and Seung [5] devise two algorithms to find $\mathbf{W}$ and $\mathbf{H}$: For the least squares error and the KL divergence they show that the following recursive updates converge to a local minimum:

$$\text{Least squares error :} \quad \mathbf{W} \leftarrow \mathbf{W} \bullet \frac{\mathbf{VH}^T}{\mathbf{WHH}^T}, \quad \mathbf{H} \leftarrow \mathbf{H} \bullet \frac{\mathbf{W}^T\mathbf{V}}{\mathbf{W}^T\mathbf{WH}},$$

$$\text{KL divergence :} \quad \mathbf{W} \leftarrow \mathbf{W} \bullet \frac{\frac{\mathbf{V}}{\mathbf{WH}}\mathbf{H}^T}{\mathbf{1}\cdot\mathbf{H}^T}, \quad \mathbf{H} \leftarrow \mathbf{H} \bullet \frac{\mathbf{W}^T\frac{\mathbf{V}}{\mathbf{WH}}}{\mathbf{W}^T\cdot\mathbf{1}},$$

where $A \bullet B$ denotes element-wise multiplication, $\frac{A}{B}$ denotes element-wise division, and $\mathbf{1}$ is a matrix with all elements unity. These algorithms can be derived by minimizing the cost function using gradient descent and choosing the stepsize appropriately to yield simple multiplicative updates.

2.1 NMF2D

We extend the NMF model to be a 2-dimensional convolution of $\mathbf{W}^\tau$ which depends on time, τ, and $\mathbf{H}^\phi$ which depends on pitch, ϕ. This forms the non-negative matrix factor 2-D deconvolution (NMF2D) model:

$$\mathbf{V} \approx \mathbf{\Lambda} = \sum_{\tau} \sum_{\phi} \overset{\downarrow\phi}{\mathbf{W}^{\tau}} \overset{\rightarrow\tau}{\mathbf{H}^{\phi}}, \tag{1}$$

where $\downarrow \phi$ denotes the downward shift operator which moves each element in the matrix ϕ rows down, and $\rightarrow \tau$ denotes the right shift operator which moves each element in the matrix τ columns to the right, i.e.:

$$\mathbf{A} = \begin{pmatrix} 1\,2\,3 \\ 4\,5\,6 \\ 7\,8\,9 \end{pmatrix}, \quad \overset{\downarrow 2}{\mathbf{A}} = \begin{pmatrix} 0\,0\,0 \\ 0\,0\,0 \\ 1\,2\,3 \end{pmatrix}, \quad \overset{\rightarrow 1}{\mathbf{A}} = \begin{pmatrix} 0\,1\,2 \\ 0\,4\,5 \\ 0\,7\,8 \end{pmatrix}.$$

We note that the NMFD model introduced by Smaragdis [1] is a special case of the NMF2D model where $\phi = \{0\}$.

The NMF2D model can be used to factorize a log-frequency magnitude spectrogram of polyphonic music into factors corresponding to individual instruments: If the matrix $\mathbf{V}$ is a log-frequency spectrogram representation of a piece of polyphonic music, the columns of $\mathbf{W}^{\tau}$ correspond to the time-frequency signature of each instrument, and the rows of $\mathbf{H}^{\phi}$ correspond to the time-pitch signature of each instrument, i.e. which notes are played by the instrument at what time. In other words, the convolution in time, τ, accounts for the temporal structure of each instrument, and the convolution in pitch, ϕ, accounts for each instrument playing different tones.

2.2 Least Squares NMF2D

We now derive a set of recursive update steps for computing $\mathbf{W}^{\tau}$ and $\mathbf{H}^{\phi}$ based on gradient descent with multiplicative updates. We consider the least squares (LS) cost function which corresponds to maximizing the likelihood of a gaussian noise model:

$$C_{LS} = ||\mathbf{V} - \mathbf{\Lambda}||_f^2 = \sum_{i} \sum_{j} (\mathbf{V}_{i,j} - \mathbf{\Lambda}_{i,j})^2. \tag{2}$$

A given element in $\mathbf{\Lambda}$, defined in equation (1), is given by:

$$\mathbf{\Lambda}_{i,j} = \sum_{\tau} \sum_{\phi} \sum_{d} \mathbf{W}^{\tau}_{i-\phi,d} \mathbf{H}^{\phi}_{d,j-\tau}. \tag{3}$$

In the following we will need the derivative of a given element $\mathbf{\Lambda}_{i,j}$ with respect to a given element $\mathbf{W}^{\tau}_{k,d}$:

$$\frac{\partial \mathbf{\Lambda}_{i,j}}{\partial \mathbf{W}^{\tau}_{k,d}} = \frac{\partial}{\partial \mathbf{W}^{\tau}_{k,d}} \left(\sum_{\tau} \sum_{\phi} \sum_{d} \mathbf{W}^{\tau}_{i-\phi,d} \mathbf{H}^{\phi}_{d,j-\tau} \right) \tag{4}$$

$$= \frac{\partial}{\partial \mathbf{W}^{\tau}_{k,d}} \left(\sum_{\phi} \mathbf{W}^{\tau}_{i-\phi,d} \mathbf{H}^{\phi}_{d,j-\tau} \right) \tag{5}$$

$$= \begin{cases} \mathbf{H}^{\phi}_{d,j-\tau} & \phi = i-k \\ 0 & otherwise. \end{cases} \tag{6}$$

Differentiating C_{LS} with respect to a given element in $\mathbf{W}^\tau$ gives:

$$\frac{\partial C_{LS}}{\partial \mathbf{W}^{\tau}_{k,d}} = \frac{\partial}{\partial \mathbf{W}^{\tau}_{k,d}} \sum_i \sum_j (\mathbf{V}_{i,j} - \mathbf{\Lambda}_{i,j})^2 \tag{7}$$

$$= -2 \sum_i \sum_j (\mathbf{V}_{i,j} - \mathbf{\Lambda}_{i,j}) \frac{\partial \mathbf{\Lambda}_{i,j}}{\partial \mathbf{W}^{\tau}_{k,d}} \tag{8}$$

$$= -2 \sum_\phi \sum_j (\mathbf{V}_{\phi+k,j} - \mathbf{\Lambda}_{\phi+k,j}) \mathbf{H}^{\phi}_{d,j-\tau}. \tag{9}$$

The recursive update steps for the gradient descent are given by:

$$\mathbf{W}^{\tau}_{k,d} \leftarrow \mathbf{W}^{\tau}_{k,d} - \eta \frac{\partial C_{LS}}{\partial \mathbf{W}^{\tau}_{k,d}}. \tag{10}$$

Similar to the approach of Lee and Seung [5], we choose the step size η so that the first term in (10) is canceled:

$$\eta = \frac{\mathbf{W}^{\tau}_{k,d}}{-2 \sum_\phi \sum_j \mathbf{\Lambda}_{\phi+k,j} \mathbf{H}^{\phi}_{d,j-\tau}}, \tag{11}$$

which gives us the following simple multiplicative updates:

$$\mathbf{W}^{\tau}_{k,d} \leftarrow \mathbf{W}^{\tau}_{k,d} \frac{\sum_\phi \sum_j \mathbf{V}_{\phi+k,j} \mathbf{H}^{\phi}_{d,j-\tau}}{\sum_\phi \sum_j \mathbf{\Lambda}_{\phi+k,j} \mathbf{H}^{\phi}_{d,j-\tau}}. \tag{12}$$

By noticing that transposing equation (1) interchanges the order of $\mathbf{W}^\tau$ and $\mathbf{H}^\phi$ in the model, the updates for $\mathbf{H}^\phi$ can easily be found. In matrix notation the updates can be written as:

$$\mathbf{W}^\tau \leftarrow \mathbf{W}^\tau \bullet \frac{\sum_\phi \overset{\uparrow\phi}{\mathbf{V}} \overset{\rightarrow\tau}{\mathbf{H}^\phi}{}^T}{\sum_\phi \overset{\uparrow\phi}{\mathbf{\Lambda}} \overset{\rightarrow\tau}{\mathbf{H}^\phi}{}^T} \qquad \mathbf{H}^\phi \leftarrow \mathbf{H}^\phi \bullet \frac{\sum_\tau \overset{\downarrow\phi}{\mathbf{W}^\tau}{}^T \overset{\leftarrow\tau}{\mathbf{V}}}{\sum_\tau \overset{\downarrow\phi}{\mathbf{W}^\tau}{}^T \overset{\leftarrow\tau}{\mathbf{\Lambda}}}. \tag{13}$$

2.3 Kullback-Leibler NMF2D

We can also find similar equations for computing the NMF2D based on the Kullback-Leibler (KL) divergence:

$$C_{KL} = \sum_i \sum_j \mathbf{V}_{i,j} \log \frac{\mathbf{V}_{i,j}}{\mathbf{\Lambda}_{i,j}} - \mathbf{V}_{i,j} + \mathbf{\Lambda}_{i,j}. \tag{14}$$

Minimizing the KL divergence corresponds to assuming a multinomial noise model. By taking similar steps as in the derivation of the LS updates we find the following recursive updates for the KL cost function:

$$\mathbf{W}^\tau \leftarrow \mathbf{W}^\tau \bullet \frac{\sum_\phi \overset{\uparrow\phi}{\left(\frac{\mathbf{V}}{\mathbf{\Lambda}}\right)} \overset{\rightarrow\tau}{\mathbf{H}^\phi}{}^T}{\sum_\phi \mathbf{1} \cdot \overset{\rightarrow\tau}{\mathbf{H}^\phi}{}^T} \qquad \mathbf{H}^\phi \leftarrow \mathbf{H}^\phi \bullet \frac{\sum_\tau \overset{\downarrow\phi}{\mathbf{W}^\tau}{}^T \overset{\leftarrow\tau}{\left(\frac{\mathbf{V}}{\mathbf{\Lambda}}\right)}}{\sum_\tau \overset{\downarrow\phi}{\mathbf{W}^\tau}{}^T \cdot \mathbf{1}}. \tag{15}$$

3 Experimental Results

In order to demonstrate our NMF2D algorithm, we have analyzed a 4 second piece of computer generated polyphonic music containing a trumpet and a piano. For comparison we have also analyzed the same piece of music by the NMFD algorithm [1]. For both algorithms we used the least squares cost function. The score of the piece of music is shown in Fig. 1. The trumpet and the piano play a different short melodic passage each consisting of three distinct notes. We generated the music at a sample rate of 16 kHz and analyzed it by the short time Fourier transform with a 2048 point Hanning windowed FFT and 50% overlap. This gave us 63 FFT frames. We grouped the spectrogram bins into 175 logarithmically spaced frequency bins in the range of 50 Hz to 8 kHz with 24 bins per octave, which correponds to twice the resolution of the equal tempered musical scale. Then, we performed the NMF2D and NMFD factorization of the log-frequency magnitude spectrogram. The parameters of the two models were selected so that both methods were able to model the music almost perfectly:

For the NMF2D we used two factors, $d = 2$, since we seek to separate two instruments. The NMF2D method requires at least as many factors as the number of individual instruments. We empirically chose to use seven convolutive components in time, $\tau = \{0, \ldots, 6\}$, corresponding to approximatly 45 ms, which we found to capture the temporal structure of the instruments well. The pitch of the notes played in the music span three whole notes. Consequently, we chose to use 12 convolutive components in pitch, i.e. $\phi = \{0, \ldots, 11\}$.

For the NMFD we used six factors, $d = 6$, corresponding to the total number of different tones played by the two instruments. The NMFD method requires at least as many factors as the number of distinctive instrument notes. Similar to the experiment with NMF2D we used seven convolutive components in time. For the experiment with NMFD we used our formulation of the NMF2D algorithm with $\phi = \{0\}$, since the NMFD is a special case of the NMF2D algorithm.

The results of the experiments with NMFD and NMF2D are shown in Fig. 2 and Fig. 3 respectively. As we expected, the NMFD algorithm factorized each individual note from each instrument into a separate component, whereas the NMF2D algorithm factorized each instrument into a separate component.

We used the NMF2D factorization to reconstruct the individual instruments separately. First, we reconstructed the spectrogram of each individual instrument

Fig. 1. Score of the piece of music used in the experiments. The music consists of a trumpet and a piano which play different short melodic passages each consisting of three distinct notes.

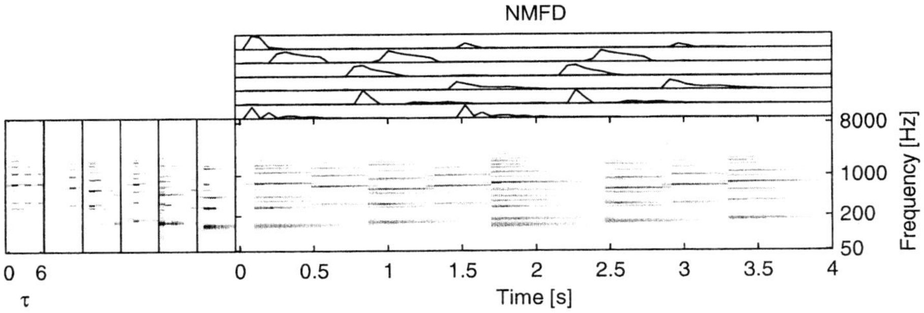

Fig. 2. Factorization of the piece of music using NMFD. The six time-frequency plots on the left are $\mathbf{W}^{\tau}$ for each factor, i.e. the time-frequency signature of the distincts tone played by the two instruments. The six plots on the top are the rows of $\mathbf{H}$ showing how the individual instrument notes are placed in time. The factors have been manually sorted so that the first three corresponds to the trumpet and the last three correspond to the piano.

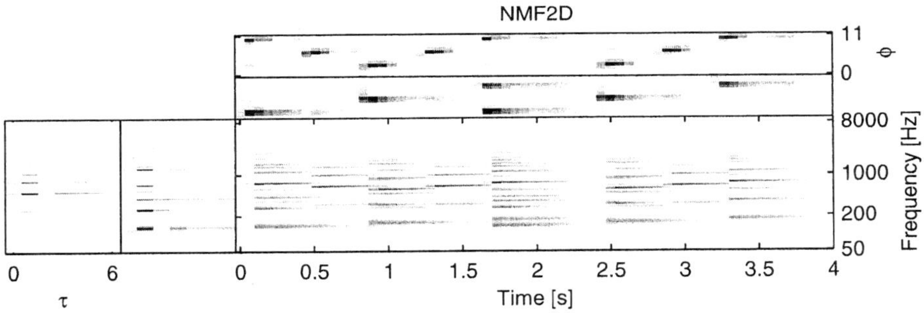

Fig. 3. Factorization of the piece of music using NMF2D. The two time-frequency plots on the left are $\mathbf{W}^{\tau}$ for each factor, i.e. the time-frequency signature of the two instruments. The two time-pitch plots on the top are $\mathbf{H}^{\phi}$ for each factor showing how the two instrument notes are placed in time and pitch.

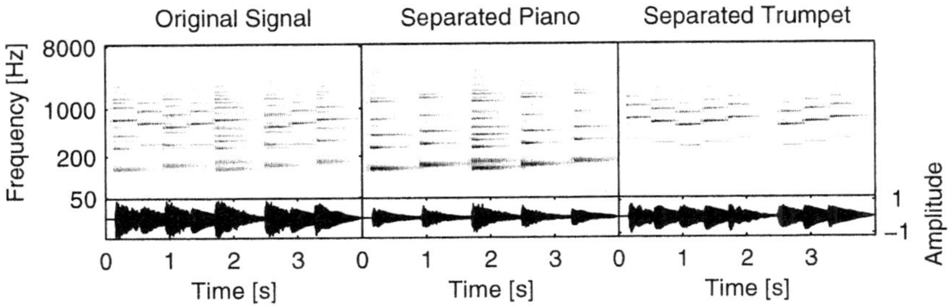

Fig. 4. Single channel source separation using NMF2D. The plots show the log-frequency spectrogram and the waveform of the music and the separated instruments.

by computing equation (3) for each specific value of d. These reconstructed spectrograms could be used directly to reconstruct the instruments, but since the log-frequency spectrograms are computed with a relatively low frequency resolution we reconstructed the signals using spectrogram masking: We mapped the reconstructed log-frequency spectrograms back into the linear-frequency spectrogram domain and computed a spectrogram mask for each instrument which assigned each spectrogram bin to the instrument with the highest power at that bin. We filtered the original spectrogram based on the masks, and computed the inverse filtered spectrogram using the original phase. The separation of the two instruments in the music is shown in Fig. 4. Informal listening test indicated, that the NMF2D algorithm was able to separate the two instruments very well.

4 Discussion

In the previous section we compared the proposed NMF2D algorithm with NMFD. Both the NMF2D and the NMFD algorithm can be used to separate the instruments in polyphonic music. However, since in NMFD the notes of the individual instruments are spread over a number of factors, these must first be grouped manually or by other means. The advantage of the NMF2D algorithm is, that it implicitly solves the problem of grouping notes.

If the assumption holds, that all notes for an instrument is an identical pitch shifted time-frequency signature, the NMF2D model will give better estimates of these signatures, because more examples (different notes) are used to compute each time-frequency signature. Even when this assumption does not hold, it migth still hold in a region of notes for an instrument. Furthermore, the NMF2D algorithm might be able to explain the spectral differences between two notes of different pitch by the 2-D convolution of the time-frequency signature.

Both the NMFD and NMF2D models almost perfectly explained the variation in the spectrogram. However, the number of free parameters in the two models is quite different. If the dimensionality of the spectrogram is $I \times J$, and n_τ, n_ϕ denote the number of convolutive lags in time and pitch, NMFD has $(n_\tau I + J) \cdot d = (7 \cdot 175 + 63) \cdot 6 = 7728$ parameters whereas NMF2D has $(n_\tau I + n_\phi J) \cdot d = (7 \cdot 175 + 12 \cdot 63) \cdot 2 = 3962$ parameters. Consequently, the NMF2D was more restricted making the NMF2D the best model from an Occam's razor point of view.

Admittedly, the simple computer generated piece of music analyzed in this paper favors the NMF2D algorithm since each instrument key is almost a simple spectral shift of the same time-frequency signature. However, when we analyze real music signals the NMF2D also gives very good results. Demonstrations of the algorithm for different music signals can be found at *www.intelligentsound.org.*

It is worth noting, that while we had problems making the NMFD algorithm converge in some situations when using the updates given by Smaragdis [1], the updates devised in this paper to our knowledge always converge.

In the experiments above we used the NMF2D based on least squares. However, using the algorithm based on minimizing the KL divergence gave similar results. It is also worth mentioning that the NMF2D analysis is computationally inexpensive; the results in the previous section took approximatly 20 seconds to compute on a 2 GHz Pentium 4 computer.

It is our belief that the NMF2D algorithm can be useful in a wide range of areas including computational auditory scene analysis, music information retrieval, audio coding, automatic music transcription, and image analysis.

References

1. Smaragdis, P.: Non-negative matrix factor deconvolution; extraction of multiple sound sources from monophonic inputs. Lecture Notes in Computer Science **3195** (2004) 494–499
2. Paatero, P., Tapper, U.: Positive matrix factorization: A non-negative factor model with optimal utilization of error estimates of data values. Environmetrics **5**(2) (1994) 111–126
3. Lee, D., Seung, H.: Learning the parts of objects by non-negative matrix factorization. Nature **401**(6755) (1999) 788–91
4. Donoho, D., Stodden, V.: When does non-negative matrix factorization give a correct decomposition into parts? NIPS (2003)
5. Lee, D.D., Seung, H.S.: Algorithms for non-negative matrix factorization. In: NIPS. (2000) 556–562
6. Helén, M., Virtanen, T.: Separation of drums from polyphonic music using non-negative matrix factorization and support vector machine. In: 13th European Signal Processing Conference. (2005)
7. Smaragdis, P., Brown, J.C.: Non-negative matrix factorization for polyphonic music transcription. IEEE Workshop on Applications of Signal Processing to Audio and Acoustics (WASPAA) (2003) 177–180
8. Wang, B., Plumbley, M.D.: Musical audio stream separation by non-negative matrix factorization. In: Proceedings of the DMRN Summer Conference. (2005)
9. Virtanen, T.: Separation of sound sources by convolutive sparse coding. SAPA (2004)

Speech Enhancement in Short-Wave Channel Based on ICA in Empirical Mode Decomposition Domain

Li-Ran Shen[1], Xue-Yao Li[1], Qing-Bo Yin[1, 2], and Hui-Qiang Wang[1]

[1] College of Computer Science And Technology,
Harbin Engineering University,
No.145 Nantong Street, Nangang District, Harbin, China
[2] Div. of Electronic, Computer and Telecommunication Engineering,
Pukyong National University, Busan, Korea

Abstract. It is well known that the non-stationary noise is the most difficult to be removed in speech enhancement. In this paper a novel speech enhancement algorithm based on the empirical mode decomposition (EMD) and then ICA is proposed to suppress the non-stationary noise. The noisy speech is decomposed into components by the EMD and ICA-based vector space, and the components are processed and reconstructed, respectively, by distinguishing between voiced speech and unvoiced speech. There are no requirements of noise whitening and SNR pre-calculating. Experiments show that the proposed method performs well suppressing of the non-stationary noise in short-wave channel for speech enhancement.

1 Introduction

In short-wave channel communication there is a great deal of interferential noise existing in the surrounding environment, transmitting media, electronic communication device and other speakers' sound, etc. common in most practical situations. In general, the addition of noise reduces intelligibility and degrades the performance of digital voice processors used for applications such as speech compression and recognition. Therefore, the problem of removing the uncorrelated noise component from the noisy speech signal, i.e., speech enhancement, has received considerable attention. In speech communication over the short-wave channel the purpose is to elevate the objective quality of speech signal and the intelligibility of noisy speech in order to reduce the listener fatigue. There have been numerous studies on the enhancement of the noisy speech signal. Many different types of speech enhancement algorithms have been proposed and tested [1–4, 6]. Spectral subtraction is a traditional method of speech enhancement [6]. The major drawback of this method is the remaining musical noise. Additionally a drawback of speech enhancement methods is the distortion of the useful signal. The resolution is the compromise between signal distortion and residual noise. Though this problem is well known, the study results indicate that both of these cannot be minimized simultaneously. Minimum mean square error (MMSE) [3] estimates on

J. Rosca et al. (Eds.): ICA 2006, LNCS 3889, pp. 708 – 713, 2006.

speech spectrum have been proposed. And Ephraimand Van Trees proposed a signal-subspace-based spectral domain algorithm, which controls the energy of residual noise in a certain threshold while minimizing the signal distortion. Hence the probability of noise perception can be minimized. The drawback of this method is that it deals only with white noise. EMD theory is the newly developed time–frequency analysis technology and is especially of interest in non-stationary signals such as water, sonar, seismic signal, etc [7-10].

In this paper we use EMD technique to produce an observed signal matrix from single short-wave channel signal. And then based on ICA the observed signal matrix was decomposed into signal subspace and noise subspace. Reconstruct speech signal using signal subspace to achieve speech enhancement.

2 Empirical Mode Decomposition

The empirical mode decomposition(EMD) was first introduced by Huang et al. [8]. The principle of this technique is to decompose adaptively a given signal $x(t)$ into oscillating components. These components are called intrinsic mode functions (IMFs) and are obtained from the signal x by means of an algorithm, called sifting. It is a fully data driven method.

The algorithm to create IMFs is elegant and simple. Firstly, the local extremes in the time series data $X(t)$ are identified, and then all the local maxims are connected by a cubic spline line $U_X(t)$, known as the upper envelope of the data set. Then, we repeat the procedure for the local minima to produce the lower envelope, $L_X(t)$.

Their mean $m_1(t)$ is given by:

$$m_1(t) = \frac{L_X(t) + U_X(t)}{2} \tag{1}$$

It is a running mean. We note that both envelopes should cover by construction all the data between them.

Then we subtract the running mean $m_1(t)$, from the original data $X(t)$, and we get the first component, $h_1(t)$, i.e.:

$$h_1(t) = X(t) - m_1(t) \tag{2}$$

To check if $h_1(t)$ is an IMF, we demand the following conditions: (i) $h_1(t)$ should be free of riding waves i.e. the. Rest component should not display under-shots or over-shots riding on the data and producing local extremes without zero crossing. (ii) To display symmetry of the upper and lower envelops with respect to zero. (iii) Obviously the number of zero crossing and extremes should be the same in both functions.

The sifting process has to be repeated as many times as it is required to reduce the extracted signal to an IMF. In the subsequent sifting process steps, $h_1(t)$ is treated as the data; then:

$$h_{11}(t) = h_1(t) - m_{11}(t) \tag{3}$$

If the function $h_{11}(t)$, does not satisfy criteria (i)–(iii), then the sifting process continues up to k times, h1k, until some acceptable tolerance is reached:

$$h_{1k}(t) = h_{1(k-1)}(t) - m_{1k}(t) \tag{4}$$

The resulting time series is the first IMF, and then it is designated as $C_1(t) = h_{1k}$ the first IMF component from the data contains the highest oscillation frequencies found in the original data $X(t)$.

The first IMF is subtracted from the original data, and this difference, is called a residue $r_1(t)$ by:

$$r_1(t) = X(t) - C_1(t) \tag{5}$$

The residue $r_1(t)$ is taken as if it was the original data and we apply to it again the sifting process. The process of finding more intrinsic modes $C_j(t)$ continues until the last mode is found. The final residue will be a constant or a monotonic function; in this last case it will be the general trend of the data.

$$X(t) = \sum_{j=1}^{n} C_j(t) + r_n(t) \tag{6}$$

Thus, one achieves a decomposition of the data into n-empirical IMF modes, plus a residue, $r_n(t)$, which can be either the mean trend or a constant. We must point out that this method do not requires a mean or zero reference, and only needs the locations of the local extremes.

3 Speech Enhancement Based on ICA

3.1 Infomax Algorithm

Informax algorithm is to maximize the network entropy. The network entropy was defined as the inter-information between input and output.

$$I(\mathbf{Y},\mathbf{X}) = H(\mathbf{Y}) - H(\mathbf{Y}|\mathbf{X}) \tag{7}$$

And then

$$\frac{\partial I(Y,X)}{\partial W} = \frac{\partial H(Y)}{\partial Y} \tag{8}$$

If the relationship between the input and the output is $Y = \tanh(WX)$,and then

$$\Delta W = \left[W^T\right]^{-1} - 2\tanh(WX)X^T \tag{9}$$

To fast convergence and simply computation, the left of the equation (9) was multiplied by $W^T W$.

$$\Delta W \propto [W - 2\tanh(WX)(WX)^T]W \tag{10}$$

So it can be used using natural grade method to solve the optimization problem in equation (9).

3.2 De-noising Method

Firstly using the EMD method to form the observed data $C_i(t)$. And then using ICA decomposes the observed space $C_i(t)$ into signal subspace and noise subspace. So we can reconstruct speech signal using signal subspace.

4 Experiments and Analysis

4.1 The Source of Experiment Data

The experiment data come from two parts. The first one is standard noise database and speech database. It is used to test the proposed method using different noise types and different SNRs. The second come from real short-wave speech signal records on the spot.

4.2 The Results of Experiments and Analysis

Using the noise database NOISEX92, add different type noise to the same pure speech signal. The SNR is 5dB. The enhanced SNR shown in table 1. Because the different of noise, the enhancement effect are different. From the table 1, we can see the proposed method can effectively reduce the pink and white noise.

The second database comes from the real records on the spot, which include many kinds of languages such as China, English, Japanese, Russian and so on. Each signal length of 8 frequency bands is 5 minutes, and the sample rate is 11025Hz. And the estimated SNR is 0.Two methods are used to test. The first one is the proposed method, and the other is the classical method spectral subtraction.

Table 1. The enhancement results of diffrent noise with different SNR

Noise	Pink	Factory	Airforce	Babble	White
–5	3.1	1.7	1.3	–1.2	5.2
0	9.2	5.4	4.5	2.5	12.7
5	15.2	11.9	11.2	8.2	18.0

Table 2. The tests result of real short-wave speech signal

Frequency band	Proposed method	SS
1	5.2	3.1
2	8.6	6.3
3	7.2	8.2
4	10	9.1
5	9.7	6.2
6	3.2	5.3
7	11.4	10.1
8	5.9	6.2

From table 2 we can see the proposed method can efficiently remove the noise. Because the different noise in different frequency band, the effects of enhancement are different also. In some band the enhancement effects of the proposed method are over performed the traditional method SS. And in the other frequency band the enhancement effects almost equal to the method SS. But it is worth to point out that not like SS the proposed method didn't produce music noise.

5 Conclusions

In this paper, a novel method for speech enhancement was proposed. Using EMD to form an observed data matrix. The ICA can decompose the matrix into signal subspace and noise subspace. The primary experiments show that the proposed method can efficiently remove the non-stationary noise. It is important for the short-wave communication, because it can elevate the objective quality of speech signal and the intelligibility of noisy speech in order to reduce the listener fatigue.

Acknowledgment

This work herein was supported by National Nature Science Foundation of China, under project No.60475016.

References

1. Martin R.:Noise Power Spectral Density Estimation Based on Optimal Smoothing and Minimum Statistics.IEEE Trans on Speech and Audio Processing,Vol.9(2001)504-512
2. Ephraim. Y,Malah. D.: Speech enhancement using a minimum mean-square error short-time spectral amplitude estimator. IEEE Transactions on Acoustics, Speech and Signal Processing, Vol.32 (1984) 1109-1121
3. Zheng W.T, Cao Z.H.: Speech enhancement based on MMSE-STSA estimation and residual noise reduction, 1991 IEEE Region 10 International Conference on EC3-Energy, Computer, Communication and Control Systems, Vol.3 (1991).265 –268
4. Liu Zhibin, Xu Naiping.: Speech enhancement based on minimum mean-square error short-time spectral estimation and its realization, IEEE International conference on intelligent processing system, Vol.28 (1997)1794-1797.
5. Lim ,J.S, Oppenheim. A.V.: Enhancement and bandwidth compression of noisy speech. Proc. of the IEEE, Vo1.67 (1979): 1586-1604.
6. Goh.Z, Tan.K and Tan.T.: Postprocessing method for suppressing musical noise generated by spectral subtraction. IEEE Trans. Speech Audio Procs, Vol. 6(1998) 287-292.
7. He. C and Zweig.Z.: Adaptive two-band spectral subtraction with multi-window spectral estimation. ICASSP, Vol.2 (1999) 793-796.
8. Huang. N.E.: The Empirical Mode Decomposition and the Hilbert Spectrum for Nonlinear and Non-stationary Time Series Analysis, J. Proc. R. Soc. Lond. A, Vol.454 (1998) 903-995
9. Huang. W,Shen.Z,Huang.N.E,Fung. Y.C.: Engineering Analysis of Biological Variables: an Example of Blood Pressure over 1 Day, Proc. Natl. Acad. Sci. USA, Vol.95 (1998) 4816-4821
10. Huang. W,Shen.Z,Huang.N.E,Fung.: Nonlinear Indicial Response of Complex Nonstationary Oscillations as Pulmonary Pretension Responding to Step Hypoxia, Proc. Natl. Acad. Sci, USA, Vol.96(1999)1833-1839

Robust Preprocessing of Gene Expression Microarrays for Independent Component Analysis

Pedro Gómez Vilda[1], Francisco Díaz[1], Rafael Martínez[1], Raul Malutan[1], Victoria Rodellar[1], and Carlos G. Puntonet[2]

[1] Universidad Politécnica de Madrid, Campus de Montegancedo, s/n, 28660, Boadilla del Monte, Madrid, Spain
[2] University of Granada, C/ Periodista Daniel Saucedo, s/n, 18071, Granada, Spain
pedro@pino.datsi.fi.upm.es

Abstract. Oligonucleotide Microarrays are useful tools in Genetic Research as they provide parallel scanning mechanisms to detect the presence of genes using test probes composed of controlled segments of gene code built by masking techniques. The detection of each gene depends on the multichannel differential expression of perfectly matched segments (PM) against mismatched ones (MM). This methodology, devised to robustify the detection process poses some interesting problems under the point of view of Genomic Signal Processing, as test probe expressions are not in good agreement with the proportionality assumption in most of the genes explored. These cases may be influenced by unexpected hybridization dynamics, and are worth of being studied with a double objective: gain insight into hybridization dynamics in microarrays, and to improve microarray production and processing as well. Recently Independent Component Analysis has been proposed to process microarrays. This promising technique requires the pre-processing of the microarray contents. The present work proposes the de-correlation of test probes based on probe structure rather than the classical "blind" whitening techniques currently used in ICA. Results confirm that this methodology may provide the correct alignment of the PM-MM pairs maintaining the underlying information present in the probe sets.

1 Introduction

The development of new techniques in Genetic Expression Microarrays has experienced a great push forward in the recent years. Microarray techniques have been successfully applied to almost every aspect of biomedical research because they open the possibility to do massive tests on genome patterns [10], [11]. DNA microarrays make use of the hybridization properties of nucleic acids to monitor DNA or RNA abundance on a genomic scale. Nevertheless there are several factors which render this methodology subject to further improvements regarding reliability:

- The dynamics of the hybridization process underlying genomic expression is complex as thermodynamic factors affecting molecular interaction are present and their influence must be taken into account.
- The microtechniques used in producing microarrays are not 100% reliable, and the results are corrupted by different kinds of noise, errors and crosstalk.
- Microarray scanning to images is prone to distortion and noise corruption.

J. Rosca et al. (Eds.): ICA 2006, LNCS 3889, pp. 714 – 721, 2006.

- The quality of the data available in the microarrays depend on how the experts in Genetics have conceived the process of gene testing (how mRNA has been segmented and deployed in the tests).

Preprocessing to detect the presence of undesired or uncontrolled effects on expression data can be afforded from a wider point of view using ICA based on Higher Order Statistics (HOS) using well-known algorithms [9], like FastICA. In this paper a whitening technique based in the de-correlation of test probes (Perfect Match from MisMatch probe pairs) is proposed. The paper focuses in modeling the hybridization process under a dynamical point of view. The next sections concentrate in the application of classical FastICA to data from the Latin Square Database [8], followed by conclusions drawn from this study.

2 Detection of Gene Expression Levels in Microarrays

The main types of tests using gene expression microarrays are serial analysis of gene expression (SAGE) [1], short oligonucleotide arrays [2], and spotted cDNA arrays [3]. The present paper is concentrated in the second type where hundreds of thousands of oligonucleotides are synthesized *in situ* on small glass *chips* by means of photochemical reaction and mask technology. Each gene or portion of a gene is represented by 11 to 20 test pairs of 25 oligonucleotides each. One of the samples in each pair is a complementary replica of the segment to be matched (Perfect Match – PM), where the other sample has been altered changing one of the central bases by its complementary, composing a mismatch (MM) probe. The difference in hybridization between the probes in the pairs, as well as their intensity ratios, mark specific target sequences. Differential estimation algorithms as MAS 4.0 and 5.0, MBEI or RMA [4], [5], [6] evaluate the *expression signal* for the probe set. In the present work a simple model of the hybridization process reliability is used, stating that the amounts of hybridized material for the perfect ($x_{i,k}^p$) and mismatch ($x_{i,k}^m$) probe pairs k corresponding to gene i may be given as

$$\begin{aligned} x_{i,k}^p &= \rho(s_{i,k}, x, y) p_t(s_{i,k} \mid z_{i,k}^p) \\ x_{i,k}^m &= \rho(s_{i,k}, x, y) p_t(s_{i,k} \mid z_{i,k}^m) \end{aligned} \tag{1}$$

$\rho(s_{i,k}, x, y)$ being the surface density of segment k in gene i for point (x,y) and $p_t(s_{i,k} \mid z_{i,k}^p)$ and $p_t(s_{i,k} \mid z_{i,k}^m)$ the respective hybridization probabilities for segment $s_{i,k}$ on the test segments $z_{i,k}^p$ (for PM) and $z_{i,k}^m$ (for MM), t being time. Assuming that hybridization thermodynamics are the same, probabilities will be proportional

$$\eta_{i,k} = \frac{p_t(s_{i,k} \mid z_{i,k}^p)}{p_t(s_{i,k} \mid z_{i,k}^m)} = \frac{\tau_{i,k}^m}{\tau_{i,k}^p} \tag{2}$$

where $\tau_{i,k}^{p/m} > 0$ are the time constants for PM or MM pairs of segments i, k, related to the *hybridization temperature* of each segment. Probe pairs could be considered proportional in a strict sense if the proportionality parameter for segment k is almost the same independently of gene segments

$$\eta_{i,k} = \eta_{j,k}; \quad \forall i, j \tag{3}$$

In general, time constants $\left\{\tau_{i,k}^{p,m}\right\}$ associated to a given gene i will not be equal, segments with small time constants will be saturated, while those showing large time constants will be unsaturated, therefore hypothesis (3) will not be fulfilled in these cases. Each probe set can be seen as composed by two vectors $\boldsymbol{x}_i^p=\left\{x_{i,k}^p\right\}$ and $\boldsymbol{x}_i^m=\left\{x_{i,k}^m\right\}$ corresponding to the PM and MM probe set. Classically in microarray processing it should be expected that strict proportionality (3) among segments holds at least up to a certain degree, and in such case the expression results for the corresponding gene k could be trusted. These cases are designated as *reliably expressed probe sets*. But in many cases this assumption can not be granted, as there are some PM-MM pairs where strict proportionality does not match that of others within the same test probe. These cases may be referred to as *unreliably expressed probe sets*. The question is how to measure the reliability of a probe set expression. This can be done (see 0) through the proportionality parameter [12] of gene i as

$$\lambda_i=\frac{\left\|\boldsymbol{x}_i^m\right\|\cos\beta_i}{\left\|\boldsymbol{x}_i^p\right\|}=\frac{\left\langle\boldsymbol{x}_i^m,\boldsymbol{x}_i^p\right\rangle}{\left\|\boldsymbol{x}_i^p\right\|^2}=\frac{\sum_{k=1}^{K}p_t(s_{i,k}\mid z_{i,k}^p)p_t(s_{i,k}\mid z_{i,k}^m)}{\sum_{k=1}^{K}p_t^2(s_{i,k}\mid z_{i,k}^p)}\tag{4}$$

This parameter is independent of the density distribution, therefore is not affected by the overall gene expression. Besides, the proportionality parameter is directly related to the projection of the vectors of conditioned probabilities for the perfect and the mismatch test probes, and therefore to the hybridization dynamic constants.

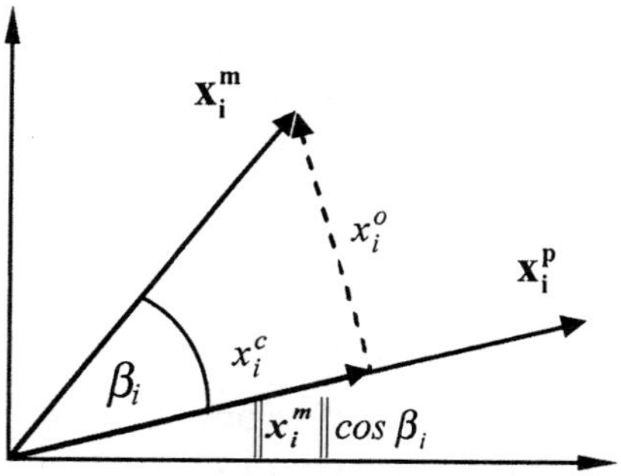

Fig. 1. Geometrical relations among PM and MM vectors

The diagram in Figure 1 may help to understand the meaning of the proportionality parameter and the effects of bad alignment between PM and MM where $\mathbf{x}_i^c$ and $\mathbf{x}_i^o$ are the co-linear and orthogonal components of $\mathbf{x}_i^m$ with respect to $\mathbf{x}_i^p$ and β_i is the angle between them. If the homogeneity hypothesis (3) holds, a high degree of co-linearity between the PM and MM vectors would be expected, the orthogonal component being small, otherwise one should conclude that the estimated expression levels are the result of underlying unknown processes. Orthogonality may be defined as

$$\gamma_i=1-\cos^2\beta_i\tag{5}$$

which ranges between *0* (co-linear vectors), and *1* (orthogonal ones). Once this model is established projection methods [7] could be used to pre-process microarray information prior to component separation and classification. If the values of γ_i are closer to 1 this will mean that the two vectors are not correlated at all, and in that case independent components can be estimated using ICA. Data alignment may help in improving the reliability of estimations in probe sets with large γ_i. Data alignment and orthogonalization may be easily derived from the projection model in Figure 1. In fact, the co-linear and orthogonal components of the MM vector will be given by

$$\mathbf{x}_i^c = \lambda_i \mathbf{x}_i^p = \frac{\langle \mathbf{x}_i^m, \mathbf{x}_i^p \rangle}{\|\mathbf{x}_i^p\|^2} \tag{6}$$

$$\mathbf{x}_i^o = \mathbf{x}_i^m - \lambda_i \mathbf{x}_i^p = \mathbf{x}_i^m - \frac{\langle \mathbf{x}_i^m, \mathbf{x}_i^p \rangle}{\|\mathbf{x}_i^p\|^2} \tag{7}$$

3 A Case of Study

The methodology described has been tested using a microarray (12-13-02-U133A-Mer-Latin-Square-Expt1-R1) from the Latin Square set of experiments [8]. A single microarray contains information of PM-MM probe sets from 22300 gene positions. The information contained in the microarray was converted on a string of gene probe sets arranged linearly. The values of λ_i and γ_i for each probe set were evaluated by the application of (4) and (5). One of the objectives of the present work is to design methods to pinpoint unreliable probe sets, which will be corrected using ICA. The selection of probe sets relies on estimating the orthogonality parameter γ_i given in (5). The estimation results for the whole set of genes in the microarray are presented in Figure 2, together with their statistical distribution. It may be seen that most of the genes have values of γ_i not fulfilling the hypothesis of co-linearity, therefore the hybridization process do not respond to condition (3). The results in Table 1 show the distribution of genes vs. the values for the co-linearity coefficient γ_i.

Table 1. Distribution of genes by the value of γ_i

Value of γ_i	Number of genes	Percent on total
<0.05	1605	7.19
$0.05 \leq \gamma_i < 0.1$	3315	14.86
$0.1 \leq \gamma_i < 0.5$	16547	74.20
≥ 0.5	833	3.73

The number of unreliable gene expression tests ($\gamma_i > 0.1$) is quite large (77.93%), thus implying that many probe tests may have been affected by corrupting hybridization

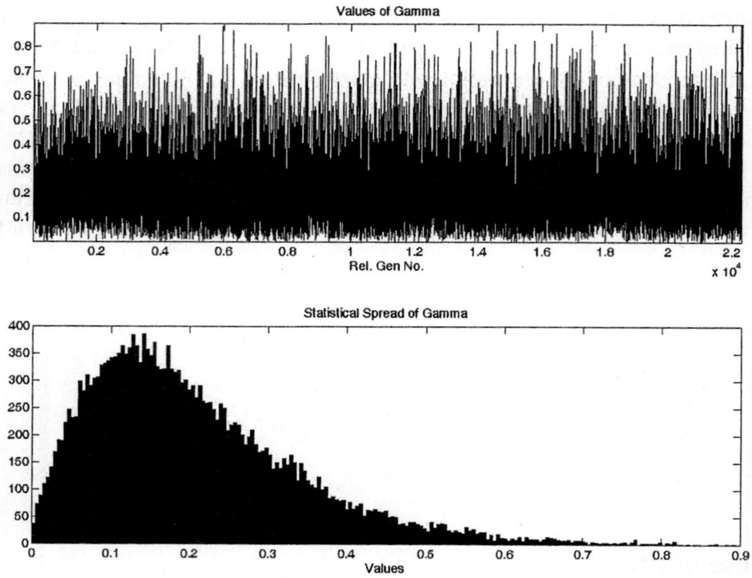

Fig. 2. Results of estimating γ_i for a practical case versus relative gene number in microarray. Top: plot from the evaluation of (5). Bottom: Histogram of γ_i.

side effects. These observations allow concluding that on one side results from reliable probe sets should be taken into account independently of their level of expression, as they are in good agreement with the coherence model. On the other hand, results from unreliably expressed probe sets should be treated to improve reliability, and studied to highlight the reasons for anomalous behavior in their underlying hybridization dynamics.

Unreliably expressed probe sets may be re-aligned using ICA assuming that x_i^p and x_i^m have been previously orthogonalized using (6) and (7) and normalized into x_i^c and x_i^o constituting the vector of observations

$$\boldsymbol{x} = \left\lfloor x_i^c , x_i^o \right\rfloor^T \tag{8}$$

such that each observed variable could be described in terms of two independent components s_i^1 and s_i^2

$$\begin{aligned} x_i^c &= a_{c1} s_i^1 + a_{c2} s_i^2 \\ x_i^o &= a_{o1} s_i^1 + a_{o2} s_i^2 \end{aligned} \tag{9}$$

where the coefficients of the mixing matrix are unknown and have be estimated as well as the independent components s using only the observed data. The statistically independent components s_i^1 and s_i^2 with non-gaussian distributions were estimated using FastICA [13], [14] as well as the mixing matrix $\boldsymbol{A}$ and its inverse $\boldsymbol{W} = \boldsymbol{A}^{-1}$

$$y_i^1 = w_{c1}x_i^c + w_{o1}x_i^o$$
$$y_i^2 = w_{c2}x_i^c + w_{o2}x_i^o \qquad (10)$$

which will be used to re-estimate the PM and MM probe sets

$$\hat{x}_i^p = y_i^1 / \lambda_i \qquad (11)$$

$$\hat{x}_i^m = y_i^2 + y_i^1 \qquad (12)$$

4 Results and Discussion

The microarray under study was separated into a PM and an MM vector. Preprocessing consisted in the application of (6) and (7) probe by probe (for the 22,300 probe tests) to generate two new vectors of co-linear and orthogonal components. After normalization these were taken for the analysis using Fast-ICA independently of their values for γ. The results of the analysis of genes 212058_at and 212062_at are given in Figure 3 as an example. The first probe set (top templates corresponding to gene 212062_at) in which most of the probe pairs behave accordingly with (3) is reliably expressed and shows high co-linearity (γ= 0.0221). The second one (bottom templates corresponding to gene 212058_at) is an unreliably expressed probe test (γ= 0.259) due mainly to probes 2, 5, and 6 which present rather different proportionality ratios, possibly due to perturbations in some underlying process or in too dissimilar hybridization dynamics. The re-estimated PM and MM components, show an improvement of co-linearity in cases with large γ, whereas this value deteriorates in cases where co-linearity was large initially, as expected from the orthogonality hypothesis used to process expression vectors prior to the application of ICA. Well aligned vectors may be seen as produced by a single underlying process, therefore strictly they do not need to be re-aligned, whereas mis-aligned vectors may be the result of several underlying independent processes and the application of ICA to these cases improves substantially the results. Most of the probe sets in the microarray (21,765 out of 23,000) were composed by 11 probe pairs. To evaluate the effects of ICA on the statistical distributions of co-linear and orthogonal components, the kurtosis of each expression probe after orthogonalization was compared against its corresponding position in the respective independent component, as given in Table 2.

For example the kurtosis of the 4th position of the co-linear component is 98.9, whereas the respective position in the independent component is 134, indicating that ICA improved the non-gaussianity of this position. This is not observed in all the positions as well-aligned probe sets were not removed from the data analyzed.

Table 2. Kurtosis of co-linear, orthogonal and independent components

i	1	2	3	4	5	6	7	8	9	10	11
x_{ic}	*122*	*88.9*	*99.7*	*98.9*	*76.9*	*108*	*96.8*	*79.5*	*67.6*	*69.8*	*74.5*
x_{io}	*45.2*	*43.9*	*53.8*	*65.3*	*87.3*	*59.5*	*66.9*	*113*	*53.4*	*46.3*	*51.6*
y_{ic}	*113*	*95*	*116*	*134*	*108*	*106*	*117*	*99.9*	*94.3*	*90.5*	*96.3*
y_{io}	*60.4*	*47.1*	*81*	*78.1*	*68.8*	*45.7*	*36*	*82.5*	*40.8*	*33.9*	*37.7*

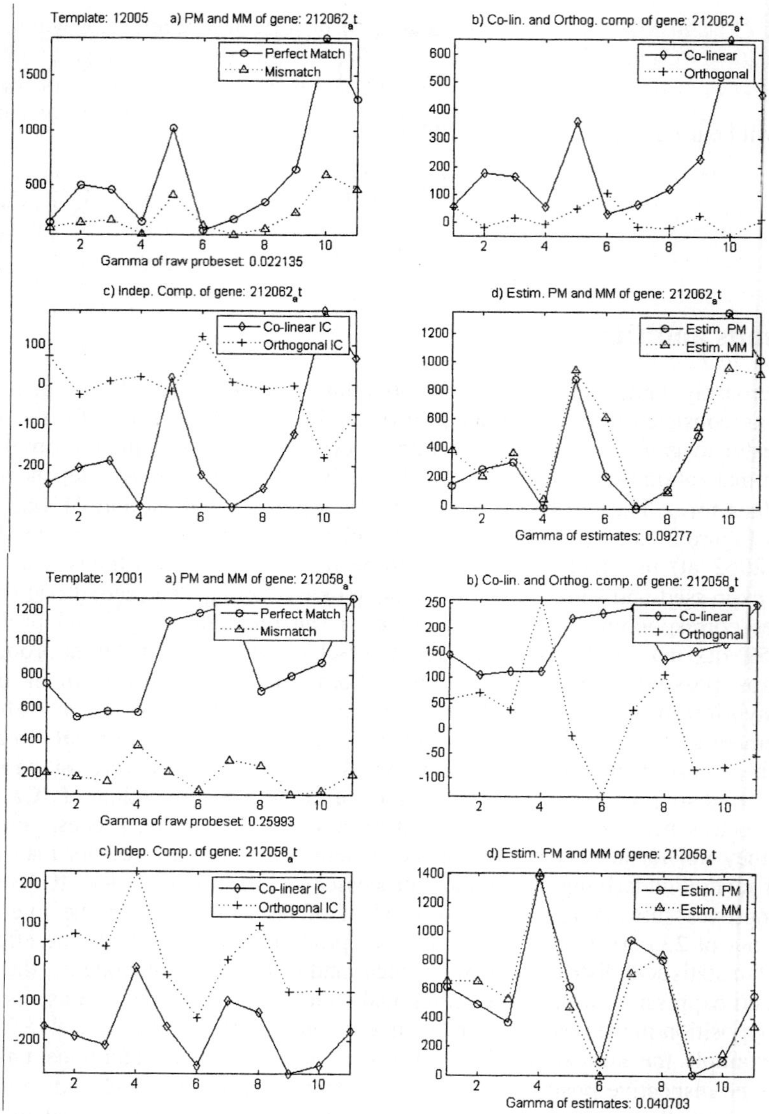

Fig. 3. a) Four top templates: Results for a reliable probe set. b) Four bottom templates: Results for an unreliable probe set. Plates a) original PM-MM probes, b) co-linear and orthogonal components after the application of (6) and (7), c) independent components from (10), and d) re-estimated PM and MM from (11) and (12).

5 Conclusions

In this work a probabilistic model has been used to justify reliable hybridization of gene probe sets in microarrays based on proportionality and correlation. Correlation between PM and MM samples may be used to provide information on which probe set results were produced from normal hybridization processes, in contrast with those which may be produced by corrupted hybridization. This can help in improving the estimation reliability of microarray data prior to their use in clustering and pattern recognition. The number of unreliable gene probe sets found in a particular microarray may be quite large, thus meaning that many probe tests may have been affected by corruption processes. These probe sets may be re-aligned by detecting their independent components by ICA, and re-estimating the PM-MM pairs from the independent components found. The results show that re-alignment improve the reliability of genes affected by underlying processes related with the independent components.

Acknowledgments

This research is being carried out under grant Nos. TIC2002-2273 and TIC2003-08756 from Programa de las Tecnologías de la Información y las Comunicaciones, Ministry of Education and Science, Spain.

References

[1] D. Porter et al., "A SAGE view of breast tumor progression", Cancer Research, Vol. 61, 1 August, 2001, pp. 5967 - 5702
[2] D. Amaratunga, and J. Cabrera, "Exploration and analysis of DNA microarray and protein array data", Ed. Wiley Interscience, Hobooken, N.J., 2004, pp. 35-36
[3] Moore, S., "Making Chips", IEEE SPECTRUM, March, 2001; pp. 54-60
[4] R. A. Irizarry, B. Hobbs,F. Collin, Y. D. Beazer-Barclay, K. J. Antonellis, U. Scherf, and T. P. Speed, "Exploration, normalization and summaries of high density oligonucleotidearray probe level data", Biostatistics. Vol. 4, Number 2 (2003), pp. 249-264
[5] C. Li,. and W. H. Wong, "Model-based analysis of oligonucleotide arrays: expression index computation and outlier detection", Proc. Nat. Acad. Sci., Vol. 98 (2002), pp. 31-36
[6] F. Naef, D. A. Lim, N. Patil, and M. Magnasco, "From features to expresion: High denstiy oligonucleotide arrays revised", Proc. DIMACS Workshop on Analysis of Gene Expression Data, (2001)
[7] D. E. Catlin, "Estimation, control, and discrete Kalman filter", Ed. Springer-Verlag New York Inc., 1989, pp. 238-254
[8] http://www.affymetrix.com/index.affx
[9] Cichocki, A., Amari, S. I., (2002). Adaptive Blind Signal and Image Processing, John Wiley & Sons
[10] Schmidt, U., Begley, C. G. (2003). "Cancer diagnosis and microarrays", The International Journal of Biochemistry and Cell Biology, Vol. 35, pp. 119-124.
[11] Whitchurch, A. K. (2002). "Gene Expression Microarrays", IEEE Potentials, February-March 2002, pp. 30-34.
[12] Lukac, R., Plataniotis, K. N., Smolka, B., Venetsanopoulos, A. N., "A Multichannel Order-Statistic Technique for cDNA Microarray Image Processing", IEEE Trans. on Nanobioscience, Vol. 3, No. 4, December 2004.
[13] Hyvärinen, A., "HUT-CIS: The FastICA package for MATLAB", http://www.cis.hut.fi/projects/ica.
[14] Hyvärinen, A. "Fast and Robust Fixed-Point Algorithm for Independent Component Analysis", *IEEE transactions on Neural Networks*, vol. 10, no. 3, May 1999.

Single-Channel Mixture Decomposition Using Bayesian Harmonic Models

Emmanuel Vincent and Mark D. Plumbley

Electronic Engineering Department, Queen Mary, University of London,
Mile End Road, London E1 4NS, United Kingdom
emmanuel.vincent@elec.qmul.ac.uk

Abstract. We consider the source separation problem for single-channel music signals. After a brief review of existing methods, we focus on decomposing a mixture into components made of harmonic sinusoidal partials. We address this problem in the Bayesian framework by building a probabilistic model of the mixture combining generic priors for harmonicity, spectral envelope, note duration and continuity. Experiments suggest that the derived blind decomposition method leads to better separation results than nonnegative matrix factorization for certain mixtures.

1 Introduction

1.1 Constrained Specific Models and Unconstrained Generic Models

Single-channel musical source separation is the problem of extracting the source signals $(s_j(t))_{1 \leq j \leq J}$ underlying a music signal $x(t) = \sum_{j=1}^{J} s_j(t)$. This problem can be addressed by building appropriate models of the sources. The source models proposed in the literature rely on different amounts of prior information.

Some methods exploit constrained source models representing the sources in a specific mixture with a good accuracy. For example, methods based on sparse coding with a fixed dictionary [1] or on factorial hidden Markov models [2] typically assume that the source models can be learnt on segments of the mixture where only one source is present. These methods provide very good separation results, given the difficulty of the problem, but until now they rely on knowing the instruments present in the mixture and performing a manual segmentation. Other methods based on Computational Auditory Scene Analysis (CASA) with instrument templates [3] or on hybrid source models [4] rely on instrument-specific timbre properties learnt on a database of isolated notes. These methods also perform satisfyingly, but they cannot be applied when some of the instruments present in the mixture are not part of the learning database.

By contrast, other methods rely on unconstrained generic source models applicable to a large range of mixtures. For example, Nonnegative Matrix Factorization (NMF) decomposes the mixture short-term magnitude spectrum into a sum of components modeled by a fixed magnitude spectrum and a time-varying

J. Rosca et al. (Eds.): ICA 2006, LNCS 3889, pp. 722–730, 2006.

gain, assuming no constraints about the spectra and the gains except positivity [5]. Source separation can then be achieved by clustering the components into sources, provided each component belongs to a single source. Good results based on automatic clustering have been reported for the separation of vocals [6] or drums [7] from real mixtures. Other studies using a manual clustering have shown that NMF can be used to separate real mixtures of non-percussive instruments [8]. However the NMF source model is not adapted to certain types of mixtures, such as those involving notes with time-varying fundamental frequency, instruments with similar spectral envelope or instruments playing synchronously.

1.2 Harmonicity as a Precise Generic Model

In this paper, we assume that each musical note is a near-periodic signal containing harmonic sinusoidal partials. Harmonicity means that at each instant the frequencies of the partials are multiples of a single fundamental frequency. This assumption is true for sustained instruments such as bowed strings and winds and approximately true for many other instruments. It is false for drums, human voice and other noisy or transient sounds. Harmonicity can thus be seen as a precise generic model: it gives more information about the sources than the NMF model while being valid for a large range of mixtures. In the following, we call *harmonic component* a set of harmonic partials having common onset and offset times and we address the problem of Harmonic Component Extraction (HCE), that is the decomposition of a mixture into such components. We do not discuss the difficult issue of clustering the estimated components into sources.

Most existing HCE methods consist in performing a polyphonic pitch tracking, that is transcribing the fundamental frequencies of the notes present in the mixture, and then estimating the amplitudes and phases of their harmonics. Methods exploiting harmonicity only [9] are insufficient for source separation. Indeed harmonicity does not provide enough information to segregate partials from different sources overlapping at the same frequency. Other methods have used complementary assumptions of spectral continuity [10,11] and temporal continuity [12,10] to this aim. Since polyphonic pitch tracking is a difficult problem for which no current algorithm provides a perfect solution, the separation performance of these methods was mostly evaluated based on prior knowledge of the fundamental frequencies and few quantitative results were reported.

In the following, we recast the problem of estimating harmonic components in the Bayesian framework. We model the mixture signal as a sum of harmonic components whose parameters are governed by probabilistic priors and we estimate the number of components and their parameters using a Maximum A Posteriori (MAP) criterion. This can be seen as a coherent approach where polyphonic pitch tracking and estimation of the amplitudes and phases of the partials are performed using the same model. The proposed model is inspired by Bayesian harmonic models introduced previously in the literature for polyphonic pitch transcription [13] but it includes several modifications. Most importantly, we design a perceptually motivated residual prior and we learn the parameters of other priors on a database of isolated notes rather than setting them manually to arbitrary values. When this learning database is large, the resulting model is generic. We have also used this model recently for object coding purposes [14].

The rest of the paper is structured as follows. Section 2 presents the generative model of the mixture and the associated inference algorithm. Section 3 compares the performance of the proposed method with NMF on a few test mixtures. Finally Section 4 discusses some future research directions.

2 Bayesian Inference of Harmonic Components

2.1 Signal Model

The proposed model is expressed in the time domain. Let $x_n(t)$ be the n-th frame of the mixture signal $x(t)$ defined by $x_n(t) = w(t)x(nS+t)$ where $w(t)$ is a Hanning window of length W and S is the stepsize. We develop $x_n(t)$ as

$$x_n(t) = \sum_{c \in C_n} s_{cn}(t) + e_n(t), \tag{1}$$

where $(s_{cn}(t))_{c \in C_n}$ are the harmonic components present in this frame and $e_n(t)$ is the residual. We define each harmonic component, which generally spans several time frames, by

$$s_{cn}(t) = w(t) \sum_{m=1}^{M_c} a_{cmn} \cos(2\pi m f_{cn} t + \phi_{cmn}), \tag{2}$$

where f_{cn} is its fundamental frequency and (a_{cmn}, ϕ_{cmn}) are the time-varying amplitude and phase of its m-th partial in the n-th frame.

2.2 Frequency, Amplitude and Spectral Envelope Priors

We associate each component with a latent fundamental frequency F_c belonging to the MIDI scale, which is the discrete 1/12 octave scale used for western musical scores. We constrain the number of partials M_c of the c-th component to

$$M_c = \min((F_{\max}/F_c), M_{\max}), \tag{3}$$

where $F_{\max}$ is the Nyquist frequency and $M_{\max}$ is set to 60. On each time frame, we model the fundamental frequency by a log-Gaussian prior

$$P(\log f_{cn}) = \mathcal{N}(\log f_{cn}; \log F_c, \sigma^f), \tag{4}$$

where $\mathcal{N}(\cdot; \mu, \sigma)$ is the univariate Gaussian density of mean μ and standard deviation σ. In order to help estimate the amplitudes of the partials when partials from several notes overlap at the same frequency, we describe the amplitudes as the product of a fixed normalized spectral envelope $(\mu^a_{F_c m})_{1 \le m \le M_c}$, a latent log-Gaussian amplitude factor r_{cn} and a log-Gaussian residual, that is

$$P(\log a_{cmn} | r_{cn}) = \mathcal{N}(\log a_{cmn}; \log(r_{cn} \mu^a_{F_c m}), \sigma^a_{F_c}), \tag{5}$$

$$P(\log r_{cn}) = \mathcal{N}(\log r_{cn}; \mu^r_{F_c}, \sigma^r_{F_c}). \tag{6}$$

Finally we assume that the phases of the partials are uniformly distributed

$$P(\phi_{cmn}) = 1/(2\pi). \tag{7}$$

2.3 Duration and Continuity Priors

Perceptually annoying discontinuities may appear in the extracted source signals when the model parameters are estimated on each time frame separately. Thus we add duration and continuity priors on the parameters. We associate each point on the MIDI scale with a binary activity state in each frame determining whether a harmonic component with the corresponding latent frequency F_c is being played or not in that frame, with the constraint that different instruments cannot play notes with the same latent frequency at the same time. We assume that the sequences of activity states for different points on the MIDI scale are independent, and we model each sequence by a two-state Markov prior. We also set temporal continuity priors on the frequencies and amplitudes of the partials

$$P(\log f_{cn}|f_{c,n-1}) = \mathcal{N}(\log f_{cn}; \log f_{c,n-1}, \sigma^{f'}), \tag{8}$$

$$P(\log a_{cmn}|a_{cm,n-1}) = \mathcal{N}(\log a_{cmn}; \log a_{cm,n-1}, \sigma^{a'}_{F_c m}), \tag{9}$$

$$P(\log r_{cn}|r_{c,n-1}) = \mathcal{N}(\log r_{cn}; \log r_{c,n-1}, \sigma^{r'}_{F_c}). \tag{10}$$

The global prior on amplitudes and frequencies is then defined up to a multiplicative constant by multiplying these priors with the local priors defined above.

2.4 Perceptually Motivated Residual Prior

The role of the prior on the residual is to ensure that the largest possible number of notes present in the mixture are extracted using a given number of components. The standard Gaussian prior measures the distortion between the mixture signal and the model according to the energy of the residual. This often results in several components being used to represent high-energy notes, while low energy parts of the mixture such as low energy notes, onsets and reverberation are not transcribed despite their perceptual significance. We design instead a weighted Gaussian prior inspired from the distortion measures proposed in [15,16] which give a larger weight to perceptually significant low energy parts.

The proposed prior models the first stages of auditory processing. The incoming sound first passes through the outer and the middle ear and is split by the cochlea into several frequency subbands called auditory bands. The energy in each auditory band is then transformed nonlinearly into a loudness value taking into account masking phenomena. More precisely, we measure the power of the residual in the b-th auditory band by $\tilde{E}_{nb} = \sum_{f=0}^{W/2} v_{bf} g_f |E_{nf}|^2$, where $(E_{nf})_{0 \leq f \leq W-1}$ are the Fourier transform coefficients of $e_n(t)$, $(v_{bf})_{0 \leq f \leq W/2}$ are coefficients modeling the frequency spread of that band and $(g_f)_{0 \leq f \leq W/2}$ is the frequency response of the outer and middle ear. We measure similarly the power of the mixture signal in that band by $\tilde{X}_{nb} = \sum_{f=0}^{W/2} v_{bf} g_f |X_{nf}|^2$. Then we define the distortion due to the residual on the n-th frame by $L_n = \sum_{b=1}^{B} \tilde{E}_{nb} \tilde{X}_{nb}^{-0.75}$. It can be shown that this distortion is approximately equal to the perceived loudness of the residual on that frame [16]. We derive the residual prior from the distortion by $P(e_n) \propto \exp(-L_n/(2\sigma^{e\,2}))$. This prior can also be expressed as

$$P(E_{nf}) = \mathcal{N}(E_{nf}; 0, \sigma^e \gamma_{nf}^{-1/2}) \quad (11)$$

where

$$\gamma_{nf} = \sum_{b=1}^{B} v_{bf} g_f \left(\sum_{f=0}^{W/2} v_{bf} g_f |X_{nf}|^2 \right)^{-0.75} . \quad (12)$$

2.5 Approximate Inference of Harmonic Components

The signal model and the parameter priors define together a probabilistic generative model of the mixture signal that is used to infer the MAP values of the activity states and the frequency, amplitude and phase parameters representing a given mixture. Due to the complexity of the model, exact inference is intractable. We therefore use a three-step approximate inference procedure instead. First we estimate the MAP activity states and the corresponding MAP parameters on each time frame separately, then we refine the estimation of the states by adding the duration priors, and finally we refine the estimation of the parameters by keeping the states fixed and adding the continuity priors. More details about these steps are given in [14]. Each harmonic component is then directly synthesized from the corresponding parameters.

3 Evaluation

3.1 Training, Performance Measure and Optimal Clustering

We evaluate the proposed HCE method on test mixtures sampled at 22.05 kHz. Hyper-parameters of the generative model are set to the same values for all test mixtures: σ^f, $(\mu^a_{F_c m})$, $(\sigma^a_{F_c})$, $(\mu^r_{F_c})$, $(\sigma^r_{F_c})$, $\sigma^{f'}$, $(\sigma^{a'}_{F_c m})$ and $(\sigma^{r'}_{F_c})$ are learnt on part of the RWC[1] Musical Instrument Database whereas σ^e and the Markov transition probabilities are set manually. The frame parameters are set to $W = 1024$ (46 ms) and $S = 512$ (23 ms) and discrete fundamental frequencies span the range between MIDI 36 (65 Hz) and MIDI 100 (2640 Hz).

For comparison purposes, we also evaluate NMF on the same test mixtures. We write the NMF generative model as $|X_{nf}| = \sum_{c=1}^{C} p_{cf} q_{cn} + E_{nf}$, where $(p_{cf})_{0 \le f \le W/2}$ and $(q_{cn})_{0 \le n \le N-1}$ are the fixed spectrum and time-varying amplitude of the c-th nonnegative component respectively. We assume that these quantities are positive and that the residual E_{nf} follows the weighted Gaussian prior above. The total number of spectra C is fixed manually and the spectra and time-varying amplitudes are estimated using multiplicative update rules.

[1] `http://staff.aist.go.jp/m.goto/RWC-MDB/`

Source signals including several spectra are then synthesized by inverse Fourier transform and overlap-add using the phase spectrum of the mixture signal. This algorithm is similar to the weighted NMF algorithm introduced in [16], except the definition of the time-frequency weights (γ_{nf}) is modified by taking into account overlap between auditory bands.

For evaluation purposes, we partition components produced by HCE or NMF into source clusters based on prior knowledge of the true sources. We define the optimal clusters as those which maximize the overall source separation performance and we compute them using a beam search procedure. This "oracle" clustering is not feasible in realistic situations, however it allows the measurement of the best source separation quality potentially achievable.

The source separation performance is measured locally for each estimated source j around each time frame n using a local phase-blind Signal-to-Distortion Ratio (SDR) in decibels (dB) defined by

$$\mathrm{SDR}_{jn} = 10 \log_{10} \left(\frac{\sum_{l=0}^{W'-1} w'(l)^2 |S_{j,n+l,f}|^2}{\sum_{l=0}^{W'-1} w'(l)^2 (|\hat{S}_{j,n+l,f}| - |S_{j,n+l,f}|)^2} \right), \tag{13}$$

where $w'(l)$ is a Hanning window of length $W' = 12$ frames and $(\hat{S}_{jnf})$ and (S_{jnf}) are the short-term Fourier transforms of the j-th estimated source and the j-th true source respectively. The overall performance is measured by a global SDR defined as the median of local SDRs for all sources and all time frames. We believe that this performance measure accounts better for subjective effects than the standard time-domain SDR. Indeed the ear is approximately phase-blind and the error perceived at a given time depends only on the power of the target signal at that time, not on its total energy. However the actual subjective performance is better assessed by listening to the estimated source signals.

3.2 Results

We consider two sets of test mixtures: ten mixtures of two sources using real sources from the SQAM database[2], and ten MIDI-synthesized mixtures from the RWC Classical Music and Music Genre Databases containing two to five sources. We set the number of nonnegative components of NMF to be the same as the number of harmonic components estimated by HCE. This allows a rather fair comparison of the two methods, since in a blind context the difficulty of component clustering would depend on the number of components. We also separate MIDI-synthesized mixtures by HCE using knowledge of the note activity states. All the mixture signals and some of the estimated source signals are available for listening on `http://www.elec.qmul.ac.uk/people/emmanuelv/ICA06/`.

Table 1 shows that the global SDR achieved by HCE is on average 3 dB higher than NMF on mixtures of real sources and 6 dB higher on MIDI-synthesized mixtures. Informal listening tests suggest that the estimation errors made by the two

[2] `http://www.ebu.ch/en/technical/publications/tech3000_series/tech3253/`

Table 1. Comparison of the separation performance achieved by HCE and NMF

Separation method	Global SDR on various mixtures of real sources (dB)									
HCE	12.9	13.3	13.8	10.9	19.3	17.3	14.6	15.2	14.7	11.9
NMF	10.3	11.6	12.6	7.2	14.0	11.8	10.9	11.9	13.0	10.4
Separation method	Global SDR on various MIDI-synthesized mixtures (dB)									
HCE with true score	14.5	29.3	10.8	13.0	10.4	3.0	12.3	9.4	17.7	8.8
HCE	15.4	29.3	7.2	11.9	10.7	5.5	11.5	3.2	17.0	5.1
NMF	6.9	7.4	5.7	7.0	1.4	3.5	3.4	2.6	14.5	3.3

methods are very different. As expected, NMF often fails to separate synchronized notes in MIDI-synthesized mixtures because these notes have the same temporal evolution. This results in strong interference or in continuous artifacts. More surprisingly, NMF also produces artifacts on mixtures of real sources which are not synchronized. By contrast, HCE generally produces fewer artifacts, but some interference appears locally due to simultaneous or successive notes with the same frequency being fused into a single component, or to harmonic partials from different sources being transcribed as part of the same component.

The knowledge of the note activity states does not substantially improve the performance of HCE for seven out of ten MIDI-synthesized mixtures[3]. It is interesting to note that the number of notes estimated by HCE on MIDI-synthesized mixtures is on average 2.5 times larger than the actual number of notes being played. Most of the spurious notes have short duration and are due to the system trying to represent non-harmonic parts of the signal using harmonic components, which does not seem to affect the separation performance.

Other experiments suggested that the performance of NMF decreases when more components are allowed and does not change significantly when initializing the NMF basis spectra by the spectra of the harmonic components estimated by HCE. Thus the limited performance of NMF on the test mixtures seems to be the effect of the model itself rather than algorithmic issues.

4 Conclusion

In this paper, we address the blind source separation problem for single-channel musical mixtures where the notes are near-periodic signals containing harmonic sinusoidal partials. The proposed method, which exploits harmonicity and other generic source priors, performs better than NMF on various test mixtures. This suggests that the NMF model is not sufficiently constrained to ensure that typical audio source properties hold for the separated sources and that more precise generic source models can help separation without needing specific information about a particular mixture.

[3] For some mixtures the estimated note activity states lead to a better SDR than the true states because the perceptual weights used for decomposition are not taken into account for evaluation. In practice, the subjective performance of HCE using the true note activity states is always larger or equal to that of blind HCE.

The main limitation of HCE is that it cannot deal with mixtures containing voice or drum instruments. This limitation could be addressed using a three-component generative model including probabilistic models for wideband noise components and transient components, in the spirit of the CASA system proposed in [12]. The proposed model could also be improved by adding slightly inharmonic components to represent instruments such as piano or guitar or by performing automatic adaptation of the probabilistic priors to the mixture to increase their precision and help reduce separation errors.

References

1. Benaroya, L., McDonagh, L., Bimbot, F., Gribonval, R.: Non negative sparse representation for Wiener based source separation with a single sensor. In: Proc. IEEE Int. Conf. on Acoustics, Speech and Signal Processing (ICASSP). (2004) VI–613–616
2. Ozerov, A., Philippe, P., Gribonval, R., Bimbot, F.: One microphone singing voice separation using source-adapted models. In: Proc. IEEE Workshop on Applications of Signal Processing to Audio and Acoustics (WASPAA). (2005) 90–93
3. Kinoshita, T., Sakai, S., Tanaka, H.: Musical sound source identification based on frequency component adaptation. In: Proc. Int. Joint Conf. on Artificial Intelligence (IJCAI) Workshop on Computational Auditory Scene Analysis. (1999) 18–24
4. Vincent, E.: Musical source separation using time-frequency source priors. IEEE Trans. on Speech and Audio Processing **14(1)** (2006) To appear.
5. Smaragdis, P., Brown, J.C.: Non-negative matrix factorization for polyphonic music transcription. In: Proc. IEEE Workshop on Applications of Signal Processing to Audio and Acoustics (WASPAA). (2003) 177–180
6. Vembu, S., Baumann, S.: Separation of vocals from polyphonic audio recordings. In: Proc. Int. Conf. on Music Information Retrieval (ISMIR). (2005) 337–344
7. Helén, M., Virtanen, T.: Separation of drums from polyphonic music using non-negative matrix factorization and support vector machine. In: Proc. European Signal Processing Conf. (EUSIPCO). (2005)
8. Wang, B., Plumbley, M.D.: Musical audio stream separation by non-negative matrix factorization. In: Proc. UK Digital Music Research Network (DMRN) Summer Conf. (2005)
9. Gribonval, R., Bacry, E.: Harmonic decomposition of audio signals with matching pursuit. IEEE Trans. on Signal Processing **51** (2003) 101–111
10. Virtanen, T.: Algorithm for the separation of harmonic sounds with time-frequency smoothness constraint. In: Proc. Int. Conf. on Digital Audio Effects (DAFx). (2003) 35–40
11. Every, M.R., Szymanski, J.E.: Separation of synchronous pitched notes by spectral filtering of harmonics. In: IEEE Trans. on Speech and Audio Processing. (2006) To appear.
12. Ellis, D.P.W.: Prediction-driven computational auditory scene analysis. PhD thesis, Dept. of Electrical Engineering and Computer Science, MIT (1996)
13. Davy, M., Godsill, S.: Bayesian harmonic models for musical pitch estimation and analysis. Technical Report CUED/F-INFENG/TR.431, Cambridge University (2002)

14. Vincent, E., Plumbley, M.D.: A prototype system for object coding of musical audio. In: Proc. IEEE Workshop on Applications of Signal Processing to Audio and Acoustics (WASPAA). (2005) 239–242
15. van de Par, S., Kohlrausch, A., Charestan, G., Heusdens, R.: A new psychoacoustical masking model for audio coding applications. In: Proc. IEEE Int. Conf. on Acoustics, Speech and Signal Processing (ICASSP). (2002) II–1805–1808
16. Virtanen, T.: Separation of sound sources by convolutive sparse coding. In: Proc. ISCA Tutorial and Research Workshop on Statistical and Perceptual Audio Processing (SAPA). (2004)

Enhancement of Source Independence for Blind Source Separation*

Kun Zhang and Lai-Wan Chan

Department of Computer Science and Engineering,
The Chinese University of Hong Kong,
Shatin, Hong Kong
{kzhang, lwchan}@cse.cuhk.edu.hk

Abstract. When exploiting independent component analysis (ICA) to perform blind source separation (BSS), it is assumed that sources are mutually independent. However, in practice, the latent sources are usually dependent to some extent. Fortunately, if the sources are the same type of natural signals, they may be mutually independent in some frequency band, and dependent in other band. It is possible to make them mutually independent by temporal-filtering. In this paper we investigate ways to find the optimal filter for enhancing source independence in two scenarios. If none of the sources is known, we propose to adaptively estimate the filter and the de-mixing matrix simultaneously by minimizing the mutual information between outputs. Consequently the learned filter makes the filtered sources as independent as possible and the learned de-mixing matrix successfully separates the mixtures. If some source signals are available, we can estimate the filter more reliably by making the filtered sources as independent as possible. After that, with temporal-filtering as preprocessing, we can successfully perform BSS using ICA. Experiments on separating speech signals and images are presented.

1 Introduction

Independent Component Analysis (ICA) aims at extracting independent components given only observed data that are assumed to be mixtures of some independent sources [7]. Nowadays ICA is a popular method for blind source separation (BSS), since in many situations the hidden factors underlying the observations are statistically independent, so that they can be revealed by ICA.

However, in some real-world situations the independence property of sources may not hold, especially in biomedical signal processing and image processing, and therefore the standard ICA can not give the expected results. Some extended data models have been developed to relax the independence assumption in the standard ICA model, such as multidimensional ICA, independent subspace analysis, topographic ICA, and tree-dependent component analysis.

---

* This work was partially supported by a grant from the Research rants Council of the Hong Kong Special Administration Region, China.

J. Rosca et al. (Eds.): ICA 2006, LNCS 3889, pp. 731–738, 2006.

In this paper we consider the case in which the sources are dependent, but their sub-components in some frequency bands are mutually independent, as described by the subband decomposition ICA (SDICA) model [5,8,9]. Natural signals of the same type, such as speeches and images, may follow this model. In order to exploit ICA to successfully separate mixtures of such sources, we need to apply a filter to extract the source independent sub-components, or to enhance the source independence. Once the filtered sources are mutually independent, the mixing matrix in the data generation procedure can be estimated by applying ICA to the filtered observations. We should mention that in [3], temporal filtering is exploited for sparsification of images.

Traditionally the filter for enhancing source independence is determined either by some *a priori* information [4,5], or by exploiting some additional assumptions [6,8]. For example, the innovation processes of the hidden source signals, which are defined as errors of the best predictions for sources and usually obtained by temporal filtering sources, are believed to be more independent from each other in general [6]. The assumption that at least two groups of sub-components are statistically independent is incorporated in [8], so that the source independent sub-components and the true mixing matrix can be identified. These assumptions may be helpful in practice, but they do not always hold.

Recently, we have investigated the feasibility of estimating this filter and the mixing matrix in an adaptive manner and given an adaptive method for this task, namely band-selective ICA (BS-ICA) [9]. In BS-ICA, the filter and de-mixing matrix are learned simultaneously by minimizing mutual information between outputs, but their estimation interferes with each other, and may cause difficulty in estimation and some local optima. Therefore, here we consider estimating the optimal filter when some source signals, which are of the same type as those to be separated, are available. This can be considered as a type of semi-blind source separation: the filter for enhancing source independence is estimated from a small number of the sources signals which are known in advance; after applying this learned filter to the observations, the standard ICA is used to perform BSS. This is inspired by the fact that the source dependent sub-components usually concentrate on the same frequency band for natural signals of the same type. Consequently the temporal filter which removes dependence between the known sources also enhances independence amongst all sources of this type. Also, in this paper we report the experimental results on noisy speech separation, as well as separating images of human faces, to illustrate the behavior of our methods.

2 SDICA Model

The SDICA model assumes that the source signals can be dependent, however only some of their narrow-band sub-components are independent [5]. Without loss of generality, we can merge all independent sub-components and all dependent ones, respectively. Consequently we can represent the sources as

$$s_i(t) = s_{i,I}(t) + s_{i,D}(t) \tag{1}$$

where $s_{i,I}(t)$ denotes the independent sub-component and $s_{i,D}(t)$ denotes the dependent sub-component. More precisely, the source independent sub-components are assumed to be spatially independent stochastic sequences [9], i.e. they are mutually independent not only for the same time index, but also for different time indices. Note that $s_{i,I}(t)$ and $s_{i,D}(t)$ have different frequency bands.

The observations are still generated from the sources s_i in the same way as in the basic ICA model:

$$\mathbf{x} = \mathbf{A}\mathbf{s} \tag{2}$$

where $\mathbf{s} = [s_1, s_2, ..., s_n]^T$ and the observed data $\mathbf{x} = [x_1, x_2, ..., x_n]^T$.

As the components of $\mathbf{s}$ are not mutually independent, the linear transformation in equation 2 could not be determined by traditional ICA. In order to separate SDICA mixtures, we first need to apply a filter $h(t)$ to filter out the source dependent sub-components, i.e. $h(t)*s_i(t)$ are mutually independent. The filtered observations are then $\mathbf{z}(t) = h(t) * \mathbf{x}(t) = \mathbf{A}[h(t) * \mathbf{s}(t)]$. $\mathbf{z}(t) = [z_1(t), ..., z_n(t)]^T$ can then be considered as mixtures of independent sources. By applying linear ICA algorithms to $\mathbf{z}(t)$:

$$\mathbf{y}(t) = \mathbf{W}\mathbf{z}(t) = \mathbf{W}\mathbf{A}[h(t) * \mathbf{x}(t)] \tag{3}$$

we can obtain the de-mixing matrix $\mathbf{W}$, which is associated with the mixing matrix $\mathbf{A}$.

3 To Adaptively Estimate the Temporal Filter

3.1 When Only Observations Are Known

Now let us consider how to estimate the optimal filter which makes the filtered sources as independent as possible. We first try to separate sources described in equation 1 in a totally blind manner, i.e. given only the observations $x_i(t)$, the filter attenuating source dependent sub-components and the de-mixing matrix $\mathbf{A}$ are estimated from data simultaneously [9].

The separability of the SDICA model has been investigated [9]. It was shown that under weak conditions, the outputs of the SDICA separation system, given in equation 3, are spatially independent stochastic sequences if and only if $h(t)$ filters out the dependent sub-components $s_{i,D}(t)$ and $\mathbf{WA}$ is a generalized permutation matrix. However, it is difficult to enforce that $y_i(t)$ are spatially independent stochastic sequences. Furthermore, it was given that under certain additional conditions, $y_i(t)$ are instantaneously independent if and only if $h(t)$ filters out the dependent sub-components $s_{i,D}(t)$ and $\mathbf{WA}$ is a generalized permutation matrix. As a consequence, we can estimate the filter $h(t)$ and the separation matrix $\mathbf{W}$ in the SDICA separation system by making $y_i(t)$ as independent as possible, as BS-ICA does [9].

In BS-ICA $h(t)$ and $\mathbf{W}$ are learned by minimizing $I(\mathbf{y})$, the mutual information amongst $y_i(t)$. Let the filter for enhancing source independence be a causal finite impulse response (FIR) filter, $h(t) = [h_0, h_1, ..., h_L]$. The gradient of $I(\mathbf{y})$

with respect to h_p is

$$\frac{\partial I(\mathbf{y})}{\partial h_p} = \sum_{i=1}^{n} \frac{\partial H(y_i)}{\partial h_p} - \frac{\partial H(\mathbf{y})}{\partial h_p}$$
$$= -E_t\Big\{\boldsymbol{\psi}_{\mathbf{y}}^T(t)\cdot\mathbf{W}\cdot\mathbf{x}(t-p)\Big\} + E_t\Big\{\boldsymbol{\varphi}_{\mathbf{y}}^T(t)\cdot\mathbf{W}\cdot\mathbf{x}(t-p)\Big\}$$
$$= E_t\Big\{\boldsymbol{\beta}_{\mathbf{y}}^T(t)\cdot\mathbf{W}\cdot\mathbf{x}(t-p)\Big\} \tag{4}$$

where $\boldsymbol{\psi}_{\mathbf{y}}(\mathbf{y}) = [\psi_{y_1}(y_1), ..., \psi_{y_n}(y_n)]^T$ is called the marginal score function (MSF) [2], with $\psi_{y_i}(u)$, the score function of the random variable y_i, defined as $\psi_{y_i}(u) = (\log p_{y_i}(u))' = \frac{p'_{y_i}(u)}{p_{y_i}(u)}$. $\boldsymbol{\varphi}_{\mathbf{y}}(\mathbf{y}) = [\varphi_1(\mathbf{y}), ..., \varphi_n(\mathbf{y})]^T$ is called the joint score function (JSF), and its i-th element is $\varphi_i(\mathbf{y}) = \frac{\partial \log p_{\mathbf{y}}(\mathbf{y})}{\partial y_i} = \frac{\frac{\partial}{\partial y_i} p_{\mathbf{y}}(\mathbf{y})}{p_{\mathbf{y}}(\mathbf{y})}$. $\boldsymbol{\beta}_{\mathbf{y}}(\mathbf{y}) = \boldsymbol{\varphi}_{\mathbf{y}}(\mathbf{y}) - \boldsymbol{\psi}_{\mathbf{y}}(\mathbf{y})$ is defined as the score function difference (SFD) [2].

$\mathbf{W}$ just aims at minimizing the mutual information amongst outputs given $z_i(t) = h(t) * x_i(t)$ as inputs, so its learning rule is the same as in the basic ICA model. Here we adopt the natural gradient algorithm [1]: $\triangle\mathbf{W} \propto (I + E\{\boldsymbol{\psi}_{\mathbf{y}}(\mathbf{y})\mathbf{y}^T\})\mathbf{W}$.

3.2 When Some Sources Are Known

In the above method, the estimation of $h(t)$ and $\mathbf{W}$ interferes with each other and may cause some local optima, as illustrated by the first experiment in [9]. Therefore, we further consider the case in which we can obtain some source signals of the same type as those to be separated. These known sources can actually be used to determine the filter fast and reliably. The learned filter in this way is believed to be capable of enhancing statistical independence between all sources of this type. For example, in order to separate speeches of many channels in a noisy environment (which may produce source dependent sub-components), we can record two individual speeches in this environment and learn the filter by making the filtered version of these two speeches as independent as possible. In separating images, such as images of human faces [6], the filter which makes two filtered source images as independent as possible is believed to enhance independence between all such images.

Suppose there are some available sources, denoted by $\mathbf{s}^K = [s_1^K(t), ..., s_k^K(t)]^T$. When $s_i^{K\prime}(t) = h(t) * s_i^K(t)$ are independent from each other, the filter $h(t)$ filters out the dependent sub-components $s_{i,D}^K(t)$. The filter $h(t)$ can then be estimated by making $s_i^{K\prime}(t)$ as independent as possible. Mathematically, this can be achieved by minimizing $I(s_1^{K\prime}, ..., s_k^{K\prime})$. Let $\mathbf{s}^{K\prime} = [s_1^{K\prime}(t), ..., s_k^{K\prime}(t)]^T$. The gradient of $I(s_1^{K\prime}, ..., s_k^{K\prime})$ with respect to the coefficients of $h(t)$ is

$$\frac{\partial I(s_1^{K\prime}, ..., s_k^{K\prime})}{\partial h_p} = \sum_{i=1}^{k} \frac{\partial H(s_i^{K\prime})}{\partial h_p} - \frac{\partial H(s_1^{K\prime}, ..., s_k^{K\prime})}{\partial h_p}$$
$$= E_t\Big\{\boldsymbol{\beta}_{\mathbf{s}^{K\prime}}^T(t)\cdot\mathbf{s}^K(t-p)\Big\} \tag{5}$$

$h(t)$ can then be adjusted with the gradient-based method. After estimating $h(t)$, We can apply $h(t)$ to the observations to obtain $\mathbf{z}(t) = h(t) * \mathbf{x}(t) = \mathbf{A}[h(t) * \mathbf{s}(t)]$. The mixing matrix $\mathbf{A}$ can be recovered by applying linear ICA to $\mathbf{z}(t)$.

3.3 For High-Dimensional Data: Pairwise Score Function Difference

Due to the curse of dimensionality, it is difficult to estimate SFD in high-dimensional spaces. Fortunately, under certain conditions, we can then achieve mutual independence by minimizing the sum of the pairwise mutual information [9]. Consequently, the SFD in equations 4 and 5 is replaced by the pairwise score function difference (PSFD) $\gamma_{\mathbf{y}}(t) = [\gamma_1(t), ..., \gamma_n(t)]^T$, whose components are defined as $\gamma_i = \sum_{j=1, j\neq i}^{n} \beta_{y_i}(y_i, y_j)$.

4 Experiments

Two experiments are given for illustrative purpose. The first is devoted to separating images of human faces [6]. The second is to separate noisy speech signals. We use the Amari performance index P_{err} to assess the quality of the de-mixing matrix $\mathbf{W}$ for separating observations generated by the mixing matrix $\mathbf{A}$ [1]: $P_{err} = \frac{1}{n(n-1)} \sum_{i=1}^{n} \left\{ \left(\sum_{j=1}^{n} \frac{|p_{ij}|}{\max_k |p_{ik}|} - 1 \right) + \left(\sum_{j=1}^{n} \frac{|p_{ji}|}{\max_k |p_{ki}|} - 1 \right) \right\}$ where $p_{ij} = [\mathbf{WA}]_{ij}$. The smaller P_{err} is, the better $\mathbf{W}$ is.

4.1 Separating Images of Human Faces

The face images, represented as 1-dimensional signals, are apparently dependent, so it is very hard to separate them with the ICA technique [6]. Here the four original images of human faces are the same as in [6], as shown in Figure 1 (a).

They are mixed with a random chosen non-singular matrix. Four methods for separating these images were tested. The first is to directly apply linear ICA algorithms to the mixtures. We chose FastICA with the tanh nonlinearity and in the symmetric estimation mode and the natural gradient method with the score function estimated from data. The second is to apply ICA algorithms to the innovation processes of the mixtures [6]. We used a 10-th order auto-regressive model to estimate the innovation processes from the observations. The third is to assume there are two source images known and to learn $h(t)$ from them according to equation 5. After that $h(t)$ is applied to all mixtures and linear ICA is performed on the filtered mixtures. The last one is to apply BS-ICA to the observed mixtures. In the last two methods, the filter length is 11.

The performance index obtained is summarized in Table 1, from which we can see the result by directly applying ICA to the mixtures (column 2) is poor, and the last three methods can all recover the sources successfully. In particular, BS-ICA (the last column) gives almost the perfect result, and the filter $h(t)$ learned from two sources which are assumed known makes the filtered version of all sources approximately independent, as seen from the good performance in columns 4 $\sim$ 9. This verifies the success of the two methods we proposed. For

Table 1. The Amari performance index P_{err} of separating images of human faces by four different methods . Numbers in parenthesis in columns 4 $\sim$ 9 are the sequential numbers of the sources which are assumed to be known

Method	ICA on mixtures	ICA on Innovations	Sources assumed known:						BS-ICA
			(1,2)	(1,3)	(1,4)	(2,3)	(2,4)	(3,4)	
FastICA	0.592	0.132	0.108	0.095	0.114	0.105	0.099	0.116	0.043
Natural gradient	0.289	0.061	0.043	0.056	0.047	0.048	0.054	0.051	

Fig. 1. Separating mixtures of images of human faces. (a) Original images. Since the recovered images produced by our methods are almost the same as the original one, they are not plotted for saving space.(b) The outputs of BS-ICA, which are a filtered version of the source independent sub-components of the images

illustration, we plot the outputs of BS-ICA, which provide a filtered version of the source independent sub-components; see Figure 1 (b). We can see that the recovered source independent sub-components highlight the special features of each image. Consequently they are almost independent from each other.

The learned $h(t)$, as well as its frequency response magnitude, was given in [9] (due to space limitation, it is not plotted here). It is interesting that $h(t)$ attenuates both the low frequency part and the high frequency part of the sources, which is different from the filter $h_{AR}(t)$ for producing innovation processes of the mixtures.

4.2 Separating Speech Signals

In this experiment we recorded six speeches in a normal office room as the original source signals, as shown in Figure 2 (left). The sample rate is 22050 Hz, and to reduce the number of samples, we resample the signals with the ratio 1/10. These speeches are given by the same speaker reading an essay in Chinese. There is a little fan on the table, which may cause the source dependent sub-components.

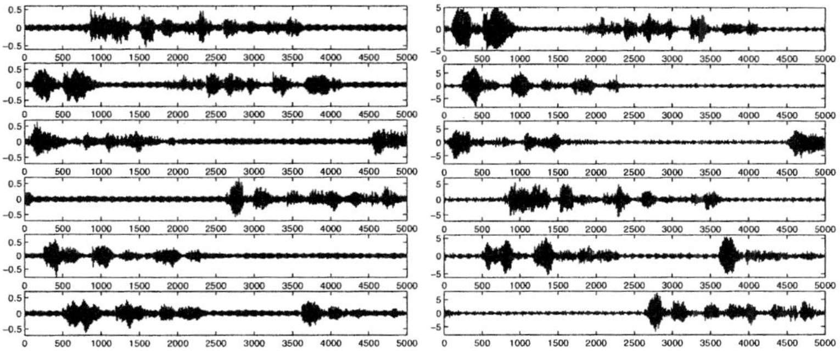

Fig. 2. Separating noisy speech signals. Left: original speeches. Right: outputs of BS-ICA. We can see that the noise effect caused by the fan has been greatly attenuated

Table 2. The Amari performance index P_{err} of separating noisy speeches by the four different methods

Method	ICA on mixtures	ICA on Innovations	With two sources known: (1,2)	(1,4)	(2,5)	(3,4)	(4,5)	(5,6)	BS-ICA
FastICA	0.124	0.021	0.025	0.026	0.043	0.024	0.030	0.021	0.036
Natural gradient	0.142	0.012	0.013	0.013	0.023	0.012	0.020	0.011	

We mixed the speeches with a randomly chosen non-singular matrix. The performance index obtained by the four different methods mentioned above is given in Table 2. In the method exploiting innovation processes we used the 22-th order auto-regressive model to estimate the innovation processes. The filter length in the two methods we proposed is 23. The table shows that all the methods except for directly applying ICA to observed mixtures (column 2) successfully recover the mixing matrix with good performance. The good performance of exploiting innovation processes is because the source dependent sub-components concentrate on a very narrow frequency band such that they are greatly attenuated by the filter producing innovation processes. In addition, we found that the innovation processes are much more non-Gaussian than the original sources, which also improves the separation performance. However, the filter producing innovation processes, which is actually used to whiten the observations, gives no knowledge about the frequency band of the source independent sub-components.

The outputs of BS-ICA, as an estimate of the source independent sub-components, are shown in Figure 2 (right). From Figure 2 we can see that the noise caused by the fan, as the source dependent sub-components, has been attenuated significantly by $h(t)$. To reduce the random effect, we repeated this experiment for 20 runs, and in each run the mixing matrix was chosen randomly. We found that in this experiment BS-ICA may converge to a local optimum if $\mathbf{W}$ is initialized as the identity matrix. However, with $\mathbf{W}$ initialized by the result of FastICA, BS-ICA always converges to the desired target. Therefore, it is recommended to initialize $\mathbf{W}$ in BS-ICA with the result of linear ICA.

5 Conclusion

Source independence is a precondition when applying independent component analysis for blind source separation. However, usually natural signals are not completely independent. They may be independent in some frequency band and dependent in other band. In this paper we considered methods to enhance source independence by using temporal-filtering. When only the observations are available, band-selective ICA can be adopted to adaptively estimate the separation matrix as well as the temporal filter. If we could find some source signals of the same type as those to be separated, the filter can be learned more reliably by making the filtered version of these sources as independent as possible. After applying this filter to observations, we can use ICA to do source separation. Experimental results on separating images of human faces as well as noisy speeches have shown the validity and good behavior of the proposed methods.

References

1. S. Amari, A. Cichocki, and H. H. Yang. A new learning algorithm for blind signal separation. In *Advances in Neural Information Processing Systems*, 1996.
2. M. Babaie-Zadeh, C. Jutten, and K. Nayebi. Separating convolutive mixtures by mutual information minimization. In *Proc. IWANN*, volume 2, pages 834–842, Granada, Spain, June 2001.
3. M. M. Bronstein, A. M. Bronstein, M. Zibulevsky, and Y. Y. Zeevi. Blind deconvolution of images using optimal sparse representations. *IEEE Trans. on Image Processing*, 14(6):726–736, 2005.
4. A. Cichocki, S. Amari, K. Siwek, and T. Tanaka. *ICALAB Toolboxes for Signal and Image Processing*, 2003. http://www.bsp.brain.riken.jp/ICALAB/.
5. A. Cichocki and P. Georgiev. Blind source separation algorithms with matrix constraints. *IEICE Transactions on Information and Systems, Special Session on Independent Component Analysis and Blind Source Separation*, E86-A(1):522–531, March 2003.
6. A. Hyvarinen. Independent component analysis for time-dependent stochastic processes. In *Proc. Int. Conf. on Artificial Neural Networks (ICANN'98)*, pages 541–546, Skovde, Sweden, 1998.
7. A. Hyvärinen, J. Karhunen, and E. Oja. *Independent Component Analysis*. John Wiley & Sons, Inc, 2001.
8. T. Tanaka and A. Cichocki. Subband decomposition independent component analysis and new performance criteria. In *Proc. IEEE Int. Conf. on Acoustics, Speech and Signal Processing (ICASSP'04)*, volume 5, pages 541–544, 2004.
9. K. Zhang and L. W. Chan. An adaptive method for subband decomposition ICA. *Neural Computation*, 18(1):191–223, 2006.

Speech Enhancement Using ICA with EMD-Based Reference

Yongrui Zheng [1], Qiuhua Lin [1], Fuliang Yin [1], and Hualou Liang [2]

[1] School of Electronic and Information Engineering,
Dalian University of Technology, Dalian 116023, China
qhlin@dlut.edu.cn
[2] School of Health Information Sciences,
The University of Texas at Houston, Houston, TX 77030, USA

Abstract. Different from the traditional ICA that recovers all the source signals simultaneously, the ICA with reference (ICA-R) extracts only some desired source signals from the mixtures of source signals by incorporating some *a priori* information into the separation process. This paper applies ICA-R to extracting a target speech signal from its noisy linear mixtures by constructing a proper reference signal with the empirical mode decomposition (EMD). Specifically, EMD is used to obtain an approximate envelope of the power spectrum of the desired speech, which is quite different from the power spectra of the environmental noises. The results of computer simulations and performance analyses demonstrate the efficiency of the proposed method.

1 Introduction

Independent Component Analysis (ICA) aims to recover a set of unknown mutually independent source signals from their observed linear mixtures without knowing the mixing coefficients [1]. It has been applied to communications, biomedical engineering, etc. [2]-[4]. The problem of speech separation has also received attention from researchers investigating ICA [5]-[7]. A recent variation, constrained ICA, also referred to as the ICA with reference (ICA-R) [8], [9], was proposed to allow *a priori* information about the desired source signals to be provided in the separation process. This differs from most approaches to ICA where the algorithms recover simultaneously all the source signals; the ICA-R extracts only the desired source signal which is the closest one, in some sense, to a properly constructed reference signal based on prior knowledge. Clearly the reference signal is important for ICA-R to extract the desired signal, but it does not need to be exactly the same as the desired source signal.

To have a proper and general reference signal for ICA-R to extract a target speech signal from its noisy linear mixtures, the power spectrum of the speech signal can be utilized to provide the prior information of the speech signal. The reason is that the speech power spectrum contains the common characteristics of the speech signals, and it is usually different from those of its environmental noises. For example, the power spectrum of the speech signal is not continuous, and its main power is regularly distributed within several given frequency bins. In practice, an approximate envelope of the speech power spectrum can be used as a reference signal.

J. Rosca et al. (Eds.): ICA 2006, LNCS 3889, pp. 739 – 746, 2006.

Empirical mode decomposition (EMD) is a general nonlinear and non-stationary signal processing method. It decomposes a signal into a finite and often small number of intrinsic mode functions (IMFs), and can be used as a filter by reconstructing the original signal with partial IMFs. In this paper, EMD is used as a low-pass filter to obtain the envelope of the speech power spectrum. Compared with the conventional low-pass filtering, EMD is able to track the brief and sharp edges of the speech power spectrum, which is very important for accurate extraction of the desired speech signal. The main advantage of the proposed method is that the speech signal is significantly enhanced by including prior information, i.e. the knowledge of the speech power spectra, into the ICA separation process, which would greatly facilitate many applications such as automatic speech recognition and speaker identification.

2 ICA with Reference

Suppose that there exist M independent source signals $\mathbf{s}(t)=[s_1(t),\ldots,s_M(t)]^T$ and N observed mixtures of the source signals $\mathbf{x}(t)=[x_1(t),\ldots,x_N(t)]^T$ (usually it is assumed that $N \geq M$). The linear ICA assumes that these mixtures are linear, instantaneous, and noiseless, i.e,

$$\mathbf{x}(t)=\mathbf{A}\mathbf{s}(t) \tag{1}$$

where $\mathbf{A}$ is a $N \times M$ mixing matrix that contains the mixing coefficients. The goal of the classical ICA is to find a $M \times N$ demixing matrix $\mathbf{W}$ such that M output signals

$$\mathbf{y}(t)=\mathbf{W}\mathbf{x}(t)=\mathbf{W}\mathbf{A}\mathbf{s}(t)=\mathbf{P}\mathbf{D}\mathbf{s}(t)\,. \tag{2}$$

where $\mathbf{P}\in \mathbf{R}^{M\times M}$ is a permutation matrix and $\mathbf{D}\in \mathbf{R}^{M\times M}$ is a diagonal scaling matrix.

Instead of separating all M number of independent sources from N mixed signals, ICA-R extracts L ($L< M$) number of desired sources from N mixed signals by incorporating some *a priori* information into the ICA learning algorithm as reference signals. These reference signals, denoted by $r(t)=[r_1(t),\ldots,r_L(t)]^T$, carry some information of the desired sources but not identical to the corresponding desired signals [8], [9]. In the following, we briefly describe the one-unit ICA-R. It finds one weight vector $\mathbf{w}$, i.e., one row of the demixing matrix $\mathbf{W}$, so that the output signal $y(t)=\mathbf{w}^T\mathbf{x}(t)$ recovers one desired source $s^*(t)$ by using $r(t)$ as the reference signal [8], [9]. For simplicity, the time index t is omitted in the equations below.

The flexible and reliable approximation of the negentropy $J(y)$ introduced by Hyvärinen in [10] is defined as the contrast function of one-unit ICA-R [8], [9]:

$$J(y)\approx \rho[E\{G(y)\}-E\{G(v)\}]^2 \tag{3}$$

where ρ is a positive constant, v is a Gaussian variable having zero mean and unit variance, $G(\cdot)$ can be any non-quadratic function.

The closeness between the ICA-R output y and the reference signal r is measured by $\varepsilon(y,r)$, which has a minimal value when $y=\mathbf{PD}s^*$. A threshold ξ is used to distinguish the desired source s^* from other source signals such that $g(\mathbf{w}) = \varepsilon(\mathbf{w}^T\mathbf{x}, r) - \xi \le 0$ is satisfied only when $y = \mathbf{PD}s^*$ among all source signals. Treating $g(\mathbf{w})$ as feasible constraint to the contrast function in (3), the problem of one-unit ICA-R can be modeled in the framework of constrained independent component analysis [9] as follows:

$$\begin{aligned} &\text{maximize} \quad J(y) \approx \rho[E\{G(y)\} - E\{G(v)\}]^2 \\ &\text{subject to} \quad g(\mathbf{w}) \le 0,\ h(\mathbf{w}) = E\{y^2\} - 1 = 0. \end{aligned} \tag{4}$$

where $h(\mathbf{w})$ is the equality constraint used to ensure the contrast function $J(y)$ and the weight vector $\mathbf{w}$ are bounded. In [9], a Newton-like learning algorithm is derived by finding the maximum of an augmented Lagrangian function corresponding to (4):

$$\mathbf{w}_{k+1} = \mathbf{w}_k - \eta \mathbf{R}_{\mathbf{XX}}^{-1} L'_{\mathbf{w}_k} / \delta(\mathbf{w}_k) \tag{5}$$

where k is the iteration index, η is the learning rate, $\mathbf{R}_{\mathbf{XX}}$ is the covariance matrix of the input mixtures $\mathbf{x}$,

$$L'_{\mathbf{w}_k} = \rho E\{\mathbf{x}G'_y(y)\} - 0.5\mu E\{\mathbf{x}g'_y(\mathbf{w}_k)\} \tag{6}$$

$$\delta(\mathbf{w}_k) = \rho E\{G''_{y^2}(y)\} - 0.5\mu E\{g''_{y^2}(\mathbf{w}_k)\} \tag{7}$$

where $G'_y(y)$ and $G''_{y^2}(y)$ are the first and the second derivatives of $G(y)$ with respect to y, and $g'_y(\mathbf{w}_k)$ and $g''_{y^2}(\mathbf{w}_k)$ are those of $g(\mathbf{w}_k)$. μ and λ are the Lagrange multipliers learned by the gradient-ascent method:

$$\mu_{k+1} = \max\{0, \mu_k + \gamma g(\mathbf{w}_k)\}, \tag{8}$$

$$\lambda_{k+1} = \lambda_k + \gamma h(\mathbf{w}_k). \tag{9}$$

where γ is the scalar penalty parameter.

3 Empirical Mode Decomposition (EMD)

EMD is a general nonlinear, non-stationary signal processing method. It has been applied to many fields such as ocean waves [11] and biomedical engineering [12]. The major advantage of EMD is that the basis functions are derived directly from the signal itself. Hence the analysis is adaptive, in contrast to Fourier analysis, where the basis functions are linear combinations of fixed sinusoids.

The principle of EMD is to decompose a signal into a sum of oscillatory functions, namely intrinsic mode functions (IMFs). An IMF is defined as any function: (1) having the same numbers of extrema and zero-crossings or differ at most by one; and (2) symmetric with respect to local zero mean. With these two requirements, the meaningfully instantaneous frequency of an IMF can be well defined. Specifically, the first condition is similar to the narrow-band requirement, whereas the second condition modifies a global requirement to a local one by using the local mean of the envelopes defined by the local maxima and the local minima, and is necessary to ensure that the instantaneous frequency will not have unwanted fluctuations as induced by asymmetric waveforms. To make use of EMD, the signal must have at least two extrema – one maximum and one minimum to be successfully decomposed into IMFs [12].

Given these two definitive requirements of an IMF, the sifting process for extracting IMFs from a given signal $z(t), t = 1,\ldots,T$ is described as follows:

(1) Extract the local minima and maxima of $z(t)$.

(2) Interpolate the local maxima and the local minima with cubic spline to form upper envelope $z_{\text{up}}(t)$ and lower envelope $z_{\text{low}}(t)$ of $z(t)$.

(3) Calculate the point-by-point mean from upper and lower envelopes, $m(t) = (z_{\text{up}}(t) + z_{low}(t))/2$.

(4) Extract the detail, $d(t) = z(t) - m(t)$. Check the properties of $d(t)$:

(5) If $d(t)$ do not meets the above-defined two conditions, replace $z(t)$ with $d(t)$. Repeat the procedure from Step 1 to 5 until it satisfies some stopping criterion.

(6) If $d(t)$ meets the above-defined two conditions, an IMF is derived and replace $z(t)$ with the residual $r(t) = z(t) - d(t)$; Repeat the above procedure until $r(t)$ has at least two extrema, else the decomposition is finished.

At the end of this process, the signal $z(t)$ can be expressed as follows:

$$z(t) = \sum_{j=1}^{N} c_j(t) + r_N(t) \tag{10}$$

where N is the number of IMFs, $r_N(t)$ denotes the final residue which can be interpreted as the DC component of the signal, and $c_j(t)$ are nearly orthogonal to each other, and all have nearly zero means. Due to this iterative procedure, none of the sifted IMFs is derived in closed analytical form [12].

As a result, the signal is decomposed into N IMFs, each with distinct time scale. More specifically, the first IMF has the smallest time scale which corresponds to the fastest time variation of the signal. As the decomposition process proceeds, the time scale increases, and hence, the mean frequency of the mode decreases. Therefore, EMD can be used as a filter by reconstructing the original signal with partial IMFs.

4 EMD-Based Reference Signal

A proper and available reference signal is the most important thing for ICA-R to extract a speech signal from its noisy linear mixtures. As mentioned above, the speech power spectrum is largely different from those of its environmental noises in that speech power is regularly distributed within several given frequency bins. Therefore, the approximate envelope of the power spectrum of the desired speech can be used as a reference signal, and can be well obtained by EMD. As an example, Fig. 1 (b) shows the power spectrum of a speech signal in Fig. 1(a). Its ten IMF components C_1, $C_2 \ldots C_{10}$ obtained by EMD are shown in Fig.2. We can find that the time scale of C_1, C_2, ... , C_{10} increases gradually, i.e., their mean frequency decreases gradually. Thus, by summing IMFs with lower frequency, a rough envelope of the speech power spectrum can be formed. The synthesized signals obtained by summing partial IMFs from C_4 to C_{10}, C_3 to C_{10}, and C_2 to C_{10}, are shown in Fig. 3(a)-(c), respectively. It is easy to

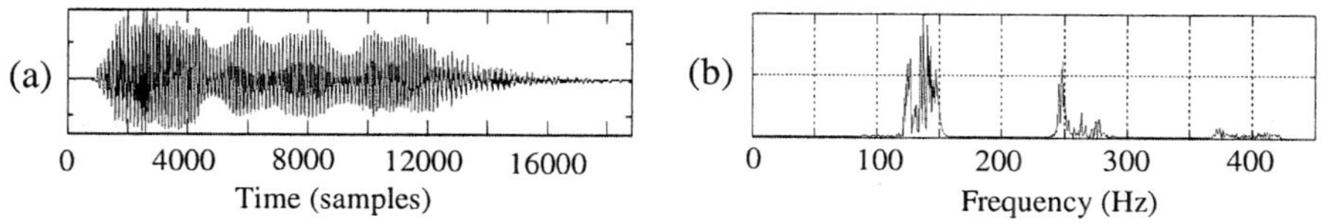

Fig. 1. A speech signal and its power spectrum. (a) The waveform of a speech signal. (b) The corresponding power spectrum of (a).

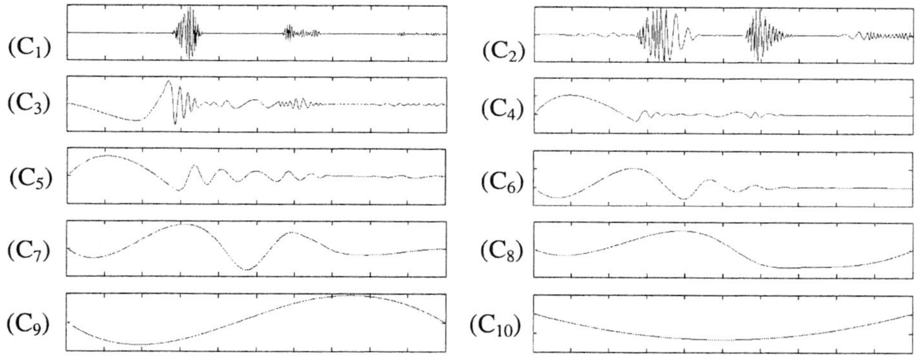

Fig. 2. Ten IMF components C_1, C_2, ... , C_{10} of speech power spectrum in Fig.1(b) obtained by EMD method

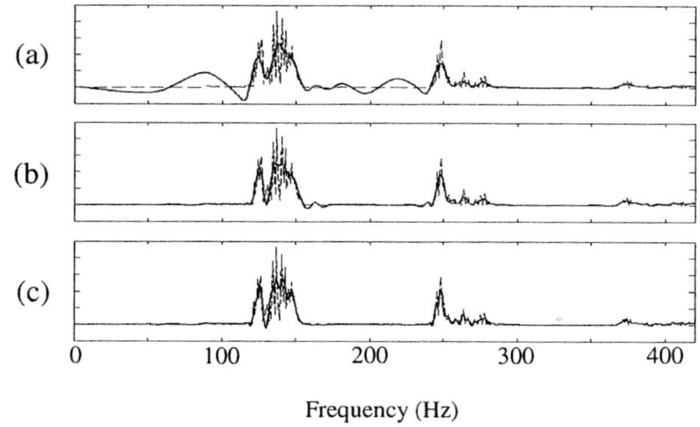

Fig. 3. Several synthesized signals of partial IMFs (solid line) compared with the power spectrum of the speech signal (dash line in each sub-figure). (a) Synthesized signal obtained by summing IMFs from C_4 to C_{10}. (b) Synthesized signal obtained by summing IMFs from C_3 to C_{10}. (c) Synthesized signal obtained by summing IMFs from C_2 to C_{10}.

find that the synthesized signals in Fig. 3(b) and Fig. 3(c) can efficiently approximate the original power spectrum in Fig. 1(b). For the proposed method, EMD-based reference signal needs not to be accurate power spectrum. Therefore, smaller number of IMFs such as C_3 to C_{10} in Fig. 3(b) are enough for constructing the reference signal.

5 Experimental Results

To illustrate the efficiency of the proposed method, extensive simulations are carried out. One experiment is to enhance a desired speech signal shown in Fig. 1(a). Three interrupted noises are noise bursts, random noise, and telephone trill, respectively, as shown in Fig. 4(a)-(c). The four noisy mixtures of the speech signal and three noises obtained by a random mixing matrix are shown in Fig. 4(d)-(g).

In the ICA-R algorithm, we use $G(y) = \log\cosh(y)$, which is a good general-purpose function [10]. The closeness between the ICA-R output signal y and the corresponding reference signal r is defined as the mean square error in frequency domain $\varepsilon(Y,r) = E\{(Y- r)^2\}$, where Y is the power spectrum of y, r is obtained by adding the larger time-scale IMFs from C_3 to C_{10}, as shown in Fig. 3(b). The threshold ξ is critical to the convergence of the algorithm of ICA-R. It may be initialized with a small value to avoid the algorithm going to a local optimum, and then is gradually increased to converge at the global maximum [9]. By performing ICA-R with the EMD-based reference signal (see Fig. 3(b)) on the four noisy speech mixtures in Fig. 4(d)-(g), the desired speech signal is extracted, as shown in Fig. 4(h). Comparing Fig. 4(h) and Fig. 1(a), we can see that the extracted speech signal by ICA-R is of good quality.

To quantitatively assess the performance of the proposed method, individual performance index (*IPI*) is defined as follows [9]:

$$IPI = \left(\sum_{j=1}^{M} \frac{|p_j|}{\max_k |p_k|}\right) - 1, \quad k = 1,\ldots,M \tag{11}$$

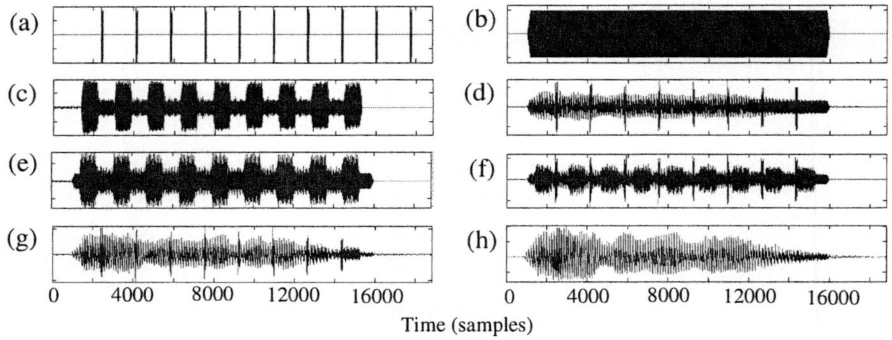

Fig. 4. Speech enhancement experiment with ICA-R. (a) Noise bursts. (b) Random noise. (c) Telephone trill. (d)-(g) Four noisy mixtures of the speech signal and three noises in (a)-(c). (h) Extracted speech signal by ICA-R.

where p_j denotes the j element of the global vector $\mathbf{P}=\mathbf{w}^{\mathrm{T}}\mathbf{A}$. The value of *IPI* is within the range of [0, 1], e.g., *IPI*=0 if $\mathbf{w}$ is perfectly estimated. The accuracy of the recovered signal y^* compared to the desired source signal s^* is measured by the signal to noise ratio (*SNR*) in dB:

$$SNR(\mathrm{dB}) = 10\log_{10}(\frac{\sigma^2}{mse}) \tag{12}$$

where σ^2 is the variance (power) of the desired speech signal, *mse* denotes the mean square error between the desired speech signal and the recovered one, that is, *mse* is the noise power.

The *IPI*, *mse*, and *SNR* (dB) are then computed on the enhanced speech signal in Fig. 4(h) by the ICA-R algorithm, which are 0.099, 0.0044, and 23.57, respectively. These results show that the desired weight vector $\mathbf{w}$ is well estimated by ICA-R, the noise power is very small, and the target speech is enhanced with very high *SNR* index. Therefore, ICA-R is capable of extracting the desired speech signal from the noisy mixtures with very high quality by using the reference signal; an approximate envelope of the speech power spectrum can be used as the reference signal; and the reference signal can be properly constructed with EMD method.

6 Conclusion

This paper presents an application of ICA-R to extracting a speech signal from its noisy linear mixtures while the reference signal is properly constructed by EMD. The EMD-based reference signal is obtained by adding the larger time-scale IMFs to approximate the envelope of the power spectrum of the desired speech signal, which is quite different from the power spectra of the environmental noises. The results of the computer simulations and the performance analyses show that the proposed method is effective.

Acknowledgement

This work was supported by the National Natural Science Foundation of China under Grant No. 60402013, No. 60372082, and No. 60172073.

References

1. Comon, P.: Independent component analysis, a new concept? Signal Processing, vol. 36, no. 3 (1994) 287–314.
2. Cardoso, J. F.: Blind signal separation: statistical principles. Proc. of the IEEE, vol. 86, no. 10 (1998) 2009–2025.
3. Amari, S., Cichocki, A.: Adaptive blind signal processing - Neural network approaches. Proc. of the IEEE, vol. 86, no. 10 (1998) 2026–2048.
4. Hyvärinen, A., Oja, E.: Independent component analysis: algorithms and applications. Neural Networks, vol. 13, no. 4-5 (2000) 411–430.

5. Barros, A. K., Rutkowski, T., Itakura, F., Ohnishi, N.: Estimation of speech embedded in a reverberant and noisy environment by independent component analysis and wavelets. IEEE Transactions on Neural Networks, vol. 13, no. 4 (2002) 888–893.
6. Visser, E., Lee, T. W.: Speech enhancement using blind source separation and two-channel energy based speaker detection. Proceedings of the IEEE International Conference on Acoustics, Speech, and Signal Processing (ICASSP '03), vol. 1 (2003) 884–887.
7. Lee, J. H., Jung, H. Y., Lee, T. W., Lee, S. Y.: Speech enhancement with MAP estimation and ICA-based speech features. Electronics Letters, vol. 36, no. 17 (2000) 1506–1507.
8. Lu, W., Rajapakse, J. C.: Constrained Independent Component Analysis. Advance in Neural Information Processing Systems, MIT Press, (2000) 570-576.
9. Lu, W., Rajapakse, J. C.: ICA with reference. Proc. Third Int. Conf. on ICA and Blind Source Separation (ICA 2001), (2001) 120–125.
10. Hyvärinen, A.: New approximations of differential entropy for independent component analysis and projection pursuit. In Advances in Neural Information Processing Systems 10 (NIPS*97), (1997) 273–279.
11. Huang, N. E., Shen, Z., Long, S. R., Wu, M. C., Shih, H. H., Zheng, Q., Yen, N-C., Tung, C. C., Liu, H. H.: The empirical mode decomposition and the Hilbert spectrum for nonlinear and nonstationary time series analysis. Proc. Royal Soc. Lond. A, vol. 454, (1998) 903–995.
12. Liang, H. L., Lin, Q. H., and Chen, J. D. Z.: Application of the empirical mode decomposition to the analysis of esophageal manometric data in gastroesophageal reflux disease. IEEE Transactions on Biomedical Engineering, vol. 52, no. 10 (2005) 1692–1701.

Zero-Entropy Minimization for Blind Extraction of Bounded Sources (BEBS)

Frédéric Vrins[1], Deniz Erdogmus[2], Christian Jutten[3], and Michel Verleysen[1,*]

[1] Machine Learning Group, Université catholique de Louvain, Louvain-la-Neuve, Belgium
[2] Dep. of CSEE, OGI Oregon, Health and Science University, Portland, Oregon, USA
[3] Laboratoire des Images et des Signaux, Institut National Polytechnique de Grenoble (INPG), France
{vrins, verleysen}@dice.ucl.ac.be, derdogmus@ieee.org, christian.jutten@inpg.fr

Abstract. Renyi's entropy can be used as a cost function for blind source separation (BSS). Previous works have emphasized the advantage of setting Renyi's exponent to a value different from one in the context of BSS. In this paper, we focus on zero-order Renyi's entropy minimization for the blind extraction of bounded sources (BEBS). We point out the advantage of choosing the *extended* zero-order Renyi's entropy as a cost function in the context of BEBS, when the sources have non-convex supports.

1 Introduction

Shannon's entropy is a powerful quantity in information theory and signal processing; it can be used e.g. in blind source separation (BSS) applications. Shannon's entropy can be seen as a particular case of *Renyi's entropy*, defined as [1]:

$$h_r[f_X] = \begin{cases} \frac{1}{1-r}\log\left\{\int f_X^r(x)dx\right\} & \text{for } r \in \{[0,1)\cup(1,\infty)\} \\ -\mathrm{E}\left\{\log f_X(x)\right\} & \text{for } r = 1 \end{cases} . \tag{1}$$

The above integrals are evaluated on the support $\Omega(X)$ of the probability distribution (pdf) f_X. The first-order Renyi's entropy $h_1[f_X]$ corresponds to Shannon's entropy; function $h_r[f_X]$ is continuous in r.

Previous works have already emphasized that advantages can be taken by considering the general form of Renyi's entropy rather than Shannon's in the BSS context [2]. For instance, it is interesting to set $r = 2$ in specific cases. Using kernel density estimates leads to a simple estimator for $h_2[.]$.

This paper points out that in particular situations, e.g. when dealing with the blind extraction of bounded sources (BEBS) application, zero-Renyi's entropy (Renyi's entropy with $r = 0$) should be preferred to other Renyi's entropies.

Renyi's entropy with $r = 0$ is a very specific case; it simply reduces to the logarithm of the support volume of $\Omega(X)$: $h_0[f_X] = \log \mathrm{Vol}[\Omega(X)]$ [3]. In the BEBS

* Michel Verleysen is Research Director of the Belgian F.N.R.S.

J. Rosca et al. (Eds.): ICA 2006, LNCS 3889, pp. 747–754, 2006.

context, it can be shown that the global minimum of the output zero-Renyi's entropy is reached when the output is proportional to the source with the lowest support measure [4], under the whiteness constraint. The other sources can then be iteratively extracted, minimizing the output zero-Renyi's entropy in directions orthogonal to the previously extracted signals. A similar conclusion has been independently drawn in [5], where it is also shown that the output support convex hull volume has a local minimum when the output is proportional to one of the sources. The main advantage in considering zero-Renyi's entropy is that, under mild conditions, this cost function is free of local minima. Hence gradient-based methods yield the optimal solution of the BEBS problem. When the sources have strongly multimodal pdfs, this property is not shared by the most popular information-theoretic cost functions, like e.g. mutual information, maximum-likelihood and Shannon's marginal entropy (see [6,7,8,9]).

This contribution aims at analyzing the condition for which the "spurious minima-free" property of zero-Renyi's entropy $h_0[f_X]$ holds in the context of BEBS. First, it is shown that the output zero-Renyi's entropy has no spurious minimum in the BEBS application when the volume of the non-convex part of the sources support is zero. Second, it is shown that the support $\Omega[.]$ should be replaced by its convex hull $\overline{\Omega}[.]$ in Renyi's entropy definition (1), in order to avoid spurious minima when the source supports have non-convex parts having a strictly positive volume measure. These two last claims are based on the Brunn-Minkowski inequality.

The following of the paper is organized as follows. The impact of choosing the support pdf or its convex hull when computing Renyi's entropy is first analyzed in Section 2. Section 3 recalls the Brunn-Minkowski inequality. The latter is used to discuss the existence of spurious zero-Renyi's entropy minima depending of the convexity of the source supports in Section 4. The theoretical results are illustrated on a simple example in Section 5.

2 Support, Convex Hull and Renyi's Entropy

The density f_X of a one-dimensional bounded random variable (r.v.) X satisfies $f_X(x) = 0$ for all $x > \sup(X)$ and $x < \inf(X)$. The support of the density is defined as the set where the r.v. *lives* [10]: $\Omega(X) \triangleq \{x : f_X(x) > 0\}$. Another viewpoint is e.g. to consider that the r.v. lives for x such that $0 < F_X(x) < 1$, where F_X is the cumulative distribution of X. Therefore, an extended definition of the support could be : $\overline{\Omega}(X) \triangleq \{x \in [\inf\{x : f_X(x) > 0\}, \sup\{x : f_X(x) > 0\}]\}$. Then, $\overline{\Omega}(X)$ can be seen as the *closed bounded convex hull* of $\Omega(X)$, and obviously: $\Omega(X) \subseteq \overline{\Omega}(X)$.

Let us abuse notation by writing $h_{r,\Omega(X)}[f_X]$ for $h_r[f_X]$. Consider the slightly modified Renyi's entropy (called in the following *extended Renyi's entropy*), defined as $h_{r,\overline{\Omega}(X)}[f_X]$: f_X^r is now integrated on the set $\overline{\Omega}(X)$ rather than on the support $\Omega(X)$ in eq. (1). For $r \neq 0$, one gets $h_{r,\Omega(X)}[f_X] = h_{r,\overline{\Omega}(X)}[f_X]$, because $0^r = 0$ for $r \neq 0$ and $0 \log 0 = 0$ by convention [10]. Conversely, $h_{0,\overline{\Omega}(X)}[f_X] > h_{0,\Omega(X)}[f_X]$ if $\mathrm{Vol}[\overline{\Omega}(X) \setminus \Omega(X)] > 0$ (the support contains 'holes' with non-zero volume measure). Indeed, consider the Lebesgue measure $\mu[.]$, which is the standard way of assigning a volume to subsets of the Euclidean space. Let us assume that $\Omega(X)$ can be written as the union of I disjoint intervals $\Omega_i(X)$ of strictly positive volume. Using the properties of Lebesgue measure, zero-Renyi's entropy becomes : $h_{0,\Omega(X)}[f_X] = \log \sum_{i=1}^{I} \mu[\Omega_i(X)]$. This quantity is strictly lower than $h_{0,\overline{\Omega}(X)}[f_X] = \log \mu[\overline{\Omega}(X)]$ if $\mu[\overline{\Omega}(X) \setminus \Omega(X)] > 0$. In summary, we have:

$$\begin{cases} h_{r,\Omega(X)}[f_X] = h_{r,\overline{\Omega}(X)}[f_X] \text{ for } r \neq 0. \\ \lim_{r\to 0} h_{r,\Omega(X)}[f_X] = h_{0,\Omega(X)}[f_X] \leq h_{0,\overline{\Omega}(X)}[f_X] \end{cases} . \tag{2}$$

The $r = 0$ case is thus very specific when considering Renyi's entropies; for other values of r, $h_{r,\Omega(X)}[f_X] = h_{r,\overline{\Omega}(X)}[f_X]$. The $r = 0$ value is also the only one for which $h_{r,\overline{\Omega}(X)}[f_X]$ can be not continuous in r. The impact of choosing $h_{0,\overline{\Omega}(X)}[f_X]$ rather than $h_{0,\Omega(X)}[f_X]$ as BEBS cost function is analyzed in Section 4. The study is based on Brunn-Minkowski's inequality [11], which is introduced below.

3 Brunn-Minkowski Revisited

The following theorem presents the original Brunn-Minkowski inequality [11].

Theorem 1 (Brunn-Minkowski Inequality). *If $\mathcal{X}$ and $\mathcal{Y}$ are two compact convex sets with nonempty interiors (i.e. mesurable) in $\mathbb{R}^n$, then for any $s, t > 0$:*

$$\mathrm{Vol}^{1/n}[s\mathcal{X} + t\mathcal{Y}] \geq s\mathrm{Vol}^{1/n}[\mathcal{X}] + t\mathrm{Vol}^{1/n}[\mathcal{Y}] . \tag{3}$$

The operator Vol[.] *stands for* volume. *The operator "+" means that $\mathcal{X} + \mathcal{Y} = \{x + y : x \in \mathcal{X}, y \in \mathcal{Y}\}$. The equality holds when $\mathcal{X}$ and $\mathcal{Y}$ are equal up to translation and dilatation (i.e. when they are* homothetic*).*

As explained in the previous section, we use the Lebesgue measure $\mu[.]$ as the volume Vol[.] operator. Obviously, one has $\mu[\overline{\Omega}(X)] \geq \mu[\Omega(X)] \geq 0$.

Inequality (3) has been extended in [10,12] to non-convex bodies; in this case however, to the authors knowledge, the *strict equality* and *strict inequality* cases were not discussed in the literature. Therefore, the following lemma, which is an extension of the Brunn-Minkowski theorem in the $n = 1$ case, states sufficient conditions so that the strict equality holds (the proof is relegated to the appendix).

Lemma 1. *Suppose that $\Omega(X) = \cup_{i=1}^{I} \Omega_i(X)$ with $\mu[\Omega_i(X)] > 0$ and $\Omega(Y) = \cup_{j=1}^{J} \Omega_j(Y)$ with $\mu[\Omega_j(Y)] > 0$, with $\Omega(X) \subset \mathbb{R}$, $\Omega(Y) \subset \mathbb{R}$. Then:*

$$\mu[\Omega(X+Y)] \geq \mu[\Omega(X)] + \mu[\Omega(Y)] ,$$

with equality if and only if $\mu[\overline{\Omega}(X) \setminus \Omega(X)] = \mu[\overline{\Omega}(Y) \setminus \Omega(Y)] = 0$.

4 Zero-Renyi's *vs* Extended Zero-Renyi's Entropy for BEBS

Consider the linear instantaneous BEBS application, and let $S_1, S_2, \cdots, S_K$ be the independent source signals. If we focus on the extraction of a single output Z, we can write $Z = \sum_{i=1}^K \mathbf{c}(i) S_i$, where $\mathbf{c}$ is the vector of the transfer weights between Z and the S_i. The vector $\mathbf{c}$ is the row of the transfer matrix $\mathbf{C}$ associated to the output Z. The latter matrix is obtained by left-multiplying the unknown mixing matrix by the unmixing matrix that has to be estimated. The unmixing matrix row associated to $\mathbf{c}$ can be blindly found by minimizing $h_{0,\Omega(Z)}[f_Z]$, under a fixed-norm constraint to avoid that $\mathrm{var}(Z)$ diverges (see [4,5]).

The following subsections discuss the impact of minimizing zero-Renyi's entropy $h_{0,\Omega(Z)}[f_Z]$ or its extended definition $h_{0,\overline{\Omega}(Z)}[f_Z]$ for the BEBS application.

4.1 Convex Supports

If the sources have convex supports $\Omega(S_i)$ (Theorem 1), or if $\mu[\overline{\Omega}(S_i) \backslash \Omega(S_i)] = 0$ (Lemma 1) for all $1 \leq i \leq K$, then both approaches are identical: $\mu[\Omega(Z)] = \mu[\overline{\Omega}(Z)]$. Brunn-Minkowski equality holds, and the following relation comes: $\mu[\Omega(Z)] = \sum_{i=1}^K |\mathbf{c}(i)|.\mu[\Omega(S_i)]$. It is known that in the $K = 2$ case, we can freely parametrize $\mathbf{c}$ by a single angle: $\mathbf{c}$ can be written as $[\sin\theta, \cos\theta]$, where θ is the transfer angle. This parametrization of $\mathbf{c}$ forces the vector to have a unit Euclidean norm. In this case $\mu[\Omega(Z)] = \mu[\overline{\Omega}(Z)]$ is concave w.r.t. θ in each quadrant [5]. Since $\log f$ is concave if f is concave, $\log \mu[\Omega(Z)] = \log \mu[\overline{\Omega}(Z)]$ is also concave. In other words, the minima of $\mu[\Omega(Z)]$ w.r.t. θ can only occur at $\theta \in \{k\pi/2 | k \in \mathbb{Z}\}$: all the minima of $h_{0,\Omega(Z)}[f_Z]$ are non-mixing (corresponding to non-spurious solutions of the BEBS problem). This last result holds for higher dimensions, i.e. for $K \geq 2$ (see [5] for more details).

4.2 Non-convex Supports

In the non-convex situation, Brunn-Minkowski equality holds for the set $\overline{\Omega}(.)$ (by Theorem 1):

$$\mu[\overline{\Omega}(Z)] = \sum_{i=1}^K |\mathbf{c}(i)|.\mu[\overline{\Omega}(S_i)] \ . \tag{4}$$

It can be shown that all the minima of the above quantity w.r.t. vector $\mathbf{c}$ are relevant; as in the convex-support case, they all correspond to non-spurious solutions of the BEBS problem [5]. By contrast, the strict Brunn-Minkowski inequality holds when a source has a support $\Omega(S_i)$ such that $\mu[\overline{\Omega}(S_i) \setminus \Omega(S_i)] > 0$. Lemma 1 gives $\mu[\Omega(Z)] > \sum_{i=1}^K |\mathbf{c}(i)|.\mu[\Omega(S_i)]$. In this case, there is no more guarantee that $\mu[\Omega(Z)]$ does not have mixing minima when a source has a non-convex support. The next section will presents simulation results showing on a simple example that spurious minima of $\mu[\Omega(Z)]$ may exist.

As a conclusion, the best integration domain for evaluating Renyi's entropy for the blind separation of bounded sources seems to be $\overline{\Omega}(Z)$, the convex hull of the output support $\Omega(Z)$. Remark that contrarily to $h_{r,\Omega(Z)}[f_Z]$, $h_{r,\overline{\Omega}(Z)}[f_Z]$ is not rigourously speaking a Renyi's entropy. Nevertheless, while $h_{0,\Omega(Z)}[f_Z]$ is the log of the volume of $\Omega(Z)$, *extended* zero-Renyi's entropy $h_{0,\overline{\Omega}(Z)}[f_Z]$ is the log of the volume of $\Omega(Z)$'s *convex hull.*

In the BEBS application, the output volume must be estimated directly from Z, since neither $\mathbf{c}$, nor the $\mu[\Omega(S_i)]$ are known. Therefore evaluating zero-Renyi's entropy requires the estimation of $\mu[\Omega(Z)]$ and computing extended zero-Renyi's

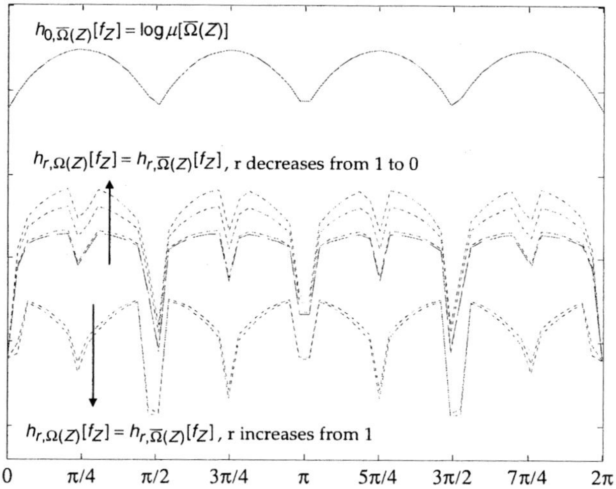

(a) $\xi = 8$, $\nu = 1$: the source supports $\Omega(S_i)$ are non-convex

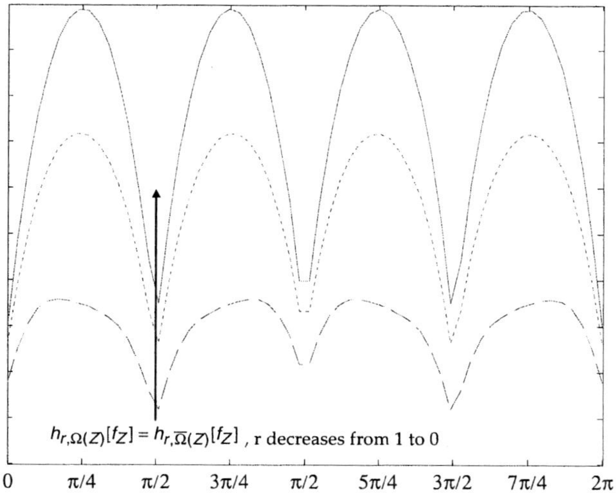

(b) $\xi = 0.5$, $\nu = 1$: the source supports $\Omega(S_i)$ are convex

Fig. 1. Extended zero-Renyi (–), Shannon (- -), and r-Renyi entropies with $r \neq \{0, 1\}$ (..)

entropy requires the estimation of $\mu[\overline{\Omega}(Z)]$. In [5], the support of $\overline{\Omega}(Z)$ is approximated by $\max(\hat{Z}) - \min(\hat{Z})$ ($\hat{Z}$ is the set of observations of Z), which is also a good approximation of $\mu[\Omega(Z)]$ (i.e. of $\exp\{h_{0,\Omega(Z)}[f_Z]\}$) when the source supports are convex.

5 Example

Let p_{S_1} and p_{S_2} be two densities of independent random variables $S_i = U_i + D_i$ where U_1 and U_2 are independent uniform variables taking non-zero values in $[-\nu, \nu]$ ($\nu > 0$) and D_1, D_2 are independent discrete random variables taking values $[\alpha, 1-\alpha]$ at $\{-\xi, \xi\}$ ($\xi > 0$). Suppose further that $\xi > \nu$. Then, both sources S_i have the same density p_S:

$$p_S(s) = \begin{cases} \frac{\alpha}{2\nu} & \text{for } x \in [-\xi-\nu, -\xi+\nu] \\ \frac{1-\alpha}{2\nu} & \text{for } x \in [\xi-\nu, \xi+\nu] \\ 0 & \text{elsewhere.} \end{cases} \tag{5}$$

It results that $\Omega(S_i) = \{x \in [-\xi-\nu, -\xi+\nu] \cup [\xi-\nu, \xi+\nu]\}$ and $\overline{\Omega}(S_i) = \{x \in [-\xi-\nu, \xi+\nu]\}$, which implies $\mu[\Omega(S_i)] = 4\nu$ and $\mu[\overline{\Omega}(S_i)] = 2\xi + 2\nu$. By Lemma 1, we have $\mu[\overline{\Omega}(S_1+S_2)] = \mu[\overline{\Omega}(S_1)] + \mu[\overline{\Omega}(S_2)]$ and $\mu[\Omega(S_1+S_2)] > \mu[\Omega(S_1)] + \mu[\Omega(S_2)]$.

Let us note $Z = \cos\theta S_1 + \sin\theta S_2$. Equation (4) guarantees that $\mu[\overline{\Omega}(Z)]$ is concave with respect to θ. By contrast, according to Section 4.2, there is no guarantee that $\mu[\Omega(Z)]$ has no minima for $\theta \notin \{k\pi/2 | k \in \mathbb{Z}\}$.

Figure 1 illustrates the effect of the source support convexity on $h_r[f_Z]$ w.r.t. θ for various values of r in the above example. Note that the omitted scales of vertical axes are common to all curves. We can observe that $h_r[f_Z]$ has spurious minima regardless of r; there exist local minima of the zero-entropy criterion for which Z is not proportional to one of the sources. By contrast, when considering the extended zero-Renyi's entropy $h_{r,\overline{\Omega}(Z)}[f_Z]$, no spurious minimum exists: all the $h_{r,\overline{\Omega}(Z)}[f_Z]$ local minima correspond to $Z = \pm S_i$. Note that in Figure 1 (a), $h_{r_1}[f_Z] < h_{r_2}[f_Z]$ if $r_1 > r_2$. This result can be theoretically proven by Hölder's inequality: Renyi's entropy $h_r[f]$ is decreasing in r, and strictly decreasing unless f is a uniform density [13].

6 Conclusion and Perspectives

This paper focusses on zero-Renyi's entropy for blind extraction of bounded sources. Theoretical results show that if the sources have convex supports, both zero-Renyi and extended zero-Renyi's entropies are free of spurious minima.

However, this is no more true when the source support contains "holes" of positive volume. In this case, simulation results seem to indicate that the order of Renyi's entropy (i.e. parameter r) has no influence on the existence of local spurious minima, see Figure 1 (a). Nevertheless, when considering the extended zero-Renyi's entropy, Brunn-Minkowski inequality shows that this cost function is free of spurious minima, when the support is correctly estimated.

Finally, new perspectives for Renyi entropy-based BSS/BEBS algorithms arise from the results presented in this paper. Despite the "no spurious minima" property of the extended zero-Renyi's entropy which is not shared by Shannon's one, both the output support volume and Shannon's entropy BEBS can be used in *deflation* algorithms for source separation. Indeed, it is known that the *Entropy Power inequality* shows that Shannon's entropy can be used in deflation procedures for BSS. On the other hand, this paper shows that *Brunn-Minkowski inequality* justifies the use of zero-Renyi's entropy for the sequential extraction of bounded sources. Conversely, to the authors knowledge, there is no proof to date justifying the use of Renyi's entropy for $r \neq 0$ and $r \neq 1$ in *deflation* BSS/BEBS algorithms. It is thus intriguing to remark that the two aforementioned information-theoretic inequalities are closely related [12]. By contrast, the sum of output Renyi's entropies can be seen as a cost function for *symmetric* BSS (all sources are extracted simultaneously), as explained in [2]. As it is known that values of r different from 0 and 1 are also interesting in specific BSS applications, future work should then study deflation methods based on general Renyi's entropy definition (of order $r \in \mathbb{R}^+$).

Acknowledgment

The authors would like to thank Prof. Erwin Lutwak from Polytechnic University (Brooklyn, NY, USA) for discussion on Hölder's inequality.

References

1. Principe, J.C Xu, D. & Fisher III, J.W. (2000) "Information-Theoretic Learning." In *Unsup. Adapt. Filtering* **I**: 265–319, Haykin, S. (edt), Wiley, New York.
2. Erdogmus, D., Hild, K.E. & Principe, J.C. (2002) Blind Source Separation Using Renyi's α-marginal Entropies. *Neurocomputing*, **49**: 25–38.
3. Guleryuz, O.G., Lutwak, E., Yang, D., & Zhang, G. (2000) Information-Theoretic Inequalities for Contoured Probability Distributions. *IEEE Trans. Info. Theory*, **48**(8): 2377–2383.
4. Cruces, S. & Duran, I. (2004) The Minimum Support Criterion for Blind Source Extraction: a Limiting Case of the Strengthened Young's Inequality. In proc. ICA, *Int. Conf. on Ind. Comp. Anal. and Blind Sig. Sep.*: 57–64, Granada (Spain).
5. Vrins, F., Jutten, C. & Verleysen, M. (2005) SWM : a Class of Convex Contrasts for Source Separation. In proc. ICASSP, *IEEE Int. Conf. on Acoustics, Speech and Sig. Process.*: V.161-V.164, Philadelphia (USA).
6. Vrins, F. & Verleysen, M . (2005) On the Entropy Minimization of a Linear Mixture of Variables for Source Separation. *Sig. Process.*, **85**(5): 1029–1044, Elsevier.
7. Learned-Miller, E.G. & Fisher III, J.W. (2003) ICA Using Spacings Estimates of Entropy. *Journal of Machine Learning Research*, **4**: 1271–1295, MIT press.
8. Pham, D.T. & Vrins, F. (2005) Local Minima of Information-Theoretic Criteria in Blind Source Separation. *IEEE Sig. Process. Lett.*, **12**(11): 788-791.

9. Cardoso, J.-C. (2000) "Entropic Contrast for Source Separation: Geometry & Stability." In *Unsup. Adapt. Filtering* **I**: 265–319, Haykin, S. (edt), Wiley, New York.
10. Cover, Th. & Thomas, J. A. (1991) "Elements of Information Theory." Wiley, New-York.
11. Gardner, R.J. (2002) The Brunn-Minkowski Inequality. *Bull. of Am. Math. Soc.*, **39**(3): 355–405.
12. Costa, M. & Cover Th. (1984) On the Similarity of the Entropy Power Inequality and the Brunn-Minkowski Inequality. *IEEE Trans. Info. Theory*, **30**(6): 837–839.
13. Lutwak, E., Yang, D., & Zhang, G. (2005) Cramér-Rao and Moment-Entropy Inequalities for Renyi Entropy and Generalized Fisher Information. *IEEE Trans. Info. Theory*, **51**(2): 473–478.

Appendix. Proof of Lemma 1

Suppose that $\mu[\Omega(X)] = \mu[\overline{\Omega}(X)] > 0$ and $\mu[\Omega(Y)] = \mu[\overline{\Omega}(Y)] > 0$. This means that $\mu[\overline{\Omega}(X)\setminus\Omega(X)] = \mu[\overline{\Omega}(Y)\setminus\Omega(Y)] = 0$. Therefore, the sets $\Omega(X)$ and $\Omega(Y)$ can be expressed as

$$\begin{cases} \Omega(X) = [\inf X, \sup X] \setminus \cup_{i=1}^{I'}\{x_i\} \\ \Omega(Y) \;= [\inf Y, \sup Y] \setminus \cup_{j=1}^{J'}\{y_j\} \end{cases} \tag{6}$$

where x_i, y_i are isolated points. Then,

$$\begin{aligned} \mu[\Omega(X+Y)] &= \mu[\overline{\Omega}(X+Y)] \\ &= (\sup X + \sup Y) - (\inf X + \inf Y) \\ &= \mu[\Omega(X)] + \mu[\Omega(Y)] \ , \end{aligned}$$

which yields the first result of the Lemma.

To prove the second claim, suppose that $X^\star = \cup_{i=1}^{I-1}[X_i^m, X_i^M] \setminus \cup_{i'=1}^{I'}\{x_{i'}\}$, $Y^\star = \cup_{j=1}^{J-1}[Y_j^m, Y_j^M] \setminus \cup_{j'=1}^{J'}\{y_{j'}\}$ and $X = X^\star \cup [X_I^m, X_I^M] \setminus \cup_{i^\star=1}^{I^\star}\{x_{i^\star}\}$, $Y = Y^\star \cup [Y_J^m, Y_J^M] \setminus \cup_{j^\star=1}^{J^\star}\{y_{j^\star}\}$ where $X_i^m < X_i^M < X_{i+1}^m$, $Y_i^m < Y_i^M < Y_{i+1}^m$ and $X_I^m = X_{I-1}^M + \epsilon$, $\epsilon > 0$. Then, if we note $\Delta_X \triangleq X_I^M - X_I^m$ and $\Delta_Y \triangleq Y_J^M - Y_J^m$, we have:

$$\mu[\Omega(X+Y)] \geq \mu[\Omega(X^\star+Y)] + \left\{(Y_J^M + X_I^M) - \max(X_{I-1}^M + Y_J^M, Y_J^m + X_I^m)\right\} \ ,$$

where the term into brackets is a lower bound of the sub-volume of $\Omega(X+Y)$ due to the interval $[X_I^m, X_I^M]$; it can be rewritten as $\min\{\Delta_X + \epsilon, \Delta_X + \Delta_Y\}$. Finally, having the Brunn-Minkowski inequality in mind, one gets:

$$\begin{aligned} \mu[\Omega(X+Y)] &\geq \mu[\Omega(X^\star+Y)] + \min\{\Delta_X + \epsilon, \Delta_X + \Delta_Y\} \\ &\geq \mu[\Omega(X)] - \Delta_X + \mu[\Omega(Y)] + \min\{\Delta_X + \epsilon, \Delta_X + \Delta_Y\} \\ &> \mu[\Omega(X)] + \mu[\Omega(Y)] \ . \end{aligned}$$

On the Identifiability Testing in Blind Source Separation Using Resampling Technique

Abdeldjalil Aïssa-El-Bey, Karim Abed-Meraim, and Yves Grenier

ENST, TSI department, 46 rue Barrault 75634, Paris Cedex 13, France
{elbey, abed, grenier}@tsi.enst.fr

Abstract. This paper focuses on the second order identifiability problem of blind source separation and its testing. We present first necessary and sufficient conditions for the identifiability and partial identifiability using a finite set of correlation matrices. These conditions depend on the autocorrelation fonction of the unknown sources. However, it is shown here that they can be tested directly from the observation through the decorrelator output. This issue is of prime importance to decide whether the sources have been well separated or else if further treatments are needed. We then propose an identifiability testing based on resampling (jackknife) technique, that is validated by simulation results.

1 Introduction

Blind source separation (BSS) of instantaneous mixtures has attracted so far a lot of attention due to its many potential applications [1] and its mathematical tractability that lead to several nice and simple BSS solutions [1, 2, 5, 13]. The underlaying model is given by:

$$\mathbf{x}(t) = \mathbf{y}(t) + \mathbf{w}(t) = \mathbf{A}\mathbf{s}(t) + \mathbf{w}(t)$$

where $\mathbf{s}(t) = [s_1(t), \cdots, s_m(t)]^T$ is the $m \times 1$ complex *source vector*, $\mathbf{w}(t) = [w_1(t), \cdots, w_n(t)]^T$ is the $n \times 1$ complex *noise vector*, $\mathbf{A}$ is the $n \times m$ full column rank *mixing matrix* (i.e., $n \geq m$), and the superscript T denotes the transpose operator. The source signal vector $\mathbf{s}(t)$, is assumed to be a multivariate stationary complex stochastic process.

In this paper we consider only the second order BSS methods and hence the component processes $s_i(t)$, $1 \leq i \leq m$ are assumed to be temporally coherent and mutually uncorrelated, with zero mean and second order moments:

$$\mathbf{S}(\tau) \stackrel{\text{def}}{=} E\left(\mathbf{s}(t+\tau)\mathbf{s}^\star(t)\right) = \text{diag}[\rho_1(\tau), \cdots, \rho_m(\tau)]$$

where $\rho_i(\tau) \stackrel{\text{def}}{=} E(s_i(t+\tau)s_i^*(t))$, the expectation operator is E, and the superscripts $*$ and $\star$ denote the conjugate of a complex number and the complex conjugate transpose of a vector, respectively. The additive noise $\mathbf{w}(t)$ is modeled as a white stationary zero-mean complex random process. In that case, the source separation is achieved by decorrelating the signals at different time lags.

J. Rosca et al. (Eds.): ICA 2006, LNCS 3889, pp. 755–764, 2006.

This is made possible under certain identifiability conditions that have been developed in [3] and recalled briefly in this paper.

Although the previous conditions are expressed in terms of the autocorrelation coefficient of the unknown source signals, we propose here a solution to test them directly out of the received data using the jackknife (resampling) technique.

2 Second Order Identifiability

In [4], Tong *et al.* have shown that the sources are blindly separable based on (the whole set) of second order statistics only if they have different spectral density functions. In practice we achieve the BSS using only a finite set of correlation matrices. Therefore, the preview identifiability result was generalized to that case in [5, 3] leading to the necessary and sufficient identifiability conditions given by the following theorem:

Theorem 1. *Let $\tau_1 < \tau_2 < \cdots < \tau_K$ be $K \geq 1$ time lags, and define $\boldsymbol{\rho}_i = [\rho_i(\tau_1), \rho_i(\tau_2), \cdots, \rho_i(\tau_K)]$ and $\tilde{\boldsymbol{\rho}}_i = [\Re(\boldsymbol{\rho}_i), \Im(\boldsymbol{\rho}_i)]$ where $\Re(x)$ and $\Im(x)$ denote the real part and imaginary part of x, respectively. Taking advantage of the indetermination, we assume without loss of generality that the sources are scaled such that $\|\boldsymbol{\rho}_i\| = \|\tilde{\boldsymbol{\rho}}_i\| = 1$, for all i* [1]*. Then, BSS can be achieved using the output correlation matrices at time lags $\tau_1, \tau_2, \cdots, \tau_K$ if and only if for all $1 \leq i \neq j \leq m$:*

$$\tilde{\boldsymbol{\rho}}_i \text{ and } \tilde{\boldsymbol{\rho}}_j \text{ are (pairwise) linearly independent} \tag{1}$$

Interestingly, we can see from condition (1) that BSS can be achieved from only one correlation matrix $\mathbf{R}_x(k) \stackrel{\text{def}}{=} E(\mathbf{x}(t+k)\mathbf{x}^\star(t))$ provided that the vectors $[\Re(\rho_i(k)), \Im(\rho_i(k)]$ and $[\Re(\rho_j(k)), \Im(\rho_j(k)]$ are pairwise linearly independent for all $i \neq j$.

Note also that, from (1), BSS can be achieved if at most one temporally white source signal exists. In contrast, recall that when using higher order statistics, BSS can only be achieved if at most one Gaussian source signal exists.

Under the condition of Theorem 1, the BSS can be achieved by decorrelation according to the following result:

Theorem 2. *Let $\tau_1, \tau_2, \cdots, \tau_K$ be K time lags and $\mathbf{z}(t) = [z_1(t), \cdots, z_m(t)]^T$ be an $m \times 1$ vector given by $\mathbf{z}(t) = \mathbf{B}\mathbf{x}(t)$. Define $r_{ij}(k) \stackrel{\text{def}}{=} E(z_i(t+k)z_j^*(t))$. If the identifiability condition holds, then $\mathbf{B}$ is a separating matrix (i.e. $\mathbf{B}\mathbf{y}(t) = \mathbf{P}\boldsymbol{\Lambda}\mathbf{s}(t)$ for a given permutation matrix $\mathbf{P}$ and a non-singular diagonal matrix $\boldsymbol{\Lambda}$) if and only if*

$$r_{ij}(k) = 0 \quad \text{and} \quad \sum_{k=\tau_1}^{\tau_K} |r_{ii}(k)| > 0 \tag{2}$$

for all $1 \leq i \neq j \leq m$ and $k = \tau_1, \tau_2, \cdots, \tau_K$.

[1] We implicitly assume here that $\boldsymbol{\rho}_i \neq 0$, otherwise the source signal could not be detected (and a fortiori could not be estimated) from the considered set of correlation matrices. This hypothesis will be held in the sequel.

Note that, if one of the time lags is zero, the result of Theorem 2 holds only under the noiseless assumption. In that case, we can replace the condition $\sum_{k=\tau_1}^{\tau_K} |r_{ii}(k)| > 0$ by $r_{ii}(0) > 0$, for $i = 1, \cdots, m$. On the other hand, if all the time lags are non-zero and if the noise is temporally white (but can be spatially colored with unknown spatial covariance matrix) then the above result holds without the noiseless assumption.

Based on Theorem 2, we can define different objective (contrast) functions for signal decorrelation. In [6], the following criterion[2] was used

$$G(\mathbf{z}) = \sum_{k=\tau_1}^{\tau_K} \log |\text{diag}(\mathbf{R}_z(k))| - \log |\mathbf{R}_z(k)| \tag{3}$$

where diag($\mathbf{A}$) is the diagonal matrix obtained by zeroing the off diagonal entries of $\mathbf{A}$. Another criterion used in [7] is

$$G(\mathbf{z}) = \sum_{k=\tau_1}^{\tau_K} \sum_{1 \leq i < j \leq m} |r_{ij}(k)|^2 + \sum_{i=1}^{m} | \sum_{k=\tau_1}^{\tau_K} |r_{ii}(k)| - 1|^2 \tag{4}$$

Equations (3) and (4) are non-negative functions which are zero if and only if $\mathbf{R}_z(k) = E(\mathbf{z}(n+k)\mathbf{z}^\star(n))$ are diagonal for $k = \tau_1, \cdots, \tau_K$ or equivalently if (2) is met.

3 Partial Identifiability

It is generally believed that when the identifiability conditions are not met, the BSS cannot be achieved. This is only half of the truth as it is possible to partially separate the sources in the sense that we can extract those which satisfy the identifiability conditions. More precisely, the sources can be separated in blocks each of them containing a mixture of sources that are not separable using the considered set of statistics. For example, consider a mixture of 3 sources such that $\tilde{\boldsymbol{\rho}}_1 = \tilde{\boldsymbol{\rho}}_2$ while $\tilde{\boldsymbol{\rho}}_1$ and $\tilde{\boldsymbol{\rho}}_3$ are linearly independent. In that case, source s_3 can be extracted while sources s_1 and s_2 cannot. In other words, by decorrelating the observed signal at the considered time lags, one obtain 3 signals one of them being s_3 (up to a scalar constant) and the two others are linear mixtures of s_1 and s_2.

This result can be mathematically formulated as follows: assume there are d distinct groups of sources each of them containing d_i source signals with same (up to a sign) correlation vector $\tilde{\boldsymbol{\rho}}_i$, $i = 1, \cdots, d$ (clearly, $m = d_1 + \cdots + d_d$). The correlation vectors $\tilde{\boldsymbol{\rho}}_1, \cdots, \tilde{\boldsymbol{\rho}}_d$ are pairwise linearly independent. We write $\mathbf{s}(t) = [\mathbf{s}_1^T(t), \cdots, \mathbf{s}_d^T(t)]^T$ where each sub-vector $\mathbf{s}_i(t)$ contains the d_i source signals with correlation vector $\tilde{\boldsymbol{\rho}}_i$.

[2] In that paper, only the case where $\tau_1 = 0$ was considered.

Theorem 3. *Let* $\mathbf{z}(t) = \mathbf{B}\mathbf{x}(t)$ *be an* $m \times 1$ *random vector satisfying equation (2) for all* $1 \leq i \neq j \leq m$ *and* $k = \tau_1, \cdots, \tau_K$. *Then, there exists a permutation matrix* $\mathbf{P}$ *such that* $\overline{\mathbf{z}}(t) \stackrel{\text{def}}{=} \mathbf{P}\mathbf{z}(t) = [\overline{\mathbf{z}}_1^T(t), \cdots, \overline{\mathbf{z}}_d^T(t)]^T$ *where* $\overline{\mathbf{z}}_i(t) = \mathbf{U}_i\mathbf{s}_i(t)$, $\mathbf{U}_i$ *being a* $d_i \times d_i$ *non-singular matrix. Moreover, sources belonging to the same group, i.e., having same (up to a sign) correlation vector* $\tilde{\boldsymbol{\rho}}_i$ *can not be separated using only the correlation matrices* $\mathbf{R}_x(k)$, $k = \tau_1, \cdots, \tau_K$.

This result (see [3])shows that when some of the sources have same (up to a sign) correlation vectors then the best that can be done is to separate them per blocks and this can be achieved by decorrelation. However, this result would be useless if we cannot check the linear dependency of the correlation vectors $\tilde{\boldsymbol{\rho}}_i$ and partition the signals per groups (as shown above) according to their correlation vectors. This leads us to the important problem of testing the identifiability condition that is discussed next.

4 Testing of Identifiability Condition

4.1 Theoretical Result

The necessary and sufficient identifiability condition (1) depends on the correlation coefficients of the source signals. The latter being unknown, it is therefore impossible to *a priori* check whether the sources are 'separable' or not from a given set of output correlation matrices. However, it is possible to check *a posteriori* whether the sources have been 'separated' or not. We have the following result [3]:

Theorem 4. *Let* $\tau_1 < \tau_2 < \cdots < \tau_K$ *be* K *distinct time lags and* $\mathbf{z}(t) = \mathbf{B}\mathbf{x}(t)$. *Assume that* $\mathbf{B}$ *is a matrix such that* $\mathbf{z}(t)$ *satisfies*[3] *equation (2) for all* $1 \leq i \neq j \leq m$ *and* $k = \tau_1, \cdots, \tau_K$. *Then there exists a permutation matrix* $\mathbf{P}$ *such that for* $k = \tau_1, \cdots, \tau_K$.

$$E(\mathbf{z}(t+k)\mathbf{z}^\star(t)) = \mathbf{P}^T\mathbf{S}(k)\mathbf{P}$$

In other words the entries of $\overline{\mathbf{z}}(t) \stackrel{\text{def}}{=} \mathbf{P}\mathbf{z}(t)$ *have the same correlation coefficients as those of* $\mathbf{s}(t)$ *at time lags* $\tau_1, \cdots, \tau_K$, *i.e.* $E(\overline{z}_i(t+k)\overline{z}_i^\star(t)) = \rho_i(k)$ *for* $k = \tau_1, \cdots, \tau_K$ *and* $i = 1, \cdots, m$.

From Theorem 4, the existence of condition (1) can be checked by using the approximate correlation coefficients $r_{ii}(k) \stackrel{\text{def}}{=} E(z_i(t+k)z_i^\star(t))$. It is important to point out that even if equation (2) holds, it does not mean that the source signals have been separated. Three situations may happen:

1. For all pairs (i,j), $\tilde{\boldsymbol{\rho}}_i$ and $\tilde{\boldsymbol{\rho}}_j$ (computed from $\mathbf{z}(t)$) are pairwise linearly independent. Then we are sure that the sources have been separated and that $\mathbf{z}(t) = \mathbf{s}(t)$ up to the inherent indeterminacies of the BSS problem. In fact, the testing of the identifiability condition is equivalent to pairwise

[3] Because of the inherent indetermination of the BSS problem, we assume without loss of generality that the exact and estimated sources are similarly scaled, i.e., $\|\tilde{\boldsymbol{\rho}}_i\| = 1$.

testing the angles between $\tilde{\boldsymbol{\rho}}_i$ and $\tilde{\boldsymbol{\rho}}_j$ for all $1 \leq i \neq j \leq m$. The larger the angle between $\tilde{\boldsymbol{\rho}}_i$ and $\tilde{\boldsymbol{\rho}}_j$, the better the quality of source separation (see performance analysis in [5]).
2. For all pairs (i, j), $\tilde{\boldsymbol{\rho}}_i$ and $\tilde{\boldsymbol{\rho}}_j$ are linearly dependent. Thus the sources haven't been separated and $\mathbf{z}(t)$ is still a linear combination of $\mathbf{s}(t)$.
3. A few pairs (i, j) out of all pairs satisfy $\tilde{\boldsymbol{\rho}}_i$ and $\tilde{\boldsymbol{\rho}}_j$ linearly dependent. Therefore the sources have been separated in blocks.

Now, having only one signal realization at hand, we propose to use a resampling technique to evaluate the statistics needed for the testing.

4.2 Testing Using Resampling Techniques

Note that in practice the source correlation coefficients are calculated from noisy finite sample data. Due to the joint effects of noise and finite sample size, it is impossible to obtain the exact source correlation coefficients to test the identifiability condition. The identifiability condition should be tested using a certain threshold α, i.e., decide that $\tilde{\boldsymbol{\rho}}_i$ and $\tilde{\boldsymbol{\rho}}_j$ are linearly independent if $||\tilde{\boldsymbol{\rho}}_i\tilde{\boldsymbol{\rho}}_j^T| - 1| > \alpha$.

To find α we use the fact that the estimation error of $\tilde{\boldsymbol{\rho}}_i\tilde{\boldsymbol{\rho}}_j^T$ is asymptotically gaussian[4] and hence one can build the confidence interval of such a variable according to its variance. This algorithm can be summarized as follows:

1. Estimate a demixing matrix $\mathbf{B}$ and $\mathbf{z}(t) \stackrel{\text{def}}{=} \mathbf{Bx}(t)$ using an existing second order decorrelation algorithm (e.g. SOBI [5]).
2. For each component $z_i(t)$, estimate the corresponding normalized vector $\widetilde{\boldsymbol{\rho}}_i$.
3. Calculate the scalar product $\hat{R}(i, j) = |\tilde{\boldsymbol{\rho}}_i\tilde{\boldsymbol{\rho}}_j^T|$ for each pair (i, j).
4. Estimate $\hat{\sigma}_{(i,j)}$ the standard deviation of $\hat{R}(i, j)$ using resampling technique (see Section 5).
5. Choose $\alpha_{(i,j)}$ according to the confidence interval. e.g. to have a confidence interval equal to 99.7% we choose $\alpha_{(i,j)} = 3\hat{\sigma}_{(i,j)}$, and compare $|\hat{R}(i, j) - 1|$ to $\alpha_{(i,j)}$ to test whether sources i and j have been separated or not.

5 Resampling Techniques: The Jackknife

In many signal processing applications one is interested in forming estimates of a certain number of unknown parameters of a random process, using a set of sample values. Further, one is interested in finding the sampling distribution of the estimators, so that the respective means, variances, and cumulants can be calculated, or in making some kind of probability statements with respect to the unknown true values of the parameters.

The bootstrap [8] was introduced by Efron [9] as an approach to calculate confidence intervals for parameters in circumstances where standard methods cannot be applied. The bootstrap has subsequently been used to solve many other problems that would be too complicated for traditional statistical analysis.

---

[4] More precisely, one can prove that the estimation error $\sqrt{T}\delta(\tilde{\boldsymbol{\rho}}_i\tilde{\boldsymbol{\rho}}_j^T)$ is asymptotically, i.e. for large sample size T, gaussian with zero mean and finite variance.

In simple words, the bootstrap does with a computer what the experimenter would do in practice, i.e. if it were possible: he or she would repeat the experiment. With the bootstrap, the observations are randomly reassigned, and the estimates recomputed. These assignments and recomputations are done hundreds or thousands of times and treated as repeated experiments.

The jackknife [10] is another resampling technique for estimating the standard deviation. As an alternative to the bootstrap, the jackknife method can be thought of as drawing n samples of size $n-1$ each without replacement from the original sample of size n [10].

Suppose we are given the sample $\mathcal{X} = \{X_1, X_2, \dots, X_n\}$ and an estimate, $\hat{\vartheta}$, from $\mathcal{X}$. The jackknife method is based on the sample delete-one observation at a time,

$$\mathcal{X}^{(i)} = \{X_1, X_2, \dots, X_{i-1}, X_{i+1}, \dots, X_n\}$$

for $i = 1, 2, \dots, n$, called the jackknife sample. This i^{th} jackknife sample consists of the data set with the ith observation removed. For each ith jackknife sample, we calculate the ith jackknife estimate, $\hat{\vartheta}^{(i)}$ of ϑ, $i = 1, 2, \dots, n$. The jackknife estimate of the standard deviation of $\hat{\vartheta}$ is defined by

$$\hat{\sigma} = \sqrt{\frac{n-1}{n}\sum_{i=1}^{n}\left(\hat{\vartheta}^{(i)} - \frac{1}{n}\sum_{j=1}^{n}\hat{\vartheta}^{(j)}\right)^2}$$

The jackknife is computationally less expensive if n is less than the number of replicates used by the bootstrap for standard deviation estimation because it requires computation of $\hat{\vartheta}$ only for the n jackknife data sets. For example, if $L = 25$ resamples are necessary for standard deviation estimation with the bootstrap, and the sample size is $n = 10$, then clearly the jackknife would be computationally less expensive than the bootstrap. In order to test the separability of the estimated signals, we have used a jackknife method to estimate the variance of the scalar product quantities $R(i,j)$ for $i, j = 1, 2, \dots, m$. This is done according to the following steps:

1. From each signal $\mathbf{z}_i = [z_i(0), \dots, z_i(T-1)]^T$, generate T vectors such as $\mathbf{z}_i^{(j)} = [z_i(0), \dots, z_i(j-1), z_i(j+1), \dots, z_i(T-1)]^T$ and $j = 0, 1, \dots, T-1$.
2. For each vector $\mathbf{z}_i^{(j)}$, estimate the corresponding vector $\widetilde{\boldsymbol{\rho}}_i^{(j)}$.
3. Estimate $\hat{\mathbf{R}}$ such as its $(i,j)^{th}$ entry is

$$\hat{R}(i,j) = \frac{1}{T}\sum_{k=0}^{T-1}\frac{\langle \widetilde{\boldsymbol{\rho}}_i^{(k)}, \widetilde{\boldsymbol{\rho}}_j^{(k)} \rangle}{\|\widetilde{\boldsymbol{\rho}}_i^{(k)}\|\|\widetilde{\boldsymbol{\rho}}_j^{(k)}\|}$$

where $\langle \cdot, \cdot \rangle$ denotes the scalar product and $\| \cdot \|$ is the euclidian norm.
4. Estimate the standard deviation of $\hat{R}(i,j)$ by

$$\hat{\sigma}_{(i,j)} = \sqrt{\frac{T-1}{T}\sum_{k=0}^{T-1}\left(\frac{\langle \widetilde{\boldsymbol{\rho}}_i^{(k)}, \widetilde{\boldsymbol{\rho}}_j^{(k)} \rangle}{\|\widetilde{\boldsymbol{\rho}}_i^{(k)}\|\|\widetilde{\boldsymbol{\rho}}_j^{(k)}\|} - \frac{1}{T}\sum_{l=0}^{T-1}\frac{\langle \widetilde{\boldsymbol{\rho}}_i^{(l)}, \widetilde{\boldsymbol{\rho}}_j^{(l)} \rangle}{\|\widetilde{\boldsymbol{\rho}}_i^{(l)}\|\|\widetilde{\boldsymbol{\rho}}_j^{(l)}\|}\right)^2}$$

6 Discussion

Some useful comments are provided here to get more insight onto the considered testing method and its potential applications and extensions.

- The asymptotic performance analysis of SOBI derived in [5], shows that the separation performance of two sources s_i and s_j depends on the angle between their respective correlation vectors $\widetilde{\boldsymbol{\rho}}_i$ and $\widetilde{\boldsymbol{\rho}}_j$. Hence, measuring this angle gives a hint on the interference rejection level of the two considered sources.
  As a consequence, one can use the measure of this angle not only to test the separability of the two sources but also to guarantee a target (minimum) separation quality. Choosing the threshold $\alpha_{(i,j)}$ accordingly is an important issue currently under investigation.
- The testing method can be incorporated into a two stage separation procedure where the first stage consists in a second order decorrelation method (e.g. SOBI). The second stage would be an HOS-based separation method applied only when the testing indicates a failure of separation at the first step.
- In many practical situations, one might be interested by only one or few source signals. This is the case for example in the interference mitigation problem in [11] or in the power plants monitoring applications [12]. In this situation, the partial identifiability result is of high interest as it proves that the desired source signal can still be extracted even if a complete source separation cannot be achieved.
- We believe that similar testing procedure can be used for HOS-based BSS methods, at least for those like JADE [13], that are based on 4th order decorrelation. This would be the focus of future research work.

7 Simulation Results

We present in this section some simulation results to illustrate the performance of our testing method. In the simulated environment we consider uniform linear array with $n = 2$ sensors receiving the signals from $m = 2$ unit-power first order autoregressive sources (with coefficients $a_1 = 0.95e^{j0.5}$ and $a_2 = 0.5e^{j0.7}$) in the presence of stationary complex temporally white noise. The considered sources are separable according to the identifiability result, i.e. their respective correlation vectors $\widetilde{\boldsymbol{\rho}}_1$ and $\widetilde{\boldsymbol{\rho}}_2$ are linearly independent. The time lags (delays) implicitly involved are $\tau_0, \cdots, \tau_9$ (i.e., $K = 10$). The signal to noise ratio (SNR) is defined as $\text{SNR} = -10\log_{10}\sigma_n^2$, where σ_n^2 is the noise variance. We use SOBI algorithm [5] to obtain the decorrelated sources. The statistics in the curves are evaluated over 2000 Monte-Carlo runs. We present first in figure 1(a) a simulation example where we compare the rate of success of the testing procedure (success means that we decide the 2 sources have been separated) to detect the sources separability for different sample sizes versus the SNR in dB. The confidence interval is fixed to $\beta = 99.7\%$. One can observe from this figure that the performance of

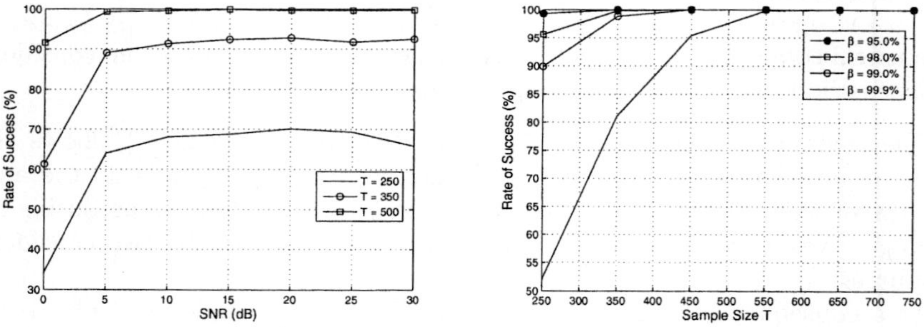

(a) (b)

Fig. 1. (a) Rate of success versus SNR for 2 autoregressive sources and 2 sensors and $\beta = 99.7\%$: comparison of the performance of our testing algorithm for different sample sizes T; (b) Rate of success versus sample size T for 2 autoregressive sources and 2 sensors and SNR=25dB: comparison of the performance of our algorithm for different confidence interval β

the testing procedure degrades significantly for a small sample size due to the increased estimation errors and the fact that we use the asymptotic normality of considered statistics. In figure 1(b), we present a simulation example where we compare the rate of success according to the sample size for different confidence intervals. The SNR is set to 25dB. Clearly, the lower the confidence interval is, the higher is the rate of success of the testing procedure. Also, as observed in figure 1, the rate of success increases rapidly when increasing the sample size. In figure 2(a), we present a simulation example where we plot the rate of success versus the confidence interval β for different sample sizes and for SNR=25dB. This plot shows somehow the evolution of the rate of success w.r.t. the 'false alarm rate' and confirms the results of the two previous figures.

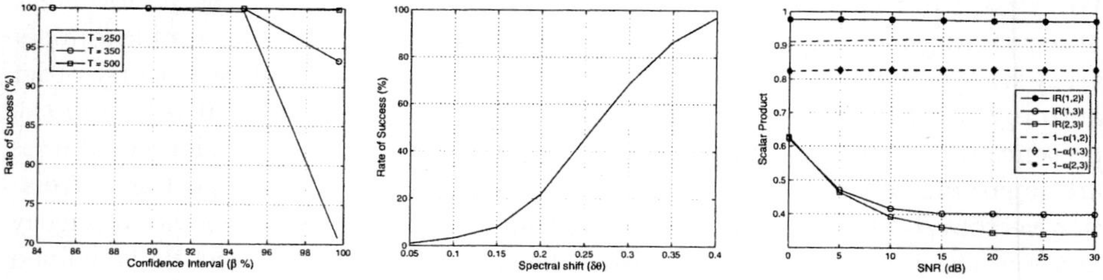

(a) (b) (c)

Fig. 2. (a)Rate of false alarm versus confidence interval β for 2 autoregressive sources and 2 sensors and SNR=25dB: comparison of the performance of our algorithm for different sample size T; (b)Rate of success versus spectral shift $\delta\theta$ for 2 autoregressive sources and 5 sensors and SNR=25dB; (c)Average values of the $|R(i,j)|$ and thresholds $1-\alpha_{(i,j)}$ versus SNR for 3 sources and 3 sensors : 2 sources are complex white gaussian processes and the third one is an autoregressive signal

The simulation example presented in figure 2(b) assumes two source signals with parameters $a_1 = 0.5e^{j0.5}$ and $a_2 = 0.5e^{j(0.5+\delta\theta)}$, where $\delta\theta$ represents the spectral overlap of the two sources. The number of sensors is $n = 5$, the sample size is $T = 1000$ and the SNR=30dB. Figure 2(b) shows the rate of success versus the spectral shift $\delta\theta$. As we can see, small values of $\delta\theta$ lead to high rates of 'non-separability' decision by our testing procedure. Indeed, when $\delta\theta$ is close to zero the two vectors $\widetilde{\boldsymbol{\rho}}_1$ and $\widetilde{\boldsymbol{\rho}}_2$ are close to 'linear dependency'. That means that the separation quality of the two sources is poor in that case which explains the observed testing results. In the last figure, we assume there exist three sources. The first two sources are complex white gaussian processes (hence $\widetilde{\boldsymbol{\rho}}_1 = \widetilde{\boldsymbol{\rho}}_2$) and the third one is an autoregressive signal with coefficient $a_3 = 0.95e^{j0.5}$. The plots in figure 2(c) compares the average values of scalar products for $\widetilde{\boldsymbol{\rho}}_i$ and $\widetilde{\boldsymbol{\rho}}_j$ $(i, j = 1, 2, 3)$ with their corresponding threshold values $1 - \alpha_{(i,j)}$ versus the SNR. The sample size is fixed to $T = 500$ and the number of sensors is $n = 3$. This example illustrates the situation where two of the sources (here sources 1 and 2) cannot be separated (this is confirmed by the testing result) while the third one is extracted correctly (the plots show clearly that $R(1, 3) < 1 - \alpha_{(1,3)}$ and $R(2, 3) < 1 - \alpha_{(2,3)}$).

8 Conclusion

This paper introduces a new method for testing the second order identifiability condition of the blind source separation problem. In simple words, this testing allows us to 'blindly' check, out of the observation, whether the unknown sources have been correctly separated or not. To evaluate the statistics needed for the testing procedure we used the jackknife (resampling) technique. The simulation results illustrate and assess the effectiveness of this testing procedure at least for moderate and large sample sizes.

References

1. A.K. Nandi (editor), "Blind estimation using higher-order statistics." *Kluwer Acadimic Publishers*, Boston 1999.
2. D. T. Pham and J. Cardoso, "Blind source separation of instantaneous mixtures of nonstationary sources," *IEEE Trans. SP.*, Vol. 49, no. 9, pp. 1837–1848, 2001.
3. K. Abed-Meraim, Y. Xiang, Y. Hua, "Generalized second order identifiability condition and relevant testing technique", *In Proc. ICASSP*, Vol. 5, 2989–2992, 2000.
4. L. Tong, R. Liu, V.C. Soon, Y.H. Huang, "Indeterminacy and identifiability of blind identification", *IEEE-T-CS*, Vol. 38, pp. 499–509, 1991.
5. A. Belouchrani, K. Abed-Meraim, J.F. Cardoso, E. Moulines, "Blind source separation using second order statistics", *IEEE Trans. SP*, pp. 434–444, 1997.
6. M. Kawamoto, K. Matsuoka, and M. Oya, "Blind separation of sources using temporal correlation of the observed signal", *IEICE Tr. on Fundamentals of Electronics Communications and Computing* E80A(4):695-704, Apr. 1997.
7. K. Abed-Meraim, Y. Hua, and A. Belouchrani, "A general framework for blind source separation using second order statistics", *Eighth IEEE Digital Signal Processing Workshop, Utah, USA, (CD-ROM)*, Aug. 1998.

8. A.M. Zoubir and B. Boashash, "The bootstrap and its application in signal processing", *IEEE Signal Processing Magazine*, Vol. 15, pp. 56–76, 1998.
9. B. Efron, "The jackknife, the bootstrap and other resampling plans", *CBMS Monograph 38, Society for Industrial and Applied Mathematics*, Philadelphia, 1982.
10. R.G. Miller, "The jackknife - A review", *Biometrika*, Vol. 61, pp. 1–15, 1974.
11. A. Belouchrani, M.G. Amin, M.G. Chenshu Wang "Interference mitigation in spread spectrum communications using blind source separation," *Asilomar Conference*, pp. 718–719, 1996
12. G. D'Urso, P. Prieur, C. Vincent "Blind identification methods applied to Electricite de France's civil works and power plants monitoring," *Higher-Order Statistics, 1997. Proceedings of the IEEE Signal Processing Workshop*, pp. 82–86, 1997.
13. J.F. Cardoso and A. Souloumiac, "A.Blind beamforming for non-Gaussian signals", *Radar and Signal Processing, IEE Proceedings F*, Vol. 140, pp. 362–370, 1993.

On a Sparse Component Analysis Approach to Blind Source Separation

Chunqi Chang[1], Peter C.W. Fung[1,2], and Yeung Sam Hung[1]

[1] Department of Electrical and Electronic Engineering,
The University of Hong Kong, Pokfulam Road, Hong Kong, P.R. China
{cqchang, yshung}@eee.hku.hk
[2] Division of Medical Physics, Department of Medicine,
The University of Hong Kong, Pokfulam Road, Hong Kong, P.R. China
hrspfcw@hkucc.hku.hk

Abstract. Blind source separation has found applications in various areas including biomedical signal processing and genomic signal processing. Often, blind source separation is performed via independent component analysis (ICA) under the assumption of mutual independence among source signals. However, in bio-signal and genomic signal processing, the assumption of independence is often untrue, and the performance of the ICA approach is not so good. Much effort has been devoted to searching alternative approaches to blind source separation without the independence assumption. In this paper we present a sparse component analysis method, which exploits the sparseness of the source signals and makes the separated signals as sparse as possible according to a properly defined sparsity function, to reliably extract source signals from their mixtures. Some related theoretical and practical issues are investigated, with support and validation by simulation results.

1 Introduction

Information is always hidden in measurements. Very often the details of the system that hides the wanted information are not fully available to us. Instead, only the nature of the system and/or the hidden information is partially known. One of the major tasks of signal processing is to extract such hidden information from the measurements only. This task is challenging but when the system that hides the information is a linear mixing system, the problem can be modeled as a blind source separation problem [1]. Blind source separation has many applications in various areas including, for example, acoustics, imaging, communication, and biomedical engineering. It has been well established in the past decade that if the sources are non-Gaussian and statistically independent of each other, the Darmois-Skitovich Theorem [2] ensures that the independent component analysis (ICA) [3], which makes the separated signals as independent of each other as possible, is a solution to the blind source separation problem. However, in most biological and genomic environments, the assumption of independence cannot be reasonably satisfied [4,5]. Alternatively, if the sources are

J. Rosca et al. (Eds.): ICA 2006, LNCS 3889, pp. 765–772, 2006.

statistically uncorrelated over a set of time lags, it is proven that making the separated signals as uncorrelated (over a certain set of time lags) with each other as possible solves the problem [6], and various methods have been developed [7,8]. Such approaches are based on second order statistics and may have promising applications in biomedical signal processing [9,10]. However, both the assumption of independence and the assumption of uncorrelatedness are very strong assumptions that are not easy to be satisfied for naturally occurred signals such as biological signals and genomic signals. Therefore alternative approaches with weaker assumptions are highly desirable. In this direction, some methods have been developed to make use of the non-negativity of the mixing matrix and/or the signals [11,12,13], or to make use of the the sparsity of the signals [14,15]. An alternative way to utilize the sparsity of the source signals has been developed by the authors of this paper and has been successfully applied to EPR spectra decomposition [16,17]. Our approach is to make the separated signals as sparse as possible, and it has been proven that such approach can give the correct solution of source separation if the sources are reasonably sparse [17]. In this paper we introduce our sparse component analysis (SCA) approach and investigate some related theoretical and practical issues.

2 Blind Source Separation

The blind source separation problem is to extract a number of M unknown source signals from a number of N known linear mixtures and at the same time estimate the unknown mixing matrix, using the following model:

$$\mathbf{x}(k) \triangleq \begin{bmatrix} x_1(k) \\ x_2(k) \\ \vdots \\ x_N(k) \end{bmatrix} = \mathbf{As}(k) \triangleq \begin{bmatrix} a_{11} & a_{12} & \cdots & a_{1M} \\ a_{21} & a_{22} & \cdots & a_{2M} \\ \vdots & \vdots & \ddots & \vdots \\ a_{N1} & a_{N2} & \cdots & a_{NM} \end{bmatrix} \begin{bmatrix} s_1(k) \\ s_2(k) \\ \vdots \\ s_M(k) \end{bmatrix}, \tag{1}$$

where the unknown N by M matrix $\mathbf{A}$ is called the mixing matrix, $\mathbf{x}(k)$ the mixtures and $\mathbf{s}(k)$ the unknown sources, $k = 1, \cdots, K$, and K is the length of the signals. Our task is to find a separating matrix $\mathbf{B}$ so that $\hat{\mathbf{s}}(k)$ is an estimate of the source vector $\mathbf{s}(k)$, with preserved waveforms but possibly undetermined scales and orders. In the following section we introduce a sparse component analysis (SCA) approach to this problem.

3 A Sparse Component Analysis Approach

3.1 Sparsity Measure of Sparse Signals

A sparse signal has some peaks and relatively flat area in between the peaks. Typical measures to quantify the degree of sparseness of a signal are L_0 norm and L_1 norm, but L_0 norm is sensitive to noise and L_1 norm is not scale invariant.

In [17] we define the sparsity function of a signal $\mathbf{s}$ as the ratio of its second order statistic to its absolute first order statistic, specifically as

$$S(\mathbf{s}) = \frac{\sqrt{1/K \sum_{k=1}^{K} s^2(k)}}{1/K \sum_{k=1}^{K} |s(k)|} = \sqrt{K}\frac{\sqrt{\sum_{k=1}^{K} s^2(k)}}{\sum_{k=1}^{K} |s(k)|} = \sqrt{K}\frac{\|\mathbf{s}\|_2}{\|\mathbf{s}\|_1}, \tag{2}$$

where $\|\mathbf{s}\|_2$ and $\|\mathbf{s}\|_1$ represent L_2 norm and L_1 norm of $\mathbf{s}$ respectively. It can be shown that $S(\mathbf{s})$ is scale invariant and is bounded between 1 (when the signal is flat) and $\sqrt{K}$ (when the signal contains one only pulse) [17]. The sparsity will not change much with the length of the signal if it is stationary and long enough to provide a stable statistic.

3.2 Blind Source Separation by Sparse Component Analysis

Theorem 1. *Let $\Omega = \{k|s_1(k) = 0, s_2(k) \neq 0\}$ be the support where $s_2(k)$ is not overlapping with $s_1(k)$. Then, $J(t) = S(\mathbf{s}_1 + t\mathbf{s}_2)$ has a local maximum at $t = 0$ if $|e_\rho| < \frac{\|\mathbf{s}_1\|_2}{\sqrt{K}} S(\mathbf{s}_1) \sum_{k\in\Omega} |s_2(k)|$, where $e_\rho = \sum_{k=1}^{K} s_1(k)s_2(k) - S(\mathbf{s}_1) \sum_{k=1}^{K} s_1'(k)s_2(k)$ and $s_1'(k) = \frac{\|\mathbf{s}_1\|_2}{\sqrt{K}}\mathrm{sgn}(s_1(k))$.*

Proof. The theorem has been proven in [17], here we summarize the proof for completeness. $J_0(t) = \frac{1}{K}J^2(t) = \frac{\sum_{k=1}^{K}[s_1(k)+ts_2(k)]^2}{\left[\sum_{k=1}^{K}|s_1(k)+ts_2(k)|\right]^2} = \frac{\|\mathbf{s}_1+t\mathbf{s}_2\|_2^2}{\|\mathbf{s}_1+t\mathbf{s}_2\|_1^2}$ will have the same maxima as $J(t)$, thus in the following we analyze the property of $J_0(t)$.

$$\begin{aligned}
\textbf{Let } g(t) &= \frac{1}{2}\|\mathbf{s}_1 + t\mathbf{s}_2\|_1^3 \frac{\partial J_0}{\partial t} \\
&= \frac{1}{2}\|\mathbf{s}_1 + t\mathbf{s}_2\|_1 \frac{\partial}{\partial t}\|\mathbf{s}_1 + t\mathbf{s}_2\|_2^2 - \frac{1}{2}\frac{\|\mathbf{s}_1 + t\mathbf{s}_2\|_2^2}{\|\mathbf{s}_1 + t\mathbf{s}_2\|_1}\frac{\partial}{\partial t}\|\mathbf{s}_1 + t\mathbf{s}_2\|_1^2 \\
&= \|\mathbf{s}_1 + t\mathbf{s}_2\|_1 [\sum_{k=1}^{K} s_1(k)s_2(k) + t\sum_{k=1}^{K} s_1^2(k)] \\
&\quad - \|\mathbf{s}_1 + t\mathbf{s}_2\|_2^2 \sum_{k=1}^{K} \mathrm{sgn}\,(s_1(k) + ts_2(k))s_2(k)
\end{aligned}$$

If $|e_\rho| < \frac{\|\mathbf{s}_1\|_2}{\sqrt{K}} S(\mathbf{s}_1) \sum_{k\in\Omega} |s_2(k)|$, that is $\|\mathbf{s}_1\|_2^2 \sum_{k\in\Omega} |s_2(k)| > |e_\rho| \|\mathbf{s}_1\|_1$, since

$$\begin{aligned}
g(t \to 0) \to \|\mathbf{s}_1\|_1 &[\sum_{k=1}^{K} s_1(k)s_2(k) - S(\mathbf{s}_1)\sum_{k=1}^{K} s_1'(k)s_2(k)] \\
&- \|\mathbf{s}_1\|_2^2 \sum_{k\in\Omega} \mathrm{sgn}(ts_2(k))s_2(k) \\
&= e_\rho \|\mathbf{s}_1\|_1 - \|\mathbf{s}_1\|_2^2 \sum_{k\in\Omega} \mathrm{sgn}(ts_2(k))s_2(k),
\end{aligned}$$

$$g(t \to 0^-) \to e_\rho \|\mathbf{s}_1\|_1 + \|\mathbf{s}_1\|_2^2 \sum_{k\in\Omega} |s_2(k)| > 0,$$

$$g(t \to 0^+) \to e_\rho \|\mathbf{s}_1\|_1 - \|\mathbf{s}_1\|_2^2 \sum_{k\in\Omega} |s_2(k)| < 0.$$

Therefore $J(t)$ has a local maximum at $t = 0$. ■

For a system with two inputs ($M = 2$) and two outputs ($N = 2$), we can extract the sources as

$$z(k) = x_1(k) + cx_2(k). \tag{3}$$

Due to the model Eq. (1) the above can be represented by the sources as

$$z(k) = \alpha(s_1(k) + t_2 s_2(k)) = \beta(s_2(k) + t_1 s_1(k)). \tag{4}$$

Then Theorem 1 says that under the condition in the theorem a linear combination of the mixtures that is locally most sparse, according to the sparsity measure defined by Eq. (2), will be an estimation of one of the source signals. This estimation makes the separated signals as (locally) sparse as possible, so we will refer to it as sparse component analysis (SCA).

The following procedure summarizes the SCA approach to blind source separation for a 2×2 system:

(1) Find c_1 and c_2 where $J(c) = S(x_1(k) + cx_2(k))$ achieves its local maxima;
(2) Estimate the source signals as $\hat{s}_m(k) = x_1(k) + c_m x_2(k)$, $m = 1, 2$.

4 Theoretical and Practical Issues

4.1 On the Condition of SCA Approach to Blind Source Separation

It is obvious that if the two sources $s_1(k)$ and $s_2(k)$ are uncorrelated and $\Omega \neq \phi$, with $s_1(k)s_2(k) = 0$ for any k as a special case, then the condition of Theorem 1 is satisfied. To investigate the general case where the sources are correlated, we need the following lemma.

Lemma 1. *Let $s_1(k)$ and $s_2(k)$ be two random time series with zero means, $s'(k) = \sqrt{m_2(s)}\text{sgn}(s(k))$, $m_2(s) = E\{s^2(k)\}$, $m_1(s) = E\{|s(k)|\}$, $S(s) = \frac{\sqrt{m_2(s)}}{m_1(s)}$ be the sparsity of $s(k)$, and $\rho(s_1, s_2) = E\{s_1(k)s_2(k)\}/\sqrt{m_2(s_1)m_2(s_2)}$ be the correlation coefficient between $s_1(k)$ and $s_2(k)$. Relate $s_1(k)$ and $s_2(k)$ with linear regression*

$$\frac{s_2(k)}{\sqrt{m_2(s_2)}} = \frac{s_1(k)}{\sqrt{m_2(s_1)}}\rho(s_1, s_2) + z(k)\sqrt{1 - \rho^2(s_1, s_2)}, \tag{5}$$

where $z(k)$ is the normalized regression residue uncorrelated to $s_1(k)$, with zero mean and unit standard deviation. then

$$\lim_{K\to\infty} e'_\rho = S(s_1)\sqrt{1 - \rho^2(s_1, s_2)}E\{\text{sgn}(s_1(k))z(k)\}, \tag{6}$$

where $e'_\rho \equiv \frac{e_\rho/K}{\sqrt{m_2(s_1)m_2(s_2)}}$ is the normalized version of e_ρ defined in Theorem 1.

Proof. It is obvious that $m_2(s_1') = m_2(s_1)$, then

$$\begin{aligned}
&\rho(s_1', s_2) = E\{s_1'(k)s_2(k)\}/\sqrt{m_2(s_1)m_2(s_2)} \\
&= \rho(s_1, s_2)E\{s_1'(k)s_1(k)\}/m_2(s_1) + \sqrt{1-\rho^2(s_1, s_2)}E\{s_1'(k)z(k)\}/\sqrt{m_2(s_1)} \\
&= \rho(s_1, s_2)E\{\text{sgn}(s_1(k))s_1(k)\}/\sqrt{m_2(s_1)} + \sqrt{1-\rho^2(s_1, s_2)}E\{\text{sgn}(s_1(k))z(k)\} \\
&= \rho(s_1, s_2)m_1(s_1)/\sqrt{m_2(s_1)} + \sqrt{1-\rho^2(s_1, s_2)}E\{\text{sgn}(s_1(k))z(k)\} \\
&= \rho(s_1, s_2)/S(s_1) + \sqrt{1-\rho^2(s_1, s_2)}E\{\text{sgn}(s_1(k))z(k)\}. \qquad (7)
\end{aligned}$$

Therefore, $e'_\rho = \frac{\frac{1}{K}\sum_{k=1}^{K} s_1(k)s_2(k)}{\sqrt{m_2(\mathbf{s}_1)m_2(\mathbf{s}_2)}} - S(\mathbf{s}_1)\frac{\frac{1}{K}\sum_{k=1}^{K} s_1'(k)s_2(k))}{\sqrt{m_2(\mathbf{s}_1)m_2(\mathbf{s}_2)}}$, $\lim_{K\to\infty} e'_\rho = \rho(s_1, s_2) - S(s_1)\rho(s_1', s_2) = S(s_1)\sqrt{1-\rho^2(s_1, s_2)}E\{\text{sgn}(s_1(k))z(k)\}$. ■

The condition in Theorem 1 can be rewritten as

$$\left|e'_\rho\right| < \frac{S(\mathbf{s}_1)\sum_{k\in\Omega}|s_2(k)|}{K\sqrt{m_2(\mathbf{s}_2)}} = \frac{S(\mathbf{s}_1)}{S(\mathbf{s}_2)}\frac{\sum_{k\in\Omega}|s_2(k)|}{\sum_{k=1}^{K}|s_2(k)|}. \qquad (8)$$

If both $s_1(k)$ and $s_2(k)$ are Gaussian, $z(k)$ is also Gaussian and independent of $s_1(k)$, then $E\{\text{sgn}(s_1(k))z(k)\} = 0$, and $\lim_{K\to\infty} e'_\rho = 0$ due to Lemma 1, therefore the condition Eq.(8) could be easily satisfied if the signals are long enough. In practical situations, e'_ρ might not be zero since we have only a finite number of data points, and the sources may not be Gaussian. If the sources are Gaussian, e'_ρ should be a small number when K is not too small. In other situations, simulation results in Section 5 show that $\left|e'_\rho\right|$ is close to 0. Consequently, the condition Eq.(8) can be satisfied if there is enough portion of the source $s_1(k)$ with large values not overlapping with the other source $s_2(k)$.

4.2 On Systems with More Than Two Sources

For a system with more than two sources, assuming $N \geq M$, the same idea of sparse component analysis can also apply. To extract a source signal, let

$$z(k) = \sum\nolimits_{n=1}^{N} c_n x_n(k). \qquad (9)$$

According to Eq.(1), $z(k)$ can be represented by the original sources as

$$z(k) = \alpha_m(s_m(k) + t\sum\nolimits_{i\neq m}\beta_i s_i(k)). \qquad (10)$$

Therefore, to extract the source signal $s_m(k)$, the combination of all other source signal can be regarded as a virtual source $\tilde{s}(k) = \sum_{i\neq m}\beta_i s_i(k)$, then the result on systems with two sources can be applied. That is, the local maxima of $S(\mathbf{c}) = S(\sum_{n=1}^{N} c_n x_n(k))$ will give the estimates of the source signals. However, one potential problem is that in the worst case the virtual signal $\tilde{s}(k)$ might be far less sparse, thus making the condition Eq.(8) more difficult to be satisfied.

4.3 Comparison with Other Sparse Methods

There has been quite a lot development in source separation methods utilizing the sparseness of the source signals, as reviewed in [14]. Most of them are geometrical methods that estimate the mixing matrix $\mathbf{A}$ through locating the clustered lines in the scatter plot of the mixtures, and then estimate the source signals through L_1 norm minimization constrained by the model Eq.(1) with $\mathbf{A}$ fixed. A special interest is on systems with less number of mixtures than sources ($N < M$). However, these approaches normally require the source signals to be significantly sparse, since as demonstrated in [16] the lines in the scatter plot are not so obvious if the source signals are not very sparse, which is common in practice. Nevertheless, this two-stage approach can deal with the situation in which the sources are overlapped to some degree, as shown in [18].

Our approach is different from all the existed ones in that we achieve source separation by directly making the separated signals as sparse as possible, where the sparseness is measured by a sparsity function whose reciprocal is a normalized L_1 norm. The normalization makes this sparsity measure scale invariant, which is of crucial importance for our SCA approach to source separation.

4.4 Signal Sparsification

Since it has been demonstrated that most ICA algorithms work better for sparser sources and if the sources are significantly sparse there may be solutions for systems with less number of mixtures than sources, it is desirable to sparsify the signals by some kinds of sparsification transforms, such as wavelet transform, that give a sparse representation of the original signals [14,18,19]. In this paper we have demonstrated that if the source signals are reasonably sparse then we can extract source signals from their linear mixtures even if the source signals are not independent of each other. Therefore, signal sparsification is also very important for our SCA approach to source separation, since the more sparse the source signals, the easier the condition Eq.(8) can be satisfied.

5 Simulation Results

In Section 4.1 the condition on the source signals to validate the SCA approach to blind source separation has been discussed. For correlated non-Gaussian sources, the condition depends on the parameter e'_ρ defined in Theorem 1 and Lemma 1. Here we investigate e'_ρ for sources with a variety of statistical distributions and it's variation over the whole range of correlation coefficients between the two sources. A number of six cases are studied: (1) both with normal distribution, (2) both with exponential distribution, (3) both with uniform distribution, (4) both with lognormal distribution, (5) $s_1(k)$ normal while $s_2(k)$ exponential, and (6) $s_1(k)$ exponential while $s_2(k)$ normal. Results are shown in Figure 1, in which 1000 Monte Carlo simulations for each case have been performed to generate the plots. It can be seen that for all cases e'_ρ is always relatively small even when the two sources are highly correlated.

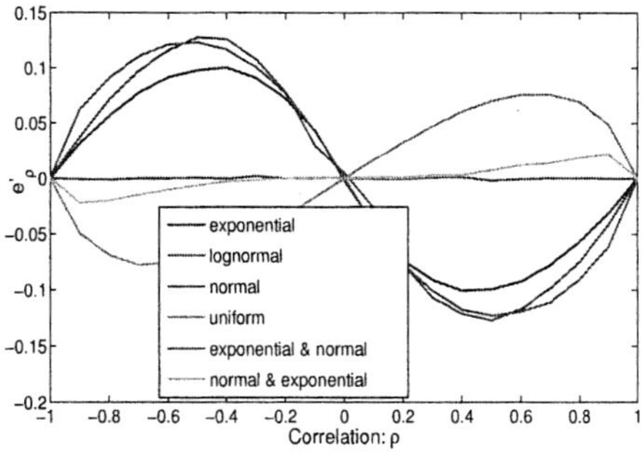

Fig. 1. Simulation results for e'_ρ

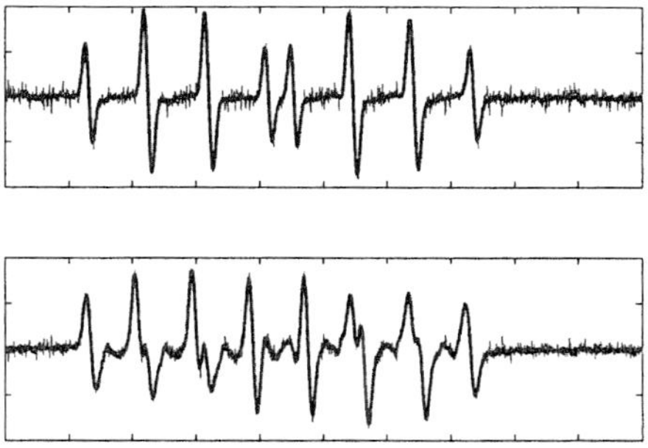

Fig. 2. Separate EPR spectra by SCA

The SCA approach has been successfully applied to EPR spectra decomposition, and the results on both simulated and real data can be found in [17]. Here we show a result for a simulation with noisy signals. Pure EPR spectra of two free radicals hydroxyl and superoxide are used as sources for the simulation, and noises are added to the linear mixtures to a level of SNR=20dB. The SCA separation results are shown in Figure 2. Bold lines are the original sources, overlayed with the corresponding estimates. Our SCA approach performs very well for this case although the correlation coefficient between these two sources is as high as 0.3454.

6 Conclusions

If the source signals are independently non-Gaussian then the ICA approach which makes the separated signals as independent as possible will give a solution for the blind source separation problem. If the source signals (not necessarily non-Gaussian) are uncorrelated over a set of time lags then an uncorrelated component analysis (UCA) approach which makes the separated signals as uncorrelated as possible will also solve the source separation problem [6]. In this paper, we have shown that if the source signals (not necessarily uncorrelated) are reasonably sparse then a sparse component analysis (SCA) approach that makes the separated signals as sparse as possible can serve as a solution. Some related theoretical and practical issues have been investigated, with support from computer simulations.

Acknowledgements

This work is supported in part by Hong Kong RGC grants under HKU7180/03E and N_HKU703/03, and the University of Hong Kong Small Project Funding under 200507176052.

References

1. Jutten, C., Herault, J.: Blind separation of sources, part i: An adaptive algorithm based on neuromimetic architechture. Signal Processing **24** (1991) 1–10
2. Kagan, A.M., Linnik, J.V., Rao, C.R.: Characterization problems in Mathematical Statistics. Wiley (1973)
3. Comon, P.: Independent component analysis, a new concept. Signal Processing **36** (1994) 287–314
4. Lee, S.I., Batzoglou, S.: Application of independent component analysis to microarrays. Genome Biology **4** (2003) Art.R76
5. Ren, J.Y., Chang, C.Q., Fung, P.C.W., Shen, J.G., Chan, F.H.Y.: Free radical EPR spectroscopy analysis using blind source separation. Journal of Magnetic Resonance **166** (2004) 82–91
6. Chang, C.Q., Yau, S.F., Kwok, P., Chan, F.H.Y., Lam, F.K.: Uncorrelated component analysis for blind source separation. Circuits Systems and Signal Processing **18** (1999) 225–239
7. Tong, L., Liu, R.W., Soon, V., Huang, Y.F.: Indeterminacy and identifiability of blind identification. IEEE Transactions on Circuits and Systems **38** (1991) 499–509
8. Chang, C.Q., Ding, Z., Yau, S.F., Chan, F.H.Y.: A matrix-pencil approach to blind separation of colored nonstationary signals. IEEE Transactions on Signal Processing **48** (2000) 900–907
9. Tang, A.C., Liu, J.Y., Sutherland, M.T.: Recovery of correlated neuronal sources from EEG: The good and bad ways of using SOBI. NeuroImage **28**(2) (2005) 507–519
10. Ting, K.H., Fung, P.C.W., Chang, C.Q., Chan, F.H.Y.: Automatic correction of artifact from single-trial event related potentials by blind source separation using second order statistics only. Medical Engineering & Physics (2006) in press
11. Lee, D.D., Seung, H.S.: Learning the parts of objects with non-negative matrix factorization. Nature **405**(6755) (1999) 788–791
12. Donoho, D., Stodden, V.: When does non-negative matrix factorization give a correct decomposition into parts? In: Advances in Neural Information Processing 16 (Proc. NIPS2003), MIT Press (2004)
13. Plumbley, M.D.: Algorithms for non-negative independent component analysis. IEEE Trans. on Neural Network **14** (2003) 534–543
14. O'Grady, P.D., Pearlmutter, B.A., Rickard, S.T.: Survey of sparse and non-sparse methods in source separation. International Journal of Imaging Systems and Technology **15**(1) (2005) 18–33
15. Georgiev, P., Theis, F., Cichocki, A.: Sparse component analysis and blind source separation of underdetermined mixtures. IEEE Trans. on Neural Networks **16**(4) (2005) 992–996
16. Chang, C.Q., Ren, J.Y., Fung, P.C.W., Chan, F.H.Y.: A sparse component analysis approach to EPR spectra decomposition. In: Proc. Int. Symp. on Nonlinear Theory and its Applications (NOLTA2004), Fukuoka, Japan (2004)
17. Chang, C.Q., Ren, J.Y., Fung, P.C.W., Hung, Y.S., Shen, J.G., Chan, F.H.Y.: Novel sparse component analysis approach to free radical EPR spectra decomposition. Journal of Magnetic Resonance **175** (2005) 242–255
18. Li, Y., Cichocki, A., Amari, S.: Analysis of sparse representation and blind source separation. Neural Computation **16**(6) (2004) 1193–1234
19. Donoho, D., Elad, M.: Optimally sparse representation in general (nonorthogonal) dictionaries via l^1 minimization. Proc. Nat. Acad. Sci. USA **100**(5) (2003) 2197–2202

Post-nonlinear Underdetermined ICA by Bayesian Statistics

Chen Wei, Li Chin Khor, Wai Lok Woo, and Satnam Singh Dlay

School of Electrical, Electronic and Computer Engineering,
University of Newcastle, Newcastle upon Tyne, NE1 7RU, United Kingdom
{Chen.Wei, L.C.Khor, W.L.Woo, S.S.Dlay}@ncl.ac.uk

Abstract. The problem of nonlinear signal separation and underdetermined signal separation has received increasing attention in the research of blind signal separation. Few of them can solve the situation where nonlinear and underdetermined characteristics exist simultaneously. In this paper, a new learning algorithm based on Bayesian statistics is proposed to solve the problem of the blind separation of nonlinear and underdetermined mixtures. This paper addresses the Blind Signal Separation (BSS) of post-nonlinearly mixed signals where the number of observations is less than the number of sources. Formal derivation shows that the source signals, mixing matrix and nonlinear functions can be estimated through an iterative technique based on alternate optimization. Simulations have been carried out to demonstrate the effectiveness of the proposed algorithm in separating signals under nonlinear and underdetermined conditions.

1 Introduction

Independent Component Analysis (ICA) has received increasing attention in recent years. The classical mixture model is represented as

$$\mathbf{Y} = \mathbf{M}\mathbf{X} + \boldsymbol{\varepsilon} \tag{1}$$

where $\mathbf{Y}$, $\mathbf{X}$ $\boldsymbol{\varepsilon}$ are mixtures, sources and noises respectively. The mixing matrix $\mathbf{M}$ is assumed to be a square $M \times M$ matrix. Many solutions have been proposed to recover the source signals [1, 2, 3]. But, none of these methods can handle the case where the mixture is nonlinear and underdetermined.

ICA with underdetermined mixture is one of the issues addressed in ICA where the dimension of the mixing matrix $\mathbf{M}$ is $N \times M$ where $N < M$. There are a number of methodologies proposed to estimate the sources and reviewed in [4, 5, 6]. To address underdetermined BSS, a Maximum a Posteriori (MAP) probability approach can be applied, which is expressed as follows

$$\left(\hat{\mathbf{X}},\hat{\mathbf{M}}\right) = \arg\max_{\mathbf{X},\mathbf{M}} P(\mathbf{X},\mathbf{M}|\mathbf{Y}) \propto \arg\max_{\mathbf{X},\mathbf{M}} P(\mathbf{X}|\mathbf{M},\mathbf{Y})P(\mathbf{M}|\mathbf{Y}) \tag{2}$$

Maximizing the joint probability of $\mathbf{X}$ and $\mathbf{M}$ in (2) can be complicated. In this paper, we propose to use an iterative optimization approach to maximize the joint probability as follows:

J. Rosca et al. (Eds.): ICA 2006, LNCS 3889, pp. 773 – 780, 2006.

$$\hat{\mathbf{X}} = \arg\max_{\mathbf{X}} P(\mathbf{X}|\mathbf{M},\mathbf{Y}) \tag{3}$$

$$\hat{\mathbf{M}} = \arg\max_{\mathbf{M}} P(\mathbf{M}|\mathbf{Y}) = \arg\max_{\mathbf{M}} \int P(\mathbf{Y} \mid \mathbf{M},\mathbf{X})P(\mathbf{M})P(\mathbf{X})d\mathbf{X} \tag{4}$$

The main contribution of this paper is to address the issue of post-nonlinear and underdetermined mixtures in BSS and propose an algorithm that resolves both issues. The structure of this paper is organized as follows: the post-nonlinear underdetermined mixing model is firstly proposed in section 2; Section 3 and 4 presents the estimation of the mixing matrix and source signals; Section 5 presents an effective method to minimize the nonlinearity mismatch. Section 6 presents simulation results and analysis to verify the effectiveness of the proposed algorithm.

2 Post-nonlinear Underdetermined Model and Generalized Gaussian Distribution (GGD) Model

The post-nonlinear underdetermined mixing model is composed of a linear mixing matrix $\mathbf{M}$ with dimension $N \times M$ where $N < M$ and a layer of nonlinear distortion function $\{f_n\}_{n=1}^{N}$ [7]. The model can be expressed as

$$\mathbf{Y} = F(\mathbf{MX}) + \boldsymbol{\varepsilon} \tag{5}$$

A generative approach is adopted in the estimation of source signals. Following (3), the expression below is derived:

$$P(\mathbf{X}|\mathbf{Y},\mathbf{M}) \propto P(\mathbf{Y}|\mathbf{X},\mathbf{M})P(\mathbf{X} \mid \mathbf{M}) \propto \ln P(\mathbf{Y}|\mathbf{X},\mathbf{M}) + \ln P(\mathbf{X}) \tag{6}$$

To incorporate the prior knowledge of $P(\mathbf{X})$, we utilize the GGD model [8] which is defined as follows:

$$g(u;p,\lambda) = \frac{p}{2\lambda\Gamma(1/p)} e^{-(|u|/\lambda)^p} \qquad g(.) = \begin{cases} \text{super-gaussian, } 0<p<2 \\ \text{gaussian, } \quad p=2 \\ \text{sub-gaussian, } p>2 \end{cases} \tag{7}$$

where $\Gamma(.)$ is the standard gamma function, λ is the proportional to the standard deviation, p controls the shape of distribution and therefore the kurtosis of the signal.

3 Estimation of Mixing Matrix

The marginal likelihood $P(\mathbf{M}|\mathbf{Y})$ is expressed as follows

$$\begin{aligned} P(\mathbf{M}|\mathbf{Y}) &= \int P(\mathbf{M},\mathbf{X} \mid \mathbf{Y})d\mathbf{X} \\ &\propto \int P(\mathbf{Y} \mid \mathbf{M},\mathbf{X})P(\mathbf{M})P(\mathbf{X})d\mathbf{X} \\ &\propto \ln P(\mathbf{M}) + \ln \int P(\mathbf{Y} \mid \mathbf{M},\mathbf{X})P(\mathbf{X})d\mathbf{X} \end{aligned} \tag{8}$$

As equation (8) is computationally intractable due to the underdetermined mixture, we obtain an approximation by using the Gaussian integral as follows:

$$\int P(\mathbf{Y} \mid \mathbf{M},\hat{\mathbf{X}})P(\hat{\mathbf{X}})d\hat{\mathbf{X}} \approx (2\pi)^{k/2} P(\mathbf{Y} \mid \mathbf{M},\hat{\mathbf{X}})P(\hat{\mathbf{X}}) \det\left(-\Lambda\left(\ln\left(P(\mathbf{Y} \mid \mathbf{M},\hat{\mathbf{X}})P(\hat{\mathbf{X}})\right)\right)\right)^{-1/} \tag{9}$$

where Λ represents the Hessian function around the estimated $\hat{\mathbf{X}}$. Details of the estimation of $\hat{\mathbf{X}}$ is laid out in section 4.

So it is possible to build a cost function when substituting (9) into (8) and the cost function is shown as follows

$$\mathbf{M} = \arg\max_{\mathbf{M}} \left(\ln P(\mathbf{M}) + \ln P(\hat{\mathbf{X}}) + \ln P(\mathbf{Y} \mid \mathbf{M}, \hat{\mathbf{X}}) - \frac{1}{2} \ln \det H(\hat{\mathbf{X}}) \right) \tag{10}$$

where $H(\hat{\mathbf{X}}) = -\Lambda\left(\ln\left(P(\hat{\mathbf{X}})P(\mathbf{Y} \mid \mathbf{M}, \hat{\mathbf{X}})\right)\right)$ is the Hessian matrix. Due to the limitation of space, we only introduce the derivation of $\frac{\partial \ln \det H(\hat{\mathbf{X}})}{\partial \mathbf{M}}$ in the following part

From (10), we know that

$$\begin{aligned} H(\hat{\mathbf{X}}) &= -\Lambda\left(\ln\left(P(\hat{\mathbf{X}})P(\mathbf{Y} \mid \mathbf{M}, \hat{\mathbf{X}})\right)\right) \\ &= -\Lambda \ln P(\mathbf{Y} \mid \mathbf{M}, \hat{\mathbf{X}}) - \Lambda \ln P(\hat{\mathbf{X}}) = \xi + \theta \end{aligned} \tag{11}$$

where $\xi = -\Lambda \ln P(\mathbf{Y} \mid \mathbf{M}, \hat{\mathbf{X}})$ and $\theta = -\Lambda \ln P(\hat{\mathbf{X}})$.

Based on the chain rule,

$$\frac{\partial \ln \det H(\hat{\mathbf{X}})}{\partial \mathbf{M}} = \frac{1}{\det H(\hat{\mathbf{X}})} \sum_{l=1}^{M} \sum_{k=1}^{M} \frac{\partial \det H(\hat{\mathbf{X}})}{\partial h_{kl}(\hat{\mathbf{X}})} \frac{\partial h_{kl}(\hat{\mathbf{X}})}{\partial m_{ij}} \tag{12}$$

and the following identity [9]

$$\frac{\partial \det H(\hat{\mathbf{X}})}{\partial h_{kl}(\hat{\mathbf{X}})} = (\det H(\hat{\mathbf{X}})) h_{lk}^{-1}(\hat{\mathbf{X}}) \tag{13}$$

substitute (13) into (12), this leads to

$$\frac{\partial \ln \det H(\hat{\mathbf{X}})}{\partial \mathbf{M}} = \sum_{l=1}^{M} \sum_{k=1}^{M} h_{lk}^{-1}(\hat{\mathbf{X}}) \frac{\partial h_{kl}(\hat{\mathbf{X}})}{\partial m_{ij}} = \sum_{l=1}^{M} \sum_{k=1}^{M} h_{lk}^{-1}(\hat{\mathbf{X}}) \frac{\partial \xi_{kl}}{\partial m_{ij}} + \sum_{l=1}^{M} \sum_{k=1}^{M} h_{lk}^{-1}(\hat{\mathbf{X}}) \frac{\partial \theta_{kl}}{\partial m_{ij}} \tag{14}$$

Theorem 1: $\frac{\partial \ln \det H(\hat{\mathbf{X}})}{\partial \mathbf{M}}$ can be calculated as follows

$$\frac{\partial \ln \det H(\hat{\mathbf{X}})}{\partial \mathbf{M}} = 2\mathbf{W}^T + [\eta_1' \eta_1^{-1} \quad \cdots \quad \eta_N' \eta_N^{-1}]^T \hat{\mathbf{X}}^T \tag{15}$$

where $\eta_n' = \left(y_n - f_n(\sum_{m=1}^{M} m_{nm} \hat{x}_m) \right) f_n'''(\sum_{m=1}^{M} m_{nm} \hat{x}_m) - 3 f_n'(\sum_{m=1}^{M} m_{nm} \hat{x}_m) f_n''(\sum_{m=1}^{M} m_{nm} \hat{x}_m)$ and $\eta_n = \left(y_n - f_n(\sum_{m=1}^{M} m_{nm} \hat{x}_m) \right) f_n'(\sum_{m=1}^{M} m_{nm} \hat{x}_m) - f_n'(\sum_{m=1}^{M} m_{nm} \hat{x}_m)^2$

The final expression for the learning rule of $\mathbf{M}$ can be expressed as:

$$\Delta \mathbf{M} = \lambda_{\mathbf{M}_P} |\mathbf{M}|^{P_M - 2} \circ \mathbf{M} - \mathbf{W}^T \left(\frac{\partial \ln P(\hat{\mathbf{X}})}{\partial \hat{\mathbf{X}}} + I \right) \hat{\mathbf{X}}^T - \frac{1}{2} [\eta_1' \eta_1^{-1} \quad \cdots \quad \eta_N' \eta_N^{-1}]^T \hat{\mathbf{X}}^T = \mathbf{L} + \mathbf{N} \tag{16}$$

where we define $\mathbf{L} = \lambda_{\mathbf{M}_p} |\mathbf{M}|^{p_{\mathbf{M}}-2} \circ \mathbf{M} - \mathbf{W}^T \left(\frac{\partial \ln P(\hat{\mathbf{X}})}{\partial \hat{\mathbf{X}}} \hat{\mathbf{X}}^T + I \right)$ as the linear component and $\mathbf{N} = -\frac{1}{2} [\eta_1' \eta_1^{-1} \ \cdots \ \eta_N' \eta_N^{-1}]^T \hat{\mathbf{X}}^T$ as the nonlinear component. Since the nonlinear mixing processing is unknown, a mismatch in the nonlinear function will occur. In the following sections, we present a learning algorithm to estimate the source signals and introduce a systematic method to minimize this nonlinear mismatch.

4 Estimation of Source Signal

The GGD model can be incorporated into (6) to form the cost function of the proposed generative network:

$$\hat{\mathbf{X}} = \arg\max_{\mathbf{X}} J(\mathbf{X}) \tag{17}$$

where

$$J(\mathbf{X}) = -\frac{1}{\lambda_\varepsilon^2} \|\mathbf{Y} - F(\mathbf{MX})\|^2 - \frac{1}{\lambda_x^{p_x}} \sum_{m=1}^{M} |x_m|^{p_x} \tag{18}$$

The derivative of the cost function with respect to $\mathbf{X}$ simply becomes

$$\nabla_{\mathbf{X}} J(\mathbf{X}) = \frac{2}{\lambda_\varepsilon^2} \mathbf{M}^T diag(F'(\mathbf{MX}))(\mathbf{Y} - F(\mathbf{MX})) - \frac{p_x}{\lambda_x^{p_x}} diag\left(|x_m|\right)^{(p_x-2)} \mathbf{X} \tag{19}$$

It can be easily inferred that the optimal solution of $\mathbf{X}$ satisfies the condition $\nabla_{\mathbf{X}} J(\mathbf{X}) = 0$. The adaptation of $\mathbf{X}$ can be expressed as the gradient-based learning algorithm:

$$\mathbf{X}(t+1) = \mathbf{X}(t) + \mu_{\mathbf{X}} \nabla_{\mathbf{X}} J(\mathbf{X}) \tag{20}$$

where $\mu_{\mathbf{X}}$ is the learning rate of the source estimation.

5 Minimization of Nonlinear Mismatch

Due to lack of knowledge about the nonlinear function, the initial estimation of nonlinear function in (16) may not match the true function. In this case, a self-adaptive algorithm is necessary to approximate the nonlinear function as similar as possible to the true nonlinear function. It has been demonstrated in the Universal Approximation Theorem [10] that for every continuous function $f_n(.)$, there always exists a Multilayer Perceptron [11]which can uniformly approximate $f_n(.)$ in the form of

$$\delta(\mathbf{U}, \mathbf{A}^{(1)}, \mathbf{A}^{(2)}, \boldsymbol{\beta}) = \mathbf{A}^{(1)} \tanh(\mathbf{A}^{(2)} \mathbf{U} + \boldsymbol{\beta}) \tag{21}$$

As long as the nonlinear functions $f_n(.)$ are wrongly specified, the estimation of $\mathbf{X}$ and $\mathbf{M}$ will consequently degrade. Since the function $f_n(.)$ is a one-to-one mapping, the MLP indeed performs non-mixing nonlinear mapping. Thus, this allows us to formulate a least square error criterion to minimize the mismatch between the true observed signal and the estimated observed signal as follows:

$$\begin{aligned}\{\mathbf{A}_n^{(1)},\mathbf{A}_n^{(2)},\beta_n\} &= \underset{\mathbf{A}_n^{(1)},\mathbf{A}_n^{(2)},\beta_n}{\arg\min}\left\| y_n - \mathbf{A}_n^{(1)}\tanh(\mathbf{A}_n^{(2)}\sum_{m=1}^{M} m_{nm}x_m + \beta_n)\right\|^2 \\ &= \underset{\mathbf{A}_n^{(1)},\mathbf{A}_n^{(2)},\beta_n}{\arg\min}\left\| y_n - \mathbf{A}_n^{(1)}\tanh(\mathbf{A}_n^{(2)T}u_n + \beta_n{}^T)\right\|^2\end{aligned} \quad (22)$$

The parameters in (22) can be estimated from the following:

$$\hat{\mathbf{A}}_n^{(1)} = -2\left(y_n - \mathbf{A}_n^{(1)}\tanh(\mathbf{A}_n^{(2)T}u_n + \beta_n{}^T)\right)\tanh(\mathbf{A}_n^{(2)}u_n + \beta_n) \quad (23)$$

$$\begin{bmatrix}\hat{\mathbf{A}}_n^{(2)} \\ \hat{\beta}_n\end{bmatrix} = \left[-2\mathbf{A}_n^{(1)}\left(y_n - \mathbf{A}_n^{(1)}\tanh(\mathbf{A}_n^{(2)T}u_n + \beta_n{}^T)\right)\mathrm{diag}\left(\mathrm{sech}^2(\mathbf{A}_n^{(2)T}u_n + \beta_n{}^T)\right)\right]\begin{bmatrix}u_n \\ 1\end{bmatrix} \quad (24)$$

Once the coefficients of Multilayer Perceptron converge, the new estimated nonlinear function is substituted into (16) and (19) to obtain refined estimates of $\hat{\mathbf{X}}$ and $\hat{\mathbf{M}}$.

6 Results and Discussion

In this section, two different experiments are carried out to evaluate the algorithm's effectiveness in BSS of nonlinear underdetermined signals and compare the performance of the proposed algorithm against the FOCUSS algorithm [10]. As a basis for comparison, a performance index is adopted to assess the results. It can be as:

$$\mathrm{P} = 2\left(1 - \frac{1}{M}\sum_{i=1}^{M}|\rho_i|\right) \qquad \rho_i = \frac{E\left[(x_i - E[x_i])^*(\hat{x}_i - E[\hat{x}_i])\right]}{\sqrt{E\left[|x_i - E[x_i]|^2\right]E\left[|\hat{x}_i - E[\hat{x}_i]|^2\right]}} \quad (25)$$

where ρ_i is the normalized cross-correlation, '$*$' and '$|\cdot|$' denote the complex conjugate and absolute operation respectively.

For all experiments presented in this paper, the initial value of $\mathbf{M}$ is selected randomly and the initial value of $\hat{\mathbf{X}}$ is obtained from the following relationship

$$\hat{\mathbf{X}} = \mathbf{M}^{+}\mathbf{Y} \quad (26)$$

where $\mathbf{M}^{+}$ is the pseudoinverse of mixing matrix $\mathbf{M}$. The signals are perturbed with Gaussian noise at the sensors distributed and is used to perturb the sensors.

6.1 Performance Improvement Compared with Linear Algorithm

In this experiment, three audio waves correspond to the source signals shown in Fig.1 (top). The source signals are transformed into two mixtures through (5) and depicted in Fig.1 (middle). The mixing matrix is randomly generated from a Gaussian distribution. The actual post-nonlinear process is set to $\tanh(.)$ and the estimated nonlinear process is assumed to be identical to the true nonlinear distortion function. Fig.1 (bottom) shows the recovered source signals by the proposed algorithm under SNR=20dB. Fig.1 (top) and Fig.1 (bottom) clearly demonstrates the close resemblance between the original source signals and the recovered sources signals.

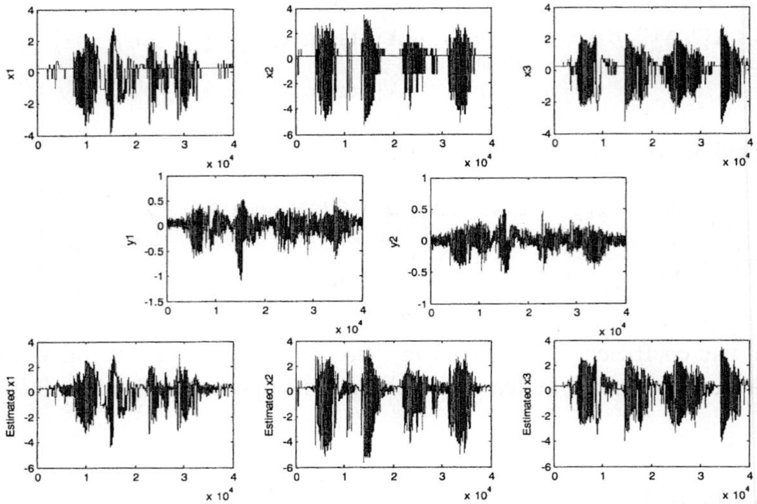

Fig. 1. Nonlinear underdetermined mixing using three speech signals. Three source signals (top); two mixtures (middle); three estimated source signals (bottom).

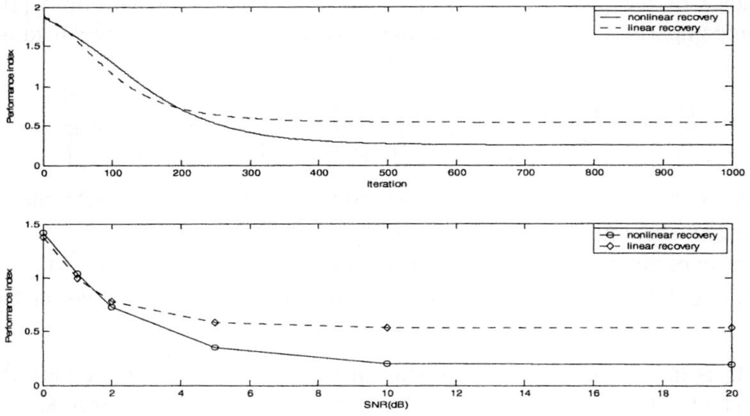

Fig. 2. Performance index comparison under SNR=20dB (top); Performance index comparison as a function of SNR (bottom)

To demonstrate the incompetence of a linear BSS algorithm approach and the significance in the proposed algorithm of post-nonlinear underdetermined mixtures, we compare the effectiveness of the proposed algorithm with the well-known FOCUSS algorithm [10]. Fig.2. (top) depicts the performance index under fixed SNR=20dB. It is seen that both of the performance index converges to a small fixed value after 500 iterations. However, performance of the proposed algorithm surpasses FOCUSS algorithm by over 100% under SNR=20dB. In the case of varying degrees of noise, when SNR<5dB the presence of noise dominates and affects the performance of both algorithms. However, a significant improvement in the performance of the proposed algorithm is observed as SNR increases and the rate in accuracy exceeds the FOCUSS algorithm.

6.2 Nonlinear Mismatch

In reality, the problem of nonlinear mismatch is inevitable in the signal separation process. Any mismatch in the nonlinear function will lead to a decrease in performance. A self-adaptive Multilayer Perceptron is employed to test the performance of the proposed algorithm in approximating the true nonlinear function.

The accuracy of the proposed algorithm in estimating the following three different nonlinear functions are measured independently

$$F(\mathbf{U}) = \left[u_1 + u_1^{1/3},\quad u_1 + u_2^5,\quad \tanh(u_3)\right]^T, \text{ where } u_i = \left(\sum_{m=1}^{M} m_{im} x_m\right)$$

Fig.3 shows the true and estimated nonlinear function for each of the three functions. Results show that, in each channel, the estimated Multilayer Perceptron converges very close to the true nonlinear function. Although minor nonlinear mismatch still exists in some cases, this mismatch is negligible compared to the case when the hypothetical nonlinear function are selected arbitrarily.

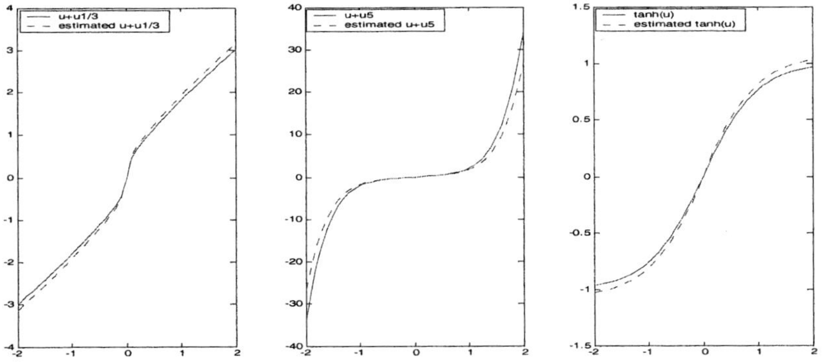

Fig. 3. Comparison between the true nonlinear function and the estimated nonlinear function by Multilayer Perceptron

7 Conclusion and Future Work

The main contribution of this paper is to present a novel algorithm which successfully recovers the source signals from a set of blind nonlinear underdetermined mixtures. The algorithm is derived from a Bayesian framework and addresses simultaneously the problem of nonlinearity and underdetermined mixtures in BSS. Simulation results demonstrate the efficacy of the proposed algorithm in separating blind nonlinear underdetermined mixtures. However, a few challenges still remains to be solved. One of them is the high computational complexity incurred by iterative process of the parameter update. An effective solution is to remove the parameters that become trivial after a few iterations. Another challenge lies in finding a general model for all kinds of distributions as the GGD model can only model unimodal distributions. A possible alternative would be the Gaussian Mixture Model (GMM) which can asymptotically accommodate any continuous distributions.

References

1. Cardoso, J.F.. Blind signal separation: Statistical principles in Proceedings of the IEEE, Vol. 86. (1998) 2009-2025.
2. Amari, S., Cichocki, A.. Adaptive blind signal processing - Neural network approaches in Proceedings of the IEEE, Vol. 86. (1998) 2026-2048.
3. Amari, S., Hyvarinen, A., Lee, S.Y., Lee, T.W., Sanchez A, V.D., Blind signal separation and independent component analysis in Neurocomputing, Vol. 49 (2002) 1-5.
4. Daives, M. and Mitianoudis, N.: 'Simple mixture model for sparse overcomplete ICA', IEE Proceedings on Vision, Image & Signal Processing (Special Section on Nonlinear and Non-Gaussian Signal Processing), Vol. 151, (2004), pp. 35-43.
5. Takigawa, I., Kudo, M. and Toyama, J.: 'Performance Analysis of Minimum l_1 -Norm Solutions for Underdetermined Source Separation', IEEE Transactions on Signal Processing, Vol. 52, (2004), pp. 582-591
6. Woo, W.L., and Dlay, S.S.: 'Neural Network Approach to Blind Signal Separation of Mono-nonlinearly Mixed Signals', IEEE Trans. on Circuits and System – Part 1, Vol. 28, (2005), pp. 1236-1247.
7. Girolami, M.: 'A Variational Method for learning Sparse and Overcomplete representations', Neural Computation, Vol. 13, (2001) pp. 2517-2532
8. Wei, C., Woo, W.L., Dlay, S.S.. A FOCUSS-Based Algorithm for Nonlinear Overcomplete Independent Component Analysis in WSEAS Transactions on Information Science and Applications, Vol. 6 (2004) 1688-1693.
9. Lewicki, M.S., Sejnowski, T.J., Learning Overcomplete Representation in Neural Computation, Vol. 12 (2000) 337-365
10. Haykin, S., Neural Networks: A comprehensive Foundation, Prentice Hall (1999)
11. Woo, W.L. and Sali, S.: 'General Multilayer Perceptron Demixer Scheme for Nonlinear Blind Signal Separation', IEE Proceedings on Vision, Image and Signal Processing, Vol. 149, (2002), pp. 253-262.
12. Rao, B.D., Kreutz-Delgado, K.. An Affine Scaling Methodology for Best Basis Selection, IEEE Transactions on Signal Processing, Vol. 47 (1999) 187-200

Relationships Between the FastICA Algorithm and the Rayleigh Quotient Iteration

Scott C. Douglas

Department of Electrical Engineering,
Southern Methodist University,
Dallas, Texas 75275 USA

Abstract. The FastICA algorithm is a popular procedure for independent component analysis and blind source separation. Recently, several of its convergence properties have been elucidated, including its average convergence performance and its finite-sample behavior. In this paper, we examine the kurtosis-based algorithm version for two-source mixtures with equal-kurtosis sources, proving that the single-unit FastICA algorithm has dynamical behavior that is identical to the Newton-based Rayleigh Quotient Iteration for finding an eigenvector of a symmetric matrix. We also derive a bound on the average inter-channel interference indicating that the initial convergence rate of FastICA is linear with a rate of (1/3). A simulation indicates its convergence performance.

1 Introduction

The FastICA algorithm of Hyvarinen and Oja [1] is a popular procedure for independent component analysis (ICA) and blind source separation. The technique is simple to set up, converges quickly, and provides good separation behavior in a variety of contexts. Moreover, when a fourth-moment or kurtosis-based contrast function is used within the algorithm, convergence is globally cubic about a separating solution for the linear ICA model with non-Gaussian sources [2]. The technique has become popular for a number of problems in signal analysis.

Various studies of the convergence and identification behavior of the FastICA have been made, including stationary point analyses in the two-source and m-source mixing cases [3,4], its average convergence performance [4,5,6], and its finite-sample behavior at convergence [7]. In this paper, we add to this knowledge about the convergence of the FastICA algorithm by studying the algorithm for mixtures of two sources with equal kurtoses, with the goal of providing additional theoretical insight into the algorithm's behavior. In this situation, we prove that

- the single-unit FastICA algorithm has dynamical behavior that is mathematically-identical to the Newton-based Rayleigh Quotient Iteration for finding a minimum eigenvalue of a symmetric matrix, and
- convergence of the average inter-channel interference is bounded above by a function that converges linearly with rate (1/3) or 4.77dB per iteration.

J. Rosca et al. (Eds.): ICA 2006, LNCS 3889, pp. 781–789, 2006.

Thus, our results support previous observations made in the ICA literature which indicated linear (exponential) convergence of FastICA with a kurtosis contrast [6], and it connects the algorithm with a well-known eigenvector search procedure. A simulation verifies the derived performance bound.

2 Kurtosis-Based FastICA for Two-Source Mixtures

We briefly introduce the FastICA algorithm for two-source mixtures so that notation can be defined; complete descriptions of the algorithm are in [1,2]. Let $\mathbf{s}(k) = [s_1(k)\ s_2(k)]^T$, where $s_1(k)$ and $s_2(k)$ are zero-mean, unit variance, non-Gaussian, and statistically-independent at time k, such that their normalized kurtoses are $\kappa_i = E\{s_i^4(k)\} - 3$, $i \in \{1,2\}$. Let $\mathbf{x}(k) = \mathbf{A}\mathbf{s}(k)$ contain a linear mixture of these sources, The single-unit FastICA procedure first determines a whitening transformation $\mathbf{P}$ such that $\mathbf{v}(k) = \mathbf{P}\mathbf{x}(k)$ contains unit-power uncorrelated signals. A weight vector $\mathbf{w}_t = [w_{1,t}\ w_{2,t}]^T$ is then adjusted such that

$$y_t(k) = \mathbf{w}_t^T \mathbf{v}(k) \tag{1}$$

is an estimate of one extracted source. The adjustment procedure is

$$\widetilde{\mathbf{w}}_t = E\{y_t^3(k)\mathbf{v}(k)\} - E\{y_t^2(k)\}\mathbf{w}_t, \qquad \mathbf{w}_{t+1} = \frac{\widetilde{\mathbf{w}}_t}{\sqrt{\widetilde{\mathbf{w}}_t^T \widetilde{\mathbf{w}}_t}} \tag{2}$$

if a kurtosis-based cost function is used. An extension allows one to use other cost functions [2]. Sampled data averages are used to compute all expectations.

For analysis, consider the behavior of FastICA in the combined coefficient vector $\mathbf{c}_t = \mathbf{A}^T\mathbf{P}^T\mathbf{w}_t$, in which case $y(k) = \mathbf{c}^T\mathbf{s}(k)$. Furthermore, for two-source mixtures, we introduce an intrinsic parametrization for $\mathbf{c}_t = [c_{1,t}\ c_{2,t}]^T$ given by

$$\mathbf{c}_t = [\cos(\theta_t)\ \sin(\theta_t)]^T. \tag{3}$$

Then, letting the number of data measurements tend to infinity, an equivalent expression for FastICA in this case is [4]

$$c_{1,t+1} = \frac{\kappa_1 c_{1,t}^3}{\sqrt{\kappa_1^2 c_{1,t}^6 + \kappa_2^2 c_{2,t}^6}}, \quad c_{2,t+1} = \frac{\kappa_2 c_{2,t}^3}{\sqrt{\kappa_1^2 c_{1,t}^6 + \kappa_2^2 c_{2,t}^6}}, \tag{4}$$

which can be represented even more compactly using (3) as

$$\tan(\theta_{t+1}) = \frac{\kappa_2}{\kappa_1}\tan^3(\theta_t). \tag{5}$$

When $|c_{2,t}| \leq |c_{1,t}|$, the ratio $c_{2,t}/c_{1,t} = \tan(\theta_t)$ is related to the inter-channel interference (ICI) at time t as

$$ICI_t = \frac{c_{2,t}^2}{c_{1,t}^2} = \tan^2(\theta_t), \tag{6}$$

which represents a useful performance factor for the algorithm. In this paper, we shall assume equal kurtosis sources, such that $\kappa_2/\kappa_1 = 1$. Equal-magnitude kurtosis sources could be handled with additional notational changes. We shall also restrict our study to the case $|c_{2,t}| \leq |c_{1,t}|$ or $|\theta_t| \leq \pi/4$, as all other convergence regions follow from the four-fold symmetry of the parameter space.

3 The Kurtosis Contrast, Newton's Method, and the Rayleigh Quotient Iteration

The single-unit kurtosis-based FastICA algorithm can be derived as an *approximate* Newton's method for minimizing the cost [1,2]

$$\mathcal{J}(\mathbf{w})) = -\left|E\{y^4(k)\} - 3\left(E\{y^2(k)\}\right)^2\right| \tag{7}$$

under a unit power constraint on $y(k)$ given by $E\{y^2(k)\} = \mathbf{w}^T\mathbf{w} = \mathbf{c}^T\mathbf{c} = 1$. In the two-source case, this constraint is exactly maintained by (3). We can express the kurtosis contrast for a two source mixture with equal-kurtosis sources in the vicinity of $\theta = 0$ as

$$\mathcal{J}(\mathbf{c}) = -|\kappa|\left[c_1^4 + c_2^4\right] \quad \text{or} \quad \mathcal{J}(\theta) = -|\kappa|\left[\cos^4(\theta) + \sin^4(\theta)\right]. \tag{8}$$

Our pursuit of knowledge about the FastICA procedure can move in either a constructive or an analytical fashion, and we choose the former approach first. Ignoring our ability to represent iterative algorithms using measured data, what are some good approaches that could be used to minimie $\mathcal{J}(\theta)$ with respect to θ? Clearly, Newton's method is of interest given the convexity and evenness of $\mathcal{J}(\theta)$ at $\theta = \{0, \pi/2, \pi, 3\pi/2\}$, as Newton-based methods converge cubically under such conditions. The one-dimensional gradient and Hessian of $\mathcal{J}(\theta)$ are

$$\frac{\partial \mathcal{J}(\theta)}{\partial \theta} = |\kappa|\sin(4\theta) \quad \text{and} \quad \frac{\partial^2 \mathcal{J}(\theta)}{\partial \theta^2} = 4|\kappa|\cos(4\theta), \tag{9}$$

such that Newton's method for adapting θ is

$$\theta_{t+1} = \theta_t - \frac{1}{4}\tan(4\theta_t). \tag{10}$$

Near $\theta_t = 0$, the algorithm is indeed cubically-convergent, as

$$\theta_{t+1} = -\frac{16}{3}\theta_t^3 + \mathcal{O}(\theta_t^5). \tag{11}$$

Eqn. (10) has appeared before in the analysis of the Rayleigh quotient iteration (RQI) for finding the eigenvector of a symmetric matrix [8], in which the cost function in the transformed eigenvector $\mathbf{c}_t$ and eigenvalues $\{\lambda_1, \lambda_2\}$ is

$$\overline{\mathcal{J}}(\mathbf{c}) = \lambda_1 c_1^2 + \lambda_2 c_2^2 \quad \text{or} \quad \overline{\mathcal{J}}(\phi) = \lambda_1\cos^2(\phi) + \lambda_2\sin^2(\phi), \tag{12}$$

Newton's method for minimizing $\overline{\mathcal{J}}(\phi)$ is

$$\phi_{t+1} = \phi_t - \frac{1}{2}\tan(2\phi_t) \;\approx\; -\frac{4}{3}\phi_t^3 + \mathcal{O}(\phi_t^5), \tag{13}$$

which is identical in form to (10) for $\theta_t = 2\phi_t$. That the two costs in (8) and (12) would produce the same Newton iteration is remarkable, but our interest here is in realizable algorithms, such as FastICA for blind source separation. The following theorem, proven in the Appendix, illuminates an important relationship between FastICA and the RQI.

Theorem 1. *In the two-source equal-kurtosis case as the number of measurements tends to infinity, the FastICA algorithm for minimizing (8) in the intrinsic parametrization variable θ_t is*

$$\theta_{t+1} = -\theta_t + \arctan\left(\frac{1}{2}\tan(2\theta_t)\right) \;=\; \theta_t^3 + \mathcal{O}(\theta_t^5), \tag{14}$$

which is identical in form to the Rayleigh Quotient Iteration for minimizing (12) in the intrinsic parametrization variable ϕ_t, as given by

$$\phi_{t+1} = \phi_t - \arctan\left(\frac{1}{2}\tan(2\phi_t)\right) \;=\; -\phi_t^3 - \mathcal{O}(\phi_t^5), \tag{15}$$

where $\theta_t = (-1)^t\phi_t$. Moreover, both the RQI and FastICA are approximate step-and-project Newton algorithms in $\mathbf{c}_t$ employing the tangent form of the Newton update within the intrinsic parameter space θ_t or ϕ_t.

The above theorem states that in the two-source case, the weight vector $\mathbf{w}_t$ for the FastICA algorithm evolves identically to that for the RQI applied to a symmetric matrix $\boldsymbol{\Gamma}\mathrm{diag}\{\lambda_1, \lambda_2\}\boldsymbol{\Gamma}^T$ when $\boldsymbol{\Gamma} = \mathbf{PA}$ is orthonormal, except for the sign changes associated with the alternating update directions in the RQI. We have verified this fact numerically to the machine precision limits of MATLAB by running each algorithm its respective task for the same $\boldsymbol{\Gamma}$ separation and eigenvector matrix, respectively. That the FastICA procedure shares the same evolutionary behavior of RQI is informative, as RQI is a well-known and well-studied procedure in the numerical linear algebra community [8,9]. The RQI is considered one of the best procedures for its task due to its local cubic convergence. This link means that convergence results for RQI can potentially be applied to the FastICA algorithm, and vice versa.

To better see the geometrical relationships of the various algorithms, Figure 1 illustrates a single iteration of each algorithm in both two-dimensional $\mathbf{c}$-space as well as one-dimensional angular space. Point O corresponds to the point on the unit circle at angle θ_t of the FastICA algorithm. Point P is the negative of this angle at $-\theta_t$, which we will set equal to ϕ_t for comparison with the RQI. Vector OD is the the component of the Newton update direction for minimizing the kurtosis-based cost in (8) in the tangent space at point O or angle θ_t. Vector PE is the component of the Newton update direction for minimizing the Rayleigh quotient cost in (12) in the tangent space at angle ϕ_t. Point A is reached by the update in (10), in which the arclength OA is equal to the linear distance OD.

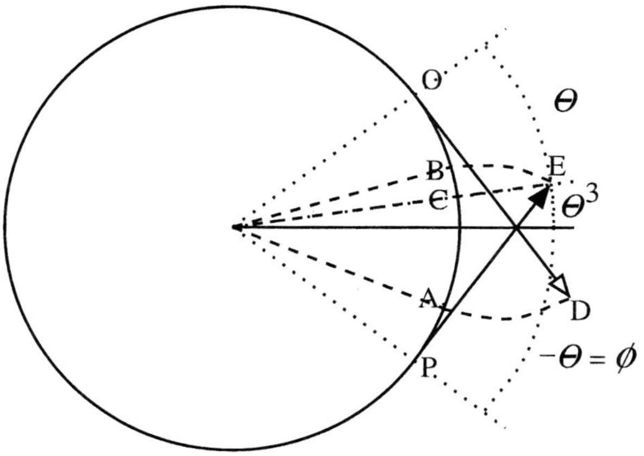

Fig. 1. Geometry of the FastICA algorithm and the Rayleigh Quotient Iteration - see text for label descriptions

Point B is reached by the update in (13), in which the arclength PB is equal to the linear distance PE. Point C is the result of both FastICA and RQI in their respective problems, which is obtained by projecting the point E to the unit circle. In all cases, the angle magnitude is reduced in proportion to the cube of the original angle θ_t or ϕ_t for small angles: Point A is at an angle of approximately $-\frac{16}{3}\theta_t^3$, Point B is at an angle of approximately $\frac{4}{3}\theta_t^3 = -\frac{4}{3}\phi_t^3$, and point C is at an angle of approximately θ_t^3.

In practice, the kurtoses of the two sources are not equal; even so, the locally-cubic convergence of the FastICA algorithm is maintained. From (5),

$$\theta_{t+1} = \arctan\left(\frac{\kappa_2}{\kappa_1}\tan^3(\theta_t)\right) = \frac{\kappa_2}{\kappa_1}\theta_t^3 + \mathcal{O}(\theta_t^5). \tag{16}$$

Convergence of θ_t to zero remains cubic. The κ_2/κ_1 factor in (16) does not significantly alter the algorithm's local convergence behavior. In fact, if $\kappa_2/\kappa_1 < 0$, the update's oscillatory behavior about $\theta = 0$ is identical to that in the RQI.

4 A Bound on the Average ICI for FastICA

The FastICA algorithm appears to converge quickly in many contexts. In [6], the average behavior of the inter-channel interference (ICI) for kurtosis-based FastICA on general m-dimensional mixtures was observed in simulations to be exponential with rate (1/3). Recent analytical work has verified this convergence property under a range of initial conditions on $\mathbf{w}_0$ [5,6]. The goal of this section is to use the simplicity of the two-source FastICA algorithm analysis to verify this property under general initial conditions for $\mathbf{w}_0$.

When $\kappa_1 = \kappa_2$, we can use (5) to write the evolutionary equation for FastICA in a remarkably simple form for the ICI at time t:

$$ICI_t = ICI_{t-1}^3 = (ICI_0)^{3^t}. \quad (17)$$

This scalar evolutionary equation for the ICI is cubically-convergent *globally* so long as the saddle point $ICI_0 = 1$ occurs with zero finite probability. Convergence depends on ICI_0, which for our analysis is assumed to have an unknown scalar p.d.f. $p_0(u)$ over the range $0 \leq u \leq 1$.

The average ICI, denoted as $E\{ICI_t\}$, is the ensemble average of the ICI values at iteration t that one would obtain by running FastICA on the same data set with different initial conditions as characterized by the p.d.f of ICI_0. Here, infinite data has been assumed. The following bound characterizes the value of $E\{ICI_t\}$ for weak assumptions on the p.d.f. of the initial ICI, the proof of which is in the Appendix.

Theorem 2. *Let ICI_0 be arbitrarily-distributed on $[0, ICI_{max}]$ with distribution $p_0(u)$, where $0 < ICI_{max} \leq 1$, subject to the additional condition that the probability density of ICI_0 has no point masses, or equivalently, the cumulative distribution function of ICI_0 is continuous over the interval $[0,1]$. Define $K = \max_{0 \leq u \leq ICI_{max}} u p_0(u)$. Then, an upper bound on the average ICI of the FastICA algorithm at iteration t for a linear mixture and infinite data in the two-source case is*

$$E\{ICI_t\} \leq \left(\frac{1}{3}\right)^t (ICI_{max})^{3^t} K. \quad (18)$$

Theorem 2 states that for reasonable distribution assumptions on the initial ICI, the average ICI at time t is bounded by a function consisting of the product of a linear-converging term and a cubically-converging term. Cubic convergence is what the deterministic analysis in [1] describes for kurtosis-based FastICA, and it is ultimately attained under stochastic initial conditions of the separation system vector if the initial distribution of the inter-channel interference is bounded away from unity. It may take a number of iterations, however, before this cubically-converging term dominates the expression. During the initial convergence period, the bound is linear with rate (1/3), as observed in simulations in [4]. Moreover, if the uncertainty about the mixing system prevents one from bounding ICI_0 away from unity – a likely situation – then the bound predicts *only linear convergence.*

Finite data records prevent one from attaining $\lim_{t\to\infty} E\{ICI_t\} = 0$ in practice. Experience show that linear convergence of FastICA with rate (1/3) is typically observed from the multiple-unit kurtosis-based FastICA algorithm applied to finite-length data sets. The performance "floor" due to finite measurements prevents one from observing the eventual cubic convergence of the FastICA procedure. The above bound indicates why linear convergence behavior is observed.

To verify the above behavior, the following simulations were carried out. The FastICA procedure was applied to 10000 different realizations of $N = 1000$ snapshots of mixtures of Unif-$[-\sqrt{3}, \sqrt{3}]$-distributed sources. The initial coefficient

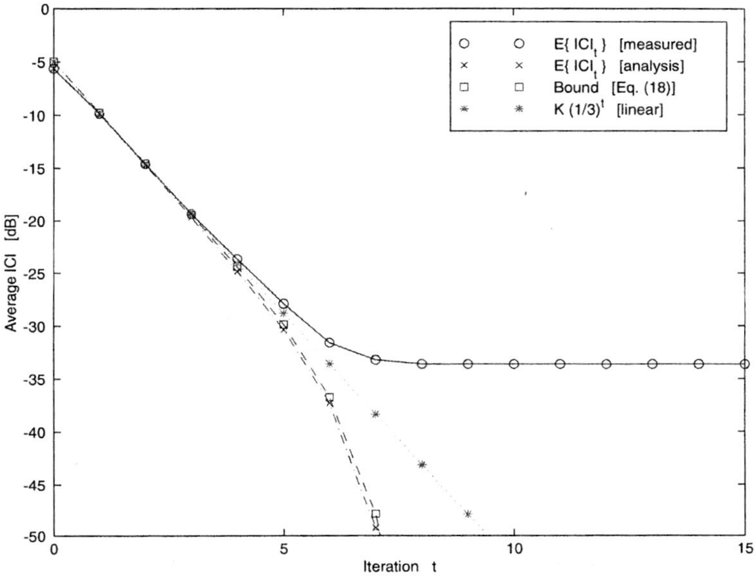

Fig. 2. Convergence of $E\{ICI_t\}$ from simulations and from predictions

vector $\mathbf{w}_0$ for each realization was randomly and uniformly-selected from an arbitrary point on the unit circle satisfying $ICI_0 < ICI_{max} = 0.999$. Ensemble averages were then used to compute $E\{ICI_t\}$ for comparison with the average behavior as predicted by the bound in (17), where $K \approx 1/\pi$ [5]. As a check, the random values of ICI_0 were used to compute an estimate $E\{\widehat{ICI}_t\}$ of $E\{ICI_t\}$ using ensemble averages of (17), which assumes infinite data ($N \rightarrow \infty$).

Figure 2 shows the evolutions of the various measures of inter-channel interference, in which the bound in (18) is seen to accurately predict both $E\{ICI_t\}$ and $E\{\widehat{ICI}_t\}$ for small t. Eqn. (18) closely follows the average behavior of (17) for larger t, but the measured $E\{ICI_t\}$ continues to converge linearly with rate $(1/3)$ until performance limits due to prewhitening and finite-sample effects are reached. In essence, the FastICA algorithm does not achieve cubic convergence on average despite having cubic convergence in a deterministic setting.

5 Conclusions

In this paper, we illustrate an important connection between the popular FastICA algorithm for independent component analysis and the Rayleigh Quotient Iteration in numerical linear algebra. We also derive a bound on the evolution of the average inter-channel interference for the FastICA algorithm for equal-kurtosis two-source mixtures which predicts linear convergence of the algorithm initially. Simulations show that the average ICI in FastICA typically converges linearly despite having cubic convergence in a deterministic setting.

References

1. A. Hyvärinen and E. Oja, "A fast fixed-point algorithm for independent component analysis," *Neural Computation,* vol. 9, no. 7, pp. 1483-1492, Oct. 1997.
2. A. Hyvarinen, J. Karhunen, and E. Oja, *Independent Component Analysis* (New York: Wiley, 2001).
3. E. Oja, "Convergence of the symmetrical FastICA algorithm," *Proc. 9th Int. Conf. Neural Inform. Processing,* Singapore, vol. 3, pp. 1368-1372, Nov. 2002.
4. S.C. Douglas, "On the convergence behavior of the FastICA algorithm," *Proc. Fourth Symp. Indep. Compon. Anal. Blind Signal Separation,* Kyoto, Japan, pp. 409-414, Apr. 2003.
5. S.C. Douglas, "A statistical convergence analysis of the FastICA algorithm for two-source mixtures," *Proc. 39th Asilomar Conf. Signals, Syst., Comput.*, Pacific Grove, CA, Oct. 2005.
6. S.C. Douglas, Z. Yuan, and E. Oja, "Average convergence behavior of the FastICA algorithm for blind source separation," to appear in *Proc. 6th Int. Conf. Indep. Compon. Anal. Blind Source Separation,* Charleston, SC, Mar. 2006.
7. Z. Koldovsky, P. Tichavsky, and E. Oja, "Cramer-Rao lower bound for linear independent component analysis," *Proc. IEEE Int. Conf. Acoust., Speech, Signal Processing,* Philadelphia, PA, vol. 3, pp. 581-584, Mar. 2005.
8. A. Edelman, T.A. Arias, and S.T. Smith, "The geometry of algorithms with orthogonality constraints," *SIAM J. Matrix Anal., Appl.* vol. 20, pp 303-353, Apr. 1999.
9. G.H. Golub and C.F. Van Loan, *Matrix Computations,* 3rd. ed. (Baltimore: Johns Hopkins, 1996).

Appendix

Proof of Theorem 1. From (5), we can write the update of the single-unit FastICA procedure applied to mixtures of two equal-kurtosis sources in the variable θ_t as

$$\theta_{t+1} = \arctan\left(\tan^3(\theta_t)\right). \tag{19}$$

Consider an expression for $\tan^3(\theta)$ of the form

$$\tan^3(\theta) = \tan(\alpha - \theta). \tag{20}$$

Applying the tan difference formula, we obtain the relationship

$$\tan^3(\theta) = \frac{\tan(\alpha) - \tan(\theta)}{1 + \tan(\alpha)\tan(\theta)}, \tag{21}$$

or $$\tan(\alpha)(1 - \tan^2(\theta)) = \tan(\theta). \tag{22}$$

Assume first that $\theta \neq \pi/4$, in which case

$$\tan(\alpha) = \frac{\tan(\theta)}{1 - \tan^2(\theta)} = \frac{1}{2}\tan(2\theta). \tag{23}$$

Now, considering the case that $\theta = \pi/4$ and the solution for $\tan(\alpha)$ in (23) , the left-hand-side of (22) can be evaluated using L'Hopital's Rule as

$$\lim_{\theta \to \pi/4} \frac{1}{2} \tan(2\theta)(1 - \tan^2(\theta)) = \lim_{\theta \to \pi/4} \frac{1}{2} \sin^2(2\theta)[\tan^3(\theta) + \tan(\theta)] = 1. \quad (24)$$

Thus, we have $\alpha = \arctan(0.5\tan(2\theta))$ for all θ, such that

$$\tan^3(\theta) = \tan\left(\arctan\left(\frac{1}{2}\tan(2\theta)\right) - \theta\right). \quad (25)$$

Setting $\theta = \theta_t$ and taking the arc-tangent of both sides of (25), the result follows.

Proof of Theorem 2. Let $p_0(u)$ denote the p.d.f. of ICI_0. Then, we have

$$E\{ICI_t\} = \int_0^{ICI_{max}} u^{3^t} p_0(u)du = \int_0^{ICI_{max}} u^{3^t - 1} \cdot u p_0(u)du. \quad (26)$$

Using the Holder inequality, we have

$$E\{ICI_t\} \le \left(\int_0^{ICI_{max}} u^{r(3^t-1)}\right)^{\frac{1}{r}} \left(\int_0^{ICI_{max}} u^s p_0^s(u)du\right)^{\frac{1}{s}} \quad (27)$$

$$\le \left(\frac{1}{r(3^t - 1) + 1}\right)^{\frac{1}{r}} (ICI_{max})^{3^t - 1 + \frac{1}{r}} \left(\int_0^{ICI_{max}} u^s p_0^s(u)du\right)^{\frac{1}{s}} \quad (28)$$

where $1/r + 1/s = 1$. Letting $s \to \infty$ and $r \to 1$, we have the inequality in (18).

Average Convergence Behavior of the FastICA Algorithm for Blind Source Separation

Scott C. Douglas[1], Zhijian Yuan[2], and Erkki Oja[2]

[1] Department of Electrical Engineering,
Southern Methodist University,
Dallas, Texas 75275, USA
[2] Neural Networks Research Centre,
Helsinki University of Technology,
Espoo 02015, Finland

Abstract. The FastICA algorithm is a popular procedure for independent component analysis and blind source separation. In this paper, we analyze the average convergence behavior of the single-unit FastICA algorithm with kurtosis contrast for general m-source noiseless mixtures. We prove that this algorithm causes the average inter-channel interference (ICI) to converge exponentially with a rate of (1/3) or -4.77dB at each iteration, independent of the source mixture kurtoses. Explicit expressions for the average ICI for the three- and four-source mixture cases are also derived, along with an exact expression for the average ICI in a particular situation. Simulations verify the accuracy of the analysis.

1 Introduction

The FastICA algorithm is a popular procedure for independent component analysis and blind source separation. The technique is simple to implement and converges quickly when applied to mixtures of independent non-Gaussian sources. The algorithm's convergence speed is locally-quadratic, and it is cubic when a kurtosis-based cost is employed [1, 2]. This cubic convergence behavior can be described using the analytical expressions for the evolution of the combined system coefficient vector $\mathbf{c}_t = [c_{1,t} \; \cdots \; c_{m,t}]^T$ for infinite data measurements, as given by

$$c_{i,t+1} = \frac{\kappa_i c_{i,t}^3}{\sqrt{\sum_{j=1}^{m} \kappa_j^2 c_{j,t}^6}}, \tag{1}$$

where κ_i is the ith source kurtosis. The vector $\mathbf{c}_t$ corresponds to the weight vector $\mathbf{w}_t$ of the single-unit FastICA algorithm in a transformed coordinate system where the independent components are explicitly included.

The FastICA algorithm's convergence behavior depends on the initial point of the algorithm, represented by $\mathbf{w}_0$ or $\mathbf{c}_0$. As this point is usually chosen fully at random in lack of any prior knowledge of the mixtures, an interesting question arises: What is the *average* convergence behavior of the algorithm across

J. Rosca et al. (Eds.): ICA 2006, LNCS 3889, pp. 790–798, 2006.

a distribution of initial points? Consider the performance metric known as the *inter-channel interference* (ICI) defined for the m-source case as

$$ICI_t^{(m)} = \frac{\sum_{i=1}^{m} c_{i,t}^2 - \max_{1\le i\le m} c_{i,t}^2}{\max_{1\le i\le m} c_{i,t}^2}. \tag{2}$$

Recently, an interesting observation about the FastICA algorithm with kurtosis contrast was made [3]: For a random initial $\mathbf{w}_0$ or $\mathbf{c}_0$, convergence of the average ICI appears to follow the "(1/3)rd Rule" given by

$$E\{ICI_t^{(m)}\} = \left(\frac{1}{3}\right) E\{ICI_{t-1}^{(m)}\}, \tag{3}$$

over almost the entire convergence period. Additional work has shown that this convergence behavior can be proven for the single-unit FastICA algorithm applied to simple two-source mixtures [4, 5], but it is not clear if such behavior extends to the general m-source mixture case.

In this paper, we prove that the FastICA algorithm with kurtosis contrast indeed obeys the "(1/3)rd Rule" for general m-source mixtures over a large portion of the convergence period. Our analysis employs a norm-constrained Gaussian prior for the initial separation system vector. Moreover, explicit expressions for the average ICI in the three- and four-source mixture cases are provided, and simulations are used to verify the analytical performance predictions.

2 Average Behavior of the FastICA Algorithm for a Three-Source Mixture

Before presenting the general m-source performance analysis, we introduce the analytical tools used in our derivations for a three-source separation task. The unconstrained (*e.g.* non-normalized) combined system vector at iteration t is

$$\mathbf{c}_t = [\kappa_1^{\frac{q}{2}} x^{\frac{p}{2}} \quad \kappa_2^{\frac{q}{2}} y^{\frac{p}{2}} \quad \kappa_3^{\frac{q}{2}} z^{\frac{p}{2}}]^T, \tag{4}$$

where $p = 2(3^t)$ and $q = 3^t - 1$. Employing this choice within (2) results in the general ICI expression derived in [1]. At time $t = 0$, we have $\mathbf{c}_0 = [x \quad y \quad z]^T$, where x, y, and z are random variables with some assumed probability density function (p.d.f.). A reasonable joint p.d.f. choice for $\{x, y, z\}$ would give a uniform prior for the direction of $\mathbf{c}_0$. We can induce such a p.d.f by letting x, y, and z be zero mean, uncorrelated, and jointly Gaussian. We can then express $ICI_t^{(3)}$ using ratios of powers of x, y, and z *without normalization*, and the resulting expectations can be evaluated without trigonometric functions.

The portion of the average ICI at iteration t in which the first kurtosis component is being extracted, such that the first element of (4) is the largest, is

$$E\{ICI_{1,t}^{(3)}\} = \frac{8}{(2\pi)^{3/2}} \int_0^{\infty} \frac{e^{-\frac{x^2}{2}}}{x^p} dx \left[\int_0^{ax} \left(\left(\frac{y}{a}\right)^p + \left(\frac{z}{b}\right)^p\right) e^{-\frac{y^2}{2}} dy\right] \int_0^{bx} e^{-\frac{z^2}{2}} dz. \tag{5}$$

$$\text{where} \qquad a = \left(\frac{\kappa_1}{\kappa_2}\right)^{\frac{q}{p}} \text{ and } b = \left(\frac{\kappa_1}{\kappa_3}\right)^{\frac{q}{p}}. \tag{6}$$

The integral in brackets on the right-hand side of (5) can be approximated as

$$\int_0^{ax} \left(\left(\frac{y}{a}\right)^p + \left(\frac{z}{b}\right)^p\right) e^{-\frac{y^2}{2}} dy \approx \left[e^{-\frac{a^2x^2}{2}} \int_0^{ax} \left(\frac{y}{a}\right)^p dy\right] + \left(\frac{z}{b}\right)^p \int_0^{ax} e^{-\frac{y^2}{2}} dy \tag{7}$$

$$= \left[ae^{-\frac{a^2x^2}{2}} \frac{x^{p+1}}{p+1}\right] + \left(\frac{z}{b}\right)^p \int_0^{ax} e^{-\frac{y^2}{2}} dy. \tag{8}$$

Substituting (8) into (5), we obtain

$$E\{ICI_{1,t}^{(3)}\}$$
$$= \frac{8}{(2\pi)^{3/2}(p+1)} \left[\int_0^{\infty} axe^{-\frac{x^2(1+a^2)}{2}} \int_0^{bx} e^{-\frac{z^2}{2}} dzdx + \int_0^{\infty} bxe^{-\frac{x^2(1+b^2)}{2}} \int_0^{ax} e^{-\frac{y^2}{2}} dydx\right]. \tag{9}$$

We can evaluate the integrals within brackets on the right-hand-side of (9) as

$$\int_0^{\infty} xe^{-\frac{x^2(1+a^2)}{2}} \left[\int_0^{bx} e^{-\frac{z^2}{2}} dz\right] dx = \frac{b}{1+a^2}\sqrt{\frac{\pi}{2(1+a^2+b^2)}} \tag{10}$$

$$\text{Therefore,} \qquad E\{ICI_{1,t}^{(3)}\} = \frac{2}{\pi(p+1)} \frac{1}{\sqrt{1+a^2+b^2}} \left[\frac{b}{a^{-1}+a} + \frac{a}{b^{-1}+b}\right]. \tag{11}$$

Now, as t increases, we have

$$\lim_{t\to\infty} \frac{q}{p} = \frac{1}{2}, \quad \lim_{t\to\infty} a = \sqrt{\frac{\kappa_1}{\kappa_2}}, \quad \lim_{t\to\infty} b = \sqrt{\frac{\kappa_1}{\kappa_3}}, \quad \text{and} \quad 2(3^t) \gg 1. \tag{12}$$

Substituting these results into (11), we obtain

$$E\{ICI_{1,t}^{(3)}\} = \frac{1}{\pi}\left(\frac{1}{3}\right)^t \frac{1}{\sqrt{\kappa_1\kappa_2+\kappa_1\kappa_3+\kappa_2\kappa_3}} \left[\frac{\kappa_1\kappa_2}{\kappa_1+\kappa_2} + \frac{\kappa_1\kappa_3}{\kappa_1+\kappa_3}\right]. \tag{13}$$

Invoking symmetry for the terms $E\{ICI_{2,t}^{(3)}\}$ and $E\{ICI_{3,t}^{(3)}\}$, we find an approximate expression for the average ICI to be

$$E\{ICI_t^{(3)}\} = \sum_{n=1}^{3} E\{ICI_{n,t}^{(3)}\} = g_3(\kappa_1, \kappa_2, \kappa_3)\left(\frac{1}{3}\right)^t, \tag{14}$$

$$g_3(\kappa_1,\kappa_2,\kappa_3) = \frac{2}{\pi} \frac{1}{\sqrt{\kappa_1\kappa_2+\kappa_1\kappa_3+\kappa_2\kappa_3}} \left[\frac{\kappa_1\kappa_2}{\kappa_1+\kappa_2} + \frac{\kappa_1\kappa_3}{\kappa_1+\kappa_3} + \frac{\kappa_2\kappa_3}{\kappa_2+\kappa_3}\right] \tag{15}$$

Eqn. (14) states that the average ICI for arbitrary three-source mixtures asymptotically obeys the "(1/3)rd Rule" in (3). Numerical evaluations of this expression show that it is extremely accurate in predicting the average ICI during the algorithm's convergence period. Moreover, across all source kurtosis combinations, the maximum value of $E\{ICI_t^{(3)}\}$ occurs when $\kappa_1 = \kappa_2 = \kappa_3$, for which

$$E\{ICI_t^{(3)}\} = \frac{\sqrt{3}}{\pi}\left(\frac{1}{3}\right)^t. \tag{16}$$

3 Average Behavior of the FastICA Algorithm for General m-Source Mixtures

Eqn. (14) provides evidence that the kurtosis-based FastICA algorithm has exponential convergence with a rate that is independent of the source distributions. Can this result be extended to m-source mixtures? And how accurate is the approximation used in (8)? The following theorem addresses these issues, the proof of which is outlined in the Appendix.

Theorem 1. *Assume that a single-unit FastICA algorithm with kurtosis contrast is applied to an m-source noiseless mixture with infinite data, and that the initial combined system coefficient vector $\mathbf{c}_0$ is uniformly-distributed on the m-dimensional unit hypersphere. Then, the average ICI at iteration t is*

$$E\{ICI_t^{(m)}\} = g_m(\kappa_1, \cdots, \kappa_m)\left(\frac{1}{3}\right)^t + R(t, \kappa_1, \cdots, \kappa_m), \tag{17}$$

where the m-dimensional function $g_m(\cdot)$ does not depend on t and $R(t, \cdot)$ decreases to zero faster than $(1/3)^t$ as $t \to \infty$.

The above theorem indicates that the "(1/3)rd Rule" holds in general for the single-unit FastICA procedure with kurtosis contrast. The convergence rate of the algorithm does not depend on the source kurtoses, which only affect the overall magnitude of the average ICI during the convergence period. This result explains why the FastICA algorithm can be called "fast" – the average convergence speed for the ICI is linear with a constant rate in all source scenarios.

The methodology used to derive the above theorem can in theory be used to find an asymptotic expression for the average ICI in (17) by determining an explicit expression for $g_m(\kappa_1, \cdots, \kappa_m)$ for any m. For $m = 2$, see [4], or set $\kappa_3 = 0$ in (14). For $m = 4$, one can show that

$$\begin{aligned} g_4(\kappa_1, \kappa_2, \kappa_3, \kappa_4) = \frac{4}{\pi^2}\,(&h_{1234} + h_{1243} + h_{1324} + h_{1342} + h_{1423} + h_{1432} \\ &+ h_{2314} + h_{2341} + h_{2413} + h_{2431} + h_{3412} + h_{3421}) \end{aligned} \tag{18}$$

$$h_{ijkl} = \frac{\sqrt{\kappa_i \kappa_j}}{\sqrt{\kappa_i} + \sqrt{\kappa_j}}\left[\frac{\sqrt{\kappa_k^{-1}}}{\sqrt{\kappa_i^{-1} + \kappa_j^{-1} + \kappa_k^{-1}}} \arctan\left(\frac{\sqrt{\kappa_l^{-1}}}{\sqrt{\kappa_i^{-1} + \kappa_j^{-1} + \kappa_k^{-1}}}\right)\right], \tag{19}$$

which reduces to (14) when $\kappa_4 = 0$. When $\kappa_1 = \kappa_2 = \kappa_3 = \kappa_4$, we have

$$E\{ICI_t^{(4)}\} = \frac{4}{\pi\sqrt{3}}\left(\frac{1}{3}\right)^t, \tag{20}$$

which is 4/3 times larger than the maximum ICI in the three-source case with $\kappa_1 = \kappa_2 = \kappa_3$ and $4/\sqrt{3} = 2.31$ times larger than the maximum ICI in the two source case with $\kappa_1 = \kappa_2$. For $m > 4$, the integrals become difficult to evaluate.

4 An Exact Expression for the Average ICI for m-Source Mixtures in a Particular Situation

Given our reliance on a uniform distribution for the direction of $\mathbf{c}_0$, one might wonder whether the "(1/3)rd Rule" for FastICA requires this assumption. The following analysis suggests that this behavior likely holds in other contexts.

Suppose the elements of $\mathbf{c}_0 = [c_{1,0} \ \cdots \ c_{m,0}]^T$ are uniformly-distributed on the interval $[0, 1]$. Of course, $\mathbf{c}_0$ is normally of unit length, but as scaling doesn't matter, we choose a scaled version of $\mathbf{c}_0$ instead. Assuming $c_{i,t} \geq 0$ does not change the value of $ICI_t^{(m)}$, either. When projected onto the unit hypersphere, this distribution tends to concentrate probability in the $[\pm 1 \ \pm 1 \ \cdots \ \pm 1]^T$ directions of m-dimensional space, making convergence somewhat more challenging for the algorithm. Moreover, we shall assume that $\kappa_i = \kappa_j$ for all i and j. Under this situation, the value of $E\{ICI_t^{(m)}\}$ is easy to compute.

Theorem 2. *For the situation above, the average ICI at iteration t is exactly*

$$E\{ICI_t^{(m)}\} = \frac{m-1}{2(3^t)+1}. \tag{21}$$

Proof: The proof relies on the facts that (a) ordering of the coefficients within the update relations does not matter in the convergence analysis, and (b) the order

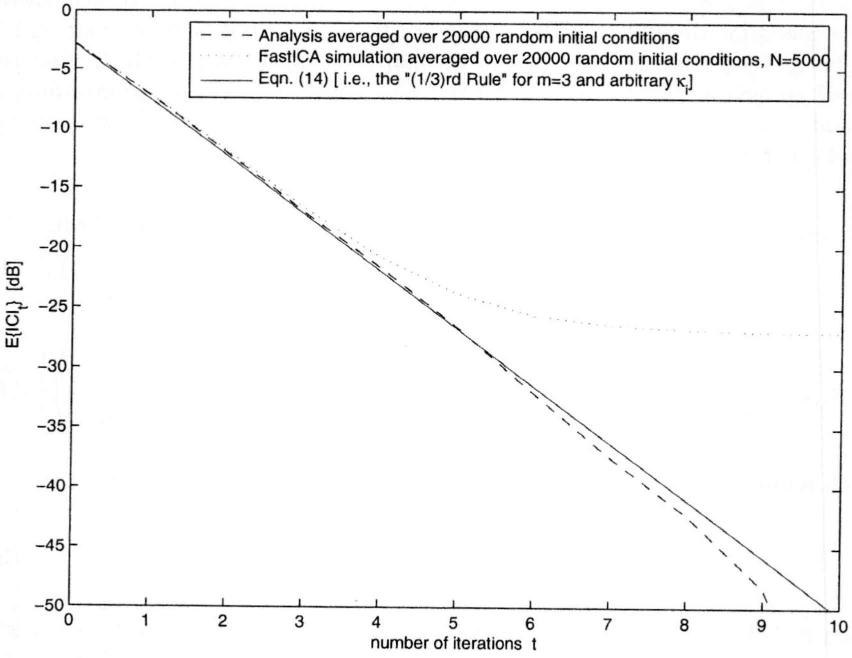

Fig. 1. Evolutions of the average ICI as determined by various methods, $m = 3$

statistics of i.i.d. Unif-[0, 1]-distributed random variables are jointly uniform over the integration volume [6]. Hence, the average ICI is given by

$$E\{ICI_t^{(m)}\} = m! \int_0^1 \int_0^{c_1} \int_0^{c_2} \cdots \int_0^{c_{m-1}} \frac{\sum_{j=2}^{m} c_j^{2(3^t)}}{c_1^{2(3^t)}} dc_m \, dc_{m-1} \ldots dc_1 \quad (22)$$

which easily integrates to (21). Simulations corroborate this exact result.

Theorem 2 is not meant as a replacement for the more-general result in Theorem 1. Rather, it shows that the average exponential convergence behavior of the kurtosis-based FastICA algorithm holds for at least one other distribution of $\mathbf{c}_0$ than a uniform angular distribution. It has been our experience that (3) predicts the average behavior of the original single-unit FastICA algorithm with kurtosis contrast quite well, and to date, all theoretical results concerning the convergence performance of this algorithm reflect (3) in one form or another.

5 Simulations

To verify our theoretical results, simulations in MATLAB were carried out. Three and four-source mixtures have been generated, in which the sources are zero-mean unit-variance binary ($|\kappa| = 2$), uniform ($|\kappa| = 6/5$), and/or Laplacian

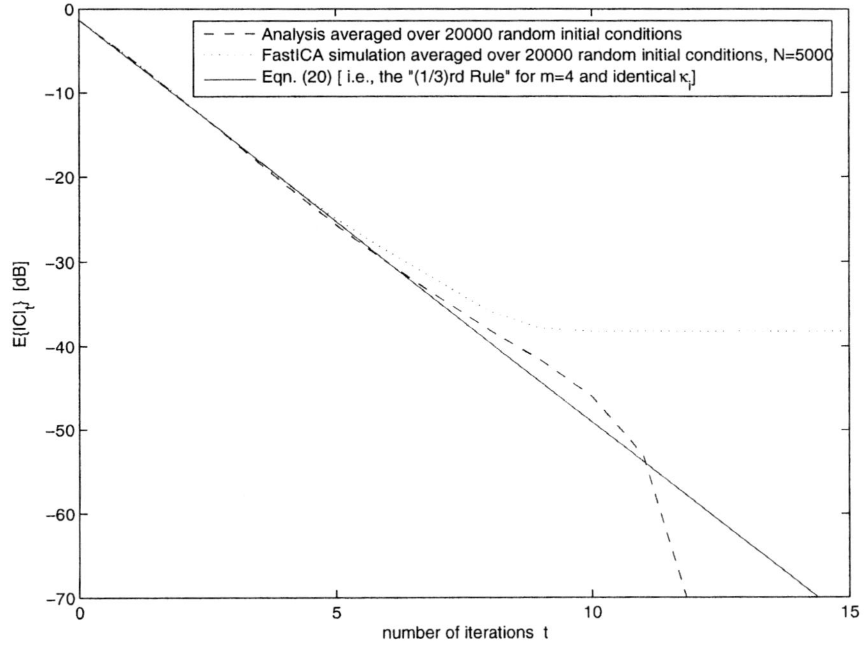

Fig. 2. Evolutions of the average ICI as determined by various methods, $m = 4$

($|\kappa| = 3$) distributed. MATLAB's `randn` function was used to generate $M = 20000$ different initial coefficient vectors $\mathbf{c}_0$ for the numerical simulations. $N = 5000$ snapshots were used to evaluate the FastICA algorithm on sampled data.

Figure 1 shows plots of the averaged value of $ICI_t^{(3)}$ as predicted by simulations of the analytical convergence expressions of the FastICA algorithm, as determined by the FastICA algorithm on sampled data, and calculated from (14) for a three-source separation task with binary, uniform, and Laplacian sources. All of the curves agree quite well up to iteration $k = 5$. For $k > 5$, our simulation method for estimating $E\{ICI_t^{(3)}\}$ using non-uniform sampling of the unit sphere via the `randn` function is not accurate enough to verify (14). The scaling factor $g_3(2, 6/5, 3) = 131\sqrt{3}/(140\pi)$ is correct for the "(1/3)rd Rule" in this case.

Figure 2 shows plots of the averaged value of $ICI_t^{(4)}$ as predicted by analytical simulations, actual performance, and the prediction in (20) of FastICA convergence behavior for $m = 4$ binary sources. The scaling factor of $4/(\sqrt{3}\pi)$ is correct for the "(1/3)rd Rule" in the four-equal-kurtosis-source case.

6 Conclusions

In this paper, we analyze the average convergence behavior of the single-unit FastICA algorithm with kurtosis contrast on m-source mixtures, showing that its behavior is exponential with rate (1/3). Accurate expressions for $m = 3$ and $m = 4$-source mixtures are provided, and simulations verify the analyses.

References

1. A. Hyvärinen and E. Oja, "A fast fixed-point algorithm for independent component analysis," *Neural Computation,* vol. 9, no. 7, pp. 1483-1492, Oct. 1997.
2. A. Hyvarinen, J. Karhunen, and E. Oja, *Independent Component Analysis* (New York: Wiley, 2001).
3. S.C. Douglas, "On the convergence behavior of the FastICA algorithm," *Proc. Fourth Symp. Indep. Compon. Anal. Blind Signal Separation,* Kyoto, Japan, pp. 409-414, Apr. 2003.
4. S.C. Douglas, "A statistical convergence analysis of the FastICA algorithm for two-source mixtures," *Proc. 39th Asilomar Conf. Signals, Syst., Comput.*, Pacific Grove, CA, Oct. 2005.
5. S.C. Douglas, "Relationships between the FastICA algorithm and the Rayleigh Quotient Iteration," to appear in *Proc. 6th Int. Conf. Indep. Compon. Anal. Blind Source Separation,* Charleston, SC, Mar. 2006.
6. H.A. David, *Order Statistics,* 2nd ed. (New York: Wiley, 1980).

Appendix: Proof of Theorem 1

Consider a single-unit m-source FastICA algorithm with cubic nonlinearity. The combined system coefficient vector at iteration t is $\mathbf{c}_t = [\kappa_1^{\frac{q}{2}} x_1^{\frac{p}{2}} \; \cdots \; \kappa_m^{\frac{q}{2}} x_m^{\frac{p}{2}}]^T$, where $p = 2(3^t)$ and $q = 3^t - 1$. Consider the portion of the ICI at iteration t in which the first kurtosis component is being extracted, as given by

$$E\{ICI_{1,t}^{(m)}\} = \frac{2^m}{(2\pi)^{m/2}} \sum_{i=2}^{m} \int_0^\infty dx_1 \int_0^{a_2 x_1} dx_2 \cdots \int_0^{a_m x_1} dx_m \frac{(\frac{x_i}{a_i})^p}{x_1^p} \exp(-\frac{\sum_{i=1}^m x_i^2}{2}) \tag{23}$$

where

$$a_i = \left(\frac{\kappa_1}{\kappa_i}\right)^{\frac{q}{p}}, \quad i \in \{2, 3, \ldots, m\}. \tag{24}$$

Using the transformation $\widetilde{x}_i = x_i/x_1$ for $i \in \{2, \ldots, m\}$, (23) becomes

$$E\{ICI_{1,t}^{(m)}\} = \frac{2^m}{(2\pi)^{m/2}} \sum_{i=2}^{m} \int_0^{a_2} d\widetilde{x}_2 \cdots \int_0^{a_m} d\widetilde{x}_m \int_0^\infty \left(\frac{\widetilde{x}_i}{a_i}\right)^p x_1^{m-1} \exp\left(-\frac{x_1^2}{2}(1 + \sum_{i=2}^m \widetilde{x}_i^2)\right) dx_1. \tag{25}$$

The most inside integral in (25) can be calculated as

$$\int_0^\infty x_1^{m-1} \exp\left(-\frac{x_1^2}{2}(1 + \sum_{i=2}^m x_i^2)\right) dx_1 = \begin{cases} \frac{(m-2)!!\sqrt{\pi}}{\sqrt{2}(1+\sum_{i=2}^m x_i^2)^{m/2}} & \text{m is odd} \\ \frac{[\frac{1}{2}(m-2)]!2^{(m-2)/2}}{(1+\sum_{i=2}^m x_i^2)^{m/2}} & \text{m is even} \end{cases}$$

where $(m-2)!! = 1 \cdot 3 \cdot 5 \cdot 7 \cdots (m-2)$. When $k \to \infty$, $(\frac{x_i}{a_i})^p \to 0$ over the interval $0 \le x_i < a_i$, and $a_i \to \sqrt{\frac{\kappa_1}{\kappa_i}} \equiv b_{1i}$. We can then approximate the integral

$$\int_0^{a_i} \left(\frac{x_i}{a_i}\right)^p \frac{1}{(1 + \sum_{j=2}^m x_j^2)^{m/2}} dx_i \approx \frac{1}{(1+b_{1i}^2+\sum_{j=2,j\neq i}^m x_j^2)^{m/2}} \frac{b_{1i}}{p}. \tag{26}$$

Noting that $p = 2(3^t)$, we have

$$E\{ICI_{1,t}^{(m)}\} \approx \begin{cases} \frac{1}{2(3)^t} \frac{2^m}{(2\pi)^{m/2}} \sum_{i=2}^m \int_0^{b_{12}} dx_2 \cdots \int_0^{b_{1m}} dx_m \frac{(m-2)!!\sqrt{\pi}}{\sqrt{2}(1+b_{1i}^2+\sum_{j=2,j\neq i}^m x_j^2)^{m/2}} & m \text{ is odd} \\ \frac{1}{2(3)^t} \frac{2^m}{(2\pi)^{m/2}} \sum_{i=2}^m \int_0^{b_{12}} dx_2 \cdots \int_0^{b_{1m}} dx_m \frac{[\frac{1}{2}(m-2)]!2^{(m-2)/2}}{(1+b_{1i}^2+\sum_{j=2,j\neq i}^m x_j^2)^{m/2}} & m \text{ is even} \end{cases}$$

Similarly, by invoking symmetry, we get

$$E\{ICI_{n,t}^{(m)}\} \approx \begin{cases} \frac{1}{2(3)^t} \frac{2^m}{(2\pi)^{m/2}} \sum_{i=1,i\neq n}^m \int_0^{b_{n1}} dx_1 \cdots \int_0^{b_{nm}} dx_m \frac{(m-2)!!\sqrt{\pi}}{\sqrt{2}(1+(b_{ni})^2+\sum_{j=1,j\neq n,j\neq i}^m x_j^2)^{m/2}} \\ \qquad m \text{ is odd} \\ \frac{1}{2(3)^t} \frac{2^m}{(2\pi)^{m/2}} \sum_{i=1,i\neq n}^m \int_0^{b_{n1}} dx_1 \cdots \int_0^{b_{nm}} dx_m \frac{[\frac{1}{2}(m-2)]!2^{(m-2)/2}}{(1+(b_{ni})^2+\sum_{j=1,j\neq n,j\neq i}^m x_j^2)^{m/2}} \\ \qquad m \text{ is even} \end{cases}$$

for $n \in \{2, \ldots, m\}$ and

$$b_{ni} = \sqrt{\frac{\kappa_n}{\kappa_i}}. \tag{27}$$

Finally, the average ICI is $E\{ICI_t^{(m)}\} = \sum_{n=1}^m E\{ICI_{n,t}^{(m)}\}$, which results in (17) with

$$g(\kappa_1, \cdots, \kappa_m) = \begin{cases} \frac{2^{m-1}}{(2\pi)^{m/2}} \sum_{n=1}^m \sum_{i=1, i\neq n}^m \int_0^{b_{n1}} dx_1 \cdots \int_0^{b_{nm}} dx_m \frac{(m-2)!!\sqrt{\pi}}{\sqrt{2}(1+(b_{ni})^2 + \sum_{j=1, j\neq n, j\neq i}^m x_j^2)^{m/2}} & m \text{ is odd} \\ \frac{2^{m-1}}{(2\pi)^{m/2}} \sum_{n=1}^m \sum_{i=1, i\neq n}^m \int_0^{b_{n1}} dx_1 \cdots \int_0^{b_{nm}} dx_m \frac{[\frac{1}{2}(m-2)]! 2^{(m-2)/2}}{(1+(b_{ni})^2 + \sum_{j=1, j\neq n, j\neq i}^m x_j^2)^{m/2}} & m \text{ is even} \end{cases}$$

Additional calculations show that the error introduced in (17) by (26) is

$$|R(t, \kappa_1, \cdots, \kappa_m)| \leq c(m) \left(\frac{1}{3}\right)^t \sum_{n=1}^m \left[(m-1) \prod_{i=2}^m b_{ni} \left(1 - \prod_{i=2}^m b_{ni}^{-\frac{2}{p}}\right) \right.$$
$$\left. + \sum_{i=2}^m \left((1 - \frac{1}{\sqrt{p}})^{p+1} b_{ni}^{1-\frac{2}{p}} + \frac{b_{ni}}{p} |1 - \frac{p}{p+1} b_{ni}^{-\frac{4}{p}}| \right) \prod_{i=2}^m b_{ni}^{1-\frac{2}{p}} \right] \tag{28}$$

Since $1 - \prod_{i=2}^m b_{ni}^{-\frac{2}{p}} \to 0$ and $(1 - \frac{1}{\sqrt{p}})^{p+1} \to 0$ as $t \to \infty$, it is easy to see that

$$\lim_{t\to\infty} \frac{R(t, \kappa_1, \cdots, \kappa_m)}{\left(\frac{1}{3}\right)^t} = 0. \tag{29}$$

Multivariate Scale Mixture of Gaussians Modeling

Torbjørn Eltoft[1], Taesu Kim[2], and Te-Won Lee[2]

[1] Department of Physics, University of Tromsø & Norut IT, Tromsø, Norway
torbjorn.eltoft@phys.uit.no
[2] Institute of Neural Computation, UCSD, CA
taesu@ucsd.edu, tewon@ucsd.edu

Abstract. In this paper, we present an approach to generate a class of multivariate probability models, which are referred to as scale mixture of Gaussians models. They are constructed as normal variance mixture models, in which the covariance matrix involves a stochastic scale factor with a given prior distribution. We limit the presentation here to the multivariate K (MK) model, which results if we apply a Γ distribution for the scale factor. We then discuss how the parameter of the model can be estimated in an iterative procedure, and include a 2-D case study, where we compare the ability of the MK model to represent real data to corresponding abilities of the multivariate Laplace and the multivariate NIG models.

1 Introduction

In many real world data sets involving multivariate observations, the data have an empirical distribution which is highly peaked at zero (or the mean vector), and which asymptotically falls off more slowly than the Gaussian distribution as the distance from zero increases. We denote these distributions *sparse distributions.* Sparse distributions are appropriate for representing the statistics of speech and image data, especially when observed in a transform domain like the wavelet or DFT domain [1]. For example, the overcomplete wavelet transform coefficients of images are found to have sparse distributions, a property that has been extensively exploited in coding and denoising [2, 3]. Several authors have studied the statistics of DFT-transformed speech signals, and found the coefficients to have sparse, heavy-tailed distributions [4, 5]. Sparse distributions are also frequently encountered in various machine learning areas, like e.g. blind source separation and independent component analysis (ICA) [6].

Multivariate observations, which are mutually correlated and have higher-order dependencies, have frequently been represented using mixture of Gaussians models. These are convenient in many respects, they have a closed form probability density function (pdf), and the parameters can easily be obtained using the EM algorithm. Recently, yet an other class of mixture models, the so-called scale mixture of Gaussians models, have emerged as a powerful set of distributions for modeling statistical dependencies in multivariate data [1]. The

J. Rosca et al. (Eds.): ICA 2006, LNCS 3889, pp. 799–806, 2006.

multivariate Normal Inverse Gaussian distribution (NIG) [7], is an example of this kind of models. The 1-D NIG pdf has been applied to model economical time series, and it has also been successfully applied in some engineering problems [8]. The model parameters of this model is also estimated using an EM- type of algorithm.

In this paper, we present a general approach to formulate multivariate distributions with probability density functions given in closed form. We use the multivariate K model, which is a multivariate scale mixture of Gaussians model with a Γ distributed scale factor, as a case study. Note that the multivariate Laplace (ML) [9] and the multivariate NIG (MNIG) [10] are other examples, created using respectively the exponential and the Inverse Gaussian distributions as priors for the scale factor. We show how we can estimate the model parameters from data, and test the ability of the MK, the ML and the MNIG to represent the DFT-coefficients of a speech signal in a log-likelihood cross-validation test.

2 A General Scheme for Generating Scale Mixture of Gaussians Models

In [11] it was shown that if the probability density function (pdf) of some random variable Y, $p_Y(y)$, is symmetric about zero, and the derivatives of $p_Y(y)$ satisfy

$$\left(-\frac{d}{dy}\right)^k p_Y(y) \geq 0 \quad \text{for } y > 0, \tag{1}$$

then there exist independent variables X and Z, with X being a standard normal variable, such that

$$Y = \sqrt{Z}X. \tag{2}$$

The variable Z is allowed to take on only positive values. A random variable Y, which can be expressed as in (2), is referred to as a *normal variance mixture model*, or a *scale mixture of Gaussians*. Of course, if the mean of Y should be non-zero, (2) may be modified by adding a scalar μ corresponding to the actual mean value. Now, let $p_Z(z)$ be the probability density function (pdf) of Z. Then, the marginal pdf of Y is obtained by averaging over Z, as in (3) below:

$$p_Y(y) = \int_0^\infty \frac{1}{\sqrt{2\pi z}} \exp(-\frac{(y-\mu)^2}{2z})\, p_Z(z)\, dz. \tag{3}$$

The multidimensional extension of the generative model described above, is straight forward. Let $\boldsymbol{X}$ be a d-dimensional, zero mean Gaussian variable with covariance matrix equal to the identity matrix. Let furthermore, $\boldsymbol{\Gamma} \in \mathcal{R}^{d\times d}$ be a positive definite matrix with determinant $\det \boldsymbol{\Gamma} = 1$, and let Z be a scalar random variable with pdf $p_Z(z)$, which can attain only positive values. We now generate a new variable $\boldsymbol{Y}$ as a *multivariate scale mixture of Gaussians* according to

$$\boldsymbol{Y} = \boldsymbol{\mu} + \sqrt{Z}\boldsymbol{\Gamma}^{\frac{1}{2}}\boldsymbol{X}, \tag{4}$$

where $\boldsymbol{\mu}$ is the mean vector. The matrix $\boldsymbol{\Gamma}$ defines the internal covariance structure of the variables of $\boldsymbol{Y}$. For this reason we will refer to this matrix as the covariance structure matrix. To obtain the marginal pdf of $\boldsymbol{Y}$, we have to perform an integration similar to the one in (3) over the prior distribution $p_Z(z)$. The integral which must be computed is accordingly given as

$$p_{\boldsymbol{Y}}(\boldsymbol{y}) = \int_0^\infty p_{\boldsymbol{Y}|Z}(\boldsymbol{y}|Z=z) p_Z(z)\,dz$$
$$= \int_0^\infty \frac{1}{(2\pi\, z)^{\frac{d}{2}}} \exp[-\frac{1}{2z}(\boldsymbol{y}-\boldsymbol{\mu})^t \boldsymbol{\Gamma}^{-1}(\boldsymbol{y}-\boldsymbol{\mu})] p_Z(z)\,dz. \quad (5)$$

In the following, we define

$$q(\boldsymbol{y}) = (\boldsymbol{y}-\boldsymbol{\mu})^t \boldsymbol{\Gamma}^{-1}(\boldsymbol{y}-\boldsymbol{\mu}), \quad (6)$$

and show that choosing the Γ distribution for $p_Z(z)$ leads to the multivariate K model, which is a sparse multivariate models with pdfs given in closed form.

2.1 The Multivariate K Distribution

The K distribution, which was first introduced in the middle of the 1970's, is a model which has been extensively applied to model the envelope statistics of various signals, e.g. ultrasound echoes [12] and microwave radar backscatter [13, 14]. This pdf model can be derived from a 2-D version of the model in (4), with $\boldsymbol{\mu} = \boldsymbol{0}$, and a Γ distributed variance Z. Using the Γ distribution for Z, i.e.

$$p_Z(z;\alpha,\lambda) = \frac{\lambda^{\alpha+1}\, z^\alpha}{\Gamma(\alpha+1)} \exp(-\lambda z), \quad (7)$$

one finds that the general d-dimensional pdf for $\boldsymbol{y}$ is given as

$$p_{\boldsymbol{Y}}(\boldsymbol{y}) = \frac{2}{(2\pi)^{\frac{d}{2}}} \frac{\lambda^{\alpha+1}}{\Gamma(\alpha+1)} \left(\sqrt{\frac{q(\boldsymbol{y})}{2\lambda}}\right)^{\alpha+1-\frac{d}{2}} K_{\alpha+1-\frac{d}{2}}(\sqrt{2\lambda q(\boldsymbol{y})}), \quad (8)$$

where $K_m(x)$ denotes the modified Bessel function of the second kind and order m, evaluated at x. We note that this multivariate scale mixture of Gaussians model is given in terms of the parameter set $\{\alpha, \lambda, \boldsymbol{\mu}, \boldsymbol{\Gamma}\}$. α is a scalar shape parameter, and λ is a scalar scale parameter. $\boldsymbol{\mu}$ is a d-dimensional location vector, and $\boldsymbol{\Gamma}$ is a positive definite matrix, determining the covariance structure of the model. We will use the notation $\boldsymbol{Y} \sim \mathrm{MK}\{\alpha, \lambda, \boldsymbol{\mu}, \boldsymbol{\Gamma}\}$ to denote that $\boldsymbol{Y}$ is multidimensional K distributed with parameters α, λ, $\boldsymbol{\mu}$, and $\boldsymbol{\Gamma}$.

The moment generating function of the multivariate K distribution is given by

$$M_Y(\boldsymbol{\omega}) = E\{e^{\boldsymbol{\omega}^t \boldsymbol{Y}}\} = \frac{\lambda^{\alpha+1}\, e^{\boldsymbol{\mu}^t \boldsymbol{\omega}}}{(\lambda - \frac{\boldsymbol{\omega}^t \boldsymbol{\Gamma} \boldsymbol{\omega}}{2})^{\alpha+1}}. \quad (9)$$

It is also easy to find that

$$\mathcal{E}\{\boldsymbol{Y}\} = \mu_{\boldsymbol{Y}} = \boldsymbol{\mu} \tag{10}$$

$$\mathcal{E}\{(\boldsymbol{Y}-\boldsymbol{\mu})(\boldsymbol{Y}-\boldsymbol{\mu})^t\} = \boldsymbol{\Sigma}_{\boldsymbol{Y}} = \frac{\alpha+1}{\lambda}\boldsymbol{\Gamma}. \tag{11}$$

Now, let $\boldsymbol{V} = \boldsymbol{A}\boldsymbol{Y} + \boldsymbol{b}$ be an arbitrary linear transformation of a multivariate K variable $\boldsymbol{Y}$ with parameters $\{\alpha, \lambda, \boldsymbol{\mu}, \boldsymbol{\Gamma}\}$, where $\boldsymbol{A}$ is a $d \times d$ real valued matrix. The transformed variable $\boldsymbol{V}$ can then be shown to be another multivariate K distributed random vector, with parameters $\{\tilde{\alpha}, \tilde{\lambda}, \tilde{\boldsymbol{\mu}}, \tilde{\boldsymbol{\Gamma}}\}$, where

$$\tilde{\alpha} = \alpha \tag{12}$$

$$\tilde{\lambda} = \lambda\,|\mathrm{det}\boldsymbol{A}|^{\frac{1}{d}} \tag{13}$$

$$\tilde{\boldsymbol{\mu}} = \boldsymbol{A}\boldsymbol{\mu} + \boldsymbol{b} \tag{14}$$

$$\tilde{\boldsymbol{\Gamma}} = \boldsymbol{A}\boldsymbol{\Gamma}\boldsymbol{A}^t|\mathrm{det}\boldsymbol{A}|^{-\frac{2}{d}}. \tag{15}$$

2.2 Some Properties of the Multivariate K Model

We observe that for the model presented above $p_{\boldsymbol{Y}}(\boldsymbol{y})$ is dependent on $\boldsymbol{y}$ through $q(\boldsymbol{y})$. If $\boldsymbol{\Gamma}$ is diagonal, two components Y_i and Y_j of $\boldsymbol{Y}$ will be uncorrelated, but they are not statistically independent. However, the joint distribution conditioned on Z will factorize, hence $Y_i|Z$ and $Y_j|Z$ are independent.

Let us for the moment assume that $\boldsymbol{\mu} = \boldsymbol{0}$, and let $||\boldsymbol{y}||_{\boldsymbol{\Gamma}^{-1}} = \sqrt{\boldsymbol{y}^t\boldsymbol{\Gamma}^{-1}\boldsymbol{y}}$ denote the weighted $\boldsymbol{\Gamma}^{-1}$-norm of $\boldsymbol{y}$. Noting that the Bessel function behaves as

$$K_d(x) \sim \sqrt{\frac{\pi}{2x}}\exp(-x), \quad \text{when } |x| \to \infty,$$

we find that

$$p_{\boldsymbol{Y}}(\boldsymbol{y}) \sim F\ \exp(-\sqrt{\frac{2}{\lambda}}||\boldsymbol{y}||_{\boldsymbol{\Gamma}^{-1}}) \text{ for large } ||\boldsymbol{y}||_{\boldsymbol{\Gamma}^{-1}}, \tag{16}$$

where F is an algebraic expression. Hence, the model presented above have an asymptotic behavior, which is a combination of an algebraic and an exponential term. It is more heavy-tailed than the Normal distribution.

3 Parameter Estimation

The model we have used to generate the multivariate K distribution involves the latent variable Z, which means that the parameters of the pdfs may be estimated using an iterative procedure. At the outset we note that the corresponding a posteriori probability density function, $p_{Z|\boldsymbol{Y}}(z|\boldsymbol{y})$, is a so-called Generalized Inverse Gaussian (GIG) distribution. This model has a pdf given as

$$p_Z(z;\theta,\delta,\gamma) = (\frac{\gamma}{\delta})^\theta \frac{1}{2K_\theta(\delta\gamma)} z^{\theta-1}\ \exp(-\frac{1}{2}(\frac{\delta^2}{z} + \gamma^2 z)), \tag{17}$$

and its kth-order moments are

$$\mu_Z^{(k)} = E\{Z^k\} = \left(\frac{\delta}{\gamma}\right)^k \frac{K_{\theta+k}(\delta\gamma)}{K_\theta(\delta\gamma)}. \tag{18}$$

Equation (10) suggests that $\boldsymbol{\mu}$ can be estimated as the first order moment of the sample set $\mathcal{Y} = \{\boldsymbol{y}_1, \boldsymbol{y}_2, \cdots, \boldsymbol{y}_N\}$, i.e., $\hat{\boldsymbol{\mu}} = \frac{1}{N}\sum_{i=1}^N \boldsymbol{y}_i$, whereas according to (11), $\boldsymbol{\Gamma}$ can be estimated as the sample covariance matrix divided by an estimate of $\eta = \frac{\alpha+1}{\lambda}$. Let

$$\hat{\boldsymbol{R}} = \frac{1}{N}\sum_{i=1}^N (\boldsymbol{y}_i - \hat{\boldsymbol{\mu}})(\boldsymbol{y}_i - \hat{\boldsymbol{\mu}})^t \tag{19}$$

denote the sample covariance matrix. Using the fact that $\det \boldsymbol{\Gamma} = 1$, an estimate of $\eta = \frac{\alpha+1}{\lambda}$ would be given as

$$\hat{\eta} = \det \hat{\boldsymbol{R}}^{\frac{1}{d}} \tag{20}$$

and accordingly, an estimate of $\boldsymbol{\Gamma}$ is

$$\hat{\boldsymbol{\Gamma}} = \frac{1}{\hat{\eta}}\hat{\boldsymbol{R}}. \tag{21}$$

The a posteriori pdf associated with the MK model can be shown to be a GIG$\{\alpha - \frac{d}{2} + 1, q(\boldsymbol{y}), \sqrt{2\lambda}\}$. Hence, using (18) it follows that, for a given observation $\boldsymbol{y}_i$, we get

$$\xi_i = \mathcal{E}\{\frac{1}{Z}|\boldsymbol{y}_i\} = \sqrt{\frac{2\lambda}{q(\boldsymbol{y}_i)}}\frac{K_{\alpha-\frac{d}{2}}(\sqrt{2\lambda q(\boldsymbol{y}_i)})}{K_{\alpha-\frac{d}{2}+1}(\sqrt{2\lambda q(\boldsymbol{y}_i)})}. \tag{22}$$

When Z is Γ distributed with a pdf as in (7), its k-order moments are

$$E\{Z^k\} = \frac{\Gamma(\alpha+1+k)}{\Gamma(\alpha+1)\lambda^k}. \tag{23}$$

Given N observations, we define

$$\bar{\xi} = \frac{1}{N}\sum_{i=1}^N \xi_i. \tag{24}$$

Regarding $\bar{\xi}$ as estimates for $\mathcal{E}\{\frac{1}{Z}\}$, estimates for α and λ may be obtained as

$$\hat{\alpha} = \frac{1}{\hat{\eta}\,\bar{\xi} - 1}, \tag{25}$$

$$\hat{\lambda} = \hat{\alpha}\,\bar{\xi}. \tag{26}$$

3.1 Iterative Procedure

(i): Set $l = 0$. Select some initial estimates for the parameters of the prior distribution. We suggest to use $\alpha_0 = 1$ and $\lambda_0 = 1$.

(ii): Estimate $\hat{\boldsymbol{\mu}} = \frac{1}{N}\sum_{i=1} \boldsymbol{y}_i$, and $\hat{\boldsymbol{R}} = \frac{1}{N}\sum_{i=1}^{N}(\boldsymbol{y}_i - \hat{\boldsymbol{\mu}})(\boldsymbol{y}_i - \hat{\boldsymbol{\mu}})^t$. Get $\hat{\eta} = \frac{1}{\det \hat{\boldsymbol{R}}^{\frac{1}{d}}}$, and $\Gamma = \frac{\hat{\boldsymbol{R}}}{\hat{\eta}}$.

(iii): Calculate ξ_i using (22) and $\bar{\xi}$ using (24).

(iv): Set $l = l + 1$. Get new estimates α_l and λ_l using (25) and (26).

(v): Repeat *(iii)* and *(iv)* until convergence.

4 Log-Likelihood Ratio Tests

In this section we apply the multivariate K distribution to model DFT coefficients of a speech signal. The signal is sampled at 8 kHz, and the 128-point DFT coefficients are calculated using a Hanning analysis. In order to be able to visualize the proposed model, we will here only discuss the 2-D model referring to the real and imaginary components of a single complex DFT component, i.e $\boldsymbol{f}_k = (\text{real}(F_k), \text{imag}(F_k))$, where F_k is the kth component.

We include a cross-validation log-likelihood test, in which we compare the goodness of fit of the MK model to the multivariate Laplace (ML) model [9], which assumes an exponential distributed scale factor, and the multivariate NIG (MNIG) model [10], which assumes an Inverse Gaussian model for this factor. The pdf of the ML model is given as

$$p_{\boldsymbol{Y}}(\boldsymbol{y}) = \frac{1}{(2\pi)^{\frac{d}{2}}} \frac{2}{\lambda} \frac{K_{\frac{d}{2}-1}(\sqrt{\frac{2}{\lambda} q(\boldsymbol{y})})}{\left(\sqrt{\frac{\lambda}{2} q(\boldsymbol{y})}\right)^{\frac{d}{2}-1}}, \tag{27}$$

and the pdf of the MNIG model is

$$p_{\boldsymbol{Y}}(\boldsymbol{y}) = 2\,\delta e^{\delta\gamma} \left(\frac{\gamma}{2\pi\sqrt{\delta^2 + q(\boldsymbol{y})}}\right)^{\frac{d+1}{2}} K_{\frac{d+1}{2}}(\gamma\sqrt{\delta^2 + q(\boldsymbol{y})}). \tag{28}$$

In the case of the MNIG model, we assume that the bias parameter $\beta = \mathbf{0}$. The parameters of the ML and MNIG models are estimated in procedures similar to the one described above for the MK model. In addition to these models, a traditional mixture of Gaussians model consisting of 5 mixtures, is also included in the test. The latter pdf is given as

$$p_{\boldsymbol{F}}(\boldsymbol{f}) = \sum_{m=1}^{M} \pi_m \mathcal{N}(\boldsymbol{\mu}_m, \boldsymbol{\Sigma}_m), \tag{29}$$

where $\mathcal{N}(\boldsymbol{\mu}_m, \boldsymbol{\Sigma}_m)$ is a 2-D Gaussian pdf, with mean $\boldsymbol{\mu}_m$ and $\boldsymbol{\Sigma}_m$. The parameters of this model is estimated from data using the EM-algorithm. Fig. 1

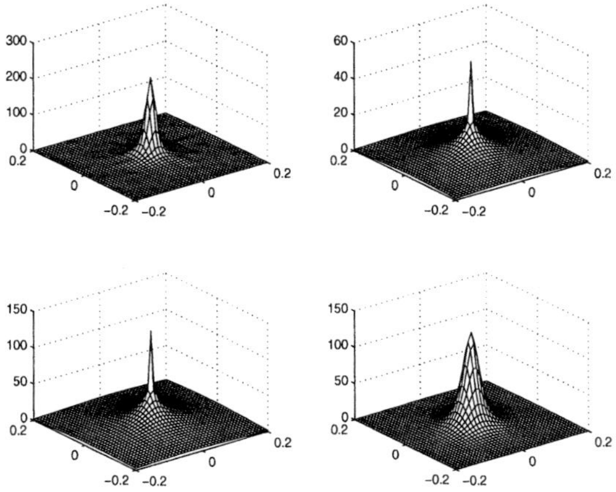

Fig. 1. An example of the 2-D pdfs fitted to the real and imaginary components of a DFT coefficient of a speech signal. *Upper left*: Parzen estimate. *Upper right*: Multivariate K. *Lower left*: Multivariate Laplace. *Lower right*: Multivariate NIG.

Table 1. Results of log-likelihood comparisons

Model	ML	MK	MNIG	Mix. of Gaussians (5)
$no.of wins$	47	0	0	16

shows in the upper left corner a 2-D Parzen estimate of the pdf of frequency component $k = 17$, and in the other panels the estimated pdfs corresponding to the multivariate scale mixture of Gaussian models. We observe that the ML and MK models have pdfs which are quite similar, and peaky at origin. The similarity is because the estimated α for the MK model is small. The shape of the multivariate NIG seems to fit the shape of the Parzen estimate better in this example. Table 1 displays the results of 10-fold cross-validation log-likelihood tests for 63 frequency components. We note that the ML model has the highest log-likelihood value in most of the cases. The mixture of Gaussians model wins 16 times, whereas the MK and MNIG have no wins. Hence, in the log-likelihood sense, these tests, which by no means are very extensive, indicate that the ML model is the best model for this data set.

5 Conclusion

In this paper, we have described an approach for generating multivariate probability models, which can represent data which are sparsely distributed, and where the various components are mutually dependent. The multivariate pdf is constructed as a scale mixture of Gaussians model, using a specific prior distribution for the scale factor. We have calculated closed form expression for the case that the prior pdf is the Γ distribution. Similar models using the exponential and the Inverse Gaussian distributions for the scale factor have been presented elsewhere. Although the expressions for the pdfs may seem complicated, it has been shown that the parameters of the models can be efficiently estimated from data. Hence, we consider these models useful in problems requiring a sparse, multivariate model, like e.g. in speech processing, image and signal enhancement, independent component analysis and blind source separation.

References

1. Simoncelli, E.P.: Modeling the joint statistics of images in the wavelet domain. In: Proc. SPIE 44th Annual Meeting, Denver (1999)
2. Portilla, J., Strela, V., Wainwright, M.J., Simoncelli, E.P.: Image denoising using scale mixture of Gaussians in the wavelet domain. IEEE Trans. Image Proc. **12**(11) (2003) 1338–1351
3. Solbo, S., Eltoft, T.: Homomorphic wavelet-based statistical despeckling of SAR images. IEEE Trans. Geosci. Remote Sensing **42**(4) (2004) 711–721
4. Martin, R., Breithaupt, C.: Speech enhancement in the DFT domain using Laplacian speech priors. In: Proc. IWAENC'03, Kyoto (2003)
5. Kim, T., Attias, H.T., Lee, S.Y., Lee, T.W.: Blind source separation exploiting higher-order frequency dependencies. IEEE Trans. Speech Audio Processing **14**(1) (2006)
6. Park, H.J., Lee, T.W.: Modeling nonlinear dependencies in natural images using mixture of Laplacian distribution. Advances in Neural Computation Systems (2004)
7. Barndorff-Nielsen, O.E.: Normal Inverse Gaussian Distributions and Stochastic Volatility Modelling. Scand. J. Statist. **24** (1997) 1–13
8. Jenssen, R., Øigård, T.A., Eltoft, T., Hanssen, A.: Sparse code shrinkage using the Normal Inverse Gaussian Density Model. In: Proceedings of ICA'2001, San Diego (2001)
9. Eltoft, T., Kim, T., Lee, T.W.: A multivariate Laplace distribution. Accetpted for publication in "IEEE Signal Processing Letters" (2005)
10. Øigård, T.A., Hanssen, A., Hansen, R.E., Godtliebsen, F.: EM-estimation and modeling of heavy-tailed processes with the multivariate Normal Inverse Gaussian distribution. Signal Processing **85**(8) (2005) 1655–1673
11. Andrews, A.F., Mallows, C.L.: Scale mixtures of Normal Distributions. Journal of the Royal Statistical Society. Series B **36**(1) (1974) 99–102
12. Shankar, P.M.: A model for ultrasonic scattering from tissues based on K distribution,. Phys. Med. Biol. **40** (1995) 1633–1649
13. Jakeman, E., Pusey, P.N.: A model for non-Rayleigh sea echo. IEEE Trans. Antennas Propag. **31** (1976) 490–498
14. Eltoft, T., Høgda, K.A.: Non-Gaussian signal statistics in ocean SAR imagery. IEEE Trans. Geosci. Remote Sensing **36**(2) (1998) 562–575

Sparse Deflations in Blind Signal Separation

Pando Georgiev[1], Danielle Nuzillard[2], and Anca Ralescu[1]

[1] ECECS Department, University of Cincinnati, ML 0030,
Cincinnati, Ohio 45221-0030, USA
pgeorgie@ececs.uc.edu, aralescu@ececs.uc.edu
[2] CReSTIC, Université de Reims - Chamapgne Ardenne (URCA),
Moulin de la Housse B.P. 1039, 51687 REIMS Cedex 2, France
danielle.nuzillard@univ-reims.fr

Abstract. We present a new deflation procedure for blind signal separation based on sparsity. It allows, under mild sparsity assumptions, to separate mixtures which could not be separated by ICA methods. We present a new algorithm for sparse deflations and apply it for sparse blind signal separation of mixtures of signals with bounded support. Relations to signals from High Performance Liquid Chromatography in chemistry are discussed and computer simulation examples are presented.

1 Introduction

The goal of the Blind Signal Separation (BSS) is to recover underlying source signals of some given set of observations obtained by an unknown linear mixture of the sources:

$$\mathbf{X} = \mathbf{A}\mathbf{S}. \tag{1}$$

Here $\mathbf{S}$ ($n \times N$-dimensional matrix) is called the *source matrix*, $\mathbf{X}$ the *mixtures* and $\mathbf{A}$ ($m \times n$-matrix) the *mixing matrix*. We speak of *complete*, *overcomplete* or *undercomplete* BSS if $m = n$, $m < n$ or $m > n$ respectively.

BSS has potential applications in many different fields such as medical and biological data analysis, communications, audio and image processing, etc. In order to decompose the data set, different assumptions on the sources have to be made. The most common assumption nowadays is statistical independence of the sources, which leads to the field of *Independent Component Analysis* (ICA), see for instance [1], [10] and references therein. ICA is very successful in the linear *complete* case, when as many signals as underlying sources are observed, and the mixing matrix is non-singular.

One of the first ideas using sparsity in inverse problems, and in particular in BSS, is to apply l_1-norm minimization (basis pursuit method) (see [6], [5]), but its applicability is limited. Another idea is to preprocess the data and to use sparsity properties of the signals in different domains: time-frequency, wavelet domain, etc. Some of these ideas are applied in recent papers on sparse BSS: [2, 3], [7–9], [15].

Recently, it is shown in [12] that using a new approach based on sparsity alone (*Sparse Component Analysis, SCA*), we can still detect both mixing matrix and sources uniquely (except for trivial indeterminacies) even when the

J. Rosca et al. (Eds.): ICA 2006, LNCS 3889, pp. 807–814, 2006.

mixing matrix is singular [11], if the sources are sufficiently sparse. Algorithms for reconstructing the mixing matrix and the sources are proposed and algebraic identifiability conditions on the source matrix are presented.

N. Delfosse N. and P. Loubaton [4] were first to use a deflation technique to BSS using independent sources. In this paper we develop another deflation procedure useful even if two columns of the mixed matrix can be estimated only in each step of the procedure. This is possible under mild sparsity assumptions on the sources (see Fig. 1, first row, for an idea of these assumptions) and doesn't need pre-whitening.

2 Motivation

A motivation of the present paper is the processing of the data matrices that are generated by a High Performance Liquid Chromatography (HPLC) apparatus when coupled to a Diode Array Detector (DAD). HPLC is widely used by pharmaceutical companies to identify and quantify the components of complex mixtures of chemical compounds, often isomeric and diastereoisomeric mixtures (compounds with one and same molecular weight, but different structure in 3D).

Each compound is characterized by the retention time of its concentration at the outlet of the chromatographic system (or concentration profile) and by its ultra-violet (UV) spectrum. UV spectra usually display bump-shaped pickes, as shown in Fig 1. Therefore HPLC-DAD data are stored as a $\mathbf{X}(m \times N)$ matrix where the m rows represent m spectra measured at different times, and where each spectrum is made of UV absorbances measured for N different wavelengths. Considering m components with a mixture, $\mathbf{A}$ the $m \times n$ matrix of compound concentration profiles and $\mathbf{S}$ the $n \times N$ matrix of compound UV spectra,

$$\mathbf{X} = \mathbf{AS} + \mathbf{E}, \tag{2}$$

where $\mathbf{E}$ is the detector noise. If there were never overlapping of concentration profiles, compound identification would always be easy. However, the resolution power of chromatography is often not ideal and the analysis of X matrices is necessary to obtain reliable information on mixtures in these difficult cases, especially when neither reference concentration profiles nor reference UV spectra are available.

3 New Deflation Procedure

We present a new deflation procedure for sparse blind signal separation, which allows to recover underlying signals from their linear mixture only under mild sparsity assumptions.

Consider the following sparse properties of the source matrix $\mathbf{S}$.

(i) there exist at least two indices i_1 and i_2, and two sets of indices $J_1 \subset \{1, ..., N\}$, $J_2 \subset \{1, ..., N\}$ such that

$$\mathbf{S}(i,j) = 0 \text{ and } \mathbf{S}(i_k,j) \neq 0 \text{ whenever } k \in \{1,2\}, i \neq i_k \text{ and } j \in J_k, \tag{3}$$

(i.e.s_{i_k} is uniquely present $|J_k|$ times);

(ii) The set $\mathcal{S} = \left\{ \pm \frac{\mathbf{s}_j}{\|\mathbf{s}_j\|} : j \in \{1, ..., N\} \setminus \{J_k \cup J_2\}, \mathbf{s}_j \neq \mathbf{0} \right\}$ has less than $\min_{i \in \{1,2\}} |J_i|$ equal elements[1].

Theorem 1. *Assume that the assumptions (i) and (ii) above are satisfied and the columns with indices i_1 and i_2 of the mixing matrix $\mathbf{A}$ are different and non-zero. Then they are identifiable up to scaling and sign.*

Proof.(Sketch) Normalize the nonzero columns of $\mathbf{X}$ and cluster them in clusters according to the rule: $\mathbf{x}$ and $\mathbf{y}$ belong to one and the same cluster if and only if either $\mathbf{x} = \mathbf{y}$ or $\mathbf{x} = -\mathbf{y}$[2]. By the assumptions (i) and (ii), the first two clusters with maximal number of elements will represent the columns i_1, i_2 (normalized) of the matrix $\mathbf{A}$ up to sign. ∎

The sparsity assumption under which we can perform BSS is:

(H) *The mixing matrix $\mathbf{A}$ is square, non-singular and any submatrix obtained after removing any k rows of $\mathbf{S}$ ($k = 1, ..., n-2$) satisfies assumptions (i) and (ii).*

Note that assumption **H** could be satisfied for dependent sources (see Fig. 1 first row) – this is the main advantage of our method. The method doesn't need pre-whitening (in fact pre-whitening cannot be performed in the case of dependent sources).

Below we describe the new deflation procedure under assumption **H**.

Without loss of generality we can assume that the first column $\mathbf{a}_1$ of $\mathbf{A}$ is identified and that $a_{11} \neq 0$ (otherwise we can change the notations so that this assumption to be satisfied). Consider the following matrix:

$$\mathbf{W} = \begin{pmatrix} 1 & 0 & 0 \ldots & 0 \\ -a_{21} & a_{11} & 0 \ldots & 0 \\ . & . & . \ldots & . \\ -a_{n1} & 0 & 0 \ldots & a_{11} \end{pmatrix}.$$

Putting $\mathbf{B} = \mathbf{WA}$ we find

$$\mathbf{B} = \begin{pmatrix} a_{11} & a_{12} & \ldots & a_{1n} \\ 0 & b_{22} & \ldots & b_{2n} \\ . & . & \ldots & . \\ 0 & b_{n2} & \ldots & b_{nn} \end{pmatrix},$$

where $b_{ij} = a_{11}a_{ij} - a_{i1}a_{1j}$, $i, j \geq 2$. If we define a new data matrix $\mathbf{Y}$ as $\mathbf{Y} = \mathbf{WX} = \mathbf{WAS}$, we have $\mathbf{Y} = \begin{pmatrix} \mathbf{x}_1 \\ \mathbf{X}_1 \end{pmatrix}$, where

$$\mathbf{X}_1 = \mathbf{B}_1 \mathbf{S}_1, \tag{4}$$

[1] See Fig. 1 (first row) for some idea of the conditions (i) and (ii): $i_1 = 1$ (left bump), $i_2 = 3$ (right bump), $J_1 = \{100, ..., 200\}$, $J_2 = \{380, ..., 480\}$. In practice, it is enough these conditions to be satisfied approximately, i.e. up to some threshold $\varepsilon > 0$ (not stated here).

[2] If and only if either $\|\mathbf{x} - \mathbf{y}\| < \varepsilon$ or $\|\mathbf{x} + \mathbf{y}\| < \varepsilon$ up to some $\varepsilon > 0$ (in the Matlab code below).

$\mathbf{S}_1$ is a submatrix of the source matrix $\mathbf{S}$ which does not contain the first row of $\mathbf{S}$ and $\mathbf{B}_1$ is the submatrix of $\mathbf{B}$ which does not contain the first row and the first column of $\mathbf{B}$. So, (4) represents our deflation procedure: it does not contain the first source. Assuming that conditions (i) and (ii) are satisfied for $\mathbf{S}_1$ and $\mathbf{X}_1$ (by assumption $\mathbf{H}$), we can identify a column of $\mathbf{B}_1$, deflate the corresponding source, and so on. In the k-th step we obtain a system

$$\mathbf{X}_k = \mathbf{B}_k \mathbf{S}_k, \tag{5}$$

with dimension $(n-k) \times (n-k)$ of the mixing matrix $\mathbf{B}_k$. When k became equal to $n-2$, we can identify the matrix $\mathbf{B}_{n-2}$ (with dimension (2×2)). By inverting $\mathbf{B}_{n-2}$, we can recover two sources. In each step of the construction we can identify two columns (by Theorem 1), which allows us in the end to identify all the sources.

The algorithm in fact is *recursive*, which acts on the nodes of a tree representing the current mixing system $\mathbf{Y} = \mathbf{B}\tilde{\mathbf{S}}$. From each node there are only two branches starting, and they correspond to the two columns of $\mathbf{B}$, which can be identified thanks to condition ($\mathbf{A}$). Therefore, each branch will correspond to the mixing model in which the corresponding row of $\tilde{\mathbf{S}}$ is deflated.

We present a Matlab program of a simple *iterative* algorithm, which works perfectly for the case $m = n \leq 4$ (if $m = n < 4$, the results contain repeating sources).

```
function S=sd(X,eps)
% S is an output matrix, whose rows are some of the estimated sources (maximum 4)
% X is the input matrix, whose rows are  mixtures of unknown sources
% A below is a matrix with two columns, which are estimates of two columns of
% the current mixing matrix
S=[]; [n,N]=size(X); X0=X; for k=1:2; Y=X0;
    for i=1:n-1;
        X=Y;
        [m,N]=size(X);
        A=two_rows_identification(X,eps);
        A1=A(1,k)*eye(m-1);
        A2=[-A(2:m,k),A1];
        W=[eye(1,m);A2];
        Z=W*X;
        Y=Z(2:m,:);
    end
    S2=inv(A)*X;
    S=[S;S2];
end
```

The function *two_rows_identification* is a Matlab code made according to the proof of Theorem 1 with some *eps* describing the accuracy of the clustering step (see footnote 4).

4 Computer Simulation Examples

Example 1. Consider a typical case of HPLC. Simulated source signals are shown in Fig.1, first row. Their mixtures with a randomly generated matrix

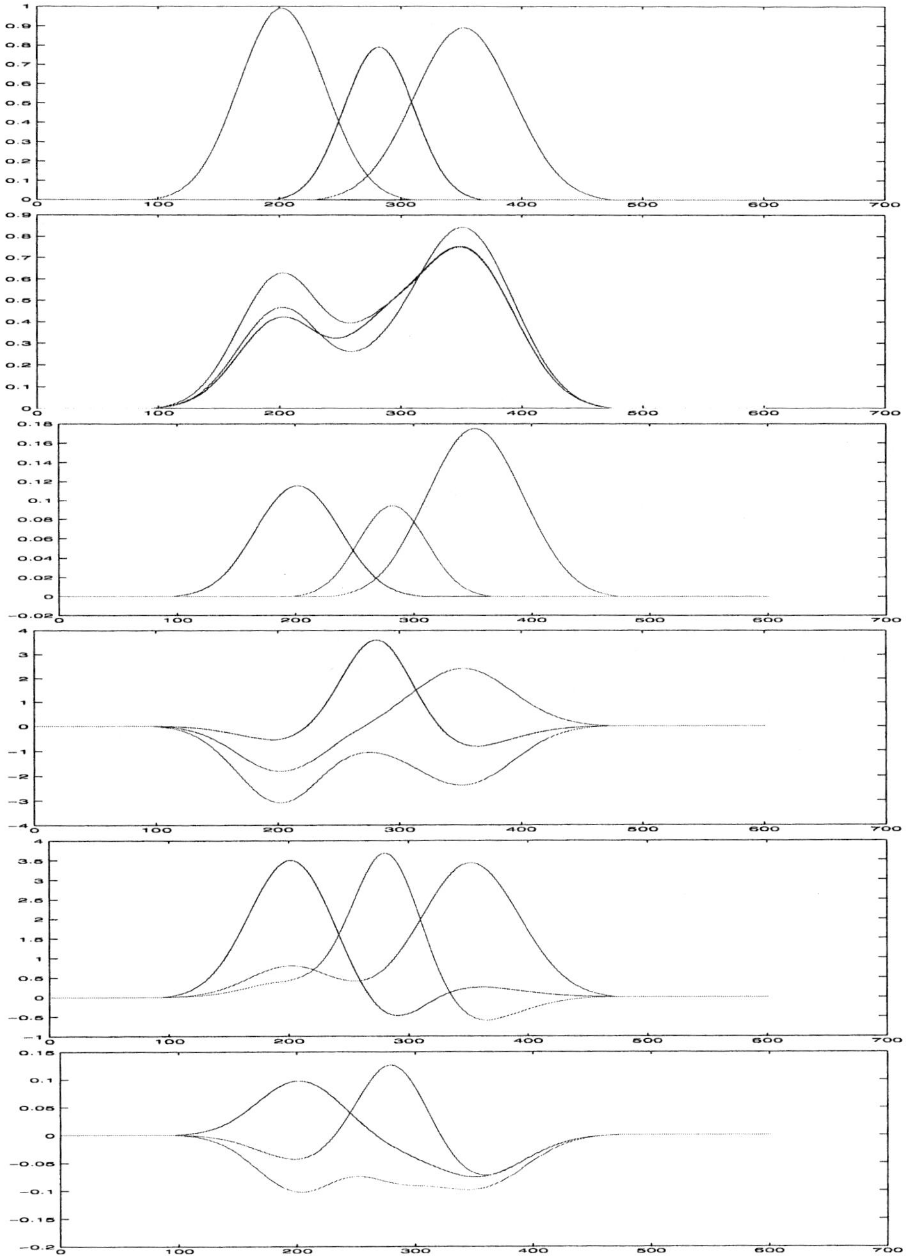

Fig. 1. Example 1: first row – simulated source signals; second row – mixed signals (resembling HPLC picture); third row – reconstructed source signals by our algorithm (sparse deflations); forth row – result obtained by JADE; fifth row – result obtained by Fast ICA; sixth row – result obtained by SOBI

(3 × 3) are shown in Fig. 1, second row – resembling a typical picture from real chromatograms. Applying our algorithm (with $eps = 10^{-5}$), we reconstruct perfectly the original sources (reconstructed sources are shown in Fig. 1, third row). In the next rows of Fig. 1 the results obtained by applying the algorithms JADE, Fast ICA and SOBI are shown – we see that they cannot separate the presented mixture. This is due to the fact that the simulated sources are not independent.

Example 2. Consider a picture to which we have added the same Gaussian noise matrix with two different multipliers to produce two mixed (noisy) images (shown in Fig. 2). Our algorithm succeeded to remove the noise, due to wavelet preprocessing with dwt2, whose diagonal coefficients (shown in Fig. 3) allow to extract one column of the mixing matrix (corresponding to the noise), and subsequently to remove the noise from the picture of interest.

Example 3. Consider now four independent Gaussian signals, which are shown in Fig. 4, first four rows. The sources are generated by the random generator in Matlab, and after that we erased a little each source in order to satisfy assumption (**H**). The mixed signals with a randomly generated matrix (4 × 4) are shown

Fig. 2. Example 2: noised picture (with two unknown proportions of unknown Gaussian noise) and denoised picture

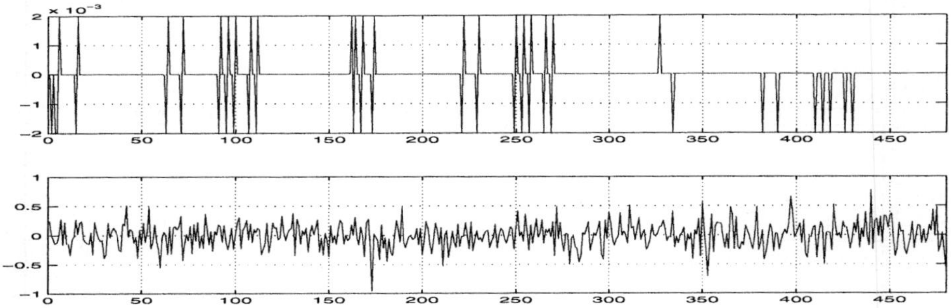

Fig. 3. Example 2: Diagonal wavelet coefficients of dwt2 of the picture and of the noise. The sparse form of those of the picture allow to recover one column of the mixing matrix (corresponding to the noise) and to isolate the picture of interest, applying one step of our algorithm (see (4)).

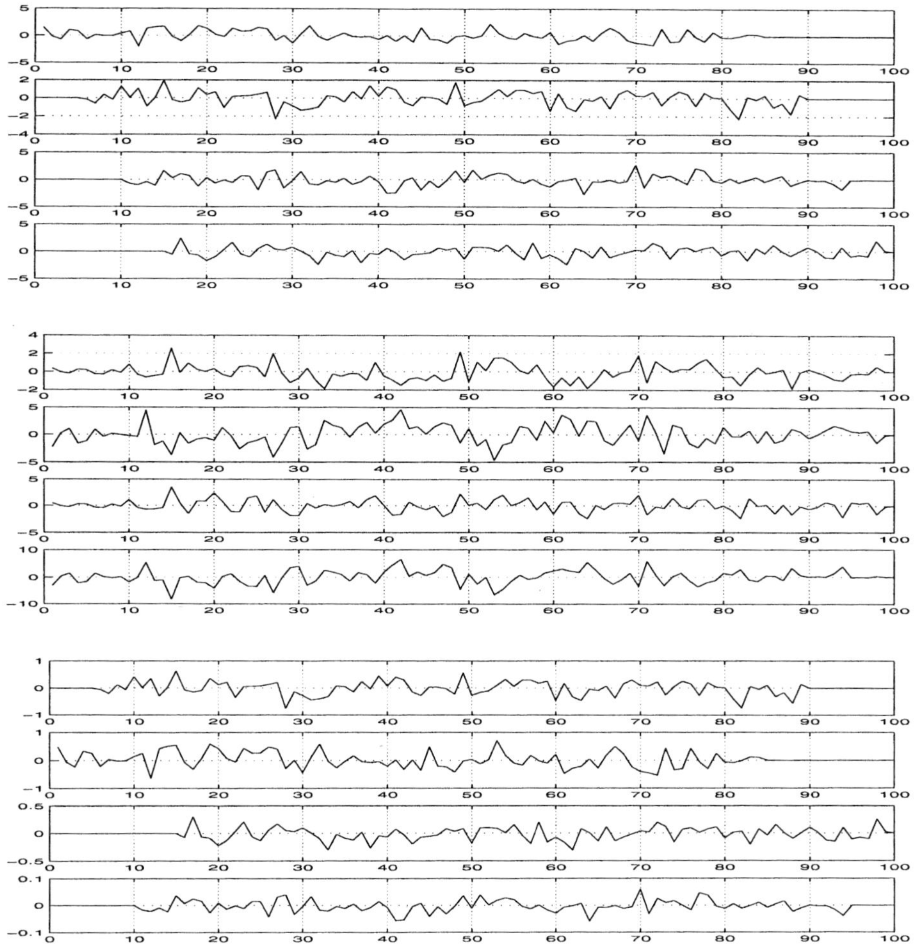

Fig. 4. Example 3: first four rows – slightly sparsified Gaussian signals; the next four rows – their mixture with a random matrix; the last four rows – the reconstructed signals using our algorithm for sparse deflations

in Fig. 4, the middle four rows. Reconstructed sources by our algorithm (shown in Fig. 4, last four rows) are perfect copies (up to permutation and scaling) of the original sources.

5 Conclusion

We presented a new deflation procedure for the BSS problem under mild sparsity assumptions. This procedure allows to solve BSS problems for sources which could be dependent or Gaussian, as well to denoise pictures after wavelet preprocessing. The main motivation is from High Performance Liquid Chromatography (HPLC) in chemistry and we hope that our method could be applied for separation of other sources too (for instance, several delays of a given bump). We presented three computer simulation examples, one of which is a typical HPLC case, the second one is a new denoising procedure, and the third one is constructed by Gaussian sources after small sparsification. The ICA methods cannot separate the mixtures in the first and third examples.

References

1. A. Cichocki and S. Amari. *Adaptive Blind Signal and Image Processing.* John Wiley, Chichester, 2002.
2. A.M. Bronstein, M.M. Bronstein, M. Zibulevsky and Y.Y.Zeevi, "Blind Deconvolution of Images using Optimal Sparse Representations", IEEE Trans. on Image Processing, to appear.
3. Bofill P., Zibulevsky, M., Underdetermined Blind Source Separation using Sparse Representations, Signal Processing, Vol.81, No 11 (2001), pp.2353-2362.
4. Delfosse N. and Loubaton P. "Adaptive blind separation of independent sources: a deflation approach". *Signal Processing*, 45: 59 - 83, 1995.
5. D. Donoho and M. Elad, "Optimally sparse representation in general (nonorthogonal) dictionaries via l^1 minimization", *Proc. Nat. Acad. Sci.*, vol.100, no.5, pp. 2197–2202, 2003.
6. S. Chen, D. Donoho and M. Saunders, " Atomic decomposition by basis pursuit", *SIAM J. Sci. Comput.*, Vol. 20, no. 1, pp. 33–61, 1998.
7. P. Kisilev, M. Zibulevsky, Y. Zeevi, "A multiscale framework for blind separation of linearly mixed signals", Source The Journal of Machine Learning Research, Vol. 4, Issue 7-8, 2004, pp. 1339 - 1364.
8. Y. Li, A. Cichocki, S. Amari, "Analysis of Sparse Representation and Blind Source Separation", Neural Computation, Vol. 16, 2004, pp. 1193-1234.
9. Y. Li, S. I. Amari, A. Cichocki and D. W. C. Ho, "Underdetermined blind source separation based on sparse representation", IEEE Trans. on Signal Processing, accepted as regular paper, to appear in 2006.
10. A. Hyvärinen, J. Karhunen and E. Oja, *Independent Component Analysis*, John Wiley & Sons, 2001.
11. P. Georgiev and F. Theis, "Blind Source Separation of Linear Mixtures with Singular Matrices", Lecture Notes in Computer Science, Springer-Verlag Heidelberg, Vol. 3195, 2004, pp. 121 – 128.
12. P. Georgiev, F. Theis and A. Cichocki, "Sparse Component Analysis and Blind Source Separation of Underdetermined Mixtures", *IEEE Trans. of Neural Networks*, Vol. 16, No. 4, July 2005, 992–996.
13. P. O'Grady, B. Pearlmutter, S. Rickard, "Survey of Sparse and Non-Sparse Methods in Source Separation," International Journal of Imaging Systems and Technology, special issue on Blind Source Separation and Deconvolution in Imaging and Image Processing, Vol. 15, Issue 1, pages 18-33, July 2005.
14. M. Zibulevsky, and B. A. Pearlmutter, "Blind source separation by sparse decomposition in a signal dictionary", *Neural Comput.*, Vol. 13, no. 4, pp. 863–882, 2001.
15. M. Zibulevsky, Y.Y. Zeevi, "Extraction of a single source from multichannel data using sparse decomposition", Neurocomputing 49, (2002), pp 163-173.

Global Analysis of Log Likelihood Criterion

Gen Hori

Brain Science Institute, RIKEN, Saitama 351-0198, Japan
hori@brain.riken.jp

Abstract. This paper introduces and investigates a gradient flow of the log likelihood function restricted on the isospectral submanifold and proves that the flow globally converges to diagonal matrices for almost all positive definite initial matrices. This result shows that the log likelihood function does not have any spurious stable fixed point and ensures the global convergence of ICA algorithms based on the log likelihood function.

1 Introduction

The log likelihood function

$$\phi(A) = \log \det \operatorname{diag}(A) - \log \det A$$

has been derived based on the maximum likelihood estimation of a covariance matrix and used as a cost function in developing algorithms for ICA and joint diagonalization (see Flury[4] and Pham[8]). Comparing to other diagonality criteria, the log likelihood function is based on the orthodox theory of statistical inference and has a favorable property of "scale invariance". The purpose of this paper is to show that the log likelihood function does not have any spurious stable fixed point and ensure the global convergence of ICA algorithms based on the log likelihood function. To this end, we introduce a gradient flow of the log likelihood function restricted on the isospectral submanifold (defined in Section 3) and prove that the gradient flow globally converges to diagonal matrices for almost all positive definite initial matrices.

The rest of the paper is organized as follows. Section 2 derives the log likelihood function for reader's convenience. Section 3 surveys two previous examples of isospectral flows. Section 4 derives the log likelihood flow and proves its global convergence to diagonal matrices. Section 5 contains concluding remarks.

2 Log Likelihood Function

Suppose that we have ν independent samples $X_1, \ldots, X_\nu$ from an n-dimensional normal distribution $N(\mu, \Sigma)$ and denote the sample mean and the (unbiased) sample covariance by $\bar{X}$ and S respectively,

$$\bar{X} = \frac{1}{\nu} \sum_{i=1}^{\nu} X_i, \quad S = \frac{1}{\nu - 1} \sum_{i=1}^{\nu} (X_i - \bar{X})(X_i - \bar{X})^T,$$

J. Rosca et al. (Eds.): ICA 2006, LNCS 3889, pp. 815–822, 2006.

where X_i's are column vectors and X^T denotes the transpose of X, then $\bar{X}$ and S are distributed according to a normal distribution and a Wishart distribution respectively,

$$\bar{X} \sim N(\mu, \Sigma/\nu), \quad (\nu - 1)S \sim W_{\nu-1}(\Sigma).$$

We consider the estimation of Σ from S by the maximum likelihood method.

The probability density function of the n-dimensional Wishart distribution $W_\nu(\Sigma)$ with the degree of freedom ν and the variance Σ is

$$f_\nu(X;\Sigma) = \frac{1}{c_\nu} \frac{(\det X)^{(\nu-n-1)/2}}{(\det \Sigma)^{\nu/2}} \exp(-\frac{1}{2}\mathrm{tr}(\Sigma^{-1}X)),$$

$$c_\nu = 2^{n\nu/2}\pi^{n(n-1)/4} \prod_{i=1}^{n} \Gamma(\frac{\nu - i + 1}{2})$$

where $\det A$ and $\mathrm{tr}A$ denote the determinant and the trace of a matrix A and $\Gamma(x)$ is the gamma function. Because $(\nu - 1)S \sim W_{\nu-1}(\Sigma)$, we have

$$f_{\nu-1}(\,(\nu-1)S\,;\,\Sigma\,) = \frac{1}{c_{\nu-1}} \frac{(\det((\nu-1)S))^{(\nu-n-2)/2}}{(\det \Sigma)^{(\nu-1)/2}} \exp(-\frac{1}{2}\mathrm{tr}(\Sigma^{-1}(\nu-1)S))$$

which gives the log likelihood function as

$$l(\Sigma) = \log f_{\nu-1}(\,(\nu-1)S\,;\,\Sigma\,) = -\frac{\nu-1}{2}(\log\det\Sigma + \mathrm{tr}(\Sigma^{-1}S)) + \mathrm{const.}$$

The maximization of the log likelihood function results in the minimization of the following function where $\nu > 1$,

$$\tilde{l}(\Sigma) = \log\det\Sigma + \mathrm{tr}(\Sigma^{-1}S).$$

Here we consider the eigendecomposition $\Sigma = U\Lambda U^T$ where U is an orthogonal matrix and $\Lambda = \mathrm{diag}(\lambda_1, \dots, \lambda_n)$ is a real diagonal matrix with diagonal elements $\lambda_1, \dots, \lambda_n$. Using $\mathrm{tr}(AB) = \mathrm{tr}(BA)$, we have

$$\tilde{l}(U\Lambda U^T) = \log\det\Lambda + \mathrm{tr}(\Lambda^{-1}U^TSU) = \sum_{i=1}^{n}(\log\lambda_i + \frac{(U^TSU)_{ii}}{\lambda_i}) \qquad (1)$$

where $(U^TSU)_{ii}$ denotes the (i,i)-th element of U^TSU. The differentiation of the right-hand side with respect to λ_i shows that $\tilde{l}(U\Lambda U^T)$ takes its minimum when $\lambda_i = (U^TSU)_{ii}$ holds, that is, for any fixed U, Λ which minimizes $\tilde{l}(U\Lambda U^T)$ is given by $\Lambda = \mathrm{diag}(U^TSU)$, where $\mathrm{diag}(A)$ denotes a diagonal matrix obtained by replacing all the non-diagonal elements of A by zeros. Substituting this in (1) reduces the problem to a minimization with respect to U,

$$\tilde{l}(U\mathrm{diag}(U^TSU)U^T) = \log\det\mathrm{diag}(U^TSU) + \mathrm{tr}(\mathrm{diag}(U^TSU)^{-1}U^TSU).$$

The second term of the right-hand side can be omitted because it is constant. We subtract a constant $\log\det(U^TSU)$ instead and define a cost function

$$\varphi(U) = \log\det\mathrm{diag}(U^TSU) - \log\det(U^TSU) \qquad (2)$$

so that we can easily see that it takes its minimum when U^TSU is diagonal and the minimum value is zero, which is guaranteed by the Hadamard inequality

$$\det \operatorname{diag}(A) \geq \det A \text{ (the equality holds} \Leftrightarrow A \text{ is diagonal).}$$

Substituting $A = U^TSU$ in (2) yields the log likelihood function

$$\phi(A) = \log\det\operatorname{diag}(A) - \log\det A.$$

3 Isospectral Gradient Flows

Isospectral flows are flows defined on the isospectral submanifold

$$\Omega(\Lambda, SO(n)) = \{U\Lambda U^T | U \in SO(n)\}, \quad \Lambda = \operatorname{diag}(\lambda_1, \lambda_2, \ldots, \lambda_n) \tag{3}$$

where Λ is a real diagonal matrix and $SO(n)$ is a matrix Lie group of $n \times n$ real orthogonal matrices. Note that $\Omega(\Lambda, SO(n))$ lies in the set of $n \times n$ real symmetric matrices. Consider a real-valued function $\phi(A)$ defined on the set of $n \times n$ real matrices and suppose that

$$\frac{d\phi}{dA} = \begin{pmatrix} \frac{\partial\phi(A)}{\partial a_{11}} & \cdots & \frac{\partial\phi(A)}{\partial a_{1n}} \\ \vdots & \ddots & \vdots \\ \frac{\partial\phi(A)}{\partial a_{n1}} & \cdots & \frac{\partial\phi(A)}{\partial a_{nn}} \end{pmatrix}$$

is symmetric for an arbitrary symmetric matrix A, then it is known that the gradient flow of the function $\phi(A)$ restricted on the isospectral submanifold (3) is given by

$$\frac{dA}{dt} = -[A, [A, \frac{d\phi}{dA}]], \tag{4}$$

where $[A, B] = AB - BA$ is a commutator product and

$$[A, \frac{d\phi}{dA}] = 0 \tag{5}$$

holds at the fixed points of the flow[2][5]. Brockett[1] introduced an isospectral gradient flow

$$\frac{dA}{dt} = [A, [A, N]] \tag{6}$$

where N is a constant real diagonal matrix with distinct diagonal elements, which is derived by substituting

$$\phi(A) = -\operatorname{tr}(NA) = -\sum_i n_i a_{ii}, \quad N = \operatorname{diag}(n_1, n_2, \ldots, n_n)$$

in (4). Brockett[1] proved that the isospectral gradient flow (6) converges to a diagonal matrix for almost all real symmetric initial matrices. All the fixed points of the flow are diagonal matrices among which one is stable and all the others unstable. Chu and Driessel[3] introduced an isospectral gradient flow

$$\frac{dA}{dt} = [A, [A, \mathrm{diag}(A)]] \tag{7}$$

which is derived by substituting

$$\phi(A) = \frac{1}{2}\mathbf{off}(A) = \frac{1}{2}\sum_{i \neq j} {a_{ij}}^2$$

in (4). Chu and Driessel[3] proved that the isospectral gradient flow (7) converges to a diagonal matrix for almost all real symmetric initial matrices. All the diagonal matrices are the stable fixed points of the flow. The flow has non-diagonal fixed points as well but they are proven to be unstable.

4 Log Likelihood Flow

This section derives and investigates the isospectral gradient flow of the log likelihood function

$$\phi(A) = \log\det\mathrm{diag}(A) - \log\det A. \tag{8}$$

Because the second term of (8) is invariant on the isospectral submanifold, the isospectral gradient flow of $\phi(A)$ is equivalent to one of $\tilde{\phi}(A) = \log\det\mathrm{diag}(A)$ for which we have

$$\frac{d\tilde{\phi}}{dA} = \mathrm{diag}(A)^{-1} = \mathrm{diag}(1/a_{11}, 1/a_{22}, \ldots, 1/a_{nn}).$$

By substituting this in (4), we obtain the log likelihood flow

$$\frac{dA}{dt} = -[A, [A, \mathrm{diag}(A)^{-1}]]. \tag{9}$$

The rest of the section proves that the log likelihood flow (9) globally converges to a diagonal matrix for almost all positive definite initial matrix. The log likelihood flow (9) evolves on the isospectral submanifold

$$\Omega(\Lambda, SO(n)) = \{U\Lambda U^T | U \in SO(n)\}, \quad \Lambda = \mathrm{diag}(\lambda_1, \lambda_2, \ldots, \lambda_n)$$

where Λ is a real diagonal matrix and $\lambda_1, \lambda_2, \ldots, \lambda_n > 0$ holds for the case where the initial matrix is positive definite. In this case, all the diagonal elements of a matrix $A \in \Omega(\Lambda, SO(n))$ are always positive because the i-th diagonal element of A is calculated as follows,

$$a_{ii} = \sum_{j=1}^{n} \lambda_i {u_{ij}}^2 > 0.$$

Therefore it is guaranteed that the diagonal elements of $\mathrm{diag}(A)^{-1}$ never diverge if the initial matrix of (9) is positive definite.

Theorem 1. *The log likelihood flow (9) converges to a diagonal matrix for almost all positive definite real symmetric initial matrices.*

Proof. First of all, we note that the isospectral submanifold $\Omega(\Lambda, SO(n))$ is compact because $U \mapsto U\Lambda U^T$ is a continuous map from $SO(n)$ to $\Omega(\Lambda, SO(n))$ and $SO(n)$ is compact. Because the log likelihood flow is a gradient flow defined on a compact set, it always converges to one of its critical points. From (5), the critical points of (9) satisfy the following condition,

$$[A, \mathrm{diag}(A)^{-1}] = 0.$$

Because the (i, j)-th element of $[A, \mathrm{diag}(A)^{-1}]$ is calculated as $a_{ij}(1/a_{jj} - 1/a_{ii})$, the condition is equivalent to the following,

$$\text{for any index pair } (i, j),\ a_{ij} = 0 \text{ or } a_{ii} = a_{jj} \text{ holds.}$$

Therefore a critical point of the log likelihood flow (9) is i) a diagonal matrix or ii) a non-diagonal matrix with non-zero non-diagonal element $a_{ij} \neq 0$ for which $a_{ii} = a_{jj} > 0$ holds. To complete the proof, it is enough to show that the former is stable and the latter is unstable as proven in the following Lemma 1 and Lemma 2 respectively. □

Lemma 1. *The diagonal critical points of the log likelihood flow (9) are stable.*

Proof. We show that the Hessian of the log likelihood function is positive definite at the diagonal critical points. Let E_{ij} denote an $n \times n$ matrix whose (i, j)-th element is one and all the other elements are zeros. Suppose that X_{ij} is a skew symmetric matrix defined as

$$X_{ij} = E_{ji} - E_{ij} \quad (i < j),$$

then a plane rotation is expressed in an exponential of X_{ij},

$$e^{\theta X_{ij}} = \begin{pmatrix} 1 & & & & & \\ & \ddots & & & & \\ & & \cos\theta & & -\sin\theta & \\ & & & \ddots & & \\ & & \sin\theta & & \cos\theta & \\ & & & & & \ddots & \\ & & & & & & 1 \end{pmatrix}.$$

We introduce a parameterized orthogonal matrix

$$\begin{aligned} U(\boldsymbol{\theta}) = {} & e^{\theta_{12}X_{12}} \cdot e^{\theta_{13}X_{13}} \cdots e^{\theta_{1n}X_{1n}} \\ & \cdot e^{\theta_{23}X_{23}} \cdots e^{\theta_{2n}X_{2n}} \\ & \cdots \\ & \cdot e^{\theta_{n-1,n}X_{n-1,n}} \end{aligned}$$

and a local coordinate system on $SO(n)$ around the identity matrix

$$\boldsymbol{\theta} = (\theta_{12}, \theta_{13}, \ldots, \theta_{1n}, \;\; \theta_{23}, \ldots, \theta_{2n}, \;\; \ldots, \;\; \theta_{n-1,n})$$

in which $\boldsymbol{\theta} = 0$ corresponds to the identity matrix. We have

$$\left.\frac{\partial}{\partial \theta_{ij}} U(\boldsymbol{\theta})\right|_{\theta=0} = X_{ij}.$$

To parameterize the isospectral submanifold around a diagonal point $\Lambda = \mathrm{diag}(\lambda_1, \lambda_2, \ldots, \lambda_n)$, we put

$$A(\boldsymbol{\theta}) = U(\boldsymbol{\theta})^T \Lambda U(\boldsymbol{\theta}).$$

The first and second order derivatives of $A(\boldsymbol{\theta})$ at Λ are calculated as

$$\left.\frac{\partial}{\partial \theta_{ij}} A(\boldsymbol{\theta})\right|_{\theta=0} = [\Lambda, X_{ij}] = -(\lambda_i - \lambda_j)(E_{ij} + E_{ji}), \tag{10}$$

$$\left.\frac{\partial}{\partial \theta_{ij}} \frac{\partial}{\partial \theta_{kl}} A(\boldsymbol{\theta})\right|_{\theta=0} = [[\Lambda, X_{ij}], X_{kl}] = \begin{cases} -2(\lambda_i - \lambda_j)(E_{ii} - E_{jj}) & (i = k, j = l) \\ (\lambda_i - \lambda_j)(E_{jl} + E_{lj}) & (i = k, j \neq l) \\ -(\lambda_i - \lambda_j)(E_{ik} + E_{ki}) & (i \neq k, j = l) \\ (\lambda_i - \lambda_j)(E_{il} + E_{li}) & (k = j, i < l) \\ 0 & (\text{otherwise}) \end{cases}, \tag{11}$$

and then the derivatives of diagonal elements of $A(\boldsymbol{\theta})$ are calculated as

$$\left.\frac{\partial a_{pp}(\boldsymbol{\theta})}{\partial \theta_{ij}}\right|_{\theta=0} = 0,$$

$$\left.\frac{\partial^2 a_{pp}(\boldsymbol{\theta})}{\partial \theta_{ij} \partial \theta_{kl}}\right|_{\theta=0} = \begin{cases} -2(\lambda_i - \lambda_j) & (i = k, j = l, p = i) \\ 2(\lambda_i - \lambda_j) & (i = k, j = l, p = j) \\ 0 & (\text{otherwise}) \end{cases}.$$

Note that the order of X_{ij} and X_{kl} appeared in (11) is the same as the order appeared in the definition of $U(\boldsymbol{\theta})$ and does not depend on the order of differentiation.

The Hessian of the log likelihood function (with the invariant second term dropped)

$$\tilde{\phi}(A) = \log \det \mathrm{diag}(A) = \sum_{p=1}^{n} \log a_{pp}$$

at a diagonal point Λ is calculated as follows,

$$\left.\frac{\partial}{\partial \theta_{ij}} \frac{\partial}{\partial \theta_{kl}} \tilde{\phi}(A(\boldsymbol{\theta}))\right|_{\theta=0} = \sum_{p=1}^{n} \left.\frac{\partial}{\partial \theta_{ij}} \frac{\partial}{\partial \theta_{kl}} \log a_{pp}(\boldsymbol{\theta})\right|_{\theta=0}$$

$$= \sum_{p=1}^{n} \left.\frac{a_{pp} \dfrac{\partial^2 a_{pp}}{\partial \theta_{ij} \partial \theta_{kl}} - \dfrac{\partial a_{pp}}{\partial \theta_{ij}} \dfrac{\partial a_{pp}}{\partial \theta_{kl}}}{a_{pp}{}^2}\right|_{\theta=0}$$

$$= \begin{cases} \dfrac{-2(\lambda_i - \lambda_j)}{\lambda_i} + \dfrac{2(\lambda_i - \lambda_j)}{\lambda_j} = 2\dfrac{(\lambda_i - \lambda_j)^2}{\lambda_i \lambda_j} & (i = k, j = l) \\ 0 & (\text{otherwise}) \end{cases}.$$

The Hessian at a diagonal point Λ is already diagonalized. Because all the diagonal elements of $\Lambda = \text{diag}(\lambda_1, \lambda_2, \ldots, \lambda_n)$ are positive where the initial matrix of the log likelihood flow is positive definite, we see that the Hessian of the log likelihood function at a diagonal point is positive definite and therefore the diagonal critical points of the log likelihood flow (9) are stable. □

Lemma 2. *The non-diagonal critical points of the log likelihood flow (9) are unstable.*

Proof. A non-diagonal critical point A_0 of the log likelihood flow (9) has at least one non-zero non-diagonal element $a_{ij} \neq 0$ for which $a_{ii} = a_{jj} > 0$ holds. We parameterize the isospectral submanifold around the non-diagonal critical point A_0 as

$$A(\boldsymbol{\theta}) = U(\boldsymbol{\theta})^T A_0 U(\boldsymbol{\theta}),$$

and show that the second order derivative of $\tilde{\phi}(A(\boldsymbol{\theta}))$ along the direction of θ_{ij} is negative.

The first and second order derivatives of $A(\boldsymbol{\theta})$ at A_0 are given as

$$\left.\frac{\partial}{\partial \theta_{ij}} A(\boldsymbol{\theta})\right|_{\theta=0} = [A_0, X_{ij}],$$

$$\left.\frac{\partial^2}{\partial {\theta_{ij}}^2} A(\boldsymbol{\theta})\right|_{\theta=0} = [[A_0, X_{ij}], X_{ij}]$$

and then the derivatives of diagonal elements of $A(\boldsymbol{\theta})$ are calculated as

$$\left.\frac{\partial a_{pp}(\boldsymbol{\theta})}{\partial \theta_{ij}}\right|_{\theta=0} = \begin{cases} a_{ij} + a_{ji} = 2a_{ij} & (p = i) \\ -(a_{ij} + a_{ji}) = -2a_{ij} & (p = j) \\ 0 & (\text{otherwise}) \end{cases},$$

$$\left.\frac{\partial^2 a_{pp}(\boldsymbol{\theta})}{\partial {\theta_{ij}}^2}\right|_{\theta=0} = \begin{cases} -2(a_{ii} - a_{jj}) = 0 & (p = i) \\ 2(a_{ii} - a_{jj}) = 0 & (p = j) \\ 0 & (\text{otherwise}) \end{cases}$$

where $a_{ij} = a_{ji}$ and $a_{ii} = a_{jj}$ are assumed. Therefore the second order derivative of $\tilde{\phi}(A(\boldsymbol{\theta}))$ along the direction of θ_{ij} is calculated as

$$\left.\frac{\partial^2}{\partial {\theta_{ij}}^2} \tilde{\phi}(A(\boldsymbol{\theta}))\right|_{\theta=0} = \sum_{p=1}^{n} \left.\frac{\partial^2}{\partial {\theta_{ij}}^2} \log a_{pp}(\boldsymbol{\theta})\right|_{\theta=0}$$

$$= \sum_{p=1}^{n} \left.\frac{a_{pp} \dfrac{\partial^2 a_{pp}}{\partial {\theta_{ij}}^2} - \left(\dfrac{\partial a_{pp}}{\partial \theta_{ij}}\right)^2}{{a_{pp}}^2}\right|_{\theta=0} = \frac{-(2a_{ij})^2}{{a_{ii}}^2} + \frac{-(-2a_{ij})^2}{{a_{jj}}^2} = -8\left(\frac{a_{ij}}{a_{ii}}\right)^2 < 0$$

where $a_{ij} \neq 0$ and $a_{ii} = a_{jj} > 0$ hold. This completes the proof. □

5 Concluding Remarks

We have shown that the log likelihood flow (9) globally converges to diagonal matrices if the initial matrix is positive definite excepting the case where the initial matrix is one of the unstable fixed points. This is equivalent to that the log likelihood function $\phi(A)$ does not have any spurious stable fixed point on the isospectral submanifold $\Omega(\Lambda, SO(n))$ as well as the function $\varphi(U)$ (the log likelihood's dependency on the unitary diagonalizer U of the estimation of the covariance Σ) does not have any spurious stable fixed point on $SO(n)$. This ensures the global convergence of the gradient-based ICA algorithms which optimizes the log likelihood function and gives basis for developing and understanding the ICA algorithms which optimizes with the method including the conjugate gradient method and the Newton method under unitary constraints (see Manton[7]).

References

1. Brockett, R.W. : Dynamical systems that sort lists, diagonalize matrices and solve linear programming problems. Linear Algebra Appl., vol. 146, pp. 79–91 (1991)
2. Brockett, R.W. : Differential geometry and the design of gradient algorithms. In *Differential Geometry: Partial Differential Equations on Manifolds* (eds. R.Green and S-T Yau), Amer. Math. Soc., Providence, pp. 69–92 (1993)
3. Chu, M.T. and Driessel, K.R. : The projected gradient method for least squares matrix approximations with spectral constraints. SIAM J. Numer. Anal, vol. 27, pp. 1050–1060 (1990)
4. Flury, B.N. : Common principal components in k groups. J. Amer. Statist. Assoc., vol. 79, pp. 892-897 (1984)
5. Hori, G. : Isospectral gradient flows for non-symmetric eigenvalue problem. Japan J. Indust. Appl. Math., vol. 17, pp. 27–42 (2000)
6. Hori, G. : A new approach to joint diagonalization. Proc. Intl. Workshop Independent Component Analysis Blind Signal Separation, pp. 151–155 (2000)
7. Manton, J. H. : Optimisation algorithms exploiting unitary constraints. IEEE Trans. Signal Processing, vol. 50, pp. 635–650 (2002)
8. Pham, D.T. : Joint approximate diagonalization of positive definite Hermitian matrices. SIAM J. Matrix Anal. Appl., vol. 22, pp. 1136-1152 (2001)

A Comparison of Linear ICA and Local Linear ICA for Mutual Information Based Feature Ranking

Tian Lan[1], Yonghong Huang[2], and Deniz Erdogmus[1,2]

[1] BME Department, OGI, Oregon Health & Science University, Portland, OR, USA
lantian@bme.ogi.edu
[2] CSEE Department, OGI, Oregon Health & Science University, Portland, OR, USA
{huang, deniz}@csee.ogi.edu

Abstract. Feature selection and dimensionality reduction is important for high dimensional signal processing and pattern recognition problems. Feature selection can be achieved by filter approach, in which certain criteria must be optimized. By using mutual information (MI) between feature vectors and class labels as the criterion, we proposed an ICA-MI framework for feature selection. In this paper, we will compare the linear ICA and local linear ICA for the accuracy of MI estimation, and study the bias-variance trade-off on feature projections and ranking.

1 Introduction

Recent trends in multi-sensor signal processing coupled with multidimensional statistical feature extraction techniques for pattern recognition leads to extremely high dimensional classification problems, EEG-based pattern recognition problems being one such scenario. Dimensionality reduction and feature selection, therefore, becomes crucial for accurate and robust classifier design. Techniques based on mutual information maximization between features and class labels has attracted increasing attention, because this approach can find out the most relevant features, therefore (i) reduces the computational load in real-time system; (ii) can eliminate irrelevant or noisy features, hence increases the robustness of the system; (iii) is a filter approach, which is independent of the design of classifier, and is more flexible.

The MI based method for feature selection is motivated by lower and upper bounds in information theory [1,2]. The average probability of error has been shown to be related to MI between the feature vectors and the class labels. Fano's and Hellman & Raviv's bounds demonstrate that probability of error is bounded from below and above by quantities that depend on the Shannon MI between these variables. Specifically, Hellman & Raviv showed that the upper bound on Bayes error is given by $(H_S(C)-I_S(\mathbf{Y},C))/2$, where $H_S(C)$ is the Shannon entropy of the a priori probabilities of the classes and $I_S(\mathbf{Y},C)$ is the Shannon MI between the continuous-valued feature vector and the discrete-valued class label. Maximizing this MI reduces the upper bound as well as Fano's lower bound, therefore, forces the probability of error to decrease.

J. Rosca et al. (Eds.): ICA 2006, LNCS 3889, pp. 823 – 830, 2006.

Estimating MI requires the knowledge of the joint pdf of the data in the feature space. This is an especially data consuming estimation problem, and if possible must be avoided. Utilizing individual mutual information of the features with the class labels will surely lead to suboptimal selections, since features are generally mutually dependent and information redundancies cannot be captured with such an approach. Several MI-based methods have been developed for feature selection in the past years [3-8]. Unfortunately, all of these methods failed to solve the particularly difficult high dimensional situation – partly because of the curse of dimensionality that is particularly severe for MI estimation.

In practice, MI must be estimated non-parametrically from the training samples. Although this is a challenging problem for multiple continuous-valued random variables, in classification, the discrete-valued class labels simplify the problem to estimating joint entropy of continuous random vectors. Further simplification is possible if the components of the random vectors are independent or made independent – then the joint entropy becomes the sum of marginal entropies, which are easier to estimate. Thus, the joint mutual information of a feature vector with the class labels is equal to the sum of marginal mutual information of each individual feature with the class labels, provided that the features are independent. In previous work, we exploited this fact and proposed a framework using ICA transformation and sample-spacing estimator to estimate the mutual information between features and class labels [9]. This framework is superior because it is open to diverse algorithms, i.e. each component, including ICA transformation and entropy estimator can be replaced by any qualified algorithm/alternative. Applying linear ICA to an arbitrary feature vector has the drawback that in nonlinear classification problems, the linear ICA model possibly fails, thus estimated MI values are inaccurate. For such situations, nonlinear ICA methods become necessary, and we focus particularly on local linear ICA for this purpose.

In this paper, we will investigate the use of linear and local linear ICA for mutual information estimation. We will compute the estimation bias arising from the possibility that linear ICA might not achieve perfect independence, and study the bias-variance trade-off on feature projections and ranking.

2 Problem Formulation and Asymptotic Analysis

Consider a group of nonlinearly distributed, n-dimensional feature vectors: $\mathbf{x}=[x_1,x_2,...,x_n]^T$. Dimensionality reduction on such a feature vector has to be done to improve the generalization capability of the following classifier without compromising accuracy. The information inequalities mentioned above indicate that the subspace projection should be carried out in a manner that maintains as much mutual information with the class labels as possible. The subspace projection can be achieved by linear/nonlinear projections, as well as feature selection (the latter is a special case of linear projections with binary matrix entries – 0 or 1).

Projection approach: The goal is to determine linear or nonlinear projections that jointly maximize their mutual information with the class labels. Specifically, if $\mathbf{y}=\mathbf{g}(\mathbf{x})$, then we must determine $\mathbf{g}(.)$ such that $I_S(\mathbf{Y};C)$ is maximal. If $\mathbf{g}$ is a solution to the nonlinear ICA problem given mixture **x**, then the best m-dimensional nonlinear projection for this NICA solution is the subset $y_1,\ldots,y_m$ such that $I_S(Y_1;C)>I_S(Y_2;C)>\ldots>I_S(Y_n;C)$. Since there are infinitely many solutions to the NICA problem, additional constraints on the form of **g** must be imposed. These constraints are typically imposed as model order limitations for parametric nonlinear projections (such as a neural network) or simply as the utilization of a linear projection. For further discussions, we will focus on feature selection for simplicity.

Feature selection: Given a high dimensional feature vector **x**, our goal is to find the best m dimensional subset of features (in terms of maximum MI with C). This is a combinatorial search problem, and often m is not defined *a priori*. An alternative strategy is to rank the features and pick the top m features from this ranking. Given previously ranked d-1 features $x_{(1)},\ldots, x_{(d-1)}$ the dth feature is the one that maximizes the joint MI: $I_S(x_{(1)},\ldots, x_{(d-1)},x_{(?)};C)$. The joint mutual information takes into account any redundancies in the new feature with the previously ranked d-1 features. This ranking procedure requires the repeated evaluation of d-dimensional MI values. The following procedure is utilized for this purpose.

We first apply a suitable clustering algorithm to segment the data into p partitions: $\mathbf{x}^{(1)}, \mathbf{x}^{(2)}, \ldots, \mathbf{x}^{(p)}$. We assume that within each partition $\mathbf{x}^{(i)}$, the data is d dimensional, and distributed in accordance with the linear ICA model. We apply the linear ICA transformation on each partition C+1 times to get feature vectors: $\mathbf{y}^{(i|c)}$ and $\mathbf{y}^{(i)}$ for each partition, where c denotes class labels and $\mathbf{y}^{(i|c)}$ are the independent components of data in cluster i from class c only, $\mathbf{y}^{(i)}$ are the independent components of data in cluster i regardless of class labels. As a result of the linear ICA transformation, we have:

$$\begin{aligned} H_S(\mathbf{x}^{(i)}) &= H_S(\mathbf{y}^{(i)}) - \log|\mathbf{W}^{i}| \\ H_S(\mathbf{x}^{(i|c)}) &= H_S(\mathbf{y}^{(i|c)}) - \log|\mathbf{W}^{i|c}| \end{aligned} \tag{1}$$

where $i = 1,\ldots,p$, and $\mathbf{W}^{i}$ and $\mathbf{W}^{i|c}$ are the corresponding ICA separation matrices. If linear ICA works perfectly, then the joint entropy of $\mathbf{y}^{(i|c)}$ and $\mathbf{y}^{(i)}$ reduces to the sum of marginal entropies. However, this is not guaranteed, therefore, the residual mutual information will remain as an estimation bias. In practice, we have an imperfect ICA solution and

$$\begin{aligned} H_S(\mathbf{x}^{(i)}) &= \sum_{l=1}^{d} H_S(y_l^{(i)}) - \log|\mathbf{W}^{i}| - I_S(\mathbf{y}^{(i)}) \\ H_S(\mathbf{x}^{(i|c)}) &= \sum_{l=1}^{d} H_S(y_l^{(i|c)}) - \log|\mathbf{W}^{i|c}| - I_S(\mathbf{y}^{(i|c)}) \end{aligned} \tag{2}$$

Mutual information satisfies the following additivity property for any partition (q_i denoting the probability mass of the corresponding partition):

$$I_S(\mathbf{x};C) = \sum_i q_i I_S(\mathbf{x}^{(i)};C) \tag{3}$$

The mutual information within each partition can be expressed as a linear combination of entropy values as follows:

$$I_S(\mathbf{x}^{(i)};C) = H_S(\mathbf{x}^{(i)}) - \sum_c p_{ic} H_S(\mathbf{x}^{(i)}|c) \tag{4}$$

where p_{ic} denotes the probability mass of class c in partition i. Substituting (2) in (4)

$$I_S(\mathbf{x}^{(i)};C) = \left(\sum_{l=1}^{d} H_S(y_l^{(i)}) - \sum_c p_{ic} \sum_{l=1}^{d} H_S(y_l^{(i|c)}) \right) - \left(\log|\mathbf{W}^i| - \sum_c p_{ic} \log|\mathbf{W}^{i|c}| \right) - \left(I_S(\mathbf{y}^{(i)}) - \sum_c p_{ic} I_S(\mathbf{y}^{(i|c)}) \right) \tag{5}$$

The last parenthesis in (5) shows the estimation bias one makes when estimating the MI within each partition if it is assumed that the local linear ICA solution in that partition achieved perfect separation. Over all partitions, the total estimation bias (estimated MI minus the actual MI) is averaged as follows:

$$Bias = \sum_i q_i \left(I_S(\mathbf{y}^{(i)}) - \sum_c p_{ic} I_S(\mathbf{y}^{(i|c)}) \right) \tag{6}$$

Note that asymptotically as the number of partitions approach infinity, one could utilize a grid partitioning structure within which the probability distributions would be uniform, thus local linear ICA would achieve perfect separation within each infinitesimal hypercube. However, in practice, one cannot utilize infinitely many partitions given a finite number of samples. Note that the analysis above also holds for the case where linear ICA is employed directly on the whole dataset without any partitions.

3 Empirical Study

We have employed the feature ranking method described above to benchmark datasets. Partitions are identified via K-means clustering, local linear ICA solutions are determined using joint diagonalization of second and fourth order cumulants [10], and marginal entropies are estimated using sample spacing estimators [11].

3.1 Experiments on a Synthetic Dataset

This dataset consists of four dimensional feature vectors: x_i (i=1,…,4), where x_1 and x_2 are nonlinearly related (Fig. 1 - left), x_3 and x_4 are independent from the first two features and are Gaussian distributed with different mean and variance (Fig. 1 - right). There are two classes in this dataset (represented as blue/red or different grayscale levels in print). These two classes are separable in the x_1 and x_2 plane, but overlapping in the x_3 and x_4 plane. It is clear that this dataset can be well classified only using x_1 and x_2, while x_3 and x_4 provides redundant and insufficient information for perfect classification. From Fig. 1 we can see that x_2 has less overlap compared with x_1, while x_3 has less overlap than x_4. So ideally, the feature ranking in descending order of

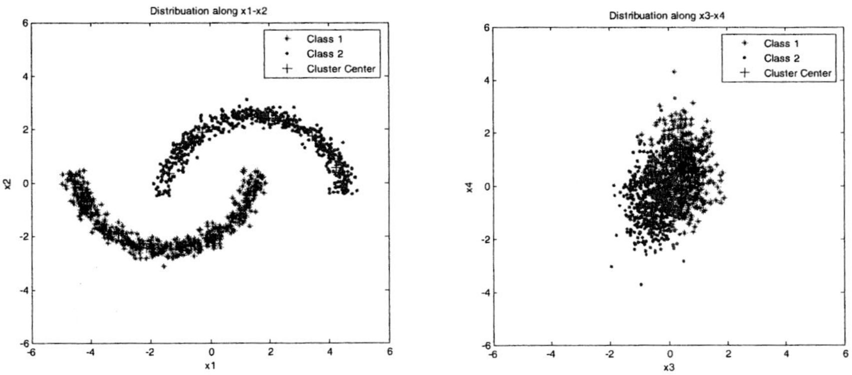

Fig. 1. Four-dimensional Synthetic dataset and corresponding cluster centers. Left: distribution of x_1 and x_2; Right: distribution of x_3 and x_4.

importance in terms of classification rate should be x_2, x_1, x_3, x_4. In our experiments, we choose the sample size as 1000 used 20 partitions. The '+' in Fig.1 represents the partition centers. We also apply linear ICA without any partitioning. The linear ICA approach finds the ranking to be x_2, x_1, x_4, x_3, while the local linear ICA approach with 20 partitions finds the expected *correct* ranking.

3.2 Experiments on the Iris Dataset

In this experiment, we applied linear and local linear ICA (with 2 partitions) approaches to the ranking of the features for the Iris dataset from the UCI database [12]. Due to the small sample size, 10 Monte Carlo rankings with randomly selected training (used for ranking) and test sets are utilized, each consisting of 50% of the available samples. For each ranked subset, a Gaussian Mixture Model (GMM) based Bayesian classifier is employed. The frequency of rankings and classification accuracy are shown in Table 1. and Fig. 2. Since both methods agree on the fourth feature as the top one, pairwise scatter plots of this feature with the remaining features are shown in Fig. 3 for visual comparison. Feature 3 seems to yield a more compact class distribution, while features 1 and 4 seem to have less overlapping samples. Still, it is difficult to judge and we rely on the GMM performances on the testing set for the final comparison. The classification accuracy in Fig. 2. shows that local linear ICA yields better performance than linear ICA in Iris data.

Table 1. Feature ranking frequencies on the Iris dataset

Methods	Ranking indices
Linear ICA	4 3 2 1 (10)
Local linear ICA	4 1 2 3 (5)
	4 2 3 1 (3)
	4 2 1 3 (2)

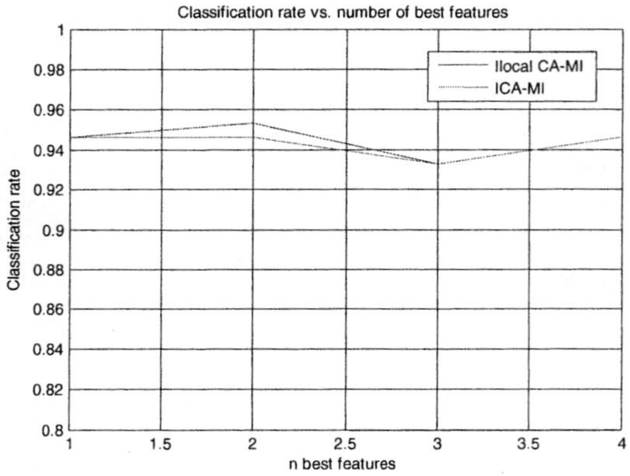

Fig. 2. Classification accuracy for Iris data by linear ICA-MI and Local linear ICA-MI methods. The classification accuracy is the average over 10 Monte Carlo simulations.

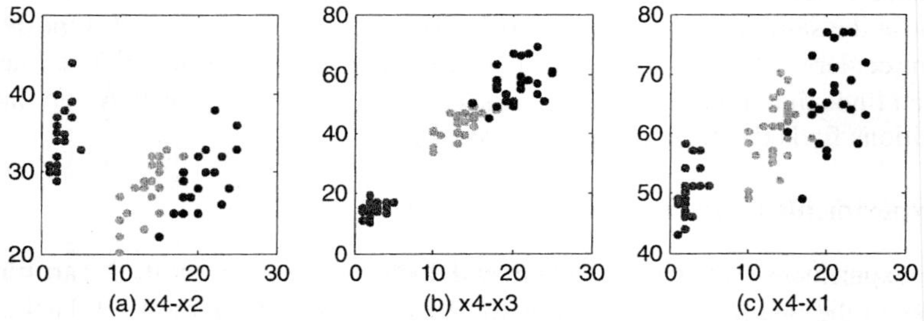

Fig. 3. Combinational distribution of 2 feature vectors of Iris dataset. Left: distribution of x_4 and x_2; Middle: distribution of x_4 and x_3; Right: distribution of x_4 and x_1.

3.3 Experiments on the Wisconsin Breast Cancer Dataset

The two methods are applied to this benchmark dataset, which has higher dimensionality than the previous two case studies. Local linear ICA approach uses 2 partitions

Table 2. Feature Ranking results on Wisconsin Breast Cancer dataset for different ICA-MI methods in 10 Monte Carlo simulations. The frequency of different ranking of 10 Monte Carlo simulations are shown inside the bracket.

Methods	Ranking indices
Linear ICA	3 2 9 4 5 6 7 8 1 (9)
	3 2 9 4 5 8 7 6 1 (1)
Local linear ICA	3 1 2 4 5 6 7 8 9 (4)
	3 4 6 8 7 1 9 2 5 (3)
	3 1 4 5 9 6 8 2 7 (3)

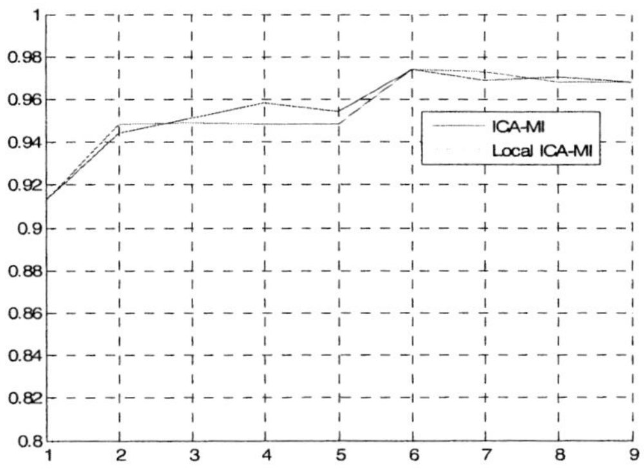

Fig. 4. Classification accuracy for Wisconsin Breast Cancer data by different ICA-MI methods. The classification accuracy is the average over 10 Monte Carlo simulations.

and the Monte Carlo ranking approach is employed as before. The ranking and classification accuracy are shown in Table 2. and Fig. 4. Local linear ICA also exhibit better performance than linear ICA. Consider the number of data samples and the dimensions, if we partition data into more segments, the performance degrades due to the lack of data for reliable linear ICA transformation within each partition.

4 Conclusions

Feature projections and feature selection are important preprocessing procedures in contemporary pattern recognition problems with extremely high dimensional feature vectors. Mutual information maximization provides a suitable *filter* methodology with proven optimality properties regarding the minimization of bounds for the probability of error one would attain when features selected based on this criteria are utilized.

In this paper, we analyzed the finite sample bias of a local linear ICA based mutual information estimation scheme that can be conveniently used for ranking features for subset selection. Experimental evaluation of the proposed method using 1 and more partitions in localization have revealed that as expected, more accurate results are obtained when large sample sets are available for MI evaluation. The sample size must increase appropriately with increasing data dimensionality; otherwise, the estimates are prone to breaking down at higher dimensional estimations, yielding unreliable rankings after a few dimensions. In very high dimensional and small data size situations, simply assuming a single partition and employing linear ICA rather than local linear ICA might lead to more robust ranking and selection results, though will be based on more biased MI estimates. The bias-variance trade-off will be the determining factor in the choice of the number of partitions for local linear ICA.

Acknowledgments

This work was supported by DARPA under contract DAAD-16-03-C-0054 and by NSF under grant ECS-0524835.

References

1. R. M. Fano, *Transmission of Information: A Statistical Theory of Communications.* Wiley, New York, 1961.
2. M.E. Hellman, J. Raviv, "Probability of Error, Equivocation and the Chernoff Bound," IEEE Transactions on Information Theory, vol. 16, pp. 368-372, 1970.
3. K. Torkkola, "Feature Extraction by Non-Parametric Mutual Information Maximization," Journal of Machine Learning Research, vol. 3, pp. 1415-1438, 2003.
4. R. Battiti, "Using Mutual Information for Selecting Features in Supervised Neural Networks learning," IEEE Trans. Neural Networks, vol. 5, no. 4, pp. 537-550, 1994.
5. A. Ai-ani, M. Deriche, "An Optimal Feature Selection Technique Using the Concept of Mutual Information," Proceedings of ISSPA, pp. 477-480, 2001.
6. N. Kwak, C-H. Choi, "Input Feature Selection for Classification Problems," IEEE Transactions on Neural Networks, vol. 13, no. 1, pp. 143-159, 2002.
7. H.H. Yang, J. Moody, "Feature Selection Based on Joint Mutual Information," in Advances in Intelligent Data Analysis and Computational Intelligent Methods and Application, 1999.
8. H.H. Yang, J. Moody, "Data Visualization and Feature Selection: New Algorithms for Nongaussian Data," Advances in NIPS, pp. 687-693, 2000.
9. T. Lan, D. Erdogmus, A. Adami, M. Pavel, "Feature Selection by Independent Component Analysis and Mutual Information Maximization in EEG Signal Classification," Proceedings of IJCNN'05, Montreal, Canada, pp. 3011-3016, Aug. 2005.
10. L. Parra, P. Sajda, "Blind Source Separation via Generalized Eigenvalue Decomposition," Journal of Machine Learning Research, vol. 4, pp. 1261-1269, 2003.
11. E.G. Learned-Miller, J.W. Fisher III, "ICA Using Spacings Estimates of Entropy," Journal of Machine Learning Research, vol. 4, pp. 1271-1295, 2003.
12. http://www.ics.uci.edu/~mlearn/MLRepository.html

Analysis of Source Sparsity and Recoverability for SCA Based Blind Source Separation

Yuanqing Li[1,*], Andrzej Cichocki[2], Shun-ichi Amari[3], and Cuntai Guan[1]

[1] Institute for Infocomm Research, Singapore 119613
[2] Laboratory for Advanced Brain Signal Processing
[3] Laboratory for Mathematical Neuroscience,
RIKEN Brain Science Institute,
Wako shi, Saitama, 3510198, Japan
yqli2@i2r.a-star.edu.sg

Abstract. One (of) important application of sparse component analysis (SCA) is in underdetermined blind source separation (BSS). Within a probability framework, this paper focuses on recoverability problem of underdetermined BSS based on a two-stage SCA approach. We consider a general case in which both sources and mixing matrix are randomly drawn. First, we present a recoverability probability estimate under the condition that the nonzero entry number of a source column vector is fixed. Next, we define the sparsity degree of a signal, and establish the relationship between the sparsity degree of sources and recoverability probability. Finally, we explain how to use the relationship to guarantee the performance of BSS. Several simulation results have demonstrated the validity of the probability estimation approach.

1 Introduction

Sparse component analysis (SCA) of signals has received a great deal of attention in recent years (e.g., [1]-[5], etc). An important application of the sparse representation is in underdetermined blind source separation (BSS), where the number of sources is greater than the number of observations. Until now, Independent Component Analysis (ICA) approach has been commonly used to solve BSS problems. However, ICA approach generally can not recover all sources in the underdetermined case [6]. Based on sparsity of sources, a two-stage clustering-then-l^1-optimization approach was proposed for underdetermined BSS in [3] etc. In this approach, the mixing matrix and the sources were estimated separately. Recently, in [7], we analyzed the two-stage SCA approach and its application in BSS.

First, we present the model and explain the two-stage SCA approach. Generally, instantaneous linear mixtures of sources can be modeled by,

$$\mathbf{X} = \mathbf{AS}, \tag{1}$$

* Corresponding author.

J. Rosca et al. (Eds.): ICA 2006, LNCS 3889, pp. 831–837, 2006.

where the unknown mixing matrix $\mathbf{A} \in R^{n\times m}$ $(n < m)$, the matrix $\mathbf{S} = [\mathbf{s}(1), \cdots, \mathbf{s}(K)] \in R^{m\times K}$ contains m unknown sources, and the only observable $\mathbf{X} = [\mathbf{x}(1), \cdots, \mathbf{x}(K)] \in R^{n\times K}$ is a data matrix containing n mixtures of the sources.

In the two-stage SCA approach, the mixing matrix $\mathbf{A}$ is first estimated, and the source matrix $\mathbf{S}$ is then estimated by solving the following set of K optimization problems $(j = 1, \cdots, K)$,

$$\min \sum_{i=1}^{m} |s_i(j)|, \text{ s. t. } \mathbf{A}\mathbf{s}(j) = \mathbf{x}(j). \quad (2)$$

In [3, 7] etc., clustering-type algorithms are used to estimate the mixing matrix $\mathbf{A}$. Besides, a new algorithm for estimating the mixing matrix can been found in [8]. When the mixing matrix is correctly (or sufficiently precisely) estimated, we need to discuss the recoverability problem, which can be rephrased as a question. How is it possible for the 1-norm solution of (2) to be equal to the true source vector? By our previous studies [7, 8], we found that the higher the sparsity of sources, the higher probability of recoverability. Especially in [8], we obtained a recoverability probability estimate under the condition that the mixing matrix is estimated. In [9], we considered a more general case where both of the sources and mixing matrix are randomly drawn. Under the condition that the nonzero entry number of the source vector is fixed, we presented a recoverability probability estimate.

Another important problem may naturally arise: when sources are not sufficiently sparse, how to guarantee a satisfying performance of BSS? We will discuss this problem in this paper through first defining sparsity degree of sources and then establishing a relationship between the sparsity degree of sources and recoverability probability.

2 Estimation of Recoverability Probability

In this section, we mainly discuss the recoverability probability estimation. For simplification, discussion in the following sections will be based on the following optimization problem, which can be seen a representative of those in (2):

$$(P_1) \min \sum_{i=1}^{m} |s_i|, \; s. \; t. \; \mathbf{A}\mathbf{s} = \mathbf{x}^*,$$

where $\mathbf{x}^* = \mathbf{A}\mathbf{s}^*$, $\mathbf{s}^* \in R^m$ is a true source vector. Hereafter, the solution of (P_1) is denoted as $\mathbf{s}_1$.

When all entries of $\mathbf{A} \in R^{n\times m}$ are drawn from a distribution (e.g., a uniform distribution valued in $[-1, 1]$), the recoverability probability then depends on the sensor number n, the source number m, and the nonzero entry number l of $\mathbf{s}^*$. Hence we denote the recoverability probability as

$$P(n, m, l) = P(\mathbf{s}_1 = \mathbf{s}^* / ||\mathbf{s}^*||_0 = l). \quad (3)$$

Before estimating the probability in (3), we need the following assumption.

Assumption 1: All nonzero entries of the source vector $\mathbf{s}^*$ take either positive or negative sign with equal probability.

Define a set of sign vector with l nonzero entries, $T_l = \{\mathbf{t} = [t_1, \cdots, t_m]^T | t_k \in \{-1, 0, 1\}, \sum |t_k| = l\}$. Note that there are $2^l C_m^l$ vectors in T_l. The following lemma is a simplified version of (23) in [8].

Lemma 1. *For a given basis matrix A, suppose that there are q_l sign column vectors in T_l that can be recovered by solving (P_1'). Then we have*

$$P(\mathbf{s}^0 = \mathbf{s}^{(1)}; ||\mathbf{s}^{(0)}||_0 = l, \mathbf{A}) = \frac{q_l}{2^l C_m^l}, \quad (4)$$

where $l = 1, 2, \cdots, n$.

From Lemma 1, we can see that the recoverability probability in (4) depends only on the sign patterns of of source vectors. The theoretical analysis can be seen in several existing references e.g. [8].

The probability $P(n, m, l)$ can be expressed by the following integral,

$$P(n, m, l) = \int\limits_{||A||_\infty \leq 1} P(\mathbf{s}^* = \mathbf{s}_1 / ||\mathbf{s}^*||_0 = l, \mathbf{A}) P(\mathbf{A}) d\mathbf{A}. \quad (5)$$

Furthermore, we can have the following approximation of the integral in (5),

$$P(n, m, l) \approx \frac{1}{k_0} \sum_{q=1}^{k_0} P(\mathbf{s}^* = \mathbf{s}_1 / ||\mathbf{s}^*||_0 = l, \mathbf{A}_q), \quad (6)$$

where $\mathbf{A}_1, \cdots, \mathbf{A}_{k_0}$ are k_0 random samples of $\mathbf{A}$, k_0 is a small positive integer (5 in our simulation examples). The probability $P(\mathbf{s}^* = \mathbf{s}_1 / ||\mathbf{s}^*||_0 = l, \mathbf{A}_q)$ are calculated according to (4). The theoretical analysis of (6) is presented in [9] and omitted here.

From (6), we can see that $P(n, m, l)$ can be estimated by taking a small number of random samples of mixing matrices $\mathbf{A}$. It is based on the conclusion proved in [9] that the recoverability probability $P(\mathbf{s}^0 = \mathbf{s}^{(1)}; ||\mathbf{s}^{(0)}||_0 = l, \mathbf{A})$ is close to be a constant for most of mixing matrix $\mathbf{A}$ especially when m is large.

3 Sparsity Degree and Recoverability

In this section, we first define a sparsity degree of a signal, then establish the relationship between sparsity degree and recoverability.

Definition 1: For a signal sequence $\mathbf{s} = (s_1, s_2, \cdots)$, suppose that the probability $P(s_k = 0)$ is invariant w.r.t. k. Then the probability $P(s_k = 0)$ is called its sparsity degree.

Suppose that the sparsity degrees of sources are α. Thus for any source vector $\mathbf{s}^* \in R^m$, we have

$$P(s_k^* = 0) = \alpha, \;\; P(s_k^* \neq 0) = 1 - \alpha, \;\; k = 1, \cdots, m. \tag{7}$$

It follows from (7) that the probability that $\mathbf{s}^*$ has exact j nonzero entries is $C_m^j(1-\alpha)^j\alpha^{(m-j)}$. Thus, we have,

$$\begin{aligned} P(\mathbf{s}^* = \mathbf{s}_1/\alpha) &= \sum_{j=0}^{m} P(\mathbf{s}^* = \mathbf{s}_1/||\mathbf{s}^*||_0 = j)P(||\mathbf{s}^*||_0 = j) \\ &= \sum_{j=0}^{m} C_m^j(1-\alpha)^j\alpha^{(m-j)}P(n,m,j), \end{aligned} \tag{8}$$

where the probability estimates $P(n,m,j)$ $(j = 0, \cdots, m)$ can be calculated by (6).

(8) reflects the relationship between recoverability probability and sparsity degree of sources. In real world applications, to guarantee a good performance of BSS, we can first set a recoverability probability constraint p_0 (e.g., 0.95). Using (8), we can determine a corresponding sparsity degree constraint by searching an α in $\{0.01, 0.02, \cdots, 1\}$ such that $P(\mathbf{s}^* = \mathbf{s}_1/\alpha)$ is just larger than p_0. Hereafter, this sparsity degree constraint of sources is denoted as $\alpha_s(n,m,p_0)$, which is related to sensor number n and source number m. Furthermore, we can estimate the corresponding sparsity degree constraint of mixtures, which is denoted as $\alpha_x(n,m,p_0)$. For the m sources, suppose that events $\{s_i(j) = 0\}$, $i = 1, \cdots, m$ are independent. Considering the linear mixture model (1), we have

$$\alpha_x(n,m,p_0) = (\alpha_s(n,m,_0))^m. \tag{9}$$

Under the recoverability probability constraint of 0.95, Table 1 shows the sparsity degrees constraints $\alpha_x(n,m,0.95)$ for mixtures for $n = 6, \cdots, 14$, and $m = n+1, \cdots, 15$.

Finally, we can apply the results shown in Table 1 in BSS. When an $n \times m$ dimensional mixing matrix is estimated, we apply a wavelet packets transformation (WPT) to all mixtures such that their average sparsity degree is larger than

Table 1. Sparsity degree constraints $\alpha_x(n,m,0.95)$ for mixtures corresponding to recoverability probability constraint of 0.95

$n \backslash m$	7	8	9	10	11	12	13	14	15
6	0.15	0.20	0.32	0.35	0.40	0.42	0.39	0.49	0.54
7		0.07	0.17	0.20	0.25	0.28	0.34	0.36	0.40
8			0.04	0.07	0.13	0.16	0.22	0.27	0.29
9				0.02	0.04	0.07	0.12	0.17	0.21
10					0.01	0.02	0.06	0.07	0.09
11						0.004	0.01	0.03	0.05
12							0.002	0.01	0.02
13								0.0005	0.003
14									0.0002

$\alpha_x(n, m, p_0)$ in the time frequency domain. This can be achieved by choosing a suitable level number of WPT (see Example 2). By this way, we can make sure the recoverability probability constrain satisfied.

4 Simulation Examples

In this section, two simulation examples are presented to illustrate our approach and results.

Example 1. In this example, we demonstrate the validity of probability estimates in (6) and (8) using simulations. Here every mixing matrix $\mathbf{A} \in R^{7\times 9}$ is taken according to the uniform distribution in $[-1, 1]$. For each l $(l = 1, \cdots, 9)$, we first estimate the probabilities $P(7, 9, l)$ by solving 3000 linear programming problems, each of which is formed by a pair of random mixing matrix and source vector. Note that each source vector has exactly l nonzero entries drawn from a uniform distribution in $[-0.5, 0.5]$ with their indices also taken randomly. Suppose that n_l source vectors can be recovered, we obtain the ratio $\bar{p}_l = \frac{n_l}{3000}$ that reflects the true probability $P(7, 9, l)$ on recoverability. All $\bar{p}_l, l = 1, \cdots, 9$, are depicted by "$o$" and the dashed curve in the first subplot of Fig. 1.

Next, we estimate $P(7, 9, l)$ $(l = 1, \cdots, 9)$ using (4) and (6), where k_0 in (6) is taken as 5. These 9 probability estimates are depicted by "$*$" and the solid curve in the first subplot of Fig. 1. The two curves in this subplot match each other very well, which demonstrates the validity of recoverability probability estimate of $P(n, m, l)$ in (6).

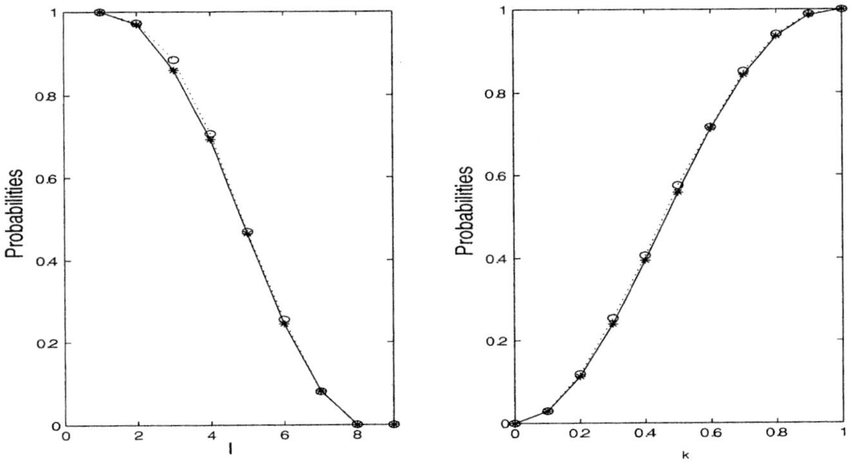

Fig. 1. Curves of estimated probabilities and true probabilities in Example 1. The left and right subplots demonstrate validity of the probability estimates in (6) and (8), respectively.

In the following, we confirm the probability estimate (8) by simulation.

For $\alpha_k = (j-1)*0.1$, $(k = 1, \cdots, 11)$, we first calculate the probabilities $P(\mathbf{s}^* = \mathbf{s}_1/\alpha_k)$ according to (8), where $n = 7, m = 9$. Note that $P(n, m, j)$ $(j = 0, \cdots, 9)$ have been obtained previously. Next, for each k $(k = 1, \cdots, 11)$, we solve 3000 linear programming problems, each of which is formed by a pair of random 7×9 dimensional mixing matrix and source vector. Note that all source vectors are drawn from the distribution in (7). Suppose that m_k source vectors are recovered, hence we obtain the ratio $\tilde{P}(\mathbf{s}_1 = \mathbf{s}^*/\alpha_k) = \frac{m_k}{3000}$, which reflects the true probability $P(\mathbf{s}_1 = \mathbf{s}^*)$ under the sparsity degrees α_k.

In the second subplot of Fig. 1, $P(\mathbf{s}^* = \mathbf{s}_1/\alpha_k)$ are depicted by " $*$ " and the solid curve, while $\tilde{P}(\mathbf{s}^* = \mathbf{s}_1/\alpha_k)$ are depicted by "o" and the dashed curve. These two curves fit very well, virtually overlapping. Thus, the relationship between sparsity degree of sources and recoverability probability shown in (8) is demonstrated by simulations.

Example 2. In this example, we explain how to use the sparsity degree constraints shown in Table 1 to guarantee a satisfying performance of BSS. Generally, real world sources are not sufficiently sparse, we can not directly solve the linear programming to separate sources. For producing sparsificaiton, we apply a WPT (Daubechies WPT here) to all mixtures. BSS is then performed in the time-frequency domain. Using estimated time-frequency coefficients and a inverse WPT, we can reconstruct sources in the time domain.

We now consider 8 speech sources, and their 6 mixtures. We apply a WPT with k levels to each speech source, and each speech mixture. The sparsity degree of all WPT coefficients is then estimated. Note that if the absolute value of a time frequency coefficient of a speech source (mixture) is less than $0.02M$ $(0.004M)$, where M is the maximum of the absolute values of all time frequency coefficients, then this coefficient is taken as zero. Table 2 shows average sparsity degrees of the 8 speech sources, and the 6 speech mixtures in the time frequency domain under 10 WPTs with level number from 1 to 10.

As an example, we now consider a blind source separation problem with 8 unknown speech sources and 6 unknown mixtures. From Table 1, the sparsity degree of mixtures should be larger than 0.2 such that recoverability probability is larger than 0.95. From Table 2, we find that 7 level WPT can produce sufficient sparsity.

In practical blind source separation, it is impossible to directly estimate the sparsity degree of sources in the analyzed domain. However, it the sparsity degree of mixtures can be estimated, hence the sparsity degree of sources can be indirectly estimated using (9).

Table 2. Average sparsity degrees of 8 speech sources and their 6 mixtures in the time frequency domain (the first row refers to the level numbers of WPT)

	1	2	3	4	5	6	7	8	9	10
Speech source	0.55	0.64	0.71	0.74	0.75	0.78	0.82	0.83	0.83	0.84
mixture	0.09	0.13	0.18	0.18	0.19	0.20	0.22	0.24	0.27	0.32

5 Concluding Remarks

In this paper, the recoverability problem was discussed when a two-stage SCA approach was used for BSS. For general case of random mixing matrix and sources, we presented the estimate of recoverability probability which depends on the numbers of sensors, sources and nonzero entries of each source column vector. Next, we defined the sparsity degree of a signal sequence, and established a relationship between sparsity degrees of sources and recoverability probability. To guarantee the performance of BSS, we need to apply a WPT to all mixtures for producing sparsification. Based on the results in this paper, we can first set a recoverability probability constraint (e.g., 0.95), and determine a corresponding sparsity degree constraint of mixtures such that the probability constraint is satisfied. Next, we choose a level number of WPT according to the sparsity degree constraint and perform BSS in the time frequency domain. Finally, the sources in the time domain can be reconstructed by corresponding inverse WPT.

References

1. Lewicki, M.S., Sejnowski, T.J.: Learning overcomplete representations. Neural Computation **12** (2000) 337–365
2. Donoho, D.L., Elad, M.: Maximal sparsity representation via l^1 minimization. the Proc. Nat. Aca. Sci. **100** (2003) 2197–2202
3. Zibulevsky, M., Pearlmutter, B.A.: Blind Source Separation by Sparse Decomposition. Neural Computations **13** (2001) 863–882
4. Lee, T.W., Lewicki, M.S., Girolami, M., Sejnowski, T.J.: Blind source separation of more sources than mixtures using overcomplete representations. IEEE Signal Processing Letter **6** (1999) 87–90
5. Girolami, M.: A variational method for learning sparse and overcomplete representations. Neural Computation **13** (2001) 2517–1532
6. Li, Y., Wang, J.: Sequential blind extraction of linearly mixed sources. IEEE Trans. On Signal Processing **50** (2002) 997–1007
7. Li, Y.Q., Cichocki A., Amari, S.: Analysis of Sparse representation and blind source separation. Neural Computation **16** (2004) 1193–1234
8. Li, Y.Q., Cichocki, A., Amari, S., Ho, D.W.C., Xie, S.L.: Underdetermined blind source separation based on sparse representation. accepted by IEEE Trans. on Signal Processing
9. Li, Y.Q., Amari, S., Cichocki, A., Guan, C.T.: Probability Estimation for Recoverability Analysis of Blind Source Separation Based on Sparse Representation. submitted to IEEE Trans. on Information Theory (revised).

Analysis of Feasible Solutions of the ICA Problem Under the One-Bit-Matching Condition

Jinwen Ma[1,3], Zhe Chen[2], and Shun-ichi Amari[1]

[1] Laboratory of Mathematical Neuroscience
[2] Laboratory of Advanced Brain Signal Processing,
RIKEN Brain Science Institute, Wako-shi, Saitama 351-0198, Japan
[3] Department of Information Science, School of Mathematical Sciences and LMAM,
Peking University, Beijing 100871, P.R. China
{jwma, zhechen, amari}@brain.riken.jp

Abstract. The one-bit-matching conjecture for independent component analysis (ICA) has been widely believed in the ICA community. Theoretically, it has been proved that under certain regular assumptions, the global maximum of a simplified objective function derived from the maximum likelihood or minimum mutual information criterion under the one-bit-matching condition corresponds to a feasible solution of the ICA problem, and also that all the local maxima of the objective function correspond to the feasible solutions of the ICA problem in the two-source square mixing setting. This paper further studies the one-bit-matching conjecture along this direction, and we prove that under the one-bit-matching condition there always exist many local maxima of the objective function that correspond to the stable feasible solutions of the ICA problem in the general case; moreover, in ceratin cases there also exist some local minima of the objective function that correspond to the stable feasible solutions of the ICA problem with mixed super- and sub-Gaussian sources.

1 Introduction

Independent component analysis (ICA) is a powerful tool for blind signal processing and has remained as an intense research subject in the literature. One of important application of ICA is used for blind source separation where the source signals are assumed to be independent and non-Gaussian. In particular, consider a conventional ICA problem which assumes an instantaneous linear mixing model: $\mathbf{x} = \mathbf{A}\mathbf{s}$, where $\mathbf{A} \in \mathbb{R}^{m\times n}$ denotes the mixing matrix; $\mathbf{s} \in \mathbb{R}^n$ and $\mathbf{x} \in \mathbb{R}^m$ correspond to the n-dimensional source vector and m-dimensional mixture vector, respectively. The goal of ICA is to seek a demixing matrix, $\mathbf{W} \in \mathbb{R}^{n\times m}$, applied to the mixture vector $\mathbf{x}$:

$$\mathbf{y} = \mathbf{W}\mathbf{x} = \mathbf{W}(\mathbf{A}\mathbf{s}) = (\mathbf{W}\mathbf{A})\mathbf{s} \tag{1}$$

where $\mathbf{y} \in \mathbb{R}^n$ corresponds to the unmixed signal vector. When the sources in $\mathbf{s}$ are statistically independent, it is hoped that the recovered $\mathbf{y}$ is also componentwise independent, that is,

J. Rosca et al. (Eds.): ICA 2006, LNCS 3889, pp. 838–845, 2006.

$$q(\mathbf{y}) = \prod_{i=1}^{n} q(y_i), \tag{2}$$

where $q(\cdot)$ denotes the probability density function. Generally and unless stated otherwise, it is assumed in the paper that $m = n$ and the square mixing matrix $\mathbf{A}$ is invertible.

The study on the ICA problem can be traced back to Tong, Inouye & Liu [1] who showed that $\mathbf{y}$ recovers the sources $\mathbf{s}$ up to scaling and permutation ambiguity when y_i $(i = 1, \ldots, n)$ become componentwise independent and at most one of them is Gaussian. Later on, Comon [2] further formalized the problem under the name ICA. Since then, the ICA problem has been widely studied from different perspectives by many researchers (e.g., [3]-[7]). In particular, one of essential goal to exploit the independence in parallel is to minimize the following objective function, or the so-called "minimum mutual information (MMI)":

$$D(\mathbf{W}) = -H(\mathbf{y}) - \sum_{i=1}^{n} \int p_{\mathbf{W}}(y_i; \mathbf{W}) \log p_i(y_i) dy_i, \tag{3}$$

where $H(\mathbf{y}) = -\int p(\mathbf{y}) \log p(\mathbf{y}) d\mathbf{y}$ represents the entropy of $\mathbf{y}$, $p_i(y_i)$ denotes the predetermined model probability density function (pdf) that is implemented to approximate the marginal pdf of $\mathbf{y}$, and $p_{\mathbf{W}}(y_i; \mathbf{W})$ denotes the joint probability distribution on $\mathbf{y} = \mathbf{W}\mathbf{x}$. In the literature, how to choose the model pdf's is an important issue for the ICA problem. It is known that, with each model pdf $p_i(y_i)$ predefined, this MMI method works only in the cases where the components of $\mathbf{y}$ are either all super-Gaussians [4] or all sub-Gaussians [5].

For the cases where sources contain both super-Gaussian and sub-Gaussian signals in an unknown manner, it was suggested that each model pdf $p_i(y_i)$ should be flexibly adjustable and be learned together with demixing matrix $\mathbf{W}$. In fact, the learning of $p_i(y_i)$ can be done by adapting the parameters in a finite mixture of sigmoid functions that learns the cumulative distribution function (cdf) of each source [8], or by learning a mixture of parametric pdf's [9]. On the other hand, it has also been found that a rough estimate of each source pdf or cdf may be sufficient for source separation. These observations motivated the proposal of the so-called *one-bit-matching conjecture* [10], which can be basically stated as "*all the sources can be separated as long as there is a one-to-one same-sign-correspondence between the kurtosis signs of all source pdf's and the kurtosis signs of all model pdf's*".

The one-bit-matching conjecture was widely believed in the ICA community since there have been many experimental studies supporting this claim (e.g., [11]-[14]). Moreover, some new ICA algorithms were already established in light of this conjecture. However, a complete understanding of the one-bit-matching conjecture requires a theoretical proof for it. In the literature, a mathematical proof [15] was given for the case involving only two sub-Gaussian sources, but the result cannot be extended to a model either with more than two sources, or with mixed sub- and super-Gaussian sources. Recently, Liu, Chiu and Xu [16] have proved that under the assumption of zero skewness for the model pdf's, the one-bit-matching condition guarantees a feasible solution of the ICA problem by *globally* maximizing the simplified objective function (to be defined later in Section 2) derived from Eq.(3). However, this result is rather restrictive in that it is generally difficult to obtain a feasible solution of the ICA problem by searching the global maximum of the objective function. As a matter of fact, it is more significant to study the local separation property of the ICA problem under the one-bit-matching condition that the sources can be separated by *locally* maximizing that objective function in the same setting. Along this direction, Ma, Liu & Xu [17] already proved that all the local maxima of the formulated objective function correspond to the feasible solutions of the ICA problem in the two-source mixing setting.

In this paper, we further investigate the formulated objective function in the general case. Specifically, we prove that there always exist many local maxima of the objective function that correspond to the stable feasible solutions of the ICA problem (i.e., the stable solutions of a local searching algorithm on the objective function) in the general case under the one-bit-matching condition. Moreover, in ceratin situation under the one-bit-matching condition, there also exist some local minima of the objective function that correspond to the stable feasible solutions of the ICA problem with mixed super- and sub-Gaussian sources. That is, the successful separation can be obtained via locally minimizing the objective function under the one-bit-matching condition in such a case with mixed super- and sub-Gaussian sources.

The rest of the paper is structured as follows. We first formulate the objective function and introduce a leema in section 2. Section 3 presents the main results of two theorems. We conclude briefly in section 4.

2 The Objective Function and a Lemma

For discussion simplicity, we assume that the source, mixed, and recovered signals are all whitened and thus $\mathbf{W}$ and $\mathbf{A}$ are both orthonormal. When the skewness and kurtosis statistics are considered and when the non-Gaussian sources have nonzero kurtosis statistics, under the zero skewness assumption for all the model pdf's, the objective function derived from Eq.(3) can be simplified as follows [16]:

$$J(\mathbf{R}) = \sum_{i=1}^{n}\sum_{j=1}^{n} r_{ij}^4 \nu_j^s k_i^m, \tag{4}$$

where $\mathbf{R} = (r_{ij})_{n\times n} = \mathbf{WA}$ is an orthonormal matrix to be estimated (the reason that we optimize $\mathbf{R}$ instead of $\mathbf{W}$ is for convenience of analysis); ν_j^s denotes the kurtosis of the source s_j, and k_i^m is a constant with the same algebraic sign as the kurtosis ν_i^m of the model pdf.

For the purpose of clarity, we define a matrix $\mathbf{K}$ by

$$\mathbf{K} = (k_{ij})_{n\times n}, \quad k_{ij} = \nu_j^s k_i^m. \tag{5}$$

By that we may rewrite (4) as

$$J(\mathbf{R}) = \sum_{i=1}^{n}\sum_{j=1}^{n} r_{ij}^4 \nu_j^s k_i^m = \sum_{i=1}^{n}\sum_{j=1}^{n} r_{ij}^4 k_{ij}. \tag{6}$$

Under the one-bit-matching condition, with the help of certain permutation we can always obtain $k_1^m \geq \cdots \geq k_p^m > 0 > k_{p+1}^m \geq \cdots \geq k_n^m$ and $\nu_1^s \geq \cdots \geq \nu_p^s > 0 > \nu_{p+1}^s \geq \cdots \geq \nu_n^s$, which will be considered as the one-bit-matching condition in this paper.

It has been proved in [16] that the global maximization of Eq.(6) under the one-bit-matching condition can only be approachable by setting $\mathbf{R}$ as an identity matrix up to certain permutation and sign indeterminacy. That is, the global maximization of Eq.(6) will recover the original sources up to sign and permutation indeterminacies if the one-bit-matching condition is satisfied. In the two-source mixing case, i.e., $n = 2$, it has been further proved in [17] that the local maxima of $J(\mathbf{R})$ are also only reachable by the permutation matrices up to sign indeterminacy under the one-bit-matching condition. In the following, we will prove that there exist many local maxima of $J(\mathbf{R})$ that correspond to the stable feasible solutions of the ICA problem. Moreover, in certain cases where both super- and sub-Gaussian sources coexist, some minima of $J(\mathbf{R})$ also correspond to the stable feasible solutions of ICA problem. Before doing so, we introduce one lemma as follows.

Lemma 1. *Suppose that $F(\mathbf{x})$ ($\mathbf{x} \in \mathbb{R}^m$) is a twice differentiable scalar function under the following constraints:*

$$C_i(\mathbf{x}) = 0, \qquad i = 1, 2, \cdots, k. \tag{7}$$

Construct a Lagrange function with a Lagrange multiplier set $\boldsymbol{\lambda} = \{\lambda_1, \lambda_2, \cdots, \lambda_k\}$, i.e., $L(\mathbf{x}, \boldsymbol{\lambda}) = F(\mathbf{x}) + \sum_{i=1}^{k} \lambda_i C_i(\mathbf{x})$, and assume that $(\mathbf{x}^*, \boldsymbol{\lambda}^*)$ is a solution of the system of the equalities that all the derivatives of $L(\mathbf{x}, \boldsymbol{\lambda})$ with respect to the variables of $\mathbf{x}$ and the Lagrange multipliers λ_i are equal to zeros. It is also assumed that these $\nabla C_i(\mathbf{x}^*)$ are linearly independent. If for any nonzero vector $q \neq 0$ under the constraints $\mathbf{q}^T \nabla_{\mathbf{x}} C_i(\mathbf{x}^*) = 0$ for $i = 1, 2, \cdots, k$, we have

$$\mathbf{q}^T \nabla_{\mathbf{x}}^2 L(\mathbf{x}^*, \boldsymbol{\lambda}^*) \mathbf{q} < 0 \ \ (or > 0), \tag{8}$$

then $\mathbf{x}^*$ is a local maximum (or local minimum) of $F(\mathbf{x})$ under the constraints.

Lemma 1 is a well-known mathematical result in optimization theory; its proof can be found in [18].

3 The Main Results

With the above background, we are ready to investigate the local maximization of objective function $J(\mathbf{R})$ defined in (6), where $\mathbf{R}$ is a permutation matrix up to sign indeterminacy (namely, as a special orthonormal matrix). We consider the general optimization problem of maximizing $J(\mathbf{R})$ with a fixed matrix $\mathbf{K}$ and $\mathbf{R}\mathbf{R}^T = \mathbf{I}$.

In order to solve this constrained optimization problem, we introduce a set of Lagrange multipliers $\boldsymbol{\lambda} = \{\lambda_{ij} : i \leq j\}$ and construct the Lagrange objective function:

$$L(\mathbf{R}, \boldsymbol{\lambda}) = \sum_{i=1}^{n}\sum_{j=1}^{n} r_{ij}^4 k_{ij} + \sum_{i=1}^{n}\sum_{j=i}^{n} \lambda_{ij}\Big(\sum_{l=1}^{n} r_{li}r_{lj} - \delta_{ij}\Big), \tag{9}$$

where δ_{ij} denotes the Kronecker function. By derivation, we have

$$\frac{\partial L(\mathbf{R}, \boldsymbol{\lambda})}{\partial r_{ij}} = 4k_{ij}r_{ij}^3 + \sum_{l=1}^{j-1} r_{il}\lambda_{lj} + 2\lambda_{jj}r_{ij} + \sum_{l=j+1}^{n} r_{il}\lambda_{jl}; \tag{10}$$

$$\frac{\partial L(\mathbf{R}, \boldsymbol{\lambda})}{\partial \lambda_{ij}} = \sum_{l=1}^{n} r_{li}r_{lj} - \delta_{ij}. \tag{11}$$

Given $\boldsymbol{\lambda}$, we define a new matrix $\mathbf{U} = (u_{ij})_{n\times n}$ as

$$u_{ij} = \begin{cases} \lambda_{ij}, & \text{if } i < j; \\ \lambda_{ji}, & \text{if } i > j; \\ 2\lambda_{ii} & \text{if } j = i. \end{cases} \tag{12}$$

In light of (10) and (12), we have

$$\frac{\partial L(\mathbf{R}, \boldsymbol{\lambda})}{\partial r_{ij}} = 4k_{ij}r_{ij}^3 + \sum_{l=1}^{n} r_{il}u_{lj}. \tag{13}$$

Note that $\mathbf{U}$ is symmetric in that $\mathbf{U}^T = \mathbf{U}$. Setting the derivatives of (10) and (11) to zeros yields

$$-4(k_{ij}r_{ij}^3)_{n\times n} = \mathbf{R}\mathbf{U}. \tag{14}$$

For clarity, we further define a new matrix $\mathbf{B}$ by

$$\mathbf{B} = (k_{ij}r_{ij}^3)_{n\times n}.$$

By virtue of the symmetry of $\mathbf{U}$, we have

$$\mathbf{R}^T\mathbf{B} = \mathbf{B}^T\mathbf{R}, \quad \text{or} \quad \mathbf{B} = \mathbf{R}\mathbf{B}^T\mathbf{R}, \tag{15}$$

which is essentially the condition for matrix $\mathbf{R}$ to be a *critical point* of the objective function (6) under the orthonormality constraint; in fact, it is equivalent to the condition that the gradient of $J(\mathbf{R})$ on the Stiefel manifold is zero [19].

Moreover, it can be easily shown that all the permutation matrices up to sign indeterminacy satisfy Eq.(15). That is, these permutation matrices will be the local maxima, minima, and saddle points of the objective function $J(\mathbf{R})$. In the following, we will study the circumstances when a permutation matrix (up to sign indeterminacy) corresponds to a local maximum, local minimum, or saddle point of $J(\mathbf{R})$. The main results are summarized into two theorems.

Theorem 1. *If* $\mathbf{R}^*$ *is a permutation matrix up to sign indeterminacy and* $k_{ij} > 0$ *at all the positions where* $|r^*_{ij}| = 1$*, it corresponds to a local maximum of the objective function* $J(\mathbf{R})$.

Proof: For convenience, we vectorize the $n \times n$ matrix $\mathbf{R}$ into an $n^2 \times 1$ vector

$$vec[\mathbf{R}] = [r_{11}, r_{21}, \cdots, r_{n1}, r_{12}, r_{22}, \cdots, r_{n2}, \cdots, r_{1n}, r_{2n}, \cdots, r_{nn}]^T \in \mathbb{R}^{n^2}.$$

Correspondingly, we may also construct a nonzero $n^2 \times 1$ vector $\mathbf{q}$

$$\mathbf{q} = [q_{11}, q_{21}, \cdots, q_{n1}, q_{12}, q_{22}, \cdots, q_{n2}, \cdots, q_{1n}, q_{2n}, \cdots, q_{nn}]^T \in \mathbb{R}^{n^2}.$$

Taking the derivative of Eq.(13) yields

$$\frac{\partial^2 L(\mathbf{R}, \boldsymbol{\lambda})}{\partial r_{ij} \partial r_{i'j'}} = \delta_{(i,j),(i',j')}[12 k_{ij} r_{ij}^2 + u_{jj}], \tag{16}$$

where $\delta_{(i,j),(i',j')}$ denotes the Kronecker function such that it equals to 1 if $(i', j') = (i, j)$ (namely, $i = i'$ and $j = j'$) and zero otherwise. It follows from Eq.(14) that

$$\mathbf{U} = -4\mathbf{R}^T \mathbf{B}. \tag{17}$$

When $\mathbf{R} = \mathbf{R}^*$ is a permutation matrix up to sign indeterminacy, $\mathbf{U}^*$ (associated with $\boldsymbol{\lambda}^*$) will be a diagonal matrix. By the condition that $k_{ij} > 0$ at *every* $|r^*_{ij}| = 1$, it follows that $u^*_{jj} = -4k_{ij} < 0$ for each j. Moreover, it can be readily verified that all $\nabla_{\mathbf{R}} C_{ij}(\mathbf{R}^*)$ are linearly independent, where we define $C_{ij}(\mathbf{R}) = \sum_{l=1}^{n} r_{li} r_{lj} - \delta_{ij}$ for $i \le j$.

In light of Eq.(16), we infer that $\nabla^2_{\mathbf{R}} L(\mathbf{R}^*, \boldsymbol{\lambda}^*)$ is a diagonal matrix. Furthermore, its diagonal elements are *negative* except those ones corresponding to $|r^*_{ij}| = 1$. However, the q_{ij} associated with $|r^*_{ij}| = 1$ will be constrained to zeros under the condition $\mathbf{q}^T \nabla_{\mathbf{R}} C_{jj}(\mathbf{R}^*) = 0$ for any nonzero vector $\mathbf{q}$. Thus, with all the constraints $\mathbf{q}^T \nabla_{\mathbf{R}} C_{ij}(\mathbf{R}^*) = 0$, we always have $\mathbf{q}^T \nabla^2_{\mathbf{R}} L(\mathbf{R}^*, \boldsymbol{\lambda}^*) \mathbf{q} < 0$ for any nonzero vector $\mathbf{q}$. It then follows from Lemma 1 that $\mathbf{R}^*$ is a local maximum of $J(\mathbf{R})$. Thus far the proof is completed. □

Remark 1. According to the one-bit-matching condition, the matrix $\mathbf{K} = (k_{ij})_{n \times n}$ can be divided into the following four blocks:

$$\mathbf{K} = \begin{pmatrix} \mathbf{K}_{11} & \mathbf{K}_{12} \\ \mathbf{K}_{21} & \mathbf{K}_{22} \end{pmatrix},$$

where $\mathbf{K}_{11}$ and $\mathbf{K}_{22}$ are, respectively, the upper left $p \times p$ submatrix and the lower right $(n-p) \times (n-p)$ submatrix of $\mathbf{R}$, with all their elements being positive; while $\mathbf{R}_{12}$ and $\mathbf{R}_{21}$ are, respectively, the upper right $p \times (n-p)$ submatrix and the lower left $(n-p) \times p$ submatrix of $\mathbf{R}$, with all their elements being negative. Thus for a permutation matrix, if its nonzero elements are all in the submatrices $\mathbf{K}_{11}$ and $\mathbf{K}_{22}$, their corresponding k_{ij} are all positive. Therefore, these permutation matrices (up to sign indeterminacy) are all local maxima of $J(\mathbf{R})$. Clearly, there are $p!(n-p)!$ such permutation matrices. For $0 \leq p \leq n$, the number of these permutation matrices is fairly large. Therefore, there always exists many local maxima of $J(\mathbf{R})$ that correspond to the stable feasible solutions of the ICA problem. In other words, the ICA problem has many stable feasible solutions under the one-bit-matching condition via locally maximizing the objective function $J(\mathbf{R})$.

In the similar context, we can prove the following theorem.

Theorem 2. *If* $\mathbf{R}^*$ *is a permutation matrix up to sign indeterminacy and* $k_{ij} < 0$ *at all the positions where* $|r^*_{ij}| = 1$*, it corresponds to a local minimum of the objective function* $J(\mathbf{R})$.

Remark 2. According to Theorem 2 and under the one-bit-matching condition, if the nonzero elements of a permutation matrix are all in the submatrices $\mathbf{K}_{12}$ and $\mathbf{K}_{21}$, it is a local minimum of $J(\mathbf{R})$. That is, it is possible that the local minimum of the objective function can be a feasible solution of the ICA problem, which actually explains why a local gradient-descent search of the objective function can also lead to a feasible solution of the ICA problem in certain scenarios. However, this kind of permutation matrix can only exist in the special case where $n = 2p$ (i.e., half super-Gaussian and half sub-Gaussian).

Moreover, since the condition (8) is also necessary for a local optimum solution (maximum or minimum) of the constrained function we can conclude that if the numbers of positive and negative k_{ij} at the positions where $|r^*_{ij}| = 1$ are both greater than 1, $\mathbf{R}^*$ will be a saddle point of the objective function $J(\mathbf{R})$. Clearly, such a permutation matrix generally exists and also corresponds to a feasible solution of the ICA problem with mixed super- and sub-Gaussian sources; however, this solution is always unstable.

To sum up the above results, we have established that under the one-bit-matching condition, there always exist many stable feasible solutions of the ICA problem via locally maximizing the objective function (6); in the meanwhile, there may exist some unstable feasible solutions of the ICA problem; in addition, there may exist local minima of $J(\mathbf{R})$ that correspond to the stable feasible solutions in the cases of mixed super- and sub-Gaussian sources.

4 Conclusion

In this paper, we have analyzed the feasible solutions of the ICA problem under the one-bit-matching condition. By mathematical analysis, we have proved that there always exist many stable feasible solutions of the ICA problem under the one-bit-matching condition. In the meanwhile, under the one-bit-matching condition, there may exist some unstable feasible solutions of the ICA problem; moreover, there may exist local minima of $J(\mathbf{R})$ corresponding to the stable feasible solutions of the ICA problem with mixed super- and sub-Gaussian sources.

References

1. Tong L., Inouye Y., and Liu R.: Waveform-preserving blind estimation of multiple independent sources. IEEE Trans. Signal Processing, 41(7)(1993) 2461-2470
2. Comon P.: Independent component analysis–a new concept?. Signal Processing, 36(1994) 287-314
3. Delfosse N. and Loubaton P.: Adaptive blind separation of independent sources: A deflation approach. Signal Processing, 45(1995) 59-83
4. Bell A. and Sejnowski T.: An information-maximization approach to blind separation and blind deconvolution. Neural Computation, 7(1995) 1129-1159
5. Amari S. I., Cichocki A., and Yang H.: A new learning algorithm for blind separation of sources. Advances in Neural Information Processing Systems, 8(1996) 757-763.
6. Oja E.: ICA learning rules: stationarity,stability, and sigmoids. In C. Fyfe (ed.), Proc. of Int.ICSC Workshop on Independence and Artificial Neural Networks, pp. 97-103, 1998.
7. Cardoso J. F.: Infomax and maximum likelihood for source separation. IEEE Signal Processing Letters, 4(1999) 112-114
8. Xu L., Cheung C. C., Yang H., and Amari S. I.: Independent component analysis by the information-theoretic approach with mixture of density. Proc. 1997 IEEE Int. Joint Conf. Neural Networks, III, pp. 1821-1826, 1997.
9. Xu L., Cheung C. C., and Amari S. I.: Learned parametric mixture based ICA algorithm. Neurocomputing, 22(1998) 69-80
10. Xu L., Cheung C. C., and Amari S. I.: Furthere results on nonlinearity and separtion capability of a liner mixture ICA method and learned LPM. In C. Fyfe (Ed.) *Proc. I&ANN'98*, pp. 39-45, 1998.
11. Girolami M.: An alternative perspective on adaptive independent component analysis algorithms. Neural Computation, 10(1998) 2103-2114
12. Everson R. and Roberts S.: Independent component analysis: A flexible nonlinearity and decorrelating manifold approach. Neural Computation, 11(1999) 1957-1983
13. Welling M. and Weber M.: A constrained EM algorithm for independent component analysis. Neural Computation, 13(2001) 677-689
14. Gao D., Ma J., and Cheng Q.: An alternative switching criterion for independent component analysis (ICA). Neurocomputing, 68(2005) 267-272
15. Cheung C. C. and Xu L.: Some global and local convergence analysis on the information-theoretic independent component analysis approach. Neurocomputing, 30(2000) 79-102
16. Liu Z. Y., Chiu K. C., and Xu L.: One-bit-matching conjecture for independent component analysis. Neural Computation, 16(2004) 383-399
17. Ma J. Liu Z., and Xu L.: A further result on the ICA one-bit-matching conjecture. Neural Computation, 17(2005) 331-334
18. Himmelblau D. M.: Applied Nonlinear Programming. McGraw-Hill Book Company, New York, 1972.
19. Edelman A., Arias T. A., and Smith S. T.: The geometry of algorithms with orthogonality constraints. SIAM J. Matrix Anal. Appl., 20(1998) 303-353

Kernel Principal Components Are Maximum Entropy Projections*

António R.C. Paiva, Jian-Wu Xu, and José C. Príncipe

Computational NeuroEngineering Laboratory,
Dept. of Electrical and Computer Engineering,
University of Florida, Gainesville, FL 32611, USA
{arpaiva, jianwu, principe}@cnel.ufl.edu

Abstract. Principal Component Analysis (PCA) is a very well known statistical tool. KERNEL PCA is a nonlinear extension to PCA based on the kernel paradigm. In this paper we characterize the projections found by KERNEL PCA from a information theoretic perspective. We prove that KERNEL PCA provides optimum entropy projections in the input space when the Gaussian kernel is used for the mapping and a sample estimate of Renyi's entropy based on the Parzen window method is employed. The information theoretic interpretation motivates the choice and specifices the kernel used for the transformation to feature space.

Keywords: Kernel PCA, information-theoretic learning, entropy projections.

1 Introduction

Many real world problems deal with a very high number of signals not all equally important for the application. Therefore, a simplification of the problem is often desirable, and sometimes imperative. The goal is to obtain a smaller number of projections that describes the data and minimize the loss of information in the projection. A very well known statistical tool for data projection is Principal Component Analysis (PCA) [1]. PCA searches for the projections of maximum variance. If the process that generated the data is Gaussian this projection is optimum. This is because Gaussian processes are totally described by their mean and variance. The same is not true, however, for other data distributions.

PCA can be formulated in terms of inner (or dot) products. Following a recent trend, a kernel-based extension named KERNEL PCA was proposed by Schölkopf et al. [2, 3]. In fact, it has been pointed out that any algorithm that can be formulated using only dot products can be immediately *kernelized*, yielding an easily trackable nonlinear formulation. KERNEL PCA performs PCA in feature space. It has been verified, that by selecting the kernel appropriately, it is possible to find a projection in the input space that is more descriptive of the data, even

* This work was supported in part by Fundação para a Ciência e a Tecnologia (FCT) grant SFRH/BD/18217/2004 and NSF grant ECS-0300340.

J. Rosca et al. (Eds.): ICA 2006, LNCS 3889, pp. 846–853, 2006.

if the data is described by a non-Gaussian distribution. Recently, Williams [4] pointed out that KERNEL PCA algorithm can be interpreted as a form of multidimensional scaling provided that the kernel function $\kappa(\mathbf{x}, \mathbf{y})$ is isotropic, i.e. it depends only on $\| \mathbf{x} - \mathbf{y} \|$. This connection provides a metric multidimensional scaling algorithm to solve KERNEL PCA instead of a eigendecomposition of the Gram matrix. Bengio et al. [5] pointed out the link between KERNEL PCA and spectral embedding. The direct relation resides in a more general learning problem: learning the principal eigenfunctions of operators defined from a kernel and the unknown data-generating density function.

In this paper we take an information-theoretic perspective to KERNEL PCA. We show a direct connection between KERNEL PCA and maximization of entropy, and prove mathematically why this happens. As Bach and Jordan [6] pointed out, this insight is also highly valuable to ICA, since ICA can be viewed as a generalization of PCA, one that depends on high order moments. Although a relation between KERNEL PCA and ICA is not made here, the demonstration we make inherently connects both concepts.

2 Kernel PCA

Let $\mathbf{x}_i$, $i = 1, \ldots, M$ be a set of M sample vectors in a N-dimensional (input) space, and $\mathbf{\Phi}(\cdot) : \mathbb{R}^N \longrightarrow \mathcal{F}$ be the mapping to the feature space. KERNEL PCA is simply PCA applied in feature space. Hence, the goal of KERNEL PCA is to find variance maximizing projections of the vectors $\mathbf{\Phi}(\mathbf{x}_i)$. If the vectors $\mathbf{\Phi}(\mathbf{x}_i)$, $i = 1, \ldots, M$ have zero mean KERNEL PCA can be stated as the following optimization problem: we wish to maximize the cost function

$$J(\mathbf{w}) = E\left\{(\mathbf{w}^T \mathbf{\Phi}(\mathbf{x}))^2\right\}. \tag{1}$$

Because the above equation depends on the norm of the projection vector, the Lagrange multiplier method is used to force the vectors to unit norm. Thus, the following cost function is maximized instead

$$\begin{aligned} J(\mathbf{w}) &= E\left\{(\mathbf{w}^T \mathbf{\Phi}(\mathbf{x}))^2\right\} - \lambda(\mathbf{w}^T \mathbf{w} - 1) \\ &= \mathbf{w}^T E\left\{\mathbf{\Phi}(\mathbf{x})\mathbf{\Phi}(\mathbf{x})^T\right\} \mathbf{w} - \lambda(\mathbf{w}^T \mathbf{w} - 1). \end{aligned} \tag{2}$$

Notice that $\mathbf{C} = E\left\{\mathbf{\Phi}(\mathbf{x})\mathbf{\Phi}(\mathbf{x})^T\right\}$ is the covariance matrix of the vectors in the feature space. The solution of (2) is found by solving

$$\mathbf{C}\mathbf{w} = \lambda \mathbf{w}. \tag{3}$$

As for PCA, the solutions to this equation are well known to be the eigenvectors and eigenvalues of the covariance matrix, although in this situation computed in feature space. Solving this problem directly in feature space is very complicated. Fortunately, this equation can be restated in terms of dot products, for which a solution can be easily found, as we shown next.

As all solutions $\mathbf{w}$ of (3) for which $\lambda \geq 0$ lie in the span of the transformed vectors we can write,

$$\mathbf{w} = \sum_{i=1}^{M} \alpha_i \mathbf{\Phi}(\mathbf{x}_i). \quad (4)$$

Also, the covariance matrix of the transformed vectors can be estimated from the vectors as

$$C = \frac{1}{M} \sum_{i=1}^{M} \mathbf{\Phi}(\mathbf{x}_i)\mathbf{\Phi}(\mathbf{x}_i)^T. \quad (5)$$

Returning to the problem of the eigendecomposition of the covariance matrix of the feature vectors, we have that (3) is equivalent to

$$\langle \mathbf{\Phi}(\mathbf{x}_k), \mathbf{C}\mathbf{w} \rangle = \lambda \langle \mathbf{\Phi}(\mathbf{x}_k), \mathbf{w} \rangle, \text{for all } k = 1, \ldots, M. \quad (6)$$

Then, substituting (4) and (5) yields

$$\frac{1}{M} \sum_{i=1}^{M} \mathbf{\Phi}^T(\mathbf{x}_k)\mathbf{\Phi}(\mathbf{x}_i) \sum_{j=1}^{M} \alpha_j \mathbf{\Phi}^T(\mathbf{x}_i)\mathbf{\Phi}(\mathbf{x}_j) = \lambda \sum_{j=1}^{M} \alpha_j \mathbf{\Phi}^T(\mathbf{x}_k)\mathbf{\Phi}(\mathbf{x}_j),$$
$$\text{for all } k = 1, \ldots, M. \quad (7)$$

Defining the Gram matrix $\mathbf{K}$, as $\mathbf{K}_{ij} = \kappa(\mathbf{x}_i, \mathbf{x}_j) = \langle \mathbf{\Phi}(\mathbf{x}_i), \mathbf{\Phi}(\mathbf{x}_j) \rangle$, $i, j = 1, \ldots, M$, we can rewrite (7) in matrix form as

$$\mathbf{K}^2 \alpha = M \lambda \mathbf{K} \alpha. \quad (8)$$

where $\alpha = [\alpha_1, \ldots, \alpha_M]^T$. This equation has solutions found by the eigendecomposition of $\mathbf{K}$ but, most important of all, is that tells us that the eigenvectors of the Gram Matrix are the coefficients the decomposition of the eigenvectors of $\mathbf{C}$. Consequently, the projection of a feature vector is

$$\langle \mathbf{\Phi}(\mathbf{x}), \mathbf{w}_j \rangle = \sum_{i=1}^{M} \alpha_i^j \langle \mathbf{\Phi}(\mathbf{x}), \mathbf{\Phi}(\mathbf{x}_i) \rangle, \quad (9)$$

where $\mathbf{w}_j$ denotes the j-th *positive* eigenvector of $\mathbf{C}$.

3 Information Theoretic Concepts

In this section we briefly introduce some of the information theoretic core concepts needed to later establish its connection to KERNEL PCA.

The key information measure in information theoretic applications is Rényi's quadratic entropy [7], defined for the pdf, $f(\mathbf{x})$, of a random variable $\mathbf{X}$ as

$$H(\mathbf{x}) = -\log \int_{-\infty}^{\infty} f^2(\mathbf{x}) dx = -\log E\{f(\mathbf{x})\}. \quad (10)$$

The argument of the logarithm,

$$V(\mathbf{x}) = \int_{-\infty}^{\infty} f^2(\mathbf{x})dx = E\left\{f(\mathbf{x})\right\}, \tag{11}$$

is what it is called the *information potential* (IP), so named due to a similarity with the potential energy field in physics [8]. Notice that the information potential depends directly on the pdf of $\mathbf{X}$, which is normally unknown. Luckily, we can circumvent the explicit estimation of the pdf because entropy is a "moment" of the pdf. In fact, using the Parzen window method [9], written as

$$\hat{f}(\mathbf{x}) = \frac{1}{N}\sum_{i=1}^{N} \kappa_{\sigma/\sqrt{2}}(\mathbf{x}, \mathbf{x}_i), \tag{12}$$

where $\kappa_{\sigma/\sqrt{2}}(\mathbf{x}, \mathbf{x}_i)$ is the estimation kernel, commonly taken as a Gaussian,with bandwidth $\sigma/\sqrt{2}$, although other kernels may be used [9]. This kernel must be a valid pdf, i.e. be positive and integrate to one. Then, substituting this estimator in the IP we do not need to explicitly compute the integral because the integral of a product of Gaussians is a Gaussian (with twice the variance), yielding directly

$$\hat{V}(\mathbf{x}) = \frac{1}{N^2}\sum_{i=1}^{N}\sum_{j=1}^{N} \kappa_{\sigma}(\mathbf{x}_i, \mathbf{x}_j). \tag{13}$$

Although the information potential as given by the previous equation is an approximation, this is only to the extent of the error in the pdf estimation. In other words, if $\hat{f}(\mathbf{x})$ from (12) equals the true pdf then the estimator given by (13) also has no error.

For any Mercer kernel, one can employ *Mercer's theorem*,

$$\kappa_{\sigma}(\mathbf{x}_i, \mathbf{x}_j) = \langle \mathbf{\Phi}(\mathbf{x}_i), \mathbf{\Phi}(\mathbf{x}_j) \rangle, \tag{14}$$

to rewrite the information potential of (13) as [10]

$$\hat{V}(\mathbf{x}) = \frac{1}{N^2}\sum_{i=1}^{N}\sum_{j=1}^{N} \langle \mathbf{\Phi}(\mathbf{x}_i), \mathbf{\Phi}(\mathbf{x}_j) \rangle = \left\langle \frac{1}{N}\sum_{i=1}^{N} \mathbf{\Phi}(\mathbf{x}_i), \frac{1}{N}\sum_{j=1}^{N} \mathbf{\Phi}(\mathbf{x}_j) \right\rangle = \| \mu_{\mathbf{\Phi}} \|^2, \tag{15}$$

where $\mu_{\mathbf{\Phi}}$ is the mean of the vectors in feature space. That is, the information potential is the squared norm of the mean vector of the data in kernel space. This equation shows exactly the duality existing between the information potential and second order statistics computed in feature space on the transformed data.

Finally, we remark that extremization (maximization or minimization) of $H(\mathbf{x})$ can be alternatively achieved by extremizing the information potential in the opposite direction, because of the minus signs in (10) and the fact that the logarithm is a monotonic function. Hence, if we wish to maximize the entropy we can simply minimize the information potential. Conversely, maximizing the information potential yields minimum entropy.

4 Characterization of Kernel PCA Projections in Input Space

In section 2 explained the fundamentals of KERNEL PCA. The most important point was to explicitly formulate KERNEL PCA as a tool for finding projections of maximum variance in feature space, as (2) states. On the other hand, (15) shows that a relationship between second order statistics in the feature space and quadratic Renyi's entropy in the input space exists.

Let us analyze in detail what is the meaning of the variance of the feature vectors. The variance of the feature vectors is

$$\mathrm{var}(\mathbf{\Phi}(\mathbf{x})) = E\left\{\mathbf{\Phi}(\mathbf{x})^T\mathbf{\Phi}(\mathbf{x})\right\} - E\left\{\mathbf{\Phi}(\mathbf{x})\right\}^T E\left\{\mathbf{\Phi}(\mathbf{x})\right\}. \tag{16}$$

Expressing the inner product as a kernel operation and using identity (15),

$$\mathrm{var}(\mathbf{\Phi}(\mathbf{x})) = E\left\{\kappa(\mathbf{x},\mathbf{x})\right\} - V(\mathbf{x}). \tag{17}$$

The quantity $E\left\{\kappa(\mathbf{x},\mathbf{x})\right\}$ is the information potential at the origin, $V(0)$. This is a constant value representing the zero entropy situation, for which the maximum value of the information potential is achieved. From (17), maximizing the variance of the feature vectors corresponds therefore to the minimization of the information potential, $V(\mathbf{x})$, in the input space.

The fact that KERNEL PCA finds projections that minimize the information potential in input space, together with the remarks made in section 3 on the relationship between the information potential and entropy prove the statement that kernel principal components are maximum entropy projections. Since entropy is associated with information [11], maximum entropy projections are the directions more informative to explore for machine learning algorithms. Furthermore, notice that at no point in our proof of this connection an assumption of a specific kernel was made, other than it has to be able to accurately provide an estimation to the input sample vectors pdf.

5 Example

In this section we illustrate what was just proved in the previous section. We will use a small example, in which the goal is to obtain the maximum informative projection of a mixture of two Gaussian distributions. The overall pdf is specified by

$$p(\mathbf{x}) = \frac{1}{2}\left(N(\mathbf{x},\mu_1,\mathbf{\Sigma}_1) + N(\mathbf{x},\mu_2,\mathbf{\Sigma}_2)\right), \tag{18}$$

where $N(\mathbf{x},\mu,\mathbf{\Sigma})$ is a Gaussian distribution with mean μ and covariance matrix $\mathbf{\Sigma}$. In this case,

$$\mu_1 = \begin{bmatrix}-1\\-1\end{bmatrix}, \quad \mathbf{\Sigma}_1 = \begin{bmatrix}1 & 0\\0 & 0.1\end{bmatrix}, \quad \mu_2 = \begin{bmatrix}1\\1\end{bmatrix}, \quad \mathbf{\Sigma}_2 = \begin{bmatrix}0.1 & 0\\0 & 1\end{bmatrix}.$$

For reference, we show the contours of constant projection (constant eigenvalue) of standard PCA in Fig. 1(a). Recall that the projection is made along a line orthogonal to the contours. The contours for KERNEL PCA using a Gaussian kernel are a little more difficult to construct, since KERNEL PCA has as many principal directions as the size of the Gram matrix. In exploratory data analysis, what matters are the directions in the input space, and it is not clear how they are related. In this case we decided to plot in Fig. 1(b)-(c) the direction corresponding to the maximum eigenvalue in kernel space, using a kernel size (variance) $\sigma^2 = 1$ and $\sigma^2 = 10$, respectively. Note how the contours bend

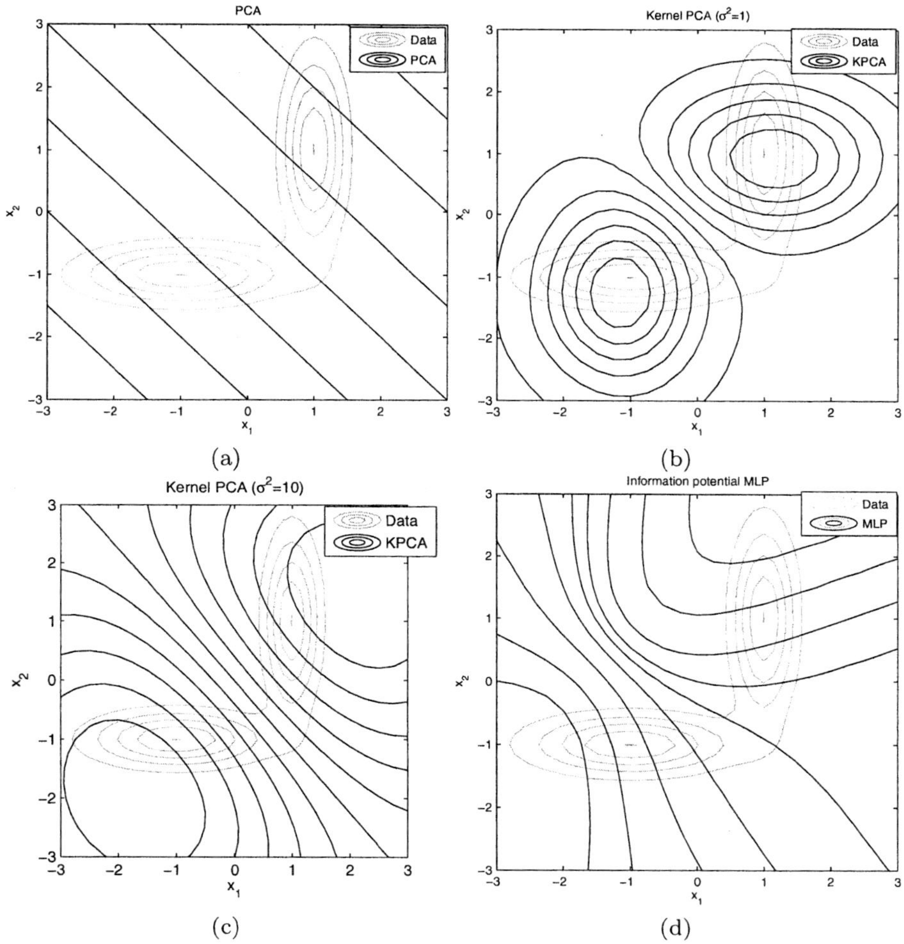

Fig. 1. Contours of constant projection and the pdf for the example of Sec. 5. From (a) to (d), the contours are for standard PCA, KERNEL PCA with $\sigma^2 = 1$, KERNEL PCA with $\sigma^2 = 10$, and MLP output.

themselves and wrap around the distribution to make the projection as uniform as possible. Also, when the kernel size is increased (Fig. 1(c)) the contours tend to those of standard PCA (Fig. 1(a)).

With our information theoretic interpretation there is another alternative to create the maximum entropy direction that uses always the input space dimension. In fact, we can train with backpropagation a MLP with architecture 2-4-1 (2 inputs, 4 hidden PEs and 1 output PE) and instead of using the conventional MSE criterion, substitute it for the maximum entropy cost [8]. The MLP was trained for 200 epochs to minimize the information potential of the outputs, as evaluated by (13), with a kernel size of 0.2. The nonlinearity used at the PEs is the hyperbolic tangent function. The contours of the surface generated by the neural network are shown in Fig. 1(d). The contours are obviously different from the ones for Kernel PCA since the basis functions are different and the method uses gradient descent learning, but is remarkable how they bend so that a projection to a line orthogonal to these contours would have maximum entropy. Although in this example we are only interested in the first projection, the neural network framework can also be used to obtain as many projections as needed up to the dimensionality of the space by using concepts of orthogonalizing the outputs [12].

6 Conclusions

Kernel PCA was proposed as an nonlinear extension of PCA. Despite this simple motivation, in this paper we prove that the principal components determined by Kernel PCA provides optimum entropy mappings when the Gaussian kernel is used both for the mapping and in Parzen window pdf estimation method. The use of the Gaussian kernel is not restrictive since the same result holds for any Mercer theorem, although the connection between the pdf estimation and IP becomes more difficult to express. This motivates the choice for the kernel and, considering the implicit pdf estimation, how to select its parameters.

The main contribution of this work is the understanding of the underlying properties of the projections found by Kernel PCA in feature space. This insight becomes especially important if we intend to use Kernel PCA as a data exploratory tool. Notice how the projections of Kernel PCA and maximum entropy achieve fundamentally the same result, although they are different due to the differences in the basis functions used (Gaussians in kernel methods, ridge functions in the MLP). This is a very interesting result given that Kernel PCA has an analytical solution, while the MLP requires adaptation. Yet, the Kernel PCA looses the intuition of the meaning of PCA in the input space. Indeed, Schölkopf et al. [2] mention about the possibility of finding more eigenvectors than the dimension of the input space which is clearly misleading in data analysis. The maximum entropy projection brings the insight that effectively Kernel PCA is projecting the data in informative directions using local bases. Therefore, Kernel PCA will require many such projections to cover the full data space. However, it is still not clear how to distinguish a minor component from a major component since the bases are local.

References

1. Diamantaras, K.I., Kung, S.Y.: Principal Component Neural Networks: Theory and Applications. John Wiley & Sons (1996)
2. Schölkopf, B., Smola, A., Müller, K.R.: Nonlinear component analysis as a kernel eigenvalue problem. Neural Computation **10**(5) (1998) 1299–1319
3. Schölkopf, B., Smola, A., Müller, K.R.: Kernel principal component analysis. In Gerstner, W., Germond, A., Hasler, M., Nicoud, J.D., eds.: Proc. Artificial Neural Networks ICANN'97, Berlin, Springer Lecture Notes in Computer Science, Vol. 1327 (1997) 583–588
4. Williams, C.K.I.: On a connection between kernel pca and metric multidimensional scaling. Machine Learning **46**(1–3) (2002) 11–19
5. Bengio, Y., Delalleau, O., Roux, N.L., Paiement, J.F., Vincent, P., Ouimet, M.: Learning eigenfunctions links spectral embedding and kernel pca. Neural Computation **16**(10) (2004) 2197–2219
6. Bach, F.R., Jordan, M.I.: Kernel independent component analysis. Journal of Machine Learning Research **3** (2002) 1–48
7. Rényi, A.: On measures of entropy and information. In: Selected paper of Alfréd Rényi. Volume 2. Akademiai Kiado, Budapest, Hungary (1976) 565–580
8. Príncipe, J.C., Xu, D., Fisher, J.W.: Information theoretic learning. In Haykin, S., ed.: Unsupervised Adaptive Filtering. Volume 2. John Wiley & Sons (2000) 265–319
9. Parzen, E.: On the estimation of a probability density function and the mode. The Annals of Mathematical Statistics **33**(2) (1962) 1065–1076
10. Jenssen, R., Erdogmus, D., Príncipe, J.C., Eltoft, T.: Towards a unification of information theoretic learning and kernel methods. In: Proc. MLSP'04, São Luís, Brazil (2004)
11. Shannon, C.E.: A mathematical theory of communication. The Bell System Technical Journal **27** (1948) 379–423, 623–656
12. Sanger, T.D.: Optimal unsupervised learning in a single layer linear feedforward neural network. Neural Networks **2**(7) (1989) 459–473

Super-Gaussian Mixture Source Model for ICA

Jason A. Palmer[1,2], Kenneth Kreutz-Delgado[1], and Scott Makeig[2]

[1] Department of Electrical and Computer Engineering,
University of California San Diego, La Jolla, CA 92093
{japalmer, kreutz}@ece.ucsd.edu
[2] Swartz Center for Computational Neuroscience,
University of California San Diego, La Jolla, CA 92093
scott@sccn.ucsd.edu

Abstract. We propose an extension of the mixture of factor (or independent component) analyzers model to include strongly super-gaussian mixture source densities. This allows greater economy in representation of densities with (multiple) peaked modes or heavy tails than using several Gaussians to represent these features. We derive an EM algorithm to find the maximum likelihood estimate of the model, and show that it converges globally to a local optimum of the actual non-gaussian mixture model without needing any approximations. This extends considerably the class of source densities that can be used in exact estimation, and shows that in a sense super-gaussian densities are as natural as Gaussian densities. We also derive an adaptive Generalized Gaussian algorithm that learns the shape parameters of Generalized Gaussian mixture components. Experiments verify the validity of the algorithm.

1 Introduction

We propose an extension of the mixture of factor [2], or independent component [6] analyzers model that enlarges the flexibility of the source density mixture model while maintaining mixtures of strongly super-gaussian densities. Mixture model source densities allow one to model skewed and multi-modal densities, and optimization of these models is subject to convergence to local optima, the mixture model is a generalization of the unimodal model and may be built up by starting with uni- or bi-modal source models, then adding components and monitoring the change in likelihood [8,3,6].

Variational Gaussian mixture models, proposed in [8,2,6,5], are ultimately mixtures of Student's t distributions after the random variance is integrated out [19,3]. In [12] a mixture generalization of the Infomax algorithm is proposed in which a mixture model is employed over sets basis vectors but not for the source component density models. The means are updated by gradient descent or by a heuristic approximate EM update. In [16] a variance mixture of Laplacians model is employed over the source densities, in which the Laplacian components in each mixture have the same mean, but differing variances. An EM algorithm is derived by exploiting the closed form solution of the M-step for the variance parameters. In [17] a mixture of Logistic source density model is estimated by gradient descent.

The property of strongly super-gaussian densities that we use, namely log-convexity in x^2, has been exploited previously by Jaakkola [10,11] in graphical models, and Girolami [9] for ICA using the Laplacian density. The model we propose extends the work

J. Rosca et al. (Eds.): ICA 2006, LNCS 3889, pp. 854–861, 2006.

in [9] in applying more generally to the (large) class of strongly super-gaussian densities, as well as mixtures of these densities. We also take the approach of [3] in allowing the scale of the sources vary (actually a necessity in the mixture case) and fixing the scale of the de-mixing filters to unity by an appropriate transformation at each iteration in order to avoid the scale ambiguity inherent in factor analysis models.

The proposed model generalizes all of these algorithms, including Gaussian, Laplacian, Logistic, as well as Generalized Gaussian, Student's t, and any mixture combination of these densities. The key to the algorithm is the definition of an appropriate class of densities, and showing that the "complete log likelihood" that arises in the EM algorithm can be guaranteed to increase as a result of an appropriate parameter update, which thus guarantees increase in the true likelihood. It is thus a "Generalized EM" (GEM) algorithm [7]. For a given number of mixture components, the EM algorithm estimates the location (mode) and scale parameters of the mixture component.

Using the natural gradient [1] to update the un-mixing matrices (the inverses of the basis matrices), we can further guarantee (in principle) increase of the likelihood. Furthermore, it is possible, for densities that are parameterized besides the location and scale parameters such that all densities in a range of the additional parameter are strongly super-gaussian, e.g. Generalized Gaussian shape parameters less than 2, to update these parameters according to the gradient of the complete log likelihood, remaining within the GEM framework and guaranteeing increase in the data likelihood under the model. The un-mixing matrices and any other shape parameters will require a step size to be specified in advance, but the mixture component locations and scales will be updated in closed form. In the Gaussian case, the algorithm reduces to the classical EM algorithm for Gaussian mixtures.

The practical situation in which we shall be interested is the analysis of EEG/MEG, the characteristics of which are a large number of channels and data points, and mildly skewed, occasionally multi-modal source densities. The large number of channels constrains the algorithm to be scalable. This along with the large number of data points suggests the natural gradient maximum likelihood approach, which is scalable and asymptotically efficient. The large amount of data also dictates that we limit computational and storage overhead to only what is necessary or actually beneficial, rather than doing Bayesian MAP estimation of all parameters as in the variational Bayes algorithms [3,6]. Also for computational reasons we consider only noiseless mixtures of complete bases so that inverses exist.

In §2 we define strongly super-gaussian densities and mixtures of these densities. In §3-5 we derive the EM algorithm for density estimation. In §6 we introduce an adaptive generalized Gaussian algorithm. §7 contains experimental verification of the theory.

2 Strongly Super-Gaussian Mixtures

Definition 1. *A symmetric probability density $p(x)$ is **strongly super-gaussian** if $g(x) \equiv -\log p(\sqrt{x})$ is concave on $(0, \infty)$, and **strongly sub-gaussian** if $g(x)$ is convex.*

An equivalent definition is given in [4], where the authors define $p(x) = \exp(-f(x))$ to be super-gaussian (sub-gaussian) if $f'(x)/x$ is increasing (decreasing) on $(0, \infty)$.

This condition is equivalent to $f(x) = g(x^2)$ with g concave, i.e. g' decreasing, where $g'(x^2) = f'(x)/x$.

In [15] we have discussed these densities in some detail, and derived relationships between them and the hyperprior representation used in the evidence framework [13] and the Variational Bayes framework [2]. Here we limit consideration to strongly super-gaussian mixture densities. If $p(s)$ is strongly super-gaussian, we have $f(s) \equiv g(s^2)$, with g concave on $(0, \infty)$. This implies that, $\forall t$,

$$f(t) - f(s) \;=\; g(t^2) - g(s^2) \;\le\; g'(s^2)(t^2 - s^2) \;=\; \frac{1}{2}\frac{f'(s)}{s}(t^2 - s^2) \qquad (1)$$

Examples of densities satisfying this criterion include: (i) Generalized Gaussian $\propto \exp(-|x|^\beta)$, $0 < \beta \le 2$, (ii) Logistic $\propto 1/\cosh^2(x/2)$, (iii) Student's $t \propto (1 + x^2/\nu)^{-(\nu+1)/2}$, $\nu > 0$, and (iv) symmetric α-stable densities (having characteristic function $\exp(-|\omega|^\alpha)$, $0 < \alpha \le 2$). The property of being strongly sub- or super-gaussian is independent of scale.

Mixture densities have the form,

$$p(s) = \sum_{j=1}^{m} \alpha_j \, p_j\!\left(\frac{s - \mu_j}{\sigma_j}\right), \quad \sum_j \alpha_j = 1\,,\; \sigma_j > 0$$

The probability density of the j_ith mixture component of the ith source is denoted $p_{ij_i}(s_{ij_i})$, with mode μ_{ij_i}, and scale σ_{ij_i}.

3 The EM Algorithm

We follow the framework of [18,14] in deriving the EM algorithm, which was originally derived rigorously in [7]. The log likelihood of the data decomposes as follows,

$$\begin{aligned}\log p(\mathbf{x};\theta) &= \int q(\mathbf{z}|\mathbf{x};\theta') \log \frac{p(\mathbf{z},\mathbf{x};\theta)}{q(\mathbf{z}|\mathbf{x};\theta')}\, d\mathbf{z} \,+\, D\big(q(\mathbf{z}|\mathbf{x};\theta') \,\|\, p(\mathbf{z}|\mathbf{x};\theta)\big)\\ &\equiv -F(q;\theta) \,+\, D(q\|p_\theta)\end{aligned}$$

where q is an arbitrary density and D is the Kullback-Leibler divergence. The term $F(q;\theta)$ is commonly called the *variational free energy* [18,14]. This representation is useful if $F(q;\theta)$ can easily be minimized with respect to θ. Since the KL divergence is non-negative, we have,

$$-\log p(\mathbf{x};\theta) = \min_q F(q;\theta)$$

where equality is obtained if and only if $q(\mathbf{z}|\mathbf{x};\theta') = p(\mathbf{z}|\mathbf{x};\theta)$. The EM algorithm at the lth iteration, given q^l and θ^l, performs coordinate descent in q and θ,

$$\theta^{l+1} = \min_\theta F\big(q^l;\theta\big)\,, \quad q^{l+1} = p\big(\mathbf{z}|\mathbf{x};\theta^{l+1}\big)$$

This algorithm is guaranteed to increase the likelihood since,

$$-\log p(\mathbf{x};\theta^l) = F(q^l;\theta^l) \ge F(q^l;\theta^{l+1}) \ge F(q^{l+1};\theta^{l+1}) = -\log p(\mathbf{x};\theta^{l+1})$$

Note however, that it is not necessary to actually minimize F to guarantee that the likelihood increases. It is enough simply to guarantee that $F(q^l;\theta^l) \geq F(q^l;\theta^{l+1})$, i.e. to guarantee that F decreases as a result of updating θ. This leads to the Generalized EM (GEM) algorithm [7], and is the approach we follow here. We maintain the global convergence (to a local optimum) property of the EM algorithm however by guaranteeing a decrease in F by an efficient closed form update for the source density parameters.

4 ICA with Strongly Super-Gaussian Mixture Sources

Let the data $\mathbf{x}_k$, $k = 1,\ldots,N$ be given, and consider the model,

$$\mathbf{x}_k = \mathbf{A}\mathbf{s}_k$$

where $\mathbf{A} \in \mathbb{R}^{n\times n}$ is non-singular, and the sources are independent mixtures of independent strongly super-gaussian random variables s_{ij_i}, $j_i = 1,\ldots,m_i$, where we allow the number of source mixture components m_i to differ for different sources.

The source mixture model is equivalent to a scenario in which for each source s_i, a mixture component j_i is drawn from the discrete probability distribution $P[j_i = j] = \alpha_{ij}$, $1 \leq j \leq m_i$, then s_i is drawn from the mixture component density p_{ij_i}. We define j_{ik} to be the index chosen for the ith source in the kth sample.

We wish to estimate the parameters $\mathbf{W} = \mathbf{A}^{-1}$ and the parameters of the source mixtures, so we have,

$$\theta = \{\mathbf{w}_i, \alpha_{ij_i}, \mu_{ij_i}, \sigma_{ij_i}\}, \quad i = 1,\ldots,n\,,\; j_i = 1,\ldots,m_i$$

where $\mathbf{w}_i$ is the ith column of $\mathbf{W}^T$. We define $\mathbf{X} = [\mathbf{x}_1 \cdots \mathbf{x}_N]$.

To use the EM algorithm, we define the random variables z_{ij_ik} as follows,

$$z_{ij_ik} = \begin{cases} 1, & j_{ik} = j_i \\ 0, & \text{otherwise} \end{cases}$$

Let $\mathbf{Z} = \{z_{ij_ik}\}$. Then we have,

$$p(\mathbf{X};\theta) = \sum_{\mathbf{Z}} \prod_{k=1}^{N} |\det \mathbf{W}| \prod_{i=1}^{n} \prod_{j_i=1}^{m_i} \alpha_{ij_i}^{z_{ij_ik}} \left[\frac{1}{\sigma_{ij_i}} p_{ij_i}\!\left(\frac{\mathbf{w}_i^T\mathbf{x}_k - \mu_{ij_i}}{\sigma_{ij_i}}\right)\right]^{z_{ij_ik}}$$

For the variational free energy, F, we have,

$$F(q;\theta) = \sum_{k=1}^{N}\sum_{i=1}^{n}\sum_{j_i=1}^{m_i} \hat{z}_{ij_ik}\left[-\log\alpha_{ij_i} - \log\sigma_{ij_i} + f_{ij_i}\!\left(\frac{\mathbf{w}_i^T\mathbf{x}_k - \mu_{ij_i}}{\sigma_{ij_i}}\right)\right] - N\log|\det\mathbf{W}| \quad (2)$$

where q is the discrete distribution defining the expectation $\hat{z}_{ij_ik} = E[z_{ij_ik}|\mathbf{x}_k]$, and where we define $f_{ij_i} = -\log p_{ij_i}$.

Let us define,

$$y_{ij_ik}^l \equiv \frac{\mathbf{w}_i^{l\,T}\mathbf{x}_k - \mu_{ij_i}^l}{\sigma_{ij_i}^l} \tag{3}$$

The $\hat{z}_{ij_ik}^l = P[z_{ij_ik} = 1|\mathbf{x}_k;\theta^l]$ are determined as in the usual Gaussian EM algorithm,

$$\hat{z}_{ij_ik}^l = \frac{p(\mathbf{x}_k|z_{ij_ik} = 1;\theta^l)P[z_{ij_ik} = 1;\theta^l]}{\sum_{j_i'=1}^{m_i} p(\mathbf{x}_k|z_{ij_i'k} = 1;\theta^l)P[z_{ij_i'k} = 1;\theta^l]} = \frac{p_{ij_i}(y_{ij_ik}^l)\,\alpha_{ij_i}^l/\sigma_{ij_i}^l}{\sum_{j_i'=1}^{m_i} p_{ij_i'}(y_{ij_i'k}^l)\,\alpha_{ij_i'}^l/\sigma_{ij_i'}^l} \tag{4}$$

as are the optimal α_{ij_i},

$$\alpha_{ij_i}^{l+1} = \frac{\sum_{k=1}^N \hat{z}_{ij_ik}^l}{\sum_{j_i'=1}^{m_i}\sum_{k=1}^N \hat{z}_{ij_i'k}^l} = \frac{1}{N}\sum_{k=1}^N \hat{z}_{ij_ik}^l$$

Now, since the p_{ij_i} are strongly super-gaussian, we can use the inequality (1) to replace $f_{ij_i}(y_{ij_ik})$ in (2) by $\big(f_{ij_i}'(y_{ij_ik}^l)/2y_{ij_ik}^l\big)\big(y_{ij_ik}^2 - y_{ij_ik}^{l\,2}\big)$. Defining,

$$\xi_{ij_ik}^l \equiv \frac{f_{ij_i}'(y_{ij_ik}^l)}{y_{ij_ik}^l} \tag{5}$$

we replace F by,

$$\tilde{F}(q;\theta) = \sum_{k=1}^N\sum_{i=1}^n\sum_{j_i=1}^{m_i} \hat{z}_{ij_ik}\left[-\log\alpha_{ij_i} - \log\sigma_{ij_i} + \frac{\xi_{ij_ik}^l}{2}\left(\frac{\mathbf{w}_i^T\mathbf{x}_k - \mu_{ij_i}}{\sigma_{ij_i}}\right)^2\right] - N\log|\det\mathbf{W}|$$

Minimizing $\tilde{F}$ with respect to μ_{ij_i} and σ_{ij_i} guarantees, using the inequality (1), that,

$$F(q;\theta^{l+1}) - F(q;\theta^l) \;\le\; \tilde{F}(q;\theta^{l+1}) - \tilde{F}(q;\theta^l) \;\le\; 0$$

and thus that $F(q;\theta)$ is decreased as required by the EM algorithm.

As in the Gaussian case, the optimal value of μ_{ij_i} does not depend on σ_{ij_i}, and we can optimize with respect to μ_{ij_i}, then optimize with respect to σ_{ij_i} given μ_{ij_i}, and guarantee an overall increase in the likelihood. The updates, using the definitions (3), (4) and (5), are found to be,

$$\mu_{ij}^{l+1} = \frac{\sum_{k=1}^N \hat{z}_{ijk}^l\xi_{ijk}^l\mathbf{w}_i^{l\,T}\mathbf{x}_k}{\sum_{k=1}^N \hat{z}_{ijk}^l\xi_{ijk}^l}, \quad \sigma_{ij}^{l+1} = \left(\frac{\sum_{k=1}^N \hat{z}_{ijk}^l\xi_{ijk}^l(\mathbf{w}_i^{l\,T}\mathbf{x}_k - \mu_{ij}^{l+1})^2}{\sum_{k=1}^N \hat{z}_{ijk}^l}\right)^{1/2} \tag{6}$$

We adapt $\mathbf{W}$ according to the natural gradient of F (equivalently of $\tilde{F}$). Defining the vector $\mathbf{u}_k^l$ such that,

$$\big[\mathbf{u}_k^l\big]_i \equiv \sum_{j_i=1}^{m_i} \hat{z}_{ij_ik}^l f_{ij_i}'(y_{ij_ik}^l)/\sigma_{ij_i}^l \tag{7}$$

we have,

$$\Delta\mathbf{W} = \left(\frac{1}{N}\sum_{k=1}^N \mathbf{u}_k^l\mathbf{x}_k^T\mathbf{W}^{l\,T} - \mathbf{I}\right)\mathbf{W}^l \tag{8}$$

5 Full ICA Mixture Model with Super-Gaussian Mixture Sources

We now consider the case where the data are generated by a mixture of mixing matrices,

$$p(\mathbf{x}_k;\theta) = \sum_{h=1}^{M} \gamma_h p(\mathbf{x}_k;\theta_h)\,, \quad \sum_{h=1}^{M} \gamma_h = 1,\ \gamma_h > 0$$

where now we have,

$$\theta = \{\gamma_h, \mathbf{w}_{hi}, \alpha_{hij}, \mu_{hij}, \sigma_{hij}\}\,,\ h = 1,\ldots,M,\ i = 1,\ldots,n,\ j = 1,\ldots,m_{hi}$$

The EM algorithm for the full mixture model is derived similarly to the case of source mixtures. Due to space constraints the details are omitted.

6 Adaptive Generalized Gaussian Mixture Model

We can obtain further flexibility in the source model by adapting mixtures of a parameterized family of strongly super-gaussian densities. In this section we consider the case of Generalized Gaussian mixtures,

$$p(s_{ij_i};\mu_{ij_i},\sigma_{ij_i},\beta_{ij_i}) = \frac{1}{2\,\sigma_{ij_i}\Gamma\left(1+\frac{1}{\beta_{ij_i}}\right)} \exp\left(-\left|\frac{s_{ij_i}-\mu_{ij_i}}{\sigma_{ij_i}}\right|^{\beta_{ij_i}}\right)$$

The parameters β_{ij_i} are adapted by scaled gradient descent. The gradient of F with respect to β_{ij_i} is,

$$\frac{dF}{d\beta_{ij_i}} = \sum_{k=1}^{N} \hat{z}_{ij_ik} \left[|y_{ij_ik}|^{\beta_{ij_i}} \log|y_{ij_ik}| - \frac{1}{\beta_{ij_i}^2} \Psi\left(1+\frac{1}{\beta_{ij_i}}\right)\right]$$

We have found that scaling this by $\beta_{ij_i}^2 / \left(\Psi(1+\frac{1}{\beta_{ij_i}})\sum_{k=1}^{N}\hat{z}_{ij_ik}\right)$, which is positive, leads to faster convergence. The update is then,

$$\Delta\beta_{ij_i} = \frac{\beta_{ij_i}^2 \sum_{k=1}^{N} \hat{z}_{ij_ik}\,|y_{ij_ik}|^{\beta_{ij_i}} \log|y_{ij_ik}|}{\Psi\left(1+\frac{1}{\beta_{ij_i}}\right)\sum_{k=1}^{N}\hat{z}_{ij_ik}} - 1$$

7 Experiments

We verified the convergence of the algorithm with toy data generated from Generalized Gaussian mixtures with randomly generated parameters. Below we show an example of a super-gaussian mixture that was learned by the adaptive Generalized Gaussian mixture algorithm, including the shape parameter update, on a real EEG separation problem. Five mixture components per were used per source. The shape parameters were initialized to 1.5, the location and scale parameters were randomly initialized. The data was sphered and the unmixing matrix initialized to identity.

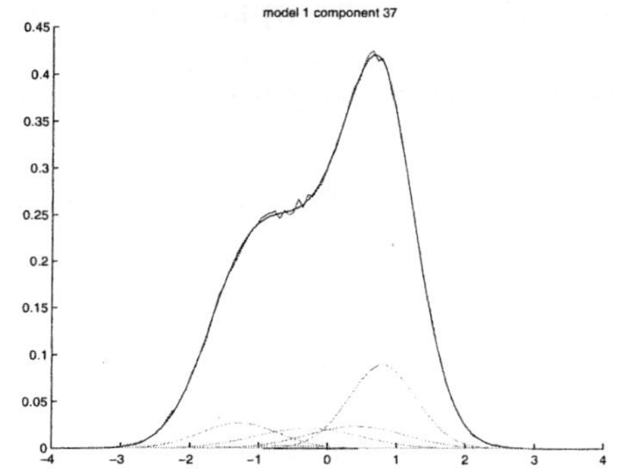

Fig. 1. Example of adaptive convergence of super-gaussian mixture model

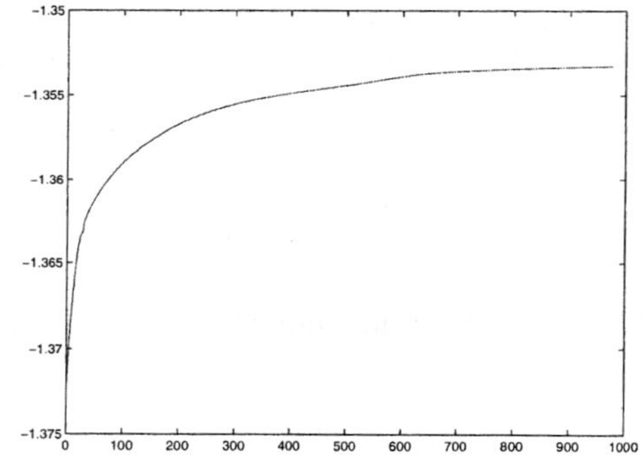

Fig. 2. Log likelihood is monotonically increasing

References

1. S.-I. Amari. Natural gradient works efficiently in learning. *Neural Computation*, 10(2):251–276, 1998.
2. H. Attias. Independent factor analysis. *Neural Computation*, 11:803–851, 1999.
3. H. Attias. A variational Bayesian framework for graphical models. In *Advances in Neural Information Processing Systems 12*. MIT Press, 2000.
4. A. Benveniste, M. Goursat, and G. Ruget. Robust identification of a nonminimum phase system. *IEEE Transactions on Automatic Control*, 25(3):385–399, 1980.
5. K. Chan, T.-W. Lee, and T. J. Sejnowski. Variational learning of clusters of undercomplete nonsymmetric independent components. *Journal of Machine Learning Research*, 3:99–114, 2002.
6. R. A. Choudrey and S. J. Roberts. Variational mixture of Bayesian independent component analysers. *Neural Computation*, 15(1):213–252, 2002.
7. A. P. Dempster, N. M. Laird, and D. B. Rubin. Maximum likelihood from incomplete data via the EM algorithm. *Journal of the Royal Statistical Society, Series B*, 39:1–38, 1977.

8. Z. Ghahramani and M. J. Beal. Variational inference for Bayesian mixtures of factor analysers. In *Advances in Neural Information Processing Systems 12*. MIT Press, 2000.
9. M. Girolami. A variational method for learning sparse and overcomplete representations. *Neural Computation*, 13:2517–2532, 2001.
10. T. S. Jaakkola. *Variational Methods for Inference and Estimation in Graphical Models*. PhD thesis, Massachusetts Institute of Technology, 1997.
11. T. S. Jaakkola and M. I. Jordan. A variational approach to Bayesian logistic regression models and their extensions. In *Proceedings of the 1997 Conference on Artificial Intelligence and Statistics*, 1997.
12. T.-W. Lee, M. S. Lewicki, and T. J. Sejnowski. ICA mixture models for unsupervised classification of non-gaussian classes and automatic context switching in blind signal separation. *IEEE Trans. Pattern Analysis and Machine Intelligence*, 22(10):1078–1089, 2000.
13. D. J. C. Mackay. Comparison of approximate methods for handling hyperparameters. *Neural Computation*, 11(5):1035–1068, 1999.
14. R. M. Neal and G. E. Hinton. A view of the EM algorithm that justifies incremental, sparse, and other variants. In M. I. Jordan, editor, *Learning in Graphical Models*, pages 355–368. Kluwer, 1998.
15. J. A. Palmer, K. Kreutz-Delgado, D. P. Wipf, and B. D. Rao. Variational EM algorithms for non-gaussian latent variable models. In *Advances in Neural Information Processing Systems*. MIT Press, 2005. Available at http://dsp.ucsd.edu/∼japalmer/.
16. H.-J. Park and T.-W. Lee. Modeling nonlinear dependencies in natural images using mixture of Laplacian distribution. In L. K. Saul, Y. Weiss, and L. Bottou, editors, *Advances in Neural Information Processing Systems 14*, Cambridge, MA, 2004. MIT Press.
17. B. A. Pearlmutter and L. C. Parra. Maximum likelihood blind source separation: A context-sensitive generalization of ICA. In M. Mozer, M. I. Jordan, and T. Petsche, editors, *Advances in Neural Information Processing Systems*. MIT Press, 1996.
18. L. K. Saul, T. S. Jaakkola, and M. I. Jordan. Mean field theory for sigmoid belief networks. *Journal of Artificial Intelligence Research*, 4:61–76, 1996.
19. M. E. Tipping. Sparse Bayesian learning and the Relevance Vector Machine. *Journal of Machine Learning Research*, 1:211–244, 2001.

Instantaneous MISO Separation of BPSK Sources

Maciej Pedzisz and Ali Mansour

École Nationale Supérieure d'Ingénieurs (ENSIETA),
Laboratoire "Extraction et Exploitation de l'Information,
en Environnements Incertains" (E^3I^2), Brest, France
pedzisma@ensieta.fr, mansour@ensieta.fr

Abstract. We present a new approach for signal separation from an undetermined instantaneous mixture of two BPSK (Binary Phase Shift Keying) signals in AWGN (Additive White Gaussian Noise) channel. The method uses frequency diversity of the mixture (frequency shift between carriers) and the fact that signals of interest are binary variables. We compare separation results of our method to a theoretical BER (Bit Error Rate) of unmixed signals, which reveal algorithm's performance for different communications scenarios.

1 Introduction

The problem of blind source separation (BSS) has been intensively studied in the literature and many effective solutions have been proposed in the case of instantaneous mixtures (memoryless channel) [1-4] and convolutive mixtures (channel effects can be considered as a linear filter) [5-9]. Most of the proposed algorithms deal with an undercomplete case (the number of sensors is equal or greater to the number of sources). For more sources than mixtures [10-14], the BSS problem is said to be overcomplete (undetermined) and is ill-posed.

In general, separation of overcomplete mixtures is still a real challenge for the scientific community. Even though the methods of identifying instantaneous mixing coefficients for n sources have been developed [15], they need at least 2 sensors. The same assumptions limit the method of separating undetermined mixtures proposed in [16] (two or more sensors).

In this contribution, the case of one sensor and two sources (undetermined problem) is addressed. The mixture is considered to be an instantaneous (memoryless channel) and the sources to be linearly modulated digital signals. Such a scenario can be found in a cellular phone reception, satellite transmissions, as well as in military communications (eg. signal interception, jamming or countermeasure). We present a new blind separation algorithm adopted to deal with BPSK signals, closely distributed in a frequency domain, as well as a method of identifying mixing coefficients. Experimental results reveal the performance of the proposed method.

J. Rosca et al. (Eds.): ICA 2006, LNCS 3889, pp. 862–867, 2006.

2 Signal Model

Let us consider a linear, instantaneous mixture $x(t)$ of two BPSK-type signals

$$x(t) = a_1 s_1(t) \mathrm{e}^{i(\omega_1 t + \varphi_1)} + a_2 s_2(t) \mathrm{e}^{i(\omega_2 t + \varphi_2)} \tag{1}$$

where a_k are unknown mixing coefficients, ω_k are carrier frequencies, φ_k are equivalent phases (sum of carrier and mixing coefficients phases), and $s_k(t)$ are equiprobable, i.i.d. random sequences during symbol period T_k[1]:

$$s_k(t) \in \{+1, -1\}, \quad \text{for} \quad t \in [l, (l+1)T_k] \tag{2}$$

We assume that source signals $s_k(t)$ are independent of each other and carrier frequencies ω_k are distinct and can be estimated [17, 18]. We consider scenario with small (compared to baud rates) frequency shifts, that any separation method based on signal filtering [19, 20] can't be applied. In a particular case where both signals are at the same frequency, the separation can be achieved using algorithm proposed in [21].

3 Theoretical Development

The basic idea of our algorithm consists of using frequency diversity of the mixture and the fact that signals of interest are binary variables. Assuming that carrier frequencies ω_k are already known, the mixing coefficients can be estimated using auxiliary signals defined as

$$Z_k(t) = x(t) \mathrm{e}^{-i\omega_k t} = a_k s_k(t) \mathrm{e}^{i\varphi_k} + a_l s_l(t) \mathrm{e}^{i((\omega_l - \omega_k)t + \varphi_l)} \tag{3}$$

for $k, l \in \{1, 2\}, k \neq l$, and mean values of its squares

$$\begin{aligned} \mathcal{E}\{Z_k^2(t)\} &= a_k^2 \mathcal{E}\{s_k^2(t)\} \mathrm{e}^{i2\varphi_k} + a_l^2 \mathcal{E}\{s_l^2(t)\} \mathcal{E}\left\{\mathrm{e}^{i2((\omega_l - \omega_k)t + \varphi_l)}\right\} \\ &\quad + 2a_k a_l \mathcal{E}\{s_k(t) s_l(t)\} \mathcal{E}\left\{\mathrm{e}^{i((\omega_l - \omega_k)t + \varphi_k + \varphi_l)}\right\} \end{aligned} \tag{4}$$

Using the fact that source signals are independent, and assuming that observation time is big enough, one has

$$\mathcal{E}\{s_k(t) s_l(t)\} = 0, \quad \text{and} \quad \mathcal{E}\left\{\mathrm{e}^{i2((\omega_l - \omega_k)t + \varphi_l)}\right\} = 0 \tag{5}$$

thus equation (4) becomes

$$\mathcal{E}\{Z_k^2(t)\} = a_k^2 \mathcal{E}\{s_k^2(t)\} \mathrm{e}^{i2\varphi_k} \tag{6}$$

For considered BPSK-type signals $\mathcal{E}\left\{s_k^2(t)\right\} = 1$, the mixing coefficients can be estimated as

$$\hat{a}_k = \sqrt{\left|\mathcal{E}\left\{Z_k^2(t)\right\}\right|}, \quad \text{and} \quad \hat{\varphi}_k = \tfrac{1}{2} \arg\left[\mathcal{E}\left\{Z_k^2(t)\right\}\right] \tag{7}$$

[1] The case of general, linear digital modulations ($s_k(t) \in \mathbb{C}$), as well as convolutive mixtures (channel effects taken into considerations) are our current topics of interest.

One should pay attention to the sign ambiguity which occurs when equivalent phases $|\varphi_k|$ are bigger than π, i.e.

$$\arg\left[e^{i2(\varphi_k+m\pi)}\right] = \arg\left[e^{i2\varphi_k}\right], \quad s_k(t)e^{i(\omega_k t+\varphi_k+m\pi)} = (-1)^m s_k(t)e^{i(\omega_k t+\varphi_k)}$$

for $m \in \mathbb{Z}$, thus the opposite sign signals can be observed (i.e. $\hat{s}_k = -s_k$). This ambiguity can be eliminated only during the modulation stage, e.g. applying differential modulation as DPSK (Differential Phase Shift Keying) instead of absolute BPSK type.

Once we estimated mixing parameters, the separation problem can be simplified to the solution of a linear system of equations. Let $\alpha = (\omega_1 - \omega_2)t + \hat{\varphi}_1 - \hat{\varphi}_2$, then one can find auxiliary signals $X_k(t)$ as

$$X_k(t) = x(t)e^{-i(\omega_k t+\hat{\varphi}_k)} = a_k s_k(t) + a_l s_l(t)e^{-i\alpha} \tag{8}$$

and the estimators of the original signals $s_k(t)$ by

$$\hat{s}_1(t) = \tfrac{1}{\hat{a}_1}\left[X_1(t) - \hat{a}_2 s_2(t)e^{-i\alpha}\right], \quad \hat{s}_2(t) = \tfrac{1}{\hat{a}_2}\left[X_2(t) - \hat{a}_1 s_1(t)e^{i\alpha}\right] \tag{9}$$

For BPSK-type signals ($s_k(t) = \pm 1$), previous equations become

$$\hat{s}_1(t) = \tfrac{1}{\hat{a}_1}\left[X_1(t) \pm \hat{a}_2 e^{-i\alpha}\right], \quad \hat{s}_2(t) = \tfrac{1}{\hat{a}_2}\left[X_2(t) \pm \hat{a}_1 e^{i\alpha}\right] \tag{10}$$

To eliminate the sign ambiguity, we propose a "solution selector" based on the minimization of the following instantaneous objective function

$$Q(t,\epsilon_1,\epsilon_2) = \left|x(t) - \hat{a}_1\hat{s}_1(t,\epsilon_1)e^{i(\omega_1 t+\hat{\varphi}_1)} - \hat{a}_2\hat{s}_2(t,\epsilon_2)e^{i(\omega_2 t+\hat{\varphi}_2)}\right|^2 \tag{11}$$

where

$$\hat{s}_k(t,\epsilon_k) = \Re\left\{\tfrac{1}{\hat{a}_k}\left[X_k(t) + \epsilon_k \hat{a}_l e^{i((\omega_l-\omega_k)t+\hat{\varphi}_l-\hat{\varphi}_k)}\right]\right\} \tag{12}$$

and $(\epsilon_1,\epsilon_2) \in \{(+1,+1),(+1,-1),(-1,+1),(-1,-1)\}$.

4 Experimental Results

To corroborate the effectiveness of the proposed algorithm in various communications scenarios, extensive simulations were conducted on linear mixtures of two BPSK signals in AWGN channel. As a measure of performance, we have chosen the mean value of BERs calculated for each of the separated signals

$$\mathrm{BER} = \frac{N_{e_1} + N_{e_2}}{2N} \tag{13}$$

versus Signal to Noise Ratio (SNR) calculated as

$$\mathrm{SNR} = 10\log\left[\frac{P_{s_1} + P_{s_2}}{P_n}\right] \tag{14}$$

where N_{e_1} and N_{e_2} are numbers of erroneous symbols in the demodulated (separated) signals, N is a total number of symbols used in each trial ($N \approx 10^7$), P_{s_1} and P_{s_2} are powers of the source signals, and P_n is a power of the additive gaussian noise calculated in the sampling frequency band $[-F_s/2, +F_s/2]$ ($F_s = 8$ kHz). The initial phases (φ_k) were randomly chosen from the range $[-\pi/2, \pi/2]$ and the SNR was varying from 0 dB to 30 dB with a 2 dB step.

In all experiments, we have compared BERs of the separated signals with mean value of BERs (solid bold line) calculated for each original BPSK signal (assuming that all parameters needed for demodulation are known).

In the first experiment, we have verified the shape of BER curves for different ratios between amplitudes $\eta = \min\left[\frac{a_1}{a_2}, \frac{a_2}{a_1}\right]$. This ratio was fixed to be $\eta \in \{0.1, 0.4, 0.7, 1\}$. For each trial, signals were composed of 256 symbols, 5 samples per symbol, 40000 different realisations, and difference between carrier frequencies $|f_1 - f_2|$ was fixed to be 20 Hz at sampling frequency of 8 kHz. Corresponding results are presented in the figure 1, for known as well as estimated coefficients $(a_1, a_2, \varphi_1, \varphi_2)$. BERs for separated signals are the lines with markers, and the solid bold lines without markers correspond to the theoretical BERs for only one BPSK signal. One should pay attention to the following facts: the best results are obtained when $\eta \in \{0.4, 0.7\}$, the method used to estimate mixing coefficients plays an important role especially for small ratios between amplitudes ($\eta \approx 0.1$), the worst results are obtained for $a_1 = a_2$.

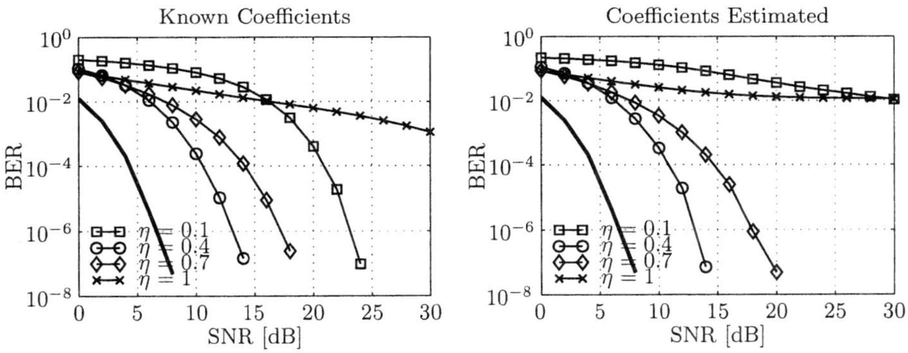

Fig. 1. BER versus SNR for different values of η

Other simulations were conducted to verify the behavior of the algorithm for different number of samples per symbol $N_{\text{samp}} \in \{5, 10, 15, 20\}$ (or equivalently Baud Rates for fixed F_s). Simulation results are shown in the figure 2 for $\eta = 0.5$ and $|f_1 - f_2| = 20$ Hz. It is evident that regardless of method used to estimate the mixing coefficients, increasing N_{samp} for the same total number of transmitted symbols, improves the performance of the algorithm (which is also true for any demodulation-detection system working on only one signal [22]).

The influence of the frequency shifts ($|f_1 - f_2|$ changing from 20 to 200 Hz), as well as the total number of emitted symbols (N_{symb} changing from 256 to 2048)

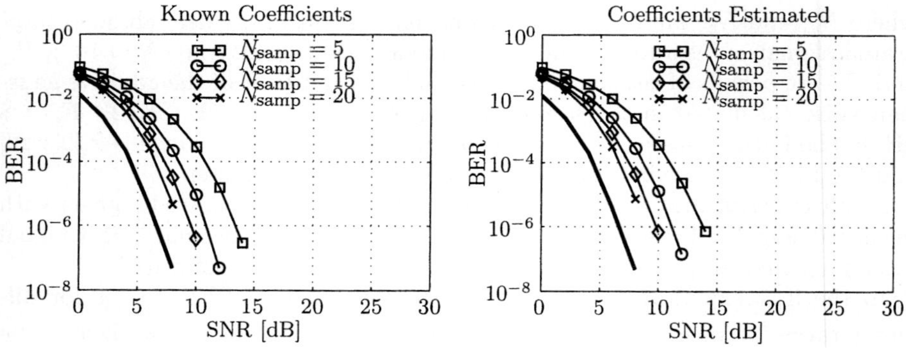

Fig. 2. BER versus SNR for different values of N_{samp}

on the performance of our algorithm has been also verified. Corresponding results reveal that our algorithm is invariant to signals' placement in the frequency domain (assuming that carrier frequencies are distinct and can be estimated), and to the number of available symbols (even when mixing coefficients have to be estimated).

5 Conclusion

Our new algorithm is targeted signal separation of undetermined instantaneous mixtures of two BPSK signals, closely distributed in a frequency domain. Using only one observation, we show a new solution for separation as well as for estimation of mixing coefficients. Experimental results reveal the robustness of our method to the number of samples per symbol (Baud Rates), total number of available symbols (possibility of working with small packets in "quasi real-time" applications), as well as to the shift between the carrier frequencies (even for overlapping bands). The separation algorithm is very robust to the ratio between amplitudes (excepted $a_1 = a_2$) and to the method used to estimate the mixing coefficients (excepted $\eta < 0.1$ or $\eta = 1$). All simulations have shown that experimental BERs are sufficiently close to the theoretical ones, which makes the proposed method of great interest in practice.

Methods of improving estimators of the mixing coefficients, as well as the possibility of using presented ideas to separate convolutive mixtures are currently investigated. Further researches will be conducted to generalise described methods for the mixtures of other types of linear modulations.

References

1. S.I. Amari, J.F. Cardoso, *Blind Source Separation – Semiparametric Statistical Approach*, IEEE Transactions on Signal Processing, vol. 45, no. 11, pp. 2692-2700, November 1997.
2. J.F. Cardoso, P. Comon, *Tensor-Based Independent Component Analysis*, in Signal Processing V, pp. 673-676, 1990.

3. J.F. Cardoso, B. Laheld, *Equivariant Adaptive Source Separation*, IEEE Transactions on Signal Processing, vol. 44, no. 12, pp. 3017-3030, December 1996.
4. A. Mansour, N. Ohnishi, C.G. Puntonet, *Blind Multiuser Separation of Instantaneous Mixture Algorithm Based on Geometrical Concepts*, Elsevier Signal Processing, vol. 82, no. 8, pp. 1155-1175, 2002.
5. N. Delfosse, P. Loubaton, *Adaptive Blind Separation of Convolutive Mixtures*, in ICASSP 1996, pp. 2940-2943, Atlanta, USA, May 1996.
6. A. Mansour, C. Jutten, P. Loubaton, *An Adaptive Subspace Algorithm for Blind Separation of Independent Sources in Convolutive Mixture*, IEEE Transactions on Signal Processing, vol. 48, no. 2, pp. 111-114, February 2000.
7. A. Mansour, *A Mutually Referenced Blind Multiuser Separation of Convolutive Mixture Algorithm*, Elsevier Signal Processing, vol. 81, no. 11, pp. 2253-2266, November 2001.
8. P. Comon, L. Rota, *Blind Separation of Independent Sources From Convolutive Mixtures*, IEICE Transactions on Fundamentals of Electronics Communications and Computer Sciences, vol. E86-A, no. 3, March 2003.
9. J. Antoni, F. Guillet, M. El Badaoui, F. Bonnardot, *Blind Separation of Convolved Cyclostationary Processes*, Elsevier Signal Processing, vol. 85, pp. 51-66, 2005.
10. P. Comon, O. Grellier, *Non-Linear Inversion of Undetermined Mixtures*, in ICA 1999, pp. 461-465, Aussois, France, January 1999.
11. L. De Lathauwer, B. Moor, J. Vandewalle, *ICA Techniques for More Sources Than Sensors*, IEEE Signal Processing Workshop, Higher-Order Statistics, pp. 116-120, 1999.
12. T. Lee, M.S. Lewicki, M. Girolami, T.J. Sejnowski, *Blind Source Separation of More Sources Than Mixtures Using Overcomplete Representations*, IEEE Signal Processing Letters, vol. 6, no. 4, pp. 87-90, 1999.
13. P. Bofill, M. Zibulevsky, *Blind Separation of More Sources Than Mixtures Using Sparsity of Their Short-Time Fourier Transform*, in ICA 2000, pp. 87-92, Helsinki, Finland, 2000.
14. L. Nguyen, A. Belouchrani, K. Abed-Meraim, B. Boashash, *Separating More Sources Than Sensors Using Time-Frequency Distributions*, in ISSPA 2001, vol. II, pp. 583-586, 2001.
15. A. Taleb, *An Algorithm for the Blind Identification of N Independent Signals with 2 Sensors*, in ISSPA 01, Kuala-Lampur, Malaysia, August 13-16, 2001.
16. P. Comon, O. Grellier, *Analytical Blind Identification of a SISO Communication Channel*, in 10th IEEE Signal Processing Workshop on Statistical Signal and Array Processing, pp. 206-210, Pocono Manor Inn, Pennsylvania, USA, August 14-16, 2000.
17. S.L. Marple, *Digital Spectral Analysis*, Prentice-Hall, 1987.
18. M. Pedzisz, A. Mansour, *HOS-Based Multi-Component Frequency Estimation*, in EUSIPCO 2005, Antalya, Turkey, 4-8 September 2005.
19. L. Benaroya, F. Bimbot, *Wiener Based Source Separation with HMM/GMM Using a Single Sensor*, in Report GdR ISIS, Paris, France, 12 June 2003.
20. J.R. Hopgood, P.J.W. Rayner, *Single Channel Nonstationary Stochastic Signal Separation Using Linear Time Varying Filters*, IEEE Transactions on Signal Processing, vol. 51, no. 11, pp. 1739-1752, July 2003.
21. K.I. Diamantaras, *Blind Separation of Multiply Binary Sources Using a Single Linear Mixture*, in ICASSP 2000, Istanbul, Turkey, June 2000.
22. J.G. Proakis, *Digital Communications*, McGraw-Hill, 4th edition, 2001.

Blind Partial Separation of Instantaneous Mixtures of Sources

Dinh-Tuan Antoine Pham

Laboratoire de Modélisation et Calcul, BP 53, 38041 Grenoble Cedex, France
Dinh-Tuan.Pham@imag.fr

Abstract. We introduce a general criterion for blindly extracting a subset of sources in instantaneous mixtures. We derive the corresponding estimation equations and generalize them based on arbitrary nonlinear separating functions. A quasi-Newton algorithm for minimizing the criterion is presented, which reduces to the FastICA algorithm in the case when only one source is extracted. The asymptotic distribution of the estimator is obtained and a simulation example is provided.

1 Introduction

Blind source separation (BSS) has attracted much attention recently, as it has many useful applications. The simplest and most widely used BSS model assumes that the observations are linear mixtures of independent sources with the same number of sources as the number of mixtures: $\mathbf{X} = \mathbf{AS}$ where $\mathbf{X}$ and $\mathbf{S}$ represent the observation and the source vectors, both of a same dimension K, and $\mathbf{A}$ is an invertible matrix. The aim is to extract the sources from their mixtures, without relying on any specific knowledge about them and quite a few good algorithms have been proposed for this task. In many applications (biomedical for ex.) however, the number K of mixtures can be very large and therefore one may be interested in extracting only a small number of (interesting) sources. In such case, many sources would be nearly Gaussian and since BSS algorithms rely on the non Gausianity, these sources would not be reliably extracted. In fact, in BSS problem with very large number of mixtures, one routinely discards most of the extracted sources and only retain some of them.

To extract only a small number of sources, one may of course proceed sequentially by extracting them one by one, using, for example, the (one-unit) FastICA algorithm [1]. However, such procedure entails a loss of performance as the accuracy of an extracted source is affected by the inaccuracy of the previously extracted ones, since the former is constrained to be uncorrelated with the later. Further, as it will be shown later, even for the first extracted source, the performance on the FastICA (with the optimal choice of the nonlinearity) is still less than extracting all sources simultaneously (through an optimal algorithm) and then retaining only one (adequately chosen) source. However, there is no loss of performance if one extracts simultaneously only $m < K$ sources, provided that the remaining $K - m$ are Gaussian.

In this paper we shall develop a class of algorithms for extracting only $m < K$ sources. For $m = 1$, this class contains the one-unit FastICA algorithm, and for $m = K$, it contains the quasi-maximum likelihood algorithm in [2] and the mutual information based algorithm in [3].

J. Rosca et al. (Eds.): ICA 2006, LNCS 3889, pp. 868–875, 2006.

In the sequel, we shall assume, for simplicity, that the sources have zero means. If they are not, one just centers them, which amounts to centering the observed vector $\mathbf{X}$.

2 Estimation Method

For the full extraction of sources, that is for the case $m = K$, the mutual information approach leads to the criterion [3]:

$$\sum_{i=1}^{K} H(Y_i) - \log \det \mathbf{B} \tag{1}$$

(to be minimized with respect to $\mathbf{B}$) where Y_i are the components of $\mathbf{Y} = \mathbf{BX}$ and $H(Y)$ denotes the Shannon differential entropy of the random variable Y: $H(Y) = -\mathrm{E}[\log p_Y(Y)]$, p_Y denoting the density of Y and E denoting the expectation operator [4]. This criterion can be written up to an additive constant as $\sum_{i=1}^{K} H(Y_i) - (1/2)\log\det(\mathbf{BCB}^T)$ where $\mathbf{C} = \mathrm{E}(\mathbf{XX}^T)$ denotes the covariance matrix of $\mathbf{X}$. The nice thing is that it involves only the statistical properties of the variables $Y_1, ..., Y_K$, as $\mathbf{BCB}^T$ represents the covariance matrix of the vector $\mathbf{Y}$. Thus one can easily extend it to the case where only $m < K$ sources are sought. More precisely, we will consider the criterion

$$C(\mathbf{B}) = \sum_{i=1}^{m} H(Y_i) - \frac{1}{2}\log\det(\mathbf{BCB}^T), \tag{2}$$

in which $\mathbf{B}$ is a $m \times K$ matrix and $Y_1, \ldots, Y_m$ are the components of $\mathbf{Y} = \mathbf{BX}$.

It has been shown in [5] that in the case where $m = K$, one can generalize the criterion (1) by replacing $H(Y_i)$ by $\log Q(Y_i)$ where is Q is a class II superadditive functional. Recall that [6] a functional Q of the distribution of a random variable Y, is said to be of class II if it is scale equi-variant[1], in the sense that $Q(aY) = |a|Q(Y)$ for any real number a, and it said to be superadditive if;

$$Q^2(Y + Z) \geq Q^2(Y) + Q^2(Z) \tag{3}$$

for any pair of independent random variables Y, Z. It is proved in [5] that this generalized criterion is still a contrast, in the sense that it can attain its minimum if and only if each of the $Y_1, \ldots, Y_K$ is proportional to a different source. We can show that this result carries to the case $m < K$, but the proof is omitted due to lack of space. Thus, in (2), one may take $H = \log Q$ where Q is a class II superadditive functional. Note that the exponential of the entropy functional has this property [6].

The superadditivity condition is quite strong because (3) must be satisfied for *any* pair of independent random variables Y, Z, but actually it is enough that this holds for random variables which are linear mixtures of sources. Thus (2) may still be a valid criterion if H is only a class II functional. In fact, for such functional, the point $\mathbf{B}$ for

[1] The definition of class II in [6] also requires that Q be translation invariant, but since we are working with zero-mean random variables, we drop this requirement.

which the components of $\mathbf{Y}$ are proportional to distinct sources, is still a stationary point of the criterion. Indeed, the gradient of the criterion (2) can be seen to be

$$\mathrm{E}[\psi_{\mathbf{Y}}(\mathbf{Y})\mathbf{X}^T] - (\mathbf{BCB}^T)^{-1}\mathbf{BC} \tag{4}$$

where $\psi_{\mathbf{Y}}(\mathbf{Y})$ is the vector with components $\psi_{Y_1}(Y_1), \ldots \psi_{Y_m}(Y_m)$ and ψ_Y denotes the "coordinate free" derivative of the functional H, defined by the condition

$$\lim_{\epsilon \to 0}[H(Y + \epsilon Z) - H(Y)]/\epsilon = \mathrm{E}[\psi_Y(Y)Z] \tag{5}$$

for any random variable Z. (For H the entropy functional, this is the score function [3].) Setting the gradient (4) to zero yields the estimation equation (for the stationary point of the criterion), which can be seen to be equivalent to

$$\mathrm{E}[\psi_{\mathbf{Y}}(\mathbf{Y})\mathbf{S}^T] - [\mathrm{E}(\mathbf{YY}^T)]^{-1}\mathrm{E}(\mathbf{YS}^T) = \mathbf{0}, \tag{6}$$

since $\mathbf{X} = \mathbf{AS}$. Note that if Y_i is proportional to a source S_{π_i},

$$\mathrm{E}[\psi_{Y_i}(Y_i)S_j] = \begin{cases} \mathrm{E}[\psi_{Y_i}(Y_i)Y_i]\mathrm{E}(Y_i S_{\pi_i})/\mathrm{E}(Y_i^2), & j = \pi_i \\ 0, & j \neq \pi_i \end{cases}$$

Thus, provided that $\mathrm{E}[\psi_{Y_i}(Y_i)Y_i] = 1$, equation (6) is satisfied as soon as $Y_1, \ldots, Y_m$ are proportional to distinct sources. On the other hand, since Q is of class II, $H(Y + \epsilon Y) = H(Y) + \log(1 + \epsilon)$, which by (5) yields immediately $\mathrm{E}[\psi_Y(Y)Y] = 1$.

A simple example of class II functional is $Q(Y) = \exp\{\mathrm{E}[G(Y/\sigma_Y)]\}\sigma_Y$, where G is some fixed function and $\sigma_Y = [\mathrm{E}(Y^2)]^{1/2}$. This functional yields, in the case $m = 1$, the same criterion as in the FastICA algorithm. Indeed, in the case $m = 1$ and with $H = \mathrm{E}[G(Y/\sigma_Y)] + \log \sigma_Y$, (2) becomes

$$C(\mathbf{b}) = \mathrm{E}[G(Y/\sigma_Y)], \qquad Y = \mathbf{bX},$$

where we have used the symbol $\mathbf{b}$ in place of $\mathbf{B}$ to emphasize that it is a row vector. The corresponding function ψ_Y is then given by

$$\psi_Y(y) = g(y/\sigma_Y)/\sigma_Y + \{1 - \mathrm{E}[g(Y/\sigma_Y)Y/\sigma_Y]\}y/\sigma_Y^2 \tag{7}$$

where g is the derivative of G.

In practice the (theoretical) criterion C would be replaced by the empirical criterion $\hat{C}$, defined as in (2) but with H replaced by an estimate $\hat{H}$ and $\mathbf{C}$ replaced by the sample covariance matrix $\hat{\mathbf{C}}$ of $\mathbf{X}$. The gradient of $\hat{C}$ is still given by (4) but with $\mathbf{C}$ replaced by $\hat{\mathbf{C}}$ and ψ_Y replaced by

$$\hat{\psi}_{Y_i}[Y_i(t)] = n\partial\hat{H}(Y_i)/\partial Y_i(t), \qquad t = 1, \ldots, n, \tag{8}$$

$\mathbf{Y}(t) = \mathbf{BX}(t)$ and $\mathbf{X}(1), \ldots, X(n)$ being the observed sample [3]. In the case $H(Y) = \mathrm{E}[G(Y/\sigma_Y)] + \log \sigma_Y$, its estimator $\hat{H}$ is naturally defined by the same expression but with E replaced by the sample average operator $\hat{\mathrm{E}}$ and σ_Y^2 replaced by $\hat{\sigma}_Y^2 = \hat{\mathrm{E}}(Y^2)$. The function $\hat{\psi}_Y$ is again given by (7) but with E replaced by $\hat{\mathrm{E}}$ and σ_Y replaced by $\hat{\sigma}_Y$.

The above argument shows that one can even start with the system of estimating equations obtained by equating (4) to zero, with ψ_{Y_i} being *arbitrary* functions (depending on the distribution of Y_i) subjected to the only condition that $\mathrm{E}[\psi_{Y_i}(Y_i)Y_i] = 1$. In practice, one would replace ψ_{Y_i} by some estimate $\hat{\psi}_{Y_i}$, E by $\hat{\mathrm{E}}$ and $\mathbf{C}$ by $\hat{\mathbf{C}}$, which results in the empirical estimating equation

$$\hat{\mathrm{E}}[\hat{\psi}_{\mathbf{Y}}(\mathbf{Y})\mathbf{X}^T] - (\mathbf{B}\hat{\mathbf{C}}\mathbf{B}^T)^{-1}\mathbf{B}\mathbf{C} = \mathbf{0}, \qquad \mathbf{Y} = \mathbf{B}\mathbf{X}. \tag{9}$$

We only require $\hat{\psi}_{Y_i}$ to satisfy $\hat{\mathrm{E}}[\hat{\psi}_{Y_i}(Y_i)Y_i] = 1$, which holds automatically if it is given by (8) and $\hat{H}$ is scale equi-variant, in the sense that $\hat{H}(\alpha Y) = \hat{H}(Y) + \log|\alpha|$.

3 The Quasi Newton Algorithm

In this section, we develop the quasi-Newton algorithm for solving (9). In the Newton algorithm, one replaces $\mathbf{B}$ in the right hand side of (9) by $\mathbf{B} + \delta\mathbf{B}$ and linearizes the result with respect to $\delta\mathbf{B}$. Here $\mathbf{B}$ denotes a current estimate and the new estimate is obtained by adding to it the solution $\delta\mathbf{B}$ of the linearized equations. In the quasi Newton algorithm, the system matrix of the linearized equations is further approximated.

Instead of working with $\delta\mathbf{B}$, it is much more convenient to work with its coefficients in a basis which contains the rows of $\mathbf{B}$ as its basis vectors. Thus we shall complete $\mathbf{B}$ to a square matrix $\bar{\mathbf{B}}$ by adding $K - m$ rows, which are chosen to be orthogonal to the rows of $\mathbf{B}$ and among themselves, in the sense of the metric $\hat{\mathbf{C}}$. More precisely, the matrix $\bar{\mathbf{B}}$ satisfies

$$\bar{\mathbf{B}}\hat{\mathbf{C}}\bar{\mathbf{B}}^T = \begin{bmatrix} \mathbf{B}\hat{\mathbf{C}}\mathbf{B}^T & \mathbf{0} \\ \mathbf{0} & \mathbf{I} \end{bmatrix}. \tag{10}$$

Let $\mathcal{E}_{ij}$, $i = 1, \dots, m, j = 1, \dots, K$, be the element of the matrix $\delta\mathbf{B}\bar{\mathbf{B}}^{-1}$, then $\delta\mathbf{Y} = \delta\mathbf{B}\mathbf{X}$ has components $\delta Y_i = \sum_{j=1}^{K} \mathcal{E}_{ij} Y_j$, where Y_j denote the components of $\bar{\mathbf{B}}\mathbf{X}$ (or of $\mathbf{Y}$ if $j \le m$). Thus, $\hat{\mathrm{E}}[\hat{\psi}_{Y_i+\delta Y_i}(Y_i + \delta Y_i)\mathbf{X}^T]$ is linearized as

$$\hat{\mathrm{E}}[\hat{\psi}_{Y_i}(Y_i)\mathbf{X}^T] + \sum_{j=1}^{K} \hat{\mathrm{E}}\{[\hat{\psi}'_{Y_i}(Y_i)Y_j + \dot{\hat{\psi}}_{Y_i;Y_j}(Y_i)]\mathbf{X}^T\}\mathcal{E}_{ij}$$

where ψ'_{Y_i} is the derivative of $\hat{\psi}_{Y_i}$ and $\dot{\hat{\psi}}_{Y_i;Y_j}$ is the derivative of $\psi_{Y_i+hY_j}$ with respect to h at $h = 0$.

We shall replace the last term in the above expression by an appropriate approximation. To this end, we shall assume that $\mathbf{B}$ is close to the solution so that the extracted sources $Y_1, \dots, Y_m$ are nearly proportional to $S_{\pi_1}, \dots, S_{\pi_m}$ for some distinct set of indexes $\pi_1, \dots, \pi_m$ in $\{1, \dots, K\}$. Since the $Y_{m+1}, \dots, Y_K$, by construction, have zero sample correlation with $Y_1, \dots, Y_m$, they would be nearly uncorrelated with $S_{\pi_1}, \dots, S_{\pi_m}$ and hence must be nearly linear combinations of the sources other than $S_{\pi_1}, \dots, S_{\pi_m}$. Thus we may treat the $Y_1, \dots, Y_m$ as independent among themselves and $(Y_{m+1}, \dots, Y_K)$ as independent of $(Y_1, \dots, Y_m)$. Further, we shall approximate $\hat{\mathrm{E}}$ by the expectation operator E and vice versa and regard $\hat{\psi}_{Y_i}$ as a fixed (non random) function. With such approximation

$$\sum_{j=1}^{K} \hat{\mathrm{E}}\{[\hat{\psi}'_{Y_i}(Y_i)Y_j + \dot{\hat{\psi}}_{Y_i;Y_j}(Y_i)]Y_k\}\mathcal{E}_{ij} \approx \begin{cases} \hat{\mathrm{E}}[\hat{\psi}'_{Y_i}(Y_i)]\hat{\mathrm{E}}(Y_k^2)\mathcal{E}_{ik} & k \neq i \\ \hat{\mathrm{E}}[\hat{\psi}'_{Y_i}(Y_i)Y_i^2 + \dot{\hat{\psi}}_{Y_i;Y_j}(Y_i)Y_i]\mathcal{E}_{ii} & k = i \end{cases}$$

But $\hat{\mathrm{E}}[(Y_i + hY_i)\hat{\psi}_{Y_i+hY_i}(Y_i + hY_i) = 1$, hence by taking the derivative with respect to h and letting $h = 0$, one gets $\hat{\mathrm{E}}[\hat{\psi}'_{Y_i}(Y_i)Y_i^2 + \dot{\hat{\psi}}_{Y_i;Y_j}(Y_i)Y_i] = -1$. Therefore, the linearization of $\hat{\mathrm{E}}[\hat{\psi}_{\mathbf{Y}+\delta\mathbf{Y}}(\mathbf{Y} + \delta\mathbf{Y})\mathbf{X}^T]$ is approximately $\hat{\mathrm{E}}[\psi_{\mathbf{Y}}(\mathbf{Y})\mathbf{X}^T] + \boldsymbol{\Delta}\bar{\mathbf{B}}^{-1T}$ where $\boldsymbol{\Delta}$ is a $m \times K$ matrix with general element

$$\Delta_{ij} = \begin{cases} \hat{\mathrm{E}}[\hat{\psi}'_{Y_i}(Y_i)]\hat{\mathrm{E}}(Y_j^2)\mathcal{E}_{ij}, & j \neq i \\ -\mathcal{E}_{ii}, & j = i \end{cases} \tag{11}$$

On the other hand, the linearization of $[(\mathbf{B} + \delta\mathbf{B})\hat{\mathbf{C}}(\mathbf{B} + \delta\mathbf{B})^T]^{-1}(\mathbf{B} + \delta\mathbf{B})\hat{\mathbf{C}}$ with respect to $\delta\mathbf{B}$ is

$$\begin{aligned}&(\mathbf{B}\hat{\mathbf{C}}\mathbf{B}^T)^{-1}\mathbf{B}\hat{\mathbf{C}} + \\ &\qquad (\mathbf{B}\hat{\mathbf{C}}\mathbf{B}^T)^{-1}\delta\mathbf{B}\hat{\mathbf{C}} - (\mathbf{B}\hat{\mathbf{C}}\mathbf{B}^T)^{-1}(\delta\mathbf{B}\hat{\mathbf{C}}\mathbf{B}^T + \mathbf{B}\hat{\mathbf{C}}\delta\mathbf{B}^T)(\mathbf{B}\hat{\mathbf{C}}\mathbf{B}^T)^{-1}\mathbf{B}\hat{\mathbf{C}}\end{aligned} \tag{12}$$

Multiplying the above expression by $\bar{\mathbf{B}}^T$ and using (10) yields

$$[\mathbf{I}\ \ \mathbf{0}] + (\mathbf{B}\hat{\mathbf{C}}\mathbf{B}^T)^{-1}[\mathbf{0}\ \ \mathcal{E}^c] - [\mathcal{E}^T\ \ \mathbf{0}]$$

where $\mathcal{E}$ and $\mathcal{E}^c$ are the matrices formed by the first m columns and by the last $K - m$ columns of $\mathbf{B}^{-1}\delta\mathbf{B}$, respectively. Note that the off diagonal elements of $\mathbf{B}\mathbf{C}\mathbf{B}^T$ nearly vanish since the $Y_1, \ldots, Y_m$ are nearly independent, hence one may replaces $\mathbf{B}\mathbf{C}\mathbf{B}^T$ by diag$(\mathbf{B}\mathbf{C}\mathbf{B}^T)$, where diag denotes the operator with builds a diagonal matrix from the diagonal elements of its argument.

Finally, $\hat{\mathrm{E}}[\hat{\psi}_{\mathbf{Y}+\delta\mathbf{Y}}(\mathbf{Y} + \delta\mathbf{Y})\mathbf{X}^T] - [(\mathbf{B} + \delta\mathbf{B})\hat{\mathbf{C}}(\mathbf{B} + \delta\mathbf{B})^T]^{-1}(\mathbf{B} + \delta\mathbf{B})\hat{\mathbf{C}}$ can be approximately linearized as

$$\mathrm{E}[\hat{\psi}_{\mathbf{Y}}(\mathbf{Y})\mathbf{X}^T] + \{\boldsymbol{\Delta} - [\mathbf{I}\ \ \mathbf{0}] + [\mathcal{E}^T\ \ \mathbf{0}] - \mathrm{diag}(\mathbf{B}\hat{\mathbf{C}}\mathbf{B}^T)^{-1}[\mathbf{0}\ \ \mathcal{E}^c]\}\bar{\mathbf{B}}^{-1T}$$

Equating this expression to zero yields, after a multiplication by $\bar{\mathbf{B}}^T$,

$$\hat{\mathrm{E}}[\psi_{\mathbf{Y}}(\mathbf{Y})(\bar{\mathbf{B}}\mathbf{X})^T] - [\mathbf{I}\ \ \mathbf{0}] + \boldsymbol{\Delta} + [\mathcal{E}^T\ \ \mathbf{0}] - \mathrm{diag}(\mathbf{B}\hat{\mathbf{C}}\mathbf{B}^T)^{-1}[\mathbf{0}\ \ \mathcal{E}^c] = \mathbf{0}.$$

This equation can be written explicitly as, noting that their i, i elements already yield the identity $0 = 0$ and that the diagonal elements of $\mathbf{B}\hat{\mathbf{C}}\mathbf{B}^T$ equal $\hat{\mathrm{E}}(Y_1^2), \ldots, \hat{\mathrm{E}}(Y_m^2)$ and $\hat{\mathrm{E}}(Y_{m+1}^2) = \cdots = \hat{\mathrm{E}}(Y_K^2) = 1$,

$$\mathrm{E}[\hat{\psi}_{Y_i}(Y_i)Y_j] + \hat{\mathrm{E}}[\hat{\psi}'_{Y_i}(Y_i)]\hat{\mathrm{E}}(Y_j^2)\mathcal{E}_{ij} + \mathcal{E}_{ji} = 0, \quad 1 \leq i \neq j \leq m \tag{13}$$

$$\hat{\mathrm{E}}[\hat{\psi}_{Y_i}(Y_i)Y_j] + \{\hat{\mathrm{E}}[\hat{\psi}'_{Y_i}(Y_i)] - 1/\hat{\mathrm{E}}(Y_i^2)\}\mathcal{E}_{ij} = 0, \quad 1 \leq i \leq m,\ m < j \leq K. \tag{14}$$

These equations can be solved explicitly for $\mathcal{E}_{ij}$ and then the new value of $\mathbf{B}$ is given by $\mathbf{B} + \mathcal{E}\mathbf{B} + \mathcal{E}^c\mathbf{B}^c$ where $\mathbf{B}^c$ is the matrix formed by the last $K - m$ rows of $\bar{\mathbf{B}}$.

It should be noted that the matrix $\mathbf{B}^c$ is not unique as one can pre-multiply it by any orthogonal matrix of size $K - m$ without affecting (10). Thus the matrix $\mathcal{E}^c$ is also not unique. However, the product $\mathcal{E}^c\mathbf{B}^c$ is. Indeed, by (14), $\mathcal{E}^c = -\mathbf{D}_\mathbf{Y}^{-1}\hat{\mathrm{E}}[\hat{\psi}_Y(\mathbf{Y})(\mathbf{B}^c\mathbf{X})^T]$ where $\mathbf{D}_\mathbf{Y}$ is the diagonal matrix with diagonal elements $\hat{\mathrm{E}}[\hat{\psi}'_{Y_i}(Y_i)] - 1/\hat{\mathrm{E}}(Y_i^2)$. Hence $\mathcal{E}^c\mathbf{B}^c = -\mathbf{D}_\mathbf{Y}^{-1}\hat{\mathrm{E}}[\hat{\psi}_Y(\mathbf{Y})\mathbf{X}^T]\mathbf{B}^{cT}\mathbf{B}^c$. But by (10),

$$\hat{\mathbf{C}}^{-1} = \bar{\mathbf{B}}^T \begin{bmatrix} (\mathbf{B}\hat{\mathbf{C}}\mathbf{B}^T)^{-1} & \mathbf{0} \\ \mathbf{0} & \mathbf{I} \end{bmatrix} \bar{\mathbf{B}} = \mathbf{B}^T(\mathbf{B}\hat{\mathbf{C}}\mathbf{B}^T)^{-1}\mathbf{B} + \mathbf{B}^{cT}\mathbf{B}^c.$$

yielding $\mathbf{B}^{cT}\mathbf{B}^c = \hat{\mathbf{C}}^{-1} - \mathbf{B}^T(\mathbf{B}\mathbf{C}\mathbf{B}^T)^{-1}\mathbf{B}$. Therefore, one can rewrite the algorithm in a form independent of the choice of $\mathbf{B}^c$ as

$$\mathbf{B} \leftarrow \mathbf{B} + \mathcal{E}\mathbf{B} + \mathbf{D}_\mathbf{Y}^{-1}\{\hat{\mathrm{E}}[\hat{\psi}_\mathbf{Y}(\mathbf{Y})\mathbf{Y}^T](\mathbf{B}\hat{\mathbf{C}}\mathbf{B}^T)^{-1}\mathbf{B} - \hat{\mathrm{E}}[\hat{\psi}_Y(\mathbf{Y})\mathbf{X}^T]\hat{\mathbf{C}}^{-1}\},$$

$\mathcal{E}$ being the $m \times m$ matrix with zero diagonal and off diagonal elements solution of (13).

Note: In the case where $m = 1$ and the extracted source is normalized to have unit sample variance, the algorithm becomes:

$$\mathbf{b} \leftarrow \mathbf{b} + \frac{\mathbf{b} - \hat{\mathrm{E}}[\hat{\psi}_Y(\mathbf{Y})\mathbf{X}^T]\hat{\mathbf{C}}^{-1}}{\hat{\mathrm{E}}[\hat{\psi}'_Y(Y)] - 1} = \frac{\hat{\mathrm{E}}[\hat{\psi}'_Y(Y)]\mathbf{b} - \mathrm{E}[\hat{\psi}_Y(Y)\mathbf{X}^T]\hat{\mathbf{C}}^{-1}}{\hat{\mathrm{E}}[\hat{\psi}'_Y(Y)] - 1}.$$

The new $\mathbf{b}$ is not normalized (but is nearly so), therefore one has to renormalize it and thus the denominator in the last right side is irrelevant. In the case where ψ_Y is given by (7) with σ_Y replaced by $\hat{\sigma}_Y = [\hat{\mathrm{E}}(Y^2)]^{1/2} = 1$, the numerator takes the same form but with $\hat{\psi}_Y$ replaced by g. One is thus led to the fixed point FastICA algorithm [1].

4 Asymptotic Distribution of the Estimator

Consider the asymptotic distribution of the estimator $\hat{\mathbf{B}}$, solution of the estimating equations (9). We shall assume that this estimator converge (as the sample size n goes to infinity) to an unmixing solution, that is a matrix $\mathbf{B}$, with rows proportional to distinct rows of $\mathbf{A}^{-1}$. Let $\delta\mathbf{B} = \hat{\mathbf{B}} - \mathbf{B}$, we may repeat the same calculations as in previous section. However, we now complete $\mathbf{B}$ to $\bar{\mathbf{B}}$ in a slightly different way: the last $K - m$ rows of $\bar{\mathbf{B}}$ are chosen so that (10) holds with the true covariance matrix $\mathbf{C}$ in place of $\hat{\mathbf{C}}$. By the same argument as in previous section,

$$\hat{\mathrm{E}}[\hat{\psi}_{\mathbf{Y}+\delta\mathbf{Y}}(\mathbf{Y} + \delta\mathbf{Y})\mathbf{X}^T] \approx \hat{\mathrm{E}}[\hat{\psi}_\mathbf{Y}(\mathbf{Y})\mathbf{X}^T] + \mathbf{\Delta}\bar{\mathbf{B}}^{-1}$$

where $\mathbf{\Delta}$ is defined as before by (11). We shall made further approximation by replacing $\hat{\psi}_\mathbf{Y}$ in the above right hand side by $\psi_\mathbf{Y}$ and $\hat{\mathrm{E}}$ and $\hat{\psi}'_{Y_i}$ in (11) by E and ψ'_{Y_i}. On the other hand, $[\mathbf{B} + \delta\mathbf{B})\hat{\mathbf{C}}(\mathbf{B} + \delta\mathbf{B})^T]^{-1}(\mathbf{B} + \delta\mathbf{B})\hat{\mathbf{C}}$ may be linearized with respect to $\delta\mathbf{B}$ as (12) as before. But since $\delta\mathbf{B}$ is small and $\hat{\mathbf{C}}$ converges to $\mathbf{C}$, one can replace, in the last two term in (12), $\hat{\mathbf{C}}$ by $\mathbf{C}$. Then by the same argument as in previous section and noting that $\bar{\mathbf{B}}$ satisfies (10) with $\mathbf{C}$ in place of $\hat{\mathbf{C}}$, the resulting expression can be written as

$$\{[\mathbf{I} \;\; (\mathbf{B}\hat{\mathbf{C}}\mathbf{B}^T)^{-1}\mathbf{B}\hat{\mathbf{C}}\mathbf{B}^{cT}] + (\mathbf{B}\mathbf{C}\mathbf{B}^T)^{-1}[\mathbf{0} \;\; \mathcal{E}^c] - [\mathcal{E}^T \;\; \mathbf{0}]\}\bar{\mathbf{B}}^{T\,-1}$$

Note that $\mathbf{BCB}^T = \mathrm{diag}(\mathbf{BCB}^T)$ since the Y_i are uncorrelated. Further, since $\mathbf{B\hat{C}B}^{cT} \to \mathbf{0}$, one may replace $(\mathbf{B\hat{C}B}^T)^{-1}\mathbf{B\hat{C}B}^{cT}$ by $[\mathrm{diag}(\mathbf{BCB}^T)]^{-1}\mathbf{B\hat{C}B}^{cT}$, which is the matrix with general element $\hat{\mathrm{E}}(Y_iY_{m+j})/\mathrm{E}(Y_i^2)$. Then by the same argument as before, the elements $\mathcal{E}_{ij}$ of $\delta\mathbf{B}\bar{\mathbf{B}}^{-1}$ can be seen to be approximatively the solution of

$$\hat{\mathrm{E}}[\psi_{Y_i}(Y_i)Y_j] + \mathrm{E}[\psi'_{Y_i}(Y_i)]\sigma^2_{Y_j}\mathcal{E}_{ij} + \mathcal{E}_{ji} = 0, \quad 1 \le i \ne j \le m$$
$$\hat{\mathrm{E}}\{[\psi_{Y_i}(Y_i) - \sigma_{Y_i}^{-2}Y_i]Y_j\} + \{\mathrm{E}[\psi'_{Y_i}(Y_i)] - \sigma_{Y_i}^{-2}\}\mathcal{E}_{ij} = 0, \quad 1 \le i \le m,\ m < j \le K.$$

where $\sigma^2_{Y_i} = \mathrm{E}(Y_i^2)$. The solution is

$$\begin{bmatrix} \mathcal{E}_{ij} \\ \mathcal{E}_{ji} \end{bmatrix} = -\begin{bmatrix} \mathrm{E}[\psi'_{Y_i}(Y_i)]\sigma^2_{Y_j} & 1 \\ 1 & \mathrm{E}[\psi'_{Y_j}(Y_j)]\sigma^2_{Y_i} \end{bmatrix}^{-1} \begin{bmatrix} \hat{\mathrm{E}}[\psi_{Y_i}(Y_i)Y_j \\ \hat{\mathrm{E}}[\psi_{Y_i}(Y_i)Y_j] \end{bmatrix}, \quad 1 \le i < j \le m,$$
$$\mathcal{E}_{ij} = -\frac{\hat{\mathrm{E}}\{[\psi_{Y_i}(Y_i) - \sigma_{Y_i}^{-2}Y_i]Y_j\}}{\mathrm{E}[\psi'_{Y_i}(Y_i)] - \sigma_{Y_i}^{-2}}, \quad 1 \le i \le m, m < j \le K.$$

One then can show, using the Central Limit Theorem, that the vectors $[\mathcal{E}_{ij}\ \ \mathcal{E}_{ji}]^T, 1 \le i < j \le m$ and the random variables $\mathcal{E}_{ij}, 1 \le i \le m, m < j \le K$ are asymptotically independently normally distributed, with covariance matrices

$$\frac{1}{n}\begin{bmatrix} \sigma_{Y_i}/\sigma_{Y_j} & 0 \\ 0 & \sigma_{Y_j}/\sigma_{Y_i} \end{bmatrix} \begin{bmatrix} \lambda_i^{-1} & 1 \\ 1 & \lambda_j^{-1} \end{bmatrix}^{-1} \begin{bmatrix} \rho_i^{-2} & 1 \\ 1 & \rho_j^{-2} \end{bmatrix} \begin{bmatrix} \lambda_i^{-1} & 1 \\ 1 & \lambda_j^{-1} \end{bmatrix}^{-1} \begin{bmatrix} \sigma_{Y_i}/\sigma_{Y_j} & 0 \\ 0 & \sigma_{Y_j}/\sigma_{Y_i} \end{bmatrix}$$

and variances $(\sigma^2_{Y_i}/n)(\rho_i^{-2} - 1)/(\lambda_i^{-1} - 1)^2$, where n is the sample size and

$$\rho_i = \frac{1}{\sigma_{Y_i}\sqrt{\mathrm{E}[\psi_i^2(Y_i)]}} = \mathrm{corr}\{Y_i, \psi_i(Y_i)\}, \qquad \lambda_i = \frac{1}{\sigma^2_{Y_i}\mathrm{E}[\psi'_i(Y_i)]}.$$

One can prove that the asymptotic variance is smallest when ψ_{Y_i} is the score function of Y_i, in this case $\lambda_i = \rho_i$ and the asymptotic variance of $\mathcal{E}_{ij}$ equals $(\sigma^2_{Y_i}/\sigma^2_{Y_i})\rho_j^{-2}/(\rho_i^{-2}\rho_j^{-2} - 1)$ if $1 \le j \le m$ and $\sigma^2_{Y_i}/(\rho_i^{-2} - 1)$ if $m < j \le K$. Thus, assuming that the extracted sources are normalized to have unit variance, there is a loss of accuracy with respect to the case where all sources are extracted, since $1/(\rho_i^{-2} - 1) > \rho_j^{-2}/(\rho_i^{-2}\rho_j^{-2} - 1)$ for $\rho_j^2 < 1$. But the loss could be negligible if the $\rho_j, m < j \le K$ are close to 1, that is if the non extracted sources are nearly Gaussian. This would not be the case if only one source is extracted since it is unlikely that all the remaining sources are nearly Gaussian.

5 An Example of Simulation

In a simulation experiment, we have generated 10 source signals of length $n = 1000$: the first is a sinusoid, the second is a sequence of uniform random variables, the third is a sequence of bilateral exponential variables and the remaining are sequences of Gaussian variables. All sources have zero mean and unit variance.

As it can be shown, our algorithm is "transformation invariant" in the sense that its behavior when applying to a mixtures with mixing matrix $\mathbf{A}$ and starting with a matrix

$\mathbf{B}$ is the same as when applying to unmixed sources and starting with the global matrix $\mathbf{G} = \mathbf{BA}$. Thus we shall apply our algorithm to the unmixed sources with a starting matrix $\mathbf{G}$ with elements randomly generated as independent standard normal variates. The following table shows the initial value of $\mathbf{G}$ and the final value produced by the algorithm after convergence.

Table 1. Initial and final matrices $\mathbf{G}$

Initial matrix G										
	1	2	3	4	5	6	7	8	9	10
1	0.7007	0.7669	1.0257	-0.6238	0.9284	1.0447	0.0076	-0.0686	1.5620	0.4070
2	-0.8775	0.4997	1.0876	0.1395	-0.0442	-0.6111	-0.2117	1.8387	-0.9778	-0.6222
3	0.6501	-1.4355	0.1399	0.3051	-0.8784	2.3058	0.0912	-1.1623	1.0585	1.0601
Final matrix G										
	1	2	3	4	5	6	7	8	9	10
1	-0.0376	0.1344	3.7040	0.0184	0.0814	0.0712	-0.0597	0.1758	-0.1045	-0.0260
2	8.4342	0.0354	0.1003	-0.0941	-0.0667	-0.0337	0.1271	-0.1267	0.0704	-0.1243
3	0.0524	-6.4488	0.0060	-0.1386	-0.0276	0.1422	0.0513	0.0122	0.0581	0.1500

One can see from the above table that the algorithm have correctly extracted the first three sources (but in the order third, first, second). However, we have observed that depending on the starting value, the algorithm may extract only two non Gaussian source and the other is a mixture of the Gaussian sources. The problem is that the algorithm may be stuck with a local minimum of the criterion; and it may be shown that any point $\mathbf{B}$ for which the random variable $Y_1, \ldots, Y_m$ are independent and at most one of them can be Gaussian, is a local minimum point of the criterion (2). Thus the algorithm may not produce the most interesting sources but only some sources and possibly a mixture of Gaussian sources in the case where there are several Gaussian sources. We currently investigate ways to avoid this problem.

References

1. Hyvärinen, A.: Fast and robust fixed-point algorithms for independent component analysis. IEEE Trans. Neural Networks **10** (1999) 626–634
2. Pham, D.T., Garat, P.: Blind separation of mixtures of independent sources through a quasi maximum likelihood approach. IEEE Trans. Signal Processing **45** (1997) 1712–1725
3. Pham, D.T.: Fast algorithms for mutual information based independent component analysis. IEEE Trans. on Signal Processing **52** (2004) 2690–2700
4. Cover, T.M., Thomas, J.A.: Elements of Information Theory. Wiley, New-York (1991)
5. Pham, D.T.: Contrast functions for blind seperation and deconvolution of sources. In Lee, T.W., Jung, T.P., Makeig, S., Sejnowski, T.J., eds.: Proceeding of ICA 2001 Conference, San-Diego, USA (2001) 37–42
6. Huber, P.J.: Projection pursuit. Ann. Statist. **13** (1985) 435–475

Contrast Functions for Blind Source Separation Based on Time-Frequency Information-Theory

Mohamed Sahmoudi[1], Moeness G. Amin[1], K. Abed-Meraim[2], and A. Belouchrani[3]

[1] Center for Advanced Communications,
Villanova University, Villanova, PA 19085, USA
{mohamed.sahmoudi, moeness.amin}@villanova.edu
[2] TSI, Telecom Paris, 37 rue Dareau, 75014, Paris cedex
abed@tsi.enst.fr
[3] Ecole Polytechnique, EE Dept., 10 Av. Hassen Badi, Algiers, Algeria
adel.belouchrani@enp.edu.dz

Abstract. This paper introduces new contrast functions for blind separation of sources with different time-frequency signatures. Two contrast functions based on the Kullback-Leibler and Jensen-Rényi divergences in the time-frequency (T-F) plane are introduced. Two iterative algorithms are proposed for the proposed contrasts optimization and source separation. One algorithm consists of spatial whitening and gradient-Jacobi maximization, combining Givens rotations and stochastic gradient. The second algorithm uses a quasi-Newton technique.

1 Introduction

Blind source separation (BSS) methods exploit the knowledge that the sources typically satisfy certain statistical or deterministic properties. In principle, BSS methods can be categorized into the following five classes:

- HOS-approaches: Higher-order statistics (HOS)-based methods assume that the sources are statistically independent. This is always enforced using the fourth order moments or cumulants of the mixtures. This class fails if more than one source is Gaussian.
- SOS-approaches: Second-order statistics (SOS)-based methods use some temporal structures less restrictive than the statistical independence. SOS approaches deal with uncorrelated sources and fails when sources have close power spectral shapes and when source are i.i.d.[1]
- NS-SOS-approaches: This class exploits the nonstationarity (NS) and SOS when sources have time-varying variances.
- FLOS-approaches: Recently introduced for separation of impulsive sources. For this king of signals, sources have heavy-tailed probability density functions (pdf), and fractional low order statistics (FLOS) can be used [12, 14].

[1] Abbreviation of independent and identically distributed.

J. Rosca et al. (Eds.): ICA 2006, LNCS 3889, pp. 876–884, 2006.

- STF-approaches: Space-time-frequency methodes refers to methods which exploit various diversity of signals, typically, time, frequency, and time-frequency structure [3, 4, 10, 15]. This is the class of diversity we analyse and we use in this contribution.

In this paper we expand on the last class and build on the nonstationarity properties in the time-frequency domain. Specifically, we propose a class of contrast functions for BSS in the time-frequency plane using information theoretic divergence measures. Two different schemes are proposed. In the first scheme, we minimize the T-F mutual information between the marginal TFD signal energies and the cross-TFD energies (joint TFD) as a useful extension of the FastICA mutual information based algorithm in the time domain [9]. In the second scheme, we maximize the divergence measures between the TFD energy support of the observation components as a natural extension to the minimum support contrasts proposed in the time domain [13]. To achieve BSS, we optimize the proposed cost functions using a two stage whitening-gradient algorithm and an iterative Quasi-Newton technique.

2 Problem Formulation

Consider m nonstationay signals for which $n \geq m$ noisy linear combinations are observed,

$$\mathbf{x}(t) = \mathbf{A}\mathbf{s}(t) + \mathbf{w}(t) \tag{1}$$

where $\mathbf{s}(t) = [s_1(t), \cdots, s_m(t)]^T$ is the $m \times 1$ complex sources vector, $\mathbf{w}(t)$ is the $n \times 1$ complex noise vector and $\mathbf{A}$ is a $n \times m$ full rank *mixing matrix*. The source signals $s_i(t), i = 1, \cdots, m$ are assumed to be nonstationary, complex stochastic processes *with different time-frequency signatures*. The additive noise $\mathbf{w}(t)$ is modelled as a stationary zero-mean complex random process. The goal of BSS is to find a $m \times n$ separating matrix $\mathbf{B}$ such that the output vector $\mathbf{z}(t) = \mathbf{B}\mathbf{x}(t)$ is an estimate of the source signals. There are two inherent ambiguities in the underlying problem. First, the original labeling of the sources cannot be ascertained, and second, exchanging a fixed scalar factor between a source signal and the corresponding column of $\mathbf{A}$ does not affect the observations. Accordingly, $\mathbf{B}$ is commonly determined up to a permutation and scaling multiplication of its columns. That is, $\mathbf{B}$ is referred to as a separating matrix if $\mathbf{z}(t) = \mathbf{B}\mathbf{x}(t) = \mathbf{P}\mathbf{\Lambda}\mathbf{s}(t)$, where $\mathbf{P}$ is a permutation matrix and $\mathbf{\Lambda}$ a non-singular diagonal matrix. Similarly, blind identification of $\mathbf{A}$ amounts to the determination of a matrix equal to $\mathbf{A}$ up to a permutation matrix and a non-singular diagonal matrix.

3 Measuring Time-Frequency Information

3.1 Time-Frequency Distributions

The discrete-time form of the Cohen's class of time-frequency distributions (TFD), for signal $x(t)$, is given by [5]

$$D_{xx}(t,f) = \sum_{l=-\infty}^{\infty} \sum_{m=-\infty}^{\infty} \phi(m,l) x(t+m+l) x^*(t+m-l) e^{-j4\pi fl} \quad (2)$$

where t and f represent the time index and the frequency index, respectively. The kernel $\phi(m,l)$ characterizes the TFD and is a function of both the time and lag variables. The cross-TFD of two signals $x_1(t)$ and $x_2(t)$ is defined by

$$D_{x_1x_2}(t,f) = \sum_{l=-\infty}^{\infty} \sum_{m=-\infty}^{\infty} \phi(m,l) x_1(t+m+l) x_2^*(t+m-l) e^{-j4\pi fl} \quad (3)$$

The main obstacle in considering the information theoretic in the T-F plane is that the TFDs are not always positive. If positive TFDs are desired, then the kernel must be signal-dependent [5]. Alternatively, one can use the Copulas theory-based non iterative method for constructing positive TFDs, as recently proposed in [7]. Therefore, in this paper we consider only positive TFDs (e.g. spectrograms). In addition, in order to ensure that the TFD behaves like a pdf, we need to normalize the TFD by their energy, $E_D = \int\int D(t,f) dtdf$.

3.2 Information-Theoretic on the Time-Frequency Plane

Recently, there has been an interest in applying information theoretic measures to the time-frequency plane in order to quantify signal information [6, 1]. Using energy density instead of probability density function, divergence measures for the probability distributions has been adapted to the TFD as follows [1]:

1. Time-frequency Kullback-Leibler divergence : This standard distance measure belongs to the class of Csiszar's ϕ-divergence with $\phi(t) = -\log(t)$. It can be employed in the time-frequency plane to quantify complexity between two TFD D_1 and D_2, as

$$\mathcal{K}(D_1, D_2) = \int\int D_1(t,f) \log \frac{D_1(t,f)}{D_2(t,f)} dtdf \quad (4)$$

2. Time-frequency Jensen-Rényi divergence: For two positive TFDs D_1 and D_2, Jensen-Rényi divergence can be defined as

$$\mathcal{J}_\alpha(D_1, D_2) = H_\alpha\left(\sqrt{D_1 D_2}\right) - \frac{H_\alpha(D_1) + H_\alpha(D_2)}{2} \quad (5)$$

where H_α is the αth T-F Rényi entropy defined as

$$H_\alpha(D) = \frac{1}{1-\alpha} \log_2 \int\int \left(\frac{D(t,f)}{\int\int D(t,f) dtdf} \right)^\alpha dtdf \quad (6)$$

where $\alpha > 0, \alpha \neq 1$. H_α converges to the Shannon entropy as $\alpha \to 1$.
3. Time-frequency divergence properties: Some of the most desirable properties of the time-frequency divergences are given in the following proposition [1].

Proposition 1.

- *The divergence measures are always positive :*

$$0 \leq \mathcal{K}(D_1, D_2), \mathcal{J}_\alpha(D_1, D_2) \leq \infty$$

- *The lower bound is reached if and only if $D_1 = D_2$.*
- *The upper bound is reached if and only if $Supp(D_1) \cap Supp(D_2) = \emptyset$.*

To separate nonstationary mixtures, the last property will be exploited to provide the mixed signals with nearly[2] disjoint supports in the TF plane through divergence measure maximizations.

4 TFD Information Based Contrast Functions

It is shown in [4] that the necessary condition to achieve BSS in the T-F domain is given in the following proposition.

Proposition 2. *Let $(t_1, f_1), \cdots, (t_K, f_K)$ are time-frequency points corresponding to auto-terms, i.e. energy concentration in the time-frequency plan and $d_i = [D_{s_is_i}(t_1, f_1), \cdots, D_{s_is_i(t_k,f_k)}]$ for $i = 1, \cdots, m$. Then, BSS can be achieved if and only if d_i and d_j are linearly independent for $i \neq j$.*

4.1 Minimum T-F Mutual Information Contrast Function

If we replace the marginal pdfs, used in the time domain, by the TFD signal energy $D_x(t, f)$ and $D_y(t, f)$, and the joint density function by the joint energy distribution defined by the cross-spectrograms of the two signals, i.e.,

$$D_{(x,y)}(t, f) \stackrel{\text{def}}{=} STFT_x(t, f) STFT_y^*(t, f) \tag{7}$$

where $STFT_x(t, f) = \int h(\tau - t)x(\tau)e^{-j2\pi f\tau}d\tau$ with $h(t)$ being the data window. Since the time marginal of $D_{(x,y)}(t, f)$ leads to the cross energy of $x(t)$ and $y(t)$, i.e. $\int D_{(x,y)}(t, f)df = x(t)y^*(t)$, it can be viewed as a joint energy distribution of the two signals. Then, T-F mutual information between $x(t)$ and $y(t)$ can be defined as

$$\mathcal{I}(D_x, D_y) \stackrel{\text{def}}{=} \mathcal{K}(|D_{(x,y)}|, D_x D_y) = \int\int |D_{(x,y)}(t, f)| \log \frac{|D_{(x,y)}(t, f)|}{D_x(t, f)D_y(t, f)} dtdf \tag{8}$$

We replaced $D_{(x,y)}(t, f)$ by its absolute value because it can be complex-valued. When the two signals $x(t)$ and $y(t)$ are separated in the T-F plane, we have $|D_{(x,y)}(t, f)| = 0, \forall\, t, f$, which implies that the T-F mutual information between $x(t)$ and $y(t)$ is equal to zero. This is analogous to the case where independent

[2] We relax the well known assumption that sources must be of disjoint T-F supports and we need just that sources are with linearly independent T-F signature vectors.

random variables have zero mutual information. It allows us to introduce a mutual information BSS framework in the T-F domain. In this paper, we define the following T-F mutual information contrast function $\mathcal{C}_1$ as,

$$\begin{aligned}
\mathcal{C}_1(\mathbf{z}) &= \sum_{1\le i\ne j\le m} \mathcal{I}(D_{z_i}, D_{z_j}) \\
&= \sum_{1\le i\ne j\le m} \sum_{k=1}^{K} \left|D_{(z_i,z_j)}(t_k, f_k)\right| \log \frac{\left|D_{(z_i,z_j)}(t_k, f_k)\right|}{D_{z_i}(t_k, f_k) D_{z_j}(t_k, f_k)} \\
&= \sum_{1\le i\ne j\le m} \sum_{k=1}^{K} \left|D_{(z_i,z_j)}(t_k, f_k)\right| \left[\log\left|D_{(z_i,z_j)}(t_k, f_k)\right| - 2\log D_{z_i}(t_k, f_k)\right] \quad (9)
\end{aligned}$$

Equation (9) implies that minimizing the T-F mutual information corresponds to minimizing the cross-terms and maximizing the auto-terms between data components, which should be achieved for proper source estimation. A separating matrix $\mathbf{B}$ can be computed as

$$\mathbf{B} = \arg\min \mathcal{C}_1(\mathbf{z}) \quad (10)$$

4.2 Maximum T-F Divergence Contrast Function

Considering the source probability density function, we have discussed the minimum support contrast function for source separation in [13]. In this section, we extend this idea to the time-frequency domain using the above time-frequency support properties. From property 1, for two signals $x(t)$ and $y(t)$, the divergence between D_x and D_y assumes a maximum value when the two signals are well-separated in the time-frequency plane, or equivalently when their TFD support are disjoint. This leads to the well known minimum support criterion for source extraction in the time domain (e.g., see [13] and references therein). Thus, we propose to maximize the divergence between the mixture components to achieve source separation.

Using the Jensen-Rényi divergence, we can define a second contrast function as

$$\begin{aligned}
\mathcal{C}_2(\mathbf{z}) &= \sum_{1\le i\ne j\le m} \mathcal{J}_\alpha(D_{z_i}, D_{z_j}) \\
&= \frac{1}{1-\alpha} \sum_{1\le i\ne j\le m} \log_2 \left(\frac{\sum_{k=1}^{K} \sqrt{D_{z_i}(t_k, f_k) D_{z_j}(t_k, f_k)}^{\alpha}}{\sqrt{\sum_{k=1}^{K} D_{z_i}^{\alpha}(t_k, f_k)} \sqrt{\sum_{k=1}^{K} D_{z_j}^{\alpha}(t_k, f_k)}} \right) \quad (11)
\end{aligned}$$

The maximization of the above criterion leads to signals with energy almost disjoint in the T-F plane. Since the log is a monotonous function, contrast $\mathcal{C}_2$ can be further simplified as

$$\mathcal{C}_2(\mathbf{z}) = \sum_{1\le i\ne j\le m} \left(\frac{\sum_{k=1}^{K} \sqrt{D_{z_i}(t_k, f_k) D_{z_j}(t_k, f_k)}^{\alpha}}{\sqrt{\sum_{k=1}^{K} D_{z_i}^{\alpha}(t_k, f_k)} \sqrt{\sum_{k=1}^{K} D_{z_j}^{\alpha}(t_k, f_k)}} \right) \quad (12)$$

Then, a separating matrix $\mathbf{B}$ can be computed as

$$\mathbf{B} = \arg\max \mathcal{C}_2(\mathbf{z}) \text{ if } \alpha < 1 \quad \text{and} \quad \mathbf{B} = \arg\min \mathcal{C}_2(\mathbf{z}) \text{ if } \alpha > 1 \tag{13}$$

4.3 Jacobi-Gradient Algorithm

To derive a nice and fast algorithm, we have proposed in [11] to combine the Jacobi-like decomposition of Givens rotations and the Gradient algorithm using a numerical computation for achieving the BSS in two stages. Firstly, the observations are pre-processed (whitening) such that the sensor vector is transformed to a white vector and at the same time the dimensionality is reduced from n to m, and then an $m \times m$ orthogonal matrix is determined as a product of Givens rotations. The so called Jacobi-Gradient algorithm can be summarized as follows:

Step 0. *Whitening (e.g., see [9, 12]).*
Step 1. *Initialize Givens angles randomly.*
Step 2. *Compute, the considered sources T-F distributions used in the contrast function:* $D_{z_i}(t_k, f_k)$ *and* $D_{(z_i,z_j)}(t_k, f_k)$ *for* $i, j = 1, \cdots, m$ *and* $k = 1, \cdots, K$.
Step 3. *Calculate the gradient of the cost function with respect to the Givens angles. The gradient is* $\frac{\partial \mathcal{C}(\mathbf{B})}{\partial \Theta}$.
Step 4. *Update the Givens angles using gradient ascent*

$$\Theta(k+1) = \Theta(k) - \eta \frac{\partial \mathcal{C}(\mathbf{B})}{\partial \Theta}$$

Step 5. *Go back to step 3 and continue until convergence.*

4.4 Iterative Quasi Newton Algorithm

Similar to [4], a block technique can be implemented for the contrasts optimization using the received samples and consists of searching solutions of equations (10) and (13) iteratively in the form

$$\mathbf{B}^{(p+1)} = (\mathbf{I} + \varepsilon^{(p)})\mathbf{B}^{(p)} \tag{14}$$

and thus

$$\mathbf{z}^{(p+1)}(t) = (\mathbf{I} + \varepsilon^{(p)})\mathbf{z}^{(p)}(t) \tag{15}$$

At the pth step, a matrix $\varepsilon^{(p)}$ is computed from a local linearization of the criterion. It is an approximate Newton technique with the benifit that $\varepsilon^{(p)}$ can simply computed, without need to Hessian inversion, under that $\mathbf{B}^{(p)}$ is close to a separating matrix.

Thus, applying the two proposed algorithms to both contrast functions $\mathcal{C}_1$ and $\mathcal{C}_2$, we will have four new different algorithms.

5 Simulation Results

5.1 First Experiment: Separation of Two Crossing Linear FM Signals

In this experiment, we consider a linear mixture of two linear frequency modulated nonstationary signals (chirp) with random Gaussian amplitudes. The model consists of two noisy mixture observations ($n = m = 2$) with number of samples $K = 1024$ and SNR = 10 dB. The plots of the spectrograms TFD of the two mixtures and the separated signals are shown in Fig. 1 using the first Jacobi-gradient algorithm and the T-F mutual information contrast $\mathcal{C}_1$. It is clear that the proposed procedure works well for noisy random Gaussian sources with overlapping TFD signatures. This specific example is chosen to test the algorithm when the mixed signals overlap in the time-frequency domain and when more than one source is Gaussian.

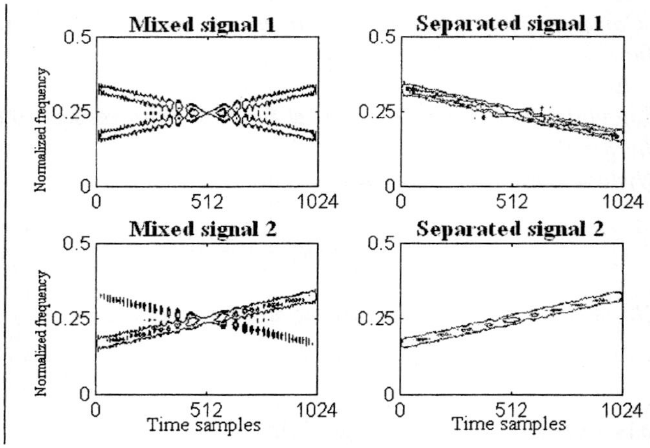

Fig. 1. Example of T-F mutual information separation

5.2 Second Experiment: Performance Comparison with FastICA

We characterize the performance of each algorithm in terms of signal rejection level. When $\mathbf{C} \stackrel{\text{def}}{=} \mathbf{BA}$, the i-th estimated source is $\hat{s_i}(t) = z_i(t) = \sum_{j=1}^{m} \mathbf{C}_{ij} s_j(t)$ which contains the j-th source signal at level $|\mathbf{C}_{ij}|^2/|\mathbf{C}_{ii}|^2$. In this case, the averaged rejection level is given by $I_{perf} = \frac{1}{m}\sum_{i=1}^{m} I_i = \frac{1}{m}\sum_{i=1}^{m}\sum_{j\neq i}\frac{|\mathbf{C}_{ij}|^2}{|\mathbf{C}_{ii}|^2}$. Here, the same default settings of the first experiment are used. We evaluate the performance of the proposed maximum T-F divergence contrast $\mathcal{C}_2$ and compare it with that of the time domain algorithm FastICA [9]. Fig. 2 presents the mean rejection level (I_{perf}) versus the signal to noise ratio (SNR) of each algorithm. It is evident from this figure that in this case the proposed T-F mutual information

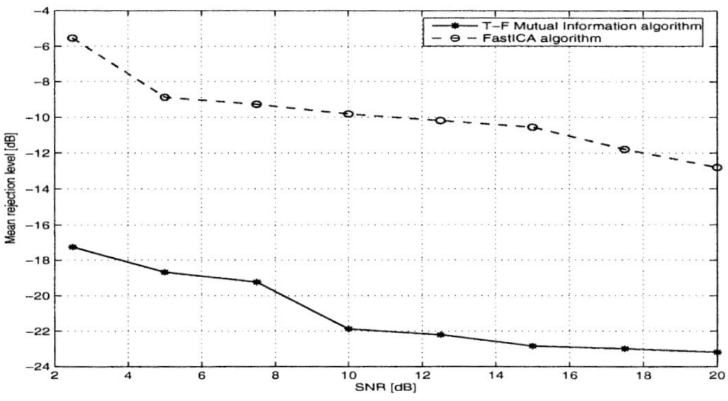

Fig. 2. Mean rejection level versus SNR. $T = 1000$

algorithm outperforms the classic FastICA. This is can be explain by the fact that the FastICA assumptions are not fitted in this example because we have more than one Gaussian source in the mixture. We note also that the proposed method can separate non independents sources contrary to the ICA methods which fails in that case.

6 Discussion and Conclusion

In this contribution, new contrast functions for BSS has been introduced that are based on the minimization of the T-F mutual information and maximization of the T-F Jensen-Rényi divergence measure. Two efficient implementation using a Jacobi-gradient algorithm and a block quasi-Newton algorithm are proposed for the contrast optimization. Unlike to the existing BSS methods, the proposed method allows the separation of Gaussian sources with identical spectral shape, provided that the sources have different time frequency signatures. Contrary to several existing time-frequency -based BSS methods, our approach don't suffer from the problem of overlapped source components. More studies and analysis need to be done like on the effect of the time-frequency kernel and the choice of the time-frequency points.

References

1. S. Aviyente, "Information processing on the time-frequency plane", *in Proc. of the IEEE Conference ICASSP'2005*, Philadelphia, USA, 2005.
2. Z. Shan and S. Aviyente, "Information-Theoretic Nonstationary Source Separation", *in Proc. of ICA'2006, Charleston USA, March 2006.*
3. A. Belouchrani and M. Amin, "Blind source separation based on time-frequency signal representations", IEEE Trans. on Signal Processing, 46(11), Nov. 1998.
4. A. Belouchrani and K. Abed-Meraim, "Contrast Functions for Blind Source Separation based on Time frequency distributions", *in ISCCSP'04; IEEE International Symposium on Control, Communications and Signal Processing*, Tunisia, 2004.

5. B. Boashash Editor, *Time Frequency Signal Analysis and Processing: a comprehensive book*, Elseiver, 2003.
6. R. G. Baraniuk, P. Flandrin, A. J. E. M. Janssen, and O. J. J. Michel, "Measuring Time-Frequency Information Content Using the Renyi Entropies", *IEEE Trans. on Signal Processing*, Vol. 47, No. 4, 2001.
7. M. Davy and A. Doucet, "Copulas: A new Insight into Positive Time-Frequency Distributions", IEEE Signal Processing Letters, Vol. 10, No. 7, July 2003.
8. K. E. Hild II, D. Erdogmus, and J.C. Principe, "Blind Source Separation using Renyi's mutual information", IEEE Signal Proc. Letters, Vol. 8, pp: 174-176, 2001.
9. A. Hyvarinen et al, *Independent Component Analysis* , Wiley, 2001.
10. S. Rickard, R. Balan and J. Rosca, "Blind source separation based on space-time-frequency diversity", *in Proc. of ICA'03*, Nara, Japan, April 2003.
11. M. Sahmoudi and K. Abed-Meraim, "Investigations on Contrast Functions for Blind Source Separation Based on Non-Gaussianity and Sparsity Measures", in Proc. of the IEEE Conf. ISSPA'2005, Sydney, Australia, Aug. 2005.
12. M. Sahmoudi, K. Abed-Meraim and M. Benidir, "Blind Separation of Impulsive Alpha-Stable Sources Using Minimum Dispersion Criterion", *in IEEE Signal Processing Letters*, Vol. 12, No. 4, April 2005.
13. Mohamed Sahmoudi, "Generalized Contrast Functions for Blind Separation of Overdetermined Linear Mixtures with Unknown Number of Sources", *in Proc. of SSP'05; the IEEE Statistical Signal Processing Workshop*, France, July 2005.
14. M. Sahmoudi, H. Boumaraf and D.-T. Pham, "Blind separation of convolutive mixtures using nonstationarity and fractional lower order statistics (FLOS) with application to audio sources", *Submitted to IEEE SAM2006*, Waltham, Massachusetts USA, July 2006.
15. Y. Zhang, M. G. Amin, "Blind separation of sources based on their time-frequency signatures", *in Proc. of ICASSP'2000.*

Information–Theoretic Nonstationary Source Separation

Zeyong Shan and Selin Aviyente

Department of Electrical and Computer Engineering,
Michigan State University, East Lansing, MI 48824, USA
{shanzeyo, aviyente}@egr.msu.edu

Abstract. Blind source separation aims at recovering the original source signals given only observations of their mixtures. Some common approaches to the source separation problem include second or higher order statistics based methods, and independent component analysis. Most of these methods are developed in the time domain, and thus inherently assume the stationarity of the underlying signals. Since most real life signals of interest are non–stationary, there have been efforts to perform source separation in the time–frequency domain. In this paper, we propose a new approach for source separation on the time–frequency plane using an information–theoretic cost function. Jensen–Rényi divergence, as adapted to time–frequency distributions, is introduced as an effective cost function to extract sources that are disjoint on the time–frequency plane. The sources are extracted through a series of Givens rotations and the optimal rotation angle is found using the steepest descent algorithm. The proposed method is applied to several example signals to illustrate its effectiveness and the performance is quantified through simulations.

1 Introduction

Blind source separation (BSS) is an important and fundamental problem in signal processing with a broad range of applications. A number of BSS algorithms have been proposed based on the instantaneous mixture model, in which the observed signals are linear combinations of the source signals and no time delays are involved in the mixtures. Among these methods, the most common ones are second order statistics based methods [1], and information–theoretic approaches which utilize cost functions such as mutual information or divergence measures, e.g. independent component analysis (ICA) [2,3]. These methods in general assume a certain structure for the underlying source signals. For example, higher–order statistics based methods assume non–Gaussian and i.i.d. source signals, whereas ICA assumes the independence of the source signals.

Most real life signals are non–stationary, and thus do not obey the underlying assumption of stationarity that is embedded in the current methods. For this reason, recently various methods have been introduced to exploit the non–stationarity of the source signals to solve the separation problem. Researchers have resorted to the powerful tool of time–frequency signal representations. For

J. Rosca et al. (Eds.): ICA 2006, LNCS 3889, pp. 885–892, 2006.

non–stationary signals, a blind separation approach using a spatial time-frequency distribution is proposed in [4]. The separation is achieved by joint diagonalization of the auto–terms in the spatial time-frequency distributions.

In this paper, we introduce a new approach to the source separation problem combining time–frequency representations with information–theoretic measures. An information–theoretic criterion, Jensen–Rényi divergence as adapted to the time–frequency distributions, is used as the objective function to separate the sources. The underlying sources are assumed to be disjoint on the time–frequency plane and it is shown that this new cost function achieves its maximum when the signals are disjoint. With the assumption that the source signals are disjoint on the time–frequency plane, signal separation is performed through a rotation transformation using a steepest descent algorithm.

2 Background on Time-Frequency Distributions and Information Measures

A time-frequency distribution (TFD), $X(t,\omega)$, from Cohen's class can be expressed as [1] [5]:

$$X(t,\omega)=\int\int\int \phi(\theta,\tau)s(u+\frac{\tau}{2})s^*(u-\frac{\tau}{2})e^{j(\theta u-\theta t-\omega\tau)}du\,d\theta\,d\tau, \qquad (1)$$

where $\phi(\theta,\tau)$ is the kernel function and s is the signal. Some of the most desired properties of TFDs are the energy preservation and the marginals. They are satisfied when $\phi(\theta,0)=\phi(0,\tau)=1\ \ \forall\tau,\theta$ and are given as follows:

$$\int\int X(t,\omega)\,dt\,d\omega=\int|s(t)|^2\,dt=\int|S(\omega)|^2\,d\omega,$$
$$\int X(t,\omega)\,d\omega=|s(t)|^2\ ,\int X(t,\omega)\,dt=|S(\omega)|^2. \qquad (2)$$

The formulas given above evoke an analogy between a TFD and the probability density function (pdf) of a two–dimensional random variable. This analogy has inspired the adaptation of information–theoretic measures such as entropy to the time–frequency plane. Although entropy measures have proven to be useful in quantifying the complexity of individual signals, they cannot be used directly to quantify the difference between signals. For this reason, well–known divergence measures from information theory have been adapted to the time–frequency plane [6]. One such distance measure is the Jensen–Rényi divergence based on the Jensen difference. For time–frequency distributions, Jensen–Rényi divergence can be defined as:

$$G_{12}^{\alpha}(X_1,X_2)=H_\alpha(\sqrt{X_1X_2})-\frac{H_\alpha(X_1)+H_\alpha(X_2)}{2}, \qquad (3)$$

[1] All integrals are from $-\infty$ to ∞ unless otherwise stated.

where H_α represents Rényi entropy defined on the time–frequency plane as:

$$H_\alpha(X) = \frac{1}{1-\alpha}\log_2 \int\!\!\int \left(\frac{X(t,\omega)}{\int\!\!\int X(u,v)du\,dv} \right)^\alpha dt\,d\omega \quad (\alpha > 0). \tag{4}$$

3 Problem Formulation and Method

3.1 Problem Statement in the Time–Frequency Domain

In this paper, we consider the problem of determining the source signals when the number of observed mixtures is equal to or greater than the number of the source signals. Assume that the M mixtures, $\{s_1(t), s_2(t), \cdots, s_M(t)\}$, of the N non-stationary complex source signals are given ($M \geq N$). Each mixture, $s_i(t)$, is first transformed to the time-frequency plane as:

$$X_i(n,\omega;\psi) = \sum_m \sum_l \psi(n-l,m) s_i\left(l + \frac{m}{2}\right) s_i^*\left(l - \frac{m}{2}\right) e^{-j\omega m}. \tag{5}$$

The time-frequency distribution corresponding to each mixture is vectorized and a matrix of time–frequency distributions is formed:

$$\mathbf{X} = \begin{bmatrix} \mathbf{X}_1 \\ \mathbf{X}_2 \\ \vdots \\ \mathbf{X}_M \end{bmatrix} = \begin{bmatrix} X_1(1) & \cdots & X_1(Q) \\ X_2(1) & \cdots & X_2(Q) \\ & \vdots & \\ X_M(1) & \cdots & X_M(Q) \end{bmatrix}, \tag{6}$$

where $\mathbf{X}_i$ is a vector of length $Q = K \times L$ points, K and L are the number of time and frequency points, respectively. The signals to be separated on the time–frequency plane are defined as:

$$\mathbf{Y} = \begin{bmatrix} \mathbf{Y}_1 \\ \mathbf{Y}_2 \\ \vdots \\ \mathbf{Y}_N \end{bmatrix} = \begin{bmatrix} Y_1(1) & \cdots & Y_1(Q) \\ Y_2(1) & \cdots & Y_2(Q) \\ & \vdots & \\ Y_N(1) & \cdots & Y_N(Q) \end{bmatrix}. \tag{7}$$

In order to make the following discussions simpler, we concentrate on the case where $M = N$. The discussions can be generalized for $M > N$ as illustrated through an example in Sect. 4.

The sources $\mathbf{Y}$ are extracted by applying a rotation transform $\mathbf{R}(\theta)$ in N–dimensions:

$$\mathbf{Y} = \mathbf{R}(\theta)\mathbf{X}. \tag{8}$$

Rotation matrix is used for extracting the sources since any unitary transform can be written in terms of rotation matrices and it provides a convenient parametrization of the problem. The rotation angle θ is adapted to maximize the following cost function:

$$G_\alpha \triangleq \sum_{i=1}^{N-1} \sum_{j=i+1}^{N} \left[H_\alpha(\sqrt{\mathbf{Y}_i \mathbf{Y}_j}) - \frac{H_\alpha(\mathbf{Y}_i) + H_\alpha(\mathbf{Y}_j)}{2} \right]. \tag{9}$$

Maximizing this cost function will ensure that the extracted components do not overlap with each other on the time–frequency plane.

3.2 Cost Function and Rotation

The Jensen–Rényi divergence between two time–frequency distributions is defined as:

$$G_{ij}^\alpha = H_\alpha(\sqrt{\mathbf{Y}_i \mathbf{Y}_j}) - \frac{H_\alpha(\mathbf{Y}_i) + H_\alpha(\mathbf{Y}_j)}{2}. \tag{10}$$

This expression can be further simplified as:

$$G_{ij}^\alpha = \frac{1}{1-\alpha} \log \left[\frac{\sum_{k=1}^{Q} \left(\sqrt{Y_i(k) Y_j(k)} \right)^\alpha}{\sqrt{\left(\sum_{k=1}^{Q} Y_i^\alpha(k) \right) \left(\sum_{k=1}^{Q} Y_j^\alpha(k) \right)}} \right], \tag{11}$$

which represents the ratio of the energy of the overlap between the two TFDs to the product of the energy of the individual TFDs. Let

$$J_{ij}^\alpha = \frac{\sum_{k=1}^{Q} \left(\sqrt{Y_i(k) Y_j(k)} \right)^\alpha}{\sqrt{\left(\sum_{k=1}^{Q} Y_i^\alpha(k) \right) \left(\sum_{k=1}^{Q} Y_j^\alpha(k) \right)}}, \tag{12}$$

and

$$J_\alpha = \sum_{i=1}^{N-1} \sum_{j=i+1}^{N} J_{ij}^\alpha. \tag{13}$$

Since log is a monotonous function, maximizing G_α is equivalent to minimizing J_α for $\alpha > 1$, or maximizing J_α for $\alpha < 1$. This means that we can equivalently use J_α as our cost function. In this paper, we will consider orders of $\alpha > 1$. The results are similar for $\alpha < 1$. One special case of $\alpha > 1$ is the quadratic one when $\alpha = 2$. When $\alpha = 2$, the cost function J_α simplifies to:

$$J_2 = \sum_{i=1}^{N-1} \sum_{j=i+1}^{N} \left[\frac{\sum_{k=1}^{Q} Y_i(k) Y_j(k)}{\sqrt{\left(\sum_{k=1}^{Q} Y_i^2(k) \right) \left(\sum_{k=1}^{Q} Y_j^2(k) \right)}} \right]. \tag{14}$$

In this paper, we will use $\alpha = 2$ since the Rényi entropy will be well–defined for this order even when the distributions are non-positive.

In N–dimensional space, the simplest rotation is in the two–dimensional plane. If a rotation is through an angle θ_{ab} in the $a - b$ plane, then the rotation matrix $\mathbf{R}_{ab}(\theta_{ab})$ equals the $N \times N$ identity matrix $\mathbf{I}_N$ except that the elements

$I_N(a,a), I_N(a,b), I_N(b,a)$, and $I_N(b,b)$ are replaced by $\cos(\theta_{ab}), \sin(\theta_{ab}), -\sin(\theta_{ab})$, and $\cos(\theta_{ab})$, respectively, where $I_N(a,b)$ is the element of $\mathbf{I}_N$ located at the ath row and bth column. From [7], we know that any N–dimensional rotation matrix can be written as the product of $N(N-1)/2$ two–dimensional–plane N–dimensional rotation matrices,which is:

$$\mathbf{R}(\theta) = \mathbf{R}_{12}(\theta_{12}) \cdots \mathbf{R}_{ab}(\theta_{ab}) \cdots \mathbf{R}_{(N-1)N}(\theta_{(N-1)N}), \tag{15}$$

where $\theta = [\theta_{12}, \cdots, \theta_{ab}, \cdots, \theta_{(N-1)N}]^T$, and $a < b$.

3.3 Proposed Algorithm

The goal of the proposed algorithm is to determine the optimal rotation transform such that the total pairwise divergence measure is maximized to achieve signal separation. We use the gradient adaptation algorithm also known as the steepest descent to update the rotation angles.

The overall update equation for stochastic gradient descent is:

$$\theta(n+1) = \theta(n) - \mu \frac{\partial J_2}{\partial \theta}, \tag{16}$$

where μ is the step size parameter. The gradient of the cost function J_2 with respect to the rotation angle θ_{ab} is derived as:

$$\frac{\partial J_2}{\partial \theta_{ab}} = \sum_{i=1}^{N-1} \sum_{j=i+1}^{N} \frac{\partial J_{ij}^2}{\partial \theta_{ab}}, \tag{17}$$

where

$$\frac{\partial J_{ij}^2}{\partial \theta_{ab}} = \frac{\sum_{k=1}^{Q}\left(\frac{\partial \mathbf{R}_i}{\partial \theta_{ab}}\mathbf{X}(k) Y_j(k) + Y_i(k)\frac{\partial \mathbf{R}_j}{\partial \theta_{ab}}\mathbf{X}(k)\right)}{\sqrt{\left(\sum_{k=1}^{Q} Y_i^2(k)\right)\left(\sum_{k=1}^{Q} Y_j^2(k)\right)}} - \frac{\sum_{k=1}^{Q} Y_i(k) Y_j(k)}{\left(\sqrt{\left(\sum_{k=1}^{Q} Y_i^2(k)\right)\left(\sum_{k=1}^{Q} Y_j^2(k)\right)}\right)^3} \times$$
$$\left[\left(\sum_{k=1}^{Q} Y_i(k)\frac{\partial \mathbf{R}_i}{\partial \theta_{ab}}\mathbf{X}(k)\right)\left(\sum_{k=1}^{Q} Y_j^2(k)\right) + \left(\sum_{k=1}^{Q} Y_i^2(k)\right)\left(\sum_{k=1}^{Q} Y_j(k)\frac{\partial \mathbf{R}_j}{\partial \theta_{ab}}\mathbf{X}(k)\right)\right], \tag{18}$$

where $\mathbf{R}_i$ is the ith row of $\mathbf{R}(\theta)$, and $\mathbf{X}(k)$ is the kth column of $\mathbf{X}$.

4 Experimental Results and Analysis

In order to evaluate the effectiveness of the proposed method, we consider the following source separation examples. The sources are assumed to be approximately disjoint on the time–frequency plane.

Example 1: Separation of two gabor logon signals and performance comparison with FastICA

In this example, the set of observed signals are linear combinations of two gabor logons. The first gabor logon is centered at the time sample point 50 and normalized frequency of 0.7, and the second gabor logon is centered at the time sample point 150 and normalized frequency of −0.7. Jensen–Rényi divergence with order $\alpha = 2$ is used as the cost function to ensure that the divergence is well–defined. In this example, we assume two mixtures of these two source signals. Each combination is transformed to the time–frequency domain using a binomial kernel [5] with $K = 50$ time samples and $L = 64$ frequency samples. Each TFD is vectorized to form a TFD observation matrix of size 2×3200. Fig. 1 (A) shows clearly that these two gabor logon signals can be separated from their mixtures through an optimal rotation under the divergence criterion.

In order to evaluate the performance of the proposed approach, we compare the MSE of the proposed algorithm with FastICA in the time–frequency domain by adding white Gaussian noise over a SNR range of 2–18 dB. We use 100 Monte Carlo simulations for each noise level. It is evident from Fig. 1 (B) that the proposed method has smaller MSE compared to FastICA. The difference in performance is due to the fact that the given sources are not necessarily independent and thus do not fit the assumptions underlying ICA.

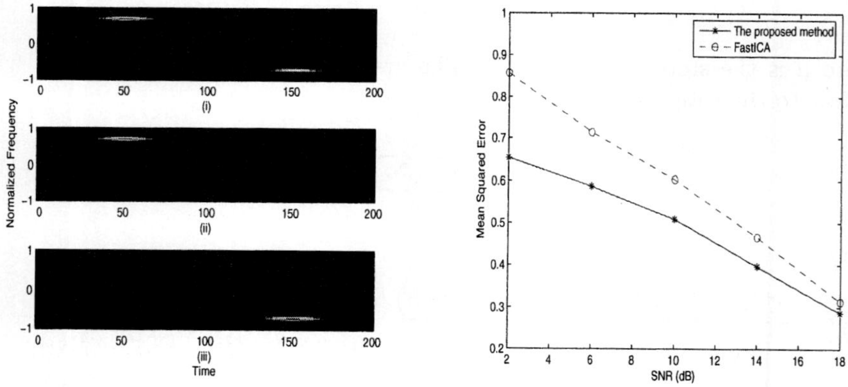

(A) The mixture and the extraction of two gabor logons: (i) the mixture, (ii) and (iii) the extracted source signals

(B) Error performance of the proposed method and FastICA versus SNR

Fig. 1. Source separation and performance comparison

Example 2: Separation of two crossing chirp signals

In this example, we consider the separation of two signals overlapping in the time–frequency domain. A mixture of two linear chirp signals is used for source separation. One of the chirp signals has an initial normalized frequency of -0.8 and its instantaneous frequency increases to a normalized frequency of 0.8. The other one has an initial normalized frequency of 0.8 and its instantaneous frequency decreases to a normalized frequency of -0.8. Obviously, these two chirp signals overlap with each other in both the time and frequency domains. Typical time domain or frequency domain separation methods can not be used to

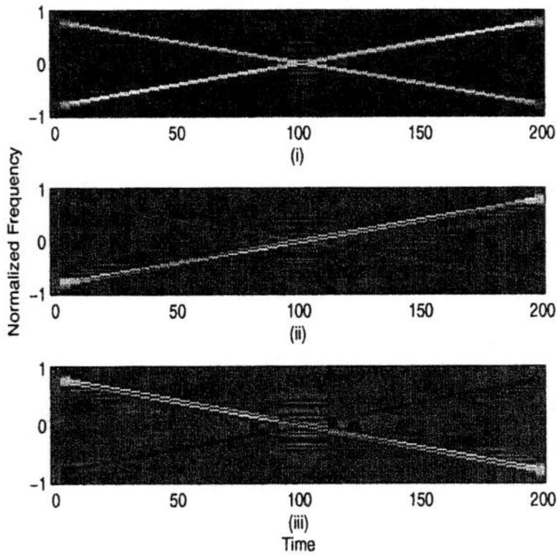

Fig. 2. The mixture and the separation of two crossing chirp signals: (i) the mixture, (ii) and (iii) the separated signals

perfectly recover them. Fig. 2 shows that using the proposed approach, we can successfully separate these two chirp signals from their mixtures.

Example 3: Number of mixtures greater than the number of sources

In this example, we consider a more general situation where the number of mixtures is larger than the number of sources. For M mixtures and N sources $(M > N)$, we construct a new $N \times M$ rotation matrix as follows:

$$\mathbf{R}_{NM}(\theta) = \mathbf{I}_{NM}\mathbf{R}_M(\theta), \tag{19}$$

where $\mathbf{R}_M(\theta)$ is an $M \times M$ rotation matrix given by (15), and $\mathbf{I}_{NM}$ is an $N \times M$ matrix with elements equal to 1 if $i = j$, 0 otherwise, where i, j represent the row and column indices, respectively. The source signals are one of the two gabor

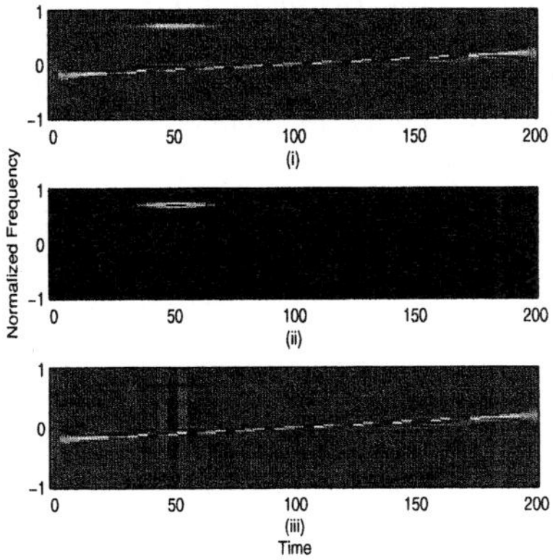

Fig. 3. Separating a chirp and a gabor logon from their three mixtures: (i) the mixture, (ii) the extracted gabor logon, (iii) the extracted chirp

logons in Example 1 and a chirp signal with an initial normalized frequency of -0.2 and a final normalized frequency of 0.2. We use the proposed approach with the new rotation matrix to extract these two signals from their three mixtures. It is indicated in Fig. 3 that the source signals can still be effectively extracted when the number of mixtures is greater than the number of sources.

5 Conclusions

In this paper, a new approach is presented for the separation of non–stationary signals on the time–frequency plane using an information-theoretic cost function. The proposed algorithm performs an N–dimensional rotation to separate the source signals. Using Jensen– Rényi divergence as the cost function, a steepest descent algorithm is implemented to update the rotation angles. The results illustrate that maximizing the divergence on the time–frequency plane can separate sources that are disjoint in the time–frequency domain.

Future work includes investigation of the effect of order α in the Jensen–Rényi divergence on the performance of the source separation algorithm, and extending the algorithm to a more challenging case, i.e., the number of mixtures is smaller than the number of sources. Another area of future work is using signal synthesis methods to transform the extracted sources from the time–frequency domain to the time domain.

References

1. A. Belouchrani, K. Abed-Meraim anf J-F. Cardoso, and E. Moulines, "A blind source separation technique using second order statistics," *IEEE Trans. on Signal Processing*, vol. 45, no. 2, pp. 434–444, 1997.
2. A. Hyvarinen, J. Karhunen, and E. Oja, *Independent Component Analysis*, John Wiley and Sons, 2001.
3. K.E. Hild II, D. Erdogmus, and J. C. Principe, "Blind source separation using Rényi's mutual informaiton," *IEEE Signal Processing Lett.*, vol. 8, pp. 174–176, 2001.
4. A. Belouchrani and M. G. Amin, "Blind source separation based on time–frequency signal representations," *IEEE Trans. on Signal Processing*, vol. 46, pp. 2888–2897, 1998.
5. L. Cohen, *Time–Frequency Analysis*, Prentice Hall, New Jersey, 1995.
6. R. G. Baraniuk, P. Flandrin, A. J. E. M. Janssen, and O. Michel, "Measuring time–frequency information content using the Rényi entropies," *IEEE Trans. on Info. Theory*, vol. 47, no. 4, pp. 1391–1409, May 2001.
7. F. D. Murnaghan, *The Unitary and Rotation Groups*, Spartan Books, Washington D.C., 1962.

Local Convergence Analysis of FastICA

Hao Shen[1,2] and Knut Hüper[1,2]

[1] Department of Information Engineering,
Research School of Information Sciences and Engineering,
The Australian National University, Canberra ACT 0200, Australia
[2] Systems Engineering and Complex Systems Research Program,
National ICT Australia, Canberra Research Laboratory,
Locked Bag 8001, Canberra ACT 2612, Australia
Hao.Shen@rsise.anu.edu.au, Knut.Hueper@nicta.com.au

Abstract. The FastICA algorithm can be considered as a selfmap on a manifold. It turns out that FastICA is a scalar shifted version of an algorithm recently proposed. We put these algorithms into a dynamical system framework. The local convergence properties are investigated subject to an ideal ICA model. The analysis is very similar to the well-known case in numerical linear algebra when studying power iterations versus Rayleigh quotient iteration.

1 Introduction

Blind Source Separation (BSS) is a challenging problem in Statistical Signal Processing. The applications of BSS arise in numerous disciplines, such as image processing, bioinformatics, noise cancellation, dimensionality reduction, and so on. Since the influential paper [1] the problem can now be solved successfully by Independent Component Analysis (ICA). Recently, several efficient ICA algorithms have been developed by researchers from various communities. The FastICA algorithm is a prominent ICA algorithm proposed by the Finnish school, see [2]. In this paper, we investigate the local convergence property of FastICA from a dynamical systems point of view.

We consider the standard noiseless linear instantaneous ICA model, $M = A \cdot S$, where $S \in \mathbb{R}^{m \times n}$ represents n samples of m sources with $m \ll n$. The invertible matrix $A \in \mathbb{R}^{m \times m}$ is a mixing matrix and $M \in \mathbb{R}^{m \times n}$ are the observed mixtures. The source signals S are assumed to be unknown, having zero mean with unit variance, being statistically independent, and at most one being Gaussian.

After a pre-whitening process, the whitened demixing model can be formulated as $Z = X^\top \cdot W$, where $W = V \cdot M \in \mathbb{R}^{m \times n}$ is the whitened observation, the invertible matrix $V \in \mathbb{R}^{m \times m}$ is the whitening matrix, the orthogonal matrix $X \in \mathbb{R}^{m \times m}$ is a new parameterisation of the problem as the demixing matrix in the whitened model, and $Z \in \mathbb{R}^{m \times n}$ is the recovered signal. Let denote $x \in S^{m-1}$ one of the columns of $X = [x_1, x_2, \ldots, x_m]$ and $w \in \mathbb{R}^m$ one of the columns of W, respectively.

J. Rosca et al. (Eds.): ICA 2006, LNCS 3889, pp. 893–900, 2006.

The FastICA algorithm was proposed to find the extrema of a generic contrast function of one-unit ICA approach [2]

$$\begin{aligned} f : S^{m-1} &\to \mathbb{R}, \\ x &\mapsto \mathbb{E}[G(x^\top w)], \end{aligned} \tag{1}$$

where $G : \mathbb{R} \to \mathbb{R}$ is an at least twice differentiable even function. It approximates some statistical properties over the sampled data. Under certain weak assumptions, the original sources signals can be recovered via computing extrema of f. The FastICA algorithm can be formulated as a self map

$$\begin{aligned} \psi : S^{m-1} &\to S^{m-1}, \\ x &\mapsto \frac{\mathbb{E}[G'(x^\top w)w] - \mathbb{E}[G''(x^\top w)]x}{\|\mathbb{E}[G'(x^\top w)w] - \mathbb{E}[G''(x^\top w)]x\|} \end{aligned} \tag{2}$$

where G', G'' are the first and second derivatives of G. This algorithm enjoys an heuristical interpretation as an approximated Newton method, see [2,3]. We will study the local convergence properties of the algorithm from a dynamical systems point of view using calculus on manifolds.

The whole analysis in this paper is subject to the ideal model. In Section 2, we characterise the critical points of a generic ICA contrast function. The local quadratic convergence of the FastICA algorithm is reexamined from a dynamical systems point of view in Section 3. Our conclusion follows in Section 4.

Note that in this paper all computations are performed using coordinate functions of the vector space $\mathbb{R}^m$ which is the embedding space of the unit sphere S^{m-1}.

2 Critical Point Analysis

As claimed by Hyvärinen *et al.* [2], the true independent sources can be reconstructed among the extrema of the function f defined in (1). In this section, we characterise the critical points of the function f.

For our analysis it is useful to find the right coordinate system to simplify calculations. Without loss of generality, we might assume $A = I_m$, i.e. $W = S$.

By the chain rule the first derivative of f can be computed as

$$\mathrm{D}\, f(x) \circ \xi = \xi^\top \cdot \mathbb{E}[G'(x^\top s)\, s], \tag{3}$$

where $\xi \in T_x S^{m-1}$ is an arbitrary tangent element. Recall that the tangent space of the unit sphere at point $x \in S^{m-1}$ is defined as

$$T_x S^{m-1} := \{\xi \in \mathbb{R}^m | x^\top \xi = 0\}. \tag{4}$$

Critical points of f are therefore characterised as solutions of

$$\xi^\top \cdot \mathbb{E}[G'(x^\top s)\, s] = 0 \tag{5}$$

for all $\xi \in T_x S^{m-1}$. Due to the geometry of S^{m-1}, equation (5) is equivalent to saying, that

$$\mathbb{E}[G'(x^\top s)\, s] = \gamma\, x, \tag{6}$$

with γ a real parameter.

Let X be an orthogonal matrix, such that x_i is the i-th column of X. By left multiplication of Equ. (6) by $X^\top$, we get

$$\mathbb{E}[G'(x_i^\top s)\, X^\top s] = \gamma\, e_i, \tag{7}$$

where e_i is i-th standard basis vector in $\mathbb{R}^m$. We therefore have proven

Lemma 2.1. *Let $X = [x_1, x_2, \ldots, x_m]$ be an orthogonal matrix with column $x_i \in S^{m-1}$, i.e. $\|x_i\| = 1$ for all $i = 1, \ldots, m$. An $x_i \in S^{m-1}$ is a critical point of the function f defined by* (1) *if and only if*

$$\mathbb{E}[G'(x_i^\top s)\, X^\top s] = \gamma\, e_i \tag{8}$$

holds. □

Obviously, the critical point set structure of f highly depends on the properties of the cost function G. In other words, the quality of the separation via an optimisation approach depends on the nature of G.

We rewrite the critical point condition (7) in an equivalent way as

$$\begin{gathered} \mathbb{E}\left[\frac{G'(x_i^\top s)}{x_i^\top s} \cdot x_j^\top s \cdot s^\top x_i\right] = \gamma \delta_{ij} \\ \Longleftrightarrow \\ x_j^\top \cdot \mathbb{E}\left[\frac{G'(x_i^\top s)}{x_i^\top s} \cdot s\, s^\top\right] \cdot x_i = \gamma \delta_{ij}. \end{gathered} \tag{9}$$

It is wellknown that

$$\mathbb{E}\left[\frac{G'(x_i^\top s)}{x_i^\top s} \cdot s\, s^\top\right]$$

can be made positive definite by requiring the function G to be even and convex.

Under the above assumptions it is easily seen that perfect separations are attained at the standard basis vectors $\{e_1, e_2, \ldots, e_m\}$. Trivially, the standard basis vectors fulfil the critical point condition in Lemma 2.1.

We will now show that the second derivative of f at a critical point corresponding to a perfect recovery, i.e., evaluated at a standard basis vector, is either positive or negative definite.

Consider the smooth extension of f to $\mathbb{R}^m$, i.e.

$$\begin{aligned} \hat{f} : \mathbb{R}^m &\to \mathbb{R}, \\ x &\mapsto \mathbb{E}[G(x^\top w)] \end{aligned} \tag{10}$$

where $\hat{f}\big|_{S^{m-1}} = f$. The Hessian at a critical point e_i can be computed from the following quadratic form on $\mathbb{R}^m \times \mathbb{R}^m$

$$Q_{e_i}(h,h) := \frac{\mathrm{d}^2}{\mathrm{d}\,\varepsilon^2}\hat{f}(x+\varepsilon h)\bigg|_{\varepsilon=0,x=e_i} = h^\top \cdot \left(\mathbb{E}[G''(e_i^\top s)\, s\, s^\top] - \mathbb{E}[G'(e_i^\top s)\, e_i^\top s]\, I_m\right) \cdot h. \tag{11}$$

By polarisation of the quadratic form $Q_{e_i}(h,h)$ and projecting the corresponding linear map $Q_{e_i}(h)$ evaluated at $h \in \mathbb{R}^m$ to the tangent space $T_{e_i}S^{m-1}$ we get an explicit discription of the Hessian of f at e_i.

Lemma 2.2. *By abuse of notation let s_i denote the i-th entry of a column s of the matrix S. Then the Hessian operator at a critical point e_i acts on tangent vectors simply by scalar multiplication, the scalar being equal to*

$$\mathbb{E}[G''(s_i)] - \mathbb{E}[G'(s_i)\, s_i].$$

Proof. This is a straightforward calculation. □

3 FastICA as a Dynamical System on S^{m-1} and Projective Space

In a suitable coordinate system, i.e., $W = S$, we consider FastICA as the map

$$\begin{aligned} \psi : S^{m-1} &\to S^{m-1}, \\ x &\mapsto \frac{\mathbb{E}[G'(x^\top s)s] - \mathbb{E}[G''(x^\top s)]x}{\|\mathbb{E}[G'(x^\top s)s] - \mathbb{E}[G''(x^\top s)]x\|}. \end{aligned} \tag{12}$$

To analyse this mapping in full generality seems to be hopeless. Nevertheless, for certain classes of functions which have been used in signal processing applications it can be shown that it is at least locally well defined and smooth.

In this paper we will assume that at a critical point e_i of the function f the denominator in (12) does not vanish. Then there exists a neighborhood of e_i in S^{m-1} where this holds as well.

We will now compute the derivative of the algorithmic mapping at such a point e_i. As it is well known the iterative mapping ψ considered as a dynamical system on the sphere might cause alternating signs in the systems coefficient vector. It is therefore convenient to take a different point of view as has been done in the literature so far. It is easily seen that the cost function f maps two antipodes on the sphere into one and the same point in $\mathbb{R}$ provided the smooth contrast function G is even. Consequently, f defines a smooth function on projective space. Moreover, we can consider the algorithmic map ψ as a selfmapping on projective space as well.

The span of any standard basis vector is a fixed point of the iteration map as seen by the following Lemma.

Lemma 3.1. *Consider* ψ *as a selfmapping on projective space. Then* span(e_i) *is a fixed point of the mapping* ψ.

Proof. We have to check if e_i solves the equation

$$x = \pm \frac{\mathbb{E}[G'(x^\top s)s] - \mathbb{E}[G''(x^\top s)]x}{\|\mathbb{E}[G'(x^\top s)s] - \mathbb{E}[G''(x^\top s)]x\|}. \tag{13}$$

According to the critical point condition (6) the numerator in (13), evaluated at $x = e_i$, is equal to a multiple of e_i. The result follows. □

The next theorem characterises the local behavior of the mapping ψ considered as a selfmap of projective space around such a fixed point span(e_i).

Theorem 3.1. *The derivative of* ψ *evaluated at* span(e_i) *is equal to zero.*

Proof. Note that locally S^{m-1} and projective space are diffeomorphic. To prove the theorem we therefore can consider ψ as a selfmapping of the sphere. We have to study the linear map

$$\mathrm{D}\,\psi(x) : T_x S^{m-1} \to T_{\psi(x)} S^{m-1} \tag{14}$$

at e_i, i.e. the linear map $\mathrm{D}\,\psi(e_i)$ assigns to an arbitrary tangent element $\xi \in T_{e_i} S^{m-1}$ the value $\mathrm{D}\,\psi(e_i)\xi$. Let denote the numerator of $\psi(x)$ by

$$\begin{aligned} F &: S^{m-1} \to \mathbb{R}^m, \\ F(x) &:= \mathbb{E}[G'(x^\top s)s] - \mathbb{E}[G''(x^\top s)]x. \end{aligned} \tag{15}$$

One computes

$$\begin{aligned} \mathrm{D}\,\psi(e_i)\xi &= \frac{\mathrm{d}}{\mathrm{d}\,\varepsilon}\psi(x+\varepsilon\xi)\Big|_{\varepsilon=0,x=e_i} \\ &= \frac{1}{\|F(e_i)\|}\underbrace{\left(\mathrm{id} - \frac{F(e_i)}{\|F(e_i)\|}\frac{F(e_i)^\top}{\|F(e_i)\|}\right)}_{=:P(e_i)}\mathrm{D}\,F(e_i)\xi. \end{aligned} \tag{16}$$

Note that $P(e_i)$ is an orthogonal projection operator. Moreover, by the critical point condition,

(i) $P(e_i)$ projects onto the complement of the span of e_i and
(ii) $\mathrm{D}\,F(e_i)\xi$ is equal to a scalar multiple of e_i.

The result follows. □

Corollary 3.1. *FastICA considered as the map* ψ *is locally quadratically convergent to a fixed point* x_*.

Proof. Using $x_{k+1} = \psi(x_k)$ and the fixed point condition $\psi(x_*) = x_*$, the result follows by the Taylor-type argument

$$\|\psi(x_k) - x_*\| \leq \sup_{y \in \overline{\mathcal{N}(x_*)}} \| \mathrm{D}^2 \psi(y)\| \cdot \|x_k - x_*\|^2$$

with $\overline{\mathcal{N}(x_*)}$ being the closure of a sufficiently small open neighborhood of x_*, as the first derivative of ψ at x_* vanishes. □

It is worthwhile to notice that the above analysis is similar to the analysis recently proposed in connection with the Rayleigh Quotient Iteration (RQI), see [4] for details. The analogy between FastICA and RQI goes even a bit further. We will now give FastICA an interpretation of a shifted version of a simpler algorithm recently proposed, see [5]. Moreover, we will show that the *scalar shift strategy* which is used in FastICA is in some sense a very clever way to ensure local quadratic convergence.

As an aside note that in numerical linear algebra it is wellknown that power iterations to compute an eigenvector of a real symmetric matrix can be significantly accelerated by incorporating scalar shift techniques. These ideas lead to the concept of RQI. Eventually, this fact is one of the reasons why the celebrated QR-algorithm is one of the most efficient algorithms in numerical linear algebra.

More recently in [5] to some extent certain fixed point algorithms for ICA were analysed. Our observation is that the algorithm which appears as equation (3) in [5] is a non-shifted version of FastICA, i.e., one might study the *zero-shift*-map

$$\begin{aligned} \phi : S^{m-1} &\to S^{m-1}, \\ x &\mapsto \frac{\mathbb{E}[G'(x^\top s)s]}{\|\mathbb{E}[G'(x^\top s)s]\|}, \end{aligned} \tag{17}$$

from a dynamical systems point of view, as well. Or more generally, one might study

$$\begin{aligned} \eta : S^{m-1} &\to S^{m-1}, \\ x &\mapsto \frac{\mathbb{E}[G'(x^\top s)s] - \rho(x)x}{\|\mathbb{E}[G'(x^\top s)s] - \rho(x)x\|}, \end{aligned} \tag{18}$$

where $\rho : S^{m-1} \to \mathbb{R}$ is a scalar valued map, the *shift-map*.

Assume that we have specified the shift-map such that e_i is a fixed point of η. We might study the local convergence properties of η in the spirit of our treatment above.

Denote the numerator in (18) by

$$\begin{aligned} H : S^{m-1} &\to \mathbb{R}^m, \\ H(x) &:= \mathbb{E}[G'(x^\top s)s] - \rho(x)x. \end{aligned} \tag{19}$$

The derivative is as $\mathrm{D}\,H(x) : T_x S^{m-1} \to \mathbb{R}^m$. Therefore, one computes

$$\begin{aligned}\mathrm{D}\,H(e_i)\xi &= \frac{\mathrm{d}}{\mathrm{d}\,\varepsilon}\eta(x+\varepsilon\xi)\Big|_{\varepsilon=0,x=e_i} \\ &= \underbrace{\frac{1}{\|H(e_i)\|}\left(\mathrm{id} - \frac{H(e_i)}{\|H(e_i)\|}\frac{H(e_i)^\top}{\|H(e_i)\|}\right)}_{=:P(e_i)} \mathrm{D}\,H(e_i)\xi. \end{aligned} \tag{20}$$

Again it is easily seen that the orthogonal projection operator $P(e_i)$ in (20) projects on the orthogonal complement of the span of e_i in $\mathbb{R}^m$, i.e. into the tangent space $T_{e_i}S^{m-1}$. Consequently,

$$\mathrm{D}\,\eta(e_i) = 0 \quad \Longleftrightarrow \quad \mathrm{D}\,H(e_i)\xi = \lambda e_i, \quad \lambda \in \mathbb{R}. \tag{21}$$

Now by the chain rule

$$\mathrm{D}\,H(e_i)\xi = \mathrm{D}\,\mathbb{E}[G'(e_i^\top s)s]\xi - (\mathrm{D}\,\rho(e_i)\xi)e_i - \rho(e_i)\xi. \tag{22}$$

Because $(\mathrm{D}\,\rho(e_i)\xi)e_i$ is already a multiple of e_i it remains to specify under which conditions on the scalar shift ρ the following holds true

$$\mathrm{D}\,\mathbb{E}[G'(e_i^\top s)s]\xi - \rho(e_i)\xi = \lambda e_i. \tag{23}$$

From the computations above, see Lemma 2.2, we already know that

$$\mathrm{D}\,\mathbb{E}[G'(e_i^\top s)s]\xi = \mathbb{E}[G''(s_i)]\xi. \tag{24}$$

Finally, the only possibility to ensure that $\mathrm{D}\,\eta(e_i)$ vanishes happens if

$$\rho(e_i) = \mathbb{E}[G''(s_i)]. \tag{25}$$

4 Conclusion and Outlook

To conclude we might state the following.

In a framework where we assume smoothness conditions on even contrast functions and the scalar shift strategy, the only scalar shift strategies which ensure local quadratical convergence are those which respect (25). This happens for instance in FastICA. One might ask the question how to modify the scalar shift strategy in FastICA to ensure third or even higher order local convergence properties. One necessary requirement is that such a modified shift strategy has to be equal to the one used in FastICA when evaluated at the standard basis vectors.

Another completely different approach to accelerate convergence is to use nonscalar shifts. For work in this direction we refer to [6,7] and a forthcoming journal paper.

Acknowledgment

National ICT Australia is funded by the Australian Government's Department of Communications, Information Technology and the Arts and the Australian Research Council through Backing Australia's Ability and the ICT Centre of Excellence Program.

References

1. Comon, P.: Independent component analysis, a new concept? Signal Processing **36** (1994) 287–314
2. Hyvärinen, A., Karhunen, J., Oja, E.: Independent Component Analysis. Wiley, New York (2001)
3. Hyvärinen, A.: Fast and robust fixed-point algorithms for independent component analysis. IEEE Transactions on Neural Networks **10** (1999) 626–634
4. Hüper, K.: A Dynamical System Approach to Matrix Eigenvalue Algorithms. In: Mathematical Systems Theory in Biology, Communications, Computation. Volume 134 of The IMA Volumes in Mathematics and its Applications. Springer, New York (2003) 257–274
5. Regalia, P., Kofidis, E.: Monotonic convergence of fixed-point algorithms for ICA. IEEE Transactions on Neural Networks **14** (2003) 943–949
6. Hüper, K., Shen, H., Seghouane, A.-K.: Local convergence properties of FastICA and some generalisations. IEEE-ICASSP 2006, Toulouse, France; accepted
7. Hüper, K., Shen, H., Seghouane, A.-K.: Geometric optimisation and FastICA algorithms. Contribution to Mini-Symposium: Geometric Optimisation in System and Control I/II The 17th International Symposium on Mathematical Theory of Networks and Systems, Kyoto, Japan, 2006; accepted

Testing Significance of Mixing and Demixing Coefficients in ICA

Shohei Shimizu[1,2], Aapo Hyvärinen[2], Yutaka Kano[1], Patrik O. Hoyer[2], and Antti J. Kerminen[2]

[1] Graduate School of Engineering Science, Osaka University, Japan
[2] Helsinki Institute for Information Technology, Basic Research Unit, Finland
http://www.cs.helsinki.fi/hiit_bru/index_neuro.html

Abstract. Independent component analysis (ICA) has been extensively studied since it was originated in the field of signal processing. However, almost all the researches have focused on estimation and paid little attention to testing. In this paper, we discuss testing significance of mixing and demixing coefficients in ICA. We propose test statistics to examine significance of these coefficients statistically. A simulation experiment implies the good performance of our testing procedure. A real example in psychometrics, which is a new application area of ICA, is also presented.

1 Introduction

Independent component analysis (ICA) [1] is a multivariate analysis technique that aims at recovering linearly mixed unobserved multidimensional independent signals from the mixed observable variables. Let $\boldsymbol{x}$ be an m-dimensional observed vector. The ICA model for $\boldsymbol{x}$ is written as

$$\boldsymbol{x} = \mathbf{A}\boldsymbol{s}, \tag{1}$$

where $\mathbf{A}$ is called a mixing matrix and $\boldsymbol{s}$ is an n-dimensional vector of independent components with zero mean and unit variance. Typically, the number of observed variables m is assumed to equal that of latent variables n, that is, $m = n$, which we assume in the following. The main goal of ICA is to estimate the mixing matrix $\mathbf{A}$ or the demixing matrix $\mathbf{B}^T = \mathbf{A}^{-1}$. (Some authors use $\mathbf{B} = \mathbf{A}^{-1}$ *without* the transpose [1].)

The ICA has been extensively studied since identification conditions for the model were provided in [2]. However, almost all the researches have focused on estimation [3,4,5], e.g., consistency, stability, robustness and asymptotic variance [6,7,8], and have not paid very much attention to testing. In this paper, we discuss testing of significance of mixing and demixing coefficients a_{ij} and b_{ij}. Such a test of significance is an important process in psychometrics for example [9].

The paper is structured as follows. First, in Section 2, we briefly review asymptotic variance of ICA and provide asymptotic covariance matrices of mixing and demixing coefficients estimated by ICA based on non-gaussianity maximization with constraints of orthogonality, e.g., FastICA [5]. In Section 3, we derive test

J. Rosca et al. (Eds.): ICA 2006, LNCS 3889, pp. 901–908, 2006.

statistics to evaluate the magnitude of significance of these coefficients using the asymptotic variances. We also consider multiple comparison procedures since we usually test significance of more than one coefficient. In Sections 4 and 5, we conduct a simulation study and provide a real data example to study how the test statistics work empirically. We conclude the paper in Section 6.

2 Asymptotic Variance of ICA

Several authors studied asymptotic variance of ICA [7,8,10,11], where the theory of estimating functions [12] was often used. Let us consider a semiparametric model $p(\boldsymbol{x}|\boldsymbol{\theta})$, where $\boldsymbol{\theta}$ is a r-dimensional parameter vector of interest. Note that the density function $p(\boldsymbol{x}|\boldsymbol{\theta})$ is unknown. Let us denote by $\boldsymbol{\theta}_0$ the true parameter vector of interest. A r-dimensional vector-valued function $\boldsymbol{f}(\boldsymbol{x}, \boldsymbol{\theta})$ is called an estimating function when it satisfies the following conditions for any $p(\boldsymbol{x}|\boldsymbol{\theta}_0)$:

$$E[\boldsymbol{f}(\boldsymbol{x}, \boldsymbol{\theta}_0)] = \mathbf{0} \tag{2}$$

$$|\det \mathbf{Q}| \neq 0, \qquad \text{where } \mathbf{Q} = E\left[\left.\frac{\partial}{\partial \boldsymbol{\theta}^T} \boldsymbol{f}(\boldsymbol{x}, \boldsymbol{\theta})\right|_{\boldsymbol{\theta}=\boldsymbol{\theta}_0}\right] \tag{3}$$

$$E[\|\boldsymbol{f}(\boldsymbol{x}, \boldsymbol{\theta}_0)\|^2] < \infty, \tag{4}$$

where the expectation E is taken over $\boldsymbol{x}$ with respect to $p(\boldsymbol{x}|\boldsymbol{\theta}_0)$.

Let $\boldsymbol{x}(1), \cdots, \boldsymbol{x}(N)$ be a random sample from $p(\boldsymbol{x}|\boldsymbol{\theta}_0)$. Then an estimator $\hat{\boldsymbol{\theta}}$ is obtained by solving the estimating equation:

$$\sum_{i=1}^{N} \boldsymbol{f}(\boldsymbol{x}(i), \boldsymbol{\theta}) = \mathbf{0}. \tag{5}$$

Under some regularity conditions including identification conditions for $\boldsymbol{\theta}$, the estimator $\hat{\boldsymbol{\theta}}$ is consistent when N goes to infinity and asymptotically distributes according to the gaussian distribution $N(\boldsymbol{\theta}_0, \mathbf{G})$, and

$$\mathbf{G} = \frac{1}{N}\mathbf{Q}^{-1} E[\boldsymbol{f}(\boldsymbol{x}, \boldsymbol{\theta}_0)\boldsymbol{f}^T(\boldsymbol{x}, \boldsymbol{\theta}_0)]\mathbf{Q}^{-T}. \tag{6}$$

In [7], an estimating function for (quasi-) maximum likelihood estimation was derived. In [13], an estimating function for JADE [4] was provided, and an estimating function for ICA based on non-gaussianity maximization with orthogonality (uncorrelatedness) constraints including FastICA [5] was also introduced.

In this paper, we restrict ourselves to testing mixing and demixing coefficients estimated by FastICA. In FastICA, we first center the data to make its mean zero and whiten the data by computing a matrix $\mathbf{V}$ such that the covariance matrix of $\boldsymbol{z} = \mathbf{V}\boldsymbol{x}$ is the identity matrix. After that, we find an orthogonal matrix $\mathbf{W}$ so that components of $\mathbf{W}^T\boldsymbol{z} = \mathbf{W}^T\mathbf{V}\boldsymbol{x}$ have maximum non-gaussianity. Then we obtain estimates of $\mathbf{A}$ and $\mathbf{B}$ by $\mathbf{A} = \mathbf{V}^{-1}\mathbf{W}$ and $\mathbf{B} = \mathbf{V}^T\mathbf{W}$.

Let us consider the following function:

$$\mathbf{F}(\boldsymbol{x}, \mathbf{W}) = \boldsymbol{y}\boldsymbol{y}^T - \mathbf{I} + \boldsymbol{y}\boldsymbol{g}^T(\boldsymbol{y}) - \boldsymbol{g}(\boldsymbol{y})\boldsymbol{y}^T, \tag{7}$$

where $\boldsymbol{y} = \mathbf{B}^T\boldsymbol{x} = \mathbf{W}^T\mathbf{V}\boldsymbol{x} = \mathbf{W}^T\boldsymbol{z}$ and $g(u)$ is the nonlinearity. The estimating function for FastICA is obtained as $\boldsymbol{f} = \mathrm{vec}(\mathbf{F})$ taking $\boldsymbol{\theta} = \mathrm{vec}(\mathbf{W})$ [13], where $\mathrm{vec}(\cdot)$ denotes the vectorization operator which creates a column vector from a matrix by stacking its columns.

According to the estimating function theory, we obtain the asymptotic covariance matrix of $\mathrm{vec}(\mathbf{W})$ by (6) (see the Appendix for the complete formula). Here we assume that the variance in the estimate of $\mathbf{V}$ is negligible with respect to the variance in $\mathbf{W}$, which is validated empirically in the simulation below. Then we obtain the asymptotic covariance matrix of $\mathrm{vec}(\mathbf{A})$ and $\mathrm{vec}(\mathbf{B})$ as follows:

$$\begin{aligned}\mathrm{acov}\{\mathrm{vec}(\mathbf{A})\} &= \mathrm{acov}\{\mathrm{vec}(\mathbf{V}^{-1}\mathbf{W})\} \\ &= (\mathbf{I}\otimes\mathbf{V}^{-1})\mathrm{acov}\{\mathrm{vec}(\mathbf{W})\}(\mathbf{I}\otimes\mathbf{V}^{-1})^T \qquad (8)\\ \mathrm{acov}\{\mathrm{vec}(\mathbf{B})\} &= \mathrm{acov}\{\mathrm{vec}(\mathbf{V}^T\mathbf{W})\} \\ &= (\mathbf{I}\otimes\mathbf{V}^{T})\mathrm{acov}\{\mathrm{vec}(\mathbf{W})\}(\mathbf{I}\otimes\mathbf{V}^{T})^T, \qquad (9)\end{aligned}$$

where $\otimes$ denotes the Kronecker product. Given an $m \times n$ matrix $\mathbf{T}$ and $p \times q$ matrix $\mathbf{U}$, the Kronecker product of $\mathbf{T}$ and $\mathbf{U}$ is the following $mp \times nq$ matrix

$$\mathbf{T}\otimes\mathbf{U} := \begin{bmatrix} t_{11}\mathbf{U} & \cdots & t_{1n}\mathbf{U} \\ \vdots & \ddots & \vdots \\ t_{m1}\mathbf{U} & \cdots & t_{mn}\mathbf{U}\end{bmatrix}. \qquad (10)$$

Matlab codes to compute $\mathrm{acov}\{\mathrm{vec}(\mathbf{A})\}$ and $\mathrm{acov}\{\mathrm{vec}(\mathbf{B})\}$ are available online at the webpage: `http://chobi.sigmath.es.osaka-u.ac.jp/~shimizu/acov/`

3 Testing Significance of Mixing and Demixing Coefficients

3.1 Wald Statistics

In this paper, we would like to test if mixing or demixing coefficients are zero or not. Such tests are related to the fundamental question typically posed in empirical sciences: Does the independent component s_j have a statistically significant effect on the observed variable x_i? Here, the null and alternative hypotheses H_0 and H_1 are as follows:

$$H_0 : a_{ij} = 0 \quad \text{versus} \quad H_1 : a_{ij} \neq 0 \qquad (11)$$

or

$$H_0 : b_{ij} = 0 \quad \text{versus} \quad H_1 : b_{ij} \neq 0. \qquad (12)$$

One can use the following Wald statistics

$$\frac{\hat{a}_{ij}^2}{\mathrm{avar}(\hat{a}_{ij})} \text{ or } \frac{\hat{b}_{ij}^2}{\mathrm{avar}(\hat{b}_{ij})} \qquad (13)$$

to test significance of a_{ij} and b_{ij}, where $\mathrm{avar}(\hat{a}_{ij})$ and $\mathrm{avar}(\hat{b}_{ij})$ denote the asymptotic variances of $\hat{a}_{ij}$ and $\hat{b}_{ij}$ computed by (8) and (9). The Wald statistics can be used to test the null hypothesis H_0. Under H_0, the Wald statistic asymptotically approximates to a chi-square variate with one degree of freedom [9]. Then we can obtain the probability of having a value of the Wald statistic larger than or equal to the empirical one computed from data. We reject H_0 if the probability is smaller than a significance level, and otherwise we accept H_0. Acceptance of H_0 implies that the assumption $a_{ij} = 0$ (or $b_{ij} = 0$) fits data. Rejection of H_0 suggests that the assumption is in error so that H_1 holds [9].

3.2 Multiple Comparison

Usually, mixing and demixing matrices have more than one element. In many cases, we need to perform more than one testing simultaneously to find out if all or all of a set of the coefficients are significantly large in an absolute value sense. Although a given significance level may be appropriate for each individual testing, it is not for the set of all the testing. We are bound to have a lot of spurious significance if we just repeat testing without any corrections. Suppose we repeat testing 1,000 times at significance level 5%. Assume all the null hypotheses are true. Nevertheless, we can always expect that approx. 50 null models are rejected. However, we should not reject the null models. We have to control the probability of having at least one spurious false positive. In such a case, we should employ multiple comparison procedures. A simple and basic method is the Bonferroni correction, where we simply divide a significance level by the number of testing to obtain the significance level for individual testing. See [14] for details. We employ the Bonferroni correction in the simulation and real data analysis below.

4 Simulation

We conducted simulations in an attempt to confirm the theoretical results above. We employed FastICA, where the hyperbolic tangent function was taken as the nonlinearity and the symmetric orthogonalization was applied.

The simulation consisted of 10,000 replications. We employed the following mixing matrix that was lower triangular:

$$\mathbf{A} = \begin{bmatrix} 1 & 0 & 0 \\ 0.5 & 1 & 0 \\ 0.65 & 0.7 & 1 \end{bmatrix}, \tag{14}$$

and then the demixing matrix $\mathbf{B}^T = \mathbf{A}^{-1}$ was

$$\mathbf{B}^T = \begin{bmatrix} 1 & 0 & 0 \\ -0.5 & 1 & 0 \\ -0.3 & -0.7 & 1 \end{bmatrix}. \tag{15}$$

In each replication, we generated independent sources and created observed signals following the ICA model (1). First, we created three independent components s_i with the sample size $N = 300, 500, 1000$ where their components were independently distributed according to the Laplace distribution. The independent components were normalized to have zero means and unit variances.

The FastICA was then applied to the data, and estimates of mixing and demixing coefficients a_{ij} and b_{ij} were obtained. Then we computed Wald statistics for these coefficients and tested the null hypotheses of the coefficients being zero as described above. The significance level was set at 5%. We computed how many null hypotheses on the coefficients in these matrices were rejected to know if the chi-square approximation worked for finite sample sizes. We also counted the numbers of cases where at least one of null hypotheses on the coefficients with zero values (here, elements in strictly upper triangular parts of $\mathbf{A}$ and $\mathbf{B}^T$) was rejected to know if the Bonferroni correction was effective.

The results are shown in Tables 1 and 2. First in Table 1, we shall examine the empirical significance levels (number of rejections) for a_{12}, a_{13} and a_{23}, and b_{12}, b_{13} and b_{23} that had zero values. Overall, we would say that the numbers of rejections of the null models were very close to the theoretically expected number 500. Second, Table 1 allows us to examine the statistical power of the test for the other coefficients that had non-zero values. The power of 0.99 (9,900 rejections) was achieved for all the conditions other than when testing b_{31} with $N = 300$. Thus, Table 1 showed that the Wald statistics were well approximated by the chi-square distribution, and the power of test was quite good.

Next in Table 2, we examine the numbers of cases where at least one of null hypotheses on the coefficients with zero values to study the performance of the Bonferroni correction for multiple comparison discussed in Section 3.2. Overall, we would say that the numbers of rejections with the Bonferroni correction were rather close to the theoretically expected number 500 for all the conditions, though the null models were rejected a bit less often than the theoretically expected number 500 when testing b_{ij}. On the other hand, the numbers

Table 1. Numbers of rejected null hypotheses with significance level 5% (10,000 replications)

	a_{11}	a_{21}	a_{31}	a_{22}	a_{32}	a_{33}	a_{12}	a_{13}	a_{23}
N =									
300	9,999	9,914	9,931	9,995	9,946	9,984	467	499	478
500	10,000	9,997	9,995	10,000	9,997	10,000	506	473	475
1,000	10,000	10,000	9,996	10,000	10,000	10,000	488	468	477

	b_{11}	b_{21}	b_{31}	b_{22}	b_{32}	b_{33}	b_{12}	b_{13}	b_{23}
N =									
300	9,994	9,903	9,159	9,990	9,962	9,999	406	452	422
500	10,000	9,994	9,893	10,000	9,998	10,000	442	454	464
1,000	10,000	10,000	9,999	10,000	10,000	10,000	428	456	507

Note: N is sample size in estimation.

Table 2. Numbers of cases where at least one of null hypotheses on coefficients with zero values was rejected (10,000 replications)

	Testing a_{ij}			Testing b_{ij}		
$N=$	300	500	1,000	300	500	1,000
Bonferroni correction	497	471	447	383	408	411
No corrections	1,307	1,344	1,326	1,139	1,222	1,249

Note: N is sample size in estimation.

of rejections with no Bonferroni correction were much larger than the theoretically expected number for all the conditions. Thus, Table 2 showed that the Bonferroni correction was effective and should be applied to real data analyses.

5 Example with Real Data

Questionnaire data about criminal psychology were analyzed as an example. The survey was conducted with university students in Japan [15]. The sample size was 222. Observed variables were standardized to have zero mean and unit variance. The labels of the variables x_1 and x_2 are "x_1: Sum of scores of items that ask subjective evaluation on frequency of your criminal opportunities when you went to high school" and "x_2: Sum of scores of items that ask subjective evaluation on frequency of your criminal behavior when you went to high school". A Kolmogorov-Smirnov test showed that all variables could not be assumed to come from the gaussian distribution (significance level 1%). Thus, ICA should be applicable to this kind of non-gaussian data.

We employed FastICA with the nonlinearity $g(u) = \tanh(u)$ and the symmetric orthogonalization. We set the significance level at 5% and used the Bonferroni method for multiple comparison. The estimated $\mathbf{A}$ by FastICA was

$$\begin{bmatrix} 0.93 & 0.36 \\ 0.77 & 0.64 \end{bmatrix}, \tag{16}$$

where a_{11}, a_{21} and a_{22} were significant, and a_{12} was not significant. See Table 3 for the Wald statistics. Thus, the matrix $\mathbf{A}$ could be seen to be lower triangular.

The fact that $\mathbf{A}$ is lower triangular allows us to interpret the results in terms of a causal ordering of the variables [16]. The result implied the causal order, $x_1 \to x_2$, that is, criminal opportunities at high schools $\to$ criminal behaviors at high schools. The link between the lower triangularity of $\mathbf{A}$ and the causal order can be seen as follows. For the lower triangular mixing matrix, x_1 is essentially

Table 3. Estimates, Wald statistics and p values

	a_{11}	a_{21}	a_{12}	a_{22}
Estimates	0.93	0.77	0.36	0.64
Wald statistics	48.70	17.57	1.69	9.24
p values	0.00	0.00	0.19	0.00

equal to s_1, up to a multiplicative constant, a_{11}. On the other hand, x_2 is a function of s_1 and s_2, $a_{21}s_1 + a_{22}s_2$. Thus, x_2 is a function of x_1 and a new independent variable, s_2, that is, $(a_{21}/a_{11})x_1 + a_{22}s_2$. This indicates that x_1 may cause x_2, but x_2 cannot cause x_1. See [16] for details.

In fact, the order $x_1 \rightarrow x_2$ was reasonable to the criminal psychology theory. According to a criminal psychology theory [17], the frequency of criminal opportunities (x_1) is a typical environmental cause of the frequency of criminal behaviors (x_2) [15]. Therefore, the possible causal order from background knowledge was $x_1 \rightarrow x_2$. Thus, the causal order founded by our method would be reasonable to the background knowledge.

6 Conclusion

In this paper, we proposed Wald statistics to test significance of mixing and demixing coefficients in ICA. We conducted a small simulation experiment, which implied that our testing procedure worked well even for finite sample sizes, although more simulation studies are needed to study to what extent the result can be generalized. We also provided a real data example in psychometrics that would be a promising new area that ICA applies.

References

1. Hyvärinen, A., Karhunen, J., Oja, E.: Independent component analysis. Wiley, New York (2001)
2. Comon, P.: Independent component analysis. a new concept? Signal Processing **36** (1994) 62–83
3. Amari, S.: Natural gradient learning works efficiently in learning. Neural Computation **10** (1998) 251–276
4. Cardoso, J.F., Souloumiac, A.: Blind beamforming for non Gaussian signals. IEE Proceedings-F **140** (1993) 362–370
5. Hyvärinen, A.: Fast and robust fixed-point algorithms for independent component analysis. IEEE Trans. Neural Networks **10** (1999) 626–634
6. Amari, S., Chen, T.P., Cichocki, A.: Stability analysis of learning algorithms for blind source separation. Neural Networks **10** (1997) 1345–1351
7. Pham, D.T., Garrat, P.: Blind separation of mixture of independent sources through a quasi-maximum likelihood approach. Signal Processing **45** (1997) 1457–1482
8. Hyvärinen, A.: One-unit contrast functions for independent component analysis: A statistical analysis. In: Neural Networks for Signal Processing VII (Proceedings of IEEE Workshop on Neural Networks for Signal Processing). (1997) 388–397
9. Bollen, K.A.: Structural Equations with Latent Variables. Wiley, New York (1989)
10. Cardoso, J.F., Laheld, B.H.: Equivariant adaptive source separation. IEEE Transactions on Signal Processing **44** (1996) 3017–3030
11. Tichavský, P., Koldovský, Z., Oja, E.: Peformance analysis of the FastICA algorithm and Cramèr-Rao bounds for linear independent component analysis. IEEE Transactions on Signal Processing (2005) In press.
12. Godambe, V.P.: Estimating functions. Oxford University Press, New York (1991)

13. Kawanabe, M., Müller, K.R.: Estimating functions for blind separation when sources have variance dependencies. Journal of Machine Learning Research **6** (2005) 453–482
14. Hochberg, Y., Tamhane, A.C.: Multiple comparison procedures. John Wiley & Sons, New York (1987)
15. Murakami, N.: Research on causes of criminal and deviant behavior. Bachelor thesis, Osaka University, School of Human Sciences (2000) (In Japanese).
16. Shimizu, S., Hyvärinen, A., Kano, Y., Hoyer, P.O.: Discovery of non-gaussian linear causal models using ICA. In: Proc. the 21st Conference on Uncertainty in Artificial Intelligence (UAI-2005). (2005) 526–533
17. Gottfredson, M.R., Hirschi, T.: A general theory of crime. Stanford University Press, Stanford, CA (1990)

Appendix. A Complete Formula of acov{vec(W)}

The formula of acov{vec($\mathbf{W}$)} for FastICA is written as

$$\mathrm{acov}\{\mathrm{vec}(\mathbf{W})\} = \frac{1}{N}\mathbf{Q}^{-1}E[\mathrm{vec}\{\mathbf{F}(\boldsymbol{x},\mathbf{W})\}\mathrm{vec}\{\mathbf{F}(\boldsymbol{x},\mathbf{W})\}^T]\mathbf{Q}^{-T}. \quad (17)$$

Denote by F_{pq} the (p,q)-element of $\mathbf{F}$ and by $\mathbf{F}_q$ the q-th column of $\mathbf{F}$, respectively. We shall provide $E(F_{pq}F_{rs})$ to compute $E\{\mathrm{vec}(\mathbf{F})\mathrm{vec}(\mathbf{F})^T\}$. Denote by i,j,k,l four different subscripts. Then we have

$$\begin{aligned}
E(F_{ii}F_{ii}) &= E(s_i^4)+1,\ E(F_{ii}F_{jj}) = 2,\ E(F_{ki}F_{ij}) = -E\{g(s_k)\}E\{g(s_j)\}\\
E(F_{ki}F_{li}) &= E\{g(s_k)\}E\{g(s_l)\},\ E(F_{ki}F_{kj}) = E\{g(s_i)\}E\{g(s_j)\}\\
E(F_{ii}F_{li}) &= -E(s_i^3)E\{g(s_l)\},\ E(F_{ki}F_{ii}) = -E(s_i^3)E\{g(s_k)\}\\
E(F_{ii}F_{ij}) &= E(s_i^3)E\{g(s_j)\},\ E(F_{ji}F_{jj}) = E(s_j^3)E\{g(s_i)\}\\
E(F_{ii}F_{lj}) &= 0,\ E(F_{ki}F_{jj}) = 0,\ E(F_{ki}F_{lj}) = 0\\
E(F_{ji}F_{ij}) &= 1+2E\{s_ig(s_i)\}E\{s_jg(s_j)\} - E\{g(s_i)^2\} - E\{g(s_j)^2\}\\
E(F_{ji}F_{lj}) &= 1+E\{s_ig(s_i)\}+E\{s_jg(s_j)\}\\
&\quad +E\{s_ig(s_i)\}E\{s_jg(s_j)\} - E\{g(s_i)\}E\{g(s_l)\}\\
E(F_{ki}F_{ki}) &= 1+2E\{s_ig(s_i)\} - 2E\{s_kg(s_k)\} + E\{g(s_i)^2\}\\
&\quad +E\{g(s_k)^2\} - 2E\{s_ig(s_i)\}E\{s_kg(s_k)\}.
\end{aligned}$$

We also give $E\left\{(\partial\mathbf{F}_i)/(\partial\boldsymbol{w}_j^T)\right\}$ to compute $\mathbf{Q} = E\left[\{\partial\mathrm{vec}(\mathbf{F})\}/\{\partial\mathrm{vec}(\mathbf{W})^T\}\right]$:

$$E\left[\frac{\partial\mathbf{F}_i}{\partial\boldsymbol{w}_i^T}\right] = \begin{cases} 2\boldsymbol{w}_i^T & (i\text{–th row})\\ \left[1-E\{s_kg(s_k)\}+E\{g'(s_i)\}\right]\boldsymbol{w}_k^T & (k\text{–th row},\ k\neq i)\end{cases}$$

$$E\left[\frac{\partial\mathbf{F}_i}{\partial\boldsymbol{w}_j^T}\right] = \begin{cases} [1-E\{g'(s_j)\}+E\{s_ig(s_i)]\boldsymbol{w}_i^T & (j\text{–th row},\ j\neq i)\\ \mathbf{0}^T & (k\text{–th row},\ k\neq j)\end{cases}.$$

Cross-Entropy Optimization for Independent Process Analysis

Zoltán Szabó, Barnabás Póczos, and András Lőrincz

Department of Information Systems,
Eötvös Loránd University, Budapest, Hungary,
Research Group on Intelligent Information Systems,
Hungarian Academy of Sciences
szzoli@cs.elte.hu, pbarn@cs.elte.hu, lorincz@inf.elte.hu
http://nipg.inf.elte.hu

Abstract. We treat the problem of searching for hidden multi-dimensional independent auto-regressive processes. First, we transform the problem to Independent Subspace Analysis (ISA). Our main contribution concerns ISA. We show that under certain conditions, ISA is equivalent to a combinatorial optimization problem. For the solution of this optimization we apply the cross-entropy method. Numerical simulations indicate that the cross-entropy method can provide considerable improvements over other state-of-the-art methods.

1 Introduction

Search for independent components is in the focus of research interest. There are important applications in this field, such as blind source separation, blind source deconvolution, feature extraction and denoising. Thus, a variety of particular methods have been developed over the years. For a recent review on these approaches and for further applications, see [1] and the references therein.

Originally, Independent Component Analysis (ICA) is 1-dimensional in the sense that all sources are assumed to be independent real valued stochastic variables. The typical example of ICA is the so called *cocktail-party problem*, where there are n sound sources and n microphones and the task is to separate the original sources from the observed mixed signals. However, applications where not all, but only certain groups of the sources are independent may have high relevance in practice. In this case, independent sources can be multi-dimensional. For example, consider the following generalization of the cocktail-party problem. There could be independent groups of people talking about independent topics, or more than one group of musicians may be playing at a party. This is the Independent Subspace Analysis (ISA) extension of ICA, also called Multi-dimensional Independent Component Analysis [2]. An important application is, e.g., the processing of EEG-fMRI data [3]. However, the motivation of our work stems from the fact that most practical problems, alike to the analysis of EEG-fMRI signals, exhibit considerable temporal correlations. In such cases, one may take

J. Rosca et al. (Eds.): ICA 2006, LNCS 3889, pp. 909–916, 2006.

advantage of Independent Process Analysis (IPA) [4], a generalization of ISA for auto-regressive (AR) processes, similar to the AR generalization of the original ICA problem [5].

Efforts have been made to develop ISA algorithms [2, 3, 6, 7, 8, 9, 10]. Theoretical problems are mostly connected to entropy and mutual information estimations. Entropy estimation by Edgeworth expansion [3] has been extended to more than 2 dimensions and has been used for clustering and mutual information testing [11]. Other recent approaches search for independent subspaces via kernel methods [7], joint block diagonalization [10], k-nearest neighbor [8], and geodesic spanning trees [9].

Here, we shall explore a particular approach that tries to solve the ISA problem by ICA transformation and then searches for an optimal permutation of the ICA components. We shall investigate sufficient conditions that justify this algorithm. Different methods for solving the IPA and the related ISA problems will be compared. The paper is constructed as follows: Section 2 formulates the problem domain and suggests a novel approach for solving the related ISA task. Section 3 contains computer studies. Conclusions are also drawn here.

2 The IPA Model

We shall treat the generative model of mixed independent AR processes. Assume that we have M hidden and independent AR processes and that only the mixture of these M *components* is available for observation:

$$\boldsymbol{s}^m(t+1) = \boldsymbol{F}^m \boldsymbol{s}^m(t) + \boldsymbol{e}^m(t), \quad m = 1, \ldots, M \tag{1}$$

$$\boldsymbol{z}(t) = \boldsymbol{A}\boldsymbol{s}(t), \tag{2}$$

where $\boldsymbol{s}(t) := \left[\boldsymbol{s}^1(t); \ldots; \boldsymbol{s}^M(t)\right]$ is the vector concatenated form of the components $\boldsymbol{s}^m$, $\boldsymbol{s}^m(t), \boldsymbol{e}^m(t) \in \mathbb{R}^d$, $\boldsymbol{e}^m(t)$ is i.i.d. in t, $\boldsymbol{e}^i(t)$ is independent from $\boldsymbol{e}^j(t)$, if $i \neq j$, and $\boldsymbol{F}^m \in \mathbb{R}^{d \times d}$. The total dimension of the components is $D := d \cdot M$, $\boldsymbol{s}(t), \boldsymbol{z}(t) \in \mathbb{R}^D$ and $\boldsymbol{A} \in \mathbb{R}^{D \times D}$ is the so called *mixing matrix* that, according to our assumptions, is invertible. It is easy to see that the invertibility of $\boldsymbol{A}$ and the reduction step using innovations (see later in Section 2.1) allow, without any loss of generality, to restrict (i) to *whitened* noise process $\boldsymbol{e}(t) := \left[\boldsymbol{e}^1(t); \ldots; \boldsymbol{e}^M(t)\right]$, and (ii) to orthogonal matrix $\boldsymbol{A}$. That is,

$$E[\boldsymbol{e}(t)] = \boldsymbol{0}, E\left[\boldsymbol{e}(t)\boldsymbol{e}(t)^T\right] = \boldsymbol{I}_D, \quad \forall t \tag{3}$$

$$\boldsymbol{I}_D = \boldsymbol{A}\boldsymbol{A}^T, \tag{4}$$

where $\boldsymbol{I}_D$ is the D-dimensional identity matrix, superscript T denotes transposition and $E[\cdot]$ is the expectation value operator. The goal of the IPA problem is to estimate the original source $\boldsymbol{s}(t)$ and the unknown mixing matrix $\boldsymbol{A}$ (or its inverse $\boldsymbol{W}$, which is called the *separation matrix*) by using observations $\boldsymbol{z}(t)$ only. If $\forall \boldsymbol{F}^m = \boldsymbol{0}$ then the task reduces to the ISA task. The ICA task is recovered if both $\forall \boldsymbol{F}^m = \boldsymbol{0}$ and $d = 1$.

2.1 Uncertainties of the IPA Model

The identification of the IPA model, alike to the identification of the ICA and ISA models, is ambiguous. First, we shall reduce the IPA task to the ISA task [5, 12, 4] by means of *innovations*. The innovation of a stochastic process $\boldsymbol{u}(t)$ is the error of the optimal quadratic estimation of the process using its past, i.e.,

$$\tilde{\boldsymbol{u}}(t) := \boldsymbol{u}(t) - E[\boldsymbol{u}(t)|\boldsymbol{u}(t-1), \boldsymbol{u}(t-2), \ldots]. \quad (5)$$

It is easy to see that for an AR process, the innovation is identical to the noise that drives the process. Therefore, constructing a block-diagonal matrix $\boldsymbol{F}$ from matrices $\boldsymbol{F}^m$, the IPA model assumes the following form

$$\boldsymbol{s}(t+1) = \boldsymbol{F}\boldsymbol{s}(t) + \boldsymbol{e}(t), \quad (6)$$
$$\boldsymbol{z}(t) = \boldsymbol{A}\boldsymbol{F}\boldsymbol{A}^{-1}\boldsymbol{z}(t-1) + \boldsymbol{A}\boldsymbol{e}(t-1), \quad (7)$$
$$\tilde{\boldsymbol{z}}(t) = \boldsymbol{A}\boldsymbol{e}(t-1) = \boldsymbol{A}\tilde{\boldsymbol{s}}(t). \quad (8)$$

Thus, applying ISA to innovation $\tilde{\boldsymbol{z}}(t)$ of the observation, mixing matrix $\boldsymbol{A}$ and thus $\boldsymbol{e}(t)$ as well as $\boldsymbol{s}(t)$ can be determined.

Concerning the ISA task, if we assume that both the components and the observation are white, that is, $E[\boldsymbol{s}] = \boldsymbol{0}$, $E\left[\boldsymbol{s}\boldsymbol{s}^T\right] = \boldsymbol{I}_D$ and $E[\boldsymbol{z}] = \boldsymbol{0}$, $E\left[\boldsymbol{z}\boldsymbol{z}^T\right] = \boldsymbol{I}_D$, the ambiguity of the problem is lessened: apart from permutations, the components are determined up to orthogonal transformations within the subspaces. It also follows from the whitening assumption that mixing matrix $\boldsymbol{A}$ (and thus matrix $\boldsymbol{W} = \boldsymbol{A}^{-1}$) are orthogonal, because:

$$\boldsymbol{I}_D = E\left[\boldsymbol{z}\boldsymbol{z}^T\right] = \boldsymbol{A}E\left[\boldsymbol{s}\boldsymbol{s}^T\right]\boldsymbol{A}^T = \boldsymbol{A}\boldsymbol{I}_D\boldsymbol{A}^T = \boldsymbol{A}\boldsymbol{A}^T. \quad (9)$$

Identification ambiguities of the ISA task are detailed in [13].

2.2 Reduction of ISA to ICA and Permutation Search

Here, we shall reduce the original IPA task further. The ISA task can be seen as the minimization of mutual information between the components. That is, we should minimize cost function $J(\boldsymbol{W}) := \sum_{m=1}^{M} H(\boldsymbol{y}^m)$ in the space of $D \times D$ orthogonal matrices, where $\boldsymbol{y} = \boldsymbol{W}\boldsymbol{z}$, $\boldsymbol{y} = \left[\boldsymbol{y}^1; \ldots; \boldsymbol{y}^M\right]$, $\boldsymbol{y}^m$ $(m = 1, \ldots, M)$ are the estimated components and H is Shannon's (multi-dimensional) differential entropy (see, e.g., [4]). Now, we present our main result:

Theorem (Separation theorem for ISA). *Let us suppose, that all the* $\boldsymbol{u} = [u_1; \ldots; u_d] = \boldsymbol{s}^m$ *components of source* $\boldsymbol{s}$ *in the ISA task satisfy*

$$H\left(\sum_{i=1}^{d} w_i u_i\right) \geq \sum_{i=1}^{d} w_i^2 H(u_i), \forall \boldsymbol{w} : \sum_{i=1}^{d} w_i^2 = 1. \quad (10)$$

Now, processing observation $\boldsymbol{z}$ *by ICA, and assuming that the ICA separation matrix* $\boldsymbol{W}_{ICA}$ *is unique up to permutation and sign of the components,* then

$\boldsymbol{W}_{ICA}$ is also the separation matrix of the ISA task up to permutation and sign of the components. In other words, the $\boldsymbol{W}$ separation matrix of the ISA task assumes the following form $\boldsymbol{W} = \boldsymbol{P}\boldsymbol{W}_{ICA}$, where $\boldsymbol{P}\left(\in \mathbb{R}^{D\times D}\right)$ is a permutation matrix to be determined.

The proof of the theorem can be found in a technical report [14] because of lack of space. Sources that satisfy the conditions of the theorem are also provided in [14], where we show that elliptically symmetric sources, among others, satisfy the condition of the theorem.

In sum, the IPA model can be estimated by applying the following steps:

1. observe $\boldsymbol{z}(t)$ and estimate the AR model,
2. whiten the innovation of the AR process and perform ICA on it,
3. solve the combinatorial problem: search for the permutation of the ICA sources that minimizes the cost function J.

Thus, after estimating the AR model and performing ICA on its estimated innovation process, IPA needs only two steps: (i) estimation of multi-dimensional entropies, and (ii) optimization of the cost function J in S_D, the permutations of length D.

A recent work [4] provides an algorithm to solve the IPA task. To our best knowledge, this is the only algorithm for this task at present. This algorithm applies Jacobi rotations for any pairs of the elements received after ICA preprocessing. We shall call it the *ICA-Jacobi* method and compare it with our novel algorithm that we refer to as the *ICA-TSP* method for reasons to be explained later. For entropy estimation, we shall apply the method suggested in [4], which is the following:

2.3 Multi-dimensional Entropy Estimation by the k-Nearest Neighbor Method

Shannon's entropy can be estimated by taking the limit of Rényi's entropy, which has efficient estimations. Let f denote the probability density of d-dimensional stochastic variable $\boldsymbol{u}$. Rényi's α-entropy of variable $\boldsymbol{u}$ $(1 \neq \alpha > 0)$ is defined as:

$$H_\alpha(\boldsymbol{u}) := \frac{1}{1-\alpha}\log \int_{\mathbb{R}^d} f^\alpha(v)dv \xrightarrow{\alpha\to 1} H(\boldsymbol{u}). \tag{11}$$

Assume that we have i.i.d. samples of T elements from the distribution of $\boldsymbol{u}$: $\boldsymbol{u}(1),\ldots,\boldsymbol{u}(T)$. For each sample $\boldsymbol{u}(t)$ let us choose the k samples, which are the closest to $\boldsymbol{u}(t)$ in Euclidean norm ($\|\cdot\|$). Let this set be denoted by $\mathcal{N}_{k,t}$. Let us choose $\alpha := \frac{d-\gamma}{d}$, and thus $\alpha \to 1$ corresponds to $\gamma \to 0$. Then, under mild conditions, the Beadword-Halton-Hammersley theorem holds [15, 16]:

$$\hat{H}\left(k,\gamma\right) := \frac{1}{1-\alpha}\log\left(\frac{1}{T^\alpha}\sum_{t=1}^{T}\sum_{\boldsymbol{v}\in\mathcal{N}_{k,t}}\|\boldsymbol{v}-\boldsymbol{u}(t)\|^\gamma\right) \xrightarrow{T\to\infty} H_\alpha(\boldsymbol{u}) + c, \tag{12}$$

where c is an irrelevant constant. This entropy estimation is asymptotically unbiased and strongly consistent [15]. In the numerical studies, we shall use $\gamma = 0.01$ and $k = 3$ alike to [4].

2.4 Cross-Entropy Method for Combinatorial Optimization

The CE method has been found efficient for combinatorial optimization problems [17]. The CE technique operates as a two step procedure: First, the problem is converted to a stochastic problem and then the following two-phases are iterated (for detailed description, see [17]):

1. Generate $\boldsymbol{x}_1, \ldots, \boldsymbol{x}_N \in \mathcal{X}$ samples from a distribution family parameterized by a θ parameter and choose the *elite* of the samples. The elite is the best $\rho\%$ of the samples according to the cost function J.
2. Modify the sample generation procedure (θ) according to the elite samples. In practice, smoothing, i.e., $\theta^{new} = \beta \cdot \theta^{proposed} + (1 - \beta) \cdot \theta^{old}$ is utilized in the update of θ.

This technique will be applied in our search for permutation matrix $\boldsymbol{P}$. Our method is similar to the CE solution suggested for the Travelling Salesman Problem (TSP) (see [17]) and we call it *ICA-TSP* method. In the TSP problem, a permutation of cities is searched for. The objective is to minimize the cost of the travel. We are also searching for a permutation, but now travel cost is replaced by $J(\boldsymbol{W})$. Thus, in our case, $\mathcal{X} = S_D$ and $\boldsymbol{x}$ is an element of this permutation group. Further, CE cost J equals to $J(\boldsymbol{P}_{\boldsymbol{x}} \boldsymbol{W}_{ICA})$, where $\boldsymbol{P}_{\boldsymbol{x}}$ denotes the permutation matrix associated to $\boldsymbol{x}$. Thus, optimization concerns permutations in $\mathcal{X}$. In the present work, θ contains transition probabilities $i \rightarrow j$ ($1 \leq i, j \leq D$), called *node transition* parametrization in the literature [17].

The above iteration is stopped if there is no change in the cost (in the last L steps), or the change in parameter θ is negligibly small (smaller then ϵ).

3 Numerical Studies

3.1 Databases

Computer simulations are presented here. We defined four different databases. They were whitened and were used to drive the AR processes of Eq. (1). Then the AR processes were mixed. Given the mixture, an AR process was identified and its innovation was computed. The innovation was analyzed by ISA. We note that this reduction step using innovations based on AR estimation ('AR-trick') can also work for non-AR processes, as it was demonstrated in [4].

Three of the four computational tasks are shown in Fig. 1. In these test examples (i) dimensions D and d were varied ($D = 12, 18, 20$, $d = 2, 3, 4$), (ii) sample number T was incremented by 100 between 300 and 1500. For all tests, we averaged the results of 10 computer runs. In the fourth task $M (= 5)$ pieces of $d (= 4)$-dimensional components were used and the innovation for each d-dimensional process was created as follows: coordinates $u_i(t)$ ($i = 1, \ldots, k$), were uniform random variables on the set {0,...,k-1}, whereas u_{k+1} was set to $mod(u_1 + \ldots + u_k, k)$. In this construction, every k-element subset of $\{u_1, \ldots, u_{k+1}\}$ is made of independent variables. This database is called the *all-k-independent* problem [9]. In our simulations $d = k + 1$ was set to 4.

Numerical values of the CE parameters were chosen as $\rho = 0.05$ $\beta = 0.4$, $L = 7$, $\epsilon = 0.005$. The quality of the algorithms was measured by the generalized Amari-distance.

Generalized Amari-Distance. The optimal estimation of the IPA model provides matrix $\boldsymbol{B} := \boldsymbol{W}\boldsymbol{A}$, a permutation matrix made of $d \times d$ sized blocks. Let us decompose matrix $\boldsymbol{B} \in \mathbb{R}^{D\times D}$ into $d \times d$ blocks: $\boldsymbol{B} = \left[\boldsymbol{B}^{ij}\right]_{i,j=1,\ldots,M}$. Let $b^{i,j}$ denote the sum of the absolute values of the elements of matrix $\boldsymbol{B}^{i,j} \in \mathbb{R}^{d\times d}$. Then the normalized version of the generalized Amari-distance (see also [9, 10]) is defined as:

$$r(\boldsymbol{B}) := \frac{1}{2M(M-1)} \cdot \left(\sum_{i=1}^{M} \left(\frac{\sum_{j=1}^{M} b^{ij}}{\max_j b^{ij}} - 1 \right) + \sum_{j=1}^{M} \left(\frac{\sum_{i=1}^{M} b^{ij}}{\max_i b^{ij}} - 1 \right) \right) \tag{13}$$

For matrix $\boldsymbol{B}$ we have that $0 \leq r(\boldsymbol{B}) \leq 1$, and $r(\boldsymbol{B}) = 0$ if, and only if $\boldsymbol{B}$ is a block-permutation matrix with $d \times d$ sized blocks.

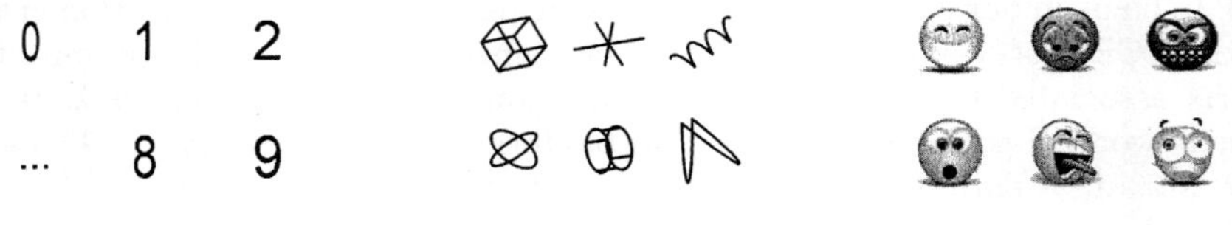

(a) numbers (b) 3D-geom (c) smiley

Fig. 1. 3 test databases: densities of $\boldsymbol{e}^m$. Each object represents a probability density. Left: *numbers*: $10 \times 2 = 20$-dimensional problem, uniform distribution on the images of numbers. Middle: *3D-geom*: $6 \times 3 = 18$-dimensional problem, uniform distribution on 3-dimensional geometric objects. Right: *smiley*: 6 basic facial expressions [18], non-uniform distribution defined in 2 dimensions, $6 \times 2 = 12$-dimensional problem.

3.2 Results and Discussion

The precision of the procedures is shown in Fig. 2 as a function of the sample number. In the ICA-Jacobi method we applied exhaustive search for all Jacobi pairs with 50 angles between $[0, \pi/2]$ several times until convergence. Still, the ICA-TSP is superior in all of the studied examples. Quantitative results are shown in Table 1. The innovations estimated by the ICA-TSP method on facial expressions are illustrated in Fig. 3.

We observed that the greedy ICA-Jacobi method seems to be similar or sometimes inferior to the global ICA-TSP, in spite of the much smaller search space available for the latter. We established rigorous conditions when the ICA-TSP is sufficient to find a global minimum, which justifies our finding. In the reduced search space of permutations, the global CE method was very efficient.

We make two notes: (1) Simulations indicate that conditions of the 'Separation Theorem' may be too restrictive. (2) For the IPA problem, the subspaces (the

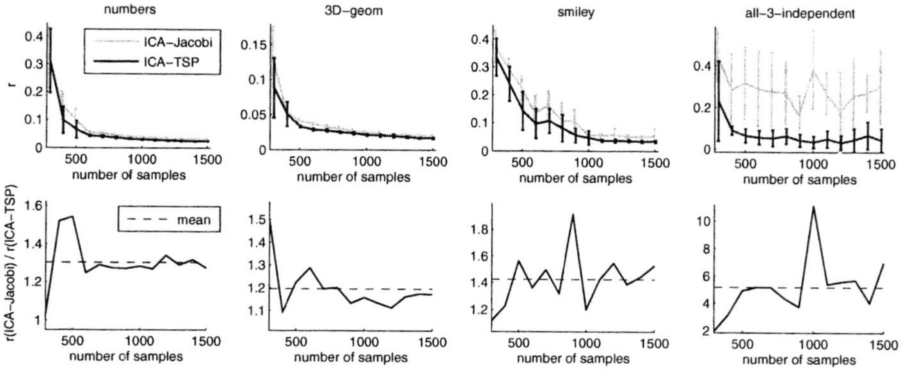

Fig. 2. Mean±standard deviation of generalized Amari-distances as a function of sample number (upper row). Gray: ICA-Jacobi, black: ICA-TSP. In the lower row, black: precision of relative estimation, dashed: average over the different sample numbers. Columns from left to right correspond to databases 'numbers', '3D-geom', 'smiley', 'all-3-independent', respectively.

Table 1. Column 1: test databases. Columns 2 and 3: average Amari-errors (in $100 \cdot r\%\pm$ standard deviation) for 1500 samples on the different databases. Column 4: precision of the ICA-TSP relative to that of ICA-Jacobi in sample domain $300-1500$.

Database	ICA-Jacobi	ICA-TSP	Improvement (min - mean - max)
numbers	3.06% (±0.22)	2.40% (±0.11)	1.03 - 1.30 - 1.54
3D-geom	1.99% (±0.17)	1.69% (±0.10)	1.09 - 1.20 - 1.50
smiley	5.26% (±2.76)	3.44% (±0.36)	1.16 - 1.43 - 1.92
all-3-independent	30.05% (±17.90)	4.31% (±5.61)	1.96 - 5.18 - 11.12

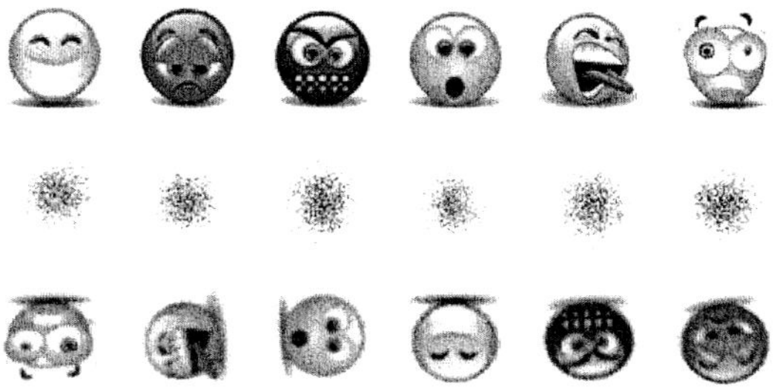

Fig. 3. Illustration of the ICA-TSP algorithm on the 'smiley' database. Upper row: density function of the sources (using 10^6 data points). Middle row: 1,500 samples of the observed mixed signals ($z(t)$). The ICA-TSP algorithm works on these data. Lower row: Estimated separated sources (recovered up to permutation and orthogonal transformation).

optimal permutation of the ICA components) may be found by transforming the observations with the learned ICA matrix followed by an AR estimation that serves to identify the predictive matrices of Eq. (1), which – under certain conditions – allows one to *list* the components of the connected subspaces [19].

References

1. Choi, S., Cichocki, A., Park, H.M., Lee, S.Y.: Blind Source Separation and Independent Component Analysis. Neural Inf. Proc. Lett. and Reviews (2005)
2. Cardoso, J.: Multidimensional Independent Component Analysis. In: ICASSP'98, Seattle, WA. (1998)
3. Akaho, S., Kiuchi, Y., Umeyama, S.: MICA: Multimodal Independent Component Analysis. In: IJCNN. (1999) 927–932
4. Póczos, B., Takács, B., Lőrincz, A.: Independent Subspace Analysis on Innovations. In: ECML, Porto, Portugal. (2005) 698–706
5. Hyvärinen, A.: Independent Component Analysis for Time-dependent Stochastic Processes. In: ICANN 1998. (1998) 541–546
6. Vollgraf, R., Obermayer, K.: Multi-Dimensional ICA to Separate Correlated Sources. In: NIPS. Volume 14. (2001) 993–1000
7. Bach, F.R., Jordan, M.I.: Finding Clusters in Independent Component Analysis. In: ICA2003. (2003) 891–896
8. Póczos, B., Lőrincz, A.: Independent Subspace Analysis Using k-Nearest Neighborhood Distances. ICANN 2005 (2005) 163–168
9. Póczos, B., Lőrincz, A.: Independent Subspace Analysis Using Geodesic Spanning Trees. In: ICML. (2005) 673–680
10. Theis, F.J.: Blind Signal Separation into Groups of Dependent Signals Using Joint Block Diagonalization. In: Proc. ISCAS 2005, Kobe, Japan (2005) 5878–5881
11. Van Hulle, M.M.: Edgeworth Approximation of Multivariate Differential Entropy. Neural Comput. **17** (2005) 1903–1910
12. Cheung, Y., Xu, L.: Dual Multivariate Auto-Regressive Modeling in State Space for Temporal Signal Separation. IEEE Tr. on Syst. Man Cyb. B **33** (2003) 386–398
13. Theis, F.J.: Uniqueness of Complex and Multidimensional Independent Component Analysis. Signal Proc. **84** (2004) 951–956
14. Szabó, Z., Póczos, B., Lőrincz, A.: Separation Theorem for Independent Subspace Analysis. Technical report, Eötvös Loránd University, Budapest (2005) http://people.inf.elte.hu/lorincz/Files/TR-ELU-NIPG-31-10-2005.pdf.
15. Yukich, J.E.: Probability Theory of Classical Euclidean Optimization Problems. Volume 1675 of Lecture Notes in Math. Springer-Verlag, Berlin (1998)
16. Costa, J.A., Hero, A.O.: Manifold Learning Using k-Nearest Neighbor Graphs. In: ICASSP, Montreal, Canada. (2004)
17. Rubinstein, R.Y., Kroese, D.P.: The Cross-Entropy Method. Springer (2004)
18. Ekman, P.: Emotion in the Human Face. Cambridge Univ. Press, New York (1982)
19. Póczos, B., Lőrincz, A.: Non-combinatorial Estimation of Independent Autoregressive Sources. (2005) submitted.

Uniqueness of Non-Gaussian Subspace Analysis

Fabian J. Theis[1] and Motoaki Kawanabe[2]

[1] Institute of Biophysics, University of Regensburg, 93040 Regensburg, Germany
[2] Fraunhofer FIRST.IDA, Kekuléstraße 7, 12439 Berlin, Germany
fabian@theis.name, nabe@first.fhg.de

Abstract. Dimension reduction provides an important tool for preprocessing large scale data sets. A possible model for dimension reduction is realized by projecting onto the non-Gaussian part of a given multivariate recording. We prove that the subspaces of such a projection are unique given that the Gaussian subspace is of maximal dimension. This result therefore guarantees that projection algorithms uniquely recover the underlying lower dimensional data signals.

An important open problem in signal processing is the task of efficient dimension reduction, i.e. the search for meaningful signals within a higher dimensional data set. Classical techniques such as principal component analysis hereby define 'meaningful' using second-order statistics (maximal variance), which may often be inadequate for signal detection, e.g. in the presence of strong noise. This contrasts to higher order models including *projection pursuit* [1, 2] or *non-Gaussian subspace analysis (NGSA)* [3, 4]. While the former extracts a single non-Gaussian independent component from the data set, the latter tries to detect a whole non-Gaussian subspace within the data, and no assumption of independence within the subspace is made.

The goal of *linear dimension reduction* can be defined as the search of a projection $\mathbf{W} \in \mathrm{Mat}(n \times d)$ of a d-dimensional random vector $\mathbf{X}$ with $n < d$ and $\mathbf{WX}$ bearing still as much information of $\mathbf{X}$ as possible. Of course this last term has to be specified in detail in terms of some distance or source model. This problem describes a special case of the larger field of model selection [1], an important tool for preprocessing and dimension reduction, used in a wide range of applications.

In the following we will use the notations $\mathrm{Gl}(n)$ and $O(n)$ to denote the group of invertible and orthogonal $n \times n$-matrices respectively. Upper-case symbols are used for both matrices and random vectors, lower case ones for scalars and vectors. Matlab-notation is employed for selecting columns and rows of matrices, so for example $\mathbf{A}(2 : n, :)$ denotes the matrix consisting of the last $(n-1)$-columns of $\mathbf{A} \in \mathrm{Mat}(n \times n)$. Random variables and vectors are defined on the probability space Ω, and the notation $X \in L_2(\Omega, \mathbb{R})$ means that the random variable X is square-integrable. Finally, we are only treating the real case here, although extensions to complex-valued random vectors along the lines of [5] are possible.

Instead of relying on second-order statistics only, higher-order statistics are used in NGSA in order to determine 'interesting' directions [3, 4]. The goal is to find a projection with maximal non-Gaussianity, removing the Gaussian part of $\mathbf{X}$. In other words, the goal is to find a projection $\mathbf{W}_N \in \mathrm{Mat}(n \times d)$ such that there exists $\mathbf{W}_G \in \mathrm{Mat}((d-n) \times d)$ with $\mathbf{W}_N\mathbf{X}$ and $\mathbf{W}_G\mathbf{X}$ being independent, and $\mathbf{W}_G\mathbf{X}$ being Gaussian.

J. Rosca et al. (Eds.): ICA 2006, LNCS 3889, pp. 917–925, 2006.

An intuitive notion of how to choose the reduced dimension n is to require that $\mathbf{W}_G\mathbf{X}$ is maximally Gaussian, and hence $\mathbf{W}_N\mathbf{X}$ non-Gaussian.

The dimension reduction problem itself can of course also be formulated within a generative model, which leads to the following linear mixing model

$$\mathbf{X} = \mathbf{A}_N\mathbf{S}_N + \mathbf{A}_G\mathbf{S}_G \tag{1}$$

such that the n-dimensional random vector $\mathbf{S}_N$ and the $(d-n)$-dimensional vector $\mathbf{S}_G$ are independent, and $\mathbf{S}_G$ Gaussian. Then $(\mathbf{A}_N, \mathbf{A}_G)^{-1} = (\mathbf{W}_N^\top, \mathbf{W}_G^\top)^\top$. This model includes the general noisy ICA model $\mathbf{X} = \mathbf{A}_N\mathbf{S}_N + \mathbf{G}$, where $\mathbf{G}$ is Gaussian and $\mathbf{S}_N$ is also assumed to be mutually independent; the dimension reduction then means projection onto the signal subspace, which might be deteriorated by the noise $\mathbf{G}$ along the subspace — the components of $\mathbf{G}$ orthogonal to the subspace will be removed. However (1) is more general in the sense that it does not assume mutual independence of $\mathbf{S}_N$, only independence of $\mathbf{S}_N$ and $\mathbf{S}_G$.

The paper is organized as follows: In the next section, we first discuss obvious indeterminacies of NGSA and possible regularizations. We then present our main result, theorem 1, and give an explicit proof in a special case. The general proof is divided up into a series of lemmas, the proofs of which are omitted due to lack of space. In section 2, some simulations are performed to validate the uniqueness result. A practical algorithm for performing NGSA essentially using the idea of separated characteristic functions from the proof is presented in the co-paper [6].

1 Uniqueness of NGSA-Based Dimension Reduction

This contribution aims at providing conditions such that the decomposition (1) is unique. More precisely, we will show under which conditions the non-Gaussian as well as the Gaussian subspace is unique.

1.1 Indeterminacies

Clearly, the matrices $\mathbf{A}_N$ and $\mathbf{A}_G$ in the decomposition (1) cannot be unique — multiplication from the right using any invertible matrix leaves the model invariant: $\mathbf{X} = \mathbf{A}_N\mathbf{S}_N + \mathbf{A}_G\mathbf{S}_G = (\mathbf{A}_N\mathbf{B}_N)(\mathbf{B}_N^{-1}\mathbf{S}_N) + (\mathbf{A}_G\mathbf{B}_G)(\mathbf{B}_G^{-1}\mathbf{S}_G)$ with $\mathbf{B}_N \in \mathrm{Gl}(n), \mathbf{B}_G \in \mathrm{Gl}(d-n)$, because $\mathbf{B}_N^{-1}\mathbf{S}_N$ and $\mathbf{B}_G^{-1}\mathbf{S}_G$ are again independent, and $\mathbf{B}_G^{-1}\mathbf{S}_G$ Gaussian.

An additional indeterminacy comes into play due to the fact that we do not want to fix the reduced dimension in advance. Given a realization of the model (1) with $n < d$, let $\mathbf{B}_G := \mathrm{Cov}(\mathbf{S}_G)^{1/2} \in \mathrm{Gl}(d-n)$. Then $\mathbf{B}_G^{-1}\mathbf{S}_G$ is decorrelated i.e. mutually independent (because of Gaussianity). By replacing $\mathbf{A}_G$ by $\mathbf{A}_G\mathbf{B}_G$, we may therefore assume that $\mathbf{S}_G$ is independent. If $\mathbf{a} := \mathbf{A}_G(:,1)$ denotes the first column of $\mathbf{A}_G$, then $\mathbf{X} = \mathbf{A}_N\mathbf{S}_N + \mathbf{A}_G\mathbf{S}_G$ can be rewritten as

$$\mathbf{X} = (\mathbf{A}_N, \mathbf{a})\begin{pmatrix}\mathbf{S}_N \\ \mathbf{S}_G(1)\end{pmatrix} + \mathbf{A}_G(:,2\!:\!d-n)\mathbf{S}_G(2\!:\!d-n) \tag{2}$$

and $(\mathbf{S}_N, \mathbf{S}_G(1))$ and $\mathbf{S}_G(2\!:\!d-n)$ are independent, with the second vector being Gaussian. In other words, without putting an additional condition of maximality onto the Gaussian part, different model realizations can be generated by simply moving random variables to the non-Gaussian part.

1.2 Uniqueness Theorem

Definition 1. $\mathbf{X} = \mathbf{AS}$ *with* $\mathbf{A} \in \mathrm{Gl}(d)$, $\mathbf{S} = (\mathbf{S}_N, \mathbf{S}_G)$ *and* $\mathbf{S}_N \in L_2(\Omega, \mathbb{R}^n)$ *is called an* n-decomposition *of* $\mathbf{X}$ *if* $\mathbf{S}_N$ *and* $\mathbf{S}_G$ *are stochastically independent and* $\mathbf{S}_G$ *is Gaussian. In this case,* $\mathbf{X}$ *is said to be* n-decomposable.

Hence an n-decomposition of $\mathbf{X}$ corresponds to the NGSA problem. If as before $\mathbf{A} = (\mathbf{A}_N, \mathbf{A}_G)$, then the n-dimensional subvectorspace $\mathrm{im}(\mathbf{A}_N) \subset \mathbb{R}^d$ is called the *non-Gaussian subspace*, and $\mathrm{im}(\mathbf{A}_G)$ the *Gaussian subspace* of the decomposition; here $\mathrm{im}(\mathbf{A})$ denotes the image of the linear map $\mathbf{A}$.

Definition 2. $\mathbf{X}$ *is denoted to be* minimally n-decomposable *if* $\mathbf{X}$ *is not* $(n-1)$*-decomposable. Then* $\dim_e(\mathbf{X}) := n$ *is called the* essential dimension of $\mathbf{X}$.

For example, the essential dimension $\dim_e(\mathbf{X})$ is zero if and only if $\mathbf{X}$ is Gaussian, whereas the essential dimension of a d-dimensional mutually independent Laplacian is d. The following theorem is the main theoretical contribution of this work. It essentially connects uniqueness of the dimension reduction model with minimality, and gives a simple characterization for it.

Theorem 1 (Uniqueness of NGSA). *Let $n < d$. Given an n-decomposition* $\mathbf{A}_N\mathbf{S}_N + \mathbf{A}_G\mathbf{S}_G$ *of the random vector* $\mathbf{X} \in L_2(\Omega, \mathbb{R}^d)$*, the following is equivalent:*

(i) The decomposition is minimal i.e. $n = \dim_e(\mathbf{X})$.
(ii) There exists no basis $\mathbf{M} \in \mathrm{Gl}(n)$ *such that* $(\mathbf{MS}_N)(1)$ *is Gaussian and independent of* $(\mathbf{MS}_N)(2:n)$.
(iii) The subspaces of the decomposition are unique i.e. another n-decomposition has the same non-Gaussian and Gaussian subspaces.

As seen in section 1.1, condition (i) is also a necessary condition for uniqueness. Together with the model assumption of existing covariance, the theorem shows that it is also sufficient.

Condition (ii) means that there exists no Gaussian independent component in the non-Gaussian part of the decomposition. The theorem proves that this is equivalent to the decomposition being minimal. Note that in (ii), it is *not* enough to require only that there exists no Gaussian component i.e. $\mathbf{v} \in \mathbb{R}^n$ such that $\mathbf{v}^\top\mathbf{S}_N$ is Gaussian. A simple counterexample is given by a two-dimensional random vector $\mathbf{S}$ with density $c\exp(-s_1^2 - (s_1^2 + s_2)^2)$ with c being a normalizing constant. Then indeed $\mathbf{S}(1) = S_1$ is Gaussian because $\int_{\mathbb{R}} c\exp(-s_1^2 - (s_1^2 + s_2)^2)ds_2 = c'\exp(-s_1^2)$, but clearly no $\mathbf{m} \in \mathbb{R}^2$ can be chosen such that $\mathbf{S}(1)$ and $\mathbf{m}^\top\mathbf{S}$ are independent. And indeed, this dependent Gaussian $\mathbf{S}(1)$ within $\mathbf{S}$ should not be removed by dimension reduction as it may contain interesting information, not being independent of the other components.

Corollary 1. *The subspaces of a* $\dim_e(\mathbf{X})$*-decomposition are unique.*

This follows from theorem 1(i)⇒(iii) in the case of $\dim_e(\mathbf{X}) < d$, and holds trivially if $\dim_e(\mathbf{X}) = d$. Also note that existence of a minimal decomposition always holds.

1.3 Proof of Theorem 1

First note that the theorem holds trivially for $n = 0$, because in this case $\mathbf{X}$ is Gaussian. So in the following let $0 < n < d$. Not (ii) $\Rightarrow$ not (i) follows similarly to equation (2), because if we had a basis in which one component of $\mathbf{S}_N$ is Gaussian and independent of the rest, then we could simply move it to the Gaussian part and reduce the order of the decomposition; hence (i) $\Rightarrow$ (ii) holds.

Also (iii) $\Rightarrow$ (i) can be easily shown: assume that the decomposition is not minimal, then we may choose an $(n-1)$-decomposition $\mathbf{X} = \mathbf{A}'_N\mathbf{S}'_N + \mathbf{A}'_G\mathbf{S}'_G$, without loss of generality $\mathbf{S}'_G \in L_2(\Omega, \mathbb{R}^{d-n+1})$ being independent (see section 1.1). The two (because $n < d$) columns $(\mathbf{a}'_G)_1$ and $(\mathbf{a}'_G)_2$ of $\mathbf{A}'_G$ are linearly independent, and similar to equation (2), we get two n-decompositions of $\mathbf{X}$ with non-Gaussian coordinates $(\mathbf{A}'_N, (\mathbf{a}'_G)_1)$ and $(\mathbf{A}'_N, (\mathbf{a}'_G)_2)$ respectively. But the two vectors $(\mathbf{a}'_G)_i$ do not lie in the image of $\mathbf{A}'_N$ and are linearly independent, so we have constructed two n-decompositions of $\mathbf{X}$ with different non-Gaussian subspaces, which contradicts (iii).

The main part of the proof now consists of showing (ii) $\Rightarrow$ (iii). Assume that (ii) holds for all n-decompositions of $\mathbf{X}$. Given two such n-decompositions, then

$$\mathbf{A}_N\mathbf{S}_N + \mathbf{A}_G\mathbf{S}_G = \mathbf{A}'_N\mathbf{S}'_N + \mathbf{A}'_G\mathbf{S}'_G.$$

By applying $(\mathbf{A}'_N, \mathbf{A}'_G)^{-1}$, we may therefore without loss of generality assume that $\mathbf{A}_N\mathbf{S}_N + \mathbf{A}_G\mathbf{S}_G = (\mathbf{S}'_N, \mathbf{S}'_G)$, or in other words that $(\mathbf{X}_N, \mathbf{X}_G)$ is also an n-decomposition of $\mathbf{X}$. Our assumption is that (ii) holds for both $\mathbf{S}$ and $\mathbf{X}$.

We introduce some notation by dividing $\mathbf{A} \in \mathrm{Gl}(d)$ into

$$\mathbf{A} = \begin{array}{|c|c|}\hline \mathbf{A}_{NN} & \mathbf{A}_{NG} \\ \hline \mathbf{A}_{GN} & \mathbf{A}_{GG} \\ \hline \end{array}$$

with $\mathbf{A}_{NN} \in \mathrm{Mat}(n \times n)$; similarly $\mathbf{S} := (\mathbf{S}_N, \mathbf{S}_G)$. Altogether we are dealing with the simple linear model $\mathbf{X} = \mathbf{AS}$, where both $\mathbf{X}$ and $\mathbf{S}$ consist of an independent Gaussian and n-dimensional non-Gaussian part. We have to show that these parts span the same subspaces respectively, which is equivalent to showing that $\mathbf{A}_{NG} = \mathbf{A}_{GN} = 0$.

Two-Dimensional Positive Density Case. For illustrative purpose we will first prove uniqueness for a two-dimensional random vector $\mathbf{X}$ with positive density $p_\mathbf{X} \in C^2(\mathbb{R}^2, \mathbb{R})$, where $C^2(.)$ denotes the algebra of twice continuously differentiable functions. By assumption $n = 1$. Note that in this simple case, we could have just used the common ICA uniqueness results as well [7, 8].

$\mathbf{X}$ is assumed to have independent Gaussian and non-Gaussian part, so its density factorizes into $p_\mathbf{X}(\mathbf{x}) = p_N(x_N)p_G(x_G)$ for $\mathbf{x} \in \mathbb{R}^2$. Let $\mathbf{B} := \mathbf{A}^{-1}$. The density of $\mathbf{S} = \mathbf{BX}$ is given by $p_\mathbf{S}(\mathbf{s}) = |\det \mathbf{B}|^{-1} p_\mathbf{X}(\mathbf{B}^{-1}\mathbf{s}) = c p_N(a_{NN}s_N + a_{NG}s_G)p_G(a_{GN}s_N + a_{GG}s_G)$ for $\mathbf{s} \in \mathbb{R}^2$, $c \neq 0$ fixed. $p_\mathbf{X}$ was assumed to be positive, then so is $p_\mathbf{S}$. $\mathbf{S}$ has also independent Gaussian and non-Gaussian components, so $\left(\partial^2/(\partial s_N \partial s_G)\right) \ln p_\mathbf{S}(\mathbf{s}) = 0$ for all $\mathbf{s} \in \mathbb{R}^2$, hence $a_{NN}a_{NG}h''_N(a_{NN}s_N + a_{NG}s_G) + a_{GN}a_{GG}h''_G(a_{GN}s_N + a_{GG}s_G) = 0$, where $h_i := \ln p_i \in C^2(\mathbb{R}^2, \mathbb{R})$. But p_G is Gaussian, so $h''_G \equiv c' \neq 0$ is constant. By setting $\mathbf{x} := \mathbf{As}$, we therefore get

$$a_{NN}a_{NG}h''_N(x_N) + a_{GN}a_{GG}c' = 0 \tag{3}$$

for all $x_N \in \mathbb{R}$, because $\mathbf{A}$ is invertible.

Now $a_{NN} \neq 0$, otherwise $X_N = a_{NG} S_G$ which contradicts (ii) for $\mathbf{X}$. If also $a_{NG} \neq 0$, then by equation (3), h''_N is constant and therefore S_N Gaussian, which again contradicts (ii), now for $\mathbf{S}$. Hence $a_{NG} = 0$. By (3), $a_{GN} a_{GG} = 0$, and again $a_{GG} \neq 0$, otherwise $X_G = a_{GN} S_N$ contradicting (ii) for $\mathbf{S}$. Hence also $a_{GN} = 0$ as was to show.

General Proof. In order to give an idea of the main proof without getting lost in details, we have divided it up into a sequence of lemmas; these will not be proven due to lack of space. The *characteristic function* of the random vector $\mathbf{X}$ is defined by $\widehat{\mathbf{X}}(\mathbf{x}) := \mathbf{E}(\exp i\mathbf{x}^\top \mathbf{X})$, and since $\mathbf{X}$ is assumed to have existing covariance, $\widehat{\mathbf{X}}$ is twice continuously differentiable. Moreover by definition $\widehat{\mathbf{AS}}(\mathbf{x}) = \widehat{\mathbf{S}}(\mathbf{A}^\top \mathbf{x})$, and the characteristic function of an independent random vector factorizes into the component characteristic functions. So instead of using $p_{\mathbf{X}}$ as in the 2-dimensional example, we use $\widehat{\mathbf{X}}$, having similar properties except for the fact that the range is now complex and that the differentiability condition can be considerably relaxed.

We will need the following lemma, which has essentially been shown in [9]; here ∇f denotes the gradient of f and $\mathbf{H}_f$ its Hessian.

Lemma 1. *Let $\mathbf{X} \in L_2(\Omega, \mathbb{R}^m)$ be a random vector. Then $\mathbf{X}$ is Gaussian with covariance $2\mathbf{C}$ if and only if it satisfies $\widehat{\mathbf{X}}\mathbf{H}_{\widehat{\mathbf{X}}} - \nabla\widehat{\mathbf{X}}(\nabla\widehat{\mathbf{X}})^\top + \mathbf{C}\widehat{\mathbf{X}}^2 \equiv 0$.*

Note that we may assume that the covariance of $\mathbf{S}$ (and hence also of $\mathbf{X}$) is positive definite — otherwise, while still keeping the model, we can simply remove the subspace of deterministic components (i.e. components of variance 0), which have to be mapped onto each other by $\mathbf{A}$. Hence we may even assume $\mathrm{Cov}(\mathbf{S}_G) = \mathbf{I}$, after whitening as described in section 1.1. This uses the fact that the basis within the Gaussian subspace is not unique. The same holds also for the non-Gaussian subspace, so we may choose any $\mathbf{B}_N \in \mathrm{Gl}(n)$ and $\mathbf{B}_G \in O(d-n)$ to get

$$\mathbf{X} = \begin{pmatrix} \mathbf{A}_{NN}\mathbf{B}_N \\ \mathbf{A}_{GN}\mathbf{B}_N \end{pmatrix} (\mathbf{B}_N^{-1}\mathbf{S}_N) + \begin{pmatrix} \mathbf{A}_{NG}\mathbf{B}_G \\ \mathbf{A}_{GG}\mathbf{B}_G \end{pmatrix} (\mathbf{B}_G^\top \mathbf{S}_G). \tag{4}$$

Here only orthogonal matrices $\mathbf{B}_G$ are allowed in order for $\mathbf{B}_G^\top \mathbf{S}_G$ to stay decorrelated, with $\mathbf{S}_G$ being decorrelated.

The next lemma uses the dimension reduction model for $\mathbf{X}$ and $\mathbf{S}$ to derive an explicit differential equation for $\hat{\mathbf{S}}_N$. The Gaussian part $\hat{\mathbf{S}}_G$ in the following lemma vanishes after application of lemma 1.

Lemma 2. *For any basis $\mathbf{B}_N \in \mathrm{Gl}(n)$, the non-Gaussian source characteristic function $\hat{\mathbf{S}}_N \in C^2(\mathbb{R}^n, \mathbb{C})$ fulfills*

$$\mathbf{A}_{NN}\mathbf{B}_N \left(\hat{\mathbf{S}}_N \mathbf{H}_{\hat{\mathbf{S}}_N} - \nabla\hat{\mathbf{S}}_N (\nabla\hat{\mathbf{S}}_N)^\top\right) \mathbf{B}_N^\top \mathbf{A}_{GN}^\top + 2\mathbf{A}_{NG}\mathbf{A}_{GG}^\top \hat{\mathbf{S}}_N^2 \equiv 0. \tag{5}$$

Lemma 3. *Let $(\mathbf{A}_{NN}, \mathbf{A}_{NG}) \in \mathrm{Mat}(n \times (n + (d-n)))$ be an arbitrary full rank matrix. If* $\operatorname{rank} \mathbf{A}_{NN} < n$, *then we may choose coordinates $\mathbf{B}_N \in \mathrm{Gl}(n)$, $\mathbf{B}_G \in O(d-n)$ and $\mathbf{M} \in \mathrm{Gl}(n)$ such that for arbitrary matrices $* \in \mathrm{Mat}((n-1) \times (n-1))$, $*' \in \mathrm{Mat}((n-1) \times (d-n-1))$:*

$$\mathbf{M}\mathbf{A}_{NN}\mathbf{B}_N = \begin{array}{|c|c|} \hline 0 & 0 \\ \hline 0 & * \\ \hline \end{array} \quad \textit{and} \quad \mathbf{M}\mathbf{A}_{NG}\mathbf{B}_G = \begin{array}{|c|c|} \hline 1 & 0 \\ \hline 0 & *' \\ \hline \end{array}$$

The basis choice from lemma 3 together with assumption (ii) can be used to prove the following fact:

Lemma 4. *The non-Gaussian transformation is invertible i.e.* $\mathbf{A}_{NN} \in \mathrm{Gl}(n)$.

The next lemma can be seen as modification of lemma 1, and indeed it can be shown similarly.

Lemma 5. *If* $\hat{\mathbf{S}}_N$ *fulfills* $\left(\hat{\mathbf{S}}_N \mathbf{H}_{\hat{\mathbf{S}}_N} - \nabla\hat{\mathbf{S}}_N(\nabla\hat{\mathbf{S}}_N)^\top\right)\mathbf{e}_1 + \hat{\mathbf{S}}_N^2\mathbf{c} \equiv 0$ *for some constant vector* $\mathbf{c} \in \mathbb{R}^n$, *then the source component* $\mathbf{S}_N(1)$ *is Gaussian and independent of* $(\mathbf{S}_N)(2:n)$.

Here more generally $\mathbf{e}_i \in \mathbb{R}^n$ denotes the i-th unit vector. Putting these lemmas together, we can finally prove theorem 1: According to lemma 4, $\mathbf{A}_{NN}$ is invertible, so multiplying equation (5) from lemma 2 by $\mathbf{B}_N^{-1}\mathbf{A}_{NN}^{-1}$ from the left yields

$$\left(\hat{\mathbf{S}}_N \mathbf{H}_{\hat{\mathbf{S}}_N} - \nabla\hat{\mathbf{S}}_N(\nabla\hat{\mathbf{S}}_N)^\top\right)\mathbf{B}_N^\top\mathbf{A}_{GN}^\top + \mathbf{C}\hat{\mathbf{S}}_N^2 \equiv 0 \tag{6}$$

for any $\mathbf{B}_N \in \mathrm{Gl}(n)$ and some fixed, real matrix $\mathbf{C} \in \mathrm{Mat}(n \times (d-n))$.

We claim that $\mathbf{A}_{GN} = 0$. If not, then there exists $\mathbf{v} \in \mathbb{R}^{d-n}$ with $\|\mathbf{A}_{GN}^\top\mathbf{v}\| = 1$. Choose $\mathbf{B}_N$ from (4) such that $\mathbf{B}_N^{-1}\mathbf{S}_N$ is decorrelated. This is invariant under left-multiplication by an orthogonal matrix, so we may moreover assume that $\mathbf{B}_N^\top\mathbf{A}_{GN}^\top\mathbf{v} = \mathbf{e}_1$. Multiplying equation (6) in turn by $\mathbf{v}$ from the right therefore shows that the vector function

$$\left(\hat{\mathbf{S}}_N \mathbf{H}_{\hat{\mathbf{S}}_N} - \nabla\hat{\mathbf{S}}_N(\nabla\hat{\mathbf{S}}_N)^\top\right)\mathbf{e}_1 + \mathbf{c}\hat{\mathbf{S}}_N^2 \equiv 0 \tag{7}$$

is zero; here $\mathbf{c} := \mathbf{C}\mathbf{v} \in \mathbb{R}$. This means that $\hat{\mathbf{S}}_N$ fulfills the condition of lemma 5, which implies that $\mathbf{S}_N(1)$ is Gaussian and independent of the rest. But this contradicts (ii) for $\mathbf{S}$, hence $\mathbf{A}_{GN} = 0$. Plugging this result into equation (5), evaluation at $\mathbf{s}_N = 0$ shows that $\mathbf{A}_{NG}\mathbf{A}_{GG}^\top = 0$. Since $\mathbf{A}_{GN} = 0$ and $\mathbf{A} \in \mathrm{Gl}(d)$, necessarily $\mathbf{A}_{GG} \in \mathrm{Gl}(d-n)$, so $\mathbf{A}_{NG} = 0$ as was to prove.

2 Simulations

In this section, we will provide experimental validation of the uniqueness result of corollary 1. In order to stay unbiased and not test a single algorithm, we have to uniformly search the parameter space for possibly equivalent model representations. The model assumptions (1) will not be perfectly fulfilled, so we introduce a measure of model deviation based on 4-th order cumulants in the following.

Let the non-Gaussian dimension n and the total dimension d be fixed. Given a random vector $\mathbf{X} = (\mathbf{X}_N, \mathbf{X}_G)$, we can without loss of generality assume that $\mathrm{Cov}(\mathbf{X}) = \mathbf{I}$. Any possible model deviation consists of (i) a deviation from the independence of $\mathbf{X}_N$ and $\mathbf{X}_G$ and (ii) a deviation from the Gaussianity of $\mathbf{X}_G$. In the case of non-vanishing kurtoses, the former can be approximated for example by

$$\delta_I(\mathbf{X}) := \frac{1}{n(d-n)d^2}\sum_{i=1}^{n}\sum_{j=n+1}^{d}\sum_{k=1}^{d}\sum_{l=1}^{d}\mathrm{cum}^2(X_i, X_j, X_k, X_l),$$

where the fourth-order cumulant tensor is defined as $\text{cum}(X_i, X_j, X_k, X_l) := E(X_i X_j X_k X_l) - E(X_i X_j)E(X_k X_l) - E(X_i X_k)E(X_j X_l) - E(X_i X_l)E(X_j X_K)$. The deviation (ii) from Gaussianity of $\mathbf{X}_G$ can simply be measured by kurtosis, which in the case of white $\mathbf{X}$ means

$$\delta_G(\mathbf{X}) := \frac{1}{d-n} \sum_{j=n+1}^{d} \left| E(X_i^4) - 3 \right| .$$

Altogether, we can therefore define a total *model deviation* as the weighted sum of the above indices; the weight in the following was chosen experimentally to approximately yield even contributions of the two measures:

$$\delta(\mathbf{X}) = 10n(d-n)\delta_I(\mathbf{X}) + \delta_G(\mathbf{X})$$

For numerical tests, we generate two different non-Gaussian source data sets, see figure 1(d) and also [4], figure 1. The first source set (I) is an n-dimensional dependent sub-Gaussian random vector given by an isotropic uniform density within the unit disc, and source set (II) a 2-dimensional dependent super- and sub-Gaussian, given by

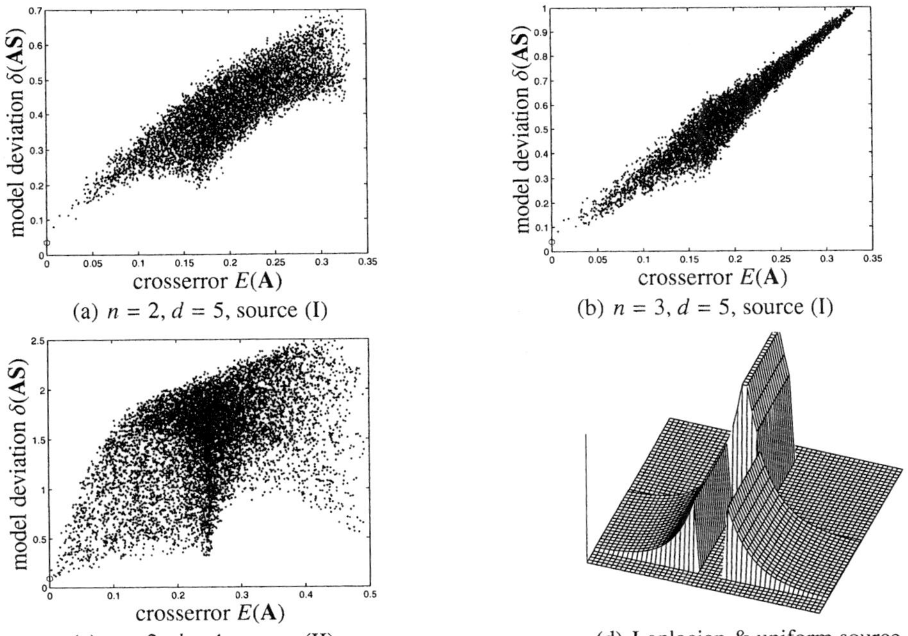

(a) $n = 2$, $d = 5$, source (I)

(b) $n = 3$, $d = 5$, source (I)

(c) $n = 2$, $d = 4$, source (II)

(d) Laplacian & uniform source

Fig. 1. (a-c): total model deviation $\delta(\mathbf{AS})$ of the transformed sources versus crosserror $E(\mathbf{A})$ of the mixing matrix for 10^4 Monte-Carlo runs. The circle $\circ$ indicates the actual source model deviation (non-zero due to finite sample sizes). (d): 2-dimensional dependent sub-Gaussian source (II).

$p(s_1, s_2) \propto \exp(-|s_1|)1_{[c(s_1),c(s_1)+1]}$, where $c(s_1) = 0$ if $|s_1| \leq \ln 2$ and $c(s_1) = -1$ otherwise. Normalization was chosen to guarantee $\mathrm{Cov}(S_N) = \mathbf{I}$ in advance.

In order to test for model violations, we have to find two representations $\mathbf{X} = \mathbf{AS}$ and $\mathbf{X} = \mathbf{A}'\mathbf{S}'$ of the same mixtures. After multiplication by $\mathbf{A}^{-1}$ we may as before assume that a single representation $\mathbf{X} = \mathbf{AS}$ is given with $\mathbf{X}$ and $\mathbf{S}$ both fulfilling the dimension reduction model (1), and we have to show that $\mathbf{A}_{NG} = \mathbf{A}_{GN} = 0$ if the decomposition is minimal (corollary 1). The latter can be tested numerically by using the so-called normalized *crosserror* $E(\mathbf{A}) := 1/(2n(d-n))\left(\|\mathbf{A}_{NG}\|_F^2 + \|\mathbf{A}_{GN}\|_F^2\right)$, where $\|.\|_F$ is some matrix norm, in our case the Frobenius norm.

In order to reduce the d^2-dimensional search space, after whitening we may assume that $\mathbf{A} \in O(d)$, so only $d(d-1)/2$ dimensions have to be searched. $O(d)$ can be uniformly sampled for example by choosing $\mathbf{B}$ with Gaussian i.i.d. coefficients and orthogonalizing $\mathbf{A} := (\mathbf{BB}^\top)^{-1/2}\mathbf{B}$. We perform 10^4 Monte-Carlo runs with random $\mathbf{A} \in O(d)$. Sources have been generated with $T = 10^4$ samples, n-dimensional non-Gaussian part (a) and (b) from above, and $(d-n)$-dimensional i.i.d. Gaussians. We measure model deviation $\delta(\mathbf{AS})$ and compare it with the deviation $E(\mathbf{A})$ from block-diagonality.

The results for varying parameters are given in figure 1(a-c). In all three cases we observe that the smaller the model deviation, the smaller also the crosserror. This gives an asymptotic confirmation of corollary 1, indicating that by random sampling no non-uniqueness realizations have been found.

3 Conclusion

By minimality of the decomposition (1), we gave a necessary condition for the uniqueness of non-Gaussian subspace analysis. Together with the assumption of existing covariance, this was already sufficient to guarantee model uniqueness. Our result allows NGSA algorithms to find the unknown, unique signal space within a noisy high-dimensional data set [6]. In practice, instantaneous mixing models are seldom found, so applications to more realistic situations, for instance to convolutive mixtures, are currently being studied.

References

1. Friedman, J., Tukey, J.: A projection pursuit algorithm for exploratory data analysis. IEEE Trans. on Computers **23** (1975) 881–890
2. Hyvärinen, A., Karhunen, J., Oja, E.: Independent component analysis. John Wiley & Sons (2001)
3. Blanchard, G., Kawanabe, M., Sugiyama, M., Spokoiny, V., Müller, K.R.: In search of non-gaussian components of a high-dimensional distribution. JMLR (2005) In revision. The preprint is available at http://www.cs.titech.ac.jp/ tr/reports/2005/TR05-0003.pdf.
4. Kawanabe, M.: Linear dimension reduction based on the fourth-order cumulant tensor. In: Proc. ICANN 2005. Volume 3697 of LNCS., Warsaw, Poland, Springer (2005) 151–156
5. Theis, F.: Uniqueness of complex and multidimensional independent component analysis. Signal Processing **84** (2004) 951–956

6. Kawanabe, M., Theis, F.: Extracting non-gaussian subspaces by characteristic functions. In: submitted to ICA 2006. (2006)
7. Comon, P.: Independent component analysis - a new concept? Signal Processing **36** (1994) 287–314
8. Theis, F.: A new concept for separability problems in blind source separation. Neural Computation **16** (2004) 1827–1850
9. Theis, F.: Multidimensional independent component analysis using characteristic functions. In: Proc. EUSIPCO 2005, Antalya, Turkey (2005)

A Maximum Likelihood Approach to Nonlinear Convolutive Blind Source Separation

Jingyi Zhang, Li Chin Khor, Wai Lok Woo, and Satnam Singh Dlay

School of Electrical, Electronic and Computer Engineering,
University of Newcastle upon Tyne Newcastle upon Tyne, NE1 7RU United Kindom
Jingyi.Zhang@ncl.ac.uk, l.c.khor@ncl.ac.uk, w.l.woo@ncl.ac.uk, s.s.dlay@ncl.ac.uk

Abstract. A novel learning algorithm for blind source separation of post-nonlinear convolutive mixtures with non-stationary sources is proposed in this paper. The proposed mixture model characterizes both convolutive mixture and post-nonlinear distortions of the sources. A novel iterative technique based on Maximum Likelihood (ML) approach is developed where the Expectation-Maximization (EM) algorithm is generalized to estimate the parameters in the proposed model. The post-nonlinear distortion is estimated by using a set of polynomials. The sufficient statistics associated with the source signals are estimated in the E-step while in the M-step, the parameters are optimized by using these statistics. In general, the nonlinear maximization in the M-step is difficult to be formulated in a closed form. However, the use of polynomial as the nonlinearity estimator facilitates the M-step tractable and can be solved via linear equations.

1 Introduction

The study of blind deconvolution so far has concentrated solely on the linear mixture [1], [2] and the existing methods only perform well when the mixture is assumed to be linear. Where nonlinear distortion in the mixture is considered, all of the previous works focused only on instantaneous mixing of signals [3]. To the best of the author's knowledge, the problem of post-nonlinear convolutive mixture of non-stationary sources has not been previously addressed. However in practical applications such as speech processing, source signals are inherently deconvolved in a real acoustic environment where signals are corrupted by noise and interferences. Furthermore, studies show that carbon-button microphones present evidence of a "phantomformant" which occurs when simple static nonlinearities were applied to speech signals. The non-uniform flux of the permanent magnet and the nonlinear response of the suspensions in the loudspeaker also contribute to the nonlinear distortions in speech signals. Therefore, an accurate representation of the mixed signals must be developed to account for the existence of the nonlinearity. The observed signal $\mathbf{x}_t$ at time t of the noisy post-nonlinear convolutive mixture can be expressed as follows:

$$\mathbf{x}_t = \mathbf{g}\left(\sum_{l=0}^{L} \mathbf{M}_l \mathbf{s}_{t-l}\right) + \mathbf{n}_t \tag{1}$$

J. Rosca et al. (Eds.): ICA 2006, LNCS 3889, pp. 926 – 933, 2006.

where vector $\mathbf{s}_t$ is the unknown non-stationary source signals, $\mathbf{g}(.)$ is the nonlinear function and $\mathbf{M}_l$ is the delayed mixing matrix. The additive noise $\mathbf{n}_t$ is assumed to be Gaussian. As in typical BSS problems, the aim is to estimate $\mathbf{s}_t$, $\mathbf{g}(.)$, $\mathbf{M}_l$ and $\mathbf{n}_t$ with only information on $\mathbf{x}_t$ available. Existing algorithms e.g. [1, 2] do not cater for both the convolutive and non-stationary properties in the mixture and hence perform poorly in solving the problem described in equation (1).

This paper presents a pioneering contribution to the noisy post-nonlinear convolutive mixing problem where we propose for the fir`st time an innovative solution to the problem. A state space model representing the post-nonlinear convolutive mixtures of non-stationary signals and a general EM framework is formulated for estimating $\mathbf{s}_t$, $\mathbf{g}(.)$, $\mathbf{M}_l$ and $\mathbf{n}_t$. The generalized EM algorithm is derived from a set of polynomials to estimate the nonlinear distortion whose coefficients are updated as part of the mixing parameters. In the proposed algorithm, the sufficient statistics associated with the source signals are inferred in the E-step and the model parameters are updated in the M-step.

2 The Model

The state space model representing the post-nonlinear convolutive mixture of non-stationary signals is constructed by two parts. First, autoregressive (AR) process is adopted to represent the temporal correlation of the non-stationary sources. The Kth order AR(K) process for source i can be modeled as

$$s_{i,t} = h_{i,t,1}s_{i,t-1} + h_{i,t,2}s_{i,t-2} + \cdots + h_{i,t,K}s_{i,t-K} + v_{i,t} \tag{2}$$

The source signal vector at time t is formed by stacking each source signal [4, 5] and can be expressed as

$$\mathbf{s}_t^{\mathrm{T}} = [\mathbf{s}_{1,t}^{\mathrm{T}} \quad \mathbf{s}_{2,t}^{\mathrm{T}} \quad \cdots \quad \mathbf{s}_{I_S,t}^{\mathrm{T}}]$$

The vector for individual source is now formed by stacking the last K signals, which can be expressed as

$$\mathbf{s}_{i,t} = [s_{i,t} \quad s_{i,t-1} \quad \cdots \quad s_{i,t-K+1}]^{\mathrm{T}}$$

Second, the convolutive mixture and nonlinearity distortion are introduced into the model. To represent the convolutive mixture, the instantaneous observation matrix is extended to the full matrix of filters, which can be expressed as follows:

$$\mathbf{M} = \begin{bmatrix} \mathbf{m}_{11} & \cdots & \mathbf{m}_{1I_S} \\ \vdots & \ddots & \vdots \\ \mathbf{m}_{I_O 1} & \cdots & \mathbf{m}_{I_O I_S} \end{bmatrix}, \quad \mathbf{m}_{ij} = \begin{bmatrix} m_{ij,1} & m_{ij,2} & \cdots & m_{ij,L} \end{bmatrix}$$

where $m_{ij,l}$ represents the lth delayed path between the sensor i and source j (L=K). Hence, the proposed post-nonlinear convolutive model is now expressed as

$$\mathbf{s}_t = \mathbf{H}_t \mathbf{s}_{t-1} + \mathbf{v}_t$$
$$\mathbf{x}_t = \mathbf{g}(\mathbf{M}\mathbf{s}_t) + \mathbf{n}_t \quad (3)$$

The matrix $\mathbf{H}_t$ is the evolution matrix. $\mathbf{W}$ and $\mathbf{R}$ are the covariance matrices of the zero mean Gaussian noise vectors $\mathbf{v}$ and $\mathbf{n}$, respectively. The prior probability distribution over initial states of the source signals $\mathbf{s}_1$ is assumed to be Gaussian with mean $\boldsymbol{\mu}$ and covariance $\boldsymbol{\Lambda}$. To satisfy the independence between the source signals, the associated parameters in (3) need to be defined in the following form:

$$\mathbf{H}_t = diag[\mathbf{H}_{1,t} \quad \mathbf{H}_{2,t} \quad \cdots \quad \mathbf{H}_{I_s,t}], \mathbf{H}_{i,t} = \begin{bmatrix} \mathbf{h}_{i,t} \\ \mathbf{I} \quad \mathbf{0} \end{bmatrix}, \mathbf{h}_{i,t} = [h_{i,t,1} \quad h_{i,t,2} \quad \cdots \quad h_{i,t,K}]$$

$$\mathbf{W} = diag[\mathbf{W}_1 \quad \mathbf{W}_2 \quad \cdots \quad \mathbf{W}_{I_s}], (\mathbf{W}_i)_{j_1 j_2} = \begin{cases} w_i & j_1 = j_2 = 1 \\ 0 & otherwise \end{cases} \quad (4)$$

$\boldsymbol{\mu}$ and $\boldsymbol{\Lambda}$ are defined in the similar way. As mentioned in [6], blind deconvolution of non-stationary signals can be achieved by segmenting them with windows in which the source signals can be assumed to be stationary. Hence, the post-nonlinear convolutive model is now expressed as:

$$\mathbf{s}_t^n = \mathbf{H}_t^n \mathbf{s}_{t-1}^n + \mathbf{v}_t^n$$
$$\mathbf{x}_t^n = \mathbf{g}(\mathbf{M}\mathbf{s}_t^n) + \mathbf{n}_t^n, \quad n = 1, 2, \ldots, N \quad (5)$$

Hence, a total number of N segments are observed. In the next section, the learning rules of the generalized EM algorithm is derived at a point where the nonlinearity is linearized by using second order Taylor Series, the Kalman recursion is then used to infer the relevant statistics in the E-step while the post-nonlinear distortion is estimated by a set of polynomials in the M-step.

3 Learning Rules

To derive the generalized EM algorithm for proposed model (5), the likelihood function is introduced as

$$L(\lambda) = \log p(\mathbf{x} \mid \lambda) = \log \int p(\mathbf{x}, \mathbf{s} \mid \lambda) ds$$

where λ denotes all the parameters in the proposed model (5). Based on Jensen's inequality

$$L(\lambda) = \log p(\mathbf{x} \mid \lambda) = \log \int p(\mathbf{x}, \mathbf{s} \mid \lambda) ds = \log \int \frac{\hat{p}(\mathbf{s})}{\hat{p}(\mathbf{s})} p(\mathbf{x}, \mathbf{s} \mid \lambda) ds$$
$$\geq \int \hat{p}(\mathbf{s}) \log \frac{p(\mathbf{x}, \mathbf{s} \mid \lambda)}{\hat{p}(\mathbf{s})} ds = \varphi_1(\lambda, \hat{p}) - \varphi_2(\hat{p}) = F(\lambda, \hat{p}) \quad (6)$$

where $\varphi_1(\lambda, \hat{p}) \equiv \int \hat{p}(\mathbf{s}) \log p(\mathbf{x}, \mathbf{s} \mid \lambda) ds$ and $\varphi_2(\hat{p}) \equiv \int \hat{p}(\mathbf{s}) \log \hat{p}(\mathbf{s}) ds$.

It is well known that in the E-step the maximization of $F(\lambda, \hat{p})$ with respect to $\hat{p}(\mathbf{s})$ is achieved when $\hat{p}(\mathbf{s})$ is chosen to be exactly the conditional distribution of $\mathbf{s}$ with the parameters obtained in the previous iteration $\hat{p}(\mathbf{s}) = p(\mathbf{s} \mid \mathbf{x}, \lambda^{[q]})$ at which point the bound becomes an equality. Then in the M-step, $\varphi_1(\lambda, \hat{p})$ is maximized with respect to λ. Each iteration is guaranteed not decrease $F(\lambda, \hat{p})$.

3.1 E-Step

The relevant statistics of the posterior distribution of the source signals $p(\mathbf{s}_t \mid \mathbf{x}_{1:\tau}, \lambda^{[q]})$ needs to be inferred and represented with the parameters obtained in the previous iteration in order to update the parameters in the proposed model. For the linear convolutive mixture model, this is achieved by using the Kalman smoother. The algorithm of Kalman smoother consists of two parts: a forward recursion named Kalman filter which uses the observation from $\mathbf{x}_1$ to $\mathbf{x}_t$ and a backward recursion which uses the observation from $\mathbf{x}_\tau$ to $\mathbf{x}_{t+1}$ where τ be the length of the observed signals. However, for the model defined by equation (5) the conditional densities are in general non-Gaussian and can lead to an intractable solution. To solve this problem, the Extended Kalman Smoother (EKS) is used in the E-step. The theory of the EKS is that the basic Kalman Smoother is applied at a linearized point of the nonlinear system. At this point, the nonlinearity is linearized by second order Taylor Series at the mean of the current filtered (not smoothed) state $\hat{\mathbf{s}}^n_{t|t-1}$. Hence, after the linearization process, the derivative matrix of the vector-valued function $\mathbf{g}$, at point $\hat{\mathbf{s}}^n_{t|t-1}$ is defined as

$$\mathbf{D}_{\hat{\mathbf{s}}^n_{t|t-1}} \equiv \left.\frac{\partial \mathbf{g}}{\partial \mathbf{s}^n_t}\right|_{\mathbf{s}^n_t = \hat{\mathbf{s}}^n_{t|t-1}} \tag{7}$$

and the model (5) can be expressed as

$$\begin{aligned} \mathbf{s}^n_t &= \mathbf{H}^n_t \mathbf{s}^n_{t-1} + \mathbf{v}^n_t \\ \mathbf{x}^n_t &= \mathbf{g}(\mathbf{M}\hat{\mathbf{s}}^n_{t|t-1}) + \mathbf{D}_{\hat{\mathbf{s}}^n_{t|t-1}} (\mathbf{s}^n_t - \hat{\mathbf{s}}^n_{t|t-1}) + \mathbf{n}^n_t \end{aligned} \tag{8}$$

Therefore, given the output the conditional distribution of the hidden states in the linearized model (8) at every instant in time is Gaussian. Hence, the basic Kalman Smoother can be applied on the model (8) to infer the associated statistics of the conditional distribution.

The inferred first order statistics is the source conditional mean $\hat{\mathbf{s}}^n_t$ for segment n, which is expressed as $\hat{\mathbf{s}}^n_t = \langle \mathbf{s}^n_t \rangle$ where $\langle . \rangle$ denotes for the integral over the source posterior $p(\mathbf{s}_t \mid \mathbf{x}_{1:\tau}, \lambda^{[q]})$. The inferred second order statistics of the hidden source signals are the autocorrelation matrix of source i for segment n denoted as $\mathbf{C}^n_{ii,tt}$ without time delay and $\mathbf{C}^n_{ii,t(t-1)}$ with time delay and expressed as follows:

$$\mathbf{C}_{ii,tt}^{n} \equiv \langle \mathbf{s}_{i,t}^{n}(\mathbf{s}_{i,t}^{n})^{\mathrm{T}} \rangle \equiv \begin{bmatrix} \mathbf{c}_{ii,1,tt}^{n} & \mathbf{c}_{ii,2,tt}^{n} & \cdots & \mathbf{c}_{ii,L,tt}^{n} \end{bmatrix}^{\mathrm{T}} \tag{9a}$$

$$\mathbf{C}_{ii,t(t-1)}^{n} \equiv \langle \mathbf{s}_{i,t}^{n}(\mathbf{s}_{i,t-1}^{n})^{\mathrm{T}} \rangle \equiv \begin{bmatrix} \mathbf{c}_{ii,1,t(t-1)}^{n} & \mathbf{c}_{ii,2,t(t-1)}^{n} & \cdots & \mathbf{c}_{ii,L,t(t-1)}^{n} \end{bmatrix}^{\mathrm{T}} \tag{9b}$$

Here, the first element in $\mathbf{c}_{ii,1,tt}^{n}$ is defined as $c_{ii,1,tt}^{n}$ and the autocorrelation matrix for $\mathbf{s}_t^n$ is defined as $\mathbf{C}_{tt}^{n}$. Because the source signals are statistical independent, for different source i and j $\mathbf{C}_{ij,tt}^{n} = \hat{\mathbf{s}}_{i,t}^{n}(\hat{\mathbf{s}}_{j,t}^{n})^{\mathrm{T}}$ .The model parameters are then estimated to maximize the likelihood in (6) in the following M-step with the relevant statistics represented in the form of model parameters obtained from the M-step of the previous iteration.

3.2 M-Step

In the M-step, $\varphi_1(\lambda)$ in (6) is maximized with respect to all the model parameters by using the relevant statistics obtained from the E-step. Represented in the form of model parameters, $\varphi_1(\lambda)$ can be expressed as:

$$\varphi_1(\lambda,\hat{p}) = -\frac{1}{2}\sum_{n=1}^{N}\Bigg[\sum_{i=1}^{I_S}\log\det\mathbf{\Lambda}_i^n + (\tau-1)\sum_{i=1}^{I_S}\log w_i^n + \tau\log\det\mathbf{R} + \sum_{i=1}^{I_S}$$
$$\langle(\mathbf{s}_{i,1}^{n}-\boldsymbol{\mu}_i^n)^{\mathrm{T}}(\mathbf{\Lambda}_i^n)^{-1}(\mathbf{s}_{i,1}^{n}-\boldsymbol{\mu}_i^n)\rangle + \sum_{t=2}^{\tau}\sum_{i=1}^{I_S}\langle\frac{1}{w_i^n}(s_{i,t}^{n}-(\mathbf{h}_{i,t}^{n})\mathbf{s}_{i,t-1}^{n})^2\rangle + \sum_{t=1}^{\tau}$$
$$\langle(\mathbf{x}_t^n-\mathbf{g}(\mathbf{M}\hat{\mathbf{s}}_{t|t-1}^{n})-\mathbf{D}_{\hat{\mathbf{s}}_{t|t-1}^{n}}(\mathbf{s}_t^n-\hat{\mathbf{s}}_{t|t-1}^{n}))^{\mathrm{T}}\mathbf{R}^{-1}(\mathbf{x}_t^n-\mathbf{g}(\mathbf{M}\hat{\mathbf{s}}_{t|t-1}^{n})-\mathbf{D}_{\hat{\mathbf{s}}_{t|t-1}^{n}}(\mathbf{s}_t^n-\hat{\mathbf{s}}_{t|t-1}^{n}))\rangle\Bigg] \tag{10}$$

For segment-wise parameters, the update equations are exactly the same as the ones for linear convolutive mixture as in [5], and the new estimator for segment n of source i is given by the following closed form equations:

$$\boldsymbol{\mu}_i^n = \hat{\mathbf{s}}_{i,1}^{n},\ \mathbf{\Lambda}_i^n = \mathbf{C}_{ii,11}^{n} - \boldsymbol{\mu}_i^n(\boldsymbol{\mu}_i^n)^{\mathrm{T}}$$

$$\mathbf{h}_{i,t}^{n} = \Big[\mathbf{c}_{ii,1,t(t-1)}^{n}\Big]\Big[\mathbf{C}_{ii,(t-1)(t-1)}^{n}\Big]^{-1},\ w_i^n = \frac{1}{\tau-1}\sum_{t=2}^{\tau}\Big(c_{ii,1,tt}^{n} - \mathbf{h}_{i,t}^{n}(\mathbf{c}_{ii,1,t(t-1)}^{n})^{\mathrm{T}}\Big) \tag{11}$$

Then $\boldsymbol{\mu}^n, \mathbf{\Lambda}^n, \mathbf{H}_t^n, \mathbf{W}^n$ can be reconstructed following the definitions in Section 2. However, the update equations for $\mathbf{M}$ and $\mathbf{R}$ which include the statistics from all observed segments are different from the ones for linear deconvolution and more complex. Because the new estimator for $\mathbf{M}$ cannot be expressed in a closed form, the update equation for $\mathbf{M}$ is derived from the gradient ascent algorithm and its elements $\mathbf{m}_{ij}$ is estimated from the following equation

$$\mathbf{m}_{ij,t+1} = \mathbf{m}_{ij,t} + \varepsilon\frac{\partial\varphi_1(\lambda,\hat{p})}{\partial\mathbf{m}_{ij}}\Big|_{\mathbf{m}_{ij}=\mathbf{m}_{ij,t}} \tag{12a}$$

$$\frac{\partial \varphi_1(\lambda,\hat{p})}{\partial \mathbf{m}_{ij}} = -\sum_{n=1}^{N}\sum_{t=1}^{\tau}\Bigg[J_1^n g_i'(\hat{\mathbf{s}}_{j,t|t-1}^n)^{\mathrm{T}} + J_2^n g_i''(\hat{\mathbf{s}}_{j,t|t-1}^n)^{\mathrm{T}} + J_3^n g_i'(\hat{\mathbf{s}}_{j,t}^n - \hat{\mathbf{s}}_{j,t|t-1}^n)^{\mathrm{T}} + \sum_{h=1}^{I_O}\sum_{k_3=1}^{I_s} g_h' r_{ih}^{-1}\mathbf{m}_{hk_3}$$
$$\left(\mathbf{C}_{k_3 j,tt}^n - \hat{\mathbf{s}}_{k_3,t}^n(\hat{\mathbf{s}}_{j,t|t-1}^n)^{\mathrm{T}} - \hat{\mathbf{s}}_{k_3,t|t-1}^n(\hat{\mathbf{s}}_{j,t}^n)^{\mathrm{T}} + \hat{\mathbf{s}}_{k_3,t|t-1}^n(\hat{\mathbf{s}}_{j,t|t-1}^n)^{\mathrm{T}}\right) g_i' \Bigg] \quad (12b)$$

$$J_1^n = -\sum_{k_1=1}^{I_o} x_{k_1,t}^n r_{ik_1}^{-1} + \sum_{k_2=1}^{I_o} g_{k_2} r_{ik_2}^{-1} + \sum_{q_2=1}^{I_o} r_{iq_2}^{-1} g_{q_2}' \mathbf{m}_{q_2}(\hat{\mathbf{s}}_t^n - \hat{\mathbf{s}}_{t|t-1}^n)$$

$$J_2^n = -\sum_{p_2=1}^{I_o} x_{p_2,t}^n r_{p_2 i}^{-1}\mathbf{m}_i(\hat{\mathbf{s}}_t^n - \hat{\mathbf{s}}_{t|t-1}^n) + \sum_{q_1=1}^{I_o} g_{q_1} r_{q_1 i}^{-1}\mathbf{m}_i(\hat{\mathbf{s}}_t^n - \hat{\mathbf{s}}_{t|t-1}^n) + \sum_{q=1}^{I_s}\sum_{p=1}^{I_o}\sum_{k=1}^{I_s} r_{ip}^{-1} g_p'$$
$$tr\{\mathbf{m}_{ik}^{\mathrm{T}}\mathbf{m}_{pq}\left(\mathbf{C}_{qk,tt}^n - \hat{\mathbf{s}}_{q,t}^n(\hat{\mathbf{s}}_{k,t|t-1}^n)^{\mathrm{T}} - \hat{\mathbf{s}}_{q,t|t-1}^n(\hat{\mathbf{s}}_{k,t}^n)^{\mathrm{T}} + \hat{\mathbf{s}}_{q,t|t-1}^n(\hat{\mathbf{s}}_{k,t|t-1}^n)^{\mathrm{T}}\right)\}$$

$$J_3^n = -\sum_{p_1=1}^{I_o} x_{p_1,t}^n r_{p_1 i}^{-1} + \sum_{q_5=1}^{I_o} g_{q_5} r_{q_5 i}^{-1} \quad (12c)$$

Where $\mathbf{m}_i = \begin{bmatrix} m_{i1} & \cdots & m_{iI_s}\end{bmatrix}$, ε is the learning rate, r_{ij}^{-1} is the ij^{th} element of the inverse matrix of $\mathbf{R}$, g_i represents $g_i\left(\sum \mathbf{m}_{ij}\hat{\mathbf{s}}_{j,t|t-1}^n\right)$, and g_i' is the first order derivative with respect to the argument $\sum \mathbf{m}_{ij}\hat{\mathbf{s}}_{j,t|t-1}^n$. The covariance matrix $\mathbf{R}$ can be estimated from the following:

$$\mathbf{R} = \tfrac{1}{N\tau}\sum_{n=1}^{N}\sum_{t=1}^{\tau}\Bigg[\mathbf{x}_t^n(\mathbf{x}_t^n)^{\mathrm{T}} - \mathbf{x}_t^n\mathbf{g}^{\mathrm{T}} - \mathbf{x}_t^n(\hat{\mathbf{s}}_t^n - \hat{\mathbf{s}}_{t|t-1}^n)^{\mathrm{T}}\mathbf{D}_{\hat{\mathbf{s}}_{t|t-1}^n}^{\mathrm{T}} - \mathbf{g}(\mathbf{x}_t^n)^{\mathrm{T}} + \mathbf{g}\mathbf{g}^{\mathrm{T}} + \mathbf{g}(\hat{\mathbf{s}}_t^n - \hat{\mathbf{s}}_{t|t-1}^n)^{\mathrm{T}}\mathbf{D}_{\hat{\mathbf{s}}_{t|t-1}^n}^{\mathrm{T}}$$
$$-\mathbf{D}_{\hat{\mathbf{s}}_{t|t-1}^n}(\hat{\mathbf{s}}_t^n - \hat{\mathbf{s}}_{t|t-1}^n)(\mathbf{x}_t^n)^{\mathrm{T}} + \mathbf{D}_{\hat{\mathbf{s}}_{t|t-1}^n}(\hat{\mathbf{s}}_t^n - \hat{\mathbf{s}}_{t|t-1}^n)\mathbf{g}^{\mathrm{T}}$$
$$+\mathbf{D}_{\hat{\mathbf{s}}_{t|t-1}^n}\left(\mathbf{C}_{tt}^n - \hat{\mathbf{s}}_t^n(\hat{\mathbf{s}}_{t|t-1}^n)^{\mathrm{T}} - \hat{\mathbf{s}}_{t|t-1}^n(\hat{\mathbf{s}}_t^n)^{\mathrm{T}} + \hat{\mathbf{s}}_{t|t-1}^n(\hat{\mathbf{s}}_{t|t-1}^n)^{\mathrm{T}}\right)\mathbf{D}_{\hat{\mathbf{s}}_{t|t-1}^n}^{\mathrm{T}}\Bigg] \quad (13)$$

where $\mathbf{g}$ represents $\mathbf{g}(\mathbf{M}\hat{\mathbf{s}}_{t|t-1}^n)$.

To estimate the nonlinearity $\mathbf{g}$, a self-adaptive algorithm is required. Only the statistics obtained from the E-step is available for the algorithm. The scalar function $g(.)$ of $\mathbf{g}$ is approximated by a set of polynomials [7, 8] defined below:

$$g_i(\mathbf{m}_i\hat{\mathbf{s}}_{t|t-1}^n) = \sum_{z=0}^{Z} a_{i,z}(\mathbf{m}_i\hat{\mathbf{s}}_{t|t-1}^n)^z = \mathbf{a}_i\mathbf{q}_i^n \quad (14)$$

where $a_{i,z}$ are the coefficients of the polynomials and $\mathbf{a}_i = \begin{bmatrix} a_{i,0} & \cdots & a_{i,Z}\end{bmatrix}$, z represents the order of expansion and $\mathbf{q}_i = \begin{bmatrix} 1 & \mathbf{m}_i\hat{\mathbf{s}}_{t|t-1}^n & \cdots & (\mathbf{m}_i\hat{\mathbf{s}}_{t|t-1}^n)^Z\end{bmatrix}^{\mathrm{T}}$. The polynomial coefficient $a_{i,z}$ can be updated as one of the model parameters by maximizing $\varphi_1(\lambda,\hat{p})$. The update equation for $a_{i,z}$ is obtained by gradient ascent algorithm with learning rate β, which can be expressed as

$$a_{i,z,t+1} = a_{i,z,t} + \beta \frac{\partial \varphi_1(\lambda, \hat{p})}{\partial a_{i,z}} |_{a_{i,z}=a_{i,z,t}} \tag{15}$$

$$\frac{\partial \varphi_1}{\partial a_{i,z}} =$$

$$\begin{cases} -\sum_{n=1}^{N}\sum_{t=1}^{\tau}\left[-\sum_{k_1=1}^{I_o} x_{k_1,t}^n r_{ik_1}^{-1} + \sum_{k_2=1}^{I_o} (\mathbf{a}_{k_2}\mathbf{q}_{k_2}^n) r_{ik_2}^{-1} + \sum_{k_3=1}^{I_o} r_{ik_3}^{-1}\left(\sum_{z=1}^{Z} z a_{k_3,z}(\mathbf{m}_{k_3}\hat{\mathbf{s}}_{t|t-1}^n)^{z-1}\right)\mathbf{m}_{k_3}(\hat{\mathbf{s}}_t^n - \hat{\mathbf{s}}_{t|t-1}^n)\right], z=0 \\ -\sum_{n=1}^{N}\sum_{t=1}^{\tau}\left[-\left(\sum_{k_1=1}^{I_o} x_{k_1,t}^n r_{ik_1}^{-1}\right)(\mathbf{m}_i\hat{\mathbf{s}}_{t|t-1}^n)^z + \left(\sum_{k_2=1}^{I_o}\left(\mathbf{a}_{k_2}\mathbf{q}_{k_2}^n\right) r_{ik_2}^{-1}\right)(\mathbf{m}_i\hat{\mathbf{s}}_{t|t-1}^n)^z - \sum_{k_3=1}^{I_o} x_{k_3,t} r_{k_3 i}^{-1}\mathbf{m}_i(\hat{\mathbf{s}}_t^n - \hat{\mathbf{s}}_{t|t-1}^n) z (\mathbf{m}_i\hat{\mathbf{s}}_{t|t-1}^n)^{z-1}\right. \\ + \sum_{k_4=1}^{I_o} r_{ik_4}^{-1}\left(\sum_{z=1}^{Z} z a_{k_4,z}(\mathbf{m}_{k_4}\hat{\mathbf{s}}_{t|t-1}^n)^{z-1}\right)\mathbf{m}_{k_4}(\hat{\mathbf{s}}_t^n - \hat{\mathbf{s}}_{t|t-1}^n)(\mathbf{m}_i\hat{\mathbf{s}}_{t|t-1}^n)^z + \sum_{k_5=1}^{I_o}\left(\mathbf{a}_{k_5}\mathbf{q}_{k_5}^n\right) r_{k_5 i}^{-1}\mathbf{m}_i(\hat{\mathbf{s}}_t^n - \hat{\mathbf{s}}_{t|t-1}^n) z(\mathbf{m}_i\hat{\mathbf{s}}_{t|t-1}^n)^{z-1} \\ \left. +\left(\sum_{k=1}^{I_o} tr\{\mathbf{m}_i^{\mathrm{T}}\mathbf{m}_k\left(\mathbf{C}_{tt}^n - \hat{\mathbf{s}}_t^n(\hat{\mathbf{s}}_{t|t-1}^n)^{\mathrm{T}} - \hat{\mathbf{s}}_{t|t-1}^n(\hat{\mathbf{s}}_t^n)^{\mathrm{T}} + \hat{\mathbf{s}}_{t|t-1}^n(\hat{\mathbf{s}}_{t|t-1}^n)^{\mathrm{T}}\right)\} r_{ik}^{-1}\left(\sum_{z=1}^{Z} z a_{k,z}(\mathbf{m}_k\hat{\mathbf{s}}_{t|t-1}^n)^{z-1}\right)\right) z(\mathbf{m}_i\hat{\mathbf{s}}_{t|t-1}^n)^{z-1}\right], z=1,\ldots,Z \end{cases} \tag{16}$$

Thus, with all model parameters estimated in the M-step the EM algorithm alternates between the E-step and M-step until if converges.

4 Results

The proposed algorithm is evaluated based on its performance in the separation of a post-nonlinear convolutive mixture of two independent speech signals with additional Gaussian noise with different signal-to-noise ratio (SNR). Both the Kth order of the AR process and L are set to 3. The full convolutive mixture matrix $\mathbf{M}$ is randomly selected and described by the following matrix:

$$\mathbf{M} = \begin{bmatrix} 0.9 & 0.1 & 0.3 & 0.1 & 0 & 0.7 \\ 0.8 & 0 & 0.2 & 1 & 0.26 & 0.1 \end{bmatrix} \tag{17}$$

The post-nonlinear distortions are selected as $g_1(\gamma_1) = \tanh(\gamma_1)$ and $g_2(\gamma_2) = \gamma_2 + \gamma_2^3$. The function $g_1(\gamma_1)$ is bounded while $g_2(\gamma_2)$ unbounded and this selection is taken merely to study the performance of the proposed algorithm under two different forms of nonlinearity. The observation signals are segmented into segments of time length $\tau = 90$. All model parameters are estimated by the proposed algorithm for each segment of each test signals with different SNR. To compare the performance of the signal separation between the proposed algorithm and the Olsson-Hansen algorithm [5] for linear convolutive mixtures, the signal to interference ratio (SIR) is adopted as a performance measure and defined as follows:

$$SIR = \frac{P_{11} + P_{22}}{P_{12} + P_{21}} \tag{18}$$

where P_{ij} is the power of the signal which is contributed the ith estimated source signal to the jth original source signal where the normalized cross-autocorrelation is used. For our evaluation, a high SIR value is desirable. The superiority of the proposed algorithm over the Olsson-Hansen algorithm is demonstrated by the significant improvements shown in the Table 1. The results proves the importance of incorporating a nonlinear model in the algorithm in cases where the observed signals have been nonlinearly distorted. The proposed algorithm is also robust under high level of noise in the separation of post-nonlinearly mixed signals.

Table 1. Performance comparisons

SIR (dB) \ SNR	10dB	20dB	30dB
Olsson-Hansen algorithm [5]	6.3	8.2	9.1
Proposed algorithm	10.2	12.6	13.5

5 Conclusions

In this paper, a novel Maximum Likelihood approach based on EM algorithm for post- nonlinear convolutive model of non-stationary signals has been proposed. The state space represented model extends the linear instantaneous mixture model to the post-nonlinear convolutive mixture. To update the model parameters, the EM algorithm is generalized where the Extended Kalman Smoother is adopted to infer the hidden source signals and a set of polynomial is utilized to estimate the post-nonlinear distortion. Experimental results show for given nonlinear data set, the proposed algorithm performs significantly better than the linear algorithm by over 50%.

References

[1] H. Attias and C.E. Schreiner: 'Blind source separation and deconvolution: the dynamic component analysis algorithm', Neural Computation, 10 (6): 1373-1424, 1998.

[2] S. Cruces-Alvarez, A. Cichochi, L. Castedo-Ribas: 'An Inversion Approach to Blind Source Separation', IEEE Trans. on Neural Networks, Vol. 11, No. 6, Nov. 2000.

[3] W.L. Woo and S.S. Dlay: 'Neural Network Approach to Blind Signal Separation of Mono-nonlinearly Mixed Signals', IEEE Transactions on Circuits and System - Part 1 2005, 52(6), 1236-1247.

[4] G. Doblinger: 'An adaptive Kalman filter for the enhancement of noisy AR signals', IEEE Int. Symp. on Circuits and Systems, Vol. 5, pp. 305-8, 1998.

[5] R.K. Olsson, L.K. Hansen: 'Probabilistic blind deconvolution for non-stationary source', 12th European Signal Processing Conference, pp. 1697-1700, 2004.

[6] L. Parra and C. Spence: 'Convolutive Blind Separation of non-stationary Sources', IEEE Trans. on Speech and Audio Processing, Vol. 8, No. 3, May 2000.

[7] W.L. Woo and S.S. Dlay: 'Nonlinear Blind Source Separation using a Hybrid RBF-FMLP Network', IEE Proceedings on Vision, Image and Signal Processing, 152(2), 173-183, 2005.

[8] W.L. Woo and L.C. Khor: 'Blind Restoration of Nonlinearly Mixed Signals using Multilayer Polynomial Neural Network', IEE Proceedings on Vision, Image and Signal Processing, 151(1), 51-61, 2002.`

On Separation of Semitransparent Dynamic Images from Static Background

Alexander M. Bronstein[1], Michael M. Bronstein[1], and Michael Zibulevsky[2]

[1] Department of Computer Science, Technion – Israel Institute of Technology, Haifa 32000, Israel
{alexbron, bronstein}@ieee.org
[2] Department of Computer Science, Technion – Israel Institute of Technology, Haifa 32000, Israel
mzib@ee.technion.ac.il

Abstract. Presented here is the problem of recovering a dynamic image superimposed on a static background. Such a problem is ill-posed and may arise e.g. in imaging through semireflective media, in separation of an illumination image from a reflectance image, in imaging with diffraction phenomena, etc. In this work we study regularization of this problem in spirit of Total Variation and general sparsifying transformations.

1 Introduction

In this paper, we consider a problem of recovering images of two objects superimposed on each other, where one of the objects is static (background) and the other one is dynamic. Such problem can arise in imaging through semireflective media, in separation of an illumination image from a reflectance image, in imaging with varying diffraction phenomena, etc. An example of semireflective layers separation is shown in Figure 1.

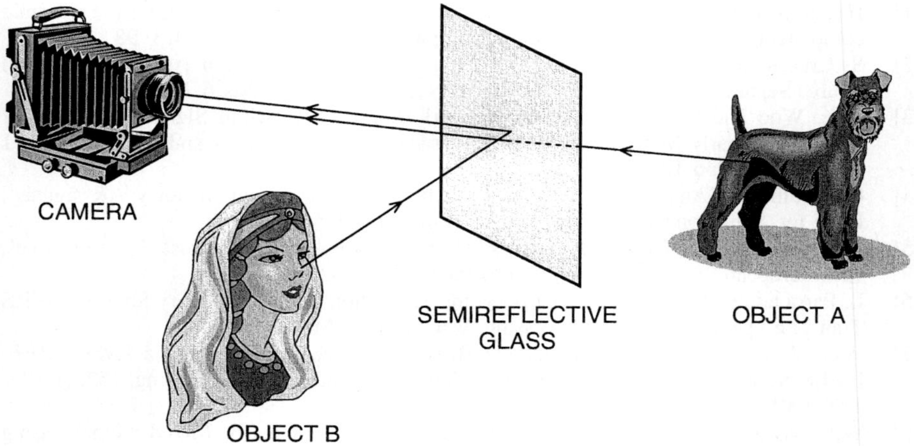

Fig. 1. Scheme of an optical setup involving a semireflector

J. Rosca et al. (Eds.): ICA 2006, LNCS 3889, pp. 934–940, 2006.

In such a setup, the camera observes a superposition of two layers: the real layer is the object (dog), seen through the glass. The virtual layer (girl) is formed by light reflected from the glass.

We assume that one of the objects is dynamic and that several time frames of the observed scene are available. Using this information, it is possible to recover both components by solving an underdetermined linear system. Since the problem is ill-posed, an appropriate regularization must be used.

Weiss [1] suggested a solution to this problem based on the observation that filtered version of an image is usually sparse, when the filter is a differentiation operator. However the suggested solution uses reconstruction in the domain of filtered images, which may suffer from noise amplification. Also explicit use of inverese filters restricts use of other (nonlinear) priors.

In our work we study more general regularizations of this problem, in spirit of Total Variation and general sparsifying transformations.

2 Regularized Separation of Layers

Let us be given a time sequence of T observations of the form

$$y^{(k)} = x^{(0)} + x^{(k)} + \xi^{(k)}, \; k = 1, ..., T, \tag{1}$$

where $x^{(0)}$ is a $M \times N$ image of the static (background) object, $x^{(k)}$ are images of the dynamic object at different times, and $\xi^{(k)}$ is additive noise, which possibly contaminates the observations. An example of such a sequence is given in Figures 3–4. We will henceforth refer to the $T+1$ images $x^{(k)}, k = 0, ..., T$ as to *sources*. The problem of recovering $T+1$ unknown images from only T observed images is ill-posed. However, plausible separation results might be obtained if the solution is restricted to some class of images, to which the sources are believed *a priori* to belong. For simplicity of presentation, we assume that both the static and the dynamic sources obey the same prior (which is in general not necessary).

Assuming that the prior can be expressed via convex penalty function $\varphi(x)$, the separation problem can be formulated as finding such $x^{(k)}$'s that obey (1) up to some allowed discrepancy due to noise and slight deviations from the linear model, and minimize $\sum_k \varphi\left(x^{(k)}\right)$. This leads to the following constrained convex minimization problem:

$$\min_{x^{(0)},...,x^{(T)}} \sum_{k=0}^{T} \varphi\left(x^{(k)}\right) \quad \text{s.t.} \quad \left\| x^{(0)} + x^{(k)} - y^{(k)} \right\|_2^2 \leq \beta \tag{2}$$
$$x^{(k)} \geq 0,$$

where β is usually chosen proportionally to the noise variance and the second constraint guarantees non-negativity of the estimated images. The latter can be reformulated as unconstrained minimization of the convex function

$$f\left(x^{(0)},...,x^{(T)}\right) = \sum_{k=0}^{T} \varphi\left(x^{(k)}\right) + \lambda_1 \sum_{k=1}^{T} \left\| x^{(0)} + x^{(k)} - y^{(k)} \right\|_2^2 + \lambda_2 \sum_{i,j,k} \psi\left(x_{ij}^{(k)}\right), \tag{3}$$

where $\psi(t)$ is a penalty on negativity and λ_1, λ_2 are parameters.

2.1 Generalized Total Variation Regularization

A powerful prior, suitable for wide classes of natural images is obtained when $\varphi(x)$ is chosen to be the *total variation* (TV) norm of the image x

$$\|x\|_{TV} = \sum_{i,j} \|\nabla x_{ij}\|_2 = \sum_{i,j} \sqrt{(\partial_x * x)_{ij}^2 + (\partial_y * x)_{ij}^2}, \tag{4}$$

where ∂_x and ∂_y are discrete derivative kernels in the x- and y-axis directions. TV has been successfully used for regularization in inverse problems, blind deconvolution, denoising, etc [2–5]. A more general form of the prior is the *generalized total variation* (GTV) [6], given by a general bivariate function of the form

$$\varphi(x) = \sum_{i,j} h\left((a * x)_{ij}, (b * x)_{ij}\right), \tag{5}$$

where a, b are some convolution kernels and e.g. $h(u,v) = \sqrt{u^2 + v^2 + \epsilon^2}$. In the latter case, the TV norm is a particular case obtained when $a = \partial_x$ and $b = \partial_y$.

2.2 Gradient and Hessian of $f\left(x^{(0)},...,x^{(T)}\right)$

Assuming the GTV prior (5), the gradient of $f\left(x^{(0)},...,x^{(T)}\right)$ from (3) is given by

$$\frac{\partial f}{\partial x_{ij}^{(m)}} = \sum_{k,l} \left(h_u\left(u_{kl}^{(m)}, v_{kl}^{(m)}\right) a_{k-i,l-j} + h_v\left(u_{kl}^{(m)}, v_{kl}^{(m)}\right) b_{k-i,l-j} \right)$$
$$+ 2\lambda_1 \left(x_{ij}^{(0)} + x_{ij}^{(m)} - y_{ij}^{(m)}\right) + \lambda_2 \psi'\left(x_{ij}^{(m)}\right) \tag{6}$$

for $m > 0$ and

$$\frac{\partial f}{\partial x_{ij}^{(0)}} = \sum_{k,l} \left(h_u\left(u_{kl}^{(0)}, v_{kl}^{(0)}\right) a_{k-i,l-j} + h_v\left(u_{kl}^{(0)}, v_{kl}^{(0)}\right) b_{k-i,l-j} \right)$$
$$+ 2\lambda_1 \sum_{k=1}^{T} \left(x_{ij}^{(0)} + x_{ij}^{(k)} - y_{ij}^{(k)}\right) + \lambda_2 \psi'\left(x_{ij}^{(0)}\right) \tag{7}$$

for $m = 0$, where $u^{(k)} = a * x^{(k)}$ and $v^{(k)} = b * x^{(k)}$. The first term accounting for the prior can be evaluated efficiently using FFT-based convolution. Since the kernels a and b are usually significantly smaller compared to the source images $x^{(k)}$, the use of the overlap-and-add (OLA) method is especially advantageous.

The Hessian of $f\left(x^{(0)},...,x^{(T)}\right)$ is given by

$$\frac{\partial^2 f}{\partial x_{ij}^{(m)} \partial x_{i'j'}^{(m')}} = \sum_{k,l} \left(h_{uu}\left(u_{kl}^{(m)}, v_{kl}^{(m)}\right) a_{k-i,l-j} a_{k-i',l-j'} \right.$$
$$+ h_{vv}\left(u_{kl}^{(m)}, v_{kl}^{(m)}\right) b_{k-i,l-j} b_{k-i',l-j'}$$
$$+ h_{uv}\left(u_{kl}^{(m)}, v_{kl}^{(m)}\right) b_{k-i,l-j} a_{k-i',l-j'}$$
$$\left. + h_{uv}\left(u_{kl}^{(m)}, v_{kl}^{(m)}\right) a_{k-i,l-j} b_{k-i',l-j'} \right) \delta_{mm'}$$
$$+ 2\lambda_1 \gamma_{mm'} \delta_{ii'} \delta_{jj'} + \lambda_2 \psi''\left(x_{ij}^{(m)}\right) \delta_{mm'} \delta_{ii'} \delta_{jj'}, \quad (8)$$

where $\delta_{mm'}$ is the Kroenecker delta and $\gamma_{mm'} = 1$ for $m = m'$ or $m = 0$ or $m' = 0$ and 0 otherwise. In the following, for notation convenience we will denote the gradient by $\nabla_{x^{(0)},...,x^{(T)}} f(x^{(0)},...,x^{(T)})$ and the Hessian by $\nabla^2_{x^{(0)},...,x^{(T)}} f(x^{(0)},...,x^{(T)})$. Parsing the $T+1$ images to MN column vector, we can represent the gradient as a $(T+1)NM$ vector and the Hessian as a $(T+1)NM \times (T+1)NM$ matrix.

The Hessian has a $(T+1) \times (T+1)$ block structure with $MN \times MN$ blocks. The first term in (8) yields a band-diagonal structure of the diagonal blocks, where the number of the diagonals and the number of bands depend on the sizes of the kernels a and b. The second and the third terms account for a constant principal diagonal in the diagonal blocks of the Hessian, whereas the second term also accounts for a constant diagonal in the first row and column blocks. Typical Hessian structure is depicted in Figure 2. The sparse structure of the Hessian is very helpful for solving efficiently the Newton system, while carrying the optimization.

2.3 Sparsity-Based Priors

Another powerful class of priors on images is related to their sparse representation using some system of basis functions, or overcomplete "dictionaries", based on wavelets, curvelets, contourlets, etc. This paradigm is already used in image denoising and in solution of some inverse problems. Assume that the original images can be represented as

$$x^{(k)} = \sum_l c_l^{(k)} \phi_l$$

or in operator form

$$x^{(k)} = \Phi c^{(k)}$$

where the coefficients $c_l^{(k)}$ are sparse. The following regularized problem can be considered:

$$\min_c \sum_{k=0}^{T} \|c^{(k)}\|_1 + \lambda_1 \sum_{k=1}^{T} \left\| \Phi c^{(0)} + \Phi c^{(k)} - y^{(k)} \right\|_2^2 + \lambda_2 \sum_{i,j,k} \psi\left((\Phi c^{(k)})_{ij} \right),$$

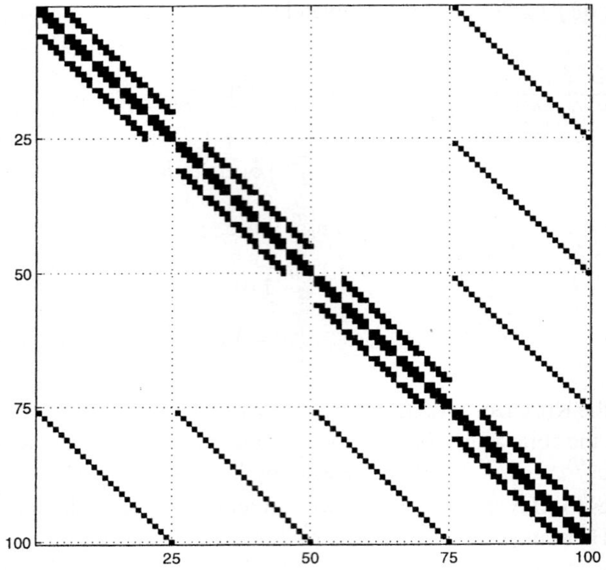

Fig. 2. Example of the Hessian sparse structure in a problem with 3 frames of size 5×5 and $a = \partial_x, b = \partial_y$. Black shows non-zero elements.

2.4 Minimization of $f\left(x^{(0)}, ..., x^{(T)}\right)$

Since the function $f\left(x^{(0)}, ..., x^{(T)}\right)$ is convex with respect to $x^{(0)}, ..., x^{(T)}$, it can be minimized using standard convex optimization technique [7]. In our case, second-order Newton-type methods appear especially appealing. Let us denote by X the $(T+1)NM$ vector of variables $x^{(0)}, ..., x^{(T)}$ in column-stack representation. In the basic Newton method, the minimization of $f(X)$ is carried out by iteratively updating X in the following manner

$$X[k+1] = X[k] + \alpha[k]d[k] \tag{9}$$

where k denotes the iteration index, $\alpha(k)$ is the step size and $d(k)$ is the Newton direction, give by the solution of the *Newton system*:

$$\nabla^2 f(X[k])d[k] = -\nabla f(X[k]). \tag{10}$$

The Newton system, in turn, can be solved iteratively to some preset degree of accuracy, e.g. using the conjugate gradients method [7], which does not require an explicit Hessian computation, but rather computation of Hessian-vector products. Such computations are very efficient due to the sparse structure of the Hessian. This version of the Newton algorithm with approximate solution of the Newton system is often referred to as *inexact* or *truncated* Newton [8] method.

Since the function is convex, global convergence is guaranteed with any initialization. Yet, selecting an initialization which is sufficiently close to the solution, e.g. using the mixture images to initialize $x^{(0)}, ..., x^{(T)}$, faster convergence can be achieved.

3 Computational Results

In this work we present computational experiments with TV prior, leaving sparsity-based priors for future study. The proposed method was tested on synthetic data created by superposition of a static image of a female face and three frames showing a running dog (Figure 3). The observed result is a sequence of 3 frames (Figure 4).

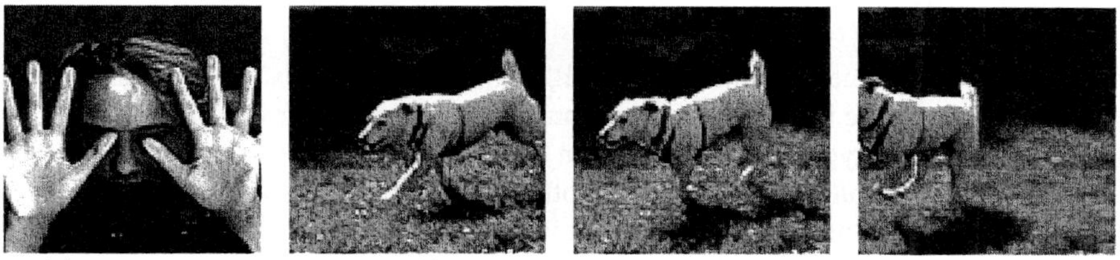

$x^{(0)}$ (background) $x^{(1)}$ $x^{(2)}$ $x^{(3)}$

Fig. 3. Source sequence. First column: static background image, second through fourth columns: three frames of the *dog* sequence.

$y(1)$ $y^{(2)}$ $y^{(3)}$

Fig. 4. Observed mixtures sequence

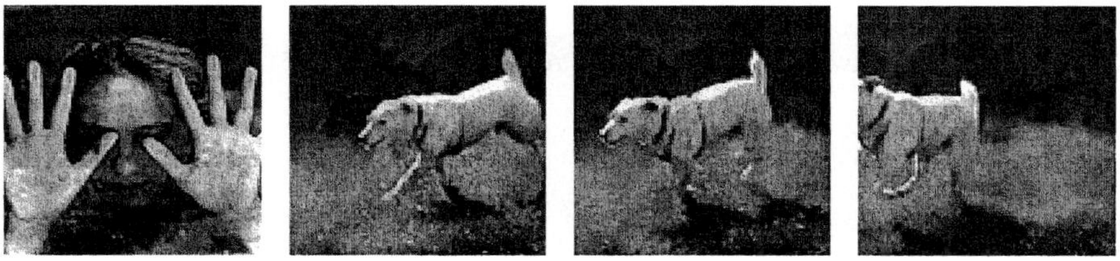

$\tilde{x}^{(0)}$ (background) $\tilde{x}^{(1)}$ $\tilde{x}^{(2)}$ $\tilde{x}^{(3)}$

Fig. 5. Unmixed sequence

The separation was carried out using the TV norm with smoothing parameter $\epsilon = 10^{-3}$ and the non-negativity penalty, with $\lambda_1 = 10^{-1}, \lambda_2 = 10^{-1}$. The mixture images $y^{(1)}, ..., y^{(3)}$ were used as the initialization for $x^{(1)}, ..., x^{(3)}$, and the image $y^{(1)}$ was used as the initialization for $x^{(0)}$. Optimization was carried out using the truncated Newton algorithm. The reconstructed images are shown in Figure 5.

Our algorithm provides a plausible separation of the background and the dynamic scene. Slight residuals of the dynamic scene are visible in the reconstructed image. These artifacts appear in regions where there is little or no motion in the dynamic scene, and thus the separation problem is ill-posed. The use of the TV prior results in a slight degradation of the texture details in the reconstructed dynamic scene.

4 Discussion

We presented an efficient solution of the ill-posed problem arising in separation of semireflective dynamic image from static background using TV prior. Further research should explore other types of priors, e.g. on coefficients of some decomposition (e.g. wavelet-type) of the images. Application to other optical problems should be considered as well. A potentially interesting application is separation of illumination and reflection components in pictures with multiple exposures [9].

References

1. Weiss, Y.: Deriving intrinsic images from image sequences. In: International Conf. Computer Vision (ICCV). (2001)
2. Rudin, L.I., Osher, S., Fatemi, E.: Nonlinear total variation based noise removal algorithms. Physica D **60** (1992) 259268
3. Blomgren, P., Chan, T.F., Mulet, P., Wong, C.: Total variation image restoration: numerical methods and extensons. In: Proc. IEEE ICIP. (1997)
4. Chan, T.F., Mulet, P.: Iterative methods for total variation image restoration. SIAM J. Num. Anal **36** (1999)
5. Chan, T.F.,Wong, C.K.: Total variation blind deconvolution. In: Proc. ONRWorkshop. (1996)
6. Bronstein, A.M., Bronstein, M.M., Zibulevsky, M., Zeevi, Y.Y.: Blind deconvolution of images using optimal sparse representations. IEEE Trans. Image Processing **14** (2005) 726736
7. Bertsekas, D.P.: Nonlinear Programming (2nd edition). Athena Scientific (1999)
8. Nash, S.: A survey of truncated-Newton methods. Journal of Computational and Applied Mathematics **124** (2000) 4559
9. Elad, M.: Retinex by two bilateral filters. In: Proc. Scale-Space. (2005)

Facial Expression Recognition by ICA with Selective Prior

Fan Chen and Kazunori Kotani

Department of Information Processing, School of Information Science,
Japan Advanced Institute of Science and Technology, Ishikawa-ken 923-1211, Japan
{chen-fan, ikko}@jaist.ac.jp

Abstract. Permutation ambiguity of the classical ICA may cause problems in feature extraction for pattern classification. To solve that, we include a selective prior for de-mixing coefficients into the classical ICA. Since the prior is constructed upon the classification information from the training data, we refer to the proposed ICA model with a selective prior as a supervised ICA. We formulate the learning rule for the supervised ICA by taking a form of the natural gradient approach, and then investigate the performance of the supervised ICA in facial expression recognition from the aspects of both the correct rate of recognition and the robustness to the number of independent components.

1 Introduction

In facial expression recognition, some facts suggest that ICA might be more effective than PCA in feature extraction. Facial expression consists of those features standing for minor, non-rigid, local variations of faces, which are usually less significant in PCA bases than those for lighting, head pose, and personal difference.[1] Further, the phase spectrum, related to higher-order statistics, contains more structural information in images that drives human perception than the power spectrum. [2] The importance of higher-order statistics in natural images to the response properties of cortical cells has been explored in Refs. [3] [4] [5], and the extraction of higher-order statistics by means of ICA was discussed in Ref.[6]. ICA has been applied to the face recognition in Ref.[2] and to the facial expression analysis in Ref.[7], where the efficiency of ICA was verified.

In the classical ICA, the derived independent components are fully exchangeable in order, i.e., permutation ambiguity, where the original order provides no information on the significance of components in discrimination. A feature selection is necessary to be performed along with the feature extraction. The selection can be applied before, after or during ICA. In Ref.[2], Best Individual Feature (BIF) selection was adopted where features were chosen according to some defined criteria individually. Methods by means of Sequential Forward Selection (SFS) and Sequential Floating Forward Selection (SFFS) were also proposed. [8] Since the selection is performed after ICA, the features are limited to be chosen from the set of the obtained independent components. To create a candidate set with enough representative features in discrimination, a large number

J. Rosca et al. (Eds.): ICA 2006, LNCS 3889, pp. 941–949, 2006.

of independent components should be learned, which may be computationally expensive. It is meaningful to search for a way to affect the selection of features before or during ICA. GEMC [9] makes a selection before ICA by heuristically replacing PCA with a discriminant analysis as the pre-processing to ICA, which still lacks a mathematical explanation. ICA in a local facial residue space is also proposed for face recognition, which can be regarded as using the pre-specified residue space to limit the selection of independent components before applying ICA. [10]

In the present paper, we consider an approach to implement the feature selection during the learning of independent components. A constraint ICA has been proposed for the analysis of EEG signals, where all components should be sparse and close to a supplied reference signal by including a correlation term. [11] In our case, we try to design a method to let those components with higher degree of class separation emerge easier than others. The classical ICA in Ref.[12] was shown to be derivable under the scheme of Maximum Log-Likelihood (MLL) estimation. [13] Instead of using the uniform prior for de-mixing coefficients in MLL, we take the Maximum *a* Posteriori (MAP) estimation. A prior defined on the degree of class separation is introduced on the de-mixing coefficients, which in turn increases the probability of the corresponding independent component to be significant in classification.

In Section 2, we will formulate the supervised ICA and give the algorithm for facial expression recognition. In Section 3, numerical experiments are made and the performance of our proposed algorithm is investigated by making comparison with the classical ICA. We also discuss on the influence of the introduced selective prior. Finally, we summarize the present paper and explain our future work.

2 Supervised Independent Component Analysis

We first formulate the supervised ICA. Let $Y = [y^{(ki)}|k \in \{1, \cdots, K\}, i \in \{1, \cdots, N_k\}]$ be the matrix of N observed samples from K classes with N_k samples in the k-th class and satisfy $N = \sum_{k=1}^{K} N_k$. The i-th sample of class k, $y^{(ki)} = [y_1^{(ki)}, \cdots, y_D^{(ki)}]^T$, is a D-dimensional vector. Provided Y as the training data set, the classical ICA assumes that these samples are generated from Q statistically independent sources. $S = [s^{(ki)}|k \in \{1, \cdots, K\}, i \in \{1, \cdots, N_k\}]$ represents the signals generated by those sources, where $s^{(ki)} = [s_1^{(ki)}, \cdots, s_Q^{(ki)}]^T$ corresponds to $y^{(ki)}$. Those signals from different sources are linearly mixed, i.e., $Y = VS$, where the D-row Q-column matrix V is for the mixing coefficients. The purpose of ICA is to search for the coefficients V that makes the sources as statistically independent as possible. If we let $W = V^{-1}$ be the inverse (or pseudo-inverse) of V, W is the de-mixing matrix and satisfies $S = WY$. For any sample y, the extracted feature in ICA will be $s = Wy$. Note that we consider the noiseless case of ICA in the present paper.

The learning rule proposed by Bell and Sejnowski[12] could also be derived by maximizing the log-likelihood criterion, i.e., $V_{\text{ICA}} = \arg\max_V \log P(Y|V)$.[13] Motivated by the reasons described in the introduction, we search for a way to make a selection of features during ICA so that those independent components with higher degree of class separation are easier to emerge than others, which is achieved by introducing a prior distribution for the coefficients. We derive the learning rule of V by maximizing the following criterion,

$$V_{\text{sICA}} = \arg\max_V \log P(V|Y) = \arg\max_V [\log P(Y|V) + \log P(V)]. \quad (1)$$

As in the classical ICA, $\log P(Y|V)$ is derived as [12]

$$\begin{aligned} \log P(Y|V) &= \log \int P(Y|V,S)P(S)dS \\ &= \log \int \prod_d \prod_k \prod_i \delta\{y_d^{(ki)} - \sum_q (V_{dq} s_q^{(ki)})\} \prod_k \prod_i \prod_q P_q\{s_q^{(ki)}\} dS \\ &= -N \log|V| + \sum_k \sum_i \sum_q \log P_q\{\sum_d [V^{-1}]_{qd} y_d^{(ki)}\} \end{aligned} \quad (2)$$

with $\delta\{x\}$ being the Dirac delta function. Without special explanations, k varies from 1 to K while i varies from 1 to N_k for the suffixes of summation here and hereafter. We define the prior as follows:

$$P(V) = P(W) = \prod_q P_w(w_q), \quad P_w(w) = \frac{1}{Z_w} \exp\{\lambda w [M_{bc}(Y) - M_{wc}(Y)] w^T\}, (3)$$

where $w_q = [w_{q1}, \cdots, w_{qD}], W = [w_1^T, \cdots, w_Q^T]^T$. Z_w is the partition function while $M_{bc}(Y)$ and $M_{wc}(Y)$ are the between-class scatter matrix and within-class scatter matrix, defined by Eq.(4).

$$M_{bc}(Y) = \frac{1}{N}\sum_k N_k ||\overline{y}^{(k)} - \overline{y}||^2, \qquad M_{wc}(Y) = \frac{1}{N}\sum_k \sum_i ||y^{(ki)} - \overline{y}^{(k)}||^2. \quad (4)$$

We define $M_s(Y) = M_{bc}(Y) - M_{wc}(Y)$ for short. $\overline{y}^{(k)}$ represents the mean vector for samples in class k and $\overline{y}$ is the mean value for all samples. λ is a hyper-parameter introduced to control the influence of the prior. For $\lambda > 0$, an independent component whose de-mixing coefficients are of larger degree of class separation will have a higher prior probability. We maximize the MAP criterion

$$\begin{aligned} &\log P(V|Y) = \log P(W|Y) \\ &= N \log|W| + \sum_k \sum_i \sum_q \log P_q\{\sum_d w_{qd} y_d^{(ki)}\} + \lambda \sum_q w_q M_s(Y) w_q^T + Const \end{aligned} \quad (5)$$

under the constraints of $||w_q|| = 1$ for all $q \in \{1, \cdots, Q\}$, by differentiating the criterion with respect to w_{qd} according to the following rule, i.e.,

$$\frac{\partial}{\partial w_{qd}} \log|W| = [W^{-1}]_{dq} = V_{dq}. \quad (6)$$

The differential reads

$$\frac{\partial \log P(Y,W)}{\partial w_{qd}} = NV_{dq} + \sum_k \sum_i \frac{P_q'(\sum_d w_{qd} y_d^{(ki)})}{P_q(\sum_d w_{qd} y_d^{(ki)})} y_d^{(ki)} + 2\lambda \sum_l w_{ql} [M_s(Y)]_{ld}. \quad (7)$$

We take $P_q(x) \propto 1/\cosh(x)$ and rewrite these differential equations for all w_{qd} into a compact form as one matrix differential which is defined as component-wise differentiation, i.e.,

$$\frac{\partial \log P(Y,W)}{\partial W} = N\{V^T - \frac{1}{N}\tanh[S]Y^T + \frac{2\lambda}{N} W M_s(Y)\}, \quad (8)$$

where $\tanh[S]$ means the calculation of tanh over all elements in matrix S. Due to the existence of inverse matrix V which is computationally expensive in the iterative learning, we adopt the natural gradient approach, proposed by Amari [14], to derive the learning rule:

$$\Delta W = \eta \left\{ \frac{\partial \log P(Y,W)}{\partial W} \right\} W^T W = N\eta\{I - \frac{1}{N}\tanh[S]S^T + \frac{2\lambda}{N} W M_s(Y) W^T\}W, (9)$$

where η is the learning rate. Comparing with the classical ICA, our supervised ICA holds a prior term for de-mixing coefficients.

When applied to facial expression recognition, the supervised ICA is performed on the PCA coefficients instead of directly on the image data X, i.e., $Y = W_{\mathrm{PCA}} X$. W_{PCA} is the matrix of PCA eigenvectors. Since all eigen-vectors that correspond to nonzero eigen-values in PCA are adopted, there is no information lost during this preprocessing. The updating rule is finally derived as follows:

$$W^{(t+1)} = W^{(t)} + N\eta\{I - \frac{1}{N}\tanh[S^{(t)}][S^{(t)}]^T + \frac{2\lambda}{N} W^{(t)} M_s(Y)[W^{(t)}]^T\}W^{(t)}. \quad (10)$$

The algorithm for the supervised ICA is summarized in Table 1 and the final bases for extracting features are computed as $W_{\mathrm{F}} = W W_{\mathrm{PCA}}$. Instead of using a Lagrange multiplier, we simply implement the constraint $||w_q|| = 1$ in Step (b) of Table 1. Exactly, the scale of w_q should not affect the sparseness of derived components in the classical ICA, i.e., scale ambiguity. As a fix-point learning algorithm, the behavior of convergence is still not fully predictable. The introduction of Step (b) requires a different learning rate and a different convergence threshold. Therefore, it is difficult to make a precise analysis on the influence of Step (b). From numerical experiments, whose data are not given in the present paper, we have found no significant differences in the recognition rate caused by applying Step (b) to the classical ICA. For the supervised ICA, the constraint $||w_q|| = 1$ is required to stabilize the influence of the prior term, which also helps improve the convergence behavior of the algorithm. Let $\widehat{X} = [x_n | n \in \{1, \cdots, \widehat{N}\}]$ be the matrix by putting all testing images into different columns and $\widehat{N}$ be the number of samples in the testing set. We define

Table 1. The learning algorithm of the supervised ICA

a) Initialize W and calculate $M_{bc}(Y) - M_{wc}(Y)$;
b) Normalize W by rows so that $\|\|w_q\|\| = 1$;
c) Calculate S from $S = WY$;
d) Calculate ΔW;
e) Update W by $W \leftarrow W + \Delta W$;
f) Calculate $\log P(Y, W)$. If the difference between two iterations is less than a threshold, exit. If not, repeat (b) to (f).

$Z = \{z_n \in \{1, \cdots, K\} | n \in \{1, \cdots, \widehat{N}\}\}$ as the true classified labels for observed data, and define a recognition rate as

$$r_c = \frac{1}{\widehat{N}} \sum_{n=1}^{\widehat{N}} \delta(z_n, z_n^*), \tag{11}$$

where $\delta(x, y)$ is the Kronecker delta. z_n^* is the estimated label value which is estimated according to the following criterion

$$z_n^* = \arg\min_k ||s_n - W\overline{y}^{(k)}||^2 \tag{12}$$

and $s_n = W_{\mathrm{F}} x_n$. A block diagram for the whole process is given in Fig. 1.

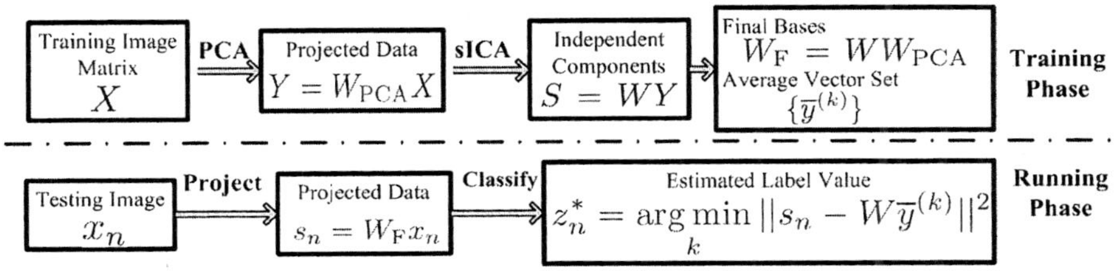

Fig. 1. A block diagram for the processing flows in both the learning phase and the running phase of facial expression recognition. All input image data will be normalized in face position and histogram-equalized as pre-processings.

3 Experiments and Discussions

In the following numerical experiments, we will focus on the comparison between the supervised ICA and the classical ICA under same conditions to investigate the effect by introducing the prior term and by changing the hyperparameter λ. We use the Japanese Female Facial Expression (JAFFE) Database [15], which includes 213 images in total. These images are aligned in face position and histogram-equalized. Some samples are given in Fig. 2. We pick up 76 images to form the training set, where at least one image from each character is

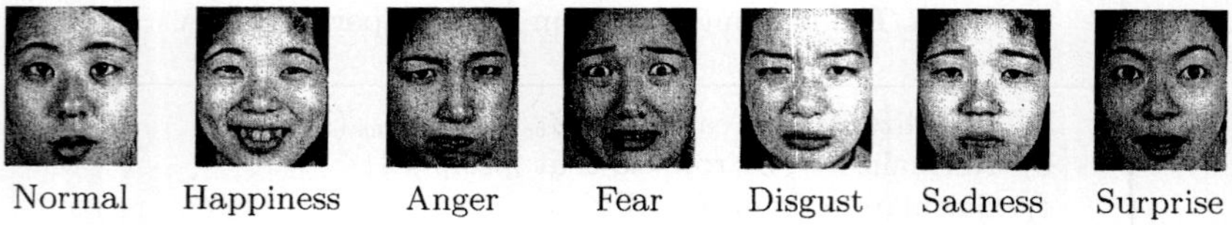

Fig. 2. Some samples in our numerical experiments from the JAFFE database

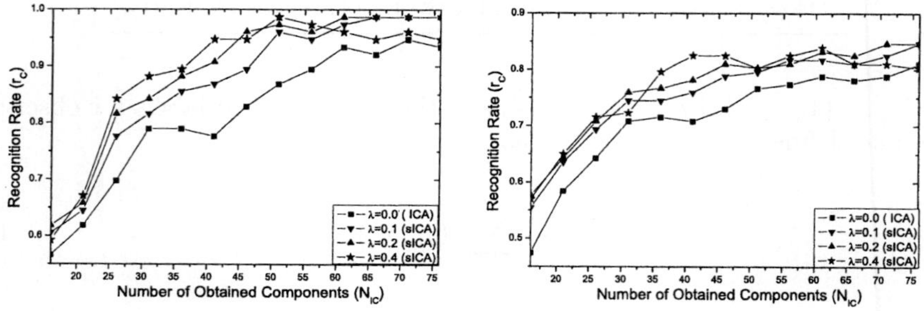

(a) (b)

Fig. 3. Recognition rate r_c is plotted as a function of the number of independent components N_{IC} under different λ for (a) the training set of samples and (b) the testing set. The supervised ICA (sICA) outperforms the classical ICA, especially for a median N_{IC}.

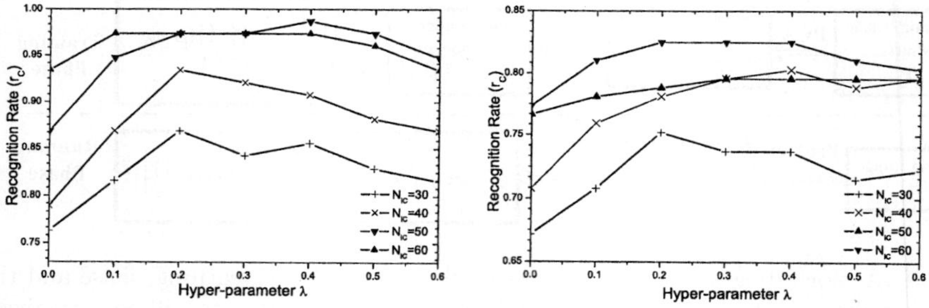

(a) (b)

Fig. 4. Recognition rate r_c is plotted as a function of λ at different N_{IC} for (a) the training set of samples and (b) the testing set. The learning rate η is set to be 0.00001 for all the cases. A properly selected λ helps improve the performance.

included. The obtained bases are tested both on the training set and a testing set consisting of the remaining 137 images. All images are resized to 32×40 pixels. Thus the dimension of x_n is 1280 and the dimension of PCA-projected data Y is $D = 75$.

We first plot the recognition rate r_c as a function of the number of independent components (N_{IC}), which is equal to Q in Section 2, for the training set and

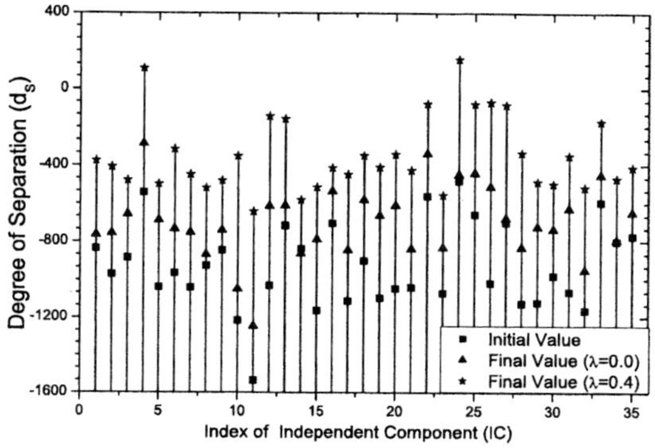

Fig. 5. Degree of class separation $d_s(w)$ is plotted for the case of $N_{\mathrm{IC}} = 35$. Each vertical bar stands for one independent component. Starting from the same initial values denoted by the rectangles, the final values of $d_s(w)$ for both the cases $\lambda = 0.0$ and $\lambda = 0.4$ are given by the triangles and the stars, respectively.

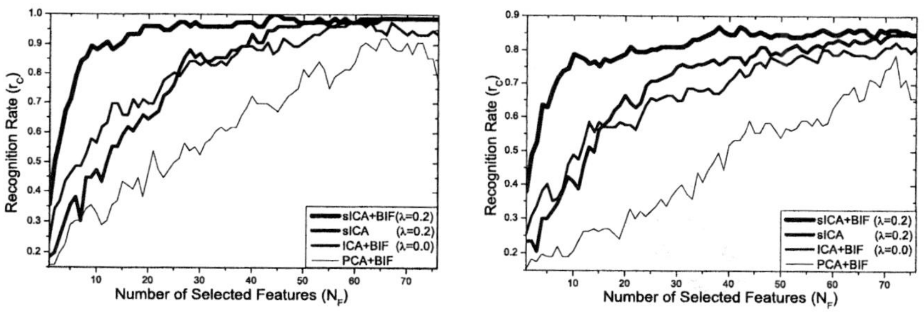

(a) (b)

Fig. 6. Recognition rate r_c is plotted as a function of N_{F}, which is the number of features we select, for sICA with BIF, sICA without BIF, the classical ICA with BIF and PCA with BIF. (a) Result for the training set. (b) Result for the testing set. We can find that since the supervised ICA provides a better set of candidate features, the supervised ICA with BIF selection has the best performance.

the testing set under different λ values in Fig. 3 (a) and (b), respectively. The learning rate η is set to be 0.00001 for all cases. We find that higher recognition rates have been achieved by including the selective prior for almost all N_{IC} values, which suggest that a better set of candidate features can be found by the supervised ICA. In Fig. 4 (a) and (b), we further investigate the transition of recognition rate as a function of λ at different N_{IC}. From both the cases of training set and testing set, the recognition rates first ascend with the increase of λ and then descend when λ gets too large and causes a heavy bias on the sparseness of the obtained independent components. A tradeoff between the sparseness and the discrimination degree should be taken to achieve the best results. We define the degree of class separation $d_s(w) = wM_s(Y)w^T$, plotted in Fig. 5 for the case of $N_{\rm IC} = 35$. Each vertical bar represents one independent component. Starting from the same initial values denoted by the rectangles, the final values of $d_s(w)$ for both the cases $\lambda = 0.0$ and $\lambda = 0.4$ are given by the triangles and the stars, respectively. For all components, the ratio of class separation increases with a large λ, which proves the effect of the prior term. As a result, those components with higher degree of class separation improve the recognition rate. For training, sICA is several times slower than ICA. The speed is same in the testing phase.

Although several methods of feature selection after the learning of ICA were proposed in Ref.[8], they are also applicable to our approach. The supervised ICA intends to search for a candidate set of features with higher degree of class separation than the classical ICA, by selecting features from which even better recognition rate can be achieved. In Fig. 6(a) and (b), we make a comparison between four methods, i.e., the supervised ICA with Best Individual Feature(BIF) selection, the supervised ICA without BIF, the classical ICA with BIF, and PCA with BIF. In the BIF selection of the present paper, all features are sorted in descendant order of their degree of class separation $d_s(w)$, and then the first $N_{\rm F}$ features are selected for classification. We note that the supervised ICA with BIF gives the best performance, which verifies the capacity of the supervised ICA in learning a better candidate set of features. We also find that the supervised ICA without BIF still outperforms the classical ICA with BIF, which confirms the robustness of the supervised ICA in recognition rate by learning those independent components with higher degree of class separation from samples when a median $N_{\rm IC}$ is used. Although BIF improves the robustness of the performance over the whole range of $N_{\rm F}$, the best recognition rate does not change much only by means of BIF selection for the same learning algorithm, as depicted in Figs. 6 (a) and (b). This result suggests that learning a candidate set of features with higher degree of class separation might be more important than performing a post selection, which is the point where the supervised ICA outperforms the classical ICA.

4 Conclusion and Future Work

In the present paper, we have proposed a supervised ICA for facial expression recognition by performing the feature selection along with the learning of ICA. A selective prior has been introduced to the classical ICA and the MAP estimation is applied to derive the learning rule. We made numerical experiments to investigate the influence of new prior term and make comparison with the classical ICA. Our method shows better performance than the classical ICA, especially in increasing the recognition rate under a median number of independent components. There are still some problems left for us to study, such as the decision of optimal λ and the design of a better learning algorithm for faster and more robust convergence. Investigation on various priors is also a part of our future work.

References

1. Nastar, C.: Face recognition using deformable matching. Face Recognition: from Theory to Applications(Wechsler, H., et al., Eds.), Springer-Verlag, New York, (1998) 206–229
2. Bartlett, M. S., Movellan, J. R., and Sejnowski, T. J.: Face recognition by independent component analysis. IEEE Trans. on Neural Networks **13** (2002) 1450–1464
3. Field, D. J.: What is the goal of sensory coding? Neural Computation **5** (1994) 559–601
4. Olshausen, B. A., and Field, D. J.: Sparse coding with an overcomplete basis set: a strategy employed by V1? Vision Research **37** (1997) 3311–3325
5. Simoncelli, E., and Olshausen, B.: Natural image statistics and neural representation. Annual Review of Neuroscience **24** (2001) 1193–1216
6. Karklin, Y., and Lewicki, M. S.: Learning higher-order structures in natural images. Network: Comput. Neural Syst. **14** (2003) 483–499
7. Bartlett, M. S., Donato, G. L., Movellan, J. R., Hager, J. C., Ekman, P., and Sejnowski, T. J.: Image representations for facial expression coding. Advances in Neural Information Processing Systems **12** (2000) 886–892
8. Ekenel, H. K., Sankur, B.: Feature selection in the independent component subspace for face recognition. Pattern Recognition Letters **25** (2004) 1377–1388
9. Eguchi, I., and Kotani, K.: Facial expression analysis by generalized eigen-space method based on class-features (GEMC). Proc. 2005 IEEE Int'l Conf. on Image Processing **1** (2005) MonAmPO3–6
10. Kim, T. K., Kim, H., Hwang, W., and Kittler, J.: Independent component analysis in a local facial residue space for face recognition. Pattern Recognition **37** (2004) 1873–1885
11. James, C. J., and Gibson, O.: Electromagnetic brain signal analysis using constrained ICA. Proc. 2nd European Medical and Biological Engineering Conference **I** (2002) 426–427
12. Bell, A. J., and Sejnowski, T. J.: An information-maximization approach to blind separation and blind deconvolution. Neural Computation **7** (1995) 1129–1159
13. MacKay, D. J. C.: Maximum likelihood and covariant algorithms for independent component analysis. Technical report, University of Cambridge (1996)
14. Amari, S.: Natural gradient works efficiently in learning. Neural Computation **10** (1998) 251–276
15. Lyons, M. J., Akamatsu, S., Kamachi, M., and Gyoba, J.: Coding facial expressions with Gabor wavelets. Proc. 3rd IEEE Int'l Conf. on Automatic Face and Gesture Recognition **1** (1998) 200–205

An Easily Computable Eight Times Overcomplete ICA Method for Image Data

Mika Inki

Neural Networks Research Centre,
Helsinki University of Technology,
P.O. Box 5400, FI-02015 HUT, Finland

Abstract. Here we present a procedure for finding an eight times overcomplete ICA description of image data using the symmetries defined by the rigid motions of a square. The procedure for estimating the basis requires only a small change to any classic ICA procedure and the data representation in this overcomplete description is unique. Coding and decoding in this description are essentially as easy as in classic ICA methods. We also show that this description is genuinely more sparse than a non-overcomplete ICA method.

1 Introduction

Independent component analysis (ICA) has been used successfully in analyzing image data, especially due to its efficient coding properties, and its links to cortical simple cell receptive fields [1,12]. In ICA the observed data is expressed as a linear transformation of latent variables that are nongaussian and independent. We can express the model as

$$\mathbf{x} = \mathbf{A}\mathbf{s} = \sum_i \mathbf{a}_i s_i, \qquad (1)$$

where $\mathbf{x} = (x_1, x_2, \ldots, x_m)$ is the vector of observed random variables, $\mathbf{s} = (s_1, s_2, \ldots, s_n)$ is the vector of latent variables called the independent components or source signals, and $\mathbf{A}$ is an unknown constant matrix, called the mixing matrix. The columns of $\mathbf{A}$ are often called features or basis vectors. A matrix that separates the original signals (or their estimates) from $\mathbf{x}$ is called a separating matrix $\mathbf{W}$, and its rows are the separation vectors $\mathbf{w}_i$. Exact conditions for the identifiability of the model were given in [4], and several methods for estimation of the classic ICA model have been proposed in the literature, see [7] for a review.

Classical methods for ICA use complete bases, i.e. $n = m$, but with image data an overcomplete description would be more appropriate. For example, by starting almost any iterative ICA algorithm from different starting points, one will obtain different bases, whose components are similar, but not identical (even after indeterminacies relating to sign, permutation and scaling are taken into account). From a different perspective, there is little reason to assume that a

J. Rosca et al. (Eds.): ICA 2006, LNCS 3889, pp. 950–957, 2006.

specific Gabor-type feature would be present in a basis, and not some spatial scaling, translation, and rotation of it.

However, an overcomplete ICA description of a data set is generally hard to estimate. The source signals are no longer recoverable, and there are thus several possible source signals for a given basis and data set [7]. Therefore appropriate additional criteria have to be used to obtain a unique description. For example, one can demand maximal sparseness of the sources [9], which (if the criterion for sparsity is chosen conveniently) may also approximately maximize the independence of the sources. Such maximally sparse descriptions can also be used to find minimal codes for the data. Nevertheless, finding such a maximally sparse overcomplete description often requires estimating the basis and source signals alternately at each iteration, which is quite demanding computationally. It has been argued that finding sparse codes is easier than sparse coding [11], and there are methods for finding overcomplete bases without estimating the sources, see e.g. [6]. However, a maximally sparse description in the (already estimated) overcomplete basis still has to be found when coding (i.e. finding the sources of) a data vector.

With image data, there are many symmetries that could basically be used to construct (or find) an overcomplete basis. One could, for example, build a basis by taking a (mother) feature and rotating, scaling, and translating it to form a basis, cf. [10]. However, the problem of finding the maximally sparse description in such a basis remains. Classic (n=m) ICA methods have the advantage that the features are orthogonal (in the whitened space) and finding the sources requires only matrix multiplication.

One could build an overcomplete basis from any classic ICA basis by using the rigid motions of a square, i.e. rotations in 90 degree increments along with their mirror images, all of which correspond to pixel-wise permutations of the basis/separation vectors (alternatively they can be understood as permutations of the data vectors). One can then choose for each data vector the permutation of basis/separation vectors that gives the sparsest sources. These eight pixel permutations are the only ones where the geometry of an image patch / basis vector is unaffected. In fact, due to the approximate rotational and mirror symmetries in images, all the permuted bases should be equally good on average. This kind of a description can be seen as an eight times overcomplete ICA method with constraints on which features can be active at the same time. Although an unconstrained description should be sparser, it should be noted that to indicate which features are used requires three bits per data vector (about 0.02 bits per pixel (bpp) for 12 by 12 windows) in this description, whereas indicating the active features for a general 8 times overcomplete basis requires 8 bits per pixel (or more than 4.2 bpp for 12 by 12 windows if a critically sampled (complete) basis is always used). Of course, the selection of the active features is also difficult in the unconstrained case, whereas in this constrained case we can test all eight possibilities.

However, this kind of a construction would include overcompleteness only as an afterthought, and wouldn't be optimal. Here we present a procedure where the rigid motions of a square are used even in estimating the basis. Although we use a complete basis for each data vector, the increase in sparsity for a minimal increase in coding cost implies that the components could be used in methods where only the most active features are retained for each data vector, in order to achieve efficient coding, cf. [2], or noise removal, cf. [5]. We will compare our method to normal ICA, and to the case when the data vectors are permuted, but not the basis.

2 Procedure

Our procedure in this paper uses FastICA, although most any ICA algorithm could be used. Starting from a random point, we (step 1) iterate FastICA with the hyperbolic tangent nonlinearity for one or more steps, increasing the sparseness of the source estimates by adjusting the basis (and separation) vectors. After this, we (step 2) estimate for each data sample which pixel permutation (rigid motion of the associated array) results in the maximally sparse description, and the vector is then permuted accordingly. In case of ties, the permutation can be chosen arbitrarily. To maximize sparsity, we chose to minimize (for each t):

$$\sum_i |y_i(t)| = \sum_i |\mathbf{w}_i \mathbf{x}(t)|. \tag{2}$$

Note that the minimization was performed for each t by calculating this value for all permutations, and picking the minimizing one. Steps 1 and 2 are repeated until no data vector changes its preferred permutation and FastICA has converged. Note that coding a data vector in this adjusted description requires only permuting the data vector by the permutation that minimizes Equation (2), and multiplying by the separating matrix.

As a technical side note, the sparsity of each sample is maximized, or in our case Equation (2) is minimized, always in a rotated version of the original whitened space. Otherwise, the basis will increasingly start to favor a certain orientation, and instead of increasing sparsity, minimizing Equation (2) divides the energy between the permutations in a less than optimal way. Therefore, the separating matrix was in all cases constrained so that it gave unit covariance sources for the original data. This was achieved using Theorem 1 in the Appendix, where in our case $\mathbf{C}_1 = \mathbf{C}$ (covariance of the original data), and additionally we know that $\mathbf{W}^T\mathbf{C}_2\mathbf{W} = \mathbf{I}$ ($\mathbf{C}_2$ is the covariance of the permuted data), so Equation (4) gives the optimal separation matrix

$$\mathbf{B}' = (\mathbf{WCW}^T)^{\frac{1}{2}}(\mathbf{CW}^T)^+, \tag{3}$$

where $^+$ denotes a pseudoinverse (in this case $\mathbf{CW}^T(\mathbf{WC}^2\mathbf{W}^T)^{-1}$) and $\mathbf{W}$ is the separating matrix given by FastICA for the permuted data. The proof presented in the Appendix expands upon the proof presented in [8].

3 Results

As data, we used 80000 12 by 12 windows sampled from natural images. The original images can be found at: **http://www.cis.hut.fi/inki/images/**. These windows (arrays) were stacked as one dimensional vectors with 144 elements. The mean values were removed from each vector. The permutations were done by transforming each data vector back to a two-dimensional matrix, which was then rotated and transposed the appropriate number of times, and finally transformed back to a vector.

In Figure 1, we have the bases associated with our experiment. As one can see, the difference between the ordinary ICA basis and the basis in our adjusted description is not visually large, but it is significant when the statistics of the associated components are compared.

In Table 1, we have the average and maximal normalized kurtosis of the separated components for a normal ICA description of the data, as well as for our adjusted description. As is readily apparent, our adjusted description produces components that are much more kurtotic, whether one compares average or maximal kurtosis. For comparison, we also calculated what the kurtosis of the new basis is when the data isn't permuted pixel-wise, and found that in such a case, the kurtosis is on average marginally lower than in the standard ICA description, as could be expected. Furthermore, we calculated the kurtosis of the components, when the basis was fixed to be the original ICA basis, but each data sample was permuted according to procedure mentioned earlier. Although the components had considerably higher kurtosis in this case (compared to the original ICA description), the description was significantly less sparse than the completely ('doubly') adjusted one.

We have also tabulated in Table 1 another measure of average sparsity, i.e. the mean of $-E\{|y_i|\}$ over all i. Note that this is related to the sparsity measure in Equation (2). Again we see that our 'doubly' adjusted description produces the sparsest components. Note also that if the components had Laplacian distributions, this value would be proportional to the negative of average marginal entropy. As the distribution is somewhat close to Laplacian, this can be seen to give a rough estimate of (the negative of) entropy. The ordering of the sparsities of the descriptions are the same in this comparison. Note that the values for the 'doubly' adjusted description are very similar (only marginally lower), if the search for the basis is started from a normal ICA solution, cf. Figure 1.

Table 1. Sparsities of the different descriptions used in this article. Boldface is used for the best (largest) values.

	Normal basis, normal data	Normal basis, Adjusted data	Adjusted basis, adjusted data	Adjusted basis, normal data
Average kurtosis	5.79	7.12	**7.94**	5.63
Maximal kurtosis	7.56	11.47	**14.83**	8.80
Average sparsity	-0.6464	-0.6268	**-0.6235**	-0.6501

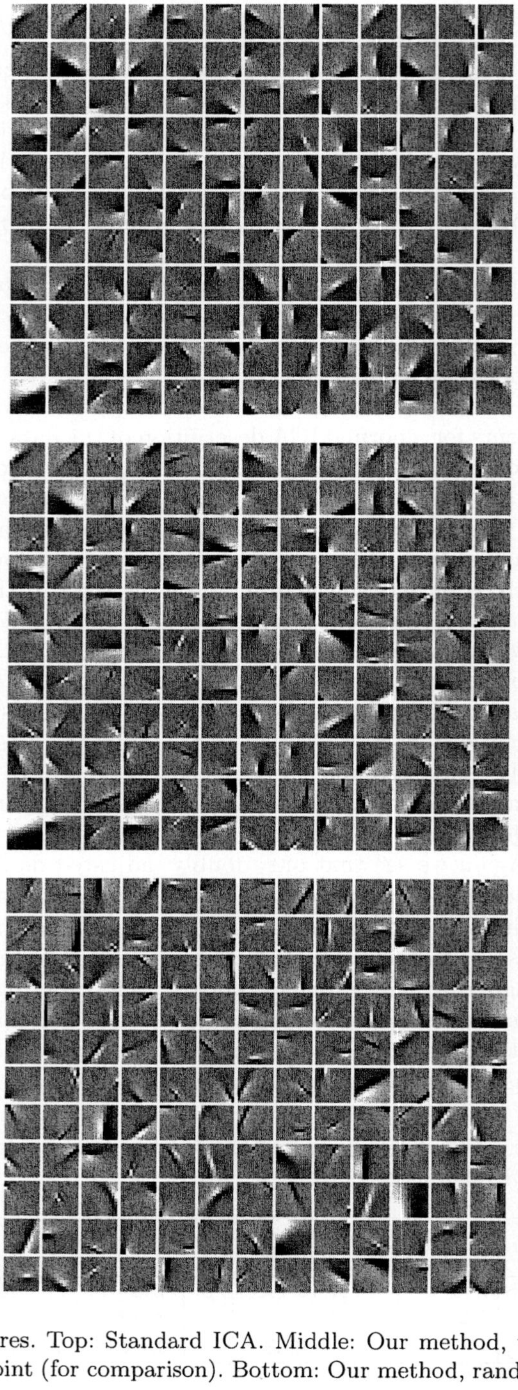

Fig. 1. Basis features. Top: Standard ICA. Middle: Our method, using the basis on top as a starting point (for comparison). Bottom: Our method, random starting point.

In Figure 2 we have the plotted the kurtosis of the components for each of the choices in Table 1. As one can see, the ordering that was apparent from Table 1 is maintained throughout all percentiles, and our description where both the data and the basis are adjusted is the best wrt. sparseness.

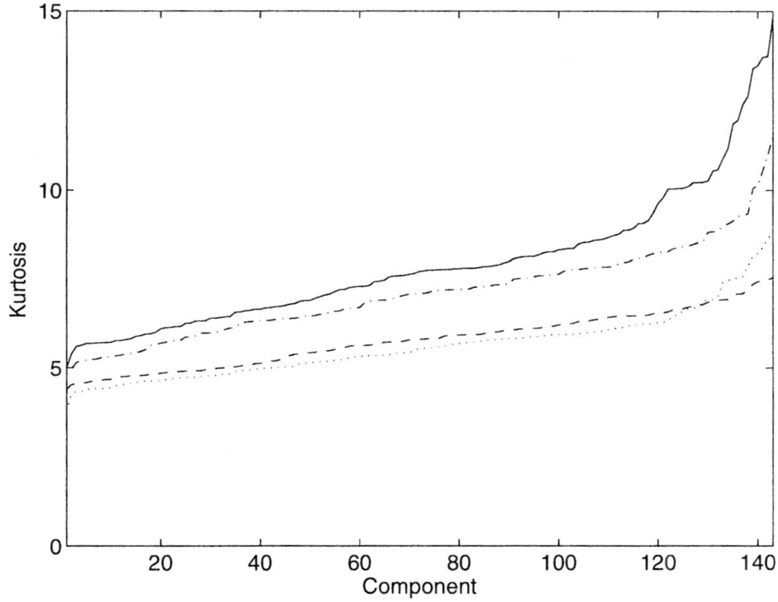

Fig. 2. Kurtosis of the components for different descriptions. Components arranged by ascending normalized kurtosis. Dashed line: Standard ICA (FastICA). Solid line: Our adjusted description. Dash-dotted: Standard ICA basis with adjusted data. Dotted line: Adjusted basis with nonadjusted data. Note that the average kurtosis in a random direction in the original data is about 2.3.

4 Conclusions

We presented here a simple method for finding an eight times overcomplete ICA description of image data. The method was based on maximizing the sparsity of a basis while also finding for each data vector the rigid motion (of the associated array) that maximizes sparsity. The description is in many ways as easy to find as a complete ICA description and, as only a single complete basis is active for any data vector, finding codes for new image patches in this description is easy. We showed that this description is genuinely sparser than a complete ICA description. This increase in sparseness should be contrasted with the minimal three bit increase in data vector coding cost over classical ICA (0.02 bits per pixel for 12 by 12 image windows).

References

1. A.J. Bell and T.J. Sejnowski. The 'independent components' of natural scenes are edge filters. *Vision Research*, 37:3327–3338, 1997.
2. R.W. Buccigrossi and E.P. Simoncelli. Image compression via joint statistical characterization in the wavelet domain. *IEEE Transactions on Image Processing*, 8(12):1688–1701, 1999.
3. B. C. Carlson and J. M. Keller. Orthogonalization procedures and the localization of Wannier functions. *Physical Review*, 105(1):102–103, 1957.
4. P. Comon. Independent component analysis—a new concept? *Signal Processing*, 36:287–314, 1994.
5. A. Hyvärinen, P. O. Hoyer, and E. Oja. Image denoising by sparse code shrinkage. In S. Haykin and B. Kosko, editors, *Intelligent Signal Processing*. IEEE Press, 2001.
6. A. Hyvärinen and M. Inki. Estimating overcomplete independent component bases from image windows. *Journal of Mathematical Imaging and Vision*, 17(2):139–152, 2002.
7. A. Hyvärinen, J. Karhunen, and E. Oja. *Independent Component Analysis*. Wiley Interscience, 2001.
8. M. Inki. *Extensions of Independent Component Analysis for Natural Image Data*. PhD thesis, Helsinki University of Technology, 2004.
9. B. A. Olshausen and D. J. Field. Sparse coding with an overcomplete basis set: A strategy employed by V1? *Vision Research*, 37:3311–3325, 1997.
10. B. A. Olshausen, P. Sallee, and M. S. Lewicki. Learning sparse image codes using a wavelet pyramid architecture. In *Advances in Neural Information Processing Systems*, volume 13, pages 887–893. MIT Press, 2001.
11. A. Pece. The problem of sparse image coding. *Journal of Mathematical Imaging and Vision*, 17(2):89–108, 2002.
12. T. Wachtler, T.-W. Lee, and T. J. Sejnowski. Chromatic structure of natural scenes. *J. Opt. Soc. Am. A*, 18(1):65–77, 2001.

Appendix

Theorem 1

Assume that we wish to find a matrix $\mathbf{B}$*, such that for a symmetric and positive definite* $\mathbf{C}_1$*,* $\mathbf{B}\mathbf{C}_1\mathbf{B}^T = \mathbf{I}$ *and the expected squared error between the separated components* $\mathbf{Bx}$ *and* $\mathbf{Wx}$ *is minimized for data with* $E\{\mathbf{x}\mathbf{x}^T\} = \mathbf{C}_2$ *positive definite. (That is,* $\mathbf{B}$ *whitens data with covariance* $\mathbf{C}_1$*, and the error is minimized for data with covariance* $\mathbf{C}_2$*, when* $E\{\mathbf{x}\} = 0$*.) Then the optimal choice for* $\mathbf{B}$ *is*

$$\mathbf{B}' = (\mathbf{W}\mathbf{C}_2\mathbf{C}_1^{-1}\mathbf{C}_2\mathbf{W}^T)^{-\frac{1}{2}}\mathbf{W}\mathbf{C}_2\mathbf{C}_1^{-1}. \tag{4}$$

Here the inverse square root of a (symmetric) matrix has all the same eigenvectors and an inverse square root has been taken of the eigenvalues.

Proof. It is easy to show that for this choice $\mathbf{B}'\mathbf{C}_1\mathbf{B}'^T = \mathbf{I}$, and any possible solution can be written as $\mathbf{B} = \mathbf{B}_2\mathbf{B}'$, where $\mathbf{B}_2$ is orthogonal. The error, R, equals:

$$\begin{aligned} R &= E\{\|\mathbf{Wx} - \mathbf{Bx}\|^2\} = \mathrm{tr}E\{(\mathbf{Wx} - \mathbf{Bx})(\mathbf{Wx} - \mathbf{Bx})^T\} \\ &= \mathrm{tr}\mathbf{W}\mathbf{C}_2\mathbf{W}^T - \mathrm{tr}\mathbf{W}\mathbf{C}_2\mathbf{B}^T - \mathrm{tr}\mathbf{B}\mathbf{C}_2\mathbf{W}^T + \mathrm{tr}\mathbf{B}\mathbf{C}_2\mathbf{B}^T \\ &= \mathrm{tr}\mathbf{W}\mathbf{C}_2\mathbf{W}^T - 2\mathrm{tr}\mathbf{B}\mathbf{C}_2\mathbf{W}^T + \mathrm{tr}\mathbf{B}\mathbf{C}_1^{\frac{1}{2}}(\mathbf{C}_1^{-\frac{1}{2}}\mathbf{C}_2\mathbf{C}_1^{-\frac{1}{2}})\mathbf{C}_1^{\frac{1}{2}}\mathbf{B}^T \end{aligned}$$

The first and third terms (traces) do not depend on $\mathbf{B}$. Note that the term $\mathbf{B}\mathbf{C}_1^{\frac{1}{2}}$ inside the third trace is an orthogonal matrix, and the trace of a matrix does not change in a similarity transformation. When $\mathbf{B} = \mathbf{B}_2\mathbf{B}'$, we obtain for the term inside the middle trace:

$$\begin{aligned}\mathbf{B}\mathbf{C}_2\mathbf{W}^T &= \mathbf{B}_2(\mathbf{W}\mathbf{C}_2\mathbf{C}_1^{-1}\mathbf{C}_2\mathbf{W}^T)^{-\frac{1}{2}}\mathbf{W}\mathbf{C}_2\mathbf{C}_1^{-1}\mathbf{C}_2\mathbf{W}^T \\ &= \mathbf{B}_2(\mathbf{W}\mathbf{C}_2\mathbf{C}_1^{-1}\mathbf{C}_2\mathbf{W}^T)^{\frac{1}{2}}\end{aligned}$$

Here the term on the last line is $\mathbf{B}_2$ multiplied by a symmetric and positive definite matrix (we assume $\mathbf{W}$ has full rank), and therefore we can write:

$$\mathrm{tr}\mathbf{B}\mathbf{C}_2\mathbf{W}^T = \mathrm{tr}\mathbf{B}_2\mathbf{E}\mathbf{D}\mathbf{E}^T = \mathrm{tr}\mathbf{E}^T\mathbf{B}_2\mathbf{E}\mathbf{D} = \mathrm{tr}\mathbf{Q}\mathbf{D} = \sum_i q_{ii}d_i$$

where $\mathbf{E}$ contains the eigenvectors of the symmetric matrix, and diagonal matrix $\mathbf{D}$ the associated eigenvalues. $\mathbf{Q} = \mathbf{E}^T\mathbf{B}_2\mathbf{E}$ is orthogonal, so $-1 \le q_{ii} \le 1$. As all the eigenvalues d_i are greater than zero, the maximum is achieved when $q_{ii} = 1 \quad \forall i$, i.e. when $\mathbf{Q} = \mathbf{I} \quad \Leftrightarrow \quad \mathbf{B}_2 = \mathbf{I}$. Thus $\mathbf{B}'$ minimizes R. □

Note that this result implies that the optimal whitening matrix (i.e. when $\mathbf{C}_1 = \mathbf{C}_2 = \mathbf{C}$ and $\mathbf{W} = \mathbf{I}$) is $\mathbf{B}' = \mathbf{C}^{-\frac{1}{2}}$, which is often used in ICA methods. Also, the optimal orthogonalization of $\mathbf{W}$ for pre-whitened data ($\mathbf{C}_1 = \mathbf{C}_2 = \mathbf{I}$) is $\mathbf{B} = (\mathbf{W}\mathbf{W}^T)^{-\frac{1}{2}}\mathbf{W}$, which is the symmetric orthogonalization [3] used in FastICA.

This proof can be used even if some eigenvalues of $\mathbf{C}_1$ and $\mathbf{C}_2$ are zero, and the associated eigenvectors span the same space. (In such a case $\mathbf{W}$ and $\mathbf{B}$ are not square.) For instance when the mean value has been removed from each image patch, there is one zero eigenvalue, $n = m - 1$. If $\mathbf{V}$ is an m by n matrix of eigenvectors spanning the nonzero space, we can transform the problem to new variables by writing $\mathbf{W}^{new} = \mathbf{W}\mathbf{V}$, $\mathbf{B}^{new} = \mathbf{B}\mathbf{V}$, $\mathbf{x}^{new} = \mathbf{V}^T\mathbf{x}$, $\mathbf{C}_1^{new} = \mathbf{V}^T\mathbf{C}_1\mathbf{V}$, $\mathbf{C}_2^{new} = \mathbf{V}^T\mathbf{C}_2\mathbf{V}$, and the proof is the same with these new variables. Equally, one can use pseudoinverses.

The InfoMin Principle for ICA and Topographic Mappings

Yoshitatsu Matsuda and Kazunori Yamaguchi

Kazunori Yamaguchi Laboratory, Department of General Systems Studies,
Graduate School of Arts and Sciences, The University of Tokyo,
3-8-1, Komaba, Meguro-ku, Tokyo, 153-8902, Japan
{matsuda, yamaguch}@graco.c.u-tokyo.ac.jp
http://www.graco.u-tokyo.ac.jp/~matsuda

Abstract. It has been well known that edge filters in the visual system can be generated by the InfoMax principle. In this paper, the "InfoMin" principle is proposed, which asserts that the information through some neighboring signals on a two-dimensional mapping must be minimized. It is shown that the standard Comon's ICA can be derived from the combination of the InfoMax principle for the whole signals and the InfoMin one for each signal under a linear model with sufficiently large noise. It is also shown that the InfoMin principle for the signals within neighboring areas can generate a topographic mapping in the same way as in topographic ICA.

1 Introduction

Independent component analysis (ICA) is a recently-developed method in the fields of signal processing and artificial neural networks, and has been shown to be quite useful for the blind separation problem [1–4]. The linear ICA is formalized as follows. Let $\boldsymbol{s}$ and $\boldsymbol{A}$ are M-dimensional source signals and an $N \times M$ mixing matrix, respectively. Then, an N-dimensional vector $\boldsymbol{x}$ of observed signals is defined as

$$\boldsymbol{x} = \boldsymbol{A}\boldsymbol{s}. \tag{1}$$

The purpose of ICA is to find out $\boldsymbol{A}$ when only the observed (mixed) signals are given. In other words, ICA blindly extracts the source signals from observed signals as follows:

$$\boldsymbol{y} = \boldsymbol{W}\boldsymbol{x}, \tag{2}$$

where $\boldsymbol{W}$ is an $M \times N$ separating matrix and $\boldsymbol{y}$ is the estimate of the source signals. This is a typical ill-conditioned problem, but ICA can solve it by assuming that the source signals are generated according to independent and non-Gaussian probability distributions. In general, the ICA algorithms find out $\boldsymbol{W}$ by maximizing a criterion (called the contrast function) such as the higher-order statistics (e.g. the kurtosis) of every component of $\boldsymbol{y}$.

It has been well known that ICA is closely related to information theory. Especially, it has been shown that the maximization of the entropy of signals (so-called

J. Rosca et al. (Eds.): ICA 2006, LNCS 3889, pp. 958–965, 2006.

"InfoMax") is equivalent to some ICA algorithms under nonlinear models [5, 3]. But, it is not available in linear models. Besides, the choice of the nonlinearity is not straightforward [6]. In this paper, we propose a new information-theoretic approach for ICA using the "InfoMin" principle, which asserts that *the information through some neighboring signals on a two-dimensional mapping must be minimized.* The InfoMin principle was proposed originally as a unifying framework for forming topographic mappings [7, 8]. We show that the standard Comon's ICA can be derived from the InfoMin and InfoMax principles under a linear model with sufficiently large noise, where the effects of the fifth and higher-order cumulants become negligible. In addition, it is shown that the InfoMin principle can form a topographic mapping in the same way as in topographic ICA [9].

This paper is organized as follows. In Section 2, the InfoMin principle is proposed. First, it is introduced intuitively from a study about the information processing in the brain in Section 2.1. Secondly, an information transfer model and the mathematical closed form of the entropy are given formally in Section 2.2. In Section 2.3, it is shown that the InfoMax principle for the whole signals is equivalent to a whitening of them and the InfoMin principle for each signal gives the standard Comon's ICA. In Section 3, under this model, a numerical experiment shows that the InfoMin principle for some large neighboring signals can generate a topographic mapping. Lastly, this paper is concluded in Section 4.

2 The InfoMin Principle

2.1 Basic Idea of the InfoMin Principle

In the information processing of the brain, it seems to be natural that the information (the entropy) of the whole neurons must be maximized. This is the InfoMax principle. On the other hand, with respect to an information processing unit (a neuron or a group of neighboring neurons), it seems to be desirable that the information to be processed is as small as possible. In other words, "the information through some neighboring neurons must be minimized," where the information is given as the entropy and each neuron corresponds to a signal. This assertion is called the "InfoMin" principle. The crucial assumption in this idea is that signals through neurons are always given with a mapping of neurons, which is a two-dimensional array where each component corresponds to a signal. Therefore, the neighboring relation of signals can be defined on the array.

2.2 Information Transfer Model

Here, the information transfer model and the mathematical closed form of the entropy are given formally.

The output signal $\boldsymbol{z}$ is given by

$$\boldsymbol{z} = \boldsymbol{y} + \boldsymbol{n}, \tag{3}$$

where $\boldsymbol{y} = \boldsymbol{W}\boldsymbol{x}$ as in Eq. (2) and $\boldsymbol{n}$ is a noise vector. Then, the entropy of $\boldsymbol{z}$, H_z, is given as

$$H_z = -\int \mathrm{d}\boldsymbol{z} P_z(\boldsymbol{z}) \log P_z(\boldsymbol{z}) \tag{4}$$

where $P_z(\boldsymbol{z})$ is the probability distribution function (pdf) of $\boldsymbol{z}$.

If $\boldsymbol{n}$ is given according to an N-dimensional non-correlated Gaussian distribution where the variances of signals are an identical and sufficiently large value Δ^2, the closed form of H_z can be derived as follows. First, by introducing $\bar{\boldsymbol{z}} = \frac{1}{\Delta}\boldsymbol{z}$ and $P_{\bar{z}} = \Delta^N P_z(\bar{\boldsymbol{z}})$, Eq. (4) is rewritten as

$$H_z = \gamma_1 H_{\bar{z}} + \gamma_2, \tag{5}$$

where γ_1 and γ_2 are constants ($\gamma_1 > 0$), and $H_{\bar{z}} = -\int P_{\bar{z}}(\bar{\boldsymbol{z}}) \log P_{\bar{z}} \mathrm{d}\bar{\boldsymbol{z}}$ ($P_{\bar{z}}$ is the pdf of $\bar{\boldsymbol{z}}$). Next, the Edgeworth expansion of $P_{\bar{z}}$ with respect to $\frac{1}{\Delta}$ is given as

$$\begin{aligned} & P_{\bar{z}}(\bar{\boldsymbol{z}}) \\ & = G(\bar{\boldsymbol{z}}) \left(1 + \frac{\sum_{i,j} \kappa_y^{ij} h_{ij}}{2\Delta^2} + \frac{\sum_{i,j,k} \kappa_y^{ijk} h_{ijk}}{6\Delta^3} + \frac{\sum_{i,j,k,l} \kappa_y^{ijkl} h_{ijkl}}{24\Delta^4}\right) \\ & + G(\bar{\boldsymbol{z}}) \frac{\sum_{i,j,k,l} \kappa_y^{ij} \kappa_y^{kl} h_{ijkl}}{8\Delta^4} + O\left(\frac{1}{\Delta^5}\right), \end{aligned} \tag{6}$$

where G is a Gaussian distribution whose covariance matrix is the $N \times N$ identity one. i, j, k, and l denote the indexes of signals. κ_y^{ij}, κ_y^{ijk}, and κ_y^{ijkl} are the second, third, and forth cumulants of $\boldsymbol{y}$, respectively. h_{ij}, h_{ijk}, and h_{ijkl} are the tensorial Hermite polynomials with G (see [10]). Then, by using the next equation for a small variable ε:

$$(1+\varepsilon)\log(1+\varepsilon) = \varepsilon + \frac{\varepsilon^2}{2} - \frac{\varepsilon^3}{6} + \frac{\varepsilon^4}{12} + O(\varepsilon^5), \tag{7}$$

the following closed mathematical form of $H_{\bar{z}}$ is obtained:

$$\begin{aligned} H_{\bar{z}} = & -\frac{1}{2 \cdot 2^2 \Delta^4} \sum_{i,j,k,l} \kappa_y^{ij} \kappa_y^{kl} \int h_{ij} h_{kl} G \mathrm{d}\bar{\boldsymbol{z}} \\ & - \frac{1}{2 \cdot 6^2 \Delta^6} \sum_{i,j,k,l,m,n} \kappa_y^{ijk} \kappa_y^{lmn} \int h_{ijk} h_{lmn} G \mathrm{d}\bar{\boldsymbol{z}} \\ & + \frac{1}{6 \cdot 2^3 \Delta^6} \sum_{i,j,k,l,m,n} \kappa_y^{ij} \kappa_y^{kl} \kappa_y^{mn} \int h_{ij} h_{kl} h_{mn} G \mathrm{d}\bar{\boldsymbol{z}} \\ & - \frac{1}{2 \cdot 24^2 \Delta^8} \sum_{i,j,k,l,m,n,o,p} \kappa_y^{ijkl} \kappa_y^{mnop} \int h_{ijkl} h_{mnop} G \mathrm{d}\bar{\boldsymbol{z}} \\ & - \frac{1}{2 \cdot 8^2 \Delta^8} \sum_{i,j,k,l,m,n,o,p} \kappa_y^{ij} \kappa_y^{kl} \kappa_y^{mn} \kappa_y^{op} \int h_{ijkl} h_{mnop} G \mathrm{d}\bar{\boldsymbol{z}} \\ & - \frac{2}{2 \cdot 24 \cdot 8 \Delta^8} \sum_{i,j,k,l,m,n,o,p} \kappa_y^{ijkl} \kappa_y^{mn} \kappa_y^{op} \int h_{ijkl} h_{mn} h_{op} G \mathrm{d}\bar{\boldsymbol{z}} \end{aligned}$$

$$+\frac{3}{6\cdot 2^2\cdot 24\Delta^8}\sum_{i,j,k,l,m,n,o,p}\kappa_y^{ijkl}\kappa_y^{mn}\kappa_y^{op}\int h_{ijkl}h_{mn}h_{op}G\mathrm{d}\bar{\boldsymbol{z}}$$

$$+\frac{3}{6\cdot 2^2\cdot 8\Delta^8}\sum_{i,j,k,l,m,n,o,p}\kappa_y^{ij}\kappa_y^{kl}\kappa_y^{mn}\kappa_y^{op}\int h_{ijkl}h_{mn}h_{op}G\mathrm{d}\bar{\boldsymbol{z}}$$

$$+\frac{3}{6\cdot 2\cdot 6^2\Delta^8}\sum_{i,j,k,l,m,n,o,p}\kappa_y^{ijk}\kappa_y^{lmn}\kappa_y^{op}\int h_{ijk}h_{lmn}h_{op}G\mathrm{d}\bar{\boldsymbol{z}}$$

$$-\frac{1}{12\cdot 2^4\Delta^8}\sum_{i,j,k,l,m,n,o,p}\kappa_y^{ij}\kappa_y^{kl}\kappa_y^{mn}\kappa_y^{op}\int h_{ij}h_{kl}h_{mn}h_{op}G\mathrm{d}\bar{\boldsymbol{z}}$$

$$+O\left(\frac{1}{\Delta^9}\right), \tag{8}$$

where several terms vanish by the property of the Hermite polynomials. This equation is transformed further as follows:

$$\begin{aligned}
H_{\bar{z}} = &-\frac{1}{4\Delta^4}\sum_{i,j}\left(\kappa_y^{ij}\right)^2 \\
&-\frac{1}{12\Delta^6}\sum_{i,j,k}\left(\kappa_y^{ijk}\right)^2+\frac{1}{48\Delta^6}\sum_{i,j,k,l,m,n}\kappa_y^{ij}\kappa_y^{kl}\kappa_y^{mn}\int h_{ij}h_{kl}h_{mn}G\mathrm{d}\bar{\boldsymbol{z}} \\
&-\frac{1}{48\Delta^8}\sum_{i,j,k,l}\left(\kappa_y^{ijkl}\right)^2+\frac{3}{16\Delta^8}\sum_{i,j,k,l}\left(\kappa_y^{ij}\right)^2\left(\kappa_y^{kl}\right)^2 \\
&+\frac{1}{144\Delta^8}\sum_{i,j,k,l,m,n,o,p}\kappa_y^{ijk}\kappa_y^{lmn}\kappa_y^{op}\int h_{ijk}h_{lmn}h_{op}G\mathrm{d}\bar{\boldsymbol{z}} \\
&-\frac{1}{192\Delta^8}\sum_{i,j,k,l,m,n,o,p}\kappa_y^{ij}\kappa_y^{kl}\kappa_y^{mn}\kappa_y^{op}\int h_{ij}h_{kl}h_{mn}h_{op}G\mathrm{d}\bar{\boldsymbol{z}} \\
&+O\left(\frac{1}{\Delta^9}\right).
\end{aligned} \tag{9}$$

Notice that the derivation of Eq. (9) requires no additional assumptions except that the noise is given according to a Gaussian distribution with sufficiently large variances. Eq. (9) shows that such large noise extinguishes the effects of the higher-order cumulants. For example, the third cumulants with the coefficient of $O\left(\frac{1}{\Delta^6}\right)$ are negligible if the effects of the second cumulants with $O\left(\frac{1}{\Delta^4}\right)$ are dominant. Conversely, higher-order cumulants are dominant only if all the lower-order ones are negligible.

2.3 Relation to ICA

First, it is shown that a whitening process is derived from the InfoMax principle for the whole signals. Under the above information transfer model, the InfoMax principle for the whole signals is equivalent to the maximization of Eq. (9) w.r.t $\boldsymbol{W}$. The first dominant term of Eq. (9) (named $H_{2\mathrm{nd}}$) is given as

$$H_{\text{2nd}} = -\alpha_{\text{2nd}} \sum_{i,j} \left(\kappa_y^{ij}\right)^2 + \beta_{\text{2nd}}, \tag{10}$$

where α_{2nd} (> 0) and β_{2nd} are constants. Now, one usual constraint is given, which requires that the variance of every signal κ_y^{ii} is a constant, e.g. 1. Then, the maximization of H_{2nd} is equivalent to a whitening where every $\kappa_y^{ij} (i \neq j)$ is 0. Because the third cumulants become dominant after this maximization of H_{2nd}, the second dominant term of Eq. (9) (named H_{3rd}) is given as

$$H_{\text{3rd}} = -\alpha_{\text{3rd}} \sum_{i,j,k} \left(\kappa_y^{ijk}\right)^2 + \beta_{\text{3rd}}. \tag{11}$$

Similarly, the third dominant term of Eq. (9) including the fourth cumulants H_{4th} is given as

$$H_{\text{4th}} = -\alpha_{\text{4th}} \left(\sum_{i,j,k,l} \left(\kappa_y^{ijkl}\right)^2 - 12 \sum_{i,j,k} \left(\kappa_y^{ijk}\right)^2 \right) + \beta_{\text{4th}}, \tag{12}$$

where α_{3rd} (> 0), α_{4th} (> 0), β_{3rd}, and β_{4th} are constants.

If $\boldsymbol{W}$ is a square matrix ($M = N$), it is shown easily that H_{3rd} and H_{4th} are constants under the whitened condition. If $M < N$, the signals with high skewness κ^{iii} need to be removed so that H_{3rd} is maximized. H_{4th} is still a constant under any $M \times M$ orthogonal transformation. The case of $M > N$ is not treated in this paper. In other words, the InfoMax principle for the whole signals requires that $\boldsymbol{y}$ is whitened and $M - N$ signals with high skewness are removed. There is no additional requirements if the terms in $O\left(\frac{1}{\Delta^9}\right)$ are negligible.

Next, in order to formalize the InfoMin principle for neighboring signals on a given array, a set of the indexes of neighboring signals is used. This set is called a neighboring filter. For example, a 4×4 neighboring filter on a two-dimensional array of signals corresponds to a set of 16 indexes of signals which are included in the area of 4×4 components in the array. Then, in this context, the InfoMin principle is the minimization of Eq. (9) in the neighboring filters. If the third cumulants are dominant, H_{3rd} is modified into

$$H_{\text{3rd}}^{\text{neigh}} = -\alpha_{\text{3rd}} \sum_{\sigma \in \Sigma} \sum_{i,j,k \in \sigma} \left(\kappa_y^{ijk}\right)^2 + \beta_{\text{3rd}}, \tag{13}$$

where Σ is the set of all neighboring filters. The InfoMin principle asserts that $H_{\text{3rd}}^{\text{neigh}}$ needs to be *minimized.* If the skewnesses are negligible (this assumption is used in quite many ICA algorithms), the InfoMin principle requires the *minimization* of

$$H_{\text{4th}}^{\text{neigh}} = -\alpha_{\text{4th}} \sum_{\sigma \in \Sigma} \sum_{i,j,k,l \in \sigma} \left(\kappa_y^{ijkl}\right)^2 + \beta_{\text{4th}}. \tag{14}$$

In the case that each neighboring filter is a minimal set including just one signal, $H_{4\text{th}}^{\text{neigh}}$ is given as

$$H_{4\text{th}}^{\text{neigh}} = -\alpha_{4\text{th}} \sum_i \left(\kappa_y^{iiii}\right)^2 + \beta_{4\text{th}}. \tag{15}$$

The minimization of Eq. (15) under the whitened condition is completely equivalent to the standard Comon's ICA in [2].

3 Formation of Topographic Mappings

Here, it is shown that the InfoMin principle can generate a topographic mapping from natural scenes. 30000 samples of natural scenes of 12 × 12 pixels were given as $\boldsymbol{x}$ ($N = 144$).

First, in order to satisfy the InfoMax principle in Section 2.3, $\boldsymbol{x}$ was whitened by PCA. Second, $H_{4\text{th}}^{\text{neigh}}$ in Eq. (14) was minimized in order to satisfy the InfoMin principle. Here, it was assumed that the third cumulants were negligible. 144 signals were placed on a 12 × 12 array. A group of 4 × 4 signals on the array is employed as a neighboring filter σ, and Σ is comprised of all possible σ over the array. In addition, because it was shown experimentally that κ_{iiii} and κ_{iijj}

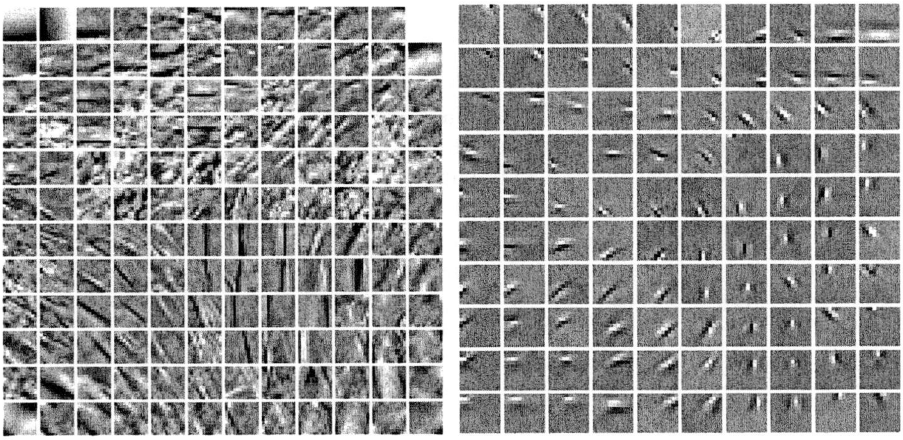

a. InfoMin b. topographic ICA

Fig. 1. Formation of a topographic mapping by the InfoMin principle: Here, natural scenes of 12 × 12 pixels were used as $\boldsymbol{x}$. They visualize the mixing matrix $\boldsymbol{A}$ (a): A topographic mapping was found by minimizing $H_{4\text{th}}^{\text{topo}}$ in Eq. (16) with 4×4 neighboring filters. (b): Topographic ICA in [9] was applied for finding a 10 × 10 topographic ICA mapping from 100 PCA-whitened components of $\boldsymbol{x}$ (with $g(u) = \tanh(u)$, a 3 × 3 neighborhood ones matrix).

were dominant in $H_{4\mathrm{th}}^{\mathrm{neigh}}$, the computational costs are reduced by employing the following approximation $H_{4\mathrm{th}}^{\mathrm{topo}}$ instead of $H_{4\mathrm{th}}^{\mathrm{neigh}}$:

$$H_{4\mathrm{th}}^{\mathrm{topo}} = -\alpha_{4\mathrm{th}} \sum_{\sigma \in \Sigma} \left(\sum_{i \in \sigma} \left(\kappa_y^{iiii}\right)^2 + 3 \sum_{i \in \sigma} \sum_{j \in \sigma, j \neq i} \left(\kappa_y^{iijj}\right)^2 \right) + \beta_{4\mathrm{th}}. \tag{16}$$

Then, $H_{4\mathrm{th}}^{\mathrm{topo}}$ was minimized by a stochastic gradient algorithm using

$$\frac{\partial H_{4\mathrm{th}}^{\mathrm{topo}}}{\partial w_{ij}} = -\alpha_{4\mathrm{th}} \sum_{\sigma \in \Sigma, \sigma \ni i} \left(8\kappa_y^{iiii} y_i^3 x_j + 24 \sum_{k \in \sigma, k \neq i} \kappa_y^{iikk} y_i y_k^2 x_j \right), \tag{17}$$

where w_{ij}, x_j, and y_i are components of $\boldsymbol{W}$, $\boldsymbol{x}$, and $\boldsymbol{y}$, respectively. κ_y^{iiii} and κ_y^{iijj} were calculated every 30000 updates. 500×30000 updates by Eq. (17) were applied for finding the optimal $\boldsymbol{W}$ under the condition that $\boldsymbol{W}$ is orthogonal.

The result of the minimization of $H_{4\mathrm{th}}^{\mathrm{topo}}$ is shown in Fig. 1-(a). It shows that the InfoMin principle could form a topographic mapping where edge detectors with similar orientation preferences are nearer. For comparison, the mapping by topographic ICA [9] is also shown in Fig. 1-(b). Topographic ICA generated distinct but short edge filters. On the other hand, the edge filters in Fig 1-(a) are rougher but longer. Because the generation of long edge filters is difficult for a simple nonlinear InfoMax approach [11], this result suggests the validity of the InfoMin principle. In addition, it is important that the proposed algorithm is based on only a simple linear model with sufficiently large noise and two general principles (InfoMax and InfoMin). On the other hand, topographic ICA is based on a complex two-layer model with nonlinearity and it requires many specific assumptions and approximations. It suggests that the InfoMin principle can give a simple and general framework for topographic mappings. Lastly, though Linsker's algorithm [12] could form an ordered map by the InfoMax principle, but it utilized only the second cumulants and formed just a quite simple map from artificial input signals.

4 Conclusion

In this paper, we proposed the "InfoMin" principle, gave its general closed forms by the Edgeworth expansion, and showed that it could generate a topographic mapping of whitened signals. The proposed method is quite simple. Because it is essentially based on the maximization and minimization of Eq. (9), it is available in any models generating $\boldsymbol{y}$. So, we are planning to apply the method to many models, for example, over-complete models ($M > N$), nonlinear models, and multilayer models.

Lastly, we are now focusing on the fact that it has been known that edge detectors are *not* formed at the maximum of cumulants. Some different nonlinearity is needed, e.g. tanh. So, the choice of nonlinearity is quite important in previous models. On the other hand, the InfoMin principle could form edge detectors by utilizing the forth cumulants. It suggests that the InfoMin principle gives a general framework for the visual processing system in the brain irrespective of nonlinearity, and we are planning to closely investigate the InfoMin principle from this perspective.

References

1. Jutten, C., Herault, J.: Blind separation of sources (part I): An adaptive algorithm based on neuromimetic architecture. Signal Processing **24** (1991) 1–10
2. Comon, P.: Independent component analysis - a new concept? Signal Processing **36** (1994) 287–314
3. Bell, A.J., Sejnowski, T.J.: An information-maximization approach to blind separation and blind deconvolution. Neural Computation **7** (1995) 1129–1159
4. Cardoso, J.F., Laheld, B.: Equivariant adaptive source separation. IEEE Transactions on Signal Processing **44** (1996) 3017–3030
5. Nadal, J.P., Parga, N.: Nonlinear neurons in the low-noise limit: a factorial code maximizes information-transfer. Network: Computation in Neural Systems **5** (1994) 565–581
6. Lee, T.W., Girolami, M., Sejnowski, T.J.: Independent component analysis using an extended infomax algorithm for mixed subgaussian and supergaussian sources. Neural Computation **11** (1999) 417–441
7. Matsuda, Y., Yamaguchi, K.: The InfoMin principle: a unifying information-based criterion for forming topographic mappings. In: ICONIP2001 Proceedings, Shanghai, China (2001) 14–19
8. Matsuda, Y., Yamaguchi, K.: The InfoMin criterion: an information theoretic unifying objective function for topographic mappings. In: Artificial Neural Network and Neural Information Processing - ICANN/ICONIP 2003. Volume 2714 of LNCS., Istanbul, Turkey, Springer-Verlag (2003) 401–408
9. Hyvärinen, A.: Complexity pursuit: separating interesting components from time series. Neural Computation **13** (2001) 883–898
10. Barndorff-Nielsen, O.E., Cox, D.R.: Asymptotic techniques for use in statistics. Chapman and Hall, London ; New York (1989)
11. Hoyer, P.O., Hyvärinen, A.: A multi-layer sparse coding network learns contour coding from natural images. Vision Research **42** (2002) 1593–1605
12. Linsker, R.: How to generate ordered maps by maximizing the mutual information between input and output signals. Neural Computation **1** (1989) 402–411

Non-negative Matrix Factorization Approach to Blind Image Deconvolution

Ivica Kopriva[1] and Danielle Nuzillard[2]

[1] Department of Electrical and Computer Engineering, The George Washington University, 801 22nd St. NW Room 615, Washington DC 20052, USA
ikopriva@gmail.com

[2] CReSTIC, Université de Reims - Chamapgne Ardenne (URCA), Moulin de la Housse, B.P. 1039, 51687 REIMS Cedex 2, France
danielle.nuzillard@univ-reims.fr

Abstract. A novel approach to single frame multichannel blind image deconvolution is formulated recently as non-negative matrix factorization (NMF) problem with sparseness constraint imposed on the unknown mixing vector. Unlike most of the blind image deconvolution algorithms, the NMF approach requires no *a priori* knowledge about the blurring kernel and original image. The experimental performance evaluation of the NMF algorithm is presented with the degraded image by the out-of-focus blur. The NMF algorithm is compared to the state-of-the-art single frame blind image deconvolution algorithm: blind Richardson-Lucy algorithm and single frame multichannel independent component analysis based algorithm. It has been demonstrated that NMF approach outperforms mentioned blind image deconvolution methods.

1 Introduction

The goal of image deconvolution is to reconstruct the original image from an observation degraded by spatially invariant blurring process and noise. Neglecting the noise term the process is modeled as a convolution of a blurring kernel $h(s,t)$ with an original source image $f(x,y)$ as:

$$g(x,y) = \sum_{s=-K}^{K} \sum_{t=-K}^{K} h(s,t) f(x+s, y+t) \tag{1}$$

where K denotes the size of the blurring kernel. If the blurring kernel is known, many so-called non-blind algorithms are available to reconstruct original image $f(x,y)$ [1]. However it is not always possible to measure or obtain information about blurring kernel, which is why blind deconvolution (BD) algorithms are important. Comprehensive comparison of BD algorithms is given in [1]. They can be divided into those that estimate the blurring kernel $h(s,t)$ first and then restore original image by some of the non-blind methods [1], and those that estimate the original image $f(x,y)$ and blurring kernel simultaneously. In order to estimate the blurring kernel, a support size has to be given or estimated. Also, quite often

J. Rosca et al. (Eds.): ICA 2006, LNCS 3889, pp. 966–973, 2006.

a priori knowledge about the nature of the blurring process is assumed to be available in order to use appropriate parametric model of the blurring process [2]. It is not always possible to know the characteristic of the blurring process. Methods that estimate blurring kernel and original image simultaneously use either statistical or deterministic *prior* on the original image, the blurring kernel and the noise [2]. This leads to a computationally expensive maximum likelihood estimation usually implemented by expectation maximization algorithm. In addition to that, exact distributions of the original image required by maximum likelihood algorithm are usually unknown. One of the most representative algorithms from this class is the blind Richardson-Lucy (R-L) algorithm. It has been originally derived for non-blind deconvolution of astronomical images in [3] and [4]. Later on, it was formulated in [5] for BD and then modified by iterative restoration algorithm in [6]. This version of blind R-L algorithm is implemented in MATLAB command '*deconvblind*'. It will be used in the section 3 for the comparison purpose during experimental performance evaluation of the NMF based blind image deconvolution method [7].

In order to overcome difficulties associated with 'standard' BD algorithms an approach was proposed in [8] based on quasi maximum likelihood with an approximate of the probability density function. It however assumed that original image has sparse or super-Gaussian distribution. This is generally not true because image distributions are mostly sub-Gaussian. To overcome that difficulty it was proposed in [8] to apply sparsifying transform to blurred image. However, design of such a transform requires knowledge of at least the typical class of images to which original image belongs. In such a case, training data can be used to design sparsifying transform. Multivariate data analysis methods such as independent component analysis (ICA) [9] might be used to solve BD problem as a blind source separation (BSS) problem. The unknown blurring process is absorbed into what is known as a mixing matrix. The advantage of the ICA approach would be that no *a priori* knowledge about the origin and size of the support of the blurring kernel is required. However, multi-channel image required by ICA is not always available. Even if it is, it would require the blurring kernel to be non-stationary, which is true for blur caused by atmospheric turbulence, but it is not true for out-of-focus blur for example. Therefore, an approach to single frame multi-channel blind deconvolution that requires minimum of *a priori* information about blurring process and original image would be of great interest.

Single frame multi-channel representation was proposed in [10]. It was based on a bank of $2-D$ Gabor filters [11] used due to their ability to realize multichannel filtering. ICA algorithms have been applied in [10] to multichannel image in order to extract the source image and two spatial derivatives along x and y directions. There is however critical condition that source image and their spatial derivatives must be statistically independent. In general this is not true as already observed in [11]. Consequently, quality of the image restoration by proposed single frame multi-channel approach depends on how well each particular image satisfies statistical independence assumption. Therefore, an extension of the ICA approach formulated in [10] is given in [7] where it has been shown that single frame multichannel BD can be formulated as NMF problem with sparseness constraint imposed on the unknown mixing vector. Consequently, no *a priori* knowledge about either the origin or the size of the blurring process is required. Because NMF is deterministic approach no *a priori* information about the statistical nature of the source image is required as well. The rest of the paper is organized as follows. We briefly introduce in section 2 blind R-L algorithm [5][6], ICA approach to single frame multichannel BD [10] and NMF approach to single frame multichannel BD with sparseness constraint [7]. Comparative experimental performance evaluation is given in section 3 for images degraded by out-of-focus blurring process. Conclusion is presented in section 4.

2 Basic Overview of the Compared Blind Image Deconvolution Algorithms

Before proceeding to description of non-bind and blind image deconvolution algorithms, we shall rewrite image observation model given in Eq.1 in the lexicographical notation:

$$g = Hf \tag{2}$$

where $g, f \epsilon \mathbb{Z}_{0+}^{MN}$, $H \epsilon \mathbb{R}_{0+}^{MN\text{x}MN}$ assuming image dimensionality of MxN pixels. Observed image vector g and original image vector f are obtained by the row stacking procedure. The matrix H is block-circulant matrix, [13], and it absorbs into itself the blurring kernel $h(s,t)$ assuming at least size of it, K, to be known.

2.1 Blind Richardson-Lucy Algorithm

The blind R-L method [5] [6] follows from the non-blind version of the R-L method [3] [4] which itself follows from Bayesian paradigm approach to statistical inference which dictates that inference about true image f should be based on conditional probability $P(f/g)$ given by the Bayes rule. The *prior* knowledge about image degradation process is incorporated in conditional probability $P(g/f)$ and *prior* probability $P(f)$. In low light level imaging such as in astronomy, microscopy and the night vision imaging, the appropriate choice for $P(g/f)$ is Poisson distribution [14]. In the high-brightness conditions the Poisson *prior* should be replaced by the Gaussian one. The R-L algorithm follows when non-informative *prior* is chosen for $P(f)$ *i.e.* $P(f) \prec$ const. The algorithm is obtained through the maximization of the log-likelihood function:

$$\widehat{f} = argmax(logP(g/f)) \tag{3}$$

The EM algorithm is employed to solve problem in Eq.3 yielding numerically efficient multiplicative iterative algorithm known as blind R-L algorithm [5]:

$$\widehat{H}_{i+1}^{(k)} = [(f^{(k-1)})^T(g\varnothing(\widehat{H}_i^{(k)}f^{(k-1)}))]\widehat{H}_i^{(k)} \tag{4}$$

$$\widehat{f}_{i+1}^{(k+1)} = [(f^{(k)})^T \bigotimes (H^T(g\varnothing(H\widehat{f}_i^{(k)} f^{(k-1)}))]\varnothing(\widehat{H}^T 1) \tag{5}$$

where index i is used to denote internal iteration of the blind R-L algorithm and k denotes main iteration index and 1 denotes a column vector with all entries equal to 1. In Eq.4, symbol $'\varnothing'$ denotes component-wise division, and in Eq.5 symbol $'\bigotimes'$ denotes component-wise multiplication.

Multiplicative update rules in Eq.4 and Eq.5, ensure positivity of both blurring kernel and reconstructed image automatically. Problem with blind R-L algorithm is that support size K of the blurring kernel must be known or estimated by some method. This knowledge is not always available *a priori*. This is especially true for non-stationary degradation process such as atmospheric turbulence where the strength of the turbulence, measured by the parameter called scintillation index, will strongly influence the size of the blur.

2.2 ICA Approach to Single Frame Multichannel BD (SFMICA)

Single frame multi-channel representation was proposed in [10]. It was based on a bank of 2-D Gabor filters [11] used due to their ability to realize multi-channel filtering and decomposing an input image into sparse images containing intensity variation over narrow range of frequency and orientation. Multichannel version of degraded image is shown to be [10]:

$$G = \begin{pmatrix} g^T \\ g_1^T \\ \cdots \\ g_L^T \end{pmatrix} \cong \begin{pmatrix} a_{01} & a_{02} & a_{03} \\ a_{11} & a_{12} & a_{13} \\ \cdots & \cdots & \cdots \\ a_{L1} & a_{L2} & a_{L3} \end{pmatrix} \begin{pmatrix} f^T \\ f_x^T \\ f_y^T \end{pmatrix} = AF \tag{6}$$

where images $g_l, l = 1, .., L$, are produced by Gabor filters, f represents source image and f_x and f_y represent spatial derivatives along x and y directions respectively.

The used Gabor filters had the following real and imaginary respectively:

$$R(x, y) = G(x, y) * cos(\frac{\pi}{\sigma}\varphi(x, y))$$
$$I(x, y) = G(x, y) * sin(\frac{\pi}{\sigma}\varphi(x, y))$$

where

$$G(x, y) = exp\left(-\frac{x^2 + y^2}{2\sigma^2}\right) \tag{7}$$

$$\varphi(x, y) = x.cos(\frac{\pi}{4}k) + y.sin(\frac{\pi}{4}k) \quad \text{with } k = 0, 1, 2, 3. \tag{8}$$

The parameter k regulates one of the four spatial orientations. The parameter $\sigma = \sqrt{2^n}$ with $n = 1, 2$ regulates one of the two spatial frequencies. Consequently, in SFMICA and later on SFMNMF BD algorithms 16 Gabor filters (8 for real and 8 for imaginary part with 4 spatial orientations and 2 spatial frequencies) were used to obtain multichannel version of the observed image.

The unknown elements a_{lm} of the mixing matrix absorb the blurring kernel assuming no *a priori* information about it including its size. The ICA algorithm has been applied in [10] to image model Eq.6 in order to extract the source image f. There is however critical condition for the source image that must hold in order to ICA algorithm to work. Image f and its spatial derivatives f_x and f_y must be statistically independent. This is in general not true as already observed in [12]. Consequently, quality of the restored image by proposed single frame multichannel approach depends on how well each particular image satisfies statistical independence assumption.

2.3 NMF Approach to Single Frame Multichannel BD (SFMNMF)

It was further shown in [7] that single frame multichannel blind deconvolution can be represented as:

$$G = \begin{pmatrix} g^T \\ g_1^T \\ \cdots \\ g_L^T \end{pmatrix} \cong \begin{pmatrix} \widetilde{a_{01}} \\ \widetilde{a_{11}} \\ \cdots \\ \widetilde{a_{L1}} \end{pmatrix} (f^T) = \tilde{a} f^T \tag{9}$$

where images $g_l, l = 1, ..., L$, were again produced by Gabor filters. Coefficients of the unknown blurring kernel were absorbed into coefficients $\widetilde{a_{lm}}$ of the unknown mixing vector $\widetilde{a}$. Image model Eq.9 suggests the existence of only one source image f in the linear image observation model. Spatially oriented Gabor filters produce images with sparse (super-Gaussian) distributions. If the source image f is sub-Gaussian, which is the case for natural images, an unknown mixing vector must be sparse. Because $\tilde{a}$ and f are non-negative, this enabled in [7] to formulate blind deconvolution problem as an NMF problem with sparseness constraint imposed on the mixing vector [15]:

$$\begin{gathered} (\widehat{\tilde{a}}, \widehat{f}) = argmin \parallel G - \widehat{\tilde{a}} \widehat{f}^T \parallel^2 \\ \text{subject to } sparseness(\tilde{a}) = S_a \end{gathered} \tag{10}$$

where 'hat' denotes estimate and the measure of sparseness S_a is number between 0 and 1, with 1 meaning that all components of vector $\tilde{a}$ are small and 0 meaning the opposite [15].

A sparseness constraint S_a must be defined for NMF algorithm. In order to obtain truly unsupervised image restoration algorithm, S_a is estimated from the multichannel image G as a ratio between number of sparse images L_s and overall number of images $L + 1$. To estimate L_s, kurtosis of each image in G is estimated. Image g_l is considered to be sparse if $\kappa(g_l) > \delta$. In our experiments we set $\delta = 0.2$.

The SFMNMF algorithm is defined without using any *a priori* information about the blurring process or original image. Because this is a deterministic approach, no assumption about statistical nature of either blur or source image is required. Only sparseness constraint must be imposed on the unknown mixing vector $\widetilde{a}$. First coefficient in $\widetilde{a}$ can initially be approximated by 1, because it represents original blurring process. The rest of the coefficients can initially be set to 0 because they correspond to sparse images. Therefore initial value of the unknown mixing vector is set to $\widetilde{a}^{(0)} = [1 \ \ 0 \ \ 0 \ \ ... \ \ 0]^T$.

The SFNMF approach to BD does not have to perform source separation due to the fact that multichannel version G of the observed image g can be approximated by the product of the unknown mixing vector and source image f as shown by Eq.9. Because there is only source image present in the observed image model, there is no need for source separation. This is the main difference with respect to the approach proposed in [10] and by Eq.6. However, it is still not clear at the moment how to apply NMF approach to BD when source image f is sparse. Because the multichannel image G is sparse and original image f is also sparse, it is not obvious in this case whether sparseness constraint must be imposed on the source image f only or it should be imposed on both source image f and unknown mixing vector.

3 Experimental Results

Fig.1 left shows blurred image obtained by digital camera in manually de-focused mode. Fig.1 right shows image reconstructed by SFNMF algorithm. Image reconstructed by SFMICA algorithm is shown in Fig.2 left, where FastICA algorithm with tanh nonlinearity was used. Fig.2 right shows image restored by the blind R-L algorithm after 5 iterations with the circular blurring kernel and radius of $R = 3$ pixels. Because the blurred image, Fig.1, was not highly de-focused blind R-L algorithm with kernel size of $R = 3$ pixels produced good result but still inferior to this produced by SFMNMF algorithm shown in Fig. 1 right. Because the size of the blurring kernel must be known *a priori* for R-L algorithm, the algorithm had to be run several times with the various values for the radius R and then the value

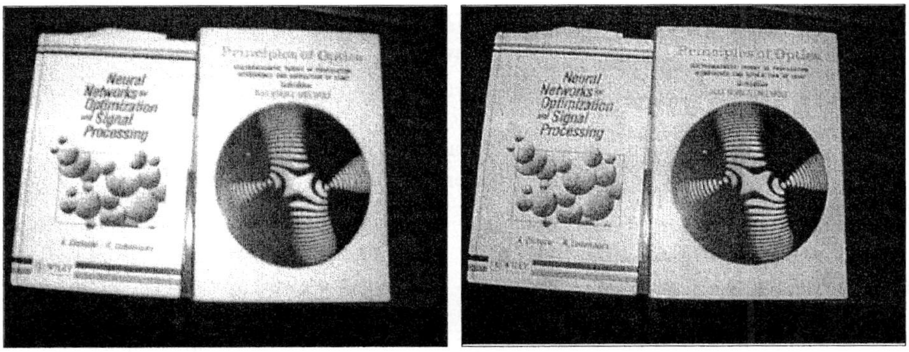

Fig. 1. (left) Image degraded by out-of-focus blur obtained by digital camera in manually de-focused mode; (right) Image reconstructed by SFMNMF algorithm

Fig. 2. (left) Image reconstructed by SFMICA algorithm; (right) Image reconstructed by blind Richardson-Lucy algorithm after 5 iterations with the circular blurring kernel with radius of $R = 3$

that corresponded to the best quality of the restored image was chosen. This was very time consuming process. In addition to that, it is known that either underestimate or overestimate of the size of the blurring kernel leads to severe distortions of the images reconstructed by blind R-L algorithm and other blind algorithms of the similar type [1]. There are no such problem with the SFMNMF algorithm. Image restored by the SFMICA algorithm has poor quality due to the fact that assumption about statistical independence between source image f and its spatial derivatives f_x and f_y does not hold. The SFMNMF algorithm eliminates all these problems due to the fact that no *a priori* knowledge about the size of the blurring kernel or statistical nature of the source image is required.

4 Conclusion

An experimental comparative performance evaluation between novel single frame multichannel blind deconvolution algorithm based on non-negative matrix factorization with sparseness constraint (SFMNMF) and other representative blind image deconvolution algorithms was presented. Image deconvolution methods were compared on image degraded by out-of-focus blur. It has been demonstrated that novel blind image deconvolution algorithm outperforms other methods. We suggest that this result is due to the characteristic of the SFMNMF algorithm which does not require any *a priori* information about the blurring kernel and original image.

References

1. M.R. Banham and A.K. Katsaggelos, "Digital Image Restoration," IEEE Signal Processing Magazine, vol. 14, no. 2, pp. 24-41, March 1997.
2. D. Kundur and D. Hatzinakos, "Blind Image Deconvolution," IEEE Signal Processing Magazine, vol. 13, no. 3, pp. 43-64, May 1996.

3. W.H. Richardson, "Bayesian-based iterative method of image restoration," J.Opt.Soc. Amer., vol. 62, pp. 55-59, 1972.
4. L.B. Lucy, "An iterative technique for rectification of observed distribution," Astron. J., vol. 79, pp.745-754, 1974.
5. D.A. Fish, A.M. Brinicombe, E.R. Pike and J.G. Walker, "Blind deconvolution by means of the Richardson-Lucy algorithm," J. Opt. Soc. Am. A, Vol. 12, No. 1, pp.58-65, January, 1995.
6. D.S.C. Biggs and M. Andrews, "Acceleration of iterative image restoration algorithms," Applied Optics, Vol. 36, No. 8, 1997.
7. I. Kopriva, "Single frame multichannel blind deconvolution by non-negative matrix factorization with sparseness constraints," Optics Letters Vol. 30, No. 23, pp. 3135-3137, December 1, 2005.
8. M.M. Bronstein, A. Bronstein, M. Zibulevsky, and Y.Y. Zeevi, "Blind Deconvolution of Images Using Optimal Sparse Representations"IEEE Tr. on Image Processing, vol. 14, No. 6, pp. 726-736, 2005.
9. A. Hyvärinen, J. Karhunen, and E. Oja, Independent Component Analysis, Wiley Interscience, 2001.
10. S. Umeyama, "Blind Deconvolution of Images by Using ICA," Electronics and Communications in Japan, Part 3, vol. 84, No. 12, 2001.
11. J. G. Daugman, "Complete Discrete 2-D Gabor Transforms by Neural Networks for Image Analysis and Compression" IEEE Tr. on Acoust., Speech and Sig. Proc., Vol. 36, pp. 1169-1179, 1988.
12. M. Numata, and N. Hamada,"Image Restoration of Multichannel Blurred Images by Independent Component Analysis", Proceedings of 2004 RISP International Workshop on Nonlinear Circuit and Signal Processing (NCSP'04),Hawaii, USA, March 5-7, 2004, pp.197-200.
13. R. L. Lagendijk and J. Biemond, Iterative Identification and Restoration of Images, KAP, 1991.
14. R. Molina, J. Nunez, F.J. Cortijo and J. Mateos, "Image Restoration in Astronomy - A Bayesian Perspective," IEEE Signal Processing Magazine, vol. 18, no. 2, pp. 11-29, March 2001.
15. P. O. Hoyer, "Non-negative matrix factorization with sparseness constraints," Journal of Machine Learning Research 5, pp. 1457-1469, 2004.

Keyword Index

Author Index

Lecture Notes in Computer Science

For information about Vols. 1–3785

please contact your bookseller or Springer

Vol. 3889: J. Rosca, D. Erdogmus, J.C. Príncipe, S. Haykin (Eds.), Independent Component Analysis and Blind Signal Separation. XXI, 980 pages. 2006.

Vol. 3884: B. Durand, W. Thomas (Eds.), STACS 2006. XIV, 714 pages. 2006.

Vol. 3881: S. Gibet, N. Courty, J.-F. Kamp (Eds.), Gesture in Human-Computer Interaction and Simulation. XIII, 344 pages. 2006. (Sublibrary LNAI).

Vol. 3879: T. Erlebach, G. Persinao (Eds.), Approximation and Online Algorithms. X, 349 pages. 2006.

Vol. 3878: A. Gelbukh (Ed.), Computational Linguistics and Intelligent Text Processing. XVII, 589 pages. 2006.

Vol. 3877: M. Detyniecki, J.M. Jose, A. Nürnberger, C. J. '. van Rijsbergen (Eds.), Adaptive Multimedia Retrieval: User, Context, and Feedback. XI, 279 pages. 2006.

Vol. 3874: R. Missaoui, J. Schmidt (Eds.), Formal Concept Analysis. X, 309 pages. 2006. (Sublibrary LNAI).

Vol. 3873: L. Maicher, J. Park (Eds.), Charting the Topic Maps Research and Applications Landscape. VIII, 281 pages. 2006. (Sublibrary LNAI).

Vol. 3872: H. Bunke, A. L. Spitz (Eds.), Document Analysis Systems VII. XIII, 630 pages. 2006.

Vol. 3870: S. Spaccapietra, P. Atzeni, W.W. Chu, T. Catarci, K.P. Sycara (Eds.), Journal on Data Semantics V. XIII, 237 pages. 2006.

Vol. 3869: S. Renals, S. Bengio (Eds.), Machine Learning for Multimodal Interaction. XIII, 490 pages. 2006.

Vol. 3868: K. Römer, H. Karl, F. Mattern (Eds.), Wireless Sensor Networks. XI, 342 pages. 2006.

Vol. 3863: M. Kohlhase (Ed.), Mathematical Knowledge Management. XI, 405 pages. 2006. (Sublibrary LNAI).

Vol. 3861: J. Dix, S.J. Hegner (Eds.), Foundations of Information and Knowledge Systems. X, 331 pages. 2006.

Vol. 3860: D. Pointcheval (Ed.), Topics in Cryptology – CT-RSA 2006. XI, 365 pages. 2006.

Vol. 3858: A. Valdes, D. Zamboni (Eds.), Recent Advances in Intrusion Detection. X, 351 pages. 2006.

Vol. 3857: M.P.C. Fossorier, H. Imai, S. Lin, A. Poli (Eds.), Applied Algebra, Algebraic Algorithms and Error-Correcting Codes. XI, 350 pages. 2006.

Vol. 3855: E. A. Emerson, K.S. Namjoshi (Eds.), Verification, Model Checking, and Abstract Interpretation. XI, 443 pages. 2005.

Vol. 3853: A.J. Ijspeert, T. Masuzawa, S. Kusumoto (Eds.), Biologically Inspired Approaches to Advanced Information Technology. XIV, 388 pages. 2006.

Vol. 3852: P.J. Narayanan, S.K. Nayar, H.-Y. Shum (Eds.), Computer Vision - ACCV 2006, Part II. XXXI, 977 pages. 2005.

Vol. 3851: P.J. Narayanan, S.K. Nayar, H.-Y. Shum (Eds.), Computer Vision - ACCV 2006, Part I. XXXI, 973 pages. 2006.

Vol. 3850: R. Freund, G. Păun, G. Rozenberg, A. Salomaa (Eds.), Membrane Computing. IX, 371 pages. 2006.

Vol. 3849: I. Bloch, A. Petrosino, A.G.B. Tettamanzi (Eds.), Fuzzy Logic and Applications. XIV, 438 pages. 2006. (Sublibrary LNAI).

Vol. 3848: J.-F. Boulicaut, L. De Raedt, H. Mannila (Eds.), Constraint-Based Mining and Inductive Databases. X, 401 pages. 2006. (Sublibrary LNAI).

Vol. 3847: K.P. Jantke, A. Lunzer, N. Spyratos, Y. Tanaka (Eds.), Federation over the Web. X, 215 pages. 2006. (Sublibrary LNAI).

Vol. 3846: H. J. van den Herik, Y. Björnsson, N.S. Netanyahu (Eds.), Computers and Games. XIV, 333 pages. 2006.

Vol. 3845: J. Farré, I. Litovsky, S. Schmitz (Eds.), Implementation and Application of Automata. XIII, 360 pages. 2006.

Vol. 3844: J.-M. Bruel (Ed.), Satellite Events at the MoDELS 2005 Conference. XIII, 360 pages. 2006.

Vol. 3843: P. Healy, N.S. Nikolov (Eds.), Graph Drawing. XVII, 536 pages. 2006.

Vol. 3842: H.T. Shen, J. Li, M. Li, J. Ni, W. Wang (Eds.), Advanced Web and Network Technologies, and Applications. XXVII, 1057 pages. 2006.

Vol. 3841: X. Zhou, J. Li, H.T. Shen, M. Kitsuregawa, Y. Zhang (Eds.), Frontiers of WWW Research and Development - APWeb 2006. XXIV, 1223 pages. 2006.

Vol. 3840: M. Li, B. Boehm, L.J. Osterweil (Eds.), Unifying the Software Process Spectrum. XVI, 522 pages. 2006.

Vol. 3839: J.-C. Filliâtre, C. Paulin-Mohring, B. Werner (Eds.), Types for Proofs and Programs. VIII, 275 pages. 2006.

Vol. 3838: A. Middeldorp, V. van Oostrom, F. van Raamsdonk, R. de Vrijer (Eds.), Processes, Terms and Cycles: Steps on the Road to Infinity. XVIII, 639 pages. 2005.

Vol. 3837: K. Cho, P. Jacquet (Eds.), Technologies for Advanced Heterogeneous Networks. IX, 307 pages. 2005.

Vol. 3836: J.-M. Pierson (Ed.), Data Management in Grids. X, 143 pages. 2006.

Vol. 3835: G. Sutcliffe, A. Voronkov (Eds.), Logic for Programming, Artificial Intelligence, and Reasoning. XIV, 744 pages. 2005. (Sublibrary LNAI).

Vol. 3834: D.G. Feitelson, E. Frachtenberg, L. Rudolph, U. Schwiegelshohn (Eds.), Job Scheduling Strategies for Parallel Processing. VIII, 283 pages. 2005.

Vol. 3833: K.-J. Li, C. Vangenot (Eds.), Web and Wireless Geographical Information Systems. XI, 309 pages. 2005.

Vol. 3832: D. Zhang, A.K. Jain (Eds.), Advances in Biometrics. XX, 796 pages. 2005.

Vol. 3831: J. Wiedermann, G. Tel, J. Pokorný, M. Bieliková, J. Štuller (Eds.), SOFSEM 2006: Theory and Practice of Computer Science. XV, 576 pages. 2006.

Vol. 3829: P. Pettersson, W. Yi (Eds.), Formal Modeling and Analysis of Timed Systems. IX, 305 pages. 2005.

Vol. 3828: X. Deng, Y. Ye (Eds.), Internet and Network Economics. XVII, 1106 pages. 2005.

Vol. 3827: X. Deng, D.-Z. Du (Eds.), Algorithms and Computation. XX, 1190 pages. 2005.

Vol. 3826: B. Benatallah, F. Casati, P. Traverso (Eds.), Service-Oriented Computing - ICSOC 2005. XVIII, 597 pages. 2005.

Vol. 3824: L.T. Yang, M. Amamiya, Z. Liu, M. Guo, F.J. Rammig (Eds.), Embedded and Ubiquitous Computing – EUC 2005. XXIII, 1204 pages. 2005.

Vol. 3823: T. Enokido, L. Yan, B. Xiao, D. Kim, Y. Dai, L.T. Yang (Eds.), Embedded and Ubiquitous Computing – EUC 2005 Workshops. XXXII, 1317 pages. 2005.

Vol. 3822: D. Feng, D. Lin, M. Yung (Eds.), Information Security and Cryptology. XII, 420 pages. 2005.

Vol. 3821: R. Ramanujam, S. Sen (Eds.), FSTTCS 2005: Foundations of Software Technology and Theoretical Computer Science. XIV, 566 pages. 2005.

Vol. 3820: L.T. Yang, X.-s. Zhou, W. Zhao, Z. Wu, Y. Zhu, M. Lin (Eds.), Embedded Software and Systems. XXVIII, 779 pages. 2005.

Vol. 3819: P. Van Hentenryck (Ed.), Practical Aspects of Declarative Languages. X, 231 pages. 2005.

Vol. 3818: S. Grumbach, L. Sui, V. Vianu (Eds.), Advances in Computer Science – ASIAN 2005. XIII, 294 pages. 2005.

Vol. 3817: M. Faundez-Zanuy, L. Janer, A. Esposito, A. Satue-Villar, J. Roure, V. Espinosa-Duro (Eds.), Nonlinear Analyses and Algorithms for Speech Processing. XII, 380 pages. 2006. (Sublibrary LNAI).

Vol. 3816: G. Chakraborty (Ed.), Distributed Computing and Internet Technology. XXI, 606 pages. 2005.

Vol. 3815: E.A. Fox, E.J. Neuhold, P. Premsmit, V. Wuwongse (Eds.), Digital Libraries: Implementing Strategies and Sharing Experiences. XVII, 529 pages. 2005.

Vol. 3814: M. Maybury, O. Stock, W. Wahlster (Eds.), Intelligent Technologies for Interactive Entertainment. XV, 342 pages. 2005. (Sublibrary LNAI).

Vol. 3813: R. Molva, G. Tsudik, D. Westhoff (Eds.), Security and Privacy in Ad-hoc and Sensor Networks. VIII, 219 pages. 2005.

Vol. 3812: C. Bussler, A. Haller (Eds.), Business Process Management Workshops. XIII, 520 pages. 2006.

Vol. 3811: C. Bussler, M.-C. Shan (Eds.), Technologies for E-Services. VIII, 127 pages. 2006.

Vol. 3810: Y.G. Desmedt, H. Wang, Y. Mu, Y. Li (Eds.), Cryptology and Network Security. XI, 349 pages. 2005.

Vol. 3809: S. Zhang, R. Jarvis (Eds.), AI 2005: Advances in Artificial Intelligence. XXVII, 1344 pages. 2005. (Sublibrary LNAI).

Vol. 3808: C. Bento, A. Cardoso, G. Dias (Eds.), Progress in Artificial Intelligence. XVIII, 704 pages. 2005. (Sublibrary LNAI).

Vol. 3807: M. Dean, Y. Guo, W. Jun, R. Kaschek, S. Krishnaswamy, Z. Pan, Q.Z. Sheng (Eds.), Web Information Systems Engineering – WISE 2005 Workshops. XV, 275 pages. 2005.

Vol. 3806: A.H. H. Ngu, M. Kitsuregawa, E.J. Neuhold, J.-Y. Chung, Q.Z. Sheng (Eds.), Web Information Systems Engineering – WISE 2005. XXI, 771 pages. 2005.

Vol. 3805: G. Subsol (Ed.), Virtual Storytelling. XII, 289 pages. 2005.

Vol. 3804: G. Bebis, R. Boyle, D. Koracin, B. Parvin (Eds.), Advances in Visual Computing. XX, 755 pages. 2005.

Vol. 3803: S. Jajodia, C. Mazumdar (Eds.), Information Systems Security. XI, 342 pages. 2005.

Vol. 3802: Y. Hao, J. Liu, Y.-P. Wang, Y.-m. Cheung, H. Yin, L. Jiao, J. Ma, Y.-C. Jiao (Eds.), Computational Intelligence and Security, Part II. XLII, 1166 pages. 2005. (Sublibrary LNAI).

Vol. 3801: Y. Hao, J. Liu, Y.-P. Wang, Y.-m. Cheung, H. Yin, L. Jiao, J. Ma, Y.-C. Jiao (Eds.), Computational Intelligence and Security, Part I. XLI, 1122 pages. 2005. (Sublibrary LNAI).

Vol. 3799: M. A. Rodríguez, I.F. Cruz, S. Levashkin, M.J. Egenhofer (Eds.), GeoSpatial Semantics. X, 259 pages. 2005.

Vol. 3798: A. Dearle, S. Eisenbach (Eds.), Component Deployment. X, 197 pages. 2005.

Vol. 3797: S. Maitra, C. E. V. Madhavan, R. Venkatesan (Eds.), Progress in Cryptology - INDOCRYPT 2005. XIV, 417 pages. 2005.

Vol. 3796: N.P. Smart (Ed.), Cryptography and Coding. XI, 461 pages. 2005.

Vol. 3795: H. Zhuge, G.C. Fox (Eds.), Grid and Cooperative Computing - GCC 2005. XXI, 1203 pages. 2005.

Vol. 3794: X. Jia, J. Wu, Y. He (Eds.), Mobile Ad-hoc and Sensor Networks. XX, 1136 pages. 2005.

Vol. 3793: T. Conte, N. Navarro, W.-m.W. Hwu, M. Valero, T. Ungerer (Eds.), High Performance Embedded Architectures and Compilers. XIII, 317 pages. 2005.

Vol. 3792: I. Richardson, P. Abrahamsson, R. Messnarz (Eds.), Software Process Improvement. VIII, 215 pages. 2005.

Vol. 3791: A. Adi, S. Stoutenburg, S. Tabet (Eds.), Rules and Rule Markup Languages for the Semantic Web. X, 225 pages. 2005.

Vol. 3790: G. Alonso (Ed.), Middleware 2005. XIII, 443 pages. 2005.

Vol. 3789: A. Gelbukh, Á. de Albornoz, H. Terashima-Marín (Eds.), MICAI 2005: Advances in Artificial Intelligence. XXVI, 1198 pages. 2005. (Sublibrary LNAI).

Vol. 3788: B. Roy (Ed.), Advances in Cryptology - ASIACRYPT 2005. XIV, 703 pages. 2005.

Vol. 3787: D. Kratsch (Ed.), Graph-Theoretic Concepts in Computer Science. XIV, 470 pages. 2005.

Vol. 3786: J. Song, T. Kwon, M. Yung (Eds.), Information Security Applications. XI, 378 pages. 2006.